CW01546306

British Warships in the Age of Sail
1817–1863

British Warships in the Age of Sail 1817–1863

Design, Construction, Careers and Fates

Rif Winfield

Seaforth

PUBLISHING

FRONTISPIECE: This lithograph after Sir Oswald Walters Brierly shows the British Baltic Fleet sailing through the Great Belt on 26 March 1854, the day before Britain declared itself at war with Russia, at the start of the campaign in the Baltic. All the ships are carefully delineated; Sir Charles Napier's flagship, the 131-gun *Duke of Wellington* (with the paddle vessels *Leopard* and *Vulture* on the starboard flank) is followed by the *St Jean d'Acre*, *Royal George*, *Princess Royal*, *Cressy* and the four screw blockships *Edinburgh*, *Hogue*, *Ajax* and *Blenheim*, as well as four screw frigates (*Impérieuse*, *Arrogant*, *Amphion* and *Tribune*) and another two paddle warships (*Valorous* and *Dragon*); the screw frigate *Dauntless* at anchor on the right provides the rounding mark.

The fleet, the first to be composed entirely of steam-powered vessels, had sailed from Spithead 13 days before. Although quickly augmented by other ships of the line, sailing as well as steam-driven ones, it soon became apparent that it lacked the support of gunboats and mortar vessels appropriate for operations in shallow coastal waters, a shortcoming that was not to be corrected until the following year's campaign. *(NMM PY8333)*

Copyright © Rif Winfield 2014

First published in Great Britain in 2014 by
Seaforth Publishing
An imprint of Pen & Sword Books Ltd
47 Church Street, Barnsley
S Yorkshire S70 2AS

www.seaforthpublishing.com
Email info@seaforthpublishing.com

British Library Cataloguing in Publication Data
A CIP data record for this book is available from the British Library

ISBN 978-1-84832-169-4

All rights reserved. No part of this publication may be reproduced or transmitted in any form or by any means, electronic or mechanical, including photocopying, recording, or any information storage and retrieval system, without prior permission in writing of both the copyright owner and the above publisher.

The right of Rif Winfield to be identified as the author of this work has been asserted by him in accordance with the Copyright, Designs and Patents Act 1988

Typeset and designed by Palindrome
Printed and bound in China

Contents

The 1825 aquatint by John Moore shows a British fleet in the harbour of Port Cornwallis on Great Andaman Island, preparing to sail for Rangoon during the First Burmese War. Alongside the frigate, the foreground shows the East India Company's paddle steam vessel *Diana*, the first known example of a steam vessel participating in a British naval operation. *(NMM D3594)*

The last fleet battle under sail took place on 20 October 1827 at Navarino (modern Pylos) during the Greek War of Independence. An allied British–French–Russian squadron attempted to overawe the Ottoman–Egyptian fleet, with a view to forcing allied terms on them. This watercolour by Lieut. William Innes Pocock (son of the famous marine painter Nicholas) shows the British contingent leading the fleet into Navarino Bay. Vice-Adm. Sir Edward Codrington (in the 84-gun *Asia*), the overall commander, hoped to avoid a battle, but the manoeuvre provoked the Turks into opening fire, and in the short but intense fighting which followed much of the Ottoman fleet was destroyed. The action embarrassed the British government, but it proved the first step to securing Greek independence. *(NMM A4916)*

Preface

This is the fourth in a series of volumes providing a guide to the construction, fitting and operation of every vessel which served in or was ordered for the (British) Royal Navy and its predecessor forces between the advent of the all-Britain monarchy in 1603 (previous naval forces pertained to England alone) and the close of the age of sail, here taken as 1863, the year in which wooden-hulled shipbuilding for the battlefleet came to an end (although wooden-hulled cruising warships were to continue to be built for a further period thereafter).

The present volume covers the period from 1817, chosen as the year closely following upon the close of the Napoleonic Wars, when the organisation and classification of British naval vessels underwent a major revision, up to 1863 when the ironclad battlefleet superseded the traditional 'wooden walls'. Earlier volumes are already available covering the periods 1603–1714, 1714–1792 and 1793–1817.

This book, like its companion volumes, gives a summary of the main technical details of each 'class' (design) of vessel built for the Navy or otherwise acquired by it, from the huge three-deckers down to the minuscule brigs, gunboats and mortar floats, together with building data for each vessel ordered to those designs. It seeks to portray those changes in the evolution of the design of each class, where these changes occurred while the vessels were under construction.

It also includes data for the surviving enemy warships taken during the Napoleonic War and added to the British Navy, and large numbers of merchant vessels purchased from civilian sources to augment British naval strength. All earlier vessels still in existence at the start of 1817 are covered.

The book also includes – as far as space allows – detailed notes on the service history for each vessel covered, with listings of their commanding officers, main deployments, actions in which they were involved (including information on enemy warships, privateers and slaving vessels in whose capture the British vessel was involved), the nature and dates of dockyard refits and major repairs – with the expenditure if known – and a variety of other relevant data. Details of the principal 'as-built' dimensions of every vessel (illustrating divergences from the design data) and of the individual shipbuilders involved, together with dates on which each vessel was ordered, laid down, launched and finished, complete this comprehensive single-volume reference source for every significant vessel of the period.

Sources and Acknowledgments

The data is this book (and its predecessors in this series) is taken almost exclusively from the archives of the British Admiralty Board, and of its subsidiary (up to 1832) the Navy Board, primarily that recorded in the National Archives at Kew and the National Maritime Museum at Greenwich. Prominent among these official records are:

ADM.32 Series – Ships' Pay Books
ADM.51 Series – Captains' Logs
ADM.53 Series – Ships' Logs
ADM.180 Series – Progress Books

There are literally tens of thousands of logbooks covering several centuries of naval vessels, so – with the best will in the world – no one is able to be fully acquainted with all of them. Furthermore, for various reasons, there are gaps in the coverage. These gaps are partially filled by reference to the Navy List and to the London Gazette (both over various dates) and by selected archives from contemporary British newspapers, notably the *Times*. Additional material on foreign-built vessels added to the Navy is taken from the archives of the country concerned where this conflicts with the (often dubious or incomplete) information held in British records. Many of these archives, both at home and abroad, are now accessible online. Further data has been taken from verifiable secondary sources where gaps exist in official British archives.

Like its predecessors, this book has been made possible by the collective work of a number of individuals, who generously supplied me with the results of their own research, offered constructive suggestions for improvements, and read drafts of the chapters and pointed out the errors that inevitably accumulate in a volume of this size. In terms of the sheer profusion of such help, first place must go to David Hepper, who invariably provided me promptly on request with vast amounts of extra research, including scrupulous trowelling through the archives listed above. Other friends to whom I must also pay my thanks include John ('Mike') Tredrea, John Houghton, Stuart Rankin, Deb Carlen, Eduard ('Ted') Sozaev, Emir Yener and Grant Walker at Annapolis. I am indebted to Andrew Lambert not only for the comprehensive overview which he provided in the 1603–1714 volume of this series, but also for his continued support and helpful comments during this project.

My thanks must go equally to those whose personal archives have been provided to me. The copious records of the late David Lyon, generously donated to me by his widow, Leo (Eleanor) Sharpston, contain the results of his many years of labour at the National Maritime Museum compiling detailed records on every vessel, particularly as regards the service histories of individual ships. The extensive lists of technical data provided by Fred Dittmar have been an invaluable source throughout this series. And I am additionally grateful to Peter Davis in Zeist, Holland, for the use of material contained on his website.

This series has relied also on the earlier research of a number of writers. David Syrett and R. L. DiNardo's *Commissioned Sea Officers of the Royal Navy 1660–1815* (Navy Records Society publication, 1994) has been a constant source of reference to verify the full names and ranks (at dates stated) of commanding officers, while David Hepper's *British Warship Losses in the Age of Sail 1650–1859* Jean Boudriot Publications, 1994) has provided reliable precise dates for ship's fates (often misquoted in many secondary sources). For the various French warships and privateers being incorporated into the British Navy, the published *répertoires* of Jacques Vichot, Frank Lecalvé and more recently of Alain Demerliac provided substantial verification of much material.

A number of museums and libraries have over the years been of vital assistance in helping my own researches and solving the frequent queries that arose, primarily the staff at the Public Records Office (now The National Archives) at Kew and the Caird Library in the National Maritime Museum at Greenwich, but also a range of museums from Rotterdam to Malta.

Inevitably much of the effort has been supplied by my publishers, and I am most grateful for the patience and careful production work of Robert Gardiner and Julian Mannering, and the scrupulous copy-editing of Paula Turner. In particular, Robert has selected and arranged use of all the illustrations which have contributed so much to the final appearance of this volume, and has shared in the work of captioning them. Finally but most essentially, I need to thank my wife Ann for her patience during the research and writing for this book, and for her constant help and encouragement.

Structure and Organisation of the Book

As with the previous volumes in this series, the first six chapters detail the rated warships of the Royal Navy, with one chapter devoted to each Rate. A major reorganisation of the rating system took place in accordance with an Admiralty Order in Council on 25 November 1816 (see below), and took effect from 1 January 1817, and these chapters reflect (even for ships built before 1817) the Rate at which they were established following that reorganisation.

These six chapters detail chronologically both the pure sailing ships and those with screw propulsion. Paddle-driven warships, however are not included in these chapters, as all paddle vessels – even those which were rated warships – are gathered together in the consolidated Chapter 11; this is because the physical layout of paddle warships meant that they were unable to fit continuous batteries of broadside guns, so that even frigates with paddle propulsion carried very many fewer guns than their pure sailing or screw-driven equivalents.

Unrated naval vessels are to be found in Chapters 7 to 14. In these chapters the pure sailing vessels are to be found in Chapters 7 to 10, and steam-driven vessels in Chapters 11 to 14. It should be remembered that throughout the early steam era, steam warships carried a full sailing rig and relied on sail power for all lengthy voyages, pending the establishment by Britain of suitable coaling stations around the globe. At the end of the period under consideration, the early ironclad warships made an appearance, beginning with the *Warrior* of 1860, and these are gathered together in a Postscript which follows Chapter 14.

Overview

The period of nearly half a century between the close of the Napoleonic Wars and the creation of the ironclad fleet can be easily divided in terms of technological evolution into three phases of almost equal length.

The years up to 1830 saw the culmination of the sailing warship; it was overseen by the post-war Tory coalition, and in design terms by Sir Robert Seppings, the last master shipwright to become Surveyor of the Navy. Its end was marked by the election of 1830, bringing in the Whigs with a new Admiralty Board under Sir James Graham, which with effect from 1 June 1832 abolished the subordinate Navy Board and replaced Seppings by a naval officer, the amateur designer William Symonds. This second period under Symonds witnessed not only major changes in hull forms and ordnance, but also the first major introduction of steam propulsion, first with paddle warships and subsequently with screw propellers.

The third phase, during which the post of Surveyor was held by Sir Baldwin Wake Walker, was initiated by the acceptance of the screw as an essential element of the future battlefleet. Unlike his predecessors, Walker was appointed more for his managerial skills (his post was redesignated Controller of the Navy in 1860, the historic title of Surveyor being abolished), with actual design responsibilities passing to his deputies, John Edye and Isaac Watts. The period witnessed not only rapid expansion prompted by the Russian War, but the final disappearance of sail as a major factor in battlefleet operations. Following this war, the construction of the first ironclad frigates, while contemporary with the last of the 'wooden-walls', foreshadowed the evolution of the modern warship over the next quarter-century. This period is deemed to have ended with the retirement of Isaac Watts as Chief Constructor for the Navy on 9 July 1863.

This half-century, the period of undisputed British naval mastery, is generally perceived to have been largely an era of (relative) peace following the close of the long-drawn-out Great War against Napoleonic France. Nevertheless the service histories of British warships listed in this volume reveal the multifarious naval involvements of the era. Apart from the Russian War (popularly called the Crimean War, but actually fought equally in the Baltic, as well as the White Sea and even the Pacific theatres), the Britain Navy was effectively the global policeman, defending the interests of the Empire and of British trading requirements in China and West Africa, New Zealand and Burma, South America and the Barbary Coast. Moreover, this was the period when it fell largely to the British Navy to combat the global slaving trade, as evidenced by the quantities of anti-slavery patrols and countless seizures of slaving vessels which are to be found in the following pages.

The Rating System

A new scale of rating warships was introduced by Admiralty Order in Council on 25 November 1816, which came into effect from 1 January 1817. Its main effect was to include the carronade armament on the upperworks (QD and Fc) in each vessel's gun rating. Previously carronades had only been included in the formal gun rating if they replaced long guns – sometimes simply theoretical long guns – in a ship's broadside. Under the new scheme all three-deckers, which now had a minimum of 100 guns (and 800 men), became First Rates. The Second Rate now included all two-deckers with a minimum of 80 guns (and 650 men), while the Third Rate started at 70 guns (and 600 men). Among the frigates, the Fourth Rate had a minimum of 50 guns (with 350 men) and the Fifth Rate 36 guns (with 280 men). The smallest frigates – soon to be redesignated corvettes – started at 35 guns (and 125 men), while among unrated vessels there were four grades of sloop, with established crews ranging from 75 up to 135 men, while even smaller vessels (brigs, schooners, cutters and bomb vessels) had either 50 or 60 men depending on size.

There was a slight alteration in rating, according to the size of the establishment, made under an Order in Council on 2 February 1842; but a completely new scale, graded according to the complement alone (instead of combining the number of guns with the number of men), was introduced by an Order in Council of 1 January 1856. The new scale differed for sailing and for steam warships.

	Steam Vessels	Sailing Vessels
First Rate	over 1,050 men	over 970 men
Second Rate	over 750 men	over 720 men
Third Rate	over 600 men	over 600 men
Fourth Rate	over 475 men	over 440 men
Fifth Rate	over 300 men	over 300 men
Sixth Rate	over 180 men	over 185 men
Sloops	over 100 men	over 80 men
Smaller	36 to 90 men	50 to 65 men

Bibliography

Archibald, Edward, *The Wooden Fighting Ship in the Royal Navy AD 897–1860*, Blandford Press, 1968, revised 1972, published 1987 (with alterations) as *The Fighting Ship of the Royal Navy 897–1984*

Brown, David K, *Before the Ironclad: Development of Ship Design, Propulsion and Armament in the Royal Navy, 1815–60*, Conway Maritime Press, London 1990

Burney, William, *Falconer's New Universal Dictionary of the Marine*, London, 1815; reprinted by Chatham Publishing, London 2006

Charnock, John, *History of Marine Architecture*, in three volumes, London 1800–1802

Colledge, James, *Ships of the Royal Navy*, 3rd edition by Greenhill Press, London 2003 (revised by Ben Warlow)

Demerliac, Alain, *Nomenclature des navires français de 1792–1799*, Editions Omega, Nice 1999

— *Nomenclature des navires français de 1800–1815*, Editions Omega, Nice 2003

Derrick, Charles, *Memoirs of the Rise and Progress of the Royal Navy*, London 1806

Douglas, Gen. Sir Howard, *A Treatise on Naval Gunnery*, John Murray, London 1855 (reprinted by Naval & Military Press, Uckfield)

Duckers, Peter, *The Crimean War at Sea: The Naval Campaigns Against Russia 1854–56*, Pen & Sword Maritime, Barnsley 2011

Evans, David, *Building the Steam Navy: Dockyards, Technology and the Creation of the Victorian Battle Fleet 1830–1906*, Conway Maritime Press/English Heritage, London 2004

Friedman, Norman, *British Cruisers of the Victorian Era*, Seaforth Publishing, Barnsley 2012

Gardiner, Robert, *The Heavy Frigate*, Conway Maritime Press, London 1994

— *Warships of the Napoleonic Era*, Chatham Publishing, London 1999

— *Frigates of the Napoleonic Wars*, Chatham Publishing, London 2000

— *The Sailing Frigate: A History in Ship Models*, Seaforth Publishing, Barnsley 2012

Gardiner, Robert (ed.), *The Line of Battle: The Sailing Warship 1650–1840*, Conway Maritime Press, 1992

— *Fleet Battle and Blockade; the French Revolutionary War 1793–1797*

— *Nelson against Napoleon: From the Nile to Copenhagen 1798–1801*

— *The Campaign of Trafalgar, 1803–1805*

— *The Victory of Seapower, 1806–1815*

— *The Naval War of 1812* (all Chatham Pictorial Histories, 1996–1998)

Haultain, Charles (ed.), *The New Navy List* (quarterly), Simpkin, Marshall and Co., London (various years)

Hepper, David, *British Warship Losses in the Age of Sail 1650–1859*, Jean Boudriot Publications, Rotherfield Sussex 1994

Howard, Dr Frank, *Sailing Ships of War 1400–1860*, Conway Maritime Press, London, 1979

Holland, A. J., *Ships of British Oak*, Newton Abbot, 1971

James, William, *The Naval History of Great Britain, from the Declaration of War by France in 1793 to the Accession of George IV*, in six volumes, Richard Bentley & Son, London, 1886

Laird Clowes, William, *The Royal Navy: A History from the Earliest Times to 1900*, Volumes 6–7, London 1898–1903, reprinted by Chatham Publishing, London, 1997–98

Lambert, Andrew, *Battleships in Transition: The Creation of the Steam Battlefleet, 1815–1860*, Conway Maritime Press, London 1984

— *The Last Sailing Battlefleet: Maintaining Naval Mastery 1815–1850*, Conway Maritime Press, London 1991

Lavery, Brian, *The Ship of the Line* (2 volumes), Conway Maritime Press, London 1983–1984

— *The Arming and Fitting of English Ships of War 1600–1815*, Conway Maritime Press, London, 1987

— *Nelson's Navy*, Conway Maritime Press, London, 1989

Lecalvé, Frank, *Liste de la Flotte de Guerre Française*, Toulon 1993

Lyon, David, *The Sailing Navy List*, Conway Maritime Press, London, 1993

Malcomson, Robert, *Warships of the Great Lakes 1754–1834*,

Chatham Publishing, London 2001

O'Byrne, William R., *A Naval Biographical Dictionary* (in 3 volumes), John Murray, London 1849

Parkinson, C. Northcote, *Britannia Rules; The Classic Age of Naval History 1793–1815*, Weidenfeld and Nicholson, 1977

Penn, Geoffrey, *'Up Funnel, Down Screw!'*, Hollis & Carter, London 1955

Preston, Antony and Major, John, *Send a Gunboat: The Victorian Navy and Supremacy at Sea 1854–1904*, (2nd edition) Conway Maritime Press, London 2007

Rodger, N. A. M., *The Command of the Ocean; A Naval History of Britain 1649–1815*, Penguin/Allen Lane, London 2004

Smyth, Adm. W. H., *Sailor's Word Book; a Dictionary of Nautical Terms*, London, 1867; reprinted by Conway Maritime Press, London, 1996

Syrett, David and DiNardo, R. L., *The Commissioned Sea Officers of the Royal Navy 1660–1815*, Navy Records Society, London, 1994

Vichot, Jacques (ed.), *Répertoire des Navires de Guerre Français*, Musée de la Marine, Paris, 1967

Ware, Chris, *The Bomb Vessel*, Conway Maritime Press, London 1994

Warlow, Lieut Cmdr Ben, *Shore Establishments of the Royal Navy*, 2nd edition, Maritime Books, Liskeard, 2000

Winfield, Rif, *British Warships in the Age of Sail, 1714–1792*, Seaforth Publishing, Barnsley, 2007

— *British Warships in the Age of Sail, 1793–1817* (2nd edition), Seaforth Publishing, Barnsley, 2007

— *The Sail and Steam Navy List: All the Ships of the Royal Navy 1815–1889* (with David Lyon), Chatham Publishing, London, 2004

Chronology of the Navy 1817 to 1863
a brief summary of naval and other events

1824   First (Anglo-)Burmese War. Capture of Rangoon by forces under Commodore Charles Grant.

1827   Battle of Navarino on 20 October: destruction of a Turkish-Egyptian fleet by a British-French-Russian fleet under Vice-Adm. Sir Edward Codrington.

1832   Abolition of the separate Navy Board and Victualling Board, with their functions being subsumed into the Board of Admiralty.

1839   Capture of Aden by Capt. Henry Smith in the *Volage* on 18 January.
       First (Anglo-) Chinese War

1840   Seizure of Chusan (off the coast of Zhejiang Province) on 5 July.
       Bombardment of St Jean d'Acre (on coast of Syria) on 3 November.

1841   Cession to Britain of Hong Kong on 26 January, and evacuation of Chusan.

1842   Capture of Shanghai on 18 June.

1845   War in the Parana River (South America): Battle of Obligado on 20 November.
       First New Zealand War.

1852   Second (Anglo-) Burmese War; capture of Rangoon on 14 April.

1854   Outbreak of War with Russia (Britain and France declared war on 27 March).
       Bombardment of Odessa on 22 April, and of Sebastopol on 17–24 October.
       Attacks on Bomarsund (off Helsinki) 23 July–16 August.
       Siege of Petropavlovsk (Kamchatka) 18–28 August.

1855   Further bombardments of Sebastopol between 9 April and 7 September.
       Allied occupation of Kertch on 24 May, followed by naval excursion into the Sea of Azov 30 May – 5 June.
       Bombardment of Sveaborg (in the Baltic) on 9–11 August.
       Bombardment of Kinburn (at the mouth of the Dneiper) on 17 October.

1856   Treaty of Paris (signed 9 March and ratified 27 April) brings Russian War to a close.
       Second (Anglo-) Chinese War; capture of Canton on 25 October by forced under Rear-Adm. Sir Michael Seymour.

1858   An Anglo-French force comprising the gunvessels *Nimrod* and *Cormorant*, together with six British and four French screw gunboats, bombarded the Taku forts at the mouth of the Peiho River on 20 May.
       Treaty of Tientsin (on 27 June) nominally ended hostilities, but the terms of peace were never implemented.

1859   The (Anglo-)Chinese War was formally renewed on 8 April 1859.
       A force of nine screw gunboats under Rear-Adm. James Hope, together with the gunvessels *Nimrod* and *Cormorant*, was repulsed in its attempts upon the Peiho River forts on 26 June.

1860   A renewed force under Rear-Adm. Hope captured the Taku forts on 21 August, subsequently occupying Tientsin and then Peking.
       Treaty of Peking (ratified 24 October) brings Chinese War to a close.

Chusan (modern Zhoushan) is the largest (194 sq. miles) island in an eponymous archipelago of 1,390 islands off the coast of Zhejiang Province, facing the port of Ningbo. It was occupied by British forces under Capt. Charles Elliot on 5 July 1840 during the First Anglo-Chinese ('Opium') War, comprising the *Wellesley* (Captain Sir John James Gordon Bremer), *Alligator*, *Conway*, *Larne*, *Algerine*, *Rattlesnake* and two HEICo. steamers, the island becoming Britain's first colonial possession off China, but it was evacuated in early 1841 in favour of Hong Kong.

A British fleet under Elliot's replacement, Colonel Sir Henry Pottinger, comprising the *Wellesley* (now with the flag of Rear-Adm. Sir William Parker), *Blenheim*, *Druid*, *Blonde*, *Modeste*, *Pylades*, *Columbine*, *Cruiser*, *Algerine*, *Rattlesnake* and the HEICo. steamers *Queen*, *Sesostris*, *Nemesis* and *Phlegethon*, recaptured Chusan on 1 October 1841, and retained it until 1846, as depicted in this watercolour by Edward Hodges Cree. The island was returned to China, but was reoccupied for a period during the Second Anglo-Chinese War. *(NMM PW6032)*

An aquatint by Lieut John Frederick Warre showing the attack and capture of Sidon on the Syrian coast in late September 1840. The Syrian campaign was the Royal Navy's first significant use of steamers in action. Three paddle warships – *Cyclops*, *Gorgon* and *Hydra* – can be seen on the left, ahead of the 84-gun *Thunderer* (acting as flagship of Commodore Charles Napier) while the bow of another steamer (the *Stromboli*) can be just made out at the far right; also participating was the 16-gun *Wasp*, together with the Austrian 49-gun *Guerriera* (under HIH the Archduke Friedrich) and the Turkish 84-gun *Mukaddime-i-Hayir*, the latter carrying the flag of Capt. Baldwin Wake Walker (subsequently the Surveyor of the Navy) who was seconded to the Ottoman Navy as their Rear-Admiral. *(NMM PY8187)*

Battles and Campaigns

1 1st Burma War Campaign, 5 March 1824–24 February 1826
Alligator, Arachne, Boadicea, Champion, Larne, Liffey, Slaney, Sophia, Tamar, Tees;
also hired steamer *Diana*.

2 Navarino, 20 October 1827
Asia 84 (flag of Vice-Adm. Sir Edward Codrington), *Genoa* 76, *Albion* 74, *Glasgow* 50, *Cambrian* 48, *Dartmouth* 42, *Talbot* 28, *Rose* 18, *Brisk* 10, *Mosquito* 10, *Philomel* 10, *Hind* (cutter) 6;
also French squadron *Breslau* 84, Scipion 80, *Trident* 74, *Sirène* 60 (flag of Contre-Adm. Henry de Rigny), *Armide* 42, *Alcyone* 10 and *Daphne* 6;
and Russian squadron *Gangoot* 84, *Azof* 74 (flag of Rear-Adm. Count de Heiden), *Yezekeyeel* 74, *Alexander Nevski* 74, *Constantin* 44, *Elena* 44, *Provornyi* 42 and *Castor* 32.

3 Aden, 19 January 1839
Cruizer, Volage
also Indian Navy ships *Coote* and *Mahé*.

4 Syrian Coast, 10 September–9 December 1840
1st Rate: *Princess Charlotte* 104 (flag of Adm. Sir Robert Stopford).
2nd Rates: *Asia* 84, *Bellerophon* 80, *Ganges* 84, *Powerful* 84 (Commodore Charles Napier), *Rodney* 92, *Thunderer* 84, *Vanguard* 80.
3rd Rates: *Benbow* 72, *Cambridge* 78, *Edinburgh* 72, *Hastings* 72, *Implacable* 74, *Revenge* 76.
5th Rates: *Castor* 36, *Pique* 36.
6th Rates: *Carysfort* 26, *Magicienne* 24, *Talbot* 26.
Sloops: *Daphne* 18, *Dido* 18, *Hazard* 18, *Wasp* 16, *Zebra* 16.
Steam paddle vessels: *Cyclops* 6, *Gorgon* 6, *Hydra* 6, *Medea* 4, *Phoenix* 4, *Stromboli* 4, *Vesuvius* 4. (Note that *Hecate* 6 also was awarded battle honours, but did not arrive in time to participate in actions.)
This list excludes the Austrian squadron under Rear-Adm. Franz Baron Bandiera, also employed in these operations.

5 China, 7 January 1841–21 July 1842
Alligator 26, *Blenheim* 72, *Conway* 26, *Cruiser* 16, *Druid* 44, *Herald* 26, *Larne* 18, *Melville* 72, *Nimrod* 20, *Pylades* 18, *Samarang* 26, *Sulphur* 8 (survey), *Volage* 26, *Wellesley* 72; (all 1841)
Algerine 10, *Blonde* 42, *Calliope* 26, *Columbine* 16, *Hyacinth* 18, *Jupiter* (transport), *Modeste* 18, *Rattlesnake* 28, *Starling* 6 (survey cutter); (all in both years)
Apollo 46, *Belleisle* 72, *Childers* 16, *Clio* 16, *Cornwallis* 72, *Dido* 18, *Endymion* 44, *Harlequin* 16, *Hazard* 18, *North Star* 26, *Plover* (survey), *Sapphire* 28, *Vixen* 6 (paddler), *Wanderer* 16, *Young Hebe* (tender to *Cornwallis*); (all 1842)

A coloured lithograph by Edward Hodges Cree depicts the destruction by the paddle sloop *Fury* (Cmdr. James Willcox) of the pirate squadron commanded by Chuiapoo in Bias Bay on 1 October 1849. The suppression of any threat to its commerce was an unabashed aim of nineteenth-century British policy, and its sharpest weapon – the Royal Navy – operated under far more robust 'rules of engagement than any modern force would countenance'. *(NMM A7547)*

John Wilson Carmichael's contemporary depiction of the bombardment of the Russian defences of Sveaborg, a group of six heavily fortified islands guarding the approaches to the then Russian base of Helsingfors (Helsinki). The channels between the islands were further protected by the guns of the Russian 120-gun *Rossiia* and 74-gun *Iezekiil'* moored broadside on. From here the Russians could dominate not only the Gulf of Bothnia but also the vital strategic route between St Petersburg and the Baltic. In the second year of the Russian ('Crimean') War, the all-steam fleet led by Rear-Admiral Richard Saunders Dundas in the 131-gun *Duke of Wellington* (depicted centrally here) and Rear-Admiral Michael Seymour in the 90-gun *Exmouth* assembled off Nargen Island (near Reval on the Estonian coast) a squadron of 22 steam warships, 16 gunboats, and 16 mortar vessels – joined by a French contingent under Rear-Admiral Charles Pénaud in the 80-gun *Tourville*.

They reached Sveaborg on 6 August 1855 and moored out of range of the Russian guns; on 9 and 10 August a continual bombardment of the defences was maintained for 45 hours until the allied ships had exhausted their ammunition, causing little destruction to the forts and batteries but considerable damage to the grenade arsenals (whose thin roofs were pierced by the mortars' plunging fire) and to the town's wooden houses, with – according to Russian records – some 62 killed and 199 wounded ashore, and 11 killed and 89 wounded aboard the *Rossiia* (on the Allied side, there were no fatalities). Although there were no strategic gains from the attack, the tactical success of the operation, in particular the achievements of the gunboat flotilla under Commodore Frederick Thomas Pelham, served to convince the Admiralty of the vital future role of the steam gunboat and led to immediate orders for mass construction – a 'Great Armament' with 80 additional gunboats (see Chapter 12) being rapidly laid down over the winter of 1855–56, as well as 4 floating batteries (Chapter 6), 34 mortar vessels and 50 mortar floats (Chapter 7). *(NMM BHC0636)*

also (no medals awarded) survey *Bentinck* 10, *Cambrian* 36, tender *Louisa*, and *Pelican* 18.

6 Battle of Vuelta de Obligado (Parana River, South America), November 1845
Paddle vessels *Gorgon* 6 and *Firebrand* 6.
Comus 18, *Philomel* 8, *Dolphin* 3 and schooner *Fanny* 1.
also 5 French vessels

7 New Zealand, 1845–47
Calliope 28, *Castor* 36, *Driver*, *Hazard* 18, *Inflexible*, *North Star* 26, *Osprey* 12 and *Racehorse* 18;
also Indian Navy ship *Elphinstone*.

8 Kua Kam (Vietnam), 20 October 1849 against Chinese pirate junks
Columbine 16 and *Fury* 6;
also HEICo.'s *Phelgethon* (Cmdr G. T. Niblett, IN).

9 Second Burmese War Campaign, 10 January 1852–30 June 1853
Hastings 72, *Hermes* 6, *Rattler* (flag of Rear-Adm. Charles John Austen), *Salamander*, *Serpent* 12 (all 1852);
Fox 42, *Spartan*, *Sphinx*, *Winchester* (all in both years)
Bittern, *Cleopatra*, *Contest*, *Styx* (all 1853)
also HEICo.'s *Feroze*, *Fire Queen*, *Mahanuddy*, *Medusa*, *Mozuffer*, *Nemesis*, *Phelgethon*, *Proserpine*, *Sesostris* and *Tenasserim*, and light river steamers *Damooda*, *Lord William Bentinck* and *Nerbudda*.

10 Baltic, 28 March–20 September 1854; 17 April– 10 December 1855
(a) Napier's fleet (sailed 11.3.1854 from Spithead):
Duke of Wellington 131 (flag of Vice-Adm. Sir Charles Napier and Rear-Adm. Michael Seymour);
Royal George 120, *St Jean d'Acre* 101 and *Princess Royal* 91; blockships *Edinburgh* 60 (flag of Rear-Adm. Henry Ducie Chads), *Hogue* 60, *Ajax* 60 and *Blenheim* 60; frigates *Impérieuse* 51, *Arrogant* 46, *Amphion* 34, *Tribune* 31, *Leopard* 18 (flag of Rear-Adm. James Hanway Plumridge), *Valorous* 16 and *Dragon* 6 (the last three named were paddle warships, all the others were screw).
(b) Other RN vessels deployed to the Baltic 1854–55 (* = as troopships):

Fort Arabat guarded the approach to the Tongue of Arabat, a tortuously thin 70-mile spit of land separating the Sivash or 'Putrid Sea' from the Sea of Azov. It was bombarded on 28 May 1855 by a squadron of gunboats and other small steamers commanded by Capt. Edmund Moubray Lyons aboard the screw corvette *Miranda*. This pen-and-ink sketch shows the fort's magazine exploding, with the Allied force comprising the *Miranda* (centre, with signal hoist at the main), twelve smaller British steamers and five French ones, at the start of a fortnight's campaign around the Sea of Azov, during which Lyons's small squadron inflicted major material and economic damage on Russia throughout the coastal areas. The drawing carries the initials O. W. B. for Oswald Walters Brierly, who was an eyewitness to many naval events of the Russian War. *(NMM PU9034)*

screw ships of the line *Royal William* 120, *St Vincent* 102*, *Algiers* 91, *Caesar* 91, *Hannibal* 91, *Exmouth* 91, *James Watt* 91, *Nile* 91, *Orion* 91, *Colossus* 81, *Cressy* 81, *Majestic* 81, *Sans Pareil* 71;

sailing ships of the line *Neptune* 120, *St George* 120, *Prince Regent* 90, *Calcutta* 84, *Monarch* 84, *Boscawen* 70 and *Cumberland* 70;

screw blockships *Cornwallis* 60, *Hastings* 60, *Hawke* 60, *Pembroke* 60 and *Russell* 60;

screw frigates *Euryalus* 51, *Dauntless* 33 and *Termagant* 24*;

paddle frigates *Retribution* 28, *Magicienne* 16, *Odin* 16, *Penelope* 16, *Centaur* 11, *Gladiator* 6* and *Vulture* 6;

screw corvettes *Pylades* 21, *Tartar* 21, *Cossack* 20, *Esk* 20, *Cruiser* 17, *Falcon* 17, *Harrier* 17, *Malacca* 17, *Archer* 14, *Miranda* 14, *Conflict* 8 and *Desperate* 8;

paddle sloops *Basilisk* 6, *Bulldog* 6, *Dragon* 6, *Driver* 6, *Geyser* 6, *Gorgon* 6, *Hecla* 6, *Merlin* 6, *Rosamond* 6, *Sphinx* 6, *Stromboli* 6*, *Firefly* 4 and *Janus* 4;

screw dispatch vessel *Wrangler* 6 (later to Black Sea);

paddle gunvessels (including ex-packets) *Cuckoo* 3, *Locust* 3, *Otter* 3, *Pigmy* 3, *Porcupine* 3 and *Zephyr* 3;

screw gunboats *Badger* 2, *Biter* 2, *Cracker* 2, *Dapper* 2, *Gleaner* 2, *Hind* 2, *Jackdaw* 2, *Lark* 2, *Magpie* 2, *Pelter* 2, *Pincher* 2, *Redwing* 2, *Ruby* 2, *Skylark* 2, *Snap* 2, *Snapper* 2, *Starling* 2, *Stork* 2, *Swinger* 2, *Thistle* 2 and *Weazel* 2;

mortar vessels *Beacon*, *Blazer*, *Carron*, *Drake*, *Grappler*, *Growler*, *Havock*, *Manly*, *Mastiff*, *Pickle*, *Porpoise*, *Prompt*, *Redbreast*, *Rocket*, *Sinbad* and *Surly*;

Steam survey vessels *Alban* 6, *Merlin* 4 and *Lightning* 3;

powder vessels *Aeolus* and *Volage*;

hospital ship *Belleisle* 6;

water tanker *Sheerness*.

In the separate White Sea operations in June 1854 a detached Squadron was deployed under Capt. Erasmus Ommanney in *Eurydice*, accompanied by the sloops *Miranda* 14 and *Brisk* 14, plus 2 French ships; they scoured the White Sea and then destroyed the port of Kola on 24 August 1854 before withdrawing home prior to the onset of winter. In May 1855 a fresh detached squadron was deployed under Capt. Thomas Baillie in *Maeander*, accompanied by the sloop *Phoenix* 8 and *Ariel* 9, plus 3 French ships (*Cléopâtre*, *Cocyte* and *Pétrel*); it left the White

Sea for home on 9 October 1855.

11 Crimea/Black Sea, 17 September 1854–9 September 1855

Screw ships of the line *Royal Albert* 121, *St Jean d'Acre* 101, *Agamemnon* 91, *Princess Royal* 91, *Algiers* 90, *Rodney* 90 and *Sans Pareil* 70;

sailing ships of the line *Britannia* 120, *Trafalgar* 120, *Queen* 116, *Hibernia* 104, *Albion* 91, *London* 90, *Vengeance* 84 and *Bellerophon* 78

screw frigate *Curacao* 31, *Tribune* 31, *Dauntless* 24 and *Highflyer* 21;

paddle frigates *Retribution* 28, *Sidon* 22, *Terrible* 21, *Leopard* 18, *Odin* 18*, *Furious* 16, *Tiger* 16 and *Valorous* 16;

sailing frigate *Arethusa* 50 and *Leander* 50;

sailing corvettes *Diamond* 28 and *Modeste* 18;

screw troopships *Apollo* 38, *Gorgon* 6, *Himalaya* —, *Transit* —;

iron screw troopships *Simoom* (iron) 18, *Vulcan* (iron) 6 and *Megaera*;

screw corvette *Desperate* 8, *Malacca* 18, *Niger* 14, *Wasp* 14

paddle sloop *Ardent* 5, *Cyclops* 6, *Firebrand* 6, *Fury* 6, *Gladiator* 6, *Inflexible* 6, *Oberon* 3, *Prometheus* 5, *Sampson* 6, *Spiteful* 6, *Spitfire* 6, *Triton* 3, *Stromboli* 6, *Vesuvius* 6, *Vulture* 6, *Firebrand* 6

paddle gunboat *Harpy* 1

paddle packet *Banshee* 2, *Caradoc* 2, *Medina* 4, Snake 6

screw storeships (unarmed) *Supply* and *Wye*;

paddle vessel *Circassian*

in addition to the vessels which also entered the Sea of Azov (listed below).

12 Sea of Azov 1855

screw ship-of-the line *Hannibal* 91;

screw sloop *Curlew* 9;

screw corvette *Miranda* 14;

screw sloop *Swallow* 9;

paddle sloops *Sphinx* 6, *Stromboli* 6, *Vesuvius* 6 and *Medina* 4;

screw dispatch/gun vessels *Arrow* 4, *Beagle* 4, *Lynx* 4, *Snake* 4, *Viper* 4 and *Wrangler* 4;

iron paddle gun vessels *Recruit* 6 and *Weser* 6;

screw gunboats *Boxer* 2, *Clinker* 2, *Cracker* 2, *Fancy* 2,

After proving their worth in the Russian War, steam gunboats were to find plenty of employment in coastal and riverine scenarios whenever Britain sought to bring its sea power to bear on the land for military or strategic ends. One successful example was the attack on the Taku forts on 20 May 1858 by an Anglo-French squadron of steam-powered small craft – four French gunboats *Mitraille*, *Fusée*, *Avalanche* and *Dragonne*, the British gunvessels *Nimrod* and *Cormorant*, and the gunboats *Slaney*, *Firm*, *Opposum*, *Leven*, *Staunch* and *Bustard*. After a bombardment which silenced the forts on both sides of the Peiho River, landing parties – seen in this contemporary lithograph in boats being towed astern of the gunboats – went ashore and destroyed the fortifications. The print was based on an eyewitness sketch by Frederick le Breton Bedwell. *(NMM PY8282)*

Grinder 2 and *Jasper* 2;
steam tenders *Danube*, *Moslem* and *Sulina*;
screw transport and supply ship *Industry*.

13 Pacific 1854–55
sail frigates *President* 50, *Winchester* 50, *Pique* 40, *Sybille* 40, *Spartan* 26, *Amphitrite* 24 and *Trincomalee* 24;
screw corvettes *Tartar* 21, *Hornet* 17, *Brisk* 14 and *Encounter* 14;
paddle sloops *Barracouta* 6, *Styx* 6 and *Virago* 6;
sail sloops *Dido* 18 and *Bittern* 12.

14 China, 1 October 1856–26 June 1858; 1 August– 24 October 1860
numerous vessels were involved in this period; those which were awarded Battle Honours were:
Acorn, Actaeon, Adventure, Amethyst, Assistance, Barracouta, Belleisle, Bittern, Calcutta, Cambrian, Camilla, Centaur, Chesapeake, Comus, Coromandel,
Cruiser, Elk, Encounter, Esk, Furious, Fury, Hesper, Highflyer, Hong Kong, Hornet, Imperieuse, Inflexible, Magicienne, Nankin, Niger, Odin, Pearl, Pique, Racehorse, Retribution, Sampson, Sans Pareil, Scout, Simoom, Sir Charles Forbes, Spartan, Sphinx, Sybille, Tribune, Urgent, Volcano, Vulcan, Watchman, Winchester;
screw gunvessels *Beagle, Cormorant, Nimrod, Pioneer, Renard, Ringdove, Roebuck, Snake, Sparrowhawk* and *Surprise;*
screw gunboats *Algerine, Banterer, Bouncer, Bustard, Clown, Cockchafer, Drake, Firm, Flamer, Forester, Grasshopper, Hardy, Haughty, Havock, Insolent, Janus, Kestrel, Lee, Leven, Opossum, Plover, Slaney, Snap, Starling, Staunch, Watchful, Weazel* and *Woodcock;*
also Indian Navy ship *Auckland*.

15 New Zealand, 1860–61
Cordelia, Iris, Niger and *Pelorus*.

Glossary and Abbreviations

The following alphabetical list of lesser-known nautical and other specialised terminology incorporates an explanation of abbreviations and definitions used in this book. Note that this brief list cannot purport to provide an explanation for all of the multitudinous terms used in maritime and naval life of the era. For more specialised terms the reader is referred to one or other of the specialised dictionaries, two of which (Falconer's and Smyth's) are rather arbitrarily selected for inclusion in the Bibliography.

Advanced ship A ship which was complete for service in all respects, with its guns on board, and more of the masting, along with all necessary stores, sails, etc., in an assigned storehouse, but not yet in commission.

Adm Admiral, RN. There were actually three ranks, in descending order these being Admiral, Vice-Admiral and Rear-Admiral. Each rank was subdivided into three grades, in descending order these being designated 'of the Red', 'of the White' and 'of the Blue'; however, I have made no attempt to distinguish the separate grades in this book. There was additionally a higher rank of Admiral of the Fleet (not subdivided), held only at the start of our period by HRH the Duke of Clarence (1811–37). A list of subsequent holders of the rank is included under 'Principal Officers of the Navy' in Appendix B.

AO Admiralty (Board) Order (usually with date of issue).

Brig A small two-masted sailing vessel (the term probably began as a shortened form of 'brigantine').

Brigantine A small two-masted vessel. By the end of the eighteenth century, they were square-rigged on both masts and the term was in practice interchangeable with 'brig'; from the 1820s, a new type of brigantine rig evolved which was fore-and-aft rigged on the mizzen while retaining square rig on the fore mast.

Capt Captain, RN. Here used only for commanding officers 'made post', i.e. having achieved the substantive rank of captain. Commanders and lieutenants who held command of unrated vessels were conventionally called 'captain' aboard their ship, but did not hold the substantive rank.

Cmdr Commander (or strictly speaking 'Master and Commander'), RN. This rank had become substantive in 1794 (until that date technically all holders prior to that date remained lieutenants); note that the present rank of 'lieutenant commander' did not become substantive until the early twentieth century.

Commodore A captain appointed to command of a squadron of vessels. Like the position of commander, at this date the term signified a temporary assignment rather than a permanent change in substantive rank. When the assignment finished , the officer concerned would revert to the substantive rank of captain. Captains who had been appointed as the commodore of a squadron were identified by flying a Broad Pendant, and they generally had a more junior captain appointed to the same ship to assume responsibility for the operation of that particular ship, so that the commodore could concentrate on overall command functions for the squadron.

Dates All dates are given in UK format, i.e. day.month.year. Abbreviations preceding these dates are:

Ord: date ordered (to be built) by the Admiralty – note that, for vessels built by a commercial shipbuilder, this is not the same as the date on which a contract was signed.

K: date the keel was laid down – note that work might in some cases have already taken place off the slipway.

L: date of launch (where known, this is also quoted for ships purchased or otherwise acquired after being built, or while building was taking place) or – for those few vessels built in a dry dock – of undocking.

C: date of completion of the fitting-out process which followed a vessel's launch – note that for vessels built by contract, this almost always took place in a Royal Dockyard.

Commissioned: a new vessel built for the Navy was generally commissioned by the first commanding officer to be appointed to that ship. Note that this usually (although not always) took place after the ship was launched, but while the ship was fitting out and consequently considerably earlier than the date of completion.

For captured vessels, a slightly different procedure is followed; while the launch date is recorded as above the overall period of construction (where known) is shown in the format *date keel laid – date of completion*.

Decks The decks on which guns were mounted are listed with the following abbreviations:

Fc *Forecastle* (deck), often abbreviated to fo'c'sle. A partial deck sheltering the forward end of the upper deck of a ship. In a flush-deck ship with no such partial deck above the upper deck, the forward end of the upper deck itself was sometimes described as the forecastle.

LD *Lower Deck*. Note this was the lowest deck to mount guns in a wooden warship of two continuous decks or more; the term gundeck was usually used in preference for this deck. In a frigate, the lower deck no longer mounted

guns, of the ports through which to fire them, although curiously the term 'gundeck' was retained until the end of the eighteenth century for this deck notwithstanding its complete absence of any guns.

MD *Middle Deck*. On a three-decker, the intervening gun-bearing deck between the lower deck and the upper deck, carrying the second full-length tier of carriage guns.

QD *Quarter Deck* (usually just quarterdeck). The partial deck extending forward from the stern, usually to about the mainmast (i.e. about halfway along the ship's length), a deck above the upper deck.

RH *Roundhouse* (deck) or poop (deck). A short additional structure above the quarterdeck at the stern of the largest ships (usually confined to three-deckers).

UD *Upper Deck*. The highest continuous deck of a vessel, excluding the superstructure or 'upper works'.

Note: other decks below the lower deck were the orlop deck and the hold itself, but they did not bear guns as they were below the waterline. The *orlop* deck was the lowest deck running the full length of a ship (or almost the full length continuous (or almost continuous) deck; in smaller vessels the orlop was non-continuous, and instead there was a series of *platforms* between the lower deck and the floor of the hold.

Note also that by mid-century, with the closing up of the waist between the fore and after superstructures (so that forecastle, spar deck and quarterdeck formed a continuous single structure, the term 'Upper Deck' was applied to this level, whereas the former upper deck (particularly on frigates) was renamed 'Main Deck' (the abbreviation 'MD' signifies this except for the three-deckers in Chapter 1, where the old designations of 'UD' and 'QD/Fc' are retained to avoid confusion)

Demonstration ship A mobilised unit used to support a diplomatic démarche, fulfilling a deterrent role, although the term is imprecise.

Dimensions Throughout the book, dimensions quoted are length x breadth x depth in hold; all measurements are in Imperial feet (ft) and inches (in). Two types of dimensions are quoted, the *design* dimensions for each class of vessel ordered to be built for the Navy, and the actual measured dimensions recorded for each individual vessel, whether built for the Navy or acquired by capture or purchase; these actual dimensions are recorded 'as built' (or in the case of an acquired vessel, as measured upon its acquisition).

Length Two *lengths* are usually quoted, separated by a comma; the first is length on the gundeck (the lower deck in larger vessels); the second is the length of the keel for tonnage (a somewhat artificial figure which is not the same as the actual length of the keel itself). Overall length is not usually quoted for vessels of the sailing era.

Breadth The *breadth* given is that used for calculating the tonnage; actually this is the breadth at the broadest part of the vessel, outside the planking but inside the wales.

Depth The depth is the perpendicular distance between the floor of the hold and the underside of the lower deck.

Draught Where known, each vessel's draught is also given in the form: *forwards/aft*. This is the light draught as at that ship's completion; clearly when fully laden each such ship would draw a considerably greater depth of water.

Dyd (Royal) Dockyard. The main bases of the British Navy, at Deptford and Woolwich on the Upper Thames, Chatham and Sheerness on the River Medway, and Portsmouth and Plymouth on the south (Channel) coast of England. Milford on the north side of Milford Haven (Pembrokeshire) had been added during the Napoleonic Wars, to be supplanted in 1815 by Pater (Pembroke Dock) on the south side of the Haven. There were also overseas dockyards at Kinsale in Ireland and at a number of other locations.

Flag As used herein as an abbreviation for *flagship*, technically a ship carrying a *flag officer* (or any one of the nine graduations of Admiral) but in common usage the officer in command of any fleet or squadron of warships.

Frigate Generally applied to any warship designed for cruising (patrol and escort duties) rather than to form part of the battlefleet, by the late eighteenth century it signified a warship with a single continuous tier of guns along its upper deck, but no guns (or gunports) below the upper deck. Technically, rated vessels of fewer than 28 guns were not classed as frigates, but instead were designated *post ships*, but most sea officers tended to apply the term in general usage to smaller Sixth Rates down to 20 guns.

Guns The *established* number of guns carried on each deck, with the calibre (weight of projectile) of each. Unless specified as carronades (usually mounted on slides), these were *long*(-barrelled) guns mounted on trucks.

HEICo. The Honourable East India Company.

Ketch In the first half of the eighteenth century, a number of unrated vessels, including many of the sloops and bombs, carried this form of two-masted rig, setting square sails on both a main and a mizzen mast as well as having fore-and-aft-rigged sails. This type of rig went out of use in the second half of the eighteenth century; when two-masted sloops of war re-appeared, most would be rigged as brigs (i.e with a fore and a main mast, instead of a main and a mizzen mast).

Lazarette (or lazaretto) A hulk adapted to be used as a floating hospital or isolation ward, often used as accommodation for men in quarantine.

Lieut Lieutenant, RN.

Men The total *established* complement of officers and ratings of the ship; the number *actually carried* at any one time naturally varied according to availability and losses.

Mld. Moulded breadth.

M/Shipwright The master shipwright was the senior naval constructor at each Royal Dockyard, and usually also at subsidiary naval dockyards. Admiralty records usually record as 'builder' only the master shipwright who was in post at that dockyard when the ship was launched and delivered to the Navy; in this book, all master shipwrights who were in post during the period of construction of the ship are attributed.

NBW Navy Board Warrant (usually with date of issue).

NMM National Maritime Museum, Greenwich.

Obusier The early French answer to the British carronade was in fact a bronze howitzer mounted on a slide carriage and firing a 36(Fr)-lb projectile. From 1804 the French introduced true carronades as a more effective short-range weapon.

Pink In seventeenth century naval usage, this applied to a ketch hull-form given a three-masted rig; later it applied to vessels retaining the ketch's narrow stern however they were rigged.

Powder hulk A vessel for storing and issuing gunpowder. There were usually under the operational control of the Ordnance Dept (a government department separate from both Navy and Army) and were for safety stationed well away from the dockyard to which they were attached.

PRO Public Records Office, Kew; subsequently renamed The National Archives.

PW Programme of Works, the annual construction programme decided upon by the Admiralty. This ran from the start of April of one year to the end of March of the next year; thus, for example, 'PW 1842' means the annual programme running from 1 April 1842 to 31 March 1843.

Rasée A vessel cut down by the removal of an entire deck, or sometimes just its quarterdeck and forecastle, to produce a lighter, usually more seaworthy vessel; most often used to give an extended life to a worn-out large warship. Often 'razee' in English texts.

Receiving ship A stationary vessel used as a floating accommodation ship for men between commissions or before they were attached to a specific vessel (in particular, newly pressed men who needed to be isolated from the shore to forestall attempts at desertion).

Sheer hulk A vessel equipped with a pair of 'sheer legs' (two large spars angled together to form an 'A' frame) to hoist masts in and out of vessels.

Ship In the more restricted usage of the sailing era, a three-masted sailing vessel, square-rigged on the fore and main masts, but usually with a fore-and-aft course on the mizzen mast, and with square topsails on all three masts.

Sloop In naval terms, an unrated vessel of war mounting fewer than 20 carriage guns, which might be either three-masted (i.e. ship-rigged) or two-masted (with one of a number of possible rigs). Note this term has no connection with the modern 'sloop rig', which indicates a single-masted fore-and-aft rig. When brigs were introduced to the RN in the late 1770s, some were initially simply rated as 'brigs', but most were subsequently re-classed as sloops. Confusingly, some small craft (even including some single-masted cutters) sometimes were re-rated as sloops either on upgrading of their ordnance or simply to allow their commanding lieutenant to receive the title (and appropriate pay) of master and commander; thus individual vessels could come and go from the sloop category depending upon who was in charge of them.

Assembled in response to growing tension with the United States during the American Civil War, the North American and West Indies Squadron is seen at Halifax about 1862. The principal vessels are all 91-gun screw battleships: *Edgar*, *Nile* (flagship of Rear-Adm. Sir Alexander Milne), *Hero* and *Agamemnon*, with the steam gunvessel *Nimble* (acting as tender to the *Nile*). Although most British warships would continue to carry some canvas for another decade or so, with all its major fighting vessels now steam-powered, for the Royal Navy this marks the end of the Age of Sail. *(NMM A.7854/N)*

Surveyors (of the Navy)   The chief designers of warships for the British Navy, and members of the Navy Board. From May 1813 the post was held jointly by Henry Peake (appointed 7 June 1806) and two new Surveyors, Joseph Tucker and Robert Seppings. Peake retired in 1822 and Tucker in 1831, leaving Seppings as the sole Surveyor; see under 'Principal Officers of the Navy' in Appendix B.

Tons   During the period covered by this book, a naval vessel's tonnage was the Builder's Measurement ('bm' throughout the book), a formula used for calculating the approximate capacity of the hull. Since the early eighteenth century, it was calculated using the formula (k x b x ½b/94), where 'k' and 'b' are respectively the keel length and the breadth of the ship outside the planking (but inside the wales); this explains why fractions of a ton are quoted in ninety-fourths. Note therefore that these are tons of volume measurement – a crude forerunner of gross tonnage – and not of weight; displacement tonnages (weight) were only a later introduction, although where relevant they are given herein for reference purposes for later steam vessels.

Tumblehome   The inward inclination of a ship's side from its broadest part near the waterline to the narrower width at upper deck level.

Waist   The area of a vessel's upper deck between the break of the forecastle (the rear of the superstructure at the forward end of the vessel) and the break of the quarterdeck) the front of the superstructure at the aft end of the vessel). The trend of naval development was for this area to be gradually encroached on over the years with gangways joining the forecastle to the quarterdeck until the waist ceased to be exposed to the elements.

Wales   These were thick strakes of timber running along the outside of a vessel's hull for structural strength, usually forming an unbroken exterior sweep running from end to end of each of the vessel's sides.

1 First Rates of 104 guns and above

The prestige ships of the British (or any) Navy were the three-deckers, carefully and lavishly built, always in Royal Dockyards and requiring many years work. Rarely commissioned in peace-time, and preserved with expensive overhauls ('refits') and rebuilding, the few First Rates served as flagships for the main fleets in the Channel and Mediterranean, but their great strength made them the strongest units of the line of battle. Nevertheless they were difficult to handle, drew much water, and the pre-war vessels had such low freeboard to their gunports that in anything of a lively sea their lower decks were liable to flooding in battle. Only with the 120-gun *Caledonia* – the first of a numerous class of capital ship – did fine sailing qualities become compatible with superior firepower, and this success led the new controller (Byam Martin) to determine that all future three-deckers should follow this model.

(A) Vessels in service or on order on 1 January 1817

At the close of 1816 there had been nine First Rates in existence – five of 120 guns, two of 112 guns and two of 100 guns; another five were under construction as follows:

In commission	In Ordinary	Building or on order
120 guns	*Hibernia* (1804)	*Britannia*
	Caledonia (1808)	*Prince Regent*
	Nelson (1814)	
	Howe (1815)	
	Saint Vincent (1815)	
112 guns	*Ville de Paris* (1795)	*Princess Charlotte*
	San Josef (1797)	*London*
100 guns		
Queen Charlotte (1810)	*Royal George* (1788)	*Trafalgar*

This situation was altered by the changes in the rating system which came into effect in January 1817, although the 120-gun ships were unaffected. While the *Ville de Paris* and *San Josef* remained as 112-gun ships, the still-building *Princess Charlotte* and *London* (most materials for their frames had been assembled) were altered from 112 guns to 110, the *Royal George* and *Queen Charlotte* from 100 guns to 108, and the *Trafalgar* to 106 guns. Most significantly, all the remaining fourteen Second Rates (see Chapter 2) were now re-classed as First Rates, which thus encompassed all twenty-eight of the Navy's three-deckers – the *Ocean* (1805) was now classed as 110 guns, the *Impregnable* (1810) as 106 guns, and the *Saint George* (1762), *Victory* (1765), *Barfleur* (1768), *Royal Sovereign* (1786), *Glory* (1788), *Prince* (1788), *Prince of Wales* (1794), *Neptune* (1797), *Temeraire* (1798),

Dreadnought (1801), *Boyne* (1810) and *Union* (1811) all as 104 guns. The *Neptune* and *Temeraire* were in need of repairs (the former was BU in 1818, while the latter was reduced to harbour service since 1813).

ROYAL GEORGE Class 104 guns. Originally built as the *Britannia*, this ship was ordered in 1751 to be built strictly to the 1745 Establishment dimensions, but allowing some variations in the design from those specified in the Establishment; on 21 May 1757 this was amended to have her built instead to the draught of the *Royal George*, the Woolwich-built vessel (which foundered 29 August 1782 at Spithead). She had been renamed in 1812 to allow the name *Britannia* to be allotted to a new *Caledonia*-class 120-gun ship.

Dimensions & tons: 178ft 0in, 144ft 6½in x 51ft 10in x 21ft 6in. 2,065⁵⁸⁄₉₄bm.

Men: 850. Guns: (originally) LD 28 x 42pdrs; MD 28 x 24pdrs; UD 28 x 12pdrs; QD 12 x 6pdrs; Fc 4 x 6pdrs. *Britannia*'s 42pdrs were replaced before 1793 by 32pdrs, and her 6pdrs by 12pdrs by AO 3.6.1790; in the 1790s she fitted 12 x 32pdr carronades in place of all 12pdrs except 2 each on QD and Fc.

Saint George Portsmouth Dyd [M/Shipwright Pierson Lock to 12.1755, Edward Allin to 5.1762, completed by Thomas Bucknall]

As built: 178ft 0in, 145ft 2in x 52ft 0½in x 21ft 6in. 2,091²⁶⁄₉₄bm. Draught 13ft 6in/18ft 6in.

Ord: 28.3.1751. K: 1.7.1751. L: 19.10.1762.

First cost: £41,729.7.1d (with extra charges up to 1764, total £45,844.2.8d).

Underwent Small Repair at Portsmouth (for £6,371.11.11d) 4–8.1772. Fitted at Portsmouth for Channel service (for £15,597.16.0d) 6.1778–4.1779.

Commissioned 8.1779 as *Britannia* under Capt. Charles Morice Pole, as flagship of Vice-Adm. George Darby 4.1779 then 6.1779 Rear-Adm. Sir John Lockhart Ross. Coppered at Portsmouth 1.1780; in 9.1780 under Capt. James Bradby, then 4.1782 Capt. Benjamin Hill; paid off into Ordinary 2.1783. Between Middling and Great Repair at Portsmouth (for £35,573) 5.1788–9.1790. Recommissioned 1.1793 under Capt. John Holloway (–1796), as flagship of Vice-Adm. William Hotham; sailed for the Mediterranean 11.5.1793; in action off Genoa 14.3.1795, then off Hyères 13.7.1795. In 1.1796 under Capt. Shuldham Peard, then 5.1796 Capt. Thomas Foley, as flagship of Vice-Adm. Hyde Parker and in 1797 of Vice-Adm. Charles Thompson; at Battle of St Vincent 14.2.1797 (1 wounded). In 3.1797 under Sir Charles Knowles, and ?6.1797 Capt. Edward Marsh; paid off 12.1797 Convalescence ship 1800. Between Small and Middling Repair at Portsmouth (for £21,739) 6.1801–1.1802. Recommissioned 4.1803 under Capt. Lord (William Carnegie, Earl of) Northesk; at blockade of Brest. In 6.1804 under Capt. Charles Bullen, as flagship of the now Rear-Adm. Northesk. In Windward column at Battle of Trafalgar 21.10.1805; had 10 killed, 42 wounded. Laid up in the Hamoaze 1806. Renamed **Princess Royal** 6.1.1812 then **Saint George** 18.1.1812. In Ordinary at Plymouth 1813, and fitted as a prison ship at Plymouth 10–12.1813; recommissioned in that role under Lieut John Cawkit 1814. Fitted as a flagship and receiving ship at Plymouth 3–6.1815; recommissioned 3.1815 under Capt. James Nash, as flagship of Adm. Sir John Duckworth at Plymouth; paid off 12.1815. Renamed *Barfleur* 2.6.1819. BU completed at Plymouth 25.2.1825.

VICTORY Class 104 guns. Design by Thomas Slade, produced 6 June 1759, approved to build 7 July 1759. The classic 100-gun ship, with an increase of 8ft in length over the 1745 Establishment, which enabled Slade's only First Rate to mount extra guns on the LD and UD (and consequently fewer on the QD and Fc).

Dimensions & tons: 186ft 0in, 151ft 3⅜in x 51ft 10in (50ft 6in mld.) x 21ft 6in. 2,162²²⁄₉₄bm.

Men: 850 (later 837; 650 in peacetime) – reduced to 738 while Second Rate. Guns: originally LD 30 x 42pdrs (replaced by 32pdrs from 5.1778 to 4.1779, and again from 28.4.1803); MD 28 x 24pdrs; UD 30 x 12pdrs; QD 10 x 6pdrs; Fc 2 x 6pdrs. All 6pdrs were replaced by 12pdrs in 1782, and these in turn were variously augmented or replaced by assorted carronades during active service. As Second Rate 98-gun from 11.1807 to 2.1817, carried LD 28 x 32pdrs; MD 30 x 18pdrs; UD 30 x 12pdrs; QD 4 x 12pdrs + 8 x 32pdr carronades; Fc 2 x 12pdrs + 2 x 32pdr carronades.

Victory Chatham Dyd [M/Shipwright John Lock to 4.1762, completed by Edward Allin]

As built: Dimensions quoted are always those of design (above), but ship actually measured 2,142bm.

Ord: 13.12.1758. K: 23.7.1759. Named 30.10.1760. L (floated out): 7.5.1765. In Ordinary at Chatham until 1778. Fitted 1765 to 1769. Fitted for sea 2–4.1778.

First cost: £57,748.1.7d to build, plus £5,426.1.5d fitting (to 1769); total £63,174.3.0d. Fitting for sea (1778) £13,296.

Commissioned 12.3.1778 under Sir John Lindsay; sailed from Chatham 13.4.1778. From 5.1778 under Rear-Adm. John Campbell and Capt. Jonathan Faulkner as flagship of Adm. Augustus Keppel, and led fleet at Battle of Ushant 27.7.1778. Paid off at Portsmouth after wartime service 2.1783. Recommissioned 10.1787 under Capt. Charles Hope; paid off at Portsmouth 12.1787. Large Repair there (for £37,523) 12.1787–4.1788, then refitted (for £6,451) to 1789. Recommissioned 5.1790 under Capt. John Knight (–12.1795), as flagship of Alexander Hood, for Spanish Armament; paid off 1.1791. Recommissioned 1.1791 as flagship of Commodore (Rear-Adm. 2.1793) Sir Hyde Parker; paid off 9.1791. Recommissioned 12.1792, still under Knight and as Parker's flagship; sailed 22.5.1793 for the Mediterranean; off Toulon 1793; Corsica operations 1794, then home; sailed for the Mediterranean 23.5.1795; in 7.1795 flagship of Rear-Adm. Robert Man; in action off Hyères 13.7.1795; in 10.1795 flagship of Vice-Adm. Robert Linzee. In 12.1795 under Capt. George Gray (–3.1797), as flagship of Adm. Sir John Jervis. In 4.1797 under Capt. Thomas Sotheby, then 6.1797 Capt. William Cuming; paid off 11.1797. Fitted as a hospital ship at Chatham 12.1797; recommissioned 12.1797 under Lieut John Rickman, as hospital ship at Chatham; paid off 10.1799. Middling Repair (later Large Repair) at Chatham (for £70,933) 2.1800–4.1803; recommissioned 4.1803 under Capt. Samuel Sutton; flagship of Vice-Adm. Lord Horatio Nelson 18–20.5.1803; sailed 25.5.1803 for the Mediterranean; took 40-gun *L'Impatiente* 29.5.1803. In 7.1803 under Capt. Thomas Masterman Hardy (–1.1806), as Nelson's flagship again; chase to West Indies, then led Weather Column at Battle of Trafalgar 21.10.1805, losing 57 dead (including Nelson),

102 wounded; paid off at Chatham 1.1806. Repairs and refit at Chatham (for £9,936) 3–5.1806; recommissioned as a Second Rate 1.1808 under Capt. John Searle; from 3.1808 flagship of Rear-Adm. Sir James Saumarez (–12.1808). From 3.1808 under Capt. Philip Dumaresque (–12.1812); evacuation of Corunna 1809; Saumarez's flagship again 4–12.1809, 3–12.1810, 4–11.1811 and 4–10.1812 (also flagship of Rear-Adm. Sir Joseph Yorke 12.1810–3.1811); paid off 11.1812. Large Repair at Portsmouth (for £79,772) 3.1814–1.1816. In Ordinary at Portsmouth 1813–23; guard ship 6.1823–1.1824; Port Admiral's flagship 1.1824–4.1830; paid off into Ordinary 4.1830.

BARFLEUR Class 104-guns (originally 98-gun Second Rate). Sir Thomas Slade design, approved 1 March 1762, based on ex-First Rate *Royal William*. Of the four ships originally built to this design, the *Princess Royal* had been broken up in 1807, and the *Formidable* in 1813, while the *Prince George* had been re-classed as a sheer hulk (see below).

Dimensions & tons: 177ft 6in, 144ft 0¾in x 50ft 3in x 21ft 0in. 1,934⁸⁷⁄₉₄bm.

Men: 750 (738 from 1794). Guns: LD 28 x 32pdrs; MD 30 x 18pdrs; UD 30 x 12pdrs; QD none (8 x 12pdrs from about 1790); Fc 2 x 9pdrs (12pdrs from about 1790).

Barfleur Chatham Dyd [M/Shipwright Edward Allin to 7.1767, completed by Joseph Harris]

As built: 177ft 8in, 144ft 0½in x 50ft 5in x 21ft 0in. 1,947⁴⁷⁄₉₄bm. Draught 14ft 3in/17ft 1in.

Ord: 19.10.1761. (named 15.11.1761) K: 22.11.1762. L: 30.7.1768. C: 1.1771 (as guard ship).

First cost: £49,222.3.1d.

Commissioned 10.1770 as guard ship at Portsmouth. Refitted as a guard ship there 4.1772 (and similarly in each of next four years); Royal Review at Spithead 22.6.1773. In dockyard hands 1777–79; Small Repair, fitted and coppered there (for £24,397.4.5d) 7.1779–5.1780; recommissioned 2.1780; sailed for West Indies 29.11.1780; sailed for England 26.4.1783 and paid off. Great Repair at Portsmouth (for £31,829.11.8d) 6.1785–10.1786; recommissioned 10.1787; paid off 9.1791. Middling to Great Repair and fitted at Portsmouth (for £51,400) 9.1792–1.1794; recommissioned 12.1793 under Capt. Cuthbert Collingwood, as flagship of Rear-Adm. George Bowyer; at Battle of Glorious First of June off Ushant 1.6.1794, losing 9 killed and 25 wounded. In 8.1794 under Capt. John Elphinston, as flagship of Rear-Adm. Sir George Elphinstone. In 3.1795 under Capt. James Dacres, as flagship of Vice-Adm. William Waldegrave; in Bridport's Action off Île Groix 23.6.1795 (no casualties); to the Mediterranean 12.1795; at Tunis in 1796; at Battle off Cape St Vincent 14.2.1797, losing 7 wounded. In 2.1799 under Capt. John Elphinston again, as flagship of Vice-Adm. Lord Keith (the former George Elphinstone) off Cadiz; to the Mediterranean 5.1799. In 8.1799 under Capt. Peter Puget, as flagship of Rear-Adm. James Whitshed; returned to the Channel. In 1800 under Capt. GeorgeStephens, as flagship (–1802) of the now Rear-Adm. Collingwood; later in 1800 under Capt. John Irwin then 10.1801 Capt. John Ommaney; paid off 5.1802. Fitted at Portsmouth (for £20,491) 1.1803–1.1805; recommissioned 11.1804 under Capt. George Martin; in Calder's Action 23.7.1805. In 10.1805 under Capt. Sir Robert Barlow, then 12.1805 Capt. Philip Durham. In 1806 under Capt. Sir Joseph Yorke (–1807), in the Channel. Recommissioned 1.1808 under Capt. Donald M'Leod, as flagship of Rear-Adm.

William Otway; sailed for Portugal 29.4.1808. In 1.1809 under Capt. Samuel Linzee, as flagship of Rear-Adm. Sir Samuel Hood at Corunna. In 2.1809 under Capt. Sir Thomas Masterman Hardy, as flagship of Vice-Adm. George Berkeley (–1812); sailed for Portugal 25.2.1809; under Cmdr John Cowan 3.1811; home in ?8.1812. In 9.1812 under Capt. Sir Edward Berry; sailed for the Mediterranean 17.11.1812. In 12.1813 under Capt. John Maitland, in the Mediterranean; laid up in Ordinary at Chatham 7.1814. BU at Chatham 9.1819.

Prince George Chatham Dyd [M/Shipwright Edward Allin to 7.1767, completed by Joseph Harris]
> As built: 177ft 6in, 143ft 10⅛ in x 50ft 6½in x 21ft 0in. 1,955 ⁴/₉₄bm. Draught 13ft 10in/17ft 7in.
>
> Ord: 11.6.1766 (and named). K: 18.5.1767. L: 31.8.1772. C: (as a guard ship) 23.1.1777.
>
> First cost: £50,043.4.3d, plus £3,580.19.7d fitting.
>
> Commissioned 11.1776 under Capt. Charles Middleton for Portsmouth & the Nore; fitted and coppered at Portsmouth (for £7,629.7.2d) 4.1780; paid off after wartime service 7.1783. Middling Repair at Chatham (for £30,625.11.2d) 7.1784–7.1785; recommissioned 10.1787 but paid off 12.1787. Middling Repair at Chatham (for £8,706) 2.1794–3.1795; recommissioned 9.1794 under Capt. James Gambier; later under Capt. Sir John Orde, as flagship of Adm. Viscount (Adam) Duncan. In 8.1795 under Capt. William Edge, for the Channel; later under Capt. James Bowen, as flagship of Rear-Adm. Hugh Christian; disabled by gale; paid off 3.1796. Fitted at Portsmouth (for £12,222) 12.1796; recommissioned 10.1796 under Capt. John Irwin, as flagship of Rear-Adm. Sir William Parker; sailed for the Mediterranean 4.1.1797; at Battle off Cape St Vincent 14.2.1797, losing 8 killed and 7 wounded. In 10.1797 under Capt. William Bowen, as flagship of Vice-Adm. Charles Thompson, in the Mediterranean. In 1798 under Irwin again, as flagship of Rear-Adm. Thomas Frederick. In 12.1798 under Capt. Joseph Bingham, as flagship of Parker again; paid off 9.1799. Fitted at Portsmouth (for £21,113) 6.1800; recommissioned 5.1800 under Capt. James Walker, for the Channel; later under Capt. John Rodd, as flagship of Rear-Adm. Sir Charles Cotton; in 1.1801 under Capt. Charles Rowley; paid off into Ordinary at Portsmouth 5.1802. Fitted at Portsmouth (for £13,395) 11.1803; recommissioned 6.1803 under Capt. Richard Curry, then 8.1803 under Capt. Joseph Yorke (–1804), for the Channel. Refitted at Portsmouth (for £11,764) 2–4.1805; recommissioned 1.1805 under Capt. George Losack (–1807), for the Channel; sailed for the West Indies 4.1.1807. In 8.1807 under Capt. Nathaniel Cochrane; paid off into Ordinary at Portsmouth 9.1807. Fitted as a sheer hulk at Portsmouth 11.1816–2.1817. BU there 24.1.1839.

DUKE Class 104 guns (originally 98-gun Second Rate). Sir John Williams's design of 1771 was the last Second Rate design produced before the American War. The 90-gun equivalent of Williams's *Royal Sovereign* (100-gun); these two designs were this Surveyor's only draughts for three-deckers. The prototype was completed early in the American War, but the following two vessels were both delayed until well after its conclusion – only the fourth of the class being expedited. *Duke* and *Glory* (with *Royal Sovereign*) were the first three-deckers to be built at Plymouth. Of the four ships built to this design, the *Saint George* was wrecked off Ringkøbing, Jutland, in a storm on Christmas Eve 1811.
> Dimensions & tons: 177ft 6in, 145ft 3in x 50ft 0in x 21ft 2in. 1,931⁴⁵/₉₄bm.

Men: 750 (738 from 1794). Guns: LD 28 x 32pdrs; MD 30 x 18pdrs; UD 30 x 12pdrs; QD none originally (8 x 6pdrs from 1782, then 8 x 12pdrs from about 1790); Fc 2 x 9pdrs (12pdrs from about 1790).
> As Third Rate, *Atlas* had (1804) LD 28 x 32pdrs; UD 30 x 18pdrs; QD 12 x 12pdrs; Fc 4 x 12pdrs.

Duke Plymouth Dyd [M/Shipwright Israel Pownoll to 2.1775, completed by John Henslow]
> As built: 177ft 6in, 145ft 2in x 50ft 2in x 21ft 9in. 1,943²⁸/₉₄bm. Draught 12ft 8in/18ft 7in.
>
> Ord: 18.6.1771. (Named 8.1771) K: 10.1772. L: 18.10.1777. C: 2.8.1778.
>
> First cost: £44,656.7.2d (+ fitting £12,300.18.5d).
>
> Commissioned 4.1778 under Capt. William Brereton; present at in Battle off Ushant 27.7.1778. Under Capt. Sir Charles Douglas, the ordnance enthusiast, who at his own cost had the *Duke* equipped with flintlocks fitted to all her carriage guns. Coppered and fitted (for £6,615.1.5d) at Portsmouth 3.1780. After other active service in the American War, paid off 6.1783. Small to Middling Repair (for £22,852.12.5d) 11.1784–6.1785. Recommissioned 3.1791 under Capt. Robert Kingsmill, then again in 8.1791 under Capt. Robert Calder, as flagship of Vice-Adm. Robert Roddam, as guard ship at Portsmouth.; from 1792 under Capt. John Knight, as flagship of Vice-Adm. Viscount (Samuel) Hood; from 1793 under Capt. Sir Andrew Snape Hammond, then under Capt. George Duff as flagship of Commodore George Murray; sailed for the Leeward Islands 24.3.1793. Led attack on the batteries at Martinique. Recommissioned 8.1796 under Capt. George Holloway, for Channel service; in 1797 flagship of Rear-Adm. Christopher Parker; involved in Spithead mutiny 5.1797; paid off 4.1798. Fitted as a lazarette at Portsmouth (for £4,696) and moored at Stangate for quarantine service 1798. Hospital ship 9.1799, but remained on Navy List until 5.1803. BU at Portsmouth 8.1843.

Glory Plymouth Dyd [M/Shipwright John Henslow to 11.1784, completed by Thomas Pollard, with fitting by Edward Sison]
> As built: 177ft 5in, 145ft 5in x 50ft 1⅛in x 21ft 2in. 1,944¹⁷/₉₄bm. Draught 13ft 3in/18ft 9in.
>
> Ord: 16.7.1774. (Named 30.8.1774) K: 7.4.1775. L: 5.7.1788. C: (9.1793–) 23.12.1793.
>
> First cost: £57,790.9.5d (+ £6,926 fitting in 1793).
>
> Commissioned 10.1793 under Capt. Francis Pender, for Howe's Fleet; later under Capt. George Duff. From 5.1794 under Capt. John Elphinston, as flagship of Rear-Adm. George Keith Elphinstone. Flagship at Glorious First of June 1.6.1794, after which Elphinstone became Viscount Keith. Under Capt. John Bourmaster from 8.1794, then Capt. Alexander Graeme from 1.1795 for Channel service. Under Cmdr John Eaton (temp.) 6.1795, then Capt. George Grey 9.1795. From 12.1795 under Capt. James Bowen, as flagship of Rear-Adm. Hugh Christian; sailed for West Indies 9.12.1795 into 'Christian's Gales'. From 4.1796 under Capt. Sir George Home, for Channel service. In 1797 under Capt. James Brine; involved in Spithead mutiny 4–5.1797 and also those in 3.1798 and 10.1798. Under Capt. Thomas Wells in 3.1799; sailed for the Mediterranean 1.6.1799 as Cotton's reinforcement. In Channel Fleet 1800–01, with her poop cut down at Plymouth 3–10.1.1801; under Capt. John Draper (acting) 6.1801. Paid off 4.1802. Fitted at Chatham (for £13,560) 2.1804–4.1805. Recommissioned 4.1804 under Capt. George Martin, for Channel service; from 8.1804 (temp.) Capt. William Champain; then 10.1804 Capt. Charles Craven, as

flagship of Vice-Adm. Sir John Orde; off Cadiz in 1805; then under Capt. Frederick Aylmer in 5.1805. From 6.1805 under Capt. Samuel Warren, as flagship of Rear-Adm. Charles Stirling, for Channel service. Participated in Calder's action off Ferrol 22.7.1805. Under Capt. William Otway from 7.1806; sailed for the Mediterranean 4.1.1807. Otway became Rear-Adm. In 10.1807, with his flag in Glory, now under Capt. Donald M'Leod; paid off 27.1.1808. Re-rated as prison ship at Chatham 27.9.1809; under Lieut Richard Simmonds 1810–11, then Lieut Robert Tyte 1812–14; paid off 6.1814 into Ordinary. Powder hulk 1815. BU ordered 1819, but not completed at Chatham until 30.7.1825.

Atlas Chatham Dyd [M/Shipwright Israel Pownoll to 4.1779, completed by Nicholas Phillips]

　As built: 177ft 7in, 145ft 8¾in x 50ft 2in x 21ft 2in. 1,950³⁷⁄₉₄bm. Draught 13ft 10in/18ft 8in.

　Ord: 5.8.1777. K: 1.10.1777. L: 13.2.1782. C: 30.3.1782.

　First cost: £50,350.7.4d (fitted).

　Commissioned 2.1782 under Capt. George Vandeput, for Howe's fleet in 7–8.1782. Sailed 11.9.1782 to assist in relief of Gibraltar; arrived about 11.10.1782, encounter with the combined fleets 20.10.1782; by 1783 under Capt. John Elphinston; paid off 3.1783. Fitted for Ordinary at Plymouth 4.1783; underwent Very Small Repair at Plymouth (for £12,611.9.5d) 9.1784–2.1785. Recommissioned 10.1787 under Capt. William Swiney; paid off 10.1787. Recommissioned 3.1795 under Capt. Edmund Dod, for Channel service; under Capt. Matthew Squire 3.1797, then Capt. Shuldham Peard 1.1799 and Capt. Theophilus Jones 4.1799 (to 1801). Underwent between Small and Middling Repair, cut down to a 74-gun Third Rate and fitted at Chatham (for £24,870) 11.1802–5.1804. Recommissioned 3.1804 under Capt. William Johnstone Hope, still for Channel service; from 1805 under Capt. William Browne; engaged in Battle of St Domingo 6.2.1806. Recommissioned 3.1806 under Capt. Samuel Pym; sailed for the Mediterranean 3.11.1806. Under Capt. James Sanders 11.1807, for operations on the Spanish coast; flagship of Rear-Adm. John Child Purvis 4.1808. Paid off into Ordinary at Plymouth 12.1810. Fitted for temporary prison ship at Portsmouth 12.1813–1.1814; fitted for a powder magazine at Portsmouth 10.1814–1.1815. BU at Portsmouth 5.1821.

ROYAL SOVEREIGN Class 104 guns. Design by John Williams, approved 21 February 1772, reverting to just 28 ports on the LD (compared with 30 in *Victory*).

　Dimensions & tons: 186ft 0in, 152ft 6in x 52ft 0in x 22ft 3in. 2,193³⁸⁄₉₄bm.

　Men: 850 (later 837); 650 in peacetime. Guns: originally LD 28 x 42pdrs (replaced by 32pdrs from 2.3.1793); MD 28 x 24pdrs; UD 30 x 12pdrs; QD 10 x 12pdrs; Fc 4 x 12pdrs. All 6pdrs were replaced by 12pdrs from 2.3.1793. The QD/Fc guns were variously augmented or replaced by assorted carronades during active service.

Royal Sovereign Plymouth Dyd [M/Shipwright Israel Pownoll to 2.1775, then John Henslow to 11.1784, completed by Thomas Pollard]

　As built: 183ft 10¼in, 150ft 9⅛in x 52ft 1in x 22ft 2½in. 2,175²⁹⁄₉₄bm. Draught 13ft 3in/18ft 8in.

　Ord: 3.2.1772. K: 1.1774. L: 11.9.1786. C: end 1787?

　First cost: £61,254.14.7d to build, plus £6,203.14.9d fitting; total £67,458.9.4d.

　Commissioned 10.1787 under Capt. James Samber; paid off 12.1787. Recommissioned 5.1790 under Capt. Richard Fisher (acting) for

Spanish Armament; sailed from Plymouth 24.8.1790; flagship of Vice-Adm. Lord Hood 9–11.1790. Recommissioned 2.1793 under Capt. Henry Nicholls as flagship of Vice-Adm. Thomas Graves, in Howe's fleet; fitted for sea (for £22,181) 3–6.1793; participated in 'Glorious First of June' Battle on 1.6.1794. In 4.1795 under Capt. John Whitby, as flagship of Cornwallis; in 'Cornwallis's Retreat' 16–17.6.1795. In 1796 under Capt. William Bedford, as flagship of Vice-Adm. Sir Alan Gardner, in the Channel; sailed for the Mediterranean 1.6.1799. In 1800 under Capt. Richard Raggett, as flagship of Vice-Adm. Sir Henry Harvey. Recommissioned 4.1803 under Capt. Richard Curry (temp). In 1.1804 under Capt. Pulteney Malcolm; sailed for the Mediterranean 2.1804. In 6.1804 under Capt. John Stuart, as flagship of Rear-Adm. Sir Richard Bickerton, off Toulon; escort to General Craig's expedition to the Mediterranean 5.1805. Later in 1805 under Capt. Edward Rotheram, as flagship of Vice-Adm. Cuthbert Collingwood; led Lee column at Battle of Trafalgar 11.10.1805, losing 47 killed and 94 wounded. Underwent between Middling and Large Repair, and fitted, at Plymouth (for £63,600) 5.1806–1.1807; recommissioned 11.1806 under Capt. Henry Garrett, as flagship of Vice-Adm. Edward Thornbrough; sailed for the Mediterranean 13.2.1807, From 6.1808 under Capt. David Colby (still Thornbrough's flag), then 4.1810 under Capt. Joseph Spear, as flagship of Rear-Adm. Francis Pickmore. In 8.1811 under Capt. John Harvey, then Capt. Robert Plampin in 12.1811. In 1812 under Capt. William Bedford, as flagship of Adm. Lord (George) Keith in the Channel (–1803); from 9.1812 under Capt. James Bissett. In 1813 under Capt. Robert Stuart Lambert, as flagship of Vice-Adm. Sir William Sidney Smith; in the Mediterranean 1814; later under Capt. Charles Thurlow Smith. In 1815 under Capt. Edward Brenton, as flagship of Rear-Adm. Sir Benjamin Hallowell; from 5.1815 under Capt. William Broughton. Laid up in Ordinary at Plymouth 7.1815. Renamed *Captain* 17.5.1825 and fitted as a receiving ship at Plymouth 6–8.1825. BU at Plymouth (for £763) 8.1841.

Revived LONDON Class 104 guns (originally 98-gun Second Rates). Of four new Second Rates begun during the American War, three were to a revival of Sir Thomas Slade design of 1759; at a time when the sole Surveyor (Williams) was increasingly incapacitated by age and ill health, it is unsurprising that Slade's well-tested designs should be revived, and no coincidence that in March 1778 a new Surveyor, Edward Hunt, was appointed to share the increased workload. The main difference from the original *London* was that the new ships were to be completed with 8 x 6pdrs added on the QD (so making them 98-gun instead of 90-gun ships); this increase was also extended to the *London*, as well as to most of the other Second Rates built since 1755. The *Impregnable* was wrecked on the Chichester Shoals off Dunnose on 18 October 1799.

　Dimensions & tons: 177ft 6in, 146ft 6in x 49ft 0in x 21ft 0in. 1,870⁹³⁄₉₄bm.

　Men: 750 (later 738). Guns: LD 28 x 32pdrs; MD 30 x 18pdrs; UD 30 x 12pdrs; QD 8 x 6pdrs (12pdrs from about 1790); Fc 2 x 9pdrs (12pdrs from about 1790).

　As Third Rate *Windsor Castle* had (1814) LD 26 x 32pdrs + 2 x 68pdrs (Millers); UD 28 x 24pdrs + 2 x 68pdrs; QD 4 x 24pdrs + 10 x 32pdr carronades; Fc 2 x 24pdrs +2 x 32pdr carronades.

Prince Woolwich Dyd [M/Shipwright John Jenner to 12.1782 [died], then Henry Peake to 12.1785, Martin Ware to 3.1787, completed by John Nelson]

　As lengthened: 194ft 6in, 163ft 6in x 49ft 0in x 21ft 0in. 2,088bm.

Draught 13ft 5in/17ft 8in.

Ord: 9.12.1779. K: 1.1.1782. L: 4.7.1788. Completed fitting 25.7.1788.

First cost: £55,041.9.11d (including fitting).

Commissioned 7.1790 under Capt. Josiah Rogers for Spanish Armament, as flagship of Sir John Jervis. Recommissioned 2.1793 under Capt. Cuthbert Collingwood, as flagship of Rear-Adm. George Bowyer in Howe's fleet. Recommissioned 10.1794; under Capt. Francis Parry from 1.1795; at Battle off Isle de Groix 23.1.1795; under Capt. Charles Powell Hamilton from 8.1795; paid off 6.1796.

Lengthened at Portsmouth by AO 2.6.1796 (with 17ft new section inserted), to 'vast improvement of her qualities as a man-of-war' (for £31,267) 5–11.1796.

Recommissioned 12.1796 under Capt. Thomas Larcom, as flagship of Rear-Adm. Sir Roger Curtis in the North Sea; sailed for the Mediterranean 2.6.1798; at blockade of Cadiz 1798; flagship of Rear-Adm. Sir Charles Cotton 1799, in the Channel. Under Capt. Samuel Sutton from 4.1799; sailed for the Mediterranean 1.6.1799; by 1800 in the Channel again, under Capt. James Walker, then Capt. William Carnegie, the Earl of Northesk, with Capt. John Loring temporarily 11–12.1801. Very Small Repair at Plymouth (for £8,668) 5–6.1802. Fitted at Plymouth (for £17,110) 4.1803–1.1804; recommissioned 4.1803 under Capt. Richard Grindall; at blockade of Brest, then in Lee column at Trafalgar 21.10.1805 (no casualties). Under Capt. William Lechmere 2.1806, in the Mediterranean, then returned to Plymouth. Laid up at Plymouth 10.1806. Under Capt. Peter Puget 12.1806, then Capt. Alexander Fraser 2.1807. In Ordinary at Plymouth 1808–13. Fitted at Plymouth as a guard ship to lie at Spithead (for £20,861) 8–10.1813. Recommissioned 8.1813 under Capt. George Fowke, as flagship of Adm. Sir Richard Bickerton; from 5.1815 under Capt. Edmund Roger, as flagship of Adm. Sir Edward Thornbrough; paid off ?9.1815. Fitted at Portsmouth as a victualling vessel and as accommodation for officers (dockyard officials) 4–6.1816. BU at Portsmouth 11.1837.

Windsor Castle Deptford Dyd [M/Shipwright Adam Hayes to 3.1785 (died), then Henry Peake to 3.1787, completed by Martin Ware]

As built: 177ft 6in, 145ft 87⁄8in x 49ft 2in x 21ft 0in. 1,87390⁄94bm. Draught 14ft 0in/18ft 2in.

Ord: 19.8.1782. (Named 21.1.1783) K: 19.8.1784. L: 31.5.1790. C: 6.6.1790.

First cost: £51,198.8.6d (including fitting).

Commissioned 7.1790 under Capt. Sir James Barclay, for Spanish Armament, as flagship of Rear-Adm. Herbert Sawyer; then paid off. Recommissioned 12.1792 under Capt. Sir Thomas Byard, as flagship of Vice-Adm. Phillips Cosby; sailed for the Mediterranean 22.4.1793. In 1794 was flagship of Rear-Adm. Robert Linzee, under (successively) Capts Edward Cooke (4.1794), William Shield (10.1794) and John Gore (11.1794) – the last following the mutiny on 11.1794. Participated in actions off Genoa 13.3.1795 and off Hyères 13.7.1795. From 12.1795 under Capt. Edward O'Bryen, as flagship of Rear-Adm. Robert Man, detached in pursuit of (Fr.) Rear-Adm. De Richery. Very Small Repair at Plymouth (£29,069) 7.1798–8.1799. Recommissioned 7.1799 under Capt. John Manley, then Capt. John Chambers White (9.1799) and Capt. Albemarle Bertie (11.1799). Recommissioned 5.1800 under Capt. James Oughton, as flagship of Vice-Adm. Sir Andrew Mitchell in the Channel; Capt. Peter Bover in temporary command 9.1800. Recommissioned 4.1803 under Capt. Philip Durham; in 6.1803 under Capt. Albemarle

Bertie, as flagship of Adm. George Montagu 1803–04. Under Capts Thomas Wells (1.1804), Samuel Osborn (4.1804), Davidge Gould (5.1804). Under Capt. Charles Boyles from 5.1805 (to 1808); participated in Battle of Finisterre 22.7.1805, Hood's action off Rochefort 24.9.1806, and Dardanelles operation in 2.1807. Laid up at Plymouth 9.1808 and paid off 11.1808 into Ordinary. Cut down to 74-gun Third Rate at Plymouth (for £47,725) 10.1813–6.1814. Fitted for sea at Plymouth (for £10,497) 12.1818–1.1819, but did not sail; used as guard ship instead, under Capt. Thomas Caulfield. Recommissioned 6.1821 (when Caulfield died) under Capt. Charles Dashwood. Paid off 1.1822 and recommissioned, still under Dashwood, for Lisbon service. Recommissioned 11.1824 under Capt. Hugh Downman. Small Repair and fitted for guard ship at Plymouth (for £27,787) 3.1825–12.1826. Recommissioned 5.1825 under Capt. Edward Durnford King. Recommissioned 5.1828 as 76-gun ship under Capt. Dunscombe Bouverie, for Mediterranean service; paid off 6.1831. Fitted as a Divisional Ship 8.1833. Depot ship at Deal 1834–38. BU at Pembroke Dock 5.1839.

ROYAL GEORGE Class 108 guns. Design by Edward Hunt, 1782, extending the length of the First Rate by a further 4ft, with the layout of guns as in *Victory*. The survivor, *Royal George*, was re-rated as 108 guns in January 1817. Her original sister ship, *Queen Charlotte*, was burnt off Livorno (Leghorn) by accident on 17 March 1800 (about 690 died), and a replacement of the same name was built in 1805–10 (see below).

Dimensions & tons: 190ft 0in, 156ft 5in x 52ft 4in x 22ft 4in. 2,27862⁄94bm.

Men: 850 (later 837); 650 in peacetime. Guns: LD 30 x 32pdrs; MD 28 x 24pdrs; UD 30 x 12pdrs; QD 10 x 12pdrs; Fc 2 x 12pdrs. The initial design included 42pdrs on the LD, and 6pdrs on the QD/Fc, but this was never carried. The QD/Fc guns were variously augmented or replaced by assorted carronades during active service.

Royal George Chatham Dyd [M/Shipwright Nicholas Phillips to 7.1790, completed by John Nelson]

As built: 190ft 0in, 156ft 2⅜in x 52ft 5½in x 22ft 4in. 2,28634⁄94bm. Draught 14ft 9in/19ft 5in.

Ord: 25.3.1782 as *Umpire* (renamed *Royal George* 11.9.1783). K: 6.1784. L: 16.9.1788. Sailed 14.4.1790 for Plymouth, where fitted for Channel service 5–7.1790.

First cost: £51,799.5.7d to build, plus £6,503.8.5d fitting at Chatham and £10,089.12.0d at Plymouth; total £68,392.6.0d.

Commissioned 5.1790 under Capt. Thomas Pringle.

Recommissioned 2.1793 under Capt. William Domett as flagship of Vice-Adm. Sir Alexander Hood, in Howe's fleet; at Battle of Glorious First of June off Ushant 1.6.1794, losing 20 killed and 72 wounded. In 8.1794 Hood became Adm. Viscount Bridport; in Bridport's Action off Île Groix 23.6.1795, losing 7 wounded. In 1796 flagship of Adm. Earl (Richard) Howe, still under Capt Domett; mutiny at Spithead 1797. In 1797 again flagship of Bridport, later flagship of Rear-Adm. Charles Pole; in attack on Spanish squadron in Basque Roads 2.7.1799. In 9.1800 under Capt. Robert Otway, as flagship of Adm. Sir Hyde Parker. In 1801 under Capt. John Child Purvis, for the Channel; paid off 4.1802. Middling Repair and fitted at Plymouth 7.1805–7.1806; recommissioned 6.1806 under Cmdr Charles Gill, for the Channel. Later in 1806 under Capt. Richard Dunn (–1808), as flagship of Vice-Adm. Sir John Duckworth; off Cadiz 1.1807. Fitted at Plymouth 1–4.1811; recommissioned 1.1811 under

Capt. John Clavell, as flagship of Rear-Adm. Sir Thomas Williams; sailed for the Mediterranean 17.11.1811. In 1812 under Capt. Andrew King, as flagship of Rear-Adm. Sir Francis Pickmore; later under Capt. Thomas Mainwaring. In 2.1813 under Capt. William Cuming; laid up at Plymouth 7.1814. BU at Plymouth 2.1822.

(1783) BOYNE Class When Sandwich was replaced as First Lord, the new Administration began the task of building new three-deckers in response to a similar effort by the French Navy. Both Keppel and Howe, who alternated the post of First Lord for some years, recognised the need to enlarge domestic designs, as the French ships were significantly larger. Edward Hunt, now effectively the sole Surveyor, produced a new design in 1783 which added 4½ft to the length of the Second Rate. Two ships were built to this design; the *Boyne* (launched 1790 at Woolwich) was burnt by accident at Spithead in 1795; the *Prince of Wales* (launched 1794 at Portsmouth as a replacement for the cancelled 74-gun ship *Bulwark*) was laid up at Portsmouth in July 1814, and completed BU at Portsmouth on 26 December 1822.

VILLE DE PARIS 112 guns. Designed by John Henslow. Originally to have been of 100 guns and to the same draught as the *Royal George*, but was altered to carry 110 guns by AO of 5 August 1788; re-rated January 1817 as 112 guns. The surprising choice of name for Britain's largest warship to be launched before 1800 is explained by her perpetuating the name of the prize taken from the French in 1782 (and lost the same year).

Dimensions & tons: 190ft 0in, 156ft 1⅛in x 53ft 0in x 22ft 4in. 2,332 24/94bm.

Men: 837. Guns: LD 30 x 32pdrs; MD 30 x 24pdrs; UD 32 x 18pdrs; QD 14 x 12pdrs; Fc 4 x 12pdrs; RH 6 x 18pdr carronades.

Ville de Paris Chatham Dyd [M/Shipwright Nicholas Phillips to 7.1790, then John Nelson to 3.1793 (died), Thomas Pollard to 6.1795, completed by Edward Sison]

As built: 190ft 2½in, 156ft 1½in x 53ft 2½in x 22ft 2½in. 2,351 22/94bm. Draught 14ft 8in/18ft 4in.

Ord: 17.1.1788. (named 26.9.1788) K: 1.7.1789. L: 7.7.1795. C: 17.9.1796.

First cost: £78,830 (including fitting).

Commissioned 10.1796 under Capt. Walter Lock; sailed for the Mediterranean 18.3.1797. In 4.1797 under Capt. Sir Robert Calder; from 6.1797 flagship of Adm. Earl St Vincent (John Jervis), in the Mediterranean to 1799. Later under Capt. George Grey, for return to England. In 3.1799 under Cmdr Walter Bathurst; to Mediterranean, but returned to England 8.1799. In 4.1800 under Capt. Sir Thomas Troubridge, as flagship of St Vincent again in the Channel. In 1801 under Capt. John Sutton, as flagship of Adm. William Cornwallis. Very Small Repair and fitted at Plymouth (for £4,031) 5–7.1802; recommissioned 4.1803 under Capt. Tristram Ricketts. Later under Capt. William Domett; took 6-gun privateer *Le Messagre* 16.8.1803. In 2.1804 under Capt. Thomas Gosselin, then under Vice-Adm. Charles Nugent (with Capt. John Whitby as 2nd Capt.), as flagship of Cornwallis again. Later under Capt. William Champain, in the Channel, as flagship of Nugent in 1805. In 2.1806 under Capt. George Aldham, then 7.1806 Capt. Henry Garrett. Middling Repair at Plymouth (for £39,289) 8.1806–7.1807; recommissioned 5.1807 under Capt. Alan Hyde Gardner, as flagship of Adm. Lord (Alan) Gardner. In 1808 under Capt. Sir Harry Neale, as flagship of Adm. Lord (James) Gambier. In 1.1809 under Capt. John Carden, at Corunna. Sailed for the Mediterranean 22.2.1809. In 4.1809 under Capt. Richard Thomas, as flagship of Vice-Adm. Lord (Cuthbert) Collingwood (died 7.3.1810). In 1811 under Capt. George Hony, as flagship of Rear-Adm. Sir Thomas Fremantle. In 1812 under Capt. George Burlton, still in the Mediterranean. Defects made good at Portsmouth (for £19,718) 12.1812–2.1813. In 1813 under Capt. Charles Jones, as flagship of the now Rear-Adm. Sir Harry Neale, in the Channel. Laid up at Plymouth 7.1814. In 1815 under Capt. Robert Jackson, as flagship of Adm. Viscount (George) Keith; paid off 8/9.1815. Re-rated at 112 guns by 2.1817. Fitted at Plymouth as a lazarette 8.1825, to lie at Milford. BU at Pembroke 5.6.1845.

***DREADNOUGHT* Class** 104 guns (originally 98-gun Second Rates). Sir John Henslow design, approved 20 March 1788. All three ships built to this design had fought at Trafalgar, and then were re-classed as 12pdr class Second Rates in 1808, and as 104-gun First Rates in January 1817. The proposed fourth ship *Ocean* (ordered in 1790 and laid down in 1792 at Woolwich) was re-ordered to a new design in 1797 (see below).

Dimensions & tons: 185ft 0in, 152ft 6⅝in x 51ft 0in (50ft 3in moulded) x 21ft 0in. 2,110 53/94bm.

Men: 738. Guns: LD 28 x 32pdrs; MD 30 x 18pdrs; UD 30 x 12pdrs; QD 8 x 6pdrs; Fc 2 x 6pdrs. By 1792 the 6pdrs had been replaced by 12pdrs.

First cost: *Dreadnought* £60,484.

Dreadnought Portsmouth Dyd [M/Shipwright George White to 3.1793, Edward Tippett to 10.1799, completed by Henry Peake]

As built: 184ft 11in, 152ft 2¾in x 51ft 2½in x 21ft 6in. 2,123 32/94bm. Draught 14ft 7in/17ft 6in.

Ord: 17.1.1788. K: 7.1788. (named 23.10.1788) L: 13.6.1801. C (sailed): 9.8.1801.

First cost: £60,484 including fitting.

Commissioned 6.1801 under Capt. James Vashon. In 4.1803 under Capt. James Bowen, then 5.1803 Capt. Edward Brace, 7.1803 Capt. William Domett, and 9.1803 Capt. John Purvis, as flagship of Adm. William Cornwallis, for the blockade of Brest. In 5.1804 under Capt. George Reynolds, then 1.1805 under Capt. Edward Rotheram as flagship of Vice-Adm. Cuthbert Collingwood, for the Channel Fleet; off Cadiz later in 1805. In 10.1805 under Capt. John Conn; in Lee column at Battle of Trafalgar; lost 7 killed, 26 wounded. Fitted at Portsmouth 9.1806–1.1807; recommissioned 12.1806 under Capt. William Lechmere, for the Channel. In 1808 under Capt. George Salt, as flagship of Rear-Adm. Thomas Sotheby. In 1810 under Capt. Valentine Collard, still Sotheby's flagship. In 8.1810 under Capt. Samuel Linzee; to Baltic 1811; paid off at Portsmouth 12.1811. Large Repair at Portsmouth (for £72,511) 8.1812–3.1814, then to Ordinary. Fitted at Portsmouth as a lazarette 9.1825, to lie at Pembroke. Fitted at Sheerness as a hospital ship 3–5.1831, then at Woolwich to 6.1831. To Greenwich as the seamen's hospital from 6.1831. BU at Woolwich 24.2–31.3.1857 (stores returned valued at £3,389).

Neptune Deptford Dyd [M/Shipwright Martin Ware until 6.1795, completed by Thomas Pollard]

As built: 184ft 9½in, 152ft 2¼in x 51ft 2in x 21ft 5½in. 2,119 29/94bm. Draught 14ft 8in/18ft 1in.

Ord: 15.2.1790. (named 24.7.1790) K: 4.1791. L: 28.1.1797. C: 12.2.1797.

First cost: £77,053 including fitting.

Commissioned 4.1797 under Capt. Henry Stanhope, for Channel service; in 8.1797 under Commodore Sir Erasmus Gower. In 3.1799 under Capt. James Vashon, then 4.1801 Capt. Edward Brace; flagship of Vice-Adm. James Gambier 1801–02. In 9.1801 under Capt. Francis Austen, then 12.1802 Capt. William O'Bryen Drury, 5.1804 Capt. Sir Thomas Williams and 6.1805 Capt. Thomas Fremantle; in Weather column at Trafalgar 21.10.1805, losing 10 killed and 34 wounded; paid off 12.1806. Fitted at Portsmouth (for £29,053) 3–11.1807; recommissioned 8.1807 under Capt. Sir Thomas Williams for the Channel station. From 1808 under Capt. Charles Dilkes, as flagship of Rear-Adm. Alexander Cochrane; sailed for the Leeward Islands 23.9.1808; took part in capture of Martinique 2.1809, and in the action with Troude's Squadron (capture of *Le d'Hautpoult*) off the Saintes 14.4.1809. Under Capt. James Wood, sailed for Leeward Islands again 14.1.1810. Fitted for Ordinary at Portsmouth 12.1810. Fitted as a temporary prison ship at Plymouth 11–12.1813, under Lieut George Lawrence. BU at Plymouth 10.1818.

Temeraire Chatham Dyd [M/Shipwright Thomas Pollard to 6.1795, completed by Edward Sison]

As built: 185ft 0in, 152ft 3$^{3}/_{8}$in x 51ft 2in x 21ft 6in. 2,120$^{58}/_{94}$bm. Draught 14ft 8in/18ft 0in.

Ord: 9.12.1790. K: 7.1793. L: 11.9.1798. C: 18.5.1799.

First cost: £73,241 including fitting.

Commissioned 3.1799 under Capt. Peter Puget. From 8.1799 under Capt. Thomas Eyles, as flagship of Rear-Adm. Sir John Borlase Warren for the Channel station. From 11.1799 under Capt. Edward Marsh, as flagship of Rear-Adm. James Whitshed for the Channel station. From 1801 under Capt. Thomas Eyles again, as flagship of Rear-Adm. George Campbell; involved in Bantry Mutiny 12.1801, then sailed for Jamaica. Fitted at Plymouth (for £16,898) 5.1803–2.1804; recommissioned 10.1803 under Capt. Eliab Harvey (–1806) for the Channel station; from 8.1804 under Capt. William H. Kelly (temp.); off Cadiz in 1805, then in Weather column at Trafalgar 21.10.1805, losing 47 killed and 76 wounded. From 12.1805 under Capt. John Larmour; paid off 1806. Fitted at Portsmouth (for £25,352) 6.1806–9.1807; recommissioned 3.1807 under Capt. Sir Charles Hamilton for the Channel station. From 1809 under Capt. Edward Sneyd Clay, as flagship of Rear-Adm. Manley Dixon in the Baltic. From 1810 under Capt. Edwin Chamberlayne; sailed for the Mediterranean 17.2.1810. From ?9.1810 under Capt. George Hony. From 3.1811 under Capt. Joseph Spear, as flagship of Rear-Adm. Francis Pickmore for the blockade of Toulon; from ?3.1812 under Capt. Samuel Hood Linzee. Fitted at Plymouth as a prison ship 11–12.1813; in Ordinary there under Lieut John Wharton from 1814. Fitted at Plymouth (for £27,733) as a receiving ship 9.1819–6.1820, to lie at Sheerness. Victualling depot 1829. Receiving ship and depot at Sheerness 8.1836. Sold to J. Beatson (for £5,530) to BU 16.8.1838 (Turner's famous painting in the National Gallery of the ship being towed to her last berth – the breaker's yard at Rotherhithe – by the tug *Monarch* dates from this time).

***OCEAN* Class** 110 guns (originally 98-gun Second Rate). Sir John Henslow design approved 21 June 1797; the *Ocean* had been originally ordered in 1790 to be of *Dreadnought* Class – but now the design lengthened to increase her capacity (to mount 18pdr on her UD instead of 12pdrs) and she was re-ordered in 1797.

Dimensions & tons: 196ft 0in, 164ft 0$^{5}/_{8}$in x 51ft 0in (50ft 3in moulded) x 21ft 6in. 2,269$^{63}/_{94}$bm.

Men: 738. Guns: LD 30 x 32pdrs; MD 32 x 18pdrs; UD 32 x 18pdrs; QD 2 x 12pdrs + 10 x 32pdr carronades; Fc 2 x 12pdrs + 2 x 32pdr carronades; RH 6 x 18pdr carronades.

Ocean Woolwich Dyd [M/Shipwright John Tovery until 7.1801, completed by Edward Sison]

As built: 196ft 6$^{1}/_{2}$in, 164ft 0$^{1}/_{8}$in x 51ft 1in x 21ft 6in. 2,276$^{49}/_{94}$bm. Draught 14ft 2in/17ft 0in.

Ord: 9.12.1790. K: 1.10.1792. Re-ordered 4.5.1797 to new design. K: 1.10.1797. L: 24.10.1805. Completed fitting 10.12.1805.

First cost: £90,076 including fitting.

Commissioned 11.1805 under Capt. Francis Pender; and sailed 1806 for the Mediterranean. In 1807 under Capt. Richard Thomas, as flagship of Collingwood in Mediterranean; paid off 7.1809. Small repair and fitted at Plymouth (for £49,622) 1.1811–4.1812; recommissioned 1.1812 under Capt. Robert Plampin, and sailed 17.4.1812 for the Mediterranean; paid off 7.1814 into Ordinary at Plymouth. Re-classed as 110-gun First Rate on 1.1.1817. Between Small and Middling Repair, and cut down (by AO 4.10.1819) to 80-gun Third Rate two-decker (for £48,753) 9.1819–7.1821. Fitted at Plymouth as a guard ship (for £11,032) 3.1824–7.1824. Fitted at Plymouth as a lazarette for Sheerness 8.1830–7.1831. Fitted as an 80-gun flagship at Sheerness in 1832. Fitted for the Captain of the Ordinary at Sheerness 9.1837–1.1838. Fitted at Chatham as coal hulk 11–12.1852, to lie at Chatham and later at Sheerness. BU completed at Chatham 11.12.1875.

***HIBERNIA* Class** 120 guns (originally 110 guns). Designed by John Henslow. Commenced building to design of *Ville de Paris*, lengthened by 11ft 2in during construction, thus providing an extra pair of gunports on each deck. Re-classed as 120 guns in January 1817, but reduced to 104 guns in 1845. This ship and all subsequent First Rates were re-armed with an all-32pdr armament under the new Establishment of Guns introduced 5 February 1839.

Dimensions & tons: 201ft 2in, 167ft 3$^{1}/_{8}$in x 53ft 0in x 22ft 4in. 2,499$^{11}/_{94}$bm.

Men: 850. Guns: LD 32 x 32pdrs; MD 32 x 24pdrs; UD 34 x 18pdrs; QD 12 x 32pdr carronades; Fc 2 x 18pdrs + 4 x 32pdr carronades; RH 6 x 18pdr carronades.

As re-armed 1845 (104 guns): LD 24 x 32pdrs (56cwt) + 4 x 68pdrs (65cwt); MD 28 x 32pdrs (48cwt) + 2 x 68pdrs (65cwt); UD 30 x 32pdrs (33cwt); QD 10 x 32pdr carronades (17cwt); Fc 6 x 32pdrs (45cwt).

Hibernia Plymouth Dyd [M/Shipwright Thomas Pollard to 4.1793, then Edward Sison to 6.1795, John Marshall to 12.1801, completed by Joseph Tucker]

As built: 201ft 2in, 167ft 4$^{1}/_{8}$in x 53ft 1in x 22ft 4in. 2,508$^{22}/_{94}$bm. Draught 16ft 4in/18ft 7in.

As re-measured by 1827: 202ft 9in, 167ft 9in x 53ft 3in x 22ft 0in. 2,530$^{4}/_{94}$bm.

Ord: 9.12.1790. K: 11.1792. L: 17.11.1804. C: 7.3.1805.

First cost: £71,139 to build, plus £17,661 fitting.

Commissioned 11.1804 under Capt. Edward Thornbrough, as flagship of Adm. Lord (Alan) Gardner, in command of the Channel Fleet from 3.4.1805. Later in 1805 under Capt. William Bedford, as flagship of Rear-Adm. John Leigh Douglas. In ?2.1806 under Capt. Edward Osborn, as flagship of Adm. Earl St Vincent (John Jervis). In 1807 under Capt. John Conn, as flagship of Gardner again. Later in 11.1807 under Capt. Charles Schomberg, as flagship of Rear-Adm. Sir William Sidney Smith, off the coast of Portugal. In 1809 under Capt. Lawrence Halsted,

The *Hibernia* was the oldest three-decker to see much post-war service as a sailing ship, but this was made possible by a massive reconstruction between 1819 and 1825 that allowed the ship to be re-armed to the contemporary standard of all-32pdrs, plus half a dozen shell guns. This is Lieut Humphrey John Julian's watercolour portrait of the ship shortening sail sometime in the 1840s. Julian was appointed on 1 March 1845 as First Lieutenant to Capt. Peter Richards in the *Hibernia*, by now reduced to 104 guns; he retained this post until 14 May 1847, when he was promoted to command the *Resistance*. (*NMM PW5999*)

as flagship of Adm. Sir Charles Cotton; later under Capt. Robert Neve, and flagship of Rear-Adm. Robert Stopford, in the Mediterranean. Refitted at Portsmouth (for £18,425) 7–9.1810; recommissioned under Capt. John Chambers White, and sailed for the Mediterranean 1.11.1810, as flagship of Rear-Adm. Sir Samuel Hood; in 4.1811 under Capt. Edward Kittoe, as flagship of Vice-Adm. Sir Richard Keats; in 1812 under Capt. Charles Smith, as flagship of Sidney Smith again, then 12.1813 under Capt. Thomas Caulfield; paid off into Ordinary at Portsmouth 10.1815. Re-rated as 120 guns in 1.1817. Large Repair at Portsmouth (for £74,302) 12.1819–10.1825. Re-armed with 100 x 32pdrs, 14 x 32pdr carronades and 6 x 68pdr shell guns about 1840–see note at end of section (C). Fitted as a flagship at Portsmouth (for £29,272) 5–7.1845; re-rated at 104 guns, and recommissioned 27.2.1845 under Capt. Peter Richards, as flagship of Vice-Adm. William Parker, for the Mediterranean; on 5.8.1847 under Cmdr Edward Codd, then 3.8.1848 under Capt. Charles Wise, still Parker's flagship in the Mediterranean; sailed for home 25.4.1849 and paid off 12.6.1849 at Plymouth. In Ordinary at Devonport 1849–55. Fitted as a receiving ship 10.1855, and sailed to Malta. Recommissioned there 13.12.1856 under Capt. Frederick Warden, as flagship of the Admiral Superintendent until 1902, first Rear-Adm. (Vice-Adm. 25.6.1858) Sir Montagu Stopford, then 27.7.1858 Rear-Adm. Henry John Codrington. On 1.4.1862 under Cmdr. Robert Beazley Harvey, still Codrington's flag, then 6.4.1863 Rear-Adm. Horation Thomas Austin, then 26.11.1864 Rear-Adm. Henry Kellett. In 5.1865 under Cmdr George Lowcay Norcock, still Kellett's flag, then 25.5.1868 flag of Rear-Adm. Edward Gennys Fanshawe. In 6.1868 under Cmdr William Luke Partridge, then 4.1870 under Cmdr. Edward Downes Panter Downes, still Fanshawe's flag, then 6.6.1870 flag of Rear-Adm. Astley Cooper Key; in 4.1872 under Cmdr. Richard Roche, still Key's flag, then 8.8.1872 flag of Rear-Adm. Sir Edward Augustus Inglefield; in 8.1875 under Cmdr. Guy Ouchterlony Twiss, still Inglefield's flag, then 22.12.1875 flag of Rear-Adm. Edward Bridges Rice;

30.5.1876 flag of William Garnham Luard. Recommissioned 1.5.1877 (further commanding officers and flag officers to 1902 will not be listed here). Sold at Malta by public auction (for £1,010) 14.10.1902.

Only one new First Rate was ordered during the early years of the Revolutionary War. The design of the *Hibernia* was stretched again, this time to 205ft, and provided 34 ports on the MD. Earl St Vincent, who became First Lord in 1801, favoured large numbers of the Common Class 74s rather than more three-deckers, and in any case believed that the latter had reached their optimum size with the 190ft of the *Ville de Paris*; nevertheless, his administration did sanction a replacement for the *Queen Charlotte* after the first vessel of that name was lost by accident in 1800.

CALEDONIA **Class** 120 guns. When ordered she was envisaged as a 100-gun ship, but not begun until 1805, and in 1806 was altered to a 120-gun design by Sir William Rule. An enlarged version of *Hibernia*, this ship on completion proved to be the most successful three-decker yet built, and was often hailed as 'faultless'. The design, which became the basis of all post-war three-deckers, was modified during construction to provide for flatter sheer but improved structural strength and sea-keeping qualities. With a foot greater freeboard than the *Queen Charlotte* (whose LD ports were only 4½ft above the waterline), she could fight all her guns in much rougher weather.

Dimensions & tons: 205ft 0in, 170ft 11in x 53ft 6in (54ft 5in oa.) x 23ft 2in. 2,602¹⁴/₉₄bm.

Men: 875. Guns: LD 32 x 32pdrs; MD 34 x 24pdrs; UD 34 x 18pdrs; QD 6 x 12pdrs + 10 x 32pdr carronades; Fc 2 x 12pdrs + 2 x 32pdr carronades; RH 6 x 18pdr carronades. In 1815 the UD 18pdrs were replaced by 24pdr Congreve guns, and in 1831 her LD 32pdrs of 64cwt were replaced by 32pdrs of 55cwt.

As re-armed 1840 (120 guns): LD 30 x 32pdrs (56cwt) + 2 x 68pdrs (60cwt); MD 32 x 32pdrs (56cwt) + 2 x 68pdrs (60cwt); UD 34 x 32pdrs (40cwt); QD 12 x 32pdr carronades (25cwt); Fc 4 x 32pdrs (25cwt).

Caledonia   Plymouth Dyd [M/Shipwright Joseph Tucker]
> As built: 205ft 0in, 170ft 9⅛in x 53ft 8in (54ft 7in oa.) x 23ft 2in. 2,616¼₉₄bm. Draught 15ft 10in/18ft 2in.
>
> Ord: 6.11.1794. (named 19.1.1796) K: 1.1805. L: 25.6.1808. C: 23.9.1808.
>
> First cost: £96,381 to build, plus £6,711 fitting.
>
> Commissioned 8.1808 under Capt. William Bedford (–1810). Flagship of Adm. Lord (James) Gambier 2–4.1809; at the Basque Roads 4.1809. Flagship of Commodore Sir Harry Neale later in 1809; in 1810 flagship of Rear-Adm. Francis Pickmore; sailed for the Mediterranean 20.2.1810 (Cadiz). From 10.1810 under Capt. Francis Austen. In 5.1811 under (temp.) Capt. Peter Heywood, in the Channel fleet; sailed for the Mediterranean again 14.6.1811, as flagship of Adm. Sir Edward Pellew. By 7.1811 under Austen again, as flagship of the now Rear-Adm. Neale, in Gambier's fleet. In 1.1813 under Capt. Jeremiah Coghlan, as flagship of Rear-Adm. Israel Pellew, in Sir Edward Pellew's fleet. In 4.1814 under Capt. E. J. Graham, then in 6.1814 Capt. Edward Sibly , still flagship of Sir Edward Pellew (now Lord Exmouth); paid off 19.8.1814 at Plymouth. Paid off 9.1814. Fitted for sea at Plymouth (for £24,205) 4–5.1815. In 5.1815 under Capt. James Brisbane, then in 7.1815 Capt. (acting) Sir Archibald Dickson. Paid off 6/7.1815 and laid up in Ordinary at Plymouth to 1830. Between Middling and Large Repair at Plymouth (for £113,839) and fitted for guard ship (for £9,511) 5.1826–5.1831. Recommissioned 1.5.1830 under Capt. Richard Curry, as flagship at Plymouth. In 1831 under Capt. Edward Curzon, as an experimental ship, then 7.11.1831 under Capt. James Hillyer, for Lisbon; paid off 5.1833, then recommissioned 17.5.1833 under Capt. Thomas Brown, as flagship of Sir Josiah Rowley, in the Mediterranean; on 31.10.1835 under Capt. George Bohun Martin, still Rowley's flag; paid off 6.9.1837 at Plymouth. Fitted as a demonstration ship at Plymouth (for £8,909) 2–6.1838. Fitted as a flagship at Plymouth (for £6,129) 11.1840. Recommissioned 27.10.1841 under Capt. Henry Eden, as flag of Adm. Sir Graham Moore; on 21.4.1842 under Capt. Alexander Milne, as flagship of Adm. Sir David Milne (his father). On 22.5.1845 under Capt. Manley Hall Dixon, then 28.1.1848, both at Plymouth; then to the Mediterranean, as flagship of Vice-Adm. Sir William Parker. Fitted as a hospital ship at Plymouth 5.1855–6.1856 ; left Plymouth 17.6.1856 under tow by Gorgon and Geyser. Renamed ***Dreadnought*** by AO 21.6.1856 (replacing former *Dreadnought* as hospital ship at Greenwich) and fitted for this at Woolwich (for £3,041) 7.1856–1.1857. Returned to Admiralty 14.7.1870, and lent to Metropolitan Asylums 6.5.1871. BU (by AO 10.11.1874) completed at Chatham 20.3.1875.

Ex-SPANISH PRIZE (1797) 112 guns. Designed by Francisco Gautier, Spain laid down ten ships of this 112-gun class between 1779 and 1794 – at Ferrol (*Purisma Concepcion*, *San Jose*, *Santa Ana*, *Salvador del Mundo* and *Reina Luisa*) and Havana (*Conde de Regla*, *Mejicano*, *Real Carlos*, *San Hermenegildo* and *Principe de Asturias*). Of the two captured off Cape St Vincent in 1797, the *San Jose* – taken by Nelson personally leading a boarding party across the already captured 80-gun *San Nicolas* ('Nelson's Patent Bridge') – was named *San Josef* in the RN; her near-sister the 112-gun *Salvador del Mundo*, built 1787 at Ferrol, was BU at Plymouth in February 1815.

San Josef (Spanish *San Jose*, built 1783 at Ferrol), 114 guns.
> Dimensions & tons: 194ft 3in, 156ft 11¼in x 54ft 3in (53ft 4in mld.) x 24ft 3½in. 2,456²⁴₉₄bm. Draught 14ft 3in/21ft 3in.

> Men: 839. Guns: LD 32 x 32pdrs; MD 32 x 24pdrs; UD 32 x 12pdrs; QD 12 x 9pdrs; Fc 6 x 9pdrs.
>
> Taken 14.2.1797 by Captain of Sir John Jervis's squadron off Cape St Vincent.
>
> From 3.1797 under Capt. Charles Lindsey, then 4.1797 under Capt. Charles Stuart. Arrived at Plymouth 5.10.1797. Registered by AO 4.12.1797. Fitted at Plymouth 6.1799–1.1801.
>
> Commissioned 12.1800 under Capt. Thomas Masterman Hardy, for the Channel; from 2.1801 under Capt. George Eyre. In 2.1802 under Capt. James Carpenter; paid off 4.1802. Recommissioned 4.1803 under Capt. Peter Spicer, then 6.1803 under Capt. John Tremayne Rodd, as flagship of Vice-Adm. Sir Charles Cotton (–4.1805), for blockade of Brest. In 1.1804 under Capt. John Dodd, then 9.1804 Capt. Tristram Roberts Ricketts, still as Cotton's flagship. Made good defects at Plymouth (for £10,259) to 8.1805; in 5.1806 under Capt. Robert Jenner Neve, then 1.1807 under Capt. John Conn; paid off 6.1807. Very Large Repair and fitted at Plymouth (for £89,308) 5.1807–6.1809; recommissioned 5.1809 under Capt. Richard Dalling Dunn, as flagship of Vice-Adm. Sir John Thomas Duckworth; in Walcheron expedition 1809. In 5.1810 under Neve again, as flagship of Adm. Sir Charles Cotton (who died 23.2.1812) for the Mediterranean. In 2.1811 under Capt. Abel Ferris, then 5.1811 under Capt. John Bowker and 7.1811 under Capt. George McKinley for return to UK. From 3.1812 under Capt. Robert Jackson, as flagship of Adm. Lord Keith 2–4.1812, with the Channel fleet. In 1.1813 under Capt. Henry Bouchier, as flagship of Rear-Adm. Edward James Foote (–4.1813). In 4.1813 under Capt. William Stewart, as flagship of Rear-Adm. Sir Richard King, for the Mediterranean fleet; in action against French fleet off Toulon 5.11.1813, with 2 wounded; to England 7.1814, then under Capt. Edward Barnard, and in 4.1815 Capt. Jeffrey Raigersfield; paid off at Plymouth into Ordinary 7.1815. Fitted for sea at Plymouth 4–8.1815, then laid up there. Fitted for Ordinary at Plymouth (for £9,796) 3–8.1822. Fitted as a flagship at Plymouth 12.1831–1.1832. Recommissioned 26.11.1831 under Capt. Richard Curry, as flagship of Sir Manley Dixon, C-in-C at Plymouth. On 1.5.1833 under Capt. Gordon Thomas Falcon, as flagship of Adm. Sir William Hargood; paid off 30.4.1836. Fitted for the Captain of the Ordinary at Plymouth 8.1836–2.1837; recommissioned 14.1.1837 under Capt. John Handcock, as gunnery training ship at Plymouth; on 4.3.1838 under Capt. Joseph Needham Tayler, as guard ship at Plymouth; on 6.8.1841 under Capt. Frederick William Burgoyne, as flagship of the Admiral Superintendent there, Rear-Adm. Sir Samuel Pym; paid off 9.1844. Recommissioned 26.11.1847 under Capt. Henry Leeke, then 28.4.1848 under Capt. Sir Thomas Maitland, as flagship of Adm. Sir William Hall Gage; finally paid off 1848. BU at Plymouth 5.1849.

IMPREGNABLE Class 106 guns (originally 98-gun Second Rate). Design by Sir William Rule, 1798.
> Dimensions & tons: 196ft 0in, 163ft 7⅜in x 51ft 0in (50ft 3in moulded) x 22ft 0in. 2,263⁵⁸₉₄bm.
>
> As rebuilt 1826: 196ft 8in, 162ft 10¼in x x 52ft 8½in x 21ft 9½in. 2,406⁵⁴₉₄bm.
>
> Men: 738. Guns: LD 28 x 32pdrs; MD 30 x 18pdrs; UD 30 x 18pdrs; QD 14 x 32pdr carronades: Fc 2 x 12pdrs + 2 x 32pdr carronades.

Impregnable (ex-*Europe*, renamed 7.3.1800) Chatham Dyd [M/ Shipwright David Polhill to 3.1803, completed by Robert Seppings]

As built: 196ft 0in, 163ft 7in x 51ft 2in x 22ft 2in. 2,278bm. Draught 14ft 3in/17ft 4in.

Ord: 13.1.1798. K: 23.2.1802. L: 1.8.1810. Completed fitting (for Ordinary) 24.10.1810.

First cost: £65,025 including fitting.

Commissioned 7.1811 under Capt. James Wilkes Maurice. Fitted for sea at Portsmouth (for £19,032) 1–4.1812; in 1812 under Capt. George MacKenzie, for service in the Channel and North Sea; flagship of Adm. William Young 1813; in 7.1813 under Capt. John Loring, then 5.1814 Capt. Charles Adam, 10.1814 Capt. Robert Hall, 3.1815 Capt. John Campbell Rowley, and 12.1815 Capt. James Pack. Recommissioned 6.1816 under Capt. Edward Brace, as flagship of Rear-Adm. David Milne; sailed 7.7.1816 for expedition to Algiers; took part in Bombardment of Algiers 27.8.1816, losing 50 killed and 160 wounded (and received 233 shot into her hull!). In 10.1816 under Capt. James Nash, as flagship of Adm. Sir John Duckworth Re-classed as 106-gun First Rate in 2.1817. In 9.1817 under Capt. Pownall Pellew, then 6.1821 Capt. Alexander Skene; paid off into Ordinary 1823. Rebuilt with circular stern at Plymouth 2.1825–4.1826. Harbour flagship at Plymouth 1839. Training ship 1862. Renamed **Kent** on 9.11.1888, then **Caledonia** on 22.2.1891. Sold to J. B. Garnham to BU 19.7.1906.

(New) BOYNE Class 104 guns (originally 98-gun Second Rates). In line with St Vincent's opinion that the *Victory* was 'a fine model for ships of 98 guns', two ships were built to Sir Thomas Slade's 1759 design for *Victory*. Both ships were intended to be 100-gun (First Rates), but were amended to 98-gun in 1806. In the 1820s both ships were intended to be raséed into 80-gun two-deckers, but this was rescinded.

Dimensions & tons: (orig.) 186ft 0in, 153ft 0¾in x 51ft 3in x 21ft 6in. 2,138⁴²⁄₉₄bm.

(as amended 1806) 186ft 0in, 151ft 3⅛in x 51ft 10in x 21ft 6in. 2,162²²⁄₉₄bm.

Men: 738. Guns: LD 28 x 32pdrs; MD 30 x 18pdrs; UD 30 x 12pdrs; QD 4 x 12pdrs + 8 x 32pdr carronades; Fc 2 x 12pdrs + 2 x 32pdr carronades.

As 76-gun ships they were intended to have 650 men and: LD 26 x 32pdr (63cwt) + 2 x 68pdr carronades; UD 28 x 32pdr (54cwt) + 2 x 68pdr carronades; QD 2 x 18pdrs + 12 x 32pdr carronades; Fc 2 x 18pdrs + 2 x 32pdr carronades. (*Boyne* in 1840 to carry 32pdrs vice 18pdrs)

First cost: *Boyne* £93,162 (fitted for sea).

Boyne Portsmouth Dyd [M/Shipwright Nicholas Diddams]

As built: 186ft 3in, 153ft 2⅞in x 51ft 5in x 22ft 1in. 2,154⁸¹⁄₉₄bm. Draught 14ft 6in/18ft 1in.

Ord: 25.6.1801 (as 100-gun). K: 4.1806. L: 3.7.1810. Completed fitting 2.4.1811.

First cost: £78,429 including fitting (later amended to £93,162).

Commissioned 1.1811 under Capt. Henry Hume Spence, as flagship of Rear-Adm. Sir Harry Burrard Neale; from 2.1811 under Capt. John M. Hanchett, and from 11.1811 under Capt. Charles Jones. From 3.1813 under Capt. George Burlton; in action against French squadron off Toulon 2.2.1814, losing 2 killed and 40 wounded; at Genoa 4.1814. Fitted as a flagship at Portsmouth (for £9,239) 8–12.1814. Recommissioned 11.1814 under Capt. Frederick L. Maitland for Mediterranean service; under Capt. James Brisbane from 3.1815, and under Capt. Air Archibald C. Dickson 5.1815. In 1816 under Capt. Edmund Boger, as flagship of Lord Exmouth. Fitted for Channel service and as a guard

ship at Portsmouth (for £2,428) 7–9.1816. Re-classed as 104-gun First Rate in 2.1817. Underwent between a Middling and a Large Repair at Portsmouth (for £44,615) 7.1818–11.1819. Cut down to a 76-gun Third Rate (two-decker) by AO 10.7.1826. Recommissioned 13.4.1832 under Capt. Thomas Hastings; fitted as a gunnery training ship at Portsmouth (for £3,340) 12.1833–3.1834; renamed **Excellent** 1.12.1834. Renamed **Queen Charlotte** 22.11.1859, and paid off 31.12.1859. BU at Portsmouth 25.6.1861.

Union Plymouth Dyd [M/Shipwright Joseph Tucker]

As built: 186ft 0⅝in, 153ft 1in x 51ft 4½in x 22ft 0in. 2,149¹⁷⁄₉₄bm. Draught 14ft 7in/17ft 8in.

Ord: 16.9.1801 (as 100-gun). K: 10.1805. L: 16.11.1811. Completed fitting 12.5.1812.

First cost: £85,601 including fitting.

Commissioned 4.1812 under Capt. Samuel Hood Linzee; sailed 19.5.1812 for the Mediterranean. From 8.1812 under Capt. William Kent (died 29.8.1812), then under Capt. Robert Rolles. Paid off at Plymouth into Ordinary 7.1814. Re-classed as 104-gun First Rate in 2.1817. Defects made good, and put into good condition in Ordinary at Plymouth (for £5,126) 11.1819–1.1820. Proposed to be cut down to a 80-gun Second Rate (two-decker) by AO 28.4.1827, amended to 76-gun Third Rate by AO 3.12.1832; but this was never completed, and order cancelled 1.1.1833. BU at Plymouth 3.1833.

QUEEN CHARLOTTE 108 guns. A late addition to the original *Royal George* Class, built to replace the former *Queen Charlotte* burnt in 1800 (see above), but with 30 guns (vice 28) on the MD, and another extra pair (12pdrs) on the QD. Re-rated (like *Royal George*) from 100 to 108 guns on 1 January 1817; by 1831 carried LD 26 x 32pdrs + 2 x 68pdrs; MD 28 x 32pdrs + 2 x 68pdrs; UD 30 x 32pdrs; QD 12 x 32pdrs + 2 x 12pdrs; Fc 2 x 32pdrs + 2 x 68pdrs.

Queen Charlotte Deptford Dyd [M/Shipwright Henry Peake to 6.1806, completed by Robert Nelson]

As built: 190ft 0½in, 156ft 2¾in x 52ft 5¾in x 22ft 4in. 2,288⁶⁹⁄₉₄bm. Draught 16ft 3in/19ft 9in.

As repaired 1831: 190ft 7in, 156ft 7¾in x 53ft 7in (52ft 8in for tonnage) x 22ft 1¾in. 2,311¹⁶⁄₉₃bm.

Ord: 9.7.1801. K: 10.1805. L: 17.5.1810. C: 18.10.1810 (for Ordinary) at Chatham.

First cost: £88,254 to build, plus £2,731 fitting.

Middling Repair at Plymouth (for £41,219) 4.1812–4.1813.

Commissioned 1.1813 under Capt. Pulteney Malcolm, as flagship of Adm. Lord Keith for Channel service; in 1.1815 under Capt. Robert Jackson (then possibly Capt. Thomas Eyles?). Fitted at Chatham (for £23,211) 8.1814–7.1815. Under Capt. Charles Inglis from 5.1815. In 10.1815 under Capt. Edmund Boger, as flagship of Sir Edward Thornbrough at Portsmouth. Defects made good and made a guard ship at Portsmouth (for £13,641) 4–7.1816; recommissioned 7.1816 under Capt. James Brisbane, as flagship of Adm. Lord Exmouth for expedition to Algiers; took part in Bombardment of Algiers 27.8.1816, losing 8 killed and 131 wounded; subsequently under Capt. William Kinghorne in 9.1816, then again under Boger as Thornbrough's flagship in 10.1816. On 15.5.1818 under Capt. Thomas Briggs, as flagship of Adm. Sir George Campbell at Portsmouth. In 2.1821 under Capt. John Baker Hay, as flagship of Adm. Sir James Whitshed at Portsmouth. Made a guard ship at Portsmouth (for £13,033) 7–10.1822. Very Large Repair at Portsmouth (for £80,718) 10.1825–1.1831, then into Ordinary. On 28.8.1845 under Capt. Henry Ducie Chads. To Sheerness to replace *Waterloo* 9.1858.

William Joy's watercolour of the *St Vincent* at Spithead during the late 1840s, with the 80-gun *Vanguard* and also (at centre) one of the royal yachts, possibly the first *Victoria & Albert*. The sailing qualities of *St Vincent* and her sisters of the *Nelson* Class did not enjoy a good reputation, and their post-war active service was limited, although this ship survived in a static training role until 1906, so was clearly well built. (*NMM PW6090*)

Commissioned 11.2.1859 as flagship at Sheerness; back to Portsmouth 11.1859. Renamed *Excellent* by AO 22.11.1859; on 31.12.1859 under Capt. Richard Strode Hewlett. Fitted as a gunnery training ship at Portsmouth 8.1860. On 4.7.1863 under Capt. Astley Cooper Key, then 3.9.1866 under Capt. Arthur William Acland Hood, 23.8.1869 under Capt. Henry Boys, 21.5.1874 under Capt. Thomas Brandreth, 9.1.1877 under Capt. Frederick Anstruther Herbert, 4.3.1880 under Capt. John Ommanney Hopkins, 21.6.1881 under Capt. William Codrington, 4.1883 under Capt. John Arbuthnot Fisher and 5.1884 under Capt. ?James Inglis. Recommissioned 7.1884 under Inglis, then 8.1884 under Fisher again; in 11.1886 under Capt. Compton Edward Domville, then 6.1890 under Capt. Hugo Lewis Pearson. Sold to J. Read, Jnr. 12.1.1892 to BU.

Apart from the replacement *Queen Charlotte*, no three-deckers were ordered for the decade after the commencement of the *Caledonia*, as construction concentrated on a sufficient quantity of the smaller battleships. But following Trafalgar it became clear that the war would be protracted – long enough for France to replace the gaps in her battlefleet, and three new First Rates were ordered. In August 1806 the Admiralty ordered the Surveyors to prepare competitive new draughts for these First Rates (and also for Second and Third Rates). It was later agreed that the Surveyors should co-operate on producing an agreed common design for each of the three Rates. While the new *Nelson* Class ships were being laid down, the *Caledonia* came into service in late 1808 and proved such a superb vessel that it was decided that future three-deckers should be built to her design, and two new First Rates were ordered to this at the start of 1812. As with the First Rates, development of Second Rate three-deckers in this period fell into two phases: firstly there was an inferior design produced in the immediate aftermath of the Battle of Trafalgar, a 'Surveyor's design' equivalent to the First Rates of the *Nelson* Class, only a single vessel being ordered to this design; then in 1812 a design was produced which reverted to well-tried schemes, and to this two Second Rates were ordered.

NELSON Class 120 guns. Designed by the Surveyors together, these were the three-decker equivalents of the *Armada* Class 74s. The design, finally approved on 1 October 1806, was modified from that of the *Caledonia*, but they proved much poorer sailers, being 'very crank'. The *Howe* was the first new warship to be built employing Seppings's new system of diagonal bracing. Although they had round bows and heavy bulwarks to protect gun crews, this design still retained the traditional square stern with quarter galleries. The *Nelson* was the only vessel launched by the end of the Napoleonic Wars, but saw no sea service; her two sisters were both complete on the stocks by May 1814, but both were left to stand on the slip to season and neither was finished for service until post-war.

Dimensions & tons: 205ft 0in, 170ft 10⅛in x 53ft 6in x 24ft 0in. 2,601⁴⁄₉₄bm.

Men: 875. Guns: LD 32 x 32pdrs; MD 34 x 24pdrs; UD 34 x 18pdrs; QD 6 x 12pdrs + 10 x 32pdr carronades; Fc 2 x 12pdrs + 2 x 32pdr carronades; RH 6 x18pdr carronades.

By 1840 had 30 x 32pdrs + 2 x 68pdr shell guns on LD, 32 x 32pdrs + 2 x 68pdr shell guns on MD (and the same on the UD), 4 x 32pdrs + 12 x 32pdr carronades on the QD, and 2 x 32pdrs + 2 carronades on the Fc.

Nelson Woolwich Dyd [M/Shipwright Edward Sison]

As built: 205ft 0¼in, 170ft 10in x 53ft 8in x 24ft 0in. 2,617¹¹⁄₉₄bm. Draught 14ft 11in/18ft 5in.

Ord: 23.11.1805. K: 12.1809. L: 4.7.1814. C: 17.8.1814 for Ordinary. First cost: £123,469 (completed for Ordinary).

Not commissioned for sea as a sailing vessel, except briefly under Capt. Thomas Burton in 1814, for fitting at Woolwich and voyage to Portsmouth. Underwent Very Large Repair, housed over fore and aft, and breadth increased by doubling while in Ordinary at Portsmouth (for £86,512) 10.1825–9.1828. Doubling removed 9.1837. Advanced ship at Portsmouth (for £15,267) 4–6.1846. Converted to screw propulsion at Portsmouth, re-armed as 90 guns 10.3.1859–7.2.1860 (see Chapter 2).

Howe Chatham Dyd [M/Shipwright Robert Seppings to 3.1813,

completed by George Parkin]
>   As built: 205ft 0in, 170ft 7in x 53ft 8¾in x 24ft 0in. 2,619³⁶⁄₉₄bm.
>   Draught 14ft 7in/18ft 5in.
>   Ord: 15.1.1806. K: 6.1808. L: 28.3.1815 (completed for Ordinary).
>   First cost: £98,105.
>   Defects made good at Sheerness (for £10,520) 9.1823–4.1824. Partial
>   Repair at Chatham (for £5,142) 10.1832–7.1833.
>   Commissioned 27.8.1835 under Capt. Alexander Ellice, and fitted
>   for flagship of Vice-Adm. Charles Elphinstone Fleeming, at
>   Sheerness (for £10,109) 7.1835–7.1836; as flagship of at the Nore
>   until 9.1840; on 26.6.1839 under Capt. Lord Clarence Edward
>   Paget, at Sheerness. Fitted for sea at Sheerness (for £5,603)
>   8–10.1840, then commissioned under Capt. Sir Watkin Owen
>   Pell, as flagship in the Mediterranean; on 21.8.1841 under Capt.
>   Robert Smart, as flagship of Rear-Adm. Sir Francis Mason, 2nd-
>   in-C Mediterranean; in 3.1843 under Capt. Thomas Forrest;
>   paid off 27.7.1843 at Sheerness. 'Advanced' (work on) the ship at
>   Sheerness (for £4,749) 10.1843–1.1844. In Ordinary at Sheerness
>   7.1843–4.1847. Fitted for sea at Sheerness (for £16,044) 3–5.1847.
>   Recommissioned 28.4.1847 under Capt. Sir James Stirling. Fitted
>   'to convey the Queen Dowager to Madeira' (for £3,042) in
>   10.1847; in Mediterranean 6.1848–7.1850, then in Ordinary at
>   Sheerness to 1853. BU completed at Sheerness 23.2.1854.

Saint Vincent Plymouth Dyd [M/Shipwright Joseph Tucker to 5.1813,
completed by Thomas Roberts]
>   As built: 204ft 11in, 170ft 6¼in x 53ft 8in x 24ft 0in. 2,612³¹⁄₉₄bm.
>   Draught 14ft 4in/18ft 9½in.
>   Ord: 15.1.1806. K: 5.1810. L: 11.3.1815 (completed for Ordinary).
>   First cost: £110,549.
>   Underwent Very Small Repair at Plymouth (for £14,256) 6–10.1823.
>   Fitted as a guard ship there (£25,818) 9.1829–5.1830.
>   Commissioned 7.9.1829 under Capt. Edward Hawker, as
>   flagship of Lord Northesk at Portsmouth; paid off 30.4.1830.
>   Recommissioned 1.5.1830 under Capt. Hyde Parker, as flagship
>   of Sir Thomas Foley, still at Portsmouth. Recommissioned
>   25.2.1831 under Capt. Humphrey Senhouse, as flagship of Vice-
>   Adm. Sir Henry Hotham, commanding the Mediterranean Fleet;
>   paid off 1834. Fitted for commission (for £3,252) 5–7.1834. Small
>   Repair and fitted as a demonstration flagship at Portsmouth
>   (for £41,016) 7.1839–8.1843. Very Small Repair and advanced at
>   Portsmouth (for £17,528) 4.1849–12.1851. Fitted as a guard ship
>   at Portsmouth (for £8,576) 3–4.1854; recommissioned 28.4.1854
>   under Capt. George Augustus Eliott, as flagship of Rear-Adm.
>   William Fanshaw Martin, and guard ship at Portsmouth (in
>   Ordinary); on 3.6.1854 under Capt. George Mansel (died
>   9.1854), as transport for French troops to the Baltic for the
>   Russian War. Recommissioned 1.4.1859 under Capt. Thomas
>   Wilson, as Reserve depot ship at Portsmouth; training ship there
>   by 6.1862. On 26.4.1866 under Cmdr Richard Carter. Sold to
>   Castle (for £5,350) 15.5.1906; arrived Falmouth to BU 23.6.1906.

TRAFALGAR **Class** 106 guns (originally 98-gun Second Rate).
Nominally 'Surveyors' (joint)' but actually Sir William Rule design,
approved 11 June 1807, modified from that for *Impregnable*. Originally intended to be built without a poop, but one was later added to
her design to make her suitable to be a flagship; also designed with
51ft breadth – increased by 1½ft during 1813 for better stability. Only
a single vessel was built, a recognition that the three-decker 98 was
already obsolete. Re-rated as a 106-gun First Rate on 1 January 1817.
>   Dimensions & tons: 196ft 0in, 162ft 9in x 52ft 6in x 22ft 8in.
>   2,386¹⁄₉₄bm.

Men: 738. Guns: LD 28 x 32pdrs; MD 30 x 18pdrs; UD 30 x 12pdrs;
>   QD 6 x 12pdrs + 8 x 32pdr carronades; Fc 2 x 12pdrs + 2 x 32pdr
>   carronades; RH 6 x 18pdr carronades.

Trafalgar Chatham Dyd [M/Shipwright George Parkin]
>   As built: 195ft 11in, 162ft 11in x 52ft 8in x 22ft 8in. 2,403⁶⁵⁄₉₄bm.
>   Draught 13ft 10in/17ft 4in.
>   Ord: 15.1.1806. K: 5.1813. L: 26.7.1820. Completed fitting (for
>   Ordinary) 7.9.1820, then sent to Sheerness where her waist and
>   forecastle were roofed over.
>   First cost: £85,082 including fitting.
>   Renamed *Camperdown* 22.2.1825. Underwent partial repair at
>   Chatham 5–12.1833, when roofing was removed, then replaced
>   over waist only, then fitted as a demonstration ship (for £6,485).
>   Roofing removed again 1.1834, then waist again roofed over
>   1.1837. Fitted as a flagship at Sheerness (for £9,156) 10.1840.
>   Commissioned 10.12.1841 under Capt. Francis Brace, as flag of
>   (his uncle) Vice-Adm. Sir Edward Brace at the Nore; paid off
>   12.1843 (following the death of the Vice-Adm. on 26.12.1843)
>   into Ordinary at Portsmouth. Fitted as a receiving ship 1855.
>   Completed fitting as a coal depot at Portsmouth (for £8,374)
>   2.1861 (–1901). Renamed *Pitt* 29.8.1882; sold to Castle and Sons,
>   Millbank (by AO 1.7.1904, for £4,025) on 15.5.1906.

On 6 January 1812 four new three-deckers were ordered, although
only one of them was laid down before the end of the war. The
Britannia and *Prince Regent* were to be First Rates of 120 guns, built
to the design of the *Caledonia*. The *Princess Charlotte* and *London*
were to be Second Rates of 98 guns, built to the lines of the *Boyne* (and
originally of *Victory*); like other Second Rate three-deckers, they were
re-classed as First Rates in 1817.

CALEDONIA **Class (1812 Orders)** 120-gun ships, continuation of
Sir William Rule's 1794 design. This highly successful design became
the post-war standard for three-deckers. In 1846–47 the *Prince Regent*
(which was crank) was raséed, turning her into a 92-gun Second Rate,
and effectively made her a unit of the *Rodney* Class (see Chapter 2).
>   Dimensions & tons: 205ft 0in, 170ft 11in x 53ft 6in x 23ft 2in.
>   2,602¹⁴⁄₉₄bm.
>   Men: 900 (820 peacetime). Guns: LD 32 x 32pdrs (56cwt); MD 34
>   x 24pdrs (49cwt); UD 34 x 24pdrs Congreve (instead of 18pdrs
>   first planned); QD 10 x 32pdr (17cwt) carronades + 6 x 12pdrs;
>   Fc 2 x 32pdr (17cwt) carronades + 2 x 12pdrs.

Britannia Plymouth Dyd [M/Shipwright Thomas Roberts to 9.1815,
completed by Edward Churchill]
>   As built: 205ft 0in, 170ft 9¼in x 53ft 8in (54ft 7in oa) x 23ft 2in.
>   2,616¹⁵⁄₉₄bm.
>   Draught 16ft 0in/18ft 7in.
>   Ord: 6.1.1812. K: 12.1813. L: 20.10.1820. C: 20.12.1820 (for
>   Ordinary).
>   First cost: £103,768 (£111,630 fitted).
>   Fitted as a guard ship (for £13,943) 12.1822.
>   Commissioned 4.10.1823 under Capt. Henry William Bruce, as
>   flagship of Adm. Sir Alexander Inglis Cochrane, C-in-C at
>   Plymouth; paid off 3.4.1824. Recommissioned 4.4.1824 under
>   Capt. Philip Pipon, as flag of Adm. Sir James Saumarez, C-in-C
>   at Plymouth. On 30.4.1827 under Capt. Edward Hawker, as flag
>   of Adm. Lord (William Carnegie, Earl of) Northesk, C-in-C at
>   Plymouth. Fitted for sea at Plymouth (for £26,172, including
>   earlier making good of defects) 9.1829. Recommissioned 10.9.1829
>   under Capt. George Burdett, as flag of Vice-Adm. Sir Pulteney
>   Malcolm, C-in-C in the Mediterranean. On 30.4.1830 under Capt.

This lithograph of *Britannia* leaving the Hamoaze (Plymouth) on 27 July 1828 celebrates an unusual, and contentious, event. The ship is flying the Royal Standard from the main truck, signifying the presence of board of HRH the Duke of Clarence (later King William IV) in his capacity as Lord High Admiral. Since the early eighteenth century the Lord High Admiral's function had been carried out by a committee, the Board of Admiralty, and the government of the day had intended the resurrection of the title to be largely honorific, but Clarence, who had served in the Navy when young, threatened to take active command of the fleet and his powers had to be reined in. The function of the cruise, which was the first seagoing commission for *Britannia*, was to try out the ship's sailing qualities, a test she passed with great credit. *(NMM X0120)*

William James Hope Johnstone, still bearing Malcolm's flag in the Mediterranean. On 27.10.1831 under Capt. Peter Rainier (–1835); at Portsmouth 1832; at Lisbon 1833, then to the Mediterranean to early 1835. Fitted as a flagship at Portsmouth (for £10,661) 2.1836. On 5.4.1836 under Capt. James Whitley Deans Dundas, as flag of Adm. Sir Philip Charles Calderwood Henderson Durham, at Portsmouth. On 2.4.1838 under Capt. Henry Dundas, as Portsmouth. On 10.11.1838 under Capt. Henry Dundas Trotter, then 19.4.1839 under Capt. Alexander Ellice and 23.11.1839 under Capt. John William Montagu, as flag of Adm. Edward Codrington, as C-in-C Portsmouth. Fitted for sea at Portsmouth (for 3,907), as flag of John Drake 1.10.1840; to the Mediterranean 1841. On 7.4.1841 under Capt. Michael Seymour, as flag of Vice-Adm. John Acworth Ommanney in the Mediterranean. In 9.1841 under Capt. George Mansel, still in the Mediterranean; home to Portsmouth at end 1842; paid off 1.1843. Fitted for commission at Portsmouth 5.1847. Fitted as a guard ship there 8.1850, then recommissioned 31.8.1850 under Capt. Richard Augustus Yates, as guard ship at Portsmouth for the Ordinary. Fitted as a flagship at Portsmouth (for £12,300) 12.1851–2.1852; recommissioned 1.12.1851 under Capt. George Goldsmith, as flag of Rear-Adm. Sir James Whitley Deans Dundas, for the Mediterranean. On 9.8.1852 under Capt. Thomas Wren Carter; at Bombardment of Sebastopol 11.10.1854; in the Mediterranean in early 1855, then paid off at Portsmouth. Fitted as a hospital ship there 7.1855. On 1.1.1859 under Capt. Robert Harris (–1.10.1862); fitted as a Training ship at Portsmouth 2.1859, then served as Cadets' Training Ship there; to Portland in same role 2.1862. From 1.10.1862 under Capt. Richard Ashmore Powell, then to Dartmouth in same role 1863. In 9.1864 under Capt. John Corbett; in 8.1871 under Capt. Fitzgerald Algernon Charles Foley (–8.1874); in 1.1875 under Capt. William Graham. To Plymouth 7.1869. Completed BU at Devonport (by AO 9.8.1869) 20.11.1869.

Prince Regent Chatham Dyd [M/Shipwright George Parkin]

As built: 205ft 0¾in, 171ft 0in x 53ft 7¾in (54ft 6¾in oa) x 23ft 2½in. 2,613⁷%₉₄bm.

Draught 15ft 9in/17ft 6in.

Ord: 6.1.1812. K: 17.7.1815. L: 12.4.1823. C: 3.9.1823 (as guard ship).

First cost: £97,715 (£119,283 fitted).

Commissioned 6.12.1822 under Capt. William Henry Webley Parry, as flagship of Vice-Adm. Sir Benjamin Hallowell at the Nore (where she remained until 1831). On 19.12.1825 under Capt. Constantine Richard Moorsom (Acting), as flag of Vice-Adm. Sir Robert Moorsom. On 24.7.1827 under Capt. George Poulett, as flag of Vice-Adm. Sir Henry Blackwood. On 6.8.1830 under Capt. James Whitley Deans Dundas; to Lisbon 1832, then paid off 2.1832. Fitted for Ordinary at Portsmouth (for £9,457) 8.1832–1.1833. In Ordinary at Portsmouth 1832–47. Repaired, cut down to 92-gun Second Rate in 3.1847, then fitted for sea (all for £53,815) from 3.1841–9.1847. Recommissioned 7.12.1847 under Capt. William Fanshawe Martin for the Mediterranean; recommissioned 21.12.1849, still under Martin (–early 1851). On 1.3.1851 under Capt. Robert Harris, with Broad Pennant of the now Commodore Martin, for Particular service. On 20.5.1852 under Capt. Frederick Hutton, as flagship of Rear-Adm. Armar Lowry Corry, with the Western Squadron (–12.1853). On 7.3.1854 under Henry Smith; to the Baltic in early 1855; paid off 16.12.1854 at Portsmouth. In Ordinary at Portsmouth 1855–60. Converted to 90-gun screw battleship 1860–61 (see Chapter 2).

PRINCESS CHARLOTTE **Class** An enlarged version of the 1801 *Boyne* Class, thereby a development of Slade's *Victory*. The design was amended on 19 June 1813 to provide a round bow. Originally ordered as 98-gun Second Rates, these ships were re-classed as 104-gun First Rates in January 1817 prior to being laid down. Broadened by 17in during construction to carry additional weight of armament; an extra (11ft 7in) section was added amidships to gain extra length – and thus speed – and additional breadth was to be obtained by diagonal framing

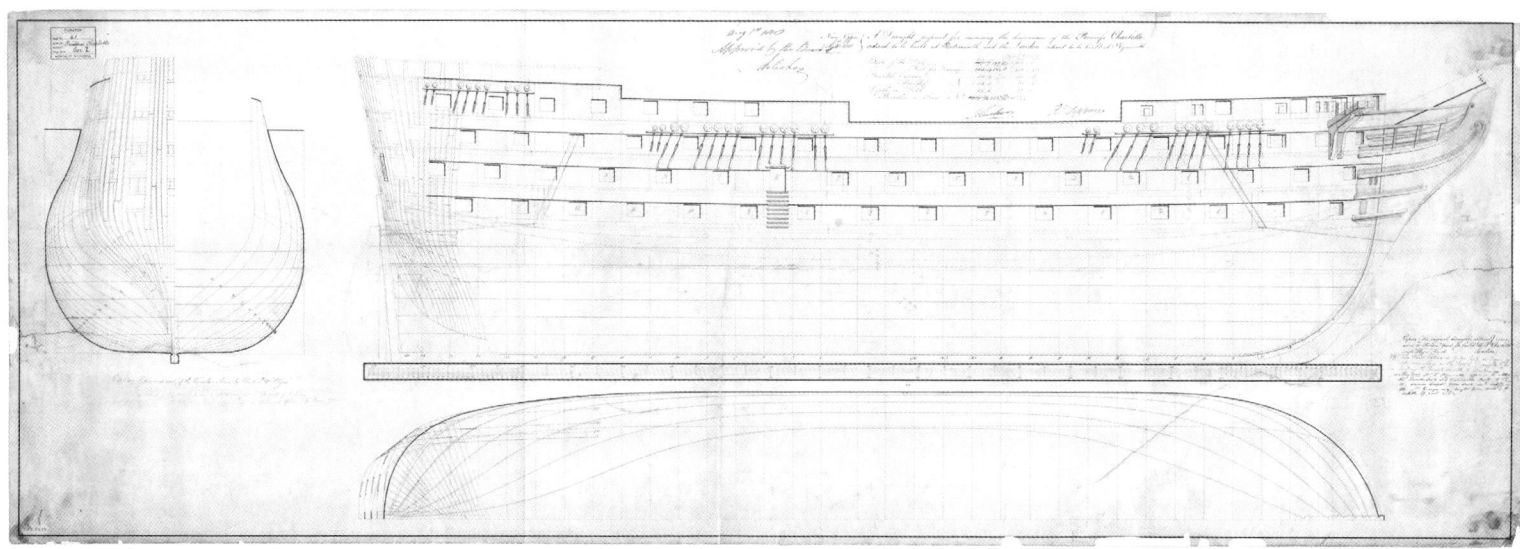

The design draught of 31 July 1818, signed by the Surveyors Joseph Tucker and Robert Seppings, for the *Princess Charlotte* and *London* (soon to be renamed *Royal Adelaide*) shows very clearly the framing of the circular sterns adopted for these three-deckers; at the same time the head was slightly altered. It notes the dimensions of the original design, but is drawn to the revised plan as decided. (*NMM J1794*)

(and doubling); they emerged as 110-gun ships and were re-rated accordingly prior to completion, when each was fitted for Ordinary (reserve) and housed over fore and aft.

>Dimensions & tons: 197ft 7in, 163ft 4in x 52ft 9in x 22ft 6in. 2,417^{44}⁄₉₄bm.

>Men: 738. Guns (as begun): LD 26 x 32pdrs (56cwt) + 2 x 8in/68pdr (65cwt) shell; MD 28 x 32pdrs (48cwt) + 2 x 8in/68pdr (65cwt) shell; UD 30 x 32pdrs (32cwt); QD 2 x 12pdrs + 12 x 32pdr carronades (17cwt); Fc 2 x 32pdr carronades (17cwt). With the extra length added during construction, a further pair of 32pdrs was added to each full deck (i.e. LD, MD and UD), making them 110-gun ships.

Princess Charlotte Portsmouth Dyd [M/Shipwright Nicholas Diddams to 1.1823, completed by John Nolloth]

>As built: 198ft 3in, 164ft 0½in x 52ft 11in (53ft 10in oa) x 22ft 6in. 2,443^{30}⁄₉₄bm.

>Draught 14ft 5½in/17ft 5½in.

>Ord: 6.1.1812. K: 11.1818. L: 11.11.1825. Remained in Ordinary until 1837.

>First cost: £89,126 including fitting.

>Fitted as a demonstration ship with housing removed (for £11,441) 7–8.1834. Fitted at Portsmouth as a flagship (for £9,919) 11.1836–5.1837.

>Commissioned 9.2.1837 under Capt. Arthur Fanshawe, as flag of Adm. Sir Robert Stopford, C-in-C in the Mediterranean; at Bombardment of Acre 3.11.1840; paid off 27.7.1841. At Portsmouth 1841–55, in Ordinary (12 guns only); converted there to an Advanced ship (for £12,557) 10.1848–11.1849. Fitted at Portsmouth as a floating barracks for Hong Kong (for £21,202) 4–7.1857. Recommissioned 23.4.1857 under Capt. George St Vincent King; on 7.2.1858 under Henry G. Thomsett, Master, as receiving ship at Hong Kong; subsequently 16.7.1761 under Capt. Matthew Stainton Nolloth, then 3.1866 under Capt. Oliver John Jones 3.1866 and then under Capt. Francis Henry Shortt. On 15.8.1873 under Commodore John Edward Parish; paid off 7.1875. Sold at Hong Kong (by AO 1.12.1874) late 1875.

Royal Adelaide (ex-*London*, renamed 10.5.1827) Plymouth Dyd [M/Shipwright Edward Churchill]

>As built: 197ft 11in, 163ft 8⅝in x 53ft 0in (53ft 11in oa) x 22ft 6in. 2,446^{19}⁄₉₄bm.

>Draught 15ft 10in/17ft 10in.

>Ord: 6.1.1812. K: 5.1819. L: 28.7.1828. Remained in Ordinary until 1835.

>First cost: £98,796 including fitting.

>Fitted at Plymouth as a flagship, with housing removed (for £20,437) 7.1835.

>Commissioned 5.9.1835 under Capt. Gordon Thomas Falcon, as flagship of Sir William Hargood, C-in-C at Plymouth; paid off 30.4.1836. Recommissioned 2.5.1836 under Capt. John Sykes, as flagship of Adm. Lord Amelius Beauclerk, C-in-C at Plymouth from 27.4.1836; on 15.8.1837 under Capt. Sir William Elliott and 19.9.1838 under Capt. Thomas White, all at Plymouth. Fitted as a demonstration ship (for £3,598) 9.1840. Flagship of Rear-Adm. Sir Thomas Pasley from 12.1857. Recommissioned 1.4.1859 under Capt. Woodford John Williams, as guard ship at Plymouth (in Ordinary); from 7.1860 under Capt. William King Hall, as depot ship for steam reserve at Plymouth; on 1.1.1862 under Capt. Charles Vesey, as flagship C-in-C at Plymouth (a role the ship then filled until 1890), with flag of Vice-Adm. (Adm., 10.11.1862) Sir Houston Stewart; on 3.11.1863 under Capt. Henry Caldwell, with flag of Vice-Adm.(Adm. 9.2.1864) Sir Charles Howe Fremantle (from 27.10.1863). Recommissioned 2.1864 under Capt. Frederick Beauchamp Paget Seymour, still with Fremantle's flag. On 26.10.1866 under Capt. George William Preedy, with flag of Adm. Sir William Fanshawe Martin. Fitted at Plymouth as a flag and receiving ship, and recommissioned 1.11.1869 under Capt. Trevenen Penrose Goode, with flag of Adm. Sir Henry John Codrington. On 1.11.1872 under Capt. Algernon Charles Fieschi Heneage, as flag of Adm. Sir Henry Keppel. On 1.10.1875 under Capt. John Ommanney Hopkins, with flag of Adm. Thomas Matthew Charles Symonds (from 1.11.1875). On 1.11.1878 under Capt. William Henry White, with flag of Adm. Arthur Farquhar (from 1.11.1878). In 11.1879 under Capt. William Burley Grant, with flag of Adm. Sir Charles Gilbert John Brydone Elliott (from 9.1.1880). Recommissioned 30.11.1880 under Capt. Richard Carter, with flag of Adm. Sir William Houston Stewart (from 1.12.1881). On 29.12.1884 under Capt. William Elrington Gordon, as flagship at Plymouth; on

30.3.1885 under Capt. William Henry Cuming, as flag of Adm. Augustus Phillimore (from 1.12.1884); on 6.4.1887 under Capt. Harry Woodfall Brent, with flag of Adm. Lord John Hay (from 25.5.1887). Later commanding officers and flag officers are left out here to save space. Fitted at Chatham as a receiving hulk (by AO 8.1.1890) to 9.1891. Sold at Chatham (by AO 26.8.1904) 4.4.1905; her figurehead was removed to the Royal Naval Barracks at Devonport by AO 23.1.1908.

The outbreak of the Anglo-American War of 1812 had led to a struggle for control of the Great Lakes system of North America and thus to a short-lived but spectacular programme which culminated in the building of three-deckers at Kingston Dyd, Ontario for service on the freshwater Lake Ontario. The end of this conflict in 1815 resulted in the first of these three-deckers – the *Saint Lawrence* – being decommissioned, while the subsequent pair – *Wolfe* and *Canada*, both still building in 1815 – were suspended in that year, although not finally cancelled until 1831–32. As the *raison d'être* for these ships vanished in 1815, they (and smaller ships on the Great Lakes) are not covered in this volume.

By 1815 the remaining smaller three-decker Second Rates (including the three old 90-gun ships already disarmed as receiving ships) were already fairly elderly and were soon hulked. The *Saint George* (ex-*Britannia*), *Victory* and *Royal Sovereign* (originally First Rates) had all been temporarily added to their ranks. However, all three-decked ships still left were re-classified as First Rates on 1 January 1817.

(B) Vessels acquired from 1 January 1817

Modified *CALEDONIA* Class Built to the draught of the *Caledonia*, but doubled with fir while building, to make her breadth (and ordnance capacity) similar to the subsequent Broadened *Caledonia* Class ships. All these early ships were later re-armed similarly to the following 'Broadened' *Caledonia* group.

Royal George (ex-*Neptune*, renamed 12.2.1823) Chatham Dyd

As built: 205ft 7½in, 171ft 0in x 54ft 6½in oa (53ft 7½in for tonnage) x 23ft 2in. 2,615 57/94bm.

Draught 15ft 8in fwd/18ft 3in aft.

Ord: 2.6.1819. K: 6.1823. D L: 22.9.1827. C: 3.12.1827 (for Ordinary), roofed over fore and aft. The doubling was removed in 2.1833.

First cost: £87,418 (fitted).

Not commissioned as a sailing ship; remained in Ordinary at Sheerness until 1850, then converted to a screw battleship (see below) in 1852–53. Cut down from a three-decker to a two-decker (89-gun Second Rate) in 1860. Sold to Castle 23.1.1875 to BU at Charlton.

Broadened *CALEDONIA* Class 120-gun ships. Broadened 1819 version of the *Caledonia* Class, to carry increased weight of guns. A further ship was ordered 29 October 1827 to this design as the *Royal Frederick*, but on 3 September 1833 she was re-ordered to the following *Queen* Class design (see Section (C).

Design dimensions & tons: 205ft 5½in, 170ft 6in x 54ft 6in x 23ft 2in. 2,693 71/94bm.

Men: 900 (820 peacetime). Guns: LD 30 x 32pdrs (56cwt) + 2 x 68pdr carronades; MD 32 x 32pdrs (55cwt) + 2 x 68pdr carronades; UD 32 x 32pdrs + 2 x 68pdr carronades; QD 16 x 32pdr carronades (25cwt); Fc 2 x 32pdrs (49cwt) + 2 x 32pdr carronades (25cwt). Later 4 x 8in/51pdr (56cwt) shell guns and 2 x 32pdrs (on the UD) replaced the 68pdr carronades (and 2 x 32pdrs on the UD), and the QD/Fc bore 6 x 32pdrs + 14 x short 32pdrs.

Saint George Plymouth Dyd

As built: 205ft 5½in, 170ft 5in x 55ft 3¼in oa (54ft 9¼in for tonnage) x 23ft 2½in. 2,719 26/94bm. Draught 13ft 11in/18ft 1in (with 60 tons ballast).

Ord: 2.6.1819. K: 5.1827. L: 27.8.1840. C: 7.1850 (as guard ship).

First cost: £76,009 to build (fitted £102,405).

Placed in Ordinary on launching, remaining at Plymouth until 1854. The ship was 'advanced' (for £17,095) 1.1843–10.1845, then fitted as a guard ship (for £9,301) 7.1850. Commissioned 31.8.1850 under Capt. Joseph Nias, as guard ship of the Ordinary at Devonport, with flag of Commodore Lord John Hay; on 8.9.1851 flagship of Commodore Michael Seymour, on same station; paid off 23.5.1853 at Plymouth. Recommissioned 23.5.1853 under Capt. John Kincome, still flag of Seymour on same station; on 6.2.1854 under Capt. Harry Eyres, for the Baltic; fitted for sea (for £8,286) 2–3.1854; deployed to Baltic during the Russian War; paid off 5.4.1856 at Plymouth. Converted to 89-gun screw battleship 1858–59 (see Chapter 2).

Neptune Portsmouth Dyd

As built: 205ft 8in, 170ft 5¾in x 55ft 6½in oa (54ft 7½in for tonnage) x 23ft 2in. 2,705 75/94bm.

Draught 14ft 10in fwd/18ft 4in aft.

Ord: 12.2.1823. K: 1.1827. L: 27.9.1832. C: 12.1832 (for Ordinary).

First cost: £73,595.

Commissioned 5.12.1851 under Capt. Richard Augustus Yates, as guard ship at Portsmouth (in Ordinary); on 27.3.1852 under Capt. Edward Hinton Scott, then 17.2.1854 under Capt. Henry Smith, in same role. On 7.3.1854 under Capt. Frederick Hutton, as flagship of Rear-Adm. Armar Lowry Corry, initially at Portsmouth, then deployed to the Baltic during the Russian War; paid off 5.1.1867. Converted to 89-gun two-decker screw battleship 1858–59 (see Chapter 2).

Royal William Pembroke Dyd

As built: 205ft 5½in, 170ft 3in x 55ft 6in oa (54ft 7in for tonnage) x 23ft 0in. 2,698 4/94bm.

Draught 15ft 1½in fwd/17ft 11½in aft.

Ord: 30.12.1823. K: 10.1825. L: 2.4.1833. C: 1834 at Plymouth (for Ordinary).

First cost: £81,255.

Commissioned 16.2.1854 under Capt. John Kingcombe, as flagship of Commodore Michael Seymour, guard ship at Plymouth (in Ordinary); for the Baltic; flagship of Rear-Adm. Sir James Hanway Plumridge 2.1855, again as guard ship at Plymouth. On 16.6.1856 under Capt. Frederick Hutton, still with Plumridge's flag, then 5.5.1857 under Capt. Woodford John Williams; paid off 31.3.1859 (replaced by *Royal Adelaide*). Converted to 89-gun two-decker screw battleship 1859–60 (see Chapter 2).

Waterloo Chatham Dyd

As built: 205ft 6in, 170ft 5⅝in x 55ft 3in oa (54ft 9in for tonnage) x 23ft 2in. 2,718 3/94bm. Draught 15ft 1in fwd/17ft 11in aft.

Ord: 5.10.1824. K: 3.1827. L: 18.6.1833 (to Sheerness 19.7.1733 for Ordinary). C: 12.1851 at Sheerness (as flagship).

First cost: £70,553 (fitted £99,898).

Commissioned 5.8.1851 at Sheerness under Capt. Montagu Stopford, intended as flagship of Vice-Adm. James Whitley Deans Dundas for the Mediterranean (but *Britannia* was substituted). In 1.1852 under Capt. George Rodney Mundy, as

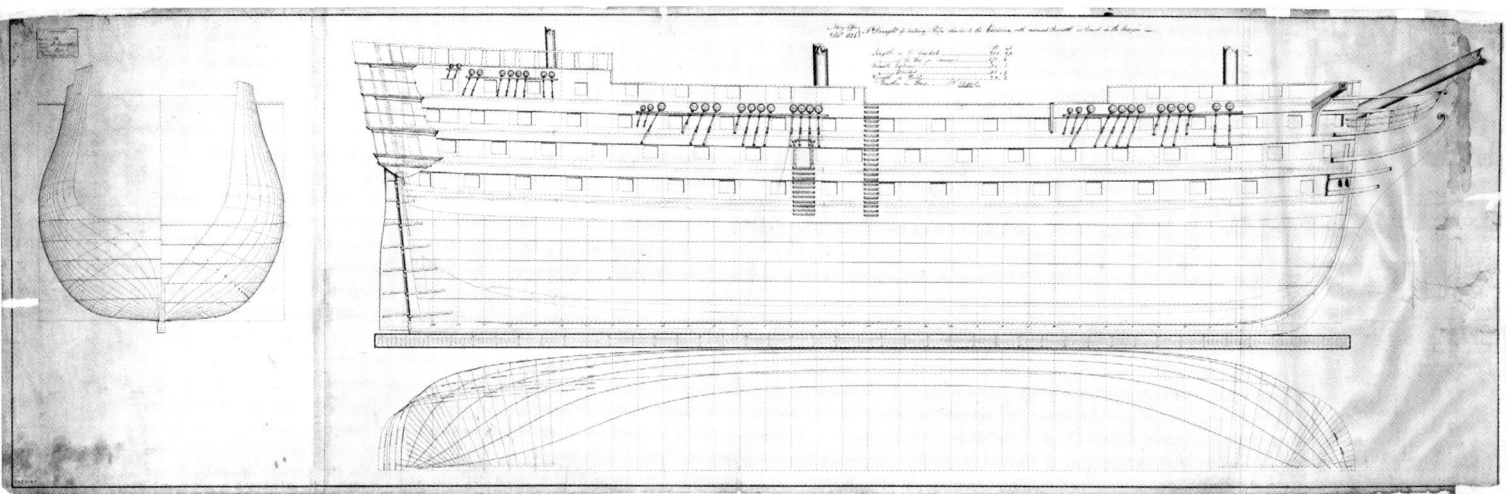

'A Draught for building Ships similar to the *Caledonia*, with increased Breadth as named in the margin, Navy Office Dec 1826.) There are no names in the margin of this copy, but the draught was used for the building of the five ships of the *St George* Class. *(NMM J1723)*

flagship of Vice-Adm. Joscelin Percy at Sheerness. On 9.12.1853 under Capt. Lord Frederick Herbert Kerr, still as flagship of Percy, later of Rear-Adm. William Gordon; paid off 28.2.1859. Converted to 89-gun screw battleship 1859 (see Chapter 2).

Trafalgar Woolwich Dyd

As built: 205ft 6in, 170ft 5in x 55ft 7½in oa (54ft 9½in for tonnage) x 23ft 2in. 2,721³²⁄₉₄bm. Draught 13ft 3in/18ft 3in (with 30 tons of ballast, 3½ tons anchors, 3 tons cables).

Ord: 22.2.1825. K: 12.1829. L: 21.6.1841. C: 22.6.1841–6.10.1842 at Sheerness (for Ordinary). Converted to an Advanced ship and stern altered at Chatham (for £17,954) 10.1844.

First cost: £75,814 (£75,797 to built, £1,283 fitting for Ordinary).

Commissioned 30.1.1845 at Sheerness under Capt. William Fanshawe Martin; fitted for sea and as flagship of Vice-Adm. Sir John Chambers White (died 4.4.1845), as C-in-C at the Nore (for £18.838) 2–5.1845; on 18.4.1845 flagship of Vice-Adm. Sir Edward Durnford King, in same role; to Experimental Squadron 1845. On 18.10.1846 under Capt. John Neale Nott, still King's flagship, for the Channel; to experimental Squadron 1846; to Channel Squadron 1847, later to Lisbon as flag of Rear-Adm. Sir Francis Augustus Collier, then 10.1847 to Mediterranean. On 11.1.1848 under Capt. Charles Hope, in the Mediterranean; paid off at Sheerness 30.6.1848. Recommissioned 26.7.1850 under Capt. Montagu Stopford; fitted for sea at Sheerness (for £9,214) 7.1850–7.1851. On 21.7.1851 under Capt. Henry Francis Grenville, to the Mediterranean; at Bombardment of Sebastopol 11.10.1854; paid off 21.7.1851 at Sheerness. Converted to 89-gun screw battleship 11.8.1858–21.3.1859 (see Chapter 2).

On 27 March 1858 it was ordered (under the 1858 Programme of Works) that the *Saint George*, *Neptune* and *Trafalgar* (plus the *Queen*, detailed below) were to be cut down a deck and converted to steam screw Second Rates, each with 90 guns and a 500hp engine. On 5 February 1859 it was similarly ordered that the *Royal William* and *Waterloo* (plus the *Rodney* and *Nelson*, detailed earlier) were to be cut down a deck and converted to steam screw Second Rates, also with 90 guns and a 500 hp engine. Their subsequent history appears in Chapter 2.

(C) Vessels acquired from November 1830

By 1830 the Admiralty's strategy had arrived at two major decisions regarding the major combatant units. The first was to base all future battlefleet requirements upon the three-decker 120-gun ship and the larger two-deckers (primarily of 90 and 80 guns). The second was to establish the 32pdr as the sole calibre for armaments, removing guns of lesser power from the broadsides of the battlefleet.

With a single exception (the *Royal Albert*), the entire history of the First Rate from 1830 onwards was identical with that of the *Queen* Class.

QUEEN Class 110 guns. At first intended to be built to the 120-gun modified *Caledonia* design, the *Royal Frederick* was ordered on 29 October 1827. However, on 2 September 1833, by which time three-quarters of the frame timbers had been cut, a fresh design was presented by William Symonds, and the following day the Admiralty instructed that the present frames should be converted, and the ship proceed as a three-decker of 110 guns, the first of a new design. A second ship to this design was ordered nine days later, also to be built at Portsmouth, and a further three weeks later two more vessels were ordered to be built at Pembroke.

On 7 May 1834 the Admiralty suspended work on the *Royal Sovereign* at Portsmouth, and on 11 December decided to construct the two Pembroke ships as two-deckers of 74 guns, albeit retaining the Symondite hull form and dimensions of a 110; they would in this form have carried: LD 28 x 8in/68pdr shell guns (65 cwt), UD 28 x 32pdrs (56cwt), QD/Fc 16 x 32pdrs (48cwt) and 2 x 32pdrs (56cwt). However, in the 1835 Navy List the *Algiers* was again shown as a 110-gun ship, while the *Victoria* was unlisted, and the 1836 List, while retaining the *Algiers* as 110 guns, reinstated the *Victoria* at 74 guns. Confusingly, the 1837 List then gave *Algiers* as the 74-gun ship, and *Victoria* as 110 guns; on 5 February 1839, the *Algiers* was restored to 110 guns. In the same year (on 12 April), the two Portsmouth ships were renamed, the first being renamed **Queen** prior to her launch on 15 May, and the second (still not begun, with construction suspended) taking over the name *Royal Frederick*.

On Boxing Day 1840 the Admiralty instructed that the *Algiers* should be built as a 90-gun Second Rate (see Chapter 2 for subsequent history). Four further 110-gun ships were subsequently ordered to the *Queen* design – the *Prince of Wales* (Portsmouth Dyd) on 14 March

The Navy's newest and largest First Rate at that time, the *Queen* was the object of a royal visit at Portsmouth by Queen Victoria and Prince Albert on 1 March 1842, the event being recorded on canvas by Lieut Robert Strickland Thomas. With the crew manning the yards, the *Queen* breaks out the Royal Standard from her masthead just as the paddle steamer alongside strikes hers, to signify the transfer of the sovereign from one ship to the other. *(NMM BHC0629)*

1842, the *Windsor Castle* (Pembroke Dyd) and a new *Royal Sovereign* (Portsmouth Dyd) both on 19 February 1844, and the *Marlborough* (Portsmouth Dyd) on 20 March 1844; however, all were – together with *Victoria* – suspended on 9 December 1844. These five ships – and the *Royal Frederick* – were re-ordered to the Modified design shown below on 29 June 1848, and completed as steam-assisted ships. Consequently, the *Queen* was the only vessel completed to the original design as a sailing three-decker.

Design dimensions & tons:

Original design: 204ft 0in, 165ft 11in x 60ft 0in oa (59ft 1¼in for tonnage) x 23ft 9in. 3,082⁹⁄₉₄bm.

Modified design: 204ft 0in, 166ft 5¼in x 60ft 0in oa (59ft 2in for tonnage) x 23ft 9in. 3,099¹⁶⁄₉₄bm.

Men: 950 (comprising 724 officers and men, 66 boys and 160 marines).

Guns: LD 28 x 32pdrs (55cwt) + 2 x 8in/68pdr (60cwt) shell; MD 28 x 32pdrs (55cwt) + 2 x 8in/68pdr (60cwt) shell; UD 32 x 32pdrs (49cwt); QD 10 x short 32pdr (40cwt); Fc 2 x short 32pdr (40cwt) + 2 x 68pdr (36cwt) carronades; RH 4 x 18pdr (10cwt) carronades.

Queen (ex-*Royal Frederick*, renamed 12.4.1839) Portsmouth Dyd

As built: 204ft 2½in, 166ft 5½in x 59ft 2½in (60ft 0½in oa) x 23ft 9in. 3,103bm.

Draught 14ft 3in fwd/18ft 11in aft.

Ord: 3.9.1833. K: 11.1833. L: 15.5.1839. C: 3.4.1840.

First cost: £70,474 (fitted for commission £77,729, + £17,408 fitting as flagship).

Commissioned 1.10.1840 under Capt. John William Montagu, as flag of Adm. Sir Robert Stopford, C-in-C in the Mediterranean; on 13.3.1841 under Capt. Henry John Codrington, as flag of Adm. Edward Codrington (his father); on 30.9.1841 under Capt. George Frederick Rich, as flag (14.10.1841) of Vice-Adm. Edward William Campbell Rich Owen, C-in-C in

the Mediterranean; in 5.1842 under (temp.) Capt. Hastings Reginald Henry, then 16.1.1843 back to Capt. Rich, still flag of Vice-Adm. Owen; on 22.4.1844 under Capt. Charles Sulivan; paid off 11.7.1844 at Portsmouth. Recommissioned 19.7.1844 under Capt. William Fanshawe Martin, as flag of Vice-Adm. John Chambers White, at Sheerness; paid off 31.1.1845. Recommissioned 16.4.1845 under Capt. Baldwin Wake Walker, as flag of Adm. Sir John West, C-in-C Devonport and 1845 Experimental Squadron; on 18.10.1845 under Capt. Sir Henry John Leeke, with 1845–46 Experimental Squadrons, then flag of Sir John West at Devonport. On 30.4.1846 under Capt. Sir James John Gordon Bremer, as flag of Sir Francis Augustus Collier in the Channel. On 4.11.1847 under Capt. Henry William Bruce, in the Mediterranean; paid off 26.5.1849 at Plymouth. Recommissioned 22.6.1849 under Capt. Charles Wise, as flagship of Vice-Adm. Sir William Parker, C-in-C in the Mediterranean; paid off at Plymouth. Recommissioned 3.7.1852 under Capt. Frederick Thomas Michell, for the Mediterranean; at Bombardment of Sebastopol 11.10.1854; on 11.7.1855 under Capt. Robert Fanshawe Stopford, in the Mediterranean (–6.9.1856); paid off 15.8.1856 at Portsmouth. Converted to 86-gun (later 74-gun) screw battleship (Second Rate) between 1858 and 5.4.1859 (undocked) (see Chapter 2 for later history).

Royal Frederick (ex-*Royal Sovereign*, renamed 12.4.1839) Portsmouth Dyd

Ord: 12.9.1833. Suspended 7.5.1834. K: 1.7.1841. Re-ordered to modified design 29.6.1848 (see below).

Algiers Pembroke Dyd

Ord: 3.10.1833. K: — . Re-ordered 26.12.1840 as a 90-gun two-decker (see Chapter 2).

Victoria Pembroke Dyd

Ord: 3.10.1833. K: 5.1844. Re-ordered to modified design 29.6.1848 (see below).

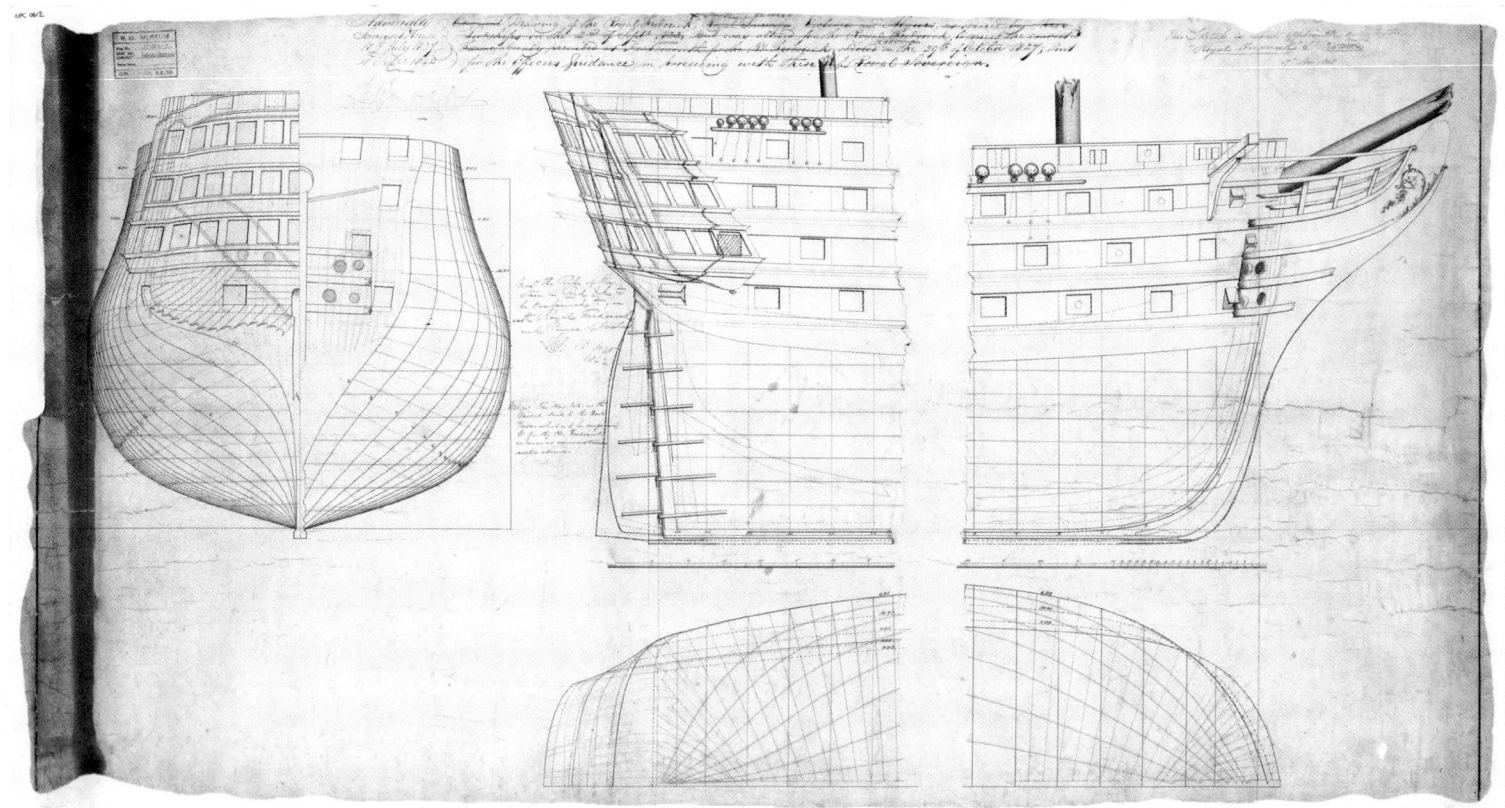

This draught, originally signed by Symonds, shows a series of planned alterations to the bow and stern of the Modified *Queen* Class, dated 18 July 1837 and 15 September 1842, the latter – signed by Peake – reducing the rake of the stern in *Royal Frederick* and *Prince of Wales*. One major criticism of Symonds's battleship designs at the time was the excessive overhang of their sterns, which rendered them vulnerable in a following sea. A later, anonymous, alteration reduced the rake still further, and a note of 17 November 1848 states 'This sketch is now applicable only to the *Royal Frederick* and *Victoria*'; the latter two ships were renamed *Frederick William* and *Windsor Castle* respectively before being completed as screw battleships to fresh designs.

Under the Establishment of Guns introduced in February 1839, all the 120-gun ships (from *Hibernia* onwards) were to mount 6 x 8in (68pdr) shell guns, 100 x 32pdrs and 14 x 32pdr carronades. These were to be mounted as follows: LD 30 x 32pdrs (of 56cwt) + 2 x 8in (of 65cwt), MD 32 x 32pdrs (of 48cwt) + 2 x 8in, UD 32 x 32pdrs (of 33cwt) + 2 x 8in, QD 4 x 32pdrs (of 45cwt) + 12 x 32pdr carronades (of 17cwt) and Fc 2 x 32pdrs (of 45cwt) + 2 x 32pdr carronades (of 17cwt).

ROYAL ALBERT 120 guns. Oliver Lang design of 1831, effectively an expanded version of the last pre-Symondite three-decker, the *Trafalgar*. The largest pure sailing battleship ever designed for the RN, and the only sailing First Rate after 1830 that was not to the *Queen* design.

> Dimensions & tons: 220ft 0in, 177ft 2¾in x 60ft 0in (60ft 10in oa) x 25ft 0in. 3,393⁷/₉₄bm.
> Design enlarged to 3,463 tons on 24.3.1851.
> Men: 1,000. Guns: LD 28 x 32pdrs (56cwt) + 4 x 8in/68pdr (60cwt) shell; MD 32 x 32pdrs (56cwt) + 2 x 8in/68pdr (60cwt) shell; UD 34 x 32pdrs (49cwt); QD/Fc 6 x 32pdrs (42cwt) + 14 x short 32pdrs (25cwt).

Royal Albert Woolwich Dyd
> Ord: 26.3.1842. K: 8.1844. L: — . Re-ordered as screw battleship 31.1.1852 (see below).

Modified *QUEEN* Class The development of the *Queen* design was complex, and the plans changed several times. Eventually all the following vessels were re-ordered as steam/screw battleships and launched as such; the following reflects their development while still projected as sailing ships.

All seven ships were first ordered to the 3,083-ton (110-gun) design, and (apart from the *Algiers*) were subsequently re-ordered as 3,186-ton ships. The *Royal Frederick*, *Algiers* and *Victoria* had all been ordered on 12 September 1833 and were shortly afterwards modified to the 3,099-ton design; the *Algiers* was re-ordered as a Second Rate in 1840, and the other two were on 29 June 1848 given a new 116-gun Establishment. The *Prince of Wales*, *Royal Sovereign* and *Windsor Castle*, which had been ordered in 1842 and 1843, were re-ordered on 29 June 1848 to the Modified design, and given a new 120-gun Establishment. The *Marlborough*, first ordered in 1843; was re-ordered on 31 March 1849 to the Modified design, and given the same 120-gun Establishment. The *Royal Sovereign* and *Marlborough* were finally amended in 1850 to provide for a LD battery entirely of 68pdrs.

Group 1
> Dimensions & tons: 204ft 0in, 165ft 11in x 60ft 0in oa (59ft 1¼in for tonnage) x 23ft 9in. 3,082⁹⁵/₉₄bm.
> Men: 950. Guns: LD 28 x 32pdrs (55cwt) + 2 x 8in/68pdr (60cwt) shell; MD 28 x 32pdrs (55cwt) + 2 x 8in/68pdr (60cwt) shell; UD 32 x 32pdrs (49cwt); QD 10 x short 32pdr (40cwt) gunnades; Fc 2 x short 32pdr (40cwt) gunnades + 2 x 68pdr (36cwt) carronades; RH 4 x 18pdr (10cwt) carronades.

Royal Frederick (ex-*Royal Sovereign*, renamed 12.4.1839) Portsmouth Dyd
> Ord: 29.6.1848. K: 7.1841 (see above). Re-ordered to complete as a screw battleship 28.2.1857.

G. H. Atkins's lithograph of the 131-gun *Duke of Wellington* off Portsmouth in 1853, before she led the British Fleet into the Baltic in 1854. Despite the speed of the conversion, and the fact that her sisters were further lengthened, the *Duke of Wellington* was regarded as a successful ship, and her performance was highly satisfactory under both steam and sail. *(NMM PY9261)*

Victoria Pembroke Dyd
  Ord: 29.6.1848. K: 5.1844 (see above). Renamed **Windsor Castle** 6.1.1855 (the ship commenced as *Windsor Castle* below having been renamed in 1852). Re-ordered to complete as a screw battleship 28.2.1857.

Group 2 3,186 ton stretched design by Edye & Watts (120 guns), approved 28.7.1848.
  Dimensions & tons: 210ft 0in, 171ft 1in x 60ft 0in oa (59ft 2in for tonnage) x 24ft 8in. 3,185⁶⁄₉₄bm.
  Men: 970. Guns: LD 30 x 8in/68pdr (68cwt) shell; MD 30 x 32pdrs (56cwt); UD 32 x 32pdrs (42cwt); QD + Fc 8 x 32pdrs (42cwt) + 14 x short 32pdrs (25cwt); RH 4 x 18pdr (10cwt) carronades.
Prince of Wales Portsmouth Dyd
  Ord: 29.6.1848. K: 10.6.1848. L: —. Re-ordered to complete as a screw battleship 9.4.1856.
Royal Sovereign Portsmouth Dyd
  Ord: 29.6.1848. K: 17.12.1848. L: —. Re-ordered to complete as a screw battleship 23.6.1854.
Windsor Castle Pembroke Dyd
  Ord: 29.6.1848. K: 5.1849. L: —. Re-ordered to complete as a screw battleship 19.1.1852 (renamed **Duke of Wellington** 1.10.1852).
Marlborough Portsmouth Dyd
  Ord: 29.6.1848. K: 1.9.1850. Suspended 1.12.1851. Re-ordered to complete as a screw battleship 30.10.1852.

(D) Vessels ordered or re-ordered as screw three-deckers (from 1850)

In 1852, two sailing First Rates of the Modified *Queen* Class, and the larger *Royal Albert*, all then building, were ordered to be completed as screw warships, and similar orders were issued for the other four of the Modified *Queen* Class in 1854 to 1857, although one of these – the *Royal Frederick* (renamed the *Frederick William*) – was instead cut down to become a screw two-decker. One older First Rate, the *Royal George* launched in 1827, was also converted to a screw three-decker, but the result was so crank that she too was quickly cut down to a two-decker. Several other older First Rates were cut down to two-

deckers when their engines and screws were fitted. Just two new three-deckers, the *Howe* and *Victoria*, were designed and built from the start as screw ships before the development of the ironclad frigate made the concept of wooden-walled battleships obsolete.

DUKE OF WELLINGTON **Class** – Conversions of 120-gun sailing First Rates (incomplete on stocks) to 131-gun screw ships. Lengthened during conversion with extra 23ft amidships and 8ft in the run. Their aft sections launched for lengthening on 7 February 1852 and 30 June 1855 respectively. The first ship, built and launched as the *Windsor Castle*, was renamed after the Duke of Wellington on 1 October 1852, the same day as the duke died. Sisters *Marlborough* and *Prince of Wales* were originally planned to be converted to the same plans, but they were further lengthened during construction (see below). The *Duke of Wellington* received the 700nhp engine (with cylinders 93⅞in diameter, 4½ft stroke) from the iron screw frigate *Simoon*, hence her brief subsequent career. *Royal Sovereign* cylinders as *Marlborough* below.
  Dimensions & tons: 240ft 7in, 201ft 11½in x 60ft 1in (59ft 3in for tons) x 24ft 8in. 3,771¹⁹⁄₉₄bm. 6,071 disp.
  Men: 1,100. Guns: LD 10 x 8in (65cwt/9ft) + 26 x 32pdrs (56cwt/9½ft); MD 6 x 8in (65cwt/9ft) + 30 x 32pdrs (56cwt/9½ft); UD 38 x 32pdrs (42cwt/8ft); QD/Fc 20 x 32pdr gunnades (25cwt/6ft) + 1 x 68pdr (95cwt/10ft) on pivot.
  Machinery: 2-cylinder horizontal single expansion (geared in *Duke of Wellington*). Single screw.
  Duke of Wellington 700nhp. 1,979ihp = 10.15kts. *Royal Sovereign* 800nhp. 2,796ihp = 12.253kts.
Duke of Wellington Pembroke Dyd/Robert Napier & Sons
  As built: 240ft 7in, 201ft 11½in x 60ft 1in (59ft 3in for tonnage) x 24ft 8in. 3,771¹⁹⁄₉₄bm. 6,071 disp. Draught 14ft 9in/20ft 3in.
  Conversion ord: 19.1.1852. Begun: 19.1.1852. L: 14.9.1852. To Portsmouth Dyd 11.10.1852 to complete. C: 27.11.1853.
  First cost: £176,918 (including £78,750 to built, £11,420 for conversion to screw, and fitting machinery £6,160 with £40,862 to Napier as contractor), plus fitting (at Portsmouth Dyd) £40,716.
  Commissioned 2.2.1853 under Commodore Henry Byam Martin (–25.4.1854), for the Western (Channel) squadron. On 4.3.1854 under Capt. George Thomas Gordon, as flag of Vice-Adm. Sir Charles Napier (with Michael Seymour as Capt. of the Fleet); sailed for the Baltic 11.3.1854. On 19.2.1853 under Capt. Henry Caldwell, as flag of Rear-Adm. Richard Saunders Douglas

(with Frederick Thomas Pelham as Capt. of the Fleet), again sailing for the Baltic; to the Mediterranean 1856; home to pay off 4.4.1857 at Portsmouth; put into 1st Division of Steam Reserve. Recommissioned 2.3.1858 under Capt. Harry Eyres, as depot ship of the Ordinary at Portsmouth; paid off 30.6.1858. Recommissioned 1.5.1863 under Capt. John Seccombe, as training ship and receiving ship; from 12.7.1863 under Capt. Charles Fellowes, then 10.9.1867 under Capt. George Hancock; from 20.12.1869 flag of Vice-Adm. Sir James Hope, as port admiral at Portsmouth; on 1.5.1872 under Capt. Henry Carr Glyn (flag of Adm. Sir George Rodney Mundy from 1.3.1872), on 1.3.1875 under Capt. Francis William Sullivan (flag of Adm. Sir George Elliot from 1.3.1875); on 16.10.1876 under Capt. Walter Cecil Carpenter (flag of Adm. Edward Gennys Fanshawe from 1.3.1878); in 11.1779 under Capt. Michael Culme-Seymour (flag of Adm. Alfred Phillipps Ryder from 27.11.1879); on 9.5.1882 under Capt. Charles John Rowley (flag of Adm. Sir Geoffrey Thomas Phipps Hornby from 28.11.1882); on 24.10.1884 under Capt. Philip Howard Colomb, then 11.1885 Capt. Edward Hobart Seymour (flag of Adm. Sir George Ommanney Willes from 28.11.1885); in 12.1887 under Capt. Robert Woodward (flag of Adm. Sir John Edmund Commerell from 20.6.1888); paid off 31.3.1888. Recommissioned 4.1900 under Capt. William Des Voeux Hamilton. Sold to Castle 12.4.1904 to BU at Charlton.

Royal Sovereign Portsmouth Dyd/Maudslay, Sons & Field

As built: 240ft 7in, 201ft 11⅛in x 60ft 0½in (59ft 2½in for tonnage) x 25ft 4in. 3,765$\frac{39}{94}$bm. Draught 14ft 5in/20ft 1in.

Conversion ord: 23.6.1854. Begun: 25.1.1855. L: 25.4.1857. C: 7.1857 (fitted for reserve only).

First cost: £178,544 (hull £113,165, machinery £49,300, fitting £16,079).

Converted to ironclad turret ship ('cupola ship') 4.1862 to 8.1864.

Commissioned 7.7.1864 under Capt. Sherard Osborn, for the Channel squadron; tender to *Excellent* from 15.10.1864. On 1.7.1865 under Capt. Frederick Anstruther Herbert, for the Channel squadron; tender to *Excellent* again from 9.10.1866. In 7.1867 under Capt. Cowper Phipps Coles, temporarily for the 1867 Naval Review; by 1.4.1869 under Capt. Arthur William Acland Hood; paid off 3.9.1869. Eventually sold 5.1885 to BU.

MARLBOROUGH – Conversion of 120-gun sailing First Rate (incomplete on stocks) to a 131-gun screw ship. On 30 October 1852 it was instructed that the *Marlborough* be completed as a screw battleship of 3,226bm. Unlike the above pair, this vessel was lengthened by an extra 5ft at the bow during her conversion. As built, she was also broader and actually measured at just over 4,000 tons bm. Problems with the launching at Portsmouth resulted in the process taking a week. After a period in the Steam Reserve at Portsmouth, her single sea-going commission was spent as the flagship of the Mediterranean Fleet.

Dimensions & tons: 245ft 6in, 206ft 11⅛in x 60ft 0in (59ft 2in for tons) x 25ft 2in. 3,853$\frac{12}{94}$bm.

Men: 1,100. Guns: LD 10 x 8in (65cwt/9ft) + 26 x 32pdrs (56cwt/9½ft); MD 6 x 8in (65cwt/9ft) + 30 x 32pdrs (56cwt/9½ft); UD 38 x 32pdrs (42cwt/8ft); QD/Fc 20 x 32pdr gunnades (25cwt/6ft) + 1 x 68pdr (95cwt/10ft) on pivot.

Machinery: 2-cylinder (82in diameter, 4ft stroke) horizontal single expansion. Single screw. 800nhp. 2,684ihp = 11.866kts.

Marlborough Portsmouth Dyd/Maudslay, Sons & Field

As (re)built: 245ft 6in, 206ft 3⅞in x 61ft 2½in oa (60ft 4½in for tonnage) x 25ft 10in. 4,000$\frac{38}{94}$bm). 6,065 disp. Draught 20ft 3in/21ft 9in.

Conversion ord: 30.10.1852. Begun: 9.6.1853. L: 31.7.1855. C: 1858.

First cost: £208,052 including hull £107,731, conversion to screw £4,498, machinery £51,120 and fitting for sea £28,333.

Commissioned 1.2.1858 under Capt. Lord Frederick Herbert Kerr, as flag of Vice-Adm. Arthur Fanshawe, C-in-C in the Mediterranean from 22.2.1858; on 3.5.1860 under Capt. William Houston Stewart, as flag of Vice-Adm. William Fanshawe Martin, C-in-C in the Mediterranean from 19.4.1860 (with Rear-Adm. Sydney Colpoys Dacres as Capt. of the Fleet); In 6.1863 under Capt. Charles Fellowes, as flag of Vice-Adm. Robert Smart, C-in-C in the Mediterranean from 20.4.1863; paid off 1.12.1864 to become a receiving ship at Portsmouth. Reduced to 98 guns by 870, and in 1875 to 74 guns. Engineers' training ship at Portsmouth 12.1878. Attached to *Asia* 1890 as receiving ship for Steam Reserve at Portsmouth. Renamed ***Vernon II*** in 3.1904. Sold to A. Butcher 10.1924 to BU, but capsized 28.11.1924 off Selsey while in tow to shipbreakers at Osea Island (Essex), with 4 drowned.

PRINCE OF WALES – Conversion of 120-gun sailing First Rate (incomplete on stocks) to 121-gun screw ship.

Dimensions & tons: 252ft 0in, 213ft 0in x 60ft 0in (59ft 2in for tonnage) x 25ft 2in. 3,966$\frac{20}{94}$bm.

Men: 1,100. Guns: LD 32 x 8in (65cwt/9ft); MD 30 x 8in (65cwt/9ft); UD 32 x 32pdrs (58cwt/9½ft); QD/Fc 26 x 32pdrs (42cwt/8ft) + 1 x 68pdr (95cwt/10ft) on pivot.

Machinery: 2-cylinder (82in diameter, 4ft stroke) horizontal single expansion, trunk. Single screw. 800nhp. 3,352ihp = 12.569kts.

Prince of Wales Portsmouth Dyd/John Penn & Son

As built: she actually measured at 60ft 2½in beam (3,994$\frac{18}{94}$bm). 6,201 disp.

Conversion ord: 9.4.1856. Begun: 27.10.1856. L: 25.1.1860 (for Ordinary).

Never fitted for sea. Engines removed in 1867. Renamed ***Britannia*** 3.3.1869 as boys' training ship at Dartmouth. Hulked 9.1909. Sold to Garnham 13.9.1914; resold to Hughes Bolckow and arrived at Blyth 7.1916 to BU.

ROYAL ALBERT – Conversion of 120-gun sailing First Rate (incomplete on stocks) to 121-gun screw ship, 1852. In 1852 the conversion of the unlaunched *Royal Albert* (on the stocks since 1844) to a screw 120-gun ship was approved, to have engines of 620nhp 'originally built for *Euphrates*'. A committee of shipwrights and engineers was summoned on 23 September 1852, and a month later concluded that the stern needed to be modified (plans approved 5 May1853) and that a 400nhp engine should be substituted for that of 620nhp. On 9 January 1854 a revised tender from Penn for 500nhp engines was accepted. So emerged the only three-decker of the last generation not to have originated as a *Queen* Class design.

Dimensions & tons: 232ft 9in, 193ft 6⅛in x 61ft 0in (60ft 2in for tons) x 24ft 2in. 3,726$\frac{26}{94}$bm. 5,517 disp.

Men: 1,050. Guns: LD 32 x 8in (65cwt/9ft); MD 32 x 32pdrs (56cwt/9½ft); UD 32 x 32pdrs (42cwt/8ft); QD/Fc 24 x 32pdrs (42cwt/8ft) + 1 x 68pdrs (95cwt/10ft).

Machinery: 2-cylinder (64½in diameter, 3ft 4in stroke) horizontal single expansion, trunk. Single screw. 500nhp. 1,801ihp = 10kts.

Royal Albert Woolwich Dyd/John Penn & Son

Conversion ord: 31.1.1852. Begun: 25.10.1852. L: 13.5.1854. C (at Portsmouth Dyd): 19.11.1854, fitted for troops.

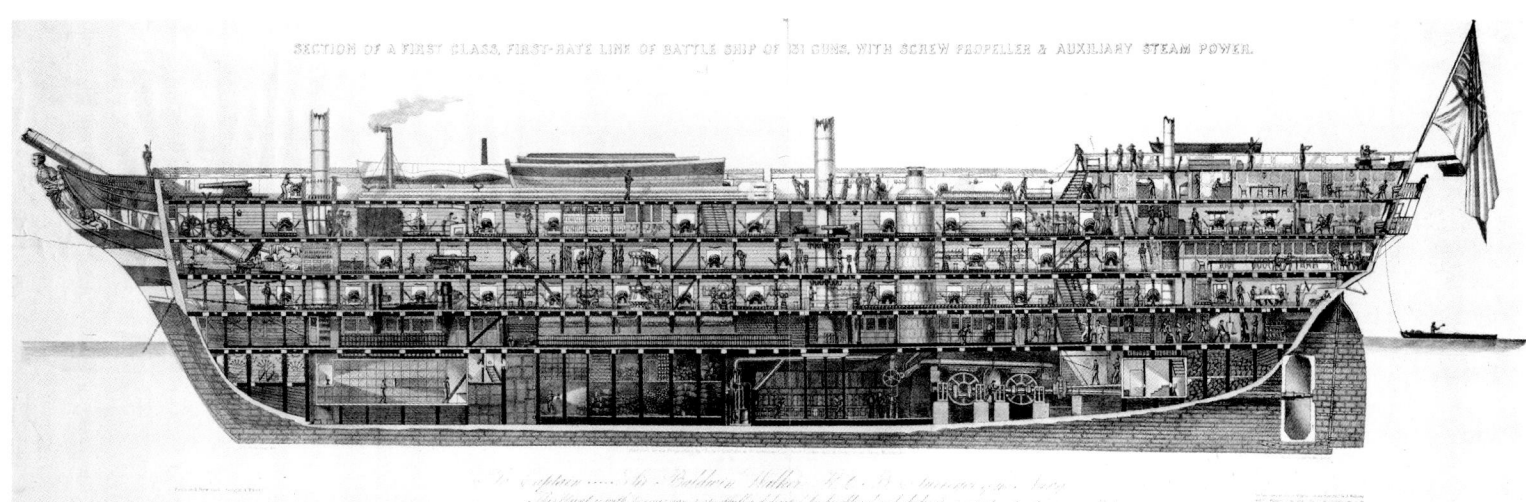

This highly detailed cutaway of the *Royal Albert* was published by Charles Lewis Pickering in 1852 – before the ship was completed – so it may not be entirely accurate, but it was dedicated 'with permission' to the Surveyor of the Navy, which suggests official support if not necessarily access to actual plans. It does provide a level of information about the interior of wooden screw battleships unobtainable even on the 'as fitted' draughts. *(NMM 9351)*

First cost: £174,345 (including hull £111,865, machinery £31,983).

Commissioned 19.6.1854 at Sheerness under Cmdr Alexander Little, then 10.1854 under Capt. Thomas Sabine Pasley; paid off 27.11.1854. Recommissioned 14.2.1855 under Capt. Robert William Mends, as flagship of Rear-Adm. Edmund Lyons, C-in-C for the Mediterranean; in 4.1857 under Capt. Francis Egerton (–20.8.1858). On 25.8.1858 under Capt. Edward Bridges Rice, as flag of Rear-Adm. Charles Howe Fremantle, with the Channel fleet; from 10.10.1860 flag of Rear-Adm. Robert Fanshawe Stopford. On 1.10.1860 under Capt. Henry James Lacon, still flag of Stopford; paid off 25.1.1861 as 'unfit for further service'. Sold to Castle 9.1884 to BU at Charlton.

ROYAL GEORGE – Conversion of 120-gun sailing First Rate (launched 1827) to a 120-gun screw ship, design of 19 April 1852. A *Caledonia* Class First Rate converted without major structural alterations. The conversion of 1852–3 left her extremely crank, and her poop and forecastle were removed in December 1854.

Men: 950. Guns: LD 8 x 8in (65cwt/9ft) + 24 x 32pdrs (56cwt/9½ft); MD 4 x 8in(65cwt/9ft) + 30 x 32pdrs (50cwt/8ft); UD 34 x 32pdrs (42cwt/8ft); QD/Fc 6 x 32pdrs (45cwt/8½ft) + 14 x 32pdr carronades (17cwt/4ft).

Machinery: 2-cylinder (58½in diameter, 3¼ft stroke) horizontal single expansion, trunk. Single screw. 400nhp. 1,417ihp = 8.65kts.

Royal George　Chatham Dyd/John Penn & Son

As built: 205ft 7½in, 171ft 0in x 54ft 6½in (53ft 7½in for tonnage) x 23ft 2in. 2,615⁵⁷/₉₄bm. 4,205 disp. Draught 15ft 8in/18ft 3in.

Conversion ord: 30.10.1852. Begun 1.11.1852. L: (undocked) 22.6.1853. C: 18.12.1853 (at Sheerness Dyd), 3.1854 (at Plymouth Dyd).

First cost: £64,236.

Commissioned 25.10.1853 at Sheerness under Capt. Henry John Codrington, sailed for the Baltic; on 13.2.1856 under Capt. Robert Spencer Robinson, as troop transport for the Crimea; paid off 29.8.1856 at Sheerness. Recommissioned 9.4.1858 under Capt. John Coghlan Fitzgerald, as guard ship of the Ordinary at Sheerness; paid off 1.7.1858. Cut down to 89-gun two-decker at Plymouth (which see) in 1860. In 4.1864 under Capt. Michael

De Courcy, as coastguard at Kingstown (replacing *Ajax*); on 3.4.1865 under Capt. Thomas Miller, in same role (but in 7.1867 to Spithead for Naval Review); in 4.1868 under Capt. Robert Jenkins, in same role; paid off 31.12.1869 (replaced at Kingstown by *Pallas*). Became receiving ship at Portsmouth. Sold to Castle 23.1.1875 to BU at Charlton.

WINDSOR CASTLE Class – Conversions of 120-gun sailing First Rates (incomplete on stocks) to 116-gun First Rates, 1857. Originally begun as *Queen* Class 110-gun sailing ships, but work was suspended and they remained on the stocks – the ship ordered as *Victoria* being renamed *Windsor Castle* on 6 January 1855. Both ships were approved in 1857 for conversion to screw three-deckers of 120 guns (without lengthening) – i.e. less heavily modified than the four *Wellington* type. But *Windsor Castle*'s poor sea-keeping qualities became apparent on trials. This resulted in her seeing no sea service, and in the conversion of *Royal Frederick* being amended by cutting her down into a two-decker. *Windsor Castle*'s guns were reduced from 116 to 102 (as below) on completion, and by 1862 to 97 guns comprising 1 x 110pdr (on pivot), 30 x 8in, 4 x 70pdrs, 6 x 40pdrs and 56 x 32pdrs.

Dimensions & tons: (original design) 216ft 9in, 178ft 8in x 55ft 7in oa (54ft 9in for tonnage) x 23ft 9¾in. 2,849⁷/₉₄bm. (final design) 204ft 0in, 166ft 5¼in x 60ft 0in (59ft 2in for tonnage) x 23ft 9in. 3,099¹⁶/₉₄bm.

Men: 930. Guns: LD 30 x 8in (65cwt/9ft); MD 30 x 32pdrs (56cwt/9½ft); UD 30 x 32pdrs (42cwt/8ft); QD/Fc 10 x 32pdrs (42cwt/8ft) + 2 x 68pdrs (95cwt/10ft).

Machinery: 2-cylinder (64½in diameter, 3ft 4in stroke) horizontal single expansion, trunk. Single screw. 500nhp. 2,052ihp = 11kts.

Windsor Castle　Pembroke Dyd/John Penn & Son

Conversion ord: 28.2.1857. Begun: 3.10.1857. L: 26.8.1858. C: 2.2.1859 at Devonport (for Ordinary).

First cost: *Windsor Castle* £117,030 including hull as sailing vessel £84,555, conversion £14,878.

Renamed *Cambridge* 1869 as gunnery training ship at Devonport; commissioned 1.1.1869 in this role under Capt. Fitzgerald Algernon Charles Foley (–1871). On 12.8.1877 under Capt. Thomas Le Hunte Ward; on 15.8.1883 under Capt. George Stanley Bosanquet (–25.8.1886); in 10.1889 under Capt. Henry John Carr; and in 8.1900 under Capt. Charles Ramsay Arbuthnot. Sold to Cox, Falmouth 24.6.1908 to BU.

Royal Frederick　Portsmouth Dyd/Maudslay, Sons & Field

Conversion ord: 28.2.1857. Re-ordered again as a screw two-decker

An early photograph showing the *Victoria* at anchor in Grand Harbour, Malta while flagship of the Mediterranean Fleet in the mid 1860s. She and her sister *Howe* were the only purpose-built steam-driven three-deckers; they were two-funnelled, although with the funnels telescoped they are barely visible above the bulwarks. *(NMM A4197)*

(see Chapter 2) on 30.4.1859, renamed *Frederick William* 28.3.1860.

VICTORIA Class New three-decker, 121-gun ships. The designs by Isaac Watts were drawn in November 1854 and approved in January 1855 – for the *Howe* on 5 January and the *Victoria* on 6 January, with the new names assigned on the same dates (the existing *Victoria* was renamed *Windsor Castle* at the same time). These were the only two-funnelled screw battleships and, apart from the French *La Bretagne* and three Russian ships, the only screw-driven three-deckers to be designed and built as such. The original 245ft design was lengthened for both ships by 15ft and the breadth increased by 1ft under AOs of 18 December 1857 and 6 March 1858.

Design dimensions & tons: 260ft 0in, 221ft 0¼in x 59ft 2in (60ft 0in oa) x 25ft 10in. 4,116³¹⁄₉₄bm. 6,959 disp.

Men: 1,000. Guns: LD 32 x 8in (65cwt); MD 30 x 8in (65cwt); UD 32 x 32pdrs (58cwt); QD/Fc 26 x 32pdrs (42cwt) + 1 x 68pdr (95cwt) pivot-mounted.

Machinery: 2-cylinder (92in diameter, 4ft stroke) horizontal single expansion (trunked in *Howe*). 8 boilers (4 fwd and 4 aft of the engines). Single screw. 1,000nhp. *Howe* 4,564ihp = 13.565kts; *Victoria* 4,403ihp = 11.797kts.

Howe Pembroke Dyd/John Penn & Son.

As completed: 260ft 0in, 219ft 10⅛in x 60ft 3in (61ft 1in oa) x 26ft 10in. 4,245³¹⁄₉₄bm.

Draught 10ft 9in fwd/20ft 9in aft.

Ord: 3.4.1854. (named 5.1.1855) K: 10.3.1856. L: 7.3.1860. C: 16.8.1860 at Devonport Dyd

First cost: total £147,465 (machinery £62,500).

Commissioned 3.5.1860 under Capt. Frederick Herbert Kerr, as flag of Vice-Adm. Houston Stewart, at Devonport. Never saw sea service, placed in reserve at Plymouth on completion. Renamed *Bulwark* 3.12.1885, then renamed *Impregnable* 27.9.1886 to become boys' training ship at Devonport. Recommissioned 9.1889 as training ship under Capt. Robert Hastings Harris (–1890). Recommissioned 1.1.1899 under Capt. A. C. B. Bromley (–1901). Recommissioned 4.1908 under Capt. Lionel Halsey, as flag of Sir Wilmot Hawksworth Fawkes; in 2.1909 under Capt. Herbert W. Savory (–1910). In 4.1911 under Capt. A. L.

Cay, as flag of Sir William Henry May; in 12.1911 under Capt. Edward Radcliffe Pears. In 3.1913 under Capt. E. P. C. Black, as flag of Sir George Le Clerk Egerton; in 10.1914 under Capt. P. Streatfield. Name reverted to *Bulwark* 12.1919. Sold to J. B. Garnham 18.2.1921 to BU.

Victoria Portsmouth Dyd/Maudslay, Sons & Field

As completed: 260ft 0in, 221ft 0in x 59ft 3in (60ft 1in oa) x 26ft 10in. 4,126⁷³⁄₉₄bm.

Draught 13ft 8in fwd/21ft 2in aft.

Ord: 3.4.1854. (named 6.1.1855) K: 1.4.1856. L: 12.11.1859. C: 20.4.1860.

First cost: total £150,578 (hull £120,593).

Retained at Portsmouth 1860–64.

Commissioned 2.11.1864 under Capt. James Graham Goodenough, as flag of Vice-Adm. Robert Smart, C-in-C for the Mediterranean. On 19.4.1866 under Capt. Alan Gardiner, as flag of Vice-Adm. Lord Clarence Edward Paget, C-in-C in the Mediterranean, based at Malta. On 10.6.1867 under Cmdr William Codrington, as flag of Adm. Sir Thomas Sabine Pasley in the Mediterranean; at Spithead Review 17.7.1867, then paid off 7.8.1867 into Ordinary at Portsmouth. Sold 31.5.1893 to BU.

Further conversions of sailing three-deckers to steam power followed these orders. On 23 June 1854 it was instructed that, following the success of the converted *Duke of Wellington*, the *Royal Sovereign* should be similarly converted to screw, with an engine of 700 or 800 hp and a new tonnage of 3,771bm. On 9 April 1856 the *Prince of Wales* was likewise ordered to be completed as a screw ship. The remaining three-deckers were not to be lengthened; on 28 February 1857 it was decided that the *Royal Frederick* and *Windsor Castle* should be fitted with 500hp engines and converted to 100-gun ships, although in the 1858 reclassifications both ships were restored to 120 guns, and on 30 April 1859 it was decided that the *Royal Frederick* should be cut down to a two-decker in the same way as the *Queen*. The *Windsor Castle* was on 24 January 1860 reduced to 116 guns. On the same day the Admiralty Board directed the Surveyor that for all future ships commissioned the First Rate was 'to comprise all ships carrying 110 guns and upwards, or the complements of such (which) consist of 1,000 men and upwards'.

2 Second Rates of 80 to 101 guns

Unlike France and Spain with their relatively extensive construction of (respectively) 118-gun and 112-gun ships, Britain built very few First Rate three-deckers. However, as a cheaper alternative the Navy Board had build a considerable number of slightly smaller (and shorter, hence slower and more leewardly) three-decker Second Rates of 90 and later 98 guns. As new First Rates of 120 guns became the norm, some of the older and smaller First Rates of 100 guns such as the *Britannia*, *Victory* and *Royal Sovereign* were reduced to 98 guns and re-classed as Second Rates. With the re-organisation of the rating system which took effect in January 1817, adding the carronades on forecastle and quarterdeck to the count of guns, all these saw increases in their gun ratings which took them over 100 guns, and all were therefore re-classed as First Rates.

The new ships which took their place in the Second Rate were the larger two-deckers – the ships of 80 or 84 guns. Over the next four decades, as structural improvements led to the ability to build longer ships, these were augmented by new two-deckers of 90 or even 100 guns (a pivot-mounted gun on the centreline added to the forecastle generally raised this to 91 and 101 respectively). However, as two-deckers and lacking the additional space for flagship facilities that three-deckers possessed, these remained classed as Second Rates.

There were recognised differences in the rôles of the smaller (80-gun) and larger (90-gun or more) Second Rate two-deckers, and it is tempting to list the developments of these classes separately. However, as steam screw propulsion was introduced from the end of the 1840s, re-ordering blurred this distinction, so this chapter deals with all Second Rates chronologically without separating the smaller and larger ships into subsections.

(A) Vessels in service or on order at 1 January 1817

The Second Rate was totally transformed by the changes in the rating system which came into force in January 1817. At the end of 1816 there had been fourteen Second Rates in existence, all three-deckers classed as 98 guns:

In commission	In Ordinary	In harbour service
98 guns		
Impregnable (1810)	*Victory* (1765)	*Saint George* (1762)
	Royal Sovereign (1786)	*Barfleur* (1768)
	Prince of Wales (1794)	*Glory* (1788)
	Dreadnought (1801)	*Prince* (1788)
	Ocean (1805)	*Neptune* (1797)
	Boyne (1810)	*Temeraire* (1798)
	Union (1811)	

Of the fourteen the *Saint George* (ex-*Britannia*), *Victory* and *Royal Sovereign* had originally been classed as 100-gun First Rates. Under the changes to the rating system which came into force in January 1817, all fourteen of these three-deckers were re-classed as First Rates, and so their details will be found in Chapter 1; in their place thirteen large two-deckers (formerly Third Rates) were newly classed as Second Rates of either 84 or 80 guns. This included the *Talavera* then under construction (she was renamed *Waterloo* in July 1817) and two ships (to the lines of the ex-French *Canopus*) on order but not yet commenced. Among the older ships, the *Foudroyant* and *Canopus* were undergoing 'Large Repairs' amounting virtually to rebuilding and giving each a much extended lease of life.

Ex-*SPANISH* Prize This very elderly vessel, even when taken in 1780, was slow and a poor sailer, while her 24pdr main armament gave her considerably less force than the newer British- and French-built vessels of similar size. Her original sister ship *Rayo* was rebuilt 1796 by Spain as a 100-gun three-decker (wrecked after Trafalgar).

Gibraltar (Spanish *Fénix*. L: 28.6.1749 at Havanna).
> Dimensions & tons: (by 1792) 178ft 10¼in, 144ft 5¾in x 53ft 3¾in x 22ft 4in. 2,184²¹⁄₉₄bm.
> Men: 650. Guns: LD 30 x 24pdrs; UD 32 x 18pdrs (quickly altered to 24pdrs by AO 11.1781); QD 12 x 9pdrs + 2 x 68pdr carronades (by 1810, 4 x 12pdr + 8 x 32pdr carronades); Fc 6 x 9pdrs (by 1810, 4 x 12pdr + 2 x 32pdr carronades).
> Taken in Rodney's Action off Cape St Vincent 16.1.1780. Named and registered by AO 20.3.1780. Fitted and coppered at Plymouth (for £16,068.5.3d) 4–8.1780.
> Commissioned 2.1780 under Capt. John Carter Allen. In action 1780–83 in West Indies and East Indies; paid off 7.1784. Fitted for Ordinary 9.1784. Underwent Middling Repair and fitted (for £36,713.0.6d) at Plymouth 2.1788–8.1790. Recommissioned 5.1790 under Capt. Samuel Goodall, for Spanish Armament. Recommissioned 5.1793 under Capt. Thomas Mackenzie; fitted at Plymouth (for £17,845) 9.1793 and fought in battle of Glorious First of June off Ushant 1.6.1794 (had 2 dead, 12 wounded). Under Capt. John Pakenham from 8.1794; sailed for the Mediterranean 23.5.1795; in action off Hyères 13.7.1795; sent home foe repairs 1.1797. Made good defects at Plymouth (for £12,818) 2–4.1797. Under Capt. William Hancock Kelly from 7.1797, sailed for the Mediterranean; involved in Warren's pursuit of Gantaume's squadron 3.1801. Under Capt. George Ryves 6.1803. Mutiny 10.1803. Paid off 7.1804. Fitted at Portsmouth (for £30,643) 7.1805–3.1806; re-classed as Second Rate 1805. Recommissioned 11.1805 under Capt. Mark Robinson, for the Mediterranean; from 1806 under Capt. William Lukin (later named Windham) and then Capt. Willougby Lake; in chase of *Le Vétéran* 26.8.1806. From 4.1807 under Capt. John Halliday, in the Channel. In 4.1809 under Capt. Henry Lidgbird Ball; action in the Basque Roads 4.1809; under (temp.) Capt. Valentine Collard 6.1809, then 1810 Capt. Robert Plampin, in the Channel. In 1.1812 under Capt. George Scott; paid off into Ordinary at Plymouth 1813. Fitted as a powder hulk at Plymouth 8–12.1813, then as a lazarette 9.1824, to lie at Milford. BU at Pembroke Dock 11.1836.

CAESAR Edward Hunt design approved 28 November 1783, the first British two-decker 80 since the 1690s.

> Dimensions & tons: 181ft 0in, 148ft 3⅛in x 50ft 3in x 22ft 11in. 1,991²⁹⁄₉₄bm.
>
> Men: 650 (later 719?). Guns: LD 30 x 32pdr; UD 32 x 24pdr; QD 14 x 9pdr; Fc 4 x 9pdr.

Caesar Plymouth Dyd [M/Shipwright Thomas Pollard to 4.1793; completed by Edward Sison]

> As built: 181ft 0in, 148ft 1in x 50ft 5in x 22ft 11in. 2,002⁷⁴⁄₉₄bm. Draught 13ft 8in/19ft 0in.
>
> Ord: 13.11.1783. K: 24.1.1786. L: 16.11.1793. C: 13.2.1794.
>
> First cost: £59,786.
>
> Commissioned 12.1793 under Capt. Anthony Molloy. Led the British column at 'Glorious First of June' (Ushant) 1.6.1794; lost 18 killed, 71 wounded; however, Molloy was dismissed his ship for failing to prosecute action to his best ability. In 8.1794 under Capt. John Whitby (?as flagship of Cornwallis), then 1.1795 under Capt. William Mitchell, 2.1795 under Capt. William Murray (temp.) and 3.1795 under Capt. Charles Nugent. Refitted at Portsmouth (for £8,221) 2–3.1796. Refitted at Plymouth (for £10,999) 11.1797–6.1798. In 1797–98 under Capt. Roddam Home; in North Sea 1797 and Channel 1798. In 3.1799 under Capt. Sir James Saumarez, in the Mediterranean. From 1.1801 under Capt. Jahleel Brenton, as flagship of the now Rear-Adm. Saumarez (–1802); led squadron in action with Linois's squadron off Algeciras 6.7.1801 (9 killed, 33 wounded), and then in action in the Gut of Gibraltar 12.7.1801. In 4.1802 under Capt. Hugh Downman; paid off 8.1802. Refitted at Plymouth (for £37,300) 7.1804–6.1805. Recommissioned 5.1805 under Capt. John Rodd; in 7.1805 under Capt. Sir Richard Strachan; attempted attack on Brest fleet in Camaret Bay 21.8.1805; Strachan led detached squadron in action with Dumanoir's squadron 4.11.1805, taking all four French vessels (*Caesar* lost 4 killed, 25 wounded). From 9.11.1805 under Capt. Thomas Shortland, as flagship of the now Rear-Adm. Strachan; from 12.1805–Summer 1806, pursuit of Leissègues and Willaumez. From 11.1.1806 under Capt. Charles Richardson (–1810), still Strachan's flagship; off the Chesapeake 1.1807; at Blockade of Rochefort 1807–08. From 1809 flagship of Rear-Adm. Robert Stopford; participated in destruction of three French 40s (*L'Italienne*, *Le Calypso* and *La Cybèle*) at Sables d'Olonne (the Basque Roads) 24.2.1809, and in the attack on the Basque Roads 12.4.1809; in Walcheren operations 1809; sailed for Portugal 13.3.1810. On 21.4.1810 under Capt. William Granger, at Lisbon; paid off 25.5.1811. In 1812 under Capt. Jeremiah Coghlan, repairing at Plymouth; to Ordinary in 1813. Fitted as an army clothing depot ship at Plymouth 12.1813–2.1814. BU there 2.1821.

FOUDROYANT Sir John Henslow design approved 16 May 1788. She underwent a 'Large Repair' in 1815–19 which amounted virtually to a rebuilding (costing as much as a new ship).

Dimensions & tons: 184ft 0in, 151ft 5⅛in x 50ft 6in x 22ft 6in. 2,054⁶¹⁄₉₄bm.

> Men: 650. Guns: LD 30 x 32pdrs; UD 32 x 24pdrs; QD 14 x 12pdrs; Fc 4 x 12pdrs (initially intended for 9pdrs on her QD and Fc, these were substituted by 12pdrs under AO 12.5.1788) + 2 x 32pdr carronades; RH 6 x 18pdr (later 4 x 24pdr) carronades. 10 of her 14 guns were replaced later by 32pdr carronades.
>
> By the 1830s she carried: LD 28 x 32pdrs + 2 x 68pdr (Miller's), UD 30 x 32pdrs + 2 x 68pdrs (Miller's), QD 4 x 32pdrs + 10 x 32pdr carronades, Fc 4 x 32pdrs + 2 x 32pdr carronades, RH nil.

Foudroyant (ex-*Superb*, renamed 16.11.1788) Plymouth Dyd [M/Shipwright Thomas Pollard to 4.1793, then Edward Sison to 6.1795, completed by John Marshall]

> As built: 184ft 8½in, 151ft 1¼in x 50ft 7¾in x 22ft 6in. 2,061⁵⁷⁄₉₄bm. Draught 14ft 3in/18ft 7in.
>
> Ord: 17.1.1788. K: 5.1789. L: 31.3.1798. C: 25.6.1798.
>
> First cost: £60,685 (£49,356 for hull, masts and yards; £11,329 for rigging and stores).
>
> Commissioned 5.1798 under Capt. James Dacres; from 25.5.1798 under Capt. Sir Thomas Byard; in Warren's action off Ireland against Bompart's squadron 11.10.1798 (9 wounded); Byard died 31.10.1798, and ship under (temp.) Cmdr William Butterfield. From 11.1798 under Capt. John Elphinstone, as flagship of Vice-Adm. Lord Keith; sailed for the Mediterranean 6.12.1798. In 4.1799 under Capt. William Brown, then from 6.1799 under Capt. Thomas Hardy, as flagship of Rear-Adm. Lord Nelson (to 6.1800). From 10.1799 under Capt. Sir Edward Berry; blockade of Malta in 1800; took French 74-gun *Le Généreux* 18.2.1800, and 80-gun *Le Guillaume Tell* 30.3.1800. In 1801 under Capt. (acting) Philip Beaver, then William Young, later Thomas Stephenson, all as flagship of Lord Keith; involved in Egypt operations. In 6.1801 under Capt. John Clarke Searle, then 9.1801 John Elphinstone. Paid off at Plymouth 7.1802. Underwent Middling Repair at Plymouth 1–11.1803. Recommissioned 6.1803 under Capt. Peter Spicer, as flagship of Sir James Dacres, from 10.1803 flagship of Sir Thomas Graves in the Channel; from 3.1804 under Capt. Peter Puget. In 10.1805 under Capt. John White, as flagship of Vice-Adm. Sir John Warren; in action against French squadron 13.3.1806, taking 74-gun *Le Marengo*. In 1.1807 under Capt. Richard Peacock, then late 1807 under Capt. Norborn Thompson, at blockade of the Tagus. In 1808 under Capt. Charles Schomberg, as flagship of Rear-Adm. Sir Sidney Smith, for the South American station. In 1810 under Capt. Richard Hancock, as flagship of Adm. Michael de Courcy; returned to England 8.1812 and paid off 11.1812. Underwent Large Repair at Plymouth (for £83,220) 1.1815–4.1819. Guard ship at Plymouth 1820. Small to Middling Repair at Plymouth (for £14,027) 10.1839–6.1840, then laid up in Ordinary. Fitted to receive Armstrong guns 3.1861, for training the Channel Squadron; gunnery training ship at Plymouth 1862–84. Attached to Cambridge 1890. Sold to J. Read 12.1.1892 then resold to German shipbreakers; later repurchased from Germany and refitted by J. R. Wheatley Cobb. Stranded on Blackpool Sands while on a fundraising and propaganda cruise 16.6.1897, and BU there; replaced by the frigate *Trincomalee* which took her name.

Ex-FRENCH PRIZES (1794–1806) All recent French 80s were built to a common specification and differed little (hence dimensions very similar). In French service, they carried 30 x 36pdrs (LD), 32 x 24pdrs (UD) and 18 x 12pdrs (QD & Fc, to which 4 x 36pdr obusiers were added). Of eight ships built to Jacques-Noël Sané's *Le Tonnant* design of 1789, four were captured during the French Revolutionary War and two more (see below) in 1805–06. The *Sans Pareil* had been converted to a sheer hulk at Plymouth in 1810, and the *Canopus*, *Tonnant*, *Malta* and *Alexandre* were also laid up in harbour by 1817; their sister *Brave* (French *Le Formidable*, built 1794–95 at Toulon) had been broken up at Plymouth in April 1816. However the *Canopus* was in excellent condition as her 'Large Repair' in 1814–16 had amounted virtually to a rebuilding (costing as much as a new ship). As such, she served as a model for a new class of 84-gun ships which took her name.

A modern model of the *Foudroyant* of 1798 depicted as first completed. One of the earliest British-designed 80-gun two-deckers, the ship survived as a training ship until 1897, a longevity which can probably be attributed to the major rebuilding she receiving between 1815 and 1819. *(NMM L3067-003)*

Sans Pareil (French *Le Sans Pareil*, built 10.1790–7.1793 at Brest. L: 8.6.1793).

> Dimensions & tons: 193ft 0in, 158ft 11¼in x 51ft 6in x 23ft 4in. 2,242²²⁄₉₄bm.
>
> Men: 738. Guns: LD 30 x 24pdrs; UD 30 x 24pdrs (Gover's); QD 2 x 24pdr + 12 x 24pdr carronades; Fc 2 x 24pdr + 4 x 24pdr carronades. This was her service armament from 1806; she had been disarmed as a sheer hulk since 1810.
>
> Taken by Lord Howe at the Glorious First of June (Ushant) 1.6.1794. Fitted at Portsmouth (for £19,051) 20.6.1794–12.4.1795. Registered in RN by AO 20.8.1794.
>
> Commissioned 3.1795 under Capt. Lord Hugh Seymour, for the Channel; Seymour made Rear-Adm. 1.6.1795, so subsequently under Capt. William Browell, as flagship of Seymour (to 1798); at Isle de Groix 23.6.1795. Refitted at Portsmouth (for £11,062) 10–11.1796, for North Sea. Refitted at Portsmouth (for £18,578) 8–11.1798, for Channel. Under Capt. David Atkins 1.1799; in attempt on Spaniards in Aix Roads 2.7.1799; under Capt. Charles Penrose 8.1799; sailed for Jamaica 20.11.1799, as flagship of Seymour (died 9.1801) then of Rear-Adm. Robert Montague; captured privateers *La Pensée* (4-gun) and *Le Sapason* (6-gun) 28.3.1800. Under Capt. James Katon 10.1801, still at Jamaica; in 8.1802 under Capt. William Essington; paid off 9.1802 and laid up at Plymouth. Fitted as a prison ship for Plymouth 1807. Converted to a sheer hulk there (for £7,484) 9–10.1810, and so served to 1838. BU 10.1842 at Plymouth.

Canopus (French *Le Franklin*, built 11.1794–3.1798 at Toulon. L: 25.6.1797).

> Dimensions & tons: 193ft 10in, 159ft 7in x 51ft 6¼in x 23ft 4½in. 2,258⁷⁷⁄₉₄bm.
>
> Men: 700. Guns: LD 32 x 32pdr; UD 32 x 18pdr; QD 2 x 18pdr + 12 x 32pdr carronades; Fc 2 x 9pdr + 4 x 32pdr carronades.
>
> Taken at Aboukir Bay 2.8.1798. Registered & named 9.12.1798. Arrived at Plymouth 17.7.1799.
>
> Commissioned 12.1798 under Capt. Bartholomew James, as flagship of Admiral Philip Affleck (died 12.1799) off Lisbon; paid off into Ordinary 8.1799. Fitted (incomplete) at Plymouth 8–11.1801, completed 1.1803; recommissioned 4.1803 under Capt. John Conn, as flagship of Rear-Adm. George Campbell; joined Nelson's fleet off Toulon 8.1803. In 2.1805 under Capt. Francis Austen, as flagship of Rear-Adm. Thomas Louis (–1807); defects made good at Plymouth (for £31,804) 6–8.1806; took

part – with rest of Duckworth's squadron – in pursuit of French squadron 1.1806 and subsequent action off San Domingo 5.2.1806 resulting in sinking or capture of five French ships of the line (losing 8 killed and 22 wounded). In 7.1806 under Capt. Thomas Shortland, sailed for the Mediterranean, still Louis's flagship; took part – again with Duckworth's squadron – in forcing of the Dardanelles 19.1.1807 and subsequent action with nine Turkish vessels sunk or taken. In 1808 under Capt. Charles Inglis (–1812), as flagship of Rear-Adm. George Martin (–1810); in action against French convoy and escorts in Gulf of Lyons 23.10.1809, driving French 80-gun *La Robuste* and 74-gun *Le Lion* ashore; flagship of Rear-Adm. Charles Boyles 1811–12; returned to England and paid off into Ordinary 2.1812. Large Repair at Plymouth (for £78,909) 3.1814–3.1816, then laid up. Fitted for sea at Plymouth (for £15,179) 5.1834. Between a Small and Middling Repair at Plymouth (for £20,816) 12.1839–5.1842. Fitted for sea at Plymouth (for £22,079) 1–5.1845. Commissioned 1.3.1845 under Capt. Fairfax Moresby, at Plymouth; for Experimental Squadron and then 12.12.1845 Channel Squadron. Laid up in Ordinary at Plymouth 5.1848. Fitted at Plymouth as a receiving ship 6–10.1862. Fitted there as a tender (to *Indus*) 2–4.1863. On 2.2.1863 under Cmdr Charles Fenton Fletcher Boughey. In 1865 under Capt. Charles Henry May, as receiving ship; on 13.5.1865 under Capt. George Le Geyt Bowyear, and on 2.9.1865 under Capt. Thomas Henry Mason. Fitted at Devonport as a mooring hulk 4–7.1869. Dismasted 4.1878. Sold to J. Pethick (for £1,750) 10.1887.

Tonnant (French *Le Tonnant*, built 11.1787–9.1790 at Toulon. L: 12.10.1789).

> Dimensions & tons: 194ft 2in, 160ft 0in x 51ft 9¼in x 23ft 3in. 2,281³⁄₉₄bm.
>
> Men: 700. Guns: LD 32 x 32pdrs; UD 32 x 18pdrs; QD 2 x 18pdrs + 14 x 32pdr carronades; Fc 4 x 32pdr carronades.
>
> Taken at Aboukir Bay 3.8.1798. Registered & named 9.12.1798. Arrived at Plymouth 17.7.1799.
>
> Commissioned 1.1799 under Capt. Loftus Bland in 1.1799; in 2.1799 under Capt. Robert Fitzgerald; laid up in Ordinary 7.1799. Middling Repair at Plymouth 12.1801–4.1803; recommissioned 3.1803 under Capt. Sir Edward Pellew; participated in Blockade of Ferrol.; in action with Calder's Squadron off Cape Ortegal against *Le Duguay-Trouin* and *La Guerrière* 2.9.1803. In 5.1804 under Capt. William Jervis, then 3.1805 Capt. Charles Tyler;

in Lee column at Trafalgar 21.10.1805, losing 26 killed and
50 wounded. Refitted at Portsmouth (for £17,890) 1–6.1806;
recommissioned 5.1806 under Capt. Thomas Browne, as flagship
of Rear-Adm. Eliab Harvey. In 7.1807 under Capt. Richard
Hancock, as flagship of Rear-Adm. Michael de Courcy. Defects
made good at Plymouth (for £18,197) 11–12.1809. In 1810 under
Capt. Sir John Gore; paid off 7.1812. Small Repair at Chatham
(for £17,978) 8–12.1813; then fitted for sea (for £19,743)
1–3.1814; recommissioned 1.1814 under Capt. Alexander Skene.
In 10.1814 under Capt. Charles Kerr, as flagship of Vice-Adm.
Alexander Cochrane on the North American station. In 11.1815
under Capt. John Tailour, for the Cork station as flagship of
Rear-Adm. Sir Benjamin Hallowell; paid off into Ordinary
11.1818. BU at Plymouth 3.1821.

Malta (French *Le Guillaume Tell*, built 9.1794–7.1796 at Toulon. L:
21.10.1795).
Dimensions & tons: 194ft 4in, 159ft 9⅜in x 51ft 7½in x 23ft 4in.
2,265⁵⁰⁄₉₄bm.
>   Men: 780. Guns: LD 30 x 32pdrs + 2 x 68pdr carronades; UD 30 x
>   24pdrs; QD 18 x 24pdrs + 8 x 24pdr carronades; Fc 2 x 12pdrs;
>   RH 2 x 68pdr carronades + 2 x 24pdr carronades.
>   Taken by *Foudroyant* and *Penelope* in the Mediterranean 30.3.1800.
>   Arrived Portsmouth 23.11.1800, and completed fitting there
>   10.7.1801.
>   Commissioned 5.1801 under Capt. Albemarle Bertie, for St Helens.
>   Following a serious fire, paid off in 4.1802. Recommissioned
>   3.1803 under Capt. Edward Buller. Under Capt. (temp.)
>   William Granger 1.1805, off Cadiz; participated in Calder's
>   Action on 22.7.1805. Flagship of Rear-Adm. Sir Thomas Louis,
>   8.1806; capture of French 44-gun *Le President*; sailed for the
>   Mediterranean 5.1.1807; participated in blockade of Cadiz
>   1807. Under Capt. William Shield 1807; blockade of Toulon
>   1808; subsequently under Capt. Robert Otway 1808; paid off
>   12.1808. Underwent Large Repair and fitted for foreign service
>   (for £82,861) at Plymouth 7.1809–12.1811. Recommissioned
>   9.1811 under Capt. Charles Paget, as flagship of Rear-Adm.
>   Sir Benjamin Hallowell; sailed for the Mediterranean 8.1.1812.
>   Under Capt. William Fahie in 1.1815. Defects made good and
>   fitted as a guard ship at Plymouth (for £30,184) 11.1815–1.1816.
>   Under Capt. Thomas Caulfield in 1.1816; paid off 7.1816 at
>   Plymouth. Fitted as an Ordinary (i.e. reserve) depot ship at
>   Plymouth (for £14,909) 10–11.1831. BU at Plymouth 8.1840.

Alexandre (French *L'Alexandre* [ex-*L'Indivisible*, renamed 2.1803],
built 5.1793–10.99 at Brest. L: 8.7.1799).
>   Dimensions & tons: 195ft 2in, 158ft 11⅜in x 51ft 4½in x 23ft 2in.
>   2,231⁴⁹⁄₉₄bm.
>   Men: 590. Guns: LD 28 x 32pdrs; UD 28 x 18pdrs; QD 4 x 12pdrs +
>   10 x 32pdr carronades; Fc 2 x 12pdrs + 2 x 32pdr carronades; RH
>   6 x 18pdr carronades.
>   Taken by Sir John Duckworth's Squadron off San Domingo
>   6.2.1806. Arrived Plymouth 11.5.1806.
>   Not commissioned. Converted to a powder hulk at Plymouth 1808.
>   Sold to Sedger at Plymouth (for £4,600) 16.5.1822.

Ex-DANISH PRIZE (1807) This Danish 90-gun ship along with
two 80-gun ships were taken by the British Navy at Copenhagen in
1807; all had carried 36pdrs (LD) and 24pdrs (UD) in Danish service.
However, of the 80-gun ships the *Neptunus* had bilged on Hveen
Island on her way to England and been destroyed, and her sister
Waldemar was broken up at Portsmouth in August 1816.

Christian VII (Danish *Christian den Syvende*, K: 20.4.1800. L:

29.7.1803. C: 1805 at Nyholm Dyd, Copenhagen; design by F. C. H.
Hohlenberg).
>   Dimensions & tons: 187ft 2¼in, 154ft 10½in x 51ft 0in x 21ft 7in.
>   2,131¹⁶⁄₉₄bm.
>   Men: 670. Guns: LD 30 x 32pdrs; UD 32 x 18pdrs; QD 4 x 12pdrs
>   + 10 x 32pdr carronades (from 10.1810–6 x 24pdr Govers + 12 x
>   32pdr carronades); Fc 2 x 12pdrs + 2 x 32pdr carronades (from
>   10.1810–2 x 24pdr Govers + 2 x 32pdr carronades); RH 4 x
>   18pdr carronades.
>   Taken at Copenhagen 7.9.1807. Arrived at Portsmouth 25.11.1807.
>   Completed fitting at Portsmouth (for £27,230) 11.9.1808.
>   Commissioned 5.1808 under Capt. Sir Joseph Yorke, for the
>   Channel. Her intended renaming as *Blenheim* in 1809 was
>   cancelled. In 5.1810 under Capt. Woodley Losack, then in 6.1810
>   under Capt. Richard Harward, as flagship of Vice-Adm. Sir
>   Edward Pellew, off Texel. Subsequently under Capt. George
>   Charles MacKenzie in 4.1811, and Capt. Edward Griffith in
>   5.1811; flagship of Adm. William Young 1811–12, in the Downs.
>   From 2.1812 under Capt. Thomas Browne, as flagship of Rear-
>   Adm. Philip Durham; in 4.1812 under Capt. Henry Lidgbird
>   Ball, still Durham's flagship, off Texel. Fitted as a lazarette at
>   Chatham (for £1,805) for Stangate Creek 12.1813–7.1814, serving
>   the quarantine service there until 1834. BU at Chatham 3.1838.

ROCHFORT Class An 1809 design by (Chevalier) Jean-Louis
Barrallier. The work was commenced on both ships on the site on the
north side of Milford Haven formerly occupied by Messrs. Jacobs
& Sons, which had gone into liquidation in 1800, when the émigré
designer Barrallier, who was supervising their construction, leased the
site on behalf of the Navy Board. Under him, the master shipwrights
responsible for the actual work were William Stone (1.7.1810 to
9.6.1813), then briefly Henry Canham (10.6–16.8.1813) and finally
Edward Churchill (from 18.8.1813) until 1815, when the entire
Establishment was transferred across to the south side of the Haven at
Pater (subsequently Pembroke Dock).
>   Dimensions & tons: 192ft 0in, 159ft 8in x 49ft 0in x 21ft 0in.
>   2,039¹⁴⁄₉₄bm.
>   Men: 640. Guns: LD 30 x 32pdrs; UD 30 x 18pdrs; QD 12 x 32pdr
>   carronades; Fc 2 x 18pdrs + 6 x 32pdr carronades; RH 6 x 18pdr
>   carronades.

Rochfort Milford Dyd [M/Shipwright see Notes above]
>   As built: 192ft 8½in, 160ft 6½in x 49ft 4½in x 21ft 10in. 2,081⁷⁷⁄₉₄bm.
>   Draught 13ft 8in/17ft 8in.
>   Ord: 1.6.1809. K: 8.1809. L: 6.4.1814. C: 7.5–15.6.1814 (for
>   Ordinary), then 8–10.1815 (for Sea) all at Plymouth Dyd
>   Commissioned 7.1815 under Capt. Sir Archibald Dickson,
>   as flagship of Rear-Adm. Sir Thomas Fremantle in the
>   Mediterranean 1816–18. Guard ship at Portsmouth 5.1818.
>   Recommissioned 25.8.1818 under Capt. Andrew Pellet Green;
>   Small Repair and fitted for sea at Portsmouth 8.1818–1.1819;
>   to the Mediterranean (still Fremantle's flagship until Fremantle
>   died 12.1819). Under Capt. Charles Marsh Schomberg 4.1820,
>   as flagship of Vice-Adm. Sir Graham Moore; sailed for the
>   Mediterranean 11.8.1820; paid off 4.1824. BU completed at
>   Chatham 20.6.1826.

Sandwich Milford Dyd [M/Shipwright see Notes above]
>   Ord: 14.10.1809. K: 12.1809. Cancelled 22.3.1811.

WATERLOO A Henry Peake design of 1809. While begun as
Talavera, this ship had two changes of name before finally coming into
service.

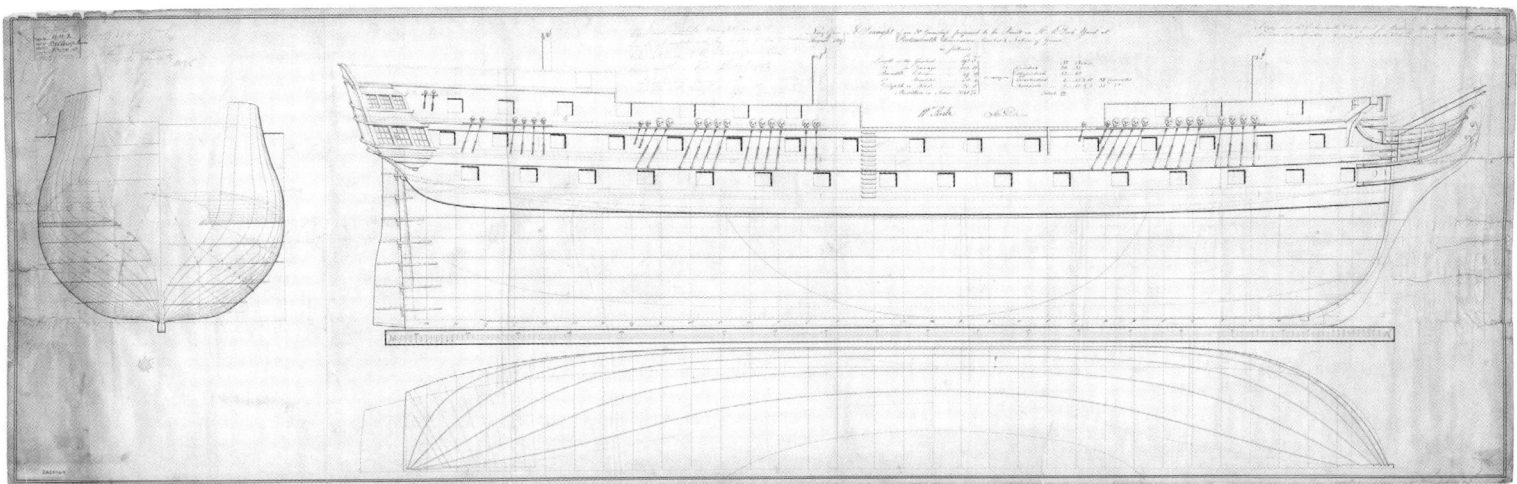

The original draught for the *Waterloo* (at that time under the name of *Talavera*) is signed by both Surveyors (Rule and Peake) and is dated August 1809, but includes subsequent alterations to raise the ship's head dated April 1813. The extra length of the 80-gun ship compared with the 74 at this time is manifest in an extra lower deck gunport (fifteen, plus a bridle port) on each broadside. (*NMM J2440*)

Dimensions & tons: 192ft 0in, 159ft 10in x 49ft 0in x 21ft 0in. 2,041²⁶/₉₄bm. 3,307 disp.

Men: 650. Guns: LD 30 x 32pdrs; UD 32 x 18pdrs; QD 4 x 12pdrs + 10 x 32pdr carronades; Fc 2 x 12pdrs + 2 x 32pdr carronades. From 1839, LD 26 x 32pdrs+ 4 x 68pdr/8in; UD 30 x 32pdrs + 2 x 68pdr/8in; QD 6 x 32pdrs + 8 x 32pdr carronades; Fc 2 x 32pdr carronades.

Waterloo (ex-*Talavera*) Portsmouth Dyd [M/Shipwright Nicholas Diddams]

As built: 192ft 0in, 159ft 7⅛in x 49ft 2½in x 21ft 0in. 2,056bm. Draught 13ft 9in/16ft 11in.

Ord: 29.8.1809. K: 11.1813. Renamed **Waterloo** by AO 23.7.1817. L: 16.10.1818. C: 28.10.1818 (for Ordinary). Roofed over for and aft 1.1819; roofing removed 12.1833. Renamed **Bellerophon** by AO 5.10.1824. Made good defects and fitted for demonstration at Portsmouth (for £8,196) 3.1833. Finally fitted for sea at Portsmouth (for £10,487) 3–7.1836.

First cost: £72,217 (fitted £85,473).

Commissioned 5.4.1836 under Capt. Samuel Jackson, as flagship of Rear-Adm. Sir Charles Paget, for the Mediterranean; on 14.4.1838 under Capt. Charles John Austen, in the Mediterranean 1838–41; operations off the Coast of Syria 1840; bombardment of Acre 3.11.1840; paid off into Ordinary 6.1841. Fitted as an Advanced ship at Portsmouth (for £15,568) 5–12.1843, then in Ordinary there until 1847. Recommissioned 23.9.1847 under Capt. Robert Lambert Baynes; fitted for sea at Portsmouth (for £11,950) 10–12.1847; troopship 1848, then to Western Squadron. On 7.11.1850 under Capt. Lord George Paulet, for the Mediterranean, then to Black Sea (bombardment of Sebastopol 11.6.1854). Fitted as a receiving ship at Portsmouth 10.1855–10.1856. Sold to J. Read, Jnr, Portsmouth to BU 12.1.1892 (sailed thence 13.4.1892).

CAMBRIDGE A 1810 design, based on Danish *Christian VII*, a prize of 1807. Later re-classed as 82 guns, but subsequently reverted to 80-gun. Launched on the day after Napoleon's abdication.

Dimensions & tons: 187ft 2in, 154ft 10½in x 50ft 9½in x 21ft 7in. 2,125²⁷/₉₄bm.

Men: 700 (630 peacetime). Guns: LD 30 x 32pdrs; UD 32 x 18pdrs; QD 4 x 12pdrs + 10 x 32pdr carronades; Fc 2 x 12pdrs + 2 x 32pdr carronades; RH 6 x 18pdr carronades. (12pdrs replaced by Congreve's 24pdrs in 1823, and then in 1839 re-armed as for *Talavera* above.)

Cambridge Deptford Dyd [M/Shipwright Robert Nelson to 7.1813, completed by William Stone]

As built: 187ft 2¼in, 154ft 10½in x 50ft 11½in x 21ft 7in. 2,139bm. Draught 14ft 11in/17ft 3in.

Ord: 16.7.1810. K: 12.1811. L: 23.6.1815. C: 12.1823 at Chatham (as guard ship).

First cost: £82,556 (+ fitting £27,389).

Commissioned 23.6.1823 under Capt. Thomas James Maling, for South America 1823–27. Between Small and Middling Repair at Chatham (for £14,904) 10.1827–10.1828. At Chatham to 1829, then Sheerness to 1834. Small Repair and fitted for sea at Sheerness (for £22,093) 1.1838–4.1840; recommissioned 31.1.1840 under Capt. Edward Barnard, for service in the Mediterranean 1840–42; operations on the Syrian coast 1840, and blockade of Alexandria then paid off into Ordinary at Plymouth 26.1.1843. Fitted as a gunnery training ship at Plymouth (for £5,373) 8.1856, and recommissioned 9.8.1856 under Capt. Richard Strode Hewlett, succeeded 3.1.1857 by Capt. Arthur William Jerningham, 1.4.1862 Capt. Leopold Heath, 20.4.1863 Capt. Charles Joseph Frederick Ewart, and 5.1867 Capt. Fitzgerald Algernon Charles Foley; paid off 1.1869 (replaced as gunnery ship by former *Windsor Castle*, which took over her name). BU (by AO 7.10.1868) completed at Plymouth 22.3.1869.

CANOPUS (or FORMIDABLE) Class Seppings's new innovation of diagonal bracing meant that longer two-deckers could now be built with less danger of hogging than before; this made the large 80-gun ship a more practical proposition, which could now replace the 74-gun ship as the 'standard' battleship. As a model, the lines of the *Canopus* (French prize *Le Franklin*, see above) was copied, and in Admiralty records this class is often labelled the *Canopus* Class; however Seppings designed new bows and stern for this class, and reduced many of the scantlings. The first vessel to this design was ordered seven weeks before Waterloo, with a second following a year later. Under the Admiralty re-classification effective in January 1817, these 80-gun Third Rates were re-classed as 84-gun Second Rates, with a further seven vessels to a modified version of this design ordered during 1817 and 1819. This class went through a series of design

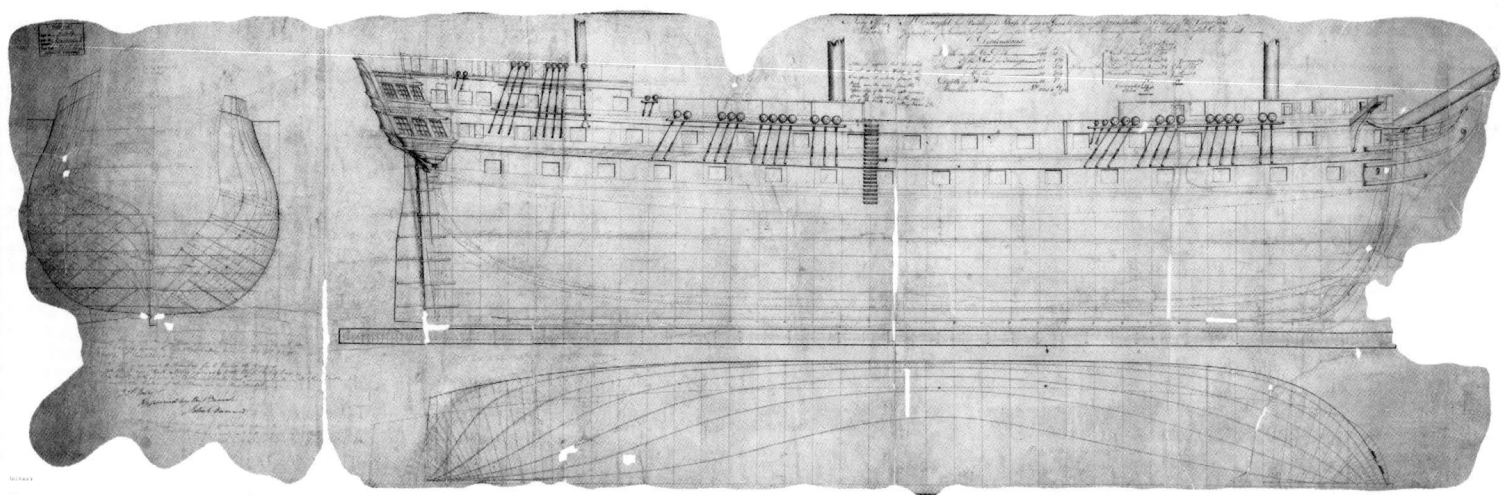

The original draught of July 1815 for the 84-gun *Formidable* notes her hull form as 'on the lines of the *Canopus*' (originally the captured French *Franklin*). The *Formidable* was built from timber seized on the stocks at Genoa in April 1814 (it had been intended for the French 74-gun *Brave*), which arrived at Chatham in May 1816. The adoption of Seppings's diagonal framing allowed increased length which in turn meant the inclusion of one more lower deck gunport, the sixteenth, on each broadside. The *Canopus* was chosen as a model because of her excellent sailing performance, notwithstanding her age, but in June 1816 the Admiralty accepted Seppings's proposals for a circular stern and by AO of 13 June 1817 directed that all new ships, down to and including Fifth Rates, should be so constructed. This made the *Formidable* one of the first three ships to be so fitted. The draught was subsequently amended in November 1816 for the *Formidable*'s intended sister *Ganges*, which was to be built in the East India Company's dockyard at Bombay. *(NMM J2310)*

modifications after the first vessels (*Ganges* and *Formidable*) were begun, later ships having a round stern. *Formidable* was built from timber captured on the stocks at Genoa 18.4.1814 (this timber being the French *Le Brave* building there). Unlike the 74s, the Bombay-built 84s did not, as sometimes erroneously recorded, bring home duplicate frames for the ships building in the UK.

Dimensions & tons: (original) 193ft 10in, 160ft 2⅛in x 51ft 5¼in x 22ft 6in. 2,254⁷⁹⁄₉₄bm.
(As modified) 195ft 4½in, 161ft 11½in x 51ft 5¼in (52ft 4½in oa) x 22ft 6in. 2,279²⁹⁄₉₄bm.
Men: 700. Guns: LD 32 x 32pdrs; UD 32 x 24pdrs; QD 4 x 24pdrs + 14 x 32pdr carronades; Fc 2 x 24pdrs + 2 x 32pdr carronades. *Vengeance* LD 28 x 32pdrs + 2 x 68pdrs (Millers); UD 28 x 24pdrs + 2 x 68pdrs (Millers); QD 16 x 32pdrs (25cwt); Fc 2 x 32pdrs (49cwt) and 4 x 32pdrs (25cwt).

Formidable Chatham Dyd [M/Shipwright George Parkin]
As built: 196ft 1½in, 162ft 0⅛in x 51ft 6½in (52ft 3½in oa) x 22ft 6in. 2,289²⁸⁄₉₄bm. Draught 13ft 8in/18ft 1in.
Ord: 8.5.1815. K: 10.1819. L: 19.5.1825. C: 5.8.1825 (for Ordinary).
First cost: £64,342 to build.
Commissioned 23.8.1841 under Capt. Edward Thomas Troubridge, fitting for the Mediterranean (–23.11.1841). On 14.12.1841 under Capt. Charles Sullivan, for the Mediterranean. On 21.4.1844 under Capt. George Frederick Rich, as flagship of Vice-Adm. Edward William Campbell Rich Owen, in the Mediterranean; paid off 1845. Recommissioned 16.6.1854 under Capt. John Jervis Tucker, as guard ship at Sheerness (in Ordinary) (–10.9.1857). On 22.9.1857 under Capt. John Coghlan Fitzgerald, still guard ship and Captain Superintendent of Sheerness Dockyard (–8.4.1858); recommissioned 1.7.1857 still under Fitzgerald (died 30.5.1859). On 3.7.1860 under Capt. William Garnham Luard, as flagship of Vice-Adm. William

James Hope Johnstone, at Sheerness. On 1.7.1863 under Capt. John Fulford, as flagship of Vice-Adm. George Robert Lambert, then Charles Talbot, at Sheerness. On 21.4.1866 under Capt. Donald McLeod Mackenzie, as flagship of Vice-Adm. Baldwin Wake Walker, at Sheerness. Lent as training ship 16.7.1869. Sold to J. B. Garnham & Son to BU 10.7.1906.

Ganges East India Company, Bombay. [M/Shipwright Jamsetjee Bomanjee to 8.1821, then Nowrajee Jamsetjee] . Teak-built.
As built: 196ft 5½in, 162ft 3½in x 51ft 5½in (52ft 2½in oa) x 22ft 6in. 2,285⁸¹⁄₉₄bm.
Ord: 4.6.1816. K: 5.1819. L: 10.11.1821. C: 12.10.1822–7.1.1823 at Portsmouth (as guard ship).
First cost: £59,865 (paid by East India Company) + £14,633 (supplied by Bombay Dyd) to build, plus £30,323 fitting at Portsmouth
Commissioned 1822 under Capt. Francis Augustus Collier at Bombay for the voyage to England; paid off 10.1822. Recommissioned 17.10.1838 under Capt. Barrington Reynolds, for the Mediterranean; operations on the coast of Syria 1840; paid off 18.4.1842. At Sheerness in Ordinary 1842–48. Recommissioned 10.1.1848 under Capt. George Thomas Gordon, at Sheerness; on 1.3.1848 under Capt. Henry Smith, for the Mediterranean; on 19.10.1849 under Cmdr Oliver John Jones, in the Mediterranean. Recommissioned 25.6.1857 under Capt. John Fulford, as flagship of Rear-Adm. Robert Lambert Baynes, for the Pacific. Training ship 5.1865. On 26.7.1866 under Cmdr Frederick William Wilson, as boys' training ship at Falmouth (–20.4.1867). On 8.8.1891 under Cmdr John Rolleston Prickett, still boys training ship; paid off 9.3.1894. Renamed **Tenedos II** on 21.6.1906, then **Indus V** on 13.8.1910, and finally **Impregnable III** on 12.10.1922. Sold to BU 31.8.1929.

(B) Vessels acquired from 1 January 1817

One 80-gun ship was still building and another two (*Formidable* and *Ganges*) had been ordered by January 1817. The additional structural strength provided by Seppings's diagonal framing meant that two-deckers could now be built with as great a length as previous three-deckers. Initially the designs for 84-gun ships derived from the lines of the *Canopus* were continued. A design for a 100-gun flush-deck two-

An anonymous watercolour showing the *Vengeance* off Malta in 1855, probably returning from service off the Crimea before hastening home to pay off at the end of a four-year commission. She was one of the later 84-gun ships to the design derived from the *Canopus*, but with their round sterns modified from the more upright design of the *Formidable* to a more aesthetically pleasing raked stern. Conversion of the frames for this ship and three sisters was begun by the Navy Board in 1817, and in 1819 two more were ordered to be built in Bombay using teak timbers, while a final ship to this design was ordered later in the same year. The ship represents a modification of the *Canopus* Class, with the lower deck armament reduced from 32 to 30 guns, although two of the latter were now 68pdrs instead of the 32pdrs in the rest of the battery.

decker was produced by the simple expedient of raséeing the design of the 120-gun three-decked *Caledonia*, but this was reduced to a 92-gun ship before completion by deleting the waist guns. From this time the Admiralty (unlike their French and American rivals) took the decision not to arm the waist on either three- or two-decker ships.

Modified *CANOPUS* Class 84 guns. Based on the lines of the *Canopus* (French prize *Le Franklin*, see above). This class had gone through a series of design modifications after the first vessels (*Ganges* and *Formidable*) were built. The upperworks saw the introduction of the lightweight (25cwt) Miller's guns rather than the former carronades.

Dimensions & tons: (modified) 195ft 4½in, 162ft 2½in x 52ft 4½in (51ft 7½in for tonnage) x 22ft 6in (Bombay ships were 196ft 1½in GD). 2,279²⁹⁄₉₄ bm.

Men: 700. Guns: LD 28 x 32pdrs + 2 x 68pdrs; UD 30 x 24pdrs + 2 x 68pdrs; QD 16 x 32pdrs (25cwt); Fc 2 x 32pdrs (49cwt) + 4 x 32pdrs (25cwt).

Powerful Chatham Dyd

As built: 196ft 1½in, 161ft 11½in x 52ft 4½in (51ft 7½in for tonnage) x 22ft 6in. 2,295⁹¹⁄₉₄ bm. Draught 13ft 2in/17ft 10in.

Ord: 23.1.1817. K: 8.1820. L: 21.6.1826. C: 4.10.1826 (for Ordinary), roofed over for and aft and moved to Sheerness. Fitted for commission (for £3,114) at Sheerness 2–3.1833, but was not commissioned. Roofing removed 12.1933. Fitted for sea at Sheerness (for £16,933) 12.1838–3.1839.

First cost: £65,926 including fitting in 1826.

Commissioned 1.1.1839 under Commodore Sir Charles Napier, for the Mediterranean; operations on the coast of Syria 1840. On 9.1.1841 under Capt. George Mansel, then 9.1841 Capt. Michael Seymore, still in the Mediterranean; paid off 6.1.1842. Fitted as an Advanced ship at Portsmouth (for £10,354) 2.1842–4.1843. At Portsmouth 1842–48. Recommissioned 25.1.1848 under

Capt. Richard Saunders Dundas; fitted for sea (for £12,207) 2–12.1848, joined Squadron of Evolution 11.1848, then 12.1848 to Mediterranean; returned Portsmouth and paid off 3.1852. Recommissioned 29.5.1854 under Capt. George Mansel again, fitted for sea (for £9,814) 5–11.1854, for Particular service; on 12.9.1854 under Capt. Thomas Lecke Massie; paid off 1858. Ordered to be converted to 81-gun screw battleship 17.1.1860, but substituted by her sister *Bombay* in 5.1860 when hull found decayed, and instead became target ship 1860. Fitted as a target ship 9–11.1862 for experiments with armour plates. BU completed at Chatham 29.8.1864 (by contract with Mr Mills).

Vengeance Pembroke Dyd

As built: 196ft 5in, 162ft 2¼in x 52ft 2¼in (51ft 5½in for tonnage) x 22ft 6in. 2,284⁷⁄₉₄ bm. Draught 13ft 4in/18ft 7in.

Ord: 24.1.1817. K: 7.1819. L: 27.7.1824. Sailed 22.8.1824 for Plymouth, where housed over from the main mast forwards (for Ordinary 1824–46). Fitted for demonstration ship at Portsmouth (for £4,938) 10–12.1840.

First cost: £60,878 to build.

Commissioned 16.10.1846 at Portsmouth under Capt. Stephen Lushington, and fitted for troops at Portsmouth (for £15,723) 10.1846–1.1847; fitted for sea at Portsmouth (for £3,536) 6–7.1847, then sailed for the Mediterranean. On 23.11.1848 under Capt. Charles Philip Yorke, Earl of Hardwicke, then 7.11.1849 under Capt. Sir Henry Martin Blackwood; paid off 10.7.1850, but recommissioned same day under Blackwood (died 7.1.1851). Refitted at Portsmouth (for £13,776) 8.1850–8.1851. On 13.1.1851 under Capt. Lord Edward Russell, for Particular service, then to the Mediterranean (–1855); at Bombardment of Sebastopol 11.10.1854. At Devonport at receiving ship (66 guns) 1855–61. Fitted as a hulk 4–7.1865, housed in 1877–8 and fitted as a masting (sheer) hulk 1878–79. Sold to Mr Scawn 10.5.1897 (for £3,410) to BU.

Monarch Deptford Dyd

As built: 196ft 1¼in, 162ft 1½in x 52ft 2⅛in oa (51ft 5⅞in for tonnage) x 22ft 6in. 2,286$\frac{27}{94}$bm. Draught 13ft 5in/17ft 8¾in.

Ord: 18.7.1817. Order transferred to Chatham Dyd on 22.2.1825 (frame sent from Deptford). K: 8.1825. L: 8.12.1832, then housed over waist. C: 10.6.1833 at Sheerness (for Ordinary). Fitted there for sea 9.1840–5.1841.

First cost: £62,591 (including fitting for Ordinary), plus £22,801 fitting for sea.

Commissioned 30.10.1840 at Sheerness under Capt. Samuel Chambers, for the Mediterranean; paid off 28.10.1843. Fitted as a guard ship at Sheerness (for £4,747) 8–10.1850. On 3.12.1850 under Capt. Michael Seymour, as guard ship at Sheerness (in Ordinary). On 29.9.1851 under Capt. Charles Hope, still at Sheerness. Defects made good and fitted at Sheerness (for £9,810) 6.1853–3.1854. Commissioned 6.2.1854 under Capt. John Elphinstone Erskine, for the Baltic during the Russian War. On 4.12.1854 under Capt. Henry Lyster, at Devonport. On 28.12.1854 under Capt. George Edwin Patey, as flag of Rear-Adm. Henry William Bruce in the Pacific; sailed 7.1.1855 from Plymouth for Valparaiso; returned 25.5.1858 to Portsmouth and paid off. Recommissioned 11.1859 under Capt. Henry Harvey, as flagship of Vice-Adm. Edward Harvey, at Sheerness. In 7.1860 under Capt. Charles Wise, still guard ship at Sheerness. Target ship 1862–4 to test resistance of armour plate to shot. Contracted 16.6.1865 with White of Cowes to BU, which completed by them 3.10.1866.

Thunderer Woolwich Dyd

As built: 196ft 1½in, 161ft 11½in x 52ft 2¼in oa (51ft 5¾in for tonnage) x 22ft 6in. 2,279bm. Draught 12ft 3in/17ft 7in.

Ord: 23.7.1817. K: 4.1823. L: 22.9.1831. C: 8.3.1833 (for Ordinary). Housed over the waist 8.1833, removed 10.1833. Completed fitting for sea at Sheerness 24.1.1834.

First cost: £72,292 (including fitting for Ordinary), plus £16,427 fitting for sea.

Commissioned 14.10.1833 under Capt. William Furlong Wise, for the Mediterranean; paid off 2.1837 at Portsmouth. Recommissioned 31.1.1840 under Capt. Maurice Frederick Fitzhardinge Berkeley, sailed 5.1840 for the Mediterranean; operations on the coast of Syria 1840; bombardment of Acre 3.11.1840. On 28 July 1841 under Capt. Daniel Pring, still in the Mediterranean; home to Plymouth 19.10.1842. Fitted at Plymouth to convey troops and sailed from Cork 28.2.1843 with troops (45th Regiment of Foot) to Cape of Good Hope, returning 25.9.1843 to Plymouth (with 87th Royal Irish Fusileers), paid off 12.10.1843 and laid up. Very Small Repair at Plymouth (for £10,049) 8.1847–1.1848, then laid up again. Fitted as a target ship (by AO 24.4.1863) 5–6.1863 for trials of armour plate. Renamed ***Comet*** on 21.4.1869, then ***Nettle*** on 9.3.1870. Sold (by AO 28.5.1900) 25.11.1901 and BU at Portchester.

Asia East India Co., Bombay. Teak-built.

As built: 196ft 4½in, 162ft 5in x 54ft 2¾in oa (51ft 5¾in for tonnage) x 22ft 6in. 2,289bm. 3,594 tons displacement.

Ord: 6 & 22.4.1819. K: 1.1822. L: 19.1.1824. C: 10.1826–1.1827 at Portsmouth.

First cost: £78,541 to build, plus £34,162 fitting at Portsmouth.

Commissioned 18.3.1824 at Bombay under Cmdr Mark John Currie, sailed from there and reached Portsmouth 10.8.1824, where laid up in Ordinary. Recommissioned 6.10.1826 under Capt. Edward Curzon, as flagship of Vice-Adm. Sir Edward Codrington, for the Mediterranean; at Battle of Navarino 20.10.1827 (losing 19 killed, 57 wounded). On 8.6.1828 under Capt. William James Hope Johnstone, as flagship of Vice-Adm. Sir Pulteney Malcolm, in the Mediterranean; paid off 28.4.1830 at Portsmouth. Fitted for sea there 1–5.1831. Recommissioned 19.12.1831 under Capt. Hyde Parker, then on same date under Capt. Peter Richards, as flagship of Rear-Adm. William Parker, for Lisbon; in 12.1831 under Capt. James Richard Dacres, as flagship of Vice-Adm. Sir Josias Rowley, at Lisbon; paid off 1834 at Sheerness. Recommissioned 18 March 1836 under Capt. William Fisher, for the Mediterranean; operations on the coast of Syria 1840; paid off 5.1841. Converted to an Advanced ship at Sheerness 3.1846, then fitted for sea there (for £9,850) 5–10.1847. Recommissioned 25.8.1847 under Capt Robert Fanshawe Stopford, as flagship of Rear-Adm. Phipps Hornby; sailed 27.11.1847 from Portsmouth for the Pacific; forced back by poor weather and defects, she put into Plymouth on the way and finally sailed 6.1.1848 for Valparaiso. Returned Portsmouth 14.5.1851 and paid off 24.5.1851 into Ordinary. On 29.4.1859 under Capt. George Thomas Gordon, as guard ship at Portsmouth (replacing *Hannibal*), in Ordinary. On 18.5.1861 under Capt. Henry Broadhead, then 17.2.1864 Capt. Henry Caldwell, as flagship of Rear-Adm. George Elliot, Admiral Superintendent at Portsmouth (ship remained in this role to 1901) and captain of the Steam Reserve; in 6.1865 flagship of Rear-Adm. George Greville Wellesley. On 24.4.1866 under Capt. William Charles Chamberlain; on 30.11.1868 under Capt. Edward Bridges Rice; in 6.1869 flagship of Rear-Adm. Astley Cooper Key, then 5.1870 of Rear-Adm. William Loring; paid off 22.12.1871. On 1.4.1876 under Capt. Charles Lodowick Darley Waddilove, as flagship of Rear-Adm. Francis Leopold McClintock. In 1877 under Capt. William Codrington, as flagship of Rear-Adm. Fitzgerald Algernon Charles Foley, then 4.1883 flagship of Rear-Adm. Frederick Anstruther Herbert. Recommissioned 1.1884 under Capt. Richard Edward Tracey; on 21.6.1884 under Capt. Henry Frederick Nicholson, as flagship of Rear-Adm. Frederick Anstruther Herbert, as guard ship of the Reserve; 8.1888 flagship of Rear-Adm. William Elrington Gordon, then 11.1889 of Rear-Adm. Albert Hastings Markham; recommissioned 10.1901. In 9.1899 flagship of Rear-Adm. Pelham Aldrich. In 6.1901 under Capt. William Wilson. Sold to Adrien Merveille, Dunkirk on 7.4.1908 to BU.

Bombay East India Co., Bombay. Teak-built.

As built: 196ft 1½in, 161ft 11½in x 52ft 2¾in oa (51ft 5¼in for tonnage) x 22ft 6in. 2,279$\frac{30}{94}$bm.

Ord: 28.4.1819. K: 5.1826. L: 17.2.1828. Sailed 26.5.1828 from Bombay (under Cmdr Alexander Campbell), arrived Plymouth 12.9.1828, laid up there in Ordinary and housed over fore and aft.

First cost: £51,176 to HEICo. (+ stores supplied by Crown £17,741).

Not commissioned as a sailing warship. Fitted as a demonstration ship (for £5,723) 1841, fitted as an Advanced ship 31.3.1846. Converted to screw battleship 5.1860–25.6.1861 by AO 3.5.1860 (see Section D).

Clarence (ex-*Goliath*, renamed 26.5.1827) Pembroke Dyd

As built: 164ft 4in, 161ft 2in x 52ft 5in oa (51ft 8in for tonnage) x 22ft 6in. 2,288$\frac{40}{94}$bm. Draught 13ft 2in/19ft 0in.

Ord: 27.5.1819. K: 8.1824. L: 25.7.1827. Arrived 17.8.1827 at Plymouth for fitting and housed over.

First cost: £64,384 (+ fitting £10,705).

Not commissioned as a sailing warship. Fitted as a demonstration ship at Plymouth (for £5,024) 10–11.1840. Lent to Liverpool R.C. Reformatory Society by AO 8.1.1864. Converted to training ship 1872. Burnt by accident in the Mersey 17.1.1884.

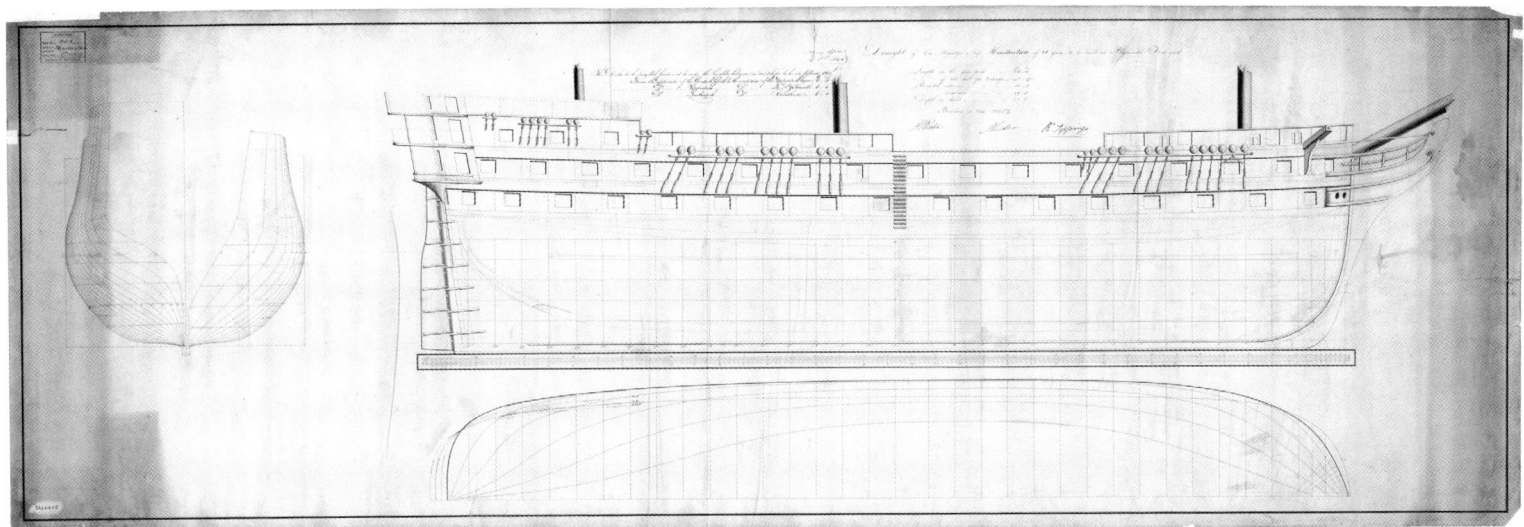

Seppings's draught of 13 October 1820 for the *Hindostan*, although signed by all three of the Surveyors, shows his characteristic round stern. Work on the ship did not commence until August 1828 as priority was given to the teak frigate *Tigris* (which was finally cancelled in 1832, and her teak materials employed in the construction of *Hindostan*), and was later suspended so that work on the 100-gun *Nile* could be prioritised, and *Hindostan* was completed with a variety of timbers including African and Danzig oak and pitch pine. By the time she was launched on a rain-soaked Monday afternoon in the summer of 1841 she was obsolete, and never saw service, lying idle in Plymouth harbour for 24 years before being taken to Dartmouth to join the cadet training ship *Britannia*. (NMM J2352)

BOSCAWEN Class 80 guns. Seppings design 1817, to 'enlarged lines of the old *Minotaur*'. A re-order of the 74-gun ship ordered in 12.1813. After cancellation in 1831, her frames were in ordered to be re-used firstly (1832) for a 50-gun frigate and then (1834) for a new 70-gun ship of this name (see Chapter 3).

> Dimensions & tons: 187ft 4½in, 153ft 8in x 50ft 0in x 21ft 6in. 2,043⁴¹⁄₉₄bm.
>
> Men: 650. Guns: LD 28 x 32pdrs + 2 x 68pdr carronades; UD 32 x 24pdrs; QD 4 x 12pdrs + 10 x 32pdr carronades; Fc 2 x 12pdrs + 2 x 32pdr carronades.

Boascawen Woolwich Dyd

> Ord: 20.11.1817. K: 1.1826. Cancelled 29.11.1832 ('the timbers to be used for *Indefatigable*).

HINDOSTAN 80 guns. Design enlarged from *Repulse* of 1803 (a Rule design of 1800). Built using the teak frames brought from Bombay aboard the *Malabar*, with material added from 1833 onwards from the cancelled frigate *Tigris*.

> Dimensions & tons: 185ft 8in, 153ft 3¼in x 50ft 8in oa (50ft 0in for tonnage) x 21ft 0in. 2,035³⁷⁄₉₄bm.
>
> Men: 700. Guns: LD 28 x 32pdr + 2 x 68pdr carronades; UD 32 x 24pdrs; QD 4 x 12pdrs + 10 x 32pdr carronades; Fc 2 x 12pdrs + 2 x 32pdr carronades.

Hindostan Plymouth Dyd

> As built: 185ft 8in, 153ft 2in x 50ft 10in oa (50ft 2in for tonnage) x 21ft 0¼in. 2,050³⁷⁄₉₄ bm. Draught 11ft 8in/16ft 4in.
>
> Ord: 24.9.1819. K: 8.1828. L: 2.8.1841. C: 9.1841 (for Ordinary).
>
> First cost: £56,717 (including fitting).
>
> Never commissioned. Designated to joined Britannia at Dartmouth as a training ship 15.5.1865; arrived at end 1865 and secured by her stern (via a covered gangway) to *Britannia*'s bow. Still cadet training ship at Dartmouth 1884; training ship for boy artificers at Portsmouth, renamed *Fisgard III* on 12.10.1905,

again renamed *Hindostan* 8.1820. Sold to J. B. Garnham & Sons 10.5.1921.

INDUS 80 guns. Design to enlarged lines of the ex-Danish *Christian VII* (taken 1807), probably an enlargement of the 74-gun *Black Prince* Class design. A re-order of the 74-gun ship ordered on 11 November 1817, using the teak frames brought from Bombay aboard the *Melville*, with material added from 1833 on.

> Dimensions & tons: 188ft 0in, 155ft 0in x 50ft 5in x 22ft 6in. 2,095⁶²⁄₉₄bm.
>
> Men: 700. Guns: LD 28 x 32pdrs + 2 x 68pdr carronades; UD 32 x 24pdrs; QD 4 x 12pdrs + 10 x 32pdr carronades; Fc 2 x 12pdrs + 2 x 32pdr carronades.

Indus Portsmouth Dyd

> As built: 188ft 8in, 155ft 2in x 51ft 2in (50ft 5in for tonnage) x 22ft 5in. 2,097⁸⁶⁄₉₄bm. Draught 12ft 0in/15ft 8in.
>
> Ord: 18.5.1820. K: 7.1824. L: 16.3.1839. C: 4.8.1841.
>
> First cost: £21,517 to build, plus £19,965 fitting for sea.
>
> Commissioned 30.10.1840 under Capt. Sir James Stirling, for the Mediterranean; paid off 6.1844. Converted to an Advanced ship at Plymouth 5–7.1844, then laid up. Recommissioned 25.11.1856 under Capt. John Charles Dalrymple Hay, as flagship of Rear-Adm. Houston Stewart, at Devonport; refitted there 12.1856–2.1857 (for £13,082). In 8.1859 under Capt. William King Hall, still Stewart's flag, to North America and West Indies. Recommissioned 14.7.1860 under Capt. Astley Cooper Key, as guard ship of the Steam Reserve at Plymouth, with flag of Rear-Adm. Thomas Sabine Pasley, as Admiral Superintendent of Devonport Dyd; on 28.11.1862 flag of Rear-Adm. Thomas Matthew Charles Symonds (–1866); remained in same role until 1890 under various captains and Admiral Superintendents. For sale to BU by AO 28.8.1897; sale suspended 4.12.1897 pending decision to use her as training ship for boys in the Thames; sold to BU (for £2,750) 11.11.1898, and removed from Devonport by breakers 4.1.1899.

RODNEY Class 92 guns. Sir Robert Seppings design of 17.11.1826. Conceived as a response to the huge American *Ohio* Class and the French *Royal Charles*, they were initially planned as 100-gun two-deckers with a fully armed spar deck (i.e. like three-deckers but without QD or Fc), based on the lines of the *Caledonia* Class. In 1828 the guns mounted on the spar deck were deleted, reducing the

A contemporary official model of *Rodney* as the design was finalised, with no guns along the wide gangways in the waist and the addition of a poop. The relentless increase in size and firepower is demonstrated by the lower deck battery; fifteen year earlier, 28 guns was the norm for two-deckers but this ship mounts 34. Apart from two 8in shell guns on each battery deck, the ships carried a homogeneous armament of 32pdrs in three different weights of gun. (*NMM F7730-001*)

ordnance by eight guns; at the same date a poop/RH was added, with 4 x 12pdr carronades mounted on it (not included in the gun rating). The *Nile* was ready for launching from No. 1 slip at Plymouth by July 1834, but the launch was postponed and finally took place five years later; she was immediately laid up in Ordinary until 1852. In March 1847 the *Prince Regent* (see Chapter 1) effectively became a fourth unit of this Class following her raséeing at Portsmouth. In 1856 during the Russian War *Rodney* and *London* were fitted as depot ships for the mortar vessels/floats, retaining lower masts but otherwise rigged as Second Class frigates.

 Dimensions & tons: 205ft 6in, 170ft 4in x 54ft 5in oa (53ft 6in for tonnage) x 23ft 0in. 2,598¼₉₄bm.
 Men: 820 (720 in peacetime). Guns: LD 32 x 32pdrs (63cwt) + 2 x 8in/68pdr (50cwt) shell; UD 32 x 32pdrs (55cwt) + 2 x 8in/68pdr (50cwt) shell; QD/Fc 24 x 32pdrs (42cwt). Later re-armed with LD 26 x 32pdrs (56cwt) + 6 x 8in/68pdr shell; UD 30 x 32pdrs (56cwt) + 4 x 8in/68pdr shell; QD/Fc 26 x 32pdrs (42cwt).
 First cost: *Rodney* £64,712 (+ fitting £23,438), *Nile* £44,532 (+ fitting £28,973), *London* £74,647 (including fitting).

Rodney Pembroke Dyd
 As built: 205 ft 8 in, 170 ft 0⅞ in x 54 ft 5½ in oa (53ft 10½ in for tonnage) x 23 ft 1 in. 2,625⁶⁹₉₄bm.
 Ord: 11.1826. K: 7.1827. L: 18.6.1833. C: 7.12.1835 at Plymouth.
 Commissioned 29.8.1835 under Capt. Hyde Parker, for the Mediterranean (–1838); paid off 12.5.1840. Recommissioned 13.5.1840 under Capt. Robert Maunsell; operations on the Syrian coast 1840; paid off 16.10.1843. Recommissioned 4.2.1845 under Capt. Edward Collier; fitted for sea at Portsmouth (for £10,935) 4–5.1845, then with the Channel Fleet, the Experimental Squadron 1845–46, to Lisbon 1847, and thence to the Mediterranean; paid off 8.3.1849. Recommissioned 6.8.1851 under Capt. Charles Graham, for the Mediterranean; Spithead Review 1851; to the Black Sea during the Russian War. From 22.11.1854 under Capt. George St Vincent King (as Graham was invalided), in the Mediterranean. On 21.7.1855 under Capt. Henry Keppel (who also commanded the Naval Brigade ashore in the Crimea during the War). On 24.1.1856 under Capt. George Knyvett Wilson; carried troops home from Black Sea; depot ship at Portsmouth 3.1856; paid off 20.8.1856. Converted to screw battleship 3.1859 to 1.1860 (see Section D).

Nile Plymouth Dyd [M/Shipwright Thomas Roberts]
 Ord: 11.1826. K: 10.1827. L: 28.6.1839. C: 12.8.1839 (for Ordinary).
 Not commissioned as sailing ship; fitted at Plymouth for Demonstration ship (for £7,579) 10.1840–3.1841; converted to screw battleship by AO 24.11.1852 (see Section D of this chapter).

London Chatham Dyd
 Ord: 11.1826. K: 10.1827. L: 28.9.1840. C: 25.2.1841 (for Ordinary) and laid up at Sheerness.
 Commissioned 28.5.1851 under Capt. George Rodney Mundy, as flagship of Vice-Adm. Josceline Percy, at Sheerness; in 1.1852 under Capt. Montagu Stopford, then 1.4.1852 under Mundy again. On 16.4.1853 under Capt. Charles Eden, at Spithead; to the Baltic for the Russian War. On 18.11.1854 under Capt. Lewis Tobias Jones (when Eden invalided), to the Mediterranean. On 13.8.1855 under Capt. Augustus Leopold Kuper, in the Mediterranean. On 24.1.1856 under Capt. William Henry Jervis; carried troops home from the Crimea; paid off 25.8.1856. Converted to screw battleship 1.1857 to 5.1858 (see Section D).

CALCUTTA 84 guns. The fourth 84-gun ship to be built at Bombay was an 1827 modification by Seppings of the *Canopus* class design with greater length of floor and waterlines filled out fore and aft, to carry a uniform 32pdr armament. Like the other Indian ships, she was built of teak.

 Dimensions & tons: 196ft 0in, 162ft 4½in x 52ft 6in oa (51ft 6in for tonnage) x 22ft 6in. 2,290⁶³₉₄bm.
 Men: 700. Guns: LD 32 x 32pdrs; UD 32 x 32pdrs; QD 16 x 32pdrs; Fc 4 x 32pdrs.
 First cost: £72,225.

Calcutta East India Co., Bombay
 Dimensions & tons: 196ft 6in, 162ft 5⅛in x 52ft 4in oa (51ft 7in for tonnage) x 22ft 6in. 2,290⁶³₉₄bm.
 Ord: 5.4.1827. K: 3.1828. (Suspended 3.12.1828, but restarted 17.9.1829). L: 14.3.1831 (actually floated out). Sailed 15.5.1831 from Bombay (under Capt. Peter Fisher), arrived Plymouth 11.10.1831 and laid up in Ordinary. Fitted for Demonstration Ship (for £7,451) 6–10.1838. Fitted for sea (for £15,451) 8–10.1840.
 First cost: £54,665 pair to HEICo., plus £17,560 worth of stores supplied by HM government.
 Commissioned 22.8.1840 under Capt. Sir Samuel Roberts, for the Mediterranean (–1842); arrived Plymouth 8.1842 and laid up after refit. At Plymouth to 1854. Recommissioned 28.4.1854 under Capt. James John Stopford, refitted there (for £9,926) 5–11.1854; to Baltic 1855 for Russian War. On 3.3.1856 under Capt. William King Hall, as flagship of Rear-Adm. Sir Michael Seymour, for China; to Canton 1856; at Fatshan Creek 1.6.1857. Fitted for Ordinary at Plymouth 7.1859, then towed to Portsmouth (by *Geyser*) 9.1863 and fitted for experimental gunnery purposes. Fitted with 4in and 5in breechloading guns 3.1887. Sold to Castle, Plymouth to BU 12.5.1908.

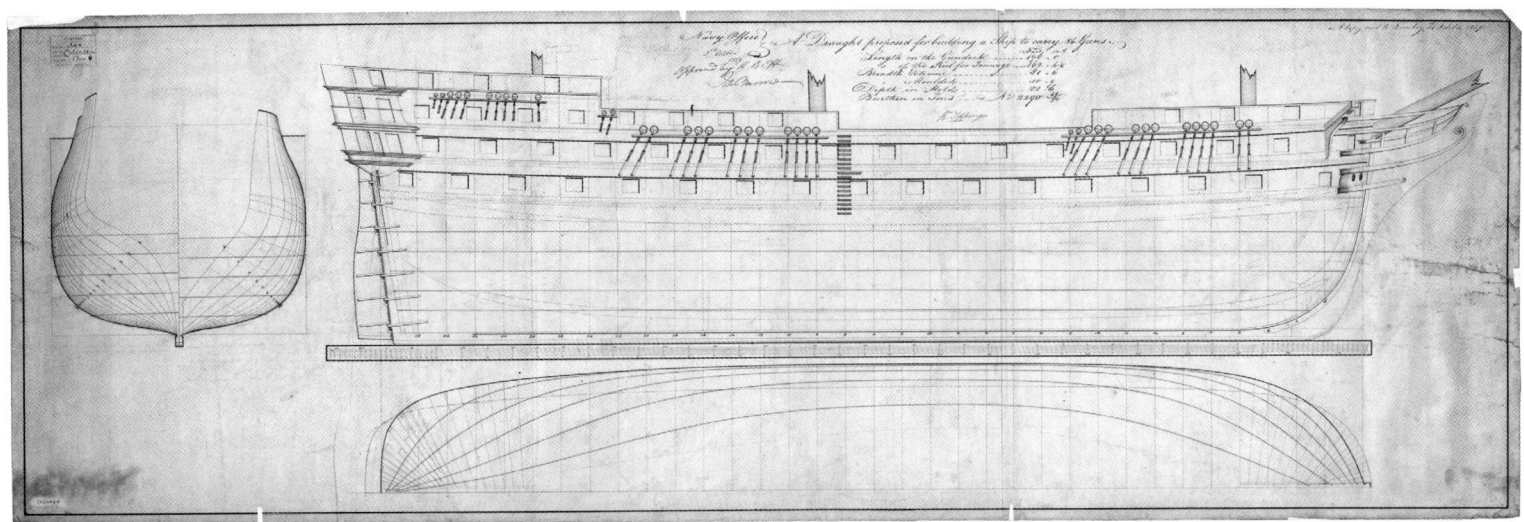

Seppings's draught of 3 October 1827 for the *Calcutta* was sent off to Bombay Dockyard aboard the *Manilla* on 31 October and the ship was laid down there in the following March. Designed for the new all-32pdr armament, the ship was more strongly built than the *Asia* and *Bombay*, being intermediate between that pair and Seppings's final 84-gun ship of 1831. (*NMM J2374*)

(C) Vessels acquired from November 1830

VANGUARD **Class** 80 guns. William Symonds design, approved 27 November 1832. Two ships were ordered in 1832, initially as 78-gun Third Rates, and two more in 1833 (although the latter pair including a rebuilding of the three-decker *Union*) to the same design. Soon after Symonds was appointed Surveyor the class was re-specified as 80-gun Second Rates, and the rebuilding of the *Union* at Devonport was cancelled by the start of 1836. Another ten were ordered to the 80-gun design in 1838–40, but two (*Albion* and *Aboukir*) were altered to a new 90-gun design (see below), another two were re-ordered as steam battleships while building, and others were converted to steam soon after their launch. A further ship, *Brunswick*, was initially ordered to this design on 19 February 1844, but was re-ordered to a new design in 1847 (see below).

Dimensions & tons: 190ft 0in, 155ft 3in x 56ft 9in oa (56ft 0in for tonnage) x 23ft 4in. 2,589^{81}/₉₄bm (the first draught was shorter on the keel – 154ft 4in – giving 2,574^{38}/₉₄bm).

Men: 645 720 (630 peacetime). Guns: LD 30 x 32pdrs (56cwt); UD 28 x 32pdrs (50cwt); QD 16 x 32pdrs (42cwt); Fc 6 x 32pdrs (42cwt). But completed with 2 LD and 2 UD guns replaced by 4 x 68pdr carronades, and 2 of the QD guns moved to the Fc.

Vanguard Pembroke Dyd

As built: 190ft 0in, 155ft 0in x 57ft 0in oa (56ft 3in for tonnage) x 23ft 4in. 2,608^{6}/₉₄bm.

Ord: 29.6.1832. K: 5.1833. L: 25.8.1835. C: 19.9.1835–5.7.1836 at Portsmouth.

First cost: £56,983 (+ fitting £20,756).

Commissioned 18.3.1836 under Capt. Duncombe Pleydell Bouverie, for the Mediterranean. On 25.1.1837 under Capt. Sir Thomas Fellowes, still in the Mediterranean. On 2.4.1840 under Capt. Sir David Dunn; refitted at Portsmouth (for £14,931) 4–9.1840, then to the Mediterranean; operations on the Syrian coast 1840; subsequently off Lisbon, then home in 1842; paid off 8.1843. Fitted for sea at Plymouth (for £7,693) 9.1844–5.1845. Recommissioned 4.2.1845 under Capt. George Wickens Willes (died 26.10.1847), for the Channel Squadron, then to

Experimental Squadron; to Lisbon 1847. On 6.11.1847 under Capt. George Frederick Rich; in the Mediterranean 1848; paid off 28.3.1849. Small Repair at Plymouth (for £20,004) 5.1849–4.1850. On 18.2.1861 under Capt. Edmund Heathcote, as guard ship at Kingstown; laid up at Sheerness 3.1862. Renamed *Ajax* 20.10.1867. BU at Chatham completed 26.6.1875.

Collingwood Pembroke Dyd

As built: 190ft 0in, 152ft 8in x 57ft 2in oa (56ft 5in for tonnage) x 23ft 4in. 2,584^{6}/₉₄bm. Draught 14ft 10in/18ft 5in (with bilgeways off and 42 tons aboard).

Ord: 29.6.1832. K: 9.1835. L: 17.8.1841. C: 13.4.1844–20.7.1844 at Portsmouth.

First cost: £55,910 (+ fitting £18,728).

Commissioned 8.5.1844 at Portsmouth under Cmdr Henry Broadhead. On 13.8.1844 under Capt. Robert Smart, as flagship of Rear-Adm. Sir George Francis Seymour, for the Pacific; paid off 20.7.1848 and laid up 1849 at Portsmouth. Converted to an Advanced ship at Portsmouth (for £8,274) 8.1848–2.1849. Converted to screw battleship 3.1860 to 7.1861 (see Section D).

Goliath Chatham Dyd

As built: 190ft 0in, 155ft 0½in x 56ft 10¼in oa (56ft 1¼in for tonnage) x 23ft 4in. 2,595^{8}/₉₄bm. Draught 14ft 0in/18ft 4in.

Ord: 9.10.1833. K: 2.1834. L: 25.7.1842. C: 7.10.1842 (for Ordinary).

First cost: £56,755 (including fitting for Ordinary).

Not commissioned as sailing ship. Converted to screw battleship 10.1856 to 11.1857 (see Section D).

Superb Pembroke Dyd

As built: 190ft 0in, 153ft 6in x 57ft 0in oa (56ft 3in for tonnage) x 23ft 4in. 2,583^{42}/₉₄bm. Draught 15ft 4½in/18ft 10in (with 42 tons on board).

Ord: 15.6.1838. K: 11.1838. L: 6.9.1842. C: as Advanced ship and for sea 7.1844–26.4.1845 at Plymouth.

First cost: £54,979 (+ fitting £25,313).

Commissioned 13.12.1844 at Portsmouth under Capt. Armar Lowry Corry, for Channel Squadron; Experimental Squadron 1845, later to Mediterranean. On 29.11.1848 under Capt. Edward Purcell; refitted at Portsmouth (for £9,929), then to the Mediterranean; paid off at Chatham 14.6.1852. Lent to the Turks 8–12.1864 as an accommodation ship in the Thames, for crews standing by two Turkish frigates building for the Ottoman Navy on that river at the time. Earmarked for use as a hospital ship for Sheerness in the event of cholera (but apparently not

A contemporary Ackermann lithograph of *Collingwood* under full sail. One of the 80-gun *Vanguard* Class, these ships were the first step by the new Surveyor, Sir William Symonds, in his intent to produce a battlefleet with significantly better sailing qualities. His work proved highly contentious, and eventually the debate was rendered irrelevant by the introduction of screw propulsion. *(NMM X0415).*

used as such). BU (by AO 3.11.1868) completed at Portsmouth 18.2.1869.

Meeanee (ex-*Madras*, renamed 19.1.1846) East India Co., Bombay.
As built: 190ft 2in, 154ft 5in x 56ft 11in oa (56ft 2in for tonnage) x 23ft 4in. 2,591¹⁴⁄₉₄bm. Draught 14ft 4in/17ft 6in (with 43¼ tons on board).
Ord: 12.3.1839 (16.9.1841). K: 4.1842. L: 11.11.1848. C: 4.4.1849 (for voyage to England under Cmdr Edward Augustus Inglefield). Arrived Chatham 1.8.1849 and laid up in Ordinary.
First cost: £87,878.
Not commissioned as sailing ship. Converted to screw battleship 12.1856 to 10.1857 (see Section D).

Centurion Pembroke Dyd
As built: 190ft 0in, 153ft 5in x 57ft 1in oa (56ft 4in for tonnage) x 23ft 4in. 2,589⁶⁄₉₄bm. Draught 15ft 3½in/18ft 10in (with 28 tons aboard).
Ord: 18.3.1839. K: 7.1839. L: 2.5.1844. C: 10.6.1844 (for Ordinary).
First cost: £57,386.
Not commissioned as sailing ship. Converted to screw battleship 9.1854 to 11.1855 (see Section D).

Colossus Pembroke Dyd
As built: 190ft 0in, 153ft 5in x 57ft 0in oa (56ft 4in for tonnage) x 23ft 4½in. 2,589⁶⁄₉₄bm. Draught 15ft 3½in/18ft 10in (with 28 tons aboard).
Ord: 18.3.1839. K: 10.1843. L: 1.6.1848. C: 3.7.1848 (sailed to Portsmouth for Ordinary).
First cost: £59,119.
Not commissioned as sailing ship. Converted to screw battleship 1.1854 to 6.1855 by AO 11.1.1854 (see Section D).

Mars Chatham Dyd
As built: 190ft 0in, 154ft 0¼in x 56ft 9½in oa (56ft 0½in for tonnage) x 23ft 4in. 2,573⁶⁷⁄₉₄bm. Draught 13ft 3in/19ft 3in.
Ord: 18.3.1839. K: 12.1839. L: 1.7.1848. C: 30.8.1848 (for Ordinary).
First cost: £57,826 (including fitting).
Not commissioned as sailing ship. Converted to screw battleship 3.1855 to 6.1856 by AO 10.8.1854 (see Section D).

Majestic Chatham Dyd
Ord: 18.3.1839. K: 2.1841. Re-ordered to complete as a screw battleship 24.11.1852 (see Section D).

Irresistible Chatham Dyd
Ord: 12.3.1840. K: 1.1.1849. Re-ordered to complete as a screw battleship 23.6.1854 (see Section D).
First cost: £35,145 as sailing vessel.

Lion Pembroke Dyd
As built: 190ft 0in, 153ft 5in x 57ft 0in oa (56ft 4in for tonnage) x 23ft 4½in. 2,589⁶⁄₉₄bm. Draught 15ft 2½in/18ft 4in.
Ord: 12.3.1840. K: 7.1840. L: 29.7.1847. C: 26.9.1847 (sailed to Plymouth for Ordinary, arriving 28.9.1847).
First cost: £59,113 (including fitting).
Not commissioned as sailing ship. Converted to screw battleship 2.1858 to 5.1859 (see Section D).

ALBION Class 90 guns. Sir William Symonds design of 13 June 1839. The first pair were originally ordered as *Vanguard* Class 80s (which see); names were allocated 20 May 1839 on this basis, but they were re-classed as 90-gun on 21 June. Another three were ordered to this 90-gun design on 12 March 1840 and were named on 14 May – although the ship first named *Prince Albert* was renamed *Princess Royal* on 26 March 1842. The. *Algiers*, first ordered 1833 as a 110-gun First Rate (see Chapter 1), was re-ordered as a 90-gun Second Rate in December 1840. *Marlborough* (at Devonport) and *St Jean d'Acre* (at Portsmouth) were first ordered 19.2.1844 to this design, but three weeks later *Marlborough* was amended to a 110-gun First Rate (see above) and the two yards interchanged. Construction of this class (excluding the nearly completed *Albion*) was suspended for a while in March 1842, and again 18 December 1844, but most re-commenced 26 March 1846.
Dimensions & tons: 204ft 0in, 165ft 11in x 60ft 0in oa (59ft 1¼in for tonnage) x 23ft 8in. 3,083¹²⁄₉₄bm.
Amended to 204ft 0in, 166ft 0in x 60ft 2¼in oa (59ft 3in for tonnage) x 23ft 8in. 3,099⁷⁄₉₄bm.
Men: 750 (peace)/820 (war). Guns: LD 4 x 68pdrs (112cwt) + 28 x 32pdrs (56cwt); UD 26 x 32pdrs (56cwt) + 6 x 8in/68pdr shell

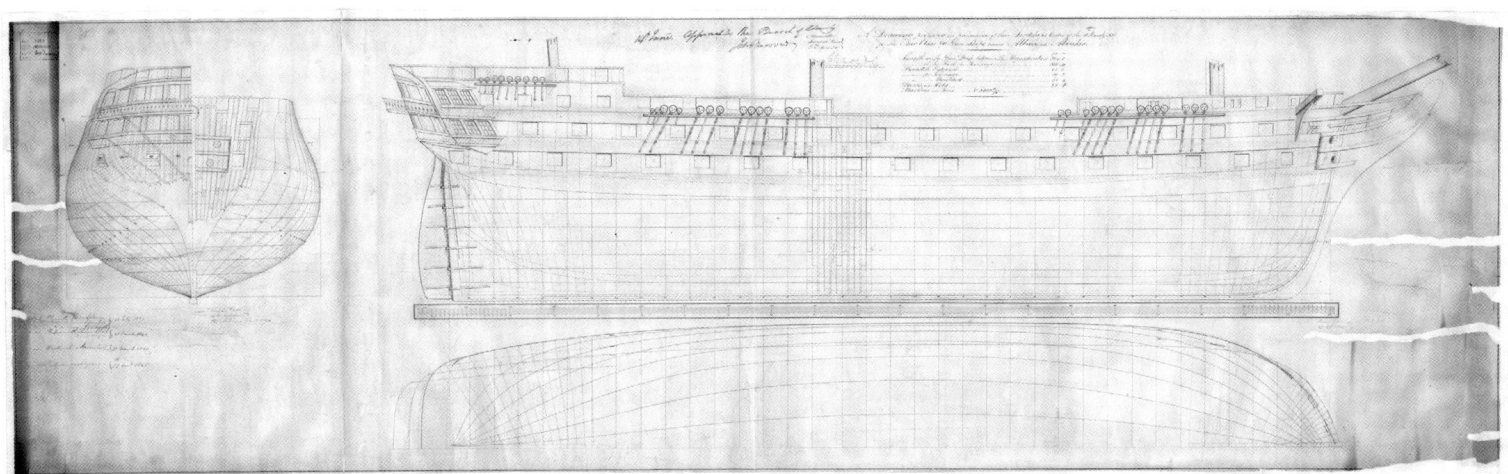

The original Symonds draught for the *Albion* Class 90-gun ships, drawn on 13 June 1839, clearly demonstrates the main features of Symonds's design philosophy – the sharp midship section, designed to produce speed, coupled with great breadth for both stability and to compensate for capacity reduced by the sharp underwater form. The draught also shows another of Symonds's innovations, the so-called elliptical stern, which was intended to combine the strength of the round stern with a more traditional appearance that pleased the seaman's (and admiral's) eye. Critics of this approach claimed Symonds's ships made poor gun-platforms because they were too 'stiff' (i.e. they rolled rapidly) and were 'uneasy' in a seaway, but the biggest disadvantage proved to be their unsuitability for the installation of steam machinery in later years.

The draught is for the *Albion* and *Aboukir*, including alterations to the design added in 1840, when the design was also approved for three new ships that were ordered in March 1840, and another in December; however all these there converted on the stocks into steam vessels. *Albion* saw service for a decade before her employment in the Black Sea during the Russian War; she was finally converted to a screw ship in 1860–61, but never saw sea service in that role. *Aboukir* was never commissioned under sail, but was converted to steam following the Russian War. *(NMM J1672)*

(65cwt); QD 16 x 32pdrs (42cwt) + 2 x 8in/68pdr shell (52cwt); Fc 8 x 32pdrs (42cwt).

Albion Plymouth Dyd
    As built: 204ft 0in, 166ft 0in x 60ft 2¼in oa (59ft 4¼in for tonnage) x 23ft 8in. 3,110⁶²⁄₉₄bm. Draught 16ft 2in/19ft 4in.
    Ord: 18.3.1839. K: 13.8.1839. L: 6.9.1842. C: 23.11.1843–23.1.1844. First cost: £77,167 (+ fitting £20,689).
    Commissioned 10.11.1843 under Capt. Nicholas Lockyer, as flagship of Adm. David Milne; on particular service (Experimental Squadron) 1844, then to Lisbon; in the Channel Squadron 1846. On 27.2.1847 under Capt. Charles Howe Fremantle (when Lockyer died), at Lisbon; to the Mediterranean 12.1847; paid off 2.4.1848 at Plymouth. Fitted for sea there (for £8,148) 6.1849–12.1850. Recommissioned 29.5.1850 under Capt. William James Hope Johnstone, for the Mediterranean. On 2.7.1852 under Capt. Stephen Lushington; to Black Sea for the Russian War (Lushington commanded the Naval Brigade from start of siege of Sebastopol on 11.10.1854 until 19.7.1855, thus ashore while *Albion* was temp. under Cmdr Henry Downing Rogers). On 7.7.1855 under Capt. James Robert Drummond. Converted to screw battleship (see Section D) 4.1860 to 5.1861, but never completed for sea. Sold to Castle & Sons to BU 8.1884.

Aboukir Plymouth Dyd
    As built: 204ft 0in, 165ft 4¾in x 60ft 1¼in oa (59ft 3¼in for tonnage) x 23ft 8½in. 3,090⁶⁰⁄₉₄bm. Draught 15ft 7in/19ft 2in (with 24 tons aboard).
    Ord: 18.3.1839. K: 8.1840. L: 4.4.1848. C: 19.4.1848 (for Ordinary). First cost: (while building as sailing 90) £80,472.
    Uncompleted as sailing ship (remained in Ordinary). Converted

to screw battleship 9.1856 to 1.1858 (see Section D). Sold to G. Lyon in Jamaica to BU 1878.

Exmouth Plymouth Dyd
    Ord: 12.3.1840. K: 13.9.1841. Converted to screw battleship on stocks (see Section D). L: 12.7.1854.

Princess Royal (ex-*Prince Albert*, renamed 26.3.1842) Portsmouth Dyd
    Ord: 12.3.1840. K: 2.1841. Re-ordered to new design 1847 (see below).

Hannibal Woolwich Dyd
    Ord: 12.3.1840. K: — . Re-ordered to new design 1847 (see below).

Algiers Plymouth Dyd
    (Re)Ord: 26.12.1840. K: 10.7.1843. Re-ordered to new design 1847 (see below).

St Jean d'Acre Plymouth Dyd (see Notes)
    Ord: 19.2.1844. K: — . Suspended (with her sisters, above) 18.12.1844 and later re-designed (see below).

AGAMEMNON 80 guns. Captain John Hayes design, 1841.
    Dimensions & tons: 195ft 0in, 161ft 0in x 54ft 8in (54ft 0in for tonnage) x 23ft 9in. 2,497²⁰⁄₉₄bm.
    Men: 750. Guns: LD 8 x 8in/68pdrs (65cwt) + 20 x 32pdrs (56cwt); UD 4 x 8in/68pdrs (65cwt) + 24 x 32pdrs (50cwt); QD/Fc 24 x 32pdrs (41cwt).

Agamemnon Woolwich Dyd
    Ord: 27.2.1841. K: — . Suspended 18.12.1844. Re-ordered as a 91-gun screw battleship 20.6.1849 (see Section D).

SANS PAREIL Class 80 guns. This ship originated in a proposal in July 1842 to rebuild the 84-gun *Sans Pareil* (the former French prize taken in 1794, see above), the Surveyor being asked whether the keel or bottom of the old ship could be used in rebuilding. On ascertaining that this was not so, the Admiralty proposed on 21 September for 'her exact lines to be taken off, and a facsimile ship to be built . . . to mount 84 guns'. On 22 April 1843 the Surveyor was ordered to prepare a sheer draught from the original lines, adapting the ports and other features to those of the 80-gun ships building. His draught was returned on 25 May with the Admiralty approving (except for minor details). Construction was suspended in October 1848 but on 15 November her completion as a steam battleship was approved.
    Dimensions & tons: 193ft 0in, 158ft 11½in x 52ft 1in (51ft 6in for tonnage) x 22ft 8in. 2,242¹⁵⁄₉₄bm.
    Men: 750. Guns (established 22.4.1843): LD 8 x 8in/68pdrs (65cwt) + 20 x 32pdrs (56cwt); UD 4 x 8in/68pdrs (65cwt) + 24 x 32pdrs (50cwt); QD/Fc 24 x 32pdrs (41cwt).

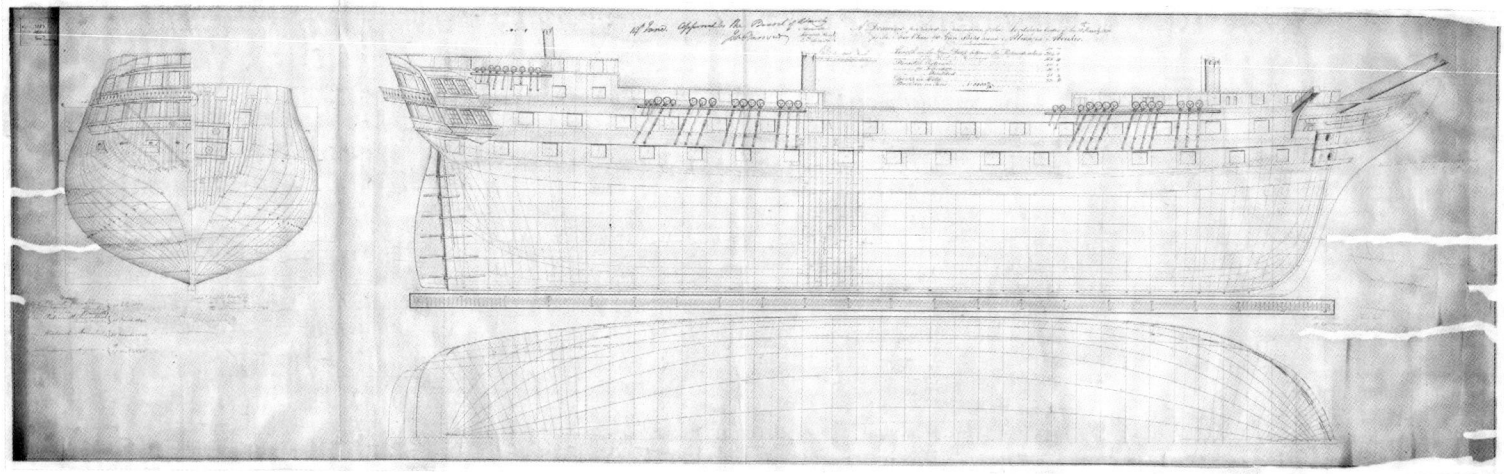

This draught of the *Hannibal* when ordered as one of the last pure sailing Second Rates shows John Edye's revised design of 30 June 1847, marking a return to less extreme hull forms, although still sharper than those of the pre-Symonds era. In September 1852 she was re-ordered as a screw vessel with a 450 nhp engine, and was completed in June 1854 and deployed to the Baltic, whence she quickly returned to the Downs conveying a large quantity of Russian prisoners taken at the capitulation of Bomarsund in August. *(NMM J1617)*

Sans Pareil Devonport Dyd

Ord: 27.2.1843. K: 1.9.1845. Suspended 2.10.1848. Re-ordered as 80-gun screw battleship 18.5.1849 and converted on stocks; however the weight (and poor performance) of the engines originally fitted meant that the guns had to be reduced to 70, so she was completed as a Third Rate [see Chapter 3].

CRESSY Class 84 guns.

In 1842 three foremen (pupils of the old School of Naval Architecture) – Samuel Read, Henry Chatfield and Augustin Creuze – formed a Committee of Naval Architecture which on 16 October 1843 completed a design for a new 80-gun ship, its lines a compromise between Seppings's flat floors of the 1820s and Symonds's sharply rising floors. The design was encouraged by Cockburn as part of his attack on the work of Symonds, who strenuously opposed the SNA design and consistently delayed it. The extra length provided room for additional LD/UD gunports. A longer alternative design was later produced but never ordered.

Dimensions & tons: 198ft 5in, 162ft 1⅜in x 55ft 0in (54ft 3in for tonnage) x 21ft 8¼in. 2,537⁸⁰⁄₉₄bm.

Men: 750. Guns: LD 6 x 8in + 24 x 32pdrs; UD 2 x 8in + 30 x 32pdrs; QD/Fc 22 x 32pdrs (6 long, 16 short).

Cressy Chatham Dyd

Ord: 19.2.1844. K: 4.1846. Re-ordered as a 91-gun screw battleship 24.11.1852 [see Section D] and converted on stocks. L: 21.7.1853.

PRINCESS ROYAL Class 90 guns.

Ordered to the *Albion* Class design, but re-ordered to a John Edye design of 1847, with breadth reduced from *Albion* design.

Dimensions & tons: 204ft 0in, 166ft 7⅜in x 58ft 0in oa (57ft 2in for tonnage) x 24ft 0in. 2,896²¹⁄₉₄bm.

Men: 750 (peace)/820 (war). Guns: LD 4 x 68pdrs (112cwt) + 28 x 32pdrs (56cwt); UD 26 x 32pdrs (56cwt) + 6 x 8in/68pdr shell (65cwt); QD 16 x 32pdrs (42cwt) + 2 x 8in/68pdr shell (52cwt); Fc 8 x 32pdrs (42cwt).

Princess Royal (ex-*Prince Albert*, renamed 26.3.1842) Portsmouth Dyd

(Re)Ord: 15.4.1847. (K: 2.1841.) Re-ordered as screw battleship [see Section D] 23.9.1852 and converted on stocks. L: 23.6.1853.

ALGIERS Class 90 guns.

Originally ordered to the *Albion* Class design, but re-ordered to a Read, Chatfield and Creuze design of 1847, lengthened by 6ft from the *Queen* design. The *Caesar* was originally ordered 24.6.1847 to this design, but subsequently re-specified as shown below.

Dimensions & tons: 210ft 0in, 170ft 0in x 60ft 0in oa (59ft 2in for tonnage) x 24ft 5in. 3,165⁴²⁄₉₄bm.

Men: 750 (peace)/820 (war). Guns: LD 32 x 68pdrs (65cwt); UD 32 x 32pdrs (56cwt); QD 16 x 32pdrs (42cwt) + 2 x 8in/68pdr shell (52cwt); Fc 8 x 32pdrs (42cwt).

Algiers Devonport Dyd

(Re)Ord: 25.4.1847. (K: 10.7.1843.) Converted to screw battleship on stocks (see Section D). L: 26.1.1854.

CAESAR Class 90 guns.

Committee of Reference design 1847, modified from *Rodney* design; approved 13.8.1847.

Dimensions & tons: 207ft 4in, 170ft 7⅜in x 56ft 0in oa (55ft 2in for tonnage) x 23ft 4in. 2,761²⁴⁄₉₄bm.

St Jean d'Acre Devonport Dyd

(Re)Ord: 25.4.1847. K: — . Finally begun as screw battleship (see Section D). L: 23.3.1853.

Caesar Pembroke Dyd (originally 28.6.1847 from Portsmouth Dyd)

(Re)Ord: 13.8.1847. K: 8.1848. Re-ordered as screw battleship 30.10.1852 and converted on stocks (see Section D). L: 8.8.1853.

HANNIBAL Class 90 guns.

Originally ordered to the *Albion* Class design, but re-ordered to a second John Edye design of 1847, lengthened from *Princess Royal* design.

Dimensions & tons: 208ft 0in, 170ft 7⅜in x 58ft 0in oa (57ft 2in for tonnage) x 24ft 0in. 2,965⁷¹⁄₉₄bm.

Hannibal Woolwich Dyd

(Re)Ord: 19.6.1847. K: 12.1848. Re-ordered as screw battleship 23.9.1852 and converted on stocks (see Section D). L: 31.1.1854.

BRUNSWICK 80 guns.

Modification (with 1ft reduced beam and fuller midships section) of the *Vanguard* (Symonds) design. The Surveyor was asked to produce new drawings on 28 September, and these were approved 14 October 1847.

Dimensions & tons: 190ft 0in, 154ft 4¾in x 55ft 9in (55ft 0in for tonnage) x 23ft 4in. 2,484²⁷⁄₉₄bm.

Men: 750. Guns: LD 8 x 8in + 20 x 32pdr; UD 4 x 8in + 24 x 32pdr; QD/Fc 24 x 32pdr.

Brunswick Pembroke Dyd

Ord: 19.2.1844. K: 8.1847. Re-ordered as 80-gun screw battleship

23.6.1854 (see Section D) and converted on stocks without lengthening. L: 1.6.1855.

ORION Class 80 guns. Originally ordered on 30 March 1848 to be 80-gun ships to the same design as the *Superb* (*Vanguard* Class). However a revised Edye & Watts design was completed on 29 June and approved 14 October 1848, for ships to be some 8ft longer and 1ft narrower than the *Superb*, with a longer floor, fuller midships section, finer run and entrance, etc. The *Hood* and *Orion* were the last sailing battleships to be laid down for the RN. However both were subsequently re-ordered as 91-gun screw battleships (see below). A third unit to this design was ordered at Pembroke Dyd on 31 March 1849, to be named *Edgar*, but before any work was done there she was re-ordered at Woolwich Dyd on 4 April 1851 as a 91-gun screw battleship (see *Agamemnon* Class below). The following data refers to these ships when ordered as pure sailing ships.

Dimensions & tons: 198ft 0in, 161ft 0¼in x 55ft 9in (55ft 1in for tonnage) x 23ft 4in. 2,600¹/₉₄bm.

Men: 720. Guns: intended to carry LD 8 x 8in/68pdr (65cwt) + 20 x 32pdrs (56cwt); UD 4 x 8in/68pdrs (65cwt) + 24 x 32pdrs (50cwt); QD/Fc 24 x 32pdrs (41cwt).

Orion Chatham Dyd
Ord: 30.3.1848. K: 1.2.1850. Re-ordered as screw battleship 30.10.1852 (see Section D).

Hood (ex-*Edgar*, renamed 29.6.1848) Chatham Dyd
Ord: 30.3.1848. K: 13.8.1849. Re-ordered as screw battleship 9.4.1856 (expenditure up to this date was £37,200) (see Section D).

(D) Vessels ordered or re-ordered as steam screw battleships (from 1847)

The official origins of steam propulsion for the battlefleet date from 1847, when in the Programme of Works for that year the Admiralty directed that a Steam Vessel should be built at Devonport. The order was placed on 25 April 1847, the Admiralty ordered that this 80-gun ship was to be named *Audacious*, and 'the Surveyor to prepare a drawing', which was to be for a screw vessel to take one of the Boulton & Watt engines previously ordered for the frigates *Vulcan* (700nhp) or *Euphrates* (620nhp) The resignation of William Symonds from the Surveyorship in June meant that the design task fell to his Chief Assistant, John Edye.

On 18 November 1847 it was ordered that this Second Rate *Audacious* should be renamed *James Watt* to honour the engineer, and fitted with the *Vulcan*'s engines; this was well before the ship was laid down. The plan drawn for this was dated 16 December; however, on the previous day Edye pointed out that the engines in question had been completed two years earlier and 'will be obsolete when the *Watt* is ready for them in four years'. It was recommended that a new engine should be ordered for the *James Watt* in two or three years, and a new frigate built to take the *Vulcan*'s engines.

On 8 January 1848 it was instructed that the 90-gun *Nile* and *London* were to be fitted with the engines of the *Simoom* (a 780nhp Napier engine) and the *Vulcan* (a 700nhp Boulton & Watt engine) respectively. Plans for the adaptations of the *Nile* and *London* were sent to the dockyards on 15 February 1849, when the Admiralty asked the Surveyor's opinion on the proposal of Hay to put these engines in the *Nile*, *London* and two 84s. On 17 February the Surveyor replied

that he strongly opposed the proposition – it would be expensive and would ruin the ships.

AGAMEMNON Class 91-guns. The Surveyor's Department (John Edye) design for the *James Watt* was submitted on 6 June 1849 and approved on 14 June. The *Agamemnon* was originally ordered as an 80-gun sailing two-decker but not laid down as such (although materials collected), see section (c) above; she was re-ordered as a screw ship on 20 June 1849. Laid down before the *James Watt*, she was the first purpose-built screw ship of the line in the British Navy. The modified final design (credited to Isaac Watts) was approved 14 June 1850. Both ships were reclassified as 91-gun on 26 March 1851.

The *Agamemnon*'s cylinders were 70¾in diameter, 3½ft stroke. In 1856/57 she was fitted to carry the (first) Atlantic telegraph cable for £12,699; she accomplished the task of laying it in company with the American steam frigate *Niagara*. The *James Watt*'s cylinders were 52¼in diameter, 3ft stroke. She required major alterations to her unsatisfactory 4-cylinder machinery (second-hand, from the iron frigate *Vulcan*) which by 1856 had cost £5,706.

Two more ships were ordered to this design on 4 April 1851, as 80-gun ships (9 days *after* the first two had been re-classed as 91 guns); they were in turn re-classed as 91 guns on 15 November 1852. The *Repulse* was renamed *Victor Emmanuel* on 7.12.1855 to honour a visit of the Italian King to the ship on that date. *Victor Emmanuel*'s and *Edgar*'s cylinders were 76¼in diameter, 3½ft stroke.

A fifth ship, the *Hero*, was ordered on 27 March 1852 to the same 80-gun design, and was re-classed as 91 guns on 2 April 1853. The design of the *Hero* was subsequently altered and she is treated separately below.

Dimensions & tons: 230 ft 0 in, 194 ft 7¼ in x 55 ft 4 in oa (54 ft 6 in for tonnage) x 24 ft 6 in. 3,074⁴⁵/₉₄bm.

Men: 860. Guns: Planned (80 guns): LD 36 x 8in (65cwt); UD 34 x 32pdrs (56cwt) + 2 x 8in (95cwt); QD/Fc 2 x 8in (95cwt) + 8 x 10in (85cwt). As completed (established 3.7.1852): LD 34 x 8in (65cwt/9ft); UD 34 x 32pdrs (56cwt/9½ft); QD/Fc 22 x 32pdrs (45cwt/8½ft) + 1 x 68pdr (95cwt/10ft).

Machinery: 2-cylinder (4-cylinder in *James Watt*) horizontal single expansion (trunked in *Agamemnon*). Single screw. 600nhp. See below for ihp/speed.

James Watt (ex-*Audacious*) Pembroke Dyd/Boulton & Watt (1,548ihp = 9.361kts)
As built: 230 ft 3 in, 194 ft 6⅛ in x 55 ft 5 in oa (54 ft 7 in for tonnage) x 24 ft 7½ in. 3,082⁷⁹/₉₄bm. Draught 13 ft 7 in/19 ft 8 in.
Ord: 25.4.1847. K: 9.1850. L: 23.4.1853. C: 27.3.1854.
First cost: £149,455 total (including £? hull, £43,875 machinery).
Commissioned 20.1.1854 at Plymouth under Capt. George Augustus Elliot; to the Baltic for the Russian War. On 23.5.1856 under Capt. Talavera Vernon Anson, at Lisbon; paid off 23.4.1857. At Plymouth 1867–59. Recommissioned 19.3.1859 under Capt. Edward Codd, for Channel Squadron; to the Mediterranean 10.1859; paid off 21.6.1862. Sold 1.1875 to Castle to BU at Charlton.

Agamemnon Woolwich Dyd/John Penn & Son. (2,268ihp = 11.243kts)
As built: 230 ft 0 in, 195 ft 2⅛ in x 55 ft 6 in oa (54 ft 8 in for tonnage) x 24 ft 6 in. 3,102⁴⁹/₉₄bm. Draught 12 ft 3 in/18 ft 8 in.
(Re-)Ord: 20.6.1849. K: 11.1849. L: 22.5.1852. C: 9.2.1853.
First cost: £141,299 total (including £74,331 hull, £32,890 machinery).
Commissioned 27.9.1852 under Capt. Sir Thomas Maitland, at Sheerness; to Portsmouth arriving 10.2.1853, at Spithead Review 1853, then to Channel Squadron. On 21.10.1853 under

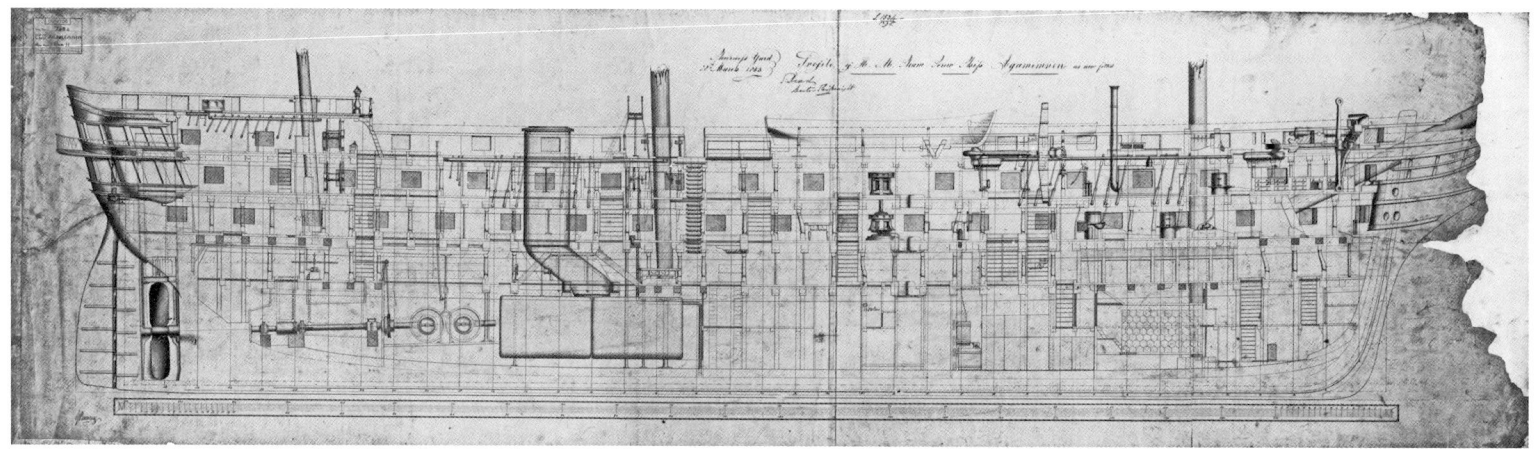

The inboard profile draught of the *Agamemnon*, the Royal Navy's first purpose-designed screw line-of-battle ship. The growing complexity of steam warships was reflected in the greater detail required on the plans, the draughtsmen responding to the difficulties of making it understandable by using more varied coloured lines and washes, an effect which – unfortunately – cannot be shown in this black-and-white reproduction. *(NMM DR7484)*

Capt. William Robert Mends, as the flag of Rear-Adm. Sir Edmund Lyons; to Black Sea for Russian War. On 26.12.1853 Capt. Thomas Matthew Charles Symonds, still Lyons's flag; at the Bombardment of Sebastopol 17.10.1854. On 27.11.1854 under Capt. Sir Thomas Sabine Pasley, still Lyons's flag, in the Black Sea. On 10.2.1856 under Capt. James John Stopford, in the Mediterranean; paid off 12.7.1856. Fitted at Portsmouth to carry the Atlantic telegraph cable (accompanied by the American steam frigate *Niagara*) 11.1856–5.1857 (for £12,699). Recommissioned 28.4.1857 under Master Commanding Cornelius Thomas Augustus Noddall, then 31.3.1858 under Capt. George William Preedy; involved in laying of first transatlantic telegraph cable 16.8.1858 (for 2 weeks); paid off 1.9.1858. Recommissioned 12.5.1859 under Capt. Thomas Hope; in the Mediterranean to end 1861, then to North American and West Indies 1.1862; paid off 18.10.1862. At Portsmouth 1862–70 (by 1865 reduced to 70 guns; by 1869 to 71 guns). Sold to W. H. Moore 12.5.1870 to BU.

Repulse Pembroke Dyd/Maudslay, Sons & Field (2,424ihp = 10.674kts)

As built: 230 ft 3 in, 195 ft 3⅛ in x 55 ft 4 in oa (54 ft 6 in for tonnage) x 24 ft 6 in. 3,085⁵⁸⁄₉₄bm. Draught 13 ft 7 in/19 ft 8 in.

Ord: 4.4.1851. K: 16.5.1853. L: 27.2.1855. Renamed *Victor Emmanuel* 4.12.1855. C: 9.9.1858.

First cost: £158,086 total (including £87,597 hull, £35,588 machinery).

Commissioned 27.7.1858 at Portsmouth under Capt. James Willcox, for the Channel Squadron; to the Mediterranean 3.1859. On 11.11.1859 under Capt. William John Cavendish Clifford, in the Mediterranean (79 guns); paid off 6.5.1862. At Portsmouth 1862–73. Recommissioned 20.11.1873 under Capt. George Henry Parkin, for the Ashanti War as hospital ship at Cape Coast Castle; then to Hong Kong 11.12.1874, replacing *Princess Charlotte* as receiving ship there. From 7.1875 under Capt. (Broad Pennant) John Edward Parish, then Capt. (Broad Pennant) Capt. George Willes Watson, then 3.1879, then Capt. (Broad Pennant) Thomas E. Smith. Recommissioned at Malta 2.1884. Receiving ship at Hong Kong again 1886, under Capt. G. D. Morant 1886, then Capt. (Broad Pennant) Edmund J. Church 12.1888 (–1890); paid off 30.9.1897. Sold in 1899.

Edgar Woolwich Dyd/Maudslay, Sons & Field (2,475ihp = 11.371kts)

As built: 230 ft 3 in, 195 ft 3⅛ in x 55 ft 4 in oa (54 ft 6 in for tonnage) x 24 ft 6 in. 3,084⁹⁹⁄₉₄bm. Draught 12 ft 10 in/19 ft 6 in.

Ord: 4.4.1851. K: 4.7.1853. L: 22.10.1858. C: 19.7.1859.

First cost: £159,284 total (including £93,846 hull, £36,540 machinery).

Commissioned 31.5.1859 under Capt. James Edward Katon, for Particular Service; as flagship of Rear-Adm. John Elphinstone Erskine, with the Channel Squadron; sailed 23.12.1860 for Lisbon; returned to Plymouth 3.4.1860. On 22.5.1861 under Capt. George Pechell Mends, still Erskine's flag; to North America and West Indies 12.1861; paid off 10.7.1862. Recommissioned 11.7.1862 under Capt. Fitzgerald Algernon Charles Foley, as flagship of Sir Sydney Colpoys Dacres, in the Mediterranean. On 14.5.1863 under Capt. Geoffrey Thomas Phipps Hornby, then 9.1865 under Capt. Thomas Brandreth, both still flag of Dacres; paid off 14.12.1865. At Portsmouth 1865–70. Lent to the Customs as a as a quarantine ship on the Motherbank, Spithead 12.12.1870 (–1890). Sold 12.4.1904 to Castle to BU at Charlton.

SAINT JEAN D'ACRE Class New two-decker 101-gun ship. Surveyor's Department (Edye & Watts) design, approved 15 February 1851. Stretched version of 91-gun *Agamemnon* Class, the order for this screw ship replaced that issued in 1844 for a 90-gun sailing Second Rate but not begun as such.

Dimensions & tons: 238ft 0in, 202ft 5in x 55ft 4in oa (54ft 6in for tonnage) x 24ft 5in. Draught 23ft 6in (fwd) 25ft 0in (aft). 3,258³²⁄₉₄bm. 5,499 disp.

Men: 900. Guns (established 17.1.1853): LD 36 x 8in (65cwt/9ft); UD 36 x 32pdrs (56cwt/9½ft); QD/Fc 28 x 32pdrs (45cwt/8½ft) + 1 x 68pdr (95cwt/10ft).

Machinery: 2-cylinder (70¼in diameter, 3½ft stroke) horizontal single expansion, trunk. Single (hoisting) screw. 600nhp. 2,136ihp = 11.199kts.

First cost: £143,708 [including hull £81,277, machinery £35,770?].

Saint Jean D'Acre Devonport Dyd/John Penn & Son

Ord: 15.2.1851. K: 6.1851. L: 23.3.1853. C: 20.9.1853.

Commissioned 21.5.1853 under Capt. Henry Keppel; joined Channel Squadron 9.1853; to Baltic 5.1854 for Russian War, home intending to pay off 11.12.1854, but instead left 2.1.1855 for Cork to embark troops and proceed to Black Sea. On 10.7.1855 under Capt. George St Vincent King, in the Mediterranean transported Earl Granville to St Petersburg 9.1856 as Head of British delegation to the Coronation of Czar

The introduction of steam propulsion did not reduce the need for expert ship-handling under sail, a point emphasised by this painting of the screw 101-gun ship *St Jean d'Arc* responding to a 'man overboard' incident in 1854. The artist, Oswald Walters Brierly, was a guest of Capt. Henry Keppel in the ship at the time. *(NMM BHC3619)*

Alexander II; paid off 21.10.1856 at Devonport. Recommissioned 4.2.1859 under Capt. Thomas Pickering Thompson, for the Mediterranean; on 26.9.1860 under Capt. Charles Gilbert John Brydone Elliot (when Thompson invalided), still in the Mediterranean; paid off 13.9.1861. Re-classed as 99-gun by 1862 and 81-gun in 1863. Sold to Castle 1.1875 (for £3.12.0d per ton), left Plymouth 19.9.1875 and BU 10.75 at Charlton.

Modified *AGAMEMNON* Class 91-gun ship. A first development of the *Agamemnon* design, the *Hero* had a finer bow, and keel lengthened by 4ft. Further development of this design produced the *Renown* Class.

Dimensions & tons: 234 ft 3 in, 199 ft 1¾ in x 55 ft 2 in oa (54 ft 4 in for tonnage) x 24 ft 6 in. Draught 21 ft 3 in (fwd), 24 ft 2 in (aft). 3,127⁸⁄₉₄bm. 4,765 disp.

Men: 860. Guns (established 3.7.1852): LD 34 x 8in (65cwt/9ft); UD 34 x 32pdrs (56cwt/9½ft); QD/Fc 22 x 32pdrs (45cwt/8½ft) + 1 x 68pdr (95cwt/10ft)

Machinery: 2-cylinder (76in diameter, 3½ft stroke) horizontal single expansion. Single screw. 600nhp. 2,662ihp = 11.85kts.

Hero Chatham Dyd/Maudslay, Sons & Field

As built: 234 ft 4 in, 199 ft 3¼ in x 55 ft 4 in oa (54 ft 6 in for tonnage) x 24 ft 6 in. Draught (light) 15 ft 3 in (fwd), 19 ft 7 in (aft). 3,148³⁹⁄₉₄bm.

Ord: 27.3.1852. K: 8.6.1854. L: 15.4.1858. C: 10.5.1859.

First cost: £142,199 [including hull £73,470, machinery £38,156]

Commissioned 29.4.1859 at Sheerness under Capt. George Nathaniel Broke (invalided 29.4.1859). On 29.4.1859 under Capt. George Henry Seymour, for the Channel Squadron. Fitted to carry the Prince of Wales to Canada and the USA; sailed on this visit 11.7.1860, returned home 11.1860. On 5.1.1861 under Capt. Alfred Phillipps Ryder; to North America and West Indies; paid off 22.11.1862. At Sheerness 1862–71 (reduced to 89 guns in 1862, 79 in 1863–69, and 71 thereafter). Sold to Castle (for £10,416) 20.6.1871 to BU at Charlton.

ALGIERS – Conversion of 90-gun Second Rate (incomplete on stocks) to 91-gun screw ship, 1852 design. Originally ordered 1847 as a 90-gun sailing Second Rate [see Section (c)], taking over frames of a ship laid down 10.7.1843. Ordered in 1852 to complete as a screw ship, for which she was lengthened by about 8ft. Her first engine (which was second-hand, from the *Megaera*, iron frigate) proved unreliable, and she was re-engined as shown below.

Dimensions & tons: (255 ft 6 in oa) 218 ft 7 in, 179 ft 9 in x 60 ft 0 in oa (59 ft 2 in for tonnage) x 24 ft 5 in. 3,347bm. 4,730 disp. Draught 24 ft 6 in (fwd), 25 ft 7 in (aft).

Men: 850 (860 after re-engining). Guns: LD 32 x 8in (65cwt); UD 32 x 32pdrs (56cwt); QD/Fc 26 x 32pdrs (42cwt) + 1 x 68pdr (95cwt).

Machinery: 4-cylinder (61in diameter, 3¼ft stroke) horizontal single expansion, geared. Single screw. 450nhp. 1,117ihp = 9kts.

First cost: £145,504 including hull as sailing vessel £79,428, machinery £21,025 (engine) + £5,505 (boilers).

Algiers Devonport Dyd/William Fairbairn & Sons

As built:

Ord: 3.8.1852. Conversion begun: 27.9.1852. L: 26.1.1854. C: 30.6.1854.

Commissioned 27.5.1854 under Capt. Charles Talbot, as troopship to the Baltic for the Russian War, and then to Black Sea. On 13.2.1856 under Commodore Henry John Codrington, in charge of a flotilla of gunboats; paid off 13.9.1856 at Portsmouth. Re-engined from 1.1857 to 3.1859 (for £52,653) with a 2-cylinder 600nhp engine by Maudslay, Sons & Field, details as for *Hero* above (2,518ihp = 12.19kts). Recommissioned 7.2.1859 under Capt. George William Douglas O'Callaghan, for the Channel Squadron; to the Mediterranean 3.1861; later in 1861 under Capt. Edward Bridges Rice, in the Mediterranean; paid off 12.12.1862 at Portsmouth. Sold to Mr D. Cooper Scott on 26.2.1870 to BU.

ORION Class 91 guns. Re-ordered 1852 and 1856 respectively as 91-gun screw battleships and were converted on the stocks at Chatham, being lengthened amidships by some 40 ft and fitted with (John Penn & Son) 600nhp engines. The following data refers to these ships following their re-orders. Although *Orion* was well regarded, she was completed in haste because of the Russian War and consequently decayed rapidly. The stern run of the *Hood* was improved (by alterations to design dated 30 December 1856) and thus differed from her sister.

Dimensions & tons: 238 ft 0 in, 200 ft 10¼ in x 55 ft 9 in (54 ft 9 in for tonnage) x 24 ft 0 in. 3,202⁴¹⁄₉₄bm.

Men: 860. Guns (established 3.4.1853): LD 34 x 8in/68pdrs (65cwt); UD 34 x 32pdrs (56cwt); QD/Fc 22 x 32pdr (45cwt) + 1 x 68pdr (95cwt).

Machinery: 2-cylinder (70¾ in diameter, 3½ ft stroke) horizontal single expansion, trunked. Single screw. 600nhp.

The *Princess Royal* photographed
at Plymouth, either in 1864
or 1867. Although based on
the *Albion* design, the ship
had been re-ordered to a John
Edye design in 1847, with the
breadth reduced by 2ft. She
was re-ordered as a 91-gun
steam battleship in 1852, when
the insertion of a new midship
section reduced the propensity
for rolling, and in general the
ship was regarded as a success.
(NMM D2166)

First cost: *Orion* £145,102, including machinery £37,242. *Hood* hull
conversion £61,174.

Orion Chatham Dyd/John Penn & Son. (2,329ihp = 11.45kts)
Re-ordered as screw battleship 30.10.1852 and converted on stocks
commencing 24.12.1852 (after part launched to lengthen ship
5.4.1853). L: 6.11.1854. Machinery fitting completed 21.12.1854.
Completed fitting for sea 29.3.1855.

Dimensions & tons: 238 ft 2 in, 202 ft 9 in x 55 ft 10 in (55 ft 2 in for
tonnage) x 24 ft 0 in. 3,281bm.

Commissioned 11.1.1855 under Capt. John Elphinsone Erskine,
for the Baltic Fleet; in 1856 to North America and West Indies;
paid off 1.10.1857. At Devonport 1857–58. Recommissioned
3.6.1858 under Capt. Edwin Clayton Tennyson D'Eyncourt,
as flag of Rear-Adm. Sir Charles Howe Fremantle for the
Channel Fleet. On 27.1.1859 under Capt. Wallace Houston, for
the Mediterranean. On 22 June 1859 under Capt. John James
Bartholomew Edward Frere. On 7.10.1861 under (acting) Capt.
William Cecil de Vere. Paid off 9.11.1861. At Sheerness 1862–66
(reduced to 89 guns by 1862). Later 79 guns. Sold to Castle
4.1867 and arrived 4.7.1867 to BU at Charlton.

Hood Chatham Dyd/John Penn & Son. (2,385ihp = 11.76kts)
Re-ordered as screw battleship 9.4.1856 and converted on stocks
commencing 5.5.1856 (after part launched to lengthen ship
21.10.1856). L: 4.5.1859. Machinery fitting and fitted for 2nd
Division of the Ordinary at Sheerness (for £7,518) by 15.6.1859.
Never fitted for sea.

Dimensions & tons: 238 ft 0 in, 206 ft 0 in x 55 ft 10½ in (55 ft 3½ in
for tonnage) x 23 ft 11½ in. 3,350bm.

Never commissioned. Remained at Sheerness 1860–70 (reduced to
54 guns by 1870). Became torpedo company barracks hulk at
Woolwich 1872. Sold 1888 (for £5,000).

PRINCESS ROYAL Class – Conversions of 90-gun Second Rates
(incomplete on stocks) to 91-gun screw ships, 1852 design. Ordered
1840 as *Albion* Class 90-gun ships (see Chapter 2), both lengthened
with new midsection inserted and converted to screw while on stocks.
Hannibal engine (71in diameter cylinders, 4ft stroke) was second-
hand, taken from *Greenock* (iron frigate) and fitted at Woolwich; she
was then fitted for sea at Chatham.

Dimensions & tons: 217ft 0in, 179ft 1¼in x 58ft 0in oa (57ft 2in
for tonnage) x 24ft 0in. Draught 23ft 6in (fwd), 26ft 6in (aft).
3,114bm. 4,540 disp.

Men: 850. Guns: LD 32 x 8in (65cwt); UD 34 x 32pdrs (56cwt);
QD/Fc 24 x 32pdrs (42cwt) + 1 x 68pdr (95cwt).

Machinery: 2-cylinder (64⅛in diameter, 3ft stroke in *Princess Royal*)
horizontal single expansion (geared in *Hannibal*). Single screw.
Hannibal 450nhp. 1,071ihp = 8.6kts. *Princess Royal* 400nhp.
1,492ihp = 11.031kts.

Hannibal Deptford Dyd/Scott, Sinclair & Co.
Dimensions & tons: 217ft 6in, 179ft 6⅝in x 58ft 1¼in oa (ca.57ft
3¼in for tonnage) x 23ft 11½in. 3,136bm.

Conversion ord: 23.9.1852. Begun: 23.9.1852. L: 31.1.1854. C:
22.6.1854.

First cost: £128,715, including hull as a sailing ship £52,729.

Commissioned 18.3.1854 under Capt. Frederick William Grey,
as commodore of fleet transporting 10,000 French troops to
the Åland Islands during the Russian War; later to the Black
Sea. On 25.1.1855 under Capt. John Charles Dalrymple Hay,
as flagship of Rear-Adm. Houston Stewart, in the Black Sea,
later to the Mediterranean; paid off 20.11.1856 at Portsmouth.
Recommissioned 1.2.1858 under Capt. George Fowler Hastings,
as guard ship at Portsmouth (in Ordinary); on 29.6.1858 under
Capt. George Thomas Gordon; replaced 29.4.1859 as guard
ship by *Asia*. On 29.4.1859 under Capt. Matthew Connolly,
as flagship of Rear-Adm. George Rodney Mundy, for the
Mediterranean; on 30.3.1860 under Capt. Arthur Farquhar, still
Mundy's flagship in the Mediterranean; paid off 24.12.1861 at
Portsmouth. On 6.2.1863 under Cmdr Thomas Curme, as depot
ship for seamen; on 1.4.1863 under Capt. John Seccombe, still
deport ship; replaced 30.4.1863 by *Duke of Wellington*. Hulked
1884. Sold to Castle 12.4.1904 to BU at Charlton.

Princess Royal (ex-*Prince Albert*, renamed 26.3.1842) Portsmouth
Dyd/Maudslay, Sons & Field
Dimensions & tons: 217ft 0in, 179ft 3½in x 58ft 1½in oa (about 57ft
3½in for tonnage) x 24ft 0in. 3,130bm.

Conversion ord: 30.10.1852. Begun: 15.11.1852. L: 23.6.1853. C:
11.3.1854.

First cost: £143,407 including machinery £24,352.

Capt. George Pechell Mends had only just arrived in the West Indies (as flag-captain of the *Edgar*) when he drew from life this illustration of the attempted salvage of the 101-gun *Conqueror*, which had grounded on a reef on the southern side of Rum Quay in the Bahamas just two weeks earlier due to a navigational error. The nearly new *Conqueror* was lost, but all 1,400 men aboard – including troops being brought to assist the French intervention in Mexico – were saved, the crew being removed by the sloop *Bulldog* which also salvaged most of her stores. The ship's remains are now designated as an underwater museum site. (*NMM PU9409*)

Commissioned 29.10.1853 under Capt. Lord Clarence Edward Paget; to the Baltic 1854 then to the Black Sea 1855, for the Russian War; at the blockade and bombardment of Sebastopol. On 13.8.1855 under Capt. Lewis Tobias Jones (Paget invalided 7.1855), in the Mediterranean; paid off 28.7.1856 at Portsmouth. Recommissioned 29.7.1856 under Capt. George Giffard, for the Mediterranean. On 6.2.1858 under Capt. Thomas Baillie, in the Mediterranean; paid off 5.12.1859 at Portsmouth. Recommissioned 21.1.1861 under Capt. Charles Fellowes, as flagship of Rear-Adm. Robert Smart, for the Channel Squadron; paid off 30.4.1861 at Plymouth. Recommissioned 12.2.1864 under Capt. William Gore Jones, as flagship of Rear-Adm. George St Vincent King, for East India and China; paid off 14.8.1867. Docked at Devonport 23.8–12.9.1872 to have copper stripped off prior to sale. Sold to Castle 7.12.1872 (for £8,500) to BU at Charlton.

CAESAR 91-gun. – Conversion of 90-gun Second Rate (a Modified *Algiers* Class ordered 1847, and incomplete on stocks) to a 91-gun screw ship. 1852 design from Committee of Reference. Note the first attempt to launch her on 21.7.1853 failed, as she stuck on the ways.
> Dimensions & tons: 207ft 4in, 170ft 6⅛in x 56ft 0¾in oa (ca.55ft 2½in for tonnage) x 23ft 4in. Draught 19ft 5in (fwd), 22ft 8in (aft). 2,767bm. 3,250 disp.
> Men: 860. Guns: LD 32 x 8in (65cwt); UD 32 x 32pdrs (56cwt); QD/Fc 26 x 32pdrs (42cwt) + 1 x 68pdr (95cwt).
> Machinery: 2-cylinder (58⅛in diameter, 3¼ft stroke) horizontal single expansion, trunked. Single screw. 400nhp. 1,420ihp = 10.274kts.
> First cost: £125,245 including hull as sailing ship £66,913.
Caesar Pembroke Dyd/John Penn & Son
> Ord: 30.10.1852. Conversion begun: 22.11.1852. L: 7.8.1853. C: 26.3.1854 at Portsmouth.
> Commissioned 21.1.1854 at Plymouth under Capt. John Robb; to Baltic during Russian War; paid off 7.3.1857. Recommissioned 14.6.1858 under Capt. Charles Frederick, for the Channel Squadron; to Central America 6.1858; to the Mediterranean 10.1859. On 2.7.1859 under Capt. Thomas Henry Mason, for the Channel Squadron; paid off 1.2.1862 and laid up at Portsmouth. Sold to C. J. Mare at Blackwall 19.4.1870.

CONQUEROR Class 101 guns. Surveyor's Department design. The first ship was ordered and named *Conqueror* in March 1852, but on 15 November it was ordered to build her as a 100-gun ship to the lines of the *Saint Jean d'Acre*, but the design was slightly lengthened (by 2ft). The second ship was ordered and named *Donegal* in December 1854 to be built on the same slip (No. 5) as the *Saint Jean d'Acre* and *Conqueror*.
> Dimensions & tons: (275 ft 0 in oa] 240 ft 0 in, 204 ft 0 in x 55 ft 4 in oa (54 ft 6 in for tonnage) x 24 ft 5 in. 3,223³⁄₉₄bm.
> Men: 930. Guns (established 26.10.1855): LD 36 x 8in (65cwt/9ft); UD 36 x 32pdrs (56cwt/9½ft); QD/Fc 28 x 32pdrs (42cwt/8ft) + 1 x 68pdr (95cwt/10ft).
> Machinery: 2-cylinder (82in diameter, 4ft stroke) horizontal single expansion, trunked. Single screw. 800nhp. See below for ihp/speed.
Conqueror Devonport Dyd/John Penn & Son. (2,812ihp = 10.806kts)
> As built: 240 ft 0 in, 204 ft 1¼ in x 55 ft 4 in oa (54 ft 6 in for tonnage) x 24 ft 5 in. 3,224⁶⁴⁄₉₄bm. Draught 13ft 11in/21ft 2in.
> Ord: 27.3.1852. K: 25.7.1853. L: 2.5.1855. C: 9.4.1856.
> First cost: £171,116 (including £91,244 for hull, and £50,919 for machinery).
> Commissioned 30.11.1855 under Capt. Thomas Matthew Charles Symonds, for the Channel Squadron. On 12.7.1856 under Capt. Hastings Reginald Yelverton, then 22.7.1859 under Capt. William John Cavendish Clifford, all in the Mediterranean. In December 1859 under Capt. James Willcox, at Plymouth, then 2.1.1860 under Capt. Edward Southwell Sotherby, with the Channel Fleet; wrecked on Rum Cay in the Bahamas 29.12.1861.
Donegal Devonport Dyd/John Penn & Son (3,103ihp = 11.77kts)
> As built: 240 ft 0 in, 204 ft 9¾ in x 55 ft 5 in oa (54 ft 7 in for tonnage) x 24 ft 5 in. 3,245⁷³⁄₉₄bm. Draught 12 ft 1 in/21 ft 0 in.
> Ord: 27.12.1854. K: 27.9.1855. L: 23.9.1858. C: 27.8.1859.
> First cost: £175,337 (including £85,783 for hull, and £49,600 for machinery, and £39,954 fitting for sea).
> Commissioned 23.6.1859 under Capt. William Fanshawe Glanville; to Liverpool (to raise men) then to Channel Squadron. On 3.11.1859 under Capt. Henry Broadhead, for the Channel Squadron. On 18.5.1851 under Capt. Sherard Osborn, in the Channel, then 11.1861 to North America and the West Indies; paid off 10.6.1862. At Devonport 1862–64 (by 1864 reduced to 81 guns). Recommissioned 1.9.1864 under Capt. James Aylmer

Dorset Paynter, then on 1.5.1867 under Capt. Edward Winterton Turnour; paid off 30.9.1870. Became torpedo and mining school ship ***Vernon I*** on 14.1.1886. Sold to Pounds, Portsmouth 18.5.1925 to BU.

ALBION **Class** 91 guns. – Conversion of the 1839-designed Symondite 90-gun sailing ships to screw propulsion, without lengthening, began in 1852 with the *Exmouth* – suspended on the stocks on 18 December 1844 ($^{3}/_{16}$ built), resumed in 1847 and nearing readiness for launch – being the first to be scheduled in 1852 for adaptation. Her sister *Aboukir*, suspended at the same date (⅜ built), had been restarted in 1847 as she was judged too advanced in her construction to be altered, and was launched a year later, but was not commissioned and was retained 'in Ordinary' and was scheduled in 1856 for adaptation to screw propulsion in 1856. The *Albion* herself (the sole ship of this design completed as a full sailing ship), was ordered to be similarly adapted in 1860, and was so modified, but never saw service as a steam battleship. *Exmouth*'s cylinders were 64in diameter, 3ft stroke; *Aboukir*'s 58⅛in diameter, 3¼ft stroke; *Albion*'s 64$^{3}/_{16}$in diameter, 3ft 8in stroke.

Dimensions & tons: (243 ft 0 in oa] 204 ft 0 in, 166 ft 0 in x 60 ft 2¼ in (59 ft 1 in for tonnage) x 23 ft 8 ft. 3,083bm. 4,382 disp.

Men: 830. Guns: LD 32 x 8in (65cwt); UD 32 x 32pdrs (56cwt); QD/Fc 26 x 32pdrs (42cwt) + 1 x 68pdr (95cwt).

Machinery: 2-cylinder horizontal single expansion (trunked in *Aboukir*). Single screw. 400nhp. See below for ihp/speed.

Exmouth Devonport Dyd/Maudslay, Sons & Field. (1,252ihp = 9.1kts)

As built: 204ft 0in, 165ft 6½in x 60ft 3in oa (59ft 5in for tonnage) x 23ft 8in. 3,108$^{57}/_{94}$bm. Draught 15ft 6in/20ft 6in (with 31 tons on board)

Ord: 30.10.1852. Conversion begun: 20.6.1853. L: 12.7.1854. C: 15.3.1855.

First cost: £146,067 (including hull as sailing ship £76,379, conversion to screw £10,257, machinery £24,620

Commissioned 18.11.1854 under Capt. Frederick Thomas Pelham. On 19.2.1855 under Capt. William King Hall, as flagship of Rear-Adm. Michael Seymour, for the Mediterranean and thence to the Baltic for the Russian War. On 20 March 1856 under Capt. Harry Eyres, for Particular service; paid off 20.6.1857 at Plymouth. From 1.2.1858 under Capt. Robert Spencer Robinson, as guard ship at Plymouth (still in Ordinary). In 5.1859 under Capt. James Jogn Stopford, for the Mediterranean; from 1.5.1960 under Capt. James Aylmer Dorset Paynter, still in the Mediterraneam; paid off 14.10.1862. Lent to Metropolitan Asylums Board 22.12.1876 as training ship for pauper boys. Sold to George Cohen 4.4.1905 and BU at Penarth.

Aboukir Devonport Dyd/John Penn & Son (1,533ihp = 9.55kts)

As built: 204ft 0in, 165ft 4¾in x 60ft 1¼in oa (59ft 3¼in for tonnage) x 23ft 8½in. 3,090$^{60}/_{94}$bm. Draught 15ft 7in/19ft 2in (with 24 tons on board).

Ord: 20.8.1856. Conversion begun: 8.10.1856. L: (undocked, completed) 1.1.1858.

First cost: Conversion cost £24,400.

Commissioned 25.5.1859 under Capt. Charles Frederick Schomberg, for the Channel Squadron. On 10.1.1860 under Capt. Douglas Curry, then 18.2.1861 under Capt. Charles Frederick Alexander Shadwell, still with Channel squadron, then to Mediterranean and West Indies 8.1862. Became receiving/depot ship (and floating battery) at Jamaica from 19.9.1862 under Capt. Peter Cracroft, as flag of Commodore Hugh Dunlop. Command of this receiving ship passed to Commodore Francis Leopold McClintock 6.9.1865, then to Commodore Augustus Phillimore 22.2.1868, Commodore Richard William Courtenay 1.12.1869, Commodore Algernon Frederick Rous de Horsey 1872 and Commodore Algernon McLennan Lyons 1875, still at Jamaica; paid off 20.7.1877. Sold 23.11.1877 in Jamaica.

Albion Devonport Dyd/Humphrys & Tennant (1,835ihp = 10.986kts)

As built: 204ft 0in, 166ft 0in x 60ft 2¼in oa (59ft 4¼in for tonnage) x 23ft 8in. 3,110$^{62}/_{94}$bm. Draught 16ft 2in/19ft 4in (with 46 tons on board).

Ord: 3.2.1860. Conversion begun: 4.4.1860. L: (undocked, 'completed') 21.5.1861.

Never commissioned or fitted for sea as screw vessel. Her machinery was only fitted in late 1861, and it was not until 21.3.1862 that she ran her trials (her only time at sea). Reduced to 86 guns in 1862 and to 76 guns in 1863. Sold to Castle 8.1884 and BU at Charlton.

RODNEY **Class** – Conversions of (1833–40 launched) 90-gun Second Rates to 91-gun screw ships. *Nile* converted without being lengthened, fitted with second-hand engines (from iron frigate *Euphrates*; reboilered and re-piped by 1855 to produce 1,247ihp = 8.2kts. The other pair were instead lengthened and fitted with new engines. *Nile*'s cylinders were 62¼in diameter, 3½ft stroke; *London*'s 70⅞in diameter, 3ft stroke; *Rodney*'s 66in diameter, 3½ft stroke.

Men: 850. Guns: LD 18 x 8in (65cwt) + 14 x 32pdrs (56cwt); UD 6 x 8in (65cwt) + 28 x 32pdrs (56cwt); QD/Fc 24 x 32pdrs (42cwt) + 1 x 68pdr (95cwt/10ft).

Machinery: 2-cylinder horizontal single expansion. Single screw. 500nhp. See below for ihp/speed.

Nile Devonport Dyd/Seaward & Capel (928ihp = 6.854kts)

Dimensions & tons: 205ft 6in, 170ft 4¾in x 54ft 3½in oa (53ft 9½in for tonnage) x 23ft 2in. 2,622$^{55}/_{94}$bm. Draught 13ft 7½in/17ft 10½in.

Conversion ord: 24.11.1852. Begun: 14.12.1852. L: (undocked) 30.1.1854. C: 4.1854.

First cost: £41,073 fitting plus machinery £38,565.

Commissioned 25.2.1854 under Capt. Henry Byam Martin, for the Baltic. On 17.7.1854 under Capt. George Rodney Mundy; returning to Plymouth 6.12.1855; at Spithead Review 23.4.1856; to North America and West Indies 6.1856, returning home 31.3.1857; paid off 20.4.1857. At Devonport 1857, fitted there 7.1857–4.1858 as a flagship for Queenstown (for £17,493). Recommissioned 1.3.1858 under Capt. Henry Chads, as flagship of Rear-Adm. Henry Ducie Chads, at Queenstown. On 7.12.1858 under Capt. Arthur Parry Eardley Wilmot Talbot, as flagship of Rear-Adm. Charles Talbot, still at Queenstown; replaced 31.12.1859 by *Sans Pareil*. On 31.12.1859 under Capt. Edward King Barnard, as flag of Rear-Adm. Sir Alexander Milne, in North America and West Indies after new boilers fitted at Devonport in 1860; paid off 23.4.1864. At Devonport in Ordinary until 22.6.1875. Renamed ***Conway*** 24.7.1876 as training ship off Rock Ferry, Liverpool, on loan to Mercantile Marine Service Association. Stranded 14.5.1953 in the Menai Strait while under tow to Birkenhead for refit. Wreck burnt by accident 31.10.1956.

London Devonport Dyd/Ravenhill, Selkeld & Co. (1,804ihp = 11.522kts)

Dimensions & tons: 213ft 3in, 178ft 0in x 54ft 3½in oa (ca.53ft 5½in for tonnage) x 22ft 7in. 2,687bm.

Conversion ord: 11.12.1856. Begun: 14.1.1857. L: (undocked) 13.5.1858. C: (fitted for sea) 21.1.1859.

First cost: £88,498 including machinery £30,396, fitting for sea £39,030.

Commissioned 12.5.1859 under Capt. Henry Chads, then 4 days later under Capt. Henry Ducie Chads, for the Mediterranean; paid off 6.2.1863. Harbour ship at Zanzibar 4.1874; from 27.7.1878 under Capt. Hamilton George Earle, then 8.6.1880 Capt. Charles James Brownrigg, and 9.12.1881 Capt. Percy Patt Luxmore. Sold to BU 1884 at Zanzibar.

Rodney Chatham Dyd/Maudslay, Sons & Field. (2,246ihp = 11.479kts)
Dimensions & tons: 214ft 4in, 181ft 11in x 54ft 0in oa (53ft 2in for tonnage) x 23ft 0in. 2,736bm. 3,126 disp.
Conversion ord: 5.2.1859. Begun: 16.3.1859. L: (undocked, completed) 11.1.1860. C: 6.1860 (fitted for reserve).
Commissioned 21.1.1867 at Sheerness under Capt. Algernon Charles Fieschi Heneage, as flagship of Vice-Adm. Henry Keppel, to China; paid off 27.4.1870 (the last wooden battleship to have been in commission). BU 2.1882.

CRESSY – Conversion of 80-gun Second Rate (incomplete on stocks), ordered 1844 as a *Vanguard* Class 80-gun ship (see Chapter 2), but converted to screw while on stocks to 80-gun screw ship, 1852.
Dimensions & tons: 198ft 5in, 162ft 1⅜in x 55ft 0in (54ft 3in for tonnage) x 21ft 8¼in. 2,537⁸/₉₄bm. 3,707 disp.
Men: 750. Guns: LD 10 x 8in/68pdrs (65cwt) + 18 x 32pdrs (56cwt); UD 4 x 8in/68pdrs (65cwt) + 24 x 32pdrs (50cwt); QD/Fc 24 x 32pdrs (42cwt).
Machinery: 2-cylinder (64in diameter, 3ft stroke) horizontal single expansion. Single screw. 400nhp. 1,076ihp = 7.206kts.
First cost: £114,693, including machinery £24,449.
Cressy Chatham Dyd/Maudslay, Sons & Field
Conversion ord: 24.11.1852. Begun: 29.11.1852. L: 21.7.1853. C: 9.3.1854 (fitted at Sheerness).
Commissioned 19.12.1853 under Capt. Richard Laird Warren, for the Mediterranean; to the Baltic 1854 for the Russian War, then to the Mediterranean; paid off 15.5.1857 at Sheerness. On 1.3.1858 under Capt. Edward Pellew Halstead, as guard ship at Sheerness (in Ordinary). Recommissioned 5.1859 under Capt. Charles Gilbert John Brydone Elliot, for the Channel Squadron; to the Mediterranean 7.1859. On 26.9.1860 under Capt. Thomas Harvey, in the Mediterranean; paid off 26.5.1861. Sold to Castle & Beech 1867 to BU at Charlton.

MAJESTIC Class – Conversion of 80-gun Third Rates, 1852 onwards. Orders were issued to convert these *Vanguard* Class 80-gun two-deckers to screw ships on dates below. *Meeanee* built of teak. *Majestic* and *Irresistible* were incomplete ships, still on stocks. *Vanguard* conversion was cancelled and *Superb* (only other member of original class) never selected for conversion. *Majestic*'s cylinders were 64in diameter, 3ft stroke (*Mars* unrecorded but probably same); *Centurion*'s were 63½in diameter, 3ft stroke; and those of Penn-engined ships were 58in diameter, 3¼ft stroke (*Colossus* fitted with engine from *Royal Albert*).
Dimensions & tons: 190ft 0in, 155ft 3in x 56ft 9in (56ft 0in for tonnage) x 23ft 4in. 2,589⁶¹/₉₄bm. 3,482 disp.
Men: 750. Guns: LD 10 x 8in (65cwt) + 18 x 32pdrs (56cwt); UD 4 x 8in (65cwt) + 24 x 32pdrs (50cwt); QD/Fc 24 x 32pdrs (42cwt).
Machinery: 2-cylinder horizontal single expansion (trunk in Penn-engined ships). Single screw. 400nhp.
Majestic Chatham Dyd/Maudslay, Sons & Field (1,192ihp = 8.783kts)
Conversion ord: 24.11.1852. Begun: 29.11.1852. L: 1.12.1853. C: 22.4.1854 (fitted at Sheerness for sea).
First cost: £118,474 (including machinery £25,310).
Commissioned 10.2.1854 under Capt. James Hope, for the Baltic

Fleet during the Russian War, later to Mediterranean; paid off 15.5.1857. Coastguard base at Liverpool 2.1860 to 8.1864, under Capt. William Robert Mends from 1.2.1860, then 1.1.1861 Capt. Edward Augustus Inglefield, then 14.4.1864 Capt. James Aylmer Dorset Paynter; paid off 5.9.1864. Sold to Marshall 4.1867 and BU 1868 at Plymouth.

Colossus Portsmouth Dyd/John Penn & Sons (1,458ihp = 9.312kts)
Conversion ord: 9.1.1854. Begun: 26.1.1854. L: (undocked) 11.6.1854. C: 27.6.1855.
First cost: conversion and fitting (including machinery) £66,209; Penn paid £25,031 for machinery.
Commissioned 15.6.1854 under Capt. Robert Spencer Robinson, for North America and West Indies; to Baltic in 1855 for Russian War; paid off 2.6.1857. Fitted for steam Reserve at Chatham 12.1857–3.1858. Fitted for 1st class Reserve at Sheerness 5.1859. Refitted for sea at Portsmouth 6.1860 then from 11.6.1860 under Capt. Francis Scott, as guard ship at Portland (to 6.1864); on 1.4.1861 under Capt. George Edward Patey, then 6.5.1862 Capt. Swynfen Thomas Carnegie, 13.4.1863 Capt. Edward Codd; crew turned over to *Frederick William* 1.7.1864. Sold to Castle & Beech 3.1867 to BU at Charlton.

Irresistible Chatham Dyd/John Penn & Sons (1,461ihp = 8.622kts)
As built: 190ft 0in, 158ft 4½in x 56ft 9in (56ft 0in for tonnage) x 23ft 4in. 2,642bm.
Conversion ord: 23.6.1854. Begun: 1.5.1855. L: 27.10.1859. C: 1.1860 (fitted at Sheerness for sea).
First cost: £37,746 as steam conversion (hull as sailing ship £35,145).
Commissioned 1.4.1864 under Capt. John Bourmaster Dickson, as guard ship at Southampton (replaced *Dauntless*). On 17.5.1865 under Capt. John Borlase; at Spithead for Review 7.1867; paid off 30.4.1868. Recommissioned 5.6.1868 at Portsmouth under Cmdr Arthur George Robertson Roe, for transport to Bermuda; paid off 30.9.1858. Depot ship 9.1868 at Bermuda (–1890). Sold 1894 at Bermuda.

Centurion Devonport Dyd/Miller, Ravenhill & Co. (1,255ihp = 8.5kts)
Conversion ord: 10.8.1854. Begun: 28.9.1854. L: (undocked, completed) 12.11.1855.
First cost: £22,425 for conversion plus £12,851 for machinery.
Commissioned 1.1.1856 under Capt. Woodford John Williams, for the Mediterranean. On 5.1856 under Capt. Edward Gennys Fanshawe, in the Mediterranean. On 28.6.1858 under Capt. George Nathaniel Broke (when Fanshaw was invalided), then 10.3.1859 under Capt. George Edwin Patey, still in the Mediterranean; paid off 7.12.1859 at Plymouth, but recommissioned next day, still under Patey. On 4.1.1860 under Capt. Henry Downing Rogers, for the Channel Squadron; paid off 30.11.1861. Sold to William Lethbridge 19.3.1870 to BU.

Mars Chatham Dyd/Maudslay, Sons & Field [never tried for speed]
Conversion ord: 10.8.1854. Begun: 7.3.1855. L: (undocked) 23.11.1855. C: 20.6.1856 (at Chatham). Fitted at Sheerness for 2nd Class Steam Reserve 6–10.1858, then fitted for sea 7–9.1859.
First cost: £17,875 at Chatham (for repair and conversion) + £25,830 at Sheerness (for machinery), + £16,548 fitting for sea.
Commissioned 20.6.1859 under Capt. James Newburgh Strange, for the Channel Squadron; to the Mediterranean 10.1860; paid off at Sheerness 7.2.1863. Lent as training ship at Woodhaven on the river Tay 13.5.1869 (to 1921). Sold to Thomas Ward of Sheffield 10.6.1929 to BU.

Goliath Chatham Dyd/John Penn & Sons (1,438ihp = 9.16kts)
Conversion ord: 20.8.1856. Begun: 16.10.1856. L: (undocked) 30.11.1857. C: 30.4.1858 (fitted at Sheerness).

First cost: £47,400 (not including machinery).

Never completed for sea as a screw ship. Ran trials 12.5.1858 (her only time at sea), then laid up at Sheerness. Lent to Forest Gate School Ships Committee at Grays as training ship for pauper boys 11.1870, under Staff Commander William Sutherland Bouchier. Burnt by accident 22.12.1875 at Grays (23 died).

Meeanee Chatham Dyd/John Penn & Sons

Conversion ord: 20.8.1856. Order moved to Sheerness Dyd/John Penn & Sons (1,456ihp = 9.701kts)

Re-ord: 26.11.1856. Begun: 6.12.1856. L: (undocked) 31.10.1857. C: 9.1858.

First cost: £20,119 for conversion

Commissioned 18.11.1862 under Capt. George Wodehouse, for the Mediterranean; paid off 1.3.1866 at Sheerness. Lent to War Dept 5.3.1867 as hospital ship at Hong Kong (commissioned for voyage to Hong Kong under tow 30.3–28.10.1868). Sold 1906 to BU.

Lion Devonport Dyd/John Penn & Sons (1,665ihp = 10.337kts)

Conversion ord: 20.8.1856. Begun: 1.2.1858. L: (undocked, completed for 2nd Division of the Steam Ordinary) 17.5.1859.

First cost: £24,200 for conversion.

Commissioned 7.9.1864 under Capt. Arthur Farquhar, as guard ship at Greenock. On 28.8.1865 under Capt. John Montagu Hayes, still at Greenock; to Londonderry 10.1867 then back to Greenock; paid off 8.9.1868. Recommissioned 1.10.1868 under Capt. John Napier, as receiving ship at Devonport; paid off 1.11.1869. Training ship at Devonport 1871. Sold 11.7.1905 at Portsmouth.

Collingwood Chatham Dyd/George Rennie & Sons (1,424ihp = 10.46kts)

Conversion ord: 9.7.1859. Begun: 23.3.1860. L: (undocked) 13.7.1861.

Never completed for sea as a screw ship; machinery fitted later. Sold to Castle 3.1867 to BU at Charlton.

Vanguard Sheerness Dyd/—

Conversion ord: 9.7.1859. Conversion cancelled during 1860 (the unconverted ship was renamed 20.10.1867 and BU 6.1875 at Chatham).

REVENGE Class 91-gun ships. First two originally ordered to the Surveyor's design for the *Agamemnon* class, but – following the slight extension with the *Hero* (lengthened by 4ft on the keel) – Walker (the Surveyor) proposed on 7 December 1854 to further lengthen the design (this time by 10½ft amidships) to improve the lines and provide space for the more powerful (800nhp) machinery; on 6 January 1855 the Admiralty approved this enlargement. Little extra speed was gained, but *Revenge* and *Renown* were always highly regarded by the Navy.

A second pair to the same design were ordered during the Russian War, but due to pressure of work in the dockyards, these were not laid down until 1858. Their lines were further modified to give them a finer stern run, but by the time they were launched in 1860 the need for armour was apparent and they were placed in Ordinary and were never completed for service.

It was initially intended that this class should have the same armament as the *Agamemnon* class, but on 18 December 1857 it was instead proposed that – as these four ships (and the *Defiance*) had ports for two more guns on the 'main' deck (as the UD was now called), two additional 32pdrs (56cwt/9½ft) should be mounted there in place of two of the lighter 32pdrs from the QD.

Dimensions & tons: 244ft 9in, 210ft 0in x 55ft 4in oa (54ft 6in for

tonnage) x 24ft 6in. 3,317 ^{78}/₉₄bm.

Men: 860. Guns: LD 34 x 8in (65cwt/9ft); UD 36 x 32pdrs (56cwt/9½ft); QD/Fc 20 x 32pdrs (45cwt/8½ft) + 1 x 68pdr (95cwt/10ft)

Machinery: 2-cylinder (82in diameter, 4ft stroke) horizontal single expansion (trunk in *Renown*). Single screw. 800nhp. See below for ihp/speed.

Revenge Pembroke Dyd/Maudslay, Sons & Field (3,028ihp = 11.53kts)

As built: 244ft 9in, 209ft 11¾in x 55ft 4½in oa (54ft 6½in for tonnage) x 24ft 6in. 3,322^{51}/₉₄bm.

Ord: 2.4.1853. K: 22.1.1854. L: 16.4.1859. Fitted for reserve 12.1859. C: 8.6.1861 at Devonport (for sea).

First cost: £178,576 until end 1859 (including £98,583 for hull and £50,800 for machinery)

Commissioned 1.5.1861 at Plymouth under Capt. Charles Fellowes, as flagship of Rear-Adm. Robert Smart, with the Channel Squadron (–14.5.1863). On 13.6.1863 under Capt. Fitzgerald Algernon Charles Foley, as 73-gun flagship of Rear-Adm. Hastings Reginald Yelverton, for the Mediterranean; paid off 26.4.1865. On 1.9.1865 under Capt. Thomas Henry Mason, as guard ship at Pembroke; on 2.9.1865 under Capt. George Le Geyt Bowyear, then 1867 under Capt. William John Samuel Pullen, in same role; paid off 11.3.1869. Recommissioned 15.4.1869 under Capt. Thomas Cochran, as flagship of Vice-Adm. Alexander Milne to the Mediterranean, where exchanged whole crew and Vice-Adm. with *Caledonia*; on 31.5.1869 thus under Capt. Henry Gardiner, as flagship of Vice-Adm. George Rodney Mundy, for voyage home. On 3.7.1869 under Capt. Richard Wells, as flagship of Vice-Adm. George Greville Wellesley, to North America and West Indies, where exchanged whole crew with *Royal Alfred*; home with new crew, paid off 27.9.1869. Recommissioned 16.11.1869 under Capt. Francis Alexander Hume, as flagship of Rear-Adm. Arthur Farquhar, to Panama, where exchanged whole crew and Rear-Adm. with *Zealous*; on 23.1.1870 thus under Capt. Richard Dawkins, as flagship of Rear-Adm. George Fowler Hastings, for voyage home; paid off 23.3.1870. Recommissioned 22.8.1872 under Capt. Benjamin Spencer Pickard as flagship of Rear-Adm. Edmund Heathcote, Port Admiral at Queenstown, Ireland (replacing *Mersey*); in 1.1874 took Vice-Adm. James Robert Drummond to the Mediterranean to assume role as C-in-C Mediterranean (also carried replacement crews for several ships on that station); returned home with Vice-Adm. Hastings Reginald Yelverton (and the superseded crews). On 17.2.1875 under Capt. Henry Rushworth Wratislaw, as flagship of Port Admiral at Queenstown. On 6.3.1878 under Capt. Charles Matthew Buckle, as flagship of Rear-Adm. Richard Vesey Hamilton in same role, then 26.2.1881 under Capt. Philip Ruffle Sharpe, then (by now reduced to 28 guns) 13.8.1881 under Cmdr William Robert Clutterbuck, then 1.3.1884 under Capt. Robert Peel Dennistoun, then 5.2.1887 under Capt. Frederick Proby Doughty, as flagship of Adm. (retired) Arthur Farquhar, then 1.8.1889 under Capt. Francis Starkie Clayton, all in same role; paid off 8.3.1890 (replaced at Queenstown by *Triumph*). Renamed **Empress** as a training ship 3.1890 for Clyde Industrial Training Ship Association. Sold 31.12.1923 and BU at Appledore.

Renown Chatham Dyd/John Penn & Son (3,751ihp = 11.43kts)

As built: 244ft 10in, 210ft 1in x 55ft 4in oa (54ft 6in for tonnage) x 24ft 6in. 3,319^{1}/₉₄bm. Draught 13ft 4in/18ft 10in.

Ord: 2.4.1853. K: 20.12.1854. L: 28.3.1857. C: 20.2.1858 at Sheerness (for sea).

First cost: £165,447 (including £82,624 for hull and £49,307 for machinery).

Commissioned 21.11.1857 under Capt. Arthur Forbes, for the Channel (briefly flagship of Rear-Adm. Charles Howe Fremantle in 8.1858); to the Mediterranean 5.1859; paid off 25.9.1861. Reduced to 89 guns, then 81 guns 1863–69, and 54 guns by 1869. Sold to the North German Confederation [Prussian] Navy 24.3.1870 and became a German training ship (until 1881); BU 1892.

Atlas Chatham Dyd/Maudslay, Sons & Field (3,732ihp = 13.022kts (unmasted, no stores)]

As built: 244ft 8in, 209ft 11⅞in x 55ft 4in oa (54ft 6in for tonnage) x 24ft 6in. 3,317⁶²⁄₉₄bm. Draught 13ft 4in/18ft 10in.

Ord: 31.3.1855. K: 23.2.1858. L: 21.7.1860. Never fitted for sea.

Never commissioned. In reserve at Sheerness from 1861, then back at Chatham from 1874. Hospital ship at Greenwich 7.1881; moved to Deptford Creek 8.1882. Lent to the Metropolitan Asylums Board 11.1884. Sold to BU 1904.

Anson Woolwich Dyd/Maudslay, Sons & Field (3,583ihp = 12.984kts)

As built: 244ft 10in, 209ft 10½in x 55ft 4in oa (54ft 6in for tonnage) x 24ft 6in. 3,315⁸⁹⁄₉₄bm. Draught 15ft 0in/18ft 6in.

Ord: 31.3.1855. K: 1.10.1858. L: 15.9.1860. Never fitted for sea.

Never commissioned. In reserve at Sheerness from 1860. Reduced to 81 guns in 1865 and 73 guns in 1875. Renamed *Algiers* 11.1.1883. On 2.9.1899 flagship of Rear-Adm. Swinton Colthurst Holland, Adm. Superintendent at Chatham Dyd. BU by Castle 4.1904 at Charlton.

BRUNSWICK – Conversion of 80-gun Second Rate (a modified *Vanguard* Class ship, originally ordered 1847 and still incomplete on stocks) to 80-gun screw ship, 1854.

Dimensions & tons: 190ft 0in, 154ft 4¾in x 55ft 9in (55ft 0in for tonnage) x 23ft 4i. 2,484²⁷⁄₉₄bm.

Men: 750. Guns: LD 10 x 8in (65cwt) + 18 x 32pdrs (56cwt); UD 4 x 8in (65cwt) + 24 x 32pdrs (50cwt); QD/Fc 24 x 32pdrs (42cwt).

Machinery: 2-cylinder (64in diameter, 3ft stroke) horizontal single expansion. Single screw. 400nhp. 1,438ihp = 7¾kts.

First cost: £122,100, including hull as sailing ship £65,687, machinery £22,791.

Brunswick Pembroke Dyd/Miller, Ravenhill & Co.

As built: 190ft 0in, 154ft 2in x 55ft 10½ in (55ft 1½in for tonnage) x 23ft 4in. 2,491⁸⁴⁄₉₄bm. 3,632 tons displacement. Draught 15ft 2in/19ft 8in (with 30½ tons on board).

Conversion ord: 23.6.1854. Begun: 7.8.1854. L: 1.6.1855. C: 27.6.1855 (at Pembroke), 2.4.1856 (at Devonport where fitted for sea).

First cost: £122,100 (including £65,687 to build, £9,865 conversion, £22,791 for machinery).

Commissioned 20.12.1855 under Capt. Hastings Reginald Yelverton, for North America and the West Indies. On 12.7.1856 under Capt. Henry Broadhead, for the Mediterranean. On 22.10.1857 under Capt Erasmus Ommanney, for the West Indies; to the Channel 1858, then the Mediterranean 1859; paid off 2.3.1860. Fitted at Plymouth for 2nd Division of the Steam Reserve 7–8.1860. Reduced to 68 guns in 1863, and then to 60 guns in 1866. Sold to Marshall, Plymouth 3.1867 to BU.

DEFIANCE Class 91-gun ship. The final derivative of Isaac Watts's original draught for the *Agamemnon* class, and the last wooden screw battleship to be launched in Britain. Originally to have been of *Revenge* Class, but the design was modified on 8 October 1858 to lengthen the vessel by a further 10ft and modify the bow. Further improvement on the 91-gun design followed, and more orders were placed, envisaging it as the standard battlefleet unit, but by the end of the decade the introduction of armoured frigates like the *Warrior* led to the swift demise of the wooden battleship, and even those vessels which were completed (the *Atlas*, *Anson* and *Defiance*) never saw service at sea.

Dimensions & tons: 254ft 9in, 219ft 11¼in x 55ft 4in oa (54ft 6in for tonnage) x 24ft 6in. 3,475³¹⁄₉₄bm. 3,958 disp.

Men: 860. Guns: LD 34 x 8in (65cwt/9ft); UD 36 x 32pdrs (56cwt/9½ft); QD/Fc 20 x 32pdrs (45cwt/8½ft) + 1 x 68pdr (95cwt/10ft)

Machinery: 2-cylinder (82in diameter, 4ft stroke) horizontal single expansion. Single screw. 800nhp. 3,550ihp = 11.834kts.

First cost: £119,442.

Defiance Pembroke Dyd/Maudslay, Sons & Field

As built: 254ft 9in, 219ft 11⅛in x 55ft 4in oa (54ft 6in for tonnage) x 24ft 6in. 3,475³¹⁄₉₄bm. Draught 15ft 0in/18ft 6in.

Ord: 31.3.1855. K: 20.9.1858. L: 27.3.1861. C: 27.6.1861.

Not commissioned until 1884. Became the torpedo & mining school ship at Plymouth 26.11.1884 (which thereafter bore her name). Commissioned 12.1884 under J. Norcock; on 12.1887 under John Durnford; in 11.1900 under J de C Hamilton and 8.1909 under H. L. D'E. Skipwith. Recommissioned 12.1910 in same role. In 8.1911 under Cmdr F. H. M. Jackson, then 7.1915 Cyril Peel and 11.1919 A. H. C. Corby. Sold to S. Castle, Millbay, Plymouth, 26.6.1931 for BU.

DUNCAN Class 101-gun ships (later 81-gun). Surveyor's Department design, dated 6 January 1857. Longer and broader version of the *Conqueror* Class, with the same armament. Both re-classed as 99-gun in 1861; further reduced to 85-gun in 1862 (with 36 x 8in; 4 x 70pdrs + 32 x 32pdrs; and 1 x 110pdr BL); 81-gun in 1863 (with 70pdrs removed). *Duncan* fitted with an iron propeller in 1861 as an experiment. In January 1867 there was a design proposed by Reed to fit her as a breastwork turret ironclad, but this was cancelled for financial reasons.

Dimensions & tons: 252 ft 0 in, 213 ft 9¼ in x 58 ft 0 in oa (57 ft 2 in for tonnage) x 25 ft 6 in. 3,715⁶²⁄₉₄bm. 5,950 disp.

Men: 930. Guns: LD 36 x 8in (65cwt/9ft); UD 36 x 32pdrs (56cwt/9½ft): QD/Fc 28 x 32pdrs (42cwt/8ft) + 1 x 68pdr (95cwt/10ft).

Machinery: 2-cylinder (82in diameter, 4ft stroke) horizontal single expansion (trunk in *Duncan*). Single screw. 800nhp. See below for ihp/speed.

First Cost: *Duncan* £132,697 (hull £100,251). *Gibraltar* £130,235.

Gibraltar Devonport Dyd/Maudslay, Sons & Field (3,448ihp = 10.898kts)

As built: 252 ft 0 in, 213 ft 10¼ in x 58 ft 1 in oa (57 ft 3 in for tonnage) x 25 ft 6 in. 3,729bm.

Ord: 31.3.1855. K: 10.1858. L: 16.8.1860. Completed fitting machinery 28.8.1860 (at Devonport 1860–63).

Commissioned 8.9.1863 under Capt. James Charles Prevost, for the Mediterranean. On 26.11.1864 under Capt. Robert Coote, still in the Mediterranean; paid off 1867. At Devonport to 1872, then lent as a training ship at Belfast 1872. Renamed *Grampian* 1889. Sold to Castle 7.1899 and BU at Charlton.

Duncan Portsmouth Dyd/John Penn & Son. (3,428ihp = 13.238kts)

As built: 252ft 4in, 213ft 9½in x 58ft 1in oa (57ft 3in for tonnage) x

The official half model of the 101-gun *Duncan*, launched in 1859. This design was the culmination of steam battleship design, the later *Bulwark* Class developed from it, although with a reduced armament, being all cancelled or completed as ironclads. As so often with the march of technology, the wooden steam battleship was replaced just as it was reaching maturity. *(NMM L0818)*

25ft 6½in. 3,727¹⁹/₉₄bm. Draught 12ft 9in/19ft 7in.

Ord: 23.12.1856. K: 2.5.1857. L: 13.2.1859. Completed fitting machinery 5.1.1860Commissioned 6.1.1864 under Capt. Robert Gibson, as flagship of Vice-Adm. James Hope, for North America and West Indies. On 15.6.1867 under Capt. George Hancock, as first reserve coastguard at Queensferry (replacing *Trafalgar*). On 10.9.1867 under Capt. Charles Fellowes; then 7.6.1869 under Capt. William Rue Rolland, in same role; paid off 28.2.1870. On 1.4.1873 under Capt. George Willes Watson, as harbour flagship at Sheerness (with officers and crew turned over to *Pembroke*). On 1.1.1875 under Capt. Charles Thomas Curme, as flagship of Vice-Adm. George Fowler Hastings, at Sheerness. In 1878 under Capt. Thomas Bridgeman Lethbridge, then 1.1.1879 under Capt. Thomas Baker Martin Sullivan, then 27.7.1781 under Capt. John D'Arcy, all at Sheerness; paid off 31.12.1881. Became naval barracks ship from 1.1.1882, still under D'Arcy (–9.1883). Renamed ***Pembroke*** 1890, for harbour service (machinery removed). Receiving ship at Chatham 1895. Renamed ***Tenedos II*** in 9.1905. Sold 11.10.1910 (for £7,525) and BU in London.

Robust Devonport Dyd/Maudslay, Sons & Field
Ord: 1.4.1857. Re-ordered as 91-gun ship during 1858.

BULWARK Class New 91-gun ships. The *Bulwark* (named on 1 April 1857) was originally ordered to the *Revenge* design, but the Surveyor's Department produced a revised design, dated 29 July 1858 and approved 12 August 1858. Of this class seven ships were completed as ironclads, while the rest were cancelled. Developed from 101-gun *Duncan* Class, with a new timbering plan. Armament as *Defiance*, except that 58cwt guns replaced 56cwt guns on MD. *Robust* originally ordered as a 101-gun ship (as *Gibraltar*), altered to 91-gun during 1858. In 1866 there was a proposal by John Henwood to fit the still incomplete *Bulwark* as an ironclad turret ship with two turrets, but this was not carried out.

Dimensions & tons: 252ft 0in, 213ft 9¼in x 58ft 0in oa (57ft 2in for tonnage) x 25ft 6in. 3,715⁶²/₉₄bm.

Men: 860. Guns: LD 34 x 8in (65cwt/9ft); UD 36 x 32pdrs (58cwt/9½ft); QD/Fc 20 x 32pdrs (45cwt/8½ft) + 1 x 68pdr (95cwt/10ft)

Machinery: 2-cylinder horizontal single expansion. Single screw. 800nhp.

Bulwark Chatham Dyd/—
Ord: 9.4.1856. K: 8.3.1859. Suspended 7.3.1861. Cancelled 22.8.1872? BU 18.3.1873.

Repulse Woolwich Dyd/—
Ord: 1.4.1857. K: 29.4.1859– —. Converted to ironclad (see Postscript).

Zealous Pembroke Dyd/—
Ord: 1.4.1857. K: 26.10.1859– Converted to ironclad (see Postscript).

Robust Devonport Dyd/Maudslay, Sons & Field
Ord: 1.4.1857. K: 31.10.1859. Suspended 12.3.1861. Cancelled 22.8.1872. BU completed 18.3.1873.

Royal Alfred Portsmouth Dyd/—
Ord: 8.4.1859. K: 1.12.1859. Began conversion to ironclad (see Postscript) 22.6.1861.

Royal Oak Chatham Dyd/—
Ord: 8.4.1859. K: 1.5.1860. Began conversion to ironclad (see Postscript) 3.6.1861.

Triumph Pembroke Dyd/—
Ord: 8.4.1859. K: 13.8.1860. Began conversion to ironclad (see Postscript) 6.6.1861.

Kent Portsmouth Dyd/—
Ord: 5.3.1860. K: 13.6.1860. Cancelled 12.12.1863 and BU on stocks.

Ocean Devonport Dyd/— ('to duplicate *Robust*')
Ord: 5.3.1860. K: 23.8.1860. Began conversion to ironclad (see Postscript) 3.6.1861.

Caledonia Woolwich Dyd/— ('to duplicate *Repulse*')
Ord: 5.3.1860. K: 1.10.1860. Began conversion to ironclad (see Postscript) 6.6.1861.

Blake Pembroke Dyd/—
Ord: 5.3.1860. K: — . Suspended 13.2.1862. Cancelled 12.12.1863 (unstarted).

Pitt Chatham Dyd/—
Ord: 5.3.1860. K: — . Suspended 13.2.1862. Cancelled 12.12.1863 (unstarted).

FREDERICK WILLIAM – Conversion of 116-gun three-decker (suspended incomplete on stocks) to 86-gun two-decker, 1859 design.

Dimensions & tons: 214ft 0in, 174ft 0½in x 60ft 0in oa (59ft 2in for tonnage) x 23ft 9in. 3,241bm. 4,502 disp.

Men: 830. Guns: LD 30 x 8in (65cwt); UD 32 x 32pdrs (56cwt); QD/Fc 22 x 32pdrs (42cwt) + 2 x 68pdrs (95cwt).

Machinery: 2-cylinder (66in diameter, 3½ft stroke) horizontal single expansion. Single screw. 500nhp. 2,276ihp = 11.777kts.

First cost: Hull as sailing vessel £84,948.

Emblematic of the radical change in technology, the *Frederick William*, one of the last conversions to a wooden screw battleship, is seen here at Queenstown (Cobh) around 1866 with the *Black Prince*, one of the first of the broadside ironclads. Suspended while building to Symonds's Modified *Caledonia* design, the *Frederick William* was completed as a steam two-decker, but – like most conversions of Symonds's designs – the ship was a poor seaboat. *(NMM A2331)*

Frederick William (ex-*Royal Frederick*, renamed 28.1.1860)
Portsmouth Dyd/Maudslay, Sons & Field
　　Ord: 28.2.1857. Conversion begun: 28.5.1859. L: 24.3.1860. C: 6.1860?
　　Commissioned 1.7.1864 under Capt. Edward Codd, as guard ship at Portland (replacing Colossus). On 7.3.1865 under Capt. Edmund Heathcote, as guard ship at Queenstown; on 31.1.1866 under Capt. John James Kennedy, in same role; paid off 31.8.1868. Renamed **Worcester** 19.10.1876 and lent to Thames Nautical Training College as training ship for merchant service at Greenhithe; returned to Admiralty 1945. Sold 7.1948 to Frary Industries to BU, but foundered 30.8.1948 in the Thames; raised 5.1953 and BU by Tennant & Horne, Grays.

QUEEN – Conversion of (19-year-old) 116-gun three-decker to 86-gun two-decker, 1858 design.
　　Dimensions & tons: 216ft 7½in, 174ft 1¾in x 60ft 0¾in oa (ca.59ft 2¾in for tonnage) x 24ft 6in. 3,249bm. 3,930 disp.
　　Men: 830. Guns: LD 30 x 8in (65cwt); UD 32 x 32pdrs (56cwt); QD/Fc 22 x 32pdrs (42cwt) + 2 x 68pdrs (95cwt).
　　Machinery: 2-cylinder (66in diameter, 3½ft stroke) horizontal single expansion. Single screw. 500nhp. 2,283ihp = 10.578kts.
Queen Sheerness Dyd/Maudslay, Sons & Field
　　Conversion ord: 10.8.1858. Begun: 23.8.1858. L: (undocked) 5.4.1859. C: 30.10.1859.
　　Commissioned 18.11.1859 under Capt. Charles Farrel Hillyar, for the Mediterranean ; sailed 23.2.1860 for Lisbon but returned to Plymouth 3.4.1860 for repairs following storm damage; landed marines to protect British Legation at Athens 7.1863; paid off 15.12.1863 at Portsmouth. Sold to Castle 1871 to BU at Charlton.

Broadened CALEDONIA Class – Conversion of (1832–41 launched) three-deckers to 89-gun Second Rates, 1858 design. Originally 120-gun First Rates of 'Broadened *Caledonia*' design, these five ships were cut

down from three to two decks while under conversion to steam ships of the line, as stopgap conversions until new construction provided.
　　Trafalgar's cylinders were 66in diameter, 3½ft stroke; *Royal William*'s were 65in diameter, 3ft stroke.
　　Men: 830. Guns: LD 32 x 8in (65cwt/9ft); UD 34 x 32pdrs (56cwt/9½ft); QD/Fc 22 x 32pdrs (42cwt/8ft) + 1 x 68pdr (95cwt/10ft).
　　Machinery: 2-cylinder (71in diameter, 3ft stroke – except as noted above) horizontal single expansion. Single screw. 500nhp. See below for individual ihp/speed.
Neptune Portsmouth Dyd/Ravenhill, Salkeld & Co. (1,910ihp = 11.24kts)
　　Dimensions & tons: 216ft 6in, 179ft 1¼in x 55ft 5in oa (54ft 6in for tonnage) x 23ft 2in. 2,830bm. 3,405 disp.
　　Conversion Ord: 10.8.1858. Begun: 10.8.1858. L: (undocked) 7.3.1859. C: 6.8.1859 (fitted for sea).
　　Commissioned 7.6.1859 under Capt. William Legge George Hoste, for the Mediterranean. On 6.12.1859 under Capt. Frederick Archibald Campbell, then 11.2.1861 under Capt. Geoffrey Thomas Phipps Hornby, still in the Mediterranean; paid off 17.12.1862. Sold to Castle 23.1.1875 to BU at Charlton.
Trafalgar Chatham Dyd/Maudslay, Sons & Field (2,275ihp = 10.9kts)
　　As converted: 216ft 0in, 183ft 3½in x 55ft 5½in oa (ca.54ft 6½in for tonnage) x 23ft 2in. 2,900bm. 3,850 disp.
　　Conversion Ord: 10.8.1858. Begun: 16.8.1858. L: (undocked) 21.3.1859. C (at Sheerness): 14.8.1859.
　　Commissioned 9.6.1859 under Capt. Edward Gennys Fanshawe, for the Channel Squadron. On 2.4.1861 under Capt. John Bourmaster Dickson, on same station. On 14.7.1862 under Capt. Thomas Baillie, to the Mediterranean. On 8 May 1863 under Capt. Thomas Henry Mason, in the Mediterranean; paid off 28.2.1864 at Sheerness. Recommissioned 1.3.1864 under Capt. Charles Frederick Schomberg, as guard ship at Leith (replacing Edinburgh); in 3.1865 under Capt. John Borlase, then 16.5.1865 under Capt. George Hancock, still at Leith; paid off 15.6.1867. Recommissioned 15.11.1867 at Sheerness under Capt. Edward King Barnard, as guard ship at Lough Swilly; paid off 20.11.1869. Recommissioned 17.8.1870 at Portsmouth under Capt. Thomas Bridgeman Lethbridge, as seagoing naval training ship; paid off 8.6.1872. Recommissioned 25.2.1873 under Cmdr Alfred

The *Trafalgar*, with *Neptune* astern, photographed around 1862, following their conversion from three-decker First Rates of the Broadened *Caledonia* Class to screw Second Rates of 89 guns. *(NMM D2164)*

Markham (–11.1.1876), and renamed as boys' training ship at Portland. On 6.2.1879 under Cmdr George Bruce Evans, in same role. In 1886 under Cmdr Algernon Charles Littleton, then 1.1890 Cmdr G. A. Primrose, and 9.1899 Cmdr T. H. M. Jerram. Sold to Castle 10.7.1906 to BU at Charlton.

Saint George Devonport Dyd/Ravenhill, Salkeld & Co. (1,730ihp = 10.93kts)

Dimensions & tons: 216ft 6in, 178ft 10½in x 55ft 4½in oa (ca. 54ft 10½in for tonnage) x 23ft 9in. 2,864bm. 4,213 disp.

Conversion Ord: 10.8.1858. Begun: 21.8.1858. L: (undocked) 19.3.1859. C: 12.7.1859.

Commissioned 6.6.1860 under Capt. Francis Egerton, as flagship of Rear-Adm. Edmund Lyons at Plymouth; to North America and West Indies 5.1861; to Channel Squadron 5.1862; to Mediterranean 11.1862; paid off 9.2.1864 at Plymouth. Recommissioned 10.2.1864 under Sydney Grenfell, as guard ship at Falmouth (replacing *Russell*). On 5 May 1864 under Capt. Edward Bridges Rice, in same role; to Portland 5.1865 (replacing *Frederick William*). On 1.5.1867 under Capt. Mathew Stainton Nolloth, in same role; cruise of the Reserve Fleet 5.1869; paid off 30.6.1869. Sold to Castle 11.1883 to BU at Charlton.

Waterloo Sheerness Dyd/Ravenhill, Salkeld & Co. (1,890ihp = 11.33kts)

Dimensions & tons: 218ft 2⅛in, 178ft 11⅛in x 55ft 4in oa (54ft 8in for tonnage) x 23ft 4½in. 2,845bm. 3,440 disp.

Conversion Ord: 5.2.1859. Begun: 1.4.1859. L: (undocked) 12.11.1859. C: 4.1860. Renamed *Conqueror* 27.2.1862.

Commissioned 24.11.1863 at Sheerness under Capt. William Garnham Luard, for China; at bombardment of Simonoseki (Japan) 9.1864; paid off 21.2.1866. Renamed *Warspite* 11.8.1876 and lent as training ship at Woolwich for Marine Society. Moved to Greenhithe 1901. Burnt 20.1.1918 in the Thames.

Royal William Devonport Dyd/Robert Napier & Sons (1,763ihp = 10.58kts)

Dimensions & tons: 216ft 9in, about 177ft x 55ft 7in oa (ca.55ft for

tonnage). 2,849bm. 3,520 disp.

Conversion ord: 5.2.1859. Begun: 21.3.1859. L: (undocked, completed) 9.2.1860.

Never commissioned as steam vessel. Became training ship and transferred to Liverpool RC Reformatory Society 11.1884; after removal of engines, arrived Liverpool 30.8.1885 and renamed ***Clarence***. Burnt by accident 26.7.1899 in the Mersey.

***NELSON* Class** – Conversion of (1814-launched) three-decker to 89-gun ship, 1859 design. Former 120-gun First Rate, cut down and lengthened along the lines of the *London*. See Chapter 1 for original details.

Dimensions & tons: 216ft 3in, 178ft 7⅛in x 54ft 6in oa (53ft 8in for tonnage) x 24ft 3in. Draught 17ft 6in (fwd), 21ft 10in (aft). 2,736bm. 3,158 disp.

Men: 850. Guns: LD 32 x 8in (65cwt); UD 34 x 32pdrs (56cwt); QD/Fc 22 x 32pdrs (42cwt) + 1 x 68pdr (95cwt).

Machinery: 2-cylinder (71in diameter, 3ft stroke) horizontal single expansion. Single screw. 500nhp. 2,102ihp = 11.533kts.

Nelson Portsmouth Dyd/Ravenhill, Salkeld & Co.

Conversion ord: 5.2.1859. Begun: 10.3.1859. L: (undocked) 7.2.1860. C: 1.9.1860.

Transferred to Colonial Government of Victoria (Australia) 7.1.1867, and re-armed as coastal defence ship with 2 x 7in, 20 x 64pdrs and 20 x 32pdrs; recommissioned 30.8.1867 under Acting Cmdr Charles Bradley Payne; sailed from Portsmouth 20.10.1867, arriving Port Philip, Victoria 2.1868; paid off 13.2.1868 at Melbourne. Raséed as a frigate 1878. Sold as a storage hulk 28.4.1898, resold as a coaling hulk for Sydney, later to Launceston and BU at Launceston 9.1928.

***POWERFUL* Class** – Conversion and 40ft extension of (1828-launched) Second Rate of *Canopus* Class to 80-gun ship. Design (for conversion) of 4.1.1860.

Dimensions & tons: 233ft 9in, 197ft 9¼in x 52ft 4½in oa (51ft 5in

A tribute to the longevity of teak construction, the *Bombay* – a ship originally launched in 1828 – was thought worthy of conversion to steam as late as 1860. She is seen here in 1861 during her conversion in dock at Chatham, before her first steam commission. *(NMM C5531K)*

for tonnage) x 23ft 1in. 2,783²⁹⁄₉₄bm.

Men: 750. Guns: LD 10 x 8in (65cwt) + 24 x 32pdrs (56cwt); UD 4 x 8in (65cwt) + 30 x 32pdrs (50cwt); QD/Fc 12 x 32pdrs (42cwt) + 1 x 68pdr (95cwt).

Machinery: 2-cylinder (64³⁄₁₆in diameter, 32in stroke) horizontal single expansion. Single screw. 400nhp. 1,688ihp = 10.157kts.

Powerful   Chatham Dyd/—

Conversion ord: 17.1.1860. Cancelled 3.5.1860 (hull found to be defective) in favour of her sister:

Bombay   Chatham Dyd/Humphrys & Tennant

As converted: 233ft 9in, 197ft 9in x 52ft 4in x 23ft 1in. 2,783bm.

Conversion ord: 3.5.1860, sailed 8.5.1860 for Chatham. Begun: 23.5.1860. C: 25.6.1861.

Commissioned 25.3.1864 under Capt. Colin Andrew Campbell, as flagship of Rear-Adm. Charles Gilbert John Brydone Elliot, for east coast of South America; burnt 14.12.1864 by accident off Montevideo, with the loss of 94 lives (formally paid off 18.2.1865).

PRINCE REGENT – Conversion of (1823-launched) three-decker, previously cut down to sailing two-decker in 1845–46, to 89-gun screw ship, 1860 design.

Dimensions & tons: 217ft 10in, 180ft 4½in x 54ft 7in oa (53ft 8in for tonnage) x 23ft 3in. 2,762bm.

Men: 830. Guns: LD 32 x 8in (65cwt); UD 34 x 32pdrs (56cwt); QD/

Fc 22 x 32pdrs (42cwt) + 1 x 68pdr (95cwt).

Machinery: 2-cylinder horizontal single expansion. Single screw. 500nhp. (never tried for ihp/speed).

Prince Regent   Portsmouth Dyd/James Watt & Co.

Conversion ord: 3.2.1860. Begun: 8.2.1860. L: (undocked, completed) 27.5.1861.

Never commissioned as a screw ship. BU completed 28.7.1873.

ROYAL GEORGE – Conversion of (1827-launched) three-decker to 89-gun screw ship, but not lengthened like other *Caledonia* Class, design dated 12.4.1860.

Dimensions & tons: see under screw three-deckers in Chapter 1.

Men: 830. Guns: LD 32 x 8in (65cwt); UD 34 x 32pdrs (56cwt); QD/Fc 22 x 32pdrs (42cwt) + 1 x 68pdr (95cwt).

Machinery: 2-cylinder (58¼in diameter, 3¼ft stroke) horizontal single expansion, trunk. Single screw. 400nhp. 1,417ihp = 8.65kts.

Reduced to 86-gun in 1861, then 78-gun in 1863 (and later to 72 guns in 1866).

Commissioned 4.1864 under Capt. Michael de Courcy, as guard ship at Kingstown (replacing *Ajax*), On 3.4.1865 under Capt. Thomas Miller, in same role; to Spithead for Royal Review 7.1867. In 1.1869 under Capt. Robert Jenkins, still at Kingstown; paid off 31.12.1869 (replaced at Kingstown by *Pallas*). Sold 23.1.1875 to Messrs. Castle to BU at Charlton.

3 Third Rates

Until 1 January 1817, the Third Rate encompassed vessels (all two-decked ships) with more than 60 but not more than 80 guns. At that date, the rating system was amended to restrict the Rate to ships carrying not less than 70 guns (or with complements of fewer than 750 but not fewer than 620 men); at the same time, recalculating the established number of guns aboard each ship (to include carronades) increased the count of ordnance on board existing 80-gun ships to 84 guns. Consequently these 84-gun ships were transferred to the Second Rate, and the last six remaining 64-gun ships (which would otherwise have become Fourth Rates) were re-classed as ancillary vessels. The Third Rate thus was left to comprise the 74-gun ships (which still numerically comprised the vast majority of the battlefleet), although some of these were re-rated as being 76 or 78 guns due to the addition of carronades in their count of ordnance. Many were subsequently to be re-classed as 72 guns in 1839.

(A) Vessels in service or on order at 1 January 1817

The Admiralty by the close of the Napoleonic War had decided to build no further small ships for the battlefleet, and there were only rare exceptions to this policy over the next forty years. Consequently, the class comprised a dwindling number of ageing ships as the existing Third Rates wore out, although a number of these acquired a new *raison d'être* in the 1820s as a dozen were raséed – cut down by a deck – to form 50-gun frigates (Fourth Rates), and right at the end of the sailing age nine of the remaining ships were converted to 60-gun 'blockships', fitted with steam engines and screws to provide a stopgap means of coastal defence. The huge total of 114 Third Rates that existed following the 1817 changes dropped to 80 in January 1820 and to 41 (of which just one was in commission) in September 1845. In September 1855 there were 16 left, excluding a single 70-gun steam battleship (*Sans Pareil*) in the Third Rate.

Following the reorganisation in January 1917, the Navy's Third Rates included 8 ships rated at 78 guns, 16 rated at 76 guns, and 90 rated at 74 guns – the latter including the only Third Rates (8 in number) still under construction. These comprised the following:

78 guns: *Spartiate*, *Superb*, *Achille*, *Kent*, *Milford* (all with 24pdrs on the upper deck); *Donegal*, *Northumberland* and *Renown* (all with 18pdrs on the upper deck).

76 guns: *Pompee* (stricken in January 1817), *Centaur* (stricken in 1819) *Bulwark*, *Conqueror*, *Mars*, *Revenge*, *Warspite* and *Chatham* (all with 24pdrs on the upper deck); *Tigre* and *Abercrombie* (stricken 1817), *Mont Blanc*, *Scipion* and *Rivoli* (stricken 1819), *Colossus*, *Genoa* and *Implacable* (all with 18pdrs on the upper deck). By January 1820 this had shrunk to 8 ships of 76 guns (3 in commission and 5 in Ordinary).

74 guns: all the remaining ships listed below in this section.

Third Rates of 74 guns: 24pdr type

***TRIUMPH* Class** The first and, for 30 years, the only 74-gun ships to mount a 24pdr armament on the upper deck; they were initially ordered on 11 January 1757 as extra units to the *Dublin* Class (the first 'true' 74s in the RN) and named on 17 March; on 30 March both were re-ordered to a 1,558-bm design, and then again re-ordered on 21 May to a new and considerably larger design copied from the lines of the French *L'Invincible* (captured 1747). By 1793 both had been re-armed with 18pdrs on the UD replacing 24pdrs. Both ships were in Ordinary in 1793, were re-commissioned for wartime service but were disarmed for harbour service before 1815.

Dimensions & tons: 171ft 3in, 139ft 0in x 49ft 3in x 21ft 3in. 1,793³⁵⁄₉₄bm.

Men: 650 (635 from 1794). Guns: LD 28 x 32pdrs; UD 30 x 24pdrs; QD 14 x 9pdrs; Fc 2 x 9pdrs.

Valiant Chatham Dyd [M/Shipwright John Lock]

As built: 171ft 2¾in, 139ft 9in x 49ft 4in x 21ft 2¾in. 1,799⁴¹⁄₉₄bm.

Ord: 21.5.1757. K: 1.2.1758. L: 10.8.1759. C: 9.10.1759.

First cost: £42,589.5.10d (including fitting).

Commissioned 8.1759 under Capt. William Brett; under Capt. Adam Duncan from 1761 until paid off 7.1764. Underwent Large Repair at Chatham (for £36,297.10.10d) 10.1771–5.1775; fitted there 1.1777. Recommissioned 11.1777–1783 (refitted and coppered at Portsmouth for £7,149.13.0d 5–8.1780). Repaired at Plymouth (for £27,446.17.11d) 2.1785–7.1786. Fitted for Channel service (for £13,361.19.2d) 5–6.1790. Recommissioned in 5.1790 under Capt. HRH the Duke of Clarence, for the Spanish Armament; then paid off 1790. Fitted at Portsmouth (for £10,301) 10.1793–3.1794; recommissioned 10.1793 under Capt. Thomas Pringle; in Howe's fleet on Glorious First of June off Ushant 1.6.1794; had 2 killed, 9 wounded; in 8.1794 under Capt. Christopher Parke; in Bridport's Action off Île Groix 23.6.1795. Under Capt. James Larcom in 7.1795, then Capt. Eliab Harvey 9.1795; sailed for Jamaica 11.8.1796. In 1797 under Capt. Edmund Crawley; (with *Thunderer*) destroyed French 44-gun *L'Harmonie* at San Domingo 17.4.1797; took 16-gun privateer *La Magicienne* 10.1797–3.1798 in West Indies. In 1799 under Capt. John Cochet. Made good defects at Portsmouth (for £11,294) 10–11.1795. Fitted as a lazarette at Chatham (for £1,506) 10–11.1799, for Stansgate Creek. Under Capt. John Bligh at Jamaica 1803–04, then under Capt. George Hope 1805. BU at Sheerness 4.1826.

Triumph Woolwich Dyd [M/Shipwright Israel Pownoll to 5.1762, completed by Joseph Harris]

As built: 171ft 3in, 138ft 8in x 49ft 9in x 21ft 3in. 1,825¹⁴⁄₉₄bm.

Ord: 21.5.1757. K: 2.1.1758. L: 3.3.1764. C: 18.4.1764.

First cost: £33,252.3.5d (including fitting).

Commissioned 1.1771 for Falkland Islands dispute. Fitted at Chatham (for £7,939.12.4d) 3.1771; fitted as a guard ship 12.1771. Fitted as a

Guard ship at Sheerness 12.1774. Fitted at Chatham (for £8,470.9.7d) 8.1778–3.1779. Wartime service in West Indies/ North America 10.1778–11.1781. Refitted and coppered (for

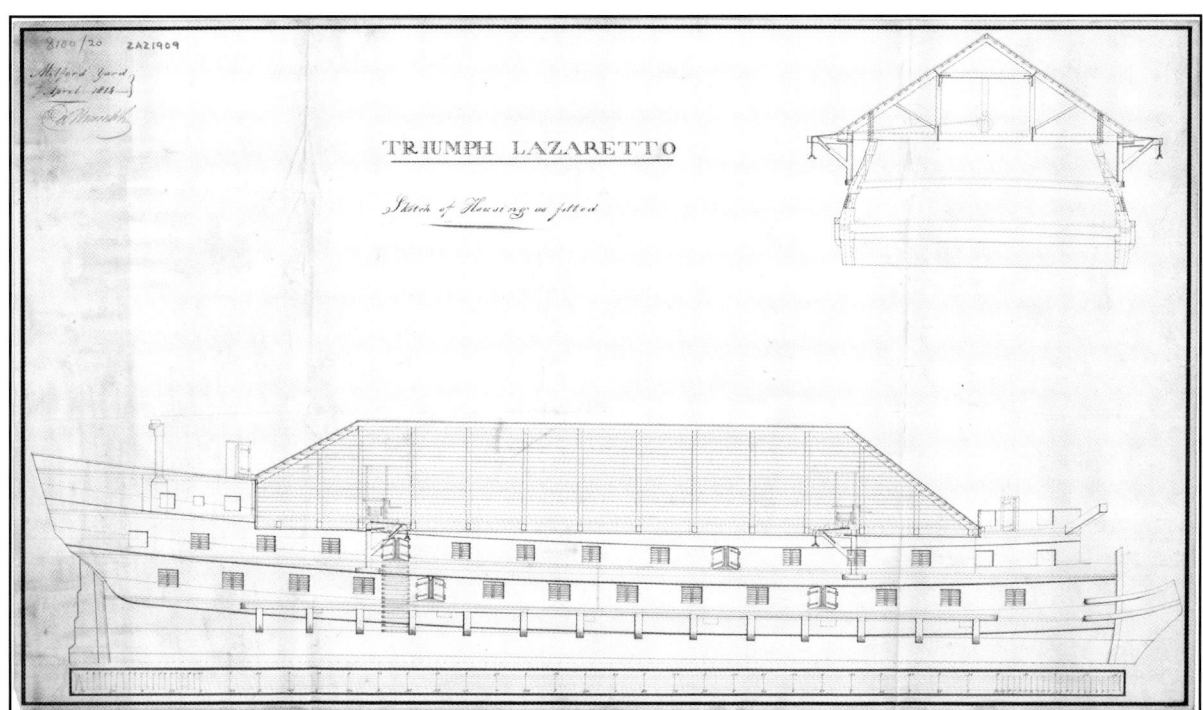

TRIUMPH LAZARETTO

Sketch of Housing as fitted

In 1813 the elderly 74-gun *Triumph* was hulked at Plymouth and fitted as a lazaretto, one of a variety of uses made of surplus battlefleet units towards the end of the Napoleonic War. This draught by Edward Churchill (assistant to the master shipwright at that dockyard from 1803 to 1813, and promoted to master shipwright there from 1815 until 1829) shows the outboard profile and a section through the hull from the waterline up, illustrating the construction of a roof over the entire waist from the forecastle to the quarterdeck. *(NMM J3031)*

£5,853.17.1d) 12.1779–2.1780. Small to Middling Repair at Chatham (for £18,321.1.6d) 5–11.1782; fitted at Portsmouth as a guard ship 7.1783. Great Repair at Portsmouth (for £46,499) 1.1792–1.1795. Recommissioned 11.1794 under Capt. Sir Erasmus Gower; in Cornwallis's retreat 16–17.6.1795. Under Capt. William Essington in 9.1797, in North Sea; at Camperdown 11.10.1797. Under Capt. Thomas Seccombe in 4.1799, flagship of Vice-Adm. Lord Collingwood from 6.1799; sailed for the Mediterranean 1.6.1799. Under Capt. Eliab Harvey in 1800, in the Channel. Under Capt. Sir Robert Barlow in 8.1801 (–1804), joined Toulon blockade. Paid off 4.1804. Fitted at Portsmouth 3–5.1805. Recommissioned 4.1805 under Capt. Henry Inman; in Battle of Finisterre 22.7.1805; in Strachan's squadron for chase of Leissègues and Willaumez. Under Capt. Sir Thomas Hardy from 5.1806 (–1809), in Strachan's squadron; to Halifax station 1807; in Beresford's squadron off Lorient 1.1809. Under Capt. Samuel Hood Linzee 1809; sailed for Portugal 22.4.1809; off Cadiz 1811. Paid off into Ordinary at Plymouth 1812. Fitted at Plymouth as a lazarette 7–10.1813 for Milford. BU at Pembroke 6.1850.

MARS Class The 1785-designed *Brunswick* had been built to much larger dimensions than the preceding 'Common' class 74s, but for the next Third Rates Sir John Henslow used a similar design to mount a heavier ordnance, with 24pdrs on the UD instead of 18pdrs; it was approved on 22 March 1788. Both ships were named by AO of 23 October 1788. Re-classed as 76-gun ships in January 1817.

> Dimensions & tons: 176ft 0in, 144ft 3in x 49ft 0in x 20ft 0in. 1,842²⁴⁄₉₄bm.
>
> Men: 640 (635 from 1794). Guns: LD 28 x 32pdrs; UD 30 x 24pdrs; QD 12 x 9pdrs; Fc 4 x 9pdrs.
>
> *Mars* later QD 2 x 24pdrs + 12 x 32pdr carronades; Fc 2 x 24pdrs + 2 x 32pdr carronades.

Mars Deptford Dyd [M/Shipwright Martin Ware]

> As built: 176ft 0in, 144ft 1in x 49ft 2in x 20ft 0½in. 1,852⁶²⁄₉₄bm. Draught 12ft 7in/17ft 5in.

Ord: 17.1.1788. K: 10.1789. L: 25.10.1794. C: 8.11.1794.
First cost: £50,270 (incl. fitting).

Commissioned 11.1794 under Capt. Charles Cotton. In Cornwallis's Retreat 16–17.6.1795. In 2.1797 under Capt. Alexander Hood; at Spithead mutiny 4.1797; took French 74-gun *L'Hercule* off Brest 21.4.1798 (losing 30 killed including Hood, 60 wounded). In 4.1798 under Capt. George Shirley, then 7.1798 Capt. John Manley. In 5.1799 under Capt. John Monkton, as flagship of Rear-Adm. George Berkeley, at blockade of Rochfort; in attack on Spanish squadron in Aix Roads 2.7.1799. In 1.1801 under Capt. Robert Lloyd, as flagship of Rear-Adm. Edward Thornbrough, at blockade of Brest (–1802). Middling Repair at Plymouth 9.1802–4.1803. Recommissioned 3.1803 under Capt. John Sutton; to blockade of Brest; (temp.) under Capt. Samuel Pym 5.1805. Under Capt. George Duff 1804, as flagship of Rear-Adm. Thomas Russell, off Ferrol. In Lee column at Trafalgar; had 29 killed (including Duff), 69 wounded; Lieut William Hennah succeeded to command. From 12.1805 under Capt. Robert Oliver; took 40-gun *Le Rhin* off Rochefort 18.7.1806. Under Capt. William Lukin 8.1806 (–1810); in Hood's Action off Rochefort 25.9.1806; in Sir Richard Keats's fleet for Copenhagen expedition 8.1807; autumn 1897 and 1808 in the Baltic. . On 9.6.1810 under Capt. John Surman Carden; sailed for Portugal 22.10.1810. On 26.11.1812 under Capt. Henry Raper, in the Baltic; paid off 25.2.1813. Laid up 12.1812. Fitted as receiving ship at Portsmouth 1813. BU there 10.1823.

Centaur Woolwich Dyd [M/Shipwright William Rule to 2.1793, completed by John Tovery]

> As built: these details quoted as identical to design (above). Draught 13ft 0in/17ft 9in.

Ord: 17.1.1788. K: 11.1790. L: 14.3.1797. C: 11.6.1797.
First cost: £59,538 (incl. fitting).

Commissioned 3.1797 under Capt. John Markham; sailed for the Mediterranean 2.6.1798; occupation of Minorca 11.1798; took 14-gun privateer *La Vierga del Rosario* 2.2.1799; (with *Cormorant*) destroyed 40-gun *El Guadaloupe* near Cape

Ovapesa 16.3.1799; capture of French squadron (40-gun *Le Junon*, 36-gun *L'Alceste*, 32-gun *La Courageuse*, 18-gun *La Salamine* and 14-gun *L'Alerte*) off Cape Secié 19.6.1799. From 3.1801 under Capt. Sir Harry Neale; and 10.1801 Capt. Arthur Legge. From 4.1802 under Capt. Bendall Littlehales, as flagship of Commodore Samuel Hood (to 1810?); recapture of St Lucia 22.6.1802 and Tobago 25.6.1802; capture of Dutch islands; took 14-gun *Hippomenes* 20.9.1802. In 1804 under Capt. Murray Maxwell (still Hood's flagship); seizure of Diamond Rock 1.1804; cut out 16-gun *Le Curieux* of Port Royal 3.2.1804; capture of Surinam 30.4–5.5.1804. In 1805 under Capts Charles Richardson, Henry Whitby (temp.) and John Talbot (12.1805). Middling Repair at Plymouth (for £41,413) 2–6.1806; in Strachan's squadron in pursuit of Leissègnes and Willaumez; boats of squadron cut out 18-gun *Le César* of the Gironde 15.7.1806; Hood's squadron took three 40-gun frigates (*La Gloire*, *L'Infatigable* and *L'Armide*) off Rochefort 24.9.1806. Under Capt. William Webly 1807, in Copenhagen expedition 8.1807; Hood made Rear-Adm. 10.1807; occupation of Madeira 12.1807; operations in Baltic 1808; (with *Implacable*) took 74-gun *Sewolod* 26.8.1808; Scheldt operations 1809; sailed for the Mediterranean 2.11.1809. Defects made good (for £15,253) 11.1812–1.1813. Under Capt. Thomas Caulfield 9.1814; in North America. Laid up at Plymouth 11.1815. BU at Plymouth 11.1819.

Up until the *Brunswick* in the mid 1780s, all 74s built for the British Navy – except for the French-derived *Carnatic* design – were all of the Small or 'Common Class' type, with similar armament and complement, and differing only fractionally in dimensions since the *Dublin* Class of 1755 (the exceptions being the *Triumph* and *Valiant* of 1757, completed to carry 24pdrs on the UD). Following criticism of the length of Common Class 74s (as well as other Rates), the Board directed both Surveyors in November 1794 to produce new draughts for 74-gun ships of 186ft length. The main aim was to increase the spacing between ports, but the new designs allowed for an extra port on each gun deck. Altogether, the Navy Board on 6 November 1794 placed orders with the Royal Dockyards for three 74-gun ships – the *Scipio* (renamed *Bulwark* in 1806 before launching) was to carry 24pdrs on her UD (see earlier section), while two 'Middling Class' designs were produced, one by each Surveyor. The competitive designs – one ship to each Surveyor's draught – were slightly reduced in scale on 6 June 1796, but still substantially exceeded the 175ft length of the Common Class 74s.

BULWARK A 'Large Class' 74, with 24pdrs on the UD. A highly successful Sir William Rule design, although she was not begun until 1804. Re-classed as a 76-gun ship in January 1817.

Dimensions & tons: 182ft 0in, 150ft 8in x 49ft 0in x 20ft 6in. 1,924¹⁹⁄₉₄bm.

Men: 640. Guns: LD 28 x 32pdrs; UD 30 x 24pdrs; QD 14 x 9pdrs (later 4 x 12pdrs + 10 x 32pdr carronades); Fc 2 x 9pdrs (later 2 x 12pdrs + 2 x 32pdr carronades); RH 6 x 18pdr carronades.

Bulwark (ex-*Scipio*, renamed 28.4.1806) Portsmouth Dyd [M/Shipwright Nicholas Diddams]

As built: 181ft 10in, 150ft 4¼in x 49ft 3in x 20ft 7in. 1,939⁸⁄₉₄bm.

Ord: 6.11.1794. K: 4.1804. L: 23.4.1807. C: 12.9.1807.

First cost: £57,240 (including fitting).

Commissioned 3.1807 under Capt. Charles Fleeming; sailed for the Mediterranean 12.10.1807; at blockade of Cadiz 1807–9. Between Middling and Large Repair and fitted at Plymouth

5.1811–1.1812. In 5.1811 under Capt. Joshua Horton, then 12.1811 under Capt. James Worth; from 5.1812 flagship of Rear-Adm. Sir Philip Durham; in the Channel in 1813. From 12.1813 under Capt. David Milne; to North America 1814. At destruction of USS corvette *Adams* in the Penobscot river 3.9.1814; took US 10-gun privateer *Harlequin* 23.10.1814. In 12.1814 under Capt. Farmery Epworth; took US 9-gun privateer *Tomahawk* 22.1.1815. Paid off 6.1815; fitted as a guard ship at Sheerness 12.1815–7.1816. Small Repair and fitted as a guard ship at Chatham 2.1819–3.1822. BU completed at Portsmouth 26.9.1826.

Earl Spencer's Board, coming into office in December 1794, decided on a programme of new large 74s, and placed orders on 30 April 1795 with commercial yards for eight new ships. Two of these were to be built to the draught of the *Triumph* Class, but with 6in higher decks, and similarly were to carry 24pdrs on the UD (*Ajax* and *Kent*), while the other six were to be of the 'Middling Class', although several of these had been refitted with 24pdrs *vice* 18pdrs by 1817. The draughts of all eight vessels were approved on 10 June 1795; all exceeded 175ft in length. A ninth vessel (*Spencer*) was ordered in September, to a design produced by the émigré French constructor, Jean-Louis Barrallier.

ACHILLE Class 'Middling Class' 74s. Derived from draught of *Le Pompée*, a French prize taken 1793 at Toulon. Re-classed as a 78-gun ships in January 1817.

Dimensions & tons: 182ft 2in, 149ft 9in x 49ft 0½in x 21ft 10½in. 1,915⁷⁹⁄₉₄bm.

Men: 640. Guns: LD 30 x 32pdrs; UD 30 x 24pdrs; QD 4 x 18pdrs+ 10 x 32pdr carronades; Fc 2 x 18pdrs + 2 x 32pdr carronades; RH 6 x 18pdr carronades. *Achille* later had 2 x 32pdr carronades replace a pair of long 32pdrs on LD.

Superb Thomas Pitcher, Northfleet

As built: 182ft 4in, 149ft 10in x 49ft 2in x 21ft 10in. 1,926⁵⁶⁄₉₄bm.

Ord: 30.4.1795. K: 8.1795. L: 19.3.1798. C: 30.6.1798 at Chatham.

First cost: Building cost unknown, £14,842 for fitting.

Commissioned 5.1798 under Capt. John Sutton; sailed for the Mediterranean ?7.1799. Under Capt. Richard Keats in 1801 (–1806); sailed for the Mediterranean 6.1801. With Saumarez's squadron off Cadiz; in action of 12.7.1801 took part in destruction of 112-gun *San Carlos* and *San Hermeneglido*, and capture of 74-gun *San Antonio*. Joined Toulon squadron in 8.1804. As flagship of Vice-Adm. John Duckworth, involved in chase to the West Indies in 1805; at battle of San Domingo 6.2.1806. Keats appointed Commodore 7.1806 off Rochefort. In 10.1807 under Capt. Donald M'Leod, as flagship of Rear-Adm. Sir Richard Keats; in Copenhagen expedition 1807. In 2.1808 under (acting) Lieut Thomas Alexander, with Strachan's squadron to the Mediterranean. In 12.1809 under Capt. Samuel Jackson in the Baltic, again Keats's flagship. Between Middling and Large Repair at Portsmouth 9.1811–11.1812. Recommissioned 9.1812 under Capt. Charles Paget, for the Channel; at Bermuda 1813; took US 6-gun privateer *Star* 9.2.1813, and took (with *Pyramus*) 6-gun privateer *Viper* 15.4.1813. Under Capt. Alexander Gordon 1814 in North America, as flagship of Rear-Adm. Henry Hotham. In 4.1815 under Capt. Humphrey Senhouse, and 9.1815 Capt. Charles Ekins. Fitted for foreign service at Plymouth 7.1816; took part in Bombardment of Algiers 27.8.1816, losing 8 killed and 84 wounded. Fitted as a guard ship at Plymouth 5.1818. Recommissioned 11.1818 under Capt. Thomas Hardy, for

South American station. Under Capt. Thomas White in 8.1818 (when Hardy made Commodore); then Capt. Adam McKenzie 6.1821 (death in 11.1823); guard ship at Plymouth 1.1822. From 10.1823 under Capt. Sir Thomas Staines, on Jamaica station; at Lisbon 1825; paid off 12.1825. BU at Portsmouth completed 17.4.1826.

Achille William Cleverley, Gravesend (note retained French spelling of name)

As built: 182ft 2in, 149ft 7in x 49ft 3in x 21ft 11in. 1,929⁸¹⁄₉₄bm.

(as rebuilt 1822: 182ft 2¾in, 150ft 11¾in x 49ft 8in x 21ft 3in. 1,981⅞⁄₉₄bm.)

Ord: 30.4.1795. K: 10.1795. L: 16.4.1798. C: 25.8.1798 at Chatham.

First cost: £38,450 to build, plus £15,165 fitting.

Commissioned 6.1798 under Capt. Henry Stanhope, for the Channel. In 4.1799 under Capt. George Murray (–1801); blockade of Cadiz. Refitted (for £6,881) 2.1800. In 4.1801 under Capt. Edward Buller (–1802). In 11.1801 under Capt. (acting) James Wallis; in 5.1802 under Capt. John Hardy. Recommissioned 4.1805 under Capt. Richard King (–1811), for the Channel. In Lee column at Trafalgar, lost 13 killed, 59 wounded. In Hood's action off Rochefort 24–25.9.1806. At blockade of Ferrol 1807. In Walcheron operation 1809. In 7.1809 under Capt. (temp.) John Hayes; sailed for the Mediterranean 18.2.1810. At blockade of Cadiz 1810. In 1.1811 under Capt. Askew Hollis (–1815), in the Adriatic; in 4.1811 under Capt. (temp.) George Dundas, then back to Hollis. Sailed for East Indies early 1814. Paid off 1815. Rebuilt with circular stern at Chatham 1817–22, then to Ordinary there; at Sheerness 1829–47. Re-rated 76 guns 1839. Sold to Castle & Beech 1.11.1865 to BU at Charlton.

CONQUEROR Class 'Middling Class' 74. Sir John Henslow design, approved 10 June 1795. By 1817 she was classed as a 76-gun ship.

Dimensions & tons: 176ft 0in, 144ft 3in x 49ft 0in x 20ft 9in. 1,842²⁴⁄₉₄bm.

Men: 590. Guns: LD 28 x 32pdrs; UD 30 x 24pdrs; QD 4 x 18pdrs + 10 x 32pdr carronades; Fc 2 x 18pdrs + 2 x 32pdr carronades; RH 6 x 18pdr carronades.

Conqueror Joseph Graham, Harwich

As built: 176ft 0in, 144ft 2in x 49ft 2in x 20ft 9in. 1,853⁶⁹⁄₉₄bm. Draught 11ft 10in/19ft 0in.

Ord: 30.4.1795. K: 10.1795. L: 23.11.1801. C: 8.5.1803 at Plymouth.

Commissioned 3.1803 under Capt. Sir Thomas Louis. In 5.1804 under Capt. Israel Pellew (–1807), in the Channel.; by 1805 in the Mediterranean, then chase to West Indies. In Weather column at Trafalgar 21.10.1805, lost 3 killed, 9 wounded, but captured Villeneuve's flagship *Le Bucentaure*. In Sir Samauel Hood's squadron off Rochefort in 1806; her boats (among others) cut out 16-gun *Le César* of the Gironde 15.7.1806. Defects made good at Plymouth (for £13,091) 4–6.1807. In Rear-Adm. Sir Sidney Smith's squadron sailing for the Tagus 15.11.1807. Later in 1807 under Capt. Edward Fellowes (–1811), in the Mediterranean 1809; with Sir Charles Cotton's squadron off Toulon 1810; action off Cape Sicié with French 40-gun frigates *L'Amélie* and *L'Adrienne* 19.7.1811. Between Small and Middling Repair at Chatham (for £23,708) 10.1812–2.1814. In 1813 under Capt. Richard Raggett (–1815); in Ordinary at Chatham. From 1816 flagship of Rear-Adm. Robert Plampin, at St Helena, under Capt. Robert Merrick Fowler (1816) and later under Capt. John Davie (1816–18). Fitted for sea at Sheerness (for £20,120) 1.1817. Recommissioned 25.1.1818 under Capt. James Wallis (temp.) and

then under Capt. Francis Stanfell (1818–20); paid off at Chatham 10.1820. BU at Chatham 7.1822, completed 2.8.1822.

AJAX Class It was approved on 10 June 1795 to build two ships by the draught of the *Triumph*, but with 6in higher decks than that ship; the design was lengthened by 11ft during construction (by AO of 19 October 1796), and the two were established by AO of 8 May 1798. The *Ajax* – built by John Randall & Co., Rotherhithe – was burnt by accident off Tenedos in the Aegean, grounded on the island and blew up on 14 February 1807 (250 died). The *Kent* was re-classed as a 78-gun ship in January 1817, and as 76-gun in 1829.

Dimensions & tons: 182ft 3in, 149ft 8⅛in x 49ft 3in x 21ft 3in. 1,931⁶²⁄₉₄bm.

Men: 690. Guns: LD 28 x 32pdrs; UD 28 x 24pdrs; QD 14 x 9pdrs (later 4 x 9pdrs + 8 x 32pdr carronades); Fc 4 x 9pdr.

Kent John Perry & Co., Blackwall

As built: 182ft 8in, 149ft 11in x 49ft 7½in x 21ft 5in. 1,963⁷³⁄₉₄bm. Draught 11ft 10in/19ft 0in.

(as rebuilt 1817–20: 184ft 2⅛in, 150ft 10½in x 50ft 0½in x 21ft 10in. 2,009⁶²⁄₉₄bm.)

Ord: 30.4.1795. K: 10.1795. L: 17.1.1798. C: 3.4.1798 at Woolwich.

First cost: £46,843 (including fitting).

Commissioned 3.1798 under Capt. William Hope, as flagship of Adm. Viscount (Adam) Duncan. To Mediterranean in 6.1800; involved in Egypt operations 1801. From 1802 under Capt. Edward O'Bryen, as flagship of Rear-Adm. Sir Richard Bickerton, for the Mediterranean; joined Nelson's fleet in 8.1801. In 8.1803 under Capt. John Stuart; paid off 1804. Repaired and refitted at Chatham (for £28,544) 5–11.1805. Recommissioned 9.1805 under Capt. Henry Garrett, as flagship of Vice-Adm. Edward Thornbrough (from 11.1805). Refitted at Portsmouth (for £15,069) 10.1809–2.1810. Paid off and laid up in Ordinary at Plymouth 1.1813. Large Repair at Plymouth, rebuilt with a circular (for £77,646) 6.1817–10.20, then laid up. Fitted as a guard ship at Plymouth (for £25,204) 2.1829–6.1831. Between Small and Middling Repair at Plymouth (for 20.704) 6.1842–3.1846. Fitted as a sheer hulk at Plymouth (for £17,507), replacing *Spartiate*, 12.1855–1.1857 (by AO 27.10.1854). BU there 1881 (by AO 12.11.1880).

MILFORD Class Jean-Louis Barrallier design, modified from his lines for *Spencer* of 1795, to which two ships were both ordered on 16 December 1796; the second ship – the *Princess Amelia*, ordered from Chatham Dockyard – was cancelled in March 1800. Re-classed as a 78-gun ship in 1817.

Dimensions & tons: 181ft 0in, 149ft 3¼in x 49ft 0in x 21ft 0in. 1,906³¹⁄₉₄bm.

Men: 590. Guns: LD 28 x 32pdrs; UD 30 x 24pdrs; QD 14 x 9pdrs; Fc 2 x 9pdrs + 2 x 32pdr carronades; RH 6 x 18pdr carronades. All 9pdrs later replaced by 32pdr carronades.

Milford Jacobs, Milford Haven. This builder 'failed' (i.e. became bankrupt), so ship was completed by 'Government' (this was the beginning of the establishment of Pembroke Dyd): Milford Dyd

As built: 181ft 1in, 149ft 4in x 49ft 1¾ x 21ft 0in. 1,918⁵¹⁄₉₄bm. Draught 13ft 6in/18ft 0in.

Ord: 16.12.1796. K: 6.1798. L: 1.4.1809. C: 19.5–30.8.1809 at Plymouth

First cost: £45,209 (including £8,290 to Jacobs) to build, plus £37,212 fitting.

Commissioned 5.1809 under Capt. Henry Bayntum, for Channel service, in 1810? as flagship of Rear-Adm. Sir Philip Durham in

The *Revenge* was one of the last of the 'Large Class' of 74-gun Third Rates, armed with 24pdr guns on their upper decks. Following the Napoleonic War, she enjoyed a longer active post-war career than most of her type, but this was due to the major reconstruction that the ship underwent between 1818 and 1823. She emerged with a new circular stern and Seppings's structural innovations, and fitted for a new role as a flagship, which she undertook with the Mediterranean Fleet from 1823 to 1827. This pen-and-wash drawing by Thomas Robinson is undated but is annotated 'Gosport', so may show the ship around 1838–39 when fitted for sea at Portsmouth. *(NMM PW8005)*

the Baltic. In later 1810 under Capt. Edward Kittoe, as flagship of Vice-Adm. Sir Richard Keats; sailed for the Mediterranean 18.8.1810. In 1812 under Capt. John Duff Markland, as flagship of Rear-Adm. Thomas Fremantle in the Mediterranean; attack on Fiume 3.7.1813; boats (with those of *Weazle*) at Rogoznica 4.8.1813; capture of Trieste 5–29.10.1813. Paid off 1814 into Ordinary. Fitted at Plymouth as a lazarette 5–6.1825, then at Milford 6–7.1825 (serving there as such until 1842). BU at Pembroke 7.1846.

REVENGE Class 'Large Class' 74s, with 24pdrs on the UD. Sir John Henslow design, approved 27 October 1796 as a 74-gun ship (the previous day's order had been for an 80-gun ship). Completed as a 74, but re-classed as a 76 in 1817.

  Dimensions & tons: 182ft 0in, 150ft 3in x 49ft 0in x 20ft 9in. 1,918⁸³⁄₉₄bm.

  Men: 590. Guns: LD 28 x 32pdrs; UD 30 x 24pdrs; QD 14 x 9pdrs (later 6 x 12pdrs + 10 x 32pdr carronades); Fc 4 x 9pdrs (later 2 x 9pdrs + 2 x 32pdr carronades); RH 6 x 18pdr carronades.

Revenge Chatham Dyd [M/Shipwright Edward Sison to 7.1801, then David Polhill to 3.1803, completed by Robert Seppings]

  As built: 181ft 11in, 150ft 0½in x 49ft 2in x 20ft 9in. 1,929²⁶⁄₉₄bm. Draught 13ft 4in/18ft 3in.

  (as rebuilt 1823: 183ft 2¼in, 150ft 5½in x 49ft 5in x 21ft 1in. 1,954³⁴⁄₉₄bm.)

  Ord: 26.10.1796. (named AO 14.2.1797) K: 8.1800. L: 13.4.1805. C: 23.6.1805.

  First cost: £58,653 (including fitting).

  Commissioned 4.1805 under Capt. Robert Moorsom, for Channel fleet. In Lee column at Trafalgar 21.10.1805; lost 28 dead, 51 wounded. Refitted at Portsmouth (for £10,687) 12.1805–2.1806. Recommissioned 1.1806 under Capt. Charles Fleeming; in 2.1806 under Capt. Sir John Gore (–1807); boat attack on shipping in

the Gironde 16.7.1806; Sir Samuel Hood's action of Rochefort 25.9.1806; sailed for the Mediterranean 26.2.1807 (for blockade of Cadiz. From 8.1808 under Capt. Charles Paget (to 1810); under Capt. John Bligh (temp.) in 1.1809; attack on shipping in Basque Roads 12.4.1809 (temp. under Capt. Alexander Robert Kerr); again under Bligh (temp.) for passage to UK 1809. In Walcheren expedition 1809. Sailed with East Indies convoy 10.6.1810. In 10.1810 under Capt. Sir John Gore; took 16-gun privateer *Le Vengeur* off Cherbourg 17.10.1810. Under Capt. J(ames or John) Nash (temp.) in 1811. Refitted at Portsmouth (for £15,110) 4–6.1811. Recommissioned 1812 under Capt. Charles Philip Butler Bateman (temp.) as flagship of Rear-Adm. Arthur Legge; sailed for the Mediterranean 18.12.1812; paid off into Ordinary 8.1814 at Chatham. Large Repair, rebuilt with a circular stern and fitted for a flag officer for foreign service at Chatham (for £76,919) 2.1818–8.1823. Recommissioned 20.3.1823 under Capt. Sir Charles Burrard, as flagship of Vice-Adm. Sir Harry Burrard Neale, C-in-C in the Mediterranean; paid off 4.1827. Fitted as a guard ship at Portsmouth (for £12,142) 6.1827–1.1828. Recommissioned 10.1830 under Capt. James Hillyar; fitted as a guard ship at Plymouth (for £18,617) 11.1830–5.1831. Recommissioned 7.11.1831 under Capt. Donald Hugh Mackay; blockade of the Dutch coast 1831; off Lisbon and back in the Mediterranean 1833; paid off 3.1834. Recommissioned 1834 under Capt. Sir William Elliott; paid off 1837. Fitted for sea and for demonstration at Portsmouth (for £17,178) 1.1838–8.1839; re-rated 76 guns 1839. Recommissioned 9.5.1840 under Capt. William Waldegrave, for the Mediterranean; at bombardment of Acre 3.11.1840; paid off 1842 at Sheerness. Surveyed 4.1849 and BU at Sheerness 10.1849.

COLOSSUS Class Sir John Henslow design, approved 13 January 1798. Re-classed as 76-gun ships in January 1817, although by then

One of a small number of wartime 74s to see active service after 1817, the *Warspite* is shown here returning to Portsmouth on 28 July 1827 after a voyage around the world. The ship had benefited from an extensive repair from 1816 to 1918, and was in good enough condition to be cut down to a frigate in 1837–40. *(NMM PU6069)*

Colossus had exchanged her 24pdrs for 18pdrs on the UD.
    Dimensions & tons: 180ft 0in, 148ft 3½in x 48ft 10in x 21ft 0in. 1,880⁸⁸/₉₄bm.
    Men: 590. Guns: LD 28 x 32pdrs; UD 30 x 24pdrs; QD 4 x 24pdrs + 10 x 32pdr carronades; Fc 2 x 24pdrs + 2 x 24pdr carronades; RH 6 x 18pdr (or 3 x 18pdr + 3 x 12pdr) carronades.
Colossus Deptford Dyd [M/Shipwright Edward Tippett to 3.1803, completed by Henry Peake]
    As built: 181ft 1½in, 148ft 4½in x 48ft 11in x 21ft 0in. 1,888⁴⁷/₉₄bm. Draught 13ft 3in/18ft 0in.
    Ord: 23.11.1797. K: 5.1799. L: 23.4.1803. C: 20.6.1803 at Woolwich.
    First cost: (not in Progress Books)
    Commissioned 3.1803 under Capt. George Martin; under (temp.) ?S. Seymour 3.1804; later under Capt. James Nicoll Morris (to 1808) as flagship of Sir Thomas Graves, in the Channel Fleet; in 1805 with Collingwood's squadron off Cadiz; in Lee column at Trafalgar 21.10.1805; had 40 dead, 160 wounded including Morris (the highest casualty rate in the British fleet) but took possession of French *Le Swiftsure* and Spanish *Bahama*. Small Repair and fitted at Portsmouth (for £18,105) 4–6.1806; recommissioned 7.1806; in Rear-Adm. Sir Richard Strachan's squadron off Rochefort 1808. Sailed for the Mediterranean 30.7.1808. Recommissioned 10.1808 under Capt. Thomas Alexander. Made good defects at Chatham (for £14,525) 2–4.1811. In Capt. Sir John Gore's squadron off Lorient 1812; captured US 12-gun privateer *Dolphin* 5.1.1812. In North Sea 1813. Laid up at Chatham 5.1814. Re-armed with 18pdr on UD by 1815. BU at Chatham 8.2.1826.
Warspite Chatham Dyd [M/Shipwright Robert Seppings]
    As built: 179ft 10in, 148ft 0¼in x 49ft 0in x 21ft 0in. 1,890³⁹/₉₄bm. Draught 13ft 7in/17ft 10in.
    Ord: 23.11.1797. K: 3.12.1805. L: 16.11.1807. C: 8.6.1808.
    First cost: £59,725 including fitting.
    Commissioned 3.1808 under Capt. Henry Blackwood (to 1813). Temporarily under Capt. William Bowles in Spring 1809 (off Cherbourg). Sailed for the Mediterranean 2.11.1809; temp. under Capt. Henry Edward Reginald Baker while in Mediterranean 1810–12. Made good defects at Chatham (for £10,772) 6–8.1812.

In the Channel Fleet in 1813; took US privateers *William Bayard* (4-gun) 12.3.1813, *Cannonier* (8-gun) 14.3.1813, and *Flash* 29.5.1813. In 11.1813 under Capt. Lord James O'Bryen, to North America with troops. Large Repair at Portsmouth (for £69,613) 1.1816–5.1818; then to Ordinary. Fitted for sea (for £20,223) 10.1825–1.1826. Made good defects at Plymouth (for £11,970) 11.1828–4.1829. Between a Middling and Large Repair, then cut down to a Fourth Rate frigate of 50 guns (by AO 19.6.1837) (see Chapter 4 for later history).

CHATHAM Design produced specifically to utilise the frames of the French *Le Royal-Hollandais*, laid down at Vlissingen (Flushing) for the French Navy to their *Le Pluton* Class design, and taken on the stocks when that port was captured on 17 August 1809. The frames were taken down, shipped to London and re-laid at Woolwich. The poor quality of the timber ensured she had only a brief life before being condemned.
    Dimensions & tons: 177ft 9in, 146ft 7⅞in x 48ft 10in x 21ft 6½. 1,860²¹/₉₄bm.
    Men: 590. Guns: LD 28 x 32pdrs; UD 28 x 24pdrs; QD 4 x 12pdrs + 10 x 32pdr carronades; Fc 2 x 12pdrs + 2 x 32pdr carronades; RH 6 x 18pdr carronades.
Chatham Woolwich Dyd [M/Shipwright Edward Sison].
    As built: 177ft 9in, 146ft 7⅞in x 48ft 10in x 21ft 6½in. 1,860²¹/₉₄bm. Draught 12ft 10in/16ft 9in.
    Ord: 1810. K: 6.1810. L: 14.2.1812. C: 25.4.1812.
    First cost: (not in Progress Books).
    Commissioned 3.1812 under Capt. Graham Moore. On 24.8.1812 under Capt. Robert Maunsell, as flagship of Rear-Adm. Matthew Henry Scott, in North Sea. In 5.1814 under Capt. David Lloyd, as a sheer hulk. Laid up at Chatham 11.1815. Sold to Joshua Crystall to BU (at Chatham?) for £5,110 on 10.9.1817.

Third Rates of 74 guns: 18pdr type

The backbone of the battlefleet, there were seventy ships of this rating in the List at the start of 1793, of which nineteen were in commission, forty-two in Ordinary (Reserve), and nine in harbour service; one further ship was building (*Minotaur*). Three of these ships were rated as 'Large Class' – the older *Triumph* and *Valiant* (see above) and the new *Brunswick*; their establishment of men and guns was the same as that of the 24pdr-armed type, and they will be found above. The other than the larger calibre guns on the UD. The other sixty-seven ships were established with a complement of 600 men (554 officers, seamen and marines; 40 servants and boys; and 6 'widows' men': increased by 2 officers but reduced by 10 servants and boys to 592 in 4.1794) and ordnance of 28 x 32pdrs, 28 x 18pdrs and 18 x 9pdrs, giving a broadside of 781 lb – to which the new carronade Establishment of 11.1794 added 2 x 32pdr type and 6 x 18pdr type, to give a 867 lb broadside.

ALBION **Class** A new design by Thomas Slade, approved 24 April 1760, based on 90-gun *Neptune* of the 1719 Establishment. Following the prototype (the *Albion* was wrecked in 1797), two sisters were ordered following the close of the Seven Years War; of these the *Grafton* was BU in 1816, but the *Alcide* survived as a receiving ship into 1817.

> Dimensions & tons: 168ft 0in, 139ft 1¼in x 47ft 3in x 18ft 10in. 1,651⁷⁵/₉₄bm.
> Men: 550. Guns: LD 28 x 32pdr; UD 30 x 18pdr; QD 16 x 9pdr; Fc 4 x 9pdr.

Alcide Deptford Dyd [M/Shipwright Adam Hayes]
> As built: 168ft 0in, 139ft 1¼in x 46ft 10⅝in x 18ft 10in. 1,651⁸/₉₄bm. Draught 11ft 7in/16ft 2in.
> Ord: 31.8.1774. K: 14.6.1776. L: 30.7.1779. C: 13.9.1779.
> First cost: £38,164.14.7d (including fitting).
> Commissioned 7.1779. Refitted and coppered at Plymouth (for £5,964.4.9d) 4–6.1780. Paid off after wartime service 7.1783. Small Repair at Portsmouth (for £15,389.2.2d) 7–12.1784. Recommissioned 10.1787 under Capt. Benjamin Caldwell. Fitted as guard ship at Portsmouth 5.1790. Recommissioned 10.1787 under Capt. Sir Andrew Snape Douglas, for Spanish Armament, and again in 8.1791, as guard ship at Portsmouth. Fitted as guard ship at Portsmouth 1.1793, and recommissioned 12.1792 under Capt. John Woodley. Refitted at Portsmouth (for £12,309) 3.1793, as flagship of Commodore Robert Linzee for Hood's fleet off Toulon; sailed for the Mediterranean 22.4.1793; in attack on Forneille 1.10.1793. Under Capt. Thomas Shivers 6.1794, as flagship of the new Rear-Adm. Linzee; under Capt. Sir Thomas Byard in 8.1794, and in 10.1794 flagship of Vice-Adm. Phillips Cosby. Paid off and laid up at Portsmouth 11.1794 in Ordinary (although listed as receiving ship from 1802). BU there 4.1817.

CANADA Design by William Bately, approved 24 April 1760. While only one ship was built in 1760–5, the design was revived for three further vessels in 1781.

> Dimensions & tons: 170ft 0in, 138ft 0in x 46ft 9in x 20ft 6in. 1,604²⁷/₉₄bm.
> Men: 550. Guns: LD 28 x 32pdr; UD 30 x 18pdr; QD 16 x 9pdr; Fc 4 x 9pdr (78 actual total).

Canada Woolwich Dyd [M/Shipwright Israel Pownoll to 5.1762, completed by Joseph Harris]
> As built: 170ft 0in, 138ft 1in x 46ft 9in x 20ft 6in. 1,605²⁴/₉₄bm. Draught 11ft 9in/18ft 3in.
> Ord: 1.12.1759. K: 1.7.1760. L: 17.9.1765. C: 11.10.1765.
> First cost: £33,546 (including fitting).
> Commissioned 2.1779. Small Repair and fitted at Chatham (for £8,932.15.11d) 10.1778–10.1779. Fitted and coppered at Plymouth (for £8,217.6.5d) 3–4.1780. Re-rated 76-gun ship,

with 2 extra 18pdrs fitted by AO 12.6.1780. Great Repair at Portsmouth (for £26,488.2.5d) 3.1783–3.1784. Recommissioned 7.1790 under Capt. Lord Hugh Conway, for Spanish Armament. Recommissioned 10.1793 under Capt. Charles Powell Hamilton; chased by Nielly's squadron 6.11.1794 (with her consort *Alexander*, which was taken). Under Capt. Sir Erasmus Gower in 1795, in Howe's fleet. From 6.1795 under Capt. George Bowen; flagship of Rear-Adm. Sir Roger Curtis 9.1795; sailed for Jamaica 24.2.1796. Under Capt. Thomas Twysden in ?9.1797, on Jamaica station. Made good defects at Plymouth (for £20,240) 1797–8. Recommissioned 11.1797 under Commodore Sir John Borlase Warren; grounded near mouth of Gironde 22.3.1798 while in chase of 36-gun *La Charente*, but later freed; action with Bompart's squadron off Ireland 12.10.1798 (had 1 man mortally wounded). Under Capt. Michael de Courcy 4.1799; in Cotton's (reinforcement) squadron to the Mediterranean 6.1799; operations off Quiberon 6.1800. Under Capt. Joseph Yorke in 4.1801. Fitted at Portsmouth (for £33,035) 5–12.1805. Recommissioned 9.1805 under Capt. John Harvey; sailed for Jamaica 28.1.1806. In Leeward Islands 1807. Paid off 1.1808. Fitted as prison ship at Chatham (for £5,364) 11.1809–3.1810. Fitted there as Powder magazine (for £138) 9–12.1814, for the Medway. Fitted as a convict ship at Sheerness 1.1826. BU at Chatham (for £1,000) 11.1834.

Modified *RAMILLIES* **Class** Design by Sir Thomas Slade, approved 13 January 1761, a still further development of his *Bellona* design. There were five contract-built ships built to this modified design (with 2in greater breadth, 2⅜in less keel than the dockyard-built vessels), of which the *Terrible* had been burnt in 1781, the *Prince of Wales* taken to pieces in 1783, the *Invincible* wrecked in 1801 and the *Russell* sold in 1811. Each was contracted to be built at a price of £16½ per ton (*Prince of Wales* at £15⅛ per ton). *Russell* and *Invincible* were (in 2.1779) the first ships of the battlefleet to be coppered.

> Dimensions & tons: 168ft 6in, 138ft 1in x 46ft 11in x 19ft 9in. 1,616⁶⁸/₉₄bm.
> Men: 550. Guns: (originally) LD 28 x 32pdr; UD 28 x 18pdr; QD 14 x 9pdr; Fc 4 x 9pdr.

Robust (John) Barnard & (John) Turner, Harwich
> As built: 168ft 8½in, 138ft 3in x 47ft 0in x 19ft 9¼in. 1,624bm.
> Ord: 16.12.1761. (contract 11.1.1762) K: 2.1762. (named 18.4.1763) L: 25.10.1764. C: 10.12.1764 by builder.
> First cost: £31,399 including fitting.
> Small Repair at Chatham (for £8,291.7.11d) 5.1774–1.1775.
> Commissioned 12.1777. Fitted at Chatham for £8,187.7.7d) 3.1778. Paid off 9.1782 after wartime service. Middling to Large Repair and coppered at Chatham (for £29,469.16.11d) 2.1783–4.1784. Commissioned 10.1787 under Capt. W Cornwallis, but paid off 12.1787. Recommissioned 7.1790 under Capt. Howland Cotton; paid off 9.1791. Very Small Repair at Chatham (for £7,292) 11.1791–4.1792. Recommissioned 1.1793 under Capt. George Elphinstone; sailed to join Hood's fleet in the Mediterranean 22.5.1793. At Toulon (Fort La Malague) 28.8.1793, under (temp.) Capt. Benjamin Hallowell; evacuation of Toulon in 12.1793. In 1794 under Capt. Christopher Parker; with Warren's squadron in 6.1795 for Quiberon operations. Under Capt. Edward Thornbrough in 8.1794; in North Sea 1796, then in Channel; mutiny at Spithead 1797. Refitted at Portsmouth (for £11,253) 10.1797. Made good defects at Portsmouth 1.1798. Took part in Warren's action with Bompart 12.10.1798. Under Capt. George Countess in 4.1799; attack on Spanish squadron in the Basque

Roads 2.7.1799. Under Capt. William Brown in 11.1800, later under Capt. John Ommanney. Under Capt. Henry Curzon in ?4.1801, later under Capt. William Ricketts as flagship of Admiral Jervis; sailed for Jamaica 4.1802. Paid off 7.1802. Fitted as a receiving ship at Portsmouth 7.1802. In Ordinary 1807–14. BU there 1.1817.

With the close of the Seven Years War, the usual cutback in state spending meant a lull in ordering. A new period of conservatism began with Anson's death, the Admiralty in early 1763 rejecting new draughts produced by the Surveyors in favour of repeat orders for existing designs. However, Slade produced two variations in 1765, and John Williams (who replaced Bately in 6.1765) a fresh design. Nineteen new 'Common Class' 74s were ordered over the next ten years, most of them to these designs. Another thirty were ordered in 1778–82 (two more were cancelled), the majority to similar designs, or indeed to revivals of earlier ones.

ROYAL OAK Class

Design by Sir John Williams, approved 18 December 1765. Less sharp bows than the typical Slade designs, but sharper in the stern above the waterline, the result having some instability but being fast in fine weather and sailing well in all but the strongest winds. *Conqueror* was BU in 1794 and the other five ships were converted to prison hulks by 1808, but the *Royal Oak* was BU in 1815 and *Sultan*, *Vengeance* and *Hector* in 1816.

Dimensions & tons: 168ft 6in, 138ft 2in x 46ft 9in x 20ft 0in. 1,606²¹⁄₉₄bm.

Men: 550. Guns: (originally) LD 28 x 32pdr; UD 28 x 18pdr; QD 14 x 9pdr; Fc 4 x 9pdr.

Bedford Woolwich Dyd [M/Shipwright William Gray to 3.1773, completed by Nicholas Phillips]

As built: 168ft 6in, 138ft 2in x 46ft 9in x 20ft 0in. 1,606²¹⁄₉₄bm. Draught 12ft 10in/18ft 0in.

Ord: 12.10.1768. K: 10.1769. L: 27.10.1775. C: 12.11.1776.

First cost: £33,011.11.1d to build, plus £5,891.13.3 fitting (as guard ship).

Commissioned 7.1776. Refitted and coppered at Plymouth (for £9,207.13.10d) 4–5.1779. Paid off 7.1783 after wartime service. Middling Repair at Portsmouth (for £22,494.6.0d) 12.1784–10.1785. Fitted as guard ship at Portsmouth 9.1787; recommissioned 6.1787 under Capt. Robert Man (–1794); Spanish Armament 1790; paid off 8.1791, then from same month under Capt. Sir Andrew Snape Hamond, as guard ship at Portsmouth, and flagship of Vice-Adm. Mark Milbank, in Evolutionary Squadron 1792. Back under Capt. Man from 1.1793, sailed for Mediterranean 22.5.1793 to join Hood's fleet at Toulon. Capture of 36-gun *La Modeste* at Genoa 17.10.1793. Flag of Sir Hyde Parker 1794. Under Capt. Davidge Gould in 1.1795; action off Genoa 13.3.1795 and off Hyères 13.6.1795. Under Capt. Augustus Montgomery, in brush with de Richery's squadron off Cape St Vincent 7.10.1795; paid off 11.1795. Fitted at Portsmouth (for £23,622) 9.1796–3.1797; recommissioned 12.1796 under Capt. Sir Thomas Byard, in Duncan's fleet at Camperdown 11.10.1797, having 30 killed, 41 wounded. Small Repair at Plymouth (for £1,363) 7–9.1799. Prison ship at Plymouth 1800–5. Large Repair and fitted for foreign service at Plymouth (for £52,317) 9.1805–10.1807. Recommissioned 9.1807 under Capt. James Walker; sailed for Rio de Janeiro 11.11.1807. In 1.1809 under Capt. Adam M'Kenzie, off Brazil. Refitted at Portsmouth (for £11,444) 9–10.1810; then Capt. James Walker again 1810–14, off Brazil. Sailed for Leeward Islands 3.4.1813.

At blockade of Flushing 1814. Paid off 1815. BU at Portsmouth 10.1817.

ALFRED Class

Design by Sir John Williams, 1772, slightly enlarged from his *Royal Oak* Class, but with significantly greater beam, and hence slightly slower ships. The prototype of the class – the *Alfred* – was BU in 1814; a fifth ship (*Edgar*) was also ordered 16.7.1774 to this design but on 25.8.1774 was altered to the Modified *Arrogant* design (*qv*).

Dimensions & tons: 169ft 0in, 138ft 5¼in x 46ft 11in x 20ft 0in. 1,620⁸²⁄₉₄bm.

Men: 550. Guns: LD 28 x 32pdrs; UD 28 x 18pdrs; QD 14 x 9pdrs; Fc 4 x 9pdrs.

Warrior Portsmouth Dyd [M/Shipwright Edward Hunt to 12.1777, Nicholas Phillips to 4.1779, completed by George White]

As built: 169ft 0in, 138ft 2¾in x 47ft 3in x 20ft 0in. 1,642bm. Draught 11ft 6in/17ft 0in.

Ord: 13 & 21.7.1773. K: 11.1773. L: 18.10.1781. C: 16.12.1781

First cost: ?

Commissioned 10.1781; paid off after wartime service 1783. Small Repair at Portsmouth 3–7.1784, then laid up there until sailed 4.11.1795. Small to Middling Repair at Chatham 9.1796–3.1797; recommissioned 3.1797 under Capt. Henry Trollope.; in 4.1797 under Capt. Henry Savage (–1799); sailed for the Mediterranean 4.4.1797. In 5.1799 under Capt. Charles Tyler (–1802), in Channel Fleet; to Baltic 1801; at Jamaica 2–6.1802. Large Repair and fitted at Plymouth (for £64,006) 1.1803–7.1804; recommissioned 5.1804 under Capt. William Bligh, for blockade of Brest. In 5.1805 under Capt. Samuel Hood Linzee; at Battle of Finisterre 22.7.1805. In 4.1806 under Capt. John Spranger (–1811); flagship in Channel 12.1806; in Mediterranean 1809–11; at occupation of Zante and Cephalonia 10.1809. Fitted at Chatham 8–11.1811; recommissioned 10.1811 under Capt. George Byng, for North Sea. Under Capt. John Rodd in 9.1814, at Jamaica; flagship of Rear-Adm. John Erskine Douglas in 1815. Laid up at Chatham 9.1815. Fitted as receiving ship at Chatham 8.1819. Fitted as Temporary quarantine ship at Chatham 7.1831. Receiving ship at Woolwich 12.1832. Fitted as convict ship there 6.1839–2.1840. BU completed at Woolwich 11.12.1857.

Alexander Deptford Dyd [M/Shipwright Adam Hayes]

As built: 169ft 0in, 138ft 5⅛in x 46ft 11in x 20ft 0in. 1,621bm. Draught 11ft 8in/16ft 9½in.

Ord: 13 & 21.7.1773. K: 6.4.1774. L: 8.10.1778. C: 6.12.1778.

First cost: £38,895.18.3d (including fitting).

Commissioned 10.1778. Fitted and coppered at Portsmouth (for £6,607.15.9d) 12.1779. Refitted and raised copper on each side at Chatham (for £8,509.12.2d) 12.1782. Paid off after wartime service 4.1783. Small Repair at Chatham (for £11,064.18.3d) 8–12.1784. Fitted at Chatham for Channel service 8.1791. Commissioned 8.1793 under Capt. Thomas West; in 5.1794 under Capt. Richard Bligh in Montagu's squadron; captured by the French (Neilly's squadron) having just finished convoy duty off the Scillies 6.11.1794, lost 40 killed and wounded; added to French Navy as *L'Alexandre*. Retaken off Lorient in Bridport's action 23.6.1795; under (acting) Alexander Wilson. Re-registered by AO 10.9.1795. Middling to Great Repair at Plymouth (for £28,847) 10.1795–4.1796. In 1796 recommissioned under Capt. Arthur Philip, for the Channel Fleet. In 1.1797 under Capt. Alexander Ball (–1800); sailed for the Mediterranean 18.3.1797; towed the dismasted *Vanguard* in a gale off Sardinia 21.5.1798; at Battle of the Nile 1.8.1798; blockade of Malta 1799. In 2.1800 under (acting) Lieut William Harrington; *Genereux*'s convoy 18.2.1800. In 2.1801 under Capt. Manley Dixon. Fitted at Portsmouth (by AO 25.4.1806) as a lazarette 6–10.1805, for the Motherbank. BU at Portsmouth 11.1819.

Montagu Chatham Dyd [M/Shipwright Israel Pownoll to 4.1779, completed by Nicholas Phillips]

As built: 169ft 0in, 138ft 4in x 47ft 1in x 19ft 11½in. 1,631¹⁶/₉₄bm. Draught 11ft 10in/16ft 11in.

Ord: 16.7.1774 (named 23.8.1774). K: 30.1.1775. L: 28.8.1779. C: 23.9.1779

First cost: £42,331.12.10d including fitting and coppering.

Commissioned 8.1779; paid off 11.1782 after wartime service. Small Repair at Portsmouth (for £11,507.10.9d) 11.1782–6.1783. Small Repair at Portsmouth (for £7,086) 3–12.1790. Recommissioned 2.1793 under Capt. James Montagu., for Howe's fleet; at Glorious First of June off Ushant 1.6.1794, lost 4 killed (including Montagu), 13 wounded; Lieut Ross Donnelly acting Cmdr Later in 1794 under Capt. William Fooks; sailed for the Leeward Islands 25.10.1794; took (with *Ganges*) 24-gun *Le Jacobin* 30.10.1794; paid off 11.1795. Small Repair and fitted (for £13,482) 11.1795–8.1796; recommissioned 4.1796 under Capt. John Knight (–1799); at Battle of Camperdown 11.10.1797; sailed for the Mediterranean 2.6.1798; in Mediterranean 1799. In 12.1799 under Capt. Charles Paterson; in 1801 under Capt. Robert Cuthbert, sailed for Jamaica. In 3.1801 under Capt. Sir Edmund Nagle, in Calder's squadron; in pursuit of Ganteaume's squadron 2.1801. Large Repair at Portsmouth (for £41,200) 1.1802–12.1805; recommissioned 3.1803 under Capt. Robert Otway (–1808), for blockade of Brest; in attempt on French fleet 21.8.1803; off Cadiz 1805; in Strachan's squadron 5–9.1806. Refitted at Portsmouth (for £7,810) 4–5.1807; flagship of Rear-Adm. George Martin, sailed for the Mediterranean 3.6.1807. In 1809 under Capt. Richard Moubray, in the Mediterranean, then in 1811 Capt. John Halliday. Small Repair and fitted for foreign service at Chatham (for £29,294) 12.1811–4.1812; in 1812 under Capt. Manley Hall Dixon, as flagship of Rear-Adm. Manley Dixon; sailed for South America 15.5.1812; in Brazil 1812–13. Under Capt. Peter Heywood in 7.1813; in North Sea 1814. Defects made good at Portsmouth (for £9,217) 9.1814–4.1815, then to Mediterranean. Paid off and laid up at Chatham 7.1816 and BU there 9.1818.

EDGAR (or Modified ARROGANT) Class

Revived design of 1774, slightly modified from Slade's (Seven Years War) original *Arrogant* design and approved 25.8.1774. Of ten ships, the *Illustrious* was a wartime loss in 1795, and the *Goliath* (cut down to a 58-gun Fourth Rate in 1812) and *Audacious* were BU in 1815 and the *Zealous* in 1816.

Dimensions & tons: 168ft 0in, 138ft 0in x 46ft 9in x 19ft 9in. 1,604²⁷/₉₄bm.

Men: 550. Guns: LD 28 x 32pdr; UD 28 x 18pdr; QD 14 x 9pdr; Fc 4 x 9pdr.

Goliath and *Vanguard* re-established 29.12.1806 with all-24pdr armament of LD 28 x 24pdrs; UD 28 x 24pdrs (Govers), QD 2 x 24pdrs (Govers) + 10 x 24pdr carronades; Fc 2 x 24pdrs (Govers) + 4 x 24pdr carronades. (*Goliath*, *Saturn* and *Elephant* as 'frigates' (actually raséed two-deckers) each had LD 28 x 32pdrs; UD 28 x 42pdr carronades + 2 x 12pdrs; 495 men.)

Retribution (ex-*Edgar*) Woolwich Dyd [M/Shipwright Nicholas Phillips to 12.1777, completed by George White]

As built: 168ft 0in, 138ft 0in x 46ft 10in x 19ft 9in. 1,609⁹³/₉₄bm. Draught 12ft 4in/17ft 9in.

Ord: 16.7.1774. K: 26.8.1776. L: 30.6.1779. C: 6.9.1779.

First cost: £42,362.0.11d (including fitting and coppering)

Commissioned 5.1779. Refitted at Portsmouth (for £8,027.1.9) 2.1782. Fitted as a guard ship at Portsmouth (for £6,028.4.9d) 4.1783. Refitted (with copper bolts) at Portsmouth for £14,243.4.7d) 9.1786–6.1787. Refitted as a guard ship at Portsmouth (for £2,601), and again in 5.1790. Refitted at Portsmouth (for £5,980) 12.1792, having recommissioned 8.1791 under Capt. Albemarle Bertie as a guard ship; then joined Howe's fleet 1793; captured privateer *Le Demourier*. Recommissioned 8.1794 under Capt. Sir Charles Knowles. Refitted at Chatham 4.1795 (for £10,791) and 9.1796 (for £11,356); recommissioned 8.1796 under Capt. John McDougall (–1799), for Channel service; under Capt. Edward Buller from 11.1799 (–1801), still in Channel. In 3.1801 under (acting) Capt. George Murray, in the Baltic; took part in Battle of Copenhagen. Under Capt. Robert Otway from 8.1801; paid off 7.1802. Refitted at Chatham 3–8.1805 (for £19,605). Recommissioned 7.1805 under Capt. John Clarke Searle; in 2.1806 under Capt. Robert Jackson, in the Downs; by 1807 flagship of Adm. Viscount (George) Keith. From 5/6.1807 under Capt. James Macnamara. Made good defects at Plymouth (for £11,439) 1–3.1808; in the Baltic 1808–9; her boats captured 18-gun *Fama* and 12-gun *Søormen* at Nyborg 11.8.1808. Refitted at Portsmouth (for £8,667) 4–6.1809. In ?3.1810 under Capt. Stephen Poyntz; by 1812 in Ordinary at Chatham. Fitted as prison ship for convicts at Chatham (for £9,179) 10–12.1813. Renamed *Retribution* 19.8.1814, as convict hulk at Sheerness to 1833. BU at Deptford 2.1835.

Vanguard Deptford Dyd [M/Shipwright Adam Hayes to 12.1785 (died), completed by Henry Peake]

As built: 168ft 0in, 137ft 8½in x 46ft 10½in x 19ft 9in. 1,609⁴¹/₉₄bm.

Ord: 9.12.1779. (named AO 17.4.1780) K: 16.10.1782. L: 6.3.1787. C: 16.3–11.5.1787 Woolwich.

First cost: £39,116.12.10 (£33,066 for hull, masts & spars), plus £2,185 fitting and coppering.

Commissioned 6.1790 under Capt. Sir Andrew Snape Hamond; fitted at Portsmouth (for £4,339) 8.1890; paid off 9.1791. Fitted (for £6,218) at Portsmouth 6.1793. Recommissioned 2.1793 under Capt. John Stanhope, for Howe's fleet; under Capt. Isaac Schomberg from 12.1793; sailed for Leedward Islands 8.3.1794. Under Capt. Charles Sawyer from 5.1794, as flagship of Commodore Charles Thompson; joined Jervis 7.6.1794. Under Capt. Simon Miller 11.1794 (–97); took 24-gun *La Perdrix* off Antigua 6.1796, and 24-gun *La Surprise* 10.10.1796; paid off 8.1797. Fitted at Chatham (for £19,121) 8.1797–2.1798. Recommissioned 12.1797 under Capt. Edward Berry, as flagship of Rear-Adm. Horatio Nelson; sailed for the Mediterranean 9.4.1798; at Battle of the Nile 1.8.1798, with 30 dead, 76 wounded; under Capt. Thomas Hardy from 4.8.1798. Under Capt. William Brown 6.1799; paid off at Portsmouth 2.1800. Fitted at Portsmouth 1–4.1801. Recommissioned 2.1801 under Capt. Sir Thomas Williams; under (temp.) Capt. Charles Inglis from 11.1801, in the Baltic. Under Capt. James Walker 4.1802; took (with others) 40-gun *La Creole* 1.7.1803, 74-gun *Le Duquesne* and 16-gun *L'Oiseau* 25.7.1803, off San Domingo, 6-gun *Le Papillon* 4.9.1803, off San Domingo, and schooner *Le Courier de Nantes* 5.9.1803. Under Capt. Lord William Fitzroy 3.1804, then Capt. Andrew Evans 4.1804, then Capt. James Newman 7.1805, off Jamaica. Paid off 11.1805. Fitted at Plymouth 1–3.1807. Recommissioned 2.1807 under Capt.

A model of a 74 ready for launching. Although it cannot be identified with any named vessel, it is a good representation of a typical 74-gun ship as they appeared around 1815 in terms of both structural details and decorative scheme. The ship has the built-up or 'round' bow introduced by Sir Robert Seppings and the berthed-up barricades with austere squared-off hances to the upperworks introduced during the war years, while the restrained and limited carved work is complemented by the simple black-and-white paint scheme that had become the norm towards the end of the conflict. *(NMM F5828-001)*

Alexander Frazer; in Copenhagen expedition 8.1807. In 2.1808 under Capt. Thomas Mainwaring, then 5.1808 under Capt. Thomas Baker, as ?flagship of Vice-Adm. Albemarle Bertie. From 1.1809 under Capt. Henry Glynn in the Baltic. Paid off 11.1811 at Plymouth. Fitted as prison ship at Plymouth 12.1812. Fitted as a Powder hulk at Plymouth 7–9.1814. BU completed there 29.9.1821.

Excellent Joseph Graham, Harwich

As built: 168ft 0in, 138ft 0in x 46ft 11in x 19ft 9in. 1,615⁷¹⁄₉₄bm.

As built: 168ft 0in, 138ft 0in x 46ft 11in x 19ft 9in. 1,615$^{71/94}$bm.

Ord: 9.8.1781. K: 3.1782. L: 27.11.1787. C: 12.1787 at Chatham.

First cost: £29,746.16.11d to build, plus £210.6.3d & £8,513.9.6d. for fitting & coppering.

Commissioned 7.1790 under Capt. John Gell, for Spanish Armament. Fitted at Chatham (for £5,432.1.7d) 9.1790. Recommissioned 9.1793 under Capt. William Clement Finch. Fitted at Portsmouth (for £7,053) 6.1794; from 5.1794 under Capt. John Samuel Smith; in 6.1794 under Capt. John Whitby, as flagship of Vice Adm. William Cornwallis in the Channel; under Capt. William Mitchell (acting) in 10.1794, then Capt. Cuthbert Collingwood from 12.1794; to the Mediterranean 7.1795. At Battle of St Vincent 14.2.1797. Flagship 9.1797–2.1798, at blockade of Cadiz; returned to UK 11.1798 and paid off. Fitted at Portsmouth (for £13,685) 6–8.1799; recommissioned 7.1799 under Capt. Robert Stopfor; took 18-gun *L'Arethuse* near Lorient 11.10.1799; cut out cutter *L'Arc* from Quiberon 20.2.1801. Under Capt. J(ohn or James) Nash, sailed for Leeward Islands 4.1802; Capt. Robert Tucker (acting) in 5.1802. Fitted at Portsmouth (for £14,898) 6–8.1803; recommissioned 7.1803 under Capt. Frank Sotheron (–1806); joined fleet off Toulon 11.1803; operations in the Bay of Naples 1806. Refitted at Chatham (for £16,267) 9.1806–3.1807; recommissioned 2.1807 under Capt. John West (–1809); sailed for the Mediterranean 6.6.1807, via Cadiz; at seizure of Fort Trinidad, Rosas Bay 1808, and destruction of convoy at Duino 28.7.1809. Between Middling and Large Repair at Portsmouth (for £54,416) 1.1812–1.1814, then to Ordinary. Ordered (by AO 11.5.1820) to be cut down to a 58-gun Fourth Rate frigate, which began 12.1825, but seemingly this conversion was never completed (see Chapter 4 for later history).

Saturn Thomas Raymond, Northam

As built: 168ft 2in, 138ft 1in x 46ft 11in x 19ft 10in. 1,616⁶⁸⁄₉₄bm.

As built: 168ft 2in, 138ft 1in x 46ft 11in x 19ft 10in. 1,616$^{68/94}$bm.

Ord: 27.12.1781. K: 8.1782. L: 22.11.1786. C: 18.12.1786–5.2.1787 (for Ordinary), 12.6.1790 (for sea) at Portsmouth.

First cost: £29,422.12.10d to build, plus £10,126.12.6 (1787 fitting) & £4,935 (fitting for sea).

Commissioned 5.1790 under Capt. Robert Linzee; paid off 9.1791. Recommissioned 10.1793 under Capt. Thomas Newnham; in 8.1794 under Capt. William Lechmere, as flagship of Rear-Adm. George Vanderput in the Downs. Under Capt. James Douglas, as flagship of Vice-Adm. Sir John Laforey; sailed for the Mediterranean 23.5.1795; in action off the Hyères 13.7.1796; in Man's squadron in 10.1795, in pursuit of de Richery. Made good defects at Plymouth (for £12,475) 2–4.1797. Under Capt. Herbert Sawyer in 8.1797, in the Channel; under Capt. Jacob Waller in 10.1797 (died 10.1798). Made good defects at Portsmouth (for £9,158) ?10.1798–1.1799; under Capt. Thomas Totty from 11.1798 (–1800). Under Capt. Charles Boyles in 1.1801, and Capt. Robert Lambert 2.1801, in Hyde Parker's fleet in the Baltic. From 8.1801 under Capt. James Brisbane , as flagship of the now Rear-Adm. Totty (died 6.1802); to the Leeward Islands in 1.1802, paid off later in 1802. Fitted at Portsmouth (for £9,135) 5–8.1805. Recommissioned 7.1805 under Capt. Lord Amelius Beauclerk (–1808), for the Mediterranean. In 1808 under Capt. Thomas Boys (acting) for blockade of Lorient; in 5.1809 under Capt. William Cumberland, in the Baltic to 1810. In Ordinary at Plymouth 1812–13. Repaired and cut down to a 58-gun Fourth Rate frigate at Plymouth (for £50,423) 4–12.1813. Recommissioned 8.1813 under Capt. James Nash; took US 10-gun privateer *Hussar* 28.5.1814. Under Capt. Thomas Brown in 11.1814; paid off into Ordinary 4.1815. Small Repair and altered at Plymouth (for £14,210) 10–12.1817, then laid up. Fitted at Plymouth as a lazarette for Milford 8–9.1825. Gunnery training ship 1830. Fitted as Harbour flag and receiving ship at Pembroke (for £2,925) 7–11.1845. By AO 1.7.1845 it was ordered to convert this very old vessel to a screw blockship, but this was cancelled 30.8.1845, and she remained a receiving ship. Fitted as a guard ship at Pembroke 10.1849–3.1850. BU (under AO 1.8.1867) completed at Pembroke 1.2.1868.

Elephant George Parsons, Bursledon
  As built: 168ft 0in, 137ft 9⅞in x 46ft 11½in x 19ft 9½in. 1,616⁴⁹/₉₄bm. Draught 12ft 6in/17ft 8in.
  Ord: 27.12.1781. K: 2.1783. L: 24.8.1786. C: 26.8–7.11.1786 at Portsmouth.
  First cost: £30,772.2.4d, plus £10,108 coppering and fitting.
  Commissioned 6.1790 under Capt. Charles Thompson, for Spanish Armament. Sailed after fitting at Portsmouth (for £4,157) 3.8.1790. Fitted for Ordinary at Portsmouth (for £190) 10.1793 (and recorded need for Middling Repair which may have occupied her for next few years, but went unrecorded). Fitted at Portsmouth (for £21,521) 8.1799–3.1800; recommissioned 12.1799 under Capt. Thomas Foley (–1801), as flagship of Rear-Adm. Sir Charles Cotton; then as flagship of Vice-Adm. Horatio Nelson 27.3–4.4.1801 for the Battle of Copenhagen, with 9 killed, 13 wounded. From 6.1801 under Capt. George Dundas; sailed for Jamaica 10.1801. Paid off 1.1805. Fitted at Chatham (for £17,695) 11.1804–7.1805. Recommissioned (still under Dundas) 5.1805, for North Sea; sailed for Leeward Islands 4.5.1806; encounter with 24-gun *Le Duguay-Trouin* 25.7.1807. Under (temp.) Cmdr George Morris in 1807, in Leeward Islands; to UK by 9.1807. Between Middling and Large Repair, and fitted at Portsmouth (for £51,290) 4.1809–9.1811. Recommissioned 7.1811 under Capt. Francis Austen; took (with *Hermes*) US 12-gun privateer *Swordfish* 28.12.1812. Cut down to a 58-gun Fourth Rate frigate at Portsmouth (for £39,407) 2.1817–3.1818, but never recommissioned. BU at Portsmouth 11.1830.

Bellerophon Edward Greaves & Co., Frindsbury
  As built: 168ft 0in, 138ft 0in x 46ft 10½in x 19ft 9in. 1,612⁷⁸/₉₄bm.
  Ord: 11.1.1782. K: 5.1782. L: 17.10.1786. C: 3.1787 for Ordinary at Chatham.
  First cost: £30,232.14.4d to builder, plus £8,376.15.2d fitting.
  Fitted for sea at Chatham (for £4,620.8.4d) 15.8.1790.
  Commissioned 19.7.1790 under Capt. Thomas Pasley for Spanish Armament, then Russian Armament; paid off 9.1791. Fitted at Chatham (for £4,164) and recommissioned 16.3.1793, still under Pasley. Under Capt. William Johnstone Hope on 9.1.1794, as flagship of the now Rear-Adm. Pasley; at Glorious First of June off Ushant 1.6.1794, lost 4 killed, 27 wounded (including Pasley, who was knighted). In 1795 under Capt. Lord (James) Cranstoun; in Cornwallis's retreat 16/17.6.1795. Made good defects at Portsmouth (for £8,103) 10–11.1795. In 9.1796 under Capt. Henry d'Esterre Darby (–1800); sailed for the Mediterranean 18.3.1797; at Battle of the Nile 1.8.1798, lost 49 killed, 148 wounded. Middling Repair and fitted at Portsmouth (for £32,608) 9.1800–8.1801; under Capt. Lord (George) Garlies in 1.1801; 12.1801 under Capt. John Loring; sailed for Jamaica 4.1802; flagship of Rear-Adm. Sir John Duckworth. Under Commodore Loring in 7.1803, squadron took 74-gun *Le Duquesne* and 16-gun *L'Oiseau* off San Domingo 25.7.1803; at surrender of French squadron at Cap François (frigates *La Surveillante*, *La Clorinde* and *La Vertu*, plus smaller) 30.11.1803. Refitted at Portsmouth (for £11,914) 9–11.1804. Under Capt. John Cooke in 4.1805; in Lee column at Battle of Trafalgar 21.10.1805, lost 27 killed (including Cooke), 123 wounded. On 21.10.1805 under Lieut William Pryce Cumby (acting), then 4.11.1805 under Capt. Edward Rotheram (–1808); repairs at Plymouth (for £18,082) 12.1805–2.1806; flagship of Rear-Adm. Albemarle Bertie in 1807. Under Capt. Samuel Warren 6.1808, as flagship of Rear-Adm. Alan Gardner in the Baltic. Under Capts Lucius Hardyman in 8.1810 and John Halstead

in 6.1811; flagship of Rear-Adm. John Ferrier 1812, in North Sea. In 2.1813 under Capt. Augustus Brine, then Capt. Edward Hawker in 3.1813, as flagship of Vice-Adm. Sir Richard Keats; sailed for Newfoundland 22.4.1813; took 16-gun privateer *Le Génie*. Under Capt. Frederick Lewis Maitland 3.1814; sailed for Newfoundland again 26.4.1814; took the surrender of Napoleon 15.7.1815. Fitted as a convict hulk at Sheerness 12.1815–12.1816. Renamed *Captivity* 5.10.1824. Fitted to navigate to Plymouth 4–6.1826; convict hulk there to 1834. Sold at Plymouth (for £4,030) 21.1.1836.

Modified *ALBION* Class These two ships were produced to a slightly modified version of Slade's 1760 design, with breadth reduced by 3in, but the *Irresistible* was BU in 1806.
  Dimensions & tons: 168ft 0in, 139ft 1in x 47ft 0in x 18ft 10in. 1,634²¹/₉₄bm.
  Men: 550. Guns: LD 28 x 32pdr; UD 28 x 18pdr; QD 14 x 9pdr; Fc 4 x 9pdr.
Fortitude John Randall & Co., Rotherhithe
  As built: 168ft 6in, 138ft 6⅛in x 47ft 3in x 18ft 9½in. 1,645¹⁹/₉₄bm. Draught 11ft 11in/17ft 1in.
  Ord: 2.2.1778. K: 4.3.1778. L: 22.3.1780. C: 3.4.1780 at Woolwich.
  First cost: £44,405.10.4d (including fitting).
  Commissioned 3.1780. Refitted at Plymouth (for £8,027.1.9d) 1–2.1782. Paid off 4.1783 after wartime service. Fitted for Ordinary at Plymouth 5.1783. Small Repair at Plymouth (for £10,261.10.7d) 3–7.1784. Recommissioned 10.1787 under Capt. Anthony Molloy; paid off 12.1787. Small Repair at Plymouth (for £8,523.2.10d) 10.1790–5.1791. Commissioned 1.1793 under Capt. William Young, for Hood's fleet in the Mediterranean; sailed for the Mediterranean 11.5.1793; attack on Mortella Town 8.1.1794. Under Capt. Thomas Taylor, action off Genoa 13.3.1795, then action off Hyères 13.7.1795; encounter with de Richery's squadron 7.10.1795 (recapture of the *Censeur*); paid off 11.1795. Prison ship at Portsmouth 10.1795, under (acting) Capt. Thomas Boys; recommissioned in same role 6.1798 under Lieut John Gourly. Powder hulk at Portsmouth 5.1802. BU at Portsmouth 3.1820.

ELIZABETH Class Revival of 1765 design by Sir Thomas Slade. Of these four ships, the *Bombay Castle* (together with the larger *Carnatic* and *Ganges*, see below) were funded by the East India Company as a contribution to the war effort. The *Bombay Castle* was wrecked in 1796, while the *Powerful* was BU in 1812 and the *Swiftsure* (in French hands 1801–05, subsequently renamed *Irresistible*) in 1816.
  Dimensions & tons: 168ft 6in, 138ft 3⅛in x 46ft 10in x 19ft 9in. 1,612⁸⁸/₉₄bm.
  Men: 550. Guns: LD 28 x 32pdr; UD 28 x 18pdr; QD 14 x 9pdr; Fc 4 x 9pdr.
Defiance John Randall & John Brent, Rotherhithe
  As built: 169ft 0in, 138ft 0in x 47ft 4in x 19ft 9in. 1,644⁵⁴/₉₄bm. Draught 12ft 2in/18ft 3in.
  Ord: 11.7.1780. K: 4.1782. L: 10.12.1783. C: 4.3.1784 at Deptford, then 6.1784 at Woolwich.
  First cost: £30,757.15.10d to build, plus fitting £8,878.
  Commissioned 8.1794 under Capt. George Keppel. Middling Repair and fitted at Chatham (for £14,925) 11.1793–4.1795. In 1795 under Capt. Sir George Home. In 3.1796 under Capt. Theophilus Jones (–1798), in the Channel. Mutiny at Spithead 1787; also ?in 1798 mutiny. Refitted at Portsmouth (for £10,270) 9–12.1798. In 1799 under Capt. Thomas Shivers, with Rear-

Adm. James Whitshed's squadron to the Mediterranean; sailed 6.5.1799 in pursuit of de Bruix's squadron. In 12.1800 under Capt. Richard Retalick, as flagship of Rear-Adm. Sir Thomas Graves; at Battle of Copenhagen 24.4.1801; paid off late 1801. Recommissioned 5.1803 under Capt. Philip Durham, for Channel Fleet; in Calder's Action off Finisterre 22.7.1805, lost 1 killed, 7 wounded; in Lee column at Trafalgar 21.10.1805, lost 17 killed, 53 wounded. Refitted at Portsmouth (for £24,311) 12.1805–4.1806. Recommissioned 3.1806 under Capt. Henry Hotham (–1810); with Rear-Adm. Robert Stopford's squadron off Rochefort 1809; at destruction of three French 40-gun frigates (*La Cybele*, *Le Calypso* and *L'Italienne*) at Sables d'Olonne 24.2.1809. In early 1809 in the Basque Roads operations. Made good defects at Plymouth (for £10,440) 3.1809–4.1810. From 8.1810 under Capt. Richard Raggett (–1813). In 1811 flagship of Rear-Adm. John Ferrier, in North Sea; in 1813 flagship of Rear-Adm. George Hope, in the Baltic. Fitted as a temporary prison ship at Chatham (for £386) 11–12.1813. In Ordinary 1814–15. BU there 5.1817.

CARNATIC Class 'Middling Class', built to the lines of the *Courageux* (ex-French prize, taken 1761), and thus noticeably larger than the standard 74-gun ships then building. The *Colossus* of this class was wrecked in 1798 and the *Minotaur* in 1810.

> Dimensions & tons: 172ft 3in, 140ft 5¼in x 47ft 9in x 20ft 9in. 1,703²¹⁄₉₄bm.
>
> Men: 640. Guns: LD 28 x 32pdrs; UD 28 x 18pdrs; QD 14 x 9pdrs; Fc 4 x 9pdrs.

Captain (ex-*Carnatic*, renamed 14.7.1815) Henry Adams & William Barnard, Deptford

> As built: 172ft 4½in, 140ft 3½in x 48ft 0in x 20ft 9½in. 1,719³⁹⁄₉₄bm. Draught 12ft 9½in/18ft 3½in.
>
> Ord: 14.7.1779. K: 3.1780. L: 21.1.1783. C: 2.2.1783 (at Deptford Dyd), then 7.1783 (at Woolwich).
>
> First cost: Fitting £13,803.19.5d (at Deptford) + £2,929.13.10d (at Wooolwich, including coppering). This ship was 'presented' (i.e. paid for) by the East India Company, the only cost to the RN being 'extra works' amounting to £112.6.7d.
>
> Commissioned 3.1783 under Capt. Anthony Molloy (–1785) as guard ship at Chatham, from 1785 at Plymouth. From 4.1786 under Capt. Peregrine Bertie (–1788). Fitted at Plymouth (for £8,743.11.3d) 2–10.1787. From 1789 under Capt. John Ford (–1792). Fitted for Channel service at Plymouth (for £5,054.11.10d) 6.1790, for Spanish Armament, from 8.1790 at flagship of Rear-Adm. John Jervis. Recommissioned 8.1791 as guard ship at Plymouth. Middling Repair and fitted at Plymouth (for £39,229) 11.1792–3.1796; recommissioned 11.1795 under Capt. Richard Grindall. From 3.1796 under Capt. Henry Jenkins (–1797), as flagship of Rear-Adm. Charles Pole; sailed for the Leeward Islands 4.1796. In 1798 under Capt. George Bowen, then 1799 under Capt. John Loring (–1800), all at Jamaica; later under Capt. Edward T. Smith. From 1.1801 under Capt. Charles Brisbane, then 6.1802 under Capt. Charles Penrose, as flagship of Rear-Adm. Robert Montagu. Fitted as temporary receiving ship at Plymouth 6–7.1805. In Ordinary there from 1812–15. Renamed *Captain* 14.7.1815. BU completed at Plymouth 30.9.1825.

Leviathan Chatham Dyd [M/Shipwright Nicholas Phillips to 7.1790, completed by John Nelson]

> As built: 172ft 3in, 140ft 4in x 47ft 10in x 20ft 9in. 1,707⁸¹⁄₉₄bm. Draught 11ft 9in/17ft 6in.

Ord: 9.12.1779. K: 5.1782. L. 9.10.1790. C: 27.8.1791 (partly fitted).

First cost: £40,810.19.8d to build, plus £2,064.16.5d fitting at Chatham.

> Commissioned 1.1793 under Capt. Hugh Conway (–1794); fitted at Sheerness to 4.1793; sailed for the Mediterranean 22.5.1793; took (with *Colossus*) privateer *Le Vrai Patriot* 7.1793; in 10.1793 under Capt. Benjamin Halowell (acting), at Toulon; refitted at Portsmouth (for £11,673) 4.1794; in Battle of Glorious First of June off Ushant 1.6.1794, losing 10 killed and 33 wounded. In 1795 under Capt. John Duckworth (–1797); sailed for Jamaica 14.5.1795; at Leogane 21.3.1796. Refitted at Plymouth (for £10,624) to 8.1797; in 8.1797 under Capt. Joseph Bingham, on Irish station when Duckworth flew Broad Pennant aboard; later under Capt. Henry Digby; sailed for the Mediterranean 2.6.1798; at capture of Minorca 11.1798. In 2.1799 under Capt. James May, as flagship of the now Rear-Adm. Duckworth (–1803). In 1800 under Capt. James Carpenter, at blockade of Cadiz; took (with *Emerald*) 36-gun *Carmen* and *Florentina* 7.4.1800; in 6.1800 under Cmdr Edward D. King; sailed for the Leeward Islands; then 1801 under Cmdr Christopher Cole and 1802 Capt. Richard Dunn (–1803), still Duckworth's flagship in the Leeward Islands; paid off 12.1803. Refitted at Portsmouth (for £22,261) 10.1803–1.1804; recommissioned 1.1804 under Capt. Henry Bayntun; sailed for the Mediterranean 26.4.1804; later at blockade of Toulon, then in chase (of Gantheaume) to the West Indies; in Weather column at Battle of Trafalgar, losing 4 killed and 22 wounded. Between Small and Middling Repair at Plymouth (for £39,026) 11.1807–8.1808; recommissioned 6.1808 under Capt. John Harvey (–1811); sailed for the Mediterranean 7.2.1809; with Martin's squadron in attack on Baudin's convoy 23.10.1809 (80-gun *La Robuste* and 74-gun *Le Lion*, run ashore and burnt near Frontignan 25.10.1809);. In 8.1811 under Capt. Patrick Campbell, in the Mediterranean; boats in attack on shipping near Fréjus 29.4.1812; destruction (with *America* and *Éclair*) of convoy near Laigueglia 9.5.1812, and (with *Imperieuse* and *Curacao*) similar convoy 27.6.1812. Made good defects at Portsmouth (for £13,396) 8–11.1813. Recommissioned 30.9.1813 under Capt. Adam Drummond; convoy escort to Jamaica 1814 and then home. In 10.1814 under Capt. Thomas Briggs, at Lisbon, Cork and then the Mediterranean again; paid off 19.7.1816. Fitted as a convict ship at Portsmouth 10.1816. Scuttled as a target ship there 10.1846. Sold to Mr Burns (for £805) 7.8.1848.

GANGES Class Edward Hunt's only 74-gun design, to which three ships were ordered (it was revived for a further pair of ships in 1801, and in modified form for a sixth vessel in 1811), but of these the *Culloden* of 1783 was BU in 1813. Fast and weatherly ships, if somewhat unstable.

> Dimensions & tons: 169ft 6in, 138ft 11¼in x 47ft 4in x 20ft 3in. 1,656⁶⁴⁄₉₄bm.
>
> Men: 590. Guns: LD 28 x 32pdrs; UD 28 x 18pdrs; QD 14 x 9pdrs; Fc 4 x 9pdrs.

Ganges John Randall, Rotherhithe

> As built: 169ft 6in, 138ft 7¾in x 47ft 8½in x 20ft 3in. 1,678⁵³⁄₉₄bm. Draught 12ft 6in/18ft 4in.
>
> Ord: 14.7.1779. K: 4.1780. L: 30.3.1782. C: 30.3–20.4.1782 (at Deptford Dyd), –26.6.1782 (at Woolwich).
>
> First cost: Fitting £11,238.5.9d (at Deptford Dyd) + £3,042.6.9d (at Woolwich, including coppering). This ship was 'presented' (i.e. paid for) by the East India Company, the only cost to the RN

being 'extra works' amounting to £279.14.3d.

Commissioned 2.1782 under Capt. Charles Feilding, for Howe's
fleet; paid off 3.1783 after wartime service, recommissioned same
month under Capt. J. Lutterell, as guard ship at Portsmouth.
Fitted as this (for £5,154.13.5d) to 8.1783. In 1784 under Capt.
Sir Roger Curtis (–1787), as flagship of Rear-Adm. Sir Francis
Drake in 10.1787. Middling to Great Repair at Portsmouth
(for £30,023) 11.1790–12.1791. Recommissioned 12.1792 under
Capt. Anthony Molloy; in chase of Vanstabel's squadron
18.11.1793. Recommissioned 1.1794 under Capt. William
Truscott; in Montagu's squadron 6.1794; sailed for the Leeward
Islands 25.10.1794; took (with *Montagu*) 24-gun *Le Jacobin*
30.10.1794. In 7.1795 under Capt. Benjamin Archer, in the
Leeward Islands. In 1796 under Capt. Lancelot Skynner; sailed
for Leeward Islands again 25.2.1796. In 4.1796 under Capt.
Robert M'Dougall, in Sir Hugh Christian's operations in the
West Indies – St Lucia 4–5.1796, and Grenada 6.1796. Refitted
at Portsmouth (for £13,899) 1.1797; still under M'Dougall,
on North Sea station to 1799; from 9.1799 under Capt. Colin
Campbell. Small Repair at Portsmouth (for £14,133) 6–9.1800;
recommissioned 8.1800 under Capt. Thomas Fremantle; at Battle
of Copenhagen 2.4.1801; under Capt. James Brisbane (acting) in
6.1801. In 10.1801 sailed for Jamaica under Capt. Joseph Baker,
then 9.1802 under Capt. George M'Kinley; paid off 7.1803 and
recommissioned under Fremantle again; paid off 11.1804. Fitted
at Portsmouth 5–6.1806; recommissioned 4.1806 under Capt.
Peter Halkett; joined Stopford's squadron in 1.1807; later in 1807
flagship of Rear-Adm. Richard Keats (–1809), for Copenhagen
expedition in 8.1807; sailed for Portugal 1.1.1808; in North Sea
station 1809. In 11.1809 under Capt. Thomas Dundas, for the
Baltic; paid off 1811. Fitted for Ordinary at Plymouth 3–4.1811.
Fitted as a prison ship at Plymouth 10.1811 (possibly lent to the
Transport Board 12.1814); under Lieut Frederick Leroux 1812–
14 then Lieut James Spratt 12.1814. BU at Plymouth 3.1816.

Tremendous William Barnard, Deptford Green
As built (originally): 170ft 4in, 139ft 3¼in x 47ft 7½in x 20ft 4in.
1,680²²⁄₉₄bm.
Ord: 1.1.1782. K: 13.8.1782. L: 30.10.1784. C: 28.1.1785 (at
Deptford Dyd),–4.1785 (coppered & fitted).
First cost: £32,366.14.7d to build, plus fitting £6,443 (to 28.1.1785) +
£4,955 (later, including coppering).
Fitted at Chatham (for £3,976.12.4d) to 10.1790, then laid up.
Commissioned 3.1793 under Capt. James Pigot, for Howe's fleet;
fitted at Sheerness (for £1,704) to 6.6.1793; refitted at Plymouth
(for £6,917) to 1.1794; in Battle of Glorious First of June off
Ushant 1.6.1794, losing 3 killed and 8 wounded. Later in 1794
under Capt. William Bentinck, then 3.1795 under Capt. William
Hope and 6.1795 under Capt. Samuel Ballard. In 1796 under
Capt. John Aylmer, as flagship of Rear-Adm. Thomas Pringle;
sailed for Cape of Good Hope 1.5.1796. In ?7.1796 under Capt.
Charles Brisbane; capture of Lucas's squadron in Saldanha Bay
17.8.1796; in 1797 under Capt. George Stephens (temp.), later
Cmdr Askew Hollis (acting); in 1797 Mutiny. In 1798 under
Capt. John Osborn, later Capt. John Searle, as flagship of Rear-
Adm. Sir Hugh Christian, on Cape of Good Hope station. Bu
1799 under Osborn again (–1806); destroyed (with *Adamant*)
36-gun *La Preneuse* off Mauritius 11.12.1809; in East Indies
1803; action against 40-gun *La Canonnière* 21.4.1806. In 1807
in Ordinary at Chatham. Large Repair (actually reconstruction)
in 1807–11 at Chatham with Seppings's diagonal frames (see
separate entry below for post-1807 history).

Modified *CULLODEN* Class The prototype *Culloden* (wrecked
1781) having been built in a Royal Dockyard, Slade's 1769 design was
altered for seven slightly smaller ships built by three Thames-side
contractors. Although all were completed during the American War,
only *Hannibal* was brought into service before 1792. Of this class, the
Hannibal was taken by the French in 1801, the *Victorious* was BU in
1803, the *Venerable* was wrecked in 1804, and *Thunderer* and *Theseus*
were BU in 1814.
Dimensions & tons: 170ft 0in, 139ft 8in x 46ft 6in x 19ft 11in.
1,652⁶⁄₉₄bm.
Men: 550. Guns: LD 28 x 32pdr; UD 28 x 18pdr; QD 14 x 9pdr; Fc
4 x 9pdr.

Terrible John & William Wells, Deptford
As built: 170ft 7in, 139ft 11in x 47ft 6in x 19ft 11in. 1,679¹⁷⁄₉₄bm.
Draught 12ft 5½in/18ft 2½in.
Ord: 13.12.1781. K: 1.1783. L: 28.3.1785. C: 28.3–12.4.1785 at
Deptford, then 2.7.1785 at Woolwich.
First cost: Coppering (at Woolwich) £1,434.
Small Repair and fitted at Chatham (for £11,449) 5.1792–2.1793.
Commissioned 12.1792 under Capt. Skeffington Lutwidge; sailed
for the Mediterranean 22.4.1793; joined Hood's Fleet there.
In 1794 under Capt. George Campbell (–1797); in Hotham's
Action off Genoa 13.3.1795 (0 killed, 6 wounded); in Hotham's
Action off Hyères 13.7.1795; in Man's squadron 10.1795,
for pursuit of de Richery. Made good defects at Plymouth
(for £12,537) 2–4.1797. In 6.1797 under Capt. John Miller; in
mutiny at Spithead 1797. In 10.1797 under Capt. Sir Richard
Bickerton; chase of Bompart's fleet 28–30.10.1798. Made good
defects at Plymouth (for £9,655) 12.1798–4.1799; in 4.1799
under Capt. Jonathan Faulknor, for the Channel. In 1799
under Capt. William Wolseley; sailed for the Mediterranean
1.6.1799; in Quiberon operations 1800. In 1.1801 under Capt.
Francis Fayerman. Small to Middling Repair at Plymouth
4.1802–12.1803; in 1803 under Cmdr (Capt. 1.1804) Lord Henry
Powlett (–1809), for the Channel; recommissioned 9.1804;
in Strachan's squadron 1806, for pursuit of Leissenges and
Willaumez; sailed for the Mediterranean 1.1.1807, via Cadiz
and Ferrol. Small Repair at Woolwich (for £31,312) 3–12.1813;
then laid up at Sheerness. Fitted as a receiving ship at Sheerness
8.1822–5.1823. Used as Coal depot at Sheerness (by AO
5.3.1829) 4.1829. BU at Deptford 3.1836.

Ramillies Randall & Brent, Rotherhithe
As built: 170ft 4in, 139ft 9in x 47ft 6in x 19ft 11½in. 1,677¹⁷⁄₉₄bm.
Draught 12ft 9½in/17ft 7in.
Ord: 19.2.1782. K: 12.1782. L: 12.7.1785. C: 18.7–31.7.1785 (at
Deptford), then 22.9.1785 at Woolwich.
First cost: Fitting (total) £9,323.3.8d.
Small Repair at Chatham (for £11,907) 8–12.1791.
Commissioned 2.1793 under Capt. Henry Harvey; in Battle of
the Glorious First of June off Ushant 1.6.1794, losing 2 killed,
7 wounded. In 8.1794 under Capt. Sir Richard Bickerton
(–1797); sailed for the Leeward Islands 20.10.1794. Refitted
at Portsmouth (for £8,671) 2.1796, for North Sea; sank lugger
Spider by collision 4.4.1796; mutiny at Spithead 4–5.1797. In
7.1797 under Capt. Bartholomew Rowley, for Channel and Irish
station. Made good defects at Plymouth (for £10,832) 3.1798.
In 10.1798 under Cmdr Henry Inman, then 2.1799 under Capt.
Richard Grindall; in Quiberon operations 1799. In 1.1801 under
Capt. John W. T. Dixon, for Parker's fleet in the Baltic; sent
to reinforce Nelson's fleet 2.4.1801. In 6.1801 under Capt. Sir
Robert Barlow, then 8.1801 Capt. Samuel Osborn; tp Channel

and Spanish coast later in year, then paid off. Fitted at Chatham (for £17,285) 7.1804–2.1805; recommissioned 12.1804 under Capt. Francis Pickmore; took (with *Illustrious*) 2-gun privateer *La Joséphine* 7.7.1805; sailed for the Leeward Islands 10.1.1807. In 4.1808 under Capt. Robert Yarker. Large Repair at Chatham (for £66,244) 7.1810–11.1812; recommissioned 10.1812 under Capt. Sir Thomas Hardy (–1814); sailed for North America; flagship of Rear-Adm. George Cockburn 1813. In 6.1815 under Capt. Charles Ogle, then 11.1815 Capt. Thomas Boys, as flagship of Rear-Adm. Sir William Hope at Leith. Fitted as a guard ship at Sheerness 6.1816. In 9.1818 under Capt. Askew Hollis, as guard ship at Portsmouth, then 8.1821 under Capt. Edward Brace, in the Downs for the Coast Blockade. Between Small and Middling Repair and fitted as a guard ship at Portsmouth (for £19,161) 5.1822–6.1823; in 5.1823 under Capt. William M'Cullock, then 11.1825 under Capt. Hugh Pigot. To Reserve for the Harbour Service by AO 3.8.1830. Fitted as a lazarette at Chatham 6.1831, for Sheerness. BU at Deptford 2.1850.

BRUNSWICK Class

The first 74 to be designed and built after the end of the American War showed a significant increase in dimensions over the wartime vessels, and carried an extra pair of 18pdrs on the UD, the first departure from the standard armament that the 'Common' Class had carried since 1755. This 'Admiralty' design was approved on 10 January 1785.

Dimensions & tons: 176ft 0in, 145ft 2in x 48ft 8in x 19ft 6in. 1,828^{72}/$_{94}$bm.

Men: 650. Guns: LD 28 x 32pdr; UD 30 x 18pdr; QD 12 x 9pdr; Fc 4 x 9pdr.

(re-established 12.1806 with an all-24pdr armament of LD 28 x 24pdrs, UD 28 x 24pdrs (Govers), QD 2 x 24pdrs (Govers) + 10 x 24pdr carronades. Fc 2 x 24pdrs (Govers) + 4 x 24pdr carronades.)

Brunswick Deptford Dyd [M/Shipwright Henry Peake to 3.1787; completed by Martin Ware]

As built: 176ft 2½in, 145ft 3in x 48ft 9in x 19ft 6in. 1,836^{1}/$_{94}$bm. Draught 13ft 0in/16ft 7in.

Ord: 7.1.1785. K: 5.1786. (named AO 14.6.1786) L: 30.4.1790. C: 17.5–18.6.1790 at Woolwich.

First cost: £43,024, plus £4,757 fitting at Woolwich.

Commissioned 5.1790 under Sir Capt. Hyde Parker, for Spanish Armament, then again under Capt. Sir Roger Curtis, for Russian Armament. Recommissioned 8.1791 as guard ship at Portsmouth.. Recommissioned 7.1793 under Capt. John Harvey as flagship of Rear-Adm. George Bowyer; in duel with 74-gun *Le Vengeur du Peuple* in Battle of Glorious First of June off Ushant 1.6.1794, with 41 killed (including Harvey) and 114 wounded; paid off and recommissioned 9.1794 under Capt. Lord Charles Fitzgerald; in Cornwallis's 'Retreat' 16/17.6.1795. In 6.1795 under Capt. William Browell (but Capt. Thomas Gosselin acting 7–9.1795), as flagship of Rear-Adm. Richard Bligh. In 1797 under Capt. William Rutherford, in the Leeward Islands. In 1798 under Lieut Hugh Cook (acting), at Jamaica. In 4.1799 under Cmdr William Chilcott, then 6.1800 under Capt. James Wallis, still at Jamaica; paid off 9.1800. Fitted at Portsmouth 2–4.1801; in 3.1801 under Capt. George Stephens. Fitted at Portsmouth, Round House taken off (for £18,818) 12.1806–3.1807; recommissioned 2.1807 under Capt. Thomas Graves; joined Gambier's fleet at Copenhagen expedition 8.1807, then Saumarez's fleet in the Baltic 1808. In Ordinary at Gillingham to 1812. Fitted as prison ship at Chatham 5–6.1812; as prison ship

under Lieut John H. Sparkes 1813–14. Fitted as a Powder hulk at Chatham 7.1814–8.1815. Lazarette at Stangate Creek 10.1825. BU at Sheerness completed 8.1826.

COURAGEUX Class

'Middling Class' 74. Sir John Henslow design, approved as modified 9 September 1796. A sister ship ordered from Portsmouth on the same date was never built or named. A third ship to this design was ordered on 15 November 1799 but was likewise not built.

Dimensions & tons: 181ft 0in, 150ft 10^{1}/$_{8}$in x 47ft 0in x 19ft 8in. 1,772^{38}/$_{94}$bm.

Men: 590. Guns: LD 28 x 32pdrs; UD 28 x 18pdrs; QD (orig. 14 x 9pdrs) 2 x 18pdrs + 12 x 32pdr carronades; Fc (orig. 4 x 9pdrs) 2 x 9pdrs + 2 x 32pdr carronades; RH 6 x 18pdr carronades.

Courageux Deptford Dyd [M/Shipwright Thomas Pollard to late 1799; completed by Edward Tippett]

As built: 181ft 0in, 150ft 9¼in x 47ft 1½in x 19ft 10in. 1,780^{93}/$_{94}$bm. Draught 13ft 1in/18ft 3in.

Ord: 6.11.1794. (named 22.5.1797) K: 10.1797. L: 26.3.1800. C: 5.4–30.5.1800 at Woolwich.

First cost: £60,701 to build, plus £3,543 fitting.

Commissioned 4.1800 under Capt. Samuel Hood; attack on Ferrol 26.8.1800. In 1801 under Capt. George Duff; in Rear-Adm. Sir Robert Calder's squadron to the West Indies 2.1801; later under Capt. Thomas Sotheby, in the Channel. Recommissioned 4.1802 under Capt. Robert Plampin, and again in 4.1803 under Capt. J(ohn or James) Hardy. In 11.1803 under Capt. Thomas Bertie, as flagship of Rear-Adm. James Dacres; sailed 4.1.1804 with convoy for West Indies, but driven back by weather. Defects made good at Plymouth (for £17,468) 2–4.1804. Under Capt. Charles Boyles in 3.1804, then under Capt. Richard Lee 1805, in the Channel. Under Capt. James Bissett 1806–7; at blockade of Cadiz 1807. Very Small Repair at Chatham (for £20,095) 3–7.1809. Recommissioned 5.1809 under Capt. Robert Plampin again; in Scheldt operations 1809. Under Capt. Adam Drummond in ?5.1810, then Capt. William Butterfield (acting) in 8.1810. Under Capt. Philip Wilkinson in 11.1810; grounded on Skerries Rocks 21.1.1811, and on Anholt Reef 13.11.1812. Fitted as a lazarette at Chatham 12.1813–2.1814. BU there 10.1832.

PLANTAGENET Class

'Middling Class' 74. Sir William Rule design (but heavily influenced by Adm. James Gambier), approved as modified 13.8.1796 and 9.9.1796. Unusually for a large 74, built without a poop.

Dimensions & tons: 181ft 0in, 151ft 3^{1}/$_{8}$in x 47ft 0in x 19ft 11in. 1,777^{28}/$_{94}$bm.

Men: 590. Guns: LD 28 x 32pdrs; UD 28 x 18pdrs; QD (orig. 14 x 9pdrs) 2 x 18pdrs + 12 x 32pdr carronades; Fc (orig. 4 x 9pdrs) 2 x 18pdrs + 2 x 32pdr carronades.

Plantagenet Woolwich Dyd [M/Shipwright John Tovery to 7.1801; completed by Edward Sison]

As built: 181ft 0in, 151ft 3^{3}/$_{8}$in x 47ft 0in x 19ft 11in. 1,777^{51}/$_{94}$bm. Draught 13ft 4in/16ft 11in.

Ord: 6.11.1794. (named 15.2.1796) K: 11.1798. L: 22.10.1801. C: 7.11.1801 (partial), then 10.4.1803.

First cost: £56,848 (including fitting)

Commissioned 3.1803 under Capt. George Hamond, for the Channel; took 4-gun privateer *Le Coureur de Terre Neuve* 24.7.1803; took (with *Rosario*) 14-gun privateer *L'Atalante* 27.7.1803, in 'the Bay' (?Biscay). In 8.1803 under Capt. Michael de Courcey. Fitted for foreign service at Plymouth 1–2.1804. In

10.1804 under Capt. Francis Pender, in the Channel. In ?3.1805 under Capt. William Bradley (–1809); took 2-gun privateer *L'Incomparable* 29.8.1807; sailed for Portugal 15.11.1807. With Sidney Smith's squadron at Lisbon, and at blockade of the Tagus ; at Corunna 1.1809, then in the Baltic 1809. Under Capt. Thomas Eyles 1810–12. Under Capt. Robert Lloyd 2.1812; sailed for North America 10.3.1813. Her boats (with others) in unsuccessful attack on US privateer *General Armstrong* at Fayal 26.9.1814. To Ordinary 1814. BU at Portsmouth 5.1817.

DRAGON Class 'Middling Class' 74. Sir William Rule design, approved 10 June 1795. One of eight 74s to five varying designs, all ordered on 30 April 1795, for which designs were approved on this date.

  Dimensions & tons: 178ft 0in, 146ft 9in x 48ft 0in x 20ft 6in. 1,798⁴⁴⁄₉₄bm.
  Men: 640. Guns: LD 28 x 32pdr; UD 30 x 18pdr; QD 12 x 9pdr; Fc 4 x 9pdr.
  (later UD 28 x 18pdrs + 2 x 32pdr carronades, QD 4 x 12pdrs + 8 x 32pdr carronades; Fc 2 x 12pdrs + 2 x 32pdr carronades; RH 3 x 18pdr carronades.)
Dragon (John & William) Wells & Co., Deptford
  As built: 178ft 0in, 146ft 6⅛in x 48ft 3in x 20ft 6in. 1,814⁷⁄₉₄bm. Draught 12ft 9in/18ft 9in.
  Ord: 30.4.1795. K: 8.1795. L: 2.4.1798. C: 2.4–11.5.1798 at Deptford Dyd
  First cost: £36,181 to build, plus £17,329 fitting.
  Commissioned 6.1798 under Capt. George Campbell. In Lord Bridport's fleet 1799; part of Sir Charles Cotton's reinforcement to the Mediterranean, sailed 1.6.1799. In 1.1801 under Capt. John Aylmer, with Warren's squadron off Cadiz; took (with *Endymion*) 16-gun *La Colombe* 18.6.1803. In 7.1803 under Capt. Edward Griffith; refitted at Portsmouth (for £12,263) 8–11.1804; escort to Craig's Force 1805; at Calder's Action off Finisterre 22.7.1805, had 4 wounded. Under Capt. Matthew Scott in the Channel 1806; off Rochefort 1.1807. Refitted at Portsmouth (for £14,370) 2–6.1807. Between Small and Middling Repair at Plymouth (for £51,887!) 4.1809–10.1810. Recommissioned 9.1810 under Capt. Thomas Forrest, as flagship of Sir ?F. la Faey; sailed for Leeward Islands 30.10.1810. Under Capt. Francis Collier from ?5.1812, then Capt. Robert Barrie in 10.1812; took part in destruction of 28-gun USS *Adams* at Hampden 3.9.1814; operations on American coast to 1815. Laid up at Plymouth 8.1815. At Portsmouth 1817, later returned to Plymouth. Fitted as a lazarette at Plymouth 8–9.1824, for Milford (–1829). Receiving ship and marine barracks 1829–42; hauled ashore and fitted as a marine barracks at Pembroke (for £1,149). Hulked and renamed *Fame* 15.7.1842. BU completed at Pembroke 23.8.1850.

NORTHUMBERLAND Class 'Middling Class' 74s. The design, derived from draught of *Impetueux* (ex-*L'Amerique*), a French prize taken in 1794, was approved 10 June 1795.

  Dimensions & tons: 182ft 0in, 150ft 1¼in x 48ft 7½in x 21ft 7in. 1,887⁷⁴⁄₉₄bm.
  Men: 640. Guns: LD 30 x 32pdrs; UD 30 x 18pdrs; QD 4 x 18pdrs + 10 x 32pdr carronades; Fc 2 x 18pdrs + 2 x 32pdr carronades; RH 6 x 18pdr carronades.
Northumberland Mrs Frances Barnard, Deptford Green
  As built: 182ft 2¼in, 150ft 3¾in x 48ft 10in x 21ft 7in. 1,906⁶⁵⁄₉₄bm. Draught 13ft 6in/19ft 3in.
  Ord: 30.4.1795. K: 10.1795. L: 2.2.1798. C: 15.3.1798 at Deptford

Dyd, 18.5.1798 at Woolwich.
  First cost: £37,456 to build, plus £13,233 fitting at Deptford and £5,414 at Woolwich..
  Commissioned 4.1798 under Capt. Edward Owen; by 8.1798 under Capt. George Martin (–1802), as flagship of Vice-Adm. Sir John Colpoys; sailed for the Mediterranean 10.1798. In Rear-Adm. Sir John Duckworth's squadron in 6.1799; blockade of Malta 1800; capture of *Le Généreux* 18.2.1800; took (with *Genereux* and *Success*) 42-gun *La Diane* off Malta 24.8.1800. Egypt operations 1801. Recommissioned 6.1803 under Capt. Alexander Cochrane, for the Channel. In 1805 under Capt. George Tobin, as flagship of now Rear-Adm. Cochrane, at blockade of Ferrol. In pursuit of Missiessy's squadron to West Indies 1805. Under Capt. (acting) John Morrison, at Battle of San Domingo 6.2.1806. Under Cmdr Joseph Spear (temp.) 6.1806. Later in 1806 under Capt. Nathaniel Cochrane (still Rear-Adm. Cochrane's flagship) in the Leeward Islands; In ?8.1807 under Capt. George Losack, at Portsmouth. In 2.1808 under Capt. William Hargood; sailed for the Mediterranean (via Cadiz) 28.2.1808. In 1810 under Capt. Henry Hotham; took 14-gun privateer *La Glaneuse* 23.11.1810. Destruction of 40-gun *L'Ariane* and *L'Andromache*, plus 16-gun *Le Mamelouk*, near Le Graul Rocks (off Lorient) 22.5.1812. In Ordinary at Chatham 1813. Large Repair at Chatham (for £54,178 plus £3,297) 9.1813–4.1815. Fitted as flagship at Chatham 7.1815; recommissioned 5.1815 under Capt. Charles Ross, as flagship of Rear-Adm. Sir George Cockburn; sailed for St Helena 8.8.1815, transporting Napoleon Bonaparte. Under Capt. James Walker in 8.1816, as flagship of Rear-Adm. Sir Charles Rowley at Sheerness. Under Capt. Sir Michael Seymour 1818, as guard ship in the Medway. Fitted as guard ship at Sheerness 1.1819; recommissioned 8.1819 under Capt. Thomas Harvey. Paid off 7.1821, but recommissioned again same month under Capt. Thomas Maling. Refitted again 4.1822. Fitted as lazarette at Sheerness 9.1826–2.1827, for Stangate Creek. BU at Deptford 7.1850.
Renown (ex-*Royal Oak*, renamed 15.2.1796) John Dudman, Deptford
  As built: 182ft 0in, 150ft 3in x 48ft 9in x 21ft 7in. 1,899³³⁄₉₄bm.
  Ord: 30.4.1795. K: 11.1795. L: 2.5.1798. C: 19.6.1798 at Deptford.
  First cost: £38,203 to build, plus £17,980 fitting.
  Commissioned 8.1798 under Capt. Albemarle Bertie, for the Channel; attack on Spanish squadron at the Basque Roads 2.7.1799. Under Capt. Thomas Eyles 11.1799, as flagship of Rear-Adm. Sir John Borlase Warren; squadron's boats took 2-gun *La Nochette* and others, near Penmarkes 11.6.1800; destroyed 20-gun *La Therese* and convoy 2.7.1800; attack on Ferrol 26.8.1800. Under Capt. John Chambers White in 10.1800 (–1804); sailed for the Mediterranean 26.11.1800. Blockade of Cadiz 2.1801. Blockade of Toulon 1803–04. Under Capt. Pulteney Malcolm later in 1804, and Capt. Sir Richard Strachan in 3.1805. Paid off 1805; recommissioned 12.1805; under Cmdr William Hellard (temp.) in 3.1806, in the Channel. Under Capt. Philip Durham 1806–9; sailed for Mediterranean 30.1.1808. Under Capt. Thomas Alexander (temp.) in ?7–10.1809, at blockade of Toulon; took 12-gun *Le Champenoite* off Toulon 4.5.1809. In Martin's squadron 10.1809, in chase of Baudin's convoy; destroyed *La Robuste* (80-gun) and *Le Lion* (74-gun) at Frontignan 26.10.1809. In Ordinary at Plymouth 1812–13. Became hospital ship at Plymouth 1.1814; later Deptford. BU at Deptford 5.1835.

SPENCER Class 'Middling Class' 74. Jean-Louis Barrallier design, approved 21 September 1795.

Dimensions & tons: 181ft 0in, 148ft 10¼in x 49ft 0in x 21ft 10in. 1,901¹/₉₄bm.

Men: 640. Guns: LD 30 x 32pdrs; UD 30 x 18pdrs; QD 4 x 18pdrs+ 10 x 32pdr carronades; Fc 2 x 9pdrs + 2 x 32pdr carronades; RH 6 x 18pdr carronades.

Spencer Balthazar & Edward Adams, Bucklers Hard

As built: 180ft 10in, 148ft 6¾in x 49ft 3in x 21ft 10in. 1,916⁶⁹/₉₄bm. Draught 13ft 9in/19ft 5in.

Ord: 19.9.1795. K: 9.1795. L: 10.5.1800. C: 12.8.1800 at Portsmouth. First cost: ? to build, plus £24,588 for fitting.

Commissioned 6.1800 under Capt. Henry d-Esterre Darby (–1802); in Battle of Algeciras 6.7.1801, and in Gut of Gibraltar 12.7.1801. Joined Calder's squadron 10.1801; sailed in chase to West Indies 12.1801. Recommissioned 5.1803 under Capt. Robert Stopford (–1807), in the Channel; joined Nelson off Toulon 8.1804; in chase to the West Indies 1805. In Duckworth's squadron 1806; at Battle of San Domingo 6.2.1806. In Copenhagen expedition 1807. In 4.1808 under Capt. John Quillam in the Channel, as flagship of the now Rear-Adm. Stopford. Large Repair at Plymouth 10.1811–3.1814. Recommissioned 1.1814 under Capt. Richard Raggett (–1815), for North America. Under Capt. William Broughton in 8.1815, as guard ship at Plymouth, from 1818 under Capt. Thomas Hardy; from 9.1818 under Capt. Samuel Rowley, as flagship of Rear-Adm. Sir Josias Rowley, at Cork. Under Capt. Sir Thomas Lavie in 12.1821. BU at Plymouth 4.1822.

Ex-FRENCH PRIZES (1793–1801)

Le TÉMÉRAIRE Class Built to Jacques-Noël Sané's design of 1782, the standard French 74-gun battleship design of the 1782–1814 era to which some ninety-eight ships were commenced (six more ordered at Toulon were cancelled unstarted) and about ninety completed, making this numerically the largest class of battleships ever built to one design. The design dimensions (in French feet) were 172 ft, 155ft x 44½ft x 22ft; compared to British-built 74s, they were rather long and were structurally weaker. Completed with 28 x 36pdr (LD) and 30 x 18pdrs (UD), they also had 16 x 8pdrs on the *gaillards* (12 QD, 4 Fc) in French service, to which 4 x 36pdr brass obusiers were later added. Of those captured by the Royal Navy during the French Revolutionary War, only the following four remained on the Navy List at the start of 1817 (and two of these were BU during that year). The *Spartiate* had undergone a major reconstruction in 1814–15, enabling her to continue in active service until 1835, and in 1817 was re-classed as a 78-gun ship with 24pdrs on her upper deck; the *Donegal* also continued in service until the 1830s.

Pompee (French *Le Pompée*, built 1.1790–2.1793 at Toulon. L: 28.5.1791).

Dimensions & tons: 182ft 2in, 148ft 7¼in x 49ft 0½in x 21ft 10½in. 1,901⁸/₉₄bm.

Men: 640. Guns: LD 30 x 32pdrs; UD 30 x 18pdrs; QD 12 x 32pdr carronades; Fc 4 x 32pdr carronades; RH 8 x 18pdr carronades.

Handed over by Royalists at Toulon 29.8.1793.

Under Lieut John Davie for passage to England. Arrived at Portsmouth 3.5.1794. Registered by AO 29.10.1794.

Commissioned 5.1795 under Capt. Charles Edmund Nugent, for the Channel. Fitted at Portsmouth (for £17,774) 2–6.1795. In 8.1795 under Capt. James Vashon. Fitted at Portsmouth (for £17,283) 12.1796–2.1797. In Mutiny at Spithead 1797. Fitted at Plymouth (for £41,229) 5–12.1805. Fitted at Chatham (for £9,177) 1–5.1808. Fitted at Plymouth as a prison ship (for

£24,599) 9.1810–1.1811. Defects made good at Portsmouth (for £15,672) 1812. Defects made good at Portsmouth (for £16,672) 9.1814–4.1815. BU at Woolwich 1.1817.

Tigre (French *Le Tigre*, built 10.1790–8.1793 at Brest. L: 8.5.1793).

Dimensions & tons: 182ft 0in, 149ft 0in x 48ft 9½in x 21ft 7⅜in. 1,886⁶⁷/₉₄bm.

Men: 640. Guns: LD 28 x 32pdrs + 2 x 68pdr carronades; UD 28 x 18pdrs + 2 x 68pdr carronades; QD 4 x 18pdrs + 10 x 32pdr carronades; Fc 2 x 18pdrs + 2 x 32pdr carronades; RH 6 x 18pdr carronades.

Taken 23.6.1795 off Île Groix by Lord Bridport. Registered by AO 2.10.1795. Fitted (for £2,340.10.11d)

Commissioned 7.1798 under Capt. Sir William Sidney Smith (–1801); sailed for the Mediterranean 29.10.1798; broad pennant at Alexandria 3.1799; at defence of Acre 15.3–20.5.1799; her boats took siege train off Mt Carmel 18.3.1799. Under (acting) Cmdr Edward Canes in early 1800. Egypt operations 1801. Under Capt. Richard Curry in 3.1802, later Capt. Robert Jackson. Large Repair and fitted at Plymouth (for £50,783) 10.1803–7.1804. Recommissioned 5.1804 under Capt. Benjamin Hallowell (–1811); under (temp.) Capt. Alexander Kerr 8.1808; with Martin's squadron in attack on Baudin's convoy 23.10.1809 (80-gun *La Robuste* and 74-gun *Le Lion*, run ashore and burnt near Frontignan 25.10.1809); at Rosas Bay 31.10.1809. Under Capt. John Halliday 8.1811 (–1815), in the Channel; sailed with convoy to St Helena 27.5.1813; paid off 8/9.1815. Powder hulk at Plymouth to 1817. BU at Plymouth 6.1817.

Spartiate (French *La Spartiate*, built 11.1794–3.1798 at Toulon. L: 24.11.1797)

Dimensions & tons: 182ft 7in, 150ft 4in x 49ft 4½in x 21ft 7in. 1,949⁴¹/₉₄bm.

Men: 640. Guns: LD 28 x 32pdrs; UD 30 x 18pdrs; QD 2 x 18pdrs + 14 x 32pdr carronades; Fc 2 x 18pdrs + 6 x 32pdr carronades.

Taken 1.8.1798 by Nelson at Aboukir Bay (Battle of the Nile). Registered by AO 9.12.1798.

Under Capt. Charles Pierrepont for passage to UK. Arrived Plymouth 17.7.1799 and laid up. Fitted there 7 .1801–4.1803.

Commissioned 3.1803 under Capt. George Murray; then under Capts John Manley (5.1803), Edward Buller and Sir Francis Laforey (3.1804–1809). Defects made good at Plymouth (for £11,238) 12.1804–1.1805. Under Laforey, joined Nelson's fleet at Barbados 4.6.1805; in Weather column at Trafalgar 21.10.1805, had 3 killed, 20 wounded. Defects made good at Plymouth (for £17,201) 1–3.1806. Blockade of Rochefort 1807, then in Strachan's squadron for pursuit of Allemand's squadron 1–2.1808. In Mediterranean 1808–9; operations in Naples Bay 6.1809; paid off 12.1809. In Ordinary at Portsmouth to 1813, then at Woolwich. Between Middling and Large Repair at Woolwich (for £59,787) 4.1814–5.1815, then laid up at Sheerness. Between Small and Middling Repair at Chatham, and fitted for a flagship for foreign service (for £36,753) 12.1822–8.1823. Recommissioned 6.1823 under Capt. Gordon Falcon as flagship of Rear-Adm. Sir George Eyre on South American station (–1825). Defects made good at Portsmouth (for £17,138) 12.1825–12.1826. Recommissioned 12.1825 under Capt. Frederick Warren (–1830); to Lisbon 1827–8, then defects made good at Portsmouth (for £14,684) 5–10.1828; to Mediterranean 1929–30; paid off 4.1830. Fitted for sea at Portsmouth 10–11.1832; recommissioned 5.10.1832 under Capt. Robert Tait, for blockade of Dutch coast; then as flagship on South American station of Rear-Adm. Sir Michael Seymour

Typical of the numerous French two-deckers captured during the Great War was the *Hoche*, taken in 1798 off Tory Island, Donegal, during the abortive attempt to aid the Irish rebellion (among those aboard taken prisoner were the expedition's commander, Commodore Bompart, and the Irish rebel leader Wolf Tone); she was commissioned in the British Navy as HMS *Donegal* in 1802. Unlike her sisters, this ship saw considerable active service during the next decade, and her excellent state of repair caused her to be brought back into service until finally paid off in the summer of 1832. This watercolour by George Pechell Mends shows the ship at the end of her active career, with typical post-war modifications of boarded-in head and solid poop barricade with the curved hancing favoured at the time. *(NMM PY0760)*

(who died 9.7.1834); paid off 1835. Fitted at Plymouth as a temporary sheer hulk 8.1842. BU at Plymouth completed 30.5.1857.

Donegal (French *Le Hoche*, ex-*Le Pégase* 1797, ex-*Le Barras* 1795, built 11.1791–2.1795 at Toulon. L: 23.3.1794)

    Dimensions & tons: 182ft 0in, 150ft 5in x 48ft 9in x 21ft 10in. 1,901⁴³/₉₄bm.

    Men: 640. Guns: LD 30 x 32pdrs; UD 30 x 18pdrs; QD 10 x 12pdrs; Fc 6 x 12pdrs + 2 x 32pdr carronades; RH 6 x 32pdr carronades.

    Taken 12.10.1798 off Ireland by Warren's squadron. Arrived Plymouth 21.12.1798. Fitted there 4–8.1801, then at Portsmouth 4–11.1802.

    Commissioned 4.1802 under Capt. Sir Richard Strachan; sailed for the Mediterranean 26.1.1803. In Campbell's brush with the French off Cape Cepet 24.5.1804. Took 40-gun *L'Amfitrite* off Cadiz 25.11.1804. Under Capt. Pulteney Malcolm 3.1805 (–1811); chase of Willaumez's fleet in the Mediterranean and West Indies. Took the Spanish 100-gun *Rayo* 23.10.1805 (she was wrecked 26.10). Battle of San Domingo 6.2.1806. Refitted at Portsmouth (for £13,039) 5–7.1806. Sailed for Lisbon 1.5.1808 (with Sir Arthur Wellesley). Under (temp.) Capt. Peter Heywood (2.1809), at blockade of Rochefort; at destruction of 3 40-gun frigates (*L'Italienne*, *Le Calypso* and *La Cybèle*) at Sables d'Olonne 27.2.1809. With Gambier's fleet at the Basque Roads. Under (temp.) Capt. Edward Brenton 7–11.1809. Took 14-gun privateer *Le Surcouf* off Cape Barfleur 6.10.1810. With other ships, attempted destruction of 40-gun frigates *L'Amazone* and *L'Eliza* at Le Hogue 15.11.1810. Paid off 1811 into Ordinary at Portsmouth. Middling Repair at Chatham (for £50,686) 11.1813–7.1815, then laid up at Sheerness. Middling Repair at Chatham (for £24,322) 7.1823–11.1824, then laid up at Sheerness. Fitted at Sheerness as a guard ship (for £21,206) 9.1829–7.1830, for the

Nore; recommissioned 9.1829 under Capt. Sir Jaleel Brenton, then 22.7.1830 under Capt. John Dick; flagship of Vice-Adm. Sir Pulteney Malcolm 5.1832, at blockade of Dutch coast; paid off Summer 1832. Fitted for flag officer at Plymouth 1.1836–10.1837; recommissioned 2.9.1837 under Capt. John Drake for Lisbon station, as flagship of Rear-Adm. Sir John Acworth Ommanney; paid off in Summer 1840. BU at Portsmouth 5.1845.

SAN ANTONIO This 74 had been ceded to France by Spain in 1800.

San Antonio (French *Le Saint Antoine*, ex-Spanish *San Antonio*, built Cartagena 1785)

    Dimensions: 174ft 10in, 139ft 8in x 47ft 10in x .... 1,700bm.

    Men: 590. Guns: LD 28 x 32pdrs; UD 28 x 18pdrs; QD 4 x 9pdrs + 10 x 32pdr carronades; Fc 2 x 9pdrs + 2 x 32pdr carronades.

    Taken 12.7.1801 by Saumarez off Gibraltar.

    From 8.1801 under Capt. George Dundas for voyage to UK. Registered 16.9.1801. Arrived at Portsmouth 8.10.1801 and laid up.

    Commissioned 10.1807 under Lieut Richard Heacock (–1812) as prison ship; from 1812 under Lieut Henry Squire (–1814). Fitted as a powder magazine at Portsmouth 5–9.1814. Sold to Mr Freake (for £2,990) 11.7.1827, then resold (Mr Freake having been declared insane) to John Small Sedger (for 2,200) 26.3.1828.

Ex-DUTCH PRIZES (1797–99) The Amsterdam Admiralty built four 74s in 1782–3 (and a fifth in 1795) to a common design measuring 179 x 48⁹/₁₁ x 22 (Dutch) feet, of which two were taken by Duncan's fleet at Camperdown and another by Mitchell's squadron off Texel. Of these the *Vryheid* was sold in 1811.

Camperdown (Dutch *Jupiter*, launched 4.1782 at Amsterdam)

    Dimensions & tons: 167ft 4in, 136ft 0in x 46ft 5in x 18ft 2½in. 1,558⁵⁴/₉₄bm.

Men: 590 (73 as prison ship). Guns: LD 28 x 32pdrs; UD 28 x 18pdrs; QD 2 x 18pdrs + 10 x 32pdr carronades; Fc 2 x 18pdrs + 4 x 32pdr carronades.

Taken 11.10.1797 at Camperdown. Registered 3.1.1798. Arrived Chatham 10.1.1798 and laid up.

Commissioned 10.1798 under Lieut Francis M'Gie as a prison ship at Chatham 1798; later under Lieut William Chilcott then Lieut Richard Hancock (1800). Fitted at Chatham as a powder magazine 4.1802. Sold to William Pannett for £1,710 to BU 10.9.1817.

Princess of Orange (Dutch *Washington*, launched 9.3.1796 at Amsterdam)

Dimensions & tons: 168ft 5in, 138ft 0¾in x 46ft 2in x 18ft 5½in. 1,565²⁵∕₉₄bm.

Men: 590. Guns: LD 28 x 32pdrs; UD 28 x 18pdrs; QD 2 x 18pdrs + 10 x 32pdr carronades; Fc 2 x 18pdrs + 4 x 32pdr carronades.

Taken in the Vlieter off Texel 30.8.1799.

Not commissioned in RN. Hulked (as 'stationary ship') 1806. Powder magazine at Chatham 1811. Sold to BU 18.4.1822.

No further 74s were begun in the final years of the century, as the prices demanded by contractors per ton were too high for the Navy Board to accept. However, a further twelve vessels were ordered from late 1799 onwards. These reflected a return to more moderate dimensions; all were 'Common Class' vessels, with lengths of 175ft and below.

FAME Class Common Class 74s. Sir John Henslow design, approved 13 December 1799.

Dimensions & tons: 175ft 0in, 144ft 0in x 47ft 8in x 20ft 6in. 1,740²⁷∕₉₄bm.

Men: 640. Guns: LD 28 x 32pdrs; UD 28 x 18pdrs; QD 2 x 18pdrs + 12 x 32pdr carronades; Fc 2 x 18pdrs + 2 x 32pdr carronades; RH 6 x 18pdr carronades.

Fame Deptford Dyd [M/Shipwright Edward Tippett to 3.1803, completed by Henry Peake]

As built: 175ft 5½in, 144ft 2in x 47ft 9in x 20ft 7½in. 1,745bm. Draught 12ft 3in/18ft 3in.

Ord: 15.10.1799. K: 22.1.1802. L: 8.10.1805. C: 8.10–22.11.1805 at Woolwich.

First cost: £63,059 (including fitting).

Commissioned 10.1805 under Capt. Graham Moore. In 5.1806 under Capt. Richard Bennet (–1809), by the Autumn in Vice-Adm. Sir John Borlase Warren's squadron in pursuit of Willaumez; sailed for the Mediterranean 28.6.1807, back in the Channel later in 1807. Seizure of Fort Trinidad, Rosas Bay 10.1808. In ?2.1810 under Capt. (acting) Phipps Hornby; by 11.1810 under Capt. Walter Bathurst (–1814), in the Mediterranean; under Capt. (acting) Abel Ferris 10–11.1811). In Ordinary at Chatham 1815. BU 9.1817 at Chatham.

Modified FAME Class Henslow modified design, approved 31 January 1800. The *Hero* of this class was a wartime loss in 1811.

Dimensions & tons: 175ft 0in, 144ft 1⅛in x 47ft 6in x 20ft 6in. 1,729⁷∕₉₄bm.

Men: 640. Guns: LD 28 x 32pdrs; UD 28 x 18pdrs; QD 2 x 18pdrs + 12 x 32pdr carronades; Fc 2 x 18pdrs + 2 x 32pdr carronades; RH 6 x 18pdr carronades.

Albion Perry, Wells & Green, Blackwall

As built: 175ft 0in, 144ft 0in x 47ft 8in x 20ft 6in. 1,740³∕₉₄bm. Draught 12ft 8in/18ft 5in.

Ord: 4.2.1800. K: 6.1800. L: 17.6.1802. C: 19.6–27.8.1802 at Woolwich, 17.3.1803 at Sheerness.

First cost: ?

Commissioned 2.1802 under Capt. John Ferrier (–1808). Flagship of Saumarez 1803, on Channel station; took (with *Minotaur* and *Thunderer*) French 40-gun *La Franchise*. Later to West Indies, where took 12-gun privateer *La Clarisse* 21.12.1803. Large Repair at Chatham (for £71,268) 12.1810–6.1813. Recommissioned 3.1813 under Capt. John Ferris Devonshire, for North American and West Indies station. In 3.1814 under Capt. Charles Ross, as flagship of Rear-Adm. George Cockburn, in North America. In 5.1815 under Capt. Philip Somerville, later Capt. James Walker. From 31.12.1815 under Capt. John Goode (–1819); in the Mediterranean 1.1816; at Bombardment of Algiers 27.8.1816, losing 3 killed and 15 wounded; in 1817 flagship of Rear-Adm. Sir Charles Penrose. Fitted as guard ship at Sheerness (for £11,635) 6.1816?. From 5.1819 under Capt. Richard Raggett, and from 6.1822 under Capt. Sir William Hoste (–1824). Fitted for sea at Portsmouth (for £6,190) 7–9.1825. Recommissioned 7.6.1825 under Capt. John Acworth Ommaney (–1828) and was at Battle of Navarino 20.10.1827, losing 10 killed and 50 wounded. In Ordinary at Portsmouth 1829, by mid-1830 used for receiving. Began fitting as receiving ship but completed as a lazarette at Portsmouth 3–7.1831; quarantine service at Leith 1832–5. BU at Deptford 6.1836.

Illustrious John Randall & John Brent, Rotherhithe

As built: 175ft 0in, 144ft 0in x 47ft 9in x 20ft 6in. 1,746⁴⁹∕₉₄bm. Draught 12ft 11in/18ft 3in.

Ord: 4.2.1800. K: 2.1801. L: 3.9.1803. C: 30.12.1803 at Woolwich.

First cost: £37,631 to builder, plus £3,246 fitting.

Commissioned 11.1803 under Capt. Sir Charles Hamilton, for the Channel. In 1805 under Capt. (acting) Michael Seymour; took (with *Ramillies*) 2-gun privateer *La Joséphine* 7.7.1805. Under Capt. William Shield 1805–7; sailed for the Mediterranean 1.1.1807. Under Capt. William Broughton 1807 (–1811); at Basque Road operation 1809, and Scheldt operations 3–4.1809. Large Repair at Portsmouth (for £68,275) 12.1813–4.1817, then laid up. Very Small Repair, and fitted for demonstration at Portsmouth (for £18,650) 11.1832–12.1837, then laid up. Fitted as flagship at Portsmouth (for £11,347) 7–10.1841. Laid up 5.1845. Fitted as an Ordinary guard ship at Portsmouth 3.1848. Partly fitted as a blockship, then used as a hospital ship 3.1853. Gunnery training ship 4.1854. Reverted to Ordinary guard ship 4.1859. Fitted for testing of Armstrong 200pdr gun 10.1863. BU completed 4.12.1868 at Portsmouth.

REPULSE Class Sir William Rule design, approved 31 January 1800.

Dimensions & tons: 174ft 0in, 143ft 2in x 47ft 4in x 20ft 0in. 1,706⁶∕₉₄bm.

Men: 590. Guns: LD 28 x 32pdrs; UD 28 x 18pdrs; QD 2 x 18pdrs + 12 x 32pdr carronades; Fc 2 x 18pdrs + 2 x 32pdr carronades; RH 6 x 18pdr carronades.

Eagle Thomas Pitcher, Northfleet

As built: 174ft 0in, 143ft 1in x 47ft 7in x 20ft 0in. 1,723²¹∕₉₄bm. Draught 13ft 3in/17ft 6in.

Ord: 4.2.1800. K: 8.1800. L: 27.2.1804. C: 3.5.1804 at Woolwich.

First cost: £37,760 to build, plus £8,967 dyd costs, plus £9,900 fitting.

Commissioned 2.1804 under Capt. David Colby (–1805), for the Leeward Islands; took 14-gun privateer *L'Empereur* 2.4.1805.

Recommissioned 11.1805 under Capt. Charles Cowley (–1813), joined Sir William Sidney Smith's squadron off Naples and Capri 5.1806. Refitted at Portsmouth (for £11,872) 2–4.1809. Sailed for the Mediterranean 20.2.1810; captured 40-gun (en flûte) La Corcyre in the Adriatic 27.11.1811; boat attack at Goro 17.9.1812 (2 gunboats taken, others burnt) ; boat attack at Goro again 29.4.1813 (5 vessels takem, another burnt); landing party destroyed battery at Farasina 11.6.1813; with Fremantle's squadron at Fiume 3.7.1813; (with *Bacchante*) captured convoy at Rovingo 2.8.1813; operations at Trieste 5–29.10.1813. Large Repair at Chatham (for £63,458) 6.1814–9.1816, then in Ordinary there. Small Repair at Chatham 10–12.1823 and cut down to fourth Rate frigate of 50 guns (by AO 26.4.1830) at Chatham 2.1830–3.1831; see Chapter 4 for later history.

Repulse (Mrs) Frances Barnard & Co., Deptford Green

As built: 174ft 0in, 142ft 11in x 47ft 8in x 20ft 0in. 1,727$\frac{23}{94}$bm. Draught 13ft 3in/17ft 6in.

Ord: 4.2.1800. K: 9.1800. L: 21.7.1803. C: 1.8–5.10.1803 at Woolwich.

First cost: ?

Commissioned 7.1803 under Capt. Arthur Legge (–1807); from 11.1803 flagship of Rear-Adm. Thomas Russell at Yarmouth; at blockade of Ferrol, action of 22.7.1805. With Warren's squadron 1806, in search for Leissègues and Willaumez; with Duckworth's squadron in the Dardanelles 1807. At Walcheron operations. In ?11.1809 under Capt. John Halliday; sailed for the Mediterranean 20.11.1809; at blockade of Toulon; rescue of *Philomel* 31.8.1810. In ?5.1811 under Capt. Richard Hussey Moubray, in the Mediterranean; in boat attack on Morgion 2.5.1813. Paid off 6.1814. Fitted for Ordinary at Plymouth 5–6.1814. BU at Plymouth 9.1820.

Sceptre John Dudman, Deptford

As built: 174ft 1¼in, 143ft 3½in x 47ft 7¼in x 20ft 0½in. 1,727$\frac{20}{94}$bm. Draught 12ft 7in/17ft 10in.

Ord: 4.2.1800. K: 12.1800. L: 11.12.1802. C: 8.4.1803 at Woolwich.

First cost: ?

Commissioned 2.1803 under Capt. Archibald Collingwood Dickson; sailed for the East Indies 9.1803; took (with *Albion*) 12-gun privateer *La Clarisse* 21.12.1803. Under Capt. Joseph Bingham 1804–9; returned to UK 1808 and paid off. Small Repair at Chatham 8.1808–6.1809. Recommissioned 3.1809, still under Bingham, for Scheldt operations; sailed for the Leeward Islands 8.11.1809. Under Capt. (temp.) Edward Dix 1–2.1810; capture of Guadeloupe 2.1810. Under Capt. Samuel Ballard 1810–11. In ?9.1811 under Capt. Sir Edward Berry, in Channel and North Sea. In 1812 under Capt. Thomas Harvey, then 1.1813 Capt. Robert Honeyman; sailed for North America 23.3.1813; later under Capt. Charles Ross, as flagship of Rear-Adm. Sir George Cockburn; took US 18-gun privateer *Anaconda* 11.7.1814. In 1814 under Capts John Devonshire, Alexander Skene (temp.) and William Waller. Laid up at Chatham 8.1814. BU at Chatham 2.1821.

Modified *CARNATIC* Class Common Class 74s. Based on *Courageux*, a French prize taken 1761, lengthened.

Dimensions & tons: 172ft 3½in, 140ft 5¼in x 47ft 9in x 20ft 9in. 1,703$\frac{21}{94}$bm.

Men: 640. Guns: LD 28 x 32pdrs; UD 28 x 18pdrs; QD 2 x 9pdrs + 12 x 32pdr carronades; Fc 2 x 9pdrs + 2 x 32pdr carronades; RH 6 x 18pdr carronades.

Aboukir Josiah & Thomas Brindley, Frindsbury

As built: 172ft 5⅝in, 140ft 2¼in x 47ft 9½in x 20ft 10in. 1,703$\frac{15}{94}$bm. Draught 13ft 11in/18ft 0in.

Ord: 16.8.1800 or 24.11.1802. K: 6.1804. L: 18.11.1807. C: 7.7.1808 at Chatham.

First cost: £41,430 to builder plus £9,022 dockyard expenses, plus fitting £22,417.

Commissioned 5.1808 under Capt. George Parker, for the Channel (earlier under Capt. Charles Cockburn, then 4.1808 under Capt. Percy Fraser. In Scheldt operations 1810, then off Texel 1810–11. In Summer 1812 under Capt. Thomas Browne, as flagship of Rear-Adm. Sir Thomas Byam Martin in the Baltic. In 1813 under Parker again; sailed for the Mediterranean 19.6.1813. In 9.1813 under Capt. Norburn Thompson in the Mediterranean; at Genoa 1814. In Ordinary at Chatham 1817–1838. Fitted as a receiving ship at Chatham (for £3,834) 7.1823–6.1824. Fitted at Chatham as a hospital ship 6.1831. Sold to J. Lachlan (for £4,250) to BU 16.8.1838.

Bombay Deptford Dyd [M/Shipwright Henry Peake to 6.1806, then Robert Nelson]

As built: 172ft 3½in, 140ft 0¼in x 47ft 9½in x 20ft 9in. 1,701$\frac{12}{94}$bm. Draught 14ft 2in/18ft 6in.

Ord: 9.7.1801 (re-affirmed 23.7.1805). K: 10.1805. L: 28.3.1808. C: 11.5.1808 at Woolwich.

First cost: £67,038 to build, plus £2,954 fitting.

Commissioned 6.1808 under Capt. William Cuming (–1811); sailed for the Mediterranean 7.2.1809. Under Capt. Norburn Thompson 1812 in the Mediterranean, subsequently Capt. George Parker. Defects made good at Portsmouth (for £23,281) 3–7.1814. In 5.1814 under Capt. Henry Bazely; flagship of Rear-Adm. Sir John Beresford in 7.1814, then of Rear-Adm. Sir Charles Penrose in 1816; paid off 7.1816. Renamed *Blake* by AO 28.4.1819. Fitted as a receiving ship at Portsmouth 12.1823. BU completed at Portsmouth 22.12.1855.

***SWIFTSURE* Class** Sir John Henslow design, approved on 11 September 1800.

Dimensions & tons: 173ft 0in, 142ft 0in x 47ft 6in x 20ft 9in. 1,704$\frac{17}{94}$bm.

Men: 590. Guns: LD 28 x 32pdrs; UD 28 x 18pdrs; QD 4 x 18pdrs + 10 x 32pdr carronades; Fc 2 x 18pdrs + 2 x 32pdr carronades; RH 6 x 18pdr carronades.

Swiftsure Balthazar & Edward Adams, Bucklers Hard

As built: 173ft 0in, 141ft 9⅝in x 47ft 10in x 20ft 9in. 1,725$\frac{73}{94}$bm. Draught 13ft 0in/18ft 6in.

Ord: 16.8.1800. K: 8.1802. L: 23.7.1804. C: 31.7–24.9.1804 at Portsmouth.

First cost: ?

Commissioned 8.1805 under Capt. Mark Robinson, for Channel; off Cadiz later in 1805; then under Capt. William Gordon Rutherford, for chase to West Indies. In Lee column at Trafalgar, had 9 killed, 8 wounded. Paid off 1807; recommissioned 1807 under Capt. John Conn (drowned 5.1810), as flagship of Vice-Adm. Sir John Borlase Warren; sailed for North America 8.12.1807. Under Capt. Charles Austen ?5.1810. Small Repair at Chatham (for £14,076) 3–9.1811; recommissioned 7.1811 under Capt. Robert Lloyd, as flagship of Vice-Adm. Herbert Sawyer, on North American station; under (temp.) Capt. Lord John Colvill 8.1811; sailed for the Mediterranean 17.11.1811. In 1812 under Capt. Temple Hardy, and 8.1812 Capt. Edward Dickson (–1814). In 1814 flagship of Vice-Adm. John Laugharne, in the Mediterranean. In 9.1814 under Capt. William Webly,

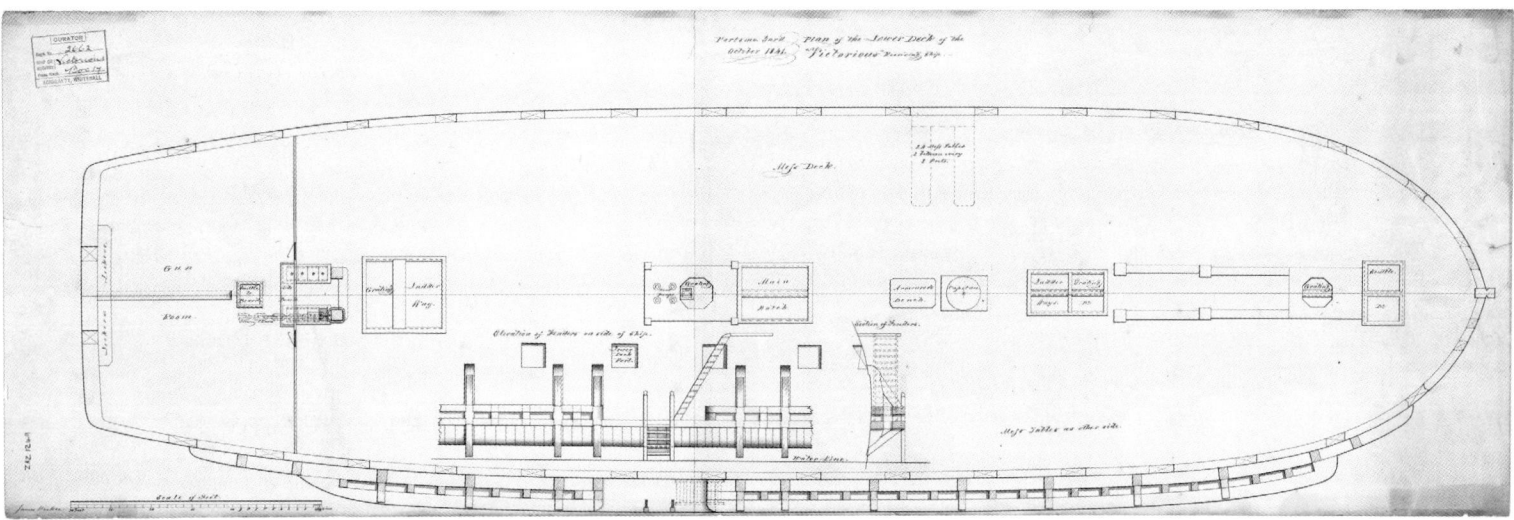

A plan view of the lower deck of the 74-gun *Victorious* as a receiving ship at Portsmouth Dockyard, dated 1841, although the ship was first so fitted in 1826. The detail shows the fenders attached to her starboard side. This kind of ancillary role was the typical fate of many wartime Third Rates after 1817. The *Victorious* and her sister *Swiftsure* were the last ships of the line to be designed by Sir John Henslow, and the last to have significant tumblehome (where the sides sloped inwards towards the upper deck). (*NMM J3055*)

in the West Indies. Paid off 8.1815. In Ordinary at Portsmouth 1816–45. Fitted as a receiving ship at Portsmouth 3–5.1819. BU by Barnard for £1,055 (at Portsmouth?) 18.10.1845.

Victorious Balthazar & Edward Adams, Bucklers Hard
As built: 173ft 2in, 142ft 1⅞in x 47ft 9in x 20ft 9in. 1,724⁶⁶/₉₄bm. Draught 12ft 8in /18ft 0in.
Ord: 7.11.1802. K: 2.1805. L: 20.10.1808. C: 7.11.1808–2.3.1809 at Portsmouth.
First cost: £41,796 (including fitting).
Commissioned 12.1808 under Capt. George Hamond; in Walcheron operations 1809. Under Capt. John Talbot from 10.1809 (–1814); sailed for the Mediterranean 20.11.1809; took (with *Weazel*) 74-gun *Le Rivoli* in the Adriatic 22.2.1812 (losing 27 men killed and 99 wounded including Talbot); temporarily under Lieut. Thomas Ladd Peake who brought the ship home; sailed for North America 20.11.1812 with Talbot resuming command; paid off 8.1814. Fitted for Ordinary at Portsmouth 3–4.1815. Fitted as a receiving ship at Portsmouth 4–5.1826. BU completed at Portsmouth 21.12.1861.

Later *GANGES* Class Two further 'Common Class' vessels were ordered to this Hunt design in 1801. Although the second of these (*Minden*) was authorised on 9 July 1801, investigations on building her at Bombay continued until 1803, when the order was confirmed.
Dimensions & tons: 169ft 6in, 138ft 11¼in x 47ft 4in (48ft 1in oa) x 20ft 3in. 1,656⁶⁴/₉₄bm.
Men: 590. Guns: LD 28 x 32pdrs; UD 28 x 18pdrs (*Invincible* had short 24pdrs); QD 4 x 12pdrs + 10 x 32pdr carronades; Fc 2 x 12pdrs + 2 x 32pdr carronades. (2 guns on LD and 2 guns on UD later replaced by 68pdr carronades.)

Invincible Woolwich Dyd [M/Shipwright Edward Sison]
Dimensions & tons: 169ft 9in, 138ft 11⅛in x 47ft 7in x 20ft 3in. 1,673⁶²/₉₄bm. Draught 13ft 2in/17ft 11in.
Ord: 25.6.1801. K: 1.1.1806. L: 15.3.1808. C: 11.4.1808.
First cost: £65,044 (including fitting).
Commissioned 3.1808 under Capt. Ross Donnelly, for North Sea; sailed for the Mediterranean 7.2.1809. In 1810 under Capt.

Charles Adam (–1813), in the Mediterranean. In Ordinary at Plymouth 1814–15. Between Small and Middling Repair at Plymouth, then housed over from main mast forward (for £48,477) 6.1814–1.1815. Small Repair at Plymouth (for £13,678) 7.1824–3.1825. Became a receiving ship and coal depot at Plymouth 1855. BU completed at Plymouth 28.1.1861.

Minden East India Co., Bombay. [M/Shipwright Jamsetjee Bomanjee] Teak-built.
Dimensions & tons: 171ft 4½in, 140ft 11½in x 47ft 11in x 20ft 3in. 1,721⁴⁷/₉₄bm.
Ord: 5.6.1803. K: 1807. L: 19.6.1810. C: 15.7.1812 at Portsmouth.
First cost: £48,260 to builder (to East India Company's charge).
Under Capt. Edward Hoare in 10.1810, for passage from India to UK.
Commissioned 1811 under Capt. William Webly, as flagship of Rear-Adm. Sir Samuel Hood (–1814); sailed for the East Indies 12.1.1811. In 1812 under Capt. Alexander Skene, in East Indies. Defects made good at Portsmouth (for £16,754) 18.5–15.7.1812. In 1813 under Capt. Joseph Prior, then 4.1814 Capt. George Henderson, still Hood's flagship. In 1.1815 under Capt. Donald Mackay; paid off 1816. Defects made good at Portsmouth (for £26,929) 2–7.1816. In 3.1816 under Capt. William Paterson (–1820), for East Indies; at Bombardment of Algiers 27.8.1816, losing 7 killed and 37 wounded; flagship of Sir Richard King from 1817; paid off 10.1820 to Ordinary (–1836). Middling Repair at Plymouth and housed over from main mast forward (for £42,284) 4.1821–11.1823. Fitted for commission (but not for sea) at Plymouth (for £5,623) 12.1833–5.1834. Fitted for sea at Plymouth (for £10,683) 3–6.1836; recommissioned for Particular service (Lisbon) 1836–8, then in Mediterranean to 1840; in 1841 to East Indies as a hospital ship. Became a seamens' hospital at Hong Kong 1842. Sold at Hong Kong 4.7.1861.

Revived *HERO* Class Five further orders to this 1800 design (see above for details) were placed following the resumption of hostilities in 1803.

Marlborough Mrs Frances Barnard, Deptford.
As built: 175ft 6in, 144ft 4⅛in x 47ft 9½in x 20ft 6in. 1,754¹⁵/₉₄bm. Draught 13ft 11in/18ft 9in.
Ord: 24.1.1805. K: 8.1805. L: 22.6.1807. C: 24.6–31.8.1807 at Woolwich.
First cost: £61,257 to builder, plus £22,391 dyd expenditure, plus

£4,223 fitting.
Commissioned 7.1807 under Capt. Graham Moore (–1810); with Sidney Smith's squadron at Lisbon 11.1807; escort of Portuguese Royal Family to Brazil; under Capt. John Phillimore 6–10.1809, for Walcheron operations; later to North Sea. In 11.1810 under Capt. George Scott, in Channel and then North Sea; later under Cmdr Francis Beaumont (temp.). In 1812 under Capt. Robert Honeyman, as flagship of Sir George Cockburn; sailed for North America 23.9.1812. From 1813 under Capt. Charles Ross, still Cockburn's flag; capture of Washington 24–25.8.1813. Large Repair at Portsmouth (for £57,119) 2.1814–2.1816, then laid up in Ordinary there. BU at Portsmouth 7.1835.

York Samuel & Daniel Brent, Rotherhithe
As built: 175ft 0in, 144ft 0⅛in x 47ft 8½in x 20ft 6in. $1,743\frac{48}{94}$bm. Draught 13ft 4in/18ft 6in.
Ord: 24.1.1805. K: 8.1805. L: 7.7.1807. C: 13.7–16.9.1807 at Woolwich.
First cost: £62,446 to build.
Commissioned 8.1807 under Capt. Robert Barton (–1812); sailed for the Leeward Islands 30.11.1807; at occupation of Madeira 26.12.1807; at capture of Martinique 2.1809; sailed for the Mediterranean 14.11.1809. In 8.1812 under Capt. Alexander Schomberg, for North Sea then the Channel; to North America with troops 1814; paid off 8.1815 into Ordinary at Plymouth. Fitted as a convict ship at Portsmouth 11.1819; served as convict hulk 1824–50. BU at Portsmouth 3.1854.

Sultan John Dudman & Co., Deptford
As built: 175ft 0in, 143ft 11¼in x 47ft 10in x 20ft 6in. $1,751\frac{72}{94}$bm. Draught 13ft 2in/18ft 0in.
Ord: 24.1.1805. K: 12.1805. L: 19.9.1807. C: 22.9–29.11.1807 at Woolwich.
First cost: £61,300 to builder, plus £22,275 Dyd expenditure, + £4,599 fitting.
Commissioned 10.1807 under Capt. Edward Griffith; sailed for the Mediterranean 18.2.1808; off Cadiz 1808; in attack on Baudin's convoy 10.1809; at destruction of 80-gun *La Robuste* and 74-gun *Le Lion* at Frontignan 25.10.1809. In 4.1810 under Capt. John West (–1814); brush with Toulon ships 19.7.1811; her boats in operation at Bastia 4.12.1811; in the Channel 1813; Bordeaux operations 1814. Large Repair at Portsmouth (for £66,067) 3.1816–9.1818. Very Small Repair and fitted for Ordinary at Portsmouth 11.1829–11.1833. Fitted as a receiving ship at Portsmouth 5–12.1861. Fitted as Target ship at Portsmouth 2–5.1862, then for trials of armour plates 8.1862. BU completed there 28.1.1864.

Hannibal Henry Adams, Bucklers Hard
As built: 176ft 0in, 145ft 2¼in x 47ft 7in x 20ft 6in. $1,748\frac{53}{94}$bm.
Ord: 24.1.1805. K: 12.1805. L: 5.1810. C: 11.5–7.9.1810 at Portsmouth.
First cost: ?
Commissioned 7.1810 under Capt. Thomas Searle, as flagship of Rear-Adm. Sir Thomas Williams, at the Scheldt; sailed for Portugal 6.10.1810. In 1811 under Capt. Andrew King, as flagship of Rear-Adm. Philip Durham, in the Baltic. In 12.1811 under Capt. Thomas Browne, then 1812 under Capt. Sir Michael Seymour (–1814), in the Channel; took 40-gun *La Sultane* off Cherbourg 26.3.1814; later sailed with convoy to West Indies; to Ordinary at Plymouth 8.1814. Fitted as a lazarette at Plymouth 7–8.1825; 1826 to Pembroke. BU at Pembroke Dock 12.1833–1.1834.

Royal Oak John Dudman & Co., Deptford
As built: 175ft 2in, 144ft 0½in x 47ft 11in x 20ft 6in. $1,759\frac{14}{94}$bm. Draught 12ft 11in/17ft 11in.
Ord: 24.1.1805. K: 6.1806. L: 4.3.1809. C: 4–16.3.1809 at Deptford, then 16.3–3.5.1809 at Woolwich.
First cost: £62,273 to build.
Commissioned 4.1809 under Capt. Lord Amelius Beauclerk, for Walcheron operations. In 8.1811 under Capt. Pulteney Malcolm. In 1.1812 under Capt. Thomas Shortland, as flagship of the new Rear-Adm. Beauclerk, on North American station. In 1813 under Capt. Edward Dix, still Beauclerk's flagship; in Chesapeake operations, then New Orleans operations. In 10.1814 under Capt. Joseph Pearce, then 1815 under Capt. Clotworthy Upton; to Ordinary at Portsmouth at end of 1815. Fitted at Portsmouth as a receiving ship 9–12.1825 for convicts at Bermuda; hulked there 1834. BU at Bermuda by AO 17.10.1850.

Revived *REPULSE* Class Five further orders were placed in 1805 to this 1800 design (see above for details); three more were built from 1812 (see below).

Magnificent Perry, Wells & Green, Blackwall
As built: 174ft 2in, 143ft 0¾in x 47ft 8½in x 20ft 0in. $1,732\frac{3}{94}$bm. Draught 13ft 3in/18ft 3in.
Ord: 24.1.1805. K: 4.1805. L: 30.8.1806. C: 9.1806–24.10.1806 at Woolwich.
First cost: : £26,172 for fitting (building cost unknown).
Commissioned 9.1806 under Capt. George Eyre (–1811); sailed for the Mediterranean (via Cadiz) 3.6.1807; in Adriatic operations 1809–10; at capture of Santa Maura 4.1810. Defects made good at Plymouth (for £24,835) 2–5.1812; recommissioned 2.1812 under Capt. Willoughby Lake, for operations on the coast of Spain; under Capt. John Hayes (temp.) for attack on Santander 8.1812; then in the Channel; paid off 1814. Fitted as a hospital ship at Portsmouth 8–12.1825, for Jamaica. Commissioned 4.1828 as a receiving ship at Rio (–1842). Sold in Jamaica to BU 10.1.1843.

Valiant Perry, Wells & Green, Blackwall
As built: 174ft 0in, 142ft 9⅝in x 47ft 6¾in x 20ft 0in. $1,718\frac{31}{94}$bm. Draught 13ft 3in/17ft 4in.
Ord: 24.1.1805. K: 4.1805. L: 24.1.1807. C: 5.2–6.3.1807 at Woolwich.
First cost: ?
Commissioned 3.1807 under Capt. Kenneth McKenzie; in 4.1807 under Capt. James Young; in Copenhagen expedition 8.1807. In 1.1808 under Capt. George Reynolds, then Capt. John Poo Beresford and in 6.1808 Capt. Thomas Briggs., 12.1808 Capt. John Hayes and 2.1809 Capt. Alexander Kerr, off Lorient. In 3.1809 under Capt. John Bligh, then in Basque Roads operations 4.1809; later under Capt. Thomas Shortland; took 14-gun *La Confiance* off Belle Île 3.2.1810 (ex-frigate *La Cannonière*, with £150,000 cargo). In 1810 under Capt. Robert Oliver (–1814); boats (with others of Neale's squadron) took two brigs and destroyed another in Basque Roads 28.9.1810; sailed for North America 14.1.1813; home in 1814. In 7.1814 under Capt. Zachary Mudge; paid off 8/9.1815 at Portsmouth. BU at Portsmouth completed 28.11.1823.

Elizabeth Perry, Wells & Green, Blackwall
As built: 174ft 0in, 143ft 1⅞in x 47ft 7in x 20ft 0in. $1,724\frac{9}{94}$bm. Draught 13ft 3in/17ft 10in.
Ord: 24.1.1805. K: 8.1805. L: 23.5.1807. C: 27.5–3.7.1807 at Woolwich.
First cost: ?
Commissioned 6.1807 under Capt. Henry Curzon (–1810); sailed

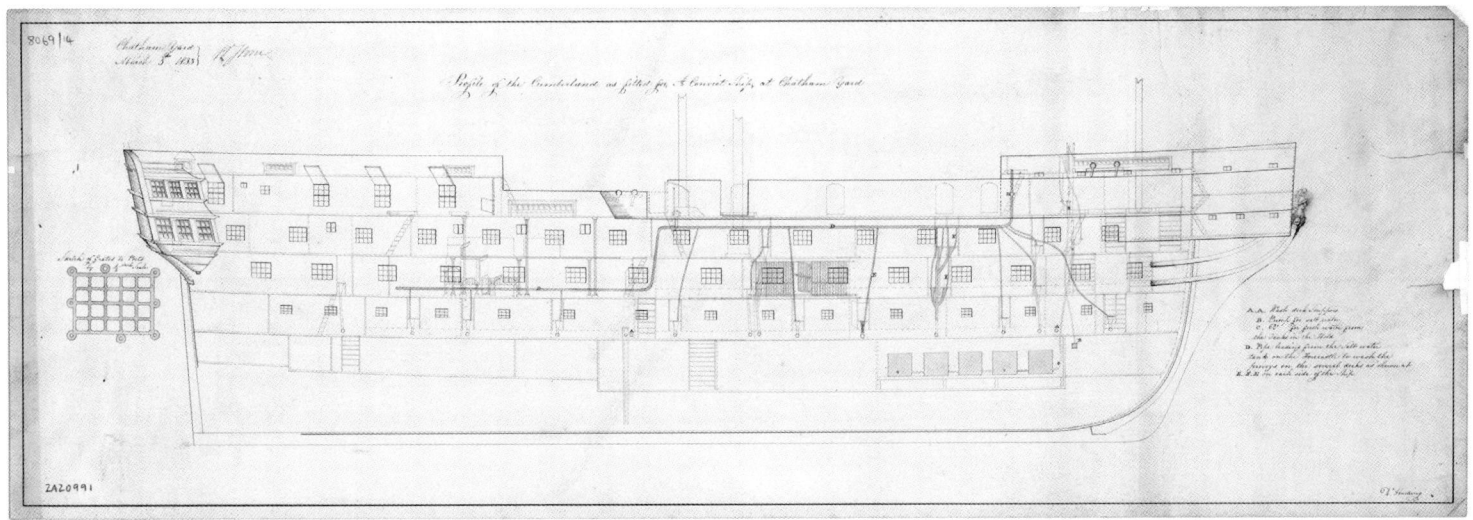

A powerful image of wartime British ports were the rows of prison hulks housing captured enemy sailors, but while many of these were no longer needed after 1815, others were converted to a similar function for civilian convicts. This inboard profile depicts the 74-gun *Cumberland*, fitted at Chatham in 1829-30 as a convict ship. The draught, signed by William Stone (master shipwright at that dockyard in 1830–39) illustrates the pipework and scuppers provided for washing the deck and flushing out the heads (toilets). The *Cumberland* belonged to the *Repulse* Class designed by Sir William Rule. *(NMM J3089)*

for Portugal 11.1807; with Sidney Smith's squadron in the Mediterranean 1807; at Corunna 1.1809, then to Brazil. In 1810 under Capt. Thomas Capel (but under acting Capt. Thomas Searle in 8.1810), later under Capt. Lord William Stuart. In 1.1811 under Capt. Edward Leveson Gower (–1813); sailed for Portugal 26.3.1811; sailed for the Mediterranean 4.8.1812. In 1814 under Capt. Gardiner Guion; her boats took 6-gun *L'Aigle* and *Chic* near Corfu 27.5.1814; paid off ?6.1814 at Woolwich. BU 8.1820 at Chatham.

Cumberland Thomas Pitcher, Northfleet
   As built: 174ft 4in, 143ft 5in x 47ft 5½in x 20ft 0in. 1,718¹⁶⁄₉₄bm. Draught 13ft 8in/17ft 7in.
   Ord: 24.1.1805. K: 8.1805. L: 19.8.1807. C: 10.1807 at Woolwich.
   First cost: £63,231 to build, plus £26,568 fitting.
   Commissioned 10.1807 under Capt. Philip Wodehouse (–1811); sailed for the Mediterranean 30.1.1808; with Strachan's squadron in the Mediterranean 2.1809; with Martin's squadron in attack on Baudin's convoy 23.10.1809 (80-gun *La Robuste* and 74-gun *Le Lion*, run ashore and burnt near Frontignan 25.10.1809); boats in attack on shipping in Rosas Bay 31.10.1809. In 7.1811 under Capt. Robert Otway for return home. Made good defects at Woolwich (for 312,040) 10–12.1811. Recommissioned 22.11.1811 under Capt. Thomas Baker (–1815), for North Sea; to North America 11.1812; made good defects at Chatham (for £16,247) 2–4.1814; to East Indies 1814–15, then paid off 2.8.1815 at Chatham. Fitted as a convict ship at Chatham (for £13,202) 10.1829–3.1830. Renamed *Fortitude* 15.11.1833. Fitted as coal depot at Chatham (for £4,097) 9.1845–8.1848. Sold to Henry Castle & Son (for £2,020) to BU at Charlton 2.1870.

Venerable Thomas Pitcher, Northfleet
   As built: 174ft 1in, 143ft 5½in x 47ft 5in (48ft 2in oa) x 20ft 0½in. 1,715⁶¹⁄₉₄bm. Draught 12ft 9in/17ft 4in.
   Ord: 24.1.1805. K: 12.1805. L: 12.4.1808. C: 1808 at builders.
   First cost: £64,318 to build, plus £26,484 fitting (all to builder).
   Commissioned 5.1808 under Capt. Sir Home Popham (–1812), as

flagship of Rear-Adm. Sir Richard Strachan; in 6.1808 under Capt. Andrew King (temp.); in Walcheron operations 1809; in 1810 flagship of Rear-Adm. Sir Thomas Williams, in the North Sea; operations on the North coast of Spain 1810; at Lequeito 20.6.1810. In 2.1812 under Capt. James Dundas, then 9.1812 under Capt. David Milne. Defects made good at Portsmouth (for £16,529) 6–7.1813. Later under Capt. James A. Worth, as flagship of Rear-Adm. Sir Philip Durham, for the Leeward Islands; took 4-gun privateer *Le Jason* 31.12.1813; took 40-gun *L'Alcmène* off Madeira 16.1.1814, and 40-gun *L'Iphigénie* 20.1.1814. In 7.1814 under Capt. William M'Culloch, later under Capt. George Pringle, then Lieut Robert Wemyss (acting), then Capt. John Thompson (acting); paid off at Portsmouth 1815. Fitted as a church ship at Portsmouth 5–10.1825. BU at Plymouth 10.1838.

BLAKE **Class** Developed from the lines of the *Courageux* (taken from the French in 1761), but lengthened by 7ft 9in on the gun deck. Timber from Holstein was used in the construction of both ships, but the *Saint Domingo* of this class was sold in April 1816, and the *Blake* in October 1816.

ARMADA **Class** Following Trafalgar, Napoleon initiated a massive programme of battlefleet construction throughout the territories controlled by France; to counter this, the Admiralty recognised the need for a similarly huge quantitative addition to the battlefleet, and begun designing a new standard class. Both Peake and Rule produced designs, but the Navy Board then instructed the two to co-operate on a joint design to utilise the best characteristics of both schemes. Thus they were designed by Peake and Rule jointly (a further development from the *Courageux*) and were known as the 'Surveyors' Class' – or more colloquially as the 'Forty Thieves' due to their alleged excessive costs and the fraudulent practices of some of their builders. In fact, they were probably as successful a design as most, but they came into service after the period of fleet actions (i.e. post-Trafalgar) and thus never acquired many battle honours. Numerically they were the largest class of battleships ever built to one design for the Royal Navy – indeed, the number was only exceeded by the French *Le Téméraire* Class (to which Sané design some 90 vessels were built), and the somewhat smaller 66-gun ships of the Russian *Slava Rossii* Class (of which some 59 examples were built). The design was approved on 1 October 1806 (*Vindictive*, which was to be built to this design, having already been on order for 8½ months).

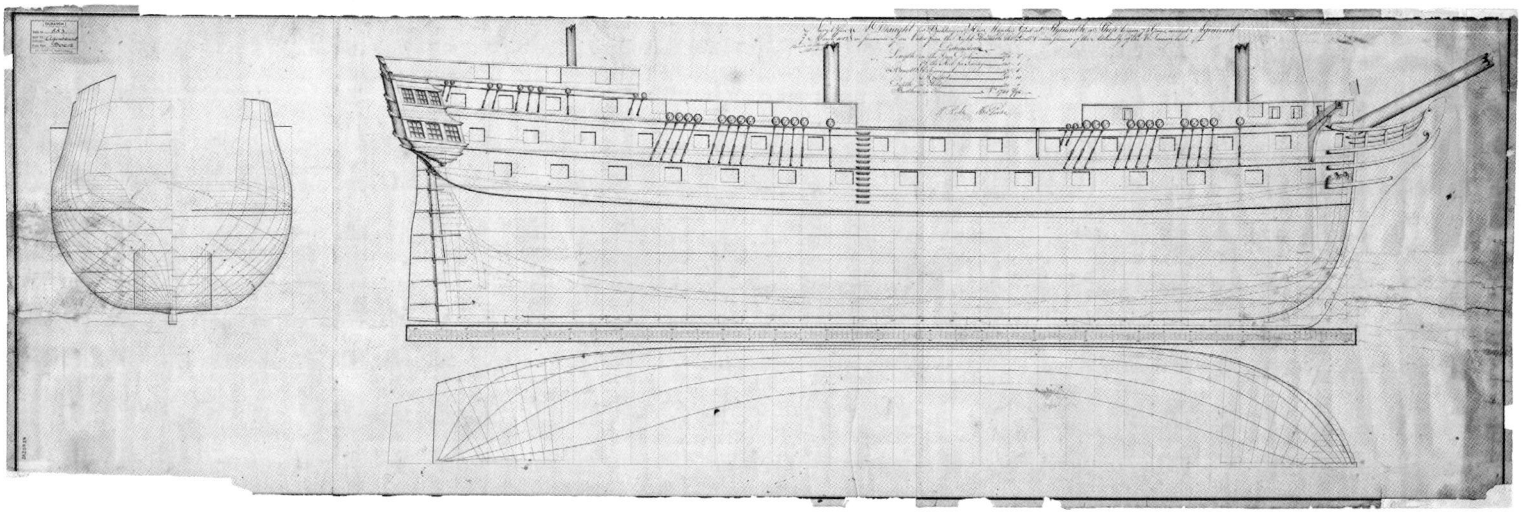

The draught for the *Agincourt*, one of the trio of *Armada* Class 74s to be ordered on 6 January 1812; the draught is dated 7 March 1812 and work on her commenced in May 1813, but was suspended as the Napoleonic War wound down. After the hull had stood unfinished for nearly a year on the stocks, her completion was ordered in March 1815 and she took to the water on Wednesday, 19 March 1817. She was housed over and stored in Ordinary for 25 years before commissioning in February 1842 as flagship for the Far East. In August 1845 and again in the summer of 1846, the *Agincourt* was engaged in a series of operations against Malay pirates in the Brunei River and at Malluda Bay, but this was her sole seagoing commission, and she returned to Devonport in September 1847 to serve as a depot ship and subsequently as a receiving ship. *(NMM J2559)*

Of the 40 ships ordered in 1806–12 which would be completed to this design, all 29 contract-built but only six of the eleven dockyard-built ships were launched by mid-1815, leaving the remaining five ships – *Pitt*, *Hercules*, *Agincourt*, *Hero* (renamed *Wellington*) and *Russell* – to be completed post-war. The 41st ship (*Boscawen*), ordered on 6 January 1812 but suspended on 2 December 1813, had been re-ordered to a new design on 14 February 1814, and was again to be re-ordered as an 80-gun ship in 1817 (see Chapter 2). A 42nd ship (*Carnatic*) ordered 30 September 1814 was to be re-ordered to a modified design in 1817 (see Section B). Two other vessels, probably but not certainly intended to be to this design were the *Akbar* (laid down 4 April 1807 at Prince of Wales Island, Penang; cancelled 12 October 1809 when the shipyard collapsed from lack of resources and manpower) and *Augusta* (laid down 1806 at Portsmouth Dyd; cancelled 1809); two more projected in 1807 but never ordered were the *Julius* (projected at Chatham Dyd; deleted 1812) and *Orford* (projected at Rio de Janeiro; deleted 1815). All survivors of this class were re-rated as 72-gun ships in 1839.

Dimensions & tons: 176ft 0in, 145ft 1in x 47ft 6in x 21ft 0in. 1,741 18/94bm.

Gloucester as rebuilt with circular stern: 177ft 1½in, 145ft 8⅛in x 48ft 1⅞in x 20ft 10in. 1,796 90/94bm.

Men: 590. Guns: LD 28 x 32pdrs; UD 28 x 18pdrs; QD 4 x 12pdrs + 10 x 32pdr carronades; Fc 2 x 12pdrs + 2 x 32pdr carronades. One pair LD guns replaced by 68pdr carronades, and one pair UD guns by 18pdr carronades after the end of the war. Most survivors reduced to 72 guns in 1839.

(As 50-gun ships; 450 men; LD 28 x 32pdrs (56cwt); QD 16 x 32pdrs (48cwt); Fc 6 x 32pdrs (48cwt).)

First (1806) Batch

Apart from the prototype, these were ordered by Thomas Grenville's Board of Admiralty.

Vindictive Portsmouth Dyd [M/Shipwright Nicholas Diddams]

As built: 176ft 2in, 145ft 2¼in x 47ft 8½in x 21ft 0in. 1,757 71/94bm. Draught 13ft 3in/17ft 8in.

Ord: 15.1.1806. K: 7.1808. L: 23.11.1813. C: 12.1813 for Ordinary.

First cost: £56,241.

Not commissioned until 9.1841. Fitted for the C-in-C of the Ordinary 12.1816. Underwent Middling Repair and cut down to a Fourth Rate frigate of 50 guns at Portsmouth 3.1828–1.1833; reduced to Fourth Rate 10.1832 (see Chapter 4 for later history).

Cressy Josiah & Thomas Brindley, Frindsbury

As built: 176ft 1in, 145ft 0⅛in x 47ft 9¾in x 21ft 1in. 1,763 63/94bm. Draught 13ft 0in/17ft 6in.

Ord: 1.10.1806. K: 3.1807. L: 7.3.1810. C: 7.3–9.5.1810 at Chatham for Ordinary, 2.4.1811 at Sheerness for Sea.

First cost: £56,855 to builder, plus fitting £9,893 at Chatham, £16,863 at Sheerness.

Commissioned 1.1811 under Capt. Charles Pater, for the Baltic; capture of Danish gunboat 5.7.1811; sailed for Leeward Islands 21.2.1813. Made good defects at Portsmouth (for £16,803) 9–11.1813. Recommissioned 8.1813 under Capt. Charles Dashwood for the North Sea, then to Leeward Islands. Paid off 5.1814 into Ordinary at Portsmouth. Ordered to be cut down to a frigate of 50 guns (by AO 20.3.1827), but conversion cancelled due to state of decay. BU at Portsmouth 12.1832.

Poictiers John King, Upnor

As built: 176ft 3in, 145ft 2⅜in x 47ft 9½in x 21ft 1in. 1,764 3/94bm. Draught 13ft 0in/17ft 5in.

Ord: 1.10.1806. K: 8.1807. L: 9.12.1809. C: 9.12.1809–4.5.1810 at Chatham.

First cost: £56,665 to builder, plus £22,458 fitting.

Commissioned 1.1810 under Capt. John Poo Beresford, for the Channel; sailed for Portugal 22.10.1810; chased Allemand's squadron off Ushant 10.1.1811. Under (Acting) Capt. Richard Jones 3.1811; Capt. Samuel Jackson (temp.) 1–7.1812, off Texel; sailed for North America 14.8.1812. Captured USS *Wasp* and retook *Frolic* 18.10.1812, then took 10-gun privateer *Herald* ?3.11.1812, 5-gun privateer *Highflyer* 9.1.1813, and (with *Maidstone* & *Nimrod*) 20-gun privateer *Yorktown* 17.7.1813. At Chatham 1814. Large Repair there (for £51,601) 4.1815–9.1817.

Fitted as guard ship at Sheerness 3.1836–9.1837, based at Chatham to 3.1848 when became Depot Ship until 1850. BU completed at Chatham 23.3.1857.

Hogue Deptford Dyd [M/Shipwright Robert Nelson]

As built: 176ft 0in, 145ft 0¼in x 47ft 7½in x 21ft 0½in. 1,749$\frac{57}{94}$bm. Draught 13ft 5in/17ft 11in.

Ord: 1.10.1806. K: 4.1808. L: 3.10.1811. C: 7.10.1811–25.1.1812 at Woolwich.

First cost: £76,715, plus fitting £1,429 at Woolwich.

Commissioned 14.12.1811 under Capt. Thomas Bladen Capel, for Texel operations; sailed for North America 14.1.1813 as flagship of Commodore William Hotham. Between Middling and Large Repair at Chatham (for £41,432 – possibly more?) 10.1814–6.1816, then to Sheerness for Ordinary. Middling Repair at Chatham (for £21,291) 11.1824–3.1826; at Chatham to 1834, then Sheerness. Small Repair at Chatham (for £2,813) 3.1839–1.1840. Converted to a screw blockship by Green, Blackwall (by AO 20.10.1845) (see Chapter 4 for later history).

Vigo Charles Ross, Rochester

As built: 176ft 9in, 145ft 6½in x 48ft 0½in x 20ft 11½in. 1,786$\frac{71}{94}$bm. Draught 12ft 7in/17ft 6in.

Ord: 20.10.1806. K: 4.1807. L: 21.2.1810. C: 20.4.1810 (for Ordinary) & 18.3.1811 (for Sea) at Chatham.

First cost: £58,330 to builder.

Commissioned 1.1811 under Capt. Richard Jones, then 6.1811 under Capt. Manley Hall Dixon, as flagship of Rear-Adm. Sir Manley Dixon, in the Baltic; on 29.1.1812 under Capt. Henry Manaton Ommanney, as flagship of Rear-Adm. James Nicoll Morris; on16.12.1812 under Capt. Thomas White, as flagship of Rear-Adm. Graham Moore; paid off 12.1813 at Portsmouth. Small Repair 3–9.1815 and laid up. Fitted for sea at Portsmouth 11.1819–3.1820; recommissioned 11.1819 under Capt. Thomas Browne, for Cape of Good Hope; flagship of Rear-Adm. Robert Lambert, at St Helena 1820–1, then of Adm. Sir James Whitshed, at Portsmouth 1822–3. Fitted as a receiving ship at Plymouth (for £1,886) 12.1826–7.1827. BU by Marshall, Plymouth 5–8.1865.

Armada Isaac Blackburn, Turnchapel, Plymouth.

As built: 176ft 0in, 145ft 0in x 47ft 7½in x 21ft 0in. 1,749$\frac{34}{94}$bm. Draught 14ft 1in/17ft 9in.

Ord: 20.10.1806. K: 2.1807. L: 22.3.1810. C: 23.3–27.9.1810 at Plymouth.

First cost: £56,890 to builder, plus £33,174 fitting.

Commissioned 8.1810 under Capt. Adam M'Kenzie, for the Texel. In 1.1812 under Capt. Charles Grant (–1814); sailed for the Mediterranean 10.3.1812; paid off into Ordinary at Plymouth 9–11.1814. Between Small and Middling Repair at Plymouth (for £51,282) 4.1815–1.1816. Powder hulk at Keyham Point 4.1844. AO to replace her by *Conquestador* 12.11.1862. Sold to Marshall, Plymouth 27.5.1863.

Vengeur Joseph Graham, Harwich

As built: 176ft 5in, 145ft 3in x 47ft 9½in x 21ft 0⅞in. 1,764$\frac{62}{94}$bm. Draught 13ft 3in/17ft 9in.

Ord: 20.10.1806. K: 7.1807. L: 19.6.1810. C: 17.7–30.10.1810 at Chatham.

First cost: ?

Commissioned 9.1810 under Capt. Thomas Brown; from 1811 flagship of Rear-Adm. Sir Joseph Yorke. From 1812 under Capt. Thomas Dundas, then Capt. James Brisbane ?8.1812; sailed for Jamaica 22.6.1813. Under Capt. Tristram Ricketts 10.1813; to North America with troops. Under Capt. Thomas Alexander 8.1815; became guard ship at Portsmouth 6.1816–5.1818, then fitted for sea there 10–12.1818; recommissioned 9.1818 under Capt. Frederick Lewis Maitland, for South America; paid off 5.1821 at Chatham. Fitted as receiving ship at Chatham 7.1823–2.1824, thence to Sheerness as receiving ship until 1838. BU at Sheerness 8.1843.

Conquestador Robert Guillaume, Northam

As built: 176ft 4in, 145ft 2⅜in x 47ft 11in x 21ft 1in. 1,773$\frac{25}{94}$bm. Draught 13ft 0in/17ft 6in.

Ord: 20.10.1806. K: 8.1807. L: 1.8.1810. C: 6.8 1810–21.3.1811 at Portsmouth.

First cost: £57,025 to builder, plus £13,291 fitting.

Commissioned 9.1810 under Capt. Lord William Stuart, for the Channel. Flagship of Adm. Lord Keith 5.1812. In West Indies 1814, then in Ordinary at Chatham 1815. Small Repair and housed over at Woolwich (for £20,783) 4.1815–4.1816; then laid up at Sheerness. Very Small Repair at Chatham (for £8,030) 9.1817–10.1818, then laid up at Sheerness. Small to Middling Repair and cut down to a Fourth Rate frigate of 50 guns (by AO 20.3.1827) 4.1827–4.1831 (see Chapter 4 for later history).

Redoubtable Woolwich Dyd [M/Shipwright Edward Sison]

As built: 176ft 5in, 145ft 6½in x 47ft 8in x 21ft 0¾in. 1,758$\frac{91}{94}$bm.

Ord: 29.12.1806. K: 4.1809. L: 26.1.1815 (completed by 9.1814, left on slip to season). To Sheerness 4.4.1815.

First cost: £76,554 (including fitting).

Not commissioned. Small Repair at Chatham (for £15,429) 7.1819–9.1820, thereafter in Ordinary at Sheerness. BU 5.1841 at Chatham.

Second (1807) Batch

These were ordered by Lord Mulgrave's Board of Admiralty.

Pitt Portsmouth Dyd [M/Shipwright Nicholas Diddams]; built to Seppings's principles.

As built: 176ft 0in, 145ft 0⅛in x 47ft 7¾in x 21ft 0in. 1,751$\frac{2}{94}$bm. Draught 13ft 0in/18ft 3in.

Ord: 17.4.1807 (but construction delayed 6 years). K: 5.1813. L: 13.4.1816 (left incomplete).

First cost: *Pitt* £78,787.

Not commissioned. Laid up after launch. Very Small Repair and fitted for Ordinary at Portsmouth (for £1,628) 7.1830–9.1832. Fitted as a coal depot and receiving ship at Portsmouth (for £4,996) 11.1852–5.1853. Refitted there as a coal depot (for £5,228) for Portland 1.1860, later back to Portsmouth. BU completed at Portsmouth 17.3.1877.

Mulgrave John King, Upnor

As built: 176ft 1in, 145ft 0⅛in x 47ft 9½in x 21ft 0in. 1,761$\frac{71}{94}$bm. Draught 12ft 10in/17ft 4in.

Ord: 23.6.1807. K: 2.1808. L: 1.1.1812. C: 22.11.1812 at Chatham.

First cost: £58,412 to builder, plus £29,971 fitting.

Commissioned 1.9.1812 under Capt. Thomas James Maling; flagship of Rear-Adm. Sir Richard King 1813; sailed for the Mediterranean 22.4.1813; paid off 1814. Small Repair at Plymouth (for £16,484) 1–3.1815. Between Middling and Large Repair at Plymouth (for £58,955) 3.1816–10.1819, then to Ordinary. Fitted as a lazarette at Pembroke (for £146) 8–9.1836. Fitted as a Powder ship at Pembroke (for £1,637) 8–9.1844. BU completed there 16.12.1854.

Ajax Perry, Wells & Green, Blackwall

As built: 176ft 0in, 144ft 11⅜in x 47ft 9½in x 21ft 0in. 1,761 (exact) bm. Draught 13ft 0in/18ft 3in.

Ord: 1.7.1807. K: 8.1807. L: 2.5.1809. C: 15.6.1809 at Woolwich.

First cost: £57,383 to builder plus £8,544 dockyard costs, plus
£27,400 fitting.

Commissioned 6.1809 under Capt. Robert Otway (–1811); sailed
for the Mediterranean 4.10.1809; in Blackwood's squadron off
Toulon 20.7.1810; in attack on Palamos 13.12.1810; took (with
Unite) French flûte *Le Dromadaire* off Elba 31.3.1811. In 4.1811
under Capt. James Brisbane; in 10.1811 under Capt. Sir Robert
Lawrie, in the Mediterranean (–1812). Defects made good
at Plymouth (for £17,064) 4–7.1813; in 5.1813 under Otway
again; at San Sebastian 8.9.1813; took French 16-gun *L'Alción*
17.3.1814. To America with troops 1814. In 10.1814 under Capt.
George Mundy, in the Mediterranean; paid off 7.1816. Large
Repair at Portsmouth (for £67,758) 10.1820–6.1829. Converted
to a screw blockship (under AO 23.10.1845) by White, Cowes
11.1845–9.1846 (see Chapter 4 for later history).

Berwick Perry, Wells & Green, Blackwall

As built: 176ft 0in, 144ft 11⅜in x 47ft 9½in x 21ft 0in. 1,761 (exact)
bm. Draught 12ft 11in/18ft 0in.

Ord: 1.7.1807. K: 10.1807. L: 11.9.1809. C: 17.11.1809 (for
Ordinary), 14.7.1810 (for sea) at Woolwich.

First cost: Unknown

Commissioned 3.1810 under Capt. James Macnamara. Chased
French 40-gun frigate *L'Amazone* (burnt to avoid capture)
off Barfleur 25.3.1811. In 10.1811 under Capt. Edward Brace
(–1815); sailed for the Mediterranean 17.11.1811; her boats (with
those of *Euryalus*) took 10-gun xebec *La Fortune* 16.5.1813. In
1815–16 under Capt. J(ames or John) Nash; laid up at Plymouth
7.1816. BU (because found to need Major Repair) at Plymouth
3.1821.

Egmont Thomas Pitcher, Northfleet

As built: 176ft 2in, 145ft 1⅛in x 47ft 9in x 21ft 0in. 1,760¹⁹⁄₉₄bm.
Draught 12ft 10in/17ft 5½in.

Ord: 13.7.1807. K: 10.1807. L: 7.3.1810. C: 20.4.1810 (at Northfleet),
7.11.1811 (at Woolwich).

First cost: £57,439 to builder, plus £11,626 dockyard expenses, plus
£18,614 fitting for sea.

Commissioned 5.1811 under Capt. Charles Austen; in 6.1811 under
Capt. Joseph Bingham (–1814) for North Sea and Channel; in
1812 flagship of Rear-Adm. Edward Griffiths in North America,
in 1813 of Rear-Adm. Sir George Hope in the North Sea, and
in 1814 of Rear-Adm. Charles Penrose for Gironde operations.
Paid off and laid up at Portsmouth 8.1814. Large Repair at
Portsmouth (for £48,208) 2.1820–10.1825, then laid up again.
Fitted as a storeship (for £24,306) by J. Samuel White 10.1861–
10.1862, then at Portsmouth to 1.1863, became a storeship at
Rio de Janeiro. Sold at Rio by public auction (for about £5,100)
2.1.1875.

Edinburgh Samuel & Daniel Brent, Rotherhithe

As built: 176ft 6in, 145ft 4⅜in x 47ft 10½in x 21ft 0½in. 1,772⁴⁵⁄₉₄bm.
Draught 13ft 0in/17ft 7in.

Ord: 13.7.1807. K: 11.1807. L: 26.1.1811. C: 7.2–7.5.1811 at
Woolwich.

First cost: £57,006 to builder, plus £21,696 & £4,326 fitting.

Commissioned 6.1811 under Capt. Robert Rolles; sailed for the
Mediterranean 8.1.1812. Under Capt. George Dundas 11.1812;
operations in Gulf of Spezia 30.4.1813; boats in operation at
Anzio 5.10.1813. Under various Captains during 1813 (Thomas
Mainwaring, Thomas Ussher, John Lampen Manley). Laid up
at Portsmouth 11.1814 (–1833). Large Repair at Portsmouth
(for £62,558) 4.1817–2.1820, then laid up. Fitted for sea at
Portsmouth (for £14,229) 2.1834; under Capt. James Dacres

10.1833–2.1837, in the Mediterranean, then 7.1837 under Capt.
William Henderson: at Bombardment of Acre 3.11.1840; paid off
at Portsmouth 7.1841. Converted to a screw blockship (under
AO 25.10.1845) at Portsmouth 8.1846–8.1852–see Chapter 4 for
later history.

Clarence Isaac Blackburn, Turnchapel, Plymouth

As built: 176ft 0¼in, 145ft 0⅞in x 47ft 7½in x 21ft 0in. 1,749bm.
Draught 14ft 0in/17ft 3in.

Ord: 13.7.1807. K: 11.1807. L: 11.4.1812. C: 12.4–9.9.1812 at
Plymouth Dyd

First cost: £58,400 to builder, plus £18,400 + £12,140 fitting.

Commissioned 7.1812 under Capt. Henry Vansittart, for North Sea
and the Channel; in 3.1814 under Capt. Frederick Warren. Laid
up 8.1814 at Portsmouth. Renamed *Centurion* and ordered to be
cut down to a Fourth Rate frigate of 50 guns (by AO 26.5.1827)
and moved 8.1827 to Chatham, but conversion cancelled due to
state of decay. Instead BU at Chatham 10.1828.

Scarborough Joseph Graham, Harwich

As built: 176ft 0in, 144ft 11⅜in x 47ft 8½in x 21ft 0½in. 1,745bm.
Draught 13ft 3in/17ft 6in.

Ord: 13.7.1807. K: 1.1808. L: 29.3.1812. C: C: 15.4.1812–18.3.1813
at Portsmouth.

First cost: £57,873 to built, plus £3,772 'yard returns', plus £25,397
fitting (total £87,042).

Commissioned 2.1813 under Capt. John Halstead, as flagship of
Rear-Adm. John Ferrier in the North Sea; on 23.12.1813 under
Capt. Charles James Johnston, as flagship of Ferrier; paid off
5.5.1814. Fitted for Ordinary at Sheerness 6.1814. Middling
Repair at Woolwich (for £24,215) 7.1816–10.1817, then laid up
at Sheerness and roofed over fore and aft. Sold at ?Deptford (for
£6,200) 3.9.1836.

Asia Josiah & Thomas Brindley, Frindsbury

As built: 175ft 7in, 144ft 4⅜in x 47ft 11in x 21ft 0in. 1,763⅜⁄₉₄bm.
Draught 12ft 9in/17ft 4in.

Ord: 13.7.1807. K: 2.1808. L: 2.12.1811. C: 9.5.1812 at Chatham.

First cost: £56,722 to builder, plus £36,347 fitting.

Commissioned 2.1812 under Capt. John Spranger; under Capt.
George Scott in 3.1812, then under Capt. Alexander Shippard
in 8.1812; sailed for Jamaica 22.4.1813. Made good defects at
Portsmouth (for £21,113) 11.1813–1.1814. Recommissioned
12.1813 under Capt. Richard Harrison Pearson, then 1.1814
under Capt. John Wainwright, as flagship of Sir Alexander
Cochrane, for North America and the West Indies; laid up at
Chatham 1816. Renamed *Alfred* 1819. Very Large Repair and cut
down to a Fourth Rate frigate of 50 guns at Chatham in 11.1822–
8.1828 (see Chapter 4 for later history).

Rodney Mrs Frances Barnard, Deptford

As built: 176ft 5in, 145ft 5⅛in x 67ft 7½in x 21ft 0¼in. 1,754bm.
Draught 13ft 5in/18ft 3in.

Ord: 13.7.1807. K: 3.1808. L: 8.12.1809. C: 10.4.1810 at Deptford &
Woolwich Dyds.

First cost: £58,330 to builder, plus £28,119 fitting.

Commissioned 2.1810 under Capt. George Burlton; sailed for
the Mediterranean 2.6.1810. In 9.1811 under Capt. Edward
Durnford King, then in 1812 under Capt. John Duff Markland,
in the Mediterranean. In Ordinary at Portsmouth 1813, then
Small Repair there (for £12,217 + £12,272) 5.1813–1.1814.
Recommissioned 12.1813 under Capt. Charles Inglis, as flagship
of Vice-Adm. Sir George Martin; paid off 8.1814. Renamed
Greenwich 17.5.1827 and ordered to be cut down to Fourth Rate
frigate of 50 guns; however, the conversion was never completed,

instead she was sold at Portsmouth (for £5,310) 8.9.1836.

Duncan John Dudman, Deptford

As built: 176ft 0in, 144ft 11¼in x 47ft 9½in x 21ft 0in. 1,761bm. Draught 13ft 0in/17ft 6in.

Ord: 13.7.1807. K: 8.1808. L: 2.12.1811. C: 13.12.1811–26.2.1812 at Woolwich.

First cost: £58,420 to builder, plus £26,632 fitting.

Commissioned 2.1812 under Capt. Robert Plampin; in 1812–14 under Capt. Robert Lambert, for North Sea; in 1814 under Capt. Thomas Smith, in the Mediterranean, then 9.1814 Capt. Samuel Chambers. Laid up at Portsmouth 6.1815. Fitted as a lazarette at Portsmouth 7.1826; to Sheerness 8.1831. Fitted as a lazarette at Chatham 2.1837, then to Stangate Creek. Returned from Customs authorities 11.1861. BU completed at Chatham 5.10.1863.

Indus John Dudman, Deptford

As built: 176ft 3½in, 145ft 4¼in x 47ft 8in x 21ft 0in. 1,756bm. Draught 13ft 0in/17ft 8in.

Ord: 31.7.1807. K: 4.1809. L: 19.12.1812. C: 29.12.1812–30.3.1813 at Woolwich.

First cost: £57,430, plus £36,246 fitting.

Commissioned 2.1813 under Capt. William Gage, for the North Sea; in 1814 in Mediterranean; laid up 8.1814 at Plymouth. Renamed *Bellona* 3.11.1818. Middling Repair at Plymouth (for £32,731) 4.1819–12.1820, then housed over from main mast forward. Fitted at Plymouth as a receiving ship (for £4,957) 11.1841–6.1842. BU (by AO 28.4.1868) completed at Plymouth 27.6.1868.

Dublin Samuel & Daniel Brent, Rotherhithe

As built: 176ft 3in, 145ft 2in x 47ft 10in x 21ft 0in. 1,766bm. Draught 12ft 11in/17ft 3in.

Ord: 31.7.1807. K: 5.1809. L: 13.2.1812. C: 4.3–25.4.1812 at Woolwich, 25.9.1812 at Sheerness.

First cost: £57,372 to builder, plus fitting £13,985 at Woolwich & £17,424 at Sheerness.

Commissioned 8.1812 under Capt. David Milne, then Capt. Richard Dunn in 9.1812 (died 6.1813) and (temp.) Capt. Robert Henderson, then Capt. Thomas Elphinstone in 1813–14, for the Channel. Laid up at Plymouth 8.1814. Middling Repair, and cut down to a Fourth Rate frigate of 50 guns at Plymouth 4.1821–12.1826 (see Chapter 4 for later history).

Stirling Castle Mrs Mary Ross, Rochester

As built: 176ft 5in, 145ft 3⅜in x 47ft 11in x 21ft 0in. 1,774bm. Draught 12ft 3in/17ft 2in.

Ord: 12.8.1807. K: 7.1808. L: 31.12.1811. C: 31.12.1811–25.2.1812 at Chatham.

First cost: £57,608 to builder, plus £18,634 fitting.

Commissioned 3.1812 under Capt. Jahleel Brenton, for the Channel. In 1813 under Capt. Augustus Brine, later under Capt. Sir Home Popham; sailed for the East Indies 20.4.1813 (to take the Viceroy, Lord Moira, to India). In 1814 under Capt. William Butterfield; paid off 11.1814. Large Repair at Plymouth (for £66,920) 1.1816–4.1819, then housed over fore and aft 'nearly down to LD ports'. Fitted as a convict ship at Plymouth 4.1839. To Portsmouth 10.1844. BU completed at Portsmouth 6.9.1861.

Medway Thomas Pitcher, Northfleet

As built: 176ft 1in, 145ft 3in x 47ft 10in x 21ft 0in. 1,768bm. Draught 12ft 2in/16ft 11in.

Ord: 19.8.1807. K: 12.1808. L: 19.11.1812. C: 29.12.1812–9.7.1813 at Sheerness.

First cost: £58,459 to builder, plus £22,638 dyd expenses, plus £5,199 fitting at Sheerness.

Commissioned 4.1813 under Capt. Augustus Brine; flagship of Vice-Adm. Charles Tyler; sailed for East Indies 31.12.1813; took 16-gun USS *Siren* 12.7.1814; laid up at Plymouth 3.1816, housed over from main mast forward. Between Small and Middling Repair at Plymouth (for £48,574) 4.1819–12.1820, then laid up there. Fitted as a convict ship for Bermuda 4–10.1847. Sold to J. D. Murphy (for £2,180) at Bermuda 2.11.1865.

America Perry & Co., Blackwall

As built: 176ft 0in, 144ft 11⅝in x 48ft 6in x 21ft 0in. 1,758bm. Draught 12ft 9in/17ft 8in.

Ord: 22.8.1807. K: 1.1808. L: 21.4.1810. C: 11.6.1810 at Woolwich, then 9.6.1811 at Sheerness.

First cost: £57,456 to builder, plus fitting £14,529 + £2,778 at Woolwich, and £16,659 at Sheerness.

Commissioned 4.1811 under Capt. Thomas Usshur; sailed for the Mediterranean 4.9.1811. Under Capt. Josias Rowley 1812–14; at capture or destruction of convoy at Laigueglia 10.5.1812; reduction of Genoa 4.1813; paid off 10.1814. Small Repair at Plymouth (for £28,820) 11.1814–4.1815, then laid up. Between Small and Middling Repair, and cut down to a Fourth Rate frigate of 50 guns (by AO 20.3.1827) at Plymouth (for £22,815 + £13,928) 3.1827–2.1835, then housed over waist and laid up (see Chapter 4 for later history).

Anson Thomas Steemson, at Paull (near Hull)

As built: 175ft 5⅛in, 144ft 6⅛in x 48ft 4¼in x 21ft 0in. 1,742bm. Draught 13ft 10in/17ft 6in.

Ord: 2.11.1807. K: 3.1808. L: 11.5.1812. C: 2.7–11.8.1812 at Portsmouth.

First cost: £56,004 to builder, plus £2,228 dyd expenses, plus £7,404 fitting.

Not commissioned until 1843. Laid up 1812 in Ordinary. Coppered 2.1815. Small Repair at Portsmouth, and fitted as a temporary lazarette (for £19,663) 9.1831. At Leith in 1834, then 1837 to Stangate Creek. Fitted at Chatham as convict ship (for £19,137) 5–8.1843, and commissioned 28.6.1843 under Lieut Francis Rogers Coghlan; then to Tasmania as a convict ship 1844. BU at Hobart 1851 (sold in 3 separate sales for £387.0.8d in 1852).

Barham Perry & Co., Blackwall

As built: 176ft 0in, 144ft 11⅜in x 47ft 9½in x 21ft 1in. 1,761bm. Draught 12ft 3in/17ft 6in.

Ord: 2.11.1807. K: 6.1808. L: 8.7.1811. C: 11.7–21.8.1811 at Woolwich (for Ordinary); 2.1812–20.5.1812 at Chatham (for sea).

First cost: £57,949.11.7d to builder, plus fitting £11,785 at Woolwich + £16,967 at Chatham.

Commissioned 2.1812 under Capt. John Spranger (from 7.1811 had been under Capt. Thomas Bladen Capel), in North Sea; at Woolwich 1813, then to Jamaica 1814; paid off 7.1814. Small Repair and housed over at Woolwich (for £22,686) 3–9.1815, then laid up at Sheerness. Small Repair and cut down to a Fourth Rate frigate of 50 guns (for £41,346) at Woolwich 2–12.1826; recommissioned 8.1826 for Jamaica (see Chapter 4). Fitted for sea at Woolwich (for £19,978) 9.1830–6.1831; recommissioned 3.1831 for the Mediterranean. BU at Deptford 3.1840.

Rippon Richard Blake & John Scott, Bursledon

As built: 176ft 4in, 145ft 1¾in x 47ft 10½in x 21ft 0½in. 1,770bm. Draught 13ft 3in/17ft 9in.

Ord: 2.11.1807. K: 10.1808. L: 8.8.1812. C: 10.8.1812–12.5.1813 at Portsmouth.

First cost: £29,448 fitting.

Commissioned 3.1813 under Capt. Sir Christopher Cole, for Channel service; present at capture (by *Scylla* and *Royalist*) of 40-gun *Weser* off Ushant 21.10.1813; to North America with troops 1814; paid off 8.1814 into Ordinary at Plymouth; roofed over 8.1816. BU at Plymouth 3.1821.

Third (1806) Batch

Again ordered by Thomas Grenville's Board of Admiralty.

Blenheim Deptford Dyd [M/Shipwright Robert Nelson]

As built: 175ft 9in, 144ft 6⅞in x 47ft 8in x 21ft 0½in. 1,747 $\frac{25}{94}$bm. Draught 13ft 9in/18ft 0in.

Ord: 4.1.1808. K: 8.1808. L: 31.5.1813. C: 10.6–22.7.1813 at Woolwich.

First cost: £76,865 to build, plus £3,215 fitting.

Commissioned 6.1813 under Capt. Samuel Warren; to the Mediterranean 1814; paid off at Chatham into Ordinary 1815, then to Sheerness. Between Middling and Large Repair at Woolwich (for £41,484) 4.1820–12.1824; at Woolwich to 1830, then to Portsmouth. Fitted as a demonstration ship at Sheerness (for £4,351) 1–6.1836. Fitted for sea at Sheerness (for £13,223) 4–7.1839; recommissioned 9.4.1839 under Capt. Humphrey Fleming Senhouse (died 14.6.1841), for the East Indies; to 1st Anglo-Chinese War 1840–1; on 14.6.1841 under Capt. Thomas Herbert, in China; paid off 28.3.1843 at Sheerness. Converted to screw blockship (by AO 20.10.1845) by Wigram, Blackwall 11.1845–3.1847 (see Chapter 4 for later history).

Pembroke Wigram, Wells & Green, Blackwall

As built: 176ft 0in, 144ft 11⅛in x 47ft 9in x 21ft 1in. 1,758 $\frac{17}{94}$bm. Draught 12ft 10in/17ft 10in.

Ord: 17.5.1808. K: 3.1809. L: 27.6.1812. C: 5.11.1812 at Woolwich.

First cost: £58,516 to build, plus £29,100 fitting.

Commissioned 9.1812 under Capt. James Brisbane; grounded on Dunnose 26.12.1812; defects made good at Portsmouth (for £13,465) 12.1812–2.1813; in the Channel 1813, then Mediterranean 1814; laid up at Portsmouth 8.1814 and paid off 9.1814. Between Middling and Large Repair at Portsmouth (for £57,316) 1.1819–2.1823. Fitted for sea at Portsmouth (for £16,604) 1–7.1836; recommissioned 25.3.1836 under Capt. Thomas Fellowes, for the Mediterranean; on 25.1.1837 under Capt. Fairfax Moresby, paid off 2.1840 into Ordinary at Portsmouth. Converted to a screw blockship at Portsmouth in 10.1854–5.1855 (see Chapter 4 for later history).

Cornwall Mrs Frances Barnard, Deptford

As built: 176ft 0⅜in, 145ft 0⅜in x 47ft 7¾in x 21ft 0in. 1,751 $\frac{25}{94}$bm. Draught 12ft 8in/17ft 10in.

Ord: 30.5.1808. K: 3.1809. L: 16.1.1812. C: 25.1–7.3.1812 at Woolwich, 8.3–9.9.1812 at Sheerness.

First cost: £58,330 to builder, plus £2,227 fitting at Woolwich & £17,814 at Sheerness.

Commissioned 7.1812 under Capt. John Broughton; on 17.2.1813 under Capt. Edward William Campbell Rich Owen (–1814), for the North Sea. To Ordinary at Chatham 1814–15. Small Repair at Woolwich (for £18,712) 11.1815–9.1816, then laid up at Sheerness. Between Small and Middling Repair and cut down to a Fourth Rate frigate of 50 guns by AO 24.2.1827 at Sheerness 3.1827–5.1830, then laid up (see Chapter 4 for later history).

Devonshire Mrs Frances Barnard, Deptford

As built: 176ft 0⅜in, 145ft 1⅜in x 47ft 6in x 21ft 0½in. 1,741 $\frac{53}{94}$bm. Draught 13ft 2in/18ft 2in.

Ord: 30.5.1808. K: 2.1810. L: 23.9.1812. C:27.9–2.12.1812 at Woolwich, then to 6.1814 at Sheerness (for Ordinary).

First cost: £58,330 to builder, plus £2,520 + £8,050 at Woolwich & £11,262 at Sheerness.

Commissioned 6.1813 under Capt. Ross Donnelly; paid off 5.1814 into Ordinary at Sheerness (–1839). Large Repair at Woolwich (for £53,889) 3.1817–2.1820. At Chatham 1842–50. Fitted as temporary hospital ship for merchant seamen 7–11.1849 and lent to the Greenwich Seamen's Hospital. To Sheerness as a prison ship for Russian PoWs at Sheerness 6.1854. School ship in 'Queenborough Swale' 1860–5. BU at Sheerness (by AO 6.10.1868) completed 5.6.1869.

Gloucester Thomas Pitcher, Northfleet

As built: 176ft 3½in, 145ft 2¼in x 47ft 10½in x 21ft 0in. 1,770 $\frac{6}{94}$bm. Draught 12ft 6in/17ft 5½in.

Ord: 11.6.1808. K: 3.1808. L: 27.2.1812. C: 20.5–11.6.1812 at Sheerness.

First cost: £62,514 for building, plus £25,232 fitting (Woolwich Dyd costs) + £111 at Sheerness.

Commissioned 4.1812 under Capt. Robert Williams (–1814), for North Sea and then Baltic; to West Indies 1814. Laid up in Ordinary at Chatham 1.1815 (–1822). Large Repair, rebuilt with circular stern and fitted for a flagship at Chatham (for £71,873) 3.1818–11.1822; recommissioned 28.3.1822 under Capt. Sir Murray Maxwell; on 25.11.1822 under Capt. Sir Edward William Campbell Rich Owen, at Jamaica; on 24.10.1823 under Capt. James Lillicrap, with broad pendant of Commodore Owen, at Jamaica; paid off 4.1824. Recommissioned 11.6.1825 under Capt. Joshua Sydney Horton; fitted at Sheerness (for £5,331) to convey the Duke of Devonshire to Russia 7.1825–5.1826. Made good defects at Sheerness (for £10,934) 8.1827–1.1828. Recommissioned 24.6.1828 under Capt. Henry Stuart at Sheerness, as guard ship in the Medway; on 22.7.1830 under Capt. Francis Holmes Coffin; to the Mediterranean 1831. Small Repair and cut down to a Fourth Rate frigate of 50 guns (by AO 28.2.1831) at Chatham 4.1831–12.1832 (see Chapter 4 for later history).

Benbow Samuel & Daniel Brent, Rotherhithe

As built: 176ft 3½in, 145ft 1⅞in x 47ft 11in x 21ft 0in. 1,772 $\frac{72}{94}$bm. Draught 12ft 6in/17ft 3in.

Ord: 11.6.1808. K: 7.1808. L: 3.2.1813. C: 6.2–17.5.1813 at Woolwich.

First cost: £57,411 plus £24,757 Deptford Dyd costs, + £6,409 fitting at Woolwich.

Commissioned 4.1813 under Capt. Richard Pearson; laid up at Portsmouth 9.1814 in Ordinary (–1839). Between Middling and Large Repair at Portsmouth (for £55,834); fitted for a demonstration ship at Portsmouth 2.1836–7.1837. Fitted for sea at Portsmouth (for £14,600) 4–6.1839; recommissioned 4.1839 under Capt. Houston Stewart, for the Mediterranean (–6.1842); at Bombardment of Acre 3.11.1840; laid up at Sheerness 5.1842. Fitted as a marine barracks at Woolwich for Sheerness 2.1848. Fitted as prison ship for Russians at Sheerness 9.1854. Fitted as Coal hulk at Chatham for Sheerness 8.1859. Sold to Castle to BU 23.11.1892.

Final batch (all at HM dockyards except *Cornwallis* at Bombay)

Defence Chatham Dyd [M/Shipwright Robert Seppings to 3.1813, completed by George Parkin]

As built: 174ft 0in, 144ft 10⅝in x 47ft 8½in (for tonnage) x 21ft 0½in. 1,754 $\frac{9}{94}$bm. Draught 13ft 3in/18ft 0in.

Ord: 23.3.1809. K: 5.1812 as *Marathon*, renamed 3.1.1815. L: 25.4.1815. C: 7.6.1815 for Ordinary.

The *Cornwallis* was one of the last of the numerous *Armada* Class 74s, and was built of teak at Bombay, which probably accounts for her long life. She spent much of the post-war era in Ordinary, but was commissioned for sea service in 1837–39 and again in 1841–44. This lithograph by Sir Oswald Walters Brierly showing her coming out of Plymouth harbour was published in 1847, but from the end of 1844 she was in Ordinary. In 1854 she was converted to a steam blockship and enjoyed active service for a further 18 months, first in the Baltic and then more peaceably off the American coast before settling down to a long career as a harbour auxiliary. (*NMM PY0786*)

First cost: £66,111 (unfitted).

Not commissioned. Never completed for sea, but was delayed at close of war and housed over from main mast forwards 10.1817; remained in Ordinary at Sheerness (at Chatham 1842–6). Fitted at Sheerness 7–9.1848 (for £2,822) then at Portsmouth 9.1848–9.1851 (for £8,960) as a convict ship for Woolwich, where based for 6 years. Damaged by accidental fire 14.7.1857, BU completed at Woolwich 21.1.1858.

Hercules Chatham Dyd [M/Shipwright Robert Seppings to 3.1813, completed by George Parkin]

As built: 176ft 1in, 145ft 1¾in x 47ft 7¼in (for tonnage) x 21ft 0in. 1,749⁵⁶/₉₄bm. Draught 13ft 6in/17ft 10in.

Ord: 16.5.1809 (repeated 6.12.1811). K: 8.1812. L: 5.9.1815. C: 17.11.1815 (for Ordinary).

First cost: £67,841.

Between Small and Middling Repair at Sheerness (for £6,631) and Chatham (for £14,720) 11.1830–1.1836; fitted at Sheerness (for £5,915) 2–7.1836.

Commissioned 25.3.1836 at Sheerness under Capt. Maurice Frederick Fitzhardinge Berkeley, for 'particular service' at Lisbon. On 16.8.1837 under Capt. John Toup Nicholas, at Lisbon. Fitted as Troopship at Plymouth (for £2,068) 1–2.1838. In Ordinary at Sheerness 1842–6. Fitted to convey timber from India at Chatham (for £4,122) 12.1847–2.1848; recommissioned under Robert Fulton, Master, as storeship 9.12.1847. Receiving ship at Malta 8.1851. Recommissioned under Cmdr Benjamin Bayntum; fitted at Chatham as Emigrant ship (for £3,485) 9–10.1852; sailed with emigrants from Scotland to Australia. Fitted as Army depot ship at Plymouth 1853. To Hong Kong as depot and receiving ship 1854. Sold there to a Chinese resident (Hop-tai-loon) to BU (for $18,000 = £3,825) 22.8.1865.

Cornwallis Bombay Dyd [M/Shipwright Jamsetjee Bomanjee] Teak-built.

As built: 177ft 5in, 145ft 1¼in x 49ft 1in oa (48ft 5in for tonnage) x 21ft 1¾in. 1,809²⁸/₉₄bm.

Ord: 25.7.1810. K: 1811. L: 12.5.1813. C: 10.1814–25.12.1814 at Portsmouth.

First cost: £44,591 to build, plus £11,266 stores send from UK, plus £20,251 fitting (+ £4,320 coppering).

Under Capt. Henry Folkes Edgell for passage to England. Arrived Portsmouth 9.6.1814, and fitted as a flagship.

Commissioned 12.1814 under Capt. John Bayley, as flagship of Rear-Adm. George Burlton (died 21.9.1815); escape of USS 20-gun *Hornet* 27–30.4.1815. In Ordinary at Plymouth 1815–1832. Fitted for commission (not for sea) at Plymouth 11.1832; fitted for sea there (for £10,102) 3–6.1836.; recommissioned 15.2.1837 under Capt. Richard Grant, as flagship of Vice-Adm. Charles Paget, for North America and West Indies; paid off 2.6.1839. Fitted as Demonstration ship 7–8.1839. Fitted for sea at Plymouth (for £10,971) 4–7.1841; recommissioned 5.4.1841 under Capt. Peter Richards, as flagship of Rear-Adm. William Parker, for the East Indies; in 1st Anglo-Chinese War 1842; paid off at end of 1844 at Plymouth. Fitted as Army depot ship 7–9.1852, intended for Hong Kong, but not sent. Converted to 60-gun screw blockship at Plymouth 10.1854–4.1855 (see Chapter 4 for later history).

Agincourt Plymouth Dyd [M/Shipwright Thomas Roberts to 9.1815; completed by Edward Churchill]

As built: 175ft 11¼, 144ft 11¼in x 48ft 4¼ (47ft 7¼in for tonnage) x 21ft 0in. 1,747⁷/₉₄bm. Draught 13ft 0in/18ft 3in.

Ord: 6.1.1812. K: 5.1813. L: 19.3.1817. C: 4.1817 (for Ordinary).

First cost: £72,850 (total incl. fitting £85,720).

Delayed at close of war and housed over from main mast forwards. Very Small Repair and fitted for Demonstration ship at Plymouth (for £14,748) 4.1833–2.1836. Old housing removed, fresh housing over waist 2.1838. Fitted for sea at Plymouth (for £4,887 + £11,085) 2–5.1842.

Commissioned 1.2.1842 at Plymouth under Capt. Henry William

Bruce; fitted for sea at Plymouth (for £4,887 + £11,085) 2–5.1842, as flagship of his brother-in-law, Rear-Adm. Thomas John Cochrane, for the East Indies. On 6.5.1845 under Capt. William James Hope Johnstone, still with Cochrane's flag in the East Indies; at Brunei 1846; paid off 4.9.1847. Fitted for Ordinary as Training ship at Plymouth 8.1847–3.1848; recommissioned 28.1.1848 under Capt. William Bowen Mends, as depot ship at Devonport; on 9.2.1850 under Capt. Lord John Hay (Broad Pennant), in same role. Renamed *Vigo* by AO 29.4.1865. Cholera hospital ship by AO 11.5.1866. Receiving ship at Plymouth 1870. Sold to Castle & Sons 10.1884 and BU at Charlton completed by 5.1.1885.

Wellington (ex-*Hero*) Deptford Dyd [M/Shipwright William Stone]
    As built: 176ft 6in, 145ft 4¾in x 48ft 5in (47ft 8in for tonnage) x 21ft 0in. 1,757¹⁹⁄₉₄bm. Draught 13ft 3in/17ft 7in.
    Ord: 6.1.1812. K: 7.1813. L: 21.9.1816. Renamed *Wellington* 4.12.1816. C: 1.10–6.12.1816 at Woolwich.
    First cost: £69,767, plus £3,018 fitting (for Ordinary at Woolwich).
    Commissioned 1.3.1848 at Sheerness under Capt. David Price (also Capt.-Superintendent of Sheerness Dyd) (*Wellington* was delayed at close of war and housed over from main mast forwards, from 12.1816). Fitted for a Divisional Ship at Sheerness 12.1840–1.1841. Ordered to be converted to screw blockship 13.8.1845, but this was cancelled 17 days later. Fitted as receiving and depot ship at Sheerness (for £5,268) 6.1848. To coastguard at Sheerness 1.1854, then fitted as a guard ship in the 'Steam Ordinary' 10.1856–11.1857. Recommissioned 25.4.1859 under Capt. Robert Spencer Robinson, as a guard ship at Plymouth (in Ordinary). On 25.5.1860 under Capt. Astley Cooper Key (this was the last time that the word 'Ordinary' was formally used to categorise Reserve status), then 14.7.1860 under Cmdr Frederick Cannon, as flagship of Rear-Adm. Thomas Sabine Pasley; paid off 15.7.1861. Lent to Liverpool Juvenile Reformatory Association as a training ship 10.5.1862, replacing former *Akbar* and taking her name; loan terminated 1.1.1908. Sold at Chatham to T. W. Ward, Ltd (for £5,025) 8.4.1908 and BU at Morecambe.

Russell Deptford Dyd [M/Shipwright William Stone]
    As built: 176ft 6in, 145ft 6in x 48ft 3¾in (47ft 6¼ for tonnage) x 21ft 0in. 1,750⁷⁴⁄₉₄bm. Draught 12ft 9in/17ft 3in.
    Ord: 6.1.1812. K: 8.1814 (housed over 9.1815 from main mast forward). L: 22.5.1822. To Chatham 13.7.1822 (for Ordinary). Housing removed 4.1834.
    First cost: £70,437.
    Commissioned 9.7.1835 at Sheerness under Capt. Sir William Henry Dillon, fitted there for sea (for £12,787) (*Russell* was delayed at close of war and housed over from main mast forwards, from 9.1815 to 4.1834), to Lisbon; in mid 1837 to the Mediterranean; paid off 1.1839 at Sheerness. Fitted for Demonstration ship at Sheerness (for £3,531) 2–6.1839. Converted to 60-gun screw blockship at Sheerness 1854–5 (see Chapter 4 for later history).

Boscawen Woolwich Dyd
    Ord: 6.1.1812. Suspended 2.12.1813, re-ordered to a new design on 14.2.1814, and was again to be re-ordered as an 80-gun ship on 20.11.1817.

Carnatic Portsmouth Dyd
    Ord: 30.9.1814. Re-ordered to a new design 10.11.1817.

Ex-FRENCH PRIZES (1803–14)

Eight French 74s were taken by Nelson's fleet at Trafalgar on 21 October 1805, but of these *Swiftsure* was a former British ship retaken (built in 1787 and captured 1795, see entry earlier), *L'Algésiras* was

retaken on 23 October by Cosmao's squadron, and the other six sank or were wrecked following the battle – *L'Achille*, *L'Aigle*, *Le Fougueux*, *L'Intrépide*, *Le Redoutable* and *Le Berwick*. However, ten further French 74s were taken by the British Navy and added to the fleet between 1803 and 1815, all but one being further units of the numerous *Le Téméraire* design.

Le TÉMÉRAIRE Class Further units of this numerous Sané design (see above) were taken from 1803 onwards. Of the nine added to the RN, the *Duquesne* was BU in 1805 and the *Brave* foundered in 1806; the *Maida* (French *Le Jupiter*) was BU in 1814 and the *Marengo* in 1816.

Implacable (French *Le Duguay-Trouin*; built 11.1794–11.1800 at Rochefort. L: 25.3.1800).
    Dimensions & tons: 181ft 6in, 148ft 11⅛in x 48ft 11in x 20ft 7in. 1,896²²⁄₉₄bm.
    Men: 640. Guns: LD 30 x 32pdr; UD 30 x 18pdr; QD 2 x 12pdr + 12 x 32pdr carronades; Fc 2 x 12pdr + 2 x 32pdr carronades.
    Taken in Strachan's Action 4.11.1805. Arrived Plymouth 10.11.1805 and laid up.
    Underwent Small Repair and fitted at Plymouth (for £34,585) 9.1807–3.1808.
    Commissioned 1.1808 under Capt. Thomas Byam Martin, for Baltic. Action with Russian *Sewolod* 26.8.1808. Corunna 1.1809. Boats (with other ships') took 6 Russian gunboats in Baro Sound 7.7.1809. From 1810 under Capt. George Cockburn, as flagship of Rear-Adm. Sir Richard Keats; sailed for the Mediterranean 17.7.1810. In 10.1811 under Capt. Joshua Watson in the Mediterranean; paid off 11.1812. Very Large Repair at Plymouth (for £66,857) 10.1812–1.1815, then to Ordinary 1815–39. Small repair at Plymouth (for £20,510) 7–12.1826. Fitted as demonstration ship at Plymouth 2.1836. Fitted for sea at Plymouth (for £12,552) 2–4.1839. Recommissioned under Capt. Edward Harvey 1.2.1839 for the Mediterranean; took part in Syrian operations 1840; paid off 31.1.1842; in Ordinary at Devonport to 1855. Fitted at Plymouth as a training ship 6–7.1855. Lent to Wheatley Cobb at Falmouth 21.3.1912 (sailed 12.9.1912 for Falmouth). Renamed *Foudroyant* 1943; paid off 1.1947. Scuttled in the Channel 2.12.1949 after a plan to preserve her at Greenwich in the dock now occupied by the *Cutty Sark* had failed.

Mont Blanc (French *Le Mont Blanc*, ex-*Le Républicain* 1796, ex-*Le Trente-et-Un Mai* 1795, ex-*Le Pyrrhus* 1793; built 7.1789–3.1793 at Rochefort. L: 19.8.1791)
    Dimensions & tons: 183ft 2in, 149ft 7⅜in x 48ft 8¼in x 20ft 6in. 1,886⁴⁴⁄₉₄bm.
    Men: 640. Guns: LD 30 x 32pdr; UD 30 x 18pdr; QD 2 x 12pdr + 12 x 32pdr carronades; Fc 2 x 12pdr + 2 x 32pdr carronades.
    Taken in Strachan's action 4.11.1805. Arrived Plymouth 10.11.1805 and laid up.
    Not commissioned. Fitted as Powder hulk at Plymouth 1811 (–1815). Sold to John Small Sedger (for £5,510) 8.3.1819.

Scipion (French *Le Scipion*, built 6.1798–9.1801 at Lorient. L: 29.3.1801)
    Dimensions & tons: 182ft 1½in, 150ft 0⅞in x 48ft 7½in x 20ft 4in. 1,887³⁹⁄₉₄bm.
    Men: 640. Guns: LD 30 x 32pdr; UD 30 x 18pdr; QD 2 x 12pdr + 12 x 32pdr carronades; Fc 2 x 12pdr + 2 x 32pdr carronades.
    Taken in Strachan's action 4.11.1805. Arrived Plymouth 10.11.1805 and laid up.
    Underwent Large Repair and fitted at Plymouth 6.1808–11.1809.

Commissioned 7.1809 under Capt. Charles Bateman, for the Channel; from 1810 flagship of Rear-Adm. Sir Robert Stopford; sailed for Cape of Good Hope 8.10.1810, and thence to East Indies. Under Capt. James Johnson 1811; capture of Java 8–9.1811. Under Capt. Henry Heathcote in 5.1812; sailed for the Mediterranean 20.7.1812; paid off into Ordinary at Portsmouth 10.1814. Began Middling Repair at Portsmouth 9.1818, but instead BU 1.1819.

Abercrombie (French *D'Hautpoult*, ex-*L'Alcide*, ex-*Le Courageux*, built 6.1803 –2.1808 at Lorient. L: 2.9.1807)

Dimensions & tons: 181ft 6in, 149ft 4¾in x 48ft 6¼in x 21ft 8in. 1,870⁷⁸/₉₄bm.

Men: 640. Guns: LD 30 x 32pdrs; UD 30 x 18pdrs; QD 4 x 12pdrs + 10 x 32pdr carronades; Fc 2 x 12pdrs + 2 x 32pdr carronades.

Taken 17.4.1809 off Puerto Rico by *Pompée*, *Castor* and *Recruit*. Arrived Plymouth 5.9.1810.

Commissioned 6.1809 under Capt. Charlier Napier at Plymouth. In 9.1809 under Capt. John Richards, and in 12.1809 under Capt. William Fahie. At capture of Guadeloupe 26.1–6.2.1810. Defects made good at Plymouth (for £16,375) 25.11.1810; sailed 30.12.1810 for Portugal. Defects made good at Portsmouth (for £13,082) 1–3.1812. In the Channel 1812–13. In Ordinary at Plymouth 1814. Sold to Mr Freake for £3,810 (at Plymouth) 30.4.1817.

Genoa (French *Le Brillant* built 2.1812–4.1815 at Genoa. L: 18.4.1815)

Dimensions & tons: 182ft 0in, 150ft 4¾in x 48ft 6¾in x 21ft 7in. 1,886⁵¹/₉₄bm.

Men: 600. Guns: LD 28 x 32pdr; UD 30 x 18pdr; QD 4 x 12pdr + 10 x 32pdr carronades; Fc 2 x 12pdr + 4 x 32pdr carronades.

Captured incomplete on the stocks at Genoa when that city surrendered on 18.4.1814. Launched there and completed for the RN. Arrived Chatham 13.10.1815?

Underwent Middling Repair and 'fitted as an English ship of war' at Chatham (for £36,839) 9.1816–2.1818. Fitted as a guard ship at Chatham (for £22,781) 6–10.1821.

Commissioned 3.10.1821 under Capt. Sir Thomas Livingstone; at Lisbon 10.1821–1825. Under Capt. William Cumberland 10.1824. Under Capt. Walter Bathurst 27.5.1825 in the Mediterranean; at Battle of Navarino 20.10.1827 (losing 26 killed including Bathurst, and 33 wounded). At Plymouth under Cmdr Dickinson in 1827; under Capt. Charles Irby in 11.1827; paid off 21.1.1828. Receiving ship at Plymouth 1833–7. BU at Plymouth 1.1838.

Le PLUTON Class A single unit building for the French Navy to a further Jacques-Noël Sané design (smaller than *Le Téméraire* Class, and indeed officially termed the '*petit modèle*') to which 25 ships were launched 1805–17 (and at least four more were begun at Venice but not completed, and another one ordered at Trieste but never begun). The design was specially conceived for production in yards situated in shallower waters. The *Rivoli* below was re-classed as 78 guns in 1817. A sister ship – building for the French Navy as *Le Royal-Hollandais* – was captured on the stocks at Vlissingen (Flushing), taken to England and completed as HMS *Chatham* (see under 24pdr type earlier).

Rivoli (French *Le Rivoli*, built 1.1807–10.1811 at Venice. L: 6.9.1810).

Dimensions & tons: 176ft 5½in, 144ft 8⅛in x 48ft 5in x 21ft 3¼in. 1,804⁴¹/₉₄bm.

Men: 590. Guns: LD 28 x 32pdr; UD 28 x 18pdr; QD 4 x 18pdr + 10 x 32pdr carronades; Fc 2 x 12pdr + 2 x 32pdr carronades; RH 6 x 18pdr carronades.

Taken by *Victorious* and *Weazel* in the Adriatic 22.2.1812. Arrived

Portsmouth 20.7.1812.

Underwent Large Repair and fitted at Portsmouth (for £51,318) 28.7.1813.

Commissioned 5.1813 under Capt. Graham Eden Hamond, for the Mediterranean. In 6.1814 under Capt. Edward Dickson; took 40-gun *La Melpomène* off Ischia 30.4.1815. From 1.1816 under Capt. Charles Ogle. Defects made good at Portsmouth and fitted as guard ship (for £13,327) 1–2.1816. In 4.1816 under Capt. Askew Hollis (and possibly in same year under Capt. Thomas Capel), as guard ship at Portsmouth. Paid off 2.1817. BU there 1.1819.

Ex-SPANISH PRIZES (1805) Five Spanish 74-gun ships had been captured by Nelson's fleet at Trafalgar in October 1805, but the *San Augustín* was burnt on Collingwood's orders and the *Monarca* was wrecked; the other three were escorted to Gibraltar and later added to the British Navy, together with another captured three months earlier by Calder's squadron. However, *Firme* (ex-*Ferme*) and *Bahama* were sold or BU in 1814, and *San Juan* (ex-*San Juan Nepomuceno*) and *Ildefonso* (ex-*San Ildefonso*) in 1816. Two earlier ships had been captured at Cape St Vincent in 1797, but the *San Ysidro* was been sold in 1814 and the *San Damaso* in 1816.

Ex-DANISH PRIZES (1807) Of the dozen Danish 74s captured by the British at Copenhagen, only two remained on the Navy List by the start of 1816. The two larger, more modern ships, *Danmark* and *Norge*, which had both carried 36pdrs on the LD in Danish service (with 18pdrs on the UD and 8pdrs on the upperworks), were re-armed for British service and rated as Middling Class 74s, but had been sold in December 1815 and March 1816 respectively. Of the nine ships of the *Prindsesse Sophia Frederica* Class that were taken, only five remained at the start of 1817 (see below). The twelfth of the prizes, the *Princess Carolina* (originally one of a pair designed by F. C. H. Hohlenberg, the newest and smallest of the Danish 74s) only carried 24pdr guns on the LD, and had been sold in February 1815.

PRINDSESSE SOPHIA FREDERICA Class The standard Danish 74-gun design by Henrik Gerner, to which eleven ships were completed, of which the nine that survived to 1807 were all captured at the fall of Copenhagen in 1807. These were vessels of average size – none of which were ever fitted for sea service in the RN. As built, they carried 24pdrs on the LD (with 18pdrs on the UD and 8pdrs on the upperworks), but were re-established in the RN (although not re-armed, as none were fitted for service) with the standard 'Common Class' armament of British-built 74s. The *Fyen*, *Kron Princen* (ex-*Kronprinds Friderich*) and *Kron Princessen* (ex-*Kronprinsesse Maria*) had been sold to BU in 1814, and the *Princess Sophia Frederica* (Danish *Prindsesse Sophia Friderich*) was BU in 1816. By 1817 only *Tree Kronen* and *Justitia* were still classed as 74s, the other three having been re-classed as receiving/victualling ships.

Tree Kronen (Danish *Tre Kroner*, K: 6.3.1788. L: 5.10.1789 at Nyholm Dyd, Copenhagen).

Dimensions & tons: 175ft 5⅛in, 143ft 4in x 47ft 10¼in x 19ft 9in. 1,745⁸⁶/₉₄bm.

Men: 590. Guns: LD 28 x 32pdrs; UD 28 x 18pdrs; QD 4 x 12pdrs + 10 x 32pdr carronades; Fc 2 x 12pdrs + 2 x 32pdr carronades; RH 6 x 18pdr carronades.

Taken at the fall of Copenhagen 7.9.1807. Arrived Portsmouth 8.11.1807 and laid up.

Not commissioned. Intended renaming as *Medway* in 1809 was cancelled. Doubled and coppered at Portsmouth 7–11.1811,

The 74-gun *Wellesley* is portrayed by William Joy off Amoy in the summer of 1841 during the First Anglo-Chinese War, shown accompanied by the sloop *Columbine* and other vessels including four HEICo. warships. Although ordered to be built at Bombay to the design of the *Black Prince* Class, the *Wellesley* was actually built of teak to the plans of the *Cornwallis*. (NMM F3467)

then laid up. Receiving ship 1817–25. Sold at Portsmouth to Mr Beatson (for £3,710) to BU 20.7.1825.

Heir Apparent Frederick (Danish *Arveprinds Friderich*, K: 6.12.1780. L: 1.6.1782 at Copenhagen)
   Dimensions & tons: 174ft 6¼in, 143ft 8½in x 47ft 10in x 19ft 6in. 1,746⁸³⁄₉₄bm.
   Men: 590. Guns: LD 28 x 32pdrs; UD 28 x 18pdrs; QD 4 x 12pdrs + 10 x 32pdr carronades; Fc 2 x 12pdrs + 2 x 32pdr carronades; RH 6 x 18pdr carronades.
   Taken at the fall of Copenhagen 7.9.1807. Arrived Portsmouth 19.11.1807 and laid up.
   Not commissioned. Intended renaming as *Cornwall* in 1809 was cancelled. Renamed *Arve Princen* in 1810. Prison ship under Lieut James Fuller 1812–14. Fitted at Portsmouth as a Victualling ship 12.1814–1.1815. Sold to Mr Freake (for £2,410) at ?Portsmouth to BU 3.4.1817.

Skiold (Danish *Skjold*, K: 23.9.1790. L: 7.12.1792 at Copenhagen)
   Dimensions & tons: 174ft 5½in, 143ft 3⅜in x 47ft 10½in x 19ft 7in. 1,747⁷⁄₉₄bm.
   Men: 590. Guns: LD 28 x 32pdrs; UD 28 x 18pdrs; QD 4 x 12pdrs + 10 x 32pdr carronades; Fc 2 x 12pdrs + 2 x 32pdr carronades; RH 6 x 18pdr carronades.
   Taken at the fall of Copenhagen 7.9.1807. Arrived Portsmouth 2.12.1807 and laid up.
   Not commissioned. Intended renaming as *Somerset* in 1809 was cancelled. Fitted as receiving ship at Portsmouth 10–12.1811. In Ordinary at Portsmouth 1812–25. Sold to John Small Sedger (for £3,200) 20.7.1825.

Odin (Danish *Odin*, K: 7.4.1787. L: 26.4.1788. C: 1789 at Copenhagen)
   Dimensions & tons: 174ft 7¼in, 143ft 2¼in x 47ft 10¼in x 19ft 7in. 1,747¹²⁄₉₄bm.
   Men: 590. Guns: LD 28 x 32pdrs; UD 28 x 18pdrs; QD 4 x 12pdrs + 10 x 32pdr carronades; Fc 2 x 12pdrs + 2 x 32pdr carronades; RH 6 x 18pdr carronades.
   Taken at the fall of Copenhagen 7.9.1807. Arrived Portsmouth 5.12.1807 and laid up.
   Not commissioned. No renaming was intended for this ship. Fitted at Portsmouth as receiving ship 12.1810–2.1811. Sold to Joshua Crystall (for £3,000) at ?Portsmouth 20.7.1825.

Justitia (Danish *Justitia*, K: 8.6.1776. L: 2.9.1777. C: 1780 at Copenhagen)
   Dimensions & tons: 174ft 3in, 144ft 2½in x 47ft 10½in x 19ft 9in. 1,758¹¹⁄₉₄bm.
   Men: 590. Guns: LD 28 x 32pdrs; UD 28 x 18pdrs; QD 4 x 12pdrs + 10 x 32pdr carronades; Fc 2 x 12pdrs + 2 x 32pdr carronades; RH 6 x 18pdr carronades.
   Taken at the fall of Copenhagen 7.9.1807. Arrived Portsmouth 5.12.1807 and laid up.
   Not commissioned. Intended renaming as *Orford* in 1809 was cancelled. Used for an experiment with Seppings's diagonal braces in 2.1817. BU at Portsmouth 3.1817.

***BLACK PRINCE* Class** 1810 'Middling Class' design based on reduced lines of the Danish 80-gun prize *Christian VII* (taken in 1807), the only foreign design that was not French to be copied for the battlefleet. The principal dimensions were virtually identical to the standard Surveyors' design (the keel was about 3in shorter). By a supplementary AO of 8 September 1813, they were to be built according to Seppings's principles of diagonal framing. However, the *Wellesley* only belonged to this Class in theory; plans for her were lost in December 1812 (with the capture of HMS *Java*) and so – with no other designs available to them – Bombay dockyard re-used the plans and the moulds of the *Cornwallis*; the *Wellesley* eventually suffered the most unusual loss of any Napoleonic warship – she was sunk by the Luftwaffe in 1940! Both Bombay-built ships (which were of teak construction) carried to the UK with them duplicate frames for use in building equivalent 74s in UK dockyards; the duplicate frame sent home in the *Wellesley* was prepared to the plans of the *Melville*, while that sent home in the *Melville* was actually used for the *Indus*. All this class were re-rated as 72-gun ships in 1839.
   Dimensions & tons: 176ft 0in, 144ft 9⅞in x 48ft 3in oa (47ft 6in for tonnage) x 21ft 0in. 1,738⁰⁄₉₄bm.
   Men: 590. Guns: LD 28 x 32pdrs; UD 28 x 18pdrs; QD 4 x 12pdrs + 10 x 32pdr carronades; Fc 2 x 12pdrs + 2 x 32pdr carronades; RH 6 x 18pdr carronades. Subsequently 4 x 68pdr carronades replaced one pair of guns on LD, and one pair on UD.

Black Prince Woolwich Dyd [M/Shipwright Henry Canham]
   As built: 176ft 1in, 144ft 10⅜in x 48ft 5in oa (47ft 8in for tonnage) x 21ft 0in. 1,750⁷⁴⁄₉₄bm. Draught 13ft 2in/16ft 10in.
   Ord: 14.8.1810. K: 7.1814. L: 30.3.1816. C: 29.8.1816 (for Ordinary, roofed over from main mast forward).
   First cost: £78,386 (+ fitting £2,296).
   Never commissioned. In Ordinary at Woolwich to 1822, and at

Sheerness 1827–49. Prison (convict) ship in 1848 for Woolwich. BU at Portsmouth completed 10.2.1855.

Wellesley East India Co., Bombay [M/Shipwright Jamsetjee Bomanjee]
>   As built: 175ft 10¾in, 144ft 11½in x 48ft 4in oa (47ft 7in for tonnage) x 21ft 0in. 1,745⁷¹⁄₉₄bm. Draught 16ft 6in/15ft 6in.
>   Ord: 6.1.1812. K: 5.1813. L: 24.2.1815. C: 2.1825 at Portsmouth (as guard ship).
>   First cost: £55,147 (including £15,630 of naval dockyard stores).
>   Commissioned 19.6.1815 under Capt. John Harper (for voyage from India); arrived Portsmouth 3.5.1816 and laid up. Recommissioned 30.3.1824 under Capt. Sir Graham Eden Hamond, for Particular Service. Fitted as guard ship at Portsmouth (for £26,597) 7.1824–2.1825. On 21.8.1825 under Capt. Gordon Thomas Falcon, as flagship of Rear-Adm. Sir George Eyre, on the South American station; paid off 1827. Recommissioned 13.2.1827 under Capt. Sir Frederick Lewis Maitland, for the Mediterranean; at Lisbon 10.1827 to mid 1828; paid off end 1829 at Portsmouth. Recommissioned 15.9.1830 under Capt. Samuel Campbell Rowley, for Particular Service; paid off 1.1832 at Plymouth. Small Repair at Plymouth (for £25,119) 2.1835–9.1837 (waist housed over in 11.1836), then fitted as a flagship at Portsmouth 9–10.1837. Recommissioned 19.6.1837 under Capt. Thomas Maitland, as flagship of Rear-Adm. Frederick Lewis Maitland (died 30.11.1839), then of Commodore Sir James John Gordon Bremer, then 7.7.1840 of Rear-Adm. Sir William Parker, then 30.11.1840 Bremer again, in the East Indies; in 1st Anglo-Chinese War, at Chusan 5.7.1840, Amoy 25/26.8.1841 and Tinghai 1.10.1841; laid up at Plymouth 7.1842. Fitted as flagship at Plymouth (for £11,157) 12.1847–3.1848. Recommissioned 6.1.1848 under Capt. George Goldsmith, as flagship of Vice-Adm. Thomas Cochrane, Earl of Dundonald, for North America and the West Indies; laid up at Chatham 6.1851. Fitted as guard ship for Ordinary at Chatham 7.1854. On 14.6.1854 under Capt. Christopher Wyvill, as guard ship at Chatham (in Ordinary); on 20.3.1856 under Goldsmith again, re-appointed 1.5.1858, then 18.4.1861 under Capt. Edward Gennys Fanshawe, in same role. Recommissioned as training ship as well as guard ship at Chatham 1862. On 9.11.1863 under Capt. William Houston Stewart; re-appointed 1.1.1866. Lent 6.4.1868 to the London School Ship Society as a reformatory, renamed *Cornwall* 18.6.1868. Sunk by the Luftwaffe in the Thames 24.9.1940.

Hawke Woolwich Dyd [M/Shipwright Henry Canham]
>   As built: 176ft 1in, 144ft 10½in x 48ft 5½in oa (47ft 8½in for tonnage) x 21ft 1½in. 1,753⁹¹⁄₉₄bm. Draught 14ft 0in/16ft 6in.
>   Ord: 6.1.1812. K: 4.1815. L: 16.3.1820. C: 25.5.1820 (for Ordinary), then laid up at Sheerness (with the waist roofed over).
>   First cost: £72,924.
>   Not commissioned until after conversion to steam in 1855. Small Repair at Chatham (for £6,713) 9.1833–9.1835. Fitted at Chatham as a Demonstration ship (for £5,889) 4.1836–1.1837. Converted to screw blockship at Chatham (for £23,365, plus £12,212 to Penn for engines) 10.1854–4.1855. Commissioned 14.2.1855 under Capt. Erasmus Ommanney, for the Baltic during the Russian War (when Ommanney was Senior British Officer in the Gulf of Riga), then to Falmouth; paid off 19.5.1856. Recommissioned 12.12.1856 under Capt. James Willcox, as guard ship at Queenstown; fitted for coastguard at Sheerness 12.1856–2.1857. In 19.7.1858 under Capt. William Crispin, later flagship of Rear-Adm. Sir Lewis Tobias Jones in same role. BU 1865.

Melville East India Co., Bombay. [M/Shipwright Jamsetjee Bomanjee]
>   As built: 176ft 2⅝in, 144ft 9in x 48ft 8in oa (47ft 11in for tonnage) x 21ft 0½in. 1,767⁷¹⁄₉₄bm.
>   Ord: 6.9.1813. K: 7.1815. L: 17.2.1817. C: 12.1826 (as guard ship) at Portsmouth.
>   First cost: £40,521 (at Bombay).
>   Commissioned 5.1817 (Capt. unknown) for voyage to UK; arrived Portsmouth 29.12.1817 and laid up in Ordinary; housed over 7.1820. Fitted as a guard ship at Portsmouth (for £16,974) 2–12.1826. Fitted as a flagship at Portsmouth (for £10,819) 12.1835–3.1836. Commissioned 19.1.1836 under Capt. Peter John Douglas, as flagship of Vice-Adm. Peter Halkett, for North America and the West Indies; paid off 8.1837. Fitted as a flagship at Portsmouth (for £6,524) 9.1837–1.1838, and commissioned 1.9.1837 under Capt. Richard Saunders Dundas, as flagship of Rear-Adm. George Elliot, for the Cape of Good Hope; later to East Indies, where served during 1st Anglo-Chinese War; paid off at end 1841. Fitted as a hospital ship for Hong Kong by J. White, Cowes (for £19,665) 9.1856–3.1857, then completed at Portsmouth (for £18,288) 5–6.1857. Reported by Commodore Short 14.9.1871 that ship very decayed and no longer fit for service. Sold for £7,565 (HK$35,600) at Hong Kong 20.10.1873 (by AO 15.10.1873).

Later *GANGES* Class 74 (later 72) guns Originally built to a Hunt design, Robert Seppings was allowed to apply his construction techniques to a reconstruction of the *Tremendous* with diagonal framing (new dimensions in consequence) and a further ship (*Minotaur*) was newbuilt to the same variant of this 1770s design. A further vessel of this design (*Boscawen*) was ordered on 2 December 1813, but was finally built as a 70-gun ship to a fresh design (see Section C).
>   Dimensions & tons: 172ft 1½in, 140ft 2in x 48ft 0in x 20ft 0in. 1,712⁵¹⁄₉₄bm.
>   Men: 550. Guns: LD 28 x 32pdr; UD 28 x 18pdr; QD 4 x 12pdrs + 10 x 32pdr carronades; Fc 2 x 12pdrs + 2 x 32pdr carronades (2 x 32pdrs on LD and 2 x 18pdrs on UD were later replaced by 4 x 68pdr carronades).

Tremendous Chatham Dyd [M/Shipwright Robert Seppings]
>   As rebuilt: 170ft 11in, 139ft 3in x 48ft 0in x 20ft 4in. 1,706⁵²⁄₉₄bm.
>   Reconstructed 2.1807–1.1811 (for £53,989 plus ?£13,700 fitting).
>   Commissioned 12.1810 under Capt. Robert Campbell (–1815); with Gore's squadron off Lorient 1811; off Texel 1812; sailed for the Mediterranean 15.8.1812; paid off 8/9.1815. Between Small and Middling Repair at Chatham (for £38,070) 4.1816–9.1819, then laid up at Sheerness. Receiving ship at Sheerness 1822–42. Renamed *Grampus* by AO 23.5.1845. Cut down to 50-gun Fourth Rate frigate and fitted for sea at Woolwich (for £27,610) 5.1844–1.1846 (see Chapter 4). Fitted as a powder depot at Portsmouth 11.1866. Lent to the War Department for the stowage of naval mines 12.1883. Sold to John Read, Portsmouth (for £1,605) 109.5.1897 (under AO 24.7.1896).

Minotaur Chatham Dyd [M/Shipwright Robert Seppings to 3.1813, completed by George Parkin]
>   As built: 170ft 11½in, 139ft 7½in x 48ft 2½in x 20ft 3¾in. 1,726⁶⁄₉₄bm. Draught 13ft 11in/18ft 0in.
>   Ord: 3.12.1811. K: 12.1812. L: 15.4.1816. C: 14.10.1816 for Ordinary, and laid up at Sheerness.
>   First cost: £63,667 (including fitting).
>   Not commissioned until 1859. Small Repair at Chatham (for £7,534) 8.1824–5.1825. Hulked as a receiving ship at Sheerness 3.1842,

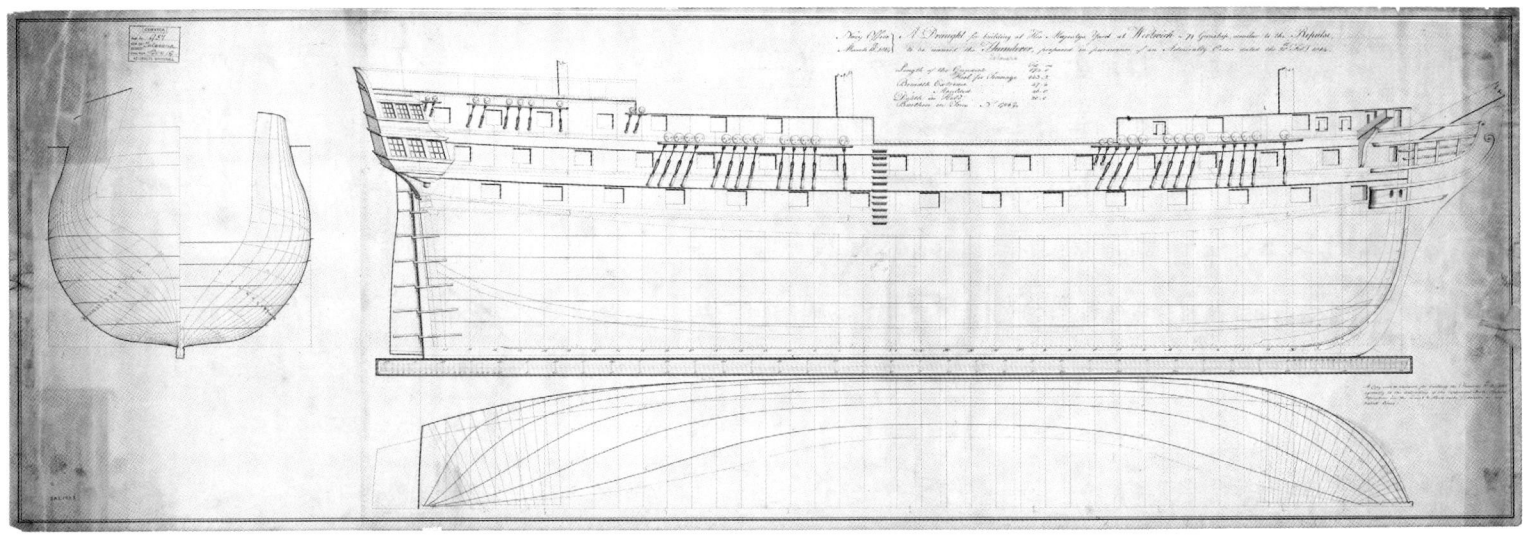

The *Talavera* was originally ordered in 1814 (under the name *Thunderer*, as shown on this sheer draught) to Rule's draught of the *Repulse* Class, but she was structurally modified while under construction to incorporate the new scheme of diagonal framing proposed by Robert Seppings, newly appointed as Joint Surveyor. *(NMM J3328)*

and fitted there as such (for £3,792) 7–11.1842. Commissioned 1.5.1859 under Capt. Edward Pellew Halstead, as guard ship of the Steam Reserve at Sheerness 1859. To Customs as a lazarette 11.1861; to Gravesend as a cholera ship 7.1866 (renamed *Hermes* 27.7.1866). BU (under AO 6.10.1868) completed at Sheerness 20.2.1869.

Further *REPULSE* Class Three further ships (*Belleisle*, *Thunderer* and *Malabar*) were ordered in 1812–15 to this revival of William Rule's 1800 design. The *Thunderer* (renamed *Talavera* while building) was modified by the introduction of Sepping's system of diagonal framing (AO 28.2.1814 'to be timbered according to Mr Seppings's principle') using oak from the Adriatic, while the *Malabar* (like all Bombay-built ships) was constructed of teak. The *Malabar* brought to the UK with her the duplicate frames intended for use for building the *Hindostan*.

 Dimensions & tons: 174ft 0in, 143ft 2in x 47ft 4in x 20ft 0in. 1,706⁸/₉₄bm.

 Men: 590, later 540 (362 officers & men, 53 boys, 125 marines).
  Guns: LD 28 x 32pdr, UD 28 x 18pdr, QD 2 x 18pdr + 12 x 32pdr carronades, Fc 2 x 18pdr + 2 x 32pdr carronades, RH 6 x 18pdr carronades.

Belleisle Pater (Pembroke) Dyd [M/Shipwright Edward Churchill]
 As built: 174ft 3in, 143ft 5in x 48ft 1in oa (47ft 4in for tonnage) x 20ft 0in. 1,709¹²/₉₄bm. Draught 13ft 7in/16ft 10in.
 Ord: 18.12.1812. K: 2.1816. L: 26.4.1819. C: 22.5–21.6.1819 at Plymouth (for Ordinary).
 First cost: £50,731 to build, plus £10,795 fitting.
 Middling Repair at Pembroke (for £8,221) 11.1832–8.1833. Fitted as a demonstration ship at Plymouth (for £5,087) 2–3.1836. Fitted there for sea (for £11,756) 4–9.1839. Commissioned 10.4.1839 under Capt. John Toup Nicholas, for the Mediterranean; paid off 1840 (as ineffective as an active man-of-war). Fitted there as a 20-gun troopship (for £4,076) 9–11.1841. Recommissioned 23.10.1841 under Capt. John Kingcome, for the East Indies; in Yangtze operations 7.1842; paid off late 1843 at Devonport. Refitted at Plymouth as a 20-gun troopship (for £8,175) 3–5.1846. Recommissioned 24.3.1846, again under Kingcome, for particular service; paid off 7.9.1848. Fitted at Plymouth

as a hospital ship for the Baltic (for £7,631) 3–5.1854. Recommissioned 13.3.1854 under Cmdr James Hosken; to the Baltic 1854; paid off 1856. Fitted at Plymouth as a hospital ship (for £22,504) for Sheerness 4–6.1857. Recommissioned 20.3.1857 under Cmdr John Rashleigh Rodd, as hospital ship for the East Indies and China; on 13.12.1858 under Cmdr Henry Maynard Bingham, in same role; paid off 1860? Lent (by AO 19.5.1866) to the Seamen's Hospital Society for cholera patients at Greenwich; returned to Sheerness (by AO 10.11.1868) on ceasing use as hospital ship. BU (by AO 15.5.1872) completed at Chatham 12.10.1872.

Talavera (ex-*Thunderer*, renamed 23.7.1817) Woolwich Dyd [M/Shipwright Edward Sison to 3.1816, completed by Henry Canham]
 As built: 174ft 0in, 143ft 2in x 48ft 3in oa (47ft 6in for tonnage) x 20ft 0in. 1,718¹⁸/₉₄bm. Draught 12ft 0in/16ft 10in.
 Ord: 28.1.1814. K: 7.1814. L: 15.10.1818. C: 17.11.1818–7.9.1819 at Chatham (for Ordinary), then roofed over and sailed to Sheerness.
 First cost: £68,409 to build (including fitting for Ordinary).
 Commissioned 15.9.1829 under Capt. Hugh Pigot, for the Preventive Service in the Downs; fitted as a guard ship at Sheerness (for £19,432) 9.1829–2.1830; on 26.10.1831 under Capt. Thomas Brown, for particular service; blockade of the Dutch coast; fitted for Lord Durham and his suite at Sheerness 6–7.1832; on 17.5.1833 under Capt. Edward Chetham, for Lisbon, then to the Mediterranean; paid off end 1834. Fitted as a demonstration ship at Plymouth (for £4,646) 3–4.1835. Fitted at Plymouth for sea (for £7,484) 3–6.1836. Recommissioned 18.3.1836 under Capt. Thomas Ball Sulivan, at Plymouth; on 26.11.1836 under Capt. William Bowen Mends; sailed 2.1837 for Lisbon, then to the Mediterranean, to Halifax and St John's (New Brunswick), then back to the Mediterranean; paid off 3.1.1840; burnt by accident at Plymouth 27.9.1840, then BU.

Malabar East India Company, Bombay Dyd [M/Shipwright Jamsetjee Bomanjee]
 As built: 174ft 3½in, 143ft 5in x 48ft 2in (47ft 5in for tonnage) x 19ft 11½in. 1,715¹⁵/₉₄bm.
 Ord: 7.3.1815. K: 4.1817. L: 28.12.1818. C: 10–11.1832 at Portsmouth (for sea).
 First cost: £56,385 to build, plus £13,624 fitting at Portsmouth.
 Sailed 5.4.1819 from Bombay; arrived Portsmouth 18.9.1819.
 Commissioned 16.10.1832 at Portsmouth under Capt. Josceline

Percy, for the Mediterranean. Recommissioned 4.1.1834 under Capt. Henry Shovell Marsham (acting from 24.12.1833), for the Mediterranean; paid off 7.1834. Recommissioned 25.7.1834 under Capt. Sir William Augustus Montagu, for Lisbon; paid off late 1837. Fitted to receive troops at Plymouth (for £5,626) 1–4.1838. Recommissioned 14.2.1838 under Capt. Edward Harvey, for North America and the West Indies; paid off early 1839. Fitted at Plymouth as a demonstration ship (for £5,333), then for sea (for £6,141) 1.1839–11.10.1841. Recommissioned 19.8.1841 under Capt. Sir George Rose Sartorius, for the Mediterranean; paid off late 1844. Fitted as a coal depot at Portsmouth (for £7,803) 10.1848. Renamed *Myrtle* by AO 30.10.1883. Sold at Portsmouth 17.7.1905.

Valiant (76 guns)
Ord at Plymouth to lines of *Bulwark*.
Canc. 1832
Dimensions & tons: 182ft 0in, 150ft 8½in x 49ft 10in (49ft 0in for tonnage) x 20ft 10in. 1,9243⁶⁶⁄₉₄bm.

There were fifteen 74s to wartime designs still under construction or on order at the close of the Hundred Days, of which six were to the numerous wartime *Armada* (or 'Surveyors') Class, four of the *Black Prince* Class (including the already-launched *Wellesley*), three of Rule's *Repulse* Class, and two to the Modified *Ganges* Class (this last pair with the new diagonal framing scheme as devised by Seppings). Under the re-classification of Rates brought in on 1 January 1817, many 74s were re-classed as 76-gun ships.

Former Third Rates of 64 guns

At the close of 1816, only six vessels of 64 guns remained on the List; this total excludes other former 64s already re-classed as ancillary vessels (hulks in harbour service), and in the re-organisation of Rates that took place on 1 January 1817, all six were re-classed as ancillary vessels, so that this once numerous type ceased to exist. As from 1817 the lower level for the Third Rate became 70 guns, any surviving ships of 62–68 guns would have been transferred to the Fourth Rate at that date. However, for continuity those former 64-gun ships which were still afloat are listed below with their original details and service histories, rather than in the chapter on Fourth Rates.

1422t	*Ardent*	1796	Prison ship 1813, BU 1824
1452t	*Argonaut*	1782	Hulked 1797, BU 1831
1366t	*Belliqueux*	1780	Hulked 1814, BU 1816
1405t	*Crown*	1782	Hulk 1806, BU 1816
1342t	*Gelderland*	1799	Sold to BU 3.1817
1409t	*Polyphemus*	1782	Hulked 1813, BU 1827
1267t	*Prince Frederick*	1796	Sold to BU 6.1817
1370t	*Ruby*	1776	Receiving ship 1813, BU 1821
1370t	*Standard*	1782	BU 1816
1317t	*Texel*	1799	Sold to BU 5.1818
1366t	*Trident*	1768	Receiving ship 1807, sold 1816
1397t	*Veteran*	1787	Prison ship 1809, BU 1816
1376t	*Vigilant*	1774	Prison ship 1799, BU 1816

Ex-FRENCH PRIZE (1761) In 1817 the Navy List still included one of the 64-gun ships taken from the French during the Seven Years War, a ship originally built for the French East India Company in 1761.
Belleisle (French East Indiaman *Le Bertin*, built 5.1760–2.1761 at Lorient. L: 29.10.1760).
Dimensions & tons: 168ft 5½in, 138ft 5½in x 45ft 0½in x 20ft 7in. 1,494²⁶⁄₉₄bm.

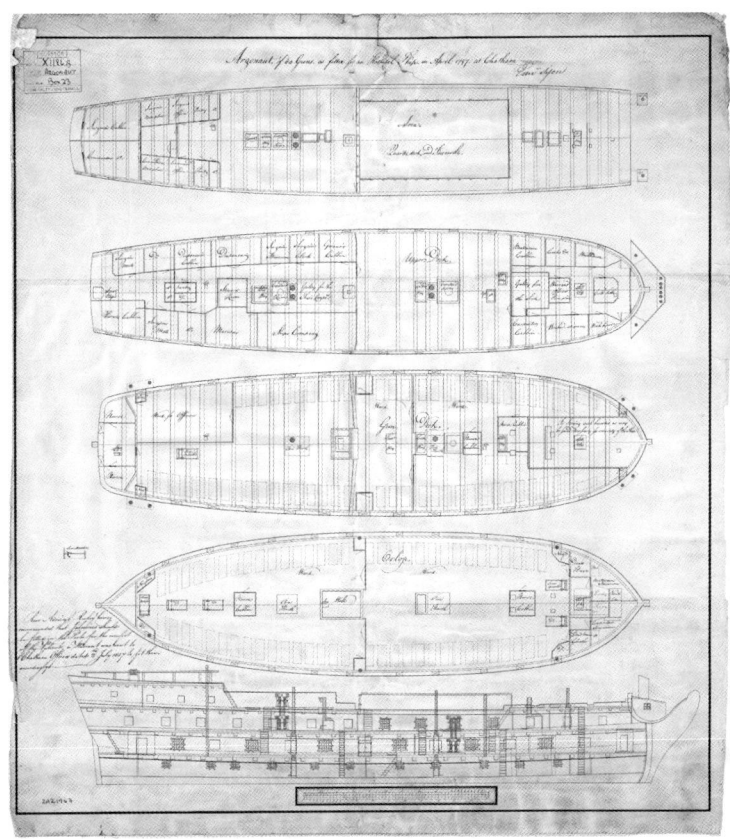

The *Argonaut* 64 was captured from the French in 1782, and this draught (dated August 1797, and signed by Edward Sison, master shipwright from 1795 to 1801) shows her conversion between the previous November and February 1797 into a hospital ship at Chatham Dockyard. She underwent a refit in 1804 and recommissioned in the same role. The draught shows plan views of the quarterdeck and forecastle, upper deck, lower deck, orlop deck and fore and aft platforms, with an inboard profile down to the waterline; it notes an amendment by Navy Board Warrant of 2 July 1817 to fit louvres (*jalousies*) to the ports to improve ventilation 'for the comfort of the patients, on the recommendation of Rear Admiral Rowley'. *(NMM J3947)*

Men: 500 (420 from 1782). Guns: Originally LD 26 x 24pdrs; UD 28 x 18pdrs; QD 8 x 9pdrs; Fc 2 x 6pdrs. From 1782. LD 24 x 68pdr carronades; UD 26 x 42pdr carronades; QD/Fc 8 x 24pdr carronades.
Taken 3.4.1761 by *Hero* and *Venus*; arrived Plymouth 5.6.1761. Purchased for the RN 3.7.1761. Named and established 8.7.1761. Commissioned 8.1761. Fitted at Plymouth (for £12,476.19.1d) 7.1761–1.1762; paid off 7.1763 after wartime service. Fitted as guard ship at Plymouth 1.1764; fitted there as troopship 5.1765 and again 3.1768. Small to Middling Repair at Plymouth (for £15,127.18.11d) 3.1771–4.1772. Fitted as guard ship at Plymouth (for £3,820.6.4d) 8.1776–1.1777. Refitted at Plymouth (for £3,186.17.10d) 12.1777–2.1778. Fitted at Plymouth for East Indies (for £9,175.5.5d) 4–6.1778, and again (for £3,347.14.2d) in 2.1779. Fitted at Chatham for North Sea with all-carronade armament (see above) and a 60-gun ship's masts and yards (for £7,397.13.2d) 3–6.1782. Paid off 3.1783 after wartime service. Fitted at Chatham as a lazarette (for £1,547.3.10d) 8–12.1784, thence to Stangate Creek where transferred to Customs and deleted from the Navy List. Sold to J. Beatson (for £2,730) 3.2.1819.

ARDENT Class Design by Thomas Slade of 1762, based on the lines of the *Fougueux* (a French prize, taken 1747). Two ships were built initially, but the design was revived in 1777 for five further ships. None remained by 1817; the name-ship, *Ardent*, had been sold in 1784; *Nassau* was wrecked in 1799 and *Agamemnon* in 1809; *Stately* was BU in 1814, *Raisonnable* in 1815, and *Belliqueux* and *Indefatigable* in 1816 – *Indefatigable* had been converted (raséed) into a Fifth Rate in 1794–5, initially becoming Edward Pellew's command, as made famous by C. S. Forester. All earlier 64s had also been disposed of before 1817, other than the elderly *Belleisle* listed above.

WORCESTER Class In November 1765 orders were issued for two new 64s to be built in the Royal Dockyards, and the two Surveyors were requested to produce competitive draughts in the usual fashion. Thomas Slade's final 64-gun draught returned to the overall length of his early 1761 design for the *Saint Albans*, but with somewhat fuller lines. The *Worcester*, the prototype of this class, was BU in 1816, but October 1768 two further vessels were ordered from the Royal Dockyards to this draught, of which the Chatham-built *Stirling Castle* was wrecked in 1780.

> Dimensions & tons: 159ft 0in, 130ft 7½in x 44ft 6in x 19ft 0in. 1,373⁷³⁄₉₄bm.
> Men: 500 (491 from 1794). Guns: LD 26 x 24pdrs; UD 26 x 18pdrs; QD 10 x 9pdrs; Fc 2 x 9pdrs (+ 2 x 24pdr carronades from 1794); RH (from 1794) 6 x 18pdr carronades.

Lion Portsmouth Dyd [M/Shipwright Thomas Bucknall to 10.1772, completed by Edward Hunt]

> As built: 159ft 0in, 130ft 4in x 44ft 8in x 19ft 0in. 1,377⁸⁶⁄₉₄bm. Draught 11ft 11in/16ft 0in.
> Ord: 12.10.1768. K: 5.1769. L: 3.9.1777. C: 7.9.1778.
> First cost: £26,220.18.10d to build, plus fitting £4,923.4.3d.
> Commissioned 5.1778; refitted and coppered (for £8,255.14.11d) at Portsmouth 12.1780–1.1781; paid off 8.1783 after wartime service. Middling Repair at Portsmouth (for £13,489) 2–9.1787. Fitted at Portsmouth (for £4,759) 7.1790; recommissioned 6.1790 under Capt. Seymour Finch, for the Spanish Armament; paid off 9.1791. Fitted at Portsmouth (for £9,460) 3–7.1792; recommissioned 5.1792 under Capt. Sir Erasmus Gower; sailed 26.9.1792 for China, with Lord Macartney's Embassy; paid off 10.1794. Fitted at Chatham (for £8,518) 10.1794–4.1795; recommissioned 2.1795 under Capt. George Palmer, for the North Sea; later under Capt. Henry Inman. In 1796 under Capt. Edmund Crawley; Nore Mutiny 1797. In 7.1797 under Capt. Charles Cobb, then 9.1797 under Capt. Manley Dixon; sailed for the Mediterranean 2.6.1798; at Blockade of Malta 1798–1800; in action against 34-gun frigates *Santa Cazilda*, *Pomona*, *Proserpine* and *Santa Dorotea* 15.7.1798 (the last-named being taken); took (with *Foudroyant* and *Penelope*) 80-gun *Le Guillaume Tell* 31.3.1800, losing 8 killed and 38 wounded. In 4.1800 under Capt. Lord William Stuart, then Capt. George Hamond 7.1800; paid off 11.1800. Fitted at Chatham (for £13,545) 2–5.1801; recommissioned 3.1801 under Capt. Henry Mitford; sailed for the East Indies 20.5.1801. Repaired by John Dudman, Deptford (for £58,124) 12.1804–12.1805; completed fitting at Deptford (for £15,509) 1.1806; recommissioned 12.1805 under Capt. Robert Rolles; sailed for the East Indies 5.1806; took 14-gun privateer *La Réciprocité* off Beachy Head 27.12.1807. In 2.1808 under Capt. Henry Heathcote; sailed for China 5.3.1808; sailed for the East Indies 18.7.1810; at reduction of Java 7.1811. In 1812 flagship of Vice-Adm. Robert Stopford at the Cape of Good Hope, then 1812–13 flagship of Rear-Adm. Charles Tyler; in this period commanded successively by Capts James Johnstone, George Douglas, Henderson Bain and John Eveleigh. Fitted as

sheer hulk at Plymouth 8.1814; to Sheerness 9.1816. Sold to John Levy & Son at Chatham (for £2,300) to BU 30.11.1837.

INTREPID Class To evaluate competitively against Slade's *Worcester* design, a slightly smaller ship was designed by John Williams, who in June 1765 had replaced William Bately when the latter retired, and this was approved 18 December 1765. Another fourteen 64s followed to this draught during Williams's tenure of office, but of these *Defiance* was wrecked in 1780 and *Repulse* in 1800, *Nonsuch* was BU in 1802 and *America* in 1807, *Anson* was wrecked in 1807, *Eagle* (renamed *Buckingham* in 1800) was BU in 1812, *Magnanime* in 1813, and *Standard* and *Vigilant* in 1816. *Anson* and *Magnanime* had been raséed into frigates in 1794–5 (re-armed like *Indefatigable*, with UD 26 x 24pdrs; QD 8 x 12pdrs + 4 x 32pdr carronades; Fc 4 x 12pdrs + 2 x 32pdr carronades).

> Dimensions & tons: 159ft 6in, 131ft 0in x 44ft 4in (43ft 8in mld.) x 19ft 0in. 1,369⁵⁹⁄₉₄bm.
> Men: 500 (491 from 1794); 310 as 44-gun frigates. Guns: LD 26 x 24pdrs; UD 26 x 18pdrs; QD 10 x 9pdrs; Fc 2 x 9pdrs (+ 2 x 24pdr carronades from 1794); RH (from 1794) 6 x 18pdr carronades. *Nonsuch* as floating battery carried 20 x 68pdr carronades (LD) and 26 long guns.

Intrepid Woolwich Dyd [M/Shipwright Joseph Harristo 7.1767, completed by William Gray]

> As built: 159ft 6in, 131ft 0in x 44ft 5in x 19ft 0in. 1,374⁶⁄₉₄bm. Draught 10ft 11½in/16ft 6in.
> Ord: 16.11.1765 (approved 18.12.1765). K: 1.1767. L: 4.12.1770. C: 31.1.1771.
> First cost: £28,281.9.9d, including fitting.
> Commissioned 1770 for Falkland Islands dispute. Fitted as a guard ship at Portsmouth 10.1771. Fitted for the East Indies (for £5,547.5.7d) 3.1772; sailed for the East Indies 16.4.1772; fitted to be laid up at Plymouth 5.1773; paid off 4.1775. Fitted for Home service at Plymouth (for £12,550.6.3d) 7.1778–4.1779; refitted and coppered at Portsmouth (for £5,071.2.4d) 5–12.1779; recommissioned 1.1779; sailed for the Leeward Islands 30.1.1780; paid off 8.1782 after wartime service. Fitted for Ordinary at Plymouth 10.1783. Middling to Great Repair at Plymouth (for £31,098) 2.1790–7.1792. Fitted at Plymouth (for £4,492) 3–7.1793; recommissioned 2.1793 under Capt. Charles Carpenter; at occupation of Toulon; sailed for Jamaica 20.5.1794; took (with *Chichester*) French 36-gun *La Sirène* off San Domingo 8.1794; took privateers in West Indies – *Le Perroux*, *La Républicain Pagest* and *Le Sans Pareil* 1–2.1795; took 26-gun *La Perçante* off Porto Plata 23.4.1796; paid off 11.1796. Fitted at Portsmouth (for £15,239) 12.1796–5.1797; recommissioned 3.1797 under Capt. Robert Parker (drowned 11.1797); sailed with convoy to Cape of Good Hope 17.6.1797. In 3.1798 under Capt. William Hargood (–1802), in the East Indies. Repaired by Daniel Brent, Rotherhithe (for £38,215) 10.1804–7.1805; completed fitting at Deptford 8.1805; recommissioned 8.1805 under Capt. Philip Wodehouse (–1807); sailed for the Mediterranean; in Sidney Smith's squadron at Naples 6.1806. In 1807 under Capt. John Laugharne, then 10.1807 Capt. Richard Worsley; sailed for the Leeward Islands 30.11.1807; with Hood's squadron at Madeira 24–26.12.1808, under Capt. Warwick Lake (acting). In 1809 under Capt. Christopher Nesham; at capture of Martinique 2.1809. Fitted as receiving ship at Plymouth 5.1810; in Ordinary there to 1815. Sold to D. Beatson (for £3,030) 26.3.1828.

Captivity (ex-*Monmouth*) Plymouth Dyd [M/Shipwright Israel Pownoll]

As built: 159ft 6in, 131ft 0in x 44ft 4in x 19ft 0in. 1,369⁵¹⁄₉₄bm.
Draught 11ft 2½in/17ft 4¾in.

Ord: 10.9.1767 (approved 22.10.1767, named 11.1767) K: 5.1768. L:
18.4.1772. C: 10.1777–9.5.1778.

First cost: £30,586.17.3d to build, plus fitting £7,426.15.1d.

Commissioned 1.1.1778; sailed for the Leeward Islands 18.6.1778;
refitted and coppered at Portsmouth (for £13,034.9.3d)
12.1779–12.1780; sailed 13.3.1781 for Porto Praya, and thence to
India; paid off 7.1784 after wartime service. Renamed *Captivity*
as prison ship at Portsmouth 20.10.1796, and fitted as such
(for £1,331) 10–11.1796. In 12.1796 under Lieut Samuel Blow
(–1799), then 1800 Lieut Emanuel Hungerford, 9.1801 Lieut
Jacob Silver and 12.1805 Lieut ? McDonald (–1806). BU at
Portsmouth 1.1818.

Ruby Woolwich Dyd [M/Shipwright William Gray to 3.1773,
completed by Nicholas Phillips]

As built: 159ft 6in, 131ft 0in x 44ft 6in x 19ft 0in. 1,369⁵⁹⁄₉₄bm.
Draught 11ft 2in/16ft 10in.

Ord: 30.11.1769 (approved 12.3.1770). K: 9.9.1772. L: 26.11.1776. C:
27.2.1778.

First cost: £26,980.12.5d to build, plus fitting £4,562.9.11d.

Commissioned 9.1777; sailed for Jamaica 24.5.1778; paid off 1.1782
after wartime service. Fitted and coppered at Portsmouth (for
£15,326.2.1d) 3–8.1782; recommissioned and sailed 11.9.1782
to relief of Gibraltar. Small to Middling Repair at Portsmouth
(for £15,038) 7.1785–3.1786. Fitted at Portsmouth (for £5,941)
3–8.1793; recommissioned 4.1793 under Capt. Sir Richard
Bickerton, for the Channel; joined Montagu's squadron in 6.1794.
Later under Capt. Henry Stanhope; sailed for the Cape of Good
Hope 12.3.1795; surrender of Dutch squadron in Saldanha Bay
17.8.1795. In 3.1796 under Capt. George Brisac, in East Indies,
then under Capt. Thomas Bertie and ?2.1797 Capt. Jacob Waller;
paid off into Ordinary at Chatham 11.1797. Repaired and fitted
at Chatham (for £24,923) 1.1798–7.1799; recommissioned 5.1799
under Capt. Alan Gardner, for the Channel. In 1800 under
Capt. Solomon Ferris; took 22-gun privateer *La Fortune* in the
South Atlantic 14.7.1800. Made good defects at Chatham (for
£6,955) 3.1801; in 4.1801 under Capt. Sir Edward Berry, for the
Baltic and North Sea; in 4.1802 under Capt. Henry Hill. Fitted
at Chatham (for £8,083) 4.1802–7.1803, then in Ordinary there.
Recommissioned 5.1803 under Capt. Francis Gardner; in 10.1803
under Capt. David Colby, then 2.1804 Capt. Charles Rowley and
7.1806 Capt. John Draper, in the North Sea; sailed for Portugal
1.1.1808. In 12.1808 under Capt. Robert Hall, for the Baltic, then
6.1809; flagship of Rear-Adm. Manley Dixon 1810. In ?6.1810
under Capt. Matthew Bradby, then 7.1810 Cmdr Thomas
White (acting). Recommissioned 10.1810 and fitted at Chatham
4–6.1811 as depot ship for Bermuda; sailed for North America
25.7.1811. In 1812 under Lieut Peter Trounce, as receiving ship at
Bermuda under Broad Pennant of Capt. Andrew Evans (–1817);
in 1813–14 under Lieut James Ward, then 1815 Lieut James
Knight. BU at Bermuda 4.1821.

Polyphemus Sheerness Dyd [M/Shipwright George White to 3.1778,
then John Jenner to 4.1779, completed by Henry Peake]

As built: 160ft 0in, 133ft 3in x 44ft 7in x 19ft 0in. 1,408⁷¹⁄₉₄bm.
Draught 11ft 0in/16ft 9in.

Ord: 1.12.1773 (approved 16.12.1773). K: 1.1776. L: 27.4.1782. C:
24.7.1782 (incl. coppering).

First cost: £37,218.5.1d including fitting & coppering.

Commissioned 4.1782; sailed for Gibraltar, and later to West Indies;
paid off 6.1783 after wartime service. Small Repair at Chatham

(for £6,371.10.4d) 12.1783–9.1784. Small Repair and fitted at
Chatham (for £8,771) 12.1793–6.1794; recommissioned 4.1794
under Capt. George Lumsdaine (–1800); capture of Dutch 74-
gun *Overyssel* 22.10.1795; refitted at Plymouth (for £5,605)
5–7.1796; flagship of Robert Kingsmill on the Irish station
1796–1800; took (with *Apollo*) 14-gun privateer *Les Deux Amis*
12.1796; capture of 40-gun *La Tortue* off Ireland 5.1.1797. Fitted
at Chatham (for £12,753) 11.1799–3.1800; in 1800–1 under
Capt. John Lawford; at Battle of Copenhagen 2.4.1801, losing
6 killed and 25 wounded; paid off into Ordinary 4.1802. Fitted
at Chatham 3–9.1804; recommissioned 7.1804 under Capt.
Robert Redmill (–1806), for the Channel fleet; took Spanish
36-gun *Santa Gertruyda* off Cape St Mary 7.12.1804; in Lee
column at Battle of Trafalgar 21.10.1805, losing 2 killed and 4
wounded; with Hood's squadron off Rochefort 1806; her boats
(with squadron's) took French 36-gun *Le César* in the Gironde
15.7.1806. Recommissioned 7.1806; in 10.1806 under Capt.
John Broughton, later under Capt. Peter Heywood, as flagship
of Rear-Adm. George Murray; in River Plate operations 1807.
In 1808 under Capt. William Cumby, as flagship of Vice-Adm.
Bartholomew Rowley; sailed for Jamaica 2.7.1808; her boats
took 3-gun *Colibri* off San Domingo 15.11.1808; at capture
of French 74-gun *Le d'Hautpoult* 17.4.1809. In 1811 under
Capt. Thomas Graves, on the Jamaica station. In 10.1811 under
Cmdr Nicholas Pateshall (temp.), then 12.1811 Capt. Cornelius
Quinton as flagship of Vice-Adm. Charles Stirling at Jamaica,
then Capt. Peter Douglas; paid off 11.1812. Fitted as powder
magazine at Chatham 3–9.1813; in Ordinary to 1815 at Chatham,
then powder ship in Medway to 1826. Fitted at Chatham for
Ordinary 2–4.1826. BU completed at Chatham 15.9.1827.

Sampson Woolwich Dyd [M/Shipwright Nicholas Phillips to end
1777, then George White to 4.1779, completed by John Jenner]

As built: 159ft 5½in, 131ft 2¼in x 44ft 5¼in x 18ft 10½in.
1,380⁵⁹⁄₉₄bm. Draught 11ft 6in/16ft 5in.

Ord: 25.7.1776. K: 20.10.1777. L: 8.5.1781. C: 29.6.1781.

First cost: £35,890.12.10d, including fitting & coppering.

Commissioned 4.1781; paid off and recommissioned 4.1783, and
fitted for guard ship at Plymouth 5–9.1783; paid off 6.1786.
Small Repair at Plymouth (for £19,255) 6–7.1792; commissioned
2.1793 under Capt. Robert Montagu; fitted at Plymouth 7.1793;
sailed with trade (i.e. convoy) for the East Indies 20.3.1794; paid
off 12.1794 and recommissioned same month. In 4.1795 under
Capt. Thomas Louis, later under Capt. William Clark; sailed
for Jamaica 23.5.1795. In 2.1796 under Capt. George Gregory,
then 5.1796 Capt. George Tripp, later Capt. Joseph Bingham;
paid off 2.1797. Recommissioned 11.1797 under Lieut William
Bevians (–1800) as prison ship at Plymouth. In 9.1801 under
Lieut John Norris; paid off 4.1802. Hulked as powder magazine
at Plymouth 8.1802. Receiving ship at Cork 10.1805 (but not
in commission). Fitted at Plymouth and re-armed as 64-gun
12.1805–1.1806; recommissioned 4.1806 under Capt. Sir Thomas
Masterman Hardy, then 7.1806 under Capt. Samuel Warren,
then Capt. William Cuming, as flagship of Rear-Adm. Charles
Stirling; sailed with convoy for Cape of Good Hope 30.8.1806,
and thence to River Plate; home in 7.1807; paid off into
Ordinary at Chatham 12.1807. Commissioned 3.1808 as prison
hulk in the Medway, under Lieut John Watherston to 1811, then
Lieut William Mounier in 1812 and Lieut Thomas Steventon in
1812–13. In 1814 in Ordinary at Chatham. Fitted as a sheer hulk
at Woolwich (for £14,147) 8–9.1814. Sold at Deptford to John
Levy, Rochester (for £1,830) 30.5.1832 to BU.

Diadem Chatham Dyd [M/Shipwright Israel Pownoll to 4.1779, completed by Nicholas Phillips]

As built: 159ft 10in, 131ft 1in x 44ft 5in x 19ft 0in. 1,375⁵³⁄₉₄bm. Draught 11ft 11in/16ft 7in.

Ord: 5.12.1777 & 11.6.1778. (named 23.7.1778) K: 2.11.1778. L: 19.12.1782. C: 19.7.1783 as guard ship.

First cost: £36,401.9.8d including fitting and coppering.

Commissioned 3.1783 as guard ship at Chatham; at Plymouth from 1784, where copper repaired (for £1,547.3.4d) to 4.1784; paid off 3.1786. Small Repair at Plymouth (for £7,706.0.3d) 1–5.1790. Fitted at Plymouth (for £5,053) 1–5.1793; recommissioned 2.1793 under Capt. Andrew Sutherland; sailed for the Mediterranean 15.10.1793; in Toulon operations 1793–4. In 1795 under Capt. Charles Taylor; in Hotham's Action off Genoa 13.3.1795, with 3 killed and 7 wounded; in Action off Hyères 13.7.1795. In 1796 under Capt. George Towry; in Nelson's squadron off Genoa 4.1796; flagship of Nelson at Leghorn 8.1796; at Battle of St Vincent 14.2.1797, with 2 killed and 7 wounded; at blockade of Cadiz 4.1797. Fitted as Troopship at Plymouth (for £7,412) 4–5.1798. Refitted at Portsmouth (for £6,733) 5–7.1799; recommissioned 1799 under Capt. John Dawson. In 1800 under Capt. Thomas Livingstone; in Quiberon operations 4–6.1800. In 1801 under Capt. John Larmour, with Keith's squadron in the Mediterranean; in landings at Aboukir Bay 8.3.1801; paid off at Woolwich 4.1802. Small to Middling Repair and fitted at Woolwich 4.1804–1.1805; recommissioned 12.1804 (as 64-gun again) under Capt. Home Popham, for the Channel; in 5.1805 under Capt. Charles Grant, then 1.1806 under Capt. Hugh Downman, as flagship of the now Rear-Adm. Popham; at capture of Cape of Good Hope and then River Plate operations; took Spanish brig *Arrogante* off Montevideo 30.7.1806. Later in 1806 under Capt. Samuel Warren, as flagship of Rear-Adm. Charles Stirling, for continued operations in River Plate area. Fitted for troops at Chatham 4–7.1810; recommissioned 6.1810 as a troopship under Capt. John Phillimore, for Lisbon; to Halifax station 1812. In 5.1813 under Capt. John Hanchett; in boat attack on Norfolk (Virginia) 22.6.1813; paid off at Plymouth 12.1814. Fitted as receiving ship at Plymouth 5–6.1815. Troopship again 1822–5, then reverted to receiving ship at Plymouth. BU there 9.1832.

INFLEXIBLE Class Designed by John Williams in 1777, based on the scaled-down lines of Slade's 74-gun *Albion* design of 1759, in turn based on the old 90-gun *Neptune* of 1730, which had been built to the 1719 Establishment. Thus the hull form was already over sixty years old by the time that the *Inflexible* took the water. She was found to be deficient in sailing qualities, and no further 64s were built to this design after 1780. Of her sisters, the *Sceptre* was wrecked in 1799 and the *Africa* was BU in 1814. The *Diadem* (above) was originally intended to be to this design, but was modified to *Intrepid* Class design in 1778.

Dimensions & tons: 159ft 6in, 131ft 0in x 44ft 6in x 18ft 0in. 1,379⁸⁄₉₄bm.

Men: 500 (491 from 1794). Guns: LD 26 x 24pdrs; UD 26 x 18pdrs; QD 10 x 9pdrs; Fc 2 x 9pdrs (+ 2 x 24pdr carronades from 1794); RH (from 1794) 6 x 18pdr carronades. (*Inflexible* re-armed 12.1806 with Gover's 24pdrs : LD 26 x 24pdrs, UD 26 x 24pdrs, QD 2 x 24pdrs + 10 x 24pdr carr., Fc 2 x 24pdr carr.)

Inflexible John Barnard, Harwich

As built: 159ft 8in, 131ft 1in x 44ft 7in x 18ft 7½in. 1,385⁸⁄₉₄bm.

Ord: 5.2.1777 & 26.2.1777. K: 4.1777. L: 7.3.1780. C: 13.6.1780 at Chatham.

First cost: £34,619.12.2d to build (£36,227.3.5d including fitting and coppering).

Commissioned 2.1780; paid off 6.1784 after wartime service. Middling Repair at Chatham (for £14,867) 4.1785–1.1786. Fitted as a storeship at Chatham (for £5,709) 9–12.1793; recommissioned 10.1793 under Capt. Solomon Ferris (–1799), for the Downs; paid off 4.1795. Fitted and re-armed as 64 at Woolwich (for £8,672) 4–7.1795; recommissioned 4.1795 for Duncan's fleet. Employed as a troopship from 3.1798; fitted as a troopship at Sheerness (for £2,543) 5–7.1799; paid off 10.1799. Fitted as a troopship at Chatham (for £6,713) 12.1799–3.1800; recommissioned 1.1800 under Capt. Benjamin Page; paid off 4.1802 into Ordinary. Fitted as troopship again at Chatham 11.1803–2.1804; recommissioned 1.1804 under Capt. Thomas Bayley, for North Sea; re-established as a 64 in 1.1805; took 4-gun privateer *L'Alerte* 12.4.1805. Fitted for sea (as a 64) at Chatham 1–6.1807; recommissioned 2.1807 under Capt. George Scott. Later under Capt. Joshua Watson for Copenhagen expedition 8.1807, then paid off. Recommissioned 8.1808 under Capt. Donald Mackay; from 10.1808 under Capt. Thomas Browne. Fitted at Chatham as a powder magazine for Halifax, Nova Scotia 5.1809 (by AO 16.4.1808), where stationed 1809–20. BU there 1820.

Dictator Robert Batson, Limehouse

As built: 159ft 4in, 130ft 8in x 44ft 8¼in x 18ft 0in. 1,387⁸⁷⁄₉₄bm. Draught 12ft 1in/15ft 11in.

Ord: 21.10.1778 (approved 13.11.1778). K: 5.1780. L: 6.1.1783. C: 18.1–30.5.1783 at Woolwich (as guard ship).

First cost: £35,383.0.8d to build, plus £2,438.11.1d fitting.

Commissioned 1.1783 as guard ship in the Medway; paid off 3.1786. Small Repair at Chatham (for £4,000) 3–7.1789. Fitted at Chatham (for £4,350) 7–10.1790; recommissioned 8.1790 under Capt. Richard Bligh, as flagship of Rear-Adm. Sir Richard King. Recommissioned 4.1791 under Capt. Thomas Tonken, as flagship of Rear-Adm. John Dalrymple and receiving ship at Blackstalks; paid off 9.1791. Fitted at Chatham (for £4,264) 9–12.1793; recommissioned 9.1793 under Capt. Edmund Dod. In 6.1793 under Capt. Nathan Brunton; sailed for West coast of Africa 5.3.1794; home late 1794 and paid off. Fitted at Portsmouth (for £9,323) 2–7.1795. In 9.1795 under Capt. Thomas Totty; sailed for Jamaica 26.2.1796; in 1797 under Capt. Thomas Western then Capt. William Rutherford. Fitted at Plymouth as a Troopship (for £10,378) 5–6.1798; in 1798 under Capt. Thomas Byam Martin, then 1799 under Capt. John Oakes Hardy (–1801), in North Sea; Egypt operations (*en flûte*); paid off 3.1802. Fitted at Chatham as a floating battery (for £6,888) 2–5.1803, for Sheerness; recommissioned 4.1803 under Capt. John Newhouse, for King's Deep; in 1804 under Capt. Charles Tinling, as guard ship in King's Deep, later under Capt. Richard Hawkins. Repaired (as 64 again) at Cox & Co., Thames (for £26,061) 10.1804–5.1805, then fitted at Deptford Dyd to 7.1805; recommissioned 6.1805 under Capt. James Macnamara, for North Sea, then under Capt. Donald Campbell in 6.1807 (–1808); in Copenhagen Expedition 8.1807; with Saumarez's fleet in the Baltic 1808. In ?3.1809 under Capt. Richard Pearson, then ?7.1810 under Capt. Robert Williams, in the Baltic, and later Capt. James Pattison Stewart (–1812). In 4.1812 under Capt. Alexander Schomberg; destroyed (with others) 40-gun *Nayaden* and took 18-gun brigs *Laaland* and *Rich* in Lyngoc Fiord 6.7.1812; her boats took Danish lugger 3.8.1812. In 8.1812 under Capt. William Hanwell; paid off 11.1812. Fitted as troopship at

Chatham 6–9.1813; in 8.1813 under Capt. George Crofton, then in 12.1813 under Cmdr Henry Dilkes Byng, to North America. In 2.1815 under (temp.) Lieut James Barnwell Tattnall, then 5.1815 under Cmdr Henry Montresor; paid off and laid up at Portsmouth 10.1815. BU there 6.1817.

CROWN Class Designed by Edward Hunt, 1779. This final British design for a 64 made few changes, but extended the overall length a further 6in to 160½ft. One notable change was the provision of a fourteenth pair of gunports on the UD, forwards in the chase position, but no additional guns were provided for these, which lay so close to the manger walls that their use must have been extremely restricted. None of this class survived into 1817 – the *Ardent* was lost in 1794, the *Scipio* was BU in 1798 and the *Crown* and *Veteran* in 1816.

Ex-FRENCH PRIZE (1782) *Le CATON* **Class** Two ships had been built at Toulon to this 1770 design by Joseph-Marie-Blaise Coulomb, both of which fell into RN hands when taken by Rodney in the Mona Channel (between Puerto Rico and San Domingo) in 1782. Her sister *Caton* (which served as a hospital ship from 1790) has been sold in 1815, as was the similar *Prothee* built in 1771–3 and taken in 1780.
Argonaut (French *Le Jason*, built 1.1778 –5.1779 at Toulon. L: 13.2.1779)

> Dimensions & tons: 166ft 0in, 136ft 8¼in x 44ft 8½in x 19ft 1in. 1,451⁷¹⁄₉₄bm.
>
> Men: 500 (491 from 1794). Guns: LD 26 x 24pdrs; UD 26 x 18pdrs; QD 10 x 9pdrs; Fc 2 x 9pdrs.
>
> Taken by Rodney's squadron in the Mona Passage 19.4.1782.
>
> First cost: £14,647.8.1d purchase price.
>
> Commissioned (by Adm. Rodney) 19.4.1782 under Capt. John Aylmer for passage home; registered 29.1.1783 (backdated to commissioning date) under original name, then renamed *Argonaut*. Small Repair at Plymouth (for £12,745.7.4d) 2–7.1783. Fitted and coppered at Plymouth (for £9,513) 4–9.1793; recommissioned 1.1793, still under Aylmer; sailed for Nova Scotia 18.5.1794. In 1795 under Capt. Alexander Ball; took (with *Oiseau*) 22-gun *L'Espérance* off the Chesapeake 8.1.1795; paid off 10.1796. Fitted at Chatham as a hospital ship (for £273) 11.1796–2.1797; recommissioned 2.1797 under Lieut Philip Hue, then under Lieut George Paul 1799, and Lieut John Stevens in 1801; paid off 4.1802. Recommissioned 8.1804 as hospital ship at Chatham under Lieut James James; paid off 7.1828 (still under James). BU there 2.1831.

PURCHASED VESSELS (1796) Five ships had been purchased on the stocks from the East India Company in early 1796 while building or being serviced by Thames-side yards; of these, *York* was lost in 1804 and *Agincourt* sold in 1814. These conversions differed from the Navy-built 64s in having a detached quarterdeck and forecastle, and being about 12ft to 14ft longer – the East India Company had designed them to carry 28 guns (18pdrs) on the LD, with one more gun port per deck on each side. This extra port, however, was not used by four of them, which were established with the same armament and men as existing 64s. The exception was *Agincourt*, allocated 28 x 24pdrs and 28 x 18pdrs to fill all her LD and UD ports; she carried only 6 x 8pdrs on the QD (vice 10 on the other ships).
Ardent (East India Company *Princess Royal*, building by Thomas Pitcher, Northfleet)

> Dimensions & tons: 173ft 3in, 144ft 0in x 43ft 0in x 19ft 10in. 1,416²⁴⁄₉₄bm.
>
> Men: 491. Guns: LD 26 x 24pdr; UD 26 x 18pdr; QD 10 x 9pdr; Fc

2 x 9pdr.

> Registered and named 14.7.1795. L: 9.4.1796. C: 26.4–7.8.1796 at Woolwich.
>
> First cost: £22,652 to builder (including coppering), plus £5,262 fitting.
>
> Commissioned 5.1796 under Capt. Richard Burges, for Duncan's fleet; in Nore mutiny, got to sea 10.6.1797; in Weather column at Battle of Camperdown 11.10.1797, lost 41 killed (incl. Burges), 107 wounded. In 11.1797 under Capt. Thomas Bertie (–1801); joined Mitchell's squadron on the Dutch coast 8/9.1799; flagship of Adm. Sir Hyde Parker 2.1801, for Copenhagen expedition; Battle of Copenhagen 2.4.1801, lost 30 killed, 64 wounded; in 6.1801 under Capt. George M'Kinley, later Capt. William Nowell. Small Repair and fitted at Chatham (for £11,829) 8.1802–4.1803; recommissioned 3.1803 under Capt. Robert Winthrop (–1805); in Pellew's squadron off Ferrol 11.1803; chased 32-gun *La Bayonnaise* (*en flûte*) 28.11.1803 – blown up by own crew to avoid capture; temp. flagship of Adm. Lord (George) Keith 2.1805, in North Sea. Recommissioned 7.1806 under Capt. George Eyre; in 9.1806 under Capt. Ross Donnelly, then Capt. Edwin Chamberlayne in 5.1807; sailed 28.9.1807 to join River Plate operations. Under Capt. William Parkinson in 2.1808. Fitted at Sheerness as guard ship 4–6.1808, to lie at Leith. Later 1808 or 1809 under Capt. John Sykes. Recommissioned 4.1808 under Capt. James Giles Vashon, as flagship of Vice-Adm. James Vashon; from 10.1808 under Capt. John Bligh. Recommissioned 4.1809 under Capt. Robert Honeyman (–1811), for the Baltic. Fitted at Chatham as Troopship 2–6.1812; recommissioned 4.1812 for Baltic under Capt. John Davie, later under Capt. George Bell. Fitted at Portsmouth as prison ship 2–5.1813, for Bermuda under Capt. John Cochet; in 5.1814 under Cmdr Sir William Burnaby (–1815). Hulk at Halifax 1817–22. BU at Bermuda 3.1824.

Monmouth (East India Company *Belmont*, building by Randall & Co., Rotherhithe)

> Dimensions & tons: 173ft 1in, 144ft 1½in x 43ft 4in x 19ft 8in. 1,439⁵¹⁄₉₄bm. Draught 10ft 7in/15ft 9in.
>
> Men: 491. Guns: LD 26 x 24pdr; UD 26 x 18pdr; QD 10 x 9pdr; Fc 2 x 9pdr.
>
> Registered and named 14.7.1795. L: 23.4.1796. C: 31.10.1796 at Deptford Dyd
>
> First cost: £24,292 to builder (including coppering), plus £5,902 fitting.
>
> Commissioned 9.1796 under Capt. Lord (William) Northesk, for North Sea; in Nore Mutiny 1797; under (acting) Cmdr James Walker in Lee column at Battle of Camperdown 11.10.1797, lost 5 killed, 22 wounded. Repairs (for £4,577) 2.1798; under Capt. Robert Deans 3.1798, in North Sea; flagship of Vice-Adm. Archibald Dickson from 1.1799. Repairs at Sheerness (for £4,873) 3.1799. Under Capt. George Hart in 4.1799 (–1805); in Mitchell's squadron in the Helder 8.1799; sailed for the Mediterranean 6.1801; flagship of Rear-Adm. Thomas Russell in North Sea 1804. Recommissioned 3.1807 under Capt. Edward Durnford King, as flagship of Rear-Adm. William O'Bryen Drury; sailed for East Indies 15.9.1807; paid off 9.1808. Recommissioned 10.1809 under Cmdr Michael Dod, as victualling vessel in the Downs; in 4.1811 under Capt. Hyde Parker, later Capt. Francis Beauman; in 1812 under Capt. William Nowell (–1813), then Capt. William Wilkinson, all as flagship of Vice-Adm. Thomas Foley as well as victualler. In Ordinary at Woolwich 1815, then sheer hulk at Deptford 6.1815;

at Woolwich 1828–33. BU at Deptford 5.1834 (although still appeared on subsequent Navy Lists for some years).

Lancaster (East India Company *Pigot*, building by Randall & Co., Rotherhithe)

Dimensions & tons: 173ft 6in, 144ft 3in x 43ft 2in x 19ft 9in. 1,429$^{69}/_{94}$bm.

Men: 491. Guns: LD 26 x 24pdr; UD 26 x 18pdr; QD 10 x 9pdr; Fc 2 x 9pdr.

L: 29.1.1797. C: 13.2–17.4.1797 at Deptford Dyd

First cost: £29,659 to builder (including coppering), plus £9,132 fitting.

Commissioned 2.1797 under Capt. John Wells; in Nore mutiny 1797 at Gravesend, returned to duty 6.6.1797; in Weather column at Battle of Camperdown 11.10.1797, lost 3 killed, 18 wounded; on Irish station 1798. Refitted (for £9,021) 4–5.1799. In 1799 under Capt. Thomas Larcom (died 4.1804), as flagship of Vice-Adm. Sir Roger Curtis; later to Cape of Good Hope and East Indies. In 1805–7 under Capt. William Fothergill, in East Indies, then in 1807 in Stirling's squadron in the River Plate. Fitted at Chatham as a receiving ship 8–9.1807, for Malta, but instead in 10.1807 in Ordinary at Chatham. Fitted as a victualler at Chatham 10–12.1808. Victualler at Plymouth in 1812, then at Sheerness 1813–15. Lent to West India Dock Co. as boys' training ship 11.3.1815. Returned to Admiralty 2.1.1832 and sold at Woolwich (for £2,410) to Joshua Crystall 30.5.1832.

Ex-DUTCH PRIZES (1796–99)

Fourteen Dutch 64s/66s/68s had been added to the RN between 1796 and 1799, and were employed as harbour vessels for many years. The largest of these, *Admiral de Vries* (built for the Frisian Admiralty in 1781 at Harlingen as the 68-gun *Admiraal Tjerk Hiddes de Vries*) was sold in 1806. In Dutch service, the first pair had been rated at 64 guns, the next four at 66 guns, and the remainder at 68 guns.

Overyssel (Dutch *Overyssel*, launched 1781 at Ferrol as Spanish *San Felipe Apostol* and purchased 21.5.1793 at Cadiz by the Dutch)

Dimensions & tons: 153ft 6⅛in, 124ft 8½in x 43ft 0in x 18ft 1¼in. 1,226$^{49}/_{94}$bm.

Men: 491. Guns: LD 26 x 24pdrs; UD 26 x 18pdrs; QD 8 x 9pdrs; Fc 4 x 9pdrs.

Taken by *Polyphemus* at Queenstown 22.10.1795. Arrived 7.5.1796 at Plymouth; registered 28.5.1796.

Commissioned 7.1796 under Capt. William Swaffield, as flagship of Adm. Joseph Peyton (–1799) in the Downs; fitted at Plymouth as flagship (for £3,074) 6–10.1796; from 11.1796 under Capt. John Young, then 1798 under Capt. John Bazely (–1801); from 1799 flagship of Vice-Adm. Skeffington Lutwidge; in Helder operations 1799; paid off into Ordinary 1.1802. Fitted at Chatham as Provision receiving ship 10–12.1805 for Sheerness. In Ordinary from 1807. Left Sheerness 8.1809 to be made a Breakwater for Harwich (by AO 17.6.1809); used at Harwich 1810. Wreck sold (for £45) 3.1.1822.

Justitia (ex-*Zealand*, ex-Dutch *Zeeland*, launched 1784 at Vlissingen)

Dimensions & tons: 156ft 8¼in, 127ft 3⅝in x 44ft 4⅜in x 18ft 4in. 1,332$^{71}/_{94}$bm.

Men: 491. Guns: LD 26 x 24pdrs; UD 26 x 18pdrs; QD 10 x 9pdrs; Fc 2 x 9pdrs.

Taken 19.1.1796 at Plymouth (having arrived there 6.2.1795). Registered 25.10.1796.

Commissioned 3.1797 under Capt. Thomas Shivers; fitted at Plymouth as guard ship (for £12,471) 5–9.1797. In 10.1797 under Capt. Thomas Parr, as guard ship at the Nore and flagship of Vice-Adm. Skeffington Lutwidge, then 1799 of Rear-Adm. Alexander Graeme. ?Troopship 6.1799. In 6.1800 under Capt. William Mitchell. Fitted as receiving ship at Sheerness (for £5,685) 5.1803. In 9.1804 under Capt. Adrian Renou (died 1805). In 1805 under Capt. Henry Lidgbird Ball, as flagship of Rear-Adm. Bartholomew Rowley, at the Nore, later under Capt. Robert Fancourt, as flagship of Rear-Adm. Thomas Wells (–1807). Fitted at Sheerness as Convict hulk for Medway 11.1809. Renamed *Justitia* 19.8.1812. Moved to Woolwich 1814. Sold to John Small Sedger (for £700) 2.11.1830.

Dordrecht (Dutch *Dordrecht*, launched 1782 by Jacob Spaans, Dort)

Dimensions & tons: 159ft 6in, 130ft 0in x 45ft 0in x 17ft 0in. 1,437bm.

Men: 491 (215 as troopship). Guns: LD 28 x 24pdrs (10 removed as troopship); UD 28 x 18pdrs (10 removed as troopship); QD 6 x 6pdrs; Fc 6 x 6pdrs.

Taken 17.8.1796 at Saldanha Bay by Elphinstone's squadron.

Commissioned 12.1796 under Capt. Charles Brisbane as guard ship at Cape of Good Hope. Under Capts John Sprat Ranier in 4.1797, Samuel Hood Linzee in 9.1797, and Brisbane again in 1.1798, then (acting) Capt. David Atkins in the Swale. Fitted as Troopship at Sheerness (for £6,350) 7–8.1799; under Capt. Robert Honeyman in 1.1800. Fitted as receiving ship at Chatham 4–9.1800; to Ordinary there 1807–14; to Sheerness 11.1814 as hulk. Sold to John Small Sedger (for £2,400) 21.5.1823 to BU.

***PRINS FREDERIK* Class** Three vessels of this Dutch 66-gun design were taken in 1796–97.

Prince Frederick (Dutch *Revolutie*, originally *Prins Frederik* 1795, launched 1777 Zwindrecht, Rotterdam).

Dimensions & tons: 156ft 9in, 129ft 4in x 42ft 11in x 16ft 3½in. 1,267$^{7}/_{94}$bm.

Men: 491 (as storeship 230). Guns: LD 24 x 24pdrs; UD 26 x 18pdrs; QD 10 x 9pdrs; Fc 2 x 9pdrs + 2 x 24pdr carronades; RH 6 x 18pdr carronades. As storeship LD nil, UD 24 x 12pdrs, QD 6 x 6pdrs, Fc 2 x 6pdrs.

Taken 17.8.1796 at Saldanha Bay by Elphinstone's squadron. Registered 11.3.1797 as storeship.

Commissioned 12.1796 under Capt. Edward Ramage as storeship; arrived Sheerness 25.1.1797. Under Cmdr John Watts in 1797. Fitted as storeship at Woolwich (for £15,567) 9.1797–5.1798; paid off 11.1797 and recommissioned 2.1798 under Capt. Charles Hare, for North Sea. In 6.1798 under Capt. John Stevens Hall. Fitted as Convalescent ship at Plymouth 4–5.1800; recommissioned 5.1800 under Lieut William Galton, for convalescents at Plymouth. Fitted at Plymouth as a hospital ship 12.1803–3.1804, for Berehaven; under Lieut Samuel Gordon (–1806), then Lieut Abdiel Orseur (1807–10), then Cmdr Peter Fisher in 10.1810. In 2.1812 under Cmdr Thomas Grove (died 1814); flagship of Vice-Adm. Sir Edward Buller at Plymouth from 1813. On 4.11.1814 under Cmdr Richard Pridham, as flagship of Rear-Adm. Sir Thomas Byam Martin; paid off 2.1815 into Ordinary at Plymouth. Sold at Plymouth to G. Bayley (for £1,200) 5.6.1817.

Wassenar (Dutch *Wassenaar*, launched 1781 at Zwindrecht, Rotterdam)

Dimensions & tons: 158ft 2in, 131ft 1¼in x 42ft 8in x 20ft 2½in. 1,269$^{48}/_{94}$bm.

Men: 491 (250 as troopship). Guns: LD 28 x 24pdrs (removed as troopship); UD 28 x 18pdrs; QD 8 x 9pdrs; Fc 2 x 9pdrs.

Taken 11.10.1797 at Camperdown by Duncan's fleet. Registered 3.1.1798. Arrived Sheerness 28.10.1797.

Commissioned 3.1798 under Capt. Charles Craven; fitted as a troopship at Sheerness (for £9,122) 2–5.1798; flagship of Adm. Joseph Peyton in the Downs 10.1798. In 1800 under Capt. John Larmour, for the Mediterranean; then Capt. Frederick Maitland in 1801, then Capt. Henry E. R. Baker in 2.1802. Fitted as Powder hulk at Chatham 9.1802; in use there to 1815. Sold to John Small Sedger (for £2,820) 13.8.1818 to BU.

Delft (Dutch *Hercules*, launched 1781 by Jacob Spaans, Dort)

Dimensions & tons: 157ft 2in, 129ft 6¼in x 42ft 10½in x 16ft 11½in. 1,266⁴³⁄₉₄bm.

Men: 491 (250 as troopship). Guns: LD 28 x 24pdrs (removed as troopship); UD 28 x 18pdrs; QD 4 x 9pdrs + 8 x 24pdr carronades; Fc 4 x 9pdrs + 2 x 24pdr carronades.

Taken 11.10.1797 at Camperdown by Duncan's fleet. Registered 3.1.1798. Arrived at Chatham 8.1.1798.

Commissioned 5.1798 under Capt. Robert Redmill; renamed by AO 30.8.1798; fitted as a troopship at Chatham (for £16,144) 2–8.1799. To Yarmouth 1801, thence to Leeward Islands. Fitted at Chatham as Powder hulk 8.1802. Fitted at Chatham as breakwater 2.1822, and sunk as breakwater at Harwich 19.9.1822.

DE RUYTER Class Eight Dutch 68-gun ships, all built at Amsterdam to this design, were taken – two at Camperdown in 1797 and the other six off Texel in 1799, but of the latter the *Verwagting* (built at Hoorn 1783) was not added to the RN. Of those added, the *De Ruyter* was wrecked in 1804, the *Gelikheid* was sold in 1814, the *Utrecht* and *Leyden* in 1815 and the *Haarlem* in 1816. A ninth vessel of this class, the original *Overyssel* (built 1783 at Amsterdam) had been lost by the Dutch in 1794 (and replaced by a vessel purchased from Spain and given the same name).

Texel (Dutch *Cerbrus*, built 1784 at Amsterdam)

Dimensions & tons: 156ft 0in, 126ft 11¼in x 44ft 2in x 17ft 1in. 1,317¹⁰⁄₉₄bm.

Men: 491 (250 as floating battery). Guns: LD 26 x 24pdrs; UD 26 x 18pdrs; QD 2 x 9pdrs.

Taken 30.8.1799 in the Vlieter off Texel by Mitchell's squadron. Arrived 4.7.1800 at Chatham.

Fitted at Chatham (for £13,260) 9.1800–1.1801.

Commissioned 11.1800 under Capt. Richard Incledon, for North Sea. In 3.1802 under Capt. Henry Garrett. Fitted as a floating battery at Chatham (for 5,887) 5.1803, possibly for defence of Leith Roads; recommissioned under Capt. George Byng 3.1803 as a floating battery for Downs station; under Cmdr Donald Campbell 9.1804, as guard ship at Leith and flagship of Rear-Adm. James Vashon (–1807); later under Capt. Robert Jackson, then Capt. James Giles Vashon. Laid up at Chatham 9.1808. Powder hulk at Chatham 1812–15. Sold to Mr Beatson at Chatham (for £2,740) 11.6.1818.

Guelderland (Dutch *Gelderland*, launched 1781 at Amsterdam)

Dimensions & tons: 157ft 2in, 128ft 3in x 44ft 4¼in x 16ft 7in. 1,342⁵⁄₉₄bm.

Men: — . Guns: — .

Taken 30.8.1799 in the Vlieter off Texel by Mitchell's squadron. Arrived Sheerness 14.4.1801 and laid up.

Not commissioned. Fitted as receiving ship at Chatham 10.1810–1.1811. Sold at Chatham (for £1,530) to Mr Freake 5.3.1817.

(B) Vessels acquired from I January 1817

It was realised by this date that, notwithstanding their sterling wartime service, the basic 74-gun concept (now 60 years old) was no longer large enough to carry the most modern armament, and virtually no further vessels of this type were acquired post-war. Surviving 74-gun vessels were reduced to 72 guns in 1839.

CARNATIC **Class** 74 (later 72) guns. The first post-war development was a modification to the *Armada* Class, with a round stern. Built using teak frames brought from Bombay aboard the *Cornwallis*. She had been originally (30 September 1814) ordered to *Armada* Class design. A second ship was ordered on 11 November 1817, initially to a similar design (*Indus*), but was re-ordered in 1820 as an 80-gun ship (see Chapter 2).

Dimensions & tons: 177ft 1in, 145ft 11in x 48ft 0in x 21ft 1¼in. 1,788²⁴⁄₉₄bm.

Men: 590. Guns: LD 28 x 32pdrs; UD 28 x 18pdrs; QD 4 x 12pdrs + 10 x 32pdr carronades; Fc 2 x 12pdrs + 2 x 32pdr carronades; RH 6 x 18pdr carronades.

First cost: £64,519.

Carnatic (ex-*Sandwich*?) Portsmouth Dyd

As built: 177ft 0in, 145ft 3¼in x 48ft 10½in oa (48ft 1½in for tonnage) x 21ft 2in. 1,789⁵⁄₉₄bm.

Re-ord: 10.11.1817. K: 1.1818. L: 21.10.1823. C: 19.11.1823 (for Ordinary). Housed over from main mast forward.

First cost: £64,519 (including fitting for Ordinary).

Never commissioned. At Portsmouth until 1859. Fitted by White, Cowes (for £11,536) 7.1859–2.1860 as a coal depot ship, then fitting completed at Portsmouth and coaling machinery installed (for £6,667) 2–8.1860. Hulked as coal depot at Portsmouth. Surveyed and reported not worth the expense of repairs estimated as necessary 8.1881. Lent as floating magazine to the War Office by AO 16.8.1886; returned to Admiralty 1.10.1891. Sold 19.2.1914 and BU in Germany.

HASTINGS 74 (later 72) guns. Built at Bombay as a speculation by the East India Co., and purchased from them by the Royal Navy on 22 June 1819 (for £32 per ton = £56,320) following her arrival in UK at Woolwich on 26 March 1819.

Dimensions & tons: 176ft 10½in, 145ft 4in x 48ft 6in oa (47ft 9in for tonnage) x 21ft 0in. 1,762⁵⁶⁄₉₄bm.

Men: 600. Guns: LD 28 x 32pdrs, UD 28 x 18pdrs, QD 4 x 12pdrs + 10 x 32pdr carronades, Fc 2 x 12pdrs + 2 x 32pdr carronades. Later re-armed with LD 26 x 32pdrs + 2 x 68pdr carronades, UD 24 x 18pdrs + 4 x 68pdr carronades, QD 4 x 12pdrs + 10 x 32pdr carronades, Fc 2 x 12pdrs + 2 x 32pdr carronades.

Registered by AO 22.6.1819; to Chatham in same month and laid up, housed over from poop forwards. Fitted for Ordinary 3–5.1827, then fitted at Chatham for sea service (for £19,060) 2.1833–4.1834.

Commissioned 7.4.1834 under Capt. Henry Shiffner, as flagship of Rear-Adm. Sir William Hall Gage, for Lisbon. On 31.1.1838 under Capt. Francis Erskine Loch, for the Mediterranean; fitted at Sheerness (for £13.624) 1–4.1838 to convey Lord Durham and his entourage to Canada; fitted at Portsmouth (for £5,047) 7–10.1838 to convey the Queen Dowager to Malta. On 4.6.1839 under Capt. John Lawrence, in the Mediterranean; operations on the coast of Syria 1840; capture of Batroun 15.9.1840; paid off 3.2.1842 at Portsmouth. Converted to Advanced ship at Portsmouth (for £9,710) 2.1842–12.1843. Recommissioned 10.4.1848 under Capt. James William Morgan; fitted for sea there

A model of the 74-gun *Hastings* that was made on board the ship during her first commission (1834–37) and presented to the Captain, Sir Henry Shiffner. A sailor-made model built by eye rather than following a draught, the proportions are not perfect but the details of the fittings and the appearance are in all probability highly authentic. *(NMM L2766-002)*

(for £12,252) 4–7.1848, as flagship of Rear-Adm. Sir Francis Augustus Collier (died 29.10.1849), for the East Indies. On 21.1.1850 under Capt. Francis William Austen, as flagship of his uncle, Rear-Adm. Charles John Austen, in the East Indies; in Second Anglo-Burmese War 9–10.1852; paid off 6.5.1853 at Portsmouth. Converted to 60-gun screw blockship 1853–5 (see Chapter 4 for later history).

Modified BULWARK Class 76 guns. A new 76 (i.e. a 'stretched' 74 with 24pdrs on the UD) was authorised by Byam Martin in 1825 using the hull form and dimensions of the 1795 Rule design, but with some variations in layout. However, little work was put into collecting timber and she was cancelled in 1831. The attempt to extend the 74 was a 'blind alley' as the lighter scantlings of a 74/76 (compared with the 84) meant they could not bear the heavier modern armament.

> Dimensions & tons: 182ft 0in, 150ft 9in x 49ft 0in x 20ft 10in. 1,925¹²⁄₉₄bm.
> Men: 640. Guns: LD 28 x 32pdrs; UD 30 x 24pdrs; QD 4 x 12pdrs + 10 x 32pdr carronades; Fc 2 x 12pdrs + 2 x 32pdr carronades.

Valiant Plymouth Dyd
> Ord: 9.6.1825. K: — . Cancelled 2.1831 (unstarted).

(C) Vessels acquired from November 1830

CUMBERLAND Class 70 guns. William Symonds design 1833, to the lines of the 74-gun *Minotaur*. While the Navy saw no role for the small two-decker in fleet actions, Symonds perceived the 70-gun ships as an enhanced trade protection warship, designed for convoy escort, and the

Cumberland was designed with this new role in mind. The second ship – *Boscawen* – may be perceived as a re-order; as mentioned above, she had originally been ordered as a 74 (of the *Armada* Class) in 1812, then re-ordered as an 80 in 1817 (and began building as such in 1826), but was cancelled on 29 November 1831 and briefly re-ordered as a 50-gun frigate in 1832 (re-named *Indefatigable*) before being finally re-ordered as a 70 (with her original name restored) in 1834; her frames were then taken down and the material converted to a Symondite 'V' hull form instead of the old 'U' form. No further ships of this rate were started, as the role was appropriated by the new 50-gun spar-deck 'frigates'.

> Dimensions & tons: 180ft 0in, 146ft 8in x 54ft 0in oa (53ft 3in for tonnage) x 22ft 4in. 2,212¹²⁄₉₄bm.
> Men: 620. Guns: Originally LD 24 x 32pdrs (55cwt) + 2 x 68pdr (60cwt) carronades; UD 26 x 32pdrs (48cwt) + 2 x 68pdr (60cwt) carronades; QD 8 x 32pdrs (25cwt); Fc 3 x 32pdrs (25cwt) + 2 x 32pdr carronades; RH 4 x 18pdrs (10cwt). Later LD 22 x 32pdrs (56cwt) + 4 x 8in/68pdr (65cwt) shell; UD 26 x 32pdrs (50cwt) + 2 x 8in/68pdr (65cwt) shell; QD 8 x 32pdrs (42cwt); Fc 4 x 32pdrs (42cwt); RH 4 x 18pdrs (10cwt) – the 18pdrs subsequently replaced by extra 4 x 32pdrs (42cwt).

Cumberland Chatham Dyd
> As built: 180ft 0in, 145ft 5⅝in x 54ft 3in oa (53ft 6in for tonnage) x 22ft 4in. 2,214³⁸⁄₉₄bm. Draught 13ft 4in/17ft 8in (with 11 tons aboard).
> Ord: 9.10.1833. K: 4.1836. L: 21.10.1842. C: 7.12.1842 (for Ordinary). Fitted as an Advanced ship (for £11,852) 8.1845–1.1846).

> First cost: £45,018 (including fitting for Ordinary).

> Commissioned 31.8.1850 under Capt. Peter Richards, as guard ship at Chatham. Recommissioned 7.1.1851 under Capt. George Henry Seymour, as flagship of his father, Vice-Adm. Sir George Francis Seymour, and fitted as a flagship at Chatham (for £17,410) 1–2.1851; to North America and West Indies, then in 1854 the Baltic for the Russian War; paid off 27.10.1854 back at Chatham. Recommissioned 1.4.1857 at Sheerness under Capt. John Bourmaster Dickson, as flagship of Rear-Adm. Provo William Parry Wallis, for the south-east coast of South America; from 23.8.1858 under Capt. Henry Downing Rogers, with flag of Rear-Adm. Stephen Lushington for same role; defects made good at Plymouth (for £13,077) 8–10.1858; paid off 5.8.1859. Recommissioned 10.59 under Capt. Edward Pellew Halsted, as guard ship of the Steam Reserve at Sheerness (replacing *Cressy*). On 10.1.1860 under Capt. Charles Frederick Schomberg; from 14.4.1863 under Capt. William King Hall, becoming a sheer hulk at Sheerness (fitted with 'masting davits' for fitting masts for small steam ships). On 18.5.1865 under Capt. George Granville Randolph, as guard ship of Steam Reserve again; in 6.1866 under Capt. George William Preedy, then 27.10.1866 under Capt. Arthur Leopold Pedro Cochrane, in same role. Became training ship in the Clyde for the Clyde Industrial Training Ship Foundation 1869; burnt by accident there 17.2.1889; wreck BU in Rosneath Bay 1889.

Boscawen Woolwich Dyd
> As built: 180ft 0in, 146ft 9⅜in x 54ft 0in oa (53ft 3in for tonnage) x 22ft 4in. 2,213⁸¹⁄₉₄bm. Draught 13ft 10in/16ft 8in.
> Ord: 3.3.1834. K: 1.1826 (to previous design as 80 gun). L: 3.4.1844. C: 26.9.1848 at Sheerness (for Ordinary). Fitted for guard ship at Chatham (for £17,314) 11.1848–6.1851.

> First cost: £56,065 (+ fitting £4,231 for Ordinary).

> Commissioned 7.1.1851 under Capt. Peter Richards, as guard ship at Chatham. Recommissioned 25.11.1853 under Capt. William

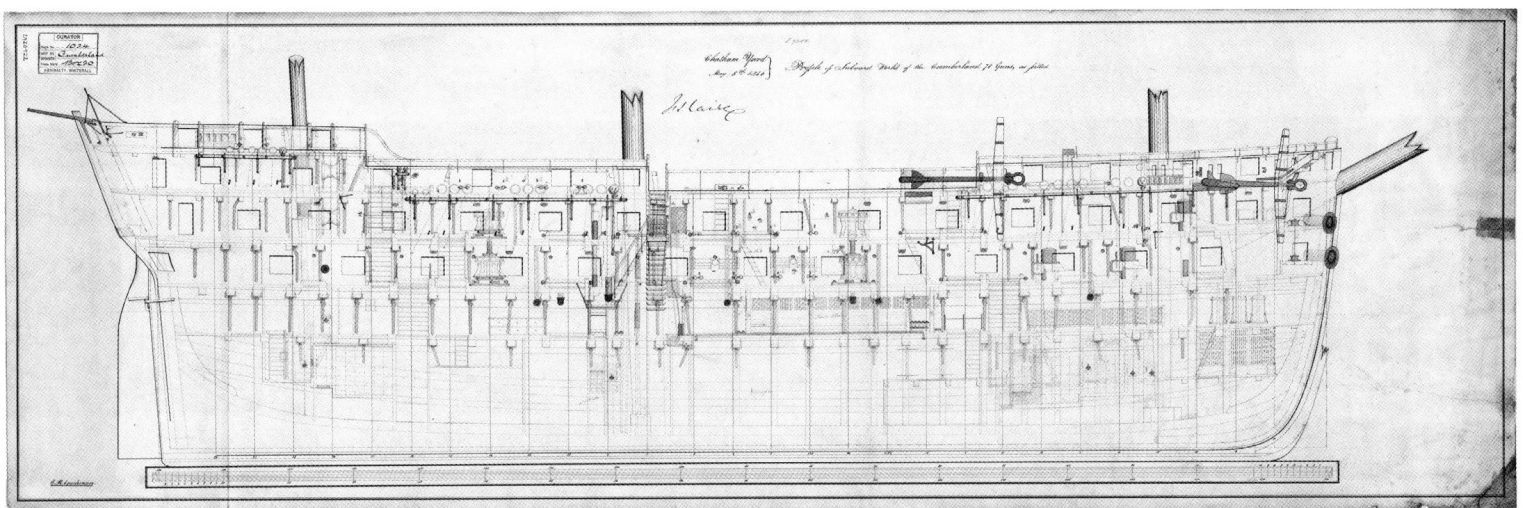

The 1842 inboard profile of the *Cumberland* as fitted out at Chatham with a priority for trade protection. It is signed by Frederick John Laire, the master shipwright at Chatham from October 1844 to May 1848, while the vessel was being fitted out as an 'Advanced' ship. *(NMM L2740)*

Fanshawe Glanville, as flagship of Rear-Adm. Arthur Fanshawe, and fitted as a flagship at Chatham (for £11,966) 12.1853–1.1854; to North America and West Indies, then in 1854 the Baltic for the Russian War. Recommissioned 1.5.1857 under Capt. Richard Ashmore Powell, as flagship of Rear-Adm. Frederick William Grey, for the Cape of Good Hope. Fitted for Ordinary at Plymouth 9.1860. Fitted at Plymouth (for £9,659) as Drill Ship for Southampton 8.1861–2.1862; on 5.3.1862 under Cmdr Frederick Thomas Chetham Strode, then 21.4.1852 under Cmdr Hubert Campion, on 1.9.1863 Cmdr George Strong Nares and 14.7.1865 Cmdr McLeod Baynes Cockroft; to Portland 10.1866; on 14.10.1867 under Cmdr Henry Fairfax, then 1869 under Cmdr James Grant, on 10.8.1872 Cmdr Marcus Augustus Stanley Hare and 21.2.1873 Cmdr Alfred Markham. Lent to Wellesley Training Establishment on the Tyne as a hulk, renamed *Wellesley* 21.3.1874; damaged by accidental fire there 3.1914; wreck was BU at Blyth.

1834 Proposed replacement of 110-gun First Rates by 74-gun Third Rates in the Building Programme

On 11 December 1834 the 1833-ordered (but unstarted) Symondite 110-gun First Rates *Victoria* and *Algiers* (see Chapter 1) were re-classed as 'special' 74s, intended to carry 68pdrs on the lower deck, but with no changes in hull designs. In the 1835 Programme, the *Algiers* was reinstated as 110-gun; in 1836 *Victoria* remained at 74-gun while *Algiers* was again listed at 110-gun; the two ships' designations were reversed in 1837; and on 5 February 1839 the *Algiers* was likewise restored to a 110-gun.

Dimensions & tons: 204ft 0in, 166ft 5¼in x 60ft 0in oa (59ft 2in for tonnage) x 23ft 9in. 3,099¹⁶/₉₄bm.

Men: — . Guns: LD 28 x 68pdr (65cwt); UD 28 x 32pdr (56cwt); QD 16 x 32pdr (48cwt); Fc 2 x 32pdr (56cwt).

IMAUM 74 (later 72) guns. (ex-Muscat & Oman *Liverpool*, built 1826 of teak construction by Bombay Dyd for the Imaum of Muscat, and presented to HM King William IV by him 9 March 1836.

Dimensions & tons: 177ft 0in, 145ft 5¾in x 48ft 4in oa (47ft 11in for tonnage) x 21ft 0in. 1,776⁶⁶/₉₄bm. 2,948 displacement.

Men: 590. Guns: LD 28 x 32pdrs, UD 28 x 18pdrs, QD 2 x 18pdrs + 12 x 32pdr carronades, Fc 2 x 18pdrs + 2 x 32pdr carronades.

Arrived at Portsmouth 23.2.1836, fitted there as receiving ship for Jamaica (for £12,989) 8.1840–9.1842.

Commanded 26.7.1842 by Commodore Henry Dilkes Byng, as receiving ship at Jamaica. On 8.9.1843 under Commodore Alexander Renton Sharpe, then 16.9.1845 under Commodore Daniel Pring (died 1.1847), 23.1.1847 Commodore George Robert Lambert, 12.2.1847 Acting Capt. William Worsfold, as flagship of Commodore Thomas Bennett, 7.2.1848 Commodore Thomas Bennett (direct), 13.3.1851 Commodore Peter McQuhae, 8.3.1854 Cmdr Samuel Morrish, 16.3.1857 Commodore Henry Kellett, 11.1859 Commodore Hugh Dunlop, then 19.5.1860 Capt. Samuel Morrish and 18.2.1862 Capt. John Barling Marsh, both with flag still of Commodore Hugh Dunlop; paid off 25.8.1862 at Jamaica (by AO 30.5.1862). BU there 1863 (by tender accepted 1862 at 15.0d per ton).

(D) Vessels ordered or re-ordered as steam screw warships (from 1848)

As the small two-decker had for many years been perceived as irrelevant to the needs of the battlefleet, the Third Rate had been in constant numerical decline since the Napoleonic era. It was certainly judged unnecessary to consider any of the 74-gun ships for fitting with steam propulsion (apart from those converted into blockships), and the only Third Rate steam warship was created simply by accident – when the 80-gun *Sans Pareil* upon being converted with engines and screw was found incapable of retaining her full armament and was re-established with 70 guns.

SANS PAREIL – Conversion of 80-gun Second Rate (suspended incomplete on stocks since 2 October 1848) to 70-gun screw ship. On 16 November 1848 the 350nhp engine intended for the frigate *Eurotas* were ordered to be appropriated for the *Sans Pareil*, and a drawing and model for altering her to a screw ship were approved on 14 December (conversion work had actually begun on 15 November). Design re-approved on 18 May 1849. The conversion to screw involved

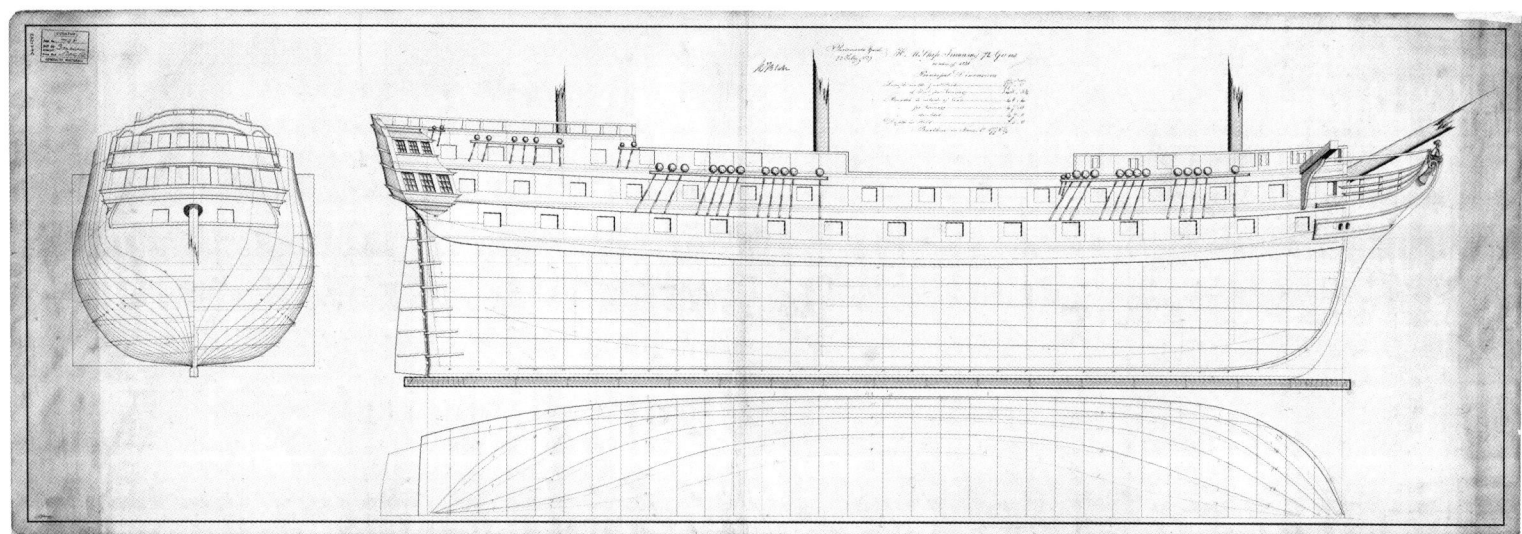

The *Imaum* was built in Bombay Dockard in 1826 (under the name *Liverpool*) for the *imam* (ruler) of Muscat and Oman. Finding her too expensive to maintain, the imam donated this teak-built ship to King William IV in 1836. The king accepted this gift, the last 74-gun ship to be acquired by the Royal Navy, renaming the ship in honour of its donor, to whom he gave a sailing yacht in exchange. Following its arrival at Portsmouth in February 1836, carrying a load of additional teak for use by the dockyard, the ship was handed over on 9 March and this draught was taken off in 1838 by Richard Blake, the master shipwright there from February 1835. The master shipwright and his assistants reported the ship to be 'wretchedly ill built of the worst materials', and she was accordingly fitted out for harbour use as a receiving ship for Jamaica, where she ended her days from 1842 to 1863. *(NMM J3011)*

lengthening *Sans Pareil* by 7ft at the stern. Initially fitted with engine intended for screw frigate (blockship) *Horatio*; the excessive weight necessitated reduction of her ordnance from 80 to 70 guns – with 10 fewer 32pdrs (25cwt) on the QD, while 45cwt 32pdrs replaced 24 of the originally planned 30 x 8in (52cwt) guns on the UD. In early 1855 this unreliable 4-cylinder engine was replaced.

Dimensions & tons: 200 ft 0 in, 165 ft 3 in x 52 ft 3 in oa (51 ft 7 ft for tonnage) x 22 ft 8 in. 2,338$^{88}/_{94}$bm (3,800 disp.). Draught 22 ft 8 in (fwd), 25 ft 8 in (aft).

Men: 700 (as 80-gun); 626 (as 70-gun). Guns: LD 30 x 32pdrs (56cwt); UD 24 x 32pdrs (45cwt) + 6 x 8in (52cwt); QD/Fc 2 x 32pdrs (56cwt) + 8 x 32pdrs (25cwt/10ft).

Machinery: 4-cylinder (43½in diameter, 2½ft stroke) horizontal single expansion, oscillating. Single screw. 350nhp. Re-engined in early 1855 with James Watt & Co. 2-cylinder (66in diameter, 3ft stroke) horizontal single expansion engine of 400nhp (1,471ihp = 9.3kts.).

First cost: £126,432, including machinery £30,888.

Sans Pareil Devonport Dyd/Boulton, Watt & Co. (622ihp = 7.05kts)

Conversion ord: 15.11.1849. Begun: 15.11.1848. L: 18.3.1851. C: 12.11.1852.

Commissioned 8.6.1852 under Capt. Sidney Colpoys Dacres, initially for Lisbon, then 1853 to Channel Squadron, finally to Black Sea for Russian War 1854; at bombardment of Sebastopol 17.10.1854. On 22.11.1854 under (acting) Capt. Leopold George Heath, still in Black Sea; returned home to repair damage and refit. Recommissioned 2.1855 under Capt. Woodford John Williams, for the Black Sea; transport of munitions (mortars) to Baltic Sea 9.1855, then home again. Recommissioned 22.1.1856 under Capt. Astley Cooper Key, as depot ship of gunboat division in Baltic; troop transport from Crimea 5.1856; to Portugal 9.1856; to India and China 3.1857 (Key commanded

the naval brigade at the capture of Canton 28.12.1857); invalided 4.1858. On 28.8.1858 under Capt. Julian Foulston Slight, still in China (during 2nd Anglo-Chinese War). On 30.6.1858 under Capt. Rochford Maguire, for voyage home from China and East Indies; paid off 15.2.1859 at Plymouth. Recommissioned 31.12.1859 under Capt. Arthur Parry Eardley Wilmot, as flagship of Rear-Adm. Charles Talbot as guard ship at Queenstown (replacing *Nile*); to Plymouth 7.1861; transported troops to Veracruz, Mexico 11.1861 (with *Donegal* and *Conqueror*). On 7.6.1862 under Capt. George Le Geyt Bowyear; transported Marines to China and invalids home; paid off 11.6.1863 at Plymouth. Reduced to 66-gun in 1866. Sold to C. Marshall (for £6,620) 8.3.1867 to BU at Plymouth.

The 70-gun *Sans Pareil* at Plymouth in June 1863 on her return from taking 300 marines to China. The ship had been ordered at Chatham in 1844 as a copy of the lines of the old French ship of that name, and slavishly duplicated the excessive tumblehome of that era. In 1848 it was intended to complete her as a 80-gun screw battleship, but the conversion of this, the first screw battleship to complete, was unsatisfactory because the unreliable second-hand machinery proved both too heavy and not powerful enough, requiring the ship to be armed with just 70 guns (and thus becoming the sole Third Rate screw battleship) and then to be re-engined in the middle of the Russian War when every steamer was urgently needed. To the right of this picture is the 91-gun *Edgar*. *(NMM D2169)*

Until 1 January 1817, the Fourth Rate encompassed vessels (all single-decked ships) with at least 48 but not more than 60 guns. Four years earlier (in January 1813) the established complements of all Fourth Rate frigates had been increased to 350 men for a 40-gun (24pdr-armed) frigate. Substantially from January 1817 the Rate was amended to comprise all ships carrying under 70 guns, but with not less than 50 guns – or whose complements were under 620, but not less than 450 men, so that ships of fewer than 50 guns were moved to the Fifth Rate. Thus both the larger frigates and smaller two-deckers were included. The majority of the Rate thus initially consisted of ships with a principal battery of 24pdr guns. From 1856 this rating was amended to restrict the Fourth Rate to ships 'the complements of which are [not more than] 600, and not less than 410 men'.

(A) Vessels in service or on order at 1 January 1817

Following the reorganisation of rates in January 1817, the Navy included in the Fourth Rate three ships rated at 60 guns, six rated at 58 guns (including two raséed from 64-gun ships), one rated at 56 guns, five rated at 54 guns and seven rated at 50 guns. For clarity this section is divided into (i) the classic 'two-deckers' and (ii) the large frigates

Fourth Rates of 50 guns (two-deckers)

Like the 64s the two-decker 50s were a fast disappearing type by 1817; their role as flagships on distant stations and other such tasks which required a ship of force but did not justify the presence of a ship of the line was being taken over by the more powerful and faster large frigates.

(Old) SALISBURY Class Sir Thomas Slade design, approved on 2 April 1766. Her half-sisters *Salisbury* and *Romney* had been wartime losses.
>   Dimensions & tons: 146ft 0in, 120ft 7⅝in x 40ft 4in x 17ft 4in. 1,043⁷⁷⁄₉₄bm.
>   Men: 350. Guns: LD 22 x 24pdrs; UD 22 x 12pdrs; QD 4 x 6pdrs; Fc 2 x 6pdrs.
Centurion John Barnard & John Turner, Harwich
>   As built: 146ft 0in, 120ft 2in x 40ft 5in x 17ft 3½in. 1,044¹¹⁄₉₄bm. Draught 10ft 8in/15ft 7in.
>   Ord: 25.12.1770. K: 5.1771. L: 22.5.1774. C: 22.6.1774–9.9.1775 at Chatham.
>   First cost: £15,023.9.11d to builder, plus £1,237.6.11d for masts (provided by the Navy), total including rigging £20,537.17.9d; plus £4,205.16.10d for fitting.
>   Commissioned 7.1775 and sailed for North America 25.10.1775; thence to West Indies; home to pay off 1780. Underwent

Middling Repair, coppered and fitted at Portsmouth (for £11,178.19.10d) 9.1780–6.1781. Recommissioned 3.1781 and sailed for North America 5.7.1781; home and paid off 10.1783. Fitted for Ordinary at Sheerness 12.1783. Great Repair at Woolwich (for £23,424) 12.1784–12.1787. Recommissioned 2.1789 under Capt. William Otway, as flagship of Rear-Adm. Philip Affleck; sailed for Jamaica 20.5.1789; home and paid off 8.1792. Fitted at Chatham (for £8,848) 8.1792–1.1793. Recommissioned 11.1792 under Capt. Samuel Osborn; sailed for Leeward Islands 26.2.1793; sailed for East Indies 28.11.1793. In action (with *Diomede*) against French 44-gun *La Prudente* and *La Cybèle*, plus 20-gun *Le Jean Bart* and 14-gun *Le Courrier*, off Mauritius 22.10.1794. Involved in capture of Ceylon 7–8.1795, and of Amboyna and Baada 2.1796. Under Capt. John Sprat Rainier from 4.1797, still in East Indies; Red Sea operations 1799–1800 (Suez), then back to Batavier 8.1800. Under Capt. James Lind (acting) in 1804; active in Lieut Phillip's defence against French 74-gun *Le Marengo*, 40-gun *L'Atalante* and 36-gun *La Sémillante* at Vizagapatam 15.8.1804. She was sent back from India in 11.1804 'as (she) will require an expensive repair if detained any longer in this Country; in her present state she may be converted by the Navy Board to some useful inferior establishment, as I know of no other mean of effectively getting rid of the White Ants onboard her, who have at times discovered themselves by serious depredations aloft' (ADM.1/174). Fitted at Chatham as a hospital ship 8.1807–1.1808 for Halifax, Nova Scotia. Recommissioned 10.1807 under Lieut Edward Webb and sailed for Halifax. Became a receiving ship and stores depot at Halifax under Capt. George Monke 11.1808. Hospital ship there 1809. As receiving ship at Halifax, under Capt. William Skipsey in 6.1813, as flagship of Rear-Adm. Edward Griffith Colpoys. Under Capt. Justice Finley from 6.1814, then Capt. David Scott from 10.1814; hulked 1817. Sunk at moorings there 21.2.1824; raised and BU 1825.

PORTLAND Class John Williams design, approved 2.4.1766. Eleven ships were built to this design, of which the following two survived into 1817.
>   Dimensions & tons: 146ft 0in, 119ft 9in x 40ft 6in x 17ft 6in. 1,044⁷⁄₉₄bm.
>   Men: 350. Guns: LD 22 x 24pdrs; UD 22 x 12pdrs; QD 4 x 6pdrs; Fc 2 x 6pdrs.
Portland Sheerness Dyd [M/Shipwright William Gray to 7.1767, completed by Edward Hunt]
>   As built: 146ft 0in, 119ft 9in x 40ft 6in x 17ft 6in. 1,044⁷⁄₉₄bm. Draught 10ft 6in/15ft 7in.
>   Ord: 18.1.1766. K: 1.1767. L: 11.4.1770. C: 10.11.1770.
>   First cost: £21,021.1.6d, plus £1,779.13.7d fitting.
>   Commissioned 9.1770 (–10.1772), recommissioned 1.1773 (–9.1774) and 1.1775 (–1782) for wartime service. Refitted and coppered at Woolwich (for £6,754.16.10d) 2–3.1779. Recommissioned 11.1797 under Lieut James Manderson as prison ship at Portsmouth; paid off 8.1800. Fitted as storeship at Portsmouth 10.1800–8.1801. Fitted at Portsmouth as a convict ship 1–5.1802,

for Langstone Harbour. Sold to Daniel List for £800 (at
Portsmouth) 19.5.1817.

Hygeia (ex-*Leander*, renamed 1813) Chatham Dyd [M/Shipwright
Israel Pownoll to 4.1779, completed by Nicholas Phillips]
>   As built: 146ft 0in, 119ft 7¾in x 40ft 8in x 17ft 5in. 1,052⁴⁹⁄₉₄bm.
>   Draught 11ft 0in/15ft 11in.
>   Ord: 21.6 and 25.7.1776. K: 1.3.1777. L: 1.7.1780. C: 21.8.1780.
>   First cost: £26,831.1.3d including fitting & coppering.
>   Commissioned 6.1780; paid off 4.1784 after wartime service.
>   Small repair at Portsmouth (for £8,466.12.5d) 6–12.1785.
>   Fitted for foreign service (for £4,465) 11.1786; recommissioned
>   8.1786 under Capt. Sir James Barclay; sailed for Nova Scotia
>   9.4.1787; flagship of Sir Hubert Sawyer 1788; paid off 9.1788.
>   Recommissioned 9.1788 under Capt. Joseph Peyton, Jnr, as
>   flagship of Rear-Adm. Joseph Peyton, Snr (–1791); sailed
>   for the Mediterranean 22.12.1788. Recommissioned 5.1795
>   under Capt. Maurice Delgano, for North Sea or Channel. In
>   11.1796 under Capt. Thomas Boulden Thompson; convoy to
>   Gibraltar 7.1.1797; attack on Santa Cruz 25.7.1797; at Battle
>   of the Nile 1.8.1798, had 0 killed, 14 wounded; taken off Crete
>   by the French *Généreux* 17.8.1798, had 35 killed, 57 wounded;
>   became French *Le Leander*, retaken at Corfu by the Russians
>   and Turks 3.3.1799 and returned to the RN. Recommissioned
>   6.1799 under Cmdr Adam Drummond in the Mediterranean; in
>   9.1799 under Capt. Michael Halliday. Refitted at Deptford (for
>   £24,962) 7.1801–6.1802. Recommissioned 5.1802 under Capt.
>   James Oughton, as flagship of Vice-Adm. Sir Andrew Mitchell
>   (–1806); sailed for Halifax 7.1802. Under Capt. Francis Fane
>   in 8.1803, then Capt. Alexander Skene 11.1803, Capt. George
>   Ralph Collier in 1804 and then Oughton again. In 11.1804 under
>   Capt. John Talbot; captured French 40-gun *La Ville de Milan*
>   and retook 32-gun *Cleopatra* on Halifax station 23.2.1805; from
>   1.1805 under Capt. William Lyall, then Capt. Henry Whitby;
>   fired on US coaster *Richard* off New York 25.4.1806. In 5.1806
>   under Capt. Salusbury Pryce Humphreys, as flagship of Vice-
>   Adm. George Berkeley, at Halifax. Under Capt. Richard Raggett
>   for passage home. Fitted as Medical depot ship at Portsmouth
>   10.1806; renamed *Hygeia* 6.5.1813. Sold to Mr Thomas for
>   £2,100 (at Portsmouth) 14.4.1817.

ANTELOPE Class Sir John Henslow design 1790
>   Dimensions & tons: 150ft 0in, 123ft 8½in x 41ft 0in x 17ft 8in.
>   1,106⁷⁄₉₄bm.
>   Men: 350. Guns: LD 22 x 24pdrs; UD 22 x 12pdrs; QD 4 x 6pdrs;
>   Fc 2 x 6pdrs.

Antelope Sheerness Dyd [M/Shipwright William Rule to 8.1790, then
Edward Sison to 5.1793, John Marshall to 6.1795, Thomas Mitchell to
7.1801, completed by Nicholas Diddams]
>   As built: 150ft 0½in, 123ft 2½in x 41ft 1¼in x 17ft 8in. 1,107²⁵⁄₉₄bm.
>   Ord: 15.2.1790. K: 6.1790. L: 10.11.1802. C: 15.3.1803.
>   First cost: £38,369 (including fitting).
>   Commissioned 11.1802 under Capt. John Melhuish; from 4.1803
>   flagship of Commodore Sir William Sidney Smith (–6.1804);
>   blockade of Ostend 1803–04; Later in 1804 under Capt. Lord
>   William Stuart, as flagship of Rear-Adm. William Domett
>   7.1804 (–8.1805), in the Downs; in 10.1804 under Capt. Sir
>   Home Popham; in attempt to destroy Fort Rouge, Calais, with
>   explosive vessels 8.12.1804. In 12.1804 under Capt. Robert
>   Plampin, in the North Sea. In 8.1805 under Capt. Henry Bazely;
>   sailed for the East Indies 1.6.1805. Under Capt. Barrington
>   Dacres 10.1805, as flagship of Commodore Sidney Smith again

(–12.1805). At Cape of Good Hope 1807. In 12.1807 under
Capt. Edward Galway; sailed for the Mediterranean 21.2.1808.
In 5.1809 under Capt. Donald M'Leod, as flagship of Vice-
Adm. John Holloway; sailed for Newfoundland 25.6.1809. In
1810 under Capt. Richard Dunn, as flagship of Vice-Adm. Sir
John Duckworth (–1812); sailed for Newfoundland 30.6.1810.
In 12.1810 under Capt. James Carpenter; sailed with convoy
for Gibraltar 30.12.1810. Under Capt. Thomas White (Acting)
2.1812, then Carpenter again; sailed for Newfoundland
22.6.1812. In 12.1812 under Capt. Edward Hawker, as flagship
of Vice-Adm. Sir Edmund Nagle on Newfoundland station.
In 1.1813 under Capt. Samuel Butcher; took Danish privateers
Kera Venner on 11.10.1813 and *Eleonara* 24.10.1813; took US
20-gun privateer *Ida* 14.8.1814. Fitted for sea at Portsmouth as a
flagship 11.1814–12.1815. Under Cmdr Richard Booth Bowden
in 5.1815; flagship of Rear-Adm. John Harvey in the Leeward
Islands 1815, as troopship. Under Capt. George Sayer in 8.1815,
still Harvey's flag; paid off 4.1819. Fitted at Chatham as convict
ship 8–11.1823, to Bermuda 1.1824. BU 7.1845.

DIOMEDE Class Sir John Henslow design, approved on 14 July
1791, modified from that of the *Antelope*. Her sister *Diomede* had been
BU in 1815.
>   Dimensions & tons: 151ft 0in, 124ft 7½in x 41ft0in x 17ft 8in.
>   1,114³¹⁄₉₄bm.
>   Men: 350. Guns: LD 22 x 24pdr; UD 22 x 12pdrs; QD 4 x 6pdrs; Fc
>   2 x 6pdrs.

Grampus (ex-*Tiger* renamed 4.3.1802) Portsmouth Dyd [M/
Shipwright George White to 3.1793, then Edward Tippett to 10.1799,
completed by Henry Peake]
>   As built: 151ft 0in, 124ft 7½in x 41ft 0in x 17ft 8in. 1,114³¹⁄₉₄bm.
>   Ord: 9.12.1790. K: 10.1792. L: 20.3.1802. C: 11.4.1803.
>   Commissioned 3.1803 under Capt. Hugh Downman; from 4.1803
>   under Capt. Thomas Caulfield (–1806); sailed 29.6.1803 for East
>   Indies. From 6.1806 under Capt. Walter Bathurst, and 10.1807
>   under Capt. James Tait, still in East Indies; home to pay off
>   end of 1809. Between Small and Middling Repair at Chatham
>   6.1809–2.1810. Recommissioned 1.1810 under Capt. William
>   Hanwell (–1812); sailed with convoy 28.4.1810 for East Indies;
>   in 11.1811 flagship of Commodore George Cockburn (–4.1812),
>   to Cadiz and thence to South America. In 6.1812 under Capt.
>   Robert Barrie, in North America; sailed for the Leeward Islands
>   2.9.1812. In 10.1812 under Capt. Francis Collier (–1815), as
>   flagship of Rear-Adm. Sir Francis Laforey in 1813; to East
>   Indies 1814–15. Laid up at Woolwich 7.1815, then at Deptford
>   1816. Troopship in 1817. Fitted at Deptford as a hospital ship
>   (by AO 13.1.1820) 2–7.1820, for the Committee for Distressed
>   Seamen in the Thames. Returned to RN 12.1831 and damaged
>   11.1832 while being hauled onto slipway at Deptford. Sold (at
>   Woolwich? for £640) to Mr Beatson in 12.1832.

PURCHASED EAST INDIAMEN (1795 and 1804) To meet the
need for small two-deckers for convoy duties during the French
wars, nine ships building as East Indiamen on the Thames had been
purchased from the East India Company in 1795 and converted to
carry 56 guns. Five of these were wartime losses, while two were
sold in 1807 and another in 1813. The remaining ship, *Glatton*, had
been experimentally fitted with a powerful all-carronade armament,
but her original 42pdr carronades on the UD were 'within a month'
replaced by those of 32pdr variety, and she was later regarded as being
vulnerable to opponents with longer range ordnance, so the LD was

re-fitted with conventional long 18pdrs in place of the former 68pdr carronades. Another two Indiamen, both built of teak at Calcutta, were purchased from the Company on 30 May 1804 and re-armed.

Glatton (ex-mercantile *Glatton*, launched 28.11.1795 by William Wells & Co., Blackwall), 60 guns.

> Dimensions & tons: 163ft 11¼in, 133ft 4¼in x 42ft 1in x 17ft 0in. 1,256²¹⁄₉₄bm.
>
> Men: 350. Guns: (orig) LD 28 x 68pdr carronades; UD 28 x 42pdr carronades; all mounted on non-recoil principal. (by 1804) LD 28 x 18pdrs; UD 28 x 32pdr carronades; Fc 2 x 32pdr carronades; RH 2 x 18pdr carronades.
>
> She made one round trip to China for the East India Company 1793–94, arriving back in England for the last time in 9.1794. Fitted by Wells & Co. (for £7,396) 15.5.1795.
>
> Commissioned 4.1795 under Capt. Henry Trollope, for North Sea; in action with French squadron in North Sea 15.7.1796. Under Capt. Charles Cobb 8.1797 (–1800), with Mitchell's squadron at surrender of Dutch fleet off Texel 28–30.8.1799. Under Capt. George Stephens 11.1800, then Capt. William Bligh 1801 for expedition to Baltic; in Battle of Copenhagen 2.4.1801; later under Capts William Nowell and William Birchall. Fitted as guard ship 'for short sea' at Sheerness (for £2,147) 7–8.1801; under Capt. John Ferris Devonshire 8.1801. Recommissioned 5.1802 as convict transport under Capt. Nathaniel Porlock; sailed to New South Wales 23.9.1802; returned home 22.9.1803. Fitted at Woolwich 11–12.1803; under James Colnett, as flagship of Rear-Adm. James Vashon. Reduced to 44-gun Fifth Rate 1804. Recommissioned 3.1806 under Capt. Thomas Seccombe (killed 1.1808); sailed for the Mediterranean 22.11.1806; her boats (with those of *Hirondelle*) took a Turkish treasure ship at Sigri harbour, on Mitylene; captured a transport off Corfu 29.11.1807; in 1.1808 under Cmdr Henry Hope (acting), then 4.1808 under Cmdr Charles Chamberlayne Irvine (acting). In 12.1808 under Capt. George Bligh; convoy home 7.1809; laid up at Sheerness 10.1809. In Baltic 1811, then under R. G. Peacock (master) at Portsmouth 1812–14. Water depot at Sheerness 1814. Fitted at Sheerness 4–6.1830 as a breakwater; sailed for Harwich 10.1830 to be expended there.

Coromandel (ex-*Malabar*, ex-mercantile *Cuvera*, launched 12.9.1798 at Calcutta), 56 guns.

> Dimensions & tons: 168ft 6in, 127ft 4in x 37ft 2in x — 935⁵⁶⁄₉₄bm.
>
> Men: 150 as storeship. Guns: LD 28 x 18pdr; UD 28 x 24pdr carronades. (as storeship UD 10 x 24pdr carronades, QD 6 x 24pdr carronades, Fc 2 x 9pdrs).
>
> A teak-built Indiaman, she made one round trip to England in 1799; used as a troop transport in India in 1801–02 and was purchased from East India Company by AO 30.5.1804. Fitted by Barnard & Co., Deptford 6.1804–24.7.1804; completed at Deptford Dyd 19.11.1804.
>
> Commissioned 7.1804 under Capt. George Byng (later Viscount Torrington); under Capt. Robert Hall, sailed for the West Indies 1805 (–1806); took (with sloop *Wolf*) privateer schooner *Le Napolèon* and destroyed similar *Le Régulator*, both off Port Azarades 2.1.1806. Under Capt. George Scott in 3.1806 and Capt. James Aycough in ?7.1806. Fitted at Woolwich 11.1806–1.1807. Recommissioned as 20-gun storeship 12.1806 under Capt. John Temple, for North Sea; sailed for the River Plate 25.6.1807. Fitted as storeship 7–8.1808; recommissioned 5.1808 under J. Henzell (master), then ?7.1809 (–1815) under F. Bradshaw (master), in the Mediterranean. Renamed *Coromandel* 7.3.1815. Fitted at Chatham 7–9.1818. Fitted at Portsmouth as Convict ship

8–10.1819, for voyage to NSW. Laid up at Portsmouth 12.1821. Fitted at Portsmouth as receiving ship 6–10.1827, for convicts at Bermuda 1828–53. BU there 12.1853 (by AO 2.5.1853).

Hindostan (ex-mercantile *Admiral Rainier*, launched 1799 by Hudson, Bacon & Co. at Calcutta), 52 guns.

> Dimensions & tons: 158ft 6in, 121ft 9in x 37ft 0in x — 886⁵⁴⁄₉₄bm.
>
> Men: 294 (141 as storeship). Guns: LD 26 x 18pdrs; UD 26 x 24pdr carronades. (as storeship UD only 20 x 24pdr carronades, + 2 x 9pdrs).
>
> A teak-built Indiaman, she made one round trip to England from India 1800; arrived in England from Bengal in 9.1803; purchased from East India Company by AO 30.5.1804.
>
> Commissioned 7.1804 under Capt. Mark Robinson; in 8.1804 under Capt. Alexander Fraser; sailed for the East Indies 2/3.1805; in action (with *Tremendous*) against French *La Canonnière* 21.4.1806. Recommissioned 12.1806 under Capt. Bendall Littlehales; repairing at Woolwich 1.1807; in 2.1807 under Capt. Thomas Bowen; sailed for the Mediterranean with convoy 28.6.1807, home at end of year. Became 22-gun storeship (by AO 11.11.1807). Recommissioned 2.12.1807 under Cmdr Lewis Hole, then 4.1808 under Cmdr Fitzowen Skinner, with squadron off Lisbon; paid off late 1808. Recommissioned as troopship 11.1808, under Cmdr John Pasco; sailed *en flûte* for New South Wales 3.5.1809; paid off 1810. Storeship again 7.1811, under Duncan Weir; in Mediterranean to 1815, then storeship at Woolwich. Renamed *Dolphin* 22.9.1819. Hulked at Woolwich as a convict ship 3.1824. Renamed *Justitia* 1831. Sold 24.10.1855.

Ex-DUTCH PRIZES (1799)

Of six Dutch two-deckers of 54 guns which had been incorporated following their capture by the RN, the *Broederschap* and (larger) *Brakel* had been disposed of during the war, the *Tromp* and *Alkmaar* were sold in 1815, and the remaining two were hulked or laid up. Neither retained any guns by this date (originally armed as follows).

> Men: 350 (215 as troopships). Guns (originally): LD 22 x 24pdrs; UD 24 x 12pdrs; QD 6 x 6pdrs; Fc 2 x 6pdrs.

Beschermer (Dutch *Beschermer*, launched 1784 at Enkhuizen).

> Dimensions & tons: 145ft 1in, 118ft 7in x 40ft 10in x 16ft 4in. 1,051⁶⁷⁄₉₄bm.
>
> Men: 350. Guns: LD 24 x 18pdrs; UD 24 x 32pdr carronades; QD 6 x 32pdr carronades; Fc 2 x 32pdr carronades. Reduced to 40 guns (20 x 24pdrs, 20 x 18pdrs), then QD 6 x 6pdrs + Fc 2 x 6pdrs added.
>
> Taken in the Texel by Vice-Adm. Andrew Mitchell's squadron 30.8.1799. Fitted as floating battery at Chatham (for £7,120) 28.6.1800–26.7.1801. Established 2.1801.
>
> Commissioned 7.1801 under Capt. Alexander Frazer; from 4.1803 under Capt. Robert Mansel, then 1.1804 under Capt. Volant Vashon Ballard. Lent to East India Dock Company as a hulk at Blackwall from 4.11.1806 to 8.1838. Sold to Joshua Crystall for £410 to BU 9.1838.

Batavier (Dutch *Batavier*, launched 1779 at Amsterdam).

> Dimensions & tons: 144ft 7in, 118ft 1⅞in x 40ft 10in x 16ft 5in. 1,047⁸⁷⁄₉₄bm.
>
> Men: 350. Guns: LD 24 x 18pdrs; UD 24 x 32pdr carronades; QD 6 x 32pdr carronades; Fc 2 x 32pdr carronades. Reduced to 40 guns (20 x 24pdrs, 20 x 18pdrs), then QD 6 x 6pdrs + Fc 2 x 6pdrs added.
>
> Taken in the Texel 30.8.1799 by Vice-Adm. Andrew Mitchell's squadron. Fitted as floating battery at Chatham (for £7,831) 14.7.1800–15.7.1801. Established 2.1801.

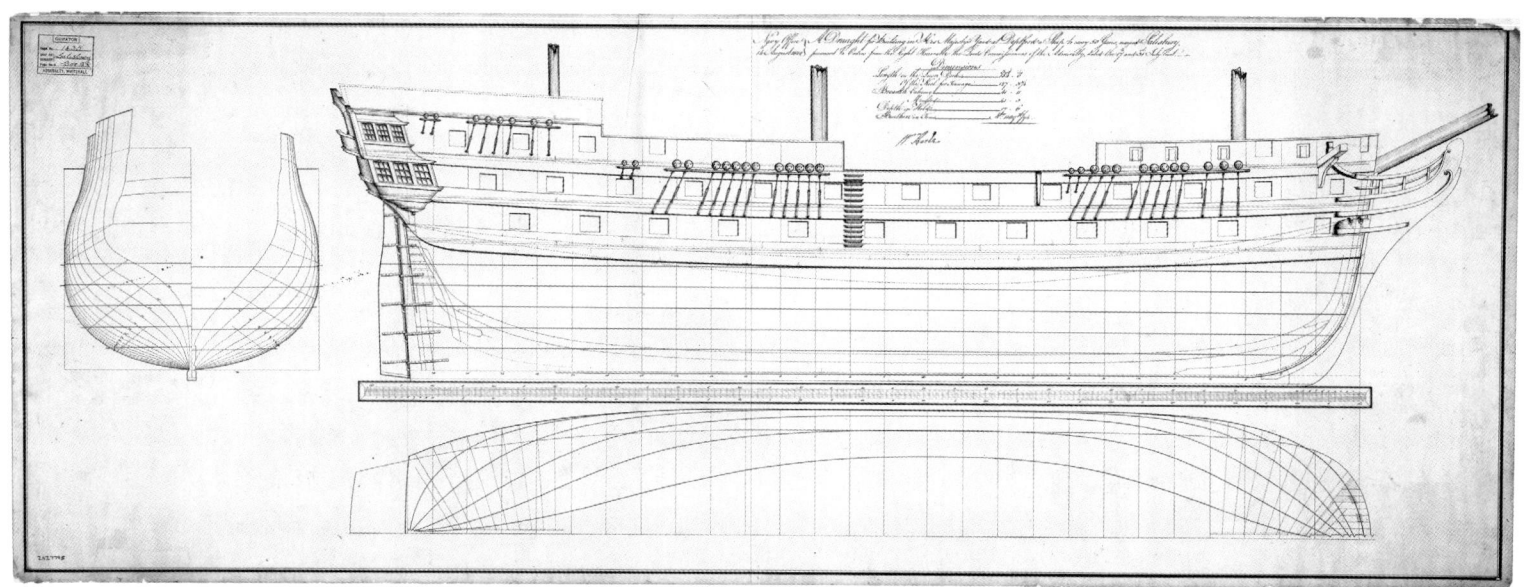

A draught of the 58-gun *Salisbury*, as designed by Sir William Rule and dated 14 August 1810. This was the final evolution of the 50-gun two-decker warship, with built-up barricades. This ship fulfilled the by now traditional role of the 50-gun ship as flagship on the lesser overseas stations. Curiously, while warships of this era were having their sterns closed in, this draught still shows an open gallery to the Great Cabin on the quarterdeck – a stubborn persistence of the privileges of flag rank. *(NMM J3507)*

Commissioned 6.1801 under Capt. William Broughton, for the Channel; from 4.1803 under Capt. Patrick Tonyn. Laid up at Chatham 8.1804–4.1809, them at Woolwich under Lieut Thomas D. Birchall as a hospital ship to 1.1817. To Blackwall to receive distressed seamen 1.1817. Fitted at Woolwich as a convict ship to lie at Sheerness 9.1817. BU at Sheerness 3.1823.

***JUPITER* Class** Sir William Rule design based on reduced lines of the 80-gun Danish prize *Christian VII*, approved on 30 June 1810.

Dimensions & tons: 154ft 0in, 127ft 5⅜in x 41ft 6in x 18ft 0in. 1,167⁴⁹⁄₉₄bm.

Men: 350. Guns: LD 22 x 24pdrs; UD 24 x 12pdrs; QD 8 x 24pdr carronades; Fc 2 x 6pdrs + 2 x 24pdr carronades.

Jupiter Plymouth Dyd [M/Shipwright Joseph Tucker to 5.1813, then Thomas Roberts]

As built: 154ft 0in, 127ft 3⅜in x 41ft 7½in x 18ft 0in. 1,173⁴⁄₉₄bm.

Ord: 30.6.1810. K: 8.1811. L: 22.11.1813. In Ordinary 1814. Began fitting for sea 3.1815, then laid up.

First cost: £36,733 to build, plus £12,494 fitting (incomplete)

Commissioned 1815 (briefly) under Cmdr Henry Meynell. Classed as troopship 11.1819. Underwent Small Repair (for £8,842) 7.1821–1.1822, then fitted (for £9,019) to carry the new Governor-General to India 5–9.1822. Recommissioned 5.1822 as troopship under Capt. George Westphal; under Capt. David Dunn from 1.1824, Capt. Sir William Wiseman 11.1824, and Capt. William Webb 12.1826, as flagship of Rear-Adm. Willoughby Lake on Halifax station from 1825; paid off 5.1827. Troopship again (30 guns) 1829. Fitted as temporary lazarette at Plymouth 8.1831. Fitted as troopship at Plymouth (for £15,671) 4–7.1832. Fitted at Woolwich to convey the Governor-General (Earl of Auckland) to India (for £2,679) and re-rated 38-gun 9.1835; recommissioned 11.1835 under Capt. Frederick Grey. Paid off 27.9.1836. Refitted at Portsmouth as troopship (for £6,158) 8–11.1837; recommissioned 23.9.1837 under Richmond Easto, Master; on 24.9.1838 under Robert Fulton, Master.

China War deployment 1839–42. On 6.2.1842 under George Hoffmeister, Master; in Vangyze operation 7.1842; returned to England and paid off 9.12.1843. Fitted as coal depot at Plymouth 4.1846. BU there (under AO 13.10.1869) completed 28.1.1870.

***SALISBURY* Class** Sir William Rule design 1810, modified from *Jupiter*. An originally intended third vessel (ordered 13 May 1811 from Woolwich Dyd) was 'not proceeded with' and never formally named, but in its place a fresh order was placed there on 10 November; this vessel (*Isis*) was converted while building to a frigate and launched in 1819 with a very different layout (see below).

Dimensions & tons: 154ft 0in, 127ft 3¾in x 41ft 11in x 17ft 6in. 1,189⁷⁶⁄₉₄bm.

Men: 350. Guns: LD 22 x 24pdrs; UD 24 x 12pdrs; QD 8 x 24pdr carronades; Fc 2 x 6pdrs + 2 x 24pdr carronades.

Salisbury Deptford Dyd [M/Shipwright Robert Nelson to 7.1813, then William Stone]

As built: 154ft 4in, 127ft 3⅜in x 42ft 1in x 17ft 6in. 1,199²⁄₉₄bbm.

Ord: 13.7.1810. K: 10.1811. L: 21.6.1814. C: 18.7–25.8.1814 at Woolwich, then 27.8.1814–22.5.1815 at Sheerness.

First cost: £45,297 (including fitting at Woolwich), plus £15,814 fitting at Sheerness.

Commissioned 3.1815 under Capt. Edward Hawker, as flagship of Vice-Adm. Sir Richard Keats on Newfoundland station. From 11.12.1815 under Capt. John Mackellar, as flagship of Rear-Adm. John Erskine Douglas on Jamaica station; under Cmdr Houston Stewart (acting Capt.) from 14.3.1817; paid off 18.4.1818. Underwent Small Repair and fitted for sea at Portsmouth (for £19,224) 5–12.1818. Under Capt. John Wilson 8.1818, as flagship of Rear-Adm. Donald Campbell (died 11.1819) on Leeward Islands station. Defects made good at Portsmouth (for £12,850) 9.1821–3.1822; under Capt. William Maude from 9.1821, as flagship of Rear-Adm. William Fahie on Halifax station. Laid up at Portsmouth 9.1824 until sold to Mr Beatson (for £2,710) to BU 12.1.1837.

Romney John Pelham, Frindsbury

As built: 154ft 9in, 127ft 8⅛in x 42ft 6in x 17ft 6in. 1,226⁶⁴⁄₉₄bm. Draught 11ft 4in/14ft 1in.

Ord: 13.5.1811. K: 8.1811. L: 24.2.1815. Laid up incomplete at Chatham.

Commissioned 3.8.1815 under Capt. John Mackellar, at Chatham; paid off 11.12.1815 (Mackellar transferred to *Salisbury*). Roofed over 12.1819, then fitted for sea (for £6,680) 7.1820. Fitted as 30-gun troopship at Sheerness (for £5,476) 7.1820–5.1822; fitted for sea at Sheerness (for £1,535) 7–8.1824; recommissioned 7.1824 under Capt. William Mingaye but never sailed. Fitted for foreign service at Chatham (for £2,414) 6.1.1825; recommissioned 17.12.1824 under Capt. Nicholas Lockyer; laid up at Plymouth 9.1827 and paid off 12.10.1827. Fitted at Plymouth as troopship (for £12,151) 1–4.1832; under Masters R. Brown 1832–3 and James Wood 11.1833–1834. Fitted at Portsmouth as a receiving ship for freed slaves (for £5,544) for Havanna 2–6.1837; recommissioned 18.5.1837 under Lieut Charles Jenkin, then from 12.8.1839 under Lieut Charles Hawkins and 18.6.1842 under Lieut Robert John Le Mesurier McClure. Sold at Jamaica 15.12.1845.

Isis Woolwich Dyd [M/Shipwright Edward Sison to 3.1816, then Henry Canham]

Ord: 10.9.1811. K: 2.1816. Converted while building to a frigate and launched 5.10.1819 with a very different layout, she appears below.

Fourth Rate Frigates

In 1793 the RN had had no 24pdr frigates, but the capture in 1794 of the very large French frigate *La Pomone* (designed as an 18pdr frigate and built by Baron Bombelle in 1782–85 at Rochefort, but by the time of her capture re-armed with 24pdrs) provided an opportunity for the Navy Board to introduce this calibre to the ranks of frigates. The *Pomone* was reduced to carry 18pdrs in 1799 and BU following the Peace of Amiens, but her design was meanwhile copied for a prototype British 24pdr frigate (*Endymion*), and five more to a slightly modified design were built to meet the urgent needs for 24pdr vessels produced by the American War of 1812.

ENDYMION Copied from design of *La Pomone* (French prize taken 1794). The prototype was built of oak, but 18 years later the design was revived for five frigates built of 'fir' (see below). Reclassified as a 50-gun Fourth Rate in January 1817, later reduced to 44-gun, finally to 38-gun in February 1839, *Endymion*'s fine qualities were such that she continued to be praised for nearly half a century.

Dimensions & tons: 159ft 2⅜in, 132ft 4¼in x 41ft 11⅜in x 12ft 4in. 1,238⁶⁷⁄₉₄bm.

Men: (orig. 300, later 340) 350. Guns: (*Endymion*) UD 26 (later 28) x 24pdrs; QD 14 (later 16) x 32pdr carronades; Fc 2 x 9pdrs + 4 x 32pdr carronades.

Endymion John Randall & Co., Rotherhithe.

Ord: 30.4.1795. K: 11.1795. L: 29.3.1797.

Commissioned 4.1797 under Capt. Sir Thomas Williams (–1800), for Channel & Irish station; took several privateers on the Irish station – 12-gun *La Revanche* 30.4.1798, 20-gun *Les Huit Amis* 10.5.1798, 6-gun *Le Brutus* and packet *San Antonio* 5.1798, 10-gun *La Sophie* 9.1798, 6-gun *La Casualidad* and 1-gun *La Prudentia* 1.1799; then 18-gun *Le Scipio* 5.1800 en passage to Mediterranean, and Spanish 4-gun *San Josef* and 2-gun *Intrepido* soon after. Under Capt. Philip Durham in 1.1801, took 14-gun privateer *La Furie* 13.4.1801. Under Capt. John Larmour in 5.1802, then Capt. Charles Paget in 4.1803; took (with *Dragon*)

16-gun *La Colombe* off Ushant 18.6.1803; took 18-gun *La Bacchante* 25.6.1803, 20-gun *L'Adour* 16.7.1804 and 16-gun privateer *Le Général Moreau* 16.8.1803. Defects made good at Plymouth (for £9,619) 12.1803–5.1804. Still under Paget, took treasure ship (ex-Vera Cruz) *Brilliante* 21.1.1805, another valuable Spanish prize 27.1.1805 and more on 4.2.1805 and 10.2.1805 (12 prizes in all). Under Capt. Edward Durnford King 4.1805; took 16-gun *La Colombe* 16.6.1805; off Cadiz 10.1805. Under Capt. Thomas Capel 1806 (–1810), in the Mediterranean, then with Duckworth's squadron in the Dardanelles 2.1807. Small Repair at Chatham (for £13,834) 5–10.1808. At Corunna? 1.1809, then on Irish station; sailed with convoy to Cape St Vincent 21.3.1809. Under Capt. Sir William Bolton 1810–11; took 14-gun privateer *Le Milan* 11.11.1810. Paid off at Plymouth 9.3.1812; between Middling and Large Repair at Plymouth (for £35,809) 5.1812–7.1813; recommissioned 18.5.1813 under Capt. Henry Hope (–1815), for North America; took US 3-gun privateer *Meteor* 18.1.1814; boats (with others) took US 15-gun privateer *Mars* 7.3.1814; took (with *Armide*) US 17-gun privateer *Herald* 15.8.1814; unsuccessful boat attack on US privateer *Prince de Neuchâtel* off Nantucket 11.10.1814; took (with consorts) the spar-deck 24pdr (nominally 44-gun) USS *President* 15.1.1815. In Ordinary at Plymouth 1815–33; Middling Repair at Plymouth (for £31,090) 1.1820–2.1822. Fitted for sea at Plymouth 6–9.1833. Recommissioned 18.6.1833 under Capt. Samuel Roberts, for the Mediterranean, then to Lisbon. Fitted for a flagship at Portsmouth 1–2.1834. Very Small Repair, and fitted as demonstration ship and for sea at Plymouth (for £15,271) 5.1837–2.1841. Recommissioned 30.10.1840 under Capt. Frederick William Grey, for the East Indies and China; Yangtze operations 7.1842; paid off late 1843. Fitted for sea at Plymouth (for £6,261) 9–12.1845. Recommissioned 8.12.1845 under Capt. George Robert Lambert, for North America and the West Indies; on 23.1.1847 under Capt. George William Conway Courtenay; on 3.8.1847 under Capt. George Fowler Hastings; paid off 18.1.1848 into Ordinary at Plymouth. Receiving ship at Devonport 1859. BU (by AO 3.6.1868) completed at Plymouth 18.6.1868.

Ex-DUTCH PRIZE (1799) The former 44-gun frigate *Vleiter* taken in 1799 was still used as a hulk into 1817.

Vleiter (Dutch *Mars*, originally built as the two-decker *Zevenwalden*, 68 – built Harlingen 1784, raséed to a frigate and renamed 1795).

Dimensions & tons: 155ft 8in, 127ft 1½in x 44ft 9½in x 16ft 9in. 1,356⁸¹⁄₉₄bm.

Men: 185 as floating battery. Guns: UD 26 x 24pdrs; QD 10 x 18pdr carronades.

Taken in the Texel 30.8.1799 by Vice-Adm. Andrew Mitchell's squadron.

Fitted at Chatham as a floating battery (for £7,177) 7.1800–4.7.1801.

Commissioned 7.1801 under Capt. William Birchall, as floating battery in the Medway; paid off 4.1802. Recommissioned 4.1803 under Capt. Adrian Renou, for the defence of the Thames; paid off 7.1804. Powder magazine at Chatham by 1806. Fitted as sheer hulk at Chatham 10.1808–1.1809 for Sheerness. BU at Sheerness 4.1817.

Ex-FRENCH PRIZE (1801)

La FORTE Class Designed by François Caro, the two vessels of this class initially carried 30 x 24pdrs and 20 x 8pdrs in French service, and so were rated 50-gun frigates, but were only 44-gun by the time of their capture (with 10 x 8pdrs and 4 x 36pdr obusiers on the QD/Fc), and were similarly rated in the British Navy. The first ship, *La Forte* taken in 1799, was wrecked in 1801. The second ship was allegedly converted on the stocks from a 74-gun ship, but this is unlikely.

Egyptienne (French *L'Egyptienne*, built 7.1798–11.1799 at Toulon. L: 17.7.1799).

Dimensions & tons: 169ft 8in, 141ft 4¾in x 43ft 8in (43ft 0in mld.) x 15ft 1in. 1,434$^{4}/_{94}$bm.

Men: 350. Guns: UD 28 x 24pdrs; QD 2 x 9pdrs + 12 x 32pdr carronades; Fc 2 x 9pdrs + 4 x 32pdr carronades.

Taken 2.9.1801 at capture of Alexandria.

Under Capt. Thomas Stephenson 1.1802, for voyage home from the Mediterranean. Arrived 13.2.1802 at Woolwich, and fitted there (for £12,625) 10–12.1802.

Commissioned 4.1803 under Capt. Charles Fleeming (–1805); took 16-gun *L'Epervier* 27.7.1803 and 14-gun privateer *La Chiffonette* 30.8.1803. Under Capt. Charles Elphinstone from 8.1805, then Capt. Charles Paget from 12.1805 (–1807); boats cut out privateer *L'Alcide* from Muros 8.3.1806; took 16-gun *L'Actéon* off Rochefort 2.10.1806; took (with *Loire*) 40-gun *La Libre* off Rochefort 24.12.1806; paid off into Ordinary 5.5.1807. Fitted as receiving ship at Plymouth after 1806; in Ordinary there 1812–15+. Sold to John Small Sedger (for £2,810) at Plymouth to BU 30.4.1817.

PURCHASED (1806) A genuine 'double-banked' or spar-decked frigate, designed by HEICo. as a long-range convoy escort.

Akbar (ex-*Cornwallis*) (orig. Bombay Marine frigate *Marquis of Cornwallis*, built 1800–01 at Bombay).

Dimensions & tons: 164ft 4½in, 140ft 4⅞in x 43ft 1¼in x 15ft 3in. 1,387$^{57}/_{94}$bm.

Men: 430. Guns: (as 'frigate') UD 30 x 24pdrs; Spar Deck 26 x 42pdr carronades. (as troopship) UD 22 x 32pdr carronades; QD 2 x 9pdrs + 8 x 32pdr carronades; Fc 2 x 9pdrs.

Purchased (for £68,630) by Admiral Sir Edward Pellew from the East India Co. at Bombay 3.1805.

Registered 13.8.1806 as *Cornwallis*, but renamed *Akbar* 2.1811.

Commissioned 'immediately' in 1805 (in Bombay). Under Cmdr (later Capt.) Charles Johnston from 2.1806. Under Capt. Christopher Cole in 1809; under Cmdr William Fisher in 12.1810; paid off 7.1812. Fitted at Woolwich as a storeship 2.1813, then as a frigate again 3–12.1813. Recommissioned 5.1813 under Capt. Archibald Dickson; under Capt. Charles Bullen from 11.1814. Flagship of Rear-Adm. Griffiths on Bermuda station 1815. Laid up at Portsmouth 12.1816, but fitted as troopship 1817. Fitted at Portsmouth as a quarantine ship 6–9.1824, for Pembroke; to Liverpool as a lazarette 9.1827. Refitted 1855 by Humble & Grayson and loaned to Liverpool Juvenile Reformatory Associaion as a detention centre for boys 7.1855. Sold 1862.

Revived *ENDYMION* Class The 1795 design was revived in 1812 for five frigates built of 'fir' (pitch pine) to meet the urgent requirement for 24pdr frigates produced by the American War of 1812; these differed from the prototype by having an extra pair of guns forward. They were all built (and coppered) by the same builder. The *Tagus* and *Eridanus* were originally ordered on 4 May 1812 as *Leda* Class 38s. All five were reclassified as 50-gun Fourth Rates in January 1817.

Dimensions & tons: 159ft 2⅜in, 132ft 4¼in x 41ft 11⅜in x 12ft 4in. 1,238$^{67}/_{94}$bm.

Men: 340 (350 as completed). Guns: UD 28 x 24pdrs; QD 16 x 32pdr carronades; Fc 2 x 9pdrs + 4 x 32pdr carronades.

Severn (ex-*Tagus*, renamed 7.1.1813) Wigram, Wells & Green, Blackwall.

As built: 159ft 2⅝in, 132ft 2in x 42ft 3in x 12ft 4in. 1,254$^{87}/_{94}$bm. Draught 9ft 9in/12ft 8in.

Ord: 4.5.1812. K: 1.1813. L: 14.6.1813. C: 1.7–11.9.1813 at Deptford Dyd.

Commissioned 6.1813 under Capt. Joseph Nourse; ?sailed for Halifax; saved convoy from 40-gun *L'Etoile* and *La Sultane* 18.1.1814; took 9-gun privateers – *Yankee Lass* 1.5.1814 and *Ind* 25.3.1815, and 4-gun privateer *Banyer* 20.12.1814; paid off 18.9.1815. Fitted at Chatham for foreign service 2–7.1816; recommissioned 21.2.1816 under Capt. Frederick Aylmer; at bombardment of Algiers 27.8.1816. In the Mediterranean in 1817, under Capts Sir James Gordon, then Robert Spencer and in 5.1817 William M'Culloch; paid off 8.6.1823 into Ordinary at Portsmouth. Sold to John Small Sedger, Rotherhithe for £3,610 (at Portsmouth) 20.7.1825.

Liffey (ex-*Eridanus*, renamed 7.1.1813) Wigram, Wells & Green, Blackwall

As built: 159ft 1½in, 131ft 8¼in x 42ft 5in x 12ft 4in. 1,260$^{23}/_{94}$bm. Draught 9ft 6in/12ft 6in.

Ord: 4.5.1812. K: 2.1813. L: 25.9.1813. C: 26.9.1813–10.6.1814 at Deptford Dyd.

Commissioned 4.1814 under Capt. John Hancock; paid off 4.1815 and laid up at Chatham 8.1815. Recommissioned 2.1816 under Capt. Sir John Louis at Chatham 1816–17. Small Repair and fitted at Chatham (for £4,682) 2.1818–1.1819; recommissioned 6.1818 under Capt. Henry Duncan (–1821), in the Mediterranean; paid off 10.1821. Recommissioned 10.1821 under Commodore Charles Grant, for East Indies; at Rangoon 11.5.1824 in First Burma War. It was instructed (by AO 15.6.1824) that she be sold in the East Indies, but this was not done. Grant died 25.7.1824, and *Liffey* was under Capt. Thomas Coe from 1825; arrived home 1.1826 and spent 18 months at Sheerness. BU at Sheerness 7.1827.

Liverpool Wigram, Wells & Green, Blackwall

As built: 159ft 2in, 132ft 1¼in x 42ft 1½in x 12ft 4in. 1,246$^{86}/_{94}$bm. Draught 9ft 7in/12ft 6in.

Ord: 26.12.1812. K: 5.1813. L: 21.2.1814. C: 21.2–30.6.1814 at Woolwich Dyd.

Commissioned 5.1814 under Capt. Arthur Farquhar, for the Cape of Good Hope; paid off 2.4.1816. Small Repair at Chatham (for £5,697) 10.1817–6.1818; recommissioned 11.2.1818 under Capt. Francis Augustus Collier, to the Persian Gulf 1820; paid off 18.1.1822. Sold at Bombay (for £3,780) 16.4.1822.

Glasgow Wigram, Wells & Green, Blackwall

As built: 159ft 2½in, 132ft 1⅞in x 42ft 4in x 12ft 4in. 1,259$^{73}/_{94}$bm. Draught 9ft 6in/12ft 5in.

Ord: 26.12.1812. K: 5.1813. L: 21.2.1814. C: 21.2–26.8.1814 at Woolwich.

First cost: £18,934 fitting.

Commissioned 7.1814 under Capt. Henry Duncan, for Home Water; paid off 9.1815. Fitted for foreign service at Chatham (for £16,704) 4–7.1816; recommissioned 2.1816 under Capt. Anthony Maitland, for the Mediterranean; at Bombardment of Algiers 27.8.1816; paid off 11.1816. Between Small and Middling Repair at Deptford (for £19,408) 3–11.1817. Recommissioned 8.1817, still under Maitland. Under Capt. Bentinck Doyle 3.1821, for the East Indies; paid off 3.1823. Fitted for foreign service at Portsmouth (for £6,314) 7–9.1823. Under Capt. James Maude 2.1825, in the Mediterranean; at Battle of Navarino 20.10.1827,

losing 2 wounded; paid off 9.1828, and BU at Chatham
24.12.1828–29.1.1829.

Forth Wigram, Wells & Green, Blackwall

As built: 159ft 3⅛in, 132ft 0¼in x 42ft 2½in x 12ft 4in. 1,251⁹⁴⁄bm.
Draught 9ft 10in/12ft 7in.

Ord: 7.1.1813. K: 2.1813. L: 14.6.1813. C: 14.7–7.9.1813 at Deptford
Dyd.

Commissioned 6.1813 under Capt. Sir William Bolton; took US
5-gun privateer *Regent* 19.9.1814. Fitted as flagship for foreign
service at Chatham 3–8.1816; recommissioned under Capt. Sir
John Louis 2.1816; paid off 7.1819. BU at Deptford 10.1819.

MAJESTIC Class Rasée 'frigates', 58-gun Fourth Rates. The success
of the large American frigates during the War of 1812 prompted the
revival of the earlier concept of cutting down a smaller two-decker
line of battle ship to produce a powerful frigate. By the end of the
Napoleonic Wars the obvious ships to use were 74s, the last of which
were ordered or building by then. The emphasis was shifting to larger
ships of the line; the substantial surplus of 74s would be used for
the next few years to produce more heavy frigates. Three ships had
been cut down in 1813; they were called 'frigates', but retained two
full gundecks, plus a 'flying forecastle' and rudimentary QD/poop,
with a single 12pdr mounted on each, on elevating carriages along
the centreline. Of these three, the *Goliath* had been taken to pieces
in June 1815 and the *Majestic* in April 1816 after stranding, but the
Saturn (formerly a 74-gun Third Rate of the Modified *Arrogant* Class)
remained in service until hulked in 1825 and not BU until 1868.

Men: 495. Guns (as Fourth Rate): UD (the former LD) 28 x 32pdrs
(56cwt); QD/Fc 2 x 12pdrs + 28 x 42pdr carronades.

Saturn Thomas Raymond, Northam, Southampton

As built: 168ft 2in, 138ft 1in x 46ft 11in x 19ft 10in. 1,616⁶⁸⁄₉₄bm.

Ord: 27.12.1781. K: 8.1782. L: 22.11.1786. C: 18.12.1786–5.2.1787
(for Ordinary), 12.6.1790 (for sea) at Portsmouth.

First cost: £29,422.12.10d to build, plus £10,126.12.6 (1787 fitting) &
£4,935 (fitting for sea).

Commissioned 5.1790 under Capt. Robert Linzee; paid off 9.1791.
Recommissioned 10.1793 under Capt. Thomas Newnham; in
8.1794 under Capt. William Lechmere, as flagship of Rear-Adm.
George Vanderput in the Downs. Under Capt. James Douglas,
as flagship of Vice-Adm. Sir John Laforey; sailed for the
Mediterranean 23.5.1795; in action off the Hyères 13.7.1796; in
Man's squadron in 10.1795, in pursuit of de Richery. Made good
defects at Plymouth (for £12,475) 2–4.1797. Under Capt. Herbert
Sawyer in 8.1797, in the Channel; under Capt. Jacob Waller
in 10.1797 (died 10.1798). Made good defects at Portsmouth
(for £9,158) 10.1798–1.1799; under Capt. Thomas Totty from
11.1798 (–1800). Under Capt. Charles Boyles in 1.1801, and
Capt. Robert Lambert 2.1801, in Hyde Parker's fleet in the
Baltic. From 8.1801 under Capt. James Brisbane , as flagship
of the now Rear-Adm. Totty (died 6.1802); to the Leeward
Islands in 1.1802, paid off later in 1802. Fitted at Portsmouth
(for £9,135) 5–8.1805. Recommissioned 7.1805 under Capt.
Lord Amelius Beauclerk (–1808), for the Mediterranean. In 1808
under Capt. Thomas Boys (acting) for blockade of Lorient; in
5.1809 under Capt. William Cumberland, in the Baltic to 1810.
In Ordinary at Plymouth 1812–13. Repaired and cut down to a
58-gun Fourth Rate frigate at Plymouth (for £50,423) 4–12.1813.
Recommissioned 8.1813 under Capt. James Nash; took US 10-
gun privateer *Hussar* 28.5.1814. Under Capt. Thomas Brown in
11.1814; paid off into Ordinary 4.1815. Small Repair and altered
at Plymouth (for £14,210) 10–12.1817, then laid up. Fitted at

Plymouth as a lazarette for Milford 8–9.1825. Gunnery training
ship 1830. Fitted as Harbour flag and receiving ship at Pembroke
(for £2,925) 7–11.1845. By AO 1.7.1845 it was ordered to
convert this very old vessel to a screw blockship, but this was
cancelled 30.8.1845, and she remained a receiving ship. Fitted
as guard ship at Pembroke 10.1849–3.1850. On 1.1.1858 under
Capt. George Ramsay, then 1.1863 under Capt. William Loring.
BU (under AO 1.8.1867) completed at Pembroke 1.2.1868.

The need for equivalents to the big American spar-decked frigates led
the Admiralty to call on 21 April 1813 for two draughts for 50-gun
frigates with two complete gundecks, designs being produced 3 days
later. Notwithstanding the armament shown below, both ships carried
an extra 2 x 24pdrs (*vice* 2 carronades) on the spar deck during their
1813–18 commission. While the flush-deck frigates ordered during the
war were designed to carry 60 or 58 guns, following the onset of peace
the Navy Board decreed in 1823 that for structural reasons the 8 guns
mounted on the gangways should be deleted, reducing the type to 52
or 50 guns. Subsequently no British 'spar-deck' frigates mounted guns
in the waist.

LEANDER Design by Sir William Rule, approved in April 1813. Built
of pitch pine. Re-rated as 60 guns in January 1817.

Dimensions & tons: 174ft 0in, 145ft 3⅛in x 44ft 10½in x 14ft 4in.
1,556³⁸⁄₉₄bm.

Men: 450. Guns: UD 30 x 24pdr; Spar Deck 26 x 42pdr carronades;
Fc 4 x 24pdr.

Leander Wigram, Wells & Green, Blackwall

As built: 174ft 0in, 145ft 1¾in x 45ft 1½in x 14ft 4in. 1,572¹⁵⁄₉₄bm.
Draught 9ft 6in/12ft 3in.

Ord: 6.5.1813. K: 6.1813. L: 10.11.1813. C: 18.2.1814 at Woolwich.

Commissioned 12.1813 under Capt. Sir George Collier. Took
US 16-gun privateer *Rattlesnake* (16-gun) 22.6.1814; chased
USS *Constitution* and recaptured the *Levant* at Porto Praya
10.3.1815. Under Capt. William Skipsey from 8.1815.
Underwent Small Repair and fitted for foreign service at
Woolwich 8.1815–2.1816. Under Capt. Edward Chetham from
5.1816, for the Mediterranean; at bombardment of Algiers
27.8.1816, firing 3,680 round shot and having 135 casualties (17
killed, 118 wounded). Flagship of Rear-Adm. Sir David Milne at
Halifax 1817. Underwent Small Repair at Portsmouth 7–11.1819.
Recommissioned 29.7.1819 under Capt. Charles Richardson,
as flagship of Rear-Adm. Sir Henry Blackwood, sailed for East
Indies; under Cmdr Henry Martin Blackwood (acting) 2–5.1822,
then Richardson returned to command; returned to pay off
31.12.1822. Receiving ship at Portsmouth 1823–30. BU there
3.1830.

NEWCASTLE Design by Jean-Louis Barrallier, approved in April
1813. Built of pitch pine. Re-rated 60 guns in January 1817.

Dimensions & tons: 177ft 0in, 150ft 2¾in x 44ft 4½in x 14ft 11in.
1,573⁴⁹⁄₉₄bm.

Men: 450 (480 originally planned). Guns: UD 30 x 24pdrs; Spar
Deck 24 x 42pdr carronades + 4 x 24pdrs [intended spar deck
armament changed several times during construction]

Newcastle Wigram, Wells & Green, Blackwall

As built: 176ft 5in, 149ft 5¾in x 44ft 8in x 15ft 1½in. 1,556bm.
Draught 10ft 8in/13ft 2in.

Ord: 6.5.1813. K: 6.1813. L: 10.11.1813. C: 23.3.1814 at Woolwich.

First cost: £39,192 to build.

Commissioned 11.1813 under Capt. Sir George Collier; from
12.1813 under Capt. Lord George Stuart; chased USS

A detail from Nicholas Pocock's painting of the *Royal Sovereign* conveying Louis XVIII from Dover to Calais on 24 April 1814, to take the French throne following Napoleon's abdication and showing the royal yacht escorted by a double-banked frigate. Both *Newcastle* and *Leander* were newly commissioned at that date, so either ship might be the intended subject, but the configuration of gunports suggests the latter. *(NMM BHC3612)*

Constitution and recaptured the *Levant* at Porto Praya 10.3.1815. Recommissioned 28.9.1815 at Woolwich under Capt. Samuel Roberts, fitted there for foreign service 10–12.1815; on 24.11.1815 under Capt. (acting, confirmed 10.4.1816) Henry Meynell, as flagship of Rear-Adm. Sir Pulteney Malcolm – at St Helena 1816 and Chatham 1817; paid off 10.9.1817. Small Repair at Chatham (for £4,658) 4–6.1818. Recommissioned 26.11.1818 under Capt. Arthur Fanshawe, fitted (– 3.1819) at Chatham as flagship of Rear-Adm. Edward Griffith Colpoys, for Halifax station; paid off 1.1822 at Portsmouth. Fitted there 4–6.1824 as a lazarette for Pembroke; to Liverpool 9.1827. Sold to John Brown (for £2,500) 12.6.1850.

ISIS 24-pounder frigate of 58 guns. Begun as a 50-gun two-decker of the *Salisbury* or Modified *Jupiter* Class (see above), but redesigned while building with 11ft lengthening and cut down to a spar-deck frigate of 58 guns.
> Dimensions & tons: 164ft 0in, 138ft 4in x 41ft 11in x 13ft 3in. 1,292⁸⁸⁄₉₄bm.
> Men: 350; 342 as 44-gun ship (243 officers & men, 39 boys, 60 marines). Guns: UD 28 x 24pdrs; QD/Fc (including spar deck) 28 x 42pdr carronades + 2 x 24pdrs; amended 3 June 1823 to complete as 50-gun with unarmed spar deck.
> First cost: £53,723.

Isis Woolwich Dyd [M/Shipwright Edward Sison to 3.1816, then Henry Canham]
> As built: 164ft 0in, 138ft 2in x 43ft 0in (42ft 4¾in for tonnage) x 13ft 2in. 1,321bm. Draught 11ft 9in/13ft 4in.
> Ord: 10.9.1811 (as a two-decker 50). K: 2.1816. L: 5.10.1819. C: 4.1820 at Chatham for Ordinary (after lengthening by 11ft). Fitted at Chatham for sea, and for a flag officer, 2.1823–4.9.1823.
> First cost: £50,523 to build, plus fitting (1823) £18,150.
> Commissioned 7.1823 under Capt. Thomas Forrest, as flagship of Sir Lawrence William Halstead, for Jamaica; on 16.9.1825 under Capt.

Hugh Patton, as flagship of Halstead, on same station. Refitted (for £11,945) at Chatham, and poop removed (by AO 1.6.1827) 6–9.1827. Recommissioned 6.1827 under Sir Thomas Staines, for the Mediterranean; destruction of pirates' base at Grabusa (off Crete) 31.1.1828. Recommissioned 10.1831 under Capt. James Polkinghorne, as flagship (5.8.1831) of Rear-Adm. Frederick Warren, for the Cape of Good Hope; on 23.11.1834 flagship of Rear-Adm. Patrick Campbell; paid off 12.1834 at Chatham. Small Repair there (for £7,343) 12.1834–6.1837. Reduced to 44-gun in 1839 (UD 26 x 32pdrs; QD 12 x 32pdrs; Fc 4 x 32pdrs + 2 x 8in shell). Fitted for sea there (for £6,371) 8–11.1841. Recommissioned 18.8.1841 under Capt. Sir John Marshall, for the Cape of Good Hope; paid off into Ordinary 1.1845. Small Repair there (for £8,717) 2–10.1851, then back to Ordinary. Fitted at Chatham as a coal depot for Sierra Leone 12.1860–3.1861. Sold there to Charles Heddle (for £1,180) 12.3.1867.

JAVA Class 24-pounder frigate of 60 guns. The prototype of a new class of 24pdr frigates was ordered in 1813 to a joint design of the three Surveyors of the Navy, approved on 9 July 1813. An alternative to the expense of building hardwood versions of the *Leander* and *Newcastle*, the new design was intended to be the minimum size that could face the American spar-decked 'super-frigates', for which Henry Peake proposed a ship of about 1,430 tons. However they were expressly not intended for the Napoleonic conflict, and even the prototype was not completed until 1826.
> Dimensions & tons: 172ft 0in, 145ft 1¼in x 43ft 4in (42ft 8in mld.) x 14ft 3in. 1,449³⁹⁄₉₄bm.
> Men: 450 (comprising 343 officers and men, 47 boys and 60 marines).
> Guns: UD 30 x 24pdrs; Spar deck originally 2 x 24pdrs + 28 x 42pdr carronades, from 6.1823 changed to 6 x 24pdrs + 16 x 42pdr carronades (with the 'spar' section left unarmed). Reduced to 50-gun in 1839.

Java Plymouth Dyd [M/Shipwright Thomas Roberts to 9.1815, completed by Edward Churchill]
> As built: 171ft 11½in, 144ft 9¾in x 43ft 6in x 14ft 3in. 1,457⁵⁶⁄₉₄bm.
> Ord: 9.7.1813. K: 3.1814. L: 16.11.1815 (then laid up). C: 12.1825–2.7.1826.
> First cost: £56,729 to build, plus £13,871 fitting (1825–26).
> Commissioned 12.1825 under Capt. John Wilson; sailed for the East Indies 3.8.1826. In 1827 under Capt. William Carroll (–1830), as flagship of Rear-Adm. William Gage; paid off 28.1.1830. Between Small and Middling Repair at Portsmouth 1.1831–8.1832, then laid up in Ordinary. Fitted for Commission at Portsmouth (for £7,822) 2–5.1846, but then laid up again. Fitted as target ship at Portsmouth 10–11.1861. BU completed at Portsmouth 22.11.1862.

SOUTHAMPTON (Modified *JAVA*) Class 50 guns. Following the prototype *Java*, the design was somewhat modified by raising the decks and adopting diagonal frames and a circular stern, and four ships were ordered to this revised design in May 1816. Another two were ordered in July 1817 and July 1818, but were suspended on the stocks and not launched until 1843; this pair lies outside the scope of this section and will be found in Section B along with two final ships subsequently ordered but never built.
> Dimensions & tons: 172ft 0in, 144ft 9in x 43ft 8in (43ft 0in mld.) x 14ft 6in. 1,468¹¹⁄₉₄bm.
> Men: 450 (comprising 343 officers and men, 47 boys and 60 marines).

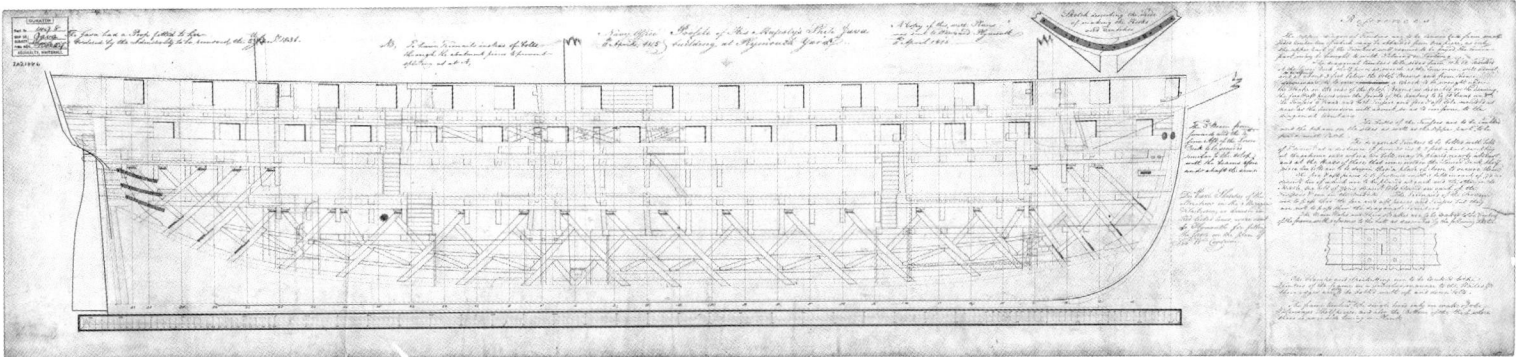

The disposition of framing for the *Java*, dated 11 April 1815 and initialled by the Surveyors Joseph Tucker and Robert Seppings. It shows the new diagonal trusses devised by Seppings. Such was the novelty of the structure that to the right of the draught is an elaborate explanation of how best to convert the timbers and precisely how they were to be fastened. As completed the ship had no upper deck guns – and indeed no barricades – in the waist. She had a poop fitted during her first commission when flagship of the East Indies station, but this was removed by AO of 1 January 1831. *(NMM J4118)*

Guns: UD 30 x 24pdrs; Spar deck originally 2 x 24pdrs + 28 x 42pdr carronades, from 6.1823 changed to 6 x 24pdrs + 16 x 42pdr carronades (with the gangways left unarmed). Reduced to 50-gun in 1839.

Southampton Deptford Dyd [M/Shipwright William Stone]

As built: 172ft 9½in, 145ft 3⅛in x 44ft 3½in oa (43ft 8½in for tonnage) x 14ft 6in. 1,474³³/₉₄bm. Draught 12ft 2in/14ft 11in.

Ord: 23.5.1816. K: 3.1817. L: 7.11.1820. C: 11.5.1821 for Ordinary, with waist housed over, then to Sheerness. Fitted 10.1828–3.1829 for sea (at Chatham).

First cost: £49,664 to build; fitting (1828–29) £22,849.

Commissioned 12.1828 under Capt. George Frederick Rich, then 1.1829 under Capt. Peter Fisher, as flagship of Rear-Adm. Sir Edward William Campbell Rich Owen, for the East Indies; on 17.4.1831 under Acting Capt. John Milligen Laws, still Owen's flagship on this station; paid off 1.1833 at Chatham. Temporarily

The heavily modified profile draught for the *Southampton* Class was originally sent to Plymouth, Deptford and Woolwich Dockyards on 16 July 1819 for the first four ships. At first, like the *Java* design from which it derived, the class was intended to be completely 'double-banked', but four waist ports along the upper deck on each side are shown to be deleted. More substantial later alterations include a modified circular stern and the replacement of the timber trusses seen in the *Java* by iron diagonals, Seppings's improved structural scheme for frigates and smaller craft. *(NMM J3436)*

housed over there 4.1833; Very Small Repair there (for £6,706) 5.1835–5.1836. Recommissioned 15.5.1840 under Capt. William Hillyar; fitted for sea (for £7,735) 5–8.1840; on 29.7.1840 flagship of Rear-Adm. Sir Edward Durnford King, for Cape of Good Hope; on 12.8.1841 under (Acting) Capt. Stephen Greville Fremantle, then 11.10.1841 Capt. Thomas Ogle, still King's flagship on same station; Vice-Adm. King from 23.11.1841; at Port Natal evacuation 24.6.1842; paid off 12.1842 at Chatham. Fitted as flagship there (for £5,545) 3–9.1838; recommissioned 3.8.1848 under Capt. Nicholas Cory, as flagship of Rear-Adm. Barrington Reynolds, for east coast of South America; paid off 1849 at Chatham. Fitted as guard ship there 9.1856–1.1857; recommissioned 2.12.1856 under Capt. Edward Philips Charlewood, as guard ship at Harwich; on 1.10.1860 under Capt. Charles Wise, as guard ship at Sheerness; paid off 15.7.1861. Fitted at Sheerness for the Humber Training Ship Association (by AO 18.4.1868) 5–6.1868. On 25.7.1879 under Capt. George Doherty Broad, as industrial school ship at Southampton. School closed 14.5.1912; sold to Hughes, Bolckow & Co., Middlesbrough (for £2,625) 26.6.1912 and handed over 10.7.1912 to BU at Blyth.

Portland (ex-*Kingston*, renamed 1817) Plymouth Dyd [M/Shipwright Edward Churchill]

As built: 172ft 9in, 145ft 5in x 44ft 3¼in oa (43ft 8¼in for tonnage) x 14ft 6in. 1,476²⁷/₉₄bm. Draught 12ft 6in/15ft 1in.

Ord: 23.5.1816. K: 8.1817. L: 8.5.1822. C: 20.8.1833 (for sea).

First cost: £51,025 to build, plus £12,006 fitting.

Commissioned 1.5.1834 under Capt. David Price, for the Mediterranean; conveyed King Otto of Greece back to his capital 14.2.1837; paid off 11.5.1838 at Plymouth. Small Repair

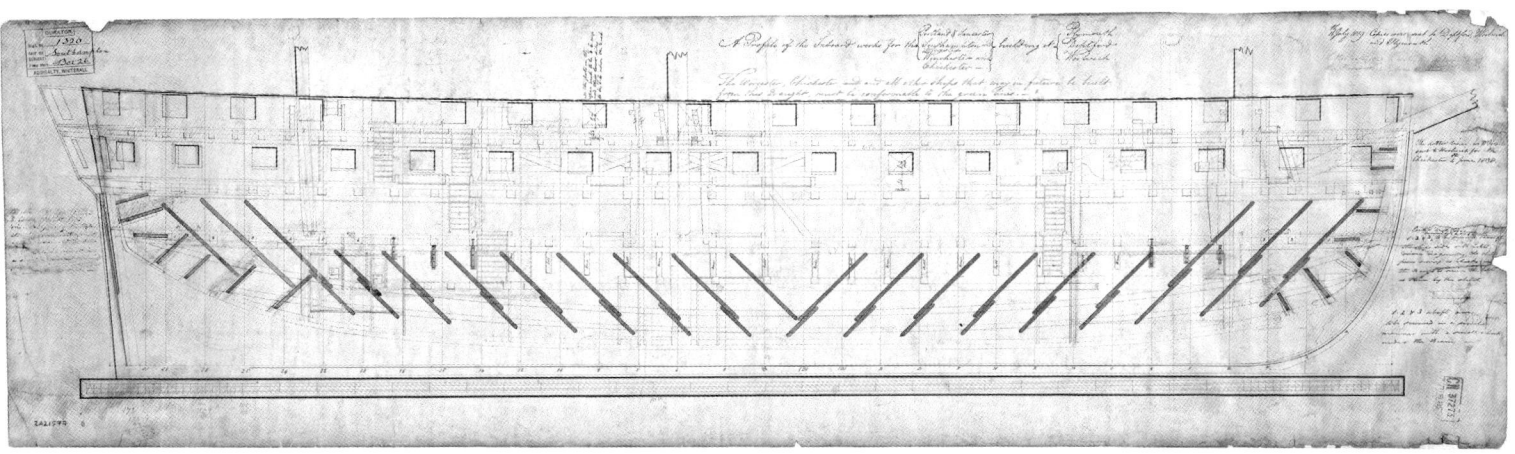

and fitted as a Demonstration ship (for £8,285) 11.1838–1.1840. Fitted as provision depot ship (for £4,360) 2–3.1846. Fitted as flagship (for £15,089) 8–11.1850. Recommissioned 26.8.1850 under Capt. Henry Chads, as flagship of Rear-Adm. Fairfax Moresby, sailed 8.11.1850 for the Pacific; paid off 11.5.1854 at Plymouth and laid up. Sold to Castle & Son 19.5.1862 (for £2,250) to BU at Charlton.

Lancaster Plymouth Dyd [M/Shipwright Edward Churchill]

As built: 172ft 9in, 145ft $4\frac{5}{8}$in x 44ft $3\frac{3}{4}$in oa (43ft $8\frac{3}{4}$in for tonnage) x 14ft 6in. $1,478\frac{74}{94}$bm. Draught 12ft 0in/15ft 2in.

Ord: 23.5.1816. K: 18.7.1818. L: 23.8.1823. C: 8.10.1823 for Ordinary (housed from main mast forward).

First cost: £47,243 (fitted for Ordinary).

Never commissioned for RN. Small Repair at Plymouth (for £7,252) 8.1837–11.1838. Fitted at Plymouth (by AO 19.6.1847) as a temporary fever hospital ship for Liverpool, to where she sailed 7.7.1847, but returned 16.8.1847 to Plymouth (still in Ordinary). Surveyed there 2–3.1863, then sold to Messrs Marshall 17.2.1864 (for £3,025) to BU.

Winchester Woolwich Dyd [M/Shipwright Henry Canham]

As built: 173ft 1in, 145ft 7in x 44ft 5in oa (43ft 10in for tonnage) x 14ft 6in. $1,487\frac{81}{94}$bm. Draught 12ft 2in/14ft 10in.

Ord: 23.5.1816. K: 11.1818. L: 21.6.1822. C: 16.9.1822 for Ordinary, 8.1829–21.1.1830 for sea (at Chatham).

First cost: £40,619 to build; fitting (1829–30) £24,838.

Commissioned 10.1829 under Capt. Charles John Austen; on 20.2.1830 flagship of Vice-Adm. Sir Edward Griffith Colpoys, for North America and the West Indies; in 1.1831 under Cmdr Robert Stuart (temp.) (Austen invalided), then on 18.5.1831 under Capt. Lord William Paget; paid off 6.1833 at Chatham. Small Repair and fitted for sea there (for £14,716 + £6,587) 9.1833–9.1834. Recommissioned 4.6.1834 under Capt. Edward Sparshott, as flagship of Rear-Adm. Sir Thomas Bladen Capel, for the East Indies; paid off 6.1838 at Chatham. Fitted as Demonstration ship then fitted for sea (for £8,901) 11.1838–6.1839. Recommissioned 18.3.1839 under Capt. John Parker, as flagship of Vice-Adm. Sir Thomas Harvey, for North America and the West Indies; paid off 1842 at Portsmouth. Fitted for sea there (for £9,660) 3–6.1842. Recommissioned 3.1842 under Capt. Charles Eden, as flagship of Rear-Adm. Josceline Percy, for the Cape of Good Hope; paid off 1846 at Portsmouth. Small Repair there (for £16,216) 2–12.1850. Fitted for sea (for £9,012) 4–5.1852. Recommissioned 16.3.1852 under Capt. Granville Gower Loch, as flagship of Rear-Adm. Fleetwood Broughton Reynolds Pellew, for the East Indies; in 2nd Burma War 1852; on 29.4.1854 under Capt. Thomas Wilson, as flagship of Rear-Adm. Sir James Stirling (from 1.1854), on same station; in 2nd Anglo-Chines War; in action against junks at Tymmoon Bay 2.11.1854, and action at Coulan Bay 13.11.1854; at Canton 10.1856; laid up 5.1857 at Chatham. Converted there as a school ship for Liverpool 8.1861, renamed *Conway* by AO 28.8.1861. Recommissioned 1.7.1862 under Cmdr Charles John Balfour, as RNR drill ship at Aberden; in 5.1865 under Cmdr William Cox Chapman, then 7.1869 Cmdr Henry Weyland Chetwynd, in same role. Returned from Liverpool to Plymouth 7.1876, and fitted as industrial training ship for the Devon & Cornwall Training Ship Society 7.1876–6.1877; renamed *Mount Edgecumbe* by AO 1.9.1876. Sold 8.4.1921 to BU.

Ex-*AMERICAN PRIZE* (1815) An American 24pdr-armed 'super-frigate' (and sister to the *Constitution* and *United States*) built to a design by Joshua Humphries, as amended by Josiah Fox. A very large spar-decked frigate, designed as 174ft $10\frac{1}{2}$in, 145ft 0in x 44ft 2in (43ft 6in moulded) x 14ft 3in, she was rated as 44-gun by the USN but carried initially 30 x 24pdrs and 22 x 12pdrs (later 32 x 24pdrs and 20 x 42pdr carronades) in US service, plus an 8in/24pdr howitzer on the QD and 5 brass 6pdrs in the tops. She was initially rated as 50 guns in the Royal Navy.

President (American *President*, launched 1.4.1800 by William Doughty & Christian Bergh, New York)

Dimensions & tons: 173ft 3in, 146ft $4\frac{3}{4}$in x 44ft 4in (43ft in moulded) x 13ft 11in. $1,533\frac{7}{94}$bm.

Men: 450 (comprising 343 officers and men, 47 boys and 60 marines).

Guns: UD 30 x 24pdrs; Spar deck 2 x 24pdrs (Fc) + 28 x 42pdr carronades.

Taken off New York 15.1.1815 by *Endymion*, *Tenedos*, *Pomone*, the rasée *Majestic* and the brig *Despatch*. Arrived Portsmouth 31.3.1815.

Not commissioned in RN. Re-rated 60 guns in January 1817. She was intended to have undergone repairs in 3.1818, but the hull was found to be too decayed, and she was BU at Portsmouth 6.1818.

(B) Vessels acquired from 1 January 1817

***SOUTHAMPTON* (Modified *JAVA*) Class** 50 guns. Two more ships to this design were ordered in 1817 and 1818. They were suspended on the stocks and not launched until 1843, being completed with a modified armament. Two further ships ordered in 1825 were never built. However, in 1844 the Admiralty, in forwarding to Symonds Fincham's plan for a steam frigate of 46 guns (the *Arrogant*, as she finally became), noted that this was to be '*Southampton* class', indicating that at least in concept the screw frigate derived from this class.

Men: 450 (395 peacetime). Guns: UD 4 x 8in (65cwt/8ft) shell + 26 x 32pdrs (50cwt/8ft); QD/Fc 4 x 32pdrs (50cwt/$8\frac{1}{2}$ft) + 16 x 32pdr carronades (25cwt/6ft).

Chichester Woolwich Dyd

As built: 173ft 1in, 145ft $2\frac{5}{8}$in x 44ft 9in oa (44ft 1in for tonnage) x 14ft 6in. $1,501\frac{10}{94}$bm. Draught 10ft 11in/14ft 6in.

Ord: 23.7.1817. K: 7.1827. L: 12.7.1843. C: 23.2.1844 (for Ordinary).

First cost: £35,035 to build.

Not commissioned as naval vessel. Towed to Sheerness 2.1844 and roofed over. Lent as training ship for the destitute boys of London to the National Refuge Society at Greenhithe 10.1866 (by AO 1.9.1866). Returned to Admiralty 8.10.1888 for disposal and sold to Castle 5.1889 for BU.

Worcester Deptford Dyd

As built: 173ft 1in, 145ft $3\frac{5}{8}$in x 44ft 3in oa (43ft 8in for tonnage) x 14ft $6\frac{1}{2}$in. $1,473\frac{67}{94}$bm. Draught 11ft 6in/14ft 6in.

Ord: 21.7.1818. K: 12.1820. L: 10.10.1843. C: 11.1843 at Sheerness (for Ordinary), then laid up.

First cost: £40,636 + £1,631, plus £2,227 fitting for Ordinary.

Not commissioned as naval vessel. Fitted at Sheerness (for £1,886) 2–5.1862 to lend as a training ship (name unchanged) at Greenhithe for Thames marine officers. Sold to Castle 8.1885 and removed 12.8.1885 from Chatham for BU at Charlton.

Liverpool Plymouth Dyd

Ord: 9.6.1825. K: — . Cancelled 5.3.1829 (unstarted).

The USS *President* was the most important prize of the naval war with the United States. Captured in 1815, she was found to be in poor condition and was quickly taken to pieces, but a replacement of the same name and to the same lines was built in 1824–29. As the prize was never commissioned in British service, this undated graphite drawing probably depicts the later ship, although the appearance of the two ships was virtually identical. *(NMM PU6144)*

Jamaica Plymouth Dyd
  Ord: 1.7.1825. K: — . Cancelled 5.3.1829 (unstarted).

PRESIDENT Class 52 guns. In May 1818 the Admiralty directed the Navy Board to build a 60-gun spar-deck frigate on the exact lines of the USS *President*, taken as a prize in 1815 but found to be in poor condition (and BU in June 1818), and to give her that (same) name. She was completed as a 52-gun ship, the design being amended to remove the 8 guns on the gangways.
  The *President* served as flagship in 1854 of the Anglo-French Squadron in the Pacific (under Rear-Adm. David Price) consisting of the *President, Pique, Amphitrite, Trincomalee* and paddle vessel *Virago*, together with the French *La Forte, L'Eurydice, L'Artémise* and *L'Obligado*, all under Rear-Adm. Auguste Febvrier-Despointes; she was involved in the attack on Petropavlosk on the Kamchatka Peninsula, in the course of which Price shot himself.
  Dimensions & tons: 174ft 10in, 143ft 6in x 45ft 1in oa (44ft 6in for tonnage) x 13ft 4in. 1,511⁴⁹⁄₉₄bm.
  Men: 450. Guns: UD 30 x 24pdrs, QD/Fc 6 x 24pdrs + 16 x 24pdr carronades; later re-armed with UD 32 x 32pdrs; QD/Fc 20 x 32pdr carronades.
  First cost: £37,637.
President Portsmouth Dyd
  As built: 173ft 5½in, 146ft 11⅛in x 44ft 10¼in oa (44ft 3¾in for tonnage) x 14ft 2in. 1,537 (1,534⁵⁶⁄₉₄ by calc.)bm. Draught 11ft 1in/16ft 6in.
  Ord: 25.5.1818. K: 6.1824. L: 20.4.1829 (and roofed over fore and aft). C: 2.1834–26.5.1834 (for sea).
  First cost: £44,781 to build, plus £12,254 fitting for sea (1834).
  Commissioned 14.2.1834 under Capt. John McKerlie, for North America; on 16.6.1834 under Capt. James Scott; flagship of Vice-Adm. Sir George Cockburn from 12.1834; flagship of Rear-Adm.

Charles Bayne Hodgson Ross from 30.8.1837, for the Pacific; on 31.10.1839 under Capt. William Broughton, still Ross's flagship, now for South America; paid off 5.1842 at Portsmouth. Recommissioned 14.8.1845 under Capt. William Pearse Stanley, as flagship of Rear-Adm. James Richard Dacres; fitted as a flagship at Portsmouth (for £10,514) 9–11.1845; to Cape of Good Hope; boats attack on Anjoxa, Mozambique in 1847; paid off 6.2.1849 at Chatham. Between a Very Small and a Small Repair at Chatham (for £4,671) 5.1850–1.1851. Refitted as a flagship at Chatham (for £10,286) 5–9.1853. Recommissioned 6.8.1853 under Capt. Richard Burridge, as flagship of Rear-Adm. David Price (suicide 31.8.1854), for the Pacific; Anglo-French squadron in the Pacific 1854 during the Russian War; on 6.12.1854 under Capt. Charles Frederick, as flagship of Rear-Adm. Henry William Bruce, in the Pacific; paid off and laid up 7.1857 at Chatham. Fitted as training ship at Chatham 8.1860–10.1861 and moved to the City Canal (London Docks), where she became RNR training ship 4.1862. On 28.6.1862 under Cmdr William Mould; (she sank in the canal 1865 but was raised); on 21.4.1866 under Cmdr Henry Wandesford Comber, then 4.1869 Cmdr Charles Gudgeon Nelson and 8.1870 Cmdr John Binney Scott; to Poplar (West India Docks) 1871; in 9.1874 under Cmdr Hardy McHardy, then 1.8.1879 under Cmdr Noel Stephen Fox Digby, 14.1.1880 Cmdr H. M. Washington, 1886 Cmdr C. Coffin, 1.1889 Cmdr Richard W. Hope, and 11.1897 Cmdr A. C. Woods. Renamed *Old President* 25.3.1903, but sold 7.7.1903.

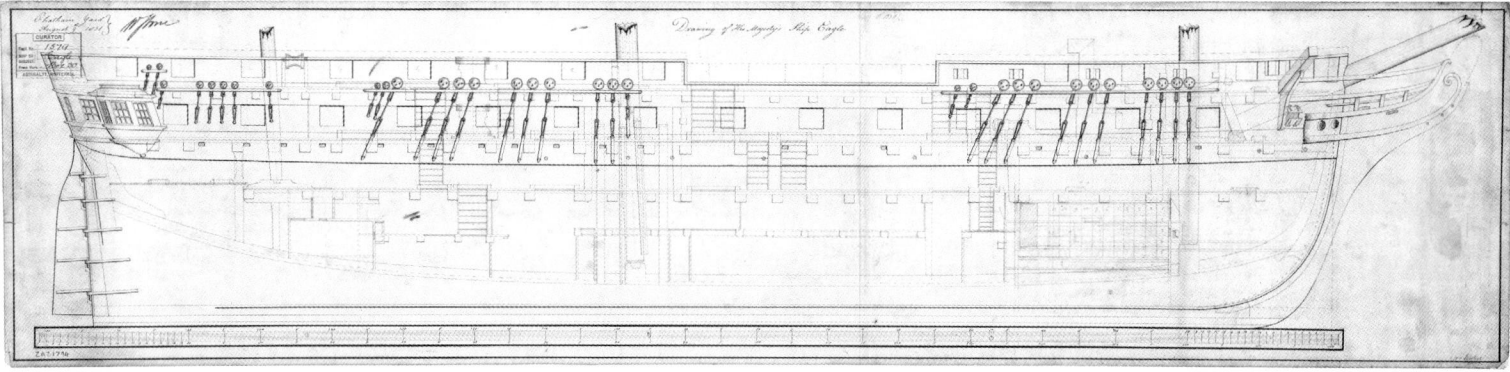

Large Frigates 50/58 guns (4th Rates) – conversions from 74-gun two-deckers

As an interim response to the appearance of the large 24-pounder US frigates, three elderly 74s had been cut down (*raséed*) in 1812–13 to produce flush-decked 58-gun two-deckers – see Section A – and two further vessels (both of the Modified *Arrogant* class of 74s) followed in the immediate post-war period, although the first of these two never put to sea and conversion of the second was cancelled. Dockyard name and order/building dates below refer to their conversions, not to the original building data, for which see Chapter 3.

Men: — . Guns: UD (the former LD) 28 x 32pdrs (56cwt); QD/Fc 2 x 12pdrs + 28 x 42pdr carronades.

Elephant Portsmouth Dyd (ex-74 of 1786)
Ord: (begun) 2.1817. C: 3.1818.
First cost (conversion): £39,407.
Never commissioned, laid up at Portsmouth. Surveyed 5.1830 and BU 11.1830.

Excellent Portsmouth Dyd (ex-74 of 1787)
Order: 11.5.1820. Begun: 12.1825. C: not completed?
Recommissioned 6.1830 under Cmdr George Smith, as gunnery training ship 1830; on 13.4.1832 under Capt. Sir Thomas Hastings; paid off 12.1834. Towed to Deptford 5.1835 and BU there 10.1835.

A further ten 74s of more recent build were cut down between 1826 and 1832 to produce 50-gun frigates. Three were ordered in May 1826, six more in early 1827 (two of which were cancelled) and another three in the following four years. All were former *Armada* Class vessels except *Eagle* (which was of the *Repulse* Class). All but the first (*Barham*) were given 'rounded' sterns on conversion. The last conversion (*Gloucester*) was ordered after the change of regime in November 1830, but before the new Surveyor was appointed, and so is included in this Section B.

Dimensions & tons: see Chapter 3 for details, except that depth in hold was reduced to about 13ft 10in.
[Note that the beam used for calculating tonnage is that measured from the outside of the planking but inside the wales; extreme or overall breadth ('oa') was 7in more in the *Armada* Class conversions (*Alfred* 6in).]
Men: 475, later 500 (393 officers and men, 47 boys, 60 marines).
Guns: UD (the former LD) 28 x 32pdrs (56cwt/9½ft); QD 16 x 32pdrs (45cwt/8½ft); Fc 6 x 32pdrs (45cwt/8½ft). 6 x 8in (65cwt/9ft) shell guns later replaced the same number of 32pdrs on the UD, and 4 more 8in replaced the same number of 32pdrs on the QD/Fc.

The former 74-gun two-decker *Eagle* of 1804 was raséed (i.e. cut down a deck to create a frigate) at Chatham between 1830 and 1831. Her lower deck guns were unaltered, but her former quarterdeck and forecastle were removed, and the upper deck amidships was opened up to form a waist between the residual after and fore sections of the upper deck – now a new quarterdeck and forecastle. This inboard profile of the reconstructed *Eagle* is signed by William Stone, Chatham's master shipwright from 1830 to 1839. Note that the raséed two-deckers had their own version of the circular stern, without the centreline 'turret' adopted on smaller frigates to protect the rudderhead. On completion, the ship was placed in Ordinary, and was not fitted for sea until 1844, when she served a single 40-month commission as flagship on the east coast of South America Station before being relegated to coastguard and later training duties. (*NMM J3040*)

Barham Woolwich Dyd (ex-74 of 1811)
Ord: 31.5.1826. L: 18.9.1826. C: 12.1826.
First cost (conversion): £41,346.
Recommissioned 30.8.1826 under Capt. Sir John Louis, as flagship as Charles Elphinstone Fleeming; paid off 12.8.1839 at Woolwich. Fitted there for sea (for £19,978) 9.1830–6.1831. Recommissioned 9.3.1831 under Capt. Hugh Pigot, for the Mediterranean, as flagship of Vice-Adm. Sir Henry Hotham. On 4.4.1835 under Capt. Armar Lowry Corry, in the Mediterranean; conveyed Earl of Durham to Istanbul; off the coast of Spain; paid off 4.1839. BU 3.1840 at Deptford.

Dublin Plymouth Dyd (ex-74 of 1812)
Ord: 31.5.1826. Commenced: 4.1821 (for repair). L: 29.12.1826. C: 12.1826. Fitted for sea 3–8.1831.
Cost of repair and conversion: £50,360 + £2,439. Fitting (1831) £20,758.
Recommissioned 4.1831 under Capt. Lord James Townshend, for the east coast of South America; paid off 1834 at Plymouth. Recommissioned 24.1.1835 under Capt. Charles Hope, as flag of Rear-Adm. Sir Graham Eden Hamond from 6.1835; in 7.1835 under Capt. Robert Tait, then 8.12.1835 under Capt. George Wickens Willes, still as flagship of Rear-Adm. Sir Graham Eden Hammond, on same station; paid off 31.5.1838. Fitted as flagship at Portsmouth (for £21,557) 9.1839–8.1841. Recommissioned 26.5.1841 at Portsmouth under Capt. John Jervis Tucker, as flagship of Rear-Adm. Richard Thomas, for the Pacific; paid off 12.4.1845 at Devonport. Receiving ship 1860; fitting hulk 1871. Sold to Castle & Sons 16.7.1884 and BU at Charlton 7.1885.

Alfred Chatham Dyd (ex-*Asia*, 74 of 1811, renamed 1819)
Ord: 31.5.1826. Commenced: 11.1822 (for repair). L: 26.8.1828. C: 8.1828. Fitted for sea 3–6.1831.
Cost of repair and conversion: £49,219.
Recommissioned 22.2.1831 under Capt. Robert Maunsell, for the Mediterranean; paid off 7.1834 at Sheerness. Fitted for sea there (for £9,923) 10.1841–2.1842. Recommissioned 1.3.1842 at Sheerness under Commodore John Brett Purvis, on east coast

of South America; paid off 6.8.1845 at Portsmouth. Fitted for commission at Portsmouth (for £9,557) 7.1845–5.1846, but then laid up in Ordinary. Fitted as gunnery trials ship (experimental 4in iron plates) at Portsmouth 7.1858; later hosted trails of the new Whitworth gun. Fitted for further armour trials there 8–11.1862. BU (by AO 10.1.1865) completed 8.5.1865 at Portsmouth.

Cornwall Sheerness Dyd (ex-74 of 1812)

Ord: 24.2.1827. Commenced 3.1827. (Re)L: 29.5.1830. C: 5.1830 for Ordinary (roofed over for and aft).

Cost of repair and conversion: £26,243.

Never recommissioned; laid up at Sheerness. Fitted 4.1859 as a Reformatory School and lent to the London School Ship Society as a reformatory school 1859. Sent to the Tyne as training ship for Newcastle Society (by AO 3.6.1868). Exchanged names with *Wellesley* on 18.6.1868; renamed *Wellesley II* on 21.5.1874. BU completed 18.1.1875 at Sheerness.

America Plymouth Dyd (ex-74 of 1810)

Ord: 20.3.1827. Commenced 3.1827. (Re)L: 26.8.1828. C: 2.1835 for Ordinary (roofed over waist).

Cost of repair and conversion: £22,815 + £13,928.

Recommissioned 22.2.1844 under Capt. John Gordon, for the Pacific; fitted for sea at Plymouth (for £11,690) 2–6.1844, then to Pacific; on 10.11.1846 under Capt. Thomas Maitland, off the coast of Portugal; paid off 20.10.1847 at Plymouth and laid up in Ordinary. Fitted as target ship there 3.1864; later to Portsmouth. Used there for experiments with 'torpedoes' (floating mines) 2.1867. BU at Portsmouth (by AO of 27.6.1867) 10.1867–6.2.1869.

Conquestador Sheerness Dyd (ex-74 of 1810)

Ord: 20.3.1827. Commenced 4.1817. (Re)L: 30.4.1831. C: 4.1831 for Ordinary (roofed over fore and aft).

Cost of repair and conversion: £17,202.

Never recommissioned; laid up at Sheerness. Lent as a powder depot to the War Office at Woolwich 12.1856, then to Purfleet. Fitted at Plymouth (by A) 11.11.1862 as a Powder depot 12.1862–2.1863. Returned by War Dept. to Naval Ordnance Dept. control 1892. Sold to Henry Scawn, Plymouth (for £1,525) 10.5.1897 to BU.

Greenwich Portsmouth Dyd (ex-*Rodney*, 74 of 1809 – renamed 17.3.1827)

Ord: 19.3.1827. (Re)L: 29.5.1830. Conversion never completed, instead sold at Portsmouth (for £5,310) 8.9.1836.

Centurion Portsmouth Dyd (ex-*Clarence*, 74 of 1812 – renamed 1826)

Ord: 19.3.1827. (Re)L: — . Conversion cancelled and BU 10.1828.

Cressy Portsmouth Dyd (ex-74 of 1810)

Ord: 19.3.1827. (Re)L: — . Conversion cancelled and BU 12.1832.

Vindictive Portsmouth Dyd (ex-74 of 1813)

Ord: 5.7.1828. Commenced 3.1828. (Re)L: 5.10.1832. C: 1.1833 for Ordinary (roofed over waist). Fitted for sea there 10.1841–15.1.1842.

Cost of Repair and conversion: £30,276 + £2,258. Fitted for sea for £19,091.

Recommissioned 30.9.1841 under Capt. John Toup Nicolas, for the East Indies, then the Pacific; paid off 26.7.1844. Recommissioned 16.1.1845 under Capt. Michael Seymour, as flagship of Vice-Adm. Francis William Austen (from 27.12.1844), for North America and the West Indies; paid off 6.1848 at Portsmouth. Fitted by J. S. White, Cowes (for £12,391) 2.1861–1.1862 as bread and coal depot ship; recommissioned 1.8.1862 under William Frederick Law, Master, as depot ship at Fernando Po; in 2.1865 under

Henry W. C. Wise, Master, as coal depot in Bight of Benin, then 12.1868 under James B. Haines, Master; for sale at Fernando Po 1871, but foundered there and wreck sold 24.11.1871.

Eagle Chatham Dyd (ex-74 of 1804)

Ord: 26.4.1830. Commenced: 3.1830. L: 1.3.1831. C: 3.1831 for Ordinary. Fitted for sea there 10.1844–2.1845.

Cost of repair and conversion: £14,243. Fitted for sea for £13,984.

Recommissioned 4.11.1844 under Capt. George Bohun Martin, as flagship of Rear-Adm. Samuel Hood Inglefield, for east coast of South America; paid off 10.3.1848. Fitted at Plymouth 10.1856–2.1857 for the coastguard; recommissioned 12.12.1856 under Capt. Henry Alexander Story, as guard ship at Falmouth; on 1.2.1858 under Capt. Edward Tatham, as guard ship at Pembroke. Fitted at Portsmouth 6–9.1860 as training ship. Recommissioned 26.8.1860 under Cmdr Frederick Thomas Chetham Strode, as training ship in Southampton Water 10.1860; paid off 6.1862 at Liverpool. Recommissioned 1.7.1862 under Cmdr John William Whyte, as RNR drill ship at Liverpool; on 16.2.1864 under Cmdr William Edward Fisher, in same role; in 7.1867 under Cmdr Edwin C. Symons, then 7.1870 under Cmdr Guy Ouchterlony Twiss, 3.1875 under Cmdr Adolphus Frederick Maximilian Meyer and 5.1880 under Cmdr Malcolm McNeile. Recommissioned 4.1881 in same role; in 1886 under Cmdr F. A. Wetherall, then 7.1889 under Cmdr G. H. Rainier, 11.1899 under Cmdr Charles Elsden Gladstone and 8.1908 under Cmdr C. W. S. Leggatt; paid off 1910. Repaired (by AO 14.10.1910) by Glover & Clayton, then lent to Mersey Division RNVR 1912. In 10.1914 under H. H. Stileman, SNO at Liverpool, then under E. G. H. Gamble, then 1.1915 under H. W. Simms. Renamed *Eaglet* 1918. Burnt by accident 1926; wreck sold to J. Hornby 4.1.1927 for BU.

Gloucester Chatham Dyd (ex-74 of 1812)

Ord: 28.2.1831. Commenced: 4.1831. L: 24.12.1831. C: 12.1832 for Ordinary (roofed over waist).

Never recommissioned. Fitted at Chatham as a receiving ship 5.1858–8.1861. For sale 7.2.1884 and sold to Messrs Castle by private agreement 3.1884 (by AO 5.1884).

(C) Vessels acquired from November 1830

VERNON **Class** 50 guns. William Symonds design 1831, based on proportions of the 80-gun *Gibraltar*, with armament and rigging specification of the rasée 50-gun *Barham*. An experimental design, she outsailed every vessel competing with her, but due to the urgency of *Vernon*'s construction, using unseasoned timber, she developed serious dry rot. A second vessel – *Indefatigable* – was ordered to this design in 1832; it was intended to build her using timbers of the incomplete 80-gun *Boscawen*, but the latter was instead re-ordered as a 70-gun ship.

Dimensions & tons: 176ft 0in, 144ft 6in x 51ft 11in oa (51ft 2¾in for tonnage) x 17ft 0in. 2,017²⁰⁄₉₄bm.

Men: 450, later 500 (393 officers and men, 47 boys, 60 marines). Guns: UD 28 x 32pdrs (56cwt/9½ft); QD 14 x 32pdrs (45cwt/8½ft); Fc 8 x 32pdrs (45cwt/8½ft); 6 x 32pdrs on the UD were later replaced by 6 x 8in/68pdr (65cwt/9ft) shell guns.

Vernon Woolwich Dyd

As built: 176ft 0in, 144ft 6⅞in x 52ft 8½in oa (52ft 0¼in for tonnage) x 17ft 1in. 2,082¹¹⁄₉₄bm. 2,388 tons displacement. Draught 13ft 6in/17ft 6½in.

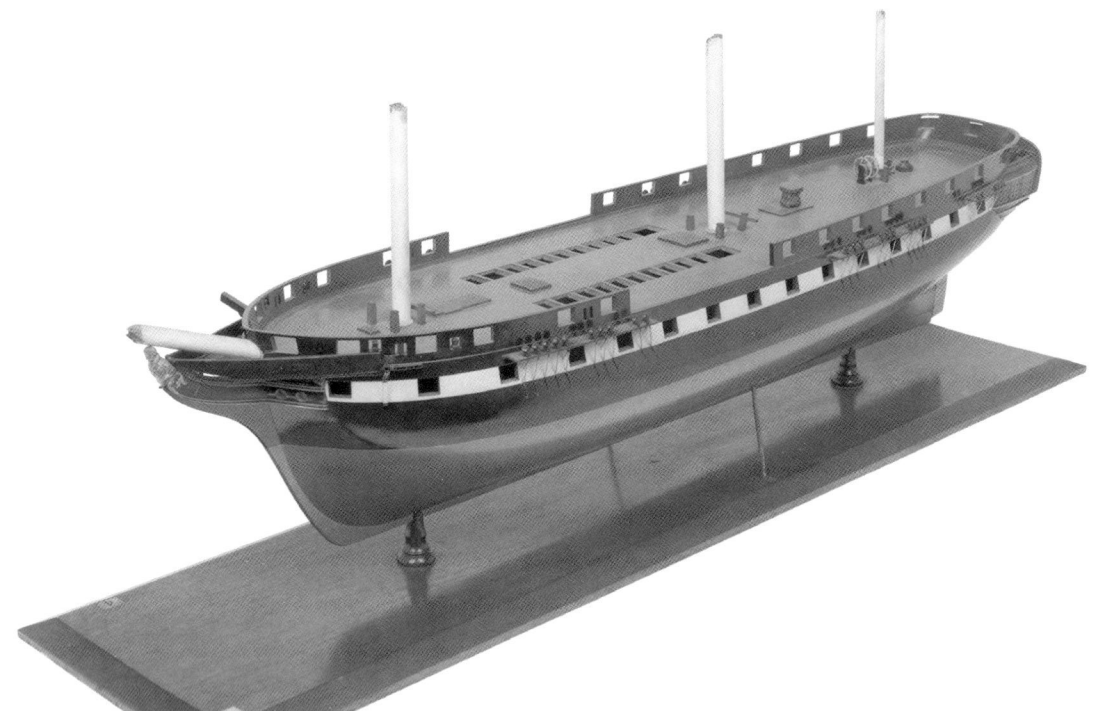

An official model of the *Vernon*, the first major warship designed by William Symonds, which was built to match the armament and masting establishment of the raséed 74s, but utilising his concept of a V-shaped hull with sharply rising floors instead of the traditional U-shape. She was built rapidly, with unprepared timbers, so that her structure proved defective and twice required major reconstruction at Sheerness, in 1837–38 and 1844–45. At over 2,000 tons burthen, she was the largest and most expensive sailing frigate ever built, and no further Fourth Rates were ordered during the rest of the 1830s (other than the raséed 74 *Warspite*). Furthermore her bluff bow above the waterline, but with fine lines below, meant that in a heavy sea, when fully laden, she was slow in rising and strained her rigging; in addition, she floated more than a foot deeper in the water than Symonds had intended. *(NMM F7788-002)*

Ord: 24.8.1831. K: 10.1831. L: 1.5.1832. C: 29.7.1832 (for sea) at Woolwich.

First cost: £48,485 (£64,161 including fitting at Woolwich).

Commissioned 1832 under Capt. Sir Francis Augustus Collier, for particular service (blockade of Dutch coast); paid off at Plymouth. Fitted for a flag officer there 12.1832–1.1833. Recommissioned 3.12.1832 under Capt. Sir George Augustus Westphal, as flagship of Vice-Adm. Sir George Cockburn for North America and the West Indies; on 27.8.1834 under Capt. John M'Kerlie (Westphal invalided); to the Mediterranean 1835; paid off 3.1837 at Sheerness. Repair there (for £12,470) 4.1837–7.1838. Fitted for sea there (for £8,444) 10.1840–3.1841; recommissioned 30.10.1840 under Capt. William Walpole, for the Mediterranean; employed as troop transport to Queenstown 4.1844; paid off 22.5.1844 at Sheerness. Fitted for sea there (for £14,737) 6.1844–6.1845; recommissioned 20.3.1845 under Capt. John Coghlan Fitzgerald, as flagship of Rear-Adm. Samuel Hood Inglefield, on east coast of South America and East India stations. Large Repair at Chatham (for £18,349) 11.1848–5.1850. Hulked 3.1863 at Chatham; coaling jetty at Portland (by AO 14.9.1867) 1870. Fitted as torpedo instruction ship at Portsmouth 4.1873; recommissioned as training ship 5.1876 under Capt. William Arthur, for the torpedo school at Portsmouth; in 6.1879 under Capt. Edward W. Gordon. Renamed *Actaeon* on 14.1.1886 and *Actaeon I* in 1906. Sold 14.9.1923.

Indefatigable Woolwich Dyd
    Ord: 29.11.1832. K: – . Cancelled 3.3.1834 (unstarted).

Raséed Third Rates 50 guns. Notwithstanding the Surveyor's aversion to the programme of raséeing the smaller – and unwanted – Third Rates, the Admiralty maintained its belief in cutting down units to make heavy 50-gun frigates, although only two such were carried out during Symonds's tenure. These two 74s of more recent build were cut down (raséed) between 1837 and 1845 to produce 50-gun frigates – *Warspite* (of *Colossus* Class) and *Grampus* (formerly the *Tremendous* of the Later *Ganges* Class).

    Dimensions & tons: see Chapter 3 for details, except that depth in hold was reduced to about 13ft 10in.
    [Note that the beam used for calculating tonnage is that measured from the outside of the planking but inside the wales; extreme or overall breadth ('oa') was 6in more in raséed *Warspite* and 9in in *Grampus*.]
    Men: 475 (445 peacetime). Guns: UD 6 x 8in/68pdr (65cwt/9ft) shell + 22 x 32pdrs (56cwt/9½ft); QD 16 x 32pdrs (45cwt/8½ft); Fc 6 x 32pdrs (45cwt/8½ft).

Warspite Portsmouth Dyd (ex-74 of 1807)
    As raséed (and re-measured): 180ft 0in, 147ft 7in x 49ft 6in x 13ft 7in. 1,885bm.
    Ord: 19.6.1837. Commenced: 11.1833 (for repair). (Re-)L: 24.8.1840. C: 9.1841–1.1842 (for sea)
    First cost: £21,300 for cut down and fitting as a demonstration ship, + £18,425 fitting for sea.
    Recommissioned 17.8.1841 under Capt. Lord John Hay, for North America and the West Indies; on 13.10.1843 under Capt. Provo William Parry Wallis, for the Mediterranean; paid off 4.1846 at Chatham. Lent to the Marine Society as a training ship 27.5.1862. Burnt by accident and scuttled 3.1.1876 off Woolwich; wreck sold to McArthur & Co. 2.2.1876 (for £2,610 except for guns and Marine Society's fittings).

Grampus Woolwich Dyd (ex-*Tremendous*, 74 of 1784 – renamed 23.5.1845)
    Ord: 17.7.1844. Commenced: 5.1844. (Re-)L (floated out): 20.8.1845. C: 1.1846 (for sea)
    First cost: £27,610 for cut down and fitting for sea.

The 50-gun frigate *Raleigh* as depicted by Robert Strickland Thomas entering Portsmouth harbour in 1850. This vessel marked the resumption of Fourth Rate building, and a variety of 50-gun frigates followed her during the 1840s. The *Raleigh* was designed by John Fincham, master shipwright at Chatham from 1839 to 1844 and a leading critic of Symonds, although he incorporated aspects of the Symondite hull form into his design. (*NMM BHC3576*)

Recommissioned 15.11.1845 under Capt. Henry Byam Martin, at Woolwich; to the Pacific; paid off 20.10.1848 at Portsmouth. Powder hulk at Portsmouth 1856. Fitted as powder depot at Portsmouth 11.1866. Lent to the War Department for the stowage of naval mines 12.1883. Sold to John Read, Portsmouth (for £1,605) 10.5.1897 (under AO 24.7.1896).

RALEIGH Class 32-pounder frigates of 50 guns. John Fincham design, approved 14 September 1842. The draught for *Severn* was altered by Fincham in January 1846 – she was the last purely sailing warship to be completed for the RN (apart from training vessels).
>   Dimensions & tons: 180ft 0in, 149ft 0in x 50ft 0in (49ft 6¼in for tonnage) x 16ft 8in. 1,943⁶⁄₉₄bm.
>   Men: 500. Guns: UD 28 x 32pdrs (56cwt); QD 14 x 32pdrs (45cwt); Fc 8 x 32pdrs (45cwt). 6 x 8in/68pdr (65cwt) shell guns later replaced 8 x 32pdrs (UD) in *Raleigh*.

Raleigh Chatham Dyd
>   As built: 180ft 0in, 148ft 9in x 50ft 1in (49ft 6in for tonnage) x 16ft 8in. 1,938⁶⁄₉₄bm. Draught 12ft 4in/17ft 3in.
>   Ord: 26.3.1842. K: 8.1842. L: 8.5.1845. C: 25.4.1846 (for sea)
>   First cost: £35,944 to build, + £23,771 fitting.
>   Commissioned 12.2.1846 under Commodore Thomas Herbert, as C-in-C on east coast of South America; on 11.1.1847 under Capt. Stephen Greville Fremantle, as flagship of Commodore Herbert; paid off and laid up 6.1850 at Portsmouth. Fitted for sea there (for £5,958) 9–11.1856. Recommissioned 17.9.1856 under Commodore Henry Keppel, for East Indies and China; in 2nd Anglo-Chinese War; wrecked 14.4.1857 on a rock off Hong Kong, and beached off Macao (no casualties).

Severn Chatham Dyd
>   As built: not measured as sailing frigate, see Section D.
>   Ord: 19.2.1844. K: 13.8.1849. L: 24.1.1856.
>   Not completed as sailing frigate; converted to screw frigate by AO 4.3.1859, see Section D.

CONSTANCE Class 50 guns. Sir William Symonds design 1843. The largest sailing frigates ever built for the RN. *Constance* had originally been ordered on 12 September 1833 as a 36-gun Fifth Rate at Portsmouth; amended on 16 February 1843 to be a 44-gun at Pembroke, finally re-ordered in March 1843 to carry the same armament as the rasées, and carrying same masts and yards. The design was approved 30 May 1843.
>   Dimensions & tons: 180ft 0in, 146ft 10¼in x 52ft 8in (52ft 2in for tonnage) x 16ft 3in. 2,125⁷⁄₉₄bm.
>   Men: 500. Guns: UD 28 x 32pdrs (56cwt); QD 14 x 32pdrs (45cwt); Fc 8 x 32pdrs (45cwt). 10 x 8in/68pdr (65cwt) shell guns later replaced 10 x 32pdrs (UD) in each.

Constance Pembroke Dyd
>   As built: 180ft 0in, 146ft 9in x 52ft 9in (52ft 3¼in for tonnage) x 16ft 4in. 2,132⁷⁄₉₄bm. Draught 14½ft/18½ft.
>   Ord: 31.3.1843. K: 10.1843. L: 12.3.1846. C: 10.4–28.6.1846 (for sea) at Plymouth Dyd
>   First cost: £41,071 to build, + £16,173 fitting.
>   Commissioned 23.4.1846 at Plymouth under Capt. Baldwin Wake Walker, for the Pacific; on 3.8.1847 under Capt. George William Conway Courtenay, on same station; paid off 9.12.1849 at Plymouth. Converted to screw frigate by AO 4.1.1860, see Section D.

Arethusa Pembroke Dyd
>   As built: 180ft 0in, 146ft 8¾in x 52ft 8¼in (52ft 2½in for tonnage) x 16ft 3⅛in. 2,127³⁄₉₄bm. Draught 14ft 6in/18ft 6in.
>   Ord: 19.2.1844. K: 30.3.1846. L: 20.6.1849. C: 2.8.1849–20.3.1850 (for sea) at Plymouth Dyd
>   First cost: £43,157 to build, + £19,384 fitting.
>   Commissioned 19.1.1850 at Plymouth under Capt. Thomas Matthew Charles Symonds, for particular service, then to the Mediterranean; on 1.1.1854 under Capt. William Robert Mends, for the Black Sea during the Russian War; on 1.7.1854 under Symonds again, still in Black Sea; paid off 15.1.1855 at Plymouth.

This drawing by Captain George Pechell Mends of sailing trials conducted off Malaga on 17 July 1852 in a 'fresh westerly wind' shows the 50-gun *Phaeton* in the process of rounding the stationary three-decker *Trafalgar* (the latter being Mends's command) which acted as the turning point in this contest. The *Phaeton*, designed by Joseph White (the Cowes shipbuilder), proved the winner in this test of speed, which comprised vessels operating entirely under sail, although several of the frigates had auxiliary steam screw propulsion. The runners-up, in order, were: the brand new screw corvette *Highflyer*, the sailing frigate *Indefatigable* (seen astern of *Trafalgar*) and the screw frigate *Arrogant*. (NMM PZ0870-001)

Converted to screw frigate by AO 4.1.1860, see Section D.

Liffey Devonport Dyd
Ord: 19.2.1844. K: – . Not begun as sailing frigate. Re-ordered 4.4.1851 as screw frigate, see Section D.

Octavia Pembroke Dyd
As built: 180ft 0in, 146ft 8¾in x 52ft 8½in (52ft 2½in for tonnage) x 16ft 3½in. 2,127³²/₉₄bm. Draught 14ft 1in/18ft 4in.
Ord: 26.3.1846. K: 9.1846. L: 18.8.1849. C: 17.10.1849 (for Ordinary) at Plymouth Dyd.
First cost: £41,645 to build.
Not completed as sailing frigate. Converted to screw frigate by AO 9.7.1859, see Section D.

LEANDER Class – 50 guns. Richard Blake design dated 4 July 1843; the designer was the master shipwright at Portsmouth (1830–35).
Dimensions & tons: 181ft 4in, 149ft 0¾in x 50ft 4in oa (49ft 10in for tonnage) x 15ft 8in. 1,969²/₉₄bm.
Men: 500. Guns: UD 22 x 32pdrs (56cwt) + 6 x 8in/68pdr (65cwt) shell; QD 14 x 32pdrs (45cwt) + 2 x 8in/68pdr (65cwt) shell; Fc 4 x 32pdrs (45cwt) + 2 x 8in/68pdr (65cwt) shell.

Leander Portsmouth Dyd
As built: 181ft 4½in, 148ft 11½in x 50ft 9in (50ft 1in for tonnage) x 15ft 8in. 1,987⁴¹/₉₄bm. Draught 13ft 8in/16ft 7in.
Ord: 8.7.1843. K: 2.1845. L: 8.3.1848. C: 3.1849–17.11.1849 (for sea)
First cost: £39,585 to build, + £16,773 fitting.
Commissioned 28.9.1849 at Portsmouth under Capt. Sidney Colpoys Dacres, for particular service (Squadron of Evolution); on 3.5.1852 under Capt. George St Vincent King, at Portsmouth;

fitted at Plymouth to convey the Earl of Clarence to New York, then to the Mediterranean and Black Sea during the Russian War; on 22.11.1854 under Capt. Swynfen Thomas Carnegie, still in Black Sea; on 6.1.1855 under Capt. William Peel, as flagship of Rear-Adm. Charles Howe Fremantle, then on 13.8.1855 under Capt. William Moorsom and on 5.1.1856 under Capt. Edward Bridges Rice, still Fremantle's flagship in the Black Sea; paid off 23.9.1856 at Chatham. Converted to screw frigate by AO 4.1.1860, see Section D.

Shannon Portsmouth Dyd
Ord: 19.2.1844. K: 8.1849. Re-ordered 4.4.1851 as screw frigate, see Section D.

Modified *CONSTANCE* Class 50 guns. For this second 1846 order (after *Octavia*), the Symonds design was modified by the Admiralty (i.e. Walker) to increase the length and reduce the breadth by 1ft, with a fuller midships section; the Surveyor was asked on 28 September 1847 to prepare new drawings, and these were approved on 14 October.
Dimensions & tons: 180ft 0in, 148ft 4⅛in x 51ft 8in oa (51ft 2in for tonnage) x 16ft 3in. 2,065⁷/₉₄ tons.
Men: 500. Guns: UD 28 x 32pdrs (56cwt); QD 14 x 32pdrs (45cwt); Fc 8 x 32pdrs (45cwt).

Sutlej Pembroke Dyd
As built: not measured as sailing frigate, see Section D.
Ord: 26.3.1846. K: 8.1847. L: 17.4.1855.
Not completed as sailing frigate; converted to screw frigate by AO 4.3.1859, see Section D.

PHAETON 50 guns. A design for a new 50-gun frigate was submitted by Joseph White, the Cowes shipbuilder. The Surveyor was asked to report on these plans on 9 January and 13 February 1845, and the name (*Arrogant*) and building yard (Deptford) were assigned on 24 February, but the name was changed to *Phaeton* five days later on 1 March 1845.
Dimensions & tons: 184ft 11in, 151ft 11⅛in x 49ft 4in oa (48ft 9in

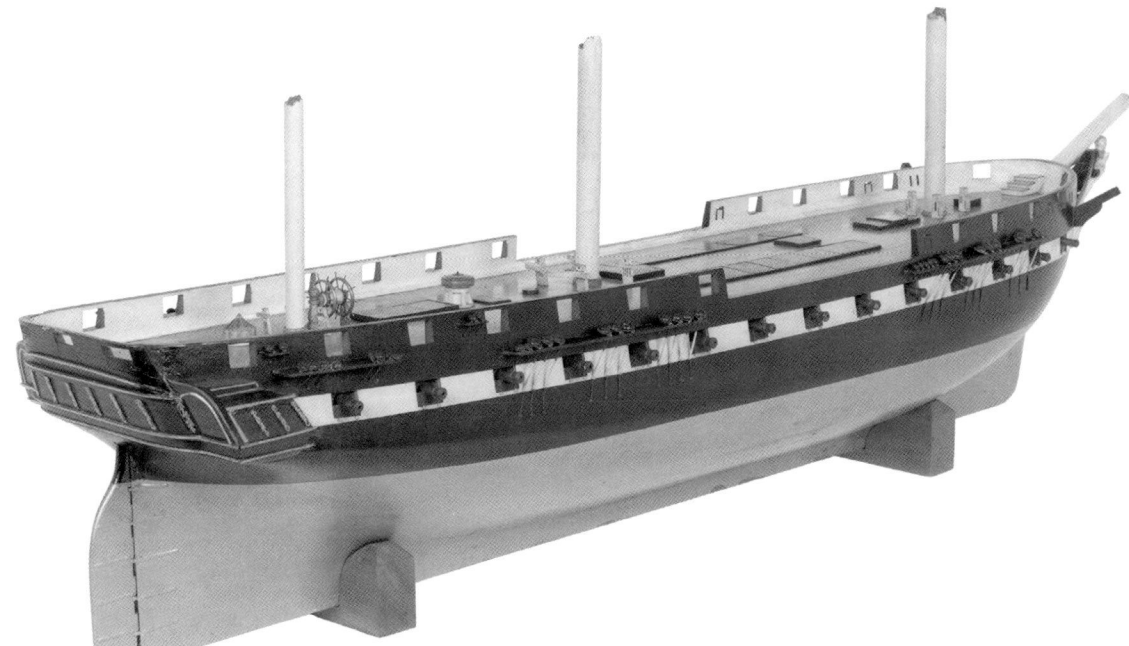

An official model of *Nankin*, one of six 50-gun frigates ordered to competing designs in 1843–46. Designed by Oliver Lang, a protégé of Seppings and a vocal critic of Symonds, the *Nankin* was the last big frigate to commission as a pure sailing ship, and unlike others of the type, she was never converted to screw propulsion. Although all these ships were built with a break in the barricades amidships, by now the waist was structurally part of a continuous weather deck, the gangways being delineated only by a double row of gratings. *(NMM F7799-003)*

for tonnage) x 15ft 10½in. 1,920⁸⁰⁄₉₄bm.

Men: 500. Guns: UD 22 x 32pdr (56cwt/9½ft) + 6 x 8in /68pdr (65cwt/9ft) shell; QD 14 x 32pdr (45cwt/8½ft) + 2 x 8in /68pdr (65cwt/9ft) shell; Fc 4 x 32pdr (45cwt/8½ft) + 2 x 8in /68pdr (65cwt/9ft) shell.

Phaeton Deptford Dyd

As built: 184ft 11in, 152ft 8½in x 49ft 5¾in (48ft 10¼in for tonnage) x 15ft 10½in. 1,942bm. Draught 11ft 7in/18ft 7½in.

First cost: £45,079 (£44,146 to build, + £833 fitting for Ordinary); £19,398 fitting for sea.

Ord: 24.2.1845. K: 1.9.1845. L: 25.11.1848. C: at Sheerness (for Ordinary) then Chatham (for sea).

Commissioned 4.12.1849 at Chatham under Capt. George Augustus Elliot, for particular service; paid off 25.1.1853 at Sheerness. Converted to screw frigate by AO 4.3.1859, see Section D.

INDEFATIGABLE Class 50 guns. A design by William Edye, the master shipwright at Devonport from 9 December 1843 to 31 December 1858 (not to be confused with the Assistant Surveyor, John Edye).

Dimensions & tons: 180ft 0in, 147ft 8½in x 51ft 6in (51ft 0in for tonnage) x 16ft 3in. 2,043⁵²⁄₉₄bm.

Men: 500. Guns: UD UD 22 x 32pdrs (56cwt) + 6 x 8in/68pdr (65cwt) shell; QD 14 x 32pdrs (45cwt) + 2 x 8in/68pdr (65cwt) shell; Fc 4 x 32pdrs (45cwt) + 2 x 8in/68pdr (65cwt) shell.

Indefatigable Devonport Dyd

As built: 180ft 0in, 147ft 11in x 51ft 6in (51ft 0in for tonnage) x 16ft 6in. 2,046⁴²⁄₉₄bm. Draught 13ft 4in/16ft 5in (with 35¼ tons on board).

Ord: 17.12.1845. K: 22.8.1846. L: 27.7.1848. C: 8–9.1848 (as Advanced ship), 11.1849–23.1.1850 (for sea).

First cost: £47,532 (+ £3,262 fitting as Advanced ship); £14,934 fitting for sea. Total £65,728.

Commissioned 7.11.1849 at Plymouth under Capt. Robert Smart, for North America and the West Indies; paid off 22.5.1852. Refitted at Plymouth (for £7,148) 5–8.1854; recommissioned 16.5.1854 under Capt. Thomas Hope, as flagship of Rear-Adm.

William James Hope Johnstone, for east coast of South America; paid off 31.10.1857 at Plymouth. Lent to the Committee for the Liverpool Training Ship (by AO 8.11.1864) 3.1.1865. Sold 26.3.1914.

Phoebe Devonport Dyd

As built: 180ft 0in, 147ft 8½in x 51ft 6in (51ft 0in for tonnage) x 16ft 3in. 2,043⁵²⁄₉₄bm. Draught 12ft 3in/15ft 9in.

Ord: 27.2.1847. K: 7.8.1848. L: 12.4.1854. C: 12.5.1854 (for Ordinary)

Not commissioned as sailing frigate. Converted to screw frigate by AO 4.3.1859, see Section D.

NANKIN 50 guns. On 13 January 1846 the Surveyor was asked to report of a design drawn up by Oliver Lang. On 26 January Chatham was instructed to build this ship on the first available slip, with the name *Nankin* allocated. However, as no slip was available at Chatham for two years, fresh instructions were given on 23 February that the frame was to be cut at Chatham and sent to Woolwich, where she was to be built on the easternmost slip.

Dimensions & tons: 185ft 0in, 149ft 10⅛in x 50ft 10in (50ft 1½in for tonnage) x 17ft 2in. 2,002⁵⁴⁄₉₄bm.

Men: 500. Guns: Guns: UD 22 x 32pdr (56cwt/9½ft) + 6 x 8in/68pdr (65cwt/9ft) shell; QD 14 x 32pdr (45cwt/8½ft) + 2 x 8in /68pdr (65cwt/9ft) shell; Fc 4 x 32pdr (45cwt/8½ft) + 2 x 8in/68pdr (65cwt/9ft) shell.

Nankin Woolwich Dyd

As built: 185ft 0in, 153ft 0⅞in x 50ft 10in (50ft 2in for tonnage) x 15ft 10½in. 2,049¹²⁄₉₄bm. 2,540 displacement. Draught 13ft 10in/15ft 10in.

Ord: 26.1.1846 and 23.2.1846 (see above). K: 6.1846. L: 16.3.1850. C (at Chatham): 17.3–12.7.1850 (for Ordinary), 10.10.1854–3.2.1855 (for sea).

First cost: £37,304 to build, + £7,291 fitting for Ordinary, + £16,066 fitting for sea; total £60,661.

Commissioned 21.9.1854 under Capt. Keith Stewart, for the East Indies; at Canton 28.10.1856; on 26.11.1857 became flagship of now Commodore Stewart; in 2nd Anglo-Chinese War; boat

action at Fatshan 1.6.1858; paid off 18.2.1859. To Pembroke as
guard ship 6.1867. In 7.1867 under Capt. Robert Hall, then 1875
under Capt. Richard Vesey Hamilton, then 8.11.1877 under
Capt. George Henry Parkin, all as receiving ship at Pembroke.
Recommissioned 4.1880 in same role; in 1886 under Capt.
Edward Kelly, then 1.1889 under Capt. Samuel Long. Sold there
(by AO 28.4.1892) 28.2.1895, and BU 1905.

Eight more 32pdr sailing frigates were ordered in 1848 (and a final unnamed pair in 1850); seven were replaced by orders for screw-driven
frigates either before or shortly after work commenced on them.

EMERALD Class 60 guns. Edye and Watts design 1848; the
Pembroke pair was originally ordered to a design by Read, Chatfield
and Creuze but on 31 May 1849 this was altered to build to same
design as other pair. All were re-ordered as screw ships on dates given
below; converted on stocks and launched as screw frigates.
>   Dimensions & tons: 185ft 0in, 152ft 2in x 52ft 0in oa (51ft 6in for
>   tonnage) x 15ft 8in. 2,146⁸¹/₉₄bm.
>   Men: 550. Guns: UD 30 x 8in /68pdr (65cwt) shell; QD/Fc 22 x
>   32pdr (45cwt) + 8 x 8in /68pdr (65cwt) shell (waist).
Emerald Deptford Dyd
>   Ord: 30.3.1848. K: 4.6.1849. L: — . Re-ordered 24.5.1854 as screw
>   frigate, see Section D.
Melpomene Pembroke Dyd
>   Ord: 30.3.1848. K: 9.1849. L: — . Re-ordered 9.4.1856 as screw
>   frigate, see Section D.
Immortalite Pembroke Dyd
>   Ord: 30.3.1848. K: 11.1849. L: — . Re-ordered 9.4.1856 as screw
>   frigate, see Section D.
Imperieuse Deptford Dyd
>   Ord: 30.3.1848. K: – . Re-ordered 14.6.1850 as screw frigate, see
>   Section D.

NARCISSUS Class 50 guns. Edye and Watts design 1848. The
Narcissus absorbed the materials collected for a Sixth Rate screw ship
of 28 guns (originally ordered as a sailing sloop of 18 guns), which was
cancelled in the same month as the new 50-gun frigate was ordered.
Two further ships were ordered from Devonport Dyd and Portsmouth
Dyd on 25 March 1850 but were neither named nor laid down. All
were cancelled as sailing frigates and re-ordered as screw frigates – the
last three about 1851 and *Narcissus* in 1857.
>   Dimensions & tons: 180ft 0in, 148ft 1⅞in x 50ft 10in oa (50ft 4in for
>   tonnage) x 15ft 10in. 1,996⁴⁹/₉₄bm.
>   Men: 500. Guns: UD 22 x 32pdr (56cwt) + 6 x 8in/68pdr (65cwt)
>   shell; QD/Fc 18 x 32pdr (45cwt) + 4 x 8in /68pdr (65cwt) shell.
Narcissus Devonport Dyd
>   Ord: 30.3.1848. K: 11.1849. L: – . Re-ordered 2.3.1857 as screw
>   frigate, see Section D.
>   First cost: £26,924 up to 1857, at point when completed taking
>   down to rebuild as a screw vessel.
Euryalus Chatham Dyd
>   Ord: 30.3.1848. K: – . Re-ordered 15.3.1851 as screw frigate, see
>   Section D.

SAN FIORENZO Class 50 guns. Read, Chatfield & Creuze design
1848, approved 31 January 1849.
>   Dimensions & tons: 187ft 4in, 155ft 4in x 50ft 6in oa (50ft 0in for
>   tonnage) x 15ft 6½in. 2,065⁵⁶/₉₄bm.
>   Men: 500. Guns: UD 28 x 8in /68pdr (65cwt) shell; QD/Fc 22 x
>   32pdr (45cwt).

San Fiorenzo Woolwich Dyd
>   Ord: 22.6.1848. K: 6.1850. L: – . Suspended 1.12.1851 and cancelled
>   4.1856.
Bacchante Portsmouth Dyd
>   Ord: 22.6.1848. K: 6.9.1856. L: – . Re-ordered 4.4.1851 as screw
>   frigate, see Section D.

(D) Vessels ordered or re-ordered as Fourth Rate steam guard ships and steam screw frigates (from 1845)

ARROGANT 46 guns. The Navy's first screw Fourth Rate frigate
was originally ordered as a sailing Fourth Rate. She was designed 'to
have steam as an auxiliary, being ordered to be masted and rigged
as a regular sailing frigate' (unlike the *Dauntless* built alongside her,
which was 'strictly a steam frigate'). The John Fincham design, dated
3 August 1844, first gave dimensions as 200ft x 45ft x 15ft 1in (=
1,861bm); to carry MD 6 x 8in (65cwt) shell + 22 x 32pdrs (56cwt); Fc
& QD 16 x 32pdrs (32cwt) + 2 x 56pdrs (85cwt). She was re-classed
as a First Class frigate when re-ordered on 11 February 1845, and
named on 4 March 1845. Originally intended to have a 360nhp Boulton
engine.
>   Dimensions & tons: 200ft 0in, 172ft 9in x 45ft 8¾in (45ft 1¾in
>   for tonnage) x 15ft 1in. 1,872²¹/₉₄bm. 2,690 disp. Draught 18ft
>   7in/20ft 3in.
>   Men: 450. Guns: MD 12 x 8in (65cwt/9ft) shell + 16 x 32pdrs
>   (56cwt/9½ft); Fc & QD 16 x 32pdrs (32cwt/6½ft) + 2 x 68pdrs
>   (95cwt/10ft).
>   Machinery: 2-cylinder (55in diameter, 36in stroke) horizontal single
>   expansion, trunked. Single screw. 360nhp. 774ihp= 8.645kts.
Arrogant Portsmouth Dyd/John Penn & Son
>   Ord: 11.2.1845. K: 9.1845. L: 5.4.1848. C: 17.6.1848–8.1.1849 at
>   Woolwich, Blackwall and Deptford.
>   First cost: £83,183.
>   Commissioned 14.3.1849 under Capt. Robert Fitzroy; to Lisbon;
>   on 15.2.1850 under Capt. Robert Spencer Robinson, at Lisbon;
>   paid off 26.9.1852 at Portsmouth. Recommissioned next day
>   under Capt. Stephen Greville Fremantle; on 24.10.1853 under
>   Capt. Hastings Reginald Yelverton; sailed to the Baltic 11.3.1854
>   with Napier's fleet; at capture of Bomarsund 16.8.1854; sailed
>   again for the Baltic 4.4.1845; in attack on Sveaborg 8.1855; on
>   20.12.1855 under Capt. Henry Lyster, for North America and
>   the West Indies; paid off 25.2.1857 at Portsmouth. Refitted
>   for coastguard there (for £21,765) 1857–58. Recommissioned
>   1.3.1858 under Capt. Leopold George Heath, for the coastguard
>   service on Southampton Water; paid off 21.8.1859. Refitted for
>   sea at Portsmouth 8.1859. Recommissioned 23.8.1859 under
>   Commodore William Edmonstone, as C-in-C on the west coast
>   of Africa; boats at Porto Novo (Niger River) 26.4.1861; at Saba
>   (Niger River) 21.2.1862; paid off 6.12.1862 at Portsmouth. Sold
>   to Castle & Beech to BU 3.1867.

The steam guard ships (often called 'blockships') constituted the
earliest attempt by the Royal Navy to apply screw propulsion to major
warships. The first vessel selected was the very elderly *Saturn*, a 74
of 1786 which in 1813 had been raséed to a 58-gun Fourth Rate (see
above for details). On 1 July 1845 her conversion at Pembroke was

The *Arrogant* was the first Fourth Rate to be built with auxiliary steam screw propulsion, and was one of the first two ships to be fitted with the innovative Penn trunk engines (the other was the smaller *Encounter*), both frigates being designed by John Fincham. As shown in this contemporary lithograph, she was masted as a regular frigate and the presence of auxiliary propulsion was only given away by the funnel. *(NMM A8115)*

ordered, and her defects instructed to be made good (23 July) with a 330nhp engine and screw fitted. The 74-gun *Wellington* (ex-*Hero*) was selected as the second conversion on 13 August 1845. However, it was reported *Saturn* was too defective to be used, and instead of the *Wellington* a programme was adopted on 30 August 1845 to modify four other *Armada* Class 74-gun ships for the defence of the Channel and Home Ports. The resultant conversions were to be labelled steam guard ships instead of the earlier term 'blockships'.

BLENHEIM **Group** Steam guard ships (converted from elderly 74s), 60 guns. Former 74-gun Third Rates of the *Armada* Class, originally launched 1809–13 (see Chapter 3 for original data). Ordered on 2 September 1845 to be converted into steam guard ships with 4 x 10in (65cwt/9ft 4in), 2 x 68pdrs, 26 x 8in, and 28 x 32pdrs. The *Blenheim* and *Hogue* had a single funnel with sterns modified and lengthened as shown below; the *Ajax* and *Edinburgh* had twin funnels and unmodified sterns. *Blenheim*'s and *Hogue*'s cylinders had 51½in diameter, 3ft stroke; *Ajax*'s and *Edinburgh*'s had 55in diameter, 2½ft stroke.

All four were dispatched to the Baltic for the Russian War under Rear-Adm. Henry Ducie Chads, sailing from Spithead on 11 March 1854 (as part of the fleet commanded by Vice-Adm. Sir Charles Napier). All participated there in the capture of Bomarsund, returning to Spithead on 16 December 1854. For the 1855 campaign, the four sailed again from Spithead on 4 April 1855 (as part of the fleet commanded by Rear-Adm. Richard Saunders Dundas).

> Men: 600. Guns: GD 28 x 32pdrs (56cwt); UD 26 x 8in/68pdr (52cwt) shell; QD/Fc 4 x 10in (67cwt) + 2 x 68pdrs (95cwt). Machinery: 4-cylinder horizontal single expansion. Single screw.

450nhp. See below for ihp.

Blenheim Sheerness Dyd (orig.)

> On 20.10.1845 the order shifted to Money Wigram & Sons, Blackwall/Seaward & Capel (938ihp = 5.816kts)
> Dimensions & tons: 181ft 2⅞in, 149ft 9½in x 48ft 6in oa (47ft 11½in for tonnage) x 21ft 3in. 1,832bm. 2,912 disp.
> Conversion ord: 2.9.1845. Begun: 12.11.1845. L: (undocked) 3.1847. C: 3–5.1847 (for sea) at Woolwich, then finally 27.10.1847.
> First cost: £31,857 (paid to Wigram), plus machinery £25,412 + £1,511 for its fitting, and fitting for sea £5,526.
> Commissioned 1.12.1847 under Capt. Richard Augustus Yates, as guard ship at Portsmouth; on 25.1.1848 under Capt. Horatio Thomas Austin, then 13.2.1850 under Capt. William Honeyman Henderson, then 14.8.1853 under Capt. Frederick Thomas Pelham, in same role; fitted for sea (for £3,977) at Portsmouth 1–3.1854; to the Baltic 3.1854; on 18.11.1854 under Capt. William Hutcheon Hall, in the Baltic 1854; again to the Baltic 4.1855; paid off 4.6.1856 at Portsmouth. Recommissioned 5.6.1856 under Pelham again, as guard ship at Portsmouth; paid off 21.11.1857 (replaced by *Hannibal*). Fitted as coastguard ship at Portsmouth (for £8,032) 2–4.1858. Recommissioned 1.2.1858 under Capt. Francis Scott, as coastguard at Portland; on 12.6.1860 under Capt. Edward Tatham, in same role at Pembroke (replaced at Portland by *Colossus*); on 1.4.1861 under Capt. Frederick Herbert Kerr, then 2.4.1864 under Capt. Thomas Henry Mason, in same role; paid off 31.8.1865 (replaced at Pembroke by *Revenge*). BU 1865.

Hogue Chatham Dyd (orig.)

> On 20.10.1845 the order shifted to Green, Blackwall/Seaward & Capel (793ihp = 7.809kts)
> Dimensions & tons: 184ft 0in, 153ft 0¼in x 48ft 4¼in oa (47ft 7¼in for tonnage) x 21ft 0½in. 1,846bm. 3,054 disp.
> Conversion ord: 2.9.1845. Begun: 1.12.1845. L: (undocked) 1846. C: 12.1846 (at Green's); 10.1847 (at Deptford); 9.1848 (at Blackwall, for machinery); 12.1848–28.7.1849 (for sea) at Sheerness.

The *Forte* is at Rio de Janeiro on 1 December 1861, bearing the flag of Vice-Admiral Richard Laird Warren, as depicted in this lithograph by Lieutenant George L. Tupman (RM Auxiliary). Designed as 50-gun frigates with a relatively low-powered (400 ihp) engine, the class was redesignated as 51-gun with the addition of a pivot-mounted 68pdr (or 95cwt) as a bow-chaser on the forecastle; this gun was replaced in 1862 by a 7in BL. With the funnel retracted, there is little to indicate that the ship has steam propulsion. By this date the barricades are flush fore and aft with no break amidships, but there are still no guns mounted on the gangways. Note the gaffs on the fore and main for what the Navy called 'Spencers' – versions of the spanker on the mizzen – and a feature of the mid-nineteenth-century rig, replacing staysails. *(NMM PY0971)*

First cost: £36,611 (paid to Green), total £81,525 fitted for sea.
Commissioned 19.6.1849 under Capt. John Macdougall, for Cork; in 10.1849 to Lisbon and thence to the Mediterranean; in 12.1850 to Queenstown again; on 1.9.1852 under Capt. William Ramsay, as guard ship at Devonport; to the Baltic 3.1854, and again in 4.1855; paid off 24.5.1856 at Plymouth. Recommissioned same date under Capt. John Fulford, as flagship of Rear-Adm. Henry Ducie Chads on 1.4.1857, at Queenstown. Fitted for coastguard service at Plymouth (for £3,680) 3–4.1858. Recommissioned 1.3.1858 under Capt. John Moore, as coastguard ship at Greenock; on 4.7.1859 under Capt. Reginald John James George Macdonald, then 1.8.1862 under Capt. Arthur Farquhar, in same role; paid off 28.2.1864. BU 1865 at Devonport.

Ajax Portsmouth Dyd (orig.)
On 23.10.1845 the order shifted to Thomas & John White, Cowes/ Maudslay, Sons & Field (846ihp = 7.147kts)
Dimensions & tons: 176ft 0in, 144ft 11½in x 48ft 6½in oa (47ft 9½in for tonnage) x 21ft 0in. 1,761bm. 2,828 disp.
Conversion ord: 2.9.1845. Begun: 17.11.1845. L: (undocked) 9.1846. Altered (at Portsmouth) 9.1846–5.1847. C: 28.9.1848 at Portsmouth. Fitted for sea 6.1850.
First cost: £15,908 (paid to Whites), plus £55,975 fitted for sea.
Commissioned 29.4.1850 under Capt. Michael Quinn, as flagship of Rear-Adm. John Brett Purvis at Queenstown (Cork); to Portsmouth 6.1853; on 10.9.1853 under Capt. Frederick Warren, for the North Sea, then to the Baltic 3.1854, and again in 4.1855; paid off 21.5.1856. Recommissioned 25.8.1856 under Capt. Robert Spencer Robinson, Superintendent of the Steam Reserve at Devonport; on 1.2.1858 under Capt. John McNeill Boyd (drowned 9.2.1861), as coastguard ship at Kingstown (Dublin); on 18.2.1861 under Capt. Edmund Heathcote, then 12.3.1862 under Capt. Michael De Courcy, in same role; paid off 21.3.1864 (replaced by *Royal George*). BU 1864.

Edinburgh Portsmouth Dyd
On 25.10.1845 the order was confirmed to remain at Portsmouth/

Maudslay, Sons & Field. (963ihp = 8.873kts)
Dimensions & tons: 176ft 6in, 145ft 4¼in x 48ft 7½in oa (47ft 10½in for tonnage) x 21ft 0½in. 1,772bm. 2,598 disp.
Conversion ord: 2.9.1845. Begun: 11.8.1846. L: (undocked) 31.12.1846. C: 19.8.1852 (as tender to *Excellent*). Fitted for sea 2–3.1854
First cost: £65,618 for conversion and fitting for sea.
Commissioned 7.2.1854 under Capt. Richard Strode Hewlett, as flagship of Rear-Adm. Henry Ducie Chads, for the Baltic 3.1854, and again in 4.1855, when took part in the attack on Sveaborg in 8.1855; paid off 9.8.1856 at Plymouth. Recommissioned same date under Capt. Edward Pellew Halstead, as guard ship at Sheerness; on 1.3.1858 under Capt. Swynfen Thomas Carnegie, as coastguard ship at Leith (replaced by *Cressy* at Sheerness); on 24.1.1859 under Capt. Edwin Claton Tennyson D'Eyncourt, then 13.2.1862 under Capt. Charles Frederick Schomberg, in same role; paid off 29.2.1864 (replaced by *Trafalgar* at Leith). Sold to Castle & Beech 11.1865 to BU at Charlton.

On 5 January 1848 it was ordered that the 50-gun sailing frigate *Worcester* be fitted as a screw frigate, using the 350nhp Boulton Watt & Co. engine originally scheduled for the *Horatio* (see Chapter 5). It was intended to convert the *Worcester* at Chatham Dyd, but on 8 April 1848 this conversion was cancelled.

IMPERIEUSE 51 (later 35) guns. Surveyor's Department design, approved 14 June 1850. The *Imperieuse*, originally ordered in March 1848 as a 60-gun sailing frigate, was re-ordered in 1850 as a screw ship. *Imperieuse* had cylinders of 55in diameter, 3ft stroke.
Dimensions & tons: 212ft 0in, 180ft 7¾in x 50ft 0½in oa (49ft 6½in for tonnage) x 16ft 8¾ft. 2,358³⁵⁄₉₄bm.
Men: 515. Guns: (original) MD 30 x 8in (65cwt) shell; UD 20 x 32pdrs (56cwt) + 1 x 68pdr (95cwt) on pivot. (by 1862) 30 x 32pdrs, 4 x 40pdr Armstrong BLs, 1 x 7in BL.
Machinery: 2-cylinder horizontal single expansion (trunked). Single

screw. 360nhp. 1,296ihp = 9.85kts.

Imperieuse Deptford Dyd/John Penn & Sons

As built: 212ft 0in, 180ft 7¼in x 50ft 0½in (49ft 6½in for tonnage) x 16ft 8¾in. 2,358³⁵⁄₉₄bm. 3,345 disp. Draught (light) 12ft 7in/17ft 11in (loaded) 21ft 2in/22ft 3in.

Ord: 14.6.1850. K: 11.1850. L: 15.9.1852. C: 30.3.1853 (for sea) at Woolwich and Chatham.

First cost: £91,833 total (hull £49,918, machinery £21,458, fitting £20,457).

Commissioned 17.12.1852 at Woolwich under Capt. Rundle Burges Watson; to Portsmouth, then to North America and the West Indies; in 1854 in the Baltic during the Russian War; led Advanced Squadron to the Baltic; to North America again 3.1856; paid off 5.2.1857 at Portsmouth. Refitted at Portsmouth (for £23,850) 3.1858–8.1859. Recommissioned 23.6.1859 under Capt. Rochford Maguire, as flagship of Rear-Adm. Lewis Tobias Jones; for East Indies and China; on 16.4.1861 under Capt. George Ommaney Willes, as flagship of Vice-Adm. James Hope, in China; Taiping rebellion; on 1.10.1862 under Acting Capt. Robert Gibson (Willes invalided), in China; paid off 21.3.1863 at Portsmouth. Sold to Castle & Beech 3.1867 to BU at Charlton.

FORTE Class 51 (later 35) guns. Slightly modified from the design of the *Imperieuse*. The *Euryalus*, originally ordered in March 1848 as a 50-gun sailing frigate, was re-ordered in 1851 as a screw ship. She had cylinders of 58⅟₁₆in diameter, 3¼ft stroke; those of the others were 63in diameter, 3ft stroke. *Bacchante*, *Liffey* and *Shannon* (all re-ordered 4 April 1851 to this design), and *Topaze* (ordered to this design on 27 March 1852), were later completed to a modified design, as were *Mersey* and *Tweed* (ordered 31 March 1855), see below. *Aurora* was given a lengthened bow in February 1860 before launching. *Euryalus* and *Chesapeake* were built in dry docks.

Dimensions & tons: 212ft 0in, 180ft 8½in x 50ft 2in oa (49ft 6in for tonnage) x 16ft 9in. 2,355⁴⁄₉₄bm.

Aurora (lengthened 15ft from original design) 227ft 0in, 195ft 8½in x 50ft 2in oa (49ft 6in for tonnage) x 16ft 9in. 2,550⁵⁄₉₄bm.

Men: 515. Guns: (original) MD 30 x 8in (65cwt) shell; UD 20 x 32pdrs (56cwt) + 1 x 68pdr (95cwt) on pivot. (by 1862) 30 x 32pdrs, 4 x 40pdr Armstrong BLs, 1 x 7in BL.

Machinery: 2-cylinder horizontal single expansion (trunked in *Euryalus*). Single screw. 400nhp. See below for ihp/speed.

Euryalus Chatham Dyd/John Penn & Sons. (1,162ihp = 9.47kts)

As built: 212ft 0in, 180ft 9in x 50ft 2in (49ft 8in for tonnage) x 16ft 9in. 2,371bm. 3,356 disp. Draught (light) 11ft 11in/18ft 4in (loaded) 20ft 7in/22ft 11in.

Ord: 15.3.1851. K: 10.1851. L: 5.10.1853. C: 5.3.1854 (for sea).

First cost: £91,357 total (hull £40,001, machinery £25,545, fitting £25,811).

Commissioned 26.12.1853 at Chatham under Capt. George Ramsay, for the Baltic during the Russian War; in 1856 to North America and the West Indies; paid off 23.4.1857 at Plymouth. Refitted at Devonport (for £21,533) 1857–58. Recommissioned 15.2.1858 under Capt. John Walter Tarleton, for the Channel Squadron, then to the Mediterranean (with HRH Prince Albert as naval cadet from 10.1858), then to Cape of Good Hope (still with Prince Alfred aboard) 1860; paid off 19.11.1860 at Portsmouth. Recommissioned 24.1.1862 under Capt. Alexander Crombie Gordon, at Portsmouth; on 11.2.1862 under Capt. John James Steven Josling, as flagship of Rear-Adm. Augustus Leopold Kuper, for East Indies and China; in Bombardment of Kagoshima during 'Anglo-Satsuma War' (Josling killed

15.8.1863); on 20.9.1863 under Capt. John Hobhouse Inglis Alexander, in East Indies and China; at Bombardment of Simonoseki (Alexander gravely wounded 24.11.1864); on 24.11.1864 under Capt. William Montagu Dowell, still flagship of Keper on same station; paid off 24.11.1864 at Portsmouth. Sold to Castle & Beech 3.1867 to BU at Charlton.

Aurora Pembroke Dyd/Maudslay, Sons & Field. (1,576ihp = 10.21kts)

As built: 212ft 0in, 180ft 10⅛in x 50ft 1in (49ft 7in for tonnage) x 16ft 4in. 2,364bm. 3,498 disp.

As lengthened (2.1860): 227ft 0in, ?196ft 3in x 50ft 2in oa (49ft 6in for tonnage) x 16ft 9in. 2,558bm.

Draught (light) 12ft 9in/18ft 10in (loaded) 21ft 7in/22ft 7in.

Ord: 4.4.1851. (named 8.6.1853) K: 5.9.1854. L: 22.6.1861. C: 1863.

Commissioned 18.11.1863 at Plymouth under Capt. Francis Leopold McClintock, for the Channel Squadron; escort in 9.1864 for Royal yacht *Osborne* conveying HRH Prince Alfred and Princess Alexander (later to become HM King Edward VII and Queen Alexander) to Copenhagen and the Baltic; to North America and West Indies 1.1865; on 1.11.1865 under Capt. Algernon Frederick Rous De Horsey, on same station; senior officer on Canadian Lakes during the Fenian disturbances; paid off 20.12.1867 at Plymouth. Recommissioned 20.2.1872 under Capt. Benjamin Spencer Pickard, as sea-going traing ship for boys; on 21.8.1872 under Capt. Sholto Douglas; to Detached Squadron 1873, then temporary flagship at Queenstown; paid off 20.4.1874. Recommissioned 15.5.1874 under Capt. William Graham, as guard ship at Greenock; on 10.8.1874 under Capt. Trevenen Penrose Coode, then 12.7.1875 Capt. Henry Duncan Grant, in same role; paid off 19.6.1877 at Plymouth. BU 12.1881.

Forte Deptford Dyd/Maudslay, Sons & Field (1,539ihp = 11.435kts)

As built: 212ft 0in, 180ft 10⅛in x 50ft 1in (49ft 7in for tonnage) x 16ft 9in. 2,364bm. 3,456 disp. Draught (light) 12ft 9in/18ft 10in (loaded) 16ft 11in/18ft 7in.

Ord: 27.3.1852. (named 8.6.1853) K: 5.5.1854. L: 29.5.1858. C: 7.4.1860.

First cost: £90,446 (hull £65,040, machinery £25,406) excluding fitting.

Commissioned 25.1.1860 under Capt. Edward Winterton Tournour, as flagship of Rear-Adm. Sir Henry Keppel, for Cape of Good Hope. In 6.1861 under Capt. Thomas Saumarez, as flagship of Vice-Adm. Richard Laird Warren, for east coast of South America; on 12.11.1862 under Capt. Arthur Mellersh, still Warren's flagship on this station; paid off 8.9.1864 at Sheerness. Recommissioned 21.8.1868 under Capt. John Hobhouse Inglis Alexander, as flagship of Commodore Sir Leopold George Heath, for the East Indies; on 6.9.1870 under Capt. Henry Fairfax, as flagship of Rear-Adm. James Horsford Cockburn, on same station; paid off 17.2.1872 at Sheerness. Became receiving ship at Chatham 1879. Coal hulk at Sheerness 3.1894. Burnt by accident at Sheerness on 23.11.1905.

Chesapeake Chatham Dyd/Maudslay, Sons & Field. (1,159ihp = 9.658kts)

As built: 212ft 0in, 181ft 2in x 50ft 2in (49ft 8in for tonnage) x 16ft 9in. 2,377bm. 3,334 disp. Draught (light) 11ft 10in/18ft 1in (loaded) 20ft 6in/22ft 10in.

Ord: 27.3.1852. (named 4.11.1853) K: 30.5.1854. L: 27.9.1855. C: 28.8.1857 (for sea) at Chatham (after machinery fitted at Sheerness).

First cost: £100,920 total (hull £56,354, machinery £25,684, fitting £18,882).

Commissioned 21.7.1857 at Chatham under Commodore Rundle

This contemporary model of a *Liffey* Class screw frigate is far more detailed than the usual official models of this era and depicts all the main deck and rigging fittings. With part of the side cut away to show the structure, it emphasizes the mixture of old and new technology in these transitional warships; on the one hand, the wooden hull, broadside armament and full rig look traditional, but on the other hand the funnel and screw propeller reveal the new power of steam propulsion. As cruising warships, frigates needed sails to meet the required endurance when deployed globally, before the creation of coaling stations around the world; and to minimise the drag effect when under canvas the two-bladed screw was fitted in a 'banjo frame' so that it could be disconnected from the shaft and hoisted up into a well in the stern. The funnel was usually fitted to telescope when steam was not in use. *(NMM F8988-004)*

Burges Watson, for the East Indies and China; on 27.12.1858 under Commodore Harry Edmond Edgell, on same station; on 16.2.1859 under Capt. George Ommaney Willes, as flagship of Rear-Adm. James Hope, on same station; in 2nd Anglo-Chinese War; on 16.4.1861 under Capt. Rochford Maguire, as flagship of Rear-Adm. Lewis Tobias Jones, for return home; paid off 17.10.1861 at Sheerness. Sold to Castle & Beech 4.1867 to BU at Charlton.

LIFFEY Class 1854 51 (later 39, then 31) guns. Surveyor's Department design. *Bacchante* (in 1848), *Liffey* and *Shannon* (both in 1844) were originally ordered as sailing frigates of 2,126bm design. *Glasgow* and *Newcastle*, originally ordered 31 March 1855 to this design, were completed to modified designs (see below).

  Dimensions & tons: (285ft oa) 235ft 1in, 203ft 10in x 50ft 1½in oa (49ft 7½in for tonnage) x 18ft 4½in. 2,651bm. 3,915 displacement.
  Men: 560. Guns: MD 30 x 8in (65cwt/9ft) shell; UD 20 x 32pdrs (56cwt/9¼ft) + 1 x 68pdr (95cwt/10ft) on pivot.
  Machinery: 2-cylinder horizontal single expansion (trunked in *Liffey* and *Shannon*). *Liffey*'s and *Shannon*'s cylinders were 70¾in diameter, 3½ft stroke; the others' cylinders were 76in diameter, 3½ft stroke (*Liverpool* 3¼ft). Single screw. 600nhp. 11.8kts.

Liffey Devonport Dyd/John Penn & Sons (1,905ihp = 11.097kts)
  As built: 235ft 0in, 203ft 10in x 50ft 1½in (49ft 7½in for tonnage) x 18ft 4½in. 2,667bm. 3,915 disp. Draught 13ft 7in/18ft 6in.
  Ord: 4.4.1851. K: 12.7.1854. L: 6.5.1856. C: 11.1.1859.
  First cost: £135,774 (hull £65,717, machinery £37,730, fitting £32,327).
  Commissioned 1.11.1858 at Portsmouth under Capt. George William Preedy, for the Channel squadron; in 7.1859 to the Mediterranean, then 3.1862 to the West Indies; paid off 17.7.1862 at Portsmouth. Recommissioned same day under Capt. George Parker, for the Mediterranean; in 3.1865 to North America and the West Indies; paid off 31.10.1865 at Plymouth. Recommissioned 26.6.1867 under Capt. Gerard John Napier, for the Naval Display; on 25.7.1867 under Capt. John Ormsby Johnson, for the south coast of Ireland; in 5.1868 escorting Prince and Princess of Wales (later Edward VII and Queen Alexandra) to Denmark; in 9.1868 to the West Indies; in 6.1869 joined Flying Squadron; on 24.11.1869 under Capt. Robert Gibson; paid off 29.11.1870. Fitted at Devonport as store hulk for Coquimbo 1877; recommissioned 1.4.1878 at Devonport, then recommissioned 17.11.1878 at Coquimbo in this role under Staff Cmdr John F. R. Aylen; in 1886 under Staff Cmdr F. K. Taylor, then 2.1890 under Staff Cmdr A. R. Wonham and 6.1901

under Staff Cmdr J. D. Moulton; paid off there 17.3.1904. Sold there as a hulk 4.1903.

Shannon Portsmouth Dyd/John Penn & Sons. (2,125ihp = 11.499kts)

As built: 235ft 1in, 203ft 10in x 50ft 1½in (49ft 7½in for tonnage) x 18ft 4½in. 2,667bm. 3,636 disp. Draught 2ft 1in/17ft 0in.

Ord: 4.4.1851. K: 1.1854. L: 24.11.1855. C: 29.12.1856 (for sea).

First cost: £127,163 (hull £62,759, machinery £37,325, fitting £27,079).

Commissioned 13.9.1856 under Capt. William Peel (died 27.4.1858 of smallpox) for the East Indies; took Lord Elgin to China; formed a Naval Brigade under Capt. Peel 14.8.1857 for Lucknow Relief Force (leaving ship under George A. Waters, Master, with 140 men), in which Peel was severely wounded and 4 VCs were awarded; on 28.4.1858 under Capt. Julian Foulton Slight (Acting); paid off 15.1.1859 at Portsmouth. Recommissioned 17.1.1862 under Capt. James Francis Ballard Wainwright, for the Channel squadron; on 18.6.1862 under Capt. Oliver John Jones, for the Mediterranean; in 1864 to North America and the West Indies; paid off 11.4.1865 at Portsmouth. Sold to Castle 31.5.1871 to BU at Charlton.

Topaze Devonport Dyd/Maudslay, Sons & Field. (2,538ihp = 11.122kts)

As built: 235ft 0in, 203ft 5⅝in x 50ft 1in (49ft 7in for tonnage) x 18ft 5in. 2,659bm. 3,915 disp. Draught 13ft 6in/18ft 3in.

Ord: 27.3.1852. K: 4.1856. L: 12.5.1858. C: 23.7.1859 (for sea).

First cost: £103,673 (hull £64,623, machinery £37,050), excluding fitting.

Commissioned 11.6.1859 under Capt. John Welbore Sunderland Spencer, for the Channel squadron; in 10.1859 to the Pacific (Spencer as Commodore on southern division of this station); paid off 31.12.1863 at Plymouth. Recommissioned 26.1.1866 under Capt. William Montagu Dowell, for the Pacific; on 23.5.1866 under Commodore Richard Ashmore Powell, in the Pacific; paid off 1.9.1969 at Plymouth. Recommissioned 14.6.1871 under Capt. Radulphus Bryce Oldfield, for the Detached Squadron; recommissioned 9.10.1872 under Capt. Edward Hardinge; on 21.7.1874 under Capt. Arthur Thomas Thrupp; paid off 22.5.1877. Recommissioned 17.7.1877 under Capt. Charles John Rowley, as coastguard at Kingstown; paid off 26.6.1878 at Plymouth (replaced at Kingstown by *Belleisle*). Sold to Castle 14.2.1884 to BU at Charlton.

Bacchante Portsmouth Dyd/Maudslay, Sons & Field. (2,490ihp = 12.074kts)

As built: 235ft 1in, 203ft 7in x 50ft 1½in (49ft 7½in for tonnage) x 18ft 4½in. 2,667bm. 3,631 disp.

Ord: 4.4.1851. K: 9.1856. L: 31.7.1859. C: 29.6.1860 (for sea).

First cost: £107,914 (hull £71,914, machinery £36,000), excluding fitting.

Commissioned 17.4.1860 at Portsmouth under Capt. Donald McLeod Mackenzie, as flagship of Rear-Adm. Sir Thomas Maitland, for the Pacific; paid off 3.8.1864 at Portsmouth. BU 10.1869 at Portsmouth.

Liverpool Devonport Dyd/Humphrys & Tennant. (1,935ihp = 10.811kts)

As built: 235ft 0in, 203ft 6in x 50ft 1in (49ft 7in for tonnage) x 18ft 5in. 2,656bm. 3,919 disp.

Ord: 31.3.1855. K: 14.11.1859. L: 30.10.1860. C: 11.12.1860.

Commissioned 30.4.1863 under Capt. Rowley Lambert, for the Channel squadron; in 1864 to North America and the West Indies; in 1865 to Channel squadron again; on 23.3.1866 under Capt. John Seccombe; paid off 10.8.1867 at Plymouth. Recommissioned 8.5.1869 under Capt. John Ommanney Hopkins, as flagship of Rear-Adm. Geoffrey Thomas Phipps Hornby, for the Flying Squadron; paid off 2.12.1870 at Plymouth. Sold to Castle 26.6.1875 to BU at Charlton.

The big Walker frigates, conceived during the Crimean War, commenced their existence through orders issued on 3 April 1854 for three 'special type frigate-corvettes' of 2,479bm and 800nhp to be built at Pembroke (*Diadem* and *Doris*) and at Deptford (*Ariadne*). On 31 March 1855 a fourth unit was ordered to this design: the *Orlando* at Pembroke. While the first pair went ahead as planned (as 32-gun ships), revised plans were approved on 2 July 1856 for the *Ariadne* (plus the *Galatea*, ordered at Woolwich as a *Jason* Class corvette on 9 April 1856) to be re-designed as corvettes with a covered deck and carrying 26 guns, while the *Orlando* (plus the *Mersey*, ordered at Chatham as a *Forte* Class frigate on 31 March 1855) to be re-designed as larger frigates carrying 36 guns.

***DIADEM* Class** Screw frigates (originally corvettes), 32 guns. Surveyor's Department (Isaac Watts) design, approved 24 January 1855. The first of 'Walker's Big Frigates'. Originally ordered (with *Ariadne*) in 1854 as 'special type (frigate-)corvettes' of 30 guns, although rated First Class frigates; *Orlando* ordered 1855 was originally to same design.

Dimensions & tons: 240ft 0in, 208ft 0¼in x 48ft 0in oa (47ft 4in for tonnage) x 16ft 7in. 2,479 42/94bm. 3,714 disp.

Men: 475. Guns: MD 20 x 10in (95cwt) MLSB shell guns (broadside); UD 10 x 32pdrs (56cwt) MLSB + 2 x 68pdrs MLSB (pivot-mounted).

Machinery: 2-cylinder (82in diameter, 4ft stroke) horizontal single expansion (trunk in *Doris*). Rectangular boilers (20 lb/sq.in). Single screw. 800nhp. 2,979ihp = 12kts (*Diadem*); 3,087ihp = 12.865kts (*Doris*).

Diadem Pembroke Dyd/Maudslay, Sons & Field.

As built: 240ft 0in, 208ft 0⅜in x 48ft 0½in oa (47ft 4½in for tonnage) x 16ft 7in. 2,483 49/94bm. Draught 11ft 0in/16ft 6in. 3,880 disp.

Ord: 3.4.1854. K: 6.1855. L: 14.10.1856. C: 3.1.1858 (for sea) at Portsmouth Dyd.

First cost: £137,479 (hull £59,140, machinery £47,252, fitting £31,087).

Commissioned 19.8.1857 under Capt. William Moorsom (invalided 10.1859); sailed 6.9.1859 to join the Channel squadron, then to West Indies; on 15.2.1858 under Capt. Peter Cracroft (acting, during Moorsom's illness); on 8.10.1859 under Capt. James Horsford Cockburn, for the Channel, them to North America and the West Indies; on 28.10.1861 under Capt. George Granville Randolph, on same station, then 1.4.1862 under Capt. Francis Scott at Portsmouth, where paid off 15.4.1862. At Portsmouth 1862–70 (by 1870 reduced to 22 guns). Sold 23.1.1875 to Castle for BU at Charlton.

Doris Pembroke Dyd/John Penn & Sons.

As built: 240ft 0in, 208ft 0⅜in x 48ft 0½in oa (47ft 4½in for tonnage) x 16ft 7in. 2,483 49/64bm. Draught 10ft 9in/15ft 10in. 3,677 disp.

Ord: 3.4.1854. K: 6.1856. L: 25.3.1857. C: 30.3.1859 (for sea) at Devonport Dyd.

First cost: £138,328 (hull £60,033, machinery £51,756, fitting £26,539).

Commissioned 7.3.1859 under Capt. Edmund Heathcote, for the Mediterranean; on 18.2.1861 under Capt. Francis Leopold McClintock, on the same station; paid off 27.12.1862 at Plymouth. Recommissioned 5.2.1866 under Capt. Charles Vesey,

The *Galatea* at Sydney in 1867 during her round-the-world cruise. One of the group generically known as 'Walker's big frigates', the ship traded a larger number of guns (notice there are only twelve gunports on the broadside) for a far heavier calibre (10in shell guns). Behind her is the screw corvette *Challenger*, the flagship on the Australia station but later fitted out for the pioneering oceanographic voyage of 1872-76 when she also circumnavigated the globe. *(Allan C. Green, via the State Library of Victoria)*

for the West Indies; in 1868 under Capt. Henry Carr Glyn, on the same station; paid off 16.6.1869 at Plymouth. Reduced to 40 men and 24 guns. Recommissioned 16.10.1872 under Capt. William Henry Edye, for the Flying Squadron; on 11.9.1874 under Capt. Edmund Robert Fremantle, on same service; paid off 4.9.1876 at Plymouth. Sold 1885 to BU.

ARIADNE Class Screw frigates (originally corvettes), 26 guns.
Surveyor's Department (Isaac Watts) design, approved 2 July 1856.
    Dimensions & tons: 280ft 0in, 245ft 8in x 50ft 0in oa (49ft 6in for tonnage) x 19ft 4in. 3,201⁷⁸⁄₉₄bm. 4,426 disp.
    Men: 450. Guns: MD 24 x 10in (85cwt) MLSB shell guns (broadside); UD 2 x 68pdrs (65cwt) MLSB (pivot-mounted, soon replaced by slide-mounted 110pdr Armstrong BLs).
    Machinery: 2-cylinder (82¹⁄₁₆in diameter, 44in stroke) horizontal single expansion (trunk in *Galatea*). Rectangular boilers (20 lb/sq. in). Single screw. 800nhp. 3,350ihp = 13.087kts (*Ariadne*); 3,061ihp = 11.796kts (*Galatea*).

Ariadne Deptford Dyd/Maudslay, Sons & Field
    As built: Dimensions unavailable. 3,214bm/4,583 disp as built.
    Ord: 3.4.1854. K: 1.8.1856. L: 4.6.1859. C: 16.12.1859 (for sea) at Woolwich and Chatham Dyds.
    First cost: £147,496 total (hull £75,980, machinery £48,600, fitting £22,916).
    Commissioned 18.11.1859 under Capt. Edward Westby Vansittart, for the Channel squadron; Fitted to carry Prince

of Wales' suite (in *Hero*) to Canada in 7–11.1860, then to North America and the West Indies; paid off 19.3.1864 at Sheerness. Recommissioned 26.11.1868 at Portsmouth under Capt. Colin Andrew Campbell (invalided 18.1.1869), as royal yacht for Prince and Princess of Wales (later Edward VII and Queen Alexandra) in the Mediterranean; on 18.1.1869 under Capt. Frederick Archibald Campbell, on same service; the Royal couple left the ship at Brindisi 5.1869; paid off 8.6.1869 at Portsmouth. Recommissioned 16.12.1871 under Capt. Walter Cecil Carpenter, as naval cadet training ship in the Mediterranean; paid off 2.9.1873 at Portsmouth. Part of *Vernon* torpedo training establishment 5.1876. Harbour training ship for naval cadets 1884. Renamed *Actaeon* 6.6.1905. Recommissioned 6.1905 in 8.1909 under Cmdr A. B. F. Dawson, then 8.1911 under Capt. Godfrey M. Payne, then 12.1912 under Capt. E. C. Villiers and 2.1919 under Capt. Hugh S. Currey. Sold 11.12.1922.

Galatea Woolwich Dyd/John Penn & Son
    As built: Dimensions unavailable. 3,227bm/4,686 disp as built.
    Ord: 9.4.1856. K: 2.2.1857. L: 14.9.1859. C: 27.12.1859 (for Reserve); 2.1862 (for sea) at Plymouth Dyd.
    First cost: £129,991 (hull £81,991, machinery £48,000) excluding fitting.
    Commissioned 19.5.1862 under Capt. Rochford Maguire, for the Channel squadron, later to North America and the West Indies; paid off 23.1.1866 at Plymouth. Recommissioned 22.1.1867 under Capt. Alfred Ernest Albert Saxe-Coburg-Gotha (HRH the Duke of Edinburgh), for world tour to South America, the Cape of Good Hope, Australia, China and Japan; paid off 2.7.1871 at Plymouth. BU 6.1883 by Castle at Charlton.

Acasta Deptford Dyd/—

The *Melpomene*, photographed about 1860, was a sailing frigate converted during construction for screw propulsion. This involved the insertion of a 52ft midship section to accommodate a 600nhp engine – the hull was cut in two and both halves launched separately, before the lengthened hull was re-assembled and re-launched as one. On trials the Penn trunk engine developed 2,171 ihp for a trials speed of 12.4 knots. *(NMM 5334)*

Ord: 3.2.1861. K: 16.4.1860. (named 21.3.1861) Cancelled 12.12.1863.
Hyperion Woolwich Dyd/—
Ord: 3.2.1861. (named 21.3.1861) K: — . Cancelled 12.12.1863.

EMERALD Class – screw frigates, converted on stocks from sailing frigates, 51 guns. Originally ordered in 1848 as 60-gun sailing frigates and laid down in 1849 (see Chapter 2); the original fourth member of class (*Imperieuse*) was begun as a screw frigate in 1850 to a different design (see previous page). The remaining three were re-ordered in 1854–56 to the above 2,852bm design, being lengthened by 52ft. *Emerald* emerged 5½in broader. *Immortalité* was further lengthened during construction by about another 14ft (by AO of 22 February 1858) which increased her to 3,050bm in design (actual tonnage as below); she was allegedly the fastest wooden warship under sail (making over 12kts). The after parts of *Melpomene* and *Immortalité* were launched for lengthening on 9 August 1856 and 28 August 1856 respectively. All three ships reduced to 35 guns in 1862–63 and then to 28 guns in 1868. Their cylinders were each 76in diameter (except *Melpomene*, 70¼in), 3½ft stroke.

Dimensions & tons: (design) 237ft 0in, 202ft 2in x 52ft 0in (51ft 6in for tonnage) x 16ft 8in. 2,852¹⁰⁄₉₄bm.
Men: 550–600. Guns: MD 30 x 8in (65cwt/9ft) shell (broadside); UD 20 x 32pdrs (56cwt/9½ft) + 1 x 68pdr (95cwt/10ft) on pivot. In 1868 MD 18 x 64pdr MLR + 4 x 7in/110pdr (6½ton), UD 6 x 64pdr MLR.
Machinery: 2-cyl. horizontal single expansion (trunk in *Melpomene*). Single (hoisting) screw. 600nhp. See below for individual ihp/speed.

Emerald Deptford Dyd/Miller, Ravenhill & Salkeld. (2,323ihp = 12kts)
Dimensions & tonnage: 237ft 0in, 202ft 3¼in x 52ft 6½in oa (52ft 0½in for tonnage) x 16ft 8in. 2,913⁸⁶⁄₉₄bm. 3,503 disp.

Conversion ord: 24.5.1854. K: (4.6.1849). Begun: 29.5.1854. L: 19.7.1856. C: 1.6.1859.
First cost: £127,933 including hull £69,722, machinery £33,860.
Commissioned 14.5.1859 at Sheerness under Capt. Arthur Cumming, for the Channel Squadron; paid off 7.11.1863 at Sheerness. Sold to Castle 2.12.1869 to BU at Charlton.

Melpomene Pembroke Dyd/John Penn & Son (2,171ihp = 12.4kts)
Dimensions & tonnage: 237ft 0in, 202ft 1½in x 52ft 1in oa (51ft 7in for tonnage) x 16ft 8in. 2,860⁷¹⁄₉₄bm. 2,741 (light) disp.
Conversion ord: 9.4.1856. K: (9.1849). Begun: 30.6.1856. L: 8.8.1857. C: 18.6.1859.
First cost: £136,213 including hull £43,835 as sailing ship plus £31,535 for conversion, machinery £37,617.
Commissioned 2.6.1859 at Portsmouth under Capt. Charles Joseph Frederick Ewart, for the Channel Squadron; to the Mediterranean 8.1860; to North America and the West Indies 1.1862; paid off 20.3.1863 at Portsmouth. Sold to Castle 23.1.1875 to BU at Charlton.

Immortalité Pembroke Dyd/Maudslay, Sons & Field. (2,366ihp = 12.3kts)
Dimensions & tonnage: 251ft 0in, 216ft 1½in x 52ft 1in oa (51ft 7in for tonnage) x 16ft 8in. 3,058⁸¹⁄₉₄bm. 3,690 disp.
Conversion ord: 9.4.1856. K: (11.1849). Begun: 14.7.1856. L: 25.10.1859. C: 7.2.1861.
First cost: hull £4,468 as sailing ship plus £39,357 for conversion, machinery £36,000.
Commissioned 17.11.1860 at Portsmouth under Capt. George Hancock, for the Mediterranean; to North America and the West Indies 8.1861; paid off 15.7.1864 at Portsmouth. Recommissioned 1.12.1870 under Capt. Francis William Sullivan; in 1871 to the Detached Squadron; on 30.10.1871 under Capt. William

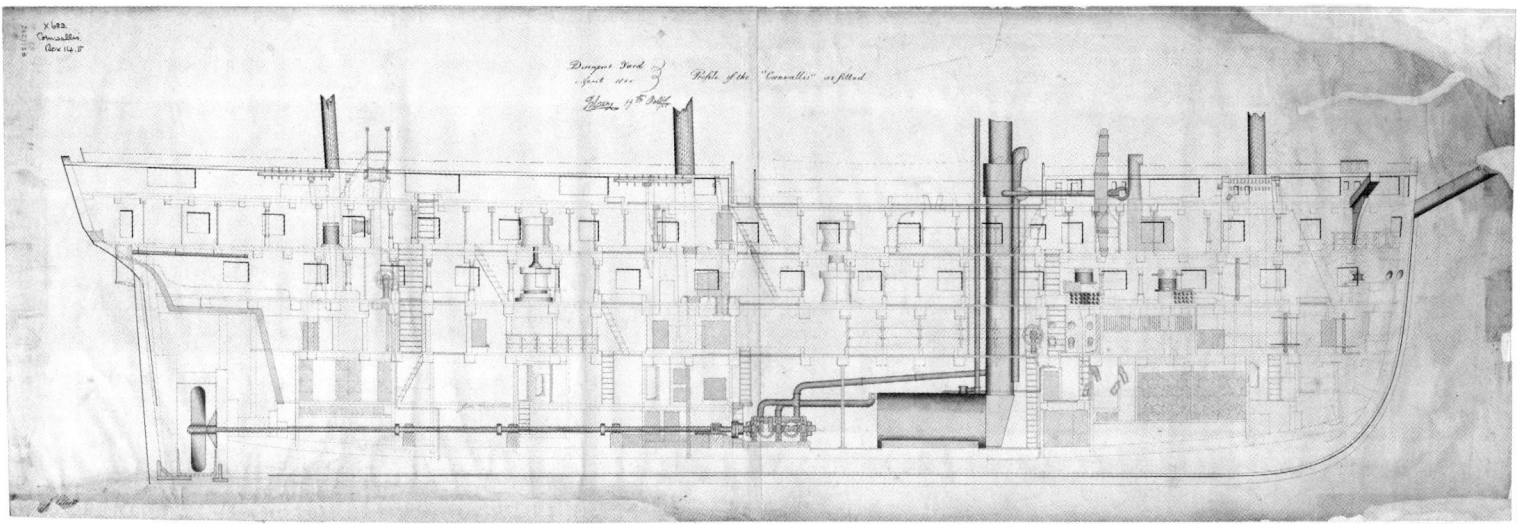

This inboard profile of the *Cornwallis* reveals how the 74-gun Third Rate of 1813 was adapted for screw propulsion into a 60-gun blockship at Plymouth between October 1854 and April 1855. The draught, signed by John Edye (the Navy's Assistant Surveyor) who had charge of design work during the tenure of the non-specialist Captain (Rear-Admiral from 1858) Sir Baldwin Wake Walker as Surveyor, shows the location of the boiler and engine in the ship's hold, with the smoke exhaust rising through three decks to a funnel located just aft of the forecastle. *(NMM J2677)*

Graham, in same role; paid off 12.10.1872 at Portsmouth. Recommissioned 13.10.1872 under Capt. Algernon M'Lennan Lyons, for the Detached Squadron; in 1874 under Capt. Francis Alexander Hume; paid off 21.5.1877 at Portsmouth. Sold 6.1883 to Castle & Sons to BU (towed away 1.9.1883).

CORNWALLIS Group Steam guard ships (converted from elderly 74s), 60 guns. On the outbreak of the Russian War, a further five vessels were similarly converted from 74-gun Third Rates, originally launched 1812–22 (see Chapter 3 for original data), used as stopgap conversions during the Crimean War, intended for use as floating batteries to attack Russian forts in the Baltic. This group were mainly refitted with second-hand materials. The cylinders in the first three (by Penn) were 30¼in diameter, 2½ft stroke; in last pair (by Maudslay) they were 30in diameter, 2½ft stroke. Four of the five (not the *Russell*) were dispatched to the Baltic for the 1855 campaign, sailing from Spithead on 4 April 1855 (as part of the fleet commanded by Rear-Adm. Richard Saunders Dundas), and the *Cornwallis*, *Pembroke* and *Hastings* took part in the 9 August 1855 attack on Sveaborg.

Men: 600. Guns: GD 24 x 32pdrs (56cwt) + 4 x 8in/68pdr (65cwt) shell guns; UD 26 x 32pdrs (50cwt); QD/Fc 4 x 10in (85cwt) + 2 x 68pdrs (95cwt).

Machinery: 2-cylinder horizontal single expansion (trunked in the Penn-engined ships). Single screw. 200nhp. See below for ihp and speed on trials.

Russell Sheerness Dyd/John Penn & Son (578ihp = 6.68kts)
Dimensions & tons: 176ft 6in, 145ft 6in x 48ft 3¾in oa (47ft 6¼in for tonnage) x 21ft 0in. 1,751bm. 2,903 disp.
Conversion ord: 1854. Begun: 7.10.1854. L: (undocked) 2.2.1855. C: 6.6.1855.
First cost: £23,935 (including machinery £11,903).
Commissioned 3.2.1855 under Capt. Francis Scott, for Portsmouth and then Queenstown (Cork); paid off 16.5.1856 at Sheerness. Fitted as coastguard ship there (for £9,613) 9.1857–3.1858. Recommissioned 1.2.1858 under Capt. Henry Alexander Story,

as coastguard ship at Falmouth; on 1.1.1859 under Capt. George Wodehouse, then 4.1.1862 under Capt. William King Hall, and 14.4.1863 under Capt. Sidney Grenfell, in same role; paid off 4.2.1864 at Plymouth (replaced at Falmouth by *Saint George*). BU 1865.

Cornwallis Devonport Dyd/John Penn & Son (787ihp = 7.188kts)
Dimensions & tons: 177ft 1in, 145ft 7¼in x 49ft 1in oa (48ft 4in for tonnage) x 21ft 1¼in. 1,809bm. 2,678 disp.
Conversion ord: 1854. Begun: 9.10.1854. L: (undocked) 8.2.1855. C: 25.4.1855.
First cost: £39,521 total (including £12,039 for machinery)
Commissioned 9.2.1855 under Capt. George Greville Wellesley, for the Baltic; then to North America and the West Indies; paid off 13.8.1856 at Plymouth. Recommissioned 2.12.1856 under Capt. Richard Ashmore Powell, as a coastguard ship on the Humber; on 20.4.1857 under Capt. George Granville Randolph, the 30.4.1861 under Capt. Sidney Grenfell, then 14.4.1863 under Capt. James Newburgh Strange, in same role; paid off 31.3.1864 (replaced at Hull by *Dauntless*). Made a jetty at Sheerness 1865. Renamed *Wildfire* as base ship 1.4.1916; paid off 15.3.1922. BU 1957 at Sheerness.

Hawke Chatham Dyd/John Penn & Son. (500ihp = 6.525kts)
Dimensions & tons: 176ft 1in, 144ft 10½in x 48ft 5½in oa (47ft 8½in for tonnage) x 21ft 1½in. 1,754bm. 2,808 disp.
Conversion ord: 1854. Begun: 10.10.1854. L: (undocked) 5.1855. C: 20.5.1855.
First cost: fitting £23,365 plus machinery £12,212.
Commissioned 14.2.1855 under Capt. Erasmus Ommanney, for the Baltic, then to Falmouth; paid off 19.5.1856 at Sheerness. Fitted for coastguard at Sheerness 12.1856–2.1857. Recommissioned 12.12.1856 under Capt. James Willcox, as coastguard ship at Queenstown; on 19.7.1858 under Capt. William Crispin, then 5.8.1861 under Capt. Thomas Harvey, as flagship of Rear-Adm. Sir Lewis Tobias Jones, then 11.2.1863 under Capt. Edward Codd, and 4.1.1864 under Capt. Edward Heathcote, in same role; paid off 24.3.1864/1965? at Plymouth. BU 1865.

Pembroke Portsmouth Dyd/Maudslay, Sons & Field. (572ihp = 7.602kts)
Dimensions & tons: 176ft 0in, 144ft 11¼in x 48ft 6in oa (47ft 9in for tonnage) x 21ft 1in. 1,758bm. 2,465 disp.
Conversion ord: 1854. Begun: 10.10.1854. L: (undocked) 3.2.1855. C: 24.5.1855.

The *Orlando*, one of a pair of frigates that constituted the longest wooden hulls ever built for the Royal Navy. Despite half a century of structural development going back to the wartime work of Sir Robert Seppings, this class was a step too far, and the weakness of their hulls ensured a short active life, made even shorter than evident by the time spent in refit – for example, the *Orlando* needed six months in dock at Bermuda to make good her defects after her first short commission on the North America station. *(NMM 9077)*

First cost: £22,751, plus machinery £11,605.

Commissioned 5.2.1855 under Capt. George Henry Seymour, for the Baltic, then to North America and the West Indies; paid off 5.8.1856 at Portsmouth. Fitted as coastguard ship at Sheerness 11.1856–1.1857. Recommissioned 3.12.1856 under Capt. John Fraser, as coastguard ship at Leith; paid off 28.2.1858 (replaced at Leith by *Edinburgh*). Recommissioned 1.3.1858 under Capt. Edward Philips Charlewood, as coastguard ship at Harwich; in 1.1861 under Capt. George Edwin Patey, then 2.4.1861 under Capt. Thomas Fisher, as flagship of Commodore Hastings Reginald Yelverton; on 6.2.1863 flagship of Commodore Swynfen Thomas Carnegie; on 28.4.1863 under Capt. John Ormsby Johnson, as flagship of Commodore Alfred Phillipps Ryder; then 27.4.1866 under Capt. John Welbore Sunderland Spencer, as flagship of Commodore (Rear-Adm. 1868) John Walter Tarleton, in same role; paid off 11.6.1869 (replaced at Harwich by *Penelope*). On 12.6.1869 under Capt. Thomas Miller, as flagship of Vice-Adm. Richard Laird Warren, C-in-C at the Nore, at Sheerness; on 24.6.1870 under Capt. John Crawford Wilson, as flagship of Vice-Adm. Charles Gilbert John Brydone Elliot; paid off 4.1873. Barrack ship at Chatham from 4.1873. Recommissioned 1.1.1875 under Capt. John Clark Soady, as Capt. Superintendent of the Steam Reserve on the Medway; paid off 31.12.1877. Recommissioned 11.1878 under Capt. Walter James Hunt-Grubbe; in 2.1879 flagship of Rear-Adm. Thomas Brandreth; on 17.12.1881 under Capt. Lord Walter Talbot Kerr, in same role; paid off 31.3.1885. Recommissioned 5.1886 under Capt. Alexander Buller, as flagship of ?Rear-Adm. William Codrington; in 11.1887 flagship of Rear-Adm. Edward Kelly; in 4.1888 flagship of Commodore Henry Frederick Stephenson. Receiving ship 1890, renamed *Forte*. Sold 1905.

Hastings Portsmouth Dyd/Maudslay, Sons & Field.(597ihp = 6.702kts)

Dimensions & tons: 176ft 10½in, 145ft 4in x 48ft 6in oa (47ft 9in for tonnage) x 21ft 0in. 1,763bm. 2,730 disp.

Conversion ord: 1854. Begun: 11.11.1854. L: (undocked) 5.2.1855. C: 12.5.1855.

First cost: £29,192, plus machinery £11,646.

Commissioned 6.2.1855 under Capt. James Crawford Caffin, for the Baltic; on 21.8.1855 under Capt. Edward Gennys Fanshawe, in the Baltic, then to Queenstown (Cork); paid off 12.5.1856 at Portsmouth. Fitted for coastguard 7.1857. Recommissioned 3.4.1857 under Capt. William Robert Mends, as coastguard ship at Liverpool 4.1857; paid off 1.2.1860 at Liverpool (replaced by *Majestic*). Recommissioned 4.2.1860 under Cmdr Charles Maxwell Luckraft, as RNR drill ship at Liverpool; on 29.1.1862 under Cmdr John William White, in same role. Recommissioned 8.10.1862 under Capt. Charles Frederick Alexander Shadwell, as flagship of Rear-Adm. Lewis Tobias Jones at Queenstown; on 24.6.1864 under Capt. John Corbett, still flag of Rear-Adm. Jones, then 3.1865 flagship of Rear-Adm. Frederick Warren, on same station; paid off 18.5.1866. Coal hulk 1870 at Devonport. Sold 9.1885.

***MERSEY* Class** Frigates, 40 guns. Surveyor's Department (?Isaac Watts) design, approved 2.7.1856. *Orlando* was originally ordered under the 1855 Programme of Works (PW) to the 2,479bm design of *Doris*, and *Mersey* to the 2,355bm design of *Tweed*; in PW 1856 both were re-ordered (with *Diadem* and *Doris*) as 26-gun 'frigate-corvettes'; on 2 July 1856 plans were approved to build the pair as 36-gun frigates.

The final pair of 'Walker's big frigates' were the longest wooden-hulled ships ever built for the Royal Navy, with an overall length of 335¾ feet. These ships proved unsuccessful, as the design exceeded the practical limits of wooden hulls by placing enormous strains on the structure and the hull flexed badly, even though they were fitted with internal iron strapping to support the hull. *Orlando* first sailed from Devonport on 13 November 1860 after being fitted, in search of *Hero* with the Prince of Wales on board returning from his visit to Canada.

Dimensions & tons: 300ft 0in, 264ft 2in x 52ft 0in oa (51ft 6in for tonnage) x 19ft 10in. 3,726⁷/₉₄bm. 5,493 and 5,385 disp.

Men: 560–600. Guns: MD 28 x 10in (84cwt/9ft 4in) MLSB shell guns (broadside); UD 12 x 68pdr MLSB (95cwt/10ft) (pivot-mounted). Orlando's ports were altered to allow her to carry 50 guns, with 38 x 8in (65cwt/9in) in lieu of 28 x 10in. *Mersey*

This watercolour portrait of
Narcissus is dated 1879 (and so is at
the end of the ship's career), but the
presence of an admiral's flag at the
mizzen suggests that it is meant to
represent the ship in her glory days
as flagship of the Detached or Flying
Squadron, for which she was most
famous. *(NMM PW5708)*

re-armed 1865 with 4 x 7in MLR replacing 8 of the 10in MLSB
on the MD, and had 12 x 7in BLR on UD in lieu of the 68pdr
MLSB. In 1868 *Orlando* was given MD 4 x 7in MLR and 30 x
64pdr MLR, UD 12 x 64pdr MLR.

Machinery: 2-cylinder (92in diameter, 4ft stroke) horizontal single
expansion (trunk in *Orlando*). Rectangular boilers. Single screw.
1,000nhp (see below for individual trial ihp/speed).

Orlando Pembroke Dyd/John Penn & Son (3,617ihp = 13kts)
As built: 300ft 0½in, 264ft 3¼in x 52ft 1in (51ft 7in for tons) x 19ft
10in. 3,740³¹⁄₉₄bm. Draught 12ft 3in/16ft 7in. 5,493 disp.
Ord: 31.3.1855. K: 11.1856. L: 12.6.1858. C: at Devonport (for
Steam Reserve).
First cost: £187,042 (hull £86,447, machinery £65,234, fitting
machinery £35,361).
Commissioned 17.12.1861 under Capt. Francis Scott, for North
America; on 1.4.1862 under Capt. George Granville Randolph,
on same station; to the Mediterranean 1863; on 16.5.1865 under
Capt. John Bourmaster Dickson, in the Mediterranean; paid off
3.1.1866 at Plymouth. Sold 15.6.1871 to C. Marshall for BU.

Mersey Chatham Dyd/Maudslay, Sons & Field (3,691ihp = 12.6kts)
As built: 300ft 9½in, 264ft 7¼in x 52ft 0in (51ft 6in for tons) x 19ft
11in. 3,732⁹⁄₉₄bm. Draught 12ft 3in/17ft 0in. 5,643 disp.
Ord: 31.3.1855. K: 26.12.1856. L: 13.8.1858. C: 21.4.1859 (for sea) at
Portsmouth.
First cost: £173,898 (hull £72,922, machinery £65,642, fitting
£35,334).
Commissioned 5.3.1859 under Capt. Henry Caldwell, for the
Channel squadron; in 1860 to North America and the West
Indies; paid off 13.8.1862 at Portsmouth. Recommissioned
1.7.1867 under Capt. John Corbett, as flag of Rear-Adm. Claude
Henry Mason Buckle, at Queenstown; on 20.9.1867 under
Capt. Richard Dunning White, then 4.1.1869 under Capt. John
Seccombe, in same role; cruise of the Reserve Fleet in 5.1869;
on 4.4.1870 under Capt. Alan Henry Gardner, still flagship

at Queenstown; paid off 15.8.1872 at Plymouth (replaced at
Queenstown by *Revenge*). Sold 23.1.1875 to Castle for BU at
Charlton.

NARCISSUS Frigate, 51 guns. Modification of Eyde & Watts design,
1857. First ordered as an 18-gun sailing corvette in 1846, re-ordered as
a sailing frigate in 1848, and laid down November 1849 (see Section C)
– on slip no. 2 at Devonport. Re-ordered, 'carefully taken down with
a view of the materials being used in the construction of a similarly
formed vessel adapted for the screw by being lengthened', and re-laid
as a screw frigate on slip no.3 in 1857.

The *Narcissus* served as flagship of the Detached (or Flying)
Squadron formed in October 1874 at Gibraltar for a world-wide
deployment to any area as needs arose. The six screw frigates
comprising the Squadron were the *Narcissus* (by then reduced to 28
guns), the *Immortalité* (also 28), *Topaze* (28), *Newcastle* (28), *Doris* (24)
and the iron screw *Raleigh* (22).

Dimensions & tons: (261ft 0in oa) 228ft 0in, 194ft 6in x 51ft 3in oa
(50ft 9in for tonnage) x 18ft 2in. 2,635bm.
Men: 540. Guns: MD 8 x 8in (65cwt/9ft) shell + 22 x 32pdrs
(56cwt/9½ft) MLSB; UD 2 x 8in (65cwt/9ft) shell + 18 x 32pdrs
(45cwt/8½ft) MLSB + 1 x 68pdr (95cwt/10ft) MLSB – on
pivoted slide. By completion the UD carried 8 x 40pdr + 1 x
7in/110pdr Armstrong BLs, and the MD 8 x 8in (65cwt/9ft) shell
+ 18 x 110pdrs x for a total of 35 guns by 1860; in 1865, 12 x
64pdr MLR guns replaced the MD 110pdrs.
Machinery: 2-cylinder (64in diameter, 3ft stroke) horizontal single
expansion. Single screw. 400nhp. 1,731ihp = 10.6kts.
First cost: Hull £26,924 as sailing ship.

Narcissus Devonport Dyd/Ravenhill, Salkeld & Co.
As built: 223ft 0in, 194ft 6in x 51ft 3in oa (50ft 9in for tonnage) x
18ft 2in. 2,664⁵⁷⁄₉₄bm. 3,548 displacement. Draught 20ft 0in (fwd),
23ft 9in (aft).
Ord: 23.3.1857. K (re-laid): 4.1857. L: 26.10.1859. C: 23.2.1861.

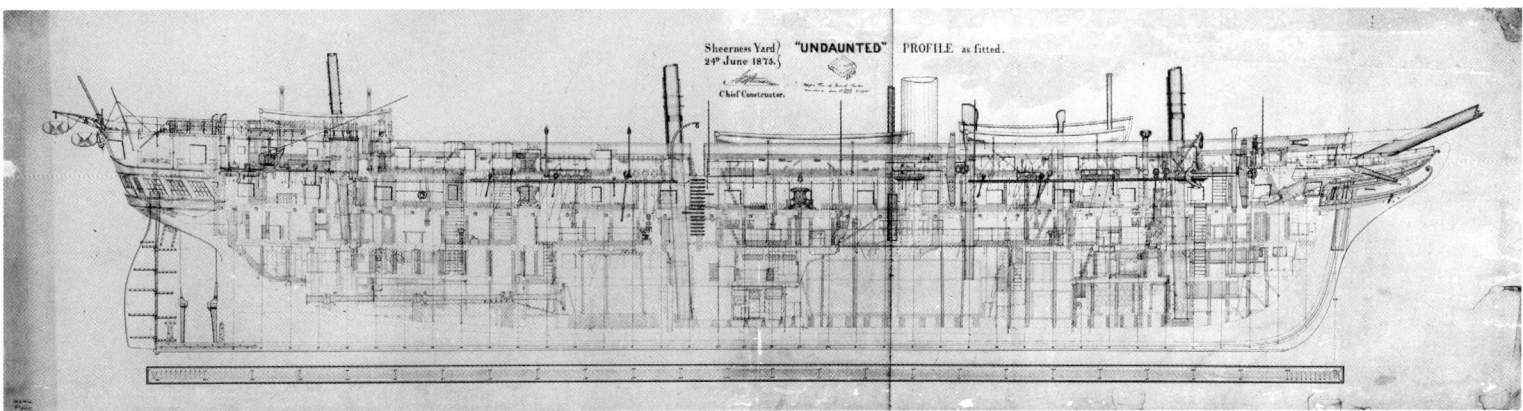

This highly detailed profile draught of the *Bristol* Class frigate *Undaunted* was made at Sheerness and is dated 24 June 1875. The draught shows in great detail how the ship was fitted out for her first commission (having spent fourteen years in Reserve). She was the last wooden frigate to serve as a flagship on a distant station – coming home from the Far East at the close of 1878 – and for this role in the East Indies a poop cabin had been added, but in most respects the internal layout and general arrangements would be common to most of the big screw frigates of this generation. (*NMM DR7026*)

Commissioned 20.12.1860 at Plymouth under Capt. Richard Hawkins Risk, as flagship of Rear-Adm. Baldwin Wake Walker, for Cape of Good Hope; on 27.1.1862 under Capt. Joseph Grant Bickford, still Walker's flagship on this station; paid off 6.8.1864 at Plymouth. Recommissioned 3.4.1865 under Capt. Colin Andrew Campbell, as flagship of Rear-Adm. Charles Gilbert John Brydone Elliot, on east coast of South America (to replace *Bombay*); on 18.5.1866 under Capt. John Crawford Wilson, as flagship of Rear-Adm. George Ramsay, on same station; paid off 17.7.1869 at Plymouth, reduced to 28 guns. Recommissioned 1.12.1870 under Capt. William Codrington, as flagship of Detached or Flying Squadron; paid off 22.5.1872 at Plymouth. Recommissioned 9.10.1872 under Capt. John Ommanney Hopkins, with the Detached Squadron, as flagship of Rear-Adm. Frederick Archibald Campbell, then 5.12.1873 flagship of Rear-Adm. Sir George Granville Randolph; on 4.7.1874 under Capt. Nathaniel Bowden-Smith, with Detached Squadron, as flagship of Rear-Adm. Rowley Lambert 31.5.1875; on 9.6.1875 under Capt. Lord Charles Thomas Montague Douglas Scott, still flagship with Detached Squadron; paid off 22.5.1877 at Plymouth. Commissioned 20.7.1877 under Capt. Henry Duncan Grant, as guard ship at Greenock; paid off 8.5.1878 at Plymouth. Reduced to 18 guns by 1880. Sold to Castle 17.2.1883 to BU at Charlton.

Post-Crimean War frigate plans were extremely confused and intentions were changed several times. Theory and experience having shown that ships designed for high speed need fine lines, Walker advanced plans on 18 December 1857 for wholesale lengthening of frigate (and other warship) designs, applying them also to ships then building that were not too far advanced. Two variants of 51-gun frigate designs were approved: a 2,651-ton, 600nhp design similar to the *Liffey* class (the *Liverpool*, *Glasgow* and *Newcastle* of the 1855 Programme, and the *Bristol* of the 1856 Programme) and a 2,355-ton, 400nhp design similar to the *Euryalus* class (the *Mersey* and *Tweed* of the 1855 Programme).

In 1857 the order for the *Newcastle* was altered to the smaller design, and in 1858 the other 600nhp ships (plus the *Ister* of the 1857 Programme) were altered to a 3,353-ton design, with the *Undaunted* and *Dryad* of the 1858 Programme also ordered to this design.

However in the 1859 Programme this 3,353-ton design was cut back to the original idea of a 3,027-ton ship, and the *Glasgow*, *Newcastle*, *Tweed*, *Bristol*, *Ister*, *Undaunted* and *Dryad* were all to be to this design (as was the newly ordered *Belvidera* from that year's fresh orders), while the *Liverpool* was now to be to the 2,651-ton design. The 1860 Programme saw a further change, with a 2,478-ton, 500nhp design being approved for the *Ister* (and for four sister ships newly ordered that year); another four to the 3,027-ton design were also ordered in 1860, and a final three in 1861, but almost all of these vessels were cancelled in 1863–64 as the Admiralty concentrated funding on its new ironclad shipbuilding programmes.

***BRISTOL* Class** Frigates, 51 (later 26/28) guns. Surveyor's Department (?Isaac Watts) design, approved 12 August 1858. Altogether fourteen ships from the 1855–61 Programmes were finally ordered to this design (another, the *Ister*, was also at one stage to this design but became the basis of a separate Class). The *Tweed* – and in February 1864 *Belvidera* – were proposed to be fitted as ironclads before their final cancellation; their frames were to be taken down by AO of 16 December 1864, which also applied to *Dryad*. In the completed ships, all cylinders were each 76in diameter (except 72¼in in *Bristol*), 3½ft stroke.

> Dimensions & tons: 250ft 0in, 214ft 7¼in x 52ft 0in oa (51ft 6in for tonnage) x 19ft 2in. 3,027⁴⁵⁄₉₄bm. 3,600 displacement.
> Men: 550. Guns: (original) MD 30 x 8in (65cwt) MLSB shell; UD 20 x 32pdrs (56cwt) MLSB + 1 x 68pdr (95cwt) MLSB on pivot (slide). In January 1868 had MD 4 x 110pdr/7in MLR + 18 x 64pdrs; UD 6 x 64pdrs (except *Bristol* which had MD 10 x 8in shell SB + 12 x 64pdrs; UD 4 x 64pdrs).
> Machinery: 2-cylinder horizontal single expansion. Single (hoisting) screw. 600nhp. See below for ihp/speed.

Newcastle Deptford Dyd/Ravenhill,Salkeld & Co. (2,354ihp = 12.416kts)
> As built: 250ft 0in, 215ft 1¼in x 52ft 0in oa (51ft 6in for tonnage) x 19ft 2in. 3,034¹⁹⁄₉₄bm. 4,020 disp. Draught 12ft 9in/18ft 6in.
> Ord: 31.3.1855. K: 6.12.1858. L: 16.10.1860. C. 1860 (for Reserve) at Sheerness; 9.1874 (for sea).
> Commissioned 21.9.1874 at Sheerness under Capt. Robert Gordon Douglas; in 1875 to the Flying Squadron; paid off 1.6.1877 at Sheerness. Recommissioned 22.7.1879 under Capt. Edward Kelly, as a training ship; paid off at Sheerness 7.9.1880 (having been the last wooden masted frigate in service). Loaned to the War Dept as a powder hulk at Devonport 1889; to Naval Ordnance Dept 8.1914. Sold 1929.

Glasgow Portsmouth Dyd/Ravenhill, Salkeld & Co. (2,020ihp = 11.548kts)

As built: 250ft 1in, 214ft 7⅛in x 52ft 1in oa (51ft 7in for tonnage) x
19ft 2in. 3,037⁷⁷⁄₉₄bm. 4,020 disp. Draught 12ft 9in/18ft 7in.
Ord: 31.3.1855. K: 12.9.1859. L: 28.3.1861. C: 1861 (for Reserve);
5.1871 (for sea).
Commissioned 24.5.1871 under Capt. Theodore Morton Jones,
as flagship of Rear-Adm. James Horsford Cockburn (died
10.2.1872) for the East Indies, then of Rear-Adm. Arthur
Cumming in same role; paid off 20.7.1875 at Portsmouth. Sold
12.1884 to BU.

Tweed Pembroke Dyd /—
Ord: 31.3.1855. K: 3.7.1860. L: – . Cancelled 16.12.1864.
First cost: £19,392 before cancellation.

Bristol Woolwich Dyd/Robert Napier & Sons (2,088ihp = 11.271kts)
As built: 250ft 0in, 214ft 7in x 52ft 0in oa (51ft 6in for tonnage) x
18ft 8in. 3,027²⁶⁄₉₄bm. 4,020 disp. Draught 13ft 5in/18ft 10in.
Ord: 9.4.1856. K: 16.9.1859. L: 12.2.1861. C: 10.1865 at Chatham.
Commissioned 4.10.1865 under Capt. Leveson Eliot Henry Somerset,
as flagship of Commodore Geoffrey Thomas Phipps Hornby,
for the west coast of Africa; paid off 25.1.1868 at Sheerness.
Recommissioned 17.2.1868 under Capt. Frederick William
Wilson, as training ship for naval cadets; to Flying Squadron 1869
(to 2.8.1869); paid off 27.1.1871 at Portsmouth. Recommissioned
28.1.1871 under Capt. Walter Cecil Carpenter, still as naval cadet
training ship; on 23.11.1871 under Capt. Philip Saumarez; paid off
15.12.1871 at Portsmouth. Sold to Castle 7.1883 to BU.

Undaunted Chatham Dyd/Ravenhill, Salkeld & Co. (2,261ihp =
11.837kts)
As built: 250ft 0in, 214ft 9¾in x 52ft 1in oa (51ft 7in for tonnage) x
18ft 10in. 3,040³¹⁄₉₄bm. 4,094 disp. Draught 12ft 6in/18ft 10in.
Ord: 27.3.1858. K: 28.5.1859. L: 1.1.1861. C: 16.7.1861 (for Reserve)
at Sheerness (after fitting machinery at Victoria Docks).
Commissioned 2.3.1875 at Sheerness under Capt. Hugh Campbell,
as flagship of Rear-Adm. Reginald John James George
Macdonald, for the East Indies; later in 1875 under Capt. Harry
Woodfall Brent, then under Capt. Nathaniel Bowden-Smith, still
Macdonald's flagship, then 2.4.1877 flagship of Rear-Adm. John
Corbett on same station; in 10.1878 under Capt. John D'Arcy;
paid off 21.12.1878 at Sheerness (having been the very last
wooden screw frigate to serve as flagship on an overseas station).
Sold to Castle 7.11.1882 to BU at Charlton.

Dryad Portsmouth Dyd/Robert Napier & Sons.
Ord: 27.3.1858. K: 2.1.1860. L: – . Cancelled 16.12.1864 (completed
taking down frames on 11.3.1865).
First cost: £15,339 before cancellation.

Belvidera Chatham Dyd/—
Ord: 8.4.1859. K: 30.4.1860. L: – . Cancelled 16.12.1864.
First cost: £18,945 before cancellation.

Pomone Chatham Dyd/—
Ord: 5.3.1860. K: 10.9.1860. Cancelled 12.12.1863.

Raleigh Portsmouth Dyd/—
Ord: 5.3.1860. K: — . Cancelled 12.12.1863 (unstarted).

Briton Portsmouth Dyd/—
Ord: 5.3.1860. K: — . Cancelled 12.12.1863 (unstarted).

Barham Portsmouth Dyd/—
Ord: 5.3.1860. K: — . Cancelled 12.12.1863 (unstarted).

Boadicea Chatham Dyd/—
Ord 2.1861. K: — . Cancelled 12.12.1863 (unstarted).

Bucephalus Portsmouth Dyd/—
Ord 2.1861. K: — . Cancelled 12.12.1863 (unstarted).

Dextrous Pembroke Dyd/—
Ord 2.1861. K: —. Cancelled 12.12.1863 (unstarted).

Modified *BRISTOL* Class Frigates, 51 guns. Stretched design of
3,353bm (lengthened to 270ft), approved 12 August 1858. Most of the
Bristol Class were initially ordered to this longer design of 3,353 tons
(and the vessels of earlier programmes retrospectively re-ordered to
this design), but all were reinstated at 3,027bm in 1859. The 1861
orders were probably also to the larger design, but in the event work on these
was never begun.

***ISTER* Class** Frigates, 21 (later 22) guns. Surveyor's Department
(?Isaac Watts) design, approved 23 February 1860. Originally ordered
as 51-gun (First Class) frigates. *Endymion* was the last wooden frigate
to be built; work on her was suspended in 1862, but recommenced in
1863; the planned gun establishment was reduced, first to 36, then to
30, and finally to 21-gun on completion. *Ister*'s engines were retained
and supplied to the corvette *Thalia* in 1869. The final two were not
laid down – the 'K' dates below are for the start of conversion of their
frames.
Dimensions & tons: 240ft 0in, 207ft 11¼in x 47ft 10in oa (47ft 4in for
tonnage) x 17ft 3in. 2,478³⁵⁄₉₄bm. 3,147 disp.
Men: 450. Guns: (30 gun design) 8 x 64pdr/6in MLR + 22 x 32pdrs
MLSB. (21 guns, *Endymion* as completed: MD 4 x 110pdr/7in
MLR + 14 x 8in MLSB (6½ ton); UD 3 x 110pdr/7in (82cwt)
BLR. In 1874 she carried MD 4 x 7in + 14 x 64pdr/6in; UD
4 x 64pdrs (2 side, 2 chase); by 1.1878 the 7in guns had been
removed.)
Machinery: 2-cylinder (65½in diameter, 3ft stroke) horizontal single
expansion. Single screw. 500nhp. *Endymion* 1,620ihp = 11¼kts.

Ister Devonport Dyd/Robert Napier & Sons.
Ord: 1.4.1857. K: 8.11.1860. L: – . Cancelled by AO 16.12.1864
(completed taking down frames 29.4.1865).
First cost: £14,603 before cancellation.

Endymion Deptford Dyd/Robert Napier & Sons.
As built: 240ft 0in, 207ft 11in x 47ft 11in oa (47ft 5in for tonnage) x
17ft 3in. 2,486⁴⁹⁄₉₄bm/3,197 disp.
Ord: 5.3.1860. K: 26.10.1860. L: 18.11.1865. C: 9.1866.
Commissioned 27.9.1866 at Sheerness under Capt. Charles Wake,
and sailed 28.10.1866 for the Mediterranean; engines repaired
at Malta 11.1867–3.1868; paid off 20.5.1869 at Portsmouth.
Recommissioned 21.5.1869 under Capt. Edward Lacy, for the
Detached (Flying) Squadron; paid off 30.11.1870 at Portsmouth.
Refitted there and re-masted (with heavier masts ex-*Mersey*)
11–12.1971. Recommissioned 24.4.1872 under Capt. Edward
Madden, as training ship for cadets in the Mediterranean; in
1.1873 to the Detached Squadron; on 18.9.1873 under Capt.
Henry Dennis Hinkley; paid off 31.7.1874 at Sheerness.
Recommissioned 18.9.1874 for coastguard on the Humber
(at Hull), still under Hinkley's command; on 29.9.1876 under
Capt. John Moresby, then 6.3.1878 under Capt. Henry Bedford
Woollcombe, in same role; paid off 31.7.1879 at Chatham
(replaced on the Humber by *Audacious*). Lent to Metropolitan
Asylums Board 5.7.1881 as an administration and hospital ship
(for smallpox) at Greenwich, later at Dartford. Sold as a hospital
hulk in 1885 and BU 1905.

Dartmouth Woolwich Dyd/—
Ord: 5.3.1860. K: 6.11.1860. L: — . Cancelled by AO 16.12.1864
(frame completed dismantling 28.2.1865).
First cost: £10,524 before cancellation.

Blonde Woolwich Dyd/—
Ord: 5.3.1860. K: 10.9.1860. L: — . Cancelled by AO 12.12.1863.
First cost: £3,600 before cancellation.

Astraea Devonport Dyd/—

The 30-gun *Ister* Class were the last wooden screw frigates to be designed before the introduction of the ironclad forced a complete reconsideration of cruiser design. Of the five ships ordered, only the *Endymion* – seen here in Australian waters – was completed, with 21 guns, to become the last wooden frigate to be built for the Royal Navy. *(Allan C. Green, via the State Library of Victoria)*

Ord: 5.3.1860. K: 21.10.1861. L: — . Cancelled by AO 12.12.1863. First cost: £250 before cancellation.

Converted Group (8 ships) 51 guns. On 4 March 1859 four of the 50-gun sailing Fourth Rate frigates were ordered to be converted to screw propulsion, as First Class screw frigates. On 9 July a fifth frigate (*Octavia*) was to be similarly converted, and this order was repeated on 4 January 1860, when three more frigates were added to the list to be converted. Each one (original designs varied) was formerly about 180ft in overall (GD) length, and they were each 'drawn asunder' in dock to lengthen them to accommodate boilers, engines and bunkerage. All were given 500hp engines except for *Phaeton* and *Leander*, which were 400hp; *Constance* was the first RN vessel driven by a compound engine (Woolf's principle), with six cylinders in triplets with a high-pressure cylinder (60in diameter; 60 lb/sq.in pressure) sited between two low-pressure ones (78in diameter; 32½ lb/sq.in; 3¼ft stroke).

The 1860 reclassifications confirmed that all First Class frigates (with complements of over 400 and up to 600 men), from the *Imperieuse* onwards, would be included in the Fourth Rate.

    Men: 525. Guns: Initially planned to carry UD 30 x 8in (65cwt/9ft) shell; QD/Fc 1 x 69pdr (95cwt) on pivot + 20 x 32pdrs (56cwt). As built UD (now 'Main deck') 8 x 8in (65cwt/9ft) shell + 22 x 32pdrs (56cwt/9½ft); QD/Fc (now 'Upper deck') 20 x 32pdrs (56cwt/8½ft) + 1 x 68pdr (95cwt/10ft) on pivot; all subsequently reduced to 35 guns by about 1860 and to 28 guns by 1868 – MD 18 x 64pdrs (64cwt) + 4 x 7in (6½ton/11ft) shell; UD 6 x 64pdrs (64cwt).

    Machinery: 2-cylinder horizontal single expansion (except *Constance*, see above). Single screw. 400 or 500hp (see below). See below for individual ihp/speed.

Phaeton Sheerness Dyd/James Watt & Co.

    Dimensions & tons: 220ft 3⅛in, 185ft 8⅛in x 49ft 10in oa (49ft 3in for tonnage) x 16ft 8½in. 2,396¹²⁄₉₄bm. 3,099 displacement. Draught 18ft 9in (fwd), 21ft 0in (aft).

    Machinery: Cylinders 64in diameter, 3ft stroke. 400nhp. 1,566ihp = 10.466kts.

    Conversion ord: 4.3.1859. Begun: 15.4.1859. After section launched 26.5.1859 to lengthen the ship. L: (undocked) 12.12.1859.

    Recommissioned 1.11.1861 at Chatham under Capt. Edward

Tatham, for North America and the West Indies; on 24.8.1863 under Capt. George Le Geyt Bowyear, in same role; paid off 24.3.1865 at Chatham. BU from 24.6.1875 at Chatham.

Severn Chatham Dyd/Maudslay, Sons & Field. (after a plan by Lang)

    Dimensions & tons: 240ft 6in, 212ft 3¼in x 50ft 1in oa (49ft 6in for tonnage) x 17ft 6in. 2,767bm. 3,536 displacement. Draught 21ft 0in (fwd), 22ft 9in (aft).

    Machinery: Cylinders 66in diameter, 3½ft stroke. 500nhp. 2,092ihp = 11.696kts.

    Conversion ord: 4.3.1859. Begun: 5.3.1859. L: (undocked) 8.2.1860. Completed fitting machinery at Sheerness Dyd 21.8.1860 and put straight into reserve.

    Commissioned 19.7.1862 at Plymouth under Commodore Frederick Byng Montresor, for the East Indies (as Senior Officer); paid off 12.6.1866 at Sheerness. BU completed 8.4.1876.

Sutlej Portsmouth Dyd/Maudslay, Sons & Field.

    Dimensions & tons: 254ft 6in, 220ft 2⅛in x 51ft 8in oa (51ft 2in for tonnage) x 17ft 2in. 3,065bm. 3,826 displacement. Draught 21ft 3in (fwd), 23ft 5in (aft).

    Machinery: Cylinders 66in diameter, 3½ft stroke. 500nhp. 2,270ihp = 11.807kts.

    Conversion ord: 4.3.1859. L: (undocked) 26.3.1860.

    Commissioned 18.9.1862 at Portsmouth under Capt. Matthew Connolly, as flagship of Rear-Adm. John Kingcome, for the Pacific; on 3.1.1864 under Capt. Trevenen Penrose Coode, as flagship of Rear-Adm. Joseph Denman from 5.1864, on same station; paid off 1.8.1867 at Portsmouth. BU from 22.2.1869 at Portsmouth.

Phoebe Devonport Dyd/Robert Napier & Sons. (after a plan by Peakes)

    Dimensions & tons: 240ft 6in, 206ft 11¼in x 51ft 9½in oa (51ft 3½in for tonnage) x 17ft 2in. 2,895⁷⁷⁄₉₄bm. 3,584 displacement. Draught 16ft 7in (fwd), 19ft 4in (aft).

    Machinery: Cylinders 65¼in diameter, 3ft stroke. 500nhp. 1,780ihp = 11.925kts.

    Conversion ord: 4.3.1859. Begun: 19.5.1959. After section launched 7.6.1859 to lengthen the ship. L: (undocked) 10.4.1860. C: 3.1861.

    Commissioned 23.9.1862 at Plymouth under Capt. Thomas Dyke Acland Fortescue (died 10.5.1865), for the Mediterranean;

Concern over the numbers of steam frigates in French service led the Admiralty in 1859 to instigate a programme to convert twelve or thirteen existing First Class sailing frigates to steam. The First Lord of the Admiralty (in the Earl of Derby's Cabinet), Sir John Pakington, intended there to be five new frigates afloat by the year's end, and another four conversions to be re-launched. Only four were eventually ordered in March 1859, with a further four ordered under Palmerston's government which took office in June, as the need for ironclad capital ships took precedence. With speeds around 11 knots or less under steam, they were already outclassed and these conversions generally saw only one active commission, and often on a remote station, before they were relegated to harbour service. The *Severn*, photographed here with the crew manning the yards, went to the East Indies as senior officer's ship between 1862 and 1866, but was otherwise unemployed until broken up in 1876. *(NMM 5333)*

on 19.5.1865 under Capt. William Rae Rolland, in the Mediterranean; paid off 29.6.1866 at Plymouth. Sold to Castle 6.4.1876 to BU.

Octavia Portsmouth Dyd/Maudslay, Sons & Field.
  Dimensions & tons: 252ft 6in, 217ft 0¼in x 52ft 10in oa (52ft 4in for tonnage) x 17ft 1in. 3,161⁵¹/₉₄bm. 3,832 displacement. Draught 20ft 10in (fwd), 23ft 10in (aft).
  Machinery: Cylinders 66in diameter, 3½ft stroke. 500nhp. 2,415ihp = 11.538kts.
  Conversion ord: 9.7.1859. Begun: 26.3.1860. After section launched 3.5.1860 to lengthen the ship. L: (undocked) 11.4.1861.
  Commissioned 14.6.1865 under Capt. Charles Farrell Hillyar, for East Indies and Cape of Good Hope; on 26.9.1865 Hillyar became Commodore as C-in-C this station; on 29.7.1867 under Commodore Leopold George Heath, in same role; on 16.9.1867 under Capt. Colin Andrew Campbell, still flagship of Heath in East Indies, then 12.5.1868 under Capt. Sidmouth De Ros Hall, with same flag; on 1869 under Capt. John Hobhouse Inglis Alexander, for voyage home; paid off 21.7.1869 at Portsmouth. BU completed 17.2.1876.

Leander Sheerness Dyd/James Watt & Co.
  Dimensions & tons: 241ft 0in, 205ft 7⅛in x 50ft 10¼in oa (50ft 2¼in for tonnage) x 16ft 0in. 2,759⁶⁹/₉₄bm. 3,539 displacement. Draught 18ft 0in (fwd), 23ft 4in (aft).
  Machinery: Cylinders 64in diameter., 3ft stroke. 400nhp. 1,568ihp = 9.703kts.
  Conversion ord: 4.1.1860. Begun: 1.1.1860. L: (undocked) 16.2.1861.
  Recommissioned 23.5.1863 at Sheerness under Commodore Thomas Harvey, for the Pacific (southern division); on 1.1.1866 under Commodore Michael De Courcy, on same station; on 16.6.1866 under Capt. William Montagu Dowell, for voyage home; paid off

17.11.1866 at Sheerness. Sold to Castle & Beech 4.1867 to BU at Charlton.

Arethusa Chatham Dyd/John Penn & Son.
  Dimensions & tonnage: 252ft 4in, 217ft 1in x 52ft 8in oa (52ft 2in for tonnage) x 17ft 1in. 3,142³¹/₉₄bm. 3,708 displacement. Draught 20ft 8in (fwd), 23ft 6in (aft).
  Machinery: Trunked engine, with surface condensers. Cylinders 80in diameter, 3½ft stroke. 500nhp. 3,165ihp = 11.704kts.
  Conversion ord: 4.1.1860. Begun: 23.4.1860. After section launched 2.6.1860 to lengthen the ship. L: (undocked) 9.8.1861.
  Recommissioned 10.6.1865 at Sheerness under Capt. Reginald John James George Macdonald, for the Mediterranean; on 1.6.1867 under Capt. Robert Coote, in the Mediterranean; paid off 7.1.1869 at Sheerness. Lent to Shaftesbury Homes as training ship for destitute boys at Greenhithe 1874 (engines removed), as accommodation ship to supplement *Chichester*; replaced 1931 by new vessel, which took her name. Sold to Castle 2.8.1933 and BU at Charlton 1934.

Constance Devonport Dyd/Randolph, Elder & Co.
  Dimensions & tons: 253ft 11in, 219ft 2in x 53ft 0in oa (52ft 6in for tonnage) x 17ft 1in. 3,212bm. 3,786 displacement. Draught 21ft 1in (fwd), 23ft 7in (aft).
  Machinery: 6-cylinder trunked compound engine, with surface condensers. 500nhp. 2,301ihp = 10.779kts.
  Conversion ord: 4.1.1860. Begun: 5.5.1860. After section launched 12.7.1860 to lengthen the ship. L: (undocked) 15.4.1862.
  Recommissioned 10.6.1865 at Plymouth under Capt. Edward King Barnard, for North America and the West Indies; on 27.10.1867 under Capt. Hugh Talbot Burgoyne, on same station; paid off 24.12.1868 at Plymouth. Sold to Castle 23.1.1875 to BU at Charlton.

5 The Fifth Rates (frigates)

Until 1 January 1817, the Fifth Rate encompassed vessels (all single-decked ships, in the sense of having one main gun deck) with at least 32 but not more than 44 guns. Four years earlier (in January 1813) the established complements of all Fifth Rate (18pdr-armed) frigates had been increased to 320 men for a 38-gun frigate, 284 for a 36-gun frigate and 270 for a 32-gun frigate. In January 1817 these groups were re-classed as 46-gun, 42-gun and 36-gun respectively, and the divisions were altered such that ships of 46 or 48 guns were transferred from the Fourth Rate, and the lower limit for Fifth Rates was placed at 36-gun ships. The most numerous categories were the ships now rated at 42 or 46 guns. From 1856 this was amended to restrict the Fifth Rate to ships 'the complements of which are [not more than] 400, and not less than 300 men'.

(A) Vessels in service or on order at 1 January 1817

Following the reorganisation in January 1817, the Navy included in the Fifth Rate 9 ships rated at 48 guns, 62 rated at 46 guns, 4 rated at 44 guns and 49 rated at 42 guns. By now the 38-gun frigate had become the standard British frigate; in January 1817 the recalculation of armament meant that these became re-rated as 46-gun ships (the 36-gun frigates largely became re-rated at 42-gun), and construction of this type continued until the start of the 1830s, when the decision to adopt a uniform 32pdr main armament for all new frigates (and larger warships) resulted in a quantitative reduction to 36 guns. In 1839 the remaining 46-gun frigates (predominantly the *Leda* Class) were re-rated as 42-gun.

24pdr-armed type

The Admiralty was always reluctant to enlarge its frigates except when forced to do so, as its strategy depended upon maximising the numbers of ships of any type rather than producing a smaller number of large individual vessels. This strategy continued into the post-1815 period (and well into the twentieth century!), as British global operational requirements were for a large quantity of minimally adequate cruisers rather than the 'super-frigate' designs of smaller navies; the only exceptions were the conversions of surplus 74-gun ships into 50-gun 'heavy frigates' (see Chapter 4).

CAMBRIAN Design by Sir John Henslow, 1795. Re-rated as 48 guns in January 1817.
  Dimensions & tons: 154ft 0in, 128ft 5¼in x 41ft 0in x 14ft 0in. 1,148³⁹⁄₉₄bm.
  Men: 320. Guns: 28 x 24pdrs; QD 8 x 9pdrs + 6 x 32pdr carronades; Fc 4 x 9pdrs + 2 x 32pdr carronades. In 1799, 6 x 9pdrs were replaced by 32pdr carronades. In 1805, the 24pdrs were replaced by 18pdrs.

Cambrian George Parsons, Bursledon
  As built: 154ft 0in, 128ft 3½in x 41ft 3in x 14ft 0in. 1,161¹⁴⁄₉₄bm. Draught 10ft 9in/15ft 10in.
  Ord: 30.4.1795. K: 9.1795. L: 13.2.1797. C: 16.6.1797 at Portsmouth.
  First cost: £17,592 to build, plus £11,974 fitting.
  Commissioned 4.1797 under Capt. Sir Thomas Williams, on Irish station and the Channel; in 5.1797 under Capt. Arthur Legge (–1801); took (with *Indefatigable* and *Childers*) 12-gun privateer *Le Vengeur* in the Channel 4.1.1798; took further privateers in the Channel – 16-gun *Le César* 27.3.1798, 16-gun *Le Pont de Lidi* 30.3.1798, *La Revanche* 19.10.1798 and 14-gun *La Cantabré* 8.12.1798; took (with *Fisgard*) 14-gun *Le Dragon* in the Channel 5.5.1800; took 14-gun privateer *L'Audacieux* on passage to St Helena 6.4.1801. In 5.1802 under Capt. William Bentley, then Capt. John Beresford 10.1804; flagship of Adm. Mark Milbank 5.1802 and of Vice-Adm. Sir Andrew Mitchell 7.1802–3.1803; took privateers – 14-gun *Maria* 13.6.1805 and 10-gun *Matilda* 3.7.1805. Refitted at Portsmouth (for £13,483) 4–6.1807; recommissioned under Capt. Richard Vincent, for the Mediterranean. In 1810 under Capt. Francis Fane; took part in disastrous attack on anchored convoy at Palamos 13.12.1810, with Fane captured and numerous casualties. On 1.1.1811 under Capt. Charles Bullen; paid off 9.12.1811. Between Middling and Large Repair at Plymouth (for £33,728) 5.1813–9.1814, then laid up. Fitted for sea at Plymouth (for £7,310) 7–10.1820; recommissioned 7.1820 under Capt. Gawen William Hamilton, for the Mediterranean; at Battle of Navarino 20.10.1827, losing 1 killed and 1 wounded. Wrecked off Grabusa 31.1.1828 (following collision with *Isis* while in combat against Greek pirate schooners).

Ex-DUTCH PRIZES (1799) One of the largest Dutch frigates, captured in 1799; the slightly smaller *Amphitrite* (taken at the same time) had been BU in 1805.

Vleiter (Dutch *Mars*, originally two-decker *Zevenwalden*, 68 – built Harlingen 1784, raséed to a frigate and renamed 1795).
  Dimensions & tons: 155ft 8in, 127ft 1½in x 44ft 9½in x 16ft 9in. 1,356⁸¹⁄₉₄bm.
  Men: 185 as floating battery. Guns: UD 26 x 24pdrs; QD 10 x 18pdr carronades.
  Taken 30.8.1799 by Mitchell's squadron in the Texel.
  Fitted at Chatham as a floating battery (for £7,177) 7.1800–4.7.1801.
  Commissioned 7.1801 under Capt. William Birchall, as floating battery in the Medway; paid off 4.1802. Recommissioned 4.1803 under Capt. Adrian Renou, for the defence of the Thames; paid off 7.1804. Powder magazine at Chatham by 1806. Fitted as sheer hulk at Chatham 10.1808–1.1809 for Sheerness. BU at Sheerness 4.1817.

Ex-FRENCH PRIZE (1801)

La FORTE Class The survivor of a pair of large frigates designed by François Caro, which initially carried 30 x 24pdrs and 20 x 8pdrs in French service, and so were rated 50-gun frigates, but were only 44-gun by the time of their capture (with 10 x 8pdrs and 4 x 36pdr obusiers on the QD/Fc), and were similarly rated in the British Navy. The *Forte* was wrecked in 1801, while the *Egyptienne* has been re-classed as a receiving ship.

Egyptienne (French *L'Egyptienne*, built 7.1798–11.1799 at Toulon. L: 17.7.1799)

    Dimensions & tons: 169ft 8in, 141ft 4¾in x 43ft 8in (43ft 0in mld.) x 15ft 1in. 1,434¼₉₄bm.

    Men: 330. Guns: UD 28 x 24pdrs; QD 2 x 9pdrs + 12 x 32pdr carronades; Fc 2 x 9pdrs + 4 x 32pdr carronades.

    Taken 2.9.1801 at capture of Alexandria.

    Under Capt. Thomas Stephenson 1.1802, for voyage home from the Mediterranean. Arrived 13.2.1802 at Woolwich, fitted there (for £12,625) 10–12.1802.

    Commissioned 4.1803 under Capt. Charles Fleeming (–1805); took 16-gun *L'Epervier* 27.7.1803 and 14-gun privateer *La Chiffonette* 30.8.1803. Under Capt. Charles Elphinstone from 8.1805, then Capt. Charles Paget from 12.1805 (–1807); boats cut out privateer *L'Alcide* from Muros 8.3.1806; took 16-gun *L'Actéon* off Rochefort 2.10.1806; took (with *Loire*) 40-gun *La Libre* off Rochefort 24.12.1806; paid off into Ordinary 5.5.1807. Fitted as Receiving ship at Plymouth after 1806; in Ordinary there 1812–15+. Sold to John Small Sedger (for £2,810) at Plymouth to BU 30.4.1817.

18pdr-armed type

By 1817 virtually all surviving Fifth Rates carried a broadside armament of 18pdr guns (the last 12pdr-armed frigates were only in harbour use, generally as hulks, by this date).

***MINERVA* Class** 46 (originally 38) guns. The sole survivor of a class of four, whose design by Sir Edward Hunt was approved by the Admiralty on 6 November 1778; the *Minerva* had been BU in 1803, and *Arethusa* and *Thetis* in 1814.

    Dimensions & tons: 141ft 0in, 117ft 0⅜in x 38ft 10in x 13ft 9in. 938⁷²⁄₉₄bm.

    Men: 320 (280 in 1780). Guns: UD 28 x 18pdrs; QD 8 x 9pdrs, 6 x 18pdr carronades; Fc 2 x 9pdrs, 4 x 18pdr carronades.

Phaeton John Smallshaw, Liverpool

    As built: 141ft 0in, 116ft 5¼in x 39ft 0½in x 13ft 10¼in. 944⁷⁄₉₄bm.

    Ord: 3.3.1780. K: 6.1780. L: 12.6.1782. Fitted and coppered at Plymouth 11.1782–27.12.1782.

    First cost: £18,986.8.1d for building, plus £7,565.18.4d for fitting.

    Commissioned 3.1782; paid off within a year. Recommissioned 4.1783; sailed for the Mediterranean 8.10.1783; paid off 1786. Recommissioned 9.1786; sailed for the Mediterranean 17.12.1786; paid off 12.1788. Great Repair at Portsmouth (for £15,788) 12.1790–1.1793; recommissioned 12.1792 under Sir Andrew Douglas for Howe's fleet; took privateers – lugger *L'Aimable Liberté* in the Channel 3.1793, and 44-gun *Le Général du Mourier* 14.4.1793 – also retook Lima ship (Gell's squadron); took 20-gun *La Prompte* off the Spanish coast 28.5.1793; took (with sloop *Weasel*) privateers – 10-gun *Le Poisson Volant* and *Le Général Washington* – in the Channel 6.1793; took (with *Latona*) 38-gun *La Blonde* off Ushant 27.11.1793; paid off 2.1794. Recommissioned 3.1794 under Capt. William Bentinck; at Battle of Glorious First of June off Ushant 1.6.1794. In 9.1794 under Capt. Robert Stopford (–1799); in Royal escort for Princess Carolina of Brunswick 1795; in Cornwallis's 'Retreat' 16–17.6.1795; took 20-gun *La Bonne Citoyenne* off Cape Finisterre 10.3.1796; took privateers – 18-gun *L'Actif* in the Channel 6.3.1797, 6-gun *Le Chasseur* in the Channel 16.9.1797, 16-gun *L'Indien* 24.9.1797 off Les Roches Bonnes (with *Unite*), *Le Découverte* 7.10.1797 (with *Unite* and *Stag*) and 12-gun *Le Hasard* 28.12.1797; retook (with *Anson*) 20-gun *Daphne* in the Bay of Biscay 29.12.1797; took more privateers – *L'Aventure* 1.1.1798 and 18-gun *La Légère* in the Channel 19.2.1798; destroyed large frigate near Cordoban lighthouse 22.3.1798; took (with *Anson*) 18-gun privateer *Le Mercure* 31.8.1798 and 36-gun *La Flore* 6.9.1798; took 16-gun privateer *Le Lévrier* in the Channel 8.10.1798; took (with *Ambuscade* and *Stag*) 20-gun *L'Hirondelle* 20.11.1798; took privateers – 18-gun *La Résolue* 24.11.1798 and (with *Stag*) 10-gun *La Ressource* 6.12.1798. In 7.1799 under Capt. James Morris; sailed for Constantinople with Lord Elgin 4.9.1799; in the Mediterranean 1800–01; her boats took 14-gun polacca at Fuengirola 27.10.1800, and (with *Naiad*'s) took *El Reposo* and destroyed *La Alcudia*, both packets at Pontevedro 16.5.1801; returned to England and paid off 3.1802. Large Repair and fitted at Deptford (for £21,545) 12.1802–8.1803; recommissioned 7.1803 under Capt. George Cockburn, for the East Indies; in action (with *Harrier*) against 40-gun *La Sémillante* off Jacinto 2.8.1805. In 10.1806 under Capt. John Wood, then 7.1808 under Capt. Fleetwood Broughton Reynolds Pellew (–1812); at Nagasaki (Japan) 1808; at reduction of Java 8.1811; with *Sir Francis Drake* at Sumenap (Madura) 8.1811; home with East Indies convoy 8.1812. Large Repair at Deptford 10.1812–4.1814, then fitted for Ordinary at Sheerness. Fitted for sea at Sheerness 11.1815–1.1816; recommissioned 10.1816 under Capt. Francis Stanfell, for St Helena (–1817). In 14.4.1818 under Capt. William Henry Dillon, to East Indies; paid off 10.1819. Recommissioned 29.10.1819 under Capt. William Augustus Montagu, for Halifax; paid off 9.1822. Recommissioned same month under Capt. Henry Evelyn Pitfield Sturt; paid off at Portsmouth 17.10.1825. Sold 11.7.1827 to Mr Freake (for £3,430), but sale cancelled ('Mr Freake having been declared insane'); re-sold to Joshua Crystall (for £2,500) to BU 26.3.1828.

***PERSEVERANCE* Class** 42 (originally 36) guns. Built to an Edward Hunt design of 1779, as a slightly shorter 36-gun equivalent of his *Minerva* Class 38s. Of this quartet, the *Leda* was lost in 1796 and the *Phoenix* in 1816, while the *Perseverance* was re-classed as a receiving ship from 1800.

    Dimensions & tons: 137ft 0in, 113ft 5½in x 38ft 0in x 13ft 5in. 871bm.

    Men: 284 (270 in 1780). Guns: UD 26 x 18pdrs; QD 8 x 6pdrs (9pdrs from 1780), 4 x 18pdr carronades; Fc 2 x 6pdrs (9pdrs from 1780), 4 x 18pdr carronades.

Perseverance John Randall & Co., Rotherhithe

    As built: 137ft 0in, 113ft 4¼in x 38ft 3in x 13ft 5in. 882¼₉₄bm. Draught 9ft 10½in/15ft 9½in.

    Ord: 3.12.1779. K: 8.1780. L: 10.4.1781. Fitted and coppered at Deptford 11.4–3.6.1781.

    First cost: £11,544.15.2d for building (+£140.2.5d dockyard exp.), plus £9,743.1.11d for fitting.

    Commissioned 3.1781 under Capt. Skeffington Lutwidge; paid off 9.1783 after wartime service. Recommissioned 10.1787 under Capt. William Young; paid off 12.1787. Fitted at Portsmouth (for £2,096) 10–12.1788; recommissioned 10.1788 under Capt. Isaac Smith; sailed for the East Indies 11.2.1789; paid off 8/9.1793. Laid up in Ordinary at Portsmouth 6.1793–1.1800, then receiving ship there until 1822. Sold to Joshua Crystall to BU (for £2,530) on 21.5.1823.

Inconstant William Barnard, Deptford Green
> As built: 137ft 9in, 114ft 2in x 38ft 3½in x 13ft 5in. 890³¹/₉₄bm.
> Draught 9ft 6in/14ft 9in.
> Ord: 8.12.1781. K: 12.1782. L: 28.10.1783. Fitted (for Ordinary)
> at Deptford 28.10.1783–22.3.1784; to Woolwich 10.1788 where
> fitted for sea 6.1790–3.11.1790.
> First cost: £16,226.0.1d including fitting for Ordinary; fitting for sea
> (1790) £6,627.
> Commissioned 8.1790 under Capt. George Wilson; paid off 9.1791.
> Fitted at Woolwich (for £7,239) 1–2.1793; recommissioned
> 1.1793 under Capt. Augustus Montgomery for Howe's fleet;
> to the West Indies 4.1793; took 14-gun *Le Curieux* in the
> West Indies 3.6.1793; returned to England 7.1793; sailed for
> the Mediterranean 21.11.1793; with Hood's fleet at Toulon. In
> 1794 under Capt. George Cockburn (acting). In 1.1795 under
> Capt. Thomas Fremantle (–1797); in action against 80-gun *Ça
> Ira* off Genoa 10.3.1795; recapture of brig *Speedy* 25.3.1795;
> with Nelson's squadron off Vado 8.1795; took 24-gun *L'Unité*
> at Boma 20.4.1796; at evacuation of Leghorn 26.6.1796; paid
> off 9.1797. Fitted as 20-gun troopship at Woolwich (for
> £7,193) 3–6.1798; recommissioned 4.1798 under Cmdr Milham
> Ponsonby; paid off 10.1799. Refitted at Woolwich (for £4,787)
> 10.1799–3.1800; re-established 4.1798 with 16 x 9pdrs and 4 x
> 6pdrs; recommissioned 12.1799 under Cmdr John Ayscough,
> for the North Sea; in Quiberon operation 6.1800; in Egypt
> operations 1801. Recommissioned 10.1802 under Capt. Richard
> Byron. In 12.1802 under Capt. Edward Dickson (–1810); fitted
> as a troopship at Chatham 10–12.1803; at capture of Goree
> 3.1804; fitted at Portsmouth (as frigate) 12.1805–2.1806; flagship
> of Vice-Adm. Sir James Saumarez in the Channel Islands
> 1806–08; large Repair and fitted at Portsmouth 9.1808–10.1809;
> allowed 12 x 32pdr carronades (QD) 6.1809; recommissioned
> 6.1809 (still under Dickson); sailed for the Cape of Good Hope
> 27.12.1809; paid off 1810. Fitted at Portsmouth 9–12.1810;
> recommissioned 10.1810 under Capt. John Quilliam, for the
> North Sea; in 12.1810 under Capt. Edward Owen (–1812). In
> 3.1814 under Capt. Sir Edward Tucker, for South America; in
> 8.1815 flagship of Capt. Sir James Yeo; home 9.1817 and docked
> for repairs, but found to be in 'a very decayed and defective
> state', and paid off 23.10.1817. BU at Portsmouth 11.1817.

ADVENTURE Class A pre-war 44-gun two-decker design by
Hunt in 1782, to which eight ships were built. Four sisters were war
losses 1804–13, while the *Chichester* was BU in June 1815 and the
Adventure in September 1816. The remaining pair only just survived
the reclassification, both in ancillary roles, being taken to pieces in
February 1817.
> Dimensions & tons: 140ft 0in, 115ft 2½in x 38ft 3in x 16ft 10in.
> 896⁵¹/₉₄bm.
> Men: 300. Guns: LD 20 x 18pdrs; UD 22 x 12pdrs; Fc 2 x 6pdrs.

Expedition Randall & Co., Rotherhithe
> Ord: 5.6.1782. K: 10.1783. L: 29.10.1784.
> Troopship 4.1798. Ballast ship at Chatham 1810. BU at Chatham
> 2.1817.

Gorgon Perry & Hankey, Blackwall
> Ord: 19.6.1782. K: 12.1782. L: 27.1.1785.
> Troopship 1787–89. Storeship 7.1793. Floating battery 1805;
> hospital ship 5.1808, in 1808 under Cmdr Robert Brown Tom,
> then 1810 under Cmdr Charles Webb, then 1811 under Cmdr
> Alexander Milner and 15.10.1813 under Cmdr Richard Bowden;
> paid off 6.1815. BU at Portsmouth 2.1817.

The outbreak of war in February 1793 saw new orders placed for six
38-gun frigates of Henslow's new *Artois* Class and four 32-gun frigates
of his *Alcmene* Class. Six 36-gun frigates were ordered in May 1794,
four to a Henslow design and two to Rule's corresponding draught.
The subsequent programme for three more 38s in November 1794
(*Active*, *Amazon* and *Leda*) was deferred and the orders to proceed
not given until April 1796. In the meantime six extra frigates were
ordered on 4 February 1795 to be built from 'fir' (pitch pine), in three
pairs based on the 38-gun, 36-gun and 32-gun designs of the previous
two years, although the 32-gun ships were completed with a 12pdr
armament and so appear thereunder. The 1795 Programme provided
for eight more frigates, mainly of new design (all ordered 30.4.1795).
Besides the 24pdr-armed *Endymion* and *Cambrian* (see below), the
Admiralty ordered three to domestic designs (the *Acasta* and *Naiad*,
and a last unit of the *Artois* Class) and three copies of French designs,
notably all three (including a 36-gun type) being of *La Junon* Class of
Toulon-built ships designed by Joseph-Marie-Blaise Coulomb.

ARTOIS Class 46 (originally 38) guns. Design by Sir John Henslow,
approved 2 March 1793. Sole survivor of six ships were originally
ordered in 1793, and three more in 1795 (of which the *Tamer* and
Clyde were built of pitch pine); *Artois* was wrecked in 1797, *Jason* in
1798, and *Apollo* and *Ethalion* in 1799, while *Tamer* was BU in 1810,
Diamond in 1812 and *Clyde* in 1814, and *Diana* was sold to the new
Dutch Navy in 1815.
> Dimensions & tons: 146ft 0in, 121ft 7⅛in x 39ft 0in x 13ft 9in.
> 983⁷⁸/₉₄bm.
> Men: 270 (later 315). Guns: UD 28 x 18pdrs; QD 2 x 9pdrs + 12 x
> 32pdr carronades; Fc 2 x 9pdrs + 2 x 32pdr carronades.

Seahorse Marmaduke Stalkartt, Rotherhithe
> As built: 146ft 4in, 121ft 10in x 39ft 3in x 13ft 9½in. 998³⁴/₉₄bm.
> Draught 10ft 3in/15ft 7in.
> Ord: 14.2.1793. K: 3.1793. L: 11.6.1794. C: 16.6–16.9.1794 at
> Deptford Dyd.
> First cost: £21,630 including fitting.
> Commissioned 7.1794 under Capt. John Peyton, for the Irish
> station. In 1796 under Capt. George Oakes (–1797, but temp.
> under Capt. Robert Oliver in 6.1796); took (with *Cerberus*)
> 6-gun privateer *Calvados* off the Irish coast 7.1796; took (with
> *Diana* and *Cerberus*) privateer *L'Indemnité* 28.8.1796; took 16-
> gun *Princesa* off Corunna 16.9.1796; sailed for the Mediterranean
> 4.1.1797. In 7.1797 under Capt. Thomas Fremantle; in
> Nelson's attack on Santa Cruz 25.7.1797. In 10.1797 under
> Capt. Edward Foote (–1802); took (with *Melampus*) 18-gun
> privateer *La Belliqueuse* off the Irish coast 16.1.1798; sailed
> for the Mediterranean 8.3.1798; took 36-gun *La Sensible* off
> Malta 27.5.1798; with Hood's squadron off Alexandria 8.1798;
> destroyed (with *Emerald*) 6-gun *L'Anémone* at Damietta
> 2.9.1798; surrender of French garrisons at Naples 6.1799;
> grounded off Leghorn 29.7.1799, then returned to UK 10.1799;
> as flagship of Rear-Adm. Sir Richard Bickerton, sailed for the
> Mediterranean 23.5.1800; in the Mediterranean 1801, then East
> Indies in 1802; home and paid off 10.1802. Fitted at Portsmouth
> 4–6.1803; recommissioned 5.1803 under Capt. Courtnay Boyle;
> sailed for the Mediterranean 1803; her boats (with *Narcissus*'s
> and *Maidstone*'s) in action at Lavandon (Hyères) 11.7.1804;
> captured store brig at San Pedro 4.5.1805. In 7.1805 under
> Capt. Robert Corbett; on Jamaica station 1806. In 2.1806 under
> Capt. John Stewart (died 10.1811); sailed for the Mediterranean
> 30.4.1807; action with 52-gun *Badere-i-Zeffee* and 26-gun *Aziz
> Fezzan* off Chgiliodrmia 5.7.1807 (taking first named, sinking

This unusually detailed model gives a good impression of what any of the surviving older frigates would look like when commissioned after the Napoleonic War. It has been identified as a *Phoebe* Class (1795) 36-gun (later 42-gun) ship, in which case *Dryad* would be the strongest candidate as she was given a major repair in 1813–15, and later refitted for sea in 1826. The model does not have a circular stern as would in all probability have been fitted if this 18pdr frigate had been reconstructed later. *(NMM L3252-002)*

other); took 2-gun privateer *La Stella di Napoleon* 8.5.1810; paid off 6.1811. Between Small and Middling Repair at Woolwich 8–10.1812; recommissioned 9.1812 under Capt. Sir James Gordon (–1815); sailed for North America 11.1812; sank 16-gun privateer *La Subtile* off Beachy Head 13.11.1813; in Potomac operations 8.1814; laid up at Plymouth 9.1815. BU at Plymouth 7.1819.

PHOEBE (or Lengthened *INCONSTANT*) Class 42 (originally 36) guns. Design by Sir John Henslow, approved 29 May 1794, as lengthened versions of Hunt's *Inconstant* design of 1779. Four ships were ordered 24 May 1794, and were named on 26 February and established 16 March 1795; however, *Doris* was wrecked (and then burnt) in 1805 and *Caroline* was BU in 1815.

Dimensions & tons: 142ft 6in, 118ft 10½in x 38ft 0in x 13ft 5in. 913¹³/₉₄bm.

Men: 264. Guns: UD 26 x 18pdr; QD 8 x 9pdr + 6 x 32pdr carronades; Fc 2 x 9pdr + 4 x 32pdr carronades.

Dryad William Barnard, Deptford
As built: 142ft 8in, 119ft 0in x 38ft 2½in x 13ft 5in. 924⁶⁴/₉₄bm. Draught 10ft 3in/14ft 10in.
Ord: 24.5.1794. K: 6.1794. L: 4.6.1795. C: 5.6–16.6.1795 (at Woolwich) –15.8.1795 (at Deptford Dyd).
First cost: £14,027 to build, plus fitting £1,174 at Woolwich + £8,542 at Deptford.
Commissioned 6.1795 under Capt. Robert Forbes (died 7.12.1795), for Irish station. In 12.1795 under Lord Amelius Beauclerk (–1798); took (while under acting Cmdr John Pulling) 14-gun cutter *L'Abeille* off the Lizard 2.5.1796; took 40-gun *La Proserpine* off Cape Clear 13.6.1796; took privateers – 9-gun *Le Vantour* 16.10.1796, 14-gun *L'Eclaire* 19.8.1797, 16-gun *La Brune* 10.10.1797 (with *Doris*), and 16-gun *Le Mars* off Cape

Clear 4.2.1798; sank 42-gun privateer *La Cornélie* 9.9.1797. In 12.1798 under Capt. Charles Mansfield (–1801); took 14-gun privateer *Le Premier Consul* 5.3.1801. In 1802 under Capt. Robert Williams. Recommissioned 9.1802 under Capt. William Domett, then in 1803 under Capt. John Giffard as flagship of Adm. Lord (Alan) Gardner on the Irish station. In 1.1805 under Capt. Adam Drummond (–1807). In 1808 under Capt. William Cumby; took 12-gun privateer *Le Rennais* 22.3.1808. In 1809 under Capt. Edward Galway (–1813); in Scheldt operations 1809; destroyed 22-gun brig off Île d'Yue 23.12.1812; to Newfoundland 1813, then home to pay off. Large Repair at Sheerness (for £25,538 plus £2,396) 7.1813–9.1815. Roofed over waist 6.1819. Began fitting for sea at Sheerness (for £6,866) 9–10.1825; completed at Chatham (for £11,280) 4.1826. Very Small Repair at Plymouth (for £20,620) 8.1829–8.1830. Laid up at Portsmouth 8.1832. Fitted as receiving ship at Portsmouth 8.1837–3.1838. BU completed there 9.2.1860.

Phoebe John Dudman, Deptford

As built: 142ft 9in, 119ft 0in x 38ft 3in x 13ft 5½in. 926⁸⁶/₉₄bm. Draught 9ft 7in/15ft 0in.

Ord: 24.5.1794. K: 6.1794. L: 24.9.1795 (coppered by builder). C: 27.10–23.12.1795 at Deptford Dyd.

First cost: £15,791 to build, plus £9,118 fitting.

Commissioned 10.1795 under Capt. Robert Barlow (–1800), for Pellew's squadron off the Irish coast; took 16-gun *L'Atalante* off the Scillies 10.1.1797, and 36-gun *La Néréide* 22.12.1797; made good defects at Plymouth (for £9,095) 10–12.1798; took (with *Revolutionnaire*) 26-gun *Le Bourdelais* and 18-gun *Le Grand Ferrailleu* 11.10.1799; took privateers – 14-gun *La Belle Garde* 21.2.1800 and 22-gun *L'Heureux* 11.3.1800. In 1.1801 under Capt. Thomas Baker; took 40-gun *L'Africaine* in the Mediterranean 19.2.1801. Fitted at Plymouth (for £2,657) 6.1801; in 6.1802 under Capt. James Shephard. Recommissioned 9.1802 under Capt. Thomas Capel; sailed for the Mediterranean 28.9.1802; in the chase to the West Indies, then at Battle of Trafalgar 21.10.1805. Refitted at Portsmouth (for £8,920) 1–4.1806; in 1.1806 under Capt. James Oswald (–1808), in North Sea and Channel; sailed for the Mediterranean 18.1.1808. In 4.1809 under Capt. Hassard Stackpole, for the Baltic. In 8.1809 under Capt. James Hillyer (–1815). Between Small and Middling Repair and fitted for sea at Plymouth (for £18,427) 3–7.1810; sailed for the East Indies 18.7.1810; at reduction of Mauritius 12.1810; in Schomberg's Action off Madagascar 20.5.1811 (40-gun *La Renommée* taken); capture of *La Néréide* at Tamatave 24.5.1811; in Java operaions 9.1811; sailed with convoy for Quebec 9.4.1812; took US 14-gun privateer *Hunter* 23.12.1812; in search for 32-gun USS *Essex*-1813–14, ending in capture of Essex-at Valparaiso 28.3.1814. Paid off 1814 and laid up at Plymouth 8.1815. Fitted as receiving ship and slop ship at Plymouth (for £1,803) 1.1823–10.1826. Sold at Plymouth to Joshua Crystall (for £1,750) 27.5.1841.

Fortunee Perry & Co., Blackwall

As built: 142ft 8in, 119ft 0in x 38ft 1½in x 13ft 5in. 921 (920¼/₉₄ by calc.) bm. Draught 9ft 10in/15ft 3½in.

Ord: 28.1.1800. K: 4.1800. L: 17.11.1800. C: 14.1.1801 at Woolwich.

Commissioned 11.1800 under Capt. Lord Amelius Beauclerk, for the Channel.; took 12-gun privateer *La Mascarade* in the Channel 5.4.1801; took (with *Trent* and cutter *Dolphin*) 10-gun privateer *Le Renard* 20.4.1801. In 8.1802 under Capt. John Clements (acting); grounded near Texel but salved. Recommissioned 12.1802 under Capt. Henry Vansittart, for

the North Sea; sailed for the West Indies 2.2.1804; took 1-gun privateers on Jamaica station – *Le Cézar* 3.4.1804 and *Le Vautour* 10.8.1804; mistaken action against British merchantman *Leander* off Hayti 9.1804; took more privateers – 7-gun *Le Grand Juge Bertolio* 18.5.1806, and 2-gun *Le Magicien* in Home waters 20.11.1807; on Irish station 1808–11; tppk (with *Saldanha*) 18-gun privateer *Le Vice-Amiral Martin* 11.10.1811. In 6.1812 under Capt. George Seymour, still on Irish station; then 1.1813 Capt. Frederick Aylmer and 5.1813 Capt. William Goate; laid up at Portsmouth 12.1813 and paid off 1.1814. Sold there to Mr Freake (for £2,740) 29.1.1818.

AMAZON Class 42 (originally 36) guns. Designed by Sir William Rule, 1794. The first pair were ordered 24 May 1794, and were named on 26 February and established 16 March 1795. The second pair were built of 'fir' (pitch pine) by AO of 4 February 1795, and were named on 20 February 1795 (before the first pair!). Of this class, *Amazon* was wrecked in 1797 and *Glenmore* was sold in 1814, while *Emerald* and *Trent* survived but the last had been re-classed as a receiving ship.

Dimensions & tons: 143ft 0in, 119ft 6in x 38ft 2in x 13ft 6in. 925⁸⁷/₉₄bm.

Men: 264. Guns: 26 x 18pdrs; QD 8 x 9pdrs + 6 x 32pdr carronades; Fc 2 x 9pdrs (12pdrs in second pair) + 2 x 32pdr carronades.

Emerald Thomas Pitcher, Northfleet

As built: 143ft 2½in, 119ft 5½in x 38ft 4in x 13ft 6in. 933⁶/₉₄bm. Draught 11ft 2½in/14ft 6in.

Ord: 24.5.1794. K: 6.1794. L: 31.7.1795, then coppered by builder. C: 12.10.1795 at Woolwich.

First cost: £14,419 (+ fitting £9,390).

Commissioned 8.1795 under Capt. Velters Berkeley; sailed for the Mediterranean 7.1.1797; pursuit of *St Trinidad* 20.2.1797; took (with *Irrestistable*) Spanish 34-gun *Ninfa* and destroyed similar *Santa Helena* off Cadiz 26.4.1797. In 1797 under Capt. Thomas Waller (–1800); in attack on Santa Cruz 25.7.1797; in 12.1797 under Capt. Lord William Proby (temp., to ?4.1798); took 8-gun privateer *Le Chasseur* Basque on the Lisbon station 12.2.1798; in Nelson's squadron, but parted company in a gale 21.5.1798; with Hood's squadron at Alexandria 8.1798; destroyed (with *Seahorse*) 4-gun *L'Anémone* 2.9.1798; attacked (with *Leviathan* and *Swiftsure*) Spanish convoy off Cadiz 5.4.1800; took (with *Leviathan*) Spanish 32-gun *El Carmen* and 34-gun *Florentine* 7.4.1800. In 12.1800 under Capt. James O'Bryen; sailed for Leeward Islands 30.3.1801; took 16-gun *L'Enfant Prodigue* off St Lucia 24.6.1803; at reduction of St Lucia, Tobago and Dutch West Indies 6–9.1803; took 10-gun privateer *Le Mozambique* at Seron (Martinique) 13.3.1804; capture of Surinam 30.4.1804. Middling Repair at Deptford 2–6.1806; recommissioned 6.1806 under Capt. John Larmour. In 1807 under Capt. Frederick Maitland (–1811); in Basque Roads operations 4.1807; took 14-gun privateer *L'Austerlitz* 14.4.1807; boats destroyed 8-gun schooner *Apropus* at Vivero 13.3.1808; took 8-gun privateer *L'Incomparable* off the Irish coast 8.10.1809; took 16-gun *Le Fanfaron* off Guadeloupe 6.11.1809; took 8-gun privateer *La Belle Etoile* in the Bay of Biscay 23.3.1810, and 18-gun privateer *L'Augusta* in the Channel 6.4.1811. Laid up 11.1811 in Ordinary at Portsmouth. Fitted as receiving ship there 4–5.1822. BU 1.1836 at Portsmouth.

Trent Woolwich Dyd [M/Shipwright John Tovery]

As built: 143ft 0in, 119ft 6in x 38ft 2in x 13ft 6in. 925⁸⁷/₉₄bm. Draught 9ft 0in/14ft 0in.

Ord: 24.1.1795. K: 3.1795. L: 24.2.1796. C: 26.5.1796 at Woolwich.

First cost: £25,915 (including fitting).

Commissioned 3.1796 under Capt. Edward Bowater (–1797), for the North Sea; in 9.1796 under Capt. John Gore (acting); took 14-gun privateer *Le Poisson Volant* off Yarmouth 27.6.1797. In 10.1797 under Capt. Richard Bagot (died 12.6.1798); sailed for the Leeward Islands 11.1797. In 6.1798 under Capt. Robert Otway, at Jamaica; took a small privateer in the West Indies 7.1798; took (with *Squirrel*) 14-gun privateer *Penada* in ?2.1799; took another Spanish privateer 3.1800. In 10.1800 under Capt. Sir Edward Hamilton (dismissed 1.1802); took (with *Fortunee* and cutter *Dolphin*) 10-gun privateer *Le Renard* off Jersey 20.4.1801. In 2.1802 under Capt. Charles Brisbane; sailed for Jamaica 13.2.1802; later under Capt. Isaac Wolley then Capt. James Katon. Fitted at Plymouth as a hospital ship 6–8.1803 for Cork, recommissioned 6.1803 under Cmdr Walter Grossett (–1810), also as flagship of Adm. Lord Alan Gardiner 1803–04, Rear-Adm. William O'Bryen Drury 1805, Lord Gardiner again 1806 and Vice-Adm. James Whitshed 1807–10. On 2.11.1810 under Cmdr Thomas Young, still as hospital ship and flagship of Vice-Ad. Edward Thornbrough to 1813; on 7.5.1814 under Cmdr George Ourry Lempriere, still hospital ship and flagship of Vice-Adm. Sir Herbert Sawyer to 1815. Laid up and re-classed as a receiving ship at Cork by AO 11.11.1815. BU at Hawlbowline 2.1823.

NAIAD 46 (originally 38) guns. Design by Sir William Rule, approved 13 August 1795. An expanded version of his *Amazon* Class.

Dimensions & tons: 147ft 0in, 122ft 8⅜in x 39ft 5in x 13ft 9in. 1,013⁹⁰⁄₉₄bm.

Men: 284 (later 315). Guns: UD 28 x 18pdrs; QD 2 x 9pdrs + 10 x 32pdr carronades; Fc 2 x 9pdrs + 4 x 32pdr carronades.

Naiad Hill & Mellish, Limehouse

As built: 147ft 0in, 122ft 6¼in x 39ft 6½in x 13ft 9in. 1,018⁹¹⁄₉₄bm. Draught 10ft 7in/15ft 10in.

Ord: 30.4.1795. K: 9.1795. L: 27.2.1797. C: 6.5.1797 at Deptford.

First cost: £24,989 including fitting.

Commissioned 3.1797 under Capt. William Pierrepoint, for the Channel; took (with *Jason*) gunvessel *L'Arrogante* 23.4.1798; took 8-gun privateer *Le Tigre* off Finisterre 11.8.1798; took (with *Magnanime*) 36-gun *La Décade* off Finisterre 24.8.1798; took 16-gun privateer *L'Heureux Hazard* in the Bay of Biscay 3.3.1799; took (with *Ethalion*, *Triton* and *Alcmene*) 34-gun *Santa Brigida* and *Thetis* off Finisterre 15.10.1799. In 12.1799 under Capt. John Murray, then 12.1800 Capt. William Henry Ricketts; cut out packets *Alcudia* and *Raposa* from Pontevedra 16.5.1800. In 6.1801 under Capt. Philip Wilkinson; grounded near Île Rhé 10.1801. Fitted at Plymouth 12.1801–11.1802; recommissioned 9.1802 under Capt. James Wallis (–1804); took 10-gun *L'Impatiente* in Bay of Biscay 29.5.1803; boats took schooner *La Providence* off Brest 4.7.1803. In 1804 under Capt. Thomas Dundas (–1808); at Battle of Trafalgar 21.10.1805; blockade of Brest 1808; made good defects at Plymouth (for £9,055) 6–8.1808; took (with *Narcissus*) privateers 16-gun *La Fanny* and 4-gun *Le Superbe* 16.12.1808; in Stopford's squadron off Rochefort 2.1809 (signalled approach of *L'Italienne*, *Le Calypso* and *La Cybèle*). In 3–10.1809 under Capt. George Cocks (acting), then 10.1809 Capt. Henry Hill and 7.1811 Capt. Philip Carteret; action with gunboats off Boulogne; took 16-gun *La Ville de Lyon* 20/21.9.1811; took two 16-gun privateers in the Channel – *Le Milan* 6.10.1811 and *Le Requin* 6.11.1811; paid off 1813 at Portsmouth. Fitted at Portsmouth to raise wreck of

merchantman *Queen Charlotte*. Large Repair at Portsmouth (for £28,862) 7.1814–4.1815. Fitted for sea at Portsmouth (for £15,058) 4–8.1823; recommissioned 4.1823 under Capt. Robert Spencer (–1826), for the Channel and Mediterranean; affair with 20-gun Algerian Tripoli; boats destroyed an Algerian brig at Bona 23.5.1824; paid off Autumn 1826. Small Repair at Portsmouth (for £12,789) 4–6.1828, then laid up. Fitted as Coal depot at Portsmouth (for £11,115) 7.1846–1.1847, for Callao, Peru; recommissioned 23.10.1846 under William Lindsay Browne, Master, then 12.1851 under Samuel Strong, Master, then 12.1856 under William Ward Dillon, Master, and 10.1861 under George Reid, Master. Sold at Callao to the Pacific Steam Navigation Co. as a coal depot (for $2,000) 2.2.1866, and BU 1898.

ACASTA 50 (originally 40) guns. Design by Sir William Rule. A much larger concept, the first British frigate to introduce 30 main guns on the UD.

Dimensions & tons: 154ft 0in, 129ft 0¼in x 40ft 6in x 14ft 3in. 1,127²²⁄₉₄bm.

Men: 320. Guns: UD 30 x 18pdrs; QD 8 x 9pdrs + 4 x 32pdr carronades; Fc 2 x 9pdrs + 4 x 32pdr carronades.

Acasta John Randall & Co., Rotherhithe

As built: 154ft 0in, 128ft 11in x 40ft 9½in x 14ft 3in. 1,141²⁄₉₄bm.

Ord: 30.4.1795. K: 9.1795. L: 13.3.1797. C: 24.6.1797 at Deptford Dyd.

Commissioned 4.1797 under Capt. Richard Lane, for the Channel; sailed for Jamaica station 2.1798; took privateers – *Santa Maria* 1.5.1798, *San Antonio* 12.5.1798, *L'Hirondelle* and another in 5.1798, *Trompe* 30.6.1798, *San Josef de Victoria* 2.7.1798, and *San Miguel y Acandoa* 13.7.1798, also *L'Actif*, *Cincinnatus* and another (name unknown) in late 1798 or early 1799. In 11.1799 under Capt. Edward Fellowes; sunk privateer *La Victoire* in early 1800; to UK with convoy 10.1801; sailed for Jamaica 2.1802. In 5.1802 under Capt. James Wood; to UK 7.1802 and refitted at Portsmouth. From 1.1803 on North Sea station, at Leith, later at the Nore. In 4.1803 under Capt. James Oswald (temp, to 6.1803); in Channel 7.1803; took privateer *L'Aventure* 2.10.1803; to West Indies 1804, returned to UK 4.1805. In 5.1805 under Capt. Richard Dunn, in Channel; in Duckworth's squadron for blockade of Cadiz 1805; action off San Domingo 6.2.1806. Under Capt. Philip Beaver 9.1806; sailed for Leeward Islands 2.1807; took 18-gun *Le Serpent* at La Guaira 17.7.1808; at capture of Martinique 30.1.1809; capture of Les Saintes Islands 14.4.1809; to UK 1810 and paid off at Plymouth. Recommissioned 3.1811 under Capt. Alexander Kerr, with Channel Fleet, then to North America 1812; took privateers *Curlew* 24.7.1812 and *Herald* 10.12.1812; distant engagement and chase of US squadron off New London 6.1813; chase of USS *Constitution* 1.1815; retook *Levant* 1.1815. To UK 7.1815, paid off 12.9.1815 to Ordinary at Chatham. BU at Woolwich 1.1.1821.

BOADICEA 46 (originally 38) guns. Design copied from *Imperieuse* (Coulomb-designed French prize taken 1793).

Dimensions & tons: 148ft 6in, 124ft 0½in x 39ft 8in x 12ft 8in. 1,038¹⁵⁄₉₄bm.

Men: 284 (later 315). Guns: UD 28 x 18pdrs; QD 14 x 32pdr carronades; Fc 2 x 9pdrs + 2 x 32pdr carronades.

Boadicea Balthazar & Edward Adams, Bucklers Hard

As built: 148ft 6in, 123ft 10½in x 39ft 11½in x 12ft 8in. 1,052¹⁄₉₄bm.

Draught 10ft 8in/15ft 3in.

Ord: 30.4.1795. K: 9.1795. (named 14.11.1795) L: 12.4.1797. C: 17.4–9.9.1797 at Portsmouth.

First cost: £15,862 to build; £30,669 including fitting.

Commissioned 6.1797 under Capt. Richard Keats (–1801); took (with *Anson*) privateers 8-gun *Le Zéphyr* 19.10.1797 and 20-gun *Le Railleur* in the Channel 17.11.1797; took more privateers – 20-gun *L'Invincible Bonaparte* 9.12.1798, 14-gun cutter *Le Milan* in Channel 20.2.1799 (with sloop *Atalante*), 14-gun *Le Requin* 8.3.1799 (it capsized next day), and 16-gun *L'Unité* 1.4.1799; in Quiberon operations 6.1800; took gunvessel *La Bombarde* off Brest 1.1801. In 3.1801 under Capt. Charles Rowley; her boats (with *Fisgard*'s and Diamond's) took 20-gun *El Neptuno* at Corunna 20.8.1801. Fitted at Plymouth 4.1802–3.1803; recommissioned 12.1802 under Capt. John Maitland (–1808); took a small privateer 6.1803; encounter with 74-gun *Le Duguay Trouin* and 40-gun *La Guerrière* off Cape Ortegal 29/30.8.1803; took 12-gun lugger *Le Vautour* off Finisterre 25.11.1803; off Rochefort 1804; sighted Dumanoir's squadron 2.11.1805, and Leissegue's and Willaumez's squadrons in the Atlantic 15.12.1805; on Irish station 1806–07. In 6.1808 under Capt. John Hatley; sailed for Cape of Good Hope 9.2.1809; occupation of St Paul, Réunion 7.1810; recapture of 38-gun *L'Africaine* off Réunion 12.9.1810; capture of 40-gun *La Vénus* and recapture of 32-gun *Ceylon*. In 2.1811 under Capt. Viscount (Ralph) Neville; laid up at Plymouth 5.1811 in Ordinary. Between Middling and Large Repair at Plymouth (for £35,433) 2.1815–8.1816, then laid up. Fitted for sea at Plymouth (for £9,387) 10.1824–1.1825; recommissioned 10.1824 under Capt. Sir James Brisbane (died 19.12.1826), for East Indies; in Burma War 1825; under Capt. John Wilson for passage home 1827; paid off 13.8.1827 at Chatham. Between Small and Middling Repair at Chatham (for £10,027) 12.1829–12.1830, then laid up. Coastguard divisional ship 1854. BU completed at Chatham 22.5.1858.

HYDRA 46 (originally 38) guns. Design copied from *Melpomene* (Coulomb-designed French prize taken 1794); from 1813 she had been re-classed as a troopship.

   Dimensions & tons: 148ft 2in, 123ft 7⅜in x 39ft 4in x 12ft 8in. 1,016²⁴⁄₉₄bm.

   Men: 284 (later 315). Guns: UD 28 x 18pdrs; QD 12 x 32pdr carronades; Fc 2 x 12pdrs + 2 x 32pdr carronades.

Hydra William Cleverley, Gravesend

   As built: 148ft 3in, 123ft 7¾in x 39ft 6½in x 12ft 8in. 1,024⁵⁹⁄₉₄bm.

   Ord: 30.4.1795. K: 11.1795. (named 15.10.1796) L: 13.3.1797. C: 25.6.1797 at Woolwich.

   First cost: £23,012 including fitting.

   Commissioned 4.1797 under Capt. Sir Francis Laforey (–1801); destroyed (with *Vesuvius* and cutter *Trial*) 36-gun *La Confiante* near Le Havre 31.5.1798; sailed for the Leeward Islands 1.1799. In 4.1801 under Capt. Charles Paget, as guard ship at Weymouth; to the Mediterranean 1802. Recommissioned 12.1802 under Capt. George Mundy (–1810); took (with cutter *Rose*) 4-gun privateer *La Phoebe* 25.3.1803; to Mediterranean 7.1803; her boats took 4-gun *Le Favori* at Le Toquet 1.8.1803; took gunbrig ?*No. 5* and lugger ?*No. 411* on 30.1.1804; close blockade of Cadiz 10.1805; chased La Meillerie's squadron and took 20-gun brig *Le Furet* 27.2.1806; took 12-gun *Argonauta* 12.4.1806; to Spanish coast; her boats took 10-gun *Principe Eugenio*, 12-gun *Bella Carolina* and 4-gun *Carmen del Rosario* at Begu 7.8.1807; returned to UK for refit 7.1808; to the Mediterranean 11.1808;

paid off and laid up at Portsmouth in Ordinary 9.1810. Fitted as Troopship at Portsmouth 5–9.1813; recommissioned 7.1813 under Cmdr Joseph Digby. In 11.1815 under Cmdr Daniel Roberts, for the West Indies; laid up at Portsmouth 11.1817. Sold to Job Cockshot (for £2,410) 13.1.1820.

AMAZON Class 46 (originally 38) guns. Design by Sir William Rule, an enlargement of his *Naiad* design, and thus a further development of the earlier *Amazon* type of 1794; her sole sister, *Hussar,* was wrecked and then burnt in 1804.

   Dimensions & tons: 150ft 0in, 125ft 7¾in x 39ft 5in x 13ft 9in. 1,038⁶⁴⁄₉₄bm.

   Men: 284 (later 300). Guns: UD 28 x 18pdrs; QD 8 x 9pdrs + 6 x 32pdr carronades (by completion only 2 x 9pdrs + 12 x 32pdr carronades); Fc 2 x 9pdrs + 2 x 32pdr carronades.

Amazon Woolwich Dyd [M/Shipwright John Tovery]

   As built: to design exactly. Draught 11ft 3in/15ft 3in.

   Ord: 27.4.1796. K: 4.1796. L: 18.5.1799. C: 5.7.1799.

   First cost: £33,972 including fitting.

   Commissioned 5.1799 under Capt. Edward Riou (killed 2.4.1801); took 18-gun privateer *Le Bougainville* in Channel 14.2.1800 (prize sunk by collision next day); at Battle of Copenhagen 2.4.1801 (Riou killed in engagement of Tre Kroner forts). In 1801 under Capt. ?Evelyn Sutton. Recommissioned 8.1802 under Capt. William Parker (–1810); took 16-gun privateer *Le Felix* 26.7.1803; brush with French fleet off Cape Capet 2.5.1804; in chase to the West Indies 1805; took 24-gun ?Spanish privateer *El Principe de la Paz* off Ushant 17.9.1805; in Warren's chase to the West Indies 1806; took 40-gun frigate *La Belle Poule* 13.3.1806; took 14-gun privateer *Le Général Pérignon* 21.1.1810. In 5.1810 under Capt. John Joyce; took French 14-gun privateer *Le Cupidon* 23.3.1811. Laid up at Plymouth 12.1811, paid off into Ordinary 1812 and BU at Plymouth 5.1817.

ACTIVE 46 (originally 38) guns. Design by Sir John Henslow, developed from his *Artois* Class.

   Dimensions & tons: 150ft 0in, 125ft 2⅞in x 39ft 9in x 13ft 9in. 1,052⁵¹⁄₉₄bm.

   Men: 284 (later 315). Guns: UD 28 x 18pdrs; QD 8 x 9pdrs + 6 x 32pdr carronades; Fc 2 x 9pdrs + 2 x 32pdr carronades.

Active Chatham Dyd [M/Shipwright Edward Sison]

   As built: 150ft 0in, 125ft 2in x 39ft 10½in x 13ft 9in. 1,058⁵⁹⁄₉₄bm. Draught 10ft 10in/15ft 3in.

   Ord: 27.4.1796. K: 7.1798. L: 14.12.1799. C: 4.2.1800.

   First cost: £28,610.

   Commissioned 12.1799 under Capt. Charles Davers; convoy for East India ships 1800; took 14-gun privateer *La Quirole* in the Channel 26.1.1801; sailed for the Mediterranean with convoy later in 1801. In 9.1801 under John Giffard (temp.), then 10.1801 under Cmdr Thomas Shortland (temp.). In 8.1804 under Capt. Richard Mowbray (–1808), for blockade of Toulon; took *La Jeune Isabelle* 1804; pursuit of Villeneuve in the Mediterranean 1805; on Irish station 1806, then with Louis's squadron in the Levant; at Dardanelles 19.2.1807; took 4-gun privateer *Les Amis* 27.4.1807; took (with *Standard*) 16-gun *Friedland* off Cape Blanco 26.3.1808. In 8.1808 under Capt. James Gordon. Fitted at Chatham 5–9.1809; recommissioned 6.1809; sailed for the Mediterranean 6.1809; at Battle of Lissa 13.3.1811 (losing 4 killed, 24 wounded); boat attack on grain convoy near Ragozmia 29.7.1811; took (with *Alceste* and *Unite*) 40-gun *La Pomone* and storeship *La Persane*; paid off 6.1812. Middling Repair at

An experiment with man-powered paddle wheels, invented by one Lieutenant (?James) Burton, was tried in the frigate *Active* (36 guns, built 1799) in 1819. There is a sketch made in Portsmouth harbour by the well-known marine artist John Christian Schekty which confirms the details of this anonymous lithograph. The experiment was not pursued, so presumably was deemed a failure, but it may have simply been overtaken by events; two years later the Navy ordered its first steam-powered paddle vessel, the *Comet*. (NMM PW7995)

Woolwich 6.1813–3.1814, then to Sheerness. In 1815 under Capt. William King, then 10.1815 under Capt. Philip Carteret (–1817); fitted for sea at Sheerness 11.1815–4.1816; in 1817 on Jamaica station. Fitted with Lieut Burton's experimental propelling machinery (man powered paddles) 1819; recommissioned 1.1819 under Capt. Sir James Gordon (–1821), on Halifax station. In 12.1821 under Capt. Andrew King 'on particular service', then 9.1824 under Capt. Robert Rodney, on Lisbon station. Fitted as receiving ship at Plymouth 10.1825–2.1826. Renamed *Argo* 15.11.1833. BU completed at Plymouth 24.10.1860.

LAVINIA 48 (originally 44) guns. Initially intended to be to *Acasta* design, but built to design by Jean-Louis Barrallier with an even longer hull for its 30-gun main battery. Barrallier acted as the overseer for Jacobs's work; when the contractor's business 'failed', he undertook to supervise completion of the work on behalf of the government, renting land (from a Mr Greville) to create what became Milford Dyd.
> Dimensions & tons: 158ft 0in, 132ft 8in x 40ft 8in x 14ft 0in. 1,166⁹²/₉₄bm.
> Men: 294 (later 340). Guns: 30 x 18pdrs (short); QD 6 x 9pdrs + 8 x 32pdr carronades; Fc 2 x 9pdrs + 4 x 32pdr carronades).

Lavinia Jacobs & Sons, Milford (completed *in situ* by 'Government' after Jacobs' bankruptcy).
> As built: 158ft 1in, 132ft 9½in x 40ft 8¾in x 14ft 0in. 1,171⁶/₉₄bm. Draught 11ft 6in/13ft 6in.
> Ord: 15.2.1797. K: 5.1798. L: 6.3.1806. C: 11.7.1806 at Plymouth.
> First cost: £6,660 to Jacobs, plus £20,267 to complete building, + £15,669 fitting at Plymouth.
> Commissioned 2.1806 under Capt. Lord William Stuart (–1810, except Capt. John Hancock acting in 1st half 1808), for the

Channel; sailed for the Mediterranean 30.1.1808; sailed with East Indies convoy 28.4.1809; Walcheren operation 1809; sailed for Portugal 17.7.1810. Under Capt. George Digby in 11.1810 (–1812), in the Mediterranean. Fitted for Ordinary at Plymouth 2.1813; between Middling and Large Repair at Plymouth (for £40,468 + £4,216) 2.1815–9.1816, then laid up there until 1835. Fitted as lazarette at Plymouth 4–7.1836, then to Liverpool; in quarantine service at Liverpool to 1850. Fitted as coal depot at Plymouth 1–4.1852; hulk at Plymouth to 1868. Order to BU rescinded 7.10.1868; sunk in collision with Hapag s.s. *Cimbria*, and by AO 1.3.1870 sold 'as she lies' 31.3.1870.

AMPHION Class 38 (originally 32) guns. Design by Sir William Rule (although his original plan was modified in favour of Gambier's preferences for a flatter sheer and more raked bow), approved 27 June 1796. Three ships were initially ordered, with a further pair being ordered in 1805 by the Barham Admiralty Board; however *Proserpine* was a war loss in 1809, while by the end of 1816 *Medusa* had been BU, and *Nereus* and *Aeolus* had been re-classed as troopships (BU February and October 1817).
> Dimensions & tons: 144ft 0in, 121ft 7½in x 37ft 6in (36ft 10in mld.) x 12ft 6in. 909⁷¹/₉₄bm.
> Men: 254. Guns: UD 26 x 18pdrs; QD 4 x 6pdrs + 4 x 24pdrs (1800 – 2 x 6pdrs + 8 x 24pdr carronades); Fc 2 x 6pdrs + 2 x 24pdr carronades.

Amphion James Betts, Mistleythorn
> As built: 144ft 0½in, 121ft 6⅞in x 37ft 7¼in x 12ft 6in. 914⁴⁹/₉₄bm.
> Ord: 11.6.1796. K: 7.1796. (named 16.12.1796) L: 19.3.1798. C: 6.7.1798 at Chatham.
> First cost: £13,958 to build, plus £7,709 fitting.

Commissioned 5.1798 under Capt. Richard Bennet (–1801); sailed for Africa and thence to Jamaica 20.12.1798; took (with *Alarm*) 24-gun Spanish privateer *Astuzzana* 25.11.1799. Refitted at Chatham (for £8,091) 11.1801–1.1802; recommissioned 5.1802 under Alexander Fraser, in the Channel. In 9.1802 under Capt. Thomas Masterman Hardy; from 5.1802 flagship of Vice-Adm. Horatio Nelson, for passage to Mediterranean. In 7.1803 Nelson transferred to *Victory*, also taking Hardy, with Capt. Samuel Sutton moving from *Victory* to take command of *Amphion*, for blockade of Toulon; took (with *Indefatigable*, *Lively* and *Medusa*) Spanish frigates *Medea*, *Fama* and *Clara* 5.10.1804, and sank fourth frigate *Mercedes*; in Orde's squadron for chase to the West Indies 1805. In 10.1805 under Capt. William Hoste (–1811); at bombardment of fortress of Cotrone 26.7.1806; boats in abortive attempt to cut out storeship *La Baleine* from Rosas 12.5.1808; at blockade of Trieste 11.1808; her boats (with *Redwing*'s) cut out French brig and a trabaccola from Melada 8.2.1809; bombardment of Pesaro (with *Spartan* and *Mercury*) 23.4.1809; boats cut out six gunboats from Cortellazzo 27.8.1809, and (with *Cerberus*'s and *Active*'s) took 14 merchant prizes at Grado 28.6.1810, burning 11 more; at Battle of Lissa 13.3.1811, with 15 killed and 47 wounded; returned to UK 4.1812 to pay off. Large Repair at Deptford 4.1812–6.1813; recommissioned 5.1813 under Capt. James Stewart (–1815), for North Sea; her boats (with *Champion*'s) in attack on French vessels in the Western Scheldt 6.3.1814. In 1815 under Capt. John Brett Purvis, for convoy from Ireland to Bermuda. In 10.1815 under Capt. William Bowles (–1818), for convoys to Brazil. In 10.1818 under Capt. William Dashwood; paid off 20.5.1819 at Woolwich. Fitted for distressed seamen at Deptford 10.1822, but then sunk as breakwater at Woolwich 11.1820. Remains sold to Joliffe & Banks (for £168) 9.1823 to raise and remove.

NARCISSUS Class 38 (originally 32) guns. Design by Sir John Henslow, approved 13 January 1798. Three vessels were completed to this design (of which the *Tartar* was a wartime loss and *Cornelia* was BU in 1814), with two more ordered in 1805 but cancelled in 1806.

Dimensions & tons: 142ft 0in, 118ft 5in x 37ft 6in x 12ft 6in. 885⁹/₉₄bm.

Men: 254. Guns: UD 26 x 18pdrs; QD 2 x 9pdrs + 8 x 24pdr carronades; Fc 2 x 9pdrs + 2 x 24pdr carronades.

Narcissus Deptford Dyd [M/Shipwright Edward Tippett]

As built: 142ft 0½in, 118ft 4⅞in x 37ft 8in x 12ft 6in. 893⁵¹/₉₄bm. Draught 10ft 2in/14ft 4in.

Ord: 23.11.1797. K: 2.1800. L: 12.5.1801. C: 2.7.1801.

First cost: £34,013 including fitting.

Commissioned 1.1801 under Capt. Percy Fraser. In 10.1801 under Capt. Ross Donnelly (–1806); sailed for the Mediterranean 2.1802; at blockade of Toulon 1802; took 16-gun *L'Alcion* off Sardinia 8.7.1803; her boats (with *Seahorse*'s and *Maidstone*'s) destroyed convoy at Lavandon (Hyères Bay) 11.7.1804; sailed for Jamaica and Cape of Good Hope 4.1805; took 4-gun privateer *Le Prudent* on the African coast 29.10.1805; destroyed 32-gun privateer *Le Napoleon* near Cape of Good Hope 24.12.1805; with Popham's squadron at Cape of Good Hope 1806; in River Plate operations 6–10.1806. In 8.1806 under Capt. Charles Malcolm; took Spanish schooner *Cantela* in the Bay of Biscay 18.8.1807; at blockade of Lorient 3.1808; took (with *Naiad*) privateers 16-gun *La Fanny* and 4-gun *Le Superbe* 16.12.1808. In 7.1809 under Capt. Frederick Aylmer; took 14-gun privateers – *Le Duguay Trouin* 19.1.1810 and *L'Aimable*

Joséphine 5.2.1810; paid off 3.1812. Between Small and Middling Repair at Plymouth 2–8.1812; recommissioned 6.1812 under Capt. John Lumley; sailed for North America 29.9.1812; took US 12-gun *Viper* on Jamaica station 17.1.1813; took US 12-gun privateer *Revenge* 30.3.1813; boats took cutter *Surveyor* in the York River 12.6.1813. In 1814 under Capt. Alexander Gordon, then 3.1815 Capt. George Crofton; laid up at Deptford 6.1816. Fitted there as a receiving ship 7.1822–10.1823, then fitted at Woolwich for the temporary receipt of convicts 10–12.1823; fitted as convict hospital ship there 7. 1824. Sold to J. Levy (for £1,490) 12.1.1837.

AIGLE Class 42 (originally 36) guns. Design by Sir John Henslow, 1798. Survivor of a class of two, of which *Resistance* had been wrecked in 1803.

Dimensions & tons: 146ft 0in, 122ft 1½in x 38ft 6in x 13ft 0in. 962⁸¹/₉₄bm.

Men: 264. Guns: UD 26 x 18pdrs; QD 4 x 9pdrs + 8 x 32pdr carronades; Fc 4 x 9pdrs + 2 x 32pdr carronades.

Aigle Balthazar & Edward Adams, Bucklers Hard.

As built: 146ft 2in, 122ft 1in x 38ft 8in x 13ft 0in. 970⁸⁴/₉₄bm (990 as corvette).

Ord: 15.9.1798. K: 11.1798. L: 23.9.1801. C: 28.9–6.10.1801 (for Ordinary) 24.3.1803 (for sea) at Portsmouth.

First cost: £14,335 to builder (after £1,000 deducted for 16 months late delivery).

Commissioned 12.1802 under Capt. George Wolfe (–1811), for the Channel; took 14-gun privateer *L'Alerte* 27.9.1803; drove ashore and destroyed 20-gun *La Charente* and 8-gun *La Joie* near mouth of Gironde 12.7.1804; in 2.1805 under Cmdr Henry Sturt (acting); action against 40-gun *L'Italienne* and 38-gun *La Sirène* off Île de Groix 22.3.1808; action in Basque Roads 4.1809; Walcheren operations 8.1809; took 18-gun privateer *Le Phoenix* 12.9.1810. In 10.1811 under Capt. Sir John Louis; sailed for the Mediterranean 20.11.1811; paid off 1813. Re-rated 42-gun 2.1817. Between Middling and Large Repair at Woolwich (for £36,427) 3.1817–5.1819, then laid up. Very Small Repair and cut down into a 24-gun Sixth Rate corvette at Chatham 3–7.1831, then laid up again. Fitted at Chatham for sea (for £9,670) 8–11.1841; recommissioned 8.1841 under Capt. Lord Clarence Paget, for the Mediterranean; paid off 30.8.1845. Fitted as Coal hulk (and receiving ship) at Woolwich 10–11.1852; to Sheerness 9.1869. Appropriated as Torpedo target 15.8.1870. Sold (by AO 3.11.1870) to BU to A. W. Howe (for £925) 26.11.1870.

ETHALION 42 (originally 36) guns. 'Admiralty' (probably Adm. James Gambier's) design, approved 5 September 1799, developed from the earlier *Triton* and likewise incorporating Gambier's preferences for minimal sheer and a wall-sided hull. With the arming of this frigate the carronade reached its zenith, in replacing all QD/Fc guns except one pair of chase 9pdrs.

Dimensions & tons: 152ft 0in, 129ft 2¼in x 38ft 0in x 13ft 0in. 992²⁵/₉₄bm.

Men: 264. Guns: UD 26 x 18pdrs; QD 14 x 32pdr carronades; Fc 2 x 9pdrs + 4 x 32pdr carronades.

Ethalion Woolwich Dyd [M/Shipwright Edward Sison]

As built: 152ft 5½in, 129ft 7¾in x 38ft 0in x 13ft 0in. 995⁷⁴/₉₄bm.

Ord: 7.9.1799. K: 5.1800. L: 29.7.1802. C: 25.10.1802 at Woolwich, then 31.10.1802–14.3.1803 at Sheerness.

First cost: £30,700 including fitting.

Commissioned 11.1802 under Capt. Charles Stuart (–1806), for

the North Sea; flagship of Adm. Lord Keith summer 1803;
recommissioned 11.1804 and sailed for Leeward Islands
20.4.1805. In 1807 under Capt. Jonas Rose, in the Leeward
Islands, later Capt. William Fahie, then 1809 Capt. Thomas
John Cochrane; capture of Martinique 2.1809; paid off 8.1810.
Between Small and Middling Repair and fitted at Portsmouth
6.1810–1.1811; recommissioned 11.1810 under Capt. Edmund
Heywood (–1813), for the Baltic. In 5.1814 under Capt. William
Dobbie, for the Irish station; paid off 9.1815 at Woolwich; later
to Deptford then back to Woolwich. Fitted as convict ship
at Woolwich 8–9.1823. Fitted as temporary receiving ship at
Woolwich 7.1824, then permanently 10.1824–3.1825. Fitted as
breakwater at Woolwich 4–5.1835, for Harwich where employed
1838–75. BU notified 6.1877.

FORTE 44 (originally 38) guns. The design was copied from
Révolutionnaire (French prize of Forfait design, taken 1794), and
approved 11 October 1802. This vessel was originally ordered on 9 July
1801 to be built at Sheerness Dyd, but this work was never started, and
the order was transferred to Woolwich in 1809.
> Dimensions & tons: 157ft 2in, 131ft 9⅞in x 40ft 5½in x 12ft 5in.
> 1,147⁶⁸⁄₉₄bm.
> Men: 284 (later 315). Guns: UD 28 x 18pdrs; QD 12 x 32pdr
> carronades; Fc 2 x 9pdrs + 2 x 32pdr carronades.

Forte Woolwich Dyd [M/Shipwright Edward Sison]
> As built: 157ft 5⅝in, 132ft 1½in x 40ft 6½in x 12ft 5in. 1,155¹¹⁄₉₄bm.
> Draught 10ft 8in/14ft 9in.
> Re-ord: 15.9.1809. K: 3.1811. L: 21.5.1814. C: 5.9.1814 (for
> Ordinary).
> First cost: £37,636 including fitting.
> In Ordinary at Sheerness 1814–17. Small Repair at Chatham (for
> £5,652) 5–10.1817. Fitted for sea at Chatham (for £14,836)
> 6–10.1820.
> Commissioned 6.1820 under Capt. Thomas John Cochrane
> (–1822), for Halifax station as flagship of Rear-Adm. William
> Fahie. Between Snall and Middling Repair and fitted for sea at
> Plymouth (for £24,415) 8.1825–8.1826; recommissioned 3.1826
> under Capt. Jeremiah Coghlan (–1830), for South America.
> Fitted for sea at Plymouth (for £10,017) 11.1832–8.1833;
> recommissioned 5.1833 under Capt. Watkin Pell (–1837), for
> North America and West Indies. Laid up at Plymouth 3.1837; to
> Deptford 3.1844. BU at Deptford 10.1844.

PERSEVERANCE Class 42 (originally 36) guns. Further vessels built
at St Vincent's insistence to Edward Hunt's design of 1783 (an earlier
four had been built during the American War), with minor changes
such as built-up forecastle barricades, but with equally poor sailing
performance. Contrary to some reports, the two Bombay-built ships
were not built for the East India Co. (Bombay Marine); for full details
see Robert Gardiner's *Frigates of the Napoleonic Wars*, p. 11, which
explains the confused re-naming history; although nominally to this
Hunt design, these two vessels differed significantly from the standard
design, *Salsette* particularly being much beamier. *Orlando* was a late
addition to the class, following an mistaken order to Chatham to build
a similar frigate following the launch of the *Iphigenia*. Of this class,
Shannon was wrecked in 1803 and *Meleager* in 1808.
> Dimensions & tons: 137ft 0in, 113ft 2½in x 38ft 0in x 13ft 5in.
> 869¹⁰⁄₉₄bm.
> Men: 260 (by 1815, 264). Guns: UD 26 x 18pdrs; QD 2 x 9pdrs + 10
> x 32pdr carronades; Fc 2 x 9pdrs + 2 x 32pdr carronades.

Tribune George Parsons, Bursledon

> As built: 137ft 1½in, 113ft 1½in x 38ft 4in x 13ft 5in. 884¹⁹⁄₉₄bm.
> Draught 10ft 4in/15ft 6in.
> Ord: 6.5.1801. K: 7.1801. L: 5.7.1803. C: 7.7–20.8.1803 at
> Portsmouth.
> Commissioned 7.1803 under Capt. George Towry. In 1804 under
> Capt. Richard Bennet (–1805); took gunbrigs (cannonnières)
> *No. 43* and *No. 47* on 30.1.1804; under Capt. Richard Curry
> (temp.) 5–6.1805. In 1806 under Capt. Thomas Baker; chase of
> *Le Vétéran* unto Baie de la Forêt 26.8.1806. In 1808 under Capt.
> George Reynolds (–1813), for the Baltic; action with 4 Danish
> brigs off Mandal (Norway) 12.5.1810; sailed for Leeward Islands
> 5.3.1811. Middling Repair at Woolwich (for £32,397) 1.1814–
> 6.1815, then laid up at Chatham. Fitted for foreign service at
> Chatham (for £15,041) 10.1818–12.1819; recommissioned 9.1818
> under Capt. Josiah Nesbit Willoughby, for the Leeward Islands;
> paid off 9.1822. Small Repair and fitted for sea at Chatham (for
> £12,248) 9.1822–3.1823; recommissioned 11.1822 under Capt.
> Gardiner Guion (–1825); on Lisbon station 1825. Between Small
> and Middling Repair and fitted for sea at Chatham (for £29,123)
> 7.1826–5.1828; recommissioned 1.1828 under Capt. John Wilson,
> for South America; from 12.1829 under Capt. John Duntze. Very
> Small Repair and cut down to a 24-gun Sixth Rate corvette at
> Chatham (for £6,111) 1.1832–3.1833; fitted for sea at Chatham
> (for £7,771) 5–9.1834; recommissioned 5.1834 under Capt. James
> Tompkinson, for the Mediterranean; on 24.5.1838 under Capt.
> Charles Williams; wrecked near Tarragona on 29.11.1839.

Iphigenia Chatham Dyd [M/Shipwright Robert Seppings]
> As built: 137ft 0in, 113ft 1¼in x 38ft 2in x 13ft 5in. 876³⁄₉₄bm.
> Draught 10ft 6in/14ft 3in.
> Ord: 9.7.1801. Cancelled 26.7.1805, but reinstated 20.1.1806. K:
> 2.1806. L: 26.4.1808. C: 24.6.1808 at Chatham.
> First cost: £26,150, including fitting.
> Commissioned 5.1808 under Capt. Henry Lambert (–1810); sailed
> for the Cape of Good Hope 28.1.1809; at capture of Réunion
> 8.7.1810; surrendered to French squadron at Mauritius 28.8.1810;
> became French *L'Iphigénie*; retaken by Bertie's squadron
> 6.12.1810. Under Capt. Thomas Caulfield for passage home; paid
> off 4.1811. Ftted at Portsmouth (for £17,402) 11.1811–2.1812;
> recommissioned 1.1812 under Capt. Lucius Curtis; sailed with
> convoy for East Indies 25.3.1812. Later under Capt. Fleetwood
> Pellew; sailed for the Mediterranean 6.12.1812. In 2.1813
> under Capt. Andrew King, in the Mediterranean; made good
> defects at Chatham (for £15,064) 6–9.1815; sailed for the East
> Indies 10.1815. In 5.1816 under Capt. John Tancock, for the
> Mediterranean. Made good defects at Plymouth (for £16,834)
> 1–6.1818; in 3.1818 under Capt. Hyde Parker, for Jamaica. Fitted
> for sea at Plymouth (for £6,720); recommissioned 6.1821 under
> Capt. Sir Robert Mends, for the African station; took slaver
> *El Conde de Ville Flor* 21.2.1822; paid off 10.11.1922. Fitted at
> Woolwich 12.1832–7.1833 and to the Marine Society as a training
> ship until 1848. BU at Deptford 5.1851.

Salsette (ex-*Pitt*, renamed 19.2.1807) Bombay Dyd [M/Shipwrights
Jamsetjee Bomanjee Wadia & Franjee Maneckjee Wadia jointly; the
latter died 12.1804]
> As built: 137ft 0in, 112ft 11in x 38ft 9in x 13ft 7in. 901⁸²⁄₉₄bm.
> Ord: 12.5.1802. K: 19.7.1803. L: 17.1.1805.
> First cost: £27,922 to East India Company (includes fitting).
> Commissioned 1805 at Bombay under Capt. James Vashon, for the
> East Indies; at blockade of Mauritius 1805–06. In 1806 under
> Capt. Walter Bathurst, then 2.1807 Capt. George Waldegrave
> (–1809). Refitted at Portsmouth (for £7,600) 1.1808–17.3.1808;

The elderly *Leda* Class frigate *Shannon*, famous for her victory over the USS *Chesapeake* during a ferocious and bloody single-ship action off Boston in 1813, survived as a storage hulk at Sheerness into the late 1850s. Technically she had been renamed *St Lawrence* in March 1844, but when Capt. George Pechell Mends painted this watercolour of her six months later he retained her former name in the inscription at the bottom, an indication of how her reputation still persisted among naval officers. *(NMM PW6175)*

later to Baltic 1808–09. In 1810 under Cmdr Henry Montresor, then ?William Bertie (drowned 12.1810), Cmdr John Hollingworth and 1811 Capt. Henry Hope; took privateers – 2-gun *La Comète* in the Mediterranean 21.4.1812, and 16-gun *Le Mercure* off Isle of Wight 14.10.1812. In 12.1812 under Capt. John Bowen (–1815); sailed for the East Indies 25.3.1813. In 1815 under Capt. Joseph Drury; paid off and laid up at Portsmouth 12.7.1816; housed over 11.1823 but remained in Prdinary; fitted as a lazarette at Portsmouth 7.1831, for Hull. Fitted as Receiving ship at Woolwich (for £867) 10.1835. To Sheerness by AO 7.9.1869. BU completed 20.3.1874 at Chatham.

Doris (ex-*Pitt*, renamed 26.8.1807, ex-*Salsette*, renamed 19.2.1807) Bombay Dyd [M/Shipwright Jamsetjee Bomanjee Wadia]
  As built: 137ft 0in, 113ft 2½in x 38ft 0in x 13ft 5in. 869¹⁹⁄₉₄bm.
  Ord: 5.6.1803. K: 25.4.1806. L: 24.3.1807.
  First cost: £39,774 to East India Company (includes fitting).
  Commissioned 1808 at Bombay under Capt. Christopher Cole (–1810), for the East Indies; in 1810 under Capt. William Lye; at capture of Mauritius 12.1810; at capture of Java 9.1811. In 1812 under Cmdr John Harper for passage to UK. Arrived Plymouth 8.11.1812. Fitted for foreign service at Plymouth (for £13,787) 12.1812–3.1813; recommissioned 1.1813 under Capt. Robert O'Brien (–1815); sailed for China 25.3.1813. In 1816 under Capt. John Allen; paid off 4.3.1816. Small Repair at Sheerness (for £4,250) 10.1817–4.1818. Fitted for sea at Sheerness (for £13,337) 3–6.1821; recommissioned 3.1821 under Capt. Thomas Graham (died 9.4.1822), for South America. On 9.4.1822 under Lieut James Henderson (acting), then 29.4.1822 Capt. Frederick Vernon, then 4.8.1823 Capt. Thomas Bourchier (Acting) and 23.3.1824 Capt. William Hope Johnstone; paid off 1.1825. Recommissioned 2.1825 under Sir John Gordon Sinclair (–1829). Sold at Valparaiso (for $5,590) because of her decayed state 4.1829.

Orlando Chatham Dyd [M/Shipwright Robert Seppings]
  As built: 137ft 0in, 113ft 1¼in x 38ft 2in x 13ft 5in. 876³³⁄₉₄bm.
    Draught 10ft 5in/14ft 5in.
  Ord: 2.5.1808. K: 3.1809. L: 20.6.1811. C: 20.7.1811 at Chatham.

Commissioned 6.1811 under Capt. John Clavell (–1815); sailed for the Mediterranean 20.11.1811. During 1815 under Cmdr Charles Orlando Bridgeman (temp.). Fitted for foreign service at Deptford 6–9.1815, then to the East Indies, still under Clavell; paid off 2.1819 at Trincomalee. Fitted as hospital ship at Trincomalee by AO 15.10.1819. Sold there (for 7,000 Rupees) 3.1824.

LEDA Class (1803–09 orders) 46 (originally 38) guns. The prototype to this 1794 Design (taken from the lines of the *Hébé*, a French prize taken in 1782) was the *Leda* launched in 1800, which had been wrecked in 1808. Eight ships were ordered in 1802–09 to the same design, the first three from private contractors, with four following from orders placed with the dockyards in September 1808 (and a fifth in 1809); however, the first ship (the *Pomone* ordered in 1802) had been wrecked in 1811.
  Dimensions & tons: 150ft 1½in, 125ft 4⅞in x 39ft 11in x 12ft 9in. 1,062⁷⁹⁄₉₄bm.
  Men: 284 (later 300). Guns: UD 28 x 18pdrs; QD 8 x 9pdrs + 6 x 32pdr carronades; Fc 2 x 9pdrs + 2 x 32pdr carronades.

Shannon Josiah & Thomas Brindley, Frindsbury
  As built: 150ft 2in x 125ft 6½in x 39ft 11⅜in x 12ft 11in. 1,065⁶²⁄₉₄bm.
    Draught 10ft 10in/14ft 9in.
  Ord: 24.10.1803. K: 8.1804. L: 5.5.1806. C: 3.8.1806 at Chatham.
  First cost: £19,484 to builder; fitting £13,564.
  Commissioned 5.1806 under Capt. Philip Broke (–1813), for the Channel; occupation of Madeira 26.12.1807; took French privateer *Le Pommereuil* 27.1.1809; sailed for North America 9.8.1811; capture (with squadron) of 16-gun USS *Nautilus* 16.7.1812; chase of USS *Constitution* 17–20.7.1812; took US 18-gun privateer *Thorn* 31.10.1812; took 16-gun privateer *L'Invincible* off Cape Ann 16.5.1813; took 36-gun USS *Chesapeake* 1.6.1813 (33 killed and 50 wounded including Broke). From 6.1813 under Cmdr Humphrey Senhouse (acting). In Ordinary at Portsmouth 1814–15. Between Middling and Large Repair at Chatham (for £26,328) 7.1815–3.1817, then to Ordinary at Chatham. Very Small Repair at Chatham (for £4,969) 5–7.1826. Fitted for sea at Chatham (for £14,746) 8–12.1828; recommissioned 9.1828 under Capt. Benjamin Clement (–1830). Fitted as receiving ship and temporary hulk at Sheerness 11.1831. Renamed *Saint Lawrence* 11.3.1844. BU at Chatham 12.11.1859.

Leonidas John Pelham, Frindsbury
As built: 150ft 1½in, 125ft 5⅞in x 39ft 11¾in x 12ft 9in. 1,066 83/94bm.
Ord: 19.7.1805. K: 11.1805. L: 4.9.1807. C: 10.12.1807 at Chatham.
First cost: £21,610 to builder; fitting £13,041.
Commissioned 9.1807 under Capt. James Dunbar; sailed for the Mediterranean 10.2.1808. In 7.1809 under Capt. Anselm Griffiths (–1812); in Mediterreanean 1809–11, then Irish station 1812; took (with *Dasher*) 14-gun privateer *La Confiance* 17.1.1812; took 14-gun privateer *La Gazelle* 16.2.1812. On 1.1.1813 under Capt. George Francis Seymour, then 2.1813 under Capt. Frederick William Aylmer, on Irish station; took US 16-gun privateer Paul Jones 23.5.1813. In 1814 under Capt. William King, at Jamaica; paid off 6.1814. Between Middling and Large Repair at Sheerness (for £30,472) 5.1815–12.1816. Housed over 11.1818 and laid up at Sheerness. Became a powder hulk at Sheerness 1872. Sold to Castle, Charlton to BU 23.11.1894.

Briton Chatham Dyd [M/Shipwright Robert Seppings]
As built: 149ft 11in, 125ft 3¾in x 40ft 3in x 12ft 8½in. 1,079 81/94bm. Draught 11ft 0in/14ft 10in.
Ord: 28.9.1808. K: 2.1810. L: 11.4.1812. C: 22.6.1812.
First cost: £34,758 including fitting.
Commissioned 4.1812 under Capt. Henry Whitby (died 6.5.1812), for the Channel. In 6.1812 under Capt. Sir Thomas Staines (–1815); took 14-gun privateer *Le Sans-Souci* 15.12.1812; took 4-gun privateer *La Melance* off Bordeaux 9.9.1813; sailed for East Indies 31.12.1813; took US 2-gun privateer *Joel Barlow* 3.7.1814; located *Bounty* mutineers on Pitcairn Island 9.1814; paid off 8.1815. Small Repair at Portsmouth (for £16,736) 8.1816–3.1817. Fitted for sea at Portsmouth (for £11,775) 12.1822–4.1823; recommissioned 11.1822 under Capt. Sir Maxwell Murray, for South American station. Fitted for the reception of an Ambassador and for sea at Woolwich (for £12,260) 9.1826–5.1827; recommissioned 5.3.1827 under Capt. William Gordon, for Home Waters; small Repair at Portsmouth (for £15,335) 11.1828–3.1829; paid off 27.4.1830. Fitted for sea at Portsmouth (for £7,268) 4–8.1830; recommissioned 28.4.1830 under Capt. John Duff Markland, on particular service; paid off and laid up at Portsmouth 1.1833. Became a convict ship at Portsmouth 1841. Target ship there 2.1860. BU at Portsmouth completed 18.9.1860.

Tenedos Chatham Dyd [M/Shipwright Robert Seppings]
As built: 150ft 0in, 125ft 1⅜in x 40ft 4in x 12ft 9½in. 1,082 58/94bm. Draught 10ft 11½in/14ft 10in.
Ord: 28.9.1809. K: 5.1810. L: 11.4.1812. C: 16.6.1812.
First cost: £36,129 including fitting.
Commissioned 4.1812 under Capt. Sir Hyde Parker; sailed for North America 28.8.1812; took (with *Curlew*) US 4-gun privateer *Enterprise* 21.5.1813; took (with *Endymion*, *Pomone* and *Majestic*) USS *President* 15.1.1815; paid off and laid up at Chatham 8.1815; to Woolwich 1819. Middling Repair at Woolwich (for £26,178) 7.1820–4.1826, then laid up again there; to Sheerness 1830 and Chatham 1839. Fitted as convict ship at Chatham (for £9,823) 12.1842–4.1843, to lie at Bermuda. Accommodation ship 4.1863. BU at Bermuda completed 20.3.1875.

Lacedemonian Portsmouth Dyd [M/Shipwright Nicholas Diddams]
As built: 150ft 1½in, 125ft 4⅞in x 39ft 11in x 12ft 9in. 1,073bm. Draught 10ft 6in/14ft 9in.
Ord: 28.9.1808. K: 5.1810. L: 21.12.1812. C: 23.2.1813.
Commissioned 21.12.1812 under Capt. Samuel Jackson (–1815); sailed for North America 2.6.1813; paid off and laid up at Portsmouth 6.1815. Small Repair intended 9.1822, but instead BU there completed 22.11.1822.

Lively (ex-*Scamander*, re-named 7.12.1812) Chatham Dyd [M/Shipwright Robert Seppings to 3.1813; then George Parkin]
As built: 150ft 1in, 125ft 1¼in x 40ft 3½in x 12ft 10in. 1,080 22/94bm. Draught 11ft 1in/14ft 6in.
Ord: 28.9.1808. K: 7.1810. L: 14.7.1813. C: 15.7.1813 (for Ordinary).
First cost: £25,248 (not fitted).
Under Capt. Frederick William Aylmer briefly in 1813 while commissioning intended, but this was cancelled. Intended sale to John Small Sedger, Rotherhithe 22.7.1819 to BU, but this was likewise cancelled. Small Repair at Chatham (for £8,026) 7–9.1821. Fitted for foreign service at Chatham (for £13,815) 12.1823–27.1.1824.
Commissioned 11.1823 under Capt. William Elliott, for Lisbon station; paid off 4.12.1826 at Plymouth. Became a receiving ship at Plymouth 4.1831 (–1860). Sold to J. & E. Marshall, Plymouth (for £1,215) to BU 28.4.1863.

Surprise Milford Dyd (following failure of Jacobs's business) [M/Shipwright William Stone]
As built: 150ft 4in, 125ft 8⅞in x 40ft 0½in x 12ft 9in. 1,072 33/94bm. Draught 11ft 3in/14ft 9in.
Ord: 10.4.1809. K: 1.1810. L: 25.7.1812. C: 9.8–1.12.1812 at Plymouth.
Commissioned 31.8.1812 under Capt. Sir Thomas John Cochrane; sailed for the Leeward Islands 19.12.1812; took US 12-gun privateer *Decatur* 16.1.1813. On 7.6.1814 under Capt. George William Henry Knight; took 9-gun privateer schooner *Yankee Lass* 1.5.1814; paid off 8.1815. Fitted as convict ship at Plymouth 5–6.1822, to lie at Cork. Sold there (for £2,010) to BU by AO 2.10.1837.

A further seven vessels to this design (*Diamond*, *Amphitrite*, *Trincomalee*, *Thetis*, *Arethusa*, *Blanche* and *Fisgard*) were ordered between 1812 and August 1815, and appear below.

***APOLLO* Class** 42 (originally 36) guns. Three vessels were originally ordered to Sir William Rule's design in 1798 and 1799, all to be built by contract. Twenty-four more ships were ordered to this design between 1803 and 1812 (the last one was later altered into a fresh 38-gun design), of which twelve orders were placed with commercial yards and an equal number with Admiralty yards, but two commercial builders 'failed' and two ships were consequently added to the dockyard output. Four ships had been wartime losses: the prototype *Apollo* in 1804, *Blanche* in 1805, *Saldanha* in 1811 and *Manilla* in 1812; the *Malacca* (which differed considerably and only belonged nominally to this class; built of a mixture of timbers, her dimensions as built were significantly different) was broken up in 1816.
Dimensions & tons: 145ft 0in, 121ft 9⅜in x 38ft 2in x 13ft 3in. 943 53/94bm.
Men: 264. Guns: UD 26 x 18pdrs; QD 2 x 9pdrs + 10 x 32pdr carronades; Fc 2 x 9pdrs + 4 x 32pdr carronades.

Euryalus Balthazar & Edward Adams, Bucklers Hard
As built: 145ft 2in, 121ft 11¾in x 38ft 2¼in x 13ft 3in. 946 16/94bm. Draught 10ft 11in/14ft 10in.
Ord: 16.8.1800. K: 10.1801. L: 6.6.1803. C: 7.6–9.8.1803 at Portsmouth.
First cost: £14,290 to builder (after £1,000 deducted from £15,568.16.0d contract price for 17 months late delivery).
Commissioned 6.1803 under Capt. Henry Blackwood, for the Channel and Irish coast; at Battle of Trafalgar 21.10.1805.

In 2.1806 under Capt. George Dundas (–1812), for the Mediterranean to 1807; her boats (with *Cruiser*'s) took Danish gunboat in the Great Belt 16.6.1808; in Walcheren operations 1809; took 14-gun privateer *L'Etoile* off Cherbourg 18.11.1809; sailed for the Mediterranean 26.4.1810; took (with *Swallow*) 2-gun privateer *L'Intrépide* off Corsica 7.6.1810. In ?5.1811 under George Waldegrave; sailed for the Mediterranean 16.4.1811. In 11.1812 under Capt. Thomas Ussher, later under Capt. Jeremiah Coghlan. In 1813 under Capt. Sir Charles Napier; at blockade of Toulon 1813; her boats (with *Berwick*'s) took 10-gun xebec *La Fortune* 16.5.1813; destroyed storeship *La Baleine* at Calvi 23.12.1813; in Potomac operations 1814; paid off 6.1815. Recommissioned 6.1815 under Capt. Thomas Huskisson (–1821). Between Small and Middling Repair at Woolwich 12.1815–11.1816, then to Chatham. Fitted for sea at Chatham 5–10.1818; to Leeward Islands 1819 (as flagship 11.1819–5.1820), then at Jamaica (flagship 6–12.1820). On 30.12.1820 under Cmdr. Isham Fleming Chapman (acting), then 6.3.1821 under Capt. Wilson Braddyll Bigland, and 22.10.1821 under Capt. Augustus William James Clifford; to the Mediterranean 1823; paid off 3.1825. Fitted as convict ship at Chatham (for £7,936) 8–11.1825. Used as a coal hulk at Sheerness 1845, then fitted as a convict ship again (for £11,104) 7.1846–2.1847, then to Gibraltar as a convict ship 1847. Renamed *Africa* 1859. Sold to A.C.Recano at Gibraltar (for £337.6.8d) 16.8.1860.

Dartmouth   Benjamin Tanner, Dartmouth
  As built: 145ft 0½in, 121ft 9⅛in x 38ft 4in x 13ft 3in. 951⁶⁶/₉₄bm. Draught 10ft 8in/13ft 11in.
  Ord: 17.3.1803. K: 7.1804 – firm failed 2.1807, so re-ordered – contract transferred to Mr Cook, Dartmouth. Re-ord: 2.6.1809. L: 28.8.1813. C: 11.9–20.9.1813 at Plymouth for Ordinary, then laid up.
  First cost: £18,000 to builder (Cook), plus £6,508 fitting.
  Between Middling and Large Repair at Plymouth (for £25,338) 9.1816–9.1819, then laid up. Fitted for sea (for £7,872) 7.1824–3.9.1824.
  Commissioned 5.1824 under Capt. James Maude, for Jamaica. In 2.1825 under Capt. Henry Dundas (–1826). Fitted for sea at Portsmouth 2–5.1827; recommissioned 2.1827 under Capt. Thomas Fellowes, for the Mediterranean; at Battle of Navarino 20.10.1827, losing 6 killed and 8 wounded; paid off 3.1830. Fitted for quarantine service at Chatham 6–7.1831, in 1834 to Leith. BU completed 2.11.1854 at Deptford.

Creole   Benjamin Tanner, Dartmouth
  As built: 144ft 11in, 121ft 5⅛in x 38ft 3⅞in x 13ft 3in. 948⁸¹/₉₄bm. Draught 9ft 11in/14ft 3in.
  Ord: 17.3.1803 – firm failed 2.1807, AO not to build 2.6.1809, then re-ordered from: Plymouth Dyd.
  Re-ord: 23.12.1810. K: 9.1811. L: 1.5.1813. C: 20.8.1813. [M/Shipwright Joseph Tucker)
  First cost: £39,430 including fitting.
  Commissioned 6.1813 under Capt. Robert Forbes. In 1814 under Capt. George MacKenzie; in action (with *Astraea*) against 40-gun *L'Etoile* and *La Sultane* 23.1.1814. Small Repair at Sheerness (for £4,342) 8.1815–3.1816. Fitted for sea at Sheerness (for $14,363) 7–10.1818; recommissioned under Capt. Sir George Collier, for the African coast; in 7.1818 under Capt. William Dashwood; to South America 1818; in 10.1818 flagship of Capt. William Bowles, off South America; home in 4.1820. Recommissioned 13.5.1820 under Capt. Adam Mackenzie, for same station; on 24.1.1821 under Capt. Thomas White, then

26.8.1822 under Capt. Frederick Spencer; paid off 13.2.1824 and laid up at Chatham. Fitted for harbour service (unspecified) there 6–7.1833. BU at Deptford 8.1833.

Semiramis   Deptford Dyd [M/Shipwright Robert Nelson]
  As built: 145ft 0⅛in, 121ft 10in x 38ft 2in x 13ft 3in. 944¹/₉₄bm. Draught 10ft 2in/14ft 4in.
  Ord: 25.3.1806. K: 4.1807. L: 25.7.1808. C: 6.9.1808.
  First cost: £35,473 including fitting.
  Commissioned 8.1808 under Capt. William Granger; sailed for Portugal 15.12.1808. In 7.1810 under Capt. George Richardson (–1814); destroyed (with *Diana*) 14-gun *Le Teazer* and 16-gun *Le Pluvier* in the Gironde 24.8.1811; on Irish station 1812; took 14-gun privateer *Le Grand Jean* 29.2.1812; sailed for the Cape of Good Hope 28.10.1812; paid off at Portsmouth 8.1814. Middling Repair there 11.1814–3.1815. Fitted for foreign service 11–12.1817; commissioned 10.1817 under Capt. Sir James Yeo, for west coast of Africa (died 8.1818); in 8.1818 under Lieut (Cmdr 9.1818) Joseph Harrison; home to pay off 9.1818. Fitted as guard ship at Portsmouth 7–11.1821, for Cork; recommissioned 1.9.1821 under Capt. Thomas Huskisson, as flagship of Lord John Colville and harbour guard ship at Cork; on 25.2.1822 under Capt. Peter Ribouleau, still Colvill's flag; on 24.4.1825 under Capt. Robert Rowley, as flagship of Vice-Adm. Robert Plampin, still at Cork. Cut down (raséed) into a 24-gun Sixth Rate corvette at Devonport 4–6.1828; recommissioned 27.5.1828 under Capt. Maurice Frederick Fitzharding Berkeley, as flagship of Rear-Adm. Sir Charles Paget at Cork; paid off 6.1831 and laid up at Plymouth 9.1831. At Sheerness in 1834. Depot ship at Plymouth 1832. BU at Plymouth 11.1844.

Owen Glendower   Thomas Steemson, Paull (near Hull)
  As built: 145ft 3in, 121ft 11⅛in x 38ft 3½in x 13ft 3in. 951³/₉₄bm.
  Ord: 1.10.1806. K: 1.1807. L: 19.11.1808. C: 3.12.1808–22.3.1809 at Chatham.
  First cost: £18,364 to builder, plus £13,102 fitting.
  Commissioned 1.1809 under Capt. William Selby (died 3.1811); at occupation of Anholt 8.5.1809; took 14-gun privateer *La Camille* in the Channel 10.3.1810; sailed with convoy for Quebec 26.4.1810; sailed with convoy for Cape of Good Hope 27.9.1810; took 16-gun privateer *L'Indomptable* off the Lizard 1.10.1810. In 3.1811 under Capt. Edward A'Court, then 7.1811 Capt. Bryan Hodgson (–1816); sailed for East Indies 3.10.1811; flagship of Sir Samuel Hood 1811–12; took US 12-gun privateer *Hyder Ally* 5.1814; paid off 5.1816 at Chatham. Large Repair at Chatham (for £42,521) 3.1817–5.1819; fitted for sea there (for £7,507) 6–10.1819; recommissioned 8.1819 under Capt. Sir Robert Spencer, for South America; paid off 9.1822. Fitted for Foreign service at Chatham (for £6,856) 10.1822–1.1823; recommissioned 11.1822 under Capt. Sir Robert Mends (died 9.1823), for the African station; later under Cmdr John Filmore (acting). Fitted for foreign service at Chatham (for £15,619) 10.1824–2.1825; recommissioned 1.9.1824 under Capt. Hood Christian, for the African station; paid off 16.8.1828. Between Small and Middling Repair at Chatham (for £18,223 including survey) 12.1828–12.1829. Fitted as convict ship at Chatham (for £9,185) 3–10.1842, for Gibraltar. Receiving ship there 1880. Sold to F. Danino at Gibraltar (for £1,035.17.8d) 1884.

Curacoa   Robert Guillaume, Northam
  As built: 145ft 2in, 121ft 11in x 38ft 4in x 13ft 4in. 952⁸⁷/₉₄bm. Draught 11ft 2in/14ft 9in.
  Ord: 1.10.1806. K: 1.1808. L: 23.9.1809. C: 26.9.1809–23.1.1810 at Portsmouth.

First cost: £18,411 to builder, plus £15,066 fitting.

Commissioned 10.1809 under Capt. John Tower (–1815), for the Channel Islands; took 14-gun privateer *La Vénus* off Land's End 9.11.1810; sailed for the Mediterranean 17.11.1811; her boats (with *Leviathan*'s, *Imperieuse*'s and *Éclair*'s) destroyed a convoy at Laigueglia 27.6.1812; paid off 1815. Large Repair at Deptford (for £28,650) 3.1817–2.1819, then laid up at Greenhithe; at Woolwich from 1822. Cut down at Chatham into a 24-gun Sixth Rate corvette (for £15,057) 2–6.1831; recommissioned 4.1831 for the East Indies (–1834). Fitted for sea at Chatham (for £9,983) 2–7.1839; recommissioned 8.4.1839 under Capt. Jenkin Jones, for South America and Pacific; on 22.2.1843 under Capt. Sir Thomas Pasley, then 12.1.1846 under Capt. William Broughton, for southeast America; paid off 4.9.1847. BU at Sheerness 3.1849.

Hotspur George Parsons, Warsash

As built: 145ft 0in, 121ft 9in x 38ft 4in x 13ft 3½in. 951¹⁵/₉₄bm. Draught 10ft 0in/13ft 6in.

Ord: 1.10.1806. K: 8.1807. L: 13.10.1810. C: 14.10.1810–9.2.1811 at Portsmouth.

Commissioned 11.1810 under Capt. Joceline Percy (–1815), for the Channel; destroyed 3 gunbrigs at Calvados 8.9.1811; took privateers – 12-gun *L'Impératrice Reine* 13.5.1813 and US 5-gun *Chesapeake* 26.10.1813; convoy to West Indies 1814; to South Amereica 1815; paid off 15.8.1816 into Ordinary at Portsmouth. BU at Sheerness 1.1821.

Havannah Wilson & Co., Liverpool (originally ordered from Fletcher, Naylor & Hassall?)

As built: 145ft 0½in, 121ft 9¼in x 38ft 3¼in x 13ft 3¾in. 948⁶⁴/₉₄bm. Draught 10ft 3in/14ft 8in.

(as repaired (1822): 146ft 1in, 122ft 3¾in x 38ft 11in x 13ft 5¼in. 964bm.)

Ord: 1.10.1806. K: 3.1808. L: 26.3.1811. C: 14.5–29.7.1811 at Plymouth.

First cost: £18,259 to builder, plus £9,333 fitting.

Commissioned 6.1811 under Capt. George Cadogan (–1813); sailed for the Mediterranean 25.12.1811; her boats cut our gunboat No. 8 on 6.1.1813, and destroyed four gunboats at Manfredonia 7.2.1813; destroyed (with *Partridge*) two gunboats 18.7.1813; capture of Zara 6.12.1813. In 1814 under Capt. James Black (acting), then Capt. Edward Sibly (temp). Recommissioned 4.1814 under Capt. Gawen Hamilton, at Portsmouth; paid off 3.7.1816 and laid up at Sheerness. Between Middling and Large Repair (for £22,126) at Sheerness 4.1819–10.1822, then laid up. Very Small Repair, and cut down to a 24-gun Sixth Rate (corvette, with 24 x 32pdr 40cwt guns) at Deptford (for £5,574) 2–6.1845, then laid up at Sheerness. Fitted for sea at Sheerness (for £11,388) 12.1847–3.1848; recommissioned 24.2.1848 under Capt. John Elphinstone Ersine, for New Zealand; paid off 18.12.1851. Repaired and refitted at Plymouth (for £8,734) 7.1852–10.1855; recommissioned 3.8.1855 under Capt. Thomas Harvey, for the Pacific; paid off 17.11.1859 at Plymouth. Fitted there as a 'ragged school ship' and lent 19.3.1860 as training ship at Cardiff. Sold (by AO 26.10.1904) to BU 1905.

Orpheus Deptford Dyd [M/Shipwright Robert Nelson]

As built: 145ft 0in, 121ft 8¾in x 38ft 3in x 13ft 4in. 947²⁸/₉₄bm. Draught 11ft 0in/14ft 3in.

Ord: 27.2.1808. K: 8.1808. L: 12.8.1809. C: 21.9.1809.

Commissioned 8.1809 under Capt. Patrick Tonyn (died 22.1.1810); sailed for the Leeward Islands 24.11.1809. Following Tonyn's death, under Capt. Charles Dilkes then Robert Preston. In 11.1810 under Capt. Hugh Pigot (–1814), in North America;

took US privateers – 8-gun *Wampoc* 28.4.1813, 20-gun *Holkar* 11.5.1813, and (with *Shelburne*) 22-gun *Frolic* 20.4.1814; convoy to East Indies 11.1814; laid up at Chatham 9.1816. BU at Chatham 8.1819.

Leda Woolwich Dyd [M/Shipwright Edward Sison]

As built: 145ft 0in, 121ft 8½in x 38ft 3in x 13ft 3in. 947²⁵/₉₄bm. Draught 10ft 6in/15ft 0in.

Ord: 23.3.1808. K: 10.1808. L: 9.11.1809. C: 8.12.1809.

Commissioned 11.1809 under Capt. George Sayer (–1815); sailed for the East Indies 9.6.1810; at reduction of Java 7–9.1811; Sayer was acting C-in-C 12.1814–6.1815 (following death of Sir Samuel Hood) and again 9.1815–1816 (following death of Sir George Burlton); paid off 28.12.1816 at Sheerness. Sold to Job Cockshott (for £2,450) 30.4.1817.

Theban George Parsons, Warsash.

As built: 144ft 10in, 121ft 5⅛in x 38ft 5in x 13ft 4in. 953⁴⁹/₉₄bm. Draught 10ft 8in/14ft 3in.

Ord: 5.1808. K: 6.1808. L: 22.12.1809. C: 23.12.1809–25.4.1810 at Portsmouth.

Commissioned 11.1809 under Capt. James Hillyar, for the Downs. In 1810 under Capt. Stephen Digby; sailed for China. In 1813 under Lieut Basil Hall (acting), then 2.5.1813 under Capt. Samuel Leslie, in East Indies; paid off 6.4.1816 at Plymouth. BU at Plymouth 5.1817.

Astraea Robert Guillaume, Northam.

As built: 145ft 4in, 122ft 1⅛in x 38ft 4½in x 13ft 4in. 956⁴⁴/₉₄bm. Draught 10ft 11in/14ft 8in.

Ord: 26.9.1808. K: 12.1808. L: 5.1810. C: 24.5–24.9.1810 at Portsmouth.

First cost: £18,552 to builder, plus £14,566 fitting.

Commissioned 6.1810 under Capt. Charles Schomberg (–1813); sailed for the Cape of Good Hope 8.10.1810; in action (with *Phoebe*, *Galatea* and *Racehorse*) against 40-gun *La Clorinde*, *La Néréide* and *La Renommée* off Madagascar 20.5.1811 (*La Renommée* taken, and *La Néréide* taken at Tamatave on 26.5.1811). In 8.1813 under Capt. John Eveleigh (acting); in action (with *Creole*) against 40-gun *La Sultane* and *L'Etoile* off Cape Verde Islands 23.1.1814 (9 killed including Eveleigh, 37 wounded). In 1.1814 under Lieut John Bulford (acting). In 7.1814 under Capt. William Black, then 12.1814 under Capt. Edward Kittoe and finally Capt. John Harvey; laid up at Plymouth 8.1815. Fitted as 8-gun depot ship at Plymouth (for £10,863) 4–8.1823, for Superintendent of packet service at Falmouth; recommissioned 4.1823 under Capt. William King (–1835); from 4.7.1835 under Capt. John Clavell, then 7.4.1837 under Capt. James Plumridge and 6.8.1841 under Capt. Alexander Ellice, in same role to 1848. From 28.1.1848 under William Yeames, Master, as storeship at Mylor; paid off 5.3.1851. BU at Plymouth 4.1851.

Belvidera Deptford Dyd [M/Shipwright Robert Nelson]

As built: 145ft 0¾in, 121ft 9¼in x 38ft 2½in x 13ft 3in. 945⁵¹/₉₄bm. Draught 10ft 9in/14ft 10in.

Ord: 28.9.1808. K: 12.1808. L: 23.12.1809. C: 16.2.1810.

First cost: £30,619 to build, plus £7,686 fitting.

Commissioned 1.1810 under Capt. Charles Dashwood. In 3.1810 under Capt. Richard Byron (–1814); sailed for the Greenland Fishery 10.5.1810; her boats (with *Nemesis*'s) took Danish 8-gun *Balder* and *Thor* and destroyed another vessel near Studtland (Norway) 22.7.1810; sailed for North America 27.10.1810; in action against USN 44-gun *President* 23.6.1812, with 9 casualties (escaped from Rodgers's squadron); with Broke's squadron in capture of USN 14-gun *Nautilus* 15.7.1812; chase of USS *Constitution* 17–20.7.1812; took US 7-gun privateer *Bunker's Hill* 21.8.1812; her boats (with others') took US 6-gun privateer *Lottery* in Chesapeake Bay 8.2.1813; took US schooner *Vixen* 3.1.1814 and US 6-gun privateer *New Zealander* 2.4.1814; paid off at Portsmouth 9.1814. Between Middling and Large Repair at Portsmouth (for £34,912) 5.1816–10.1817, then laid up. Small Repair and fitted for sea (for £24,534) 7.1829–4.1831; recommissioned 20.11.1830 under Cmdr Richard Saunders Dundas, for the Mediterranean, then 8.12.1833 under Capt. Charles Burrough Strong, for North America and the West Indies; paid off 11.1837. Small Repair and fitted as a demonstration ship at Portsmouth (for £5,443) 4.1838–11.1839. Fitted for sea there (for £5,776 + ?£1,279) 9–12.1841; recommissioned 28.8.1841 under Capt. Charles Grey, for the Mediterranean; paid off 5.3.1845. Fitted at Portsmouth as a store depot 10.1846. Fitted as receiving ship 8–11.1852 (used to 1890). Sold at Portsmouth to J. B. Garnham (for £1,800) 10.7.1906.

Galatea Deptford Dyd [M/Shipwright Robert Nelson]
As built: 145ft 0in, 121ft 8¾in x 38ft 3in x 13ft 3½in. 947³⁵⁄₉₄bm. Draught 11ft 2in/14ft 8in.
Ord: 12.5.1809. K: 8.1809. L: 31.8.1810. C: 18.10.1810.
First cost: £39,558 including fitting.
Commissioned 9.1810 under Capt. Woodley Losack (–1815); sailed for Cape of Good Hope 31.12.1810; in Schomberg's action off Tamatave 20.5.1811 (capture of *La Renommée* and *La Néréide*); chased by USS *President* and *Congress* 31.10.1812; sailed for Lisbon 1.6.1813; laid up at Portsmouth 10.1815. To Deptford 1820. Between Middling and Large Repair and fitted for sea at Deptford (for £36,187) ?11.1819–2.1826; recommissioned 8.1825 under Capt. Sir Charles Sullivan (–1829). Refitted and fitted with experimental (man-powered) propelling machinery 1–5.1829 under Capt. Charles Napier, then refitted 12.1829–2.1830 (combined cost £12,595). Fitted as receiving ship and coal depot at Plymouth for Jamaica 8–9.1836. BU there by AO 24.9.1849.

Maidstone Deptford Dyd [M/Shipwright Robert Nelson]
As built: 145ft 0in, 121ft 8¾in x 38ft 3in x 13ft 3½in. 947³⁵⁄₉₄bm. Draught 10ft 6in/14ft 1in.
Ord: 8.1.1810. K: 9.1810. L: 18.10.1811. C: 13.12.1811.
First cost: £38,083 including fitting.
Commissioned 11.1811 under Capt. George Burdett (–1813); took 2-gun privateer *Le Martinet* in the Mediterranean 4.4.1812; sailed for North America 19.6.1812; her boats (with *Spartan*'s) took US privateers in the Bay of Fundy – 1-gun *Morning Star* and *Polly* on 6.8.1812, and 6-gun *Commodore Barry* and 2-gun *Madison*, *Olive* and *Spence* on ?3.8.1812; took (with *Colibri*) US 2-gun privateer *Dolphin* off Cape Sable 13.8.1812 and (with *Spartan*) US 16-gun privateer *Rapid* on St George's Bank 17.10.1812; took further US privateers – 8-gun *Cora* 14.2.1813, and (with *Poictiers* and *Nimrod*) 20-gun *Yorktown* 17.7.1813. Later in 1813 under Capt. Alexander Gordon; took US 2-gun privateer *Black Swan* 31.10.1813; her boats (with others') took US 6-gun privateer *Lottery* 9.2.1814. On 6.8.1814 under Capt. William Skipsey; paid off 7.9.1815. Between Middling and Large Repair at Woolwich (for £26,201) 7.1816–4.1818, then fitted for Ordinary 8.1818 for Greenhithe. Fitted for the African station at Deptford (for £1,994) and Woolwich (£14,349) 11.1823–3.1824. Recommissioned 16.12.1823 under Capt. Charles Bullen, for the coast of Africa; took slavers *Avizo* 26.9.1824, *Bey* 19.5.1826, *Z* (?) 31.7.1825, *Segundo Gallego* 29.9.1825, *De Hoop* 3.1.1826, *Nicanor* 25.5.1826, *Prince of Guinea* 6.8.1826, *Paulita*

6.12.1826, and six more (*Creola*, *Venturoso*, *Tentadora*, *Carlota*, *Providencia* and *Conceicoa Paquita Rio*) off Fernando Po in 4.1827; paid off 18.8.1832 at Portsmouth. Defects made good at Portsmouth (for £13,273) 9–12.1827. Recommissioned 16.9.1827 under Skipsey again, for the coast of Africa; on 30.9.1828 under Capt. Charles Marsh Schomberg; paid off 18.8.1832 into Ordinary as Portsmouth, as receiving ship. Fitted as coal depot at Portsmouth (for £1,120) 2–3.1838. BU completed by J. White at Cowes 26.1.1867 (by contract of 16.6.1865).

Stag Deptford Dyd [M/Shipwright Robert Nelson]
As built: 145ft 0in, 121ft 8¾in x 38ft 3in x 13ft 3½in. 947³⁵⁄₉₄bm. Draught 10ft 2in/13ft 11in.
Ord: 17.10.1810. K: 1.1811. L: 26.9.1812. C: 11.1812.
Commissioned 8.1812 under Capt. Phipps Hornby (–1814); sailed for Cape of Good Hope 19.5.1813 and thence to South America; laid up at Plymouth 11.1814. To Sheerness 1.1821, scheduled for Middling Repair, but instead BU there 20.9.1821.

Magicienne Daniel List, Binstead, Isle of Wight
As built: 145ft 2in, 121ft 8⅜in x 38ft 3½in x 13ft 3in. 949⁴⁄₉₄bm. Draught 10ft 9in/13ft 10in.
Ord: 14.12.1810. K: 4.1811. L: 8.8.1812. C: 9.8–24.10.1812 at Portsmouth.
First cost: £20,194 to builder, plus £17,176 fitting.
Commissioned 8.1812 under Capt. Thomas John Cochrane; in 9.1812 under Capt. William Gordon; sailed for Portugal 8.11.1812; joined Collier's squadron on north coast of Spain; took US 7-gun privateer *Adeline* 22.3.1814; in Ordinary at Portsmouth in 1815. Small Repair and fitted for sea at Portsmouth (for £13,994) 10.1815–2.1816; recommissioned 10.1815 under Capt. John Brett Purvis, for the East Indies; paid off 1818 at Woolwich. Between Middling and Large Repair there (for £24,733) 2.1821–5.1826, then laid up at Sheerness. Cut down into a 24-gun Sixth Rate corvette (for £17,400) at Woolwich Dyd 4–11.1831 (under AO 15.4.1831); recommissioned 7.1831 under Capt. James Hanway Plumridge, for East Indies, then 4.1835 for Lisbon; paid off 28.3.1835. Recommissioned 4.1835 under Capt. George William St John Mildmay, for Lisbon and Home Waters; paid off 27.9.1838 at Portsmouth. Recommissioned 15.2.1840 under Capt. William Burnett, for the Mediterranean; in operations on Syrian coast; on 4.8.1841 under Capt. Richard Laird Warren; paid off 27.11.1843. BU at Portsmouth 3.1845.

Pallas Robert Guillaume, Northam
As built: 145ft 5in, 122ft 2⅛in x 38ft 3in x 13ft 3in. 951¹³⁄₉₄bm. Draught 10ft 8in/14ft 5in.
Ord. 19.3.1811. K: 5.1811. On his bankruptcy, the order transferred to Portsmouth Dyd [M/Shipwright Nicholas Diddams]
Ord. 10.12.1813. K: 4.1814. L: 13.4.1816. C: 27.4.1816 (for Ordinary).
First cost: £34,668 including fitting (excludes £2,200 paid to Guillaume).
Commissioned 28.8.1828 under Capt. Adolphus Fitzclarence (after 12 years in Ordinary), for conveying passengers to and from India; fitted for sea at Portsmouth (for £15,472) 10–12.1828; sailed to India and return 1829. On 15.3.1831 under Capt. Manley Hall Dixon, then 20.1.1832 under Capt. William Walpole, on West Indies station; paid off 5.1834. Fitted as coal depot at Plymouth 9–11.1836; served until 1860. Sold to Marshall (for £1,426) 11.1.1862.

Barrosa Deptford Dyd [M/Shipwright Robert Nelson]
As built: 145ft 0in, 121ft 8¾in x 38ft 3in x 13ft 3½in. 947³⁵⁄₉₄bm. Draught 10ft 9in/14ft 3in.

Just as ships of the line were cut down to make powerful frigates, so frigates could be similarly raséed to make 'corvettes' (the term was not officially adopted until later). Some of the early conversions, like *Magicienne* seen in this print, effectively became flush-decked by removing most of the quarterdeck and forecastle, leaving just short platforms fore and aft. However, the main aim of such conversions was to modernise the armament, which meant fewer but more powerful guns, and this required a more radical adjustment of the gunports. Furthermore, an exposed battery in action was considered a weakness, so the later rasée corvettes retained the frigate's complete weather deck, merely cutting away the barricades for most of its length and mounting just a handful of lighter guns on it. *(NMM PW8015)*

Ord: 4.4.1811. K: 10.1811. L: 21.10.1812. C: 10.12.1812.

Commissioned 10.1812 under Capt. William Shirreff; sailed for North America 31.1.1813; in action (with *Junon*) against 15 US gunboats in Hampton Roads 20.6.1813. In 8.1814 under Capt. John Maxwell; took US privateers – 8-gun *Engineer* 29.9.1814 and 1-gun *Highflyer* 14.11.1814. In 1815 under William M'Culloch; laid up at Portsmouth 9.1815. Fitted as slop ship at Portsmouth 8–9.1823, later receiving ship and ordnance depot (–1833). Sold to John Beatson (for £1,830) 27.5.1841.

Tartar Deptford Dyd [M/Shipwright Robert Nelson to 7.1813, completed by William Stone]

As built: 145ft 2in, 121ft 10¼in x 38ft 3¼in x 13ft 3in. 949²⁷⁄₉₄bm. Draught 10ft 4in/14ft 3in.

Ord: 6.1.1812. K: 10.1812. L: 6.4.1814. C: 6.5.1814 (for Ordinary).

First cost: £28,203 including fitting.

To Ordinary at Sheerness 5.1814. Fitted for sea at Chatham (for £14,128) 6–9.1818.

Commissioned 5.1818 under Capt. Sir George Collier (–1820), for African station; took slavers *Francisco* and *Maria* 30.1.1820,

Gazetta 2.3.1820, *Erato* 10.3.1820 and *Donna Eugenia* 23.3.1821; paid off 10.1821. In 10.1822 under Capt. Thomas Browne; fitted for sea at Plymouth (for £12,329) 11.1822–2.1823; sailed for South America 5.2.1823; paid off 2.1826 at Woolwich. Fitted as receiving ship there 11.1827–3.1830, for Sheerness and (1834–1859) at Chatham. BU completed there 30.9.1859.

Brilliant Deptford Dyd [M/Shipwright Robert Nelson to 7.1813, completed by William Stone]

As built: 145ft 6½in, 122ft 3¼in x 38ft 3½in x 13ft 3in. 953⁵¹⁄₉₄bm. Draught 10ft 9in/14ft 5in.

Ord: 11.12.1812. K: 11.1813. L: 28.12.1814. C: 5.3.1815 (for Ordinary).

First cost: £34,701 including fitting.

Housed over fore and aft 10.1814 (prior to launch). Under Arthur Stow in 1814. In Ordinary (incomplete) at Chatham to 1844.

Not commissioned until 2.1846. Middling Repair at Chatham (for £16,579; ?plus £3,076) 10.1828–1.1830. Cut down into a 24-gun 6th Rate corvette and fitted for sea at Sheerness (for £13,386) 9.1844–4.1846. Commissioned 28.2.1846 under Capt. Rundle Burges Watson, for the Cape of Good Hope; paid off 9.10.1849. Fitted for commission at Sheerness (for £3,058) 4–6.1850. Refitted for sea at Sheerness (for £6,400) 6–8.1856. Recommissioned 6.6.1856 under Capt. James Aylmer Paynter, for North America and the West Indies; paid off 21.11.1857 at Chatham. Fitted as training ship for coast volunteers at Chatham (for £1,479) 2.1860, as RNR Drill Ship at London (at Dundee from 1864, and Inverness 1880). Renamed *Briton* 8.11.1889. Lent to Army

Council as temporary accommodation ship at Inverness 9.1906. Sold to Forth Shipbreaking Co., Bo'ness (for £1,600) 12.5.1908.

Blonde Deptford Dyd

Ord: 11.12.1812. Re-ordered to new design from 1816 (see below).

***LIVELY* Class** 46 (originally 38) guns. Sir William Rule design, fifteen vessels were copied from the prototype *Lively* (ordered 1799). The finest British frigate design of the French Wars, and much prized by the Navy Board, orders continued until Rule's death in early 1813; even then, it was probably only the embarrassing loss of the *Macedonian* of this class which caused the Admiralty to terminate the programme. The prototype *Lively* was wrecked in August 1812, the *Macedonian* was captured by the US frigate *United States* on 26 October 1812, and the *Statira* was wrecked in February 1815.

Far more weatherly than the smaller *Leda* Class, and easily capable of stowing adequate provisions for Britain's long-range patrol needs. Unlike the prototype, which had been produced with separate gangways between forecastle and quarterdeck, the 1803 orders were modified (in 1805) by having the gangways integrated into the upperworks, and this step towards enclosure of the waist was taken a step further with the introduction of gratings to cover the spars between the gangways.

Dimensions & tons: 154ft 0in, 129ft 8in x 39ft 5in (38ft 9in mld) x 13ft 6in. 1,071⁵⁶⁄₉₄bm.

Men: 284 (320 from 26.1.1813). Guns: UD 28 x 18pdr; QD 2 x 9pdrs + 12 x 32pdr carronades (later ships 14 carronades and no 9pdrs); Fc 2 x 9pdrs + 2 x 32pdr carronades.

Undaunted was completed (at Capt. Maling's request) with a single-calibre armament of 28 x 24pdr Gover's guns on the UD and 18 x 24pdr carronades (14 on QD and 4 on Fc); but deficiencies in the Gover's gun resulted in subsequent re-arming as per the establishment.

Resistance Charles Ross, Rochester

As built: 154ft 1½in, 129ft 8in x 39ft 7in x 13ft 6in. 1,081bm. Draught 10ft 10in/15ft 3in.

Ord: 7.11.1803. K: 3.1804. L: 10.8.1805. C: 19.10.1805 at Chatham.

First cost: £33,065 (£20,683 to builder plus £12,382 fitting).

Commissioned 27.8.1805 under Capt. Charles Adam; took 14-gun privateer *L'Aigle* 27.12.1807 on Home station; sailed for the Mediterranean 22.6.1809; on 7.4.1810 under Cmdr John Hollinworth (Acting); on 12.7.1810 under Capt. Philip Rosenhagen, still in the Mediterranean, then 1.1813 under Capt. Fleetwood Broughton Reynolds Pellew; paid off 2.1814. Underwent Middling Repair at Plymouth (for £43,698) 5.1816–5.1818, housed over fore and aft, then laid up in Ordinary. Fitted at Plymouth as a 10-gun troopship (for £12,085) 11.1841–4.1842; recommissioned 5.3.1842 under Cmdr Charles George Edward Patey, then 21.11.1846 under Cmdr Gower Lowe (invalided 14.5.1848); on 14.5.1848 under Lieut Humphrey John Julian (Acting); paid off 14.9.1848. Recommissioned 18.4.1850 under Manser Bradshaw, Master, then 22.2.1850 under John Huntley, Master, then 4.2.1856 under John W. McIntosh Hall, and 28.2.1856 under James Thomas Russell, Master; paid off 9.10.1856 at Chatham. Completed BU there 17.4.1858.

Apollo George Parsons, Bursledon

As built: 154ft 3½in, 129ft 9⅜in x 39ft 8in x 13ft 6in. 1,085⁷⁷⁄₉₄bm. Draught 10ft 11in/15ft 3in.

Ord: 7.11.1803. K: 4.1804. L: 27.6.1805. C: 26.9.1805 at Portsmouth.

First cost: £34,601 (£19,056 to builder plus £15,545 for fitting).

Commissioned 7.1805 under Capt Edward Fellowes; sailed for the Mediterranean 25.1.1806. Under Capt Alexander

Schomberg from 10.1806; in Hallowell's Squadron at Alexandria 3.1807. Under Capt. Bridges Taylor from 1808; participated in boat attack by Hallowell's squadron on Baudin's convoy in Rosas Bay. Fitted at Portsmouth (for £12,062) 3–8.1810. Recommissioned 5.1810; sailed for the Mediterranean 29.10.1810. Took 20-gun slaver *Merinos* off Cape Corso 13.2.1812, then 6-gun privateer xebec *L'Ulysse* off Corfu 17.9.1812. Boats (with those of *Weazle*) destroyed fort at St Cataldo 21.12.1812; occupied Lagosta 18.1.1813; landing party at St Cataldo 24.4.1813; boat action with gunboats (2 taken) off Faro 27.5.1813. Occupation of Paxo 13.2.1814. Taylor drowned 1814, and *Apollo* under Capt. Edwards Lloyd Graham from 6.1814, then Cmdr Anthony Valpy (acting) from 8.1814, later Capt. William Hamilton (still in Mediterranean). In Ordinary at Portsmouth from 1815. Underwent between Middling and Large Repair at Portsmouth (for £40,467) 5.1816–12.1817. Roofed over for and aft (for £1,219.8s.6d) 6.1819. Partially fitted as a Royal yacht by AO 20.11.1821 (for £14,408) 12.1821–6.1822, but not completed. Recommissioned 12.1821 under Capt. Sir Charles Paget 12.1821; paid off 21.5.1823 into Ordinary. Fitted as troopship at Portsmouth (for £11,404) 12.1837–4.1838, with 80 men, 6 x 18pdr carronades + 2 x 9pdrs; recommissioned 15.11.1841 under Cmdr Charles Frederick, for East Indies and China; on 5.3.1842 under Cmdr Charles George Edward Patey; Yangtze Operation in 7.1842; on 3.3.1845 under Cmdr William Radcliffe, then 29.12.1848 under Cmdr James Rawstorne; paid off 21.8.1851. Fitted as storeship at Chatham (for £5,340) 8–11.1853; recommissioned 12.10.1853 under George Johnson, Master, for the Mediterranean–then troopship in Black Sea to 1855; paid off 26.6.1856. BU completed at Portsmouth 16.10.1856.

Hussar Balthesar & Edward Adams, Bucklers Hard

As built: 154ft 0in, 129ft 8in x 40ft 0½in x 13ft 6in. 1,077bm. Draught 11ft 3in/15ft 1in.

Ord: 7.11.1803. K: 3.1806. L: 23.4.1807. C: 27.6.1807 at Portsmouth.

First cost: 34,326 (£18,199 to builder plus £16,127 fitting).

Commissioned 3.1807 under Capt. Robert Lloyd; sailed for the Leeward Islands and Halifax 16.1.1808. Escorted convoy home from Jamaica in 1809. Under Capt. Alexander Skene from 8.1809; in Baltic 1810. Recommissioned ?12.1810 under Capt. James Crawford; sailed for the East Indies 16.2.1811. At reduction of Java 8–9.1811; under Capt. George Elliott in East Indies in 1813. Underwent Large Repair at Deptford (for £30,684) 6–8.1814, then in Ordinary at Woolwich 1814–15, and at Chatham 1817–22. Small Repair at Chatham (for £3,859) 6–10.1822; fitted there for sea (for £13,288) 3–7.1823. Recommissioned under Capt. George Harris 3.1823 for Jamaica station; paid off at Chatham 2.11.1826. Recommissioned 2.1827 under Capt. Edward Boxer for Halifax station (flagship of Rear-Adm. Sir Charles Ogle); paid off 7.1830. Fitted at Chatham as receiving ship (for £2,815) 5–9.1833. Fitted 3.1861 as a target for gunnery experiments at Chatham 5–6.1861; set on fire at Shoeburyness while in use as target 7.1861 and allowed to burn out.

Horatio George Parsons, Bursledon

As built: 154ft 3in, 129ft 11¼in x 39ft 8½in x 13ft 7in. 1,089⁷⁴⁄₉₄bm. Draught 10ft 6in/15ft 1in.

Ord: 15.6.1805. K: 7.1805. L: 23.4.1807. C: 28.5–4.8.1807 at Portsmouth.

First cost: £22,266 to builder, plus £15,670 fitting.

Commissioned 6.1807 under Capt. George Scott (–1812); sailed with convoy for Halifax 7.9.1807; took (with *Latona*, *Driver* and *Superieure*) 40-gun *La Junon* 10.2.1809; took 26-gun *Le Nécessité* (while en flûte) 21.2.1810; sailed to meet East Indies ships 13.4.1810; in 9.1810 under Lord George Stuart (?temp.); her boats took Danish 6-gun schooner *No. 114* and 4-gun cutter *No. 97* on the Norwegian coast 2.8.1812. In 6.1814 under Capt. William Dillon, for North America; paid off 10.1.1817. Large Repair at Deptford (for £26,680) 3.1817–8.1819, then laid up. To Woolwich 1825, then Sheerness 1827. Converted to a screw blockship 11.1845–12.1849, then engine fitted (for £19,793) in East India Docks 12.1849–6.1850. Completed as screw guard ship at Chatham 6.1850–1.1851, then fitted for sea at Sheerness 12.1851–5.1852; full cost of conversion and fitting £42,804. Recommissioned 19.1.1852 under Capt. Swinfen Thomas Carnegie, as guard ship of Steam Reserve at Sheerness; on 2.6.1853 under Cmdr Robert Jenner, then 10.2.1855 under Capt. Arthur Auckland Cochrane; to the Mediterranean and back 9.1855 as troopship. Fitted as screw mortar ship at Sheerness (for £7,188) 10.1855, with 2 x 10in/95cwt mortars; paid off 12.5.1856 at Sheerness. Sold to Castle & Beech to BU at Charlton 1.11.1865.

Spartan Charles Ross, Rochester
> As built: 154ft 2in, 129ft 6¼in x 39ft 8in x 13ft 6in. 1,084bm. Draught 10ft 4in/15ft 0in.
> Ord: 24.8.1805. K: 10.1805. L: 16.8.1806. C: 6.10.1806 at Chatham.
> Commissioned 8.1806 under Capt. George Astle; on Channel Islands station 2.1807. Under Capt. Jahleel Brenton from 1807; sailed for the Mediterranean ?3.1807. In 5.1807 escaped from French squadron (74-gun *L'Annibal*, 40-gun *La Pomone*, 38-gun *L'Incorruptible* and 14-gun *La Victorieuse*). Unsuccessful boat attack on vessel off Nice 14.5.1807; shadowed Ganteaume's squadron 4.1808; operations (with *Imperieuse*) on Spanish coast 9.1808. Temporarily under Capt. Richard Thomas in 1.1809; with *Amphion* & *Mercury* at Pesaro 23.4.1809; with *Mercury* at Cesenatico 2.5.1809. Present at surrender of Cerigo 9.1809. In action with 40-gun *La Ceres*, 30-gun *Fama*, and 8-gun *Sparviero* (taken) and *L'Achille* in Bay of Naples 3.5.1810. Boats (with those of *Success* & *Espoir*) cut our vessels at Monte Cicero 25.8.1810. Recommissioned under Capt. Edward Pelham Brenton from 9.1810; sailed for North America ?25.7.1811. Took US privateers *Active* (2 guns) off Cape Sable 16.7.1812, *Actress* (4 guns) off Cape St Mary 18.7.1812, and *Intention* (1 gun) off Anapolis 19.7.1812; her boats (with *Maidstone's*) took US 1-gun privateers *Polly* and *Morning Star* in Bay of Fundy 1.8.1812; took (with *Maidstone*) US privateers 6-gun *Commodore Barry* and 2-gun *Madison*, *Olive* and *Spence*, all on 3.8.1812, and 16-gun privateer *Rapid* on St George's Bank 17.10.1812; paid off 1813. Middling Repairs at Portsmouth 10.1813–3.1814, then fitted there 12.1814–2.1815; recommissioned 12.1814 under Capt. Phipps Hornby, for the Mediterranean; home to Portsmouth 1817. Fitted for sea at Portsmouth 1–3.1818; recommissioned 1.1818 under Capt. William Wise; at Algiers 1818, then Halifax and West Indies in 1819; paid off 1.1821. BU at Plymouth 4.1822.

Undaunted Woolwich Dyd [M/Shipwright Edward Sison]
> As built: 154ft 9in, 130ft 3¼in x 39ft 7in x 13ft 6in. 1,086bm. Draught 10ft 6in/14ft 6in.
> (This ship had originally been ordered on 7.11.1803 from Joseph Graham, Harwich, but this builder went bankrupt and the order was transferred to Woolwich.)
> Ord: 6.1.1806. K: 4.1806. L: 17.10.1807. C: 2.12.1807.
> First cost: £36,967 (including fitting).

Commissioned 27.10.1807 under Capt. Thomas James Maling, for West Indies and the Channel; sailed for Portugal 7.5.1808; took 14-gun privateer *St Joseph* in the Channel 12.2.1809. In 6.1810 under Capt. George Mackenzie; sailed with convoy for Malta 30.8.1810. In 2.1811 under Capt. Richard Thomas; sailed for the Mediterranean 8.3.1811; her boats (with *Leviathan's*) destroyed privateer near Fréjus 29.4.1812, then same day her boats (with *Volontaire's* and *Blossom's*) destroyed a convoy at the mouth of the Rhone. In 1813 under Capt. Thomas Usher; landing party destroyed works near Marseilles 18.3.1813; her boats (with *Volontaire's* and *Redwing's*)near Morgion 30.3.1813; landing party (also from *Repulse*, *Volontaire* and *Redwing*) at Morgion 2.5.1813; her boats (with *Fleet's*, *Redwing's* and *Kitt's*) destroyed batteries and shipping at Cassis 18.8.1813; embarked Napoleon at Fréjus for transport to Elba 28.4.1814. In 11.1814 under Capt. Charles Smith; paid off at Chatham 10.1815. Large Repair at Chatham (for £27,331) 9.1816–12.1818; roofed over 5.1819 (extended 9.1819). Fitted for sea at Chatham (for £17,801) 5–11.1827; recommissioned 8.1827 under Capt. Augustus Clifford (–1830), for East Indies; paid off 11.1830. Fitted for sea at Portsmouth (for £7,253) 11.1830–2.1831; recommissioned 11.1831 for Cape of Good Hope and thence to East Indies; paid off 1.834 at Portsmouth. Waist housed over 6.1834. Fitted 5–6.1859 as target for gunnery experiments at Portsmouth (with iron plates bolted to her sides) and moored in Portchester Lake, then sunk during experiments. BU completed at Portsmouth 3.12.1860.

Menelaus Plymouth Dyd [M/Shipwright Joseph Tucker]
> As built: 154ft 1in, 129ft 6¼in x 39ft 6½in x 13ft 6in. 1,077¹⁵/₉₄bm. Draught 10ft 10in/14ft 8in.
> Ord: 28.9.1808. K: 11.1808. L: 17.4.1810. C: 21.6.1810 at Plymouth.
> First cost: £38,836 (including fitting).
> Commissioned 5.1810 under Capt. Peter Parker (Bart. 21.12.1811); sailed for Cape of Good Hope 11.7.1810. At capture of Mauritius 29.11–3.12.1810. Sailed for Mediterranean 7.11.1811. Boats took 16-gun *St Joseph* off Fréjus 29.2.1812; landing party at San Stefano 13.8.1812; boats took 2-gun privateer *St Esprit* at Mignone 2.9.1812; three sloops destroyed at Porto Ercole 3.9.1812; took storeship *Fidèle* 4.9.1812; took packet *L'Hirondelle* and 6-gun privateer *Le Nouveau Phoenix* 22.3.1813. In North America 1814; landing party in Chesapeake 30.8.1814 (Parker killed). Under Capt. Edward Dix from 9.1814; paid off 10.1815. Underwent Large Repair at Sheerness & Chatham (for £34,409) 9.1815–2.1819; then laid up at Chatham. Marine Hospital at Chatham 1834. Fitted at Chatham 9–10.1836 for the Quarantine Service at Stangate Creek. To Portsmouth 10.1841 to Quarantine Service, moored at the Motherbank (Spithead). Returned to Portsmouth Dyd 1896 and sold to John Read, Jnr (for £1,157) there 10.5.1897.

Nisus Plymouth Dyd [M/Shipwright Joseph Tucker]
> As built: 154ft 0in, 129ft 6in x 39ft 6in x 13ft 6in. 1,074¹¹/₉₄bm. Draught 10ft 7in/15ft 0in.
> Ord: 28.9.1808. K: 12.1808. L: 3.4.1810. C: 15.6.1810 at Plymouth.
> Commissioned 5.1810 under Capt. Philip Beaver; sailed for Cape of Good Hope 22.6.1810. At Cape station to 1812. Beaver died 10.4.1813; thence under Capt. Charles Marsh Schomberg. Arrived Plymouth with convoy from Brazil 3.1814 and paid off 2.5.1814. In Ordinary there until 6.1822 when intended for Small Repair but instead completed BU 24.9.1822.

Crescent Woolwich Dyd [M/Shipwright Edward Sison]
> As built: 154ft 5⅛in, 130ft 1in x 39ft 7in x 13ft 6in. 1,084⁹/₉₄bm.

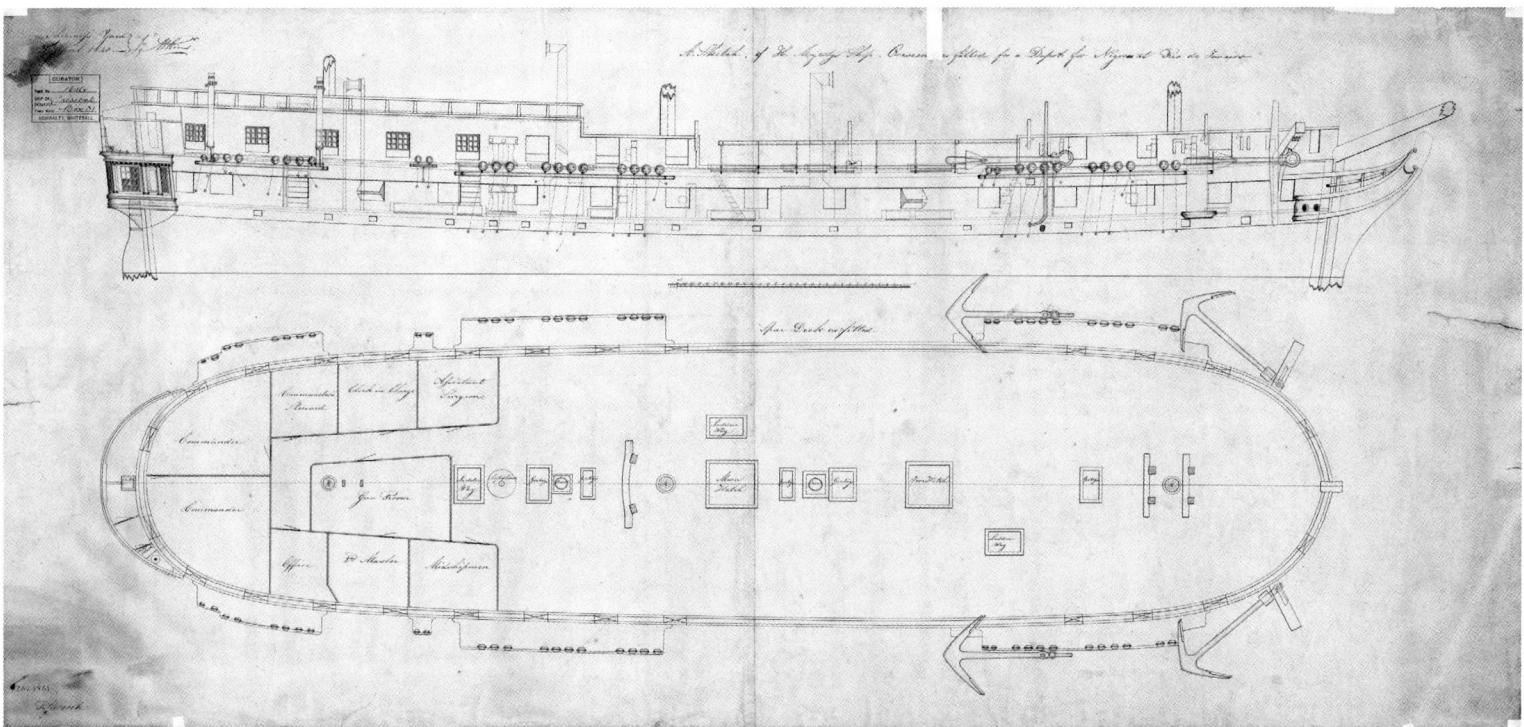

The *Lively* Class frigate *Crescent* had undergone a Major Repair following the Napoleonic War, which involved fitting a circular stern, and had been roofed over and stored in Ordinary at Sheerness. In 1839 it was decided to employ her as a floating depot for freed slaves at Rio de Janeiro, and this inboard profile, and plan of the quarterdeck, covered-in waist and forecastle, was drawn and signed on 21 April 1840 by James Atkins, the master shipwright at Sheerness from 1839 to 1848. *(NMM J3936)*

Draught 10ft 9in/14ft 10in.

Ord: 28.9.1808. K: 9.1809. L: 11.12.1810. C: 7.2.1811.

Oak from Holstein was used in some of the floor timbers and first futtocks.

First cost: £39,792 (including fitting).

Commissioned 12.1810 under Capt. John Quilliam. In Baltic 1812; sailed for Newfoundland 3.4.1813; took 14-gun US privateer *Elbredge Gerry* 16.9.1813. In North America 1814, but returned to UK and paid off 4.9.1815. Large Repair at Woolwich (for £33,881) 8.1817–9.1820. Laid up at Sheerness (roofed over) 10.1820. Fitted at Sheerness (9.1839–1.1840) as a receiving ship for freed slaves at Rio de Janeiro; on 29.11.1839 under Lieut Malachi Donellan, then 9.7.1844 under Lieut William Glassford Hemsworth, then 5.2.1847 under Lieut Thomas Charles Meheux, and 5.1849 under George Lee Bradley, Master all at Rio; paid off 31.12.1853. Sold there 1854.

Bacchante Deptford Dyd [M/Shipwright Robert Nelson]

As built: 154ft 0½in, 129ft 7½in x 39ft 6¼in x 13ft 6¼in. 1,076⁶⁹/₉₄bm. Draught 10ft 7in/14ft 6in.

Ord: 12.6.1809. K: 7.1810. L: 16.3.1811. C: 25.1.1812.

First cost: £42,145 (including fitting).

Commissioned 24.11.1811 under Cmdr David Dunn (acting); on 21.3.1812 under Capt. William Hoste; sailed for the Mediterranean 3.6.1812; took French 3-gun privateer *La Victoire* 26.7.1812; her boats took several vessels near Canale de Leme 31.8.1812, then several near Tiemite 18.9.1812; took (with boats also of *Weazle*) five 2-gun gunboats off Otranto 6.1.1813 (*L'Indomptable*, *La Diligente*, *L'Arrogante*, *La Salamino* and *La Calypso*); then captured and destroyed French 2-gun *L'Alcinous* 14.2.1813. Destroyed batteries at Carlopago 15.5.1813; boats captured convoy at Giulianovo 12.6.1813; participated in Fremantle's capture of Fiume [Rijeka] 3.7.1813; with *Eagle* at Rovigno 2.8.1813; with *Saracen* at Ragusa [Dubrovnik] 12.10.1813 and surrender of Cattaro [Kotor] 28.1.1814. Under Capt. Francis Stanfell from 3.1814. Destruction of American *John Adams* 5.9.1814. Paid off and laid up at Portsmouth 7.1815, housed over fore and aft. Underwent Middling Repair at Deptford (for £25,731) 8.1819–3.1822, then laid up at Chatham. Fitted at Chatham for Quarantine Dept (for £794) 12.1836–2.1837; to quarantine service at Stangate Creek 1837–1846. At Chatham 1850 and Deptford 1855. Fitted at Deptford as a temporary hospital 10.1856. BU completed at Chatham 20.2.1858.

Nymphe George Parsons, Warsash

As built: 154ft 2in, 129ft 10½in x 39ft 8in x 13ft 6in. 1,087⁹⁰/₉₄bm. Draught 10ft 5in/14ft 4in.

Ord: 14.12.1810. K: 1.1811 as *Nereide* (renamed 1811) L: 13.4.1812. C: 22.6.1812 at Portsmouth.

First cost: £32,538 (£23,739 to builder plus £8,799 for fitting).

Commissioned 5.1812 under Capt. Farmery Predam Epworth; sailed for North America 9.7.1812. Chased by USS *President* and *Congress* 10.10.1812. On American station, took US privateers *Montgomery* (12 guns) 5.5.1813, *Juliana Smith* (3 guns) 12.5.1813, *Thomas* (12 guns) off Nova Scotia 29.6.1813, and *Paragon* (11 guns) 14.8.1813. Under Capt. Joseph Price from 6.1814, then Capt. Hugh Pigot from 10.1814. Underwent Middling Repair at Portsmouth (for £24,611) 7.1815–5.1816, the laid up in Ordinary (housed over 12.1820). Fitted as temporary hospital ship at Portsmouth 8.1831; at Leith by 1834. Fitted at Chatham as receiving ship (for £2,744) 12.1835–5.1837; at Sheerness 1838–1855 as receiving ship. Fitted for Water Police at Sheerness 7.1861. Altered at Sheerness to a Roman Catholic chapel 1–3. 1863. Renamed *Handy* by AO 7.9.1871. BU at Chatham by AO 31.12.1874 completed 9.3.1875.

Sirius Richard Blake & John Tyson, Bursledon (the last ship to be built at Bursledon for the RN)

As built: 155ft 0⅛in, 129ft 7⅜in x 39ft 9in x 13ft 7in. 1,089$\frac{59}{94}$bm. Draught 10ft 8in/14ft 6in.

Ord: 14.12.1810. K: 9.1811. L: 11.9.1813. C: 29.9.1815 at Portsmouth (for Ordinary).

First cost: £31,990 (£23,551 to builder plus £8,439 fitting).

Not commissioned. Large Repair at Portsmouth (for £22,126) 12.1819–8.1821. Very Small Repair at Portsmouth (for £4,937) 12.1829–5.1830. Fitted with iron plates on sides for gunnery experiments there (for £1,852) 7.1860. BU at Portsmouth 26.9.1862.

Laurel John Parsons & John Rubie, Warsash

As built: 154ft 4in, 129ft 11¾in x 39ft 8in x 13ft 6in. 1,084$\frac{9}{94}$bm. Draught 10ft 7in/14ft 5in.

Ord: 21.3.1812. K: 7.1812. L: 31.5.1813. C: 13.9.1813 at Portsmouth.

First cost: £42,290 (£24,109 to builder plus £18,181 fitting).

Commissioned 7.1813 under Capt. Granville Proby. On Cape station 1814, but returned to UK and laid up at Portsmouth 11.1814. Middling Repair at Portsmouth (for £22,704) 6.1823–7.1825, then housed over. Fitted at Portsmouth as a 'lavatory' by AO 10.9.1864 (wash house 1870–1880). Sold there to Castle & Sons 11.1885.

HYPERION 32 guns. Design by Sir John Henslow, from lines of *La Magicienne* (French prize taken 1781).

Dimensions & tons: 143ft 9in, 118ft 8⅛in x 39ft 2½in x 12ft 4½in. 970$\frac{39}{94}$ bm.

Men: 254. Guns: UD 26 x 18pdrs; QD 2 x 9pdrs + 10 x 24pdr carronades; Fc 4 x 9pdrs + 2 x 24pdr carronades.

Hyperion William Gibson, Hull

As built: 143ft 9in, 118ft 7⅝in x 39ft 4½in x 12ft 4in. 978$\frac{34}{94}$bm. Draught 10ft 0in/13ft 6in.

Ord: 13.6.1805. K: 2.1806. L: 3.11.1807. C: 9.12.1807–23.4.1808 at Chatham.

First cost: £20,165 to builder.

Commissioned 1.1808 under Capt. Thomas Brodie (died 3.1811); sailed for the Mediterranean 29.6.1808; sailed for Jamaica 14.1.1810. In 5.1811 under Capt. William Cumby (–1815); sailed for Newfoundland 13.5.1812; took US 16-gun privateer *Rattlesnake* 3.6.1814. In 1.1815 under Capt. James Lillecrap (temp.); paid off 8.1815 and laid up at Portsmouth. Small Repair and fitted for sea at Portsmouth 6–11.1818; recommissioned 9.1818 under Capt. Thomas Searle, for South American station; paid off 4.1821. Recommissioned 4.1821 under Lillicrap again (–1822), for Cape of Good Hope. Fitted for the Coast Blockade at Sheerness 1–3.1825, for Newhaven; recommissioned 1.1825 under Capt. William Mingaye; paid off 5.1831. BU at Portsmouth 6.1833.

BUCEPHALUS 32 guns. Design by Sir William Rule, from lines of *La Topaze* (French prize taken 1793); however, from 1814 she was classed as a troopship.

Dimensions & tons: 150ft 0in, 126ft 6⅛in x 38ft 0in x 12ft 1in. 971$\frac{66}{94}$bm.

Men: 254. Guns: UD 26 x 18pdrs; QD 2 x 9pdrs + 10 x 24pdr carronades; Fc 2 x 9pdrs + 2 x 24pdr carronades.

Bucephalus William Rowe, Newcastle

As built: 150ft 0in, 126ft 5⅝in x 38ft 1in x 12ft 0in. 975$\frac{61}{94}$bm.

Ord: 19.6.1805. K: 8.1806. L: 3.11.1808. C: 2.3–17.6.1809 at Chatham.

First cost: ?

Commissioned 3.1809 under Capt. Charles Pelly; in Scheldt operations 1809; sailed for East Indies 15.11.1809. Under Capt. Joseph Drury ?5.1812, then Capt. Barrington Reynolds in 1813. Fitted as troopship at Woolwich 1–5.1814; under Cmdr George d'Aeth 3.1814, for New Orleans operations. Under Cmdr Amos Westropp 6.1815. Laid up at Portsmouth 8.1816. Fitted as Receiving ship at Portsmouth 6.1822. BU at Portsmouth 9.1834.

PYRAMUS 42 (originally 36) guns. Design to lines of *La Belle Poule* (French prize taken 1780).

Dimensions & tons: 140ft 0in, 115ft 11⅝in x 38ft 2in x 11ft 11in. 898$\frac{59}{94}$bm.

Men: 264. Guns: UD 26 x 18pdrs; QD 2 x 9pdrs + 10 x 32pdr carronades; Fc 2 x 9pdrs + 2 x 32pdr carronades.

Pyramus (Charles) Greensword & (Thomas) Kidwell, Itchenor, near Chichester

As built: 141ft 1in, 117ft 2⅝in x 38ft 5in x 12ft 0in. 920$\frac{18}{94}$bm. Draught 10ft 3in/12ft 9in.

Ord: 29.6.1805. K: 4.1806. Builder failed 1807; order and frames transferred to: Portsmouth Dyd.

Re-ord: 5.1808. K: 11.1808. L: 22.1.1810. C: 4.5.1810.

First cost: (at Itchenor) £10,120 to builder; (at Portsmouth) £18,842, including fitting.

Commissioned 1.1810 under Capt. Charles Dashwood (–1812), for the Baltic. In 9.1812 under Capt. James Whitley Deans Dundas, for Channel in 1813; took (with *Belle Poule*) 10-gun US privateer *Zebra* 20.4.1813; took 14-gun privateer *La Ville de l'Orient* 14.2.1814. Fitted for Ordinary at Plymouth (for £736) 4.1814; in Ordinary there 1814–1820. Between Middling and Large Repair at Plymouth (for £32,127) 3–5.1819, then laid up again. Fitted for sea at Plymouth (for £9,104) 2–6.1821; recommissioned 2.1821 under Capt. Francis Newcombe, for the West Indies; on 18.6.1825 under Capt. Robert Gambier, at Portsmouth; on 6.1826 under Capt. George Sartorius; paid off 1828. Very Small Repair at Plymouth (for £8,477) 1–4.1829, then laid up there. Fitted as receiving and convict ship for Halifax, Nova Scotia 11.1832–7.1833; at Halifax 1834–75 Sold to BU (for £1,600) there 10.11.1879.

Ex-French Prizes

The standard French 18pdr frigate (with 28 x 18pdrs on the UD) was rated at 40 guns by the French Navy (here including its Dutch and Italian dependants), but were re-rated at 38 guns when added to the British Navy. The survivors were all re-rated as 46s in January 1817. The shorter type (with only 26 x 18pdrs) were registered as 38-gun in the French Navy but became 36-gun when added to the RN; unless otherwise stated all the following were of the former type. While no enemy prizes were ever considered or treated by the Admiralty or Navy Boards as belonging to a 'class' in the modern sense of built to a common draught, the large number of these vessels captured by and added to the RN makes it convenient and useful to consider them as complete 'classes' in the same design sense as with British-built vessels.

***La JUNON* Class** (Joseph-Marie-Blaise Coulomb design of 1782). Six frigates were built to this draught, all at Toulon; all were added to the RN by the end of 1799. The first pair, *La Junon* and *La Minerve* (i), were completed as 38-gun ships, with 26 x 18drs on the UD; the last four – *L'Impérieuse*, *La Melpomène*, *La Perle* and *La Minerve* (ii)

– were completed as 40-gun ships, with an extra pair of ports forwards (chase ports) for 18pdrs. This distinction was preserved upon their registration in the British Navy, although each ship was registered with one fewer (QD) pair of secondary guns. By any standards, these were remarkably well-built ships: the *Amethyst* (ex-French *La Perle*) was wrecked in 1795, *Minerve* (ii) was retaken in 1803 and *Melpomene* was BU in 1815, but the others were repaired and lasted well for several decades of RN service.

Unite (ex-*Imperieuse*) (ex-French *L'Impérieuse*, built 2.1786–5.1788 at Toulon. L: 11.7.1787), 42 guns.

Dimensions & tons: 148ft 6in, 124ft 10in x 39ft 7in (39ft 0in mld) x 12ft 8in. 1,040³¹⁄₉₄bm.

Men: 315. Guns: UD 28 x 18pdrs; QD 8 x 9pdrs + 6 x 32pdr carronades; Fc 2 x 9pdrs + 2 x 32pdr carronades.

Taken by *Captain* off Genoa 10.10.1793. Arrived Chatham 7.12.1794 and registered 30.3.1795; fitted there (for £8,432) 6–10.1795. Renamed *Unite* 3.9.1803. Fitted for Trinity House 10.1803. Repaired by Hill & Mellish, Limehouse (for £12,144) 9.1804–4.1805.

Commissioned 4.1805 under Capt. Charles Ogle. Completed fitting at Deptford (for £9,449) 6.1805. Made good defects at Plymouth (for £14,324) 9–12.1809. Underwent between Middling and Large Repair at Deptford (for £30,536) 4.1815–10.1816. Laid up at Chatham, housed over fore and aft. Underwent between Small and Middling Repair at Chatham (for £10,042) 12.1825–4.1826. Fitted at Woolwich as a convict hospital ship 9–10.1836. Completed BU at Chatham 9.1.1858.

San Fiorenzo (ex-French *La Minerve* (i), built 1.1782–10.1782 at Toulon. L: 31.7.1782), 42 guns

Dimensions & tons: 148ft 8in, 124ft 4⅛in x 39ft 6in (38ft 11in mld.) x 13ft 3in. 1,031⁸⁶⁄₉₄bm.

Men: 274. Guns: UD 26 x 18pdrs; QD 6 x 6pdrs + 6 x 32pdr carronades; Fc 2 x 9pdrs + 2 x 32pdr carronades.

Found scuttled at San Fiorenzo, Corsica 19.2.1794 (after damage by English shore batteries on 18.2.1794); raised and added to RN.

Initially under Capt. Charles Tyler; from 7.1794 under Capt. Sir Charles Hamilton. Arrived Chatham 22.11.1794, and registered 30.5.1795; completed fitting at Chatham (for £8,459) 14.8.1795.

Commissioned 6.1795 under Capt. Sir Harry Neale (–1800), for Weymouth; took (with *Nymphe*) 40-gun *La Résistance* and 22-gun *La Constance* off Brest 9.3.1797; took 14-gun privateer *L'Unité* off Owers 3.6.1797; during Nore Mutiny, escaped to Harwich; took 14-gun privateer *Le Castor* off Scilly Isles 1.7.1797; took (with *Triton*) 14-gun privateer *La Rosée*, in the Channe; 12.1798, also took Spanish 6-gun privateer. Action with three frigates off Belleisle 9.4.1799; attempt on Spanish squadron in Aix Roads 2.7.1799. Under Capt. Charles Paterson in 1.1801, in the Mediterranean. Recommissioned 5.1802 under Capt. Joseph Bingham (–1804); sailed for the Cape of Good Hope; her boats took 2-gun chasse-maree *Le Passe-Partout* on the Malabar Coast 14.1.1804. Under Capt. Walter Bathurst 1805–06; under Capt. (acting) Henry Lambert, action with 36-gun *La Psyché* (captured) and 4-gun *L'Equivoque*, off Ganjam 13.2.1805. Under Capt. Patrick Campbell 1806–07, still in East Indies, then Capt. George Hardinge; chase and capture of 40-gun *La Pièmontaise* 6–8.3.1808 (Hardinge killed); under Capt. John Bastard, paid off later in 1808. Fitted for Baltic service at Woolwich 3–5.1809. Recommissioned 3.1809 under Capt. Henry Matson; in Walcheren operation 1809; paid off 3.1810. Fitted as troopship at Woolwich 6–10.1810; recommissioned 5.1810 at Lisbon as 22-gun troopship under Cmdr Edmund Knox. Fitted as receiving

ship at Woolwich 9.1812; then to Ordinary at Chatham. Fitted as lazarette at Sheerness 4–8.1818, for Stangate Creek. BU at Deptford 9.1837.

Andromache (ex-*Princess Charlotte*) (ex-French *La Junon*, built 2.1782–2.1783 at Toulon. L: 14.8.1782), 44 guns

Dimensions & tons: 148ft 10in, 124ft 9in x 39ft 4½in (38ft 9½in mld.) x 12ft 10in. 1,028⁷³⁄₉₄bm.

Men: 264. Guns: UD 26 x 18pdrs; QD 2 x 9pdrs + 12 x 32pdr carronades; Fc 2 x 9pdrs + 4 x 32pdr carronades.

Taken 18.6.1799 by Markham's squadron in the Mediterranean.

Commissioned 11.1799 in the Mediterranean under (temp.) Capt. Thomas Masterman Hardy, later under Capt. Thomas Stephenson then Capt. Sir Edward Berry. Arrived Woolwich 11.1.1801; completed refitting there (for £13,930) 7.1801. Middling Repair and fitted at Plymouth (for £23,850) 10.1803–7.1804; recommissioned 5.1804 under Capt. Francis Gardner; in West Indies 1804–06; took 11-gun privateer *Le Regulus* 13.12.1803; under Capt. George Tobin in 1805 (–1814), recaptured 26-gun *Cyane* off Tobago 5.10.1805. On Irish station and Channel 1807–14; took 14-gun privateer *L'Aimable Flore* 9.1.1811. Renamed *Andromache* 6.1.1812. Took 14-gun privateer *Le Sans-Souci* 15.12.1812; with Collins's squadron off St Sebastian in 1813, took 40-gun *La Trave* 23.10.1813; took schooner *Le Prospère* 20.12.1813; took US 4-gun privateer *Fair American* 18.1.1814; took 14-gun privateer *La Comète* off Bordeaux 14.3.1814; paid off 12.1814. Large Repair at Deptford (for £28,270) 8.1814–7.1815, then to Ordinary Fitted for foreign service at Chatham (for £12,378) 11.1817; recommissioned 9.1817 under Capt. William Shirreff (–1821), for Brazil. Made good defects at Portsmouth (for £14,680) 9.1821–3.1822. Recommissioned 1.1822 under Capt. Joseph Nourse (died 4.9.1824), for Cape of Good Hope; on 29.12.1824 under Capt. Constantine Richard Moorsom; paid off 3.9.1825. In Ordinary at Woolwich 1825–28. BU completed at Deptford 14.6.1828.

L'HÉBÉ Class An early (1782) design by Jacques-Noël Sané, to which six ships were built, of which three were taken by the RN; the prototype *L'Hébé* was taken by the RN in 1782 and was BU in 1811; *La Vénus* was wrecked in 1789, *La Sibylle* below was taken in 1794, *La Proserpine* taken in 1796 (renamed **Amelia**, and BU in 1816), *La Carmagnole* was wrecked in 1800 and *La Dryade* was disposed of in 1801.

Sybille (French *La Sibylle*, built 4.1790–5.1792 at Toulon. L: 30.7.1791), 48 guns

Dimensions & tons: 154ft 3in, 127ft 4¾in x 40ft 1½in x 12ft 4in. 1,090⁹¹⁄₉₄bm.

Men: 300 (later 315). Guns: UD 28 x 18pdr; QD 12 x 9pdr; Fc 4 x 9pdr. By 1799 QD had 4 x 9pdrs + 12 x 32pdr carronades, Fc 2 x 9pdrs + 2 x 32pdr carronades; later QD 8 x 9pdrs + 6 x 32pdr carronades.

Taken 17.6.1794 by *Romney* at Miconi in the Mediterranean. Fitted at Portsmouth (for £7,705) 30.11.1794–17.4.1795.

Commissioned 4.1795 under Capt. Edward Cooke (died 25.5.1799), for Sidney Smith's squadron; in the Mediterranean 1796, to Cape of Good Hope and East Indies stations 1797–99; with *Fox* at Manilla 1.1.1798; took 50-gun *La Forte* off Sand Heads 1.3.1799, with 5 dead and 17 wounded (Cooke mortally); in 3.1799 under Cmdr Joseph Turner (acting), then Capt. William Waller (acting). In 6.1799 under Capt. Charles Adam (–1803); took (with *Daedalus*, *Centurion* and *Braave*) Dutch brig 23.8.1800 (name unknown, but became HMS *Admiral Rainier*); took 36-gun *La*

Chiffonne off the Seychelles 19.8.1801; paid off 4.1803. Major Repair by Dudman, Deptford (for £22,224) 9.1804–7.1805, then fitted at Deptford Dyd (for £12.690) 8.1805; recommissioned 7.1805 under Capt. Robert Winthrop, for Channel Fleet; took French 4-gun privateer *L'Oiseau* 3.5.1807. In 8.1807 under Capt. Clotworthy Upton (–1813); in Copenhagen expedition 8.1807; took French privateers – 4-gun *Le Grand Argus* 25.1.1808 and 16-gun *L'Espiègle* 16.8.1808; made good defects at Plymouth (for £11,286) 12.1808–2.1809; to Newfoundland in Summer 1809; took more French privateers – 14-gun *L'Edouard* off the Irish coast 10.1810, 14-gun *L'Aigle* 10.5.1812, and 14-gun *Le Brestois* 5.2.1813 (after sailing 11.1812 to North America); sailed for Newfoundland again 25.3.1813. In 9.1813 under Capt. James Sanders, then 1814 under Capt. Thomas Forrest, to Greenland. Between Middling and Large Repair at Woolwich (for £30,477) 4.1815–10.1816; re-classed as 48-gun in 1.1817; fitted for foreign service at Chatham (for £12,642) 11.1817; recommissioned 9.1817 under Capt. Charles Malcolm, as flagship of Rear-Adm. Sir Home Popham at Jamaica; in 5.1819 under Capt. William Popham. In 8.1820 under Capt. Joshua Rowley, as flagship of Rear-Adm. Sir Charles Rowley at Jamaica. Made good defects at Deptford (for £15,604) 7–10.1823; recommissioned 7.1823 under Capt. Samuel Pechell, for the Mediterranean; destroyed four pirate vessels 18.6.1826; paid off 11.1826. Made good defects at Portsmouth (for £14,734) 11.1826–4.1827; recommissioned 12.1826 under Capt. Francis Collier (–1830), for the African station; took slavers *Henriquetta* 6.9.1827, *Diana* 12.10.1827, *Gertrudis* 12.1.1828, *Fanny* 19.3.1828, *Vengador* 16.5.1828, *Josephine* 4.7.1828, *Almirante* 1.2.1829, *Uniao* 6.2.1829, *Donna Barbara* 21.2.1829, *Carolina* 6.3.1829, *Hosse* 23.3.1829, *La Panchita* 29.4.1829, *Emilia* 16.8.1829, *Cristina* 11.10.1829, *Tentadora* 1.11.1829, *Nossa Senhora de Guia* 7.1.1830, *Umbelino* 15.1.1830, *Primera Rasalia* 23.1.1830 and *Manzanares* 1.4.1830; paid off 7.7.1830 at Portsmouth. Fitted at Portsmouth 7.1831 as a lazarette for Dundee. Sold to Mr Henry (for £2,460) 7.8.1833.

La REVOLUTIONNAIRE Class (Pierre-Alexandre Forfait design of 1793). Four frigates were built to this design in 1793–95, all at Le Havre – *La Seine*, *La Révolutionnaire*, *La Spartiate* and *L'Indienne* – of which the first two were taken by and added to the British Navy, but the Seine was lost in 1803.

Revolutionaire (French La Révolutionnaire, built 10.1793–7.1794 at Le Havre. L: 28.5.1794), 46 guns.
   Dimensions & tons: 157ft 2in, 131ft 9⅞in x 40ft 5½in (39ft 10in mld.) x 12ft 6in. 1,147⁶⁸⁄₉₄bm.
   Men: 280 (later 315). Guns: (by 1812) UD 28 x 18pdr; QD 8 x 9pdrs + 6 x 32pdr carronades; Fc 2 x 9pdr + 2 x 32pdr carronades. (Note: from 1803 to 1812 had carried 32pdr carronades on UD in lieu of 18pdr.)
   Taken 21.10.1794 by *Artois* and others of Pellew's squadron off Brest. Fitted at Portsmouth (for £9,014) 4.11.1794–7.5.1795.
   Commissioned 4.1795 under Capt. Francis Cole (died 18.4.1798); at Battle off Île Groix 23.6.1795; in Pellew's squadron 1796; took 38-gun *L'Unité* off Brest 12.4.1796. In 4.1798 under Capt. Thomas Twysden (died 9.1801); took privateers – 16-gun *La Victoire* 30.5.1799, *L'Hyppotine* 2.6.1799, 18-gun *Le Determiné* 29.6.1799, and (with *Phoebe*) 26-gun *Le Bordelais* and 18-gun *La Grand Ferroilleur* ?11.10.1799; chase of Bompart's stragglers 15.10.1799; took more privateers on Irish station – 14-gun *Le Coureur* 4.3.1800 and 16-gun *Le Moucheron* 16.2.1801. In 10.1801 under ?Cmdr Murray (temp), then 6.1802 under Capt.

Thomas Capel, still on Irish station. Recommissioned 4.1803 under Capt. Walter Lock; sailed for Gibraltar 5.6.1803; in 8.1803 under Capt. Robert Hall, for Channel service. Recommissioned 4.1804 under Henry Hotham; in Strachan's Action 3.11.1805. In 2.1806 under Capt. Charles Feilding, in Channel service. Large Repair at Plymouth 10.1811–12.1812; recommissioned 10.1812 under Capt. John Woolcombe (–1815); in Collier's squadron off North coast of Spain 1813, later to North America; took US 1-gun privateer *Matilda* 25.7.1813; sailed with convoy to East Indies 31.12.1813. Very Small Repair at Plymouth 2–4.1817; fitted for sea there 8.1818–1.1819; recommissioned 25.8.1818 under Capt. Fleetwood Broughton Pellew (–1822), for the Mediterranean; paid off 5.7.1822 at Plymouth. BU at Plymouth 4.10.1822.

La VIRGINIE Class (Jacques-Noël Sané design of 1793.) Eight vessels were begun to this design 1793–99 (of which *Le Zéphyr* was never launched); two were taken by the RN during the 1793–1801 period, but *La Justice* (taken 1801) was handed over by the RN to the Turks. Two more (*La Didon* and *La Volontaire*) were captured in 1805 and 1806. From 1801 onwards, four further ships were built to this Sané design, of which three were taken by and added to the RN over the period (the fourth, *L'Atalante*, was burnt in 1805). *Didon* was BU in 1811, *Surveillante* in 1814, and *Volontaire* and *Belle Poule* in 1816, by which time *Virginie* had been re-classed as a receiving ship.

Virginie (French *La Virginie*, built 11.1793–12.1794 at Brest. L: 26.7.1794)
   Dimensions & tons: 151ft 3¼in, 126ft 3¼in x 39ft 10in (39ft 3½in mld.) x 12ft 8in. 1,065⁶⁄₉₄bm.
   Men: 284 (later 315). Guns: UD 28 x 18pdr; QD 12 x 32pdr carronades; Fc 2 x 9pdr + 2 x 6pdrs + 2 x 32pdr carronades.
   Taken 22.4.1796 by *Indefatigable* in the Atlantic (SW of the Lizard). Fitted at Plymouth (for £12,360) 25.4–4.11.1796. Named and registered 25.6.1796.
   Commissioned 8.1796 under Capt. Anthony Hunt (died 8.1798); sailed for East Indies 11.1797 (carrying Lord Mornington). In ?8.1798 under Capt. George Astle; took 12-gun *Braak* and *Helena* and 8-gun *Helena Prau* in the East Indies 26.4.1799; remained in East Indies; home to pay off 2.1803. Fitted at Deptford 5–8.1803; recommissioned 7.1803 under Capt. John Poo Beresford, but paid off 8.1804. Repaired (after Survey at Sheerness 7–10.1804) by Graham, Harwich (for £9,176) 11.1804–8.1805, then fitted at Chatham 8–10.1805; recommissioned 9.1805 under Capt. Edward Brace (–1810); took Spanish 14-gun privateer *Vengador* 9.4.1806; sailed with convoy for West Indies 23.2.1807; on Irish station 7.1807; took 14-gun privateer *Le Jesus-Maria-Josef* 28.9.1807 and 36-gun *Gelderland* in the North Sea 19.5.1808; paid off 3.1810. Fitted at Plymouth as receiving ship 2.1811; in Ordinary there to ?1816, then as receiving ship 1817–25. Sold to Mr Freake (for £3,050) 8.7.1826, but retained when Freake was declared insane and re-sold (for £2,310) to BU 11.7.1827.

Rhin (French *Le Rhin*, built 6.1801–10.1802 at Toulon. L: 15.4.1802), 46 guns
   Dimensions & tons: 152ft 1in, 127ft 1½in x 39ft 11½in (39ft 4in mld.) x 13ft 0in. 1,079⁶⁄₉₄bm.
   Men: 284 (later 315). Guns: UD 28 x 18pdrs; QD 14 x 32pdr carronades; Fc 2 x 9pdrs + 2 x 32pdr carronades.
   Taken 27.7.1806 by *Mars* off Rochefort. Arrived Plymouth 8.8.1806. Repaired and fitted there (for £24,302) 3–8.1809.
   Commissioned 6.1809 under Capt. Frederick Aylmer, for the

Channel; in 7.1809 under Capt. Charles Malcolm (–1815); took privateers – 16-gun *Le Navarrois* in the Bay of Biscay 22.3.1810, 14-gun *Le San Josef* off the Lizard 29.9.1810, 14-gun *La Contesse de Montalivet* 9.10.1810, 10-gun *Le Braconnier* 2.1811, and 15-gun *La Courageuse* off Eddystone 9.11.1811; with Popham's squadron on the North coast of Spain 1812; took US 16-gun privateer *Decatur* 5.6.1812; sailed for North America 3.4.1813; paid off 1815. Large Repair at Sheerness (for £29,204) 5.1817–8.1820, then laid up (roofed over). Fitted at Chatham as a lazarette 5–10.1838, and to Sheerness. Lent to the Sub-committee for the Inspection of Shipping on the Thames as a smallpox hospital ship 9.9.1871. Sold to Castle & Sons, Charlton (for £1,250) by AO 26.5.1884.

La ROMAINE **Class** (Pierre-Alexandre Forfait design) Survivor of a large class of Forfait-designed *frégates-bombardes* (nine of this design were launched 1794–96; a further eleven ordered 1793–94 were not built, or were built to altered designs) originally to carry 20 x 24pdrs and 12 x 8pdrs, plus a 12in mortar, but by 1798 carrying 24 x 24pdrs, 14 x 8pdrs and 2 or 4 x 36pdr brass obusiers. Her sister *L'Immortalité* taken in 1798 was BU in 1806, while a third prize (*Le Libre* taken 12.1805) was never added to the RN. The 24pdrs were deemed too heavy for British use, and both *Immortalite* and *Desiree* were re-armed with 26 x 18pdrs, requiring the permanent use of the chase ports. Notwithstanding the rounded Forfait hull form, these ships had relatively limited stowage, and were chiefly restricted to operations in home waters (*Desiree* was briefly deployed to Jamaica in 1802–04 and to the East Indies in 1814–15).

Desiree (French *La Désirée*, 38 guns, built 2.1794–12.1798 at Dunkirk. L: 23.4.1796), 36 guns

> Dimensions & tons: 147ft 3in, 124ft 2¼in x 39ft 2¾in (38ft 7¾in mld.) x 11ft 9in. 1,016⁵⁹⁄₉₄bm.
>
> Men: 264. Guns: UD 26 x 18pdrs; QD 2 x 9pdrs + 8 x 32pdr carronades; Fc 2 x 9pdrs + 2 x 32pdr carronades.
>
> Cut out of Dunkirk Roads by *Dart* and her squadron 8.7.1800; arrived Sheerness 12.7.1800 and fitted there (for £10,258) 7–11.1800.
>
> Commissioned 8.1800 under Capt. Henry Inman; at Battle of Copenhagen 2.4.1801. In 1802 under Capt. Richard Dacres; sailed for Jamaica 2.1802. Later in 1802 under Capt. Samuel Linzee, then 15.10.1802 Capt. Charles Bayntun Hodgson Ross. In 2.1804 under Capt. Henry Whitby; her boats took privateer *La Jeune Adelle* on the Jamaica station 24.2.1804; on 19.4.1805 under Capt. Henry Heathcote, for passage home; paid off 1805. Large Repair at Portsmouth 9.1808–11.1809; recommissioned 8.1809 under Capt. Arthur Farquhar (–1814), for Texel operations; her boats (with others') destroyed a 6-gun privateer and took two others in the Vlie 29.5.1810; took 14-gun privateer *La Velocifere* off Texel 10.3.1811 and 16-gun privateer *Le Brave* off the Vlie 12.12.1811; Farquhar became Senior Officer at Heligoland 10.1813; operations in the Elbe 11.1813–1.1814. In 3.1814 under Capt. William Woolridge; to East Indies in 1814. In 1815 under Capt. Philip Carteret. Laid up at Sheerness 8.1815, then fitted as a slop ship 1–12.1823. Sold to Joseph Christie at Rotherhithe (for £2,020) 22.8.1832.

LOIRE A one-off design by Pierre Degay, she was rather leewardly and had relatively poor stowage capacity.

Loire (French *Le Loire*, ex-*Le Loire Inférieure*, built 4.1794–12.1797 at Nantes. L: 23.3.1796), 46 guns

> Dimensions & tons: 153ft 8in, 128ft 2⅛in x 40ft 2in x 12ft 11¼in.

1,100³¹⁄₉₄bm.

> Men: 284 (later 315). Guns: UD 28 x 18pdrs; QD 8 x 9pdrs + 4 x 32pdr carronades (from 1807, 14 x 32pdr carronades); Fc 4 x 9pdrs + 2 x 32pdr carronades (from 1807, 2 x 9pdrs + 4 x 32pdr carronades).
>
> Taken 18.10.1798 by *Anson* and *Kangaroo* of Warren's squadron off Ireland. Arrived Plymouth 27.10.1798. Registered 31.12.1798. Fitted at Plymouth 1–6.1799.
>
> Commissioned 3.1799 under Capt. James Newman (–1801); took (with consorts) 38-gun *Le Pallas* off St Malo 15.2.1800 (see next entry); took 12-gun privateer *La Françoise* on Irish station 15.5.1800. Recommissioned 10.1802 under Capt. Frederick Maitland (–1806); her boats cut out 10-gun *Le Venteux* at Île de Batz 27.6.1803 (6 wounded); took privateers – 16-gun *Le Brave* 16.3.1804, 30-gun *La Blonde* on Irish station 17.8.1804 and 7-gun Spanish *Esperanza* at Camarinas 2.6.1805 (destroyed by boats); attack on Muros 3.6.1805 (capture of 26-gun *La Confiance* and *Le Bélier*); took 30-gun Bordeaux privateer *Le Vaillant* 25.6.1805; took (with *Egyptienne*) 38-gun *La Libre* off Rochefort 24.12.1805; took 14-gun Spanish privateer *Princessa de la Paix* 22.4.1806. Middling to Large Repair at Deptford 3–12.1807; recommissioned 10.1807 under Capt. Alexander Schomberg (–1812); sailed for Greenland fisheries 1.4.1808; took 20-gun *L'Hébé* in the Bay of Biscay 5.1.1809; convoy to East Indies 1809; in Baltic 1810; under Capt. George Blamey (acting) 1.1812, then Capt. Thomas Browne 4.1812; sailed with convoy to East Indies 22.6.1812; sailed for North America 22.4.1813; under Capt. James Nash for passage home. Laid up at Plymouth 12.1814. BU there 4.1818.

PIQUE This vessel commenced building as a *frégate-bombarde* of Pierre-Alexandre Forfait's *La Romaine* Class (see above), but completed as a French 38-gun frigate to a design modified by François Pestel, and on capture was commissioned in the RN as a 36, and was re-rated as 42 guns in 1817.

Pique (ex-*Aeolus*) (French *Le Pallas*, 38 guns, built 11.1795–9.1799 at St Malo. L: ?early 1799), 42 guns

> Dimensions & tons: 146ft 8in, 123ft 1½in x 39ft 7½in (39ft 0¾in mld.) x 11ft 11in. 1,028 ²⁹⁄₂₄ bm.
>
> Men: 274. Guns: UD 26 x 18pdrs; QD 2 x 9pdrs + 10 x 32pdr carronades; Fc 2 x 9pdrs + 4 x 32pdr carronades.
>
> Taken 6.2.1800 by *Loire*, *Danae* and others off St Malo. Arrived Plymouth 2.3.1800. Registered as *Aeolus* (renamed *Pique* 1801). Fitted there 3–9.1800.
>
> Commissioned 6.1800 under Capt. James Young; sailed for the Mediterranean 1800. In 8.1802 under Capt. William Cumberland, on Irish station, later to Jamaica; took (with *Pelican*) 18-gun *Le Goéland* and a cutter at Aux Cays (San Domingo) 10.1803. In 2.1804 under Capt. Charles Ross (–1807); in Curacoa operations 1–2.1804; took 10-gun cutter *La Terreur* 18.3.1804; took (with *Diana*) 28-gun *Diligencia* 12.1804; took 18-gun *Orquijo* 8.2.1805; took privateers – Spanish 1-gun *Santa Clara* 17.3.1806 and 16-gun *Le Phaeton* and *Le Voltigeur* in the West Indies 26.3.1806; destroyed two small privateers 2.11.1806; paid off 1807. Middling to Large Repair and fitted at Woolwich 4.1809–9.1811; recommissioned 8.1811 under Capt. Anthony Maitland (–1815); sailed for the Leeward Islands 7.6.1812; took US 5-gun privateer *Hawk* 21.6.1814. Rated as a 32 until 1813. In 2.1816 under Capt. James Tait, for Jamaica, then 3.1817 under Capt. John Mackellar; paid off 12.1818. Sold to Mr Freake (for £4,250) at Deptford 22.7.1819.

***L'URANIE* Class** (Jean-François Gauthier design). Two ships were built at Basse-Indre (near Nantes) to this design, of which *L'Uranie* of 1797–1801 was burnt in French service to avoid capture on 3 February 1814.

Clorinde (French *La Clorinde*, ex-*La Havraise*, built 9.1797–6.1801 at Basse-Indre [Nantes]. L: 31.10.1800), 48 guns

Dimensions & tons: 158ft 6in, 133ft 2in x 40ft 10in (40ft 3in mld.) x 12ft 2in. 1,181$\frac{54}{94}$bm.

Men: 330 (later 315). Guns: UD 28 x 18pdr; QD 16 x 32pdr carronades; Fc 2 x 12pdr + 2 x 32pdr carronades.

Taken 30.11.1803 at the capitulation of San Domingo.

Commissioned 5.1804 at Jamaica under Capt. Robert O'Brien. Arrived at Plymouth 23.7.1804. Small to Middling Repair 11.1807–12.1808; recommissioned 10.1808 under Capt. Thomas Briggs (–1814); sailed for the East Indies 17.2.1809; took 8-gun privateer *L'Henri* in the East Indies 28.1.1810; at capture of Mauritius 12.1810. In 9.1814 under Capt. Samuel John Brooke Pechell; paid off 13.6.1815 at Plymouth. Sold to Mr Freake (for £2,500) 6.3.1817.

***L'ARMIDE* Class** (Pierre Rolland design of 1802). Fifteen frigates were ordered to this design: *L'Armide*, *La Pénélope*, *La Minerve*, *La Flore*, *L'Amphitrite*, *L'Andromède* (two ships to this name), *Le Niémen*, *L'Emeraude*, *La Circé*, *L'Alcmène*, *L'Antigone*, *La Cornélie*, *La Cléopâtre* and *La Magicienne* (six at Rochefort, four at Bordeaux, three at Cherbourg and two at Bayonne), of which three were never launched; four were taken by and added to the British Navy, but of these *Armide* and *Niemen* were BU in 1815.

Alceste (French *La Minerve*, ex-*La Ville de Milan*, built 5.1804–11.1805 at Rochefort. L: 9.9.1805)

Dimensions & tons: 152ft 5in, 128ft 8$\frac{5}{8}$in x 40ft 0$\frac{1}{2}$in x 12ft 8in. 1,097$\frac{71}{94}$bm.

Men: 284 (later 315). Guns: UD 28 x 18pdr; QD 14 x 32pdr carronades; Fc 2 x 9pdr + 2 x 32pdr carronades.

Taken 25.9.1806 by Sir Samuel Hood's squadron off Rochefort. Arrived Plymouth 26.10.1806. Fitted at Plymouth 4–8.1807.

Commissioned 3.1807 under Capt. Murray Maxwell (–1812); sailed for the Mediterranean 6.12.1807; at occupation of Madeira 25.12.1807; in action (with *Mercury* and *Grasshopper*) against about 20 gunboats off Rota 4.4.1808 (two gunboats destroyed); at blockade of Toulon 1809; attack on shipping at Fréjus 22.5.1810; with *Belle Poule* at Parenzo (Istria) 4.5.1811; in action (with *Active*) against French 40-gun *La Pomone* and *La Pauline* 28.11.1811 (*La Pomone* taken). In Ordinary at Deptford 1813–14. Fitted as Troopship at Deptford 2–7.1814; recommissioned 5.1814 under Cmdr Daniel Lawrence; sailed for North America: paid off 9.1815 at Portsmouth. Recommissioned 21.10.1815 under Maxwell again; sailed for China 9.2.1816 (with Lord Amherst); passage of the 'Bocca Tigris' (below Canton); wrecked by striking a rock in the Straits of Gaspar 18.2.1817, and burnt 22.2.1817.

Immortalite (ex-*Dunira*) (French *L'Alcmène*, built 7.1810–3.1812 at Cherbourg. L: 3.10.1811)

Dimensions & tons: 152ft 8in, 127ft 11$\frac{3}{8}$in x 39ft 10in (39ft 3in mld.) x 12ft 7$\frac{1}{2}$in. 1,079 $\frac{78}{94}$bm.

Men: 315. Guns: UD 28 x 18pdr; QD 14 x 32pdr carronades; Fc 2 x 9pdr + 2 x 32pdr carronades.

Taken 16.2.1814 by *Venerable* off Madeira. Arrived Portsmouth 18.8.1814.

Not commissioned. Incorporated as *Dunira*, but renamed *Immortalite* 8.11.1814. Never fitted for sea after arrival in Britain. Fitted at Portsmouth as a receiving ship 3.1822. Sold to W. Goldsworthy (for £1,610) 12.1.1837.

***La CONSOLANTE* Class** (François Pestel design, derived from that of Sané). Six frigates were built to this design at 'St Malo' (actually in St Servan): *La Consolante*, *La Piémontaise*, *L'Italienne* (ex-*La Sultane*), *La Bellone*, *La Néreïde* and *L'Illyrienne*, of which three were taken by and added to the British Navy – the *Piedmontaise* (ex-French *La Piémontaise*) taken in 1808 and BU in 1813, and the two below. Note two further vessels were built at Genoa to the same draught: *La Danaé* and *La Galathée*.

Men: 300 (later 315). Guns: UD 28 x 18pdr; QD 14 x 32pdr carronades; Fc 2 x 9pdr + 2 x 32pdr carronades.

Junon (French *La Bellone*, built 5.1803–7.1808 at St Malo. L: 2.1808)

Dimensions & tons: 154ft 0in, 129ft 5$\frac{1}{4}$in x 40ft 3in (39ft 7in mld.) x 12ft 5in. 1,116$\frac{44}{94}$bm.

Taken 6.12.1810 at the fall of Mauritius. Arrived at Portsmouth 20.10.1811. Completed Small Repair 5.1812.

Commissioned 3.1812 under Capt. James Sanders; sailed with convoy for East Indies 4.6.1812; sailed for North America 8.8.1812; beat off attack by 15 gunboats in Hampton Roads 20.6.1813; her boats took US gunboat *No. 121* on 29.7.1813. In 9.1813 under Capt. Clotworthy Upton, for Halifax station. In 7.6.1815 under Capt. James Haldane Tait, for the West Indies; paid off 9.8.1816. BU at Deptford 2.1817.

Madagascar (French *La Néreïde*, built 3.1806–5.1809 at St Malo. L: 12.1808)

Dimensions & tons: 154ft 6in, 129ft 3$\frac{1}{4}$in x 40ft 3in (39ft 7in mld.) x 12ft 11in. 1,113$\frac{90}{94}$bm.

Taken 26.5.1811 by Schomberg's squadron at the capture of Tamatave, Madagascar. Arrived Portsmouth 6.4.1812. Between Small and Middling Repair there 9.1812–4.1813.

Commissioned 2.1813 under Capt. Lucius Curtis, for the Channel. In 8.1814 under Capt. Bentinck Doyle, then 17.11.1815 under Capt. Sir James Alexander Gordon), then 17.12.1816 under Cmdr William Augustus Baumgardt (temp); paid off 3.2.1817 into Ordinary at Sheerness. Commenced between a Middling and Large Repair at Sheerness 2.1819, but found to be in too poor a condition to make this worthwhile, and instead BU 5.1819.

***La NYMPHE* Class** (Pierre-Augustin Lamothe design). Four frigates were built to this design: *La Nymphe*, *La Thétis*, *La Cybèle* and *La Concorde*, all at Brest, of which two were taken by the British Navy, but *La Concorde* (prize taken 4 August 1800) was not added, and *Brune* (ex-*La Thétis*) was re-classed as a troopship in 1810.

Brune (French *La Thétis*, built 10.1787–10.1788 at Brest. L: 16.6.1788)

Dimensions & tons: 153ft 9$\frac{1}{2}$in, 126ft 8$\frac{1}{2}$in x 40ft 2$\frac{5}{8}$in x 12ft 8in. 1,090$\frac{18}{94}$bm.

Men: 284 (later 315). Guns: UD 28 x 18pdr; QD 14 x 32pdr carronades; Fc 2 x 9pdr + 2 x 32pdr carronades.

Taken by *Amethyst* off Lorient 10.11.1808. Arrived Plymouth 15.11.1808 and laid up.

Fitted as 22-gun troopship at Plymouth 5–9.1810.

Commissioned 6.1811 under Cmdr George Douglas. In 4.1812 under Cmdr John Thompson, then 11.6.1812 (acting, confirmed 13.8.1812) Cmdr William Stanhope Badcock; in the Mediterranean 1812–13, then in North America 1814; paid off 8.1815. Fitted at Sheerness as a victualling depot 8.1815–7.1816; at Chatham 1817–34 (except 1827–30 at Sheerness). In 8.1836 under Cmdr Robert Scallon, at Chatham, then 4.1837 under Capt. John Clavell, as Capt. Superintendent of Chatham

Dockyard; paid off 30.6.1838. Sold to J. Levy (for £1,560) at Chatham 16.8.1838.

***L'HORTENSE* Class** (Jacques-Noël Sané design of 1802). Eight ships were built to this Sané design, of which three were taken by and added to the RN during the Napoleonic War, but *Ambuscade* (ex-French *La Pomone*) was BU in 1812, and *Daedalus* (ex-French *La Corona*) was wrecked in 1813..

Bourbonnaise (French *La Caroline*, built 5.1804–12.1806 at Antwerp. L: 15.8.1806), 46 guns

Dimensions & tons: 151ft 6in, 127ft 4⅞in x 39ft 10⅝in x 12ft 2in. 1,078$^{19}/_{94}$bm.

Men: 300 (later 315). Guns: UD 28 x 18pdr; QD 14 x 32pdr carronades; Fc 2 x 9pdr + 2 x 32pdr carronades.

Taken 21.9.1809 at the seizure of St Paul, Reunion by Rowley's squadron.

Commissioned 9.1809 at Reunion under Capt. Robert Corbett; arrived in Plymouth 16.2.1810, paid off into Ordinary and never again fitted for sea. BU at Plymouth 4.1817.

***La PALLAS* Class** (Jacques-Noël Sané design of 1805). This Sané design amounted to the standard 40-gun French frigate design of the Napoleonic Empire period, to which some sixty-plus ships were ordered. Twelve frigates of this design were taken by and added to the RN over the last few years of the war, but of these the *Laurel* (ex-French *La Fidèle*) was wrecked in January 1812, the *Java* (French *La Renommée*) was taken by the 24pdr-armed spar-deck frigate USS *Constitution* in December 1812, and the *Pomone* (French *L'Astrée*) and *Modeste* (French *La Terpsichore*) were BU in 1816. The *Weser* and *Trave* had been re-classed as troopships in 1814.

Men: 300 (later 315). Guns: UD 28 x 18pdr; QD 14 x 32pdr carronades; Fc 2 x 9pdr + 2 x 32pdr carronades.

Weser (French *La Weser*, built 4.1811–8.1812 at Amsterdam. L: 12.5.1812)

(Note that the Netherlands was annexed by France in 1810, all ships built at Amsterdam or Rotterdam therefore being entirely French rather than belonging to the puppet Kingdom of Holland.)

Dimensions & tons: 152ft 6in, 127ft 2in x 39ft 11¾in x 12ft 6in. 1,081$^{13}/_{94}$bm.

Taken 21.10.1813 by *Scylla* and *Royalist* off Ushant. Arrived Plymouth 3.11.1813.

Commissioned 3.1814 as troopship under Cmdr Thomas Ball Sulivan, for North America; fitted as Troopship at Plymouth 2–4.1814. In 3.1815 under Bartholomew Kent; on 21.10.1816 under Cmdr Daniel Lawrence; paid off 16.8.1816. Sold to Mr Bailey at Portsmouth (for £1,500) to BU 17.9.1817.

Trave (French *La Trave*, built 4.1811–8.1812 at Amsterdam. L: 12.5.1812)

Dimensions & tons: 151ft 5¼in, 126ft 5⅛in x 39ft 10½in x 12ft 4in. 1,069$^{26}/_{94}$bm.

Taken 23.10.1813 by *Andromache* in the Atlantic. Arrived Portsmouth 3.11.1813.

Commissioned 4.1814 as troopship under Cmdr Rowland Money; fitted as Troopship at Portsmouth 3–5.1814; for Potomac operations 8.1814; in New Orleans operations 12.1814–1.1815. On 28.3.1815 under Capt. John Boulton, then 13.11.1815 under Capt.. John Codd in; arrived Plymouth 7.1816 and paid off into Ordinary. Sold to Mr Holmes at Plymouth (for £2,100) to BU 7.6.1821.

Seine (French *La Cérès*, built 3.1811–1.1813 at Brest. L: 12.8.1812), 46 guns

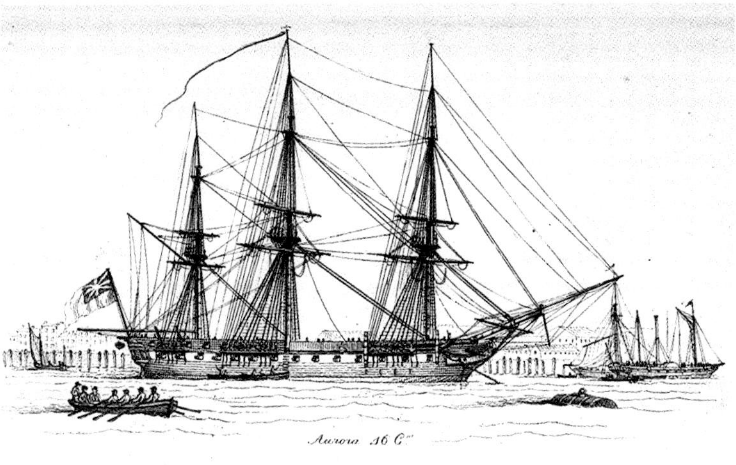

An 1826 etching by Henry Moses of the frigate *Aurora* in Portsmouth harbour prior to her departure for a deployment to the Jamaica station. The *Aurora* was the former French *Clorinde*, captured in 1814 by the British but not commissioned by her new owners until the 1820s. As the *Clorinde*, she had been one of the numerous *Pallas* Class, the standard French design of the Imperial era, but her lengthy and expensive repair adapted her to British standards, and – unusually for a French-built ship – she served two commissions on overseas stations before being relegated to harbour service. *(NMM PU7919)*

Dimensions & tons: 152ft 0in, 126ft 11½in x 39ft 10½in (39ft 3½in mld.) x 12ft 8in. 1,073$^{70}/_{94}$bm.

Taken 6.1.1814 by *Niger* and *Tagus* between Brazil and the Cape Verde Islands.

Not commissioned. Under Capt. William Bowles for passage home. Arrived Portsmouth 29.6.1814, surveyed to 19.7.1814. To Chatham 6.1817 for repair, then to Woolwich 7.1818, then Deptford 3.1819. BU at Deptford 5.1823.

Gloire (ex-*Palma*) (French *L'Iphigénie*, built 5.1809–11.1810 at Cherbourg. L: 20.5.1810), 46 guns

Dimensions & tons: 154ft 5in, 126ft 10¼in x 39ft 9in x 12ft 7½in. 1,066$^{14}/_{94}$bm.

Taken 20.1.1814 by *Venerable* and *Cyane* off Madeira.

Not commissioned. In 1814 under Capt. James Andrew Worth; at Spithead 7.1814. Incorporated as *Palma*, but renamed *Gloire* 8.11.1814. Arrived Plymouth 23.2.1814, and laid up in Ordinary. Sold to Mr Freake (for £1,750) 10.9.1817.

Aurora (French *La Clorinde*, built 11.1806–3.1809 Paimboeuf [Nantes]. L: 8.8.1808), 46 guns

Dimensions & tons: 152ft 1in, 127ft 0⅛in x 40ft 0½in x 13ft 1in. 1,083$^{14}/_{94}$bm.

Taken 26.2.1814 by *Dryad* and *Eurotas* off Ushant. Arrived Portsmouth 17.3.1814. Copper repaired (for £1,838) then laid up. Large Repair at Woolwich (for £34,594) 8.1817–10.1820, then fitted for sea at Chatham (for £14,238) 4–6.1821.

Commissioned 6.4.1821 under Capt. Henry Prescott (–1824); to Brazil 1821 thence via the west coast of South America 1811 to the Pacific; home and paid off 2.1825 then recommissioned the same month for the Lisbon station under Capt. John Maxwell (died 31.5.1826). In 2.6.1826 under Capt. Charles John Austen, for the Jamaica station; paid off 23.12.1828. Fitted as coal depot at Chatham 3.1832, for Falmouth to 1850. BU at Plymouth 5.1851.

Sultane (French *La Sultane*, built 1.1810–9.1813 at Paimboeuf [Nantes]. L: 30.5.1813), 46 guns

Dimensions & tons: 151ft 6in, 127ft 0¼in x 39ft 9in x 12ft 6in. 1,067$^{51}/_{94}$bm.

Taken 26.3.1814 by *Hannibal* off Cherbourg. Arrived Portsmouth 30.3.1814 and laid up.

Not commissioned. Fitted 6.1817 to be sent to Chatham, thence to Woolwich 7.1818, finally to Deptford where BU by AO 3.8.1819.

Topaze (French *L'Étoile*, ex-*L'Hymenée*, built 8.1811–10.1813 at Paimboeuf [Nantes]. L: 28.7.1813), 46 guns

Dimensions & tons: 151ft 5⅜in, 126ft 8⅛in x 39ft 8in x 12ft 5¼in. 1,060²¹/₉₄bm.

Men: 315. Guns: UD 28 x 18pdr; QD 14 x 32pdr carronades; Fc 2 x 9pdr + 2 x 32pdr carronades.

Taken 27.3.1814 by *Hebrus* off La Hogue. Registered & named 15.6.1814. Small Repair at Plymouth 29.3.1814–24.2.1815, then laid up. Very Small Repair and fitted for sea at Plymouth 3.1818–1.7.1818.

Commissioned 4.1818 under Capt. John Lumley (died 23.7.1821) for the East Indies; at bombardment of Mocha 26 & 30.12.1820. On 29.7.1821 under Capt. Charles Richardson, then 9.2.1822 under Capt. Armar Lowry Corry (who did not take up command but was invalided home in same month), then finally Cmdr Price Blackwood (acting) in 5.1822; paid off 10.1822. Fitted at Portsmouth as receiving ship 1–2.1823, in which role served until 1850. Became target for HMS *Excellent* 3.1850. BU at Portsmouth 12.1851.

Melpomene (French *La Melpomène*, built 3.1811–6.1812 at Toulon. L: 17.5.1812), 46 guns

Dimensions & tons: 152ft 10½in, 127ft 0⅞in x 40ft 1¼in x 12ft 6¼in. 1,087¹⁰/₉₄bm.

Men: 300. Guns: UD 28 x 18pdr; QD 14 x 32pdr carronades; Fc 2 x 9pdr + 2 x 32pdr carronades.

Taken 30.4.1815 by *Rivoli* off Ischia (the only warship prize of the Hundred Days). Arrived at Portsmouth 28.12.1815 and laid up.

Not commissioned. Never fitted for sea after arrival in Britain. Sold at Portsmouth to Mr Freake (for £2,460) 7.6.1821.

Ex-Spanish Prize (1804)

Three Spanish frigates were captured on 5 October 1804 (and another, the *Mercedes*, destroyed) in the attack upon the Plate convoy by Sir Graham Moore's squadron in the Atlantic, before the declaration of war upon Spain; of the three prizes, the *Fama* (never commissioned in the RN) was sold in 1812 and *Clara* (a 12pdr frigate, and renamed *Leocadia* in the RN) was sold in 1815. Two more Spanish frigates taken later in the year were also added to the RN but had gone before 1817.

Imperieuse (Spanish *Medea*, 40 guns, launched 1797 at Ferrol), 38 guns

Dimensions & tons: 147ft 2in, 122ft 4¼in x 40ft 1in x 12ft 0in. 1,045⁶¹/₉₄bm.

Men: 284 (later 315). Guns: UD 28 x 18pdr; QD 10 x 32pdr carronades; Fc 2 x 9pdr + 2 x 32pdr carronades.

Taken while part of the Plate fleet on 5.10.1804, before the declaration of war on Spain.

Arrived Plymouth 19.10.1804. Registered as *Iphigenia*, but renamed *Imperieuse* 3.12.1805. Large Repair 2–11.1806.

Commissioned 8.1806 under Capt. Lord Cochrane (–1809); at blockade of Basque Roads 1806; took two prizes of Sable d'Olonne 19.12.1806, and another in the mouth of the Garonne 31.12.1806; her boats destroyed a battery at Arcachon 7.1.1807; sailed for the Mediterranean 12.9.1807; took Maltese privateer *King George* off Corsica 14.10.1807; operations on the Spanish coast 2.1808; action against four gunboats near Cartagena 19.2.1808 (two sunk, one taken); boats cut out a large ship at Almeira 21.2.1808; took ten more small prizes 2–4.1808; bombardment of barracks at Ciudadella (Minorca) 13.4.1808;

defence of Fort Trinidad, Rosas 25.11–5.12.1808; took French 7-gun cutter *La Gauloise* and 5-gun *La Julie* at Galdagues Bay 30.12.1808; fireship attack on Basque roads 12.4.1809. In 6.1809 under Capt. Thomas Garth, then 9.1810 Capt. Henry Duncan; sailed for the Mediterranean 27.6.1811; took two gunboats and destroyed another off Positano 11.10.1811; destroyed (with *Thames*) ten Neapolitan gunboats off Palinuri 2.11.1811; with squadron at Laigueglia 27.6.1812, and at Anzio 5.10.1812; took French 3-gun privateer *L'Audacieux* off Bonifaccio 31.8.1813. In 1813 under Capt. Philip Dumaresq, then 1814 Capt. Joseph James; to Ordinary at Sheerness 1815. Fitted at Sheerness as a lazarette 3–5.1818, for Stangate Creek. Sold at Sheerness to John Small Sedger, Rotherhithe (for £1,705) 10.9.1838.

The 2nd Melville Board (1812)

The outbreak of war with the United States on 18 June 1812 forced the Admiralty to face the need for additional frigates, and even prior to its start Melville's new Board (installed in March) had anticipated the need by placing orders in May for six extra 36-gun ships with commercial yards. As a short-term need was envisaged, the Admiralty chose to request 'fir'-built frigates, with the Navy Board supplying red pine timber to the yards from dockyard stocks. Four more 'fir' 36-gun ships were ordered in late 1812 (and two oak-built *Apollo*s), plus eight 38-gun 'fir' ships (and four extra *Leda*s).

SCAMANDER Class 42 (originally 36) guns. A joint design by Sir William Rule and Henry Peake, but really a 'utility' softwood version of the former's *Apollo* Class 36s, with slightly reduced dimensions, a flatter sheer and reduced depth in hold, and featuring a square tuck stern and a higher head (with the hawse-holes between the check pieces) to distinguish them from the *Apollo* Class All were built by contract, with softwood 'fir' timber (actually the first seven were built of red pine, the last three of yellow pine) supplied from HM Dockyards' stores.

The first six were all begun under original names (as issued on 7 July 1812), and were renamed as shown below on 11 December 1812 – except *Eridanus* and *Tagus*, which were renamed on 26 January 1813. The seventh vessel was named *Greyhound* on 12 October, but renamed *Euphrates* on 7 December. The next (Limehouse) pair were named on 3 December, and the final ship on 7 December.

Dimensions & tons: 143ft 0in, 120ft 0¾in x 38ft 2in x 12ft 4in. 930²¹/₉₄bm.

Men: 274 (284 from 26.1.1813). Guns: UD 26 x 18pdrs; QD 12 x 32pdr carronades; Fc 2 x 9pdrs + 2 x 32pdr carronades.

Eridanus (ex-*Liffey*) Mrs Mary Ross, Rochester

As built: 143ft 3in, 120ft 0¼in x 38ft 5¼in x 12ft 4in. 945bm. Draught 9ft 3in/13ft 3in.

Ord: 4.5.1812. K: 8.1812. L: 1.5.1813. Fitted at Chatham 1.5.1813–13.7.1813.

Commissioned 3.6.1813 under Capt. Henry Prescott, for Bay of Biscay. From 4.1815 under Capt. William Paterson, and from 2.1816 under Capt. William King; paid off 25.6.1817 at Deptford. Sold there to Mr Freake (for £2,540) 29.1.1818.

Orontes (ex-*Brilliant*) Josiah & Thomas Brindley, Frindsbury

As built: 142ft 10½in, 119ft 9½in x 38ft 4¾in x 12ft 4in. 939bm. Draught 8ft 10in/12ft 10in.

Ord: 4.5.1812. K: 8.1812. L: 29.6.1813. Fitted at Chatham 29.6.1813–13.12.1813.

Commissioned 10.1813 under Capt. Nathaniel Day Cochrane for Irish station; to the West Indies 4.1814, then home 2.1815; to Cape of Good Hope and St Helena in 1816; paid off 21.3.1817.

BU at Sheerness 4.1817.

Scamander (ex-*Lively*) Josiah & Thomas Brindley, Frindsbury

As built: 143ft 0½in, 119ft 11½in x 38ft 5in x 12ft 4in. 941bm. Draught 8ft 8in/12ft 9in.

Ord: 4.5.1812. K: 8.1812. L: 13.7.1813. Fitted at Chatham 30.7.1813–24.12.1813, thence to Sheerness.

Commissioned 10.1813 under Capt. Gilbert Heathcote for Channel service; from 26.8.1815 under Capt. Sir John Louis, at Sheerness; on 15.1.1816 under Capt. Charles Sibthorpe John Hawtayne; to Leeward Islands in 4.1816; paid off 11.1818 at Portsmouth. Sold there to John Small Sedger, Rotherhithe (for £3,010) 22.7.1819.

Tagus (ex-*Severn*) Daniel List, Fishbourne, Isle of Wight

As built: 143ft 8in, 120ft 7¼in x 38ft 5½in x 12ft 4in. 949bm. Draught 9ft 2in/13ft 4in.

Ord: 4.5.1812. K: 8.1812. L: 14.7.1813. Fitted at Portsmouth 15.7.1813–9.11.1813.

Commissioned 9.1813 under Capt. Philip Pipon; to South America 1814. Captured (with *Niger*) 44-gun *La Cérès* off Cape Verde 6.1.1814. On 16.8.1815 under Capt. James Whitley Deans Dundas, for Mediterranean service (to 1.1819); paid off 21.1.1819 at Deptford; sold there to Beatson (for £2,550) on 19.4.1822.

Ister (ex-*Blonde*) William Wallis, Blackwall

As built: 143ft 6in, 120ft 1¼in x 38ft 5⅛in x 12ft 4in. 945bm. Draught 9ft 6in/13ft 6in.

Ord: 4.5.1812. K: 8.1812. L: 14.7.1813 (then coppered by builder). Fitted at Woolwich 12.8.1813–11.11.1813.

Commissioned 10.1813 under Capt. John Cramer for West Indies service. From 1815 under Capt. John Phillimore at Leith; from 8.1815 under Capt. Thomas Forrest for Mediterranean service; from 8.1816 Flagship of Rear-Adm. Sir Charles Penrose, arriving at Algiers 29.8.1816 (at close of bombardment); paid off 7.9.1818 at Portsmouth. Sold there to Thomas Beech (for £3,000) 8.3.1819.

Tigris (ex-*Forth*) John Pelham, Frindsbury

As built: 143ft 0in, 119ft 11⅞in x 38ft 3in x 12ft 4in. 934bm. Draught 8ft 7in/12ft 9in.

Ord: 4.5.1812. K: 9.1812. L: 26.6.1813. Fitted at Chatham 26.6.1813–24.12.1813.

Commissioned 10.1813 under Capt. Robert Henderson for Irish station; paid off 1814, but recommissioned and sailed with convoy 10.1814 for Leeward Islands; returned home 5.1815 and paid off 13.1.1818 at Portsmouth. Sold to Joshua Crystall for £3,020 on 11.6.1818.

Euphrates (ex-*Greyhound*) John King, Upnor.

As built: 143ft 3½in, 120ft 2¼in x 38ft 5in x 12ft 4in. 943bm. Draught 8ft 5in/12ft 8in.

Ord: 12.10.1812. K: 1.1813. L: 8.11.1813. Fitted at Chatham 8.11.1813–24.9.1814.

Commissioned 8.1814 under Capt. Robert Preston, for the Channel Squadron; paid off 23.6.1817 at Deptford. Sold there to Mr W. Thomas (for £2,679) 29.1.1818.

Hebrus John Barton, Limehouse

As built: 143ft 0in, 120ft 1⅛in x 38ft 4½in x 11ft 11¼in. 939bm. Draught 8ft 8in/12ft 10in.

Ord: 16.11.1812. K: 1.1813. L: 13.9.1813 (coppered by builder). Fitted at Deptford 27.9–18.12.1813.

Commissioned 10.1813 under Capt. Edmund Palmer; took 40-gun *L'Etoile* 27.3.1814; in Potomac operations 8.1814; to the Mediterranean 6.1816; took part in Bombardment of Algiers 27.8.1816, losing 4 killed and 15 wounded; paid off 2.11.1816 at Sheerness. Sold to Joshua Crystall (for £2,110) on 3.4.1817.

Granicus John Barton, Limehouse

As built: 143ft 0½in, 119ft 11½in x 38ft 5in x 12ft 4in. 942bm. Draught 8ft 9in/12ft 8in.

Ord: 16.11.1812. K: 1.1813. L: 25.10.1813. Fitted at Deptford 2.11.1814–31.1.1814.

Commissioned 11.1813 under Capt. William Furlong Wise; took 6-gun US privateer *Leo* 2.12.1814; to the Mediterranean 3.1816; took part in Bombardment of Algiers 27.8.1816, losing 16 killed and 42 wounded; paid off 28.10.1816 at Plymouth. Sold to Thomas Beech (for £2,100) on 3.4.1817.

Alpheus William Wallis, Blackwall

As built: 143ft 4¼in, 120ft 2⅜in x 38ft 6½in x 12ft 4in. 949bm. Draught 9ft 0in/12ft 8in.

Ord: 7.12.1812. K: 7.1813. L: 6.4.1814. Fitted at Woolwich 6.4–11.7.1814.

Commissioned 5.1814 under Capt. George Langford, for the East Indies; paid off 31.12.1816 at Chatham. Sold to Bailey & Co. (for £2,320) on 10.9.1817.

LEDA Class (1812–15 orders)

46 (originally 38) guns. After a gap of over three years, seven more of these oak-built frigates were ordered from the Royal Dockyards between 1812 and 1815. Their specification was identical to the earlier ships of this 1794 Henslow design (see above). These were not part of the emergency programme and none was completed prior to the end of the war. All were re-classed from 38-gun to 46-gun on 1 January 1817, and to 42-gun in February 1839.

Dimensions & tons: 150ft 1½in, 125ft 4⅞in x 39ft 11in x 12ft 9in. 1,062⁷⁹⁄₉₄bm.

Men: 315. Guns: UD 28 x 18pdrs (38cwt); QD 14 x 32pdr (17cwt) carronades; Fc 2 x 9pdrs (26cwt) + 2 x 32pdr (17cwt) carronades.

Diamond Chatham Dyd [M/Shipwright George Parkin]

As built: 150ft 0½in, 125ft 1¾in x 40ft 2½in x 12ft 9in. 1,076¹⁸⁄₉₄bm. Draught 11ft 2in/15ft 0in.

Ord: 30.6.1812. K: 8.1813. L: 16.1.1816, then laid up at Chatham. Fitted for sea 2.1824–24.7.1824.

First cost: unknown

Commissioned 5.1824 under Capt. Lord (William John) Napier, for South America; paid off 1.12.1826. Accidentally burnt at Portsmouth 18.2.1827, wreck docked 28.5.1827, then BU 6.1827 there.

Amphitrite Bombay Dyd [M/Shipwright Jamsetjee Bomanjee] Teak-built.

As built: 150ft 1¼in, 125ft 5in x 39ft 11¼in x 12ft 9in. 1,064⁴⁄₉₄bm. Draught 12ft 6in/13ft 6in.

Ord: 21.10.1812. K: 8.1814. L (floated out): 14.4.1816. Arrived Portsmouth 22.12.1816 and laid up in Ordinary.

First cost: £21,549.

Commissioned 29.4.1816 under Capt. James Hanway Plumridge; paid off 18.1.1817 at Portsmouth. Cut down to a 26-gun Sixth Rate corvette at Portsmouth 5.1845–2.1846, then fitted for sea 8–9.1847 at (combined) cost of £24,203. Recommissioned 14.7.1847 under Capt. Thomas Rodney Eden (died 11.1.1850), for the coast of Africa; took slavers *Triumpho de Brasil* 19.5.1848, *Secundho de Julio* 2.6.1848, *Curioso* 24.6.1848, *San Antonio* 10.9.1848 and *Josepha* 18.8.1848; to the Pacific 1.1849. On 2.1850 under Capt. John MacDougall; paid off 27.7.1850 at Portsmouth. Recommissioned 13.12.1850 under Capt. Charles Frederick, for the Pacific; on 6.12.1854 under Capt. Richard Burridge, on same station; paid off 9.6.1856 at Portsmouth. Refitted for coastguard service 10.1856–2.1857. Recommissioned 5.3.1857 under Capt. Edward Tatham, for coastguard at Milford

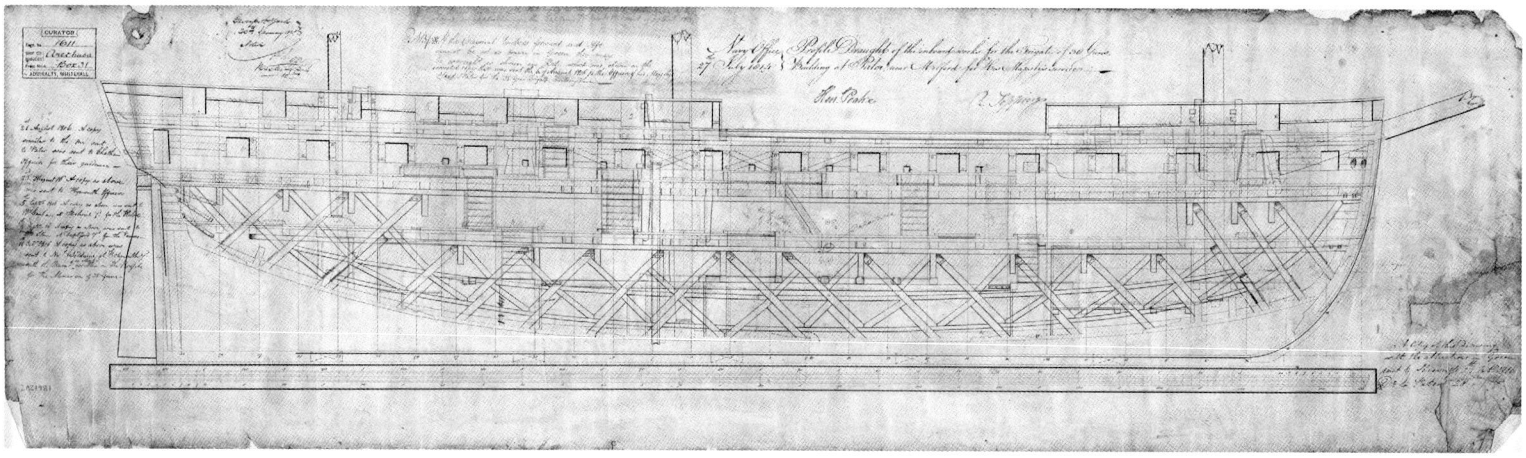

This draught signed by Surveyors Henry Peake and Robert Seppings on 27 July 1814 shows the diagonal riders devised by Seppings and other structural alterations proposed to be fitted into the last of the *Leda* Class. This inboard profile mentions the *Thetis*, ordered in 1812 at Pembroke but not laid down until December 1814, and to her name has been subsequently added the ships ordered in 1814 to 1816. *(NMM J3937)*

Haven; paid off 1.1858 at Plymouth and laid up. BU at Plymouth 30.1.1875.

Trincomalee Bombay Dyd [M/Shipwright Jamsetjee Bomanjee] Teak-built.

As built: 150ft 4$\frac{1}{2}$in, 125ft 7$\frac{1}{4}$in x 39ft 11$\frac{1}{4}$in x 12ft 9in. 1,065$\frac{59}{94}$bm. Draught 12ft 6in/13ft 8in.

Ord: 30.10.1812. K: 2.1816. L: 12.10.1817 (sailed 30.5.1818).

First cost: £23,643 (+£6,681 for stores).

Commissioned 15.10.1818 at Bombay under Cmdr Philip Henry Bridges, for voyage to UK; arrived Portsmouth 3.4.1819, paid off 27.4.1819 and laid up in Ordinary. Cut down to a 26-gun Sixth Rate corvette at Portsmouth 4.1845–9.1847, then fitted for sea at (combined) cost of £21,643. Recommissioned 10.10.1848 under Capt. Robert Laird Warren, for North America and the West Indies; paid off 16.8.1850. Recommissioned 8.1852 under Capt. Wallace Houston for the Pacific; paid off 15.9.1857. RNR drill ship 1.1861. Subsequently sold 19.5.1897, renamed *Foudroyant* and restored (still in existence in 2013, now back under her original name, at Hartlepool).

Thetis Pembroke Dyd [M/Shipwright Edward Churchill to 9.1815, then Thomas Roberts]

As built: 150ft 9in, 126ft 7$\frac{1}{8}$in x 40ft 2in x 12ft 9in. 1,086$\frac{32}{94}$bm. Draught 10ft 7in/14ft 8in.

Ord: 18.12.1812. K: 12.1814. L: 1.2.1817. Sailed 21.8.1817 for Plymouth, where fitted for Ordinary 25.8–20.9.1817.

First cost: £27,435 (+£7,699 for fitting).

Commissioned 15.3.1823 under Capt. John Phillimore, for particular service; fitted for sea at Plymouth (for £12,959) 28.2–20.8.1823; to South America 6.1826. On 8.11.1826 under Capt. Arthur Batt Bingham (died 19.8.1830); on 29.11.1830 under Capt. Samuel Burgess; wrecked off Cape Frio, Brazil 5.12.1830 (22 men drowned) – two-thirds of the cargo of bullion was recovered shortly afterwards.

Arethusa Pembroke Dyd [M/Shipwright Edward Churchill to 9.1815, then Thomas Roberts]

As built: 150ft 11in, 126ft 11in x 40ft 1in x 12ft 9in. 1,084$\frac{60}{94}$bm. Draught 10ft 7in/14ft 7in.

Ord: 22.11.1814. K: 2.1815. L: 29.7.1817. Sailed 21.8.1817 for Plymouth, where fitted for Ordinary 25.8–27.9.1817.

First cost: £25,923.

Not commissioned. Fitted at Plymouth 4–6.1836 as a lazarette for Liverpool. Renamed *Bacchus* 12.3.1844. Coal depot 1851–52. Sold to Castle & Son (for £1,450) to BU 14.8.1883.

Blanche Chatham Dyd [M/Shipwright George Parkin]

As built: 150ft 1$\frac{1}{4}$in, 125ft 1$\frac{1}{8}$in x 40ft 2in x 12ft 10in. 1,073$\frac{48}{94}$bm. Draught 11ft 3in/14ft 11in.

Ord: 29.5.1815. K: 2.1816. L: 26.5.1819. Fitted for Ordinary 7–9.1819.

First cost: £33,250.

Commissioned 2.2.1824 under Capt. William Bowes Mends, for South America; fitted for foreign service (for £14,285) 1–5.1824; paid off 25.10.1827 at Chatham. Fitted for sea at Plymouth (for £14,509) 2–6.1830. Recommissioned 26.2.1830 under Capt. Anthony Farquhar, for the West Indies; paid off 2.11.1833 at Portsmouth. Became a receiving ship at Portsmouth 10–11.1833. Sold to White, Cowes 16.6.1865, BU by 7.10.1865.

Fisgard Pembroke Dyd [M/Shipwright Thomas Roberts]

As built: 150ft 1$\frac{1}{2}$in, 125ft 4$\frac{1}{4}$in x 40ft 0$\frac{1}{2}$in x 12ft 9in. 1,069$\frac{5}{94}$bm. Draught 11ft 0in/14ft 4in.

Ord: 24.8.1815. K: 2.1817. L: 8.7.1819. Fitted for Ordinary at Plymouth 24.7–27.8.1819.

First cost: £25,429 (+£7,094 for fitting).

Commissioned 13.5.1843 under Capt. John Alexander Duntze, for the Pacific; paid off 7.10.1847 at Woolwich. Became a harbour flagship (fitted for a Commodore) at Woolwich 10.1847. Recommissioned 24.10.1847 under Commodore James John Gordon Bremer, at Woolwich; on 3.12.1850 under Commodore Henry Eden, then 31.12.1853 under Commodore John Shepherd, renewed 1.1.1858, then 20.12.1858 under Commodore James Robert Drummond, 29.6.1861 under Commodore Frederick William Erskine Nicolson, 18.2.1863 under Capt. Frederick Archibald Campbell, 1.1.1864 under Commodore Hugh Dunlop, and 6.4.1868 under Commodore William Edmonstone, all at Woolwich; paid off 1.10.1869 on closure of Woolwich Dyd. Recommissioned as a receiving and depot ship at Woolwich 2.11.1869 under Cmdr Thomas Whillier; on 6.6.1870 under Cmdr Frank Inglis, then 11.1873 under Cmdr John Palmer; moved to Greenwich 31.7.1874 as a training and depot ship; on 14.6.1876 under Cmdr William Henry Sharp; paid off 4.1878 and moved to Chatham; BU completed at Chatham 8.10.1879.

Notes: There is a question over the sequence of the first two orders at Pembroke Dyd; I have taken the view that the first frigate ordered there (1812) was the first laid down (i.e. *Thetis*), but some sources reverse *Arethusa/Thetis* orders.

***CYDNUS* Class** 46 (originally 38) guns. A 'fir'-built (actually, red pine) version of the *Leda* design vessels, ordered from 1812 onwards (which see for dimensions, guns, etc.). Like the *Scamander* Class, these softwood frigates required the inclusion of a square tuck sterm, and post-war modifications seem to have similarly raised the head, at least in the Blackwall-built vessels. The nameship of the class was BU in 1816, but is included below to complete the class listing.

Cydnus Wigram, Wells & Green, Blackwall
- As built: 150ft 1½in, 125ft 2⅜in x 40ft 3in x 12ft 9in. 1,078 82/94bm. Draught 9ft 3in/13ft 1in.
- Ord: 16.11.1812. K: 12.1812. L: 17.4.1813, then coppered by builder. Fitted at Woolwich 15.5–30.6.1813.
- Commissioned 5.1813 under Capt. Frederick William Aylmer, then later same month under Capt. Frederick Langford; captured (with *Pomone*) 14-gun US privateer *Bunker's Hill* 4.3.1814. Convoy duty to East Indies 1814. From 1815 under Capt. Robert Spencer. BU at Portsmouth 2.1816.

Eurotas Wigram, Wells & Green, Blackwall
- As built: 150ft 8¼in, 125ft 2½in x 40ft 3¼in x 12ft 9in. 1,080 8/94bm. Draught 8ft 11in/13ft 5in.
- Ord: 16.11.1812. K: 12.1812. L: 17.4.1813, then coppered by builder. Fitted at Woolwich 15.5–16.6.1813.
- Commissioned 5.1813 under Capt. John Phillimore; at capture of *La Trave* (by *Andromache*) 23.10.1813; in action with 44-gun *La Clorinde* 25.2.1814 (which was taken next day by *Dryad* and *Achates*); on 22.3.1814 under Capt. Edmund Sexton Pery Knox (as flag-captain to Rear-Adms. Charles Elphinstone Fleeming and Samuel Hood Linzee), off Cadiz and Gibraltar; paid off 21.6.1814. Recommissioned 8.4.1815 under Capt. James Lillicrap; paid off 22.1.1816. Reported 3.1816 to be in poor condition (a Hansard report quotes her as only fit 'to be sold for firewood'). BU at Deptford 8.1817.

Niger Wigram, Wells & Green, Blackwall
- As built: 150ft 0¼in, 125ft 3¼in x 40ft 0in x 12ft 9in. 1,066 49/94bm.
- Ord: 16.11.1812. K: 12.1812. L: 29.5.1813. Fitted at Woolwich 29.5–22.7.1813.
- Commissioned 6.1813 under Capt. Peter Rainier. Captured the privateer *Dart* 11.1813, and (with *Tagus*) took 40-gun French frigate *La Cérès* off Cape Verde 6.1.1814. To Brazil 1814. Recommissioned 29.8.1815 under Capt. Samuel Jackson for Halifax station; paid off and hulked at Halifax, Nova Scotia 9.1817. Sold there to BU 1820.

Meander Thomas Pitcher, Northfleet
- As built: 150ft 1½in, 125ft 5in x 40ft 0in x 12ft 9in. 1,067 35/94bm. Draught 8ft 9in/12ft 10in.
- Ord: 16.11.1812. K: 1.1813. L: 13.8.1813. Completed fitting by builder 8.12.1813. Fitted at Woolwich for Ordinary 12.1813–3.1814; and for sea there 7.1814–9.1814.
- Commissioned 7.1813, from 1814 under Capt. John Bastard. Fitted for Channel service at Woolwich, Deptford and then Woolwich again 1–6.1816; from 1816 under Cmdr Arthur Fanshawe (acting), from 24.10.1816 under Capt. Sir James Alexander Gordon. Grounded 19.12.1816 on shoal off Yarmouth and nearly lost; paid off 21.1.1817. BU at Sheerness 2.1817.

Pactolus Mrs Frances Barnard, Deptford
- As built: 150ft 2¾in, 125ft 6⅛in x 39ft 11½in x 12ft 9½in. 1,065 88/94bm. Draught 9ft 0in/14ft 2in.
- Ord: 16.11.1812. K: 1.1813. L: 14.8.1813. Coppered by builder to 28.8.1813. Fitted at Deptford 28.8–30.10.1813.
- Commissioned 14.9.1813 under Capt. Frederick William Aylmer; on 15.1.1816 under Capt. William Hugh Dobbie, for Halifax station; paid off 9.8.1817. Sold to Mr Maund for £2,790 on 29.1.1818.

Tiber Daniel List, Binstead, Isle of Wight
- As built: 150ft 1in, 125ft 4¾in x 40ft 2in x 12ft 9in. 1,076 10/94bm. Draught 8ft 8in/13ft 1in.
- Ord: 16.11.1812. K: 2.1813. L: 10.11.1813. Fitted for Ordinary at Portsmouth 23.11–6.12.1813. Fitted there 7.1814–29.9.1814; fitted for sea at Woolwich 11.1815–4.1816.
- Commissioned 23.7.1814 under Capt. James Richard Dacres (to 1817); took 7-gun US privateer Leo 8.3.1815; from 1816 Flagship of Vice-Adm. Francis Pickmore on Newfoundland station; paid off 18.9.1818. Sold to Mr Durkin, Southampton for £3,200 on 1.1820.

Araxes Thomas Pitcher, Northfleet
- As built: 150ft 6¼in, 125ft 8½in x 40ft 0in x 12ft 9½in. 1,069 81/94bm. Draught 8ft 11in/12ft 10in.
- Ord: 7.12.1812. K: 1.1813. L: 13.9.1813. Completed fitting by builder 8.12.1813. Fitted at Woolwich 9.12.1813–18.9.1814.
- Commissioned 7.1814 under Capt. George Miller Bligh; paid off 7.1816 and laid up in Ordinary at Portsmouth. Sold to Mr Manlove for £2,500 (?at Sheerness) 10.9.1817.

Tanais Mrs Mary Ross, Rochester
- As built: 150ft 8in, 126ft 0¼in x 40ft 2¼in x 12ft 10in. 1,084 79/94bm. Draught 8ft 8in/13ft 3in.
- Ord: 7.12.1812. K: 2.1813. L: 27.10.1813. Fitted at Chatham 27.10.1813–22.9.1814.
- Commissioned 8.1814 under Capt. Joseph James for Jamaica station; paid off 5.1816 and laid up in Ordinary at Sheerness. Sold at Portsmouth to Mr Beatson for £3,200 on 8.3.1819.

BLONDE 18-pounder frigate of 46 guns (rated 38-gun until 1 January 1817). Originally ordered to lines of 36-gun *Apollo* Class, but design modified and enlarged to make her a 38-gun frigate; re-ordered as such 22 January 1816. Intended to combine best features of both *Leda* and *Lively* designs, and proved fast and weatherly, but was crank and had other deficiencies which meant that the design was not repeated.
- Dimensions & tons: 155ft 0in, 130ft 10in? x 39ft 8in x 13ft 6in. 1,095bm (estimated).
- (note the keel length is actually recorded as 143ft 2in; this is clearly miscalculated and a corrected figure appears here)
- Men: 315. Guns: UD 28 x 18pdr (38cwt), QD 14 x 32pdr (17cwt) carronades, Fc 2 x 9pdr (26cwt) + 2 x 32pdr (17cwt) carronades.
- First cost: £38,266 (+ fitting £15,241).

Blonde Deptford Dyd
- As built: 155ft 1in, 130ft 11in x 40ft ?in (39ft 9½in for tonnage) x 13ft 6½in. 1,102 7/94bm. Draught 10ft 6in/13ft 6in.
- Ord: 11.12.1812. K: 3.1816 (housed over 3.1817). L: 12.1.1819. In Ordinary (reserve) at Greenhithe 13.4.1819 until 6.1924–9.9.1824 when completed and fitted at Woolwich.
- First cost: £38,266 to build, plus £15,241 fitting for sea in 1824.
- Commissioned 5.6.1824 under Capt. Lord Byron (George Anson); sailed for the Sandwich Islands (carrying the remains of the King and Queen of those Islands) 9.9.1824; paid off 15.12.1826 at Portsmouth. Fitted for sea there (for £11,962) 1–4.1828. Recommissioned 18.1.1828 under Capt. Edward Lyons, for the Mediterranean; in 11.1830 under Capt. Sir Thomas Sabine Pasley; paid off 5.1831 at Portsmouth. Defects made good and fitted for Ordinary 5–8.1832. Fitted there for sea (for £11,748) 11.1833–2.1834. Recommissioned 11.1833 under Capt. Francis Mason, for South America; paid off 11.1837 at Portsmouth. Housed over the waist, and fitted for Demonstration ship there

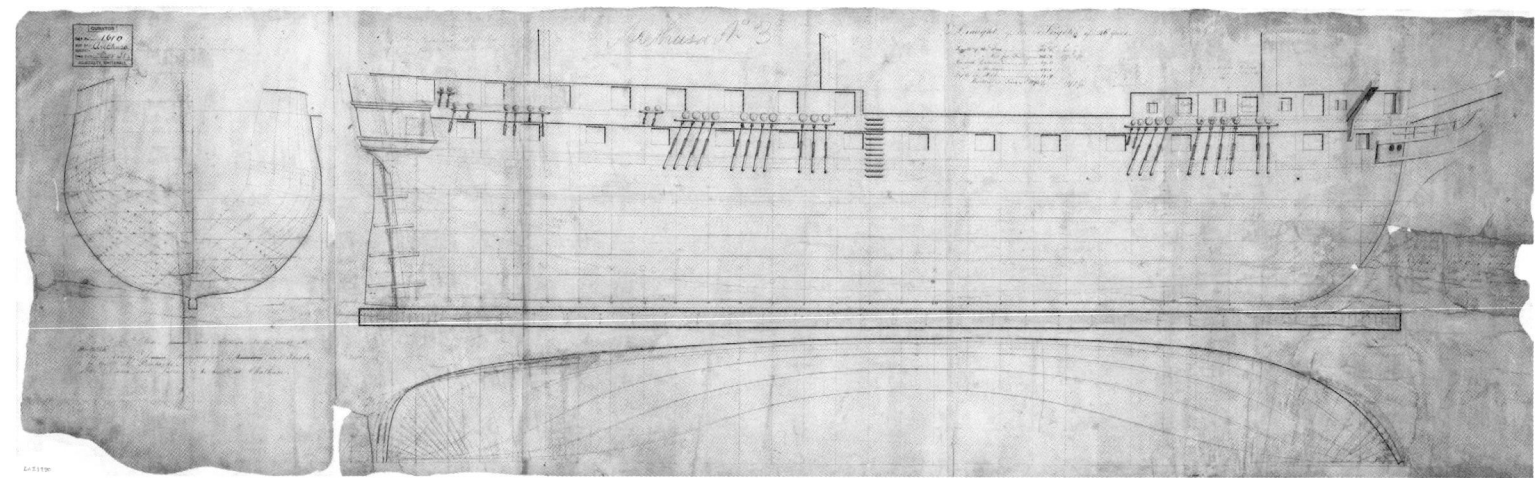

12.1837–4.1838. Fitted for sea (for £8,255 including 1837–38 working) 4.1839–2.1840. Recommissioned 6.11.1839 under Capt. Thomas Bourchier; in China War 1839–42; at Amoy 3.7.1840 and 25/26.8.1840; at Tsinghae 1.10.1841; at Woosung 16.6.1842; in Yangtse operations 7.1842; paid off 3.1843 at Portsmouth. Fitted as receiving ship there (for £2,900) 10–11.1850. Made good defects as temporary hospital ship 7.1866. Renamed *Calypso* by AO 9.3.1870. Sold to Castle & Co. 28.2.1895 (removed by purchaser 28.5.1895).

***SERINGAPATAM* Class** 18-pounder frigates of 46 guns (rated 38-gun until 1 January 1817). As a replacement for the *Lively* Class 38-gun frigates, a design based on the captured French frigate *Le Président* (a Forfait design, the prize being built in 1804 and taken in 1806) was selected for the prototype, which was ordered to be built of teak at Bombay on 6 September 1813. Compared with the *Président*, the new design had a flatter sheer (thus increasing the freeboard of the midships gunports) and a lesser tumblehome.

 Dimensions & tons: 157ft 6in, 132ft 1in x 40ft 11in (40ft 5in for tonnage) x 13ft 6in. 1,147⁶¹⁄₉₄bm.

 Men: 315. Guns: UD 28 x 18pdr (38cwt), QD 14 x 32pdr (17cwt) carronades, Fc 2 x 9pdr (26cwt) + 2 x 32pdr (17cwt) carronades; subsequently re-armed with UD 26 x 32pdrs + 2 x 8in/68pdr shell guns, QD 10 x 32pdrs, Fc 4 x 32pdrs.

 First cost: £36,749.

Seringapatam Bombay Dyd. Teak-built.

As built: 157ft 6in, 132ft 0½in x 41ft 0in (40ft 6in for tonnage) x 13ft 6in. 1,152bm.

 Ord: 21.8.1813. K: 11.1817. L: 5.9.1819.

 Commissioned 7.12.1819 under Capt. William Walpole, for passage to England (transporting frame for second ship); sailed 30.12.1819 from Bombay. Recommissioned 12.1820 under Capt. Samuel Warren; paid off 2.1824. Recommissioned 18.5.1824 under Capt. Charles Sotheby, for the Mediterranean. Recommissioned 7.2.1829 under Capt. William Waldegrave, for South America and the Pacific; paid off 28.8.1832 at Portsmouth. Recommissioned 6.2.1837 under Capt. John Leith, for North America and the West Indies; paid off 20.11.1841 at Sheerness. Became a receiving ship at the Cape of Good Hope 7.1847, then a coal hulk 1852. BU there between 6.1873 and 1883 (by AO 20.12.1871), and remains sold.

Modified *LEDA* Class 18-pounder frigates of Henslow design, modified to incorporate Seppings's circular stern and 'small-timber'

This sheer draught of 1817 for the Modified *Leda* class lists fifteen of the ships built to this design (plus the *Medusa*, which was later cancelled) but also names the final four ordered to the original *Leda* design (*Thetis*, *Arethusa*, *Blanche* and *Fisgard*) which were altered to incorporate Seppings's design for his 'small-timber' method of construction. (*NMM J3943*)

system of construction. All six were re-classed as 46-gun on 1 January 1817, and were re-rated 42-gun in 1839, but in practice only *Melampus* was brought into sea service, and the rest spent their lives in harbour auxiliary roles. Consequently the established guns and men were simply theoretical.

 Dimensions & tons: 151ft 9in, 125ft 0in x 40ft 3in x 12ft 9in. 1,077¹⁶⁄₉₄bm.

 Men: 315. Guns: UD 28 x 18pdr (38cwt), QD 14 x 32pdr (17cwt) carronades, Fc 2 x 9pdr (26cwt) + 2 x 32pdr (17cwt) carronades.

Venus Deptford Dyd

 As built: 153ft 7½in, 125ft 1⅛in x 40ft 5in (40ft 1in for tonnage) x 12ft 9½in. 1,069⁴⁄₉₄bm. Draught 10ft 4in/14ft 9in.

 Ord: 1.5.1816. K: 3.1817. L: 10.8.1820. C: 27.4.1821 (for Ordinary); housed over fore and aft at Sheerness 5.1821.

 First cost: £34,836 to build (£37,185 including fitting for Ordinary).

 Not commissioned. Fitted at Chatham 6–7.1848 then Woolwich 7–10.1848 (combined costs £2,356) and lent to the Marine Society as training hulk until 1862. Returned by the Peruvian government 31.9.1863 (after use as accommodation ship for crews of ships building on the Thames) and moved back to Chatham. Sold to Castle & Beech 7.10.1864 and BU completed 24.2.1865 at Charlton.

Melampus Pembroke Dyd

 As built: 151ft 9½in, 127ft 1⅛in x 40ft 5½in (40ft 1½in for tonnage) x 12ft 9in. 1,088⁵⁄₉₄bm. Draught 10ft 4in/14ft 7in.

 Ord: 1.5.1816. K: 8.1817. L: 10.8.1820. C: 2–23.9.1820 (for Ordinary) at Plymouth; housed over from main mast forwards.

 First cost: £23,007 to build (+£7,072 fitting for Ordinary).

 Between small and middling repair (for £10,192) 4.1832–2.1835. Became Advanced ship (for £5,837) 4.1843. Fitted for sea (for £7,693 + £1,463) 3.1845–18.5.1845.

 Commissioned 12.3.1845 under Capt. John Norman Campbell, for east coast of South America; paid off 12.1.1849 at Chatham. Fitted at Sheerness 2–3.1855 to carry the machinery for *St Jean d'Acre* to Plymouth. Fitted at Sheerness 12.1854–2.1855 as a store and receiving ship (for £3,481) for Constantinople; recommissioned 7.4.1855 under Capt. John Borlase, for the Mediterranean, though it is uncertain whether she was ever sent to Constantinople; paid off 10.1856. Fitted at Sheerness

12.1856–2.1857 (for £2,327) for the coastguard at Southampton; paid off 24.2.1858 at Portsmouth. Fitted 9–12.1866 (by AO 2.6.1866) as a Roman Catholic chapel at Portsmouth. Returned to the RN 8.1886 at Portsmouth then handed to the War Office there (by AO 16.8.1886) as an ordnance store until returned to the Admiralty 1.10.1891. To Chatham by AO 25.11.1895. Sold (by AO 16.6.1903) to Harris, Bristol 3.4.1906 to BU.

Minerva Portsmouth Dyd

As built: 151ft 5in, 126ft 6⅞in x 40ft 5in (40ft 1in for tonnage) x 12ft 9½in. 1,081$^{66}/_{94}$bm. Draught 10ft 9in/14ft 8in.

Ord: 1.5.1816. K: 10.1817. L: 13.6.1820. C: 26.6.1820 (for Ordinary); housed over fore and aft.

First cost: £33,659 (£36,080 including fitting for Ordinary).

Not commissioned. Became a Floating workshop at Portsmouth 1.1859. Hulked for harbour service 1861. Sold 28.2.1895 at Portsmouth.

Latona Chatham Dyd

As built: 150ft 1in, 125ft 0⅞in x 40ft 5½in (40ft 1½in for tonnage) x 12ft 10in. 1,071$^{10}/_{94}$bm. Draught 11ft 0in/15ft 3in.

Ord: 1.5.1816. K: 10.1818. L: 16.6.1821. C: 9.1821 (for Ordinary), roofed over fore and aft.

First cost: £29,747.

Not commissioned. Became a mooring vessel at Sheerness 1868. To Portsmouth as a powder depot 1872. Intended 1874 to become a training ship but this was not put into effect, and instead BU was completed 20.3.1875 at Chatham.

Diana Chatham Dyd

As built: 151ft 5in, 126ft 9⅛in x 40ft 5in (40ft 1in for tonnage) x 12ft 10in. 1,083$^{29}/_{94}$bm. Draught 11ft 0in/15ft 3in.

Ord: 1.5.1816. K: 2.1819. L: 8.1.1822. C: 22.2.1822 (for Ordinary), housed over from main mast forwards.

First cost: £30,159 (including fitting for Ordinary).

Not commissioned. Used as depot ship 1834–55 (approx). Fitted as receiving ship at Sheerness 9–10.1868. Offered as Training Ship for Boys at Aberdeen 11.1871, but offer declined. BU (by AO 22.3.1873) completed 9.2.1874 at Chatham.

Hebe Woolwich Dyd

As built: 151ft 9¼in, 127ft 1in x 40ft 3¼in (39ft 11¼in for tonnage) x 12ft 9in. 1,078$^{17}/_{94}$bm. Draught 10ft 6½in/14ft 5½in.

Ord: 1.5.1816. K: 5.1820. L: 14.12.1826. C: 5.1827 (for Ordinary), roofed over fore and aft.

First cost: £30,202 (including fitting for Ordinary).

Not commissioned. To Sheerness 5.1830, then Chatham 4.1837. Fitted as convict receiving ship at Woolwich (for £1,628) 2.1839. To Sheerness (by AO 21.12.1871) 6.1.1872. BU (by AO 26.2.1872) completed 31.3.1873 at Chatham.

Ex-American Prizes (1812–14)

Two American frigates captured during the War of 1812 were added as 18pdr frigates to the British Navy. The *Chesapeake* was armed as a 18pdr frigate at her capture, while the *Essex*-mounted a 12pdr main battery when originally built but was re-armed by the RN with 18pdrs and appears here under that classification. The *Chesapeake* was rated as 38-gun by the USN, but carried 48 guns when taken: 28 x 18pdrs and 20 x 32pdr carronades, and in 1817 the Navy Board re-classed her as 48-gun. The *Essex*-was built to a design of 141ft 0in (GD) x 37ft 0in (36ft 6in moulded) x 12ft 3in, and was rated at 32-gun, although carrying 46 guns when taken (40 of them 32pdr carronades, this type having replaced most of her 12pdrs in 1810, although 6 of the long guns remained – 3 on the GD and 3 on the spar deck); in 1817 she was re-classed as 42-gun.

Chesapeake Gosport Navy Yard, Norfolk, Virginia (to design by Josiah Fox), 38 guns.

Dimensions & tons: 151ft 0in, 127ft 5in x 40ft 11in (40ft 4in mld.) x 13ft 9in. 1,134$^{63}/_{94}$bm.

Men: 315. Guns: UD 28 x 18pdrs; QD 14 x 32pdr carronades; Fc 2 x 9pdrs + 2 x 32pdr carronades.

L: 20.6.1799. C: 2.12.1799. Taken by the *Shannon* off Cape Anne (Boston Bay) 1.6.1813.

Commissioned 1813 under Cmdr Alexander Gordon (Capt. 10.2.1814); later under Capt. George Burdett; arrived at Plymouth 9.10.1814 to repair defects, and on same day under Capt. Francis Newcombe; laid up there 9.1815. Sold to Joshua Holmes for £3,450 on 18.8.1819.

Essex *Enos* Briggs, Salem, Massachusetts (to design by William Hackett), 36 guns.

Dimensions & tons: 138ft 7in, 117ft 2⅞in x 37ft 3½in x 11ft 9in. 867$^{22}/_{94}$bm.

Men: 280. Guns: UD 26 x 18pdrs; QD 12 x 32pdr carronades; Fc 2 x 9pdrs + 2 x 32pdr carronades.

K: 14.4.1799. L: 30.9.1799. Taken by the *Phoebe* and *Cherub* off Valparaiso 28.3.1814.

Not commissioned. Never fitted for sea after arrival in Britain. Re-classed as a 42-gun ship in the RN from 1.1.1817. Troopship 7.6.1819. Hulked at Cork as a convict ship 10.1823. Sold for £1,230 on 6.7.1837.

(B) Vessels acquired from 1 January 1817

Modified *LEDA* Class 18-pounder frigates of 46 guns. Continuing orders were placed in 1817 to this Henslow design, two on 25 April, a third on 23 June and nineteen more on 23 July – of which three were subsequently re-ordered to a different design in 1820, and another two (*Medusa* and *Pegasus*) were eventually cancelled. Of those launched, none were brought into service at that time, but were fitted for Ordinary and laid up, most being used only for auxiliary duties, although *Daedalus* and *Amazon* were cut down to become Sixth Rate corvettes in the 1840s, and *Fox* and *Penelope* were experimentally lengthened (and were later converted to steam).

Dimensions & tons: 151ft 9in, 125ft 0in x 40ft 7in (40ft 3in for tons) x 12ft 9in. 1,077$^{16}/_{94}$bm.

Men: 300. Guns: UD 28 x 18pdrs (38cwt); QD 14 x 32pdr carronades (17cwt); Fc 2 x 9pdrs (26cwt) + 2 x 32pdr carronades (17cwt).

Nereus Pembroke Dyd

As built: 151ft 10in, 127ft 6in x 40ft 6in (40ft 2in for tons) x 12ft 9in. 1,094$^{16}/_{94}$bm. Draught 10ft 6in/14ft 6in.

Ord: 25.4.1817. K: 1.1819. L: 30.7.1821. C: 23.8–12.9.1821 at Plymouth (housed over from main mast forwards).

First cost: £23,233 to build, + £3,892 fitting in 1821. Fitted at Plymouth (for £10,557) as a coal depot 8–12.1843.

Entered service (not commissioned) 11.1843 under Francis W. Bateman, Master, as a storeship for Valparaiso (Chile). On 20.12.1851 under Alexander M. P. Mackey, Master, still at Valparaiso. In 12.1856 under John Clark Barlow, Master, at Callao. On 1.8.1863 under Charles Raguenau Pecco Forbes, Master, at Valparaiso. In 10.1866 under John P. Dillon, Master, at Valparaiso. In 8.1874 under John A. R. Petch, Master, at Coquimbo. Sold to J. L. Page, Coquimbo (for £500) 22.1.1879 as a hulk.

Hamadryad Pembroke Dyd
As built: 151ft 9¼in, 127ft 1in x 40ft 4in (40ft 0in for tons) x 12ft 9in. 1,081$^{53}/_{94}$bm. Draught 10ft 6in/15ft 2in.
Ord: 25.4.1817. K: 9.1819. L: 25.7.1823. C: 10–23.8.1823 at Plymouth (housed over from main mast forwards).
First cost: £24,679 to build.
Became a Divisional ship at Plymouth for the Ordinary 5.1847. Lent as a hospital ship for sick seamen at Cardiff (by AO 9.3.1866). Sold at Portsmouth 11.7.1905.

Amazon Deptford Dyd
As built: 151ft 5in, 126ft 7½in x 40ft 4in (40ft 0in for tons) x 12ft 9in. 1,077$^{62}/_{94}$bm. Draught 10ft 6in/15ft 3in.
Ord: 23.6.1817. K: 10.1817. L: 15.8.1821. C: 8.10.1821 (roofed over fore and aft) and taken to Sheerness.
First cost: £34,887 (including fitting for Ordinary).
Roof removed, cut down to a 26 gun corvette (6th Rate) and fitted for sea at Sheerness (for £14,226) 3.1844–4.1.1845.
Commissioned 13.11.1844 under Capt. James John Stopford; to Mediterranean 1846–47; paid off 1848. Refitted for loan to President of Liberia (for £6,543) 7–12.1848. Recommissioned 25.10.1848 under Capt. Edward Norwich Troubridge, for the East Indies (died 10.1850; on 29.9.1850 under Capt. Charles Barker, on same station; paid off at Portsmouth 5.1852. Sold to W. Lethbridge of Plymouth (for £1,820) by AO 11.9.1863.

Aeolus [*Eolus*] Deptford Dyd
As built: 151ft 9in, 127ft 0⅞in x 40ft 3in (39ft 11in for tons) x 12ft 9in. 1,076$^{91}/_{94}$bm. Draught 10ft 8in/14ft 8in.
Ord. 23.7.1817. K: 10.1818. L: 17.6.1825. C: 5.8.1825 to Woolwich, where laid up (roofed over fore and aft).
First cost: £30,119 (including fitting for Ordinary).
To Sheerness 1830.
Commissioned 17.10.1846 at Sheerness under John Thomas, Master, and fitted there as a storeship (for £7,163) 10–11.1846, for Ireland; paid off 14.10.1847 at Portsmouth. Fitted at Portsmouth (for £8,582) as a shot and shell ship 3–6.1855; on 27.8.1855 under William Lindsay Brown, Master, 'as a shell depot (ship), to attend n the Baltic Fleet' 1855–56; paid off 2.6.1856. Fitted as hospital ship (by AO 5.6.1866) at Portsmouth 6–7.1866, then lent to Southampton Board of Health; returned (by AO 30.10.1866) and moored at the Motherbank. Sold to Castle 3.1886 to BU at Charlton.

Thisbe Pembroke Dyd
As built: 151ft 9in, 127ft 0¼in x 40ft 4⅜in (40ft 0⅜in for tons) x 12ft 9in. 1,082$^{67}/_{94}$bm. Draught 10ft 0in/15ft 4in.
Ord. 23.7.1817. K: 8.1820. L: 9.9.1824. C: 5.10.1824 to Plymouth, where laid up (housed over from main mast forwards)
First cost: £26,335 to build.
Never commissioned. Employed as depot ship at Plymouth 1850–1863. Became a floating church at Cardiff 13.8.1863. Sold to W. H. Caple (for £1,005) 11.8.1892.

Cerberus Plymouth Dyd
As built: 151ft 9in, 127ft 0¼in x 40ft 3½in (39ft 11½in for tons) x 12ft 9in. 1,078$^{73}/_{94}$bm. Draught 11ft 3in/15ft 0in.
Ord. 23.7.1817. K: 11.1820. L: 30.3.1827. C: 11.4.1827 for Ordinary (housed over fore and aft).
First cost: £31,614 (including fitting for Ordinary).
Never commissioned. BU completed 10.1.1866 by J. & E. Marshall at Plymouth.

Circe Plymouth Dyd
As built: 151ft 9in, 127ft 0⅝in x 40ft 3½in (39ft 11½in for tons) x 12ft 9in. 1,079$^{4}/_{94}$bm. Draught 11ft 3in/14ft 9in.
Ord. 23.7.1817. K: 11.1820. L: 22.9.1827. C: 23.10.1827 for Ordinary (housed over fore and aft)
First cost: £31,805 to build (+ fitting for Ordinary £1,170).
Never commissioned. Became an accommodation hulk at Plymouth 1.4.1866, as tender to *Indefatigable*. Fitted as swimming bath 6–11.1885. 1916 renamed *Impregnable IV*. Sold to S. Castle, Plymouth 7.1922 for BU.

Clyde Woolwich Dyd
As built: 152ft 0in, 127ft 3⅞in x 40ft 3½in (39ft 11½in for tons) x 12ft 8½in. 1,079$^{4}/_{94}$bm. Draught 10ft 6in/14ft 7in.
Ord. 23.7.1817 (by AO 30.8.1817, of winter-felled timber). K: 1.1821. L: 9.10.1828. C: 11.1828–2.1829 for Ordinary (housed over fore and aft). To Sheerness 11.5.1830.
First cost: £32,074 (including fitting for Ordinary).
Commissioned 7.1869 as RNR Drill Ship 7.1869 under Cmdr Henry Weyland Chetwynd. Fitted at Sheerness as an RNR training ship (by AO 3.2.1870, for £3,184) for Aberdeen 7.1870. Became training ship at Aberdeen 8.1870, replacing *Winchester*. Subsequently 7.1872 under Cmdr Robert Hornby Boyle, 7.1775 under Henry D. Best, and 7.1878 under Sir William Wiseman. Recommissioned 4.1881, still under Wiseman; in 1886 under Cmdr Alfred Wilmot Warry, then 1.1889 under Cmdr E. S. Dugdale; in 1898 under Cmdr James Pipon Montgomery (–1901). Sold at Portsmouth (by AO 4.5.1904, for £1,825) 5.7.1904.

Thames Chatham Dyd
As built: 151ft 9¾in, 127ft 0⅞in x 40ft 5½in (40ft 1½in for tons) x 12ft 9½in. 1,088$^{22}/_{94}$bm. Draught 10ft 11in/15ft 4in.
Ord. 23.7.1817. K: 6.1821. L: 21.8.1823. C: 3.10.1823 for Ordinary (housed over fore and aft).
First cost: £26,401 (including fitting for Ordinary).
Never commissioned. Fitted as convict ship at Deptford (for £3,396) 10.1840–1.1841; fitted at Chatham as a convict ship for Bermuda (for £9,215) 2–4.1844. Returned to Admiralty control by Convict Dept. at Bermuda in early 1860s. Sunk at moorings there 6.6.1863 and wreck sold to J. Murphy (for £900) for BU.

Fox Portsmouth Dyd
As built: 151ft 8¼in, 126ft 11½in x 40ft 4in (40ft 0in for tons) x 12ft 8¾in. 1,080$^{47}/_{94}$bm. Draught 9ft 7in/14ft 5in.
As lengthened (1843): 159ft 4in, 133ft 7in x 40ft 4in (40ft 0in for tons) x 12ft 9in. 1,137bm.
Ord. 23.7.1817. K: 6.1821. L: 17.8.1829.
First cost: £35,215.
Lengthened at the bows 16.3–17.11.1843 (for £18,160, including fitting) in accordance with plans suggested by J. White, Cowes.
Commissioned 14.10.1843 under Capt. Sir Henry Martin Blackwood, for Tarbert (on River Shannon) as a guard ship; sailed thence 8.7.1844 for the East Indies, as Commodore; paid off 5.8.1848 at Portsmouth. Refitted there (for £11,484) 8.1849–9.1850. Recommissioned 17.7.1850 under Capt. George Robert Lambert, again for the East Indies, as Commodore; in 2nd Anglo-Burmese War; landing parties repulsed 4.2.1852; at Rangoon 10–15.4.1852; on 27.9.1852 under Capt. John Walter Tarleton, as flag of Commodore Lambert; paid off 22.5.1854 at Portsmouth. Converted to a screw frigate from 9.1855 to 3.1856 (see Section D for subsequent data).

Unicorn Chatham Dyd
As built: 151ft 9¾in, 127ft 1½in x 40ft 4½in (40ft 0½in for tons) x 12ft 9½in. 1, 084$^{16}/_{94}$bm. Draught 11ft 3in/15ft 3in.
Ord. 23.7.1817. K: 2.1822. L: 30.3.1824. C: 14.5.1824 for Ordinary (housed over for and aft).

First cost: £26,461 (fitted for Ordinary).

Never fitted for sea, or commissioned until 1874. Lent to the War Office as a powder hulk at Woolwich 1857–62, then to Ordinary at Sheerness 1862–73. Refitted at Sheerness as RNR training ship 11.1873 (re-armed with 10 guns – 1 x 9in, 1 x 6in, 4 x 64pdrs & 4 x 32pdrs) and towed to Dundee. Commissioned 1.1.1874 under Cmdr Harry Woodfall Brent, then 11.4.1874 under Cmdr William Edward Saxton Browne; then 4.1877 under Cmdr Henry Dolphin, and 13.10.1779 under Cmdr Orford Somerville Cameron. Recommissioned 4.1881 in same role; on 18.10.1882 under Cmdr Stanhope Grove, then 15.10.1885 under Cmdr William Henry George Nowell, then 18.10.1888 under Lieut Francis Avenell Brookes, then 1.8.1891 under Lieut William Frederick Glascock Clarke, then 9.11.1898 under Lieut Ernest Leigh Austen, and 1.1.1904 under Lieut Samuel Montagu Agnew. Lent to Clyde Division RNVR (by AO 15.3.1906) 1.4.1906. Renamed *Unicorn II* in 2.1939, then *Cressy* 20.11.1941, resumed *Unicorn* 14.7.1959. Sold 29.9.1968 to the Unicorn Preservation Society who fitted her out as a museum ship in Dundee Harbour, where she remains afloat under her original housing.

Daedalus Deptford Dyd

As built: 151ft 10¼in, 127ft 4½in x 40ft 3½in (39ft 11½in for tons) x 12ft 9in. 1, 081$\frac{74}{94}$bm. Draught 10ft 5in/14ft 6in.

Ord: 23.7.1817. Not begun at Deptford. Transferred to Sheerness Dyd 14.8.1821.

K: 10.1822. L: 2.5.1826. C: for Ordinary (housed over fore and aft).

First cost: £27,579 (fitted for Ordinary).

Cut down to a 20-gun Sixth Rate corvette (for £12,657 + £276) at Woolwich 3.1844–28.11.1844 and fitted for sea.

Commissioned 16.10.1844 under Capt. Peter M'Quhae, for the East Indies; boat attack on Sherif Osman's stronghold at Malluda Bay (Borneo) 19.8.1845; paid off 13.10.1848. Recommissioned 27.7.1849 under Capt. George Greville Wellesley, for the Pacific; paid off and laid up 8.1853 at Plymouth. Fitted there as a training ship for Bristol (for £5,437) 3–6.1861. Recommissioned 5.6.1861 under Cmdr William Henry Fenwick, as RNR drill ship at Bristol; in 7.1864 under Cmdr Edward Field, then 2.1869 under Cmdr Albert Henry William Battiscombe, on 4.1870 under Cmdr Charles William Manthorp, on 11.1873 under Cmdr H. B. Stewart, and in 4.1879 under Cmdr Robert Sidney Hunt, all on same station. Recommissioned 4.1881, still under Hunt, then 1886 under C. Hawkins, on 9.1888 under Lieut F. H. N. Harvey and in 1900 under Basil C. Barber, all on same station; paid off 30.9.1910 at Bristol. Sold to J. B. Garnham (for £1,010) 24.9.1911 to BU.

Pegasus Deptford Dyd

Ord: 23.7.1817: K: Not begun at Deptford. Transferred to Sheerness Dyd 17.2.1825.

K: 3.1828. Transferred to Pembroke Dyd 7.8.1830, then cancelled 10.1.1831.

Properpine Plymouth Dyd

As built: 151ft 10in, 127ft 2¾in x 40ft 3in (39ft 11in for tons) x 12ft 10in. 1, 078$\frac{27}{94}$bm. Draught 10ft 10in/14ft 10in.

Ord. 23.7.1817. K: 11.1822 (in South Dock). L: 1.12.1830 (floated out). C: for Ordinary (housed over fore and aft).

First cost: £34,342 (including fitting for Ordinary).

Never commissioned. Sold at Plymouth to Robinson.Ridley 21.1.1864 (for £2,460) to BU.

Mermaid Chatham Dyd

As built: 151ft 10½in, 127ft 2½in x 40ft 4½in (40ft 0½in for tons) x 12ft 9in. 1, 085$\frac{83}{94}$bm. Draught 11ft 2in/15ft 3in.

Ord. 23.7.1817. K: 9.1823. L: 30.7.1825. C: 30.8.1825 for Ordinary (housed over fore and aft).

First cost: £27,134 (fitted for Ordinary).

Never commissioned. Lent to the Army as a powder ship at Purfleet 1858. Returned to the RN 8.5.1863 and fitted at Woolwich as a powder depot 6.1863, then sailed to Dublin. BU completed by Mr & Mrs Kevitt 7.1874–6.1875 at Dublin.

Mercury Chatham Dyd

As built: 151ft 9½in, 127ft 1⅞in x 40ft 4⅝in (40ft 0⅝in for tons) x 12ft 9¼in. 1,085bm. Draught 12ft 2in/15ft 3in.

Ord. 23.7.1817. K: 4.1824. L: 16.11.1826. C: 15.12.1826 for Ordinary (housed over fore and aft).

First cost: £27,673 (fitted for Ordinary).

Never commissioned. Fitted at Deptford as a coal depot for Woolwich (for £3,042) 4–8.1862. By 1870 to Sheerness, still as a coal depot (–1901). Sold to Harris Brothers, Bristol 3.4.1906 (by AO 1.2.1904, for £1,375) to BU.

Penelope Pembroke Dyd

As built: 151ft 10⅛in, 127ft 1in x 40ft 6in (40ft 2in for tons) x 12ft 9½in. 1,090$\frac{55}{94}$bm. Draught 10ft 7in/13ft 1in.

Ord: 23.7.1817. K: not begun at Pembroke. Frames and order transferred to Chatham Dyd 26.5.1827.

K: 11.1827. L: 13.8.1829. C: 1.1830 for Ordinary (housed over fore and aft).

First cost: £31,476 (fitted for Ordinary).

Never commissioned as sailing frigate. Ordered 26.3.1842 to be converted to a paddle frigate (the only such conversion). Lengthened amidships at Chatham (for £18,337) 6.1842–4.1843, then engines fitted at Limehouse and Chatham 1843 as a paddle frigate (see Chapter 11).

Thalia Portsmouth Dyd

As built: 151ft 5¾in, 126ft 9¼in x 40ft 4⅞in (40ft 0⅞in for tons) x 12ft 9¼in. 1,083$\frac{79}{94}$bm. Draught 11ft 2in/15ft 6in.

Ord: 23.7.1817. K: not begun at Portsmouth. Transferred to Chatham Dyd 26.5.1827.

K: 2.1828. L: 12.1.1830. C: 10.2.1830 for Ordinary (housed over fore and aft).

First cost: £29,476 (fitted for Ordinary).

Housing removed and fitted for sea (for £12,941) 2.1834–25.8.1834.

Commissioned 5.1834 under Capt. Robert Wanchope, for the Cape of Good Hope; as flagship of Rear-Adm. Sir Patrick Campbell; paid off 1838. Refitted as a Demonstration Ship at Chatham (for £1,495) 12.1838–3.1839; fitted for sea there (for £5,658) 9–11.1841. Recommissioned 28.8.1841 under Capt. Charles Hope, for the East Indies and Pacific; paid off 11.1845 at Portsmouth into Ordinary. Fitted as Roman Catholic chapel ship at Portsmouth 11–12.1855. BU by White at Cowes (by contract 16.6.1865) completed 25.11.1867.

Nemesis Pembroke Dyd

Ord: 23.7.1817. K: — . Re-ordered to *Seringapatam* design (see below) 26.10.1820.

Statira Plymouth Dyd

Ord: 23.7.1817. K: — . Re-ordered to *Seringapatam* design (see below) 26.10.1820.

Jason Woolwich Dyd.

Ord: 23.7.1817 (or 18.7?). K: — . Re-ordered to *Seringapatam* design (see below) 26.10.1820.

Medusa Woolwich Dyd

Ord: 23.7.1817 (or 18.7?). K: not begun. Transferred to Pembroke Dyd 7.8.1830, then cancelled 22.4.1831.

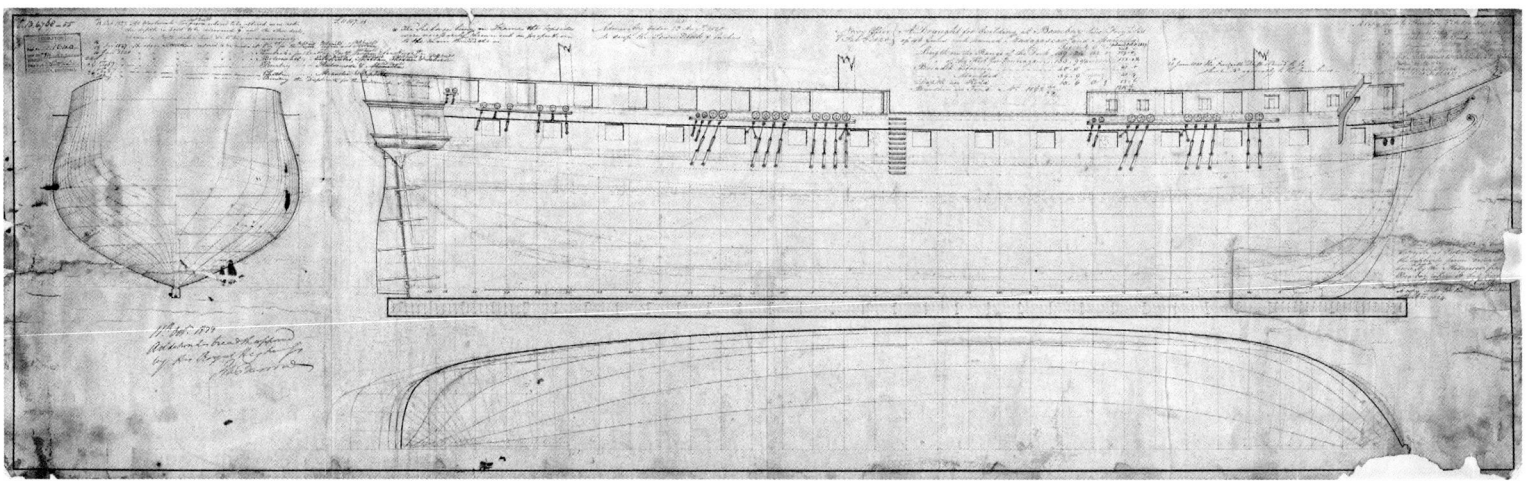

A sheer draught, dated 1 February 1820, originally intended for the Bombay-built *Madagascar* and *Manilla*, but used to record modifications to all the later ships of the class. During the latter half of the eighteenth century the Royal Navy's designers had been inspired by French hull forms, often adapting them to British requirements and occasionally simply copying them. As a source of fresh ideas this had proved largely beneficial, but it needed to be employed with intelligence. The underwater form of this class ultimately derived from the French *Président*, which was a lightly built ship (reflecting France's different operational requirements), while the *Seringapatam*, the prototype of what was to be this new class, was built of much heavier teak and accordingly floated much deeper, ruining her sailing qualities. The *Druid* group were modified as shown on this draught, with decks lowered and breadth increased slightly, but the result was still inadequate, and thus the hull was further broadened and deepened for the *Andromeda* group. A satisfactory frigate was eventually produced, but by this stage it was quite a distance from its French (Forfait) inspiration. *(NMM J3877)*

SERINGAPATAM Class Seven more ships were ordered to the original 1813 design (see under Section (a) for *Seringapatam* herself): the *Madagascar* and *Manilla* in April 1819 and the *Pique, Tigris, Jason, Nemesis* and *Statira* in October 1820 (the last three had originally been ordered in 1817 to Modified *Leda* design, see above); in 1822 the *Madagascar, Nemesis, Pique* and *Jason* were re-ordered as *Druid* Class; and in 1827 the *Manilla, Tigris* and *Statira* were re-ordered to the *Andromeda* design; but the last five were never in fact completed.

DRUID/ANDROMEDA Class 18-pounder frigates of 46 guns. A modified and slightly enlarged version of the Seringapatam design (and hence of French Président) was drawn in 1817, with the addition of circular sterns. Two vessels were ordered in 1817 (on the same day as the batch of nineteen Modified Leda Class frigates); two more were ordered in 1819 (to be built of teak at Bombay) and orders to this design continued until 1827; however in 1827 the design was revised, increasing the breadth by a further foot, and around this time most of the later orders were amended to the later version (described below as Andromeda design), without being formally re-ordered. The later ships were established to carry 44 guns vice 46 – their armament was to be 42 x 32pdr carronades and 2 x 8in shell guns.

Dimensions & tons:
Original (*Druid*) design – 159ft 0in, 133ft 9⅜in x 40ft 11in oa (40ft 5in for tonnage) x 12ft 9in. 1,162³⁸/₉₄bm.
Revised (*Andromeda*) design – 159ft 0in, 133ft 2¼in x 41ft 11in (41ft 5in for tons) x 13ft 6in. 1,215¹⁶/₉₄bm.
Men: 315. Guns: UD 28 x 18pdr (38cwt), QD 14 x 32pdr (17cwt) carronades, Fc 2 x 9pdr (26cwt) + 2 x 32pdr (17cwt) carronades; subsequently re-armed with UD 26 x 32pdrs + 2 x 8in /68pdr

shell guns, QD 10 x 32pdrs, Fc 4 x 32pdrs.

Druid Pembroke Dyd
As built: 159ft 6in, 133ft 4½in x 41ft 11in oa (40ft 7in for tonnage) x 12ft 9in. 1,168⁴²/₉₄bm. Draught 11ft 4in/15ft 4in.
Ord: 23.7.1817. K: 8.1821. L: 1.7.1825. C: 18.7–21.12.1825 at Plymouth (for sea).
First cost: £31,980 at Pembroke (incl. fitting there), + £13,054 fitting at Plymouth.
Commissioned 7.1825 under Capt. Samuel Chambers, for Jamaica; on 1.5.1828 under Capt. Williams Sandom, on same station; paid off 10.1829. Fitted for sea and for a consul (for £13,234) at Plymouth 10.1829–3.1830. Recommissioned 11.1829 under Capt. S W Hamilton, for South America; on 4.12.1832 under Capt. Samuel Roberts, for the River Tagus. Very Small Repair and fitted as a Demonstration ship at Plymouth 6.1835–3.1837. Fitted for sea at Plymouth (for £15,566 including cost of 1835–37 repairs) 3–8.1839. Recommissioned 9.4.1839 under Capt. Henry John Spencer Churchill, for the East Indies (died 2.6.1840); on 30.6.1840 under Capt. Henry Smith, on same station; in 1st Anglo-Chinese War; paid off 1843. Fitted as lazarette at Plymouth 3–6.1846, for Liverpool. Sold at Liverpool to Messrs. Marshall (by AO 13.4.1863, for £2,087) to BU.

Nemesis Pembroke Dyd
As built: 159ft 1in, 133ft 3⅛in x 41ft 11in oa (40ft 7in for tonnage) x 12ft 9in. 1,167⁴²/₉₄bm. Draught 11ft 0½in/15ft 0½in.
Ord. 23.7.1817. K: 8.1823. L.:19.8.1826. C: C: 9.1826 at Plymouth (for Ordinary, housed over fore and aft).
First cost: £26,614.
Not commissioned. Remained in Ordinary until BU by J. & E. Marshall completed 4.7.1866 at Plymouth.

Madagascar Bombay Dyd. Teak-built.
As built: 159ft 6in, 134ft 0in x 40ft 11½in oa (40ft 5½in for tonnage) x 12ft 9in. 1,166⁶⁶/₉₄bm.
Ord: 5.4.1819. K: 10.1821. L: 15.11.1822. C: 10.1828–1.1829 at Portsmouth (for sea).
First cost: £38,355 (£29,582 supplied by East India Co., + £8,773 supplied by the magazines at Bombay).
Commissioned 4.3.1823 at Bombay under Lieut Evan Nepean, for voyage to UK; sailed from Bombay 21.1.1823, and from Trincomalee 19.3.1823; arrived Portsmouth 21.9.1823, and paid off 7.10.1823 and laid up in Ordinary. Recommissioned 26.9.1828 under Sir Robert Cavendish Spencer (died 4.11.1830 at Alexandria), for the Mediterranean; on 12.1831 under Capt.

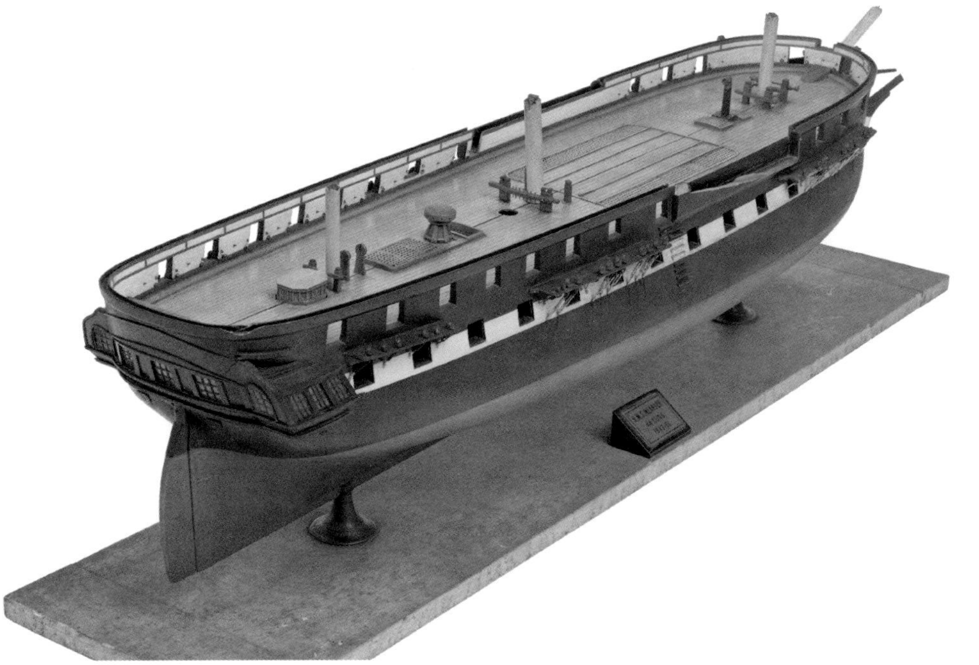

After 1815 the Admiralty embarked on a substantial programme of frigate construction to replace the large numbers of worn-out wartime-built ships, but these were not likely to be needed in peacetime, so the new ships were built slowly in the Royal Dockyards and their timbers were allowed to stand to season for many years. Knowing that they could be launched and fitted out in the same timescale as a major mobilisation, the Navy treated them as a strategic reserve. One consequence was that the design could evolve during the period of construction, which happened during the 1820s with the frigates based on the *Seringapatam* design, which altered several times between the early and the later ships.

A typical example is the *Maeander* (depicted in this model as finally completed), a ship which spent over a decade on the stocks before being launched, roofed over and placed in Ordinary; she did not go to sea until 1848. When fitted out these ships adopted the latest practice, so the circular stern of the original design was replaced by the elliptical form seen here, while the waist was almost entirely covered to form a structurally continuous spar deck. The gangways were only defined by two narrow runs of gratings and the waist hammock netting was replaced by solid boxed-in stowage – clearly seen in this model because the outer planking is broken. *(NMM F8901-003)*

Edmund Lyons; paid off 17.1.1835. Recommissioned 6.1836 under Capt. Sir John Strutt Peyton (died 20.5.1838), for North America and the West Indies; on 14.4.1838 under Capt. Provo William Parry Wallis, on same station; paid off 9.1839. Fitted as demonstration ship at Portsmouth (for £2,328) 9–12.1839; fitted there for sea (for £6,158) 10–11.1841. Recommissioned 18.8.1841 under Capt. John Foote, for the coast of Africa; paid off 1844. Fitted at Plymouth as a provision depot 8–9.1846; in 9.1846 under William J. W. Burney, Master. Fitted at Plymouth as a receiving ship for Rio de Janeiro (for £6,981) 12.1852–11.1853; to Rio 1853 as a receiving ship, then provisions ship; in 3.1855 under Cmdr James Ptolemy Thurburn, then 13.9.1859 under Capt. Richard Dunning White. Sold at Rio by public auction (for £2,304.14.9d) 5.5.1863.

Manilla Bombay Dyd. Teak-built.
Ord. 5.4.1819. K — . Cancelled 7.2.1831 (or 21.2.1831?).

Pique Plymouth Dyd
Ord: 25.10.1820. K: 6.1822. Cancelled 16.6.1832.

Tigris Plymouth Dyd (using teak frames ex-Bombay Dyd, brought to England aboard the *Seringapatam*)
Ord: 25.10.1820. K: 6.1822. Cancelled 31.8.1832 (frames BU 5.1833 and material used to complete 80-gun *Hindostan*).

Leda Pembroke Dyd
As built: 159ft 0in, 133ft 5¼in x 41ft 1½in oa (40ft 7½in for tonnage) x 12ft 9in. 1,171$^{38}/_{94}$bm. Draught 11ft 0in/15ft 0in.
Ord: 15.5.1821. K: 10.1824. L: 15.4.1828. C: 5.1828 at Plymouth (for Ordinary, housed over fore and aft).
First cost: £30,356.
Not commissioned. Remained in Ordinary until fitted as a water police ship at Plymouth 10.1864–3.1865. Sold to Mr Harris, Bristol 15.5.1906.

Hotspur Pembroke Dyd
As built: 159ft 0in, 133ft 8⅛in x 41ft 1in oa (40ft 7in for tonnage) x 12ft 9in. 1,171$^{9}/_{94}$bm. Draught 10ft 10in/15ft 0in.
Ord: 15.5.1821. K: 7.1825. L: 9.10.1828. C: 4.1829 at Plymouth (for Ordinary, housed over fore and aft).
First cost: £30,433.
Not commissioned. Remained in Ordinary until became a Roman Catholic chapel ship at Plymouth 1859. Renamed *Monmouth* 16.10.868. Sold under AO 17.9.1902.

Africaine Chatham Dyd
As built: 159ft 2½in, 133ft 10⅝in x 41ft 1in oa (40ft 7in for tonnage) x 12ft 8¾in. 1,172$^{87}/_{94}$bm. Draught 11ft 2in/14ft 7in.
Ord: 8.1.1822. K: 9.1825. L: 20.12.1827. C: 3.3.1828 (for Ordinary, housed over fore and aft).
First cost: £29,934 (including fitting for Ordinary)
Not commissioned. Remained in Ordinary at Chatham until sold (for £2,050) to Trinity House 9.5.1867 to serve as a hulk (BU in 1903).

Euphrates Plymouth Dyd (from teak frames ex-Bombay Dyd)
Ord. 22.10.1822. K: 30.6.1828. Cancelled 7.2.1831.

Seahorse Pembroke Dyd
As built: 159ft 10in, 133ft 3⅞in x 41ft 10in (41ft 4in for tons) x 13ft 3in. 1,211$^{53}/_{94}$bm. Draught 11ft 0½in/14ft 10in.
Ord: 9.1.1823. K: 11.1826. L: 22.7.1830. C: 8.1830 to Plymouth, for Ordinary (housed over fore and aft).
First cost: £29,272 (fitted for Ordinary).
Never commissioned as a sailing frigate. Ordered converted to screw frigate 2.9.1845; undocked as screw mortar frigate 3.1856– see section D).

Stag Pembroke Dyd
As built: 159ft 3in, 133ft 3¼in x 41ft 11½in (41ft 5½in for tons) x 13ft 3in. 1,218$^{49}/_{94}$bm. Draught 11ft 4in/14ft 8in.
Ord: 9.1.1823. K: 4.1828. L: 2.10.1830. C: 10.1830 to Plymouth, for Ordinary; Fitted for sea 10.1830–9.7.1831.
First cost: £31,770 (fitted for Ordinary); fitting 1830–31 for sea £16,454.
Commissioned 15.4.1831 under Capt. Edward Thomas Troubridge, as C-in-C Queenstown; on 12.10.1832 under Capt. Nicholas Lockyer, for the coast of Portugal; paid off 16.12.1835. Recommissioned 26.11.1836 under Commodore Thomas Ball Sulivan, for South America; paid off 4.1841. Fitted as Divisional ship at Plymouth 5.1844, but not recommissioned. BU completed 8.8.1866 by J. & C. Marshall, Plymouth.

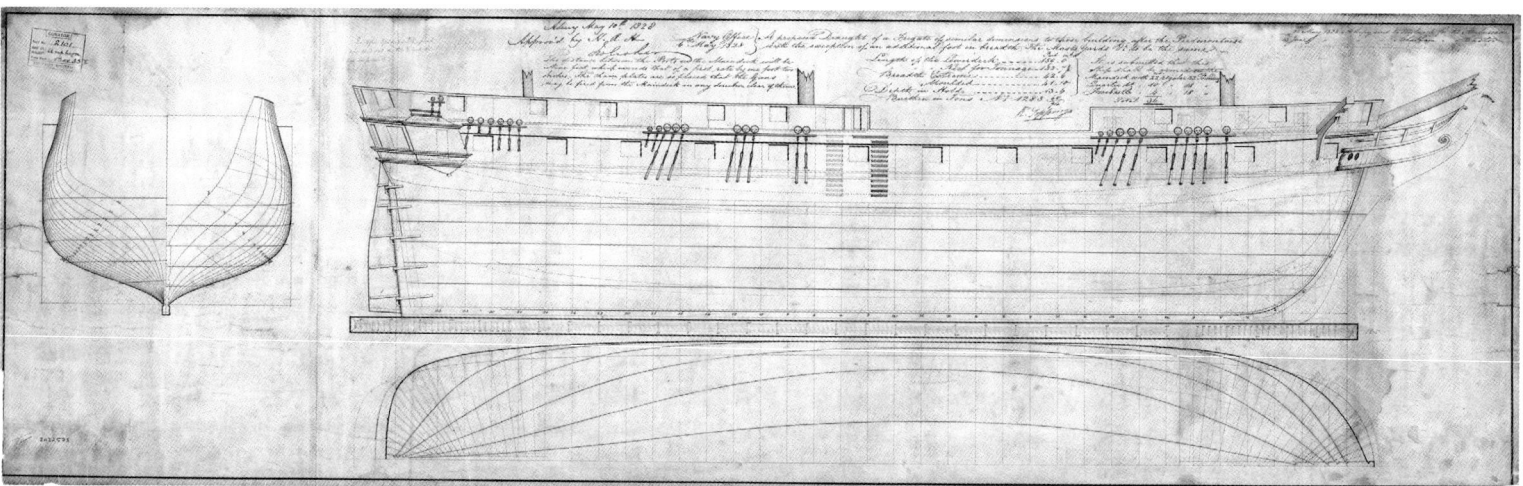

This sheer draught, dated 6 May 1828, is for the *Castor*, Seppings's last frigate design. It describes the ship as 'of similar dimensions to those building, after the *Piedmontaise* with the exceptions of an additional foot of breadth' (the preceding *Andromeda* Class, itself derived from the *Seringapatam* design of 1813, was in turn based on the French *Président* of 1804, later renamed *Piedmontaise*, albeit with less sheer and a reduced tumblehome). The midship section shows subtle but significant differences from its predecessors, notably the hollow garboards and a longer flat either side of the load waterline.

The additional breath was necessary because this was the first frigate designed specifically for a 32pdr main battery (although she also carried 8in shell guns) and the ports were arranged to allow oblique fire (a note on the draught points out that the chain plates 'are so placed that the Guns may be fired from the Maindeck in any direction clear of them'. Note also Seppings's modification of his circular stern. In service, the ship was highly regarded, not least by Admiral Sir George Cockburn, who wrote to Seppings of 'the character your *Castor* still maintains', and described her as 'an excellent ship for sea work'. *(NMM J5352)*

Eurotas Chatham Dyd
As built: 159ft 1½in, 133ft 10⅛in x 41ft 0in oa (40ft 6in for tonnage) x 12ft 9in. 1,167⁷⁷⁄₉₄bm. Draught 11ft 0in/14ft 11in.
Ord: 13.9.1824. K: 2.1827. L: 19.2.1829. C: 20.3.1829 (for Ordinary, housed over fore and aft).
First cost: £29,683 (including fitting for Ordinary)
Not commissioned as a sail frigate. Remained in Ordinary until ordered converted to a screw frigate 2.9.1845 (see Section D).

Maeander Chatham Dyd
As built: 159ft 1in, 133ft 3½in x 41ft 11in (41ft 6in for tons) x 13ft 5½in. 1,221⁷⁄₉₄bm. Draught 10ft 10in/14ft 7in.
Ord: 13.9.1824. K: 2.1829. L: 5.5.1840. C: for Ordinary.
First cost: £15,058 + £22,476 to build, + £7,923 fitting for Ordinary (total £45,457).
Commissioned 1.11.1847 under Capt. Henry Keppel, for the East Indies; fitted for sea (for £9,745) 11.1847–17.1.1848; to East Indies, later to Australia and the Pacific; paid off 1851. Recommissioned 14.7.1852 under Capt. Charles Talbot, for the Cape of Good Hope. On 30.5.1854 under Capt. Thomas Baillie, for the White Sea in the Russian War. Fitted at Plymouth for the coastguard 10.1856–1.1857. Recommissioned 2.12.1856 under Capt. James Robert Drummond, for the coastguard. Fitted at Portsmouth as a coal and storeship for Ascension Island (for £8,997 + £1,579) 5–11.1859. Recommissioned 1859 under Cmdr Malcolm MacGregor, for Ascension; on 1.11.1859 under Capt. William Farquharson Burnett; on 23.2.1861 under Capt. Frederick Lamport Barnard, then 24.12.1864 under Capt. Joseph Grant Bickford, on same service; paid off ?1.1866. Ordered to be BU there, the hull was partly stripped 10.1865, but she then foundered in a gale there 7.1870. In 1985 this wreck was investigated by RAF divers.

Spartan Portsmouth Dyd
Ord: 13.9.1824. K: — . Cancelled 7.2.1831.

Theban Portsmouth Dyd
Ord: 13.9.1824. K: — . Cancelled 7.2.1831.

Forth Pembroke Dyd
As built: 159ft 0in, 133ft 0⅜in x 42ft 2in (41ft 8in for tons) x 13ft 3in. 1,228⁴⁶⁄₉₄bm. Draught 10ft 11½in/14ft 8in.
Ord: 9.6.1825. K: 11.1828. L: 1.8.1833. C: 2.9.1833 at Plymouth for Ordinary.
First cost: £28,542 fitted for Ordinary.
Never commissioned as a sailing frigate. Ordered converted to a screw frigate at Devonport 6.9.1845; undocked as screw mortar frigate 21.1.1856 (see Section D).

Inconstant Sheerness Dyd
Ord: 9.6.1825. K: — . Cancelled 9.3.1832.

Orpheus Chatham Dyd
Ord: 9.6.1825. K: — . Cancelled 7.2.1831.

Severn Plymouth Dyd
Ord: 9.6.1825. K: — . Cancelled 7.2.1831.

Tiber Portsmouth Dyd
Ord: 9.6.1825. K: — . Cancelled 7.2.1831.

Jason Woolwich Dyd
(Re)-ord: 1826. K: — .. Cancelled 7.2.1831.

Statira Plymouth Dyd
(Re)-ord: 1826. K: 12.1823. Cancelled 31.8.1832 (frames BU 10.1832 and material used to repair *Melampus*).

Andromeda Bombay Dyd. Teak-built.
As built: 159ft 0in, 133ft 2¼in x 41ft 11in (41ft 5in for tonnage) x 13ft 6in. 1,215¹⁶⁄₉₄bm.
Ord: 5.4.1827. K: 8.1827. L: 3.1829 (or 6.1.1829?).
First cost: £38,887 to build, plus £7,516.7.10d value of stores supplied.
Commissioned 16.3.1829 under Capt. John Furneaux, for passage to England; arrived Plymouth 23.10.1829 and paid off 11.1829 (housed fore and aft). Hulked at Liverpool 1836. Fitted at Plymouth as a provision depot/storeship 9–11.1846; recommissioned 9.10.1846 under Cmdr Edmund Williams Gilbert, as storeship for Ireland; paid off 10.1847. In Ordinary at Plymouth 9.1848. Sold to Barnett & Wake (for £4,500) 24.12.1863.

An official model of the *Pique*, launched in 1834, the first Fifth Rate designed by Symonds. Like his Fourth Rate *Vernon*, this class had a bluff bow above the waterline, but fine lines below. The result was that in a heavy sea, when fully laden, they were slow in rising and strained their rigging; in addition, they floated a foot or more deeper in the water than Symonds had intended. They were fast but too lively in a seaway to make good use of the heavy 32pdr gun battery for which they were designed. *(NM F8830-002)*

CASTOR Class 32-pounder frigates of 36 guns; the first frigate designed for a 32pdr battery. Sir Robert Seppings design of 1828, derived from *Andromeda* Class (18pdr type) with further 13in of beam to mount the heavier ordnance. Two ships were ordered to this design in May 1828, but the second was not laid down until 1840 and was finally completed as a screw frigate.

> Dimensions & tons: 159ft 0in, 133ft 7⅜in x 42ft 6in x 13ft 6in. 1,283⁶⁸⁄₉₄bm. 1,808 disp.
> Men: 275. Guns: UD 22 x 32pdrs (55cwt); QD 10 x 32pdr gunnades (25cwt); Fc 4 x 32pdr gunnades (25cwt). In 1839 had – UD 18 x 32pdrs (56cwt) + 4 x 8in (50cwt) shell guns; QD and Fc as before. Two 25cwt 32pdrs were replaced by 50cwt models in 1848.

Castor Chatham Dyd

> As built: 159ft 1½in, 133ft 6⅞in x 43ft 1in (42ft 8in for tons) x 13ft 6in. 1,293³⁹⁄₉₄bm. 1,808 disp. Draught 12ft 8in/15ft 2in.
> Ord: 13.5.1828. K: 1.1830. L: 2.5.1832. C: 6.1832.
> First cost: £38,292 (fitted for sea).
> Commissioned 24.9.1832 under Capt. Lord John Hay, for the North Sea and then to the Tagus; home to UK 25.1.1837 and paid off at Chatham 8.2.1837. Recommissioned 18.4.1837 under Capt. Edward Collier, sailed 9.6.1837 for the Mediterranean; off Syria 1840; bombardment of Acre 3.11.1840; home to UK 23.7.1841 and paid off at Chatham. Recommissioned 28.4.1843 under Capt. Charles Graham; sailed 7.7.1843 for the East Indies; in New Zealand War 1845–47 (Ruapekapeka Stockade 11.1.1846); paid off at Chatham 16.11.1847. Recommissioned 7.5.1849 under Commodore Christopher Wyvill, for the Cape of Good Hope; paid off at Chatham 1853. Recommissioned 1855 under Capt. Henry Lyster, with pennant of Commodore Henry Trotter. Fitted at Chatham as training ship for coast volunteers 3–4.1860; recommissioned 16.1.1860 under Cmdr John Palmer, as RNR training ship (22-gun) at North Shields; in 1.1864 under Cmdr Charles Keats Jackson, then 4.1869 under Cmdr William A. Smith, 11.1873 under Cmdr Edward Bouverie Pusey, 7.1879 under Cmdr Edward John Jermain, 1886 under Cmdr Charles Edward Dring Wilcox and finally 1.1890 Cmdr Herbert J. G. Garbett, all as drill ship at North Shields; paid off by AO 23.11.1896. Sold at Sheerness (by AO 3.3.1902) 25.8.1902.

Amphion (ex-*Ambuscade*, renamed 31.3.1831) Woolwich Dyd

> Ord: 13.5.1828. K: 15.4.1840. Re-ordered as screw frigate 18.6.1844 (see Section D).
> First cost: £6,594 expended before the decision to convert her to a steam vessel.

(C) Vessels acquired from November 1830

All further Fifth Rates were to carry a mixed main battery of 32-pounder long guns and 68-pounder shell guns instead of the 18pdrs of their predecessors. As these long guns weighed 56cwt apiece, while the shell guns weighed 65cwt each, Symonds abandoned the Seppings *Druid/Andromeda/Castor* design to produce a much heavier vessel, adding over 6ft to the breadth.

PIQUE Class 32-pounder frigates of 36 guns. William Symonds design 1832. One of the cancelled 18pdr frigates at Plymouth, the *Pique*, was restored to the building programme as a 32pdr frigate in January 1833, and a second ship, *Flora*, ordered there two months later. Four further ships to this design were ordered in September-October, and another ship in 1834; however two of these were later replaced by orders for later designs. The class was subsequently considered for conversion to steam screw, but this was never put into effect.

> Dimensions & tons: 160ft 0in, 131ft 0in x 48ft 8in oa (48ft 3in for tonnage) x 14ft 6in. 1,622²⁰⁄₉₄bm.
> Men: 275. Guns: UD 22 x 32pdr (56cwt); QD 10 x 32pdr (25cwt); Fc 4 x 32pdr (25cwt). In 1839, 4 x 8in/68pdr (60cwt) shell guns replaced 4 x 32pdrs on UD (and Fc/QD 32pdrs swopped 25cwt type for 41cwt variety). All re-rated 40-gun in 1846–48, with UD 22 x 32pdrs (56cwt) + 6 x 8in (60cwt) shell; QD/Fc 16 x 32pdrs (42cwt).

Pique Devonport Dyd

> As built: 160ft 0in, 130ft 9½in x 48ft 10½in oa (48ft 5½in for tonnage) x 14ft 6in. 1,633⁶¹⁄₉₄bm. Draught 12ft 10in/16ft 2in.
> Ord: 14.1.1833. K: 7.1833. L: 21.7.1834. C: 2.2.1835 (for sea)
> First cost: £42,885 including fitting for sea.

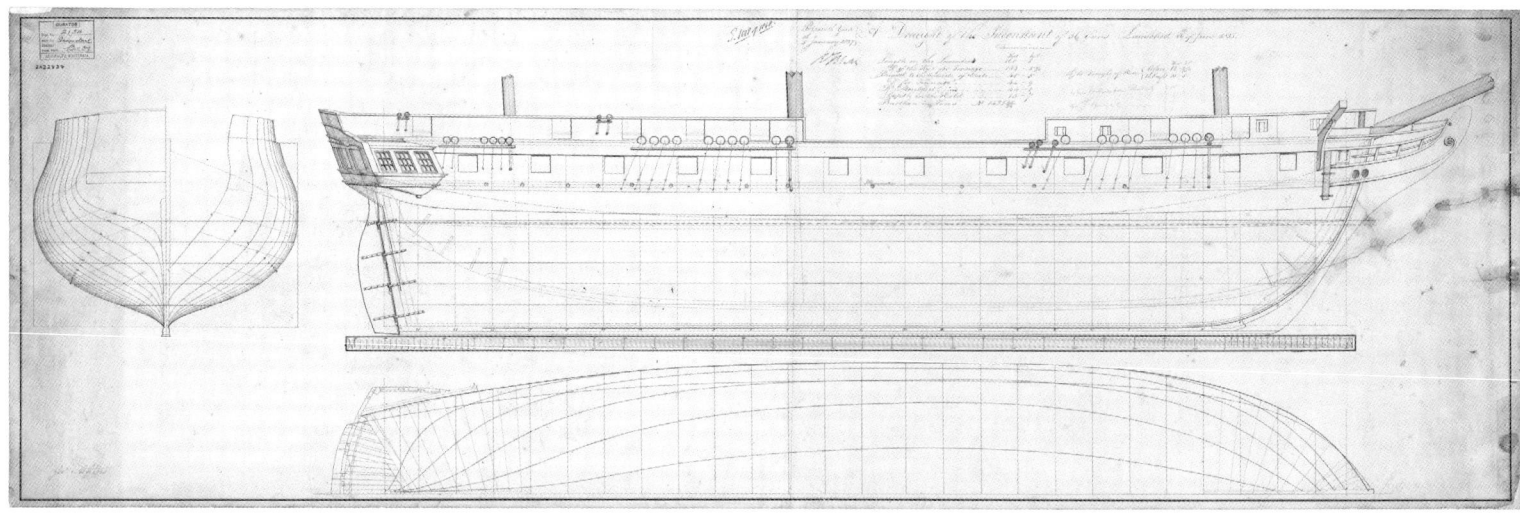

Captain John Hayes, who had spent the first five years of his working life as a shipwright apprentice at Deptford before joining the Navy, was strongly critical of the Surveyor's designs, and persuaded the Admiralty to let him design the one-off *Inconstant* to evaluate against the *Pique* Class. This 'as-built' draught is dated 5 January 1837. In the sailing trials of that year her sharply raked stem and lighter structure was judged to give superior sailing performance compared to Symonds's standard 32pdr frigate, although she carried the same armament, but the Surveyor's opposition means that no further frigates were built to Hayes's designs. *(NMM J5775)*

Commissioned 17.11.1834 under Capt. Henry John Rous, for Particular service at Lisbon 11.1834; ran ashore on Labrador coast 1835, but salved; paid off 1835. Refitted at Portsmouth (for £7,735) 10.1835–9.1836. Recommissioned 15.7.1836, still under Rous, for Lisbon station, then north Spanish coast. On 3.8.1837 under Capt. Edward Boxer (–8.1841), for North America and West Indies; to the Mediterranean 1840 (including operations on the Syrian coast; bombardment of Acre 3.11.1840); on 18.8.1841 under Capt. Richard Augustus Yates, then 14.11.1841 under Capt. Henry Forbes, in the West Indies; on 1.8.1842 under Capt. Montagu Stopford, in North America and West Indies; paid off mid 1846. Repaired at Pembroke (for £13,528) 9.1851–12.1853, then fitted for sea at Plymouth (for £10,859) to 3.1854. Recommissioned 26.12.1853 under Capt. Frederick William Erskine Nicolson, for the East Indies and Pacific; in Anglo-French squadron during Russian War and 2nd Anglo-Chinese War; at Petropavlovsk 1.9.1854. Fitted for Ordinary at Plymouth 2.1859. Fitted as receiving ship at Devonport 7–11.1862. Lent as hospital ship 25.3.1871. Loaned to Plymouth Town Council (as hospital ship) 30.3.1892. Sold to Cox, Falmouth 12.7.1910 to BU.

Flora Devonport Dyd
As built: 160ft 0in, 130ft 9⅞in x 48ft 10½in oa (48ft 5½in for tonnage) x 14ft 6in. 1,634⁴⁹/₉₄bm. Draught 13ft 1½in/16ft 7½in.
Ord: 11.3.1833. K: 12.1834. L: 11.9.1844. C: 27.9.1844 (for Ordinary).
First cost: £34,712 (+£712 fitting for Ordinary).
Remained in Ordinary at Plymouth (reduced to 10 guns) until AO 7.11.1864 ordered her to replace the storeship *Alexander* as guard ship at Ascension Island; fitted at Plymouth as storeship 3–11.1865.
Commissioned 13.10.1865 under Cmdr John Fletcher Rees, for passage to Ascension; in 1.1866 under Capt. Joseph Grant Bickford, as guard ship and storeship at Ascension; on 10.12.1866 under Capt. Arthur Wilmhurst, then 7.1869 under Capt. Edward Francis Kerby; to Simonstown (Cape of Good Hope) 12.1871

as receiving ship. On 6.8.1872 under Cmdr John Brasier Creagh, then 11.1873 under Cmdr Clayton Mitchell, then 31.7.1878 under Cmdr Henry Townley Wright, and 11.7.1881 under Cmdr Pearson Campbell Johnstone. Recommissioned 10.1881, still under Johnstone; on 11.6.1884 under Cmdr George Harvey Rainier, then 11.6.1887 under Cmdr Charles Kennedy Purvis. From 2.1889 was storeship as tender to *Penelope*. Sold there to Runciman 9.1.1891.

Constance Portsmouth Dyd
Ord: 12.9.1833. K: —. Never begun; replaced by order for 50-gun ship in 1843 (see Chapter 4).

Active Chatham Dyd (originally ordered at Pembroke)
As built: 160ft 0in, 130ft 9½in x 48ft 9in oa (48ft 4¼in for tonnage) x 14ft 6in. 1,627bm. Draught 12ft 10in/16ft 2in.
Ord: 9.10.1833. K: 8.1836. L: 19.7.1845. C: 18.8.1845 (for Ordinary).
First cost: £34,190.
No active service. Fitted as drill ship (Reserve training ship) at Chatham (for £6,347) for Sunderland 3–5.1863 (re-armed with 12 x 32pdrs (56cwt) and 8 x 8in (60cwt)). Commissioned 31.1.1863 under Cmdr Thomas Heard; to Sunderland 6.1863; in 12.1864 under Capt. Basil Sidmouth de Ros Hall; on 26.4.1866 under Cmdr George Graham Duff. Renamed *Tyne* 30.7.1867, then *Durham* 18.11.1867. On 27.4.1869 under Cmdr Willliam Alfred Gambier. Gunport and deck altered to take a 100pdr smooth-bore gun 7.1870 (with 3 x 32pdrs returned to store). Gunport fitted for a 7in gun in 11–12.1871. On 6.5.1872 under Cmdr William Henry Goold, then 3.5.1875 under Cmdr Henry Compton Best, then 10.7.1875 under Cmdr William Howorth, then 8.11.1877 under Cmdr Uvedale Corbet Singleton, then 12.4.1878 under Cmdr Mather Byle and 17.6.1880 under Cmdr John Henry Vidal. Recommissioned 4.1881, still under Vidal. On 22.6.1983 under Cmdr Henry Holford Washington, then 6.3.1888 under Lieut William J. Moore, as RNR drill ship at Leith. In 1.1900 under Reginald H. Curtis, then 22.1.1903 Lieut Cmdr Harry Thompson. Sold to Forth Shipbreaking Co. (for £4,325, by AO 15.3.1906) on 10.7.1906 and BU at Bo'ness.

Sybille Pembroke Dyd [40 guns: UD18 x 32pdrs (56cwt) + 6 x 8in (60cwt); QD 16 x 32pdrs (42cwt)]
As built: 160ft 0in, 131ft 0in x 48ft 9in oa (48ft 5in for tonnage) x 14ft 6½in. 1,633⁴¹/₉₄bm. Draught 14ft 0in/16ft 7in.
Ord: 28.10.1833. K: 12.1835. L: 15.4.1847. Arrived Plymouth 4.6.1847 and placed in Ordinary. C: 5.1853–23.8.1853 at Plymouth.

In April 1842 the Admiralty, with growing concern over the Symondite hull form, commissioned three of the graduates of the School of Naval Architecture to jointly prepare a series of standard designs for future construction. The first design prepared by the trio – known as the Chatham Committee from the venue of their deliberations – was for a 36-gun frigate, the *Thetis*, the last Fifth Rate to be completed as a sailing vessel. The draught, submitted in February 1843 and approved a month later, produced an amalgam of Sepping's flat floor and Symonds's sharply rising floor, and was strongly resented by the Surveyor. This anonymous wash drawing depicts the ship as completed. (*NMM PU8505*)

First cost: £36,021 to built, plus £14,410 fitting for sea in 1853.

Commissioned 5.1853 under Capt. Charles Gilbert John Brydone Elliot, for the East Indies and China; made Commodore Elliot 26.1.1855; in 2nd Anglo-Chinese War; at Canton 10.1856; boats at Fatshan 1.6.1857; paid off 30.4.1858 at Plymouth. Surveyed 28.2–3.3.1863. BU (by AO 3.2.1866) completed 28.4.1866 by Marshall at Plymouth.

Cambrian Pembroke Dyd

As built: 160ft 0in, 130ft 7in x 48ft 9½in oa (48ft 4½in for tonnage) x 14ft 7½in. 1,625⁴¹⁄₉₄bm. Draught 13ft 5in/15ft 10½in.

Ord: 28.10.1833. K: 8.1837. L: 5.7.1841. C: 20.7–8.11.1841 at Plymouth (for sea).

First cost: £31,153 (+ fitting £12,882).

Commissioned 20.8.1841 under Capt. Henry Ducie Chads, for the East Indies and China; in 1st Anglo-Chinese War; paid off 1845. Refitted at Portsmouth (for £13,358) 5–9.1846. Recommissioned 5.8.1847 under Capt. James Hanway Plumridge, for the East Indies; from 20.10.1847 under Capt. (acting) Henry Gahe Morris, as flag of Commodore Plumridge; paid off 1850. Repaired at Plymouth (for £18,016) 2.1851–10.1854. Refitted there (for £8,383; ?+£658) 11.1857–2.1858. Recommissioned 23.11.1857 under Capt. James Johnstone M'Cleverty, for the East Indies and China; destruction of pirate junks at Lingting Island 22.8.1856; Taku Forts 21.8.1860; paid off 25.10.1861. Mooring hulk at Devonport 5.1869. Floating factory 1880. Sold to John Read, Portsmouth 12.1.1892 to BU.

Chesapeake Chatham Dyd

Ord: 10.12.1834. K: — . Never begun; replaced by order for screw frigate in 1851 (qv).

INCONSTANT Class 32-pounder frigate of 36 guns. Captain John Hayes design of 1833, ordered to evaluate against the *Pique* class. Much lighter than Symonds's design, with frames 37in apart (*vice* 27in), she proved vastly superior, but the Surveyor's opposition meant no more were built to this design.

Dimensions & tons: 160ft 1in, 133ft 5⅜in x 45ft 5in oa (44ft 9in for tonnage) x 13ft 7in. 1,421¹⁄₉₄bm.

Men: 275. Guns: UD 18 x 32pdr (56cwt) + 4 x 8in/68pdr (65cwt) shell; QD 10 x 32pdr (25cwt); Fc 4 x 32pdr (25cwt).

Inconstant Portsmouth Dyd

Ord: 12.9.1833. K: 8.1834. L: 16.6.1836. C: 14.10.1836 (for sea).

First cost: £30,919 (£41,260 fitted for sea).

Commissioned 1.12.1836 under Capt. Daniel Pring; sailed for Lisbon 18.3.1837; later to North America (grounded in the St Lawrence 1838) and West Indies; fitted for troops at Plymouth (for £2,234) 12.1838–1.1839; later to Irish and finally Mediterranean stations. On 4.8.1841 under Capt. Frederick Thomas Michell, in the Mediterranean; paid off 30.3.1843. Recommissioned 20.5.1843 under Capt. Charles Howe Fremantle, for North America and the West Indies; in 1846 to the Mediterranean. On 28.2.1847 under Charles Wise; paid off 20.4.1847. Very small repair at Plymouth (for £9,038) 7.1847–1.1848. Recommissioned 4.12.1847 under Capt. John Shepherd, for the east coast of South America; paid off 7.12.1850. Fitted as temporary hospital hulk for emigration at Plymouth 8–9.1854. Fitted there as a receiving ship for Cork 1.1861. Sold 8.12.1862 and BU 10.1866 at Cork.

THETIS Class 32-pounder frigate of 36 guns. Design produced jointly by Samuel Read (M/Shipwright at Sheerness), Henry Chatfield (Assistant M/Shipwright at Woolwich) and Augustin Creuze (Principal Surveyor at Lloyd's), for the School of Naval Architecture, approved on 16 March 1843.

Men: 330. Guns: UD 18 x 32pdr (56cwt) + 4 x 8in/68pdr (65cwt) shell; QD 10 x 32pdr (25cwt); Fc 4 x 32pdr (25cwt).

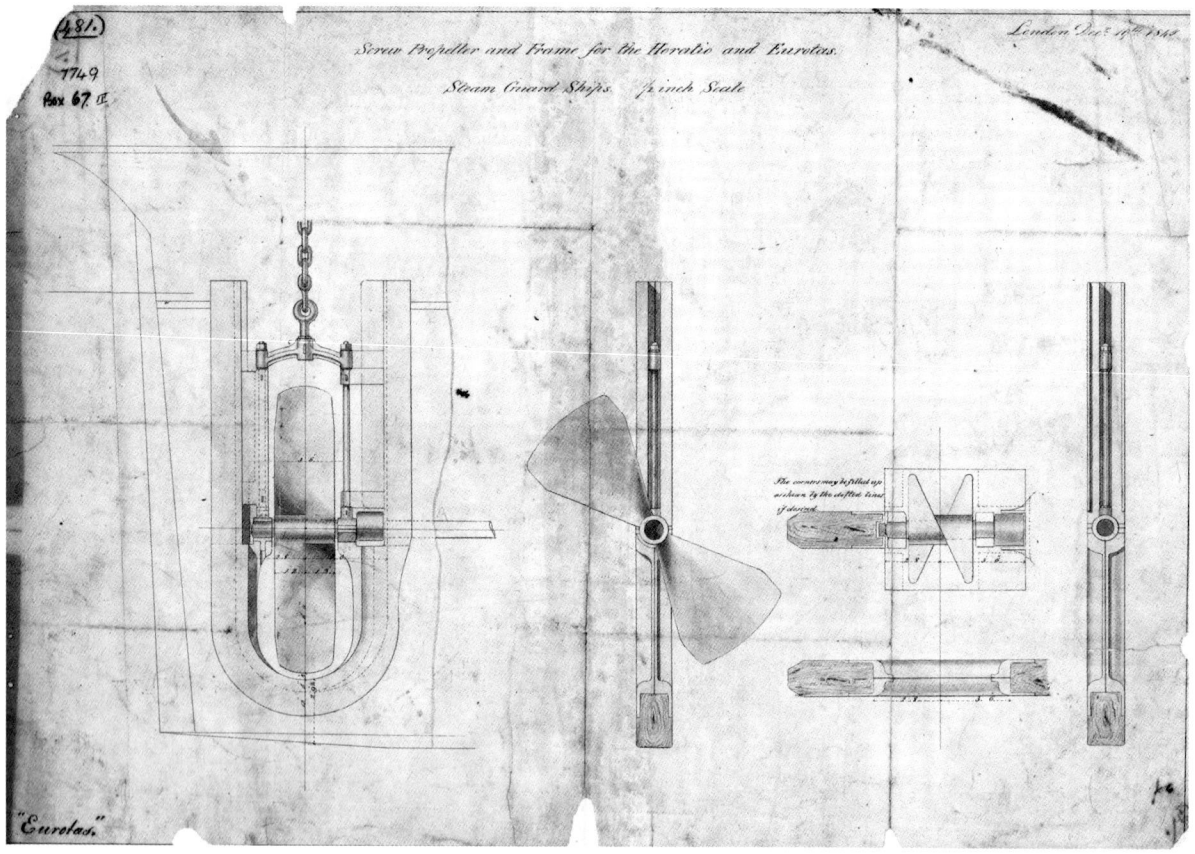

Soon after the widespread adoption of the screw, it was realised that the propeller caused significant drag when the ship was under sail, as the screw did not turn in the wake. The answer was to detach the propeller when not in use, a solution which required an elaborate lifting device called a 'banjo frame' and a watertight recess or well in the stern of the vessel into which the frame could be hoisted. This draught dated 19 December 1845 shows the propeller and frame intended for the conversion of the *Lively* Class sailing frigate *Horatio* into a screw frigate, which was completed at Sheerness by December 1849. The same conversion was intended for the *Druid* Class frigate *Eurotas*, but that conversion was deferred, and it was not until 1855 that approval was given to Sheerness to complete the latter, together with two more of this class at Devonport, as screw-driven blockships. (*NMM J0693*)

Thetis Devonport Dyd
　　As built: 164ft 7¼in, 135ft 10in x 46ft 8¾in oa (46ft 0¾in for tonnage) x 13ft 6½in. 1,533bm. 1,894 disp. Light draught 13ft 10in (fwd), 15ft 5in (aft) with 52 tons on board.
　　Ord: 23.4.1842 & 16.2.1843. K: 2.12.1844. L: 21.8.1846. C: 30.12.1846 (for sea).
　　First cost: £40,605 (£51,926 fitted for sea).
　　Commissioned 14.10.1846 under Capt. Henry John Codrington; attached to Royal Naval College; at Portsmouth 1847, later in the Mediterranean; paid off 28.8.1849. Recommissioned 7.1850 under Capt. Augustus Leopold Kuper, for the Pacific; paid off 5.2.1854. Disarmed and given to Prussian government 12.1.1855 in exchange for two steam gunboats (*Nix* and *Salamander*, which in British service became the *Weser* and *Recruit*); sailed from Plymouth 25.3.1855, and formed (with former *Musquito* and *Rover*, see under Brigs) part of the original nucleus of the Imperial German Navy.

At 1 January 1850, the Navy included 41 purely sailing Fifth Rates, consisting of 10 rated at 44 guns, 22 rated at 42 guns, 5 (*Pique, Cambrian, Flora, Active* and *Sybille*) at 40 guns, 1 (the elderly *Belvidera*) at 38 guns and 3 (*Castor, Inconstant* and *Thetis*) at 36 guns. There were 5 in commission: the 44-gun *Maeander*, the 40-gun *Cambrian*, and the three 36-gun ships; the other 36 were in Ordinary. Another 39 elderly Fifth Rates remained afloat, and on the Navy List, as fit only for harbour service.

(D) Vessels ordered or re-ordered as Fifth Rate steam screw frigates (from 1844)

AMPHION **Class** 30 (later 36) guns. Originally ordered as a sailing frigate of the *Castor* Class to a Sir Robert Seppings design of 1828 (see Section B) and converted on the stocks into a steam frigate. On 30 May 1844, Oliver Lang was ordered to report whether the space and weight required by Miller & Ravenhill for a 300hp engine on the plans of Count Rosen could be spared, and if not how much could, while still preserving her efficiency as a sailing frigate. Rosen was Ericsson's agent, and was also responsible for the machinery installed in the French frigate *La Pomone*. On 18 June a tender from Count Rosen and from Miller & Ravenhill for auxiliary engines of 300hp, with tubular boilers and an Ericsson screw, was received and accepted, and the conversion of the *Amphion* ordered. White submitted a plan on 23 August, involving the lengthening of the *Amphion* by 16ft at the bow; this alteration was suspended on 9 May 1845, as 'Mr White is about to propose a plan for firing guns from the bow in a line with the keel'; on 9 July this plan was approved, with a 32pdr (56cwt) gun to be mounted on a pivot in a line with the keel. Her figurehead was a bust of the late Captain Sir William Hoste.
　　Men: 330. Guns: UD 6 x 8in (65cwt/8ft 10in) + 14 x 32pdrs (56cwt/9½ft); QD/Fc 2 x 68pdrs (95cwt/10ft) on pivots + 2 x 32pdr carronades (25cwt/6ft). By 1856 had 36 guns, with 14 x 32pdrs (42cwt) instead of 8 x 25cwt guns.
　　Machinery: 2-cylinder (48in diameter, 4ft stroke) horizontal single expansion. Single screw (14ft diameter). 300nhp. 592ihp = 6.75kts.
Amphion Woolwich Dyd.
　　As built: 177ft 0in, 152ft 5⅜in x 43ft 2in oa (42ft 8in for tonnage) x 13ft 4½in. 1,474bm (calc. 1,476¹⁷⁄₉₄)bm. Draught (light) 10ft

10in/14ft 11in; (mean) 19ft 1½in at 2,049½ tons disp.

(re-)Ord: 18.6.1844. K: 15.4.1840. L: 14.1.1846. Machinery fitted in East India Docks 21.1–23.4.1846. C: 13.5.1847 at Woolwich (for sea).

First cost: £36,115 to build, plus £16,673 for machinery and £22,794 fitting (for sea).

Commissioned 24.10.1846 under Capt. Woodford John Williams, at Woolwich, then to Western and Experimental Squadrons; paid off 31.10.1848 at Sheerness. Recommissioned 18.12.1852 under Capt. Charles George Edward Patey, for Lisbon and then the Channel Squadron; fitted for sea at Sheerness 2–4.1853; on 8.11.1853 under Capt. Ashley Cooper Key (Patey invalided); to the Baltic 1853 for the Russian War; at Bombardment of Bomarsund 1854; at Bombardment of Sveaborg 8.1855; on 9.1.1856 under Capt. Henry Chads, for North America and the West Indies; paid off 30.1.1857 at Sheerness. Recommissioned 12.1858 under Capt. Thomas Cochran, initially as flagship of Commodore Charles Talbot at Queenstown; underwent repairs and (correcting) defects at Chatham (for 7,040) 4–6.1859. Recommissioned 22.6.1859, still under Cochran, then completed for sea at Sheerness (for £5,540) 6–9.1859; to the Mediterranean in 9.1859. On 12.7.1861 under Capt. Thomas Francis Birch (Cochran invalided), in the Mediterranean; on 12.2.1862 under Capt. Alexander Crombie Gordon, in the Mediterranean; paid off 20.12.1862 at Sheerness. Sold to W. Williams 12.10.1863 to BU.

HORATIO – Conversion of this elderly frigate to a 44-gun screw frigate/blockship (with a 350nhp Boulton & Watt engine) was approved 2 September 1845. The housing and topside fittings were removed, her copper was stripped off, and the gunports reduced from 16 to 11 per side. She was fitted with a 250nhp engine in December 1849, and was re-armed, becoming a screw mortar frigate during the Crimean War.

Horatio Sheerness Dyd/Seaward & Capel

As converted: 154ft 3in, 129ft 11¼in x 40ft 2½in oa (39ft 8½in for tonnage) x 13ft 7in. 1,090bm. Mean draught 19ft 1½in at 1,707 tons. disp.

Men: 320. Guns: As screw blockship had LD 20 x 42pdrs (66cwt/9½ft) (from 1853, 18 x 8in only); UD 2 x 56pdrs (85cwt/10ft) on pivots + 2 x 10in (86cwt/9ft 4in). As mortar frigate, carried 2 x 13in mortars (also LD 8 x 32pdrs, UD 2 x 68pdrs).

Machinery: 2-cylinder (54in diameter, 3ft stroke) horizontal single expansion, geared. 13lb/sq.in. Single screw. 250nhp. 553ihp = 8.855kts.

First cost: £42,804 (including machinery £19,793).

Begun: 11.1845. Completed 12.1849 at Chatham as screw frigate blockship.

Recommissioned 19.1.1852 at Sheerness under Capt. Swynfen Thomas Carnegie, as guard ship at Sheerness; on 2.6.1853 under Cmdr Robert Jenner. Fitted as screw mortar frigate at Sheerness 1855. Recommissioned 7.2.1855 under Cmdr Arthur Auckland Leopold Pedro Cochrane; paid off 12.5.1856 at Sheerness. Sold to Castle 1865 to BU at Charlton.

DRUID/ANDROMEDA Class 24 guns. Originally built as sailing frigates (see Section B) three ships of this class were ordered on 2 September 1845 to be converted into 44-gun steam guard ships, with 350hp engines (by Boulton & Watt in the Eurotas, and Rennie in the other pair), at the same time as the older Horatio. The original order initially included the Stag, replaced 4 days later by her sister Forth.

Unlike the *Horatio*, the three were not fitted with engines (indeed, on 31 December 1852 they were ordered to be re-registered as sail frigates) until 29 June 1855, when it was approved that they be lengthened and fitted as screw mortar vessels with 200nhp engines and two mortars each (*Eurotas* at Sheerness, *Forth* and *Seahorse* at Devonport); however, it appears that the three, although fitted with engines (cylinders 30¼in diameter, 2½ft stroke; HP engines of 60 lb/sq.in), were never re-armed.

Men: 320. Guns: UD 20 x 42pdr (66cwt/9½ft); QD/Fc 2 x 56pdrs (85cwt/10ft) + 2 x 10in (86cwt/9ft 4in).

Eurotas Chatham Dyd/John Penn & Son

Dimensions & tons: 166ft 2in, 137ft 8½in x 40ft 10in oa (ca.40ft 6in for tonnage) x 12ft 9in. 1,201bm. 1,293 disp.

Machinery: 2-cylinder horizontal, trunk (high pressure). Single screw. 200nhp. 562ihp = 7.58kts.

First cost: As frigate blockship £8,379 for hull, plus £12,192 for machinery. As mortar frigate £25,471.

Conversion ord: 4.9.1845. Begun: 12.1845. Partly completed as frigate blockship 2.1848; finally (without engine) 4.1851. Completed at Sheerness Dyd as screw mortar frigate 7.1855 to (undocked 13.2.1856) 10.4.1856.

Commissioned 24.1.1856 under Capt. William Moorsom; at Royal Naval Review 1856; paid off 27.9.1856 at Sheerness. Sold to Castle & Beech 1.11.1865 to BU at Charlton.

Seahorse Devonport Dyd/John Penn & Son

Dimensions & tons: 164ft 0in, 136ft 7½in x 42ft 2in oa (ca.41ft 7in for tonnage) x 19ft 10in. 1,257bm. 1,799 disp.

Machinery: 2-cylinder horizontal, trunk (high pressure). Single screw. 200nhp. 832ihp = 9.3kts.

First cost: As frigate blockship £7,292 for hull, plus £12,241 for machinery. As mortar frigate £30,260.

Conversion ord: 12.9.1845. Begun: 11.1845. Completed as frigate blockship 7.1847. Fitted at Plymouth as a screw mortar frigate 7.1855–4.1856 (undocked 7.3.1856).

Commissioned 1.11.1855 under Capt. Leopold George Heath; employed as troopship from Constantinople 6.1856; paid off 27.9.1856 and became a coal hulk, renamed *Lavinia* 5.5.1870. Sold 1902.

Forth Devonport Dyd/John Penn & Son

Dimensions & tons: 164ft 0in, 136ft 7½in x 42ft 2in oa (ca.41ft 7in for tonnage) x 19ft 10in. 1,258bm. 1,793 disp.

Machinery: 2-cylinder horizontal, trunk (high pressure). Single screw. 200nhp. 834ihp = 9.38kts.

First cost: As frigate blockship £8,340 for hull, plus £12,232 for machinery. As mortar frigate a further £26,006.

Conversion ord: 12.9.1845. Begun: 6.10.1845. Completed as frigate blockship 7.1847. Fitted at Plymouth as a screw mortar frigate 7.1855–4.1856 (undocked 21.1.1856).

Commissioned 9.12.1855 under Capt. Lord John Hay; at Royal Naval Review 1856; paid off 23.10.1856 and became a coal hulk, renamed *Jupiter* 12.1869. Sold to Castle 4.8.1883 to BU at Charlton.

TRIBUNE Class Second Class frigates, 31 guns. Surveyor's Department design for 'Auxiliary power screw frigate[s]', approved 1 November 1850. *Tribune*, first ordered in 1846 as a sailing frigate and re-ordered in 1850 as a screw ship, utilised the frames originally provided for the 28-gun sailing frigate *Diamond*; *Curacoa* utilised the frames prepared for (cancelled) 36-gun sailing frigates. *Tribune's* cylinders were 55in diameter, 2½ft stroke; *Curacoa's* were 57⅛in diameter, 2¾ft stroke. Both reduced to 23-gun corvettes in 1863.

Dimensions & tons: 192ft 0in, 163ft 4in x 43ft 0in oa (42ft 6in for

A contemporary model of the Second Class steam frigate *Tribune*, 31 guns. Note the early form of bridge deck over the steering wheel, necessary for the officer conning the ship to get a good view of the tall bulwarks; this is the origin of the term 'bridge' for the control position over any modern ship. *(NMM F8958-003)*

tonnage) x 12ft 11in. 1,569²⁴/₉₄bm. *Tribune* 2,220 disp. *Curacao* 2,385 disp. Draught (*Tribune*) 17ft/20¼ft.

Men: 350. Guns: MD 20 x 32pdrs (56cwt/9½ft); UD 10 x 32pdrs (42cwt/8ft) + 1 x 10in (85cwt/9ft 4in) on pivot.

Machinery: 2-cylinder horizontal single expansion. Single screw. *Tribune* 300nhp. 1,068ihp = 10.416kts. *Curacoa* 350nhp. 1,354ihp = 10.75kts.

Tribune Sheerness Dyd/Maudslay, Sons & Field
Ord: 1.11.1850. K: 4.1851. L: 21.1.1853. C: 3.8.1853 at Woolwich Dyd.
First cost: £68,304 total (hull £34,636, machinery £18,396, fitting £15,272).
Commissioned 14.5.1854 under Capt. Swinfen Thomas Carnegie, for the Baltic Fleet; later to Black Sea; at Bombardment of Sebastopol 11.10.1854; on 11.12.1854 under Capt. James Robert Drummond, in the Black Sea; in expedition to the Sea of Azov 5.1855. On 13.8.1855 under Capt. Harry Edmund Edgell; at Bombardment of Kinburn 17.10.1855; to the East Indies and China 1857; in boat action at Fatshan Creek 1.6.1857. On 18.8.1858 under Capt. Geoffrey Thomas Phipps Hornby, for the Pacific; resolution of 'Pig War' Anglo-US dispute over San Juan Island (off Vancouver); home and paid off 8.1860. Recommissioned 17.3.1862 at Portsmouth under Capt. Viscount Richard Gilford, again for the Pacific; paid off 16.5.1866 at Portsmouth. Sold to C. Marshall, Plymouth 8.1866 to BU.

Curacoa Pembroke Dyd/Maudslay, Sons & Field
Ord: 4.4. & 18.7.1851. (named 20.9.1851) K: 1.1852. L: 13.4.1854. C: 14.11.1854 at Portsmouth Dyd.
First cost: £70,608 total (hull £34,759, machinery £21,583, fitting 14,266).
Commissioned 1.9.1854 under Capt. George Fowler Hastings, for the Mediterranean, then to Black Sea; at Bombardment of Kinburn 17.10.1855; on 14.5.1857 under Capt. Arthur Forbes; paid off 9.9.1857 at Portsmouth. Refitted there (for £9,233) 1857–

58. Recommissioned 30.11.1857 under Capt. Thomas Henry Mason, for the Channel Squadron; on 2.7.1859 under Capt. Augustus Phillimore, for the east coast of South America; paid off 4.6.1862 at Portsmouth. Recommissioned 20.4.1663 under Commodore Sir William Saltonstall Wiseman; at occupation of Merimeri (New Zealand) 9.1863; boats in capture of Rangariri 20.11.1863; bombardment of Tanna and Erromanga (New Hebrides) 1865. BU completed 17.7.1869.

FOX Screw frigate, conversion from 46-gun sailing frigate, launched 1829 at Portsmouth Dyd (see above) but not completed. Conversion of this *Leda* class frigate to a screw frigate was carried out at Portsmouth Dyd (for £7,621 plus £12,103 for machinery) from September 1855 to March 1856, with a 200nhp engine being installed, but she was incomplete when work stopped, and saw no service in this role, being retained in Ordinary until March 1862, when she was finally finished and employed as a screw transport/storeship.

Dimensions & tons: 159ft 4in, 133ft 6in x 40ft 4in oa (39ft 11in for tonnage) x 12ft 9in. 1,131bm. 1,780 disp.

Men: — . Guns: (never completed as screw frigate).

Machinery: 2-cylinder (45in diameter, 2ft stroke) horizontal single expansion. Single screw. 200nhp. 741ihp = 9.3kts.

First cost: £7,621 for hull conversion, plus £12,103 machinery.

Fox Portsmouth Dyd/Miller, Ravenhill & Salkeld
Conversion ord: ?1854. Begun: 9.1855. L: (undocked as screw transport) 18.3.1856. C: 3.1862 as screw storeship.
Commissioned 22.2.1862 as a storeship under Thomas C. Pullen, Master. From late 1864 command passed to a series of Staff Commanders: from 5.12.1864 she was under Henry Augustus Moriaty, then 8.12.1867 under William S. Luke, them 18.5.1868 under Robert Barrie Batt, then 17.10.1868 under John Hillary Allard, then 25.9.1874 Joseph G. Dathan, and then 7.9.1877 under Henry D. Sarratt; paid off 11.11.1879 at Devonport. BU there 3.1882.

6 The Sixth Rates (later corvettes)

Until 1 January 1817, the Sixth Rate encompassed vessels (all single-decked ships) with at least 20 but not more than 28 guns. At that date, the divisions were altered such that ships of 34 or 32 guns were transferred from the Fifth Rate, and the lower limit for Sixth Rates was placed at 24-gun ships. In practice, as the surviving 32-gun ships were rapidly discarded, the Rate now covered ships of 28, 26 or 24 guns. From 1856 this was amended to restrict the Sixth Rate to ships of which the complements were less than 300 men, but were commanded by a captain.

(A) Vessels in service or on order at 1 January 1817

Following the reorganisation in January 1817, the Navy included five ships rated at 32 guns, thirteen rated at 28 guns (excluding the *Bittern*, rated as a tender), eighteen rated at 26 guns (excluding the *Goree*, rated as a prison ship) and three rated at 24 guns (mainly former fireships). Almost all of the 26/28-gun ships had formerly been ship sloops, 'frigate-built' with quarterdecks and forecastles, and their addition to the Sixth Rates reflected the addition of the carronades carried on these superstructures. The 24-gun ships were former fireships. In addition, a single 28-gun ship (*Atholl*) had been ordered in 1816, the first of a new class of such vessels, but for convenience she has been included with her sisters in Section B.

PORCUPINE Class 32 guns. The *Eurydice* was by 1817 the sole survivor of this Sixth Rate 24-gun class, built to a John Williams design of 1776, to which ten ships were ordered over two years, enlarged from the lines of the previous 20-gun *Sphinx* Class. Among her sisters had been the *Pandora*, lost in 1791 while returning from the hunt for the *Bounty* mutineers. The *Eurydice* was re-rated as 32 guns in January 1817, but by then actually only carried 24 guns.

Eurydice Portsmouth Dyd [M/Shipwright Nicholas Phillips to 4.1779, then George White]
- As built: 114ft 3in, 94ft 2¼in x 32ft 3in x 10ft 3in. 521²⁸⁄₉₄bm. Draught 7ft 0in/11ft 10in.
- Men (1815): 140. Guns (1815): UD 14 x 9pdrs + 8 x 18pdr carronades; QD 2 x 6pdrs.
- Ord: 24.7.1776. K: 2.1777. L: 26.3.1781. C: 3.6.1781.
- First cost: £12,391.4.0d including fitting & coppering.
- Commissioned 3.1781 under Capt. George Wilson; sailed for Leeward Islands; in Frigate Bay (St Kitts) 25/26.1.1782; at Battle of the Saintes 12.4.1782, then home with dispatches. In ?4.1782 under Capt. George Courtnay (–1785), for the Channel and Channel Islands; with Elliot's squadron in Autumn 1782; took *Les Amis* off Île de Bas 14.10.1782; paid off 1782/3; recommissioned 4.1783; sailed 10.4.1783 to East Indies; paid off 7.1785. Small Repair at Woolwich (for £2,290) 1–4.1786. Fitted for sea at Woolwich (for £3,386) 5–7.1788; recommissioned 6.1788

under Capt. George Lumsdaine; sailed for the Mediterranean 27.11.1788. Fitted by Wells & Co. (for £1,856) 2–3.1793, then at Woolwich (for £3,507) 3–6.1793; recommissioned 4.1793 under Capt. Francis Cole; escaped from 50-gun *Le Sciavola* and *Le Brutus* 8.6.1794. In 1795 under Capt. Thomas Twysden, then 1796 Capt. Richard Bennet, in cruising and convoy duties. Recommissioned 8.1796 under Capt. John Talbot (–1800); took privateers in North Sea – *Le Sphinx* on 15.12.1796, 14-gun *Le Flibustier* on 6.2.1797, *Le Voligeur* on 7.3.1797, and (with *Snake*, off Beachy Head) 14-gun *L'Hirondelle* 10.11.1799. Refitted at Portsmouth (for £4,440) 5–6.1798. Under Capt. Walter Bathurst in 1.1801 (–1802); took 14-gun privateer *Le Bougainville* in the Atlantic 8.5.1801; sailed for the East Indies 20.10.1801. Refitted at Portsmouth 6–10.1803; recommissioned 9.1803 under Capt. John Nicholas; with convoy to Quebec 16.5.1804. Under Capt. William Hoste 11.1804; in the Mediterranean 1805; took 6-gun privateer *El Mestuo La Solidade* 6.10.1805. Under Capt. Sir William Bolton 12.1805; in Channel 1806–07. In 8.1808 under (acting) Capt. David Ramsay, later under Capt. James Bradshaw; at capture of Martinique 2.1809; on North American station 1809–11; in Ordinary at Deptford 1812–14. Temporary Repair at Deptford 9.1813–6.1814; fitted for sea there 8–10.1814; recommissioned 8.1814 under Capt. Valentine Gardner; in 6.1815 under Capt. Robert Spencer on Irish Station; on 19.4.1816 under Capt. Robert Wauchope, off St Helena; paid off 20.12.1819 and laid up at Deptford; in 1821 to Woolwich. Fitted as receiving ship at Woolwich 8.1823–1.1824. BU at Deptford 3.1834.

TISIPHONE Class 24 guns. Originally built as fireships (and rated at 14 guns when used as sloops). Based on the lines of the French *Amazon* of 1745 (and thus antecedents of the succeeding *Cormorant* Class sloops). Built for the American War of Independence, the continued survival of these former fireships throughout the Napoleonic Wars indicates how solidly built they were. Of the original nine ships built to this design, four were wartime losses (*Conflagration*, *Vulcan* and *Comet* were expended as fireships, and *Incendiary* captured by the French), and *Alecto* was sold in 1802; the four survivors were re-rated as sloops in 1808, but *Tisisphone* was sold in early 1816. The *Spitfire* was reclassified as a 24-gun Sixth Rate in January 1817 (but was laid up in Ordinary), while *Pluto* was re-rated as a receiving ship and *Megaera* was again rated as a fireship.
- Dimensions & tons: 108ft 9in, 90ft 7in x 29ft 7in x 9ft 0in. 421⁶⁄₉₄bm.
- Men: 121 (originally 55 as fireship, 121 as sloop). Guns (1817): UD 16 x 32pdr carronades; QD 6 x 18pdr carronades; Fc 2 x 6pdrs.

Spitfire Stephen Teague, Ipswich
- As built: 108ft 10in, 90ft 8in x 29ft 7in x 9ft 0in. 422⁶⁄₉₄bm.
- Ord: 28.11.1780. K: 12.1780. L: 19.3.1782. C: 21.4–18.7.1782 at Sheerness (including coppering).
- First cost: £., .77.16.1d to build, plus £3,391.17.0d fitting & coppering.
- Commissioned 3.1782 but paid off in same year. Fitted for Ordinary at Sheerness 7–8.1783. Fitted at Sheerness (for £519) 5–6.1790; recommissioned 5.1790 under Cmdr Robert Watson, then paid

William Smyth painted this watercolour of the 24-gun *Blossom* off the Sandwich Islands during her lengthy explorations of the Pacific between 1825 and 1828, including the surveying of the islands lying east of Tahiti and of the Pacific coast of North America. Although *Blossom* was specifically fitted for exploration, she retained the basic layout of the quarterdecked ship sloops built during the Napoleonic Wars. (*NMM PW5964*)

off. Fitted at Sheerness (for £1,749) 4–5.1791; recommissioned 3.1791 under Cmdr Thomas Fremantle; paid off 9.1791. Recommissioned 9.1791 as sloop under Cmdr John Woodley, for Irish Sea and Channel. In 2.1793 under Cmdr Philip Durham; took privateer cutter *L'Afrique* 2.1793; took privateer *Le St Jean* and destroyed *La Marguerite* 3.1793. In 6.1793 under Cmdr James Cook (drowned 1.1794) then Cmdr John Clements, in Macbride's squadron, then ?10.1794 under Cmdr Amherst Morris. In 1796 under Cmdr Michael Seymour (–1800); took storeship *L'Allègre* off Ushant 12.1.1797, 6-gun privateer *Les Bons Amis* off Eddystone 2.4.1797, 14-gun privateer *L'Aimable Manet* in the Channel 1.5.1797, 6-gun privateer *La Trompeuse* 5.1797, 3-gun privateer *L'Incroyable* off the Lizard 15.9.1797, 14-gun transport *Le Wilding* in the Bay of Biscay 28.12.1798, 14-gun privateer *La Résolue* off Scilly 31.3.1799, 14-gun privateer *L'Heureuse Societé* in the Channel 17.4.1800 and 14-gun *L'Heureux Coureur* in the Channel 19.6.1800. Refitted at Plymouth (for £2,449) 1–2.1798. In 7.1800 under Cmdr Robert Keen (–1804); on Irish station 1802–04; recommissioned 4.1803. Middling to Large Repair and fitted at Sheerness 4.1805–4.1806; recommissioned 2.1806 under Capt. William Green; took 4-gun privateer *Les Deux Frères* in Channel 29.12.1806. Recommissioned 2.1807 as a fireship under Cmdr Henry Butt, for the Downs station. In 1808 under Cmdr John Ellis (–1814), for Leith; sailed with convoy for Quebec 23.5.1810; at Portsmouth 1811–12, and on Greenland station 1813. On 28 July 1814 under Cmdr James Robert Dalton, off West Africa; laid up at Portsmouth 28.5.1815; sold there to Mr Ranwell (for £1,205)

30.7.1825 to BU.

Pluto Joshua Stewart, Sandgate
   As built: 108ft 10½in, 90ft 7in x 29ft 9in x 9ft 0in. 426⁴²⁄₉₄bm. Draught 8ft 3in/11ft 0in.
   Ord: 4.12.1780. K: 1.1781. L: 1.2.1782. C: 24.2–31.5.1782 at Deptford (including coppering).
   First cost: £9,365.1.7d including fitting & coppering.
   Recoppered at Chatham 11.1787–1.1788. Recoppered & fitted at Chatham (for £1,398) 7–9.1790. Fitted at Portsmouth (for £938) 4.1791.
   Commissioned 3.1791 under Cmdr Robert Faulknor; paid off 9.1791. Fitted at Portsmouth (for £3,651) 11.1792–2.1793. Fitted as sloop (but not registered as such) by AO 12.1.1793; recommissioned 12.1792 under Cmdr James Nicoll Morris; sailed for Newfoundland 16.5.1793; took 12-gun *Le Lutine* off Newfoundland 25.7.1793. In 10.1793 under Cmdr Richard Raggett (–1795), then 1796–98 Cmdr Ambrose Crofton, still on Newfoundland station. On 5.3.1798 under Cmdr Henry Folkes Edgell; sailed for Newfoundland 30.7.1798; on Newfoundland station in 1801, surveying. In 4.1802 under Cmdr Robert Forbes. Fitted at Deptford 7–9.1803; recommissioned 8.1803 under Cmdr Edward Kittoe, for the Channel. In 5.1804 under Cmdr Richard Janverin (–1809), on the Downs station. Laid up at Portsmouth 2.1809, in Ordinary. Sold to Mr Warwick there (for £950) 19.7.1817 to BU.

Megaera Stephen Teague, Ipswich
   As built: 108ft 11½in, 90ft 10½in x 29ft 8in x 9ft 0in. 425⁴⁹⁄₉₄bm.
   Ord: 16.2.1782. K: 5.1782. L: 5.1783. C: 7.6.1783–10.1784 at

Chatham, then laid up there (for £1,709).

Fitted at Woolwich (for £5,202) 27.6–10.9.1793.

Commissioned 7.1793 under Cmdr Charles Mansfield. In ?10.1794 under Cmdr Henry Blackwood; at Bridport's Action off Île Groix 23.6.1795. In 9.1795 under Cmdr Samuel Ballard, then in 1796 Cmdr Archibald Dickson, in 12.1796 Cmdr John Miller, in 9.1797 Cmdr George Shirley, in 4.1798 Cmdr George White, in 11.1799 Cmdr Humphrey West, in 1800 Cmdr Peter Bover, in 9.1800 Cmdr Henry Hill, in 1.1801 Cmdr Tristram Robert Ricketts and in 3.1801 Cmdr John Newhouse; paid off 1802. Fitted at Portsmouth 10–12.1803; recommissioned 25.10.1803 under Cmdr Archibald Duff, for the North Sea; laid up at Plymouth 7.1805. Sold to James Dashin there (for £910) 3.4.1817.

PYLADES Class 24 guns. Sole survivor by 1817 of six ships ordered in February 1793 to this Sir John Henslow design as 16-gun ship sloops. Of her sisters, the *Alert* and *Ranger* had been war losses, the *Albacore* had been sold in 1802 and the *Pylades* and *Rattler* sold in late 1815. The *Peterel* was re-classed as a 24-gun Sixth Rate in 1817.

Dimensions and tons: 105ft 0in, 86ft 7½in x 28ft 0in x 13ft 6in. 361¹⁴⁄₉₄bm.

Men: 121 (from 1794). Guns: UD 16 x 24pdr carronades, QD 6 x 12pdr carronades, Fc 2 x 12pdr carronades.

Peterel John Wilson & Co., Frindsbury

As built: 105ft 1in, 86ft 7⅝in x 28ft 2in x 13ft 6in. 365⁵⁷⁄₉₄bm.

Ord: 18.2.1793. K: 5.1793. L: 4.3.1794. C: 6.1794 at Chatham.

First cost: £3,936 to build; total £7,694 including fitting.

Commissioned 4.1794 under Cmdr Stephen George Church. On 8.10.1794 under Cmdr Edward Leveson Gower, then 7.1795 under Cmdr Charles Ogle, in the Downs squadron. On 11.1.1796 under Cmdr John Temple; in 4/5.1796 in Nelson's squadron off Genoa; later under Cmdr Philip Wodehouse. In 3.1797 under Cmdr Lord (William) Proby, then Cmdr Thomas Caulfield in 8.1797, later Lieut Adam Drummond then Cmdr Henry Digby; took 12-gun privateer *Le Léopard* 30.4.1798. In 10.1798 under Cmdr Hugh Downman, then in 11.1798 Lieut George Long, with Duckworth's squadron at Minorca; captured off Minorca by four French frigates 13.11.1798; retaken next day by *Argo*. In 2.1799 under Cmdr George Jones, then Cmdr Frederick Austen in 7.1799; action with 14-gun *Le Cerf* and *La Ligurrienne*, and 6-gun *Lijoille* 20.5.1800, taking *La Ligurrienne*. In 6.1800 under Cmdr Charles Inglis, in Egypt operations. Under Cmdr John Lamborn from 4.1802 (–1809); took small privateer on the Jamaica station 23.1.1805; took 5-gun Spanish privateer *Santa Anna* off Cuba 13.5.1805. Fitted as receiving ship at Plymouth 7–8.1811, and served as such to 1825. Sold to Joshua Crystall at Plymouth (for £730) 11.7.1827.

CORMORANT Class 26 guns (before 1817 had 18 guns). A joint design by Henslow and Rule, with hull form based on the French *Amazon* of 1745. Of six ships ordered 1793, *Cormorant* and *Lark* were wartime losses and *Lynx* sold in 1813; the seventh order (in 1794), *Stork*, was sold in May 1816. These originally carried 16 x 6pdrs on the UD, but were re-armed with carronades. Of twenty-four more ships ordered 1805–06 to the same design, *Anacreon* and *Tweed* were wartime losses, and *Serpent* was cancelled in 1810; *Rosamond* was sold in 1815, *Sabrina* and *Partridge* were sold or BU in 1816. *Rosamund* and the other six ships ordered in October 1806 were sometimes designated as a separate (*Rosamund*) class, but their dimensions and hull forms were identical and in practice these differed in no significant way from the rest of the batch. All were re-classed as 20-gun Sixth Rates in 1810–

12, and most of the survivors were in January 1817 re-rated as 26-gun Sixth Rates; three exceptions (*North Star*, *Minstrel* and *Wanderer*) were re-rated as 20-gun sloops (see Chapter 7). *Ranger*, the 24th ship built as a 'one-off' stretched version of this standard class, was BU in 1814.

Dimensions & tons: 108ft 4in, 90ft 9⅛in x 29ft 7in x 9ft 0in. 422⁶¹⁄₉₄bm.

Men: 125 (121 from 1794). Guns (1817): UD 16 x 32pdr carronades, QD 6 x 18pdr carronades, Fc 2 x 6pdrs + 2 x 12pdr carronades.

1793 batch

Goree (ex-*Favourite*, renamed 1807) Randall & Brent, Rotherhithe

As built: 108ft 5in, 90ft 8¼in x 29ft 9in x 9ft 0in. 426⁸⁸⁄₉₄bm. Draught 7ft 6in/11ft 6in.

Ord: 18.2.1793. K: 4.1793. L: 1.2.1794. C: 14.5.1794 at Deptford.

First cost: £5,083 to build; total £9,404 including fitting.

Commissioned 3.1794 under Cmdr Charles White, for the Downs. In 9.1795 under Cmdr James Athol Wood; sailed for the Leeward Islands 9.1795; took 8-gun *Le Général Rigaud* 2.1796; destroyed *Le Banan* and another privateer (ex-packet *Hind*) 1796; at capture of Trinidad 2.1797. In 5.1797 under ?S Powel, then 7.1797 Cmdr James Hanson, then 1.1798 Capt. Lord (Thomas) Camelford, all in West Indies. In 5.1799 under Cmdr Joseph Westbeach (–1801); came home with trade 7.8.1799; in North Sea 1800; took three 14-gun privateers – *Le Voyageur* off Flamborough Head 15.1.1801, *L'Optimiste* in North Sea 12.3.1801, and *L'Antichristi* off Shields 17.4.1801. Small to Middling Repair, and fitted at Sheerness 5.1803–6.1804; recommissioned 5.1804 under Cmdr Charles Foote, for North Sea; at Bombardment of Le Havre 1.8.1804; took 14-gun privateer *La Raccrocheuse* 12.12.1804. Under Cmdr John Davie 12.1804; took 16-gun privateer *Le Général Blanchard* off African coast 28.12.1805; taken off the Cape Verde Islands by a French squadron 6.1.1806; retaken by *Jason* off Guiana Coast 27.1.1807, then renamed *Goree* 1807. Recommissioned 8.1807 at Antigua under Cmdr William Parkinson. In 2.1808 under Cmdr George Alfred Crofton, later Cmdr Joseph Spear; action with 16-gun *La Patinare* and *Le Pilade* off Marie Galante Island 22.4.1808; took 8-gun privateer *L'Amiral Villaret* 24.11.1808; at capture of Martinique 2.1809. Under Cmdr John Simpson in 4.1809, then on 12.12.1809 under Cmdr Henry Byng (–4.3.1813), on Halifax station (as prison ship) to 1813, then at Bermuda. On 19.7.1814 under Cmdr Constantine Richard Moorsom, then Lieut John Boulton in 6.1815 and Cmdr John Wilson on 6.6.1815; laid up at Halifax 1.5.1816. BU 1817 at Bermuda.

Hornet Marmaduke Stalkart, Rotherhithe

As built: 109ft 2in, 91ft 6⅞in x 29ft 8in x 9ft 0in. 428⁵⁵⁄₉₄bm. Draught 7ft 10in/11ft 5in.

Ord: 18.2.1793. K: 4.1793. L: 3.2.1794. C: 14.5.1794 at Deptford.

First cost: £9,039 including fitting.

Commissioned 3.1794 under Cmdr Christmas Paul; paid off 2.1795 and recommissioned under ?W Lakin, for cruising. In 1.1796 under Cmdr Robert Larkan, in Home Waters, then 11.1796 under Cmdr John Nash (–1802); refitted at Portsmouth (for £3,554) 6–7.1799, then to Leeward Islands. In 8.1802 under Lieut Robert Tucker, then Cmdr Peter Hunt in 9.1802, still in Leeward Islands. In 1804 under Cmdr John Lawrence. Fitted at Plymouth for the Military Medical Staff 9.1804–7.1805; recommissioned 6.1805 under Lieut Charles Williams (–1811), as hospital ship in the Scilly Isles; laid up at Plymouth 5.1811. Sold to Mr Bailey (for £920) 30.10.1817.

Hazard Josiah & Thomas Brindley, Frindsbury

As built: 108ft 4¾in, 90ft 6¼in x 29ft 9in x 9ft 0in. 425bm. Draught 7ft 0in/11ft 0in.

Ord: 18.2.1793. K: 5.1793. L: 3.3.1794. C: 8.6.1794 at Chatham.

First cost: £4,618 to build; £8,735 including fitting.

Commissioned 4.1794 under Cmdr John Loring. In 1795 under Cmdr Robert D. Oliver, for the Irish Sea. In 4.1796 under Cmdr Alexander Ruddach; took privateers – 14-gun *Le Terrible* off Cape Clear 16.7.1796, 22-gun *La Musette* 1.1.1797, 18-gun *Le Hardi* 1.4.1797 and 20-gun *Le Neptune* 12.8.1798 – the last three off the Irish coast. From 7.1798 under Cmdr William Butterfield (–1801). Fitted at Portsmouth 6–8.1802; recommissioned 6.1802 under Cmdr Robert Neve (–1806), for the Channel. In 2.1806 under Cmdr Charles Dilkes (–1808), still in the Channel; her boats (with consorts') took nine chasses-marees in the Pertuis Breton 27.6.1807; sailed for West Indies 16.11.1807. In 1.1809 under Cmdr Hugh Cameron; took (with *Jason* and *Cleopatra*) 40-gun *La Topaze* at Guadeloupe 22.1.1809; at capture of Martinique 2.1809; her boats (with *Pelorus*'s) destroyed a privateer at Marie Galante 17.10.1809; attack on Basseterre 18.12.1809 (Cameron killed). In 12.1809 under Cmdr William Elliott; sailed for Newfoundland 15.6.1810. In 12.1810 under Cmdr John Cookesley (–1815); sailed for Newfoundland again 23.3.1811; paid off 16.12.1816. Sold to Mr Spratley (for £1,010) at Portsmouth 30.10.1817.

1805–06 batch

Hyacinth John Preston, Great Yarmouth (later 24)

As built: 108ft 8in, 91ft 1⅝in x 29ft 7in x 9ft 0in. $424\frac{23}{94}$bm. Draught 7ft 10in/10ft 6in.

Ord: 12.7.1805. K: 11.1805. L: 30.8.1806. C: 13.9–21.11.1806 at Chatham.

Commissioned 10.1806 under Cmdr John Davie, for the North Sea; sailed for South America 15.2.1808. On 22.8.1809 under Cmdr John Carter, on South America station. Fitted at Portsmouth 5–7.1811, and re-classed as Sixth Rate (20 guns). On 24.5.1811 under Capt. Thomas Ussher; sailed for the Mediterranean 9.3.1812; her boats (with those of consorts) took 10-gun Neapolitan privateers *Intrepido* and *Napoleone* and destroyed other vessels at Malaga 29.4.1812; operations on south coast of Spain 5.1812. In 6.1812 under Capt. ?William Hamilton, then 11.1812 Capt. John Lampen Manley (acting). On 22.1.1813 under Capt. Alexander Renton Sharpe (–1818), in the Mediterranean to 1815, then to the North Sea, the Channel and finally Brazil 1817; paid off 9.1818. BU at Deptford 12.1820.

Herald Jeffrey Carver & George Corney, Littlehampton. (later 24)

As built: 108ft 10in, 90ft 11½in x 29ft 9½in x 8ft 11½in. $429\frac{38}{94}$bm. Draught 7 ft 9in/10ft 9in.

Ord: 12.7.1805. K: 12.1805. L: 27.12.1806. C: 14.1–1.4.1807 at Portsmouth.

Commissioned 3.1807 under Cmdr George Hony; sailed for the Mediterranean 18.5.1807; took 4-gun privateer *Le César* 11.11.1807. In 1.1808 under Cmdr (Capt. 8.1811) George Jackson (–1813), in the Mediterranean; re-classed as Sixth Rate (20 guns) 8.1810; sailed for Jamaica 4.7.1812. On 1.1.1813 (acting) 28.5.1813 (confirmed) under Capt. Clement Milward; at Halifax 1814; paid off 11.10.1815. BU at Chatham 9.9.1817.

Fawn Thomas Owen, Topsham (later 26)

As built: 108ft 7in, 90ft 11¾in x 29ft 7in x 9ft 0in. $423\frac{49}{94}$bm.

Ord: 19.10.1805. K: 3.1806. L: 22.4.1807. C: 7.5–3.9.1807 at Plymouth.

Commissioned 5.1807 under Cmdr Fasham Roby (died 4.1808); sailed for the Leeward Islands 11.11.1807. In 4.1808 under Cmdr Michael de Courcy. In 1809 under Cmdr George Crofton (–1812); at capture of Martinique 2.1809; took 10-gun privateer *La Téméraire* 11.10.1810; re-classed as Sixth Rate (20 guns) ?1812; sailed with Lisbon convoy 28.6.1812. In 9.1812 under Capt. Thomas Fellowes, on the Jamaica station; destroyed US privateer *Rosamond* 11.1.1813. Arrived Plymouth 9.1813, laid up and paid off 10.1813. Sold to Mr Young (for £1,450) 20.8.1818.

Acorn George Crocker, Bideford (later 26)

As built: 108ft 4¾in, 90ft 8¾in x 29ft 10in x 9ft 0in. $429\frac{49}{94}$bm. Draught 8ft/11ft 6in.

Ord: 19.10.1805. K: 3.1806. L: 30.10.1807. C: 29.2–22.5.1808 at Plymouth.

Commissioned 3.1808 under Cmdr Robert Clephane (–1811); sailed for the Mediterranean 28.5.1808; her boats (with those of *Excellent* and *Bustard*) cut out convoy at Duino (near Trieste) 28.7.1809; in the Adriatic in 1810. In 3.1811 under Capt. George Bligh (–1812); re-classed as Sixth Rate (20 guns) ?1812. Middling Repair at Woolwich 1–7.1813; recommissioned 6.1813 under Capt. George Henderson; sailed for the East Indies. In 4.1814 under Capt. Joseph Prior; paid off into Ordinary at Chatham 9.1816. BU there 5.1819.

Raccoon John Preston, Gt.Yarmouth (later 26)

As built: 108ft 4in, 90ft 8¾in x 29ft 8½in x 9ft 0in. $425\frac{88}{94}$bm. Draught 7ft 0in/10ft 8in.

Ord: 19.10.1805. K: 3.1806. L: 30.3.1808. C: 19.4–12.8.1808 at Chatham.

Commissioned 6.1808 under Cmdr James Welsh (died 11.1809); sailed for Jamaica 16.6.1809. In 11.1809 under Cmdr William Black (–1813), at Jamaica; to South America 1813–14. In 1.1815 under Lieut James Mangles (acting) at Cape of Good Hope, then 5.1815 under Cmdr John Cook Carpenter; re-classed as Sixth Rate (20 guns) 1.1817; paid off 12.1818. Fitted as convict hospital ship at Portsmouth 9–10.1819. Sold to Mr Soames (for £820) at Portsmouth 16.8.1838.

North Star Benjamin Tanner, Dartmouth, completed by John Cock after Tanner's bankruptcy (later 26)

As built: 108ft 4⅝in, 90ft 0⅛in x 30ft 1in x 9ft 0in. $433\frac{28}{94}$bm. Draught 8ft 4in/11ft 3½in.

Ord: 19.10.1805. K: 5.1806. L: 21.4.1810. C: 13.5–17.7.1810 at Plymouth.

Commissioned 6.1810 under Cmdr (Capt. 4.1811) Thomas Coe (–1815); re-classed as Sixth Rate (20 guns) 4.1811; sailed for the Leeward Islands 6.6.1813. In 11.1815 under Cmdr George Bentham, on Jamaica station; paid off 4.1816. Sold to Thomas Pitman to BU (for £1,010) 6.3.1817, but later re-sold to become mercantile *Colombo*.

Myrtle Richard Chapman, Bideford (later 24)

As built: 108ft 5in, 90ft 9¾in x 29ft 9½in x 9ft 0in. $428\frac{68}{94}$bm. Draught 7ft 3½in/11ft 6in.

Ord: 19.10.1805. K: 6.1806. L: 2.10.1807. C: 1.3–23.5.1808 at Pymouth.

Commissioned 3.1808 under Cmdr Thomas Innes; sailed for Portugal 20.5.1808. Recommissioned 9.1809 under Cmdr John Smith Cowan (also acting in 7.1808). On 3.4.1811 under Capt. Clement Sneyd; sailed for Portugal 23.11.1811; re-classed as Sixth Rate (20 guns) in 1812. On 13.6.1813 under Capt. Henry Bourchier, for Lisbon station. On 29.11.1813 under Capt. Arthur Bingham (–1815); for North Sea 1814; paid off 11.1815. At Plymouth in 1817; ordered to be repaired 5.1818, but instead BU there 6.1818.

Hesper Benjamin Tanner, Dartmouth, completed by John Cock after Tanner's bankruptcy (later 26)

As built: 108ft $3\frac{3}{4}$in, 90ft $6\frac{7}{8}$in x 29ft 8in x 9ft 0in. $424\frac{1}{94}$bm. Draught 8ft 6in/11ft 3in.

Ord: 19.10.1805. K: 6.1806. L: 3.7.1809. C: 19.7–30.9.1809 at Plymouth.

Commissioned 8.1809 Cmdr William Buchanan; in 18.8.1809 under Cmdr Edward Wallis Hoare; sailed for the East Indies 9.10.1809. In 10.1810 under Cmdr David Paterson, in 2.1812 Cmdr Charles Thurston, then 30.6.1812 Cmdr Henry Theodosius Browne Collier, then Cmdr Joseph Prior. In 8.1813 under Cmdr Charles Biddulph, then 20.9.1815 Cmdr Michael Matthews; paid off 8.1816. Sold at Mauritius (by AO 6.12.1816, for £1,660.15.0d) 8.7.1817.

Cherub John King, Dover (later 26)

As built: 108ft 4in, 90ft $9\frac{1}{2}$in x 29ft $7\frac{1}{2}$in x 9ft 0in. $423\frac{79}{94}$bm. Draught 7ft 10in/11ft 6in.

Ord: 19.11.1805. K: 1.1806. L: 27.12.1806. C: 13.1–14.6.1807 at Chatham.

Commissioned 4.1807 under Cmdr George Ravenshaw, for North Sea. Sailed for the Leeward Islands 29.2.1808. In 1809 under Cmdr (Capt., 8.1811) Thomas Tucker (–1815); at capture of Martinique 2.1809; re-classed as Sixth Rate (20 guns) 8.1811; on South American station 1813–15; with *Phoebe* took the 32-gun USS *Essex*-at Valparaiso 28.3.1814. Fitted for sea at Portsmouth 8–10.1816; recommissioned 8.1816 under Capt. William Fisher, for African station. In 10.1817 under Capt. George Willes, still on African station. Sold to Mr Holmes (for £940) 13.1.1820.

Minstrel Nicholas Bools & William Good, Bridport (later 24)

As built: 108ft 4in, 90ft $9\frac{5}{8}$in x 29ft 7in x 9ft 0in. $422\frac{65}{94}$bm.

Ord: 19.11.1805. K: 1.1806. L: 25.3.1807. C: 29.4–24.7.1807 at Plymouth.

Commissioned 3.1807 under Cmdr John Hollinworth; sailed for the Mediterranean 10.10.1807; took 10-gun Venetian schooner *Ortenzia* 16.7.1808; violation of Algerian neutrality at Bougia 9.1809. In 2.1810 under Cmdr Ralph Wormeley, then Cmdr Colin Campbell; landing party (with others ships') in attack on Palamos 13.12.1810. Re-classed as a Sixth Rate (20 guns) 1811; in 10.1811 under Capt. John Peyton. In 8.1812 under Lieut Michael Dwyer, at Benidorm; boats took ammunition ships at Valencia 29.9.1812. On 2.2.1813 under Cmdr Robert Mitford, in the Mediterranean; invalided home 8.1814. On 29.9.1814 under Capt. Francis Loch; paid off 12.1815 at Chatham. Sold to Mr Younge (for £1,010) 6.3.1817.

Wanderer James Betts, Mistleythorn (later 24)

As built: 109ft 3in, 91ft 6in x 29ft 9in x 9ft $0\frac{1}{2}$in. $430\frac{71}{94}$bm.

Ord: 19.11.1805. K: 2.1806. L: 29.9.1806. C: 25.10.1806–31.1.1807 at Chatham.

Commissioned 12.1806 under Cmdr Edward Crofton (–1811), for North Sea in 1807. In ?1810 under Cmdr William Robilliard; landing party from *Wanderer* (with consorts) at Sint Maarten 3.7.1810; re-classed as Sixth Rate (20 guns) ?1811. In 4.1811 under Capt. Francis Newcombe (–1814); sailed with Lisbon convoy ?17.6.1812; sailed for North America 28.8.1812. In ?10.1814 under Capt. John Palmer, then 12.1814 under Capt. William Dowers, in the Channel; paid off 11.1815 at Plymouth. Sold to Mr Splidt (for £1,150) 6.3.1817.

Sapphire Josiah & Thomas Brindley, King's Lynn (later 26)

As built: 108ft 5in, 90ft 11in x 29ft $8\frac{1}{2}$in x 9ft 0in. 426bm. Draught 7ft 4in/10ft 4in.

Ord: 19.11.1805. K: 2.1806. L: 11.11.1806. C: 3.12.1806–18.7.1807 at Chatham.

Commissioned 2.1807 under Cmdr George Davies (–1811); sailed for the Persian Gulf 27.10.1807; at Cape of Good Hope 1809; home in 1810; sailed for Jamaica 21.9.1810. In 11.1810 under Cmdr Joseph N. Tayler, at Jamaica. In 1812 under Cmdr Henry Haynes (–1815); took (with *Forester*) US 2-gun privateer *Mary Anne* off San Domingo 15.5.1812. In 4.1814 under Cmdr Adam Brown; paid off 9.1815 at Deptford. Small Repair at Deptford 4–8.1818; fitted for sea (and fitted with Baines Patent Perambulator) 10–11.1818; from 8.1818 under Capt. Henry Hart, for the Leeward Islands, then in 7.1820 under Capt. Alexander Montgomerie; paid off 9.1821. Sold at Chatham to Mr Manlove (for £1,510) 18.4.1822.

Blossom Robert Guillaume, Northam (later 24)

As built: 108ft $4\frac{1}{2}$in, 90ft $11\frac{1}{2}$in x 29ft $8\frac{3}{4}$in x 9ft 0in. 427bm. Draught 8ft 3in/11ft 4in.

Ord: 19.11.1805. K: 2.1806. L: 10.12.1806. C: 15.12.1806–21.4.1807 at Portsmouth.

First cost: £7,169 to builder, plus fitting £8,701.

Commissioned 12.1806 under Cmdr George Pigot, for Newfoundland station; sailed for Portugal 13.2.1808; her boats (with *Nymphe*'s) tried to cut out ex-Portuguese *La Garotta* from the Tagus 23.4.1808. In 4.1808 under Cmdr Thomas Dench, then Cmdr Henry Probyn in 10.1808 and Cmdr Francis Beaufort in 6.1809; sailed with Quebec convoy 19.6.1809. In 5.1810 under Cmdr William Stewart; sailed for the Mediterranean 18.6.1810; her boats took 4-gun privateer *Le César* off Cape Socié 4.11.1810; took 7-gun privateer *Le Jean Bart* off Naples 28.2.1812; destroyed (with boats of *Undaunted* and *Voluntaire*) a convoy in the Rhône Estuary 29.4.1812. In 1813 under Cmdr Edward Sibly; re-classed as Sixth Rate (20 guns) ?1813. In 1814 under Capt. Joshua Ricketts Rowley, then to Ordinary at Sheerness. Middling Repair at Woolwich)for £13,367) 8.1815–1.1817, then fitted for foreign service at Chatham (for £6,243) 6–8.1817; recommissioned 6.1817 under Capt. Frederick Hickey, for South America; paid off 8.1819. In 7.1820 under Capt. Francis Edward Vernon Harcourt, then 9.1822 Capt. Archibald M'Lean, still on South American station. Fitted 'for the icy seas' (as Exploration ship) at Deptford (for £5,509) and Woolwich (for £8,160) 7.1824–4.1825; recommissioned 1.1825 under Cmdr (Capt., 5.1827) Frederick William Beechey, for exploration of the Pacific North West of America; paid off 5.1828. Fitted at Woolwich as a Survey ship 4–8.1829; recommissioned 5.1829 under Richard Owen, for the Jamaica station (–1832). Fitted at Chatham as a lazarette 1.1833 for Stangate Creek, where served 1833–46. BU at Chatham 8.1848.

Egeria Nicholas Bools & William Good, Bridport (later 26)

As built: 108ft 4in, 90ft $9\frac{1}{2}$in x 29ft $7\frac{3}{4}$in x 9ft $0\frac{1}{2}$in. $424\frac{41}{94}$bm. Draught 7ft 5in/11ft 3in.

Ord: 19.11.1805. K: 6.1806. L: 31.10.1807. C: 14.2–19.6.1808 at Plymouth.

First cost: £7,278 to builder.

Commissioned 3.1808 under Cmdr Fitzowen Skinner, for the North Sea (at Leith). In 5.1808 under Cmdr Lewis Hole (–1812); took 10-gun Danish privateer *Noesois* 21.12.1808; took 6-gun cutter *Aalborg* in the Skaw 2.3.1809; took 14-gun privateer *Alvor* in the North Sea 31.12.1811. Re-classed as Sixth Rate (20 guns) 2.1810. To Ordinary at Sheerness 1813–16, including Large Repair there 6.1813–1.1814. Fitted for sea at Sheerness 2.1817; recommissioned 12.1816 under Capt. Robert Rowley, for the Newfoundland station. In 11.1819 under Capt. Henry Shiffner; paid off 1.1820. In 1.1820 under Capt. John Toup Nicholas (–1822); involved in Newcastle coal strike 1822; paid off 1823. Small Repair and fitted for sea at Plymouth 2–4.1823; recommissioned 1.1823 under Capt. Samuel Roberts (–1825). Fitted as receiving ship at Plymouth 6.1826. Lent to the breakwater department there by AO 16.12.1843, becoming an accommodation hulk 12.1845; police accommodation ship 4.1860. BU 1865.

Favorite Jabez Bailey, Ipswich (later 26)
As built: 108ft 9½in, 91ft 2⅝in x 29ft 8in x 9ft 0in. 427$^{3}/_{94}$bm. Draught 7ft 2in/11ft 1in.
Ord: 30.11.1805. K: 2.1806. L: 13.9.1806. C: 26.9–25.12.1806 at Chatham.
Commissioned 11.1806 under Cmdr John Nairne (died 7.1807), for the Channel; sailed with convoy for Africa 21.5.1807. In ?7.1807 under Lieut Frederick Hoffman (acting), on African coast and then Leeward Islands. In 1808 under Cmdr Benjamin Clement (–1811), on Jamaica station. In 8.1811 under Capt. Robert Forbes, when re-classed as a Sixth Rate (20 guns). In 4.1812 under Capt. John Maxwell; sailed for West coast of Africa 2.6.1813. In Ordinary at Woolwich 1814, under Capt. John Maples; then Small Repair at Deptford 6–10.1814; recommissioned 10.1814 under Capt. James Maude (–1817); sailed for East Indies; paid off 6.1817. Fitted for foreign service at Deptford 9–12.1817; recommissioned 9.1817 under Capt. Hercules Robinson (–1820), for Newfoundland station; paid off 2.1821. BU at Portsmouth 2.1821.

Jalouse Plymouth Dyd. (later 26) [M/Shipwright Joseph Tucker]
As built: 108ft 4½in, 90ft 9¾in x 29ft 8in x 9ft 0in. 425$^{12}/_{94}$bm. Draught 8ft 7in/10ft 9in.
Ord: 15.1.1806. K: 7.1808. L: 13.7.1809. C: 10.9.1809.
First cost: £12,923 including fitting.
Commissioned 6.1809 under Cmdr Henry Morris, for Irish station; took (with *Phoenix*) 14-gun privateer *Le Charles* 29.1.1810. In 9.1812 under Cmdr Abraham Lowe, then ?6.1814 Cmdr James Bashford, ?6.1815 Cmdr John Undrell, for West Indies, and then Cmdr Edward Hall in 1816; paid off 1.1816 at Chatham. Sold to G. T. Young at Chatham (for £1,660) 8.3.1819.

Dauntless Deptford Dyd. (later 26) [M/Shipwright Robert Nelson]
As built: 108ft 4in, 90ft 9⅝in x 29ft 7in x 9ft 0in. 422$^{66}/_{94}$bm. Draught 8ft 2in/11ft 0½in.
Ord: 25.3.1806. K: 11.1807. L: 20.12.1808. C: 14.8.1809.
Commissioned 7.1809 under Cmdr Josiah Wittman (died 1.1810), for African coast. Under Cmdr Daniel Barber from 5.1810 (–1815); sailed with trade for Archangel 15.6.1810; on Irish station 1812–13; Newfoundland in 1814. In Ordinary at Portsmouth 12.1815. Between Small and Middling Repair, and fitted for sea at Portsmouth 6.1818–2.1819; recommissioned 11.1818 under Capt. Valentine Gardner (died 11.1820), for East Indies; then 11.1820 under Cmdr John Norman Campbell (acting), then 6.1821 Capt. George Gambier, still in East Indies; paid off 10.1823 at Portsmouth. Sold to Thomas Smith (for £2,330) 27.1.1825.

***BITTERN* Class** 28 guns. Originally 18-gun ship sloops of John Henslow design, enlarged from the 1793 *Pylades* Class, and approved 29 January 1795. The prototype, *Brazen*, was first ordered (with the *Stork*, mentioned above) in late 1794 and subsequently delayed. Another four vessels were ordered 24 January 1795, and were named and established by AO of 13 October 1795. *Cerf* (ex-*Cyane*) of this batch was sold in 1809. The *Brazen*, *Plover* and *Termagant* were re-classed as 28-gun Sixth Rates in January 1817; their sister *Bittern* had been classed as a tender since 1812.
Dimensions & tons: 110ft 0in, 90ft 8⅜in x 29ft 6in x 8ft 6in. 419$^{76}/_{94}$bm.
Men: 121. Guns (1817): UD 18 x 32pdr carronades; QD 6 x 18pdr carronades; Fc 2 x 6pdrs + 2 x 18pdr carronades. As a tender, *Bittern* had 10 guns (2 x 6pdrs + 8 x 12pdr carronades).

Brazen Portsmouth Dyd. [M/Shipwright Nicholas Diddams]
As built: 110ft 3in, 90ft 9⅞in x 29ft 6½in x 8ft 6in. 421$^{57}/_{94}$bm. Draught 8ft 0in/11ft 9in.
Ord: 6.11.1794, cancelled 1799 but later reinstated. K: 15.6.1807. L: 26.5.1808. C: 25.7.1808.
Commissioned 5.1808 under Cmdr Lewis Shephard; to Jamaica in 1809. In 11.1810 under Cmdr Richard Plummer Davies, then 10.1812 under Cmdr James Stirling (–1817), on North Sea and Irish stations; sailed with convoy for Hudson's Bay 3.6.1813; to North America 1814. Fitted for foreign service at Woolwich 10–12.1815; to Jamaica 1817. Between Small and Middling Repair, and fitted for sea at Portsmouth 12.1818–1.1820; recommissioned 11.1819 under ?Capt. William Shepheard (–1822); to St Helena 1821; at Cork 1822. In 1.1823 under Capt. George Willes (–1826), for South America and West Coast of Africa. Fitted as floating church at Chatham 5–9.1827, then to Deptford where delivered to Committee of the Floating Church 10.2.1828. Returned 1846 and BU at Deptford 7.1848.

Plover James Betts, Mistleythorn
As built: 110ft 0in, 90ft 7½in x 29ft 7in x 8ft 6in. 421$^{82}/_{94}$bm. Draught 9ft 10in/10ft 10in.
Ord: 24.1.1795. K: 5.1795. L: 23.4.1796. C: 16.5–10.8.1796 at Sheerness.
First cost: £5,362 to build, plus £1,794 fitting.
Commissioned 4.1796 under Cmdr John Chesshyre (–1800), for the Downs; took (with cutter *Resolution*) 10-gun privateer *L'Erin-go-Brah* in the North Sea 28.12.1798. In 1.1800 under Cmdr Edward Galway (–1802); took 4-gun privateer *Le Messina* off Dunkirk 10.3.1800; paid off ?4.1802. Recommissioned 6.1803 under Cmdr Richard Hancock (–1806); to Newfoundland 1804 and 1805. In 9.1806 under Cmdr Philip Browne (–1810); took privateers – 14-gun L'*Eliza* in the Channel 1.1.1807, 2-gun *La Bohémienne* 30.10.1807, and 4-gun *L'Amiral Martia* 22.1.1809; in Walcheren operations 1809; took (with Lively) 16-gun privateer *L'Aurore* off Beachy Head 18.9.1809; took more privateers – 10-gun *L'Hirondelle* off Falmouth 22.10.1809, *Le Lezard* 6.11.1809 and 14-gun *Le Saratu* off Scilly Isles 10.1.1810. In 6.1810 under Cmdr Colin Campbell (–1813); destroyed three small privateers in the Channel 16.11.1810; took privateers – 10-gun *Le Ferego* off the Naze 6.7.1811 and 16-gun *Le Petit Edouard* in the North Sea 23.10.1811; in the Baltic 1812; sailed 15.1.1813 with convoy for Quebec. On 7.6.1814 under Cmdr John Skekel; with convoy to Mediterranean 1814, later to Newfoundland; paid off 30.5.1823. Loaned to the Committee for Distressed Seamen at Deptford from 10.1816. Returned to the RN in 1.1818. Sold to G. T. Young, Limehouse (for £1,520) 8.3.1819.

Termagant John Dudman, Deptford
As built: 110ft 2in, 90ft 8½in x 29ft 9in x 8ft 6in. 427$^{3}/_{94}$bm. Draught 7ft 11in/11ft 10in.
Ord: 24.1.1795. K: 5.1795. L: 23.4.1796. C: 23.4–9.6.1796 at Deptford Dyd.
First cost: £5,524 to build, plus £2,719 fitting.
Commissioned 5.1796 under Cmdr D'Arcy Preston, for the North

Sea. In ?6.1796 under Cmdr David Lloyd; took 14-gun privateer *La Victoire* off Spurn Point 20.12.1797. In ?3.1799 under Cmdr Richard Allen (died 10.1799); sailed for Halifax 4.1799. In 1800 under Cmdr William Skipsey (–1801); sailed for the Mediterranean 8.1800; took 6-gun *La Capricieuse* off Corsica 1.9.1800 and 2-gun privateer *Le Général Holtz* 4.9.1800. On 29.4.1802 under Cmdr Charles Marsh Schomberg, later under Lieut Charles Foote (acting) and ?J. Stuart/Stewart. In 8.1803 under Cmdr George Elliott, still in Mediterranean. In 1.1804 under Cmdr Robert Pettet (–1807); to Ordinary at Chatham 4.1807. Middling Repair and fitted at Chatham 2.1808–3.1809; recommissioned 10.1808 under Cmdr Henry Sturt (–1811); sailed for Halifax 3.4.1809; sailed for the Mediterranean 27.1.1810. In 11.1810 under Cmdr Richard Buck (temp.). Re-classed ?12.1811 as 20-gun Sixth Rate, under Capt. Gawen Hamilton; operations on Spanish coast 1812; took 3-gun privateer *L'Intrépide* 22.7.1812. In 1813 under Capt. John William Andrew, in the Mediterranean, then in early 1814 under Capt. John Lampen Manley; in 7.1814 under Capt. George Shaw; paid off and laid up at Chatham 10.1816. Sold to James Graham, Harwich at Chatham (for £1,460) 3.2.1819.

Bittern Bathazar & Edward Adams, Bucklers Hard
  As built: 110ft 2½in, 90ft 10⅛in x 29ft 6¼in x 8ft 6in. 422⁴⁶⁄₉₄bm. Draught 8ft 9in/12ft 3in.
  Ord: 24.1.1795. K: 6.1795. L: 7.4.1796. C: 10.4–2.7.1796 at Portsmouth.
  First cost: £5,404 to build; £12,668 including fitting.
  Commissioned 5.1796 under Cmdr Thomas Lavie; sailed for the Leeward Islands 22.2.1797; took 6-gun privateer *La Casca* off Barbados 15.3.1797; took 18-gun privateer *L'Agréable* off Tortola 13.9.1797. In 10.1797 under Cmdr Edward Kittoe (–1801); took 12-gun privateer *Le Dix Août* off Marie Galante 8.9.1798; home in 6.1800. In 9.1802 under Cmdr Robert Corbet; sailed for the Mediterranean 1.1803; took privateers there – 6-gun *La Caille* 10.9.1803, and 14-gun *L'Hirondelle* 28.4.1804. In 1.1805 under Cmdr Henry Duncan (acting), then 7.1805 Cmdr John Louis. In 1806 under Cmdr Edward Augustus Down, for the Mediterranean (Gibraltar); took privateer *Verga del Rosario* 2.8.1806. In 1808 under Cmdr (Capt. 5.1808) Thomas Ussher. In Ordinary at Plymouth to 1812. Re-classed as a 10-gun tender in 4.1812 and fitted as such at Plymouth (for £13,295) 5–9.1812; in 7.1812 under Cmdr George Augustus Hire; laid up at Plymouth 12.1815. Fitted at Plymouth for sea 'to carry new raised men' (for £4,120) 7–12.1824; recommissioned tender 12.11.1824 under Lieut Michael Dwyer, then 31.12.1825 under Lieut Robert Rochfort, at Liverpool; laid up at Plymouth 2.1828. Sold to Tibbett & Spence (for £1,930) 30.8.1833.

LAUREL Class 32 guns. In January 1805 the Admiralty ordered what were, in effect, a dozen wartime replacements for the pre-war 24-gun ships. As usual, the Navy Board commissioned competitive draughts from the two Surveyors, and split the orders between the two designs. This John Henslow design, approved on 28 March 1805, covered six vessels. Although initially rated 22-gun, they actually carried 24 carriage guns including 2 chase 6pdrs on the forecastle as well as 22 x 9pdrs on the main battery. Of the first three, the *Boreas* and *Laurestinus* (originally *Laurel*) of this class had been wartime losses, while the *Comus* had been wrecked off Newfoundland in October 1816. However the last three, all surviving into 1817, were completed with 22 x 32pdr carronades on UD and 18pdr carronades on QD/Fc. They were re-rated 32-gun frigates from 1 January 1817. Of the other six vessels built

at the same time to a similar William Rule design (the *Banterer* Class), the *Banterer* and *Cyane* had been wartime losses, and the *Daphne*, *Porcupine*, *Cossack* and *Crocodile* were sold or BU during 1816.
  Dimensions & tons: 118ft 0in, 98ft 7¼in x 31ft 6in x 10ft 3in. 526³⁹⁄₉₄ bm.
  Men: 155. Guns: Designed UD 22 x 9pdr; QD 6 x 24pdr carronades; Fc 2 x 6pdr (chase) + 2 x 24pdr carronades. But the three below were completed, and *Comus* was re-armed, with 22 x 32pdr carronades on the UD, and 18pdr (*vice* 24pdr) carronades on QD/Fc.

Garland Richard Chapman, Bideford
  As built: 118ft 1½, 98ft 8in x 31ft 8in x 10ft 3½in. 526²⁶⁄₉₄bm. Draught 8ft 4in/12ft 4in.
  Ord: 30.1.1805. K: 8.1805. L: 25.4.1807. C: 11.7–8.9.1807 at Plymouth.
  Commissioned 3.1807 under Capt. Header Whitter; sailed for the West Indies 11.11.1807. Based on Jamaica station for rest of the war, from 10.6.1808 under Capt. Rowland Bevan, then 14.3.1811 under Capt. Thomas Huskisson, and finally 10.1812 under Capt. Richard Plummer Davies; deployed to the Mediterranean 1815; paid off 2.1816. Sold to Mr Hill (for £1,550) to BU 9.5.1817.

Perseus Thomas Sutton & Co., Ringmore
  As built: 118ft 5in, 98ft 4¼in x 31ft 7in x 10ft 3¼in. 527⁷⁄₉₄bm. Draught 10ft 3in/12ft 3in.
  Ord: 30.1.1805. K: 11.1805. L: 30.11.1812. C: 22.3.1813 at Plymouth.
  First cost: £11,894 to builder (excludes fitting).
  Commissioned 2.1813 under Capt. Edward A'Court; sailed on mission to Barbary States 24.4.1813; to Newfoundland 1814 and then Halifax. In 11.1815 under Capt. Thomas Toker (–2.1817). Fitted as receiving ship for distressed seamen at Deptford in 12.1816. Fitted for the same purpose at Deptford 4–5.1818 for the Tower of London, where under command from 4.5.1818 of Lieut William Collins Barker, then 6.7.1821 under Capt. James Crouch. Paid off 1.1831, but remained stationed at the Tower. BU at Deptford 9.1850.

Volage Richard Chapman, Bideford
  As built: 118ft 2½in, 98ft 9in x 31ft 9in 10ft 3in. 529⁴⁷⁄₉₄bm. Draught 8ft 9in/12ft 3in.
  Ord: 30.1.1805. K: 1.1806. L: 23.3.1807. C: 11.7–8.9.1807 at Plymouth.
  Commissioned 5.1807 under Capt. Philip Rosenhagen; sailed for the Mediterranean 11.10.1807; took 10-gun cutter *Succes* in the Mediterranean 6.11.1807, then 16-gun *Le Requin* off Corsica 28.7.1808, 2-gun privateer *L'Annunciate* 6.9.1809, and 6-gun privateer *Le Jason* 20.9.1809. Under Capt. Phipps Hornby (from 2.1810 to 10.1811) participated in Hoste's Action at Lissa 13.3.1811 (losing 13 killed, 33 wounded). Under Capt. Arthur Bingham from ?2.1812, then 20.9.1811 under Capt. Donald Hugh Mackay; sailed for East Indies where remained for rest of war under Capts Samuel Leslie (in 1813), Joseph Drury (?2.1814), Charles Biddulph (to death in 4.1815), John Allen (nominally from 20.9.1815, but never took actual command) and John Bartholomew Hoar Curran (1816). Returned to UK to pay off, and sold to Mr Lackland (for £1,600) 29.1.1818 for mercantile use, renamed *Rochester*.

THAIS Class 24 guns. John Henslow design of 1805, virtually identical with that of *Tisiphone* Class. These six ships were ordered and classified as fireships but actually employed as sloops, and re-rated as such in 3.1808. They were almost identical with the *Cormorant* Class, but completed with spar deck. The majority were re-classed as 20-gun Sixth Rates in early 1811 or 1812. The *Comet* was sold in 1815 and the *Tartarus* and *Lightning* in 1816, and the three survivors were re-classed as 24-gun Sixth Rates on 1 January 1817.

Dimensions & tons: 108ft 9in, 90ft 7in x 29ft 7in x 9ft 0in. 421$\frac{63}{94}$bm.

Men: 121 (originally 55 as fireship, 121 as sloop). Guns (1817): UD 16 x 32pdr carronades; QD 6 x 18pdr carronades; Fc 2 x 6pdrs.

Thais Benjamin Tanner, Dartmouth

As built: 108ft 9in, 90ft 9⅝in x 29ft 10½in x 9ft 0in. 431$\frac{7}{94}$bm. Draught 8ft 1in/10ft 9in.

Ord: 1.10.1805. K: 10.1805. L: 19.8.1806. C: 16.9.1806–7.9.1807 at Plymouth.

Commissioned 6.1807 as fireship under Cmdr Isaac Ferrieres (–1810); sailed for the West Indies 9.9.1807; fitted as a sloop at Plymouth 2–4.1808; in North Sea 1809; sailed with Mediterranean convoy 23.5.1810. In 11.1810 under Cmdr Edward Scobell; sailed for Africa 2.5.1812; re-classed as Sixth Rate 1811 (Scobell Capt. 4.1811); took US 12-gun privateer *Rambler* on the African coast 31.3.1813. In 1.1814 under Capt. Henry Weir; cut down (spar deck removed) to 16-gun at Plymouth 10–12.1814; to East Indies 1815. In Ordinary at Plymouth 1817. Sold to Mr Price (for £1,400) 13.8.1818.

Prometheus Ralph Thompson, Chapel (Southampton)

As built: 109ft 6in, 91ft 4⅞in x 29ft 9½in x 9ft 0in. 431$\frac{49}{94}$bm. Draught 8ft 3in/10ft 3in.

Ord: 1.10.1805. K: 12.1805. L: 27.3.1807. C: 30.3–22.7.1807 at Portsmouth.

Commissioned 3.1807 as fireship under Cmdr Hyde Parker, for the Downs. Fitted as sloop at Sheerness 12.1807–2.1808. In ?12.1807 under Cmdr Thomas Forrest, for the Baltic; her boats (with those of *Implacable*, *Bellerophon* and *Melpomene*) attacked Russian convoy in Bäro Sound 7.7.1809 (6 gunboats and 12 other vessels taken: one gunboat sunk); her boats (with those of *Princess Caroline*, *Minotaur* and *Cerberus*) took 3 gunboats and a brig at Fredrickshamm 25.7.1809. In 7.1809 under Cmdr Hercules Robinson (–1814); destroyed 6-gun privateer *La Messilina* near Pillau 2.8.1810; sailed for Jamaica 11.5.1811; to Newfoundland 1814; took 2-gun privateer schooner *Lizard* 8.5.1814. Recommissioned 15.11.1814 under Cmdr William Bateman Dashwood, for the Mediterranean; cut down (spar deck removed) to 16-gun at Portsmouth 12.1814–3.1815; at Bombardment of Algiers 27.8.1816. On 15.11.1816 under Cmdr Constantine Richard Moorsom; paid off 8.1818. Fitted as lazarette at Portsmouth 4–5.1819, to lie at the Motherbank; by 1829 a receiving ship. Renamed *Veteran* by AO of 2.5.1839. BU at Portsmouth 8.1852.

Erebus Thomas Owen, Topsham

As built: 108ft 9in, 90ft 6in x 29ft 8in x 9ft 0in. 423$\frac{63}{94}$bm.

Ord: 1.10.1805. K: 1.1806. L. 20.8.1807. C: 16.10–11.11.1807 at Plymouth.

Commissioned 1.1808 as a sloop under Cmdr William Autridge, for the Baltic; sailed 4.1808. In 1809 under Cmdr ?Henry Withy, then 8.1812 Cmdr Henry James Lyford, but Autridge back (now Capt.) in same month. On 6.12.1813 under Cmdr John Forbes, for the North Sea. Fitted as Congreve rocket ship 1814; under Cmdr David Bartholomew 8.1814 for Potomac operation (and immortalised for 'the rockets' red glare' in 'The Star-Spangled Banner'). On 20.6.1815 under Cmdr Francis le Hunte; paid off at Deptford 4.9.1815. Sold 22.7.1819 (for £1,510) to Mr Manlove to BU.

Ex-FRENCH Prize (1809) This prize was a flush-decked corvette of the *Victorieuse* class, built to an 1804 design by François Poncet, as amended by Jacques-Noël Sané. Re-classed as a 26-gun ship in January 1817.

Ganymede (French corvette *L'Hébé*, built 6.1807–12.1808 at Bordeaux. L: 20.9.1808).

Dimensions & tons: 126ft 7in, 105ft 5⅛in x 32ft 8⅞in x 10ft 4in. 601$\frac{17}{94}$bm.

Men: 175. Guns: UD 20 x 9pdrs; QD 2 x 6pdrs; Fc 2 x 6pdrs + 2 x 18pdr carronades.

Taken by *Loire* in the Atlantic 5.1.1809.

Commissioned 9.1809 under Capt. Robert Cathcart, for the Channel Fleet; to the Mediterranean 5.1810, returning with a large convoy 10.1810; sailed 20.11.1810 with convoy for the West Indies, returning 8.1811 with another convoy. In 9.1811 under Capt. Robert Preston, who was court-martialled 10.1811 for alleged cruelty – acquitted but replaced as commander 10.1811 by Capt. John Brett Purvis, for the Spanish coast and Mediterranean; in defence of Cadiz 1812; took 7-gun privateer *Le Vanteur* 18.8.1813; at Corfu 1814; to North American station 1.1815, returning 8.1815. In 8.1815 under Capt. William McCulloch, for Channel service, in the Downs. In 5.1817 under Capt. Robert Spencer, for the Mediterranean; returned home to pay off 8.1819. Hulked at Portsmouth as a convict ship 11.1819; convict hulk in the Medway from 1820; at Woolwich from 1823. Capsized at Woolwich 19.3.1839 (when part of her moorings collapsed), raised and BU there 2.1840.

***CONWAY* Class (1813)** 28, later 26 guns. Designed by Sir William Rule and originally designated as 'sloops', this class of ten vessels were built as 20-gun post ships, re-rated as 28-gun Sixth Rates from 1 January 1817, then as 26-gun later in 1817 when their forecastle carronades were removed. *Eden* was deliberately scuttled in the Hamoaze on 9 November 1816 and raised 12 March 1817 as an experiment 'to ascertain the efficacy of sea water in curing dry rot'; *Mersey* was proposed to be similarly sunk and raised (by AO 30 March 1818), but this was never done.

Dimensions & tons: 108ft 0in, 89ft 9⅝in x 30ft 6in x 9ft 0in. 444$\frac{33}{94}$bm.

Men 155. Guns: UD 18 x 32pdr carronades; QD 6 x 12pdr carronades; Fc 2 x 6pdrs + 2 x 12pdr carronades.

Mersey William Courtney, Chester

As built: 108ft 6in, 90ft 1¼in x 30ft 8in x 9ft 0in. 450$\frac{69}{94}$bm. Draught 9ft 0in/11ft 10in.

Ord: 18.1.1813. K: 3.1813. L: 3.1814. C: 13–26.4.1814 at Plymouth (for Ordinary)

First cost: £9,720 to build.

Middling Repair at Plymouth 5.1818–3.12.1818 (under AO 7.4.1818).

Commissioned 8.9.1818 under Capt. Edward Collier, for Halifax station; paid off 30.11.1821 at Portsmouth. Fitted for sea at Portsmouth 3–8.1823; recommissioned 24.4.1823 under Capt. John Macpherson Ferguson (–1827), for South America. Recommissioned 26.12.1827 under Capt. Alexander Barclay Branch, for Jamaica station; under Capt. George William Conway Courtenay from 5.1.1829; paid off and laid up 7.1831 at Portsmouth. Receiving ship there 1834–50. BU at Portsmouth 7.1852.

Eden William Courtney, Chester

As built: 108ft 6in, 90ft 1¼in x 30ft 8in x 9ft 0in. 450$\frac{69}{94}$bm. Draught 10ft 3in/11ft 1in.

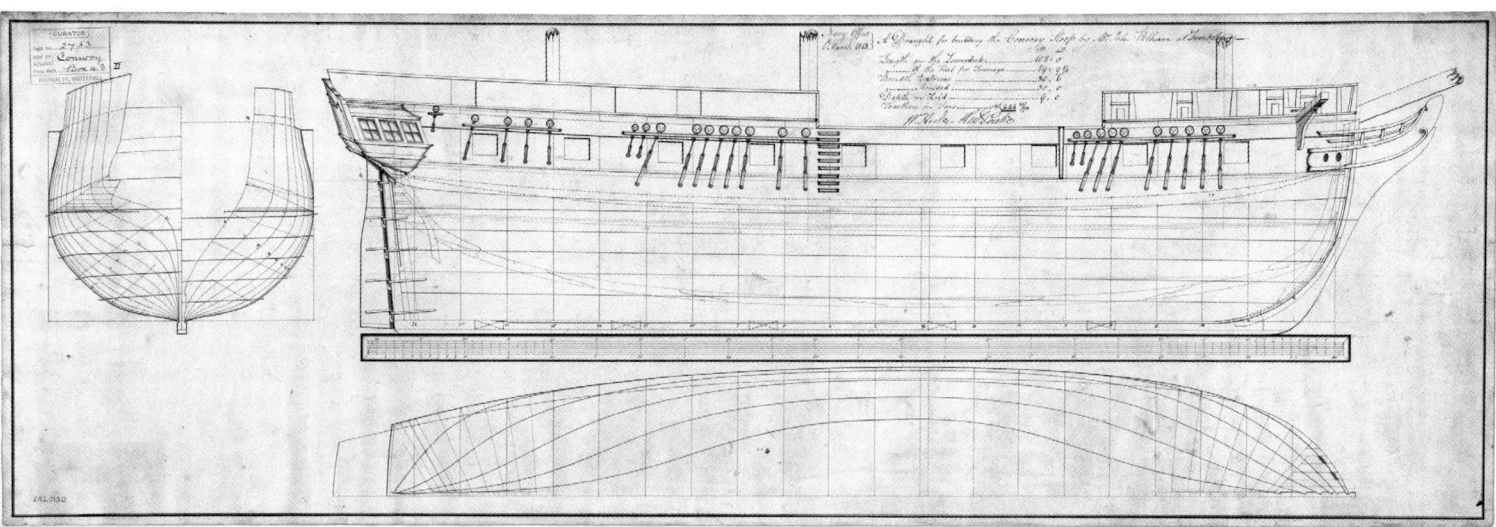

A sheer draught, dated 6 March 1813, for the Sixth Rate *Conway*, signed by the two incumbent Surveyors, Rule and Peake. In layout terms there was little to differentiate the smallest post ships from quarterdecked ship sloops, and they were reclassified a number of times. *(NMM J6118)*

Ord: 18.1.1813. K: 3.1813. L: 19.5.1814. C: 5–20.6.1814 at Plymouth (for Ordinary).

Fitted at Plymouth 2.1818–30.5.1818.

Commissioned 21.3.1818 under Capt. Francis Erskine Loch, for the East Indies; in attack on pirates at Ras-al-Khaimah in 11.1819; made good defects at Bombay 10.1820; paid off 8.1821. Very Small Repair at Deptford 8.1821–11.1822; recommissioned 31.8.1822 under Capt. John Lawrence, for the West Indies. Fitted at Woolwich for service in Africa 9.1826–7.1827; recommissioned 2.2.1827 under Capt. William Fitzwilliam Owen, for the African station; to South America 1830; at Rio 7.1830, then to the Pacific 9.1830; returned home 8.1831 to pay off. BU at Portsmouth 5.1833.

Conway John Pelham, Frindsbury

As built: 108ft 2in, 89ft 9¼in x 30ft 9in x 9ft 1in. 451⁴⁸⁄₉₄bm. Draught 8ft 4in/11ft 3in.

Ord: 18.1.1813. K: 5.1813. L: 10.3.1814. C: 10–22.3.1814 at Chatham (for Ordinary); 7.11.1814 (for sea).

Commissioned 1.10.1814 under Capt. John Tancock; sailed 1815 for East Indies. on 22.9.1816 under Capt. John Reynolds, then 12.1816 under Capt. William Hill and 4.7.1817 Capt. Edward Barnard; paid off 20.1.1820. Fitted for sea 4–7.1820; recommissioned 5.1820 under Capt. Basil Hall; sailed for South America 10.8.1820; in Pacific 1821–22; arrived home 2.1823 and paid off 3.1823 at Chatham. Sold to Edward Cohen (for £2,210) 13.10.1825.

Tamar Josiah & Thomas Brindley, Frindsbury

As built: 108ft 0½in, 89ft 7¾in x 30ft 9in x 9ft 0¾in. 450⁸³⁄₉₄bm. Draught 8ft 1in/11ft 5in.

Ord: 18.1.1813. K: 5.1813. L: 23.3.1814. C: 23.3–4.4.1814 at Chatham (for Ordinary); 5.11.1814 (for sea).

First cost: £9,571 to build, plus fitting £2,955 (for Ordinary) + £5,301 (for sea) in 1814.

Commissioned 10.1814 under Capt. Charles Sotheby. Fitted for sea at Chatham again (for £5,633) 2–4.1817; recommissioned 2.1817 under Capt. Thomas Toker, for Newfoundland station. On 20.11.1818 under acting (until 31.12.1818) Capt. John Gordon (Toker invalided 11.1818), then 12.1819 under Capt. Arthur Stow

(died 26.9.1820) at Jamaica, then Cmdr George Pechell (acting); on 22.11.1820 under Capt. Sir William Saltonstall Wiseman, at Jamaica, then in 9.1822 under Cmdr John Theed (died 11.1822), and on 25.11.1822 under Capt. Thomas Herbert; paid off 8.1823. Recommissioned 18.9.1823 under Capt. James John Gordon Bremer, for the East Indies and Northern Australia; arrived home 28.11.1827 and paid off into Ordinary at Plymouth. Coal hulk at Plymouth 3.1831 (or as early as 12.1827). Sold 3.1837.

Dee Jabez Bailey, Ipswich

As built: 108ft 1⅛in, 89ft 11in x 30ft 7in x 9ft 1in. 447³³⁄₉₄bm. Draught 8ft 8in/11ft 8in.

Ord: 18.1.1813. K: 5.1813. L: 5.5.1814. C: 10.5–29.10.1814 at Sheerness (for sea).

Commissioned 1.10.1814 under Capt. John William Andrew, for North Sea; to Hudson's Bay; paid off 24.1.1816. In 1816 under Capt. George Sartorius. In 9.1816 under Capt. Samuel Chambers, for Halifax station; paid off 12.1816. Sold to Thomas Pitman (for £1,320) 22.7.1819.

Towey Balthazar Adams, Bucklers Hard

As built: 108ft 2in, 90ft 0in x 30ft 7in x 9ft 0in. 447⁷²⁄₉₄bm. Draught 7ft 7in/12ft 5in.

Ord: 18.1.1813. K: 5.1813. L: 6.5.1814 (AO was mistakenly under the name of *Fowey*). C: 8.5–1.6.1814 at Portsmouth (for Ordinary); 9.1814–6.12.1814 (for sea).

Commissioned 10.1814 under Capt. Hew Stewart (–1816), for the East Indies. In 12.1816 under Capt. William Hill (–1817). Laid up at Plymouth 4.1819. BU completed at Plymouth 11.11.1822.

Menai Josiah & Thomas Brindley, Frindsbury

As built: 108ft 1in, 89ft 8⅞in x 30ft 8in x 9ft 0½in. 448⁸⁶⁄₉₄bm. Draught 7ft 11in/11ft 4in.

Ord: 18.1.1813. K: 6.1813. L: 5.4.1814. C: 5.4–4.5.1814 at Chatham (for Ordinary), then 10.1814–8.12.1814 (for sea).

First cost: £9,572 to build, plus fitting £3,156 (for Ordinary) + £5,589 (for sea) in 1814.

Commissioned 4.10.1814 under Capt. Watkin Owen Pell (–1815), on the Irish Station, later for North America; paid off 2.1817. Middling Repair and fitted for sea at Chatham (for £17,001) 7.1817–7.1819; recommissioned 26.4.1819 under Capt. Fairfax Moresby, for guard at St Helena; paid off 9.1823? On 21.10.1823 under Capt. Houston Stewart, for Halifax station; paid off into Ordinary at Woolwich 12.1826. Fitted for sea at Woolwich (for £4,428) 1–4.1827; recommissioned 7.1.1827 under Capt. Sir

Michael Seymour, for South America; on 12.9.1827 under Capt. Thomas Bourchier; paid off into Ordinary at Portsmouth 9.1829. Receiving ship at Portsmouth 1829–50; target ship for HMS *Excellent* 11.1852. BU at Portsmouth 4.1853.

Tyne Robert Davy, Topsham

As built: 108ft 0in, 89ft 9¼in x 30ft 6⅝in x 9ft 0¼in. $445\frac{67}{94}$bm. Draught 7ft 7in/11ft 3in.

Ord: 18.1.1813. K: 8.1813. L: 20.5.1814. C: 19.6–16.7.1814 at Portsmouth (for Ordinary): 9.11.1814 (for sea).

Commissioned 11.1814 under Capt. John Harper; sailed for the East Indies 12.11.1814. In 1815 under Capts Charles Allen and Robert Campbell and in 11.1816 under Capt. John Curran, then paid off. Recommissioned 24.6.1817 under Capt. Gordon Thomas Falcon, for South America; paid off 1820. Small Repair and fitted for sea at Portsmouth 1–5.1821; recommissioned 2.1821 under Capt. James White, for the West Indies. In 1822 under Capt. William Godfrey, then 6.5.1822 under Capt. John Edward Walcott; boats (with those of *Thracian*) took pirate schooner *Zaragozana* at Malta 31.3.1823. On 16.6.1823 under Capt. John Walter Roberts; paid off 1.1824. Sold at Portsmouth to Thomas Pitman (for £1,820) 27.1.1825.

Wye (Benjamin) Hobbs & (George) Hellyer, Redbridge (Southampton)

As built: 108ft 1⅞in, 89ft 11½in x 30ft 6¾in x 9ft 0in. $446\frac{89}{94}$bm. Draught 8ft 6in/11ft 6in.

Ord: 18.1.1813. K: 9.1813. L: 17.8.1814. C: 18–23.8.1814 at Portsmouth (for Ordinary); 10.7.1815 (for sea).

First cost: £9,720 to build.

Commissioned 5.1815 under Capt. Andrew Green, as flagship of Rear-Adm. Sir Thomas Fremantle, in the Channel Islands. In 1816 under Capt. James Lillicrap, then 2.1816 Capt. John Harper, on Halifax station; in 1818 under Capt. George Wickins Willes, then 5.1820 Capt. Peter Fisher; paid off 1823. Fitted as Convict hospital ship and floating breakwater at Sheerness 9.1825; to Chatham 4.1834; by 1850 convict hospital ship at Chatham. BU at Deptford 10.1852.

Tees William Taylor, Bideford

As built: 108ft 0in, 90ft 1in x 30ft 8½in x 9ft 0in. $451\frac{80}{94}$bm.

Ord: 18.1.1813. K: 10.1813. L: 17.5.1817. C: 6–7.1817 at Plymouth (for Ordinary); 30.5.1818 (for sea).

First cost: £9,720 to build.

Commissioned 2.1818 under Capt. George Rennie (–1821), for St Helena; sailed for the East Indies 6.1.1821. In 7.1821 under Capt. Thomas Coe; in Burma War 1824–25. In 4.1825 under Cmdr (Capt. 25.7.1825) Frederick Marryat; paid off 1.1826. Fitted as Church ship at Chatham 8–10.1826, and lent to the Church Society at Liverpool. Sank at her moorings 1872, and sold at Liverpool to Henry Robinson (for £265) 28.6.1872.

(B) Vessels acquired from 1 January 1817

Under the re-classing which took effect in 2.1817, the smallest rated ships – the Sixth Rate – comprised vessels with 24 guns and above but less than 36 guns. Typically, the post-war Sixth Rates were built as 28-gun; however, most were reduced to 26 guns in 1839. During the 1820s, a variety of designs from different sources were experimentally tried, with differing hull forms being evaluated to try and achieve a fast cruising performance. The French term 'corvette' came to be used for these vessels, which were also referred to (disparagingly) as 'donkey frigates'.

ATHOLL Class 28 guns. The prototype was ordered in late 1816, but is included in Section (b) as the first of a new species of 28-gun frigate. The design by the Surveyors of the Navy was approved on 20 May 1817. While no record of an order for the similar *Niemen* can be found before April 1818, the *Atholl*'s order in 1816 stated that she was 'to be built of larch for an experiment to be compared with the *Niemen*', indicating the intention to build the second vessel was then in existence. Certainly another six were ordered in April 1818, with a further batch of nine ordered in June 1819 (although three of these were never built to this design); three of the latter were built from teak by the East India Company. All this class were originally ordered as Sixth Rates but later re-classed as corvettes. On 9 March 1826 the last three were re-ordered as 20-gun sloops; the *Andromeda* (renamed *Nimrod* while on the stocks), constructed of African timber, was raséed on the stocks and completed as such a sloop, while the two ships at Plymouth were subsequently cancelled.

Dimensions & tons: 113ft 8in, 94ft 8¾in x 31ft 10in oa (31ft 6in for tonnage) x 8ft 9in. $499\frac{91}{94}$bm.

As Sixth Rates: Men: 175. Guns: UD 20 x 32pdr carronades; QD 6 x 18pdr carronades, Fc 2 x 9pdrs.

As Sloops: Men 130. Guns: UD 18 x 32pdrs (25cwt), 2 x 9pdrs.

Atholl Woolwich Dyd (built of larch)

As built: 113ft 10in, 94ft 10in x 31ft 11in oa (31ft 7in for tonnage) x 8ft 9in. $503\frac{16}{94}$bm. Draught 8ft 8in/11ft 3in.

Ord: 27.10.1816. K:11.1818. L: 23.11.1820. C: 9.2.1821 (for sea).

First cost: £14,589 (with fitting, £18,239).

Commissioned 20.11.1820 under Capt. Henry Bourchier, for Halifax; paid off 1822? Recommissioned 14.8.1824 under Capt. James Arthur Murray, for African station; to Pacific 1827, then to East Indies; paid off 19.10.1827. Between Very Small and Small Repair at Portsmouth (for £8978) 11.1827–7.1829; recommissioned 4.1829 under Capt. Alexander Gordon, for Africa; on 4.1831 under Cmdr Edward Webb. Fitted as troopship at Woolwich (for £7,019) 5–11.1832, under AO 25.7.1832; recommissioned 10.1832 under Master Andrew Kearley, as troopship. On 2.1838 under Master C.P. Bellamy; by 1842 reduced to 2 guns. Between Small and Middling Repair at Portsmouth, and fitted as troopship again (for £12,926) 4.1842–9.1845; recommissioned under E. J. P. Pearn. Fitted as storeship at Sheerness (for £3,438) 4.1850–1.1851; recommissioned 18.11.1850 under Cmdr William Alfred Rumbulow Pearse, for West coast of Africa. Fitted as depot ship at Portsmouth (for Greenock) 10–11.1854; recommissioned 8.11.1854 as receiving ship (4 guns) under Cmdr George Sayer Boys (died 13.4.1859); on 7.5.1859 under Cmdr Edmund Wilson. Fitted for reserve at Plymouth 9.1861; BU there 4.1863.

Niemen Woolwich Dyd (built of Baltic fir)

As built: 113ft 8in, 94ft 8½in x 31ft 11in oa (31ft 7in for tonnage) x 9ft 1in. $502\frac{48}{94}$bm. Draught 9ft 0in/11ft 6in.

Ord: 30.4.1818. K: 7.1819. L: 23.11.1820. C: 2.1821 (for sea).

First cost: £14,622 (plus £677 for masts & yards).

Commissioned 11.1820 under Capt. Edward Reynolds Sibly, for Halifax; paid off 6.1824. Recommissioned 4.6.1824 under Cmdr Provo William Parry Wallis, for Cape of Good Hope and in 1825 to Halifax; Wallis then led the first Experimental Squadron, consisting of *Niemen*, *Champion*, *Orestes*, *Pylades*, *Calliope* and *Algerine*. In 12.1826 under Cmdr W. P. Canning, then on 10.3.1827 under Capt. Charles Simeon, still at Halifax; returned

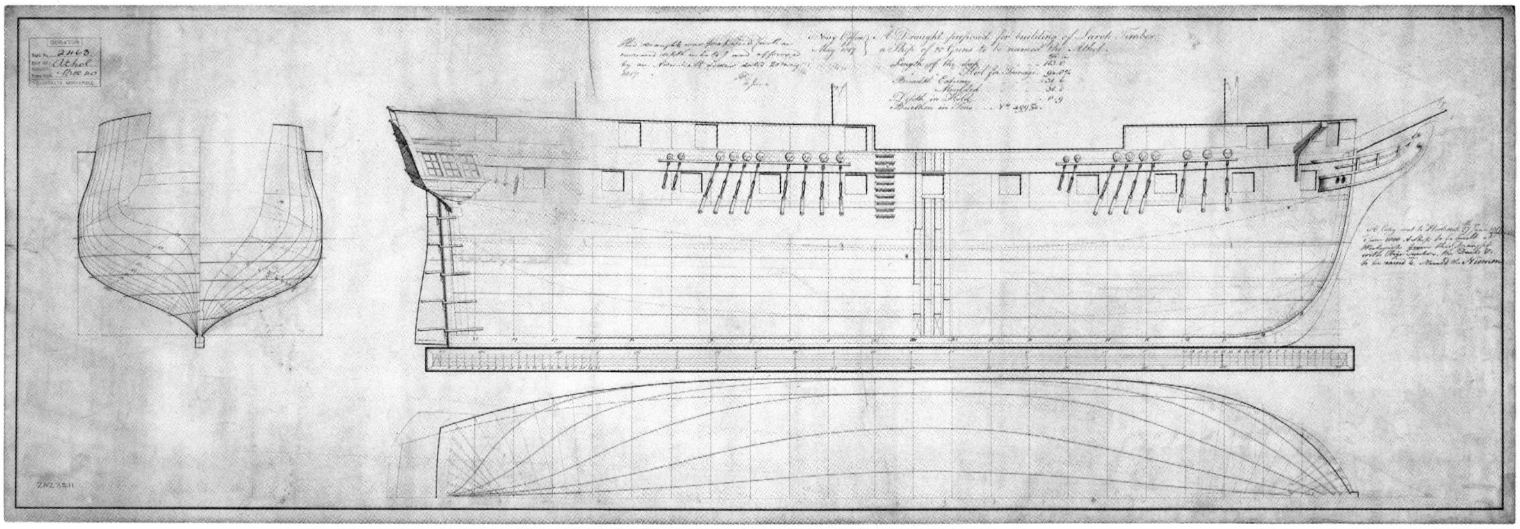

The sheer draught for the *Atholl* Class, dated May 1817. The first of a new type of 28-gun frigate, they were much sharper in hull form than previous Sixth Rates, evidence that Seppings pioneered a move away from the fuller hull forms of wartime frigates, a move that would subsequently be taken to an extreme by his successor, Symonds. After the name-ship, the rest of the class were modified slightly with increased depth in hold resulting from raising the deck by 4in. The class was the subject of comparative trials with construction using less traditional timbers – larch for *Atholl*, 'Riga timber' for *Niemen*, 'African timber' for *Nimrod*, with three other ships built of teak, and the remainder of oak. *(NMM J6605)*

home 5.1827 to pay off. Surveyed 6–8.1827 at Portsmouth, and BU completed there in 1.1828.

Ranger Portsmouth Dyd

As built: 113ft 8in, 94ft 6⅝in x 31ft 11in oa (31ft 7in for tonnage) x 8ft 9in. 501$^{64}/_{94}$bm. Draught 9ft 6in/12ft 0in.

Ord: 30.4.1818. K: 1.1819. L: 7.12.1820. C: 14.12.1820 (for Ordinary).

First cost: £20,303 (including fitting)

Commissioned 2.1822 under Capt. Peter Fisher; fitted for sea (for £6,435) 3.1822–12.6.1822. On 3.1825 under L. H. F. Thyme, for South America; arrived Portsmouth 12.6.1828 to pay off. Small Repair and fitted for sea at Portsmouth 7–10.1828. Recommissioned 2.10.1828 under Capt. William Walpole; sailed 3.1.1829 for Jamaica; returned home 4.1832 to pay off. Ordered to be reduced to corvette 23.5.1832, but found to be decayed; instead sold to J. Jackson (for £2,520) 11.1832 to BU.

Rattlesnake Chatham Dyd

As built: 113ft 9½in, 94ft 6⅞in x 31ft 11¼in oa (31ft 7¼in for tonnage) x 8ft 9¼in. 502$^{42}/_{94}$bm. Draught 9ft 8in/11ft 10in.

Ord: 30.4.1818. K: 8.1819. L: 26.3.1822. C: 23.4.1822 (for Ordinary).

First cost: £15,825 (including fitting for Ordinary).

Commissioned 28.11.1824 under Capt. Hugh Patton, after fitting for foreign service (for £8,257) 1.1824–8.5.1824; sailed for Jamaica. In 11.11.1825 under Capt. John Leith; paid off 9.1827. Fitted for sea at Woolwich (for £4,462) 9–11.1827; recommissioned 7.9.1827 under Capt. Charles Orlando Bridgeman, for the Mediterranean; at Grabusa 31.1.1828. In 5.1830 under Cmdr Sir Thomas Sabine Pasley (acting). On 4.11.1830 under Capt. Charles Graham, for South America; paid off 10.11.1833. Recommissioned 12.1834 under Capt. William Hobson, for the East Indies. Fitted as troopship at Portsmouth (for £5,134) 3–10.1839; recommissioned 1839 under Master W. Brodie; in 1st China War 1839–42. On 6.1841 under Master James Sprent (when Brodie died); in Yangtze operation 7.1842; returned home 5.1844 and laid up at Portsmouth. Refitted as a Survey vessel at Portsmouth (for £12,032) 26.5.1846–12.1846. Recommissioned 20.9.1846 under Capt. Owen Stanley (died 13.3.1850) and undocked 2 days later as Survey vessel for East Indies (New Guinea); on 7.4.1850 under Capt. James Cockburn; home to pay off 10.1850. Refitted for the Arctic at Sheerness and Portsmouth (for £6,214) 12.1852–2.1853. In 12.1852 under Capt. Henry Trollope, as Survey vessel in Bering Sea; paid off 12.1855 at Chatham. BU at Chatham completed 13.1.1860.

North Star Woolwich Dyd

As built: 113ft 9in, 94ft 9¾in x 31ft 10¼in oa (31ft 6¼in for tonnage) x 8ft 9in. 501$^{7}/_{94}$bm. Draught 9ft 9in/11ft 8in.

Ord: 30.4.1818. K: 4.1820. L: 7.2.1824. C: 12.1825–26.5.1826 (for foreign service).

First cost: £17,155 to built, plus £8,426 fitting.

Commissioned 23.12.1825 under Septimus Arabin, for African station; took several Spanish and Brazilian slavers; paid off 1829. Small Repair and fitted for sea at Portsmouth (for £11,222) 8.1829–4.1830. Recommissioned 12.12.1828 under Capt. Lord William Paget, for the West Indies and Halifax; paid off summer 1833. Recommissioned 26.3.1834 under Capt. Octavius Harcourt, for South America. In 8.3.1837 under Capt. Lord John Hay; paid off 1840. Small Repair and fitted for sea at Portsmouth (for £8,710) 10.1840–11.1841. Recommissioned 30.8.1841 under Sir James Everard Home, for the East Indies; in 1st Anglo-Chinese War; Yangtze operations 7.1842; New Zealand War 1846; at Ruapekapeki 11.1.1846; paid off 1846. Fitted as storeship (22 guns) for Arctic Seas at Sheerness and Woolwich (for £8,163) 3–5.1849; bow lines with galvanised iron for icebreaking; recommissioned 4.1849 under Cmdr James Saunders as 22-gun storeship for Arctic service; in 1852 under Cmdr William John Samuel Pullen (–1854). BU completed at Chatham 15.3.1860.

Tweed Portsmouth Dyd

As built: 113ft 8in, 94ft 7in x 31ft 10½in oa (31ft 6½in for tonnage) x 8ft 9in. 501$^{79}/_{94}$bm. Draught 9ft 6in/11ft 10½in.

Ord: 30.4.1818. K: 12.1820. L: 14.4.1823. C: 14.5.1823 (for Ordinary), then 12.1823–12.4.1824 for sea.

First cost: £15,942 (£22,183 fitted for sea).

Commissioned 28.11.1823 under Capt. Frederick Hunn, for South America, Irish and the West Indies stations. In 5.1827 under Lord H. T. S. Churchill, for the Cape of Good Hope; paid off 1.1831. Cut down (raséed) to 20-gun 'corvette' (by AO

18.3.1831), at Portsmouth 3–8.1831; recommissioned 5.1831 under Cmdr Allen Bertram, for North America and the West Indies. On 4.4.1835 under Cmdr Thomas Maitland, at Lisbon. On 17.1.1837 under Capt. Frederick Thomas Pelham; paid off autumn 1838. Reduced to sloop 1838, with 18 x 32pdr carronade and 2 x 12pdrs. Recommissioned 10.1840 under Hugh D. C. Douglas, for North America and the West Indies; paid off 1842. Fitted for sea at Plymouth (for £5,584) 6–7.1848; recommissioned 7.6.1848 under Capt. Lord Francis John Russell, for South America;. paid off into Ordinary 7.1.1852. Sold to Willson & Co. (for £1,125) 10.9.1852 to BU.

Talbot Pembroke Dyd

As built: 113ft 8in, 94ft 7¾in x 31ft 10¼in oa (31ft 6¼in for tonnage) x 8ft 9in. 500¹⁸⁄₉₄bm. Draught 9ft 6in/12ft 0in.

Ord: 30.4.1818. K: 3.1821. L: 9.10.1824. Sailed 5.11.1824 for Plymouth. C: 7.11.1824–21.12.1825.

First cost: £14,415 to build, +£9,101 fitting at Plymouth.

Commissioned 21.9.1825 under Capt. Frederick Spencer, for the Mediterranean; at Battle of Navarino 20.10.1827; arrived Portsmouth to pay off 12.1828. Defects made good at Portsmouth (for £10,727) 6.1829–7.1830. Recommissioned 4.1830 under Capt. Richard Dickinson, for the Cape of Good Hope; home to Plymouth 5.1834. On 31.5.1834 under Capt. Follett Walrond Pennell, for South America; paid off 7.1837. On 12.3.1838 under Capt. Henry John Codrington, for the Mediterranean; operations on the coast of Syria 1840; at Bombardment of Acre 3.11.1840. On 13.3.1841 under Capt. Robert Fanshawe Stopford, still in the Mediterranean; home to pay off 4.1842. Recommissioned 20.4.1842 under Capt. Sir Thomas Raikes Trigge Thompson, for South America and the Pacific; paid off 26.3.1847 at Portsmouth. Fitted as storeship at Deptford and Woolwich (for £1,956) 2–5.1854, to convey stores to Beechey Island to relieve Edward Belcher's Arctic expedition; recommissioned 1.3.1854 at Deptford under Cmdr Robert Jenkins for this role; paid off 19.1854. Lent to Ordnance Dept as (powder) depot ship at Woolwich 2.1855. On sales list by AO 11.1.1895; sold to Campbell P. Ogilvie of Suffolk 5.3.1896 to BU.

Rainbow Chatham Dyd

As built: 113ft 9½in, 94ft 7½in x 31ft 11¼in oa (31ft 7½in for tonnage) x 8ft 8½in. 502⁶⁸⁄₉₄bm. Draught 9ft 6in/11ft 10in.

Ord: 30.4.1818. K: 4.1822. L: 20.11.1823. C: 17.1.1824 (for Ordinary).

First cost: £14,294 (including fitting for Ordinary).

Commissioned 30.7.1825 under Capt. Henry John Rous, for the East Indies; fitted for foreign service (for £7,493) 8.1825–6.11.1825: paid off 25.8.1829 at Portsmouth. Very Small Repair there (for £13,000) 12.1829–11.1830. Recommissioned 23.8.1830 under Capt. Sir John Franklin, for the Mediterranean; paid off 1.1834. Recommissioned 7.2.1834 under Capt. Thomas Bennett, for North America and the West Indies; paid off 3.1838. Sold to Messrs Buck (for £1,210) 8.11.1838.

Alligator East India Co., Cochin. Teak-built.

As built: 113ft 8in, 94ft 8⅜in x 31ft 10in oa (31ft 6in for tonnage) x 9ft 0in. 499⁸⁶⁄₉₄bm.

Ord: 5.6.1819. K: 11.1819. L: 29.3.1821. Arrived Woolwich 11.10.1821. C: 4.1822–3.9.1822 (for sea).

First cost: £10,999 (including £929.8.5½d worth of stores supplied by Crown), plus £8,955 fitting at Woolwich.

Commissioned ?5.1821 at Cochin under Lieut James Wilkie, for voyage to UK; arrived Portsmouth 5.10.1821 and then to Woolwich 11.10.1821. Recommissioned 16.5.1822 at Woolwich under Capt. Thomas Alexander (died 11.11.1825), for the East Indies; Burma War 1825–26; in 11.1825 (acting) 27.4.1826 (confirmed) under Capt. Henry Ducie Chads, for the East Indies; paid off 3.1.1827. Recommissioned 3.1827 under Capt. W. P. Canning (died 24.9.1828 at Madeira), for Halifax; returned home 1.1828 for Plymouth-based home service; in 9.1828 under Lieut Julius James Farmer Newell; paid off 20.11.1828. Recommissioned 20.11.1828 under Capt. Charles Philip Yorke, for the Mediterranean; returned Portsmouth to pay off 9.1831. Recommissioned 3.9.1831 under Capt. George Robert Lambert, for the East Indies; paid off 27.8.1835. Recommissioned 12.7.1837 under Capt. James John Gordon Bremer, for Australia; visited New Holland and founded the settlement at Port Essington; during 1840 under Acting Capt. Patrick John Blake, for 1st Anglo-Chinese War; on 5.3.1840 under Acting Capt. Augustus Leopold Kuper then (?) Cmdr Samuel Perkins Pritchard, still in East Indies during 1st Anglo-Chinese War. Depot ship at Trincomalee 6.1841. Recommissioned 31.1.1842 under Richard Browne, Master, as troopship; on 27.11.1843 under Joseph King, Master, in same role. Hulked as seamen's hospital at Hong Kong 12.1846. Sold there to Ton Ping (for $HK 5,000) 30.10.1865.

Termagant East India Co., Cochin. Teak-built.

As built: 113ft 8in, 94ft 8¾in x 31ft 10in oa (31ft 6in for tonnage) x 9ft 0in. 499⁹¹⁄₉₄bm.

Ord: 5.6.1819. K: 3.1820. L: 15.11.1822. C: 10.7–20.8.1823 at Portsmouth (for Ordinary), then 23.6–16.7.1824 'for the conveyance of ambassadors'.

First cost: £10,999 to build, plus £202 fitting for Ordinary, plus £10,409 fitting in 1824.

Commissioned 30.7.1822 at Cochin under Lieut Robert Wallace Dunlop, for voyage to UK; arrived Portsmouth 7.7.1823. In 7.1823 under Capt. Lord Henry Frederick Thynne (nominally appointed from 30.7.1822), for the East Indies. Renamed *Herald* 15.5.1824, rated as yacht. Recommissioned 31.5.1824 under Cmdr Henry John Leeke; to St Petersburg, then to West Indies, home from Havana to England, them to Quebec and finally to Malta. Recommissioned 27.5.1826 under Capt. Sir Augustus William James Clifford, for the Duke of Devonshire's Embassy to Russia. In 11.1826 under Cmdr Henry Eden, then in 4.1827 under Cmdr Edward William Curry Astley and 7.4.1829 Cmdr George Berkeley Maxwell; to St Petersburg, later to Cartagena, to Quebec and then home to pay off in 1830. Fitted at Portsmouth for sea (for £3,852) 4–7.1830. On 20.11.1830 under Capt. Robert Gordon, at Portsmouth; paid off 1.1831. Fitted for sea at Portsmouth (for £7,229) 11.1837–8.1838. Recommissioned 24.5.1838 under Capt. Joseph Nias, for the East Indies and China; 1st Anglo-Chinese War; in Canton operations 2–3.1841; Shaming Creek 21.5.1841; paid off 1842 at Chatham. Very Small Repair and fitted as a surveying vessel (reduced to 8 guns) at Sheerness (for £8,558) 7.1844–6.1845. Recommissioned 8.2.1845 under Capt. Henry Kellett, for surveying in the Pacific; conducted (with *Pandora*) a survey of the British Columbia coast (to resolve Oregon boundary dispute with USA); paid off 1848. Recommissioned 18.2.1852 under Capt. Henry Mangles Denham, for surveying in Fiji. Recommissioned 1.4.1857 in same role; paid off 1859. Chapel ship at Shoreham 9.1861. Sold there to H. Castle & Son (for £1,635) by AO 28.4.1862 for BU at Charlton.

Samarang East India Co.,Cochin. Teak-built.

As built: 113ft 8in, 94ft 8¾in x 31ft 10in oa (31ft 6in for tonnage) x 9ft 0in. 499⁹¹⁄₉₄bm.

Small and economical to run, Sixth Rates saw a lot of service in the post-war years. Although sometimes contemptuously called 'donkey frigates', they provided seagoing experience for many young post captains. One of the busiest was the teak-built *Samarang* of 1822, which was reconstructed as a flush-decked ship in 1841-43, as shown in this very detailed model, for service in the East Indies. A print showing the ship heaved down at Sarawak, published in 1848, confirms the details of the model. (*NMM F8831-001*)

Ord: 5.6.1819. K: 3.1821. L: 1.1.1822. C: 2.1824–7.6.1824 at Portsmouth (for sea).

First cost: £10,999 (including £953.13.2½d worth of stores supplied by Crown), plus £8,588 fitting Portsmouth.

Commissioned 1.1.1822 at Cochin under Cmdr John Norman Campbell, for voyage to UK; arrived Portsmouth 12.7.1822 and paid off that month into Ordinary. Recommissioned 4.2.1824 under Capt. Sir William Saltonstall Wiseman., for Halifax. On 28.11.1824 under Capt. David Dunn, for the Cape of Good Hope. On 15.11.1826 under Capt. William Fanshawe Martin, for the Mediterranean; fitted for sea at Woolwich (for £7,348) 4–8.1828; fitted for sea at Portsmouth 4.1831, then paid off 6.1831. Recommissioned 3.6.1831 under Capt. Charles Henry Paget, for South America; paid off 1.1835. Recommissioned 25.10.1836 under Capt. William Broughton, for the coast of Spain (during the Carlist disputes), then to South America. On 31 October 1839 under Capt. James Scott for South America; later to East Indies; in 1st Anglo-Chinese War; Chuenpea Forts 1841; returned home 8.1841 and paid off into Ordinary. Refitted at Portsmouth (for £9,289) 12.1841–1.1843. Recommissioned 18.11.1842 under Capt. Sir Edward Belcher, for the East Indies; action against Malay pirates 3–4.6.1844; paid off 18.1.1847 at Sheerness. Fitted at Chatham (for £2,411) as guard ship for Gibraltar 1–5.1847, then in that role to 1880. Sold at Gibraltar (for £674) 10.1883.

Nimrod (ex-*Andromeda*, renamed 10.5.1827) Deptford Dyd (built of African timber)

As built: 113ft 10⅜in, 94ft 11⅛in x 31ft 10¼in oa (31ft 6¼in for tonnage) x 8ft 9¼in. 501⁶⁴⁄₉₄bm. Draught 9ft 5in/10ft 11in.

Ord: 5.6.1819. K: 10.1821. Raséed while building to a 20-gun sloop. L: 26.8.1828. C: 11.12.1828 (for sea).

First cost: £16,254 (£22,297 including fitting).

Commissioned 18.9.1828 under Cmdr Samuel Radford, for Cork; paid off 2.3.1832. Recommissioned 9.4.1832 under Cmdr Lord Edward Russell, for Lisbon; on 27.8.1833 under Cmdr John

McDougall (Russell invalided), in same role. On 21.9.1835 under Cmdr John Fraser, for North America and the West Indies; paid off 22.8.1839. Recommissioned 2.12.1839 under Cmdr Charles Anstruther Barlow, for the East Indies; on 8.6.1841 under Cmdr Joseph Pearse, for the 1st Anglo-Chinese War, then 28.6.1841 under Cmdr Frederick Henry Hastings Glasse; paid off 22.11.1844. Small Repair and fitted for sea at Plymouth (for £6,462) 6.1845–2.1846. Recommissioned 26.11.1845 under Cmdr James Richard Dacres, for African coast; died 14.2.1848 at Mozambique. On 23.5.1848 under Cmdr Thomas Belgrave, at Cape of Good Hope. Fitted at Plymouth as a Coal hulk 2.1853. Renamed *C.1* in 1895 then *C.76*. Sold to Hamley & Son 9.7.1907.

Success Pembroke Dyd

As built: 113ft 8in, 94ft 7¾in x 31ft 11⅛in oa (31ft 7⅛in for tonnage) x 8ft 9in. 503⁷⁸⁄₉₄bm. Draught 9ft 4in/12ft 1in.

Ord: 5.6.1819. K: 7.1823. L: 30.8.1825. Sailed 14.9.1825 for Plymouth. C: 16.9.1825–3.6.1826 at Plymouth.

First cost: £14,310 to build, + £9,594 fitting for sea.

Commissioned 25.1.1826 under Capt. James Stirling, for the East Indies and Australia; established settlement at Raffles Bay in Torres Strait; paid off 1.1828. Recommissioned 21.2.1828 under Capt. John Fitzgerald Studdert, for the East Indies. On 6.8.1828 under Capt. William Clarke Jervoise; paid off 11.1831. Fitted as receiving ship at Portsmouth 1–4.1833. BU there 6.1849.

Crocodile Chatham Dyd

As built: 113ft 9¾in, 94ft 8½in x 31ft 10½in oa (31ft 6½in for tonnage) x 8ft 0½in. 501¹⁷⁄₉₄bm. Draught 9ft 10in/11ft 6in.

Ord: 5.6.1819. K: 12.1823. L: 28.10.1825. C: 10.12.1825 (for Ordinary), 5.1828–27.8.1828 (for sea).

First cost: £15,151 (including fitting for Ordinary) + fitting for sea £7,805.

Commissioned 7.6.1828 under Capt. John William Montagu, for the East Indies; paid off 1832. Recommissioned 10.1837 under Capt. James Polkinghorne (died 9.1.1839), for North America and the West Indies. On 30.1.1839 under Capt. Alexander Milne; took

Spanish slaver *Mercedita*; paid off 11.1840. Recommissioned 3.1841, still under Milne; paid off 11.1841. Fitted as troopship at Portsmouth (for £5,687) 12.1841–5.1842; classed as troops 2.1842, and recommissioned 9.3.1842 at Plymouth under Thomas Elson, Master; took part of 52nd Regiment to Canada 7.1842, and brought other troops home from Canada and the West Indies in 1843; paid off 1843. Recommissioned 25.6.1845 under Capt. John Balfour Maxwell, as flagship of Rear-Adm. Hugh Pigot at Queenstown. On 12.5.1846 under Cmdr Gower Lowe, on particular service. Recommissioned 27.11.1846 at Portsmouth under Lieut Cmdr Samuel Rosser Protheroe, as receiving ship at Cork. In 4.1847 under Cmdr George Augustus Bedford, as flagship of Rear-Adm. Manley Hall Dixon. Fitted as receiving ship at Deptford for Tower of London 7–8.1850. On 14.8.1850 under Lieut Cmdr William Greet, as receiving ship off the Tower of London; and same of 16.1.1854 and 18.1.1858. Sold to Castle & Son (for £1,015) by AO 22.11.1861 for BU at Charlton.

Alarm Pembroke Dyd
    Ord: 5.6.1819. K: — . Re-ordered as *Tyne* Class 9.3.1826, then as *Conway* Class 1828 (see below).
Daphne Plymouth Dyd
    Ord: 5.6.1819. K: — . Re-ordered as 20-gun sloop 9.3.1826, but cancelled 1.11.1832 (unstarted).
Porcupine Plymouth Dyd.
    Ord: 5.6.1819. K: — . Re-ordered as 20-gun sloop 9.3.1826, but cancelled 1.11.1832 (unstarted).

VOLAGE 28 guns. Designed and built by the 'superior class of Portsmouth shipwright apprentices', and ordered just 11 days after the final batch of *Atholl* Class.
    Dimensions & tons: 113ft 10in, 96ft 8½in x 31ft 10in x 8ft 9in. 521²⁶⁄₉₄bm (516 as completed).
    Men: 175. Guns: UD 20 x 32pdr carronades, QD 6 x 18pdr carronades, Fc 2 x 9pdrs.
Volage Portsmouth Dyd.
    As built: 113ft 10in, 95ft 7¾in x 32ft 2in oa (31ft 10in for tonnage) x 8ft 9in. 515⁵¹⁄₉₄. Draught 9ft 9in/12ft 2in.
    Ord: 16.6.1819. K: 8.1819. L: 20.2.1825. C: 26.1.1826 (for sea).
    First cost: £18,909 (£24,794 including fitting).
    Commissioned 13.9.1825 under Capt. Richard Saunders Dundas, for Australia; in 1.1826 under Capt. Thomas Bourchier, then 3.1727 under Capt. Robert Tate and 12.9.1827 under Capt. Michael Seymour, on the South American station; paid off early 1829. Made good defects at Portsmouth (for £9,993) 4–9.1829. Recommissioned 9.10.1829 under Capt. Charles Abbot (Lord Colchester); to Brazil and thence home in 4.1831, then enforcing embargo off Dutch coast in Winter 1832. Fitted for sea at Portsmouth (for £5,396) 3–6.1833. Recommissioned 17.4.1833 under Capt. George Bohan Martin, for Lisbon and thence to the Mediterranean. In 1836 under Capt. P. Richardson, still in the Mediterranean, then 27.11.1837 under Capt. Henry Smith, for the East Indies; occupation of Aden 18.1.1839. On 3.6.1840 under Capt. (acting) George Elliot, then under Capt. William Warren; in Canton River 1840. On 30.8.1841 under Capt. Sir William Dickson, for North America and the West Indies; paid off 1845. Fitted as survey vessel (disarmed) at Portsmouth (for £10,167) 4.1846–4.1847; attached to Royal Naval College 1847. In 3.1849 under Capt. Thomas Graves, for the Mediterranean. Fitted as powder depot at Sheerness (for £3,624) 3–6.1855; recommissioned as storeship for the Baltic 1855, under J. C.

Hutchins. Temporary loan to War Dept. as a powder depot on the Medway 19.10.1864; returned by War Dept. 10.2.1871, then lent to them again 19.9.1871 as a floating depot for gun cotton at Upnor. BU completed at Chatham 12.12.1874.

TYNE Class 28 guns. Sir Robert Seppings design 1825. The *Alarm*, originally ordered as an *Atholl* class vessel, was re-ordered to the same design on 9 March 1826, but was again re-ordered to the *Conway* design in 1828.
    Dimensions & tons: 125ft 0in, 106ft 6in x 32ft 10in oa (32ft 6in for tonnage) x 9ft 6in. 598¹⁷⁄₉₄bm.
    Men: 175. Guns: UD 20 x 32pdr carronades, QD 6 x 18pdr carronades, Fc 2 x 9pdrs.
Tyne Woolwich Dyd
    As built: 125ft 0in, 106ft 7⅛in x 32ft 10½ oa (32ft 6½in for tonnage) x 9ft 9in. 600³⁄₉₄bm. Draught 10ft 11in/13ft 3in.
    Ord: 9.6.1825. K: 11.1825. L: 30.11.1826. C: 5.3.1827 (for sea).
    First cost: £19,663 (£26,128 including fitting).
    Commissioned 11.1826 under Capt. James Kearney White (died 2.3.1828), for Halifax. On 17.5.1828 under Capt. Sir Richard Grant, for Halifax; paid off 6.1830. Recommissioned 21.10.1830 under Capt. Charles Hope; sailed for South America 30.9.1832; paid off 1.1834. On 30.1.1834 under Capt. Henry John Chetwynd, Viscount Ingestrie, for the Mediterranean; paid off 4.1837. Recommissioned 5.9.1837 under Capt. John Townshend, in the Mediterranean; paid off end 1841. Between a Small and Middling Repair, and fitted for sea (for £8,325 + £1,476) at Portsmouth 12.1841–6.1843; recommissioned 18.4.1843 under Capt. William Nugent Glascock; paid off 1.1847. Fitted as provision depot ship at Plymouth 3.1847; fitted as a 22-gun storeship at Plymouth (for £2,101) 10–12.1850; then in Ordinary at Chatham; in 3.1855 under Cmdr Peter Wellington. Sold to Castle & Beech by AO 17.2.1862 (for £1,200) for BU.

CONWAY Class (1826) 28 guns. Sir Robert Seppings design of 13 November 1826, increased by 1ft in breadth from *Tyne* Class design. Seppings again recast the design in February 1827, which was approved in July 1828, and the *Imogene* and *Conway* were laid down to this design. The Admiralty approved further changes on 23 November 1829, and in all the breadth was increased by nearly another 1ft while building. The *Alarm* had previously been ordered to *Tyne* design, and was re-ordered in 1828 to *Conway* design.
    Dimensions & tons: original design 125ft 0in, 106ft 0in x 33ft 6in x 9ft 6in. 632⁷¹⁄₉₄bm.
    as amended 125ft 0in, 106ft 0in x 34ft 5in (34ft 0in for tonnage) x 9ft 6in. 651⁷⁴⁄₉₄bm.
    Men: 175. Guns: UD 20 x 32pdr carronades, QD 6 x 18pdr carronades, Fc 2 x 18pdrs.
Imogene (ex-*Pearl*, renamed 23.2.1826) Pembroke Dyd
    As built: 125ft 0in, 105ft 2¾in x 34ft 8in (34ft 4in for tonnage) x 10ft 0½in. 659⁷⁴⁄₉₄bm. Draught 10ft 5in/13ft 7in.
    Ord: 9.6.1825. K: 11.1829. L: 24.6.1831.
    Commissioned 7.1831 under Capt. Price Blackwood, for the East Indies and China; in the Canton River 7.9.1833; paid off 1834. Recommissioned 7.6.1836 under Capt. Henry William Bruce, on the South American station; paid off end 1839. Burnt by accident at Plymouth 27.9.1840.
Conway Chatham Dyd
    As built: 125ft 4in, 105ft 9⅜in x 34ft 5½in (34ft 0½in for tonnage) x 10ft 0½in. 652¹⁄₉₄bm. Draught 10ft 11in/13ft 9in.
    Ord: 9.6.1825. K: 12.1829. L: 2.2.1832. C: 11.5.1832 (for sea).

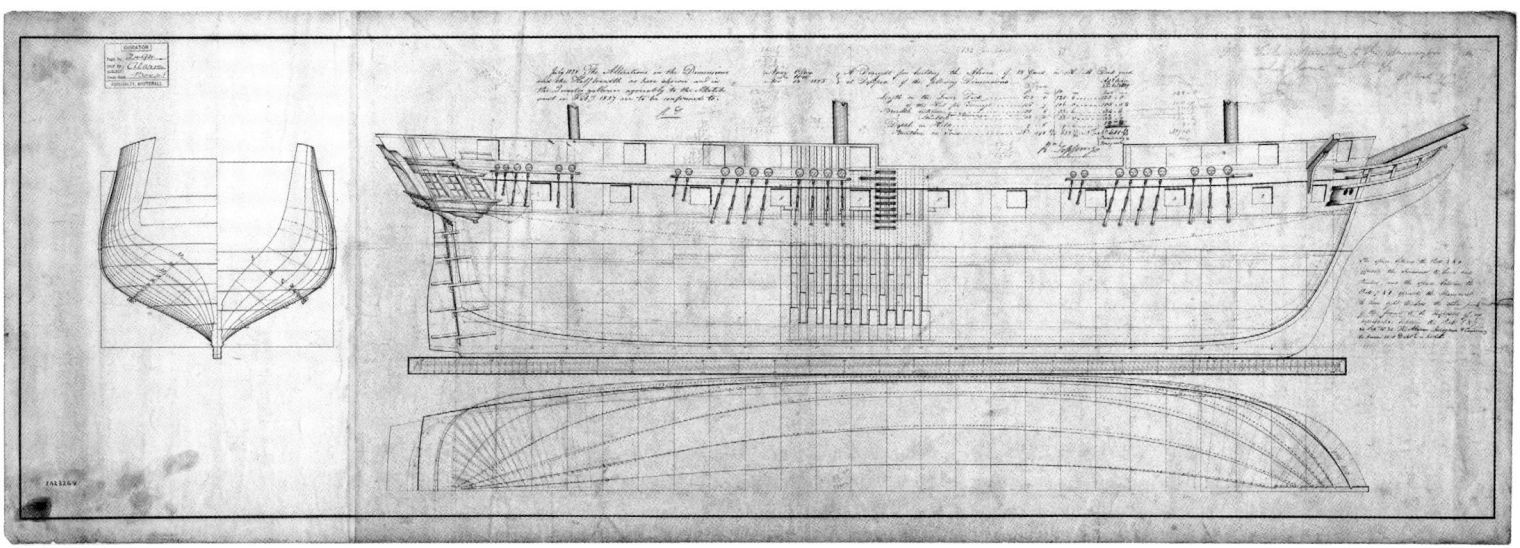

Seppings's design, dated 13 November 1826, for the proposed 28-gun *Alarm* after she was re-ordered as a sister to the *Tyne*, showing the midships framing on his 'small timber' principle. In February 1827 Seppings revised his design for the *Alarm*, along with the *Imogene* and *Conway*, but the changes were not approved until July 1828; these included a stern on the same model as the Fifth Rate *Castor*. Further alterations approved on 23 November 1829 are noted on the draught, and were applied to *Imogene* and *Conway*, but the *Alarm* was suspended in 1832 and subsequently cancelled. One look at the midship section gives the lie to the notion that Seppings was always conservative in his hull designs. *(NMM J6499)*

Commissioned 17.2.1832 under Capt. Henry Eden, for North Sea, later to Lisbon and then to South America; paid off 10.1835. Lent as training ship to Mercantile Marine Association of Liverpool 2.1859; to Aberdeen RNR 28.8.1861, renamed *Winchester*. BU at Sheerness 6.1871.

Alarm Deptford Dyd
Ord: 1825 (see above). K: 11.1826. then transferred to Pembroke Dyd. K: 1.1832. L: — . Suspended 19.6.1832; cancelled 14.9.1832 and BU from 10.1832.

CHALLENGER 28 guns. Design by Captain Lord John Hayes, a naval officer (rising to flag rank in 1851, three weeks before his death) and MP who also proved an amateur but talented naval architect.
Dimensions & tons: uncertain design details.
Men: 175. Guns: UD 20 x 32pdr carronades, QD 6 x 18pdr carronades, Fc 2 x 9pdrs.

Challenger Pembroke Dyd
As built: 125ft 7½in, 105ft 11½in x 32ft 8½in x 9ft 3¼in. 602⁹/₉₄bm. Draught 10ft 6½in/13ft 2½in.
Ord: 9.6.1825; then transferred to Portsmouth Dyd 27.6.1825. K: 11.1825. L: 14.11.1826. C: 9.3.1827 (for sea).
First cost: £26,341 (including fitting).
Commissioned 11.1826 under Capt. John Hayes, her designer, at Portsmouth. On 2.7.1827 under Capt. Lord Adolphus FitzClarence ('natural' son of King William IV), to Canada and back conveying the retiring Governor-General; then to Lisbon. On 14.9.1827 under Cmdr Joseph Harrison, at Plymouth. On 5.11.1828 under Capt. Charles Howe Fremantle, sailed 5.12.1828 for Cape of Good Hope and thence to East Indies; took formal possession of Western Australia 2.5.1829; arrived home 1.6.1833 and paid off 12.6.1833. Large refit (for £8,153) at Portsmouth 6–10.1833; recommissioned 14.6.1833 under Capt. Michael Seymour, for South America; wrecked off Mocquilla Point on the coast of Chile 19.5.1835 (all crew rescued except two).

SAPPHIRE 28 guns. Designed 1825 and built by the 'superior class of shipwright apprentices at Portsmouth'.
Dimensions & tons: 120ft 0in, 100ft 6¼in x 33ft 10in x 8ft 0in. 604³⁹/₉₄bm.
Men: 175. Guns: UD 20 x 32pdr carronades, QD 6 x 18pdr carronades, Fc 2 x 9pdrs.

Sapphire Portsmouth Dyd
As built: 119ft 0in, 100ft 7½in x 34 ft 0in oa (33ft 8in for tonnage) x 8ft 0in. 606⁶²/₉₄bm. Draught 9ft 11in/12ft 10½in.
Ord: 11.8.1825. K: 11.1825. L: 31.1.1827. C: 26.3.1827 (for sea).
First cost: £19,044 (£25,577 including fitting).
Commissioned 20.12.1826 under Capt. Henry Dundas, for South America and the Pacific; paid off 11.1830. Fitted for sea at Portsmouth (for £8,633) 11.1830–3.1831. Recommissioned 28.11.1830 under Capt. William Henry George Wellesley, for North America and the West Indies; in 9.10.1832 under Capt. George Rolle Walpole Trefusis; paid off 10.1834. Recommissioned 4.4.1835 under Capt. Richard Freeman Rowley, for the Mediterranean; paid off 8.1838 and laid up at Portsmouth. Fitted as troopship at Portsmouth (for £5,541) 10.1838–11.1839 (re-rated to 22 guns in 1842); recommissioned under George William Nembhard, Master, for East Indies and China. In 7.1840 under Gilbert H. Cole, Master, for China; in 1st China War; Yangtze operation 7.1842; under John Fittock, Master, from 28.7.1842. Receiving ship at Trincomalee 1847; sold there (for £710, by AO 24.6.1864) 5.11.1864.

ACTAEON 26 guns. Designed 1827 by the Superior Class of Shipwright Apprentices with the School of Naval Architecture.
Dimensions & tons: 121ft 6in, 100ft 4in x 34ft 1in (for tonnage) x 9in 7in. 620bm.
Men: 175. Guns: UD 20 x 32pdr gunnades, QD 4 x 32pdr carronades, Fc 2 x 9pdrs (or 2 x 32pdr carronades).
First cost: £24,319.

Actaeon Portsmouth Dyd
As built: 121ft 6½in, 100ft 4in x 34ft 5in oa (34ft 1in for tonnage) x 9in 7in. 619⁹¹/₉₄bm. Draught 10ft 2in/12ft 11½in.
Ord: 23.10.1827. K: 9.1828. L: 31.1.1831. C: 16.4.1831 (for sea).
First cost: £24,319 (including fitting).
Commissioned 25.11.1830 under Capt. Frederick William Grey,

A print by Henry Moses showing the *Challenger* sailing from Portsmouth on 31 October 1827 with dispatches for Lisbon. She had just made a rapid passage to Canada and then home, and the ship's sailing qualities were highly regarded. *(NMM PU6135)*

for the Mediterranean; Portsmouth in 1831, Lisbon 1832, then Mediterranean 1833–34; paid off 4.9.1834. Recommissioned 11.1834 under Capt. Lord Edward Russell, at Portsmouth; to South America 1836. On 14.8.1838 under Capt. Robert Russell, for South America and the Pacific; paid off 1842. Between a Small and Middling Repair and fitted for sea at Plymouth (for £10,481) 5.1843–2.1845. Recommissioned 14.12.1844 under Capt. George Mansel, for the west coast of Africa. Fitted as survey ship at Portsmouth (for £8,986) 8–11.1856. Recommissioned 1.8.1856 under Cmdr William Thornton Bate (killed 29.12.1857), for 'coast of China and Tartary'; on 30.12.1857 under Capt. Robert Jenkins (following Bate, killed ashore with Naval Brigade); on 24.9.1858 under Cmdr John Ward; paid off 19.6.1862 at Portsmouth. Handed over by AO 3.8.1866 to the Guardians of the Poor at Cowes for use as a hospital ship for cholera patients, but handed back at 30.8.1866. Made good defects for use as a hospital ship 5.1867. Lent as hulk to Cork Harbour Board 2.1870. Used for torpedo experiments, attached to HMS *Vernon* 1–3.1874. Sold to John Read at Portsmouth 2.1889 to BU.

Cut down from Frigate

The 42-gun frigate *Semiramis* (originally launched 1808) was raséed (cut down to a flush deck ship) to become a 24-gun Sixth Rate in 1828. The early service history and dimensions of this ship are recorded in Chapter 5.

Men: 210 (110 in peacetime). Guns: UD 16 x 32pdrs; QD 6 x 18pdrs; Fc 2 x 6pdrs.

Semiramis (L: 1808) Cut down 4–6.1828 at Plymouth into a 24-gun Sixth Rate.

Recommissioned 27.5.1828 under Capt. Maurice Frederick

An official model of the *Actaeon*, possibly made by the same class of apprentices from the School of Naval Architecture who designed the ship. For a number of years it became the norm for the 'Superior Class' of apprentices to design a Sixth Rate that was actually built and then trialled as a practical test of their competence. This attempt to give British ship design a more scientific foundation was undermined by the appointment as Surveyor of the amateur William Symonds, who was disinclined to give formal training much credit. *(NMM F8926-002)*

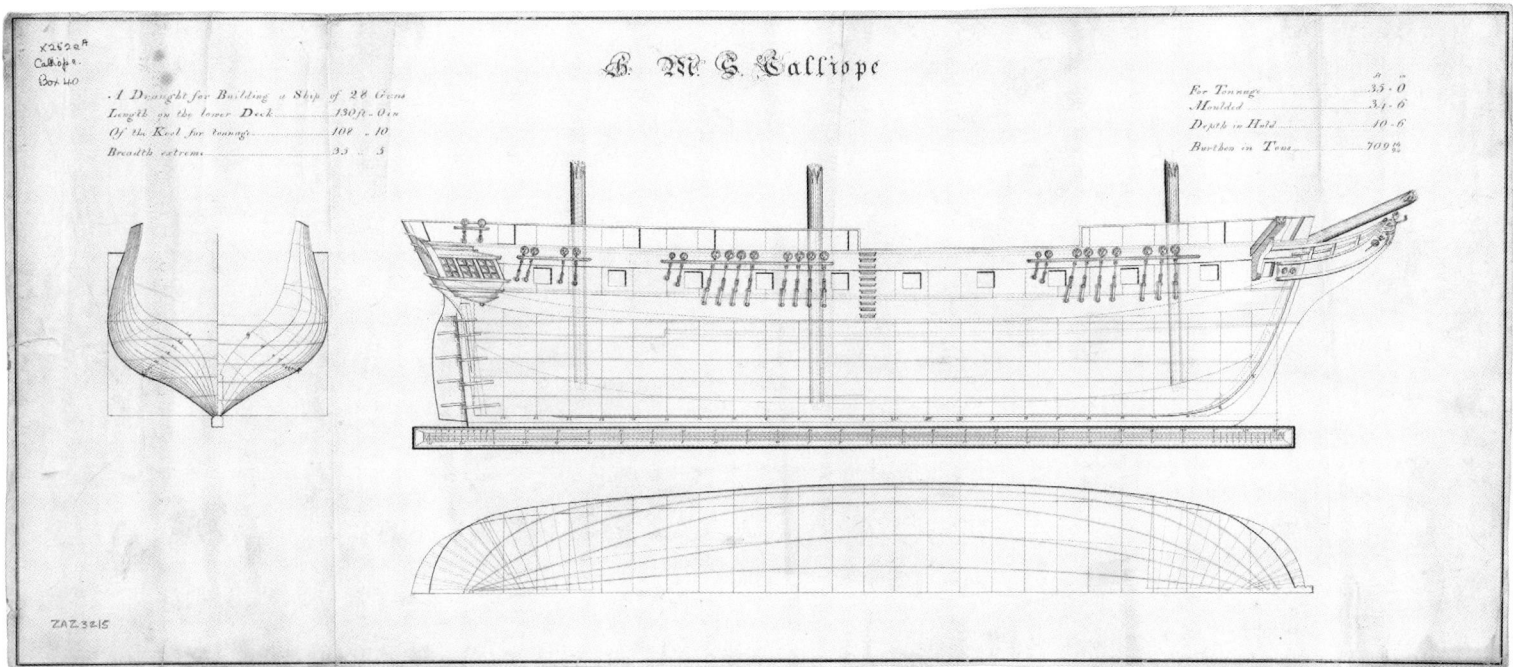

This draught of the *Calliope*, Seppings's final 28-gun design, shows considerably enlargement from the 600-ton designs of the 1820s. While the 32pdrs of her predecessors were carronades or gunnades, the ordnance on the *Calliope* and her sister *Andromache* were standard long 32pdrs, and the design has clearly been enlarged – as in the Fifth Rate *Castor* – to accommodate the heavier weapons. The very sharp midship section with exaggerated hollows proved that Seppings did not object to fine lines, but unlike his successor Symonds he did not think they were appropriate in ships of the line. *(NMM J6548)*

Fitzhardinge Berkeley, as flagship of Vice-Adm. Sir Charles Paget, at Cork; paid off 6.1831 and laid up at Portsmouth 9.1831. At Sheerness in 1834. Depot ship at Plymouth 1832. BU there 11.1844.

ANDROMACHE Class 28 guns (26 guns from 1839). Sir Robert Seppings's final design prior to his leaving office (his dismissal took effect 31 March 1832), 1830. On 25 February 1831 the Navy Board instructed Sheerness to 'use such of the materials of the *Pegasus* as are applicable to building this ship ' (*Calliope*).
  Dimensions & tons: 130ft 0in, 108ft 10in x 35ft 5in oa (35ft 0in for tonnage) x 10ft 6in. 709¹⁹⁄₉₄bm.
  Men: 175. Guns: UD 20 x 32pdrs, QD 6 x 32pdrs, Fc 2 x 32pdrs (all Dixon's 25cwt 32pdrs).

Calliope Sheerness Dyd
  As built: 130ft 2in, 109ft 3in x 35ft 5½in oa (35ft 2½in for tonnage) x 10ft 7in. 720³⁴⁄₉₄bm.
  Ord: 29.10.1830. K: 1.1831. L: 5.10.1837.
  First cost: £5,741 (cost until abolition of Navy Board), plus £12,141 completion and fitting for sea.
  Commissioned 10.11.1837 under Capt. Thomas Herbert, for the Nore; to South America 1838–39, then Pacific 1840–42; operations in 1st Anglo-Chinese War 1841–42; Chuenpee Forts action 7.1.1841, then Canton River operations. On 14.6.1841 under Capt. Augustus Leopold Kuper; Yangtze operations 7.1842; home and paid off 3.1843 at Plymouth. Very small repair and fitted for sea there (for £8,197) 2.1845–8.1848; recommissioned 18.7.1845 under Capt. Edward Stanley, for the East Indies; New Zealand War 1845; Ruapekapeki stockade 11.1.1846; home and paid off 1.1849. Between Small and Middling Repair at Plymouth (for £3,287) 6–9.1849. Recommissioned 28.11.1850 under Capt. Sir

James Everard Home, for the Australian station; fitted for sea at Plymouth (for £3,047) 11.1850–3.1851; on 2.11.1853 under Capt. John Gennys, then 20.1.1854 under Capt. John Coghlan Fitzgerald; home and paid off 6.1855. Fitted as floating chapel at Plymouth 1855 (or later); fitted as a factory ship (floating workshop) between 1861 and 1865. BU completed there 1.11.1883.

Andromache Pembroke Dyd
  As built: 130ft 0in, 108ft 8¼in x 35ft 6¾in oa (35ft 2¾in for tonnage) x 10ft 6½in. 717⁷³⁄₉₄bm. Draught 11ft 5in/13ft 7in.
  Ord: 29.10.1830. K: 8.1831. L: 27.8.1832. C: 11.10.1833–7.2.1834 at Plymouth.
  First cost: £14,845 (+ fitting £7,759).
  Commissioned 19.9.1833 under Capt. Bernard Yeoman, at Plymouth. On 2.1.1834 under Capt. Hugh Ducie Chads, for the East Indies; forcing of the Boca Tigris (China) in 9.1834; home and paid off 9.1837. Recommissioned 2.2.1838 under Capt. Robert Lambert Baynes, for North America; to West Indies 1839; at Portsmouth 1840; to Cape of Good Hope 1841–42; paid off 3.1843. Fitted at Plymouth as Provision depot ('frigate-depot') 9–11.1846, with 2 guns and 72 men; recommissioned 13.10.1846 under Thomas Johnson, Master, based on Galway to convey food and stores to Ireland for famine relief; paid off 9.1848 and laid up at Devonport. Fitted as powder depot 8–9.1854, then based at Milford. BU at Devonport (by AO 23.5.1871) completed 3.3.1875.

Vestal Chatham Dyd
  Ord: 29.10.1830. K: — Cancelled 8.12.1831 (unstarted).

Cleopatra Pembroke Dyd
  Ord: 28.3.1831. (named 4.7.1831) K: — Re-ordered to following 26-gun design 28.3.1832.

Carysfort Pembroke Dyd
  Ord: 29.6.1831. (named 4.7.1831) K: — Re-ordered to following 26-gun design 24.6.1832.

(C) Vessels acquired from November 1830

Cut down from Frigates

Three elderly 42-gun frigates surviving from the Napoleonic era were raséed to 26-gun Sixth Rate 'corvettes' in 1831: the *Curacoa* and *Aigle* at Chatham and the *Magicienne* at Woolwich; a fourth vessel – *Tribune* – followed suit at Chatham in 1832–33. The early histories and dimensions of these ships are recorded in Chapter 5. These 26-gun conversions (later 24-gun) replaced the 18pdrs by a smaller number of more powerful 32pdr guns, significantly raising firepower while reducing manning. A further six 42-gun frigates would undergo similar conversion in 1844–45 (see below).

Men: 210 (by 1839, 230). Guns: UD 12 x 32pdr (50cwt) + 6 x 8in /68pdr (65cwt); QD & Fc 2 x 56pdrs (85cwt) + 6 x 32pdrs (25cwt). By 1839 this had become UD 20 x 32pdrs (40cwt/7½ft) + 2 x 32pdrs (50cwt/9ft); QD 2 x 8in (52cwt/8ft).

Curacoa (L: 1809) Cut down 2–6.1831 (for £15,057) into a 26-gun corvette (24-gun in 1837, with 24 x 32pdrs).

Ord: 10.1.1831. Begun: 16.2.1831. Undocked 28.3.1831. C: 6.1831 (for sea).

Commissioned 2.4.1831 under Capt. David Dunn, for the East Indies; paid off summer 1835. Fitted for sea at Chatham (for £9,983) 2–7.1939; recommissioned 8.4.1839 under Capt. Jenkin Jones, for South America; paid off 30.11.1842. Recommissioned 22.2.1843 under Capt. Thomas Sabine Pasley, for east coast of South America. On 12.1.1846 under Capt. William Broughton, in same role; paid off 4.9.1847. Surveyed at Sheerness 10–11.1848 and BU there 3.1849.

Aigle (L: 1801) Cut down 3–7.1831 into a 26-gun corvette (24-gun in 1839).

Ord: 10.1.1831. Begun: 29.3.1831. Undocked 9.7.1831. C: 7.1831 (for Ordinary).

Laid up at Chatham until fitted for sea (for £9,670) 8–11.1841.

Commissioned 23.8.1841 under Capt. Lord Clarence Edward Paget, for the Mediterranean. On 15.4.1843 under acting Capt. Hastings Reginald Henry, still in the Mediterranean; laid up 8.1845. Fitted as Coal hulk (and receiving ship) at Woolwich 10–11.1852; to Sheerness by AO 9.9.1869. Torpedo target by AO 15.8.1870, and sunk by torpedoes at Sheerness. Sold 24.11.1870 (by AO 3.11.1870, for £925) to A. W. Howe to BU.

Magicienne (L: 1812) Cut down 4–11.1831 (for £17,400) into a 24-gun corvette.

Ord: 15.4.1831. Begun: 29.4.1831. Undocked 27.8.1831. C: 11.1831 (for sea).

Commissioned 18.7.1831 under Capt. James Hanway Plumridge, for the East Indies; paid off early 1835. Recommissioned 14.4.1835 under Capt. George William St John Mildmay, for Home station; at Lisbon 1837; paid off end 1838. Recommissioned 23.4.1840 under Capt. Frederick Thomas Michell, for the Mediterranean; operations on the coast of Syria 1840; on 4.8.1841 under Capt. Richard Laird Warren, still in the Mediterranean; paid off at end 1843. BU 3.1845 at Portsmouth.

Tribune (L: 1803) Cut down 6.1832–3.1833 into a 24-gun corvette.

Ord: 15.4.1831. Begun: 1.1832. Undocked 3.1833. C: 5 — 9.1834 (for sea).

Commissioned 5.1834 under Capt. James Tompkinson, for the Mediterranean; on 24.5.1838 under Capt. Charles Hamlyn Williams; wrecked 28.11.1839 near Tarragona.

VESTAL Class 26 guns. William Symonds design, 1831, replacing the Seppings design for those *Andromache* Class Sixth Rates not yet begun.

Dimensions & tons: 130ft 0in, 104ft 6in x 40ft 6in x 10ft 6in. 911⁷/₉₄bm.

Men: 210 (152 officers & men, 33 boys, 25 marines), later 240. Guns: UD 16 x 32pdrs (40cwt/7½ft) + 2 x 8in (52cwt/8ft) shell guns, QD & Fc 6 x 32pdr carronades (25cwt/6ft) + 2 x 32pdrs (42cwt/8ft).

Vestal Sheerness Dyd

As built: 130ft 0in, 105ft 9in x 40ft 7½in oa (40ft 3½in for tonnage) x 10ft 10in. 913¹⁶/₉₄bm. Draught 10ft 0in/15ft 4in.

Ord: 9.11.1831. (named 8.12.1831) K: 5.1832. L: 6.4.1833. C: 15.7.1833 (for sea)

First cost: £19,173 (£27,508 including fitting).

Commissioned 6.5.1833 under Capt. William Jones, for North America and the West Indies; paid off 9.1837. Recommissioned 27.11.1837 under Capt. Thomas Wren Carter, for same role; on 18.8.1841 under Capt. John Parker; paid off 1842. Very Small Repair, and fitted for sea at Sheerness (for £13,457 + £665) 8.1842–1.1844; recommissioned 2.11.1843 under Capt. Charles Talbot, for the East Indies; at Malluda Bay 19.8.1845; paid off 7.1847. Between a Small and Middling Repair, and fitted for sea at Sheerness (for £13,646) 6.1849–8.1852. Recommissioned 7.6.1852 under Capt. Cospatrick Baillie Hamilton, for North America and West Indies. Recommissioned 7.6.1853 under Capt. Thomas Pickering Thompson, on same service; paid off into Ordinary at Chatham 12.1856. Sold to Castle & Beech (for £1,715, by AO 17.2.1862) to BU 2.1862.

Cleopatra Pembroke Dyd

As built: 130ft 0in, 105ft 11in x 40ft 8½in oa (40ft 4½in for tonnage) x 10ft 11in. 918³⁷/₉₄bm. Draught 10ft 9in/15ft 4in.

(Re)Ord: 28.3.1832. K: 6.1832. L: 28.4.1835. Sailed 15.5.1835 for Sheerness. C: 25.5–13.9.1835 at Sheerness (for sea)

First cost: £19,060 (+ fitting £6,077 at Sheerness).

Commissioned 12.4.1835 under Capt. George Grey, to St Petersburg, then for South America, and to St Petersburg again; paid off autumn 1838. Fitted at Portsmouth and then Sheerness for the reception of an ambassador (for £618) 9–10.1838. Recommissioned 11.1840 under Capt. Alexander Milne, for the West Indies; took slaver *El Segundo Rosario* (with 284 slaves aboard) 27.1.1841. In 4.1842 under Capt. Christopher Wyvill, for North America and West Indies, subsequently the Cape of Good Hope; paid off 1847. Recommissioned 26.4.1849 under Capt. Thomas Lecke Massie, for the East Indies; paid off 4.1853 at Chatham and laid up. Accommodation ship at Blackwall 9.1857–3.1858 (for crew of Peruvian frigate *Amazon*); then returned to Chatham and laid up. Sold to Castle & Beech to BU (by AO 17.2.1862, for £1,810).

Carysfort Pembroke Dyd

As built: 130ft 0in, 105ft 10⅛in x 40ft 9⅞in oa (40ft 5⅞in for tonnage) x 11ft 0in. 922⁹³/₉₄bm. Draught 10ft 3in/15ft 0in.

(Re)Ord: 24.6.1832. K: 9.1832. L: 12.8.1836. Sailed 14.9.1836 for Sheerness. C: 21.9.1836–18.2.1837 at Sheerness (for sea).

First cost: £19,703 (+ fitting £6,401 at Sheerness).

Commissioned 21.11.1836 under Capt. Henry Byam Martin; sailed 12.3.1837 for the Mediterranean; in Syrian campaign 1840, with Stopford's fleet (capture of Acre 3.11.1840); laid up at Pembroke 1841. Recommissioned 28.12.1841 under Capt. Lord George Paulet, for South America; annexation of Hawaii 2.1843; paid off 6.1845. Recommissioned 12.12.1845 under Capt. George Henry

The *Vestal* was Symonds's replacement Sixth Rate design to supersede Seppings's *Andromache* Class. With the breadth widened by over 5ft, this design showed a further enormous leap, of some 200 tons, in the size of Sixth Rates. This lithograph was produced in 1833 by John Christian Schetky, and shows the *Vestal* off Culver Cliffs on 24 July, preparatory to her deployment for America; aboard her are HRH the Duchess of Kent and her 14-year-old daughter, the Princess Alexandrina Victoria (the future queen). Among the flotilla accompanying them are the *Emerald*, *Sylvia*, *Waterwitch*, *Louise*, *Nautilus* and several other yachts. *(NMM 9943)*

Seymour, for the Pacific; paid off at Portsmouth in Spring 1848. Surveyed 7–8.1848, then laid up. Sold to Ritherdon & Thompson (for £1,800, by AO 22.11.1861 for BU.

SPARTAN Class 26 guns. Sir William Symonds design of 12 July 1837, modified from *Vestal* design.
>Dimensions & tons: 131ft 0in, 106ft 1in x 40ft 6¼in oa (40ft 2¼in for tonnage) x 10ft 9in. 911¹³/₉₄bm.
>Modified for 3 ships (* below): keel 107ft 0in; breadth 40ft 7¼in oa (40ft 3¼in for tonnage). 923bm. 958 disp.
>Men: 190/240. Guns: originally UD 18 x 32pdrs (40cwt), QD 6 x 32pdr carronades (26cwt), Fc 2 x 32pdrs (40cwt); later UD 16 x 32pdrs (42cwt) + 2 x 8in (56cwt) shell guns, QD 6 x 32pdr carronades (25cwt), Fc 2 x 12pdrs (25cwt).

Spartan Plymouth Dyd
>As built: 131ft 0in, 106ft 10½in x 40ft 7¼in oa (40ft 2¼in for tonnage) x 10ft 9in. 918¹¹/₉₄bm. Draught 10ft 8in/15ft 9in (with c.32½ tons on board).
>Ord: 20.2.1837. K: 6.1838. L: 16.8.1841. C: 28.11.1841 (for sea).
>First cost: £19,994 to build, plus £8,879 fitting for sea.
>Commissioned 21.8.1841 under Capt. Charles Gilbert John Brydone Elliot, for North America and West Indies; paid off 27.8.1845. Recommissioned 30.5.1846 under Capt. Thomas Matthew Charles Symonds (the son of her designer), for the Mediterranean; paid off 1849. Very Small Repair and fitted for sea at Plymouth (for £1,793) 9.1849, but then laid up in Ordinary. Fitted for sea at Plymouth (for £2,092) 6–7.1852; recommissioned 7.6.1852 under Capt. Sir William Legge George Hoste, for the East Indies and China; in action 2.11.1854 against junks in Tymmoon Bay; in action 13.11.1854 against pirate stronghold at Coulan; paid off at Plymouth 1.11.1857. Sold to J. & E. Marshall (by AO 28.4.1862, for £1,410) but sale cancelled; instead sold to H.Castle & Sons (by AO 19.5.1862, for £1,875) to BU.

Iris Pembroke Dyd
>As built: 131ft 1in, 106ft 6⅛in x 40ft 6in oa (40ft 0in for tonnage) x 10ft 9in. 906⁷⁷/₉₄bm. Draught 10ft 7in/15ft 7in (with 19 tons on board).
>Ord: 20.2.1837. K: 9.1838. L: 14.7.1840. Sailed 11.8.1840 for Chatham. C: 16.8.1840–11.1.1841 at Chatham.
>First cost: £17,234 (+ £368) to build; fitting £7,110 at Chatham.
>Commissioned 3.1.1841 at Chatham under Capt. Hugh Nurse, for West coast of Africa (died 21.10.1841). On 1.11.1841 under Cmdr William Tucker; took slaver *Duquesa de Braganza* 22.1.1842. In 8.1842 under acting Capt. Thomas Rodney Eden, then 4.10.1842 under Capt. George Rodney Munday; paid off 15.8.1843. Recommissioned 18.10.1843, still under Mundy, for the East Indies; operations in the Brunei River against Borneo pirates 7.1846; paid off 8.1847. Between a Small and Middling Repair at Chatham (for £8,177), then laid up in Ordinary 1.1848. Refitted at Chatham (for £7,739) 6.1856–2.1857. Recommissioned 24.12.1856 under Commodore William Loring, for Australia; paid off 3.8.1861 at Chatham. Lent to Atlantic Telegraph Co. 1864, then to Pitcher to assist in the recovery of the *Foyle* 4.1867, then to Telegraph Construction & Maintenance Co. as a cable ship 6.1867, and sold to the latter (for £3,400) 16.10.1869.

*Juno** Pembroke Dyd
>As built: 131ft 2in, 107ft 1¼in x 40ft 7in oa (40ft 1in for tonnage) x 10ft 10in. 923³/₉₄bm. Draught 9ft 9in/15ft 6in.
>Ord: 20.2.1837. K: 4.1842. L: 1.7.1844. Sailed 4.9.1844 for Chatham. C: 24.3.1845–31.10.1845 at Chatham.
>First cost: £17,579 (+ £67 fitting at Pembroke) to build; fitting £9,281 at Sheerness.
>Commissioned 3.9.1845 under Capt. Patrick John Blake, for the Pacific station; paid off 1847. Recommissioned 10.1853 under Capt. Stephen Grenville Freemantle, for the Australian station; fitted for sea 10.1853–1.1854; paid off 11.1857 into Ordinary at Portsmouth. Fitted as Water police ship (by AO 11.9.1861) at Portsmouth 2–4.1862; renamed *Mariner* 10.1.1878 but returned to RN as seagoing training ship, renamed *Atalanta* 22.1.1878 and fitted as such as Pembroke. Lost, after sailing from Bermuda 1.2.1880, presumed foundered near Bermuda 12/16.2.1880 with all (280) hands.

*Creole** Devonport Dyd
>As built: 130ft 11in, 107ft 0in x 40ft 7¼in oa (40ft 3¼in for tonnage)

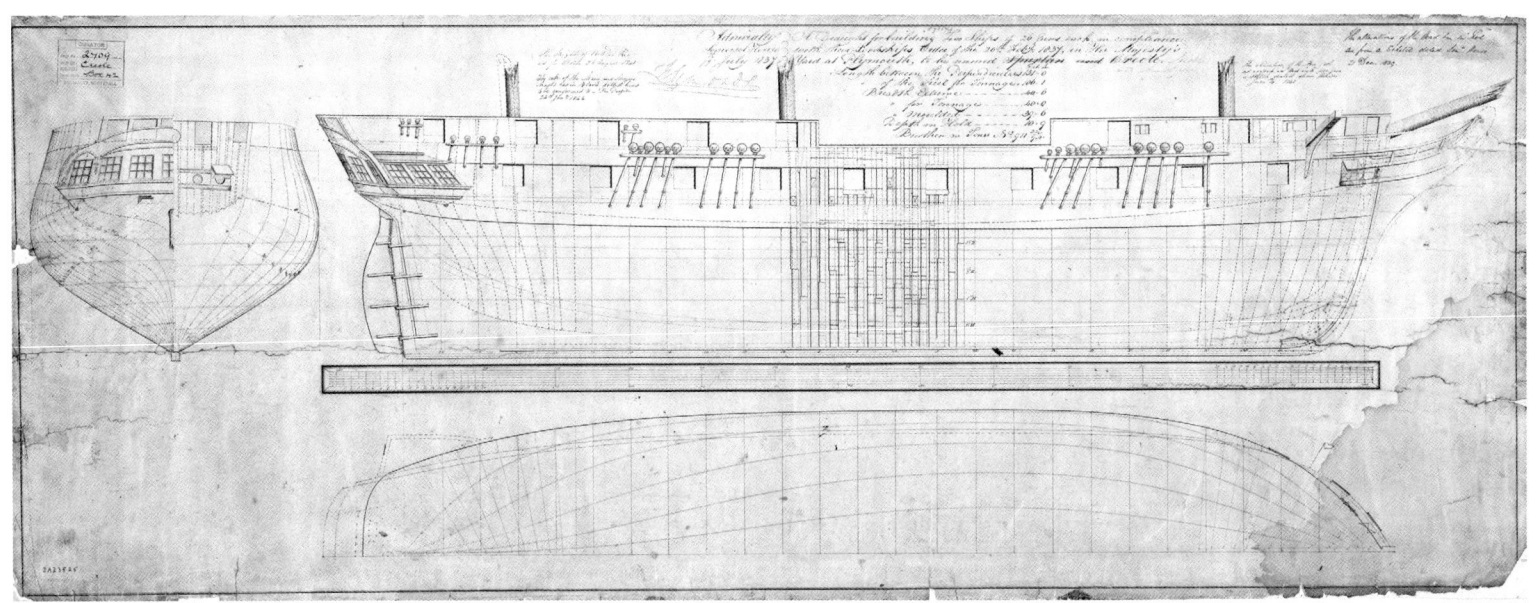

Symonds's draught, dated 12 July 1837, for the *Spartan* class. The draught bears amendments of December 1839 in respect of the *Creole*, *Amethyst* and *Niobe* consisting of alterations to the head. Further alterations to the design noted on the draught are of 4 June 1841 (to amend the fore part of the ship), 26 August 1841 (to change the height of the ports) and 24 January 1844 (to modify the rake of the main and mizzen masts). Symonds was in the habit of continuously tinkering with his designs – doubtless he regarded it as fine-tuning – and any two vessels of a class were rarely entirely identical. *(NMM J6286)*

x 10ft 8½in. 923¹⁄₉₄bm. Draught 9ft 10in/15ft 7in (with 12 tons on board).

Ord: 20.2.1837. K: 4.1842. L: 1.10.1845. C: 14.10.1845 (for Ordinary).

First cost: £20,544 (including fitting for Ordinary).

Not commissioned. Partially fitted as an Advanced ship (for £4,572) 10–12.1856. Lent to Cork Industrial Training Ship Organisation by AO 10.2.1870. BU at Devonport (by AO 22.11.1873) completed 6.3.1875.

Amethyst * Devonport Dyd

As built: 130ft 11in, 107ft 0in x 40ft 7¼in oa (40ft 3¼in for tonnage) x 10ft 10in. 923¹⁄₉₄bm. Draught 9ft 11in/15ft 7in (with 16½ tons on board).

Ord: 16.8.1839. K: 4.1843. L: 7.12.1844. C: 2.2.1845 (for Ordinary). Fitted to sea 5.1856–20.9.1856

First cost: £20,882 (+ fitting for Ordinary £424). Fitting for sea (1856) £12,627.

Commissioned 8.7.1856 under Capt. Sidney Grenfell, for East Indies and China; subsequently to the Pacific; in 2nd Anglo-Chinese War; at Battles of Fatshan and Ling-Ting; paid off 22.12.1860 at Chatham. Lent to Atlantic Telegraph Co. 1864, then to Pitcher to assist in the recovery of the *Foyle* 1866, then to Telegraph Construction & Maintenance Co. as a cable ship 24.7.1869, and sold to the latter (for £3,500) 16.10.1869.

Niobe Devonport Dyd

Ord: 16.8.1839. K: — . Re-ordered to *Diamond* Class design 28.3.1846 (see below).

Malacca Bombay Dyd. Teak-built.

Ord: 26.5.1841. K: — . Suspended 14.10.1841. Named 7.7.1843. Ordered 11.10.1848 to be build as a screw sloop (at Moulmeim) (see Chapter 7).

Alarm Sheerness Dyd

As built: 131ft 0in, 106ft 0⅜in x 40ft 6½in oa (40ft 2½in for tonnage)

x 10ft 9in. 911⁷⁶⁄₉₄bm. Draught 10ft 7½in/15ft 1in.

Ord: 20.2.1843. K: 9.1843. L: 22.4.1845. C: 4.12.1845 (for sea).

First cost: £28,259 (including fitting).

Commissioned 9.10.1845 under Capt. Charles Colville Frankland, for North America and the West Indies. On 16.10.1846 under Capt. Granville Gower Loch, on same station; at Fort Serapoqui (Nicaragua) 12.2.1848. On 21.8.1849 under Capt. George Ramsay, on same station; paid off 1852. Recommissioned 13.6.1855 under Capt. Douglas Curry, for the Pacific; paid off 8.9.1859 at Plymouth. Fitted for Ordinary there 9.1859. Fitted for coal depot at Plymouth (no derricks fitted) 8.1861; 7.1877 to Queenstown under tow of *Scotia*; landing stage at Berehaven by AO 24.8.1900; sold at Queenstown (for £65) by AO 7.7.1904.

EURYDICE Class 26 guns. The design by Rear-Admiral George Elliot (brother of the Earl of Minto, First Lord 1835–41), was begun 24 July 1841 and was approved 31 August, and the 26-gun ship ordered 4 days later, to be armed as *Andromache* Class but with scantlings and fittings similar to *Vestal* Class.

Dimensions & tons: 141ft 2in, 117ft 9¾in x 38ft 7½in oa (38ft 1½in for tonnage) x 8ft 9in. 910⁸¹⁄₉₄bm.

Men: 190. Guns: UD 18 x 32pdrs (long, 45cwt/8½ft), QD 6 x 32pdrs (short, 25cwt/6ft), Fc 2 x 32pdrs (long).

Eurydice Portsmouth Dyd

As built: 141ft 3in, 117ft 10¼in x 38ft 10in oa (38ft 4in for tonnage) x 8ft 9in. 921⁴⁶⁄₉₄bm. Draught 10ft 7in/13ft 7in.

Ord: 27.8.1841. K: 4.1842. L: 16.5.1843. C: 1.9.1843 (for sea)

First cost: £16,137 to build, plus £9,312 fitting.

Commissioned 27.6.1843 (appointed 7.6) under Capt. George Augustus Elliot (eldest son of the ship's designer), for North America and the West Indies. Recommissioned 30.5.1846, still under Capt. Elliot; from 12.10.1846 under Capt. Talavera Vernon Anson, for the Cape of Good Hope; paid off 1850. Small Repair at Portsmouth (for £2,917) 12.1850–1.1852. Recommissioned 4.4.1854 under Capt. Erasmus Ommanney; fitted for sea (for 7,057 including making good defects later in 1854) 4–5.1854; to the White Sea during the Russian Way, then 11.1854 to North America and the West Indies. On 3.1.1855 under Capt. John Walter Tarleton, on same station; paid off 12.6.1857 at Chatham. Fitted at Portsmouth as a harbour training ship for cadets 6.1861. Reconstructed and

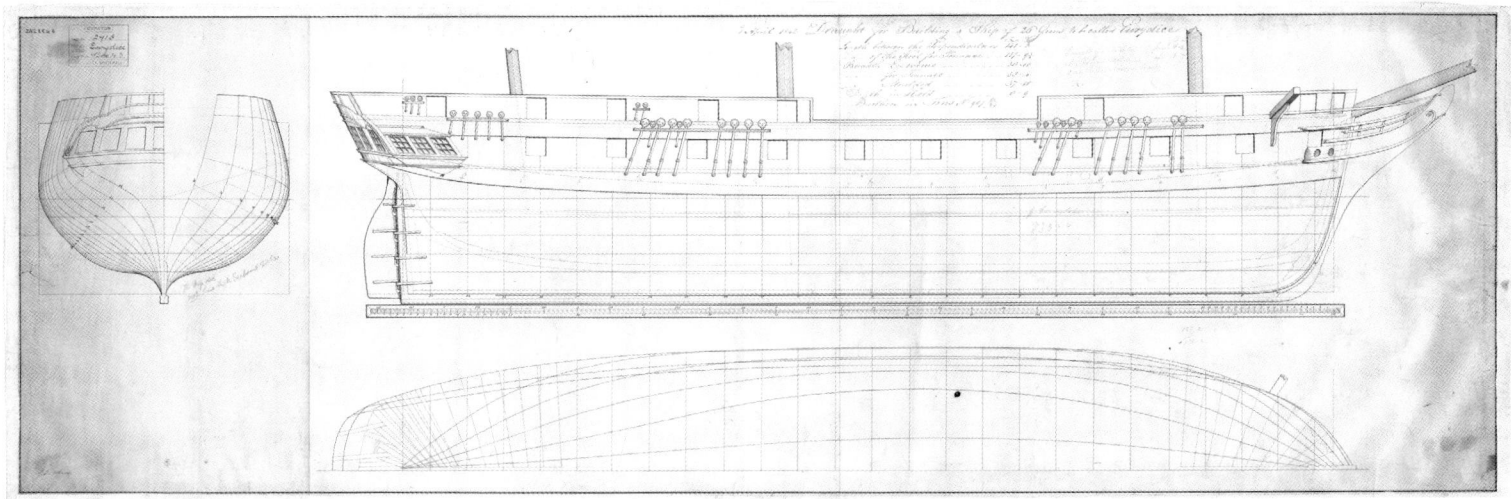

The draught of *Eurydice* dated 1 April 1842. This design by Adm. Sir George Elliot, the First Lord of the Admiralty's brother, was another of the many alternatives to Symonds's hull forms being sponsored by the Admiralty; the body plan was somewhat reminiscent of an eighteenth-century French frigate, although the hollow of the garboards was exaggerated. She was swamped in 1878 when – serving as a training ship – she was blown over in a sudden squall, submerging her lee gunports (which had been left open) and she was overwhelmed. *(NMM J6254)*

refitted as a training vessel (4 guns) at Portsmouth from 3.1875, then 3.1876 at John Whitc, Cowes (for £11,289) to 31.1.1877. Recommissioned 7.2.1877 under Capt. Marcus Augustus Stanley Hare, as seagoing training ship for Ordinary seamen; foundered in a squall off the Isle of Wight 24.3.1878 (only 5 were picked up out of 368 men aboard, and 3 of these died of exposure); wreck salvaged 1.9.1878 and towed to Portsmouth, and BU (by AO 11.9.1878) completed 11.10.1878 there.

DIAMOND Class 28 guns. Sir William Symonds design, approved 28 March 1845. Originally to have had 26 guns, and to have been built as *Spartan* Class units.

Dimensions & tons: 140ft 0in, 113ft 9¾in x 42ft 0in oa (41ft 8in for tonnage) x 11ft 1in. 1,051$\frac{1}{94}^{b}$m.

Men: 240. Guns: UD 20 x 32pdrs (long, 45cwt/8½ft), QD 1 x 68pdr/8in (56cwt) shell gun + 4 x 32pdrs (short, 25cwt/6ft), Fc 1 x 68pdr/8in (56cwt) shell gun + 2 x 32pdrs (short, 25cwt/6ft).

Diamond Sheerness Dyd
As built: 140ft 0in, 113ft 4in x 42ft 2in oa (41ft 10in for tonnage) x 11ft 0½in. 1,054$\frac{9}{94}^{b}$m. Draught 10ft 3in/14ft 6in.
Ord: 4.12.1845. K: 7.1846. L: 29.8.1848. C: 16.10.1848 (for Ordinary).
First cost: £22,903 (including £151 fitting for Ordinary); £8,287 fitting for sea (1853).
Commissioned 19.10.1853 under Capt. William Peel; fitted for sea 10.1853–26.12.1853; to the Black Sea for the Russian War. On 15.1.1855 under Capt. Gospatrick Baillie Hamilton, still in Black Sea, then the Mediterranean; paid off 18.2.1857 at Chatham and laid up. Lent to Seamen's Mission on the Tyne as training ship 4.1866, sailed to North Shields 6.1866; renamed *Joseph Straker* 13.1.1868. Returned to Admiralty 25.7.1885. Sold to Castle & Son (for £1,000) 9.1885 to BU.

Tribune Sheerness Dyd
Ord: 26.3.1846. K:— . Re-ordered as screw frigate 1.11.1850 (see Chapter 5).

Niobe Devonport Dyd
As built: 140ft 0in, 113ft 11½in x 42ft 0in oa (41ft 8in for tonnage) x 11ft 1½in. 1,052$\frac{34}{94}^{b}$m. Draught 11ft 0in/14ft 5in.
(Re)Ord: 28.3.1846. K: 5.1847. L: 18.9.1849.
First cost: £22,224.
Never commissioned in British Navy. Sold to Prussian Navy as a training ship (for £15,892) by AO 9.7.1862.

Cut down from Frigates
During the mid 1840s, another two *Apollo* Class 42-gun frigates (*Brilliant* and *Havannah*) and four *Leda* Class 42-gun (originally 46-gun) frigates were cut down and re-rated as corvettes (Sixth Rates); the early histories and dimensions of these ships are recorded in Chapter 5. As with the earlier quartet of raséed frigates, this replacement of the 18pdrs by a smaller number of more powerful 32pdr guns, significantly raised firepower while reducing manning. The expenses detailed below should be compared with the comparative costs of building 'a new frigate of equal class and armament', estimated to be £27,000 including 21,000 for materials and £5,000 for workmanship.

Amazon (L: 1821) Cut down at Sheerness (for £14,226 including fitting) into a 26-gun corvette (19-gun in 1845).
Men: 240. Guns: 26 x 32pdrs (50cwt/9ft). By 1848 possibly armed as *Daedalus*.
Begun: 27.3.1844. Undocked: 27.6.1844. C: 4.1.1845 (for sea).
Commissioned 13.11.1844 under Capt. James John Stopford, for Lisbon, then to Mediterranean 1846–47. Refitted (for 6,543) 7–12.1848 for loan to Liberia 1848 to 1852; instead recommissioned 26.10.1848 under Capt. Edward Norwich Troubridge (died 10.1850), for the East Indies. On 29.9.1850 under Capt. Charles Barker, in the East Indies; arrived back at Portsmouth 2.5.1852 and paid off. Sold to W. Lethbridge of Plymouth (by AO 11.9.1863, for £1,820) to BU.

Daedalus (L: 1826) Cut down at Woolwich (for £12,933 including fitting) into a 20-gun corvette (19-gun in 1848).
Men: 230. Guns (1848): UD 6 x 8in (52cwt/8ft) + 12 x 32pdrs (50cwt/9ft); QD 1 x 56pdr (87cwt/10ft) on pivot mounting.
Begun: 29.3.1844. Undocked: 18.5.1844. C: 28.11.1844 (for sea).
Commissioned 16.10.1844 under Capt. Peter M'Quhae, for the East Indies; in boat attack on Sherif Osman's stronghold at Malluda Bay (Borneo) 19.8.1845; paid off 1847. Recommissioned 27.7.1849 under Capt. George Grenville Wellesley, for the Pacific; paid off 8.1853 at Plymouth. Fitted at Plymouth as an RNR drill ship (for £5,437) 3–6.1861; commissioned 5.6.1861

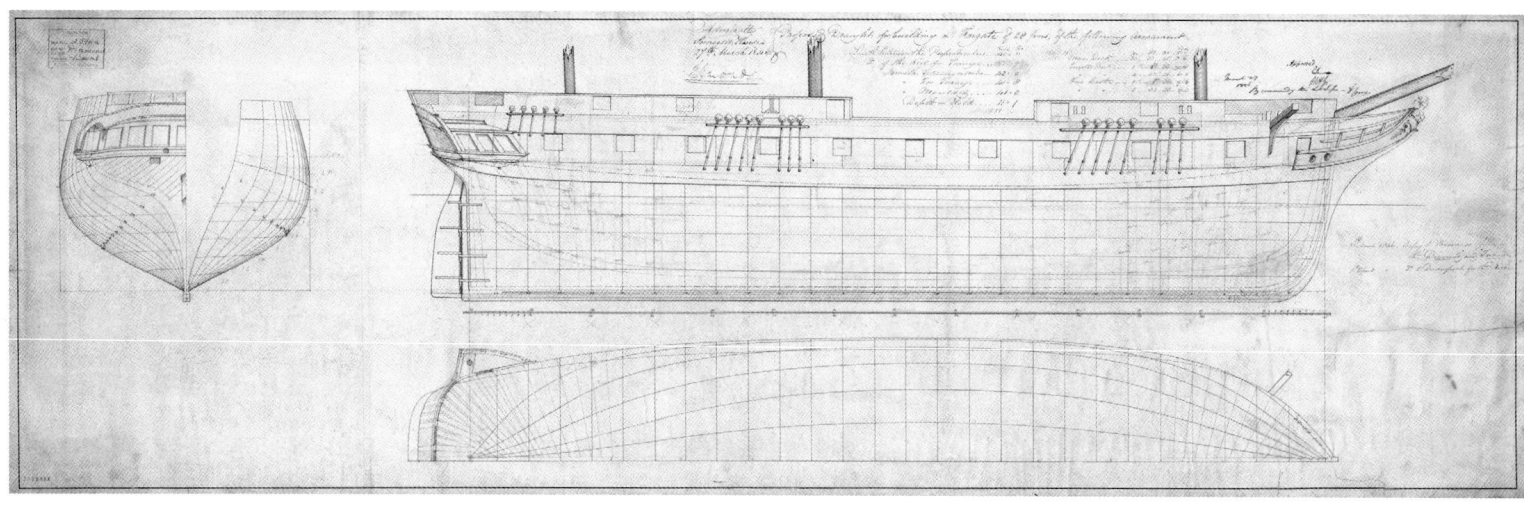

Symonds's final Sixth Rate design for the *Diamond*, the last sailing corvette, was enlarged again from the *Spartan* Class to a massive 1,051 tons. The additional 9ft of length enabled the addition of a tenth pair of 32pdrs on the upper deck (the previous classes had ten ports but only mounted nine guns on the upper deck broadside). Her sister *Niobe* was launched on Tuesday, 18 September 1849 from No. 1 slip at Devonport, and moved directly to the Ordinary; she never served in the British Navy, being finally sold to the Prussian government in 1862. *(NMM J6392)*

under Cmdr William Henry Fenwick, as RNR drill ship (16 guns) at Bristol; in 7.1864 under Edward Field, then 2.1869 under Albert Henry William Battiscombe, 4.1870 under Charles William Manthorp, 11.1673 under H. B. Stewart and 4.1879 under Robert Sidney Hunt, all in same role. Recommissioned 4.1881 under C. Hawkins, then 9.1888 under Lieut F. H. N. Harvey and 1900 under Basil C. Barber; paid off 30.9.1910. Sold to J. B. Garnham (for £1,010) 24.9.1911 for BU.

Brilliant (L: 1814) Cut down at Sheerness (for £13,386 including fitting) into a 22-gun corvette (20-gun in 1848).

Men: 230. Guns (1948): UD 6 x 8in (52cwt/8ft) + 10 x 32pdrs (50cwt/9ft); QD & Fc 2 x 56pdrs (87cwt/10ft) on pivot mountings + 2 x 32pdr carronades (25cwt/6ft).

Begun: 13.9.1844. Undocked: 25.7.1845. C: 4.1846 (for sea).

Commissioned 28.1.1846 under Capt. Rundle Burges Watson, for the Cape of Good Hope; paid off 10.1849. Fitted for sea at Sheerness (for £3,058) 4–6.1850. Recommissioned 30.5.1856 under Capt. James Aylmer Dorset Paynter, and fitted for sea at Sheerness (for £6,400) 6–8.1856; to North America and the West Indies; paid off 21.11.1757 at Chatham. Fitted at Chatham (for £1,479) as RNR drill ship (20-gun) at London in 1859. Fitted at Chatham as RNR drill ship (16-gun) 11.1861–1.1862, for Dundee; recommissioned 22.11.1861 under Cmdr Grey Skipworth. On 26.11.1864 under Cmdr James Edward Bickford, then 6.7.1868 Cmdr Francis Mowbray Trattenr, then 30.5.1870 Cmdr William Pym, all in same role; paid off 9.7.1871. Subsequently fitted at Dundee in 11.1875 for same role (8-gun) at Inverness in 1876. In 6.1880 under A. J, O'Rorke, then recommissioned 4.1881; in 1886 under Lieut E. S. Evans. Re-named *Briton* 8.11.1889. Sold to Forth Shipbreaking Co. 19.5.1908 to BU at Bo'ness.

Havannah (L: 1811) Cut down at Deptford (for £5,574, plus £11,388 for 1847–48 fitting) into a 22-gun corvette (19-gun in 1848).

Men: 230. Guns (1848): UD 6 x 8in (52cwt/8ft) + 10 x 32pdrs (50cwt/9ft); QD & Fc 1 x 56pdr (87cwt/10ft) on pivot mounting + 2 x 32pdrs (50cwt/9ft).

Begun: 8.2.1845. Undocked: 20.6.1845. C: 6.1845 (for Ordinary),

then laid up at Sheerness. Fitted there for sea 12.1847–3.1848.

Commissioned 24.2.1848 under Capt. John Elphinstone Erskine, for the East Indies; thence to Australia (where Senior Officer); paid off 1850. Repaired and refitted at Plymouth (for £8,734) 7.1852–10.1855; recommissioned 8.1855 under Capt. Thomas Harvey, at Plymouth. Arrived 7.11.1859 at Plymouth for survey; fitted there 7.1860 and lent as a 'Ragged School' training ship at Cardiff 19.3.1860. Sold (by AO 26.10.1904) 1905 to BU.

Trincomalee (L: 1817) Cut down at Portsmouth (for £21,643) into a 26-gun corvette, reduced to 24 guns by 1847.

Men: 240. Guns: UD 10 x 8in (65cwt/9ft) + 8 x 32pdrs (56cwt/9½ft); QD & Fc 2 x 56pdrs (87cwt/10ft) on pivot mountings + 4 x 32pdr carronades (25cwt/6ft).

Begun: 16.4.1845. Undocked: 29.11.1845. C: 9.1847 (for sea).

Recommissioned 20.7.1847 under Capt. William Warren, for North America and the West Indies. On 19 June 1852 under Capt. Wallace Houston, for the Pacific; with Anglo-French squadron there during Russian War. Recommissioned 19.11.1860 under Cmdr Thomas Heard, fitted 1.1861 as RNR training ship (16-gun) at Sunderland 1.1861; on 31.3.1863 under Cmdr Edward Field, in same role. In 7.1864 under Cmdr Walter James Pollard, as RNR drill ship at West Hartlepool; in 7.1867 under Edward T. Knott, then 12.1872 Richard George Kinahan and 4.1880 Herbert Franklyn Crohan, in same role. Recommissioned 4.1881. In 1886 under Frederick Edwards, as RNR drill ship at Southampton; in 5.1890 under Cmdr Reynell J. Fortescue, in same role. Sold to John Read (for £1,323) 19.5.1897; resold 19.5.1897 to Wheatley Cobb as a training ship at Falmouth where renamed *Foudroyant* 1900. Later to Portsmouth Harbour where she remained afloat as a youth training ship. In 7.1987 was loaded aboard a barge and towed up to West Hartlepool, where she underwent restoration to her original state from 1.1990 to 4.2001. (See Andrew Lambert's *Trincomalee*, Chatham Publishing, 2002.)

Amphitrite (L: 1816) Cut down at Portsmouth (for £24,203 including 1847 fitting) into a 26-gun corvette, reduced to 24 guns by 1848.

Men: 240. Guns (1948): UD 10 x 8in (65cwt/9ft) + 8 x 32pdrs (56cwt/9½ft); QD & Fc 2 x 56pdrs (87cwt/10ft) on pivot mountings + 4 x 32pdr carronades (25cwt/6ft).

Begun: 19.5.1845. Undocked: 12.4.1846. C: 4.1846 (for Ordinary), then fitted 8–9.1847 (for sea).

Recommissioned 13.7.1747 under Capt. Thomas Rodney Eden (died 11.1.1850 at Mazatlan), for west coast of Africa, then

to Pacific. On 2.1850 under Capt. John McDougall, in the
Pacific; paid off 1850 at Portsmouth. Refitted there (for £7,114)
11.1850–2.1851; recommissioned 13.12.1850 under Capt.
Charles Frederick, for the Pacific; with Anglo-French squadron
there during Russian War. On 6.12.1854 under Capt. Thomas
Burridge, in the Pacific. Fitted as guard ship at Portsmouth (for
£4,904) 10.1856–5.1857. Lent by the coastguard 14.7.1762 to the
War Office contractor for the forts at Plymouth; returned by
War Office 16.5.1871. BU completed 30.1.1875 at Devonport.

(D) Vessels ordered or re-ordered as steam screw corvettes (from 1845)

The first screw frigates ordered in 1845 were considered by Symonds to
be dubious experiments; the Surveyor wrote in October 1846 that the
prototypes *Termagant* and *Dauntless*, with their long narrow designs,
lacked the displacement needed to cater for the stores and provisions for a
sufficient crew to work their heavy deck armament. Consequently sister
ships ordered for these two were first suspended and later cancelled.

DAUNTLESS **Class** Second Class (from 31 July 1846, First Class)
frigates, 24 guns. *Dauntless* originally ordered on 19 February 1844 to
be built at Portsmouth as a paddle vessel (SV1) 'similar to *Sampson*',
but on 13 August 1844 the Surveyor was asked to prepare two
draughts, one for paddlewheels and the other as a screw vessel, and on
15 November 1844 she was ordered to be fitted for screw propulsion,
with the order switched to Deptford Dyd, but the order was restored
to Portsmouth on 12 February 1845. Designed by John Fincham
(the master shipwright at Portsmouth), whose draught was dated 6
December 1844. The *Dauntless* was lengthened by a 9½ft section at
Portsmouth Dyd in 1850. Re-armed as 33-gun frigate in 1854.

> Dimensions & tons (design): 210ft 0in, 182ft 8in x 39ft 2in oa (38ft
> 8in for tonnage) x 26ft 0in. 1,452⁵/₉₄bm.
> Men: 250. Guns: UD (now 'main deck') 18 x 32pdrs (56cwt); QD/
> Fc ('spar deck') 4 x 10in/84pdr (85cwt/9ft 4in) shell + 2 x 68pdrs
> (95cwt). Originally main deck was to carry 6 x 68pdrs (65cwt) +
> 12 x 32pdrs (56cwt).
> Machinery: 2-cylinder (84in diameter, 4ft stroke) horizontal single
> expansion, geared. Single screw (14¾ft diameter). 580nhp.
> 1,388ihp = 10.016kts.

Dauntless Portsmouth Dyd/Robert Napier & Sons
> As built: 210ft 0in, 182ft 7¾in x 39ft 9in oa (39ft 3in for tonnage) x
> 26ft 0in. 1,496⁶/₉₄bm. 2,242 disp. Draught 8ft 0in/10ft 3in (with
> 30 tons ballast aboard).
> As lengthened 1850: 219ft 6in, 192ft 1¼in x 39ft 9in (39ft 3in for
> tonnage) x 26ft 0in. 1,574⁸/₉₄bm. 2,457 disp.
> Ord: 12.2.1845. K: 4.1845. L: 5.1.1847. Fitted engines at Glasgow
> 4.1847–8.1848. C: 16.10.1850.
> First cost: £96,441 (including hull £42,365, machinery £41,032).
> Commissioned 7.8.1850 under Capt. Edward Pellew Halstead, for
> the Experimental Squadron, then 1852 to North America and the
> West Indies; paid off 19.5.1853 at Portsmouth. Recommissioned
> 28.12.1853 under Capt. Alfred Phillipps Ryder, for the Baltic
> and the Black Sea during the Russian War; in 2.1855 in
> action against Russian attacks on Eupatoria; bombardment
> of Sevastopol 4.1855; bombardment of Kinburn 10.1855;
> then to the Mediterranean; paid off 13.3.1857 at Portsmouth.
> Recommissioned 10.6.1859 under Capt. William Edmonstone;

on 12.8.1859 under Capt. John Borlase, then 23.8.1859 under
Capt. Leopold George Heath, as guard ship in Southampton
Water (or Newhaven?); on 1.1.1861 under Capt. James Willcox,
then 13.6.1862 Capt. Sherard Osborn and 14.7.1862 Capt. John
Bourmaster Dickson, in same role. On 1.4.1864 under Capt.
James Newburgh Strange, as guard ship at Hull; on 30.9.1865
under Capt. Edward Pelham Brenton Von Donop, then
19.11.1868 Capt. Charles Codrington Forsyth, in same role; paid
off 31.12.1869. On 1.1.870 tender to Humber coastguard ship
Wyvern at Hull; laid up at Plymouth 1878. Sold 1.5.1885.

Vigilant Portsmouth Dyd/—
> Ord: 26.3.1846. K: — . Suspended 9.9.1846. Cancelled 22.5.1849.

TERMAGANT **Class** Second Class (from 31 July 1846, First Class)
frigates, 24/25 guns. White design, dated 6 December 1844. Originally
to have been 208ft long, but *Termagant*'s bow was extended by White
to improve speed in June 1846. Main fitting out completed 25 August
1848. In 1854, the original engine was replaced at Portsmouth with
a 2-cylinder (62½in diameter, 3ft stroke) horizontal 400nhp engine
(1,206ihp = 10.66kts). The *Euphrates*'s design was stretched on 3 July
1846 to 215ft 7in length (2,402 tons displacement) before she was
cancelled.

> Dimensions & tons (design): 208ft 0in, 185ft 6in x 40ft 0in (39ft 4in
> for tonnage) x 26ft 4in. 1,527bm.
> Men: 250. Guns: 12 x 42pdrs (84cwt), 8 x 32pdrs (56cwt), 2 x 8in
> (112cwt), 6 x 10in (84cwt) as ordered. Completed with 24 guns
> only; in 1856 became 25-gun, with 1 x 10in (85cwt/9ft 4in) on
> pivot added.
> Machinery: 4-cylinder (62in diameter, 3½ft stroke) horizontal single
> expansion, geared. Single screw. 620nhp. 1,351ihp = 9.51kts.

Termagant Deptford Dyd/Seaward & Capel
As built: 210ft 1in, 181ft 10in x 40ft 6in (40ft 0in for tonnage) x 26ft
4in. 1,547⁴⁸/₉₄bm. 2,312 disp.
> Ord: 19.2.1845. K: 4.1845. L: 25.9.1847. C: 12.10.1854.
> First cost: Hull £34,264, machinery £35,314, fitting £36,905.
> Commissioned 30.8.1852 at Portsmouth under Capt. George
> Giffard, for particular service; paid off 6.1.1853. Re-engined at
> Portsmouth 1854. Recommissioned 14.6.1854 under Capt. Keith
> Stewart, for conveying French troops to the Baltic during the
> Russian War and returning with prisoners of war 22.9.1854. On
> 3.10.1854 under Commodore Thomas Henderson, for the West
> Indies (as Senior Officer there); on 3.8.1855 under Commodore
> Henry Kellett, in same role. On 16.3.1857 under Lieut Cmdr
> Colin Cambell Kane, for return home; paid off 19.9.1857. Fitted
> for reserve at Portsmouth 1858. Recommissioned 27.1.1859
> under Capt. Robert Hall, for the Channel Squadron until 8.1859;
> paid off 9.1.1863 at Portsmouth. Sold to Castle & Beech 3.1867
> to BU at Charlton.

Euphrates Deptford Dyd/Seaward & Capel
> Ord: 23.1.1846 & 26.3.1846. K: — . Suspended 9.9.1846. Cancelled
> 22.5.1849.

FERVENT First Class frigate, 21 guns. Oliver Lang design, 1845.
Ordered to be built as a screw steam frigate as close to the *Terrible*
paddle frigate (see Chapter 11) 'as may be consistent with the
difference between the screw and paddle wheels'. Probably intended
to be similar to *Dauntless* (which at design stage was also listed as
an identical 1,453bm). Originally named *Watt* on 4 March 1845, but
renamed *Fervent* on 22 August 1845.

> Dimensions & tons: 210ft 0in, ?175ft 0in x 40ft 0in (?39ft 6in for
> tonnage) x — 1,453bm. 1,847 disp.

Men: 250. Guns: —
Machinery: — 800nhp.

Fervent (ex-*Watt*, renamed 22.3.1845) Woolwich Dyd
Ord: 20.2.1845. K: — . Suspended 7.4.1845. Cancelled 22.5.1849.

NARCISSUS 28 guns. On 25 April 1847 it was instructed that the *Narcissus*, on order as a sailing sloop at Devonport, should be completed as a 1,200bm screw ship of 28 guns (Sixth Rate). This was cancelled in 1848, with the materials transferred to a new 50-gun sail frigate, which was given the same name.

The category of screw corvette did not formally exist in the Royal Navy until the start of 1854. Nevertheless the following vessels with their primary battery of 20 guns were clearly the lineal descendents of the sailing Sixth Rates and have been so treated in this work. In 1854 the existing *Highflyer*, and the still-building *Esk* and *Pylades* (plus the three newly ordered *Pearl* Class vessels) were re-classified from frigates. The two vessels building for Russia and sequestered in April 1854 (*Cossack* and *Tartar*) were added to this category.

The Programme of Works for 1850 had included fresh orders for four small screw 'frigates' of 1,500bm, placed on 23 March 1850 with Woolwich, Chatham, Sheerness (replacing the paddle sloop *Resolute*, cancelled on this date) and Pembroke Dyds (one ship each). While these were never built to this design, the Woolwich and Chatham orders were to materialise as the first *Pearl* Class corvettes (the Sheerness order became the *Pylades*, and the Pembroke order the frigate *Aurora*). The orders were amended to 1,161bm ('as *Highflyer*') in 1852 and to a new 1,390bm design in 1853. Corvette designs were rapidly enlarged during the 1860s, as the corvette began to take over the traditional frigate role of policing the high seas. To this end they were all built as steam auxiliaries, designed to cruise under sail.

HIGHFLYER **Class** First Class corvette, 21 guns. Surveyor's Department design, approved 30 November 1849. *Highflyer* was originally ordered 25 April 1847 at Woolwich Dyd (from 31 March 1849, Chatham Dyd) as 3rd Class screw sloop ('as *Rattler*, enlarged' – i.e. *Brisk* Class), and went through several changes before final design was approved on 30 November 1849, when building was switched to contract with Mare; she was redesignated a frigate (Fifth Rate) on 3 March 1852. A sister ship was ordered on 23 March 1850 from Deptford Dyd; but on 27 March 1852 this was replaced by an order for the sloop *Fawn*. Another to this design, the *Esk*, was built in exchange for the iron frigate *Greenock* by Scott Russell acting on behalf of the Australian Royal Mail Co. – the agreement with Scott Russell was approved on 18 August 1852 and dated 16 September 1852.

Dimensions & tons: original design 192ft 0in, 168ft 5in x 36ft 4in oa (36ft 0in for tonnage) x 22ft 8in. 1,161bm.; final design 192ft 0in, 167ft 3¾in x 36ft 4in oa (36ft 0in for tons) x 22ft 8in. Draught 15¾ft (mean). 1,153³⁹⁄₉₄bm. 1,737½ disp.

Men: 250. Guns: 20 x 32pdrs/8in (52cwt/8ft) – broadside (later altered to 18 x 8in); 1 x 68pdr/10in (84cwt/9ft 4in) on pivot.

Machinery: 2-cylinder (55³⁄₁₆in diameter, 2½ft stroke) horizontal single expansion engine, geared (*Highflyer*). 2-cylinder (50in diameter, 2¾ft stroke) inclined, oscillating engine (*Esk*). Single screw. 250nhp. *Highflyer* 702ihp = 9.399kts. *Esk* 657ihp = 9.439kts.

Highflyer C. J. Mare & Co., Blackwall/Maudslay, Sons & Field
Tonnage as completed: 1,153bm. 1,902 disp.
Ord: 25.4.1847. K: 1.1850. L: 13.8.1851. C: 10.4.1852.
First cost: £56,075 total (hull £27,105, machinery £17,431, fitting £11,539).
Commissioned 15.3.1852 at Woolwich under Capt. Henry James

Matson (died 14.12.1852), for North America and the West Indies. On 15.12.1852 under Capt. Edmund Heathcote, on same station. On 1.4.1853 under Capt. John Moore, for the Mediterranean; at Bombardment of Sebastopol 11.10.1854; still in Black Sea 1855; paid off 7.6.1856 at Portsmouth. Refitted there (for £8,308) 1856. Recommissioned 1.8.1856 under Capt. Charles Frederick Alexander Shadwell, for East Indies and China; took part in 2nd Anglo-Chinese War; in boat action at Fatshan 1.6.1857, capture of Canton 12.1857, and attack on Peiho forts 25.6.1859. On 2.1.1860 under Acting Capt. William Andrew James Heath, still in East Indies and China; paid off 31.5.1861 at Portsmouth. Recommissioned 15.12.1864 under Capt. Thomas Malcolm Sabine Pasley, for the Cape of Good Hope and thence to East Indies; paid off 31.8.1868. BU 5.1871 at Portsmouth.

Esk J. Scott Russell & Co., Millwall/J. Scott Russell & Co.
Tonnage as completed: 1,169bm. 1,900 disp.
Ord: 18.8.1852. K: 4.1853. L: 12.6.1854. C: 21.12.1854.
Commissioned 28.10.1854 at Woolwich under Capt. Thomas Francis Birch, for the Baltic; paid off 25.2.1856. Recommissioned 1.3.1856 under Capt. Sir Robert John Le Mesurier McClure, for East Indies and China; at Capture of Canton 12.1857; paid off 26.6.1861 at Portsmouth. Recommissioned 22.5.1863 under Capt. John Fane Charles Hamilton (killed 29.4.1864), for New Zealand; in assault on Gate Pah 29.4.1864. On 29.4.1864 under Cmdr George Graham Duff, then 16.7.1864 under Capt. John Proctor Luce, in New Zealand and Australia; paid off 17.10.1867 at Portsmouth. BU 1870 at Portsmouth.

PYLADES **Class** First Class corvette, 21 guns. Surveyor's Department design of 1852. Originally ordered to the lines of the *Highflyer*, but fresh plans were drawn, developed from lines of *Highflyer* but with extra 2ft of beam. The 1852 order replaced one given to Sheerness on 23 March 1850 for a 1,500-ton frigate. *Pylades* was named on 8 June 1853 (along with *Esk* and *Pearl*).

Dimensions & tons: 192ft 9in, 165ft 0½in x 38ft 4in oa (38ft 0in for tonnage) x 23ft 11in. Draught 19ft 7in. 1,267⁷⁷⁄₉₄bm. 1,956 disp.

Men: 250. Guns: 20 x 8in (42cwt) MLSB – on broadside trucks; 1 x 68pdr/10in (95cwt) MLSB – pivot-mounted at bow. Reduced to 17 guns in 1868.

Machinery: 2-cylinder (55in diameter, 3ft stroke) horizontal single expansion, trunk. Single screw. 350nhp. 1,106ihp = 10.119kts.

First cost: £68,333 (including machinery £21,684).

Pylades Sheerness Dyd/John Penn & Son
Tonnage as completed: 1,278bm. Draught 16ft 0in (fwd), 19ft 0in (aft).
Ord: 24.12.1852. K: 9.5.1853. L: 23.11.1854. C: 29.3.1855.
Commissioned 5.1.1855 at Sheerness under Capt. Edwin Clayton Tennyson d'Eyncourt, for the Baltic during the Russian War; paid off 18.11.1856 at Sheerness. Recommissioned 16.7.1857 under Capt. Michael de Courcy, for the East Indies and China, later to Pacific; paid off 30.7.1861 at Chatham. Recommissioned 2.12.1862 under Capt. Arthur William Acland Hood, for North America and the West Indies; paid off 3.9.1866 at Chatham. Recommissioned 4.12.1867 at Sheerness under Capt. Cecil William Buckley, for the Pacific; paid off 7.1870 at Gibraltar. Recommissioned 20.7.1871 under Capt. Augustus Chetham Strode, for the east coast of South America. On 20.8.1873 under Acting Capt. Arthur Richard Wright, returning from same station; paid off 31.12.1873 at Sheerness. Sold to Castle 23.1.1875 to BU at Charlton.

An undated photograph of the 21-gun First Class corvette *Pylades*. Note the row of scuttles along the lower deck, a vital addition to enhance ventilation of the crew accommodation for ships designed for frequent deployment to tropical climates. *(NMM 8948)*

COSSACK Class First Class corvettes, 20 guns. Building for Russia by Pitcher when seized on 5 April 1854 for RN, who also requisitioned the machinery building by Maudslay. The original contract price to the Russians was £46,198 (including £16,750 for machinery). For *Cossack*, the Admiralty then paid £9,591 to the contractors to complete this contract payment (including £4,187 for machinery) plus £715 for tonnage in excess of specification and £2,206 for modifications for HM service; for *Tartar*, they paid £16,607 (including £4,187 for machinery) plus £715 for excess tonnage and £1,883 for modifications.

Note the transliteration of the Russian names is 'Continental style', i.e. with the W and J being pronounced as V and I (or Y), which render the names as *Vityaz* and *Voyn*.

> Dimensions & tons: 195ft 0in, 172ft 1¾in x 38ft 6in oa (38ft 0in for tonnage) x 22ft 4in. Draught 17ft 7in (fwd), 17ft 8in (aft). 1,322⁶⁰⁄₉₄bm. 1,965 disp.
> Men: 270 Guns: 18 x 8in (60cwt/8ft 10in) MLSB – on broadside trucks, + 2 x 68pdrs (95cwt/10ft) on pivot mountings.
> Machinery: 2-cylinder (55in diameter, 2¼ft stroke) horizontal single expansion. Single screw. 250nhp. See below for individual ihp/speeds.

Cossack (ex-Russian *Witjas*) W. & H. Pitcher, Northfleet/Maudslay, Sons & Field (870ihp = 9.055kts)

A photograph of the *Cossack* at Mauritius in the late 1860s. She and her sister *Tartar* had been originally ordered for Russia's Black Sea fleet from William Pitcher in December 1852 as *Vitiaz* ('Knight') and *Voin* ('Warrior') at a cost of 290,000 silver roubles each. Work had begun by September 1853 at Northfleet, but with deteriorating relations between Britain and Russia, the latter feared possible confiscation; both had been sold to the Hamburg trading house of Henry Mersk & Co., and they were seized by the British Treasury in April 1854 following the outbreak of war. *(NMM C9357)*

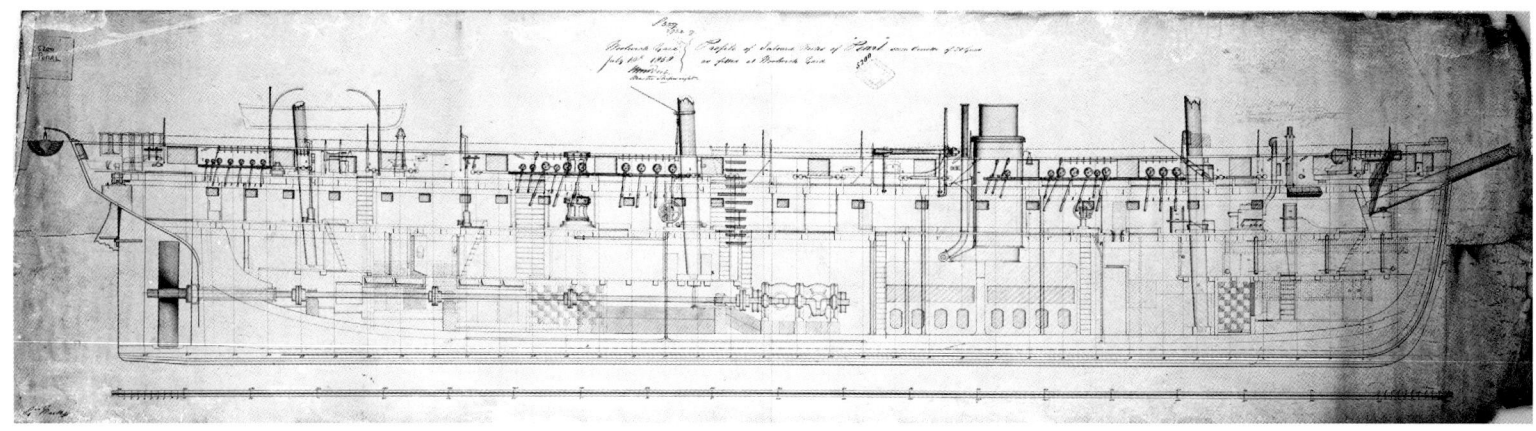

Seized 5.4.1854. L: 15.5.1854. C: 28.7.1854 (partial fitting by
  builders), then 10.10.1854 at Chatham (for sea).
First cost: see above for amount to contractors. Fitting at Chatham
  £8,611 giving total cost of £57,730.
Commissioned 19.8.1854 at Chatham under Capt. Edward Gennys
  Fanshawe, for the Baltic during the Russian War. On 21.8.1855
  under Capt. James Horsford Cockburn, in the Baltic; to
  North America and West Indies 7.1856 (chiefly at Greytown,
  Nicaragua); paid off 15.7.1857. Refitted for sea at Sheerness (for
  £12,075) 1859; recommissioned 16.6.1859 under Capt. Richard
  Moorman, for North America and the West Indies, then to
  China; paid off 22.8.1862 at Sheerness. Recommissioned 19.5.1863
  under Capt. William Rae Rolland, for the Mediterranean; on
  26.5.1865 under Capt. Richard Dunning White, still in the
  Mediterranean; paid off 23.2.1867 at Sheerness. Recommissioned
  9.12.1868 under Capt. John Edward Parish, for the East Indies;
  on 12.8.1871 under Capt. Robert Gordon Douglas, for Australia;
  paid off 18.7.1873 at Sheerness. Sold to Castle 19.5.1875 for BU at
  Charlton.
Tartar (ex-Russian *Wojn*) W. & H. Pitcher, Northfleet/Maudslay, Sons
  & Field (731ihp = 9.4kts)
    Seized 5.4.1854. L: 17.5.1854. C: 10.1854 (partial fitting by builders),
      then 26.1.1855 at Chatham (for sea).
    First cost: see above for amount to contractors. Fitting at Chatham
      £8,546 giving total cost of £57,342.
    Commissioned 30.10.1854 under Capt. Hugh Dunlop, for the
      Baltic during the Russian War, then to North America and the
      West Indies; paid off 22.11.1859 at Sheerness. Recommissioned
      15.8.1860 under Capt. John Montagu Hayes, for China; at
      bombardment of Simonoseki 5–8.9.1861; paid off 17.5.1865 at
      Sheerness. Sold 2.1866 to Castle for BU at Charlton.

PEARL Class First Class corvettes, 21 guns. Surveyor's Department
design, approved 18 November 1853. First three were ordered to a
design of 1,390 tons, enlarged to 1,462 tons later in November 1853.
The order for *Cadmus* replaced an 1834 order at Chatham for sailing
corvette *Coquette* (which was never formally cancelled). Three more
were ordered a year later, and a final four in 1855. Penn cylinders
were of 58⅛in diameter, 3¼ft stroke; Ravenhill cylinders were of
64in diameter, 3ft stroke. *Charybdis* was lent to the Canadian government
from October 1880 to August 1882 as a training ship. *Challenger* was
the vessel used for the pioneer oceanographic world voyage of 1872–76.
On 10 February 1858 the three last ships above were redesignated as a
separate class, with a second pivot-mounted 68pdr added.
    Dimensions & tons: (225ft 3in oa) 200ft 0in, 171ft 9¾in x 40ft 4in
      oa (40ft 0in for tonnage) x 23ft 11in. 1,462²²⁄₉₄bm. 2,187 tons

The inboard profile for the screw corvette *Pearl*, as fitted at Woolwich and dated 10 July
1858. These 'open battery' corvettes mounted all their guns on an exposed weather deck
(note the pivot gun shown forward), but as in eighteenth-century frigates there was a
complete unarmed deck below, indicated by a row of ventilation scuttles. This provided
plenty of berthing space and led to the description of these ships as 'troop frigates' because
they could transport soldiers when required. *(NMM DR5200)*

displacement (last three 2,306 tons).
  Men: 270. Guns: MD 20 x 8in (60cwt/8ft 10in) MLSB shell guns – on
    broadside trucks; UD ('spar deck') 1 x 68pdr/10in (95cwt/10ft)
    MLSB – pivot-mounted at bow (2 x 68pdrs in *Challenger*,
    Racoon and *Clio*). By 1868 most were reduced to 17 guns (18 in
    final trio); generally 64pdrs had replaced some or all of the 8in
    shell guns.
  Machinery: 2-cylinder horizontal single expansion, trunk in Penn-
    engined ships. Single screw. 400nhp. See below for ihp/speed.
Pearl Woolwich Dyd/John Penn & Son (1,324ihp = 11.313kts)
  Tonnage as completed: 1,469bm. 2,115 disp. Draught 16ft 2in (fwd),
    19ft 9in (aft).
  Ord: 2.4.1853. (named 8.6.1853) K: 1.1.1854. L: 13.2.1855. C:
    25.1.1856.
  First cost: £75,054 including £39,185 for hull and £25,354 for
    machinery.
  Commissioned 1.1.1856 at Woolwich under Capt. Edward Southwell
    Sotheby, for the Pacific ; in Peru during revolution of 1857;
    to India during the Mutiny (Pearl's Naval Brigade was ashore
    under Sotheby's command); paid off 16.6.1859 at Portsmouth.
    Recommissioned 12.8.1859 under Capt. William Edmonstone,
    then 23.8.1859 Capt. John Borlase, for the East Indies and
    China; in Taiping rebellion; in Bombardment of Kagoshima
    during 'Anglo-Satsuma War'; paid off 18.6.1864 at Portsmouth.
    Recommissioned 4.5.1866 under Capt. John Francis Ross, for
    China; joined Flying Squadron at Yokohama 17.4.1870; paid off
    1.12.1870 at Portsmouth. Recommissioned 16.5.1873 under Capt.
    James Graham Goodenough (died 20.8.1875), for Australia , of
    which station she was flagship from 16.10.1873. On 7.9.1875
    under Commodore Anthony Hiley Hoskins, in Australia; from
    1.1877 under Capt. Lindesay Brine, for return to UK; paid off
    5.6.1877 at Portsmouth. Sold to Castle 8.1884 to BU at Charlton.
Satellite Devonport Dyd/John Penn & Son (1,214ihp = 10.55kts)
  Tonnage as completed: 1,462bm. 2,138 disp. Draught 16ft 8in (fwd),
    19ft 6in (aft).
  Ord: 2.4.1853. (named 18.11.1853) K: 8.7.1854. L: 26.9.1855. C:
    23.12.1856.
  First cost: £74,185, including £32,535 for hull and £26,329 for
    machinery.
  Commissioned 30.9.1856 under Capt. James Charles Prevost, for

the Pacific; paid off 26.1.1861. Recommissioned 27.11.1861 under Capt. John Ormsby Johnson, for east coast South America; on 5.5.1862 under Capt. Stephen Smith Lowther Crofton (Johnson invalided), on same station; paid off 22.9.1865 at Plymouth. Recommissioned 27.10.1866 under Capt. Richard Purvis; on 14.11.1866 under Capt. Joseph Edye, for China (died 13.9.1868); on 11.11.1868 under Capt. William Henry Edye, on China station, then home with the 1869 Flying Squadron; paid off 1.12.1870 at Plymouth. BU completed 25.12.1879 at Devonport.

Cadmus Chatham Dyd/John Penn & Son (1,526ihp = 11.825kts)
Tonnage as completed: 1,461bm. 2,216 disp. Draught 16ft 11in (fwd), 20ft 1in (aft).
Ord: 2.4.1853. (named 3.4.1854) K: 10.5.1855. L: 20.5.1856. C: 4.6.1859.
First cost: £70,171, including £33,409 for hull and £26,777 for machinery.
Commissioned 11.5.1859 under Capt. Henry Shank Hillyar, for North America and the West Indies (–7.1862). On 24.11.1862 under Capt. John Francis Ross, in same role; paid off 6.5.1863 at Chatham. Recommissioned 15.12.1864 under Capt. Alexander Crombie Gordon, for same station; paid off 28.5.1868 at Chatham. Recommissioned 15.4.1869 at Sheerness under Capt. Robert Gibson, for the Flying Squadron; damaged by grounding on Salcombe Rocks 4.6.1869 and returned to Sheerness to pay off 8.6.1869 (replaced in Flying Squadron by *Barrosa*). Recommissioned 1.12.1870 at Plymouth under Capt. William Henry Whyte; to Detached Squadron 1871; to China 4.1872; paid off 26.11.1874. BU 9.1879 at Deptford.

Scout Woolwich Dyd/John Penn & Son (1,327ihp = 10.875kts)
Tonnage as completed: 1,462bm. 2,272 disp. Draught 17ft 9in (fwd), 19ft 10in (aft).
Ord: 3.4.1854. K: 10.1854. L: 31.12.1856. C: 9.7.1859.
First cost: £74,642, including £37,001 for hull and £24,373 for machinery.
Commissioned 14.6.1859 at Sheerness under Capt. John Corbett, for East Indies and China; paid off 23.4.1864. Recommissioned 17.5.1865 under Capt. Charles Henry May, then on 22.5.1865 under Capt. John Adolphus Pope Price, for the Pacific; paid off 5.5.1869 at Sheerness. Commissioned 2.3.1871 under Capt. Ralph Peter Cator, for the Pacific; paid off 2.6.1875. BU completed 6.3.1877 at Chatham.

Scylla Sheerness Dyd/John Penn & Son (1,376ihp = 10.838kts)
Tonnage as completed: 1,467bm. 2,199 disp. Draught 17ft 8in (fwd), 19ft 2in (aft).
Ord: 3.4.1854. K: 4.1.1855. L: 19.6.1856. C: 15.8.1859.
First cost: £76,023, including £41,301 for hull and £24,585 for machinery.
Commissioned 13.6.1859 at Sheerness under Capt. Rowley Lambert, for the Mediterranean; paid off 3.1.1863. Recommissioned 9.9.1863 under Capt. Samuel Gurney Cresswell; then on 24.9.1863 under Capt. Richard William Courtenay, for China; paid off 7.8.1867. Recommissioned 15.4.1869 under Capt. Frederick Anstruther Herbert, for the Flying Squadron, then to the Pacific (relieved *Charybdis* at Esquimalt on 15.5.1870); on 13.5.1871 under Capt. Charles Richard Fox Boxer, in the Pacific; paid off 15.4.1873 at Sheerness. Sold to Castle 7.11.1882 to BU at Charlton.

Charybdis Chatham Dyd/Miller, Ravenhill & Salkeld (1,363ihp = 10.081kts)
Tonnage as completed: 1,462bm. 2,231 disp. Draught 17ft 4in (fwd), 19ft 10in (aft).

Ord: 3.4.1854. K: 29.3.1856. L: 1.6.1859. C: 19.11.1860.
First cost: £43,912 for hull.
Commissioned 5.11.1860 at Sheerness under Capt. George Disney Keane, for the East Indies, then to the Pacific; on 8.6.1863 under Capt. Edward Winterton Turnour, in the Pacific; paid off 14.6.1865 at Sheerness. Reduced to 18 guns. Recommissioned 8.1.1867 under Capt. Algernon McLennan Lyons, for Australia and the Pacific; from Esquimalt to Valparaiso 28.5–14.8.1869 with the 1869–70 Flying Squadron; paid off 30.8.1871 at Sheerness. Commissioned 24.9.1873 under Capt. Thomas Edward Smith, for China; Lingie River 5.1874; on 9.2.1877 under Capt. Charles Frederick Hotham. Recommissioned 8.5.1877 at Hong Kong, still under Hotham, on China station; paid off 9.11.1880. Lent to Canadian Govt 11.1880 as a training ship; returned to RN 8.1882 at Halifax. Sold 1884 at Halifax.

Pelorus Devonport Dyd/Miller, Ravenhill & Salkeld (1,408ihp = 10.912kts)
Tonnage as completed: 1,464bm. 2,290 disp. Draught 17ft 3in (fwd), 20ft 6in (aft).
Ord: 31.3.1855. K: 25.3.1856. L: 5.2.1857. C: 10.9.1857.
First cost: £74,434, including £32,664 for hull and £24,272 for machinery.
Commissioned 16.7.1857 at Plymouth under Commodore Frederick Beauchamp Paget Seymour, for the East Indies and China; later to Australia (during the New Zealand War); paid off 12.12.1862 at Plymouth. Recommissioned 11.9.1863 under Cmdr Henry Boys, for the East Indies and China; on 23.11.1864 under Capt. William Henry Haswell, on China station; paid off 17.4.1868 at Plymouth. BU completed 3.2.1869 at Devonport.

Challenger Woolwich Dyd/John Penn & Son (1,450ihp = 10.721kts)
Tonnage as completed: 1,465bm. 2,137 disp. Draught 17ft 4in (fwd), 18ft 10in (aft).
Ord: 31.3.1855. K: 3.10.1855. L: 13.2.1858. C: 28.5.1861 at Sheerness.
First cost: £76,272 (40,069 to build, plus £24,264 for machinery (paid to Penn), £11,939 fitting at Sheerness).
Commissioned 5.5.1861 at Sheerness under Capt. John James Kennedy, for North America and the West Indies; at Veracruz 1.1862; paid off 3.2.1865 at Sheerness. Recommissioned 10.4.1866 under Commodore Rochfort Maguire, as flagship for Australia; on 12.8.1867 under Commodore Rowley Lambert; paid off 4.3.1871 at Sheerness. Fitted as survey ship 1872. Recommissioned 23.5.1872 at Sheerness under Capt. George Strong Nares, for Particular Service, Survey and Discovery (an Admiralty-Royal Society joint round-the-world oceanographic expedition). On 11.12.1874 under Capt. Frank Tourle Thomson; paid off 12.6.1876 at Sheerness. Recommissioned 26.6.1876 under Capt. William Samuel Brow, as temporary guard ship of 1st Reserve at Harwich; paid off 26.4.1878. Hulked as training ship 1880, then as a receiving ship; from 1910 an accommodation ship at Chatham. Sold to J. B. Garnham 6.1.1921 to BU.

Racoon Chatham Dyd/Miller, Ravenhill & Salkeld (1,651ihp = 11.416kts)
Tonnage as completed: 1,467bm. 2,192 disp. Draught 17ft 3in (fwd), 19ft 6in (aft).
Ord: 31.3.1855. K: 21.4.1856. L: 25.4.1857. C: 8.2.1858.
First cost: £69,586, including £29,524 for hull and £24,281 for machinery.
Commissioned 22.11.1857 at Chatham under Capt. James Aylmer Dorset Paynter, for the Channel Squadron; on 1.5.1860 under Capt. William Charles Chamberlain, for the Mediterranean; paid off 21.1.1862 at Chatham. Recommissioned 29.1.1863 under

Capt. Count Gleichen, for the West Indies; to the Mediterranean
4.1864; to Queenstown 3.1866; on 9.4.1866 under Capt. William
Armytage, at Queenstown; paid off 3.11.1866 at Plymouth.
Recommissioned same day under Capt. Richard Purvis, for Cape
of Good Hope, then to North America and West Indies; paid
off 6.5.1870 at Bermuda. Recommissioned 7.5.1870 under Capt.
Edward Henry Howard, for North America and West Indies; paid
off 2.7.1873 at Plymouth. BU completed 28.8.1877 at Devonport.

Clio Sheerness Dyd/Miller, Ravenhill & Salkeld (1,540ihp = 11.202kts)
  Tonnage as completed: 1,458bm. 2,153 disp. Draught 17ft 6in (fwd),
    18ft 10in (aft).
  Ord: 31.3.1855. K: 25.6.1856. L: 26.8.1858. C: 18.9.1859.
  First cost: £81,651, including £42,972 for hull and £24,268 for
    machinery.
  Commissioned 23.6.1859 at Sheerness under Capt. Thomas Miller,
    for the Pacific; paid off 29.8.1863 at Sheerness. Recommissioned
    6.7.1864 under Commodore Frederick Henry Stirling, for
    Australia; paid off 21.1.1874 at Sheerness. Recommissioned
    22.3.1870 under Capt. Nicolas Edward Brook Turnour, for the
    Pacific; paid off 18.7.1868 at Sheerness. Reduced to 18 guns (2 x
    64pdrs on pivots, 16 x 64pdrs on broadsides). Recommissioned
    22.3.1870 under Capt. Frederick Henry Stirling, for Australia , of
    which station she was flagship from 3.9.1870 to 16.10.1873; holed
    in Bligh Sound 1871 and beached to avoid sinking, then salved
    and repaired; sailed 16.10.1873 for Portsmouth and paid off 1874.
    Training ship 1876 for Landegfan, in the Menai Strait. Lent as a
    training ship for Boys to the North Wales Association, Bangor
    1890. Sold on 3.10.1919 and BU at Bangor.

JASON Class Open battery corvettes, 21 (later 17) guns. Surveyor's
Department design, approved 10 February 1858. The first three were
ordered in 1856 as units of the preceding *Pearl* Class, although the
Galatea order was quickly altered to that for a Fourth Rate frigate. The
next two were ordered a year later; the *Jason* and *Barrosa* were named
on the same date. Following Baldwin Walker's decision on 18 December
1857 to increase the length of all ships ordered where building was not
too far advanced, the design was enlarged to 1,623 tons on 10 February
1858 (before any vessel was laid down). Three more ships (of which the
North Star was subsequently altered to a new design) were ordered in
1858. A further enlargement to 1,702 tons of the design took place in
1859, and all but *Jason* were re-ordered to it; *Jason* followed suit in the
1860 Programme. *North Star* and *Favourite* (see below) were originally
to have been to this design.
  Referred to as 'troop frigates', all were ship-rigged, with telescopic
funnels. The Armstrong BL guns were removed in 1864 following
a series of accidents. In all, 4 x 8in guns were replaced by 4 x 40pdrs
(28cwt/8ft), then the surviving quartet were all reduced to 17 guns by
1869; *Jason* had 8 x 64pdrs and 8 x 8in shell, plus the 110pdr Armstrong
BL; *Barrosa* and *Wolverene* had 17 x 64pdrs; and *Rattlesnake* had 9 x
64pdrs (one on a pivot mounting) and 8 x 8in shell. *Wolverene*'s original
engine was never satisfactory, and Ravenhill replaced it with a new type
(also 400nhp) in 1876; now displacing 2,568 tons, she reached 11.314kts
with 1,493ihp.
  Dimensions & tons: 1858 design: 225ft 0in, 187ft 7in x 40ft 8in oa
    (40ft 4in for tonnage) x 24ft 2in. 1,623¹⁴/₉₄bm.
  1859 design: 225ft 0in, 196ft 8½in x 40ft 8in oa (40ft 4in for tonnage)
    x 24ft 2in. 1,702¹⁷/₉₄bm. 2,431 disp.
  Men: 240 (later 275). Guns: 20 x 8in (60cwt/8ft 10in) shell MLSB –
    on broadside trucks; 1 x 110pdr/7in (82cwt) Armstrong BL – on
    pivot at bow.
  Machinery: 2-cylinder (of 64in diameter, except *Orestes* 60¾in; 3ft

stroke, except *Orpheus* 2ft 8in) horizontal single expansion (with
return connecting-rods in *Wolverene*). Single (hoisting) screw.
400nhp. See below for ihp/speed.

Jason Devonport Dyd/Ravenhill, Salkeld & Co. (1,516ihp = 12.04kts)
  Tonnage as completed: 1,711bm. 2,468 disp. Draught 18ft 6in (fwd),
    20ft 3in (aft).
  Ord: 9.4.1856. (named 1.4.1857) K: 3.6.1858. L: 10.11.1859. C:
    8.4.1861 (for sea).
  First cost: Hull £41,442.
  Commissioned 20.11.1860 (appointed 16.11) at Devonport
    under Capt. Edward Pelham Brenton Von Donop, for North
    America and the West Indies; paid off 3.12.1864 at Devonport.
    Recommissioned 7.5.1866 (appointed 4.5) under Capt. Charles
    Murray Aynesley, for same station; paid off 1869 at Devonport.
    Engines removed by 1876. BU completed 26.9.1877 at
    Devonport.

Barrosa Woolwich Dyd/James Watt & Co. (1,616ihp = 11.514kts)
  Tonnage as completed: 1,702bm. 2,302 disp. Draught 16ft 10in (fwd),
    20ft 3in (aft).
  Ord: 9.4.1856. (named 1.4.1857) K: 2.8.1858. L: 10.3.1860. C:
    15.2.1861 (for Ordinary).
  First cost: Hull £42,444.
  Commissioned 24.9.1862 at Devonport under Capt. William
    Montagu Dowell, for China; at Bombardment of Simonoseki
    (Japan) 5–6.9.1864. On 23.11.1864 under Capt. Henry
    Boys, still on China station; paid off 30.1.1867 at Sheerness.
    Recommissioned 9.6.1869 under Capt. Robert Gibson, for the
    'Flying Squadron'. On 24.11.1869 under Capt. Lewis James
    Moore, with the Flying Squadron and then to China; paid off
    27.7.1873 at Sheerness. BU completed 20.1.1877 at Chatham.

Galatea Woolwich Dyd/John Penn & Son
  Ord: 9.4.1856. K: — . Re-ordered on 2.7.1856 as – and completed as
    – wooden screw frigate (which see).

Orpheus Chatham Dyd/Humphrys & Tennant (1,333ihp = 11.155kts)
  Tonnage as completed: 1,706bm. 2,365 disp. Draught 18ft 0in (fwd),
    19ft 9in (aft).
  Ord: 1.4.1857. K: 12.5.1858. L: 23.6.1860. C: — .
  Commissioned 25.10.1861 at Portsmouth under Capt. William
    Farquharson Burnett, on Australian station (Commodore from
    21.7.1862); wrecked on bar in Manukau Harbour, Auckland,
    New Zealand 7.2.1863 (189 out of 259 aboard drowned including
    Burnett).

Orestes Sheerness Dyd/Robert Napier & Sons (1,522ihp = 12.265kts)
  Tonnage as completed: 1,702bm. Draught (light) 14ft 7in (fwd), 16ft
    5in (aft).
  Ord: 1.4.1857. K: 26.8.1858. L: 18.8.1860. C: — .
  Commissioned 31.8.1861 at Sheerness under Capt. Alan Henry
    Gardner, for Cape of Good Hope; paid off 21.6.1865 at Sheerness.
    BU completed 11.1866 at Portsmouth.

Rattlesnake Chatham Dyd/Ravenhill, Salkeld & Co. (1,628ihp =
11.664kts)
  Tonnage as completed: 1,705bm. 2,450 disp. Draught 18ft 7in (fwd),
    20ft 0in (aft).
  Ord: 27.3.1858. K: 9.3.1859. L: 9.7.1861. C: 8.1862.
  Commissioned 20.8.1863 at Plymouth under Commodore Arthur
    Parry Eardley-Wilmot, for west coast of Africa; paid off 6.2.1866
    at Plymouth. Recommissioned 6.9.1867 under Commodore
    Montagu Dowell, as C-in-C at Cape of Good Hope and the west
    coast of Africa; on 16.2.1871 under Commodore John Edmund
    Commerell, with same role; in action at Chamah, West Africa
    14.8.1873 (4 men killed, Commerell among 20 wounded); paid off

A *Jason* Class corvette, the *Wolverene* is seen here in Australian waters during the late 1870s. By this date the ship had been re-armed with seventeen 64pdrs, one on a pivot mount and the rest on the broadsides. *(Allan C. Green, via the State Library of Victoria)*

2.10.1873 at Plymouth. BU 3.1882 at Devonport.

Wolverene Woolwich Dyd/Ravenhill, Salkeld & Co. (1,329ihp = 11.256kts)

Tonnage as completed: 1,703bm. 2,416 disp. Draught 17ft 11in (fwd), 20ft 4in (aft).

Ord: 27.3.1858. K: 10.4.1859. Suspended 1862, but resumed 1863. L: 29.8.1863. C: 4.1864.

Commissioned 7.5.1864 at Woolwich under Capt. Algernon Frederick Rous de Horsey, for North America and the WestIndies. On 30.12.1865 under Capt. Thomas Cochrane, on same station; paid off 27.2.1868 at Sheerness. Recommissioned 25.10.1870 under Capt. Henry Rushworth Wratislaw, for the East Indies; paid off early 1874. Recommissioned 1.8.1876 under Capt. Lindesay Brine, for voyage to Australia (to become commodore's flagship there). On 14.1.1877 under Commodore Anthony Hiley Hoskins, in Australia, then 12.9.1878 under Commodore John Crawford Wilson; paid off at Sydney but recommissioned there 1.7.1880. To New South Wales (Australia) as a training ship on 16.1.1882, paid off 16.2.1882 (handed over to the NSW colonial government) and became a sheer hulk from 2.1893; sold as such on 24.8.1923.

NORTH STAR Class Covered battery corvettes, 21 guns. Surveyor's Department design. The first two were ordered as *Jason* Class (1,702 ton) design. A new design for these two, with a covered deck, was developed from that for the *Jason* Class and approved on 23 January 1860. Six more vessels were ordered to this design in 1860 and a final pair in 1861, but all of these except *Menai* were cancelled in December 1863, and *Menai* followed a year later. The *Favorite* was ordered to be completed as an ironclad ship in 1862. A shorter design for the *North Star* was prepared in early 1865, but she was cancelled that May, and the engines scheduled for her were re-allocated to *Favorite*. The *Dido* (and later units?) were possibly intended to have been to a different design.

Dimensions & tons: (1860): 225ft 0in, 194ft 9¼in x 42ft 10in oa (42ft 4in for tonnage) x 24ft 1in. 1,856⁸³⁄₉₄bm.

1865 design (*North Star* only): 200ft 0in, 169ft 10in x 42ft 10in (42ft 4in for tonnage) x 24ft 1in. 1,619bm.

Men: 275. Guns: (presumably intended to be armed as previous class).

Machinery: 2-cylinder horizontal single expansion. Single (hoisting) screw. 400nhp.

North Star Sheerness Dyd/Humphrys & Tennant

Ord: 27.3.1858. K: 13.7.1860. Cancelled 22.5.1865; frames taken down by AO same date, done by 31.3.1866.

First cost (before cancellation): £15,575.

Favorite Deptford Dyd/Humphrys & Tennant

Ord: 8.4.1859. (named 1.10.1859) K: 23.8.1860. Re-ordered 1862 as ironclad corvette (which see).

Ontario Woolwich Dyd/—

Ord: 5.3.1860. K: 10.9.1860. Suspended 5.2.1862 and cancelled 12.12.1863.

First cost (before cancellation): £2,926.

Weymouth Sheerness Dyd/—

Ord: 5.3.1860. (named 4.10.1860) K: 18.10.1860. Cancelled 12.12.1863.

Alligator Woolwich Dyd/—

Ord: 5.3.1860. K: 1.11.1860. Suspended 5.2.1862 and cancelled 12.12.1863.

First cost (before cancellation): £4,969.

Menai Chatham Dyd/—

Ord: 5.3.1860. K: 5.1.1861. Cancelled 16.12.1864; frames taken down by AO of same date, completed by 19.8.1865.

First cost (before cancellation): £8,756.

Dido Deptford Dyd/—

Ord: 5.3.1860. K: 14.1.1861. Cancelled 12.12.1863 .

First cost (before cancellation): £4,304.

Falmouth Chatham Dyd/—
Ord: 5.3.1860. (named 4.10.1860) K: — . Cancelled 12.12.1863 (not laid down).
Nereide Woolwich Dyd/—
Ord: 3.1861. (named 21.3.1861) K: — . Cancelled 12.12.1863 (unstarted).
Ganymede Chatham Dyd/—
Ord: 3.1861. (named 21.3.1861) K: — . Cancelled 12.12.1863 (unstarted).

Until 1860, all corvettes had been built as 21-gun ships. However, in 1862 the older sloops commanded by Captains (*Archer*, *Brisk*, *Encounter*, *Malacca*, *Miranda* and *Niger*) were re-classed as corvettes – while others of similar age under Commanders (*Conflict*, *Desperate*, *Phoenix* and *Wasp*) remained sloops. When corvette-building resumed in the 1860s, the category would cover vessels with a much smaller establishment of guns.

Ironclad Floating Batteries

The technical classification of a Sixth Rate, as redefined in 1856, included all warships under the command of a captain with a complement of fewer than 300 men. Thus stated, the category included the steam/screw-propelled floating batteries, clad with plate armour, which were produced for the Russian ('Crimean') War. This type of warship was initially developed by the French, who built five iron-cased (wooden-hulled, with 110mm wrought iron armour) batteries for coastal assault – the *Dévastation*, *Tonnante*, *Lave*, *Congrève* and *Foudroyante* – along the lines advocated in 1842 by Capt. Labrousse. Ten were intended for the start of the 1855 campaign of the War against Russia, but French shipyards were only able to produce these five vessels in the time, and Britain agreed to build the other five, with flat-bottomed construction to allow them to operate in shallow waters.

AETNA **Class 1854** (Wooden-hulled) Modelled on the French batteries, of which they were a development. They were intended to carry 16 guns, but two were deleted to keep down the draught. All were named on 10 October 1854. *Trusty* was fitted in 1861 with Captain Coles's 'Shield' and used for experiments, and thus became the first ship in the world to be fitted with an armoured turret. In appearance they were double-ended, although with rudder and propeller defining the stern. Although completed as single-screw, *Meteor* (and possibly others) quickly altered to triple-screw, with wing shafts presumably driven through a belt arrangement. Note that both Mare and Green yards were listed as being at Limehouse, whereas their other building was done at Blackwall (it is not certain if this was an error). Two of this class (*Meteor* and *Glatton*) were towed to the Black Sea, but arrived a week too late for action against the Russians at Kinburn, and returned home, all four ships being present at the Naval Review at Spithead on 23 April 1856.
Dimensions & tons: 172ft 6in, 146ft 0in x 43ft 11in (43ft 6in for tonnage) x 14ft 7in. Draught 8ft 8in. 1,469bm. 1,672 disp. *Glatton* and *Trusty* had extra 15½in breadth and were thus about 1,535bm.
Men: 200. Guns: 14 x 68pdr (95cwt) MLSB. Armour: 3½–4½in iron plates.
Machinery: 2-cylinder (25½in diameter, 24in stroke) horizontal single expansion, high pressure (62 lb/sq.in). Single screw. 150nhp. 4½–5½kts.
Aetna (i) J. Scott Russell, Millwall/J. Scott Russell
Ord: 4.10.1854. K: 9.10.1854. L: — . Caught fire on slip and launched herself 3.5.1855 when nearly complete (was to have been launched 5.5.1855); her remains were BU on river bank.
Meteor C. J. Mare & Co., Limehouse/Maudslay, Sons & Field

Ord: 4.10.1854. K: 9.10.1854. L: 17.4.1855. C: 4.7.1855.
Commissioned 22.5.1855 under Capt. Frederick Beauchamp Paget Seymour, for the Black Sea; paid off 1856 at Portsmouth. BU 1861.
Thunder C. J. Mare & Co., Limehouse/Miller, Ravenhill & Co.
Ord: 4.10.1854. K: 9.10.1854. L: 17.4.1855. C: 21.7.1855.
Commissioned 9.12.1855 under Capt. George Granville Randolph; paid off 8.5.1856 at Sheerness. Remained at Sheerness in reserve; by 1870 at Chatham. BU at Chatham 6.1874.
Glatton R. & H. Green, Limehouse (No. 314)/Miller, Ravenhill & Co.
Ord: 4.10.1854. K: 9.10.1854. L: 18.4.1855. C: 3.8.1855.
Commissioned 22.5.1855 under Capt. Arthur Cumming, for the Black Sea; paid off 3.5.1856. BU 1864.
Trusty R. & H. Green, Limehouse. (No. 315)/Miller, Ravenhill & Co.
Ord: 4.10.1854. K: 9.10.1854. L: 3.5.1855. C: 13.6.1855.
Commissioned 9.12.1855 under Capt. Frederick Archibald Campbell. BU by Castle, Charlton 1864.

AETNA **Class 1855** (Wooden-hulled) Following the loss through fire of the Scott Russell hull, a replacement of the same name was quickly ordered from Chatham Dyd. This was simply a lengthened version of the previous class, with an extra pair of guns. The boilers from the earlier *Aetna* were salvaged for use in this ship.
Dimensions & tons: 186ft 0in, 157ft 9in x 43ft 11in (43ft 6in for tonnage) x 16ft 0in. Draught 6ft. 1,588bm.
Men: 200. Guns: 16 x 68pdr (95 cwt) MLSB. Armour: 4½in iron plates.
Machinery: 2-cylinder (27in diameter, 30in stroke) horizontal single expansion. 200nhp. 4kts.
Aetna (ii) Chatham Dyd/Maudslay, Sons & Field
Ord: 16.11.1855. K: 25.11.1855. L: 5.4.1856. C: — .
Never commissioned. Harbour service 1866. Burnt out at Sheerness 1873. BU 1874.

EREBUS **Class** (Iron-hulled) The British Admiralty has been sceptical about building iron-hulled warships since their disappointment with the iron-hulled paddle vessels built during the 1840s, when they learnt that the brittle forged metal plates split into more lethal shards upon cannon impact than the splinters of wood from ships of conventional build. However, by 1855 technical advances by ironmasters had produced material capable of being used as defensive plating, with the plates rolled instead of forged. The effectiveness of the three French ironclads used in the assault on Kinburn on 17 October 1855 prompted the Royal Navy to proceed with ordering from specialist contractors three iron-hulled vessels, all of which were completed for launch in little more than four months. The first iron-hulled ironclads ever built, their design was adapted from that of the modified French design for the wooden-hulled floating batteries; more graceful and sea-kindly than their wooden-hulled predecessors, they had raked *Warrior*-type bows and an extended bowsprit.
Dimensions & tons: 186ft 8½in, 156ft 2¼in x 48ft 6in x 15ft 6in. Draught 8ft 10in (fwd), 8ft 11in (aft). 1,954²⁹⁄₉₄bm.
Men: 200. Guns: 16 x 68pdr (95 cwt) MLSB. Armour: 4½in iron plates, mounted on teak of 6in in thickness.
Machinery: 2-cylinder horizontal single expansion. 200nhp. 5.5kts.
Erebus Robert Napier & Sons, Govan (No. 77)/Robert Napier & Sons
As built: 186ft 8in, 156ft 2¼in x 48ft 6in x 15ft 6in. 1,954²⁹⁄₉₄bm. 1,845 disp.
Ord: 22.12.1855. K: 1855. L: 19.4.1856. C: — .
First cost: Hull £69,724, machinery £11,500.
Never commissioned, being laid up in reserve at Portsmouth on

A contemporary model of the *Thunderbolt*, one of the iron-hulled armoured batteries built in 1856. Although produced to similar specifications, the three ships were not identical, each being completed to the shipbuilder's own fine-tuning of the concept. Although the model shows guns in every port, the ship was only established with sixteen in total; but the battery was presumably intended to operate against a fixed target on one side of the ship only, when all the ports on that broadside would be filled. *(NMM F8980-003)*

completion. Sold to Castle to BU 5.1884.

Thunderbolt Samuda Brothers, Millwall/Miller, Ravenhill & Co.

   Ord: 22.12.1855. K: 26.12.1855. L: 22.4.1856. C: — .

   First cost: Hull £67,849, machinery £11,600.

   Never commissioned; fitted at Chatham for service in Thames 1861, as tender to *Cumberland*. Floating pierhead at Chatham 13.11.1873. Renamed *Daedalus* 1916–19. Rammed and sunk by a tug 3.4.1948. Raised and BU 1949.

Terror Palmer Brothers, Jarrow (No. 52)/Robert Napier & Sons

   As built: 1,971bm. 1,844 disp.

   Ord: 22.12.1855. K: 31.12.1855. L: 28.4.1856. C: 7.5.1857 for sea, then 14.7.1857 as base ship for Bermuda.

   First cost: Hull £63,907, machinery £18,501, fitting £2,385+ as floating battery, plus £6,191 fitting for Bermuda.

   Commissioned 5.5.1857 under Capt. Frederick Hutton, as base ship for Bermuda. On 22.6.1862 under Capt. Frederick Henry Hastings Glasse, then 1.1.1865 under Capt. James Francis Ballard Wainwright, on 4.8.1869 under Capt. George Le Geyt Bowyear, on 13.9.1870 under Capt. Elphinstone D'Oyley D'Auvergne Aplin, in 9.1875 under Capt. Leveson Eliot Henry Somerset, and on 6.3.1878 under Capt. John Moresby. Recommissioned 17.6.1880 at Bermuda under Capt. Richard George Kinahan, then recommissioned again in 1.1886 under Capt. John Frederick George Grant; in 6.1899 under Capt. Thomas MacGill. Sold to Walker & Co. at Bermuda 1902.

7 Ship Sloops

Until 1 January 1817, the sloop category had included all ships (and other vessels) with fewer than 20 guns. Under the re-organisation which took effect on that date, the category was raised to include ships with 22 or 20 guns (i.e. fewer than 24); however, as carronades on the forecastle and quarter decks were now included in the count, most of the older 'frigate-built' sloops with forecastles and quarterdecks were up-rated to 24 or more guns and so became Sixth Rates. The ship sloop category thus mainly included flush-decked vessels, and all new ship sloops constructed would be of this type, predominantly carrying 18 or 20 guns.

(A) Vessels in service or on order at 1 January 1817

Following the reorganisation in January 1817, the Navy included three ships rated at 22 guns (the *Prometheus* in commission, although she is included with her class sisters in the previous chapter; and *Garland* and *Perseus* in harbour service), twenty-four rated at 20 guns (predominantly fifteen new ships of the *Cyrus* Class, plus the *Coquette*, *Minstrel*, *Wanderer*, *North Star*, *Myrmidon*, *Ariadne*, *Valorous* and the former prizes *Bonne Citoyenne* and *Florida*), and six rated at 18 guns (five of the *Dasher*/*Bermuda* Class, plus the *Rose*). Other former ship sloops (no longer armed) still existed as harbour vessels (usually re-classed as hospital, prison and receiving ships), but these are excluded below except to illustrate class connections.

***MERLIN* Class** 20 guns. The smallest surviving ship sloops of 'frigate-type' construction, i.e. with quarterdecks and forecastle; a William Rule design, approved on 29 January 1795, to which two ships were ordered in January 1795 and a fresh batch of fourteen ships to the same design in 1802 (see below). The *Merlin* was BU in 1803; four were wartime losses (*Fly*, *Martin*, *Wolf* and *Cygnet*), the *Halifax* and *Helena* were sold or BU in 1814, the *Albacore* and *Kangaroo* in 1815, and the *Brisk*, *Ariel*, *Kingfisher* and *Star* during 1816. Originally rated as 16 guns, they were re-classed as 20-gun sloops from January 1817.
> Dimensions & tons: 106ft 0in, 87ft 7in x 28ft 0in x 13ft 9in. 365³¹⁄₉₄bm.
> Men: 121. Guns: UD 14 x 32pdr carronades; QD 4 x 24pdr carronades; Fc 2 x 9pdrs.]

Pheasant John Edwards, Shoreham
> As built: 106ft 1in, 87ft 9⅞in x 28ft 3½in x 13ft 9in. 373⁸¹⁄₉₄bm. Draught 7ft 0in/10ft 3½in.
> Ord: 24.1.1795. K: 10.1795. L: 17.4.1798. C: 28.4–8.8.1798 at Portsmouth.
> First cost: £8,087 including fitting.
> Commissioned 6.1798 under Cmdr William Skipsey (–1800), for North America; sailed for Halifax 8.1798. In 1800 under Cmdr Henry Carew (–1804). In 1804 under Cmdr Robert Paul (died 1804); sailed for Jamaica 1.9.1804. In 1805 under Cmdr Robert

Henderson, in the Leeward Islands. In 1.1806 under Cmdr John Palmer (–1814); at home in 8.1806; sailed for South America 28.9.1806; in Channel Islands 1808; took privateers – 5-gun *Le Tropard* 8.5.1808, 14-gun *Le Comte de Hunebourg* 3.2.1810, and 6-gun *Le Héros* 17.6.1811. Middling Repair at Plymouth (for £11,587) 7–9.1812; to Newfoundland 1812; took (with *Warspite*) US 4-gun privateer *William Bayard* 12.3.1813; took (with *Whiting*) US 8-gun privateer *Fox* 6.5.1813; sailed for Newfoundland again 5.6.1813. In 10.1814 under Cmdr Edmund Waller, in the Channel; paid off into Ordinary at Plymouth 11.1815. Very Small Repair and fitted at Plymouth 9–12.1818; recommissioned 9.1818 under Cmdr Benedictus Kelly, for the African station. In 9.1821 under Cmdr Douglas Clavering, still on African station; paid off 2.1823. Fitted as temporary receiving ship at Woolwich 8.1823–1.1824. Sold at Deptford (for £1,250) to John Small Sedger, Rotherhithe 11.7.1827.

Otter Peter Atkinson & Co., Hull
> As built: 107ft 3in, 86ft 9in x 28ft 1½in x 13ft 7½in. 365 (exact)bm. Draught 7ft 10in/11ft 4in.
> Ord: 27.11.1802. K: 7.1803. L: 2.3.1805. C: 19.3–19.5.1805 at Sheerness.
> Commissioned 4.1805 under Cmdr John Davies, for the Downs; sailed for the Cape of Good Hope 18.8.1807. From 4.1808 under Cmdr Josiah Nesbit Willoughby, in the East Indies; boats in action at Rivière Noire (Mauritius) 14.8.1809; occupation of St Paul (Reunion) 21–28.9.1809; later under Lieut Edward Benge (acting). On 12.3.1810 under Cmdr James Tomkinson; recapture (with *Boadicea* and *Staunch*) of 38-gun *Africaine* 13.9.1810, and recapture of *Ceylon* and capture of *La Venus* 18.9.1810; on 11.10.1810 under Lieut Thomas Lamb Polden Laugharne (acting); paid off 3.1811. Fitted for Ordinary at Plymouth 4.1811. Fitted for quarantine service at Plymouth as lazarette for Pembroke 2–4.1814. Sold to J. Holmes (for £610) 6.3.1828.

Rose (Adam) Hamilton & (Thomas) Breeds, Hastings
> As built: 106ft 1½in, 87ft 8⅛in x 28ft 0½in x 13ft 9in. 366⁶⁷⁄₉₄bm.
> Ord: 27.11.1802. K: 10.1803. L: 18.5.1805. C: 24.5–3.8.1805 at Portsmouth.
> Commissioned 19.6.1805 under Cmdr Lucius Curtis. In 3.1806 under Cmdr Philip Pipon, in the Channel. On 17.9.1808 under Cmdr Thomas Mansell (–1813); in the Baltic 1808–10; paid off 4.1813. In Ordinary at Plymouth 1813–17. Sold to Thomas Pitman (for £820) at Plymouth 30.10.1817 to BU.

Ex-FRENCH PRIZE. A 'Corvette de 20 x 8', designed by Raymond-Antoine Haran. Four vessels were built to this draught at Bayonne, all of which were taken by and added to the RN in 1796–98. In effect these were flush-decked vessels, but with a long topgallant forecastle which in some vessels was armed by the British; the *Danae* had been lost (through mutiny) back to the French in 1800, the *Gayette* sold in 1808 and the *Jamaica* in 1814.

Bonne Citoyenne (French *La Bonne Citoyenne*, built at Bayonne 7.1793–5.1795. L: 9.7.1794)
> Dimensions & tons: 120ft 1in, 100ft 6¼in x 30ft 11in (30ft 6in mld.) x 8ft 7in. 511¼⁄₉₄bm.

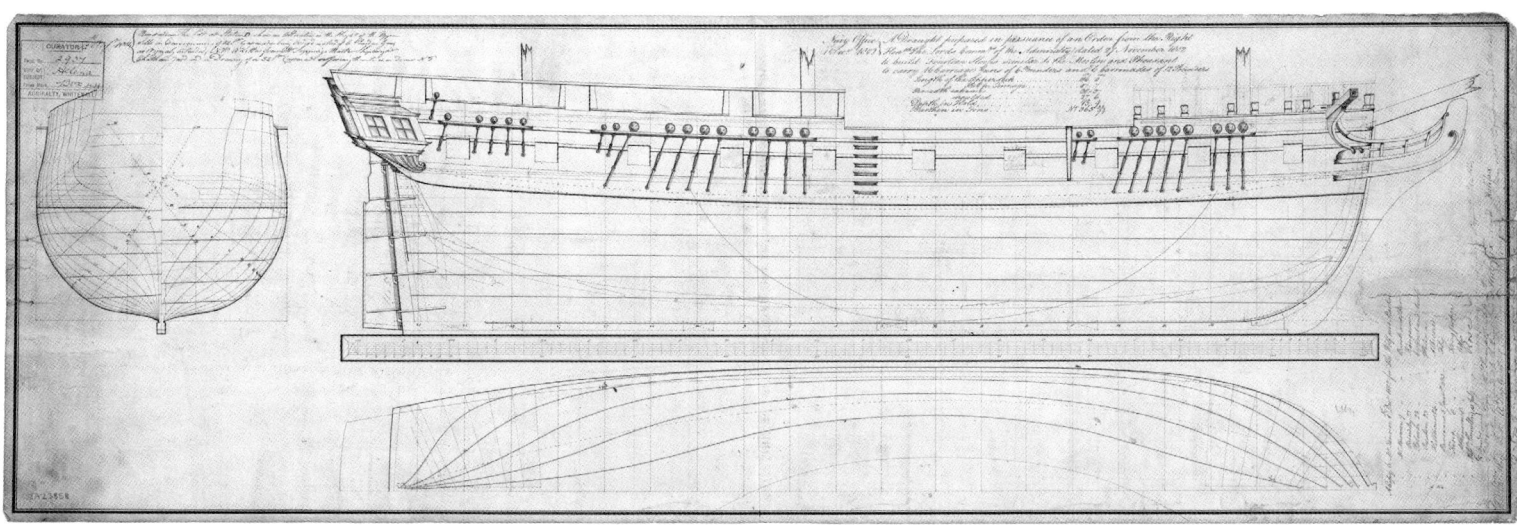

A draught dated 1 December 1802 'in pursuance of an Order . . . dated 27 November 1802 to build Fourteen Sloops similar to the Merlin and Pheasant to carry 16 Carriage Guns of 6-pounders and 6 Carronades of 12-pounders'. Unlike the first pair which are mentioned in the AO, that were built with open quarterdeck rails, the new batch had the 'berthed-up' (solid) barricades to quarterdeck and forecastle that had become standard during the French Revolutionary War. In addition, they were actually completed with 32pdr carronades on slide mountings rather than the 6pdr carriages guns in the first pair. Only thirteen of the authorised ship sloops were contracted with commercial builders in compliance with this order, the fourteenth sloop being ordered built at Halifax Dockyard in Nova Scotia (from locally cut timber) ten months later, but this last vessel had been taken to pieces in 1814, and only three of this class survived into 1817. *(NMM J4169)*

Men: 125 (later 121). Guns: UD 18 x 6pdrs, QD 2 x 32pdr carronades; later 2 x 9pdrs + 18 x 32pdr carronades.
Taken 10.3.1796 by *Phaeton* off Cape Finisterre. Arrived Portsmouth 19.3.1796. Registered and named 20.5.1796. Completed fitting there (for £4,792) 14.8.1796.
Commissioned 6.1796 under Cmdr Sir Charles Lindsay; sailed for the Mediterranean 2.1797; at Battle of Cape St Vincent 14.2.1797; at Berkeley's encounter with 130-gun *Santissima Trinidad* 20.2.1797. From 3.1797 under Capt. Lord Mark Kerr, then Cmdr Richard Retalick in 5.1797; took privateers 9-gun *Le Pluvier* and 10-gun *La Garnarde* in the Mediterranean 8.1797. In 5.1798 under Cmdr (Capt. 12.1798) Josiah Nisbet; joined Nelson's squadron in the Mediterranean in 1798 (but missed Battle of the Nile). Under Cmdr Thomas Maling 8.1799, then Lieut Archibald Duff (acting) in 9.1800 and Cmdr Robert Jackson in 10.1800; took 10-gun Spanish privateer *Vives* in the Mediterranean 31.12.1800; in Egypt operations 1801. Under Cmdr Philip Carteret in 5.1802. Paid off 1803. Middling Repair at Chatham 8.1807–6.1808; recommissioned 3.1808 under Cmdr John Thompson, for the Channel. In 5.1809 under Cmdr (Capt. 6.7.1809) William Mounsey; captured 36-gun *La Furieuse* (*en flûte*) 8.1809. In 2.1810 under Cmdr Richard James Lawrence O'Connor; sailed with convoy for Madeira 11.7.1810. In 11.1810 under Cmdr (Capt. 7.3.1811) Pitt Burnaby Greene (–1814); sailed for South America 12.3.1811; refused USS *Hornet*'s challenge to fight off San Salvador 13.12.1812; at Jamaica 1814. Under Capt. Augustus William James Clifford 23.8.1814; paid off and laid up in Ordinary at Portsmouth 1.1815. Sold to Joshua Crystall (for £1,550) 3.2.1819.

DASHER Class 18 guns. Built of Bermudan cedar. Designed, established (29 January 1796) and started as brigs, but completed as ship sloops. The possible designer was Edward Goodrich (of Goodrich & Co.), who certainly acted as contractor for these vessels, subcontracting to the individual builders indicated below (this also applied to the subsequent *Bermuda* Class).
Dimensions & tons: 107ft 0in, 86ft 10in x 29ft 6in x ?9ft 9in. 401⁸³⁄₉₄bm.
Men: 80. Guns: 2 x 9pdrs + 16 x 24pdr carronades (may have carried up to 20 carronades).

Dasher John Outerbridge & Claude McCallan, Bermuda
As built: unrecorded.
Ord: 18.1.1796. K: — . L: 1797.
Commissioned 9.1796 on the Halifax station under Cmdr John Seater. In 8.1798 under Cmdr George Tobin, for the Channel; arrived at Plymouth to make good defects 29.8.1799; paid off 10.1801. Fitted at Plymouth 11.1802–1.1803; recommissioned 11.1802 under Cmdr John Delafonds (died 9.1804); sailed for the East Indies 2.5.1803. On 20.3.1805 under Cmdr Charles Dilkes, 10.1805 Cmdr William Montagu, 9.1806 Cmdr Edward Troubridge, 4.1808 Cmdr Michael de Courcy, 14.8.1808 Cmdr Robert Festing and 1.4.1811 Lieut (Cmdr 28.11.1811) Benedictus Marwood Kelly, all in East Indies; in Java operations 8/9.1811; took (with *Leonidas*) 12-gun privateer *La Confiance* 17.1.1812; later in Ordinary at Deptford. Recommissioned 7.6.1812 under Cmdr William Henderson (–1816); sailed for the Leeward Islands 10.1812; between Small and Middling Repair at Deptford 1–8.1813; paid off 5.1816. Lent to the Committee for Distressed Seamen 19.1.1818. Fitted at Deptford 3–11.1820 for the Army in the West Indies, but not sent. Laid up at Deptford 9.1822. Fitted as convict ship 6–7.1826 at Deptford, in 1834 at Woolwich. BU at Deptford 3.1838.

Driver Nathaniel Tynes, Bermuda
As built: 107ft 0in, 83ft 9½in x 29ft 11in x 9ft 9in. 398⁸³⁄₉₄bm.
Ord: 18.1.1796. K: — . L: 1797.
Commissioned 2.1797 under Cmdr Robert Hall. In 8.1798 under Cmdr John Seater, later Cmdr Thomas Hurd; arrived Sheerness 27.1.1799. Made good defects at Sheerness (for £2,614) to 21.4.1799. In 11.1799 under Cmdr James Dunbar (–1801), for North Sea. In 7.1802 under Cmdr Francis Fane, for Channel service. Fitted at Woolwich 2–4.1803; recommissioned 2.1803 under Fane; sailed for North America 23.5.1803; in 9.1803 under Lieut William Compton (temp.). In 1804 under Cmdr William Lyall, in the Bahamas and later Halifax station. In 1.1805 under Cmdr John Carden, 1.1806 under Cmdr Robert

The suppression of the slave trade was to become an important duty for the Royal Navy early in the post-war years, which provided an opportunity for young officers to distinguish themselves at a period devoid of major naval warfare. Slave trafficking had been outlawed for British subjects by Act of 25 March 1807, with effect from 1 January 1808; in 1811 it was made punishable by transportation, and in 1824 the penalty became death, as it was equated with piracy. Spain had outlawed the practice in 1814 north of the Equator, and in 1817 this was extended to the southern hemisphere; thus the seizure of slaving vessels became a regular Royal Naval task. The sloop *Morgiana* is shown in this watercolour by Irwin Bevan on 10 December 1819 off the coast of Africa. Her gig, commanded by Midshipman William Mansell, is seen alongside the Spanish armed slaving schooner *Esperança*; Mansell and a lone marine boarded the schooner and overcame its crew; Mansell's commission as Lieut. followed in September 1821, following a warm endorsement by anti-slavery campaigner William Wilberforce. *(NMM PU9478)*

Simpson; North America 1806. In 9.1806 under Cmdr Charles Claridge, then 2.1807 Cmdr William Love (−1808); took private schooner *El Boladora* 12.6.1807. In later 1808 under Claridge again; took (with others) 40-gun *Le Junon* 10.2.1809. In 8.1809 under Cmdr George Monke, and later in 1809 under Cmdr John Lawrence, then on 12.7.1810 under Cmdr Thomas Swinnerton Dyer. Laid up at Portsmouth 12.1810. Middling Repair there 6–9.1814. In 1815 under Cmdr Lord Algernon Percy. Fitted for sea at Portsmouth 9.1815; on 22.8.1815 under Cmdr John Ross (−1817), on Leith station, then 12.12.1817 under Cmdr Charles Hope Reid; paid off 10.1821 at Portsmouth, then recommissioned 30.10.1821 under Cmdr Thomas Wolrige, for the African station; on 19.7.1822 under Cmdr Charles Bowen (−1823), for same station; paid off 1824. Fitted as coal depot at Deptford 8.1824. Convict ship 1825–31 at Deptford; by 1833 coal hulk at Woolwich. BU at Deptford 24.7.1834.

***BERMUDA* Class**, 18 guns. Built of Bermudan cedar. Modified version of *Dasher* Class of 1797, but there is a dearth of information; the only dimensions quoted are identical with the design details given below (the large depth in hold compared with *Driver* suggests they were built without a platform there). Of this class, the *Bermuda*, *Atalante* and *Sylph* had been wrecked in 1808, 1813 and 1815 respectively.

Dimensions & tons: 107ft 0in, 83ft 10⅛in x 29ft 11in x 14ft 8in. 399³¹/₉₄bm.

Men: 121. Guns: 16 x 24pdr carronades + 2 x 9pdrs (*Indian* 2 x 6pdrs).

Indian Robert Shedden, Bermuda

Ord: 23.6.1803. K: — . L: 10.1805.

Commissioned at Bermuda 10.10.1804 under Cmdr Charles John Austen, for West Indies and Halifax station; took 4-gun privateer *La Jeune Estelle* 19.6.1808. In 5.1810 under Lieut William Bowen Mends (acting); then 9.1810 under Cmdr Henry Jane; took (with *Plumper*) US 1-gun privateer *Fair Trader* in the Bay of Fundy 16.7.1812; sailed for the Leeward Islands 6.2.1813. On 8.7.1813 under Lieut (Cmdr 9.11.1813) Thomas Sykes, then 9.3.1814

under Cmdr Nicholas James Cuthbert Dunn, in North America; paid off 10.1814. In Ordinary 1814–17. Sold at Deptford to Messrs Enderby (for £1,300) 24.4.1817.

Martin David McCallan, Bermuda

Ord: 26.4.1806. K: 1806. L. 5.1809.

Commissioned at Halifax under Cmdr John Evans (−1812). In 1813 under Cmdr Humphrey Fleming Senhouse; affair with gunboats in Delaware Bay 29.7.1813; took US 6-gun privateer *Snapdragon* in Passamaquoddy Bay 30.6.1814. On 12.10.1814 under Cmdr James Arbuthnot, then 10.1816 under Cmdr Andrew Mitchell, at Cork; wrecked off Aran Islands on the West coast of Ireland 8.12.1817 (4 drowned).

Morgiana Robert Hill, Bermuda

Ord: 29.12.1808. K: 1811. L: 12.1811.

Commissioned at Bermuda 2.8.1811 under Cmdr David Scott, for North American station. On 26.5.1814 under Cmdr Vincent Newton (acting). In Ordinary at Chatham until fitted there for foreign service 11.1818–3.1819; recommissioned 17.12.1818 under Cmdr Charles Burrough Strong, for African coast, subsequently 12.8.1819 under Cmdr Alexander Sandilands and 26.5.1820 under Lieut (Cmdr 9.9.1820) William Finlaison (−1822). From 1823 at Portsmouth until sold there to Thomas Pitman (for £1,760) 27.1.1825 to BU.

***CORMORANT* Class** (20 guns). The survivors of the largest (and last) group of 'frigate-built' (i.e. with quarterdecks and forecastles, with berthed-up barricades) sloops. A joint design by Henslow and Rule, based on the lines of the French *Amazon* of 1745. Six ships were ordered in February 1793, and a seventh in 1794; all these carried a main battery of 6pdr guns. A further twenty-four ships were ordered 1805–06 to the same design, but 32pdr carronades were established (by AO of 15 June 1807) to replace the 6pdrs originally mounted in the earlier group. All were originally rated as 18 guns, but were re-classed as 20-gun Sixth Rates in 1810–12, and most of the survivors were on 1 January 1817 re-rated as 26-gun Sixth Rates (see Chapter 6); however, the three below remained briefly as 20-gun sloops, although all three were sold off on 6 March 1817.

Dimensions & tons: 108ft 4in, 90ft 9$\frac{5}{8}$in x 29ft 7in x 9ft 0in. 422$\frac{61}{94}$bm.

Men: 121. Guns: UD 16 x 32pdr carronades; QD 6 x 18pdr carronades; Fc 2 x 6pdrs + 2 x 18pdr carronades (some had no 6pdrs).

North Star Benjamin Tanner, Dartmouth, completed by John Cock after Tanner's bankruptcy

As built: 108ft 4$\frac{5}{8}$in, 90ft 0$\frac{1}{8}$in x 30ft 1in x 9ft 0in. 433$\frac{28}{94}$bm. Draught 8ft 4in/11ft 3$\frac{1}{2}$in.

Ord: 19.10.1805. K: 5.1806. L: 21.4.1810. C: 13.5–17.7.1810 at Plymouth.

Commissioned 6.1810 under Cmdr (Capt. 4.1811) Thomas Coe (–1815); re-classed as Sixth Rate (20 guns) 4.1811; sailed for the Leeward Islands 6.6.1813. On 29.11.1815 under Cmdr George Bentham, on Jamaica station; paid off 4.1816. Sold to Thomas Pitman to BU (for £1,010) 6.3.1817, but later re-sold to become mercantile *Colombo*.

Minstrel (Nicholas) Bools & (William) Good, Bridport

As built: 108ft 4in, 90ft 9$\frac{5}{8}$in x 29ft 7in x 9ft 0in. 422$\frac{65}{94}$bm.

Ord: 19.11.1805. K: 1.1806. L: 25.3.1807. C: 29.4–24.7.1807 at Plymouth.

Commissioned 3.1807 under Cmdr John Hollinworth; sailed for the Mediterranean 10.10.1807; took 10-gun Venetian schooner *Ortenzia* 16.7.1808; violation of Algerian neutrality at Bougia 9.1809. In 2.1810 under Cmdr Ralph Wormeley, then Cmdr Colin Campbell; landing party (with others ships' boats) in attack on Palamos 13.12.1810. Re-classed as a Sixth Rate (20 guns) 1811; in 10.1811 under Capt. John Peyton. In 8.1812 under Lieut Michael Dwyer, at Benidorm; boats took ammunition ships at Valencia 29.9.1812. On 2.2.1813 under Cmdr (Capt. 31.3.1813) Robert Mitford, in the Mediterranean; invalided home 8.1814. On 29.9.1814 under Capt. Francis Erskine Loch; paid off 12.1815 at Chatham. Sold to Mr Younge (for £1,010) 6.3.1817.

Wanderer James Betts, Mistleythorn

As built: 109ft 3in, 91ft 6in x 29ft 9in x 9ft 0$\frac{1}{2}$in. 430$\frac{71}{94}$bm.

Ord: 19.11.1805. K: 2.1806. L: 29.9.1806. C: 25.10.1806–31.1.1807 at Chatham.

Commissioned 12.1806 under Cmdr Edward Crofton (–1811), for North Sea in 1807. In ?1810 under Cmdr William Robilliard; landing party from *Wanderer* (with consorts) at Sint Maarten 3.7.1810; re-classed as Sixth Rate (20 guns) ?1811. In 4.1811 under Capt. Francis Newcombe (–1814); sailed with Lisbon convoy ?17.6.1812; sailed for North America 28.8.1812. In ?10.1814 under Capt. John Palmer, then 12.1814 under Capt. William Dowers, in the Channel; paid off 11.1815 at Plymouth. Sold to Mr Splidt (for £1,150) 6.3.1817.

***TALBOT* Class** Enlarged versions of the *Cormorant* Class (i.e. another *Amazon* of 1745 derivative). The two ships were built as 18-gun sloops, they were re-classed as 20-gun Sixth Rates in 1811. The *Talbot* had been sold in 1815, and the survivor *Coquette* was re-rated as a 20-gun sloop on 1 January 1817.

Dimensions & tons: 113ft 3in, 94ft 2in x 31ft 0in x 9ft 4in. 481$\frac{33}{94}$bm.

Men: 121. Guns: UD 18 x 32pdr carronades (originally planned to carry 18 x 6pdrs); QD 6 x 12pdr carronades; Fc 2 x 6pdrs + 2 x 12pdr carronades. Also mounted 1 x 12pdr on centreline.

Coquette (ex-*Queen Mab*, renamed 6.6.1807) Simon Temple, North Shields

As built: 113ft 3$\frac{1}{2}$in, 94ft 2$\frac{3}{8}$in x 31ft 1in x 9ft 5in. 484$\frac{10}{94}$bm. Draught 7ft 0in/10ft 0in.

Ord: 4.10.1805. K: 2.1806. L: 25.4.1807. C: 13.6–28.10.1807 at Chatham.

Commissioned 6.1807 under Cmdr Robert Forbes (–1810), in Rear-Adm. Richard Keats's squadron; sailed with Quebec convoy 19.4.1809. In 11.1810 under Cmdr George Hewson. Fitted at Woolwich 1–5.1812; in 3.1812 under Capt. ?Thomas Bradby, then 5.1812 Capt. John Simpson; sailed for the Leeward Islands 14.12.1812. In Ordinary at Woolwich 1814–15. Sold at ?Woolwich to Mr Ismay (for £1,090) 30.4.1817.

***HERMES/ARIADNE* Class** 26 guns. Built as post ships (Sixth Rates), reclassified on 1 January 1817 as 20-gun sloops. Originally a flush-decked design 'similar to *Bonne Citoyenne*' (French corvette prize taken 1796), to which the first pair (*Hermes* and *Myrmidon*) had been built at Milford (the former Jacobs site) on the north coast of Milford Haven. The *Hermes* had been burnt in action in 1814. The second pair were built at Pater on the south coast of the Haven (this later became Pembroke Dock) and differed slightly, being formally described as *Ariadne* Class. Under AO 6 January 1820, *Ariadne* (in 1820) and *Valorous* (in 1820–21) were modified before their first commission as 'post ships' by adding quarterdecks (called 'poops') and rudimentary forecastles, and were then re-classed as 26-gun Sixth Rates.

Dimensions & tons: 119ft 0in, 99ft 10$\frac{5}{8}$in x 30ft 11in x 8ft 7in. 507$\frac{76}{94}$bm.

Men: 135. Guns: originally 2 x 9pdrs + 18 x 32pdr carronades. As 26-gun had UD 18 x 32pdr carronades, QD 6 x 18pdr carronades, Fc 2 x 9pdrs (*Ariadne* later had a tenth pair of 32pdr carronades added on UD).

Myrmidon Milford Dyd [M/Shipwright William Stone to 9.6.1813; launched by Henry Canham]

As built: 119ft 10$\frac{1}{2}$in, 99ft 10$\frac{3}{4}$in x 30ft 11$\frac{1}{2}$in x 8ft 7$\frac{1}{2}$in. 509$\frac{25}{94}$bm. Draught 9ft 9in/10ft 3in.

Ord: 2.8.1811. K: 7.1812: L. 18.6.1813. C: 4.7.1813–6.2.1814 at Plymouth.

Commissioned 8.1813 under Capt. Valentine Gardner; in 10.1813 under Capt. Henry Bourchier, then in 1814 Capt. William Paterson and 25.4.1815 Capt. Robert Gambier (–1817); paid off 10.1815 and recommissioned, still under Gambier; in the Mediterranean 1817; paid off 19.11.1818. On 26.3.1819 under Cmdr Henry John Leeke, on African station; paid off 10.1822. BU completed 10.1.1823 at Portsmouth.

Ariadne Pater Dyd (Pembroke Dyd) [M/Shipwright Edward Churchill]

As built: 121ft 7in, 100ft 5$\frac{1}{2}$in x 30ft 11$\frac{1}{4}$in x 8ft 9in. 511$\frac{42}{94}$bm. Draught 9ft 1in/10ft 0in.

Ord. 28.11.1812. K: 4.1815. L: 10.2.1816. C: 21.3.1816 (for Ordinary).

First cost: £11,936 to build, plus £3,579 fitting.

Converted into 26-gun post ship (under AO 6.1.1820) at Plymouth 1–5.1820. Fitted for sea 3.1822–1.8.1822; total conversion and fitting cost at Plymouth £12,468 (including remedying of defects).

Commissioned 4.1823 under Capt. Robert Moorsom. In 12.1824 under Capt. Isham Chapman (dismissed 6.1826), on Cape of Good Hope station; in 2.1826 under Capt. Adolphus Fitzclarence, for Mediterranean station. In 9.1827 under Capt. Charles Leonard Irby, later same month Capt. Lewis Davies; paid off 5.1828. Fitted for sea at Plymouth (for £11,692) 8.1828–2.1829; recommissioned 10.11.1828 under Capt. Frederick Marryat, for diplomatic service at Madeira and the Western Islands; paid off

11.1830. Fitted for sea at Plymouth (for £5,820) 11.1830–1.1831; recommissioned 11.1830 under Capt. Charles Phillips (–1834), for North America and the West Indies; paid off 1835. Fitted as coal depot at Portsmouth (for £1,775) 8.1836–2.1837, to lie at Alexandria. Sold at Alexandria (for £900) by AO 23.7.1841.

Valorous Pater Dyd (Pembroke Dyd) [M/Shipwright Edward Churchill]

As built: 121ft 7¼in, 100ft 5⅝in x 31ft 0in x 8ft 9in. $513\frac{53}{94}$bm.

Ord. 28.11.1812. K: 3.1815. L: 10.2.1816. C: 11.3–26.3.1816 at Plymouth (for Ordinary).

First cost: £11,726 to build.

Converted into 26-gun post ship (under AO 6.1.1820) at Plymouth 3.1820–4.7.1821.

Commissioned 2.1821 under Capt. James Murray (–1822), for Newfoundland station. Recommissioned 8.1824 under Capt. the Earl of Huntingdon (–1825), for Jamaica station. At Chatham 1826–29. BU completed 13.8.1829 at Chatham.

Ex-AMERICAN PRIZE (1814) A flush-decked vessel, originally added as a 20-gun Sixth Rate, but re-classed as a sloop on 1 January 1817.

Florida (American *Frolic*, launched 11.9.1813 by Josiah Barker at Charlestown, Massachusetts) 20 guns.

Dimensions & tons: 119ft 5½in, 98ft 11¾in x 32ft 0in x 14ft 2in. $539\frac{11}{94}$bm.

Men: 135. Guns: 2 x 9pdr + 18 x 32pdr carronades.

Taken 20.4.1814 off Cuba by *Orpheus* and *Shelburne*. Purchase (for £8,211.1.7d) reported 24.8.1814.

Commissioned 7.6.1814 at Halifax under Capt. Nathaniel Mitchell. Arrived Woolwich 30.8.1815 and fitted for Channel service to 2.12.1815; recommissioned 9.1815 under Capt. William Elliott. In 4.1816 under Capt. Charles Sibthorpe John Hawtayne (–1819), for the North Sea; paid off 12.1818. Re-rated 22-gun sloop from 2.1817. BU at Chatham 5.1819.

***CYRUS* Class** Built as post ships (Sixth Rates), reclassified on 1 January 1817 as 20-gun sloops. They were intended as a counter to the large American sloops of the *Frolic* Class, then building. Sixteen vessels were built to this flush-decked design by Sir William Rule, approved 27.11.1812 and based on slightly reduced lines of the *Hermes* Class above. Of these, the *Tay* had been wrecked in November 1816 but is included below for completeness.

Dimensions & tons: 115ft 6in, 97ft 2in x 29ft 8in x 8ft 6in. $454\frac{80}{94}$bm.

Men: 135. Guns: UD 20 x 32pdr carronades and 2 x 6pdrs (chase).

Medina Edward Adams, Bucklers Hard

As built: 115ft 8½in, 97ft 2in x 29ft 10in x 8ft 6in. 460bm. Draught 8ft 4in/9ft 7in.

Ord: 18.11.1812. K: 1.1813. L: 13.8.1813. C: 14.8–20.12.1813 at Portsmouth.

Commissioned 18.11.1813 under Capt. Henry Bourchier (–1815); sailed for Newfoundland 1814; laid up at Chatham 12.1815. Very Small Repair and fitted for sea at Chatham 8.1820–3.1821; recommissioned 26.12.1820 under Cmdr Robert Hockings, for Mediterranean station; at Smyrna 6.1821. On 3.12.1821 under Cmdr Patrick Duff Henry Hay, then 2.1824 under Cmdr Charles Montagu Walker, in 6.1825 under Cmdr Timothy Curtis and 30.12.1826 under Cmdr William Burnaby Greene; paid off 1827. Small Repair and fitted for sea at Plymouth 7.1827–5.1828; recommissioned 15.3.1828 under Cmdr William Benjamin Suckling, for the African coast; on 2.1829 under Cmdr Edward Webb (–1831). Sold at Sheerness (for £1,610) to John Small Sedger, Rotherhithe 4.1.1832.

Cyrus William Courtney, Chester

As built: 116ft 0in, 98ft 1¼in x 29ft 10in x 8ft 6in. $464\frac{42}{94}$bm.

Ord: 18.11.1812. K: 1.1813. L: 26.8.1813. C: 7.10.1813–11.3.1814 at Plymouth.

Commissioned 10.12.1813 under Capt. Henry Hart, for the Mediterranean; on 5.6.1841 under Capt. William Carroll (–9.1818); to Irish Sea 1817. In 9.1818 under Cmdr C. W. White; paid off 1822. Sold to Bennet & Son (for £1,550) at Plymouth 23.5.1823.

Levant William Courtney, Chester

As built: 116ft 0in, 98ft 1¼in x 29ft 10in x 8ft 6in. $464\frac{42}{94}$bm. Draught 9ft 6in/10ft 0in.

Ord: 18.11.1812. K: 1.1813. L: 8.12.1813. C: 17.1–22.4.1814 at Plymouth.

Commissioned 1814 under Capt. Alexander Jones; later under Capt. George Douglas; taken (along with *Cyane*) by the *USS Constitution* off Madeira 20.2.1815: recaptured by HMS *Acasta* and others at Porto Praya 11.3.1815. In 4.1815 under Cmdr John Sheridan (acting until 13.6.1816); laid up at Chatham 27.11.1815. Intended 8.1820 to be repaired, but instead BU at Chatham completed 9.10.1820.

Esk Jabez Bailey, Ipswich

As built: 115ft 7⅝in, 97ft 3⅝in x 29ft 9in x 8ft 6in. $458\frac{7}{94}$bm. Draught 8ft 6in/9ft 8in.

Ord: 18.11.1812. K: 3.1813. L: 11.10.1813. C: 19.10.1813–14.6.1814 at Sheerness.

Commissioned 21.1.1814 under Cmdr (Capt. 6.1814) George Gustavus Lennock; paid off 9.1815 and recommissioned; Jamaica station 1816–17; paid off 12.1818. Small Repair and fitted for sea at Portsmouth 1–11.1820; recommissioned 9.8.1820 under Cmdr Edward Lloyd; sailed for Leeward Islands 8.11.1820. On 22.11.1821 under Cmdr Arthur Lee Warner, still in Leeward Islands; paid off 1824. Very Small Repair and fitted for sea at Chatham 5–7.1824; recommissioned 9.9.1824 under Cmdr William Jardine Purchas, for the Africa station; captured 9 slaving vessels between 17.7.1825 and 8.2.1827; paid off 5.1828. Fitted for Ordinary at Chatham 5–6.1828. Sold at Chatham to William Wilson (for £1,530) 8.1.1829.

Carron Edward Adams, Bucklers Hard

As built: 115ft 8½in, 97ft 2¾in x 29ft 9½in x 8ft 6⅛in. $459\frac{2}{94}$bm. Draught 7ft 11in/9ft 3in.

Ord: 18.11.1812. K: 3.1813. L: 9.11.1813. C: 11.11.1813–22.3.1814 at Portsmouth.

Commissioned 1.1814 under Capt. Robert Spencer; in unsuccessful attack on Fort Bowyer, near Mobile 15.9.1814. Under Capt. Nicholas Pateshall in 4.1815; paid off 8.1815. At Portsmouth in 1817. Recommissioned 26.5.1818 under Cmdr John Furneaux, for East Indies; wrecked 6.7.1820 near Puri, India (19 men drowned).

Tay Balthazar Adams, Bucklers Hard

As built: 115ft 8½in, 97ft 2⅝in x 29ft 10in x 8ft 6¼in. 460 ($455\frac{12}{94}$ by calc.)bm. Draught 8ft 0in/9ft 4in.

Ord: 18.11.1812. K: 4.1813. L: 26.11.1813. C: 5–12.1813 at Portsmouth (for Ordinary), 10.1814–28.11.1814 (for sea).

Commissioned 9.1814 under Capt. William Robilliard. 'Breadthened fore and main channels' at Portsmouth 1–2.1815; in 2.1815 under Capt. Robert Boyle. On 24.1.1816 under Capt. Samuel Roberts, on Jamaica station; wrecked 11.11.1816 on the Alacreanes Islands, Gulf of Mexico.

Slaney Josiah & Thomas. Brindley, Frindsbury
As built: 115ft 6½in, 96ft 10⅞in x 29ft 10⅝in x 8ft 6⅛in. $460\frac{35}{94}$bm. Draught 8ft 2in/9ft 5in.
Ord: 18.11.1812. K: 4.1813. L: 9.12.1813. C: 9–21.12.1813 at Chatham (for Ordinary), 12.1814–23.1.1815 (for sea).
Commissioned 24.8.1814 under Capt. Charles Sotheby (–27.3.1816), for Halifax, South America and later Cape of Good Hope; but from 14.12.1814 to 8.1815 under Capt. George Rose Sartorius. Fitted for sea at Plymouth 7–11.1818; recommissioned 8.1818 under Cmdr Donat O'Brien (–1820), for South America. In 3.1821 under Cmdr Henry Stanhope (–1822), still South America. Small Repair at Portsmouth 9.1822–7.1823; recommissioned 4.1823 under Cmdr Charles Mitchell, for East Indies; at Cheduba 1824; in Burma War 1824–25. In 1.1826 under Cmdr Samuel Thornton (originally appointed to her 8.4.1825), still in East Indies; paid off 1827. Fitted for sea at Woolwich 5–7.1827; recommissioned 4.5.1827 under Cmdr James Campbell, for Jamaica station; under Cmdr Henry Gossett in 6.1828, then Cmdr (Capt. 8.1829) Joseph O'Brien on 5.1.1829, and Cmdr Charles Parker (acting) on 8.9.1829; paid off 29.1.1831 after laying up 12.1830 at Bermuda; receiving ship there 1832–38. BU 1838.

Erne Robert Newman, Dartmouth
As built: 115ft 6in, 97ft 2in x 29ft 8in x 8ft 6in. 457 ($454\frac{80}{94}$ by calc.) bm. Draught 9ft 6in/10ft 0in.
Ord: 18.11.1812. K: 3.1813. L: 18.12.1813. C: 24.1–30.3.1814 at Portsmouth.
Commissioned 3.1814 under Cmdr (Capt. 6.1814) William John Napier, for the Channel. Defects made good and roundhouse added at Deptford 8–10.1815; recommissioned 9.1815 under Capt. Richard Spencer, for the Mediterranean. Fitted for sea at Plymouth 11.1817–2.1818; recommissioned 12.1817 under Cmdr Timothy Scriven; wrecked 1.6.1819 on Sal Island in the Cape Verde Islands.

Leven Jabez Bailey, Ipswich
As built: 115ft 8in, 97ft 4½in x 29ft 8⅜in x 8ft 6¾in. $456\frac{77}{94}$bm. Draught 8ft 6in/9ft 6in.
Ord: 18.11.1812. K: 3.1813. L: 23.12.1813. C: 1.1–5.3.1814 at Sheerness (for Ordinary).
In 1814 under Capt. John Tailour. Fitted for sea at Sheerness 12.1814–22.1.1815.
Commissioned 1.1815 under Capt. Buckland Bluett (–1817). Altered to a 24-gun Sixth Rate at Chatham 7–11.1818 (given a small quarterdeck and forecastle); recommissioned 8.1818 under Cmdr David Bartholomew (died 2.1821) and subsequently Capt. William F. Owen, as Survey ship for Cape Verde Islands, then African coast to 1825; fitted for sea at Woolwich 2–4.1820 and again 8.1821–1.1822. At Deptford 1826, then to Chatham 5.1827. Fitted as convict ship at Chatham 5–11.1833, to lie at Woolwich. Fitted as hulk 1841, to lie at Limehouse. BU at Deptford 7.1848.

Falmouth Richard Chapman, Bideford
As built: 115ft 6¾in, 97ft 2¼in x 29ft 8in x 8ft 6in. $454\frac{92}{94}$bm. Draught 8ft 6in/9ft 10in.
Ord: 18.11.1812. K: 4.1813. L: 8.1.1814. C: 23.3.- 3-.3.1814 at Plymouth (for Ordinary), then 12.1814–7.1815 (for sea).
Commissioned 5.1815 under Capt. George Knight; in 9.1815 under Capt. Robert Festing, for St Helena. Fitted for foreign service at Woolwich 11.1815–1.1816. In 1.1817 under Cmdr George Rich, still St Helena; then 3.1819 under Cmdr Henry Collier, for the Leeward Islands; in 9.1820 under Cmdr Edward Purcell (–1823), still Leeward Islands; paid off 7.1823. At Portsmouth in 1824. Sold to Timothy Hutchinson (for £2,260) 27.1.1825 for mercantile use, renamed *Protector*.

Cyrene Richard Chapman, Bideford
As built: 115ft 6½in, 97ft 0¾in x 29ft 11⅝in x 8ft 7½in. 457 ($463\frac{65}{94}$ by calc.)bm. Draught 8ft 8in/9ft 4in.
Ord: 18.11.1812. K: 4.1813. L: 4.6.1814. C: 10.7–30.7.1814 at Plymouth (for Ordinary); then 7.1818–12.10.1818 (for sea).
Commissioned 25.7.1818 under Cmdr Aaron Tozer, for Bermuda and Halifax; paid off 16.1.1822 At Plymouth in 1820. Recommissioned 17.1.1822 under Cmdr Percy Grace (–1825); to African coast in 2.1824, then to UK and to Mediterranean; paid off 8.1825. Fitted for sea at Deptford 9–10.1825; recommissioned 8.1825 under Cmdr Alexander Campbell, for the East Indies. Sold (by AO 20.4.1827) at Bombay for 28,500 rupees (£2,731.5.0d) 4.1828.

Bann John King, Upnor
As built: 115ft 11¾in, 97ft 3¼in x 30ft 8in x 8ft 6in. $465\frac{62}{94}$bm. Draught 8ft 6in/9ft 7in.
Ord: 18.11.1812. K: 5.1813. L: 8.1.1814 (not 8.6.1814 as some state). C: 17.2.1814 at Chatham (for Ordinary).
In 1814 under Capt. John Tancock. Doubled the bottom at line of floatation, and fitted for sea at Chatham 12.1814–23.1.1815.
Commissioned 1.1815 under Capt. Thomas Whinyates; later that year under Capt. Lewis Shepheard, then Capt. Edward Scobell and finally 14.9.1815 Capt. William Fisher; fitted for foreign service at Woolwich 9–11.1815. Fitted for sea 6–11.1818; recommissioned under Cmdr Andrew Mitchell; in 1820 under Cmdr Wilson Braddyll Bigland, for Jamaica station; paid off 6.1820 and recommissioned under Cmdr Jodrell Leigh (acting); paid off 6.1825. Sold at Chatham (for £1,050) to H. Cropman 8.1.1829.

Spey James Warwick, Eling (Southampton)
As built: 115ft 8⅜in, 97ft 4¼in x 29ft 11in x 8ft 6in. $463\frac{44}{94}$bm. Draught 8ft 7in/9ft 9in.
Ord: 18.11.1812. K: 5.1813. L: 8.1.1814. C: 6.2–24.2.1814 at Portsmouth (for Ordinary)
In 1814 under Capt. Hew Stewart. Doubled the bottom at line of floatation and fitted for sea at Portsmouth 12.1814–7.2.1815.
Commissioned 12.1814 under Capt. George Fergusson. Fitted for foreign service at Woolwich 9–12.1815; recommissioned 10.1815 under Capt. John Lake, then 11.1816 under Capt. James Arthur Murray, for Cape of Good Hope and St Helena; in 1818 under Capt. James Kearney White, then 8.1819 Capt. Frederic Noel, for the Mediterranean. In 1820 under Cmdr John Boswall, then 9.1821 Cmdr Charles Phillips; paid off 28.9.1821. Sold at Chatham to Mr Vincent (for £1,530) 18.4.1822.

Lee Josiah & Thomas Brindley, Frindsbury
As built: 115ft 7in, 96ft 11¼in x 29ft 11½in x 8ft 6in. $462\frac{73}{94}$bm.
Ord: 18.11.1812. K: 3.1813. L: 24.1.1814. C: 24.1–19.2.1814 at Chatham (for Ordinary); then 12.1814–1.1815 (for sea).
Commissioned 20.8.1814 under Capt. James John Gordon Bremer, for the Irish station; fitted with additional 6in deep false keel at Plymouth 3–4.1815. In 8.1815 under Capt. John Pasco (–1817) at Plymouth, then (as sloop) under Cmdr Stewart Blacker; paid off 9.1818. Fitted with Thomson's patent rudder at Plymouth 2.1820. BU completed at Plymouth 30.5.1822.

Hind (ex-*Barbadoes*, renamed 1813) Robert Davy, Topsham
As built: 115ft 3in, 96ft 10in x 29ft 10½in x 8ft 5¼in. $459\frac{67}{94}$bm. Draught 8ft 4in/9ft 6in.
Ord: 18.11.1812. K: 5.1813. L: 8.3.1814. C: 26.3–5.4.1814 at Plymouth (for Ordinary).
Small Repair and fitted for sea at Plymouth 3.1819–13.7.1819.

Commissioned 15.4.1819 under Cmdr Sir Charles Burrard
(–29.1.1822). Fitted with additional 6in deep false keel (by AO
5.10.1820). On 2.1822 under Cmdr H. J. Rons, then 4.1823 under
Cmdr Lord H. J. Spencer Churchill. Fitted for foreign service at
Deptford 8.1825–2.1826; recommissioned 15.9.1825 under Cmdr
John Furneaux, for the East Indies. Sold (by AO 20.4.1827) at
Bombay for 23,750 rupees (£2,276) 6.4.1829.

Larne William Bottomley, King's Lynn
>   As built: 115ft 6in, 97ft 2in x 29ft 9½in x 8ft 6in. 458⁶⁸/₉₄bm.
>
>   Ord: 18.11.1812. K: 7.1813. L: 8.3.1814. C: 9.4–5.5.1814 at Sheerness
>   (for Ordinary); then 12.1814–12.1.1815 (for sea).
>
>   Commissioned 14.12.1814 under Capt. John Fordyce Maples;
>   from 12.1.1815 under Capt. Abraham Lowe, for the Channel
>   and West Indies; on 1.5.1815 under Capt. Sir John Gordon
>   Sinclair (–8.1815, when Lowe resumed command). Fitted for
>   Channel service at Woolwich 8–11.1815; on Jamaica station
>   1817; paid off 19.1.1819. Fitted for sea at Portsmouth 5–8.1819;
>   recommissioned 24.5.1819 under Cmdr Henry Forbes, then
>   7.12.1819 under Cmdr Robert Tait, for the Mediterranean;
>   paid off 10.1822. Small Repair and fitted for sea at Portsmouth
>   11.1822–6.1823; recommissioned 31.3.1823 under Cmdr
>   Frederick Marryat (–1824), in the East Indies; occupation of
>   Rangoon 11.5.1824; in Irrawaddy operations 5–9.1824; Bassein
>   River operations 3.1825. On 15.4.1825 under (acting) Lieut
>   John Kingcombe, then 25.7.1825 under Cmdr William Burdett
>   Dobson, in New Zealand; paid off 4.4.1827. Sold to G. Bayley
>   (for £1,440) 26.3.1828 to BU.

(B) Vessels acquired from 1 January 1817

Flush-Decked Ship Sloops

From 1817, all new sloops would be flush-decked vessels. From 1819,
a new series of experimental sloops were built to varying designs, to
compare different hull forms in an effort to produce fast-sailing small
cruisers. Robert Seppings remained a strong supporter of the School of
Naval Architects established at Portsmouth (and led by its long-term
Professor, Dr James Inman), which produced a number of designs to
compare with the Surveyor's own, as well as those of other designers.
Initially these sloops all carried 18 or 20 guns, but – as ordnance
weight rose – the survivors were all reduced to 14 guns by 1848 (see
individual vessels below).

ROSE Class 18 guns. Designed (approved 29 December 1819) by
Professor Inman and built by the Superior Class of Shipwright
Apprentices at Portsmouth. This experimental vessel performed well
under sail; Inman however criticised the weight of armament as being
too heavy for her size. The Navy Board authorised four repeats to be
built to this design, but the requirement was amended instead to design
these to Professor Inman's preferred dimensions.
>   Dimensions & tons: 104ft 2½in, 85ft 10in x 29ft 10in oa (29ft 6in for
>   tonnage) x 13ft 4in. 397²⁸/₉₄bm.
>
>   Men: 125. Guns: 2 x 6pdr (bow, later 9pdr) + 16 x 32pdr carronades
>   (17cwt). Reduced to 16 guns between 1839 and 1842, and to 14
>   guns in 1848.

Rose Portsmouth Dyd
>   As built: 104ft 4in, 85ft 11⅞in x 29ft 10in oa (29ft 6in for tonnage) x

13ft 4in. 398⁴/₉₄bm.
>   Ord: 27.2.1819. K: 3.4.1820. L: 1.6.1821. C: 16.8.1821 (for sea).
>
>   First cost: £11,791 (£16,955 including fitting).
>
>   Commissioned 22.5.1821 under Cmdr Thomas Ball Clowes, for the
>   Mediterranean; on 16.5.1823 under Cmdr Henry Dundas, then
>   11.1824 Cmdr C. Abbott, and 26.1.1826 Cmdr Lewis Davies,
>   still in the Mediterranean; at Battle of Navarino 20.10.1827;
>   later on same date under Cmdr Thomas Dilke; paid off end
>   1827 or early 1828 at Portsmouth. Small Repair and fitted for
>   sea there (for £8,300) 3–9.1828; recommissioned 23.7.1828
>   under Cmdr Eaton Stannard Travers, for Teneriffe, Rio de
>   Janeiro and Cape of Good Hope, thence to Bay of Fundy and
>   Halifax station; on 19.11.1829 under Cmdr J. G. Dewar (died
>   8.1830), then on 16.8.1830 under Cmdr Edward Williams
>   Pilkington, on same station; paid off 4.1832. Small Repair and
>   fitted for sea at Sheerness (for £3,463) 5–8.1834; recommissioned
>   6.1834 under Cmdr William Barrow; sailed 17.8.1834 for the
>   East Indies. On 3.8.1838 under Cmdr Peter Christie, on the
>   Spanish and Brazilian coasts; paid off 23.11.1841 at Sheerness.
>   Recommissioned 9.3.1843 under Cmdr Henry Richard Sturt,
>   for North America and the West Indies; on 13.12.1844 under
>   Cmdr Richard Wilson Pelly, same station; paid off 21.11.1846
>   at Sheerness. Fitted for 'Relief service' there (for £688) 7.1847,
>   thence to Chatham. BU at Chatham 5.1851.

MARTIN Class 20 guns. Designed by Seppings (to the same
specification as the *Rose*), approved 25 February 1819.
>   Dimensions & tons (as built): 108ft 5in, 90ft 7¼in x 28ft 10½in x 7ft
>   9in. 401⁷⁷/₉₄bm. Draught 9ft 6in/11ft 3in.
>
>   Men: 125. Guns: 2 x 6pdr (bow) + 18 x 32pdr carronades.

Martin Portsmouth Dyd
>   Ord: 27.2.1819. K: 7.1820. L: 18.5.1821. C: 16.8.1821 (for sea).
>
>   First cost: £17,690 (including fitting).
>
>   Commissioned 22.5.1821 under Cmdr Christopher Crackenthorp
>   Askew, for the Mediterranean; on 19.7.1822 under Cmdr Henry
>   Eden; off Greece during national revolution; paid off 2.1825.
>   Recommissioned 2.1825 under Cmdr Thomas Wilson, for the
>   East Indies; lost, presumed foundered off the Cape with all (150)
>   hands in 2.1826.

COMET Class 18 guns. Longer 1823 design by Professor Inman.
Electra was originally ordered as one of the *Fly* Class (see below). First
pair renamed 1832 to avoid confusion with the steam vessels added to
the Navy List (see Chapter 11). *Electra* was reduced to 14 guns in 1846,
and the earlier pair followed suit in 1848.
>   Dimensions & tons: 113ft 3in, 92ft 10⅛in x 30ft 10in oa (30ft 6in for
>   tonnage) x 8ft 0in. 459³⁷/₉₄bm.
>
>   Men: 125. Guns: 2 x 6pdr (bow, later 9pdrs in *Comet* and 18pdrs in
>   *Electra*) + 16 x 32pdr carronades.

Comet Pembroke Dyd
>   As built: 113ft 3in, 92ft 10¼in x 30ft 11in oa (30ft 7in for tonnage) x
>   8ft 0in. 462¹⁶/₉₄bm.
>
>   Ord: 15.5.1821. K: 10.1826. L: 14.8.1828. C: 20–28.2.1829 (for sea)
>   at Plymouth.
>
>   First cost: £10,587 (+ fitting £6,927).
>
>   Commissioned 11.1828 under Cmdr Alexander Sandilands, for
>   East Indies and Pacific. Renamed *Comus* 31.10.1832, and
>   recommissioned under Cmdr William Price Hamilton, for North
>   America and the West Indies; refitted at Plymouth (for £5,015)
>   10.1832–1.1833; paid off at Portsmouth 8.1836. Recommissioned
>   28.2.1837 under Cmdr Plantagenet Pierrepont Cary, for

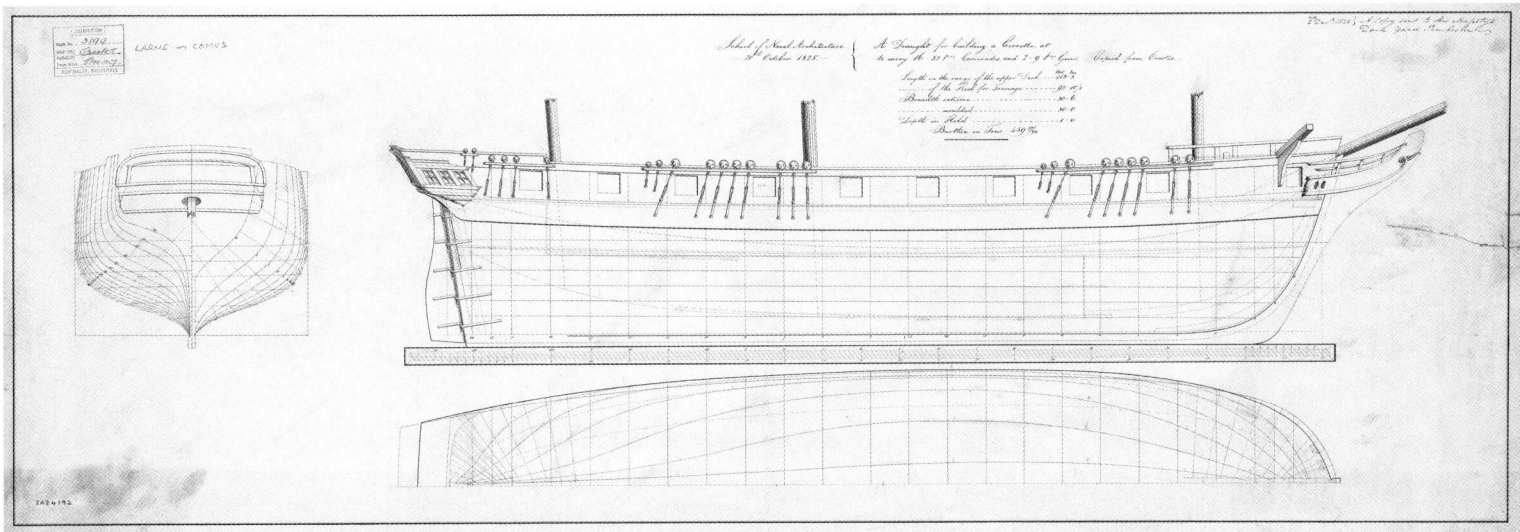

The *Comet* Class of ship sloops were built to a design by Professor James Inman of the School of Naval Architecture. Although first ordered in 1821, the *Comet* and *Lightning* were not begun until after Inman's draught had been sent to Pembroke Dock on 2 December 1825 (here dated 20 October, referred to as a 'corvette', and endorsed 'Copied from *Orestes*'); the two were renamed *Comus* and *Larne* in 1832, to avoid confusion with the steam paddle vessels of the same names which had been commissioned into the Navy. A third vessel, *Electra*, originally ordered to Inman's modified design of 1828, was built to the same draught at Portsmouth. (*NMM J6454*)

North America and West Indies; sailed 12.4.1837; on 9.5.1839 under Cmdr Evan Nepean, in the West Indies; in 1.1842 under Cmdr George Evan Davies (acting); paid off 1842 at Chatham. Between a small and middling repair there (for £11,237) 11.1842–2.1845. Recommissioned 14.12.1844 under Cmdr Thomas Sparke Thompson, for South America; on 9.11.1845 under Lieut Edward Augustus Inglefield (acting); at Battle off Punta Obligado 20.11.1845. On 17.11.1846 under Cmdr Edwin Clayton Tennyson d'Eyncourt; paid off 18.10.1848. Between a small and middling repair at Sheerness (for £4,382) 12.1850–4.1852. Recommissioned 3.5.1853 under Cmdr William Abdy Fellowes, for the East Indies; in 4.1855 under Cmdr George Blane; on 23.7.1855 under Cmdr Robert Jenkins; in 2nd Anglo-Chinese War; on 30.12.1857 under Cmdr Richard Dawkins; paid off at Chatham 7.6.1858. Completed BU there 10.5.1862.

Lightning (ex-*Orestes*, renamed 30.1.1822) Pembroke Dyd.
As built: 113ft 3in, 93ft 3in x 30ft 10½in oa (30ft 6½in for tonnage) x 8ft 0in. 462⁶³⁄₉₄bm. Draught 9ft 2in/10ft 8in.
Ord: 15.5.1821. K: 7.1828. L: 2.6.1829. C: 16.9.1829 at Plymouth.
First cost: £11,178 (+ fitting £7,291).
Commissioned 25.6.1829 under Cmdr Thomas Dickinson, for South America; recovery of $600,000 from wreck of *Thetis* off Cape Frio 1831; paid off 13.9.1832. Renamed *Larne* 12.9.1832. Recommissioned 24.9.1832 under Cmdr William Sidney Smith, for blockade of Dutch coast; stranded on the Goodwins 4.4.1833, but salved; to North America and the West Indies 1834; paid off 6.1836. Refitted at Portsmouth (for £6,371) 10.1836–5.1837. Recommissioned 9.3.1837 under Cmdr Patrick John Blake, for the East Indies; during 1839 under Cmdr Augustus Leopold Kuper (temp), then Blake again; China War 1839–41; Chusan 5.7.1840; attack on Chuenpee Forts 7.1.1841; paid off 2.7.1842 at Sheerness. Small repair and fitted for sea there (for £8,293) 3.1843–1.1844. Recommissioned 14.11.1843 under Cmdr James William Douglas Brisbane, for Africa; paid off 6.7.1847 at Sheerness. Refitted there (for £1,824) 6.1847–6.1848, then to

Ordinary. BU (by AO 18.8.1865) completed by Castle & Beech at Charlton 28.3.1866.

Electra Portsmouth Dyd
As built: 113ft 3in, 92ft 10⅛in x 30ft 10in oa (30ft 7in for tonnage) x 8ft 1in. 461⁸⁶⁄₉₄bm. Draught 8ft 1in/10ft 6in.
Ord: 1.2.1828. K: 2.1830. L: 28.9.1837. C: 17.2.1838.
First cost: £12,394 (including fitting).
Commissioned 13.11.1837 under Cmdr William Preston (–4.1839), for South America; on 7.4.1839 under Cmdr Edward Reeves Philip Mainwaring. Recommissioned 16.12.1841 under Cmdr Arthur Darley, for North America and the West Indies; paid off 1845. Recommissioned 31.1.1846 under Cmdr William Heriot Maitland (invalided 3.1847), for same station; on 17.3.1847 under Cmdr Frederick William Pleydell Bouverie. On 30.8.1852 under Cmdr William Morris, for Australia; paid off 20.3.1857. Sold to W.W. Foord (for £950) to BU 17.2.1862.

ORESTES Class 18 guns. A shorter 1823 design by Professor James Inman. After competitive trials in 1824–25 against the *Pylades* and *Champion* (see below), her masts were moved, resulting in somewhat improved performance.
Dimensions & tons: 110ft 0in, 92ft 10⅛in x 30ft 10in oa (30ft 6in for tonnage) x 7ft 6in. 459³⁷⁄₉₄bm.
Men: 125. Guns: 2 x 6pdr (bow) + 16 x 32pdr carronades. Reduced to 14 guns in 1848, with 9pdrs (17cwt) having replaced 6pdrs.

Orestes Portsmouth Dyd
As built: 109ft 11in, 92ft 10⅛in x 30ft 10in oa (30ft 6in for tonnage) x 7ft 6in. 461bm. Draught 8ft 6in/11ft 0in.
Ord: 22.5.1821. K: 4.1823. L: 31.5.1824. C: 16.9.1824 (for sea).
First cost: £12,527 (£17,741 including fitting).
Commissioned 1.6.1824 under Cmdr Henry Litchfield, first as experimental vessel, later for Halifax; on 1.5.1826 under Cmdr William Jones; paid off 1827. Recommissioned 13.2.1828 under Cmdr John Reynolds, for Cork; on 10.3.1831 under Cmdr William Nugent Glascock; at Oporto 9.1832–3.6.1833. On 28.9.1833 under Sir William Dickson, still off Lisbon. On 6.6.1834 under Cmdr Henry John Codrington, in the Mediterranean. On 28.1.1836 under Cmdr Julius James Farmer Newell; on 3.8.1838 under Cmdr Peter Sampson Hambly, in South America; paid off 12.11.1841 at Portsmouth. Recommissioned 20.8.1842 under Cmdr Swynfen Thomas Carnegie, for North America and the West Indies; to the Mediterranean 9.1843; on 22.9.1843

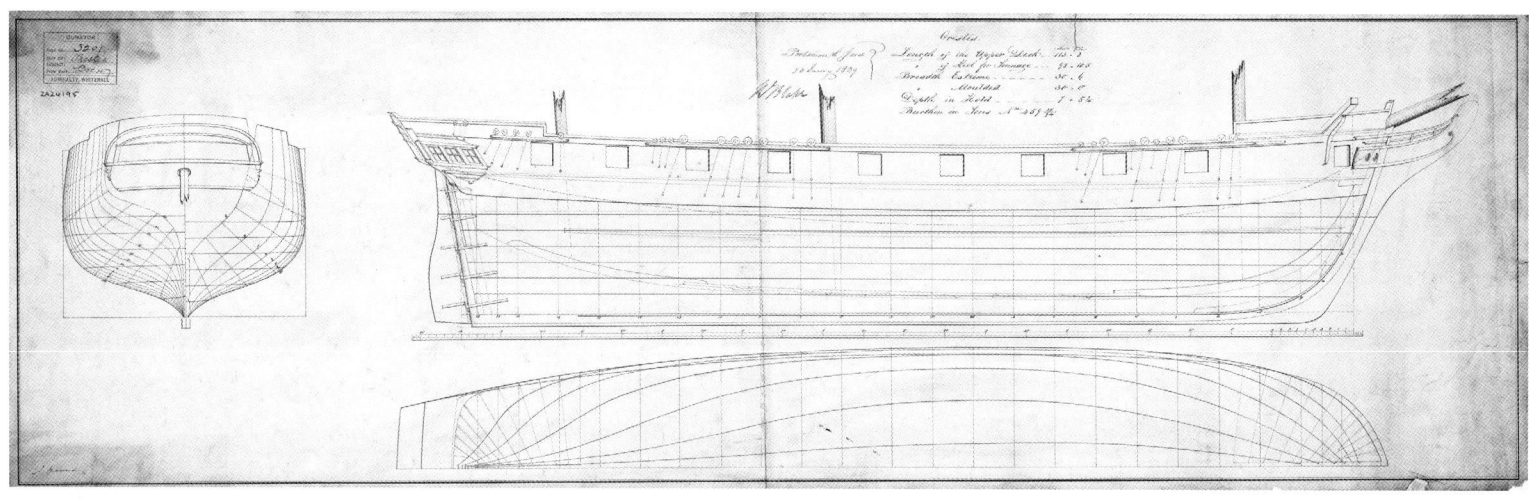

A draught of *Orestes* taken off at Portsmouth on 23 January 1839. Slightly shorter than the *Comet* Class designed at the same time (both designs by Professor Inman). *(NMM J6947)*

Cmdr Edward St Leger Cannon, for the Mediterranean; paid off 8.9.1846 at Portsmouth. Fitted there (for £7,759) 6.1847–12.1848. Recommissioned 26.10.1848 under Cmdr Henry Samuel Hawker, for the Cape of Good Hope; paid off 20.4.1852. Fitted as coal depot at Portsmouth (for £731) 11.1852; in about 1890 yard craft *C.28*. To Plymouth by 1890. Sold 1907 for commercial use, renamed *Mack* in about 1911.

PYLADES Class 18 guns. Design by Robert Seppings, to evaluate against the *Orestes* and *Champion*, and approved 20 June 1823. After her trials, a deeper keel was added and her masts were moved to improve performance.

Dimensions & tons: 110ft 0in, 90ft 1⅝in x 30ft 4in oa (30ft 0in for tonnage) x 8ft 2in. 431³⁷⁄₉₄bm.

Men: 122. Guns: 2 x 9pdr (bow) + 16 x 32pdr (17cwt) carronades.

Pylades Woolwich Dyd

As built: 110ft 1in, 90ft 2⅞in x 30ft 4¼in oa (30ft 0¼in for tonnage) x 8ft 2in. 433bm. Draught 8ft 8½in/10ft 11in.

Ord: 22.5.1821. K: 3.1823. L: 29.6.1824. C: 4.9.1824 (for sea). First cost: £15,889 including fitting.

Commissioned 1.6.1824 under Cmdr Francis Fead, for experimental squadron, then to Jamaica; on 28.6.1825 under Cmdr John Leith, then 16.9.1825 Cmdr George Vernon Jackson; paid off 1826. Recommissioned 6.3.1828 under Cmdr Patrick Duff Henry Hay, for Cork; paid off 21.5.1828. Recommissioned 22.5.1831 under Cmdr Edward Blanckley for South America; fitted for sea at Plymouth (for £3,945) 7–10.1831; paid off 6.1834. Between small and middling repair at Plymouth and fitted for sea (for £5,631) 2–10.1835. Recommissioned 4.8.1835 under Cmdr William Langford Castle, for Cape of Hood Hope; took (with *Forester*)

A watercolour by the Reverend Richard Calvert Jones of the ship sloop *Pylades*, dated precisely as 2 September 1835, shortly after recommissioning for the Cape of Good Hope station. Although described as 'flush-decked', the ship sloops of this era had an open-backed topgallant forecastle and a short platform over the captain's cabin above the upper deck, as can be seen by the position of the figures at the fore and aft extremities of the vessel. *(NMM PW6105)*

An anonymous but convincing watercolour portrait of the *Childers* at anchor. This class was a revival of a wartime design that was a ship-rigged version of the ubiquitous *Cruizer* Class brigs described in the next Chapter. The *Childers* was herself converted to a brig rig in 1834, so this illustration must pre-date that. *(NMM PW6091)*

Esperansa 1837; paid off 1838. Recommissioned 12.12.1839 under Cmdr Talavera Vernon Anson, for the East Indies; in 1st Anglo-Chinese War; at Amoy 25–26.8.1840. On 16.10.1841 under Cmdr Louis Symonds Tindal; paid off 10.1843. BU at Chatham 5.1845.

SNAKE Class 18 guns. Revival of Sir William Rule's 1796 design which produced the numerous wartime brig-sloops of the original *Cruiser* class. The former *Snake* of 1797 – ship-rigged sister to the *Cruiser* of the same year – had been sold in 1816. The new *Cruiser*, after conversion to brig, took part in the Experimental Squadron alongside the more modern brigs (see next chapter), but she outsailed all the others when before the wind.

 Dimensions & tons: 100ft 0in, 77ft 3½in x 31ft 0in oa (30ft 6in for tonnage) x 12ft 9in. 382⁴¹⁄₉₄bm. Draught 7ft 9in/11ft 11in.

 Men: 125. Guns: 2 x 6pdr (bow) + 16 x 32pdr carronades. Reduced to 16 guns in 1833, and to 12 guns in 1848.

Childers Chatham Dyd

 As built: 100ft 0in, 77ft 2½in x 31ft 1½in (30ft 7½in for tonnage) x 12ft 10in. 385¹⁸⁄₉₄bm. Draught 7ft 9in/11ft 11in.

 Ord: 10.6.1823. K: 11.1823. L: 23.8.1827. C: ?10.1827.

 First cost: £9,757.

 Commissioned 17.9.1827 under Cmdr William Morier, for the North Sea; on 24.11.1829 under Cmdr Robert Deans; off coast of Portugal 4–6.1831; paid off 17.1.1833. Converted to brig rig 1834, and recommissioned 16.5.1834 under Cmdr Henry Keppel for the Mediterranean; later to Cape of Good Hope; on 6.12.1836 under Cmdr Edward Pellew Halstead, in the East Indies; in 1st Anglo-Chinese War 1842. On 16.4.1842 under Cmdr George Greville Wellesley, in the East Indies; paid off 7.4.1844 at Portsmouth. Recommissioned 7.4.1855 under Cmdr

Victor Grant Hickey, for west coast of Africa; on 8.4.1858 under Cmdr William Swinburn; paid off 12.1858 at Chatham. Sold to Holloway 19.8.1865.

Cruiser Chatham Dyd

 As built: 100ft 0in, 77ft 3in x 31ft 0¾in oa (30ft 6¼in for tonnage) x 12ft 9½in. 383⁷⁶⁄₉₄bm.

 Ord: 10.6.1823. K: 1.1826. L: 19.1.1828. C: 6.3.1828 (for Ordinary), 10.1828–7.1.1829 (for sea).

 First cost: £9,496 (+ fitting £6,732).

 Commissioned 10.1828 under Cmdr John Edward Griffith Colpoys, for the East Indies (–1831). In 4.1831 under Cmdr (Acting) James Beckford Lewis Hay, then on 12.5.1831 under Cmdr John Parker, on same station; very small repair, converted to brig rig and fitted for sea at Sheerness 12.1832–11.1833 (for £2,112). Recommissioned 8.1833 under Cmdr John McCausland, for North America and West Indies; stranded at San Juandi (Nicaragua) 1834; paid off 1.1834. In 1837 under Cmdr William Alexander Willis, for same station. On 12.2.1838 under Cmdr Richard Henry King, for the East Indies. On 10.5.1839 under Cmdr Henry Wells Giffard, for the East Indies; occupation of Aden 18.6.1839; China War operations 1839–42; at Amoy 25/26.8.1840. On 16.10.1841 under Cmdr Joseph Pearse; at Tsinghae 1.10.1841; paid off 3.1843 at Chatham. Small repair (and ship-rigged again?) at Chatham (for £8,040) 3.1843–6.1844. Recommissioned 7.9.1844 under Cmdr Edward Gennys Fanshawe, for the East Indies; boat attack on Sherif Osman's stockade at Malluda Bay (Borneo) 19.8.1845. On 22.12.1845 under Cmdr William Maclean, then 12.11.1846 Cmdr Edward Peirse, still in East Indies; paid off 4.1848 Sold at Bombay 3.1849.

A contemporary model of the sloop *Hyacinth* of the *Favorite* Class, a lengthened version of the *Snake* Class of wartime origins. This is a sailor-made model and a little crude in detail, but accurate as to general appearance, fittings and rig. *(NMM F9232-001)*

FAVORITE Class 18 guns. Lengthened version of Rule's 1796 *Cruiser/Snake* Class design. The lengthening was done by inserting an extra 9½ft-long midships section into the design.

Dimensions & tons: 109ft 6in, 86ft 9½in x 30ft 9in oa (30ft 6in for tonnage) x 12ft 9in. 429⁴⁹⁄₉₄bm.

Men: 125. Guns: 2 x 9pdr (bow) + 16 x 32pdr carronades. Reduced to 14 guns in 1848 (1846 for *Favorite*).

Favorite Portsmouth Dyd

As built: 109ft 7in, 86ft 10¾in x 30ft 10in oa (30ft 7in for tonnage) x 12ft 11in. 432³⁹⁄₉₄bm. Draught 6ft 9in/0ft 11in.

Ord: 10.6.1823. K: 3.1825. L: 21.4.1829. C: 29.7.1829 (for sea).

First cost: £10,226 (£17,083 including fitting).

Commissioned 6.5.1829 under Cmdr Joseph Harrison, for Mediterranean and African coasts; paid off 8.1833. Recommissioned 29.8.1833 under Cmdr George Rodney Mundy, for the Mediterranean; paid off 14.2.1837 at Plymouth. Recommissioned 29.8.1837 under Cmdr Walter Croker; attack on Tongatabu ('Friendly Islands') 24.6.1840 – Croker killed; from 25.6.1840 under Cmdr Thomas Ross Sulivan; paid off 1842 at Plymouth. Recommissioned 13.3.1846 under Cmdr Alexander Murray, for African station; paid off 1847. Fitted as coal depot at Plymouth (for £1,196) 7–8.1859 (the name was spelt '*Favourite*' from about 1863); yard craft *C.3* about 1890, later *C.77*. Sold at Plymouth 17.5.1905.

Hyacinth Plymouth Dyd

As built: 109ft 8in, 86ft 11¾in x 30ft 11in oa (30ft 8in for tonnage) x 12ft 9in. 435⁵⁄₉₄bm. Draught 7ft 6in/10ft 11in.

Ord: 10.6.1823. K: 3.1826. L: 6.5.1829. C: 12.1.1830 (for sea).

First cost: £11,872 (£17,361 including fitting).

Commissioned 10.1829 under Cmdr Robert Milborne Jackson, for West Indies; on 14.3.1831 under Cmdr William Oldrey; paid off end 1832 at Portsmouth. Recommissioned 7.2.1833 under Cmdr Francis Price Blackwood, for the East Indies; paid off end 1836.

Recommissioned 18.7.1837 under Cmdr William Warren, for the East Indies; Canton (Pearl River) 11.1839; action near Maio 8.1840; 1st Anglo-Chinese War 1841. On 14.8.1841 under Cmdr George Goldsmith; 1st Anglo-Chinese War 1841–42; paid off 24.10.1842 at Sheerness. Recommissioned 7.8.1843 under Cmdr Francis Scott, for west coast of Africa; destroyed slaver in Fish Bay 13.8.1844; later to North America and West Indies; paid off 1847. Fitted at Portsmouth as coal hulk for Portland 9–10.1860. Paid off into 1st Division Reserve 11.1868. BU completed at Portsmouth 27.11.1871.

Racehorse Plymouth Dyd

As built: 109ft 8in, 86ft 11¼in x 30ft 11¼in oa (30ft 8¼in for tonnage) x 12ft 11in. 435bm. Draught 7ft 5in/10ft 11in.

Ord: 10.6.1823. K: 5.1829. L: 24.5.1830. C: 11.1830.

First cost: £11,140 (£16,807 including fitting).

Commissioned 21.9.1830 under Cmdr Charles Hamlyn Williams, for North America and the West Indies; on 31.1.1832 under Cmdr Francis Vere Cotton. Recommissioned 1.2.1834 under Cmdr Sir James Everard Home, for the West Indies; at siege of Parîa 1835; on 13.11.1837 under Cmdr Henry William Craufurd; on 29.4.1839 under Cmdr Edward Alfred John Harris, then 15.3.1841 under Cmdr, John Coghlan Fitzgerald. Recommissioned 21.12.1841 under Cmdr Edmund Peel, for same station; paid off end 1842. Small repair and fitted for sea at Plymouth (for £8,573) 6.1843–2.1845. Recommissioned 14.12.1844 under Cmdr George James Hay, for East Indies; in New Zealand War; at Ruapekapeki 11.1.1846. On 1.4.1846 under Cmdr Robert Jocelyn Otway, then on 26.6.1846 under Cmdr Edward Southwell Sotheby; paid off 1848. Recommissioned 28.11.1853 under Cmdr Edward King Barnard, for East Indies and China; action with junks 26.6.1855 and 4.7.1855; paid off 5.1858. Fitted as coal depot at Plymouth 1.1860–4.1861. Sold at Plymouth 11.5.1901.

A partner to his portrait of *Pylades* produced earlier, this watercolour sketch of September 1834 by the Rev. Richard Calvert Jones depicts the ship sloop *Champion*, a vessel designed by Capt. John Hayes. This vessel participated in sailing trials against her contemporaries *Orestes* and *Pylades*, initially proving slightly better than either, although the placings were reversed after both *Orestes* and *Pylades* had the position of their masts altered. *(NMM PW6107)*

Hazard Portsmouth Dyd

As built: 110ft 2in, 87ft 1½in x 30ft 9in oa (30ft 6in for tonnage) x 12ft 9in. 431¹⁹⁄₉₄bm. Draught 7ft 11in/10ft 7in.

Ord: 10.6.1823. K: 5.1829. L: 21.4.1837. C: 5.7.1837.

First cost: £10,500.

Commissioned 12.5.1837 under Cmdr James Wilkinson, for the Mediterranean. On 16.7.1840 under Cmdr Charles Gilbert John Brydone Elliot (son of First Lord of Admiralty); at bombardment of St Jean d'Acre 3.11.1840; paid off 25.9.1841 at Portsmouth. Recommissioned 29.9.1841 under Cmdr Charles Bell (died 8.8.1844), for the East Indies; in 1st Anglo-Chinese War 1842. On 9.8.1844 under Cmdr Francis Philip Egerton, for East Indies; paid off at Portsmouth 6.5.1847. Fitted there for experiment with Capt. Coles's shield (for £622) 12.1861–2.1862. BU completed by John White, Cowes 12.2.1866.

CHAMPION **Class** 18 guns. Designed 1823 by Captain John Hayes, and used in experimental trials against *Orestes* and *Pylades* (see above).

Dimensions & tons: 109ft 6in, 91ft 10⅛in x 30ft 10½in oa (30ft 6½in for tonnage) x 7ft 8¾in. 455⁶¹⁄₉₄bm.

Men: 125. Guns: 2 x 9pdr (bow) + 16 x 32pdr carronades. Reduced to 14 guns in 1848.

Champion Portsmouth Dyd

As built: figures as given for design. Draught 7ft 8in/10ft 10in.

Ord: 20.6.1823. K: 11.1823. L: 31.5.1824. C: 16.9.1824 (for sea).

First cost: £18,089.

Commissioned 1824 under Capt. John Hayes. In 6.1824 under Cmdr John F. Stoddart, for East Indies; in 1st Burma War 1825. On 29.11.1828 under Cmdr George Scott, for Africa and then for Halifax; on 31.1.1832 under Cmdr Charles Hamlyn Williams; paid off 23.6.1832. Recommissioned 27.6.1832 under Cmdr Arthur Duncombe, for the Mediterranean; refitted at Plymouth 12.1834. In 1836 under Cmdr Robert Fair. On 25.1.1837 under Cmdr George St Vincent King, for North America and the West

Indies; paid off end 1838. Small repair at Plymouth and fitted for sea (for £5,145) 3.1840–8.1841. Recommissioned 16.6.1841 under Cmdr Richard Byron (died 23.2.1843), for South America. Between small and middling repair at Portsmouth and fitted for sea (for £10,670) 9.1846–3.1848. Recommissioned 2.2.1848 under Cmdr John Montagu Hayes (son of Capt. Hayes), for the Pacific; paid off at Portsmouth into Ordinary 2.1.1852. Fitted as harbour police vessel at Portsmouth 9.1859. Lent to Committee on Floating Obstructions 10.2.1864, and used for explosive experiments. BU at Portsmouth 10.1867.

PEARL **Class** 20 guns. Philip Sainty design 1825. On 14 March 1825 the Admiralty ordered that she was 'to be built on such lines and principles as Mr Sainty may think proper'.

Dimensions & tons (as built): 118ft 9in, 94ft 9¼in x 33ft 7⅜in oa (33ft 3⅞in for tonnage) x 9ft 6in. 558³⁴⁄₉₄bm. Draught 7ft 6in/11ft 6in.

Men: 125. Guns: 2 x 9pdr (bow) + 18 x 32pdr carronades (Sir Alexander Dickson's type). Reduced to 18 guns in 1848.

Pearl (ex-*Nautilus*, renamed 19.3.1825) Philip Sainty, Wivenhoe.

Ord: 14.3.1825. Cancelled 11.7.1825 but re-ordered 23.2.1826. K: 7.1826. L: 17.3.1928. C: 19.3–29.6.1828 at Woolwich. Fitted there with an additional false keel and gripe 9–10.1828.

First cost: £13,392 (+ fitting £8,171).

Commissioned 4.1828 under Cmdr George Charles Blake (–1831), for Cork; later under Cmdr William Broughton, off coast of Portugal. On 22.11.1831 under Cmdr Robert Gordon, for North America and the West Indies; paid off 20.12.1834. In 1836 under Cmdr Hugh Nurse. On 17.1.1837 under Cmdr Lord Clarence Edward Paget, for Lisbon; in 1838 to West Indies and North America; took Spanish slaving schooner *Vengador* 1838; paid off 2.1.1839 at Sheerness. Recommissioned 18.12.1841 under Cmdr Richard Henry Stopford, for South America; paid off 1844. BU at Chatham 6.1851.

Because slaving was illegal, vessels involved in the trade were generally very fast and weatherly (and thus several sequestered slavers were later added to the Navy), which gave impetus to the drive for better sailing qualities in the Royal Navy's small craft charged with their capture. This oil painting by William Adolphus Knell celebrates the capture of the Spanish slaver *Vengador* in 1838 by the *Pearl*. The ship sloop is shown heaved-to, lowering a boat to take possession of the surrendered schooner astern. It is a reasonable speculation that the painting was commissioned by Lord Paget, the captain of the *Pearl* at the time. *(NMM BHC0626)*

SATELLITE Class   18 guns. Design by Robert Seppings 1825.
 Dimensions & tons: 112ft 0in, 91ft 1⅜in x 31ft 0in oa (30ft 8in for tonnage) x 13ft 10in. 455⁷¹/₉₄bm.
 Men: 125. Guns: 2 x 6pdr (bow) + 16 x 32pdr carronades. *Satellite* reduced to 14 guns in 1848.

Satellite Pembroke Dyd
 As built: 112ft 0in, 92ft 1⅜in x 31ft 2¼in oa (30ft 10¼in for tonnage) x 13ft 10in. 466⁴¹/₉₄bm. Draught 9ft 9in/11ft 9in.
 Ord: 9.6.1825. K: 6.1826. L: 3.10.1826. C: 14.2.1827 at Plymouth.
 First cost: £15,791 (including £5,434 fitting).
 Commissioned 22.11.1826 under Cmdr John Milligen Laws (–1.1831), for experimental duties, then to East Indies, then Australia and New Zealand, then back to Bay of Bengal. Refit at Plymouth (for £6,302) 5–11.1832. Recommissioned 15.9.1832 under Cmdr Robert Smart (–10.1834, then again from 2.1835), for North Sea (blockade of Dutch coast), then South America; paid off 1.1836. Refit at Plymouth and coppered (for £3,911) 7–11.1836. Recommissioned 24.10.1836 under Cmdr John Robb, for North America and West Indies; paid off 22.2.1841. Refitted and refitted at Portsmouth (for £9,658) 8.1841–7.1842. Recommissioned 2.5.1842 under Cmdr Robert Fitzgerald Gambier, for South America; on 3.12.1843 under Cmdr Robert Hibbert Bartholomew Rowley; in East Indies and South America; paid off at Sheerness 11.1847. BU at Sheerness 2.1849.

Acorn Chatham Dyd
 As built: 112ft 1in, 91ft 7⅝in x 31ft 2½in oa (30ft 10½in for tonnage) x 13ft 10in. 464⁶⁹/₉₄bm. Draught 10ft 0in/11ft 10in.
 Ord: 9.6.1825. K: 6.1826. L: 16.11.1826. C: 10.5.1827 (for sea).
 First cost: £16,656 (including fitting).
 Commissioned 11.1826 under Cmdr Alexander Ellice. In 6.1827 under Cmdr Edward Gordon; lost, presumed foundered with all (115) hands off Halifax 14.4.1828.

WOLF Class   18 guns. Design by Captain John Hayes, 1825.
 Dimensions & tons: 113ft 4½in, 91ft 8¼in x 30ft 10¼in oa (30ft 6¼in for tonnage) x 7ft 8¾in. 454⁷¹/₉₄bm.

Men: 125. Guns: 2 x 6pdr (bow) + 16 x 32pdr carronades. Reduced to 14 guns in 1848.

Wolf Portsmouth Dyd
 As built: dimensions unknown. 456bm. Draught 10ft 4in/10ft 7½in.
 Ord: 9.6.1825. K: 11.1826. L: 1.12.1826. C: 24.3.1827 (for sea).
 First cost: £13,276 (£19,816 including fitting).
 Commissioned 4.12.1826 under Cmdr George Hayes (brother of her designer), for the Mediterranean. On 2.12.1829 under Cmdr Robert Russell; stranded on the Isle of Wight 10.3.1830, but refloated (Russell dismissed the Navy, but later restored). In 10.6.1830 under Cmdr William Hamley, for the East Indies; paid off 3.1834. Recommissioned 27.5.1834 under Cmdr Edward Stanley; sailed for the East Indies 2.10.1834; in Kaffir War 1834–35; anti-piracy in Straits of Malacca 1836–38; paid off 6.1838. Between small and middling repair and fitted for sea at Plymouth (for £7,964) 6.1841–8.1842. Recommissioned 8.6.1842 under Cmdr Courtenay Osborn Hayes, for the East Indies. On 17.1.1845 under Cmdr James Alexander Gordon (died 6.1.1847), in the East Indies; on 1.1.1847 under Cmdr (Acting) Leopold George Heath. Fitted at Plymouth for a Coal hulk for Kingstown, Ireland 4–5.1848. Lent to Board of Works at Dublin 1851. Coal depot again 1860 at Queenstown. Reported unfit for further service 3.1877, so towed to Plymouth. BU completed at Devonport 5.8.1878.

SCOUT Class   18 guns. Design by Robert Seppings 1829, 'as *Satellite*' but larger. *Rover* was originally to have been built to this design.
 Dimensions & tons: 115ft 0in, 94ft 0⅜in x 31ft 4in oa (31ft 0in for tonnage) x 14ft 3in. 480⁸¹/₉₄bm.
 Men: 120/125. Guns: 2 x 9pdr (bow) + 16 x 32pdr carronades. *Scout* reduced to 14 guns in 1846.

Scout Chatham Dyd
 As built: 115ft 4½in, 94ft 7⅝in x 31ft 6in oa (31ft 2in for tonnage) x 14ft 3in. 488⁹⁹/₉₄bm. Draught 9ft 11in/10ft 10in.
 Ord: 30.1.1829. K: 10.1831. L: 15.6.1832. C: 26.10.1832 (for sea).
 Commissioned 20.7.1832 under Cmdr William Hargood, blockade of Dutch coast, then to the Mediterranean. On 10.12.1833 under Cmdr George Grey. On 21.7.1834 under Cmdr William Holt; paid off 8.1.1835. Recommissioned 4.12.1835 under Cmdr Robert Craigie, for Cape of Good Hope; took large Portuguese slaver 11.1.1837; paid off 1.1839 at Sheerness. Recommissioned 18.12.1841 under Cmdr James Robert Drummond, on the Mediterranean station; paid off 1845. Refitted at Chatham (for £5,807) 12.1845–2.1846. Recommissioned 5.1.1846 under Cmdr William Loring, for the Mediterranean, then to East Indies; on 2.2.1848 under Cmdr Frederick Erskine Johnston; destroyed two pirate junks near Chimmo Island 31.5.1848; paid off 21.9.1849. BU at Plymouth 10.1852.

Pheasant Plymouth Dyd
 Ord: 30.1.1829. K: — . Cancelled 1.11.1831 (unstarted).
Redwing Plymouth Dyd
 Ord: 30.1.1829. K: — . Cancelled 1.11.1831 (unstarted).

ROVER Class   18 guns. Originally ordered as one of the *Scout* Class (see above), but amended to new design by William Symonds, approved 10 or 11 December 1831.
 Dimensions & tons: 113ft 0in, 89ft 9½in x 35ft 4in oa (35ft 0in for tonnage) x 16ft 9in. 585¹²/₉₄bm.
 Men: 120. Guns: 2 x 32pdr (long, bow) + 16 x 32pdr (short).

Rover Chatham Dyd
 As built: 113ft 0in, 90ft 1⅜in x 35ft 5in oa (35ft 1in for tonnage) x 16ft 9in. 589⁹²/₉₄bm. Draught 9ft 3in/12ft 1in.

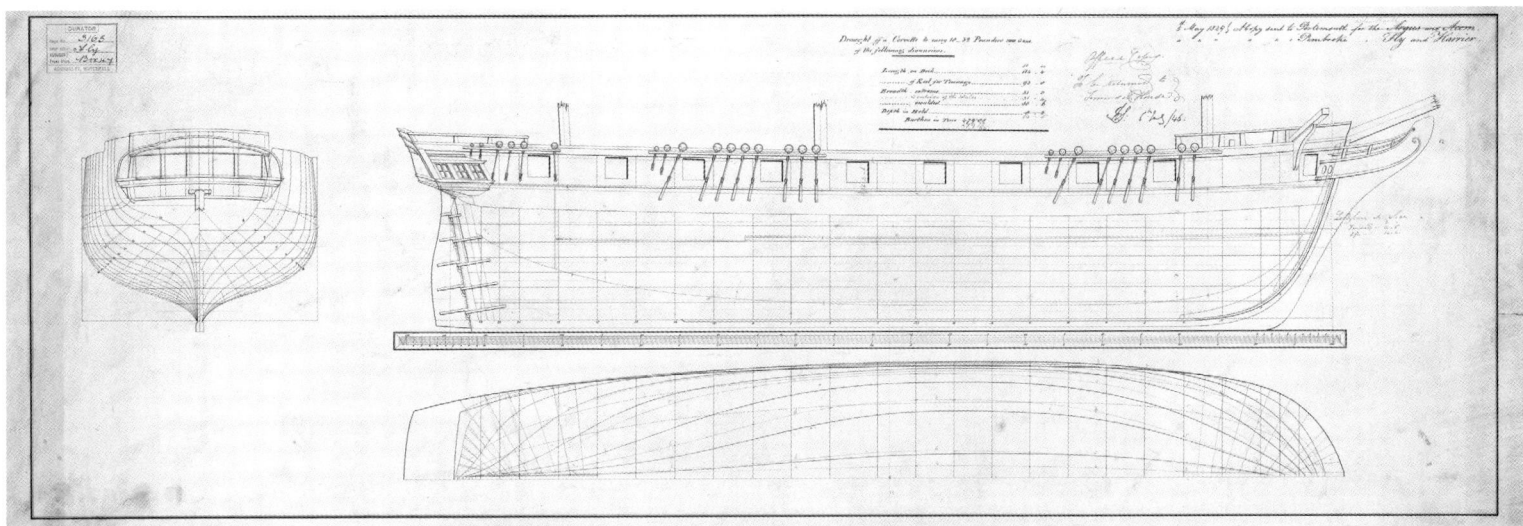

A draught, dated 5 May 1829, for the *Fly* Class ship sloops here described as 'corvettes' (as all surviving ship sloops were eventually re-designated) as designed by the School of Naval Architecture. The draught noted the intended armament as '18–32 Pounders new guns'. At this date, the class was expected to comprise four vessels – the *Fly* and *Harrier* at Portsmouth (both launched in 1831), and the *Argus* and *Acorn* at Plymouth (which were both cancelled). (*NMM J6433*)

Ord: 30.1.1829. K: 2.1832. L: 17.7.1832. C: 26.10.1832 (for sea).

Commissioned 20.7.1832 under Cmdr Sir George Young; blockade of Dutch coast, then to the Mediterranean. On 17.11.1834 under Cmdr Charles Eden, for South American station. On 27.8.1838 under Cmdr Thomas Matthew Charles Symonds (–22.2.1841), for North America and West Indies. On 13.3.1841 under Cmdr Charles Keele, for West Indies station; paid off 10.1843. BU at Chatham 9.1845.

FLY Class 18 guns. A development of Professor Inman's *Orestes* Class design. *Electra* was originally to have been built to this design.

Dimensions & tons: 114ft 4in, 93ft 8in x 31ft 4½in oa (31ft 0½in for tonnage) x 14ft 5in. 479⁷⁹⁄₉₄bm.

Men: 120. Guns: 2 x 9pdr (bow) + 16 x 32pdr carronades. *Fly* reduced to 16 guns in 1848.

Fly Pembroke Dyd

As built: 114ft 4in, 93ft 6⅛in x 31ft 7in oa (31ft 3in for tons) x 8ft 2¾in. 485⁶⁄₉₄bm. Draught 9ft 4in/11ft 5in.

Ord: 30.1.1829. K: 11.1829. L: 25.8.1831. C: 9.1831–27.1.1832 at Plymouth.

First cost: £11,761 (+ fitting £4,648).

Commissioned 17.9.1831 under Cmdr Peter M'Quhae, for North America and the West Indies; home to pay off 10.1835. Recommissioned 13.7.1836 under Cmdr Russell Eliott, for South America and the Pacific; on 12.7.1838 under Cmdr Granville Gower Loch, on same station; paid off 7.1840. Small Repair and fitted for sea at Plymouth (for £7,798) 3.1841–4.1842. Survey ship 10.1841 to 1847; recommissioned 17.11.1841 under Capt. Francis Price Blackwood, on East India station, surveying Australia; paid off 4.7.1846. Refitted and Very Small Repair at Plymouth (for £6,101) 10.1846–1.1847. Recommissioned 11.10.1847 under Cmdr Richard Aldworth Oliver, on same station; surveying New Zealand and Pacific 1847–51; paid off 12.1851. Fitted as coal depot at Plymouth 1854. On 1.7.1859 under Cmdr (Acting) John Smith Keats (–4.1861), as tender to *Cornwallis* on the Humber. Ordered to be BU by AO 6.8.1868 (but rescinded 17.10.1868); became yard craft *C.2.* in 1890. BU at Devonport 1903.

Harrier Pembroke Dyd

As built: 114ft 4in, 93ft 6⅛in x 31ft 7in oa (31ft 3in for tons) x 8ft 2½in. 485⁶⁄₉₄bm. Draught 9ft 4in/11ft 9in.

Ord: 30.1.1829. K: 11.1830. L: 8.11.1831. C: 11.1831–25.3.1832 at Plymouth.

First cost: £12,194 (+ fitting £5,387).

Commissioned 24.11.1831 under Cmdr Spencer Lambart Hunter Vassall, for the East Indies; suppression of piracy in straits of Malacca, notably at pirate strongholds of Poulo [Pulau] Arroa and Poulo [Pulau] Sujee; paid off 11.7.1835 at Portsmouth. Recommissioned 25.9.1835 under Cmdr W. H. H. Carew?, for South America; sailed 14.11.1835; paid off 18.7.1839 at Portsmouth. BU at Plymouth 3.1840.

Argus Portsmouth Dyd

Ord: 30.1.1829. K: 3.1831. Cancelled 27.4.1831 (frames taken up 6.1831).

Acorn Portsmouth Dyd

Ord: 30.1.1829. K: — . Cancelled 27.4.1831 (unstarted).

(C) Vessels acquired from November 1830

From 1834 a series of larger 18-gun cruisers were produced, but these sloops were generally described as 'corvettes' – a term adopted from French usage.

DAPHNE Class 18 guns (corvettes, open battery). William Symonds design 1834. These large and beamier vessels were classified as corvettes, while remaining within the sloop category. The second (Chatham) pair were suspended 30 November 1836; new designs were approved 20 February 1837, altering them to 20-gun corvettes with covered battery, to which they were ordered to be laid down on 6 April 1837.

Dimensions & tons: 120ft 0in, 99ft 5½in x 37ft 6in oa (37ft 2in for tonnage) x 18ft 0in. 730⁷⁄₉₄bm.

Men: 145 (later 175). Guns: 18 x 32pdrs (40cwt); *Calypso* 20 x 32pdrs (40cwt)

Dido Pembroke Dyd

As built: 120ft 0in, 99ft 0¼in x 37ft 8in oa (37ft 4in for tonnage) x 18ft 0in. 734¹⁹⁄₉₄bm. Draught 10ft 4in/14ft 6in.

Ord: 26.2.1834. K: 9.1834. L: 13.6.1836. C: 26.1.1837 at Sheerness.

First cost: £13,404 (with fitting, £20,024).

Captain (later Admiral Sir) Edward Gennys Fanshawe's pen-and-ink rendering of the *Daphne* off Moalu in the Fiji Islands on 26 September 1849, running before a light wind with her studding sails set. A double-hulled Fijian war canoe, with a high platform built across to connect the two hulls, is crossing her stern. Fanshawe noted her as 'very fine', and fast enough to sail right around the *Daphne* while the latter was under sail. *(NMM PZ4633)*

Commissioned 25.10.1836 under Capt. Lewis Davies, for the Mediterranean; operations on Syrian coast 1840; paid off 8.1841 at Sheerness. Recommissioned 30.8.1841 under Capt. Henry Keppel, for the East Indies; in China War; Yangtse operations 7.1842; operations in Borneo 1844; paid off 2.1845 at Sheerness. Recommissioned 9.5.1846 under Capt. John Balfour Maxwell, for the East Indies; paid off 1.1849 at Sheerness Very small repair (for £3,728) at Sheerness 11.1849–7.1850. Recommissioned 28.8.1851 under Capt. William Henry Anderson Morshead, for the Pacific; fitted for sea at Sheerness (for £3,027) 9–10.1851; off Russian Pacific coast during Russian War; paid off 9.1856 at Sheerness. Coal hulk there 1860–90. Sold there 3.3.1903.

Daphne Pembroke Dyd.
As built: 120ft 0in, 97ft 11½in x 37ft 8in oa (37ft 4in for tonnage) x 18ft 0in. 726²²/₉₄bm. Draught 10ft 1in/14ft 4in.
Ord: 26.2.1834. K: 12.1835. L: 6.8.1838. C: 22.9.1838–2.2.1839 at Portsmouth.
First cost: £13,515 (with fitting, £19,641).
Commissioned 15.11.1838 under Capt. John Windham Dalling, for Lisbon, then the Mediterranean; operations on Syrian coast 1840; paid off 5.1842. Recommissioned 23.7.1842 under Capt. John James Onslow, for the Pacific; grounded 31.12.1846 off St Catherines (Isle of Wight) but refloated; paid off 26.1.1847. Recommissioned 26.10.1848 under Capt. Edward Gennys Fanshawe, for the Pacific; paid off 9.8.1852. Sold to Castle & Beech 7.10.1864 and BU at Charlton completed 8.12.1865.

Calypso Chatham Dyd
As built: 120ft 1in, 99ft 3½in x 37ft 9½in oa (37ft 3½in for tonnage) x 9ft 7in. 734⁴⁵/₉₄bm. Draught 9ft 2in/13ft 8in.
Ord: 27.3.1834. K: 12.1837. L: 8.5.1845. C: 19.2.1846.
First cost: £19,050 (with fitting, £24,669).
Commissioned 12.12.1845 under Capt. Henry John Worth, for the Pacific; paid off 9.1849. Fitted for sea at Chatham 7–9.1851. Recommissioned 23.7.1851 under Capt. Arthur Forbes, for North America and the West Indies. On 23.11.1857 under Cmdr Frederick Byng Montresor, for the Pacific; refitted for sea at

Chatham 1–2.1858; beached at Tabogas Island (Bay of Panama) 2.1860; paid off 1.1862. BU completed by Castle & Beech at Charlton 29.1.1866.

Coquette Chatham Dyd
Ord: 27.3.1834. K: — . Suspended 30.11.1836. Cancelled 1851 (unstarted).

MODESTE Class 18 guns (corvette). Design by Admiral George Elliot 1837. In 1838 she was transferred from the corvette category to that for sloops, 'being stored similar to *Rover* in every respect'.
Dimensions & tons: 120ft 0in, 98ft 7in x 33ft 1in oa (32ft 9in for tonnage) x 14ft 2in. 562³⁹/₉₄bm.
Men: 120 (later 150). Guns: 2 x 32pdr (26cwt/9ft) + 16 x 32pdr (17cwt) carronades.

Modeste Woolwich Dyd
As built: 120ft 0in, 98ft 10⅛in x 33ft 2½in oa (32ft 10½in for tonnage) x 14ft 2in. 568²¹/₉₄bm. Draught 8ft 2in/10ft 0in.
Ord: 6.2.1837. K: 5.1837. L: 31.10.1837. C: 10.3.1838.
First cost: £11,095 (£15,042 including fitting).
Commissioned 21.11.1837 under Cmdr Harry Eyres, for north coast Spain, then to North America and West Indies; anti-slavery operations in Mozambique Channel, then to China for 8.1840 Amoy operations, then 5.1841 capture of Canton. On 16.10.1841 under Cmdr Rundle Burges Watson, 1st Anglo-Chinese War 1841–42, home in 1842. On 5.6.1843 under Cmdr Thomas Baillie, in the Pacific, then 27.12.1845 under Cmdr Thomas Vernon Watkins, then 11.8.1847 under Lieut Reginald John James George Macdonald (acting); paid off 15.1.1848 at Sheerness. Small repair there (for £3,184) 2–8.1848. Fitted for sea at Sheerness (for £3,126) 2.1850–10.1851. Recommissioned 28.8.1851 under Cmdr William Compton, for the Mediterranean; on 9.7.1854 under Cmdr Augustus Butler, for the Mediterranean; paid off 9.7.1856 at Sheerness. At Woolwich 1860–65. Ordered 10.2.1865 to be fitted as a coal depot (in lieu of *Terpsichore*), but was sold to Castle & Beech in 3.1866 to BU, and was BU at Charlton 18.4–6.6.1866.

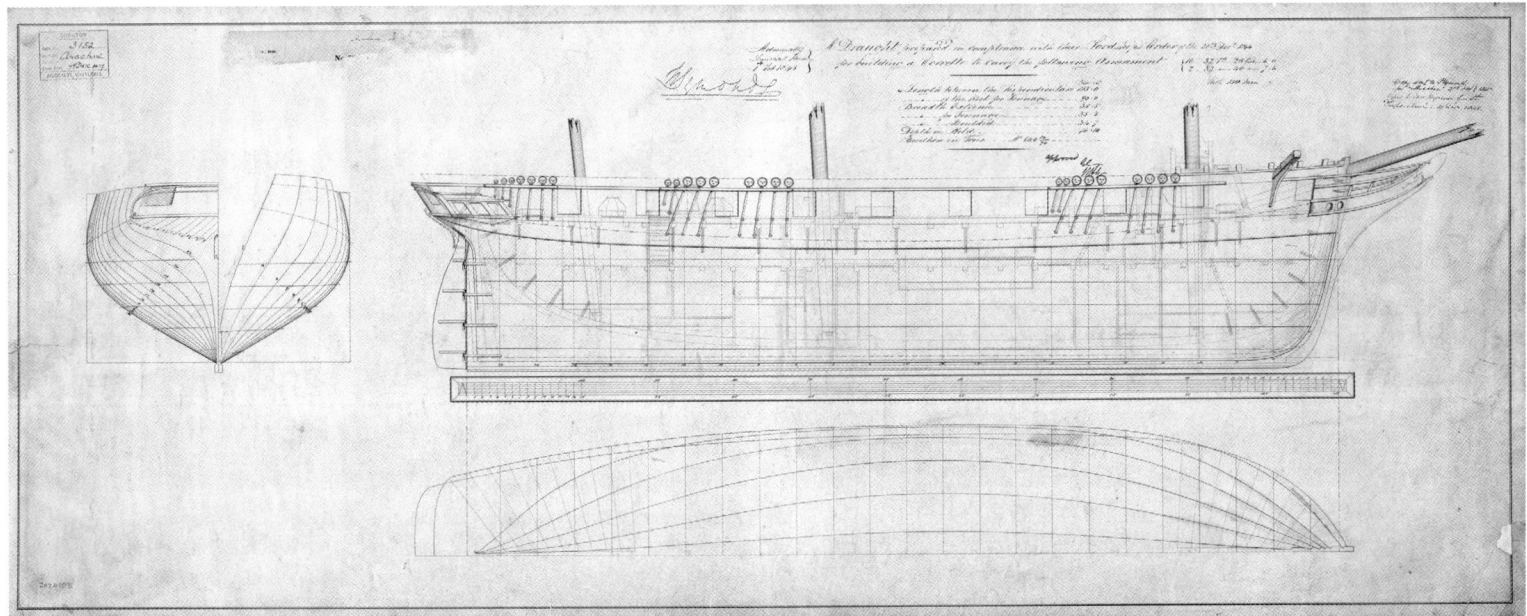

Symonds's draught of 1 February 1845 for the *Arachne*; the intended armament was sixteen 32pdr guns of 25cwt and 6ft length, and two 32pdr guns of 40cwt and 7ft 6in length. With her sister *Terpsichore*, the last sailing ship-rigged sloops to be launched. *Arachne* was not brought into service until 1855, while *Terpsichore* was never used. After a single commission on the North American station, the *Arachne* returned home to Devonport on Saturday, 17 March 1860 and was paid off 11 days later, remaining in Ordinary until sold for breaking up by contract in 1866. An intended third ship to this design, *Narcissus*, was eventually built as a 51-gun screw frigate. *(NMM J6757)*

CHALLENGER Class 18 guns (corvette). Designed 1844 by the Earl of Dundonald (Thomas Cochrane, the great frigate captain), but never built. An earlier authorisation (on 19 February 1844) was for the corvette *Camilla* to be built at Pembroke to Dundonald's plans (then stipulated 'not over 600 tons'), but on 8 July 1844 this was altered to have *Camilla* built as a brig, to be like the *Atalanta* ordered earlier that year.

> Dimensions & tons: 134ft 0in, 110ft 4¾in x 37ft 6in oa (37ft 2in for tonnage) x 17ft 6in. 810⁸⁴/₉₄bm.
>
> Men: 145 (later 175 intended). Guns: probably to have carried 2 x 32pdr (long, 7½ft) + 16 x 32pdr (short, 6ft).

Challenger Chatham Dyd
> Ord: 8.8.1844. K: — . Suspended 30.8.1845. Cancelled 22.5.1849 (unstarted).

ARACHNE Class 18 guns (corvettes). Sir William Symonds design, approved 1 February 1845 (*Arachne* named then, but not allocated to Devonport until 6 days later). *Narcissus*, originally ordered at Deptford Dyd, was named on 26 March 1846.

> Dimensions & tons: 115ft 0in, 90ft 11in x 35ft 5in oa (35ft 3in for tonnage) x 16ft 10in. 600⁸⁴/₉₄bm.
>
> Men: 150. Guns: 2 x 32pdr (long, 40cwt/7½ft) + 16 x 32pdr (short, 25cwt/6ft).

Arachne Devonport Dyd
> As built: 115ft 0in, 90ft 10in x 35ft 5½in oa (35ft 3½in for tonnage) x 16ft 10in. 601⁷²/₉₄bm. Draught 10ft 11in/12ft 3in.

Ord: 7.2.1845. K: 10.1845. L: 30.3.1847. C: 8.4.1847 (for Ordinary).
First cost: £15,419 (+ £1,034 fitting).
Channels were altered 7–8.1850; fitted for sea (for £7,663) at Plymouth 10–12.1855.
Commissioned 29.10.1855 under Cmdr Valentine Otway Inglefield, for North America and the West Indies; in 3.1859 under Cmdr John Eglinton Montgomerie, for North America and the West Indies; paid off 28.3.1860 at Devonport. Sold to J. & B. Marshall, Plymouth and BU completed 12.2.1866.

Terpsichore Money Wigram & Sons, Blackwall.
> As built: 115ft 1in, 90ft 11¼in x 35ft 5¼in oa (35ft 3¼in for tonnage) x 16ft 10in. 601⁷⁹/₉₄bm. Draught 9ft 11in/12ft 1½in.
>
> Ord: 15.12.1845. K: 1.1846. L: 18.3.1847. C: at Chatham.
> First cost: £13,484 (£11,740 to builder, plus £1,744 stores supplied by Woolwich).
>
> Never commissioned, but remained in Ordinary at Chatham; listed 1863 for disposal, but considered for use as a coal hulk there until 2.1865; used for 'torpedo' (i.e. mining) experiments, sunk 4.10.1865 and raised. BU by Castle & Beech at Charlton 6.12.1865–9.1.1866.

Narcissus Devonport Dyd.
> Ord: 23.1.1846. K: — . Order transferred to Devonport Dyd. 11.3.1846. Re-ordered as 50-gun steam/screw frigate (which see) 25.4.1847.

As at the start of 1850, the category comprised three corvettes of 18 guns (*Dido*, *Daphne* and *Calypso*), six sloops of 18 guns (*Tweed*, *Nimrod*, *Pearl*, *Modeste*, *Arachne* and *Terpsichore*) and twelve sloops reduced to 14 guns (*Rose*, *Champion*, *Orestes*, *Comus* (ex-*Comet*), *Favorite*, *Hyacinth*, *Larne* (ex-*Lightning*), *Racehorse*, *Fly*, *Scout*, *Hazard* and *Electra*). Of these, the *Nimrod* and *Tweed* were originally *Atholl* Class Sixth Rates (see Chapter 6), cut down in 1828 and 1831 respectively and completed as sloops with 130 men, 18 x 32pdr carronades (of 25cwt) and 2 x 9pdrs.

8 Brig Sloops (and other brigs)

The mass construction of two-masted sloops during the Napoleonic Wars, essentially for escort duties, had left the Royal Navy with large numbers of brig sloops, some still completing at the time the war ended. The most numerous were the seventy surviving 18-gun vessels of the *Cruizer* class, but there were also a number of the 10-gun *Cherokee*, 14-gun *Crocus* and 16-gun *Seagull* types, and a few of the even smaller gun-brigs. Their standard armament had been a pair of long guns as bow-chasers, with a battery of carronades on the broadsides. Almost all of the pre-war types that had survived the war had been deleted by 1817, as had most of the brigs taken as prizes which had been added to the Navy during the wars.

(A) Vessels in service at 1 January 1817

The renewal of the *guerre de course* between 1803 and 1814 saw increasing numbers of British merchantmen falling prey to enemy privateers and warships – 222 were captured in 1803, 387 in 1804, 507 in 1805, 519 in 1806, 559 in 1807, 469 in 1808, 571 in 1809 and 619 in 1810 – before falling to 470 in 1811, ???? in 1812, 371 in 1813 and 145 in 1814. To counter the predations of several hundred privateers, a world-wide pattern of regular convoys and (anti-corsair) cruising developed, plus the blockading of even small Continental ports. This needed many more escort and patrol vessels, and 174 extra brig sloops (plus 87 smaller gun-brigs) were ordered from a large range of shipyards between 1803 and 1813, while over fifty captured enemy brigs were likewise commissioned into the British Navy.

(i) 32pdr carronade type

In early 1795 the Admiralty identified the need for additional brig sloops to meet the urgent need for convoy duties, and – as per their usual practice – commissioned two different designs, one from each Surveyor. Five vessels to each design were ordered in March 1795, with a further three to each design following in July; eleven of these sixteen new vessels were constructed with 'fir', i.e. pine (as identified below). They were all originally to have carried 16 x 6pdr carriage guns, but by AO of 22 April 1795 were established with 16 x 32pdr carronades, with 2 x 6pdrs retained as chase weapons, so that they were classed as 18-gun sloops. However, in service this armament proved rather too heavy, and the 32pdr carronades were in most vessels replaced by 24pdr carronades.

DILIGENCE **Class** Built to a John Henslow design, approved 22 April 1795. The first five were all ordered 4 March 1795 (orders confirmed two weeks later), and were registered and named by AO on 20 June 1795. Three more, all ordered and begun in July 1795, were registered and named on 28 August. A single example, the *Harpy*, remained in the Navy by 1817. Identical numbers were ordered on the same dates to a competitive design by Henslow's co-Surveyor, William Rule; but all eight to this design – the *Albatross Class* – were gone before 1817.

> Dimensions & tons: 95ft 0in, 75ft 2½in x 28ft 0in x 12ft 0in. 313⁵⁹⁄₉₄bm.
>
> Men: 121. Guns: 16 x 32pdr carronades, plus 2 x 6pdr (bow).

Brig sloops were the Navy's 'maids-of-all-work' and huge numbers were built during the Napoleonic Wars, performing a large variety of tasks for the British Navy around the world, from convoy escort services and routine patrols to blockading minor ports. This watercolour of a typical RN brig sloop at sea off an anchorage in the early years of the nineteenth century was painted by Daniel Tandy, who spent the whole of the French Revolutionary and Napoleonic Wars as an RN lieutenant so presumably drew it from life. (*NMM PW0459*)

Harpy Thomas King, Dover
> As built: 95ft 0in, 75ft 1⅛in x 28ft 1½in x 12ft 0½in. 316¹³⁄₉₄bm.
>> Draught 7ft 3in/10ft 10in.
>
> Ord: 4 & 18.3.1795. K: 5.1795. L: 2.1796. C: 22.2–7.5.1796 at
>> Deptford.
>
> First cost: £3,683 to build, plus £3,167 fitting.
>
> Commissioned 4.1796 under Cmdr Henry Bazely (–1800), for the
>> Downs; took privateers *Le Cotentin* in the Channel 2.1797 and
>> *L'Espérance* 22.6.1797; in Popham's squadron at Ostend 5.1798;
>> in action (with *Fairy*) against 46-gun *Le Pallas* 5.2.1800 – which
>> then taken by *Loire* and *Danae*. In 4.1800 under Cmdr William
>> Birchall; retook hired cutter *Constitution* 9.1.1801; at Battle of
>> Copenhagen 2.4.1801. In 4.1801 under Cmdr Charles Boys,
>> then paid off. Recommissioned 8.1802 under Cmdr Edmund
>> Heywood (–1805), for the North Sea; took 2-gun *Penriche* off
>> Calais 12.3.1804; attack on invasion craft at Boulogne 19.7.1804;
>> attack on gunboats off Vimereux 26.8.1804; attack on invasion
>> craft near Griz Nez 24.4.1805. On 24.12.1805 under Cmdr
>> George Moubray; paid off into Ordinary at Portsmouth 1807.
>> Small Repair at Portsmouth 4–7.1809; recommissioned 5.1809
>> under Cmdr George William Blamey; in Walcheren operations
>> 1809; sailed with convoy for Halifax 13.5.1810. In 10.1810
>> under Cmdr Edward Henry A'Court; sailed for Cape of Good
>> Hope 2.1.1811. On 29.3.1811 under Cmdr Henderson Bain,
>> for Java operations in 8.1811. On 26.1.1812 under Lieut Samuel
>> Bradstreet Hore (acting), at Cape of Good Hope (–3.1813). On
>> 6.4.1813 under Cmdr Thomas Griffiths Allen (died 26.9.1814),
>> then Cmdr George Tyler on 7.2.1815; paid off 3.1816. Sold to Mr
>> Kilsby (for £710) 10.9.1817.

In December 1796, new orders were placed for four flush-decked sloops, and again these were to two differing designs by the two Surveyors. However, in order to compare the qualities of ship-rigged and brig-rigged vessels, one vessel in each design was to be completed as a ship sloop and the other as a brig. While the Henslow-designed vessels (the brig *Busy* and ship sloop *Echo*) would see no further sisters built, and the Rule-designed vessels (the brig *Cruizer* and ship sloop *Snake*) would each have a single sister ordered in the following March, Rule's *Cruizer* design would subsequently see over a hundred copies (106, to be precise) ordered between 1802 and 1815, making this easily the second largest group of sailing warships built to one design – exceeded numerically only by the smaller *Cherokee* Class brigs, most of which were actually ordered and built after 1815.

CRUIZER Class Design by Sir William Rule.
> Dimensions & tons: 100ft 0in, 77ft 3½in x 30ft 6in (30ft 0in
>> moulded) x 12ft 9in. 382⁴¹⁄₉₄bm.
>
> Men: 121. Guns: 16 x 32pdr carronades, plus 2 x 6pdrs (bow).

For convenience, these vessels are grouped below according to the Admiralty Board under which the order was placed. The contracts for these generally described them as brigantines, an indication that the terms 'brig' and brigantine' were frequently interchangeable.

Ordered by Earl Spencer's Board (2 vessels ordered)
The prototype was ordered on 19 December 1796, and was registered as a sloop under AO of 23 October 1797, and again on 16 November. A second vessel (never named) was ordered on 15 March 1797 to this design, but the order was subsequently cancelled, and no further orders for brig sloops were placed until 1802.

Cruizer Stephen Teague, Ipswich

> As built: 100ft 0in, 77ft 3in x 30ft 7in x 12ft 9in. 384³¹⁄₉₄bm. Draught
>> 7ft 6in/11ft 2in.
>
> Ord: 19.12.1796. K: 2.1797. L: 20.12.1797. Fitted at Chatham
>> 1.1.1798 to 16.3.1798.
>
> First cost: £4,588 for building plus £3,362 for fitting.
>
> Commissioned 2.1798 under Cmdr Charles Wallaston for Adm.
>> Duncan's fleet (in the North Sea); during this commission (to
>> 1800) she took a number of privateers in the North Sea – the
>> 8-gun *Le Jupiter* on 27.4.1798, the 4-gun *Le Chasseur* on
>> 19.5.1798, the 14-gun *Les Deux Frères* on 21.5.1799, the 14-
>> gun *Le Courageux* on 13.7.1799, the 14-gun *Le Persévérant*
>> on 23.3.1800, and the 14-gun *Le Flibustier* on 25.3.1800.
>> Recommissioned 1.1801 under Cmdr James Brisbane, she
>> participated in the Battle of Copenhagen (2.4.1801). From
>> 4.1801 under Cmdr John Hancock (North Sea station).
>> Recommissioned 2.1803 (still under Hancock on the North
>> Sea station), her boats, with others, cut out the schooner
>> *L'Inabordable* and brig *Le Commode* near Griz Nez on
>> 14.7.1803, and burnt the cutter *La Colombe* at Sluys on 3.3.1804.
>> The *Cruizer* captured the 17-gun privateer *Le Contre-Amiral
>> Magon* in the North Sea on 18.11.1804, and the 14-gun privateer
>> *Le Vengeur* in the Channel on 13.11.1805. Under Cmdr Pringle
>> Stoddart from 1806, she took two 16-gun privateers in the North
>> Sea – *L'Iéna* on 6.1.1807 and *Le Brave* on 26.1.1807 – and took
>> part in the bombardment of Copenhagen on 23.8.1807. Under
>> Cmdr George Charles McKenzie, her boats captured Danish
>> gunboats in the Great Belt on 16.6.1808. With Lieut Thomas
>> Wells as Acting Commander, she was in action with 20 small
>> craft off Göteborg on 1.10.1808. In 12.1808 under Cmdr Thomas
>> Richard Toker; her boats captured the 6-gun Danish privateer
>> schooner *Christianborg* off Bornholm on 31.5.1809; laid up
>> in Ordinary at Sheerness 11.1813. Sold to Job Cockshot (at
>> Sheerness) for £960 on 3.2.1819 to BU.

Ordered by St Vincent's Board (19 vessels ordered)
Of this batch, the *Leveret* was wrecked in 1807, *Ferret* in 1811 and *Belette* in 1812, and *Avon* foundered in 1814 (after recapture from the USN, while *Despatch* was disposed of in 1811, *Minorca* in 1814, *Amaranthe*, *Swallow*, *Weazle* and *Moselle* in 1815 and *Wolverene* in 1816, leaving the eight remaining vessels below.

Scorpion John King, Dover
> As built: 99ft 11½in, 77ft 2in x 30ft 7in x 12ft 9in. 383⁸⁶⁄₉₄bm.
>> Draught 6ft 0in/11ft 0in.
>
> Ord: 27.11.1802. K: 1.1803. L: 17.10.1803. C: 22.10.1803–8.1.1804 at
>> Sheerness.
>
> Commissioned 11.1803 under Cmdr George Hardinge, for the
>> Channel and Downs; boats (with those of *Beaver*) took 16-gun
>> *Atalante* in the Vlie 31.3.1804. In 4.1804 under Cmdr Philip
>> Carteret (–1807); took 16-gun *Le Bougainville* 16.2.1805; took
>> 12-gun *L'Honneur* off Dutch coast 12.4.1805; took 16-gun
>> privateer *La Glaneuse* 21.11.1805; took 10-gun privateer *Le
>> Glaneur* 3.12.1805; sailed for the Leeward Islands 28.1.1806. In
>> 1807 under Cmdr Francis Stanfell (–1809); to Leeward Islands
>> 1807, then home in 1808; sailed for the Leeward Islands 3.4.1809;
>> took 14-gun *L'Oriste* in Basseterre 12.1.1810. In 1810 under
>> Cmdr John Gore (?drowned 1812), then Cmdr Robert Giles
>> in 3.1812. Laid up at Sheerness 7.1813. Sold to G. F. Young at
>> ?Sheerness (for £1,100) 3.2.1819.

Scout Peter Atkinson & Co., Hull
> As built: 100ft 0in, 77ft 2½in x 30ft 5¾in x 12ft 9in. 381⁴⁸⁄₉₄bm.
>> Draught 6ft 6in/11ft 3in.

Ord: 27.11.1802. K: 5.1803. L: 7.8.1804. C: 19.8–24.10.1804 at
Sheerness.

Commissioned 8.1804 under Cmdr Donald Mackay, for the
Downs; joined Collingwood's fleet 23.10.1805. In 1.1806 under
Cmdr William Raitt (–1809); took Spanish privateers – 20-gun
Amiral del Tariffa 27.3.1807, 3-gun *El Determinando* 20.5.1807,
and (with *Redwing*) 3-gun *De Bon Vassallio* 13.6.1807; under
Lieut Thomas Stamp (temp), when landing party stormed
batteries 14.6.1809 and 14.7.1809. In 10.1809 under Cmdr
Alexander Sharpe (–1811); in Hallowell's squadron at Rosas
31.10–1.11.1809; destruction of *La Girafe* and *La Nourrice* in
Gulf of Sagone 1.5.1811. In 1812 under Cmdr George Hooper;
sailed for the Mediterranean 15.5.1812. In 1813 under Cmdr
Benjamin Crispin, then 12.1813 under Cmdr James Arthur
Murray (–1815), off Cadiz. Middling Repair and fitted for sea
at Deptford 2–7.1818; recommissioned 29.4.1818 under Cmdr
William Ramsden (–1820), for the Mediterranean; in 10.1821
under Cmdr John Theed, to the West Indies; in 6.1822 under
Cmdr James Wigston (–1825), in the West Indies; damaged by
stranding in Gulf of Mexico 5.1823; paid off 10.1825. Sold to
John Small Sedger to BU (for £1,010) 11.7.1827.

Musquito John Preston, Great Yarmouth

As built: 100ft 0in, 77ft 3½in x 30ft 7in x 12ft 9in. 384⁵¹⁄₉₄bm.
Draught 6ft 6in/10ft 3in.

Ord: 27.11.1802. K: 5.1803. L: 4.9.1804. C: 25.9–18.11.2804 at
Chatham.

Commissioned 10.1804 under Cmdr Samuel Jackson; took 1-gun
privateers *L'Orestes* and *Le Pylades* in the Channel 14.4.1805;
off Calais 1807. Recommissioned 10.1807 under Cmdr William
Goate, for the North Sea; took 6-gun *Sol Fuglen* in the North
Sea 25.5.1809; operations in the River Elbe 7.1809, then paid off.
Recommissioned 19.6.1811 under Cmdr Christopher Bell, then
7.2.1812 under Cmdr James Tomkinson; sailed for the Leeward
Islands 24.4.1813; in Home waters 1814; on 12.8.1815 under
Cmdr George Brine, for St Helena station; paid off at Deptford
28.11.1818. Sold to Thomas King (for £1,050) 7.5.1822.

Calypso John Dudman, Deptford

As built: 100ft 0in, 77ft 3½in x 30ft 6in x 12ft 9in. 382bm. Draught
6ft 6in/11ft 0in.

Ord: 15.10.1803. K: 3.1804. L: 2.2.1805. C: 11.2–16.5.1805 at
Deptford Dyd.

Commissioned ?8.1805 under Cmdr Matthew Forster, for the
Downs; in action against Ver Huell's convoy off Griz Nez
18.7.1805. On 22.1.1806 under Cmdr Matthew Bradby (–1810),
for the Downs and North Sea. Recommissioned 2.1810, still
under Bradby. In 6.1810 under Cmdr Henry Weir, for the Baltic
(–1812); in action (with *Dictator*, *Podargus* and *Flamer*) against
Danish squadron at Mardö (Norway) 6–7.7.1812, losing 3 dead
and 1 wounded, but Danish vessels destroyed. On 29.6.1812
under Cmdr Thomas Groube, then 6.1814 Cmdr Charles Hope
Reid, then 1816 Lieut John Sisson (acting); paid off 10.7.1816. In
Ordinary at Chatham 1817–20. BU there 3.1821.

Espoir John King, Dover

As built: 100ft 1½in, 77ft 6in x 30ft 7in x 12ft 9in. 385bm. Draught
5ft 6in/10ft 3in.

Ord: 7.11.1803. K: 2.1804. L: 22.9.1804. C: 1.10–16.12.1804 at
Sheerness.

Commissioned 10.1804 under Cmdr Joseph Edmonds. In 1.1806
under Lieut William King (acting, then later same month Cmdr
Henry Hope (–1806); at occupation of Cape of Good Hope;
sailed for the Mediterranean 29.6.1807. In 4.1808 under Lieut

William Higgs (acting); at Gibraltar, then coast of Portugal. In
1809 under Cmdr Robert Mitford (–1813), in the Mediterranean.
In 4.1813 under Lieut Higgs again (acting), then 7.1813 Cmdr
Robert Spencer. On 19.1.1814 under Cmdr Robert Russell, then
15.6.1815 under Cmdr Norwich Duff, in North America and
the West Indies; paid off 10.1816, and laid up. BU at Portsmouth
4.1821.

Surinam Obadiah Ayles, Topsham (Exeter)

As built: 100ft 1in, 77ft 4½in x 30ft 6½in x 12ft 9in. 384bm.

Ord: 7.11.1803. K: 2.1804. L: 1.1805. C: 3.3–2.5.1805 at Plymouth.

Commissioned 3.1805 under Cmdr Alexander Shippard, for the
Channel. In 2.1806 under Cmdr John Lake (–1809); took
14-gun privateer *L'Amiral Dacres* 17.11.1807; sailed for the
Leeward Islands 15.12.1808; at capture of Martinique 2.1809.
In 1810 under Cmdr Andrew Hodge, then 1811 Cmdr
Alexander Kennedy, then 1812 Cmdr John Watt (died 9.1813),
in the Leeward Islands; paid off 5.10.1813. Middling Repair
at Chatham 5.1814–10.1815, then to Ordinary at Sheerness.
Recommissioned 11.1820 under Cmdr William Godfrey, for
Jamaica. In 7.1822 under Cmdr Alfred Matthews, still on Jamaica
station, then 4.1823 Cmdr Charles Crole; paid off 2.1825. Sold at
Chatham to John Small Sedger to BU (for £1,450) 20.7.1825.

Racehorse Hamilton & Breeds, Hastings

As built: 100ft 1½in, 77ft 4¼in x 30ft 7¼in x 12ft 9½in. 385bm.

Ord: 7.11.1803. K: 6.1804. L: 17.2.1806 (345 days late!). C: ?

Commissioned 3.1806 under Cmdr Robert Forbes; sailed for the
Mediterranean 25.5.1806. In 6.1807 under Cmdr William Fisher
(–1810), for cruising; took 4-gun privateer *L'Amiral Ganteaume*
in Home aters 2.3.1808, then in Channel Islands; sailed for Cape
of Good Hope 29.9.1808. In 12.1810 under Cmdr James de
Rippe (–1813); in Schomberg's action off Madagascar 20.5.1811;
at capture of *La Néréide* at Tamatave 23.5.1811; sailed for Cape
again 7.8.1811; paid off 9.11.1813 into Ordinary at Portsmouth.
Middling Repair and fitted for sea there 2.1816–7.1818;
recommissioned 5.1818 under Cmdr George Pryse Campbell,
for the Mediterranean; on 27.1.1821 under Cmdr Charles Abbot,
then 19.2.1822 under Cmdr William Suckling; wrecked on
Langness, Isle of Man 14.12.1822 (9 drowned).

Rover Joseph Todd, Berwick

As built: 100ft 4in, 77ft 4⅜in x 30ft 7in x 12ft 9in. 384⁸¹⁄₉₄bm.
Draught 6ft 6in/11ft 5in.

Ord: 9.12.1803. K: 6.1804. L: 13.2.1808. C: 8.1808 at Leith.

Commissioned 5.1808 at Leith under Cmdr Francis Nott; at Barö
Sound, Gulf of Finland 7.7.1809. In 11.1810 under Cmdr Justice
Finley; took 14-gun privateer *Le Comte Reginaud* (ex-sloop
HMS *Vincejo*) 30.11.1811. In 1812 under Lieut Francis Loch
(acting); took US 6-gun privateer *Experiment* 21.10.1812. In
6.1814 under Cmdr James Pickard, later under Cmdr Henry
Montresor and 9.1814 Cmdr Henry Bruce, for convoys and
cruising; laid up 9.1815 at Plymouth and paid off 10.1815. Sold at
Plymouth to Adam Gordon (for £980) 26.3.1828.

Ordered by Melville's First Board (6)

All these vessels, which were – uniquely for this design – fir-built,
had gone by 1815, of which *Raven* was wrecked in 1805, *Harrier*
foundered in 1809, *Saracen* and *Elk* were BU in 1812, *Wasp* was
captured and burnt in 1814, and *Beagle* was sold in 1814.

Ordered by Barham's Board (22)

Of this batch, *Carnation* was lost in 1808 (taken by the French),
Primrose and *Foxhound* had been lost in 1809 (wrecked), *Pandora*

(wrecked), *Grasshopper* and *Alacrity* (taken by the French) in 1811, *Emulous* in 1812 (wrecked), and *Peacock* (taken by the USN) in 1813, while *Frolic* has been BU in 1813 and *Procris* sold in 1815. Four vessels (*Raleigh*, *Redwing*, *Ringdove* and *Clio*) were altered to ship rig during the 1820s.

Forester John King, Dover

As built: 100ft 0in, 77ft 2⁷⁄₈in x 30ft 7in x 12ft 9in. 384²⁶⁄₉₄bm. Draught 6ft 0in/10ft 10in.

Ord: 16.7.1805. K: 3.1806. L: 3.8.1806. C: 11.8–20.10.1806 at Sheerness.

Commissioned 9.1806 under Cmdr John Richards, for the North Sea; sailed for the Leeward Islands 29.8.1808; at capture of Martinique 2.1809. Later in 1809 under Cmdr John Watt (–1811), in the Leeward Islands. In 1812 under Cmdr Alexander Kennedy (dismissed 4.1814); sailed for Jamaica 23.3.1813; took (with *Sapphire*) US 2-gun privateer *Mary Anne* off San Domingo 15.5.1813. In 4.1814 under Cmdr William Hendry, at Jamaica; paid off 7.8.1816 at Portsmouth. Sold there to ?G. Young (for £1,130) 8.3.1819.

Mutine Henry Tucker (or Richard Chapman?), Bideford

As built: 100ft 0in, 77ft 3¼in x 30ft 8in x 12ft 10½in. 386bm. Draught 7ft 1in/11ft 6in.

Ord: 22.7.1805. K: 9.1805. L: 15.8.1806. C: 16.9.1806–11.2.1807 at Plymouth.

Commissioned 8.1806 under Cmdr Hew Stewart (–1807); in Copenhagen expedition 8.1807, then in North Sea. Recommissioned 10.1808 under Cmdr Charles Fabian (–1810); sailed for Brazil 8.11.1808. In 10.1810 under Cmdr Frederick Burgoyne, on South American station; in 11.1810 under Cmdr Nevinson de Courcy (–1814); home in 7.1812; took 16-gun privateer *L'Invincible* in the Bay of Biscay 17.4.1813. From 7.6.1814 under Cmdr James Athill, in the Leeward Islands, then 10.1814 under Cmdr James Mould; at Algiers 27.8.1816. In 10.1816 under Cmdr William Sargent, at Cork; paid off 10.1818. Sold at Plymouth to G. Young (for £1,310) 3.2.1819.

Columbine Balthazar & Edward Adams, Bucklers Hard

As built: 100ft 77ft 1¾in x 30ft 8in x 12ft 10in. 385⁸⁄₉₄bm. Draught 6ft 8in/11ft 0in.

Ord: 12.11.1805. K: 1.1806. L: 16.7.1806. C: 19.7–11.9.1806 at Portsmouth.

Commissioned 8.1806 under Cmdr James Bradshaw (–1808); sailed for Halifax 6.4.1807. In 4.1808 under Cmdr George Hills, in North America; paid off 3.1810. In 7.1810 under Cmdr James Collins; sailed for the Mediterranean 18.9.1810. In ?12.1810 under Cmdr William Shepheard, then Lieut George Westphal (acting) in 6.1811 and Cmdr Richard Muddle in 11.1811 (–1815); on the Portuguese coast 1812–13 and West Indies 1814; paid off 23.11.1815. Between Small and Middling Repair at Plymouth 11.1818–1.1820. Fitted for sea at Plymouth 4–9.1823; recommissioned 4.1823 under Cmdr Charles Abbott; wrecked on Sapienza Island (Greece) 25.1.1824.

Raleigh Francis Hurry, Newcastle

As built: 100ft 0in, 77ft 2½in x 30ft 6½in x 12ft 9in. 383⁷⁄₉₄bm. Draught 6ft 4in/11ft 3in.

Ord: 16.1.1806. K: 3.1806. L: 24.12.1806. C: 15.1–20.7.1807 at Chatham.

First cost: fitting £4,059.

Commissioned 2.1807 under Cmdr Joseph Masefield; at blockade of Rochefort. In 12.1808 under Cmdr Henry Butt; in North Sea 1809. In 2.1809 under Cmdr George Sayer; took 14-gun privateer *La Modeste* 24.2.1810; in the Baltic 1810 (–1812);

took 10-gun privateer schooner *Admiral Nils Juel* 2.11.1810. In 11.1810 under Cmdr John Sheridan (temp.), then 1.1811 Cmdr George Hooper (–1813). Repaired by Pitcher, Northfleet (for £1,415, although this may just be Dyd costs) 12.1812–2.1813; sailed for North America 18.3.1813; to Ordinary at Plymouth 1814. Between Middling and Large Repair at Plymouth (for £13,708) 12.1815–7.1817. Recommissioned 8.1818 under Cmdr William Baumgardt, for Jamaica; from 6.1820 under Cmdr George Blackman; paid off 1.1822. Small Repair and re-rigged as Ship sloop at Chatham (for £8,430) 3–7.1826; recommissioned 4.1826 under Cmdr John Dalling, for the Mediterranean; from 1.1828 under Cmdr John Dundas then 8.7.1828 under Cmdr George Haye and 4.3.1829 under Cmdr Sir William Dickson. Small Repair and fitted for sea at Woolwich (for £9,803) 3–10.1830; recommissioned 13.7.1830 under Cmdr Abraham Mills Hawkins, for the Mediterranean; paid off 31.5.1834. Recommissioned 29.7.1834 under Cmdr Michael Quinn, for the East Indies; paid off 27.10.1838. Fitted as Target ship at Sheerness 8.1839. Sold to John Small Sedger (for £810) at Sheerness 27.5.1841.

Cephalus Custance & Stone, Gt. Yarmouth

As built: 100ft 0in, 77ft 3½in x 30ft 8in x 13ft 0in. 386⁶⁰⁄₉₄bm. Draught 6ft 0in/11ft 0in.

Ord: 22.1.1806. K: 4.1806. L: 10.1.1807. C: 27.2–23.7.1807 at Chatham.

Commissioned 5.1807 under Cmdr Charles Hawtayne; sailed for the Mediterranean 10.9.1807. In 1.1808 under Cmdr Edward Harvey (–1811); took 4-gun privateer *Le Scipion* in the Mediterranean 11.1.1810. In 2.1811 under Cmdr Augustus Clifford; her boats (with *Unite*'s) at mouth of Tiber 5.7.1811; took (with *Thames*) a convoy at Porto del Infreschi 21.7.1811 (11 gunboats captured). In 7.1812 under Cmdr Edward Flin (–1814); took privateers in the Mediterranean – 8-gun *La Diligente* 22.2.1813, 10-gun *Le Jeune Thétis* off Minorca 10.6.1813, 3-gun *Le Petit Chasseur* off Sardinia 2.7.1813 and 1-gun *L'Ecureuil* off Cagliari 25.7.1813. In 6.1814 under Cmdr Henry Elton, then 7.1814 Cmdr John Furneaux; arrived Plymouth 11.9.1815 to pay off into Ordinary. BU at Plymouth completed 13.3.1830.

Redwing Matthew Warren, Brightlingsea

As built: 100ft 0in, 77ft 3½in x 30ft 6½in x 12ft 10in. 383⁴⁶⁄₉₄bm. Draught 5ft 10in/11ft 7in.

Ord: 24.1.1806. K: 3.1806. L: 30.8.1806. C: 14.9–23.11.1806 at Chatham.

Commissioned 10.1806 under Cmdr Thomas Ussher (–1808); sailed for the Mediterranean 31.1.1807; took (with *Scout*) Spanish 10-gun privateer *De Bon Vassallio* 13.6.1807; attacked Spanish convoy off Cape Trafalgar 7.5.1808 (destroyed 4 escorts and took another, took or sunk most of convoy); sailed for the Mediterranean 23.9.1808. In 8.1808 under Cmdr Edward Down (–1812); her boats (with *Amphion*'s) cut out armed brig and coaster at Melada (Dalmatia) 8.2.1809; in 1809 temp. under Lieuts Thomas Stamp and later John Nicholas; took 4-gun privateer *Le Victorieux* off Sicily 16.9.1811. In 8.1812 under Cmdr Sir John Sinclair, in the Mediterranean; her boats (with *Undaunted*'s and *Volontaire*'s) at Morgion 30.3.1813 and again (adding *Repulse*'s boats) 2.5.1813; storming of batteries at Cassis 18.8.1813. In 8.1814 under Cmdr Thomas Young; paid off 4.12.1815. Middling Repair and fitted for sea at Woolwich 8.1817–11.1818; recommissioned 8.1818 under Cmdr Frederick Hunn, for St Helena, then 8.11.1820 under Cmdr George Rolle Walpole Trefusis (–1822). Re-rigged as Ship sloop at Chatham

2–5.1824; in 2.1824 under Cmdr Adolphus Fitzclarence, at
the Nore. Paid off 1.1825 and recommissioned same month
under Cmdr Douglas Clavering; took 2 slavers off West Africa
10.1825–*Isabella* (273 slaves) and *Teresa* (199 slaves); lost,
presumed foundered with all hands on West Africa station
6.1827.

Ringdove Matthew Warren, Brightlingsea
   As built: 100ft 1½in, 77ft 4⅜in x 30ft 7in x 12ft 9in. 384⁸¹/₉₄bm.
      Draught 5ft 9in/11ft 1in.
   Ord: 27.1.1806. K: 4.1806. L: 16.10.1806. C: 25.10.1806–29.1.1807 at
      Chatham.
   Commissioned 12.1806 under Cmdr George Andrews (–1810), for
      the Baltic and North Sea; in 1808 under Lieut ?G. Peak (temp.);
      took 10-gun Danish privateer *Tordenskjold* on the Home station
      30.3.1808; sailed for the Leeward Islands 4.12.1808; at capture of
      Martinique 2.1809; in 6.1809 under Cmdr Humphrey Senhouse
      (temp.). In 1810 under Cmdr William Dowers (–1814); at
      capture of *La Loire* and *La Seine* 18.12.1809; in Leeward Islands
      1810–12; sailed for North America 2.6.1813; in Jamaica 1814.
      In 11.1814 under Cmdr James Creighton; paid off 31.8.1815.
      Large Repair at Portsmouth 1–7.1818. Forecastle and head
      housed over 8.1821. Fitted for sea at Portsmouth 12.1822–2.1823;
      recommissioned 11.1822 under Cmdr George Rich, for the West
      Indies; succeeded 7.1823 by his brother, Cmdr Edwin Rich. In
      8.1826 under Cmdr Edward Thornbrough, at Halifax, Nova
      Scotia, then 4.1827 Cmdr Charles English; paid off 28.1.1829.
      Sold to S(amuel?) Cunard & Co. (for £505) at Halifax 11.6.1829.

Sappho Jabez Bailey, Ipswich
   As built: 100ft 3in, 77ft 6½in x 30ft 6in x 12ft 9in. 383⁶⁴/₉₄bm.
   Ord: 27.1.1806. K: 4.1806. L: 15.12.1806. C: 1.1–31.3.1807 at
      Chatham.
   Commissioned 2.1807 under Cmdr George Langford, for the
      North Sea; destroyed 28-gun *Admiral Jawl* off Flamborough
      Head 3.3.1808. In 4.1808 under Cmdr William Charleton; sailed
      for Jamaica 22.6.1808. In 1810 under Cmdr Thomas Graves, at
      Jamaica. In late 1810 under Cmdr Edmund Denman, then Cmdr
      Hayes O'Grady in 1811; paid off into Ordinary at Sheerness
      4.1814. Middling Repair at Chatham 1–12.1815. Fitted for sea at
      Chatham 2–5.1818; recommissioned 2.1818 under Cmdr James
      Plumridge, for Cork; under Cmdr Henry Rous in 11.1820 and
      Cmdr William Bruce in 3.1821. In 2.1822 under Cmdr Jenkin
      Jones, for Halifax, Nova Scotia; in 4.1824 under Cmdr William
      Hotham, then 4.1825 Cmdr W. P. Canning; grounded at Halifax
      29.8.1825, re-floated but condemned by survey and paid off
      16.11.1825. Hulked at Halifax 12.1825, then BU there (under
      AO 11.11.1828) 7.1830.

Recruit Thomas Hills, Sandwich
   As built: 100ft 0in, 77ft 3½in x 30ft 6¼in x 12ft 9½in. 382⁹¹/₉₄bm.
      Draught 6ft 2in/10ft 10in.
   Ord: 27.1.1806. K: 4.1806. L: 31.8.1806. C: 3.10.1806–27.6.1807 at
      Sheerness.
   Commissioned 3.1807 under Cmdr George Acklom, then 7.1807
      under Cmdr Warwick Lake; sailed with convoy for West
      Indies 9.1807. In 8.1808 under Cmdr Charles Napier; in action
      against 18-gun *La Diligente* off Antigua 6.9.1808; at capture of
      Martinique 2.1809; in chase of *d'Hautpoult* 14–17.4.1809. In
      6.1809 under Cmdr James Murray. In 5.1810 under Cmdr John
      Cookesley, then 12.1810 under Cmdr Humphrey Senhouse;
      sailed for North America 9.11.1811. In 5.1813 under Lieut
      George Pechell (acting); drove ashore privateer *Inca* 2.11.1813.
      Under Cmdr Thomas Sykes in 2.1814, then in 1815 under Cmdr

John Lawrence; paid off into Ordinary at Plymouth 13.6.1815.
Sold there to R. Forbes (for £1,050) 7.8.1822.

Royalist Thomas Hills, Sandwich
   As built: 100ft 0in, 77ft 3½in x 30ft 6in x 12ft 9in. 382⁴²/₉₄bm.
      Draught 6ft 4in/11ft 0in.
   Ord: 27.1.1806. K: 5.1806. L: 10.1.1807. C: 1.3–30.7.1807 at
      Sheerness.
   Commissioned 5.1807 under Cmdr John Maxwell (–1810), for
      the Downs; took privateers – 6-gun Danish *Aristides* off
      Gothenburg 16.7.1808, 16-gun lugger *La Princesse* in the
      Channel 1.5.1809; 18-gun *Le Grand Napoléon* off Dungeness
      17.11.1809, 2-gun *L'Heureuse Etoile* 6.12.1809, 14-gun *Le
      Beau Marseille* 10.12.1809, 14-gun *Le François* 31.12.1809, and
      14-gun *Le Prince Eugene* in the Channel 24.2.1810. In ?7.1810
      under Cmdr George Downie (–1812); took more privateers
      – 14-gun *Le Roi de Naples* in the Channel 5.12.1810, 14-gun
      L'Aventurier off Fécamp 18.12.180, 10-gun *Le Braconnier* off
      Fécamp 3.2.1811, 14-gun *Le Rodeur* off Calais 19.12.1811, 14-
      gun *Le Furet* off Folkestone 6.1.1812, and 16-gun *Le Rusé* off
      Hythe 29.12.1812. On 1.1.1813 under Cmdr James John Gordon
      Bremer, with Collier's squadron on the north coast of Spain;
      took (with *Scylla*) 40-gun *Le Weser* off Ushant 21.10.1813; to
      North America 1814; took privateers – US 6-gun *Ned* 6.9.1814,
      and 2-gun *Antoinette* 18.12.1814; 7.1815 under Cmdr Houston
      Stewart, on Jamaica station; paid off 10.1816. Sold to W. S.
      Harper (for £1,130) 3.2.1819.

Clio James Betts, Mistleythorn
   As built: 100ft 0in, 77ft 5in x 30ft 9in x 13ft 1in. 389³¹/₉₄bm.
   Ord: 29.1.1806. K: 5.1806. L: 10.1.1807. C: 21.1–12.6.1807 at
      Chatham.
   First cost: fitting £4,213.
   Commissioned 2.1807 under Cmdr Thomas Folliott Baugh (–1810),
      for North Sea (Leith); took 6-gun Danish privateer 21.9.1808. In
      10.1810 under Lieut Matthew James Popplewell (acting), then
      under Cmdr William Farington, (–1815), for North Sea and Baltic;
      took (with *Hamadryad*) 4-gun privateer *Le Pilotin* in the Baltic
      14.10.1812; paid off 9.10.1815 at Chatham. Between Middling and
      Large Repair, and fitted for Ordinary at Chatham (for £8,183)
      11.1818–4.1820. Fitted for sea at Chatham (for £4,975) 2–5.1823;
      recommissioned 15.2.1823 under Cmdr Charles Strangways, for
      the Nore. In 4.1826 under Cmdr Robert Aitchinson, then 4.1827
      under Cmdr Robert Deans (–1829), at the Nore and the 1828–29
      at Cork. Between Very Small and Small Repair, and fitted as Ship
      sloop at Plymouth (for £9,164) 12.1829–7.1830; recommissioned
      7.5.1830 under Cmdr John James Onslow (appointed 30.4.1830),
      for South America; at re-occupation of Falklands 12.1832; paid
      off 17.6.1833. Between Small and Middling Repair, and altered to
      a 16-gun brig again and fitted for sea at Portsmouth (for £6,948)
      6.1833–6.1835; recommissioned 14.4.1835 under Cmdr William
      Richardson, for the Mediterranean; in 1838 under Cmdr Charles
      Deare, paid off 2.10.1838. Recommissioned 24.5.1839 under Cmdr
      Stephen Grenville Fremantle, for South America; to East Indies
      1840; on 6.11.1841 under Cmdr Edward Norwich Troubridge, to
      East Indies, and in Yangtse operations 7.1842; on 30.12.1842 under
      Cmdr James Fitzjames; paid off 10.10.1844. BU at Portsmouth
      3.1845.

Philomel (Nicholas) Bools & (William) Good, Bridport
   As built: 100ft 0in, 77ft 3in x 30ft 7in x 12ft 9in. 384³¹/₉₄bm. Draught
      6ft 6in/11ft 4in.
   Ord: 4.2.1806. K: 4.1806. L: 11.9.1806. C: 2.10.1806–6.2.1807 at
      Plymouth.

Commissioned 11.1806 under Cmdr George Crawley (died 3.1810); sailed for the Mediterranean 10.5.1807; at Rosas Bay 31.10.1809. In ?3.1810 under Cmdr George Davies, then Cmdr Spelman Swaine, then ?5.1810 Cmdr Gardiner Guion (–1811); in action off Porqueroiles 31.8.1810; at capture of Ithica 8.10.1810; in 2.1811 under Cmdr Augustus Clifford; in operations off Benidorm 12.8.1812. In 9.1811 under Cmdr Charles Shaw, then 7.1814 under Cmdr William Nares, 11.1814 Cmdr James Plumridge and 11.1816 Cmdr Lewis Campbell. Sold at Chatham (for £800) to Mr Manlove 30.4.1817.

Ordered by Grenville's Board (10)

Of this batch, *Magnet* was wrecked in 1809, while *Barracouta* and *Eclipse* had been sold in 1815. Two vessels (*Éclair* and *Sparrowhawk*) were altered to ship rig during the 1820s.

Derwent　Isaac Blackburn, Turnchapel (Plymouth)
> As built: 100ft …. etc. (quoted as design only) 382bm.
> Ord: 1.10.1806. K: 12.1806. L: 23.5.1807. C: 23.5–6.9.1807 at Plymouth.
> Commissioned 6.1807 under Cmdr William Goate, for the Channel. In 11.1807 under Cmdr Frederick Parker (–1809); sailed for African coast 17.11.1807; at capture of Senegal 7–13.7.1809 (Parker drowned 7.7.1809). In 8.1809 under Cmdr Joseph Tetley. Fitted at Portsmouth 1–11.1810; recommissioned 5.1810 under Cmdr George Sutton (–1813); took privateer *L'Edouard* off the Lizard 7.2.1813. In 6.1814 under Cmdr Thomas Williams; paid off 1815. Sold to Mr Young at Chatham (for £850) 7.3.1817.

Éclair　(ex-*Pelican*) Matthew Warren, Brightlingsea
> As built: 100ft …. etc (quoted as design only) 382bm.
> Ord: 1.10.1806. K: 12.1806. L: 8.7.1807. C: 19.7–21.12.1807 at Chatham.
> First cost: £5,300 to build, plus £4,707 fitting.
> Commissioned 12.1807 under Cmdr Charles Quash (–1810); sailed for the Mediterranean 20.2.1809; took 8-gun privateer *La Revanche* in the Mediterranean 10.3.1810. In 1.1811 under Cmdr John Markland, then 8.1811 Cmdr Arden Adderley and – later same month – Cmdr John Bellamy (–1815); her boats (with *America*'s) in destruction of convoy at Laiyueglia 9.5.1812, and (with *Leviathan*'s, *Imperieuse*'s and *Curacao*'s) convoy at same place 27.6.1812; squadron destroyed another convoy at Anzio 5.10.1813; paid off 9.1815 into Ordinary. Altered to a Ship sloop, and Large Repair at Deptford (for £8,504) 12.1822–5.1823; recommissioned 2.1823 under Cmdr William Hope Johnstone, for South America; in 10.1823 under Cmdr Thomas Bourchier (–1824). In 7.1827 under Cmdr Spencer Lambart Hunter Vassall; paid off 9.1827 at Plymouth. BU there 3.1831.

Nautilus　James Betts, Mistleythorn
> As built: 100ft 0in, 77ft 3¾in x 30ft 7in x 13ft 0in. 384⁶/₉₄bm. Draught 6ft 0in/11ft 0in.
> Ord: 1.10.1806. K: 1.1807. L: 5.8.1807. C: 17.8–26.11.1807 at Chatham.
> Commissioned 9.1807 under Cmdr Matthew Smith; sailed for Portugal 20.12.1807, and stationed there to 1809. In 8.1808 under Cmdr Thomas Dench (–1814); sailed for the Mediterranean 1.4.1810; took privateers – 5-gun *Le Brave* in the Mediterranean 21.7.1810, 14-gun *Le Leonide* off the Esquerques 7.2.1813, and 4-gun *La Colombe* 24.5.1813. In 11.1814 under Cmdr John Bradley (–1815). Delivered to the Committee for Distressed Seamen at Deptford 19.1.1818, but returned 3.1818. Fitted there for foreign service 10.1818–2.1819. Recommissioned 29.12.1818 under Cmdr Isham Fleming Chapman, for St Helena; to West

Indies 1820; paid off 16.5.1822. BU at Portsmouth 7.10.1823.

Pilot　Robert Guillaume, Northam
> As built: 100ft 0in, 77ft 4½in x 30ft 6⅜in x 12ft 11in. 383⁶¹/₉₄bm. Draught 6ft 6in/11ft 1in.
> Ord: 1.10.1806. K: 1.1807. L: 6.8.1807. C: 9.8.1807–28.1.1808 at Portsmouth.
> Commissioned 8.1807 under Cmdr William Walpole; took 3-gun privateer *La Princesse Pantine* in the Mediterranean 11.1808. In 7.1809 under Lieut (Cmdr 1.1810) Charles Sotheby, for the North Sea. In 1810 under Cmdr Edmund Waller, then 4.1810 Cmdr John Toup Nicholas (–1816); sailed for the Mediterranean 26.4.1810; destroyed (with *Thames* and *Weazle*) convoy at Amontea 25.7.1810 (6 Neapolitan gunboats and many transports taken); her boats took 3 settees and destroyed a fourth in Gulf of Taranto 26.5.1811; at Castella 6.9.1811, at Policastro 16.4.1812, and (with *Thames*) at Sapri 14.5.1812; took 6-gun privateer *Le Hart* off the Esquerques 4.6.1813; action against 22-gun *La Légère* off Cape Corso 17.6.1815; arrived Plymouth 7.7.1816 and laid up. Sold to Adam Gordon (for £1,010) 26.3.1828.

Sparrowhawk　Matthew Warren, Brightlingsea
> As built: 100ft 1in, 77ft 3⅞in x 30ft 7in x 12ft 7in. 384⁶/₉₄bm. Draught 5ft 9in/10ft 11in.
> Ord: 1.10.1806. K: 1.1807. L: 20.8.1807. C: 2.9–29.12.1807 at Chatham.
> First cost: £5,241 to build, plus £1,890 fitting.
> Commissioned 11.1807 under Cmdr James Pringle (–1812), for the Downs and Channel Islands; repair at Dartmouth by Cock (for £250) 2.1809; sailed for the Mediterranean 11.2.1809; took 6-gun privateer *L'Intrépide* 19.6.1809; in Roger's attack on Palamos 13.12.1809; took 2-gun privateer *L'Invincible* off Malaga 6.11.1811. In 6.1812 under Cmdr Thomas Clowes (–1814); in the Mediterranean 1813, then West Indies 1814. In 12.1814 under Cmdr Frederick Burgoyne (–1815), in the Mediterranean. Between Middling and Large Repair at Chatham (for £8,407) 5.1817–1.1819. Fitted for sea at Chatham (for £5,727) 9–11.1822; recommissioned 9.1822 under Cmdr Edward Boxer, for Halifax station; on 23.6.1823 under Cmdr Richard Saunders Dundas, at Halifax and then in Mediterranean; altered to a Ship sloop at Portsmouth (for £4,050) 12.1823–3.1824; on 17.7.1824 under Cmdr Robert Stuart, for the Mediterranean. In 11.1825 under Cmdr James Polkinghorne, for the Cape of Good Hope, then 8.1828 under Cmdr Henry Colpoys; by 4.1829 under Cmdr Thomas Gill, at Jamaica. Fitted for foreign service at Woolwich (for £9,802) 5–10.1829. Recommissioned 22.7.1830 under Cmdr Dawson Mayne, for the West Indies; paid off 5.1833. Altered to a brig again, and between Small and Middling Repair at Portsmouth (for £5,735) 6.1833–2.1834; recommissioned 9.11.1833 under Cmdr Charles Pearson, for South America; paid off at Plymouth 2.1837, but recommissioned 4.4.1837 under Cmdr John Shepherd, for South America and the Pacific; paid off 6.7.1841. Sold to Messrs Twyncham (for £880) 27.5.1841.

Zenobia　Josiah & Thomas Brindley, King's Lynn
> As built: 100ft 1½in, 77ft 4in x 30ft 7in x 12ft 9in. 384⁷/₉₄bm. Draught 6ft 0in/10ft 10in.
> Ord: 1.10.1806. K: 3.1807. L: 7.10.1807. C: 4.11.1807–29.1.1808 at Chatham.
> First cost: £5,159 to build, plus £4,757 fitting.
> Commissioned 12.1807 under Cmdr Alexander Richard Mackenzie (–1812), for the North Sea; sailed for Portugal 25.10.1810, and again 24.1.1812. In 7.1812 under Cmdr Richard Foley, then 6.1814 under Cmdr Nicholas Dobree, at Lisbon; arrived

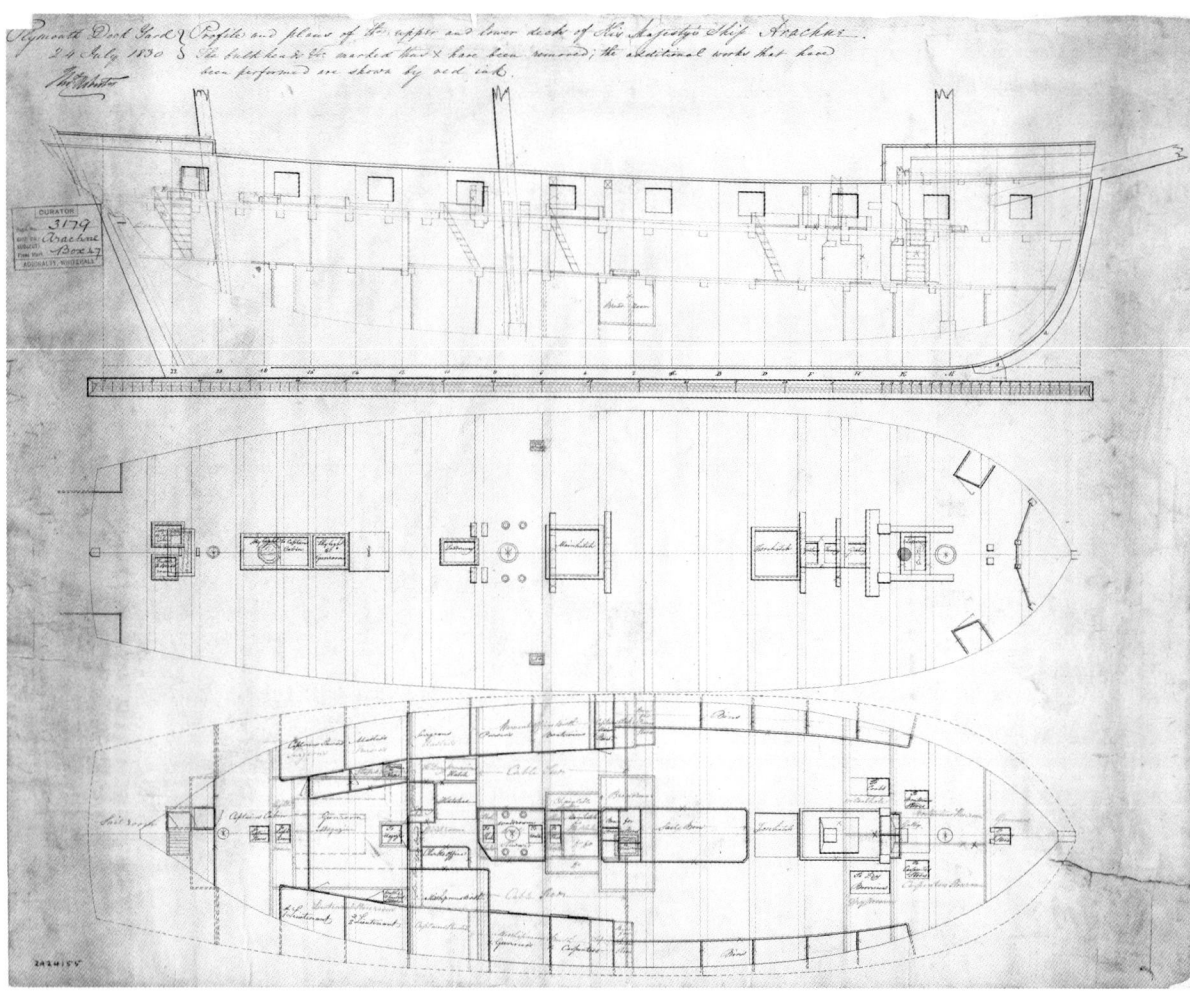

One of more than 100 *Cruiser* Class brig sloops which served the British Navy during the first half of the nineteenth century, the *Arachne* was built in Sandwich in 1808–09 and served not only during the Napoleonic conflict but continued in service for a further two decades. Like many of her sisters, she was later converted to a ship rig (in 1824) and this draught, signed on 24 July 1830 by Thomas Roberts, the master shipwright at Plymouth from 1830 to 1837, showing her profile as well as the layout of both upper and lower decks, reveals modifications made to the bulkheads, galley and companionways during repairs in 1831. *(NMM J6522)*

Plymouth 8.4.1816 and laid up. Sold to Tibbett (for £650) to BU 8.1835.

Peruvian George Parsons, Warsash
  As built: 99ft 10in, 77ft 0⅛in x 30ft 7in x 13ft 0in. 383¹³/₉₄bm. Draught 6ft 9in/11ft 1in.
  Ord: 1.10.1806. K: 9.1807. L: 26.4.1808. C: 28.4–14.7.1808 at Portsmouth.
  Commissioned 5.1808 under Cmdr Francis Douglas, for the Downs; sailed for the Leeward Islands 14.1.1810. In 11.1810 under Cmdr Francis Dickinson (died 23.4.1812), then under Lieut (Cmdr 8.1812) Amos Westropp; took US 7-gun privateer *Yankee* 24.10.1812. In 1813 under Cmdr George Kippen; took US 16-gun privateer *John* 5.2.1813; at destruction of US 28-gun *Adams* at Hampden 3.9.1814. In 10.1814 under Cmdr James Kearney White; at Cape of Good Hope 1815; laid up at Plymouth 7.1816. BU there 25.2.1830.

Ordered by Mulgrave's Board (14)
The contract price for the first two vessels was £14 per ton. Of this batch, *Persian* had been wrecked in 1813 and *Crane* lost in 1814. Five ships (*Pelorus*, *Arachne*, *Scylla*, *Thracian* and *Trinculo*) were altered to ship rig during the 1820s.

Pelorus Robert Guillaume, Northam
  As built: 100ft 0in, 77ft 3¾in x 30ft 7in x 12ft 9in. 384⁶⁰/₉₄bm. Draught 6ft 4½in/11ft 6in.
  Ord: 30.3.1807. (Contract 7.10.1807) K: 1.1808. L: 25.6.1808. C:

28.6–11.9.1808 at Portsmouth.
  First cost: £3,912 paid for building, plus £5,425 fitting.
  Commissioned 7.1808 under Cmdr James William King; sailed for the Leeward Islands 15.12.1808. In 1.1809 under Cmdr Thomas Huskinson; at capture of Martinique 1–2.1809; destroyed (with *Hazard*) a privateer schooner at St Marie, Guadeloupe 17.10.1809. In 1810 under Cmdr Alexander Kennedy, in the Leeward Islands, then ?5.1811 under Cmdr Joshua Rowley; sailed for the Mediterranean 5.4.1812. In 9.1812 under Cmdr Robert Gambier, then 1814 under Cmdr Arthur Stow; her boats (with *Endymion*'s and *Rattler*'s) took US 15-gun privateer *Mars* 7.3.1814. In 9.1814 under Cmdr John Gourly; paid off 9.1815 at Plymouth. Middling Repair at Plymouth (for £11,050), then laid up. Fitted for sea at Plymouth (for £5,570) 4–8.1823; recommissioned 4.1823 under Cmdr William Hamley, for Cork; paid off 7.1826. Altered to a ship sloop at Plymouth (for £3,882) 7–10.1826; recommissioned 20.10.1826 under Cmdr Peter Richards, for the Mediterranean; from 17.9.1828 under Cmdr Michael Quin; paid off 21.5.1830. Altered back to a brig, and between Small and Middling Repair at Portsmouth 12.1830–12.1831; recommissioned 26.9.1831 under Cmdr Richard Meredith, for the coast of Africa and Cape of Good Hope; took slavers *Segunda Teresa* 19.3.1832 and *Sutil* 17.12.1834; paid off 26.9.1835. Recommissioned 21.1.1837 under Cmdr Francis Harding; sailed 11.4.1837 for the East Indies; on 27.7.1839 under Cmdr Augustus Leopold Kuper, for Australia; paid off 6.7.1841.

Sold at Singapore 1842 (by AO 16.10.1841).

Doterel Richard Blake & John Scott, Bursledon
As built: 100ft 2in, 77ft 2¼in x 30ft 8in x 12ft 10in. 386¹¹/₉₄bm.
Draught 6ft 6in/10ft 0in.
Ord: 31.12.1807. (Contract 9.1.1808) K: 4.1808. L: 6.10.1808. C:
7.10–3.12.1808 at Portsmouth.
First cost: £5,354 to build.
Commissioned 10.1808 under Cmdr Anthony Abdy; in 3.1809
at the Basque Roads; sailed for Portugal 24.5.1809. In 10.1810
under Cmdr William Westcott Daniel, for the West Indies;
sailed for Portugal 1.1.1811; sailed for Jamaica 22.4.1813. In
12.1813 under Cmdr John Knatchbull; on North America
station 1814; paid off and laid up 8.1815 at Chatham. Fitted for
sea at Chatham 2–4.1818; recommissioned 2.1818 under Cmdr
John Gore, for Cork. In 7.1821 under Cmdr William Hendry,
for Halifax station, then 7.1822 under Cmdr Richard Hoare,
25.8.1825 under Cmdr Henry Edwards, and 17.11.1826 under
Cmdr William Alexander Baillie Hamilton. Laid up at Bermuda
by AO 10.4.1827 'in consequence of her defective state. She is
to be used as a Dwelling for the workmen employed upon the
improvements at Bermuda.' BU at Bermuda ordered 28.8.1848,
and carried out about 1855.

Arachne Thomas Hills, Sandwich
As built: 100ft 2½in, 77ft 9in x 30ft 7in x 13ft 0in. 386⁷⁷/₉₄bm.
Draught 6ft 6in/11ft 6in.
Ord: 4.8.1808. K: 9.1808. L: 18.2.1809. C: 26.3–27.8.1809 at
Sheerness (coppered).
First cost: £5,142 to build, plus £6,256 fitting & coppering.
Commissioned 5.1809 under Cmdr Samuel Chambers (–1812);
sailed for the Leeward Islands 9.5.1809. In 9.1812 under Cmdr
Charles Hope Watson, then 1813 Lieut Robert James Gordon
and 7.1814 Cmdr William Mackenzie Godfrey, still in the
Leeward Islands. Defects made good at Portsmouth (for £7,428)
1–4.1815. Between Middling and Large Repair at Portsmouth
(for £9,369) 4.1819–1.1820. Altered to a ship sloop and fitted
at Portsmouth (for £5,278) 11.1823–3.1824; recommissioned
11.1823 under Capt. Henry Ducie Chads; in 1st Anglo-Burmese
War 1824–26; from 11.1825 under Cmdr John Dawson (killed
2.12.1825), then 5.1826 under Cmdr Andrew Baird, 11.1826
under Cmdr William Pettman, 9.1827 under Cmdr George
William Conway Courtenay at Jamaica, 14.4.1828 under Cmdr
Henry Smith, and 7.1829 under Cmdr H. S. Nixon; paid
off 7.1830. Small Repair at Plymouth (for £8,671) 1–9.1831
recommissioned 7.1831 under Cmdr William Gapper Agar, for
the West Indies; on 29.4.1834 under Cmdr James Burney; paid
off 29.5.1835 at Portsmouth. Sold there to John Small Sedger (for
£1,110) 12.1.1837.

Castilian Thomas Hills, Sandwich
As built: 100ft 3in, 77ft 10in x 30ft 7in x 13ft 0in. 387²²/₉₄bm.
Draught 6ft 0in/11ft 0in.
Ord: 4.8.1808. K: 10.1808. L: 29.5.1809. C: 20.6–16.12.1809 at
Sheerness.
First cost: £5,271 to build, plus £6,032 fitting.
Commissioned 10.1809 under Cmdr Robert Tom, for the Downs.
In 10.1810 under Cmdr David Brainer, then 5.1811 under
Cmdr Edward Denman (–1815); beat off (with *Naiad*, *Rinaldo*,
Redpole and *Viper*) an attack by vessels of Boulogne Flotilla
(during visit by Napoleon), with 12-gun *La Ville de Lyons* being
taken 21.9.1811; retook (with *Bermuda*, *Rinaldo* and *Phipps*)
Apelles off Etaples 4.5.1812; on Irish station 1813–15. In 1814
under Lieut George Lloyd (acting); in action with US sloop

Wasp 1.9.1814; paid off 18.9.1815 and laid up at Deptford. BU
there 9.1829.

Charybdis Mark Richards & John Davidson, Hythe
As built: 100ft 3in, 77ft 6½in x 30ft 6¾in x 12ft 9in. 385bm. Draught
7ft 0in/11ft 0in.
Ord: 5.9.1808. K: 10.1808. L: 28.8.1809 (7½ weeks late!). C: 6.9–
1.12.1809 at Portsmouth.
Commissioned 9.1809 under Cmdr Robert Fowler; sailed for
the Leeward Islands 22.1.1810. In 6.1811 under Cmdr James
Clepham; took US 10-gun privateer *Blockade* near Saba
31.12.1811; laid up at Deptford 8.1815. Sold to Thomas Pittman
(for £1,100) 3.2.1819.

Scylla Robert Davy, Topsham
As built: 100ft 4in, 77ft 8⅛in x 30ft 6½in x 12ft 10in. 385bm.
Draught 6ft 8in/11ft 0in.
Ord: 5.9.1808. K: 12.1808. L: 29.6.1809. C: 15.8–9.12.1809 at
Plymouth.
First cost: £5,190 to build, plus £6,698 fitting.
Commissioned 9.1809 under Cmdr Arthur Atchinson (–1812);
took 11-gun *Le Canonnier* off Île de Batz 8.5.1811. In 8.1812
under Cmdr Colin M'Donald; took (with *Pheasant* and *Whiting*)
US 8-gun privateer *Fox* 6.5.1813; took (with *Royalist*) 40-gun
Le Weser off Ushant 29.9.1813. In 6.1814 under Cmdr George
Bennett Allen; paid off 30.91815 at Plymouth. Middling to
Large Repair at Plymouth (for £7,224) 2.1821–1.1822. Fitted
for sea and altered to a ship sloop at Plymouth (for £4,565)
9–12.1824; recommissioned 9.1824 under Cmdr George
Russell, for Jamaica; in 3.1826 under Cmdr William Hobson
and 1828 Cmdr Stannard Eaton Travers. Middling Repair
and fitted for sea at Portsmouth (for £8,320) 10.1829–5.1830;
recommissioned 8.5.1830 under Cmdr John Hindmarsh, for
the Mediterranean; on 3.9.1831 under Cmdr George Grey; paid
off 3.1834. Recommissioned 3.12.1834 under Cmdr Edward
John Carpenter, for North America and the West Indies; paid
off 10.4.1836. Recommissioned 26.12.1836 under Cmdr Joseph
Denman, for Lisbon station; paid off 28.3.1839. Very Small
Repair and fitted for sea at Plymouth (for £5,284) 9.1841–4.1842;
recommissioned 26.1.1842 under Cmdr Robert Sharpe, for
North America and West Indies; paid off 2.10.1845. BU at
Plymouth 1.1846.

Thracian Josiah & Thomas Brindley, Frindsbury
As built: 100ft 0in, 77ft 3½in x 30ft 6½in x 12ft 9in. 383 (384¹¹/₉₄ by
calc.)bm. Draught 6ft 6in/10ft 6in.
Ord: 30.9.1808. K: 12.1808. L: 15.7.1809. C: 20.11.1809 at Chatham.
First cost: £5,446 to build.
Commissioned 9.1809 under Cmdr James Grant. In 1810 under
Cmdr John Lawson, then 11.1810 Lieut Henry Hart (acting)
and 8.1811 Cmdr Joseph Symes; destroyed 18-gun privateer
27.12.1811. In 3.1812 under Cmdr John Carter (–1815); in Baltic
1813; took 14-gun privateer *L'Emile* off St Valery 8.2.1814; at
Jamaica 1814–15; paid off 20.10.1815. Large Repair at Chatham
3.1817–9.1818, then laid up. Altered to a ship sloop 8–10.1822;
recommissioned 6.1822 under Cmdr John Walter Roberts, for
the West Indies; her boats (with *Tyne*'s) took pirate schooner
Zaragozana 28.3.1823; in 6.1823 under Cmdr Andrew Forbes;
paid off 10.1825. BU completed at Portsmouth 6.6.1829.

Trinculo Richard Blake & John Tyson, Bursledon
As built: 100ft 5in, 77ft 8¾in x 30ft 8in x 12ft 10in. 388⁷⁸/₉₄bm.
Draught 6ft 6in/11ft 0in.
Ord: 5.11.1808. K: 10.1808. L: 15.7.1809. C: 19.7–9.12.1809 at
Portsmouth.

First cost: £5,192 to build, plus £5,524 fitting.

Commissioned 8.1809 under Cmdr Fitzowen Skinner (died
23.5.1810), for the Downs; then under Cmdr John Lamborn, and
in ?10.1810 under Cmdr Alexander Renny (–1814), on the Irish
station; paid off 16.3.1815. Between Middling and Large Repair
at Portsmouth (for £10,133) 7.1818–4.1819. Fitted for sea at
Portsmouth (for £2,751) 1–4.1823; recommissioned 1.1823 under
Cmdr Rodney Shannon, for Cork; on 3.5.1826 under Cmdr
Robert Patton, then 30.4.1827 under Cmdr Thomas Bennett.
Altered to a ship sloop at Plymouth (for £2,665) 9–11.1828; from
9.1828 under Cmdr Samuel Price (–1830), still at Cork. Between
Small and Middling Repair at Plymouth, and altered back into
a brig (for £7,359) 3.1831–6.1832; recommissioned 14.4.1832
under Cmdr James Richard Booth, for the west coast of Africa
and Cape of Good Hope; took slavers *Segundo Socorro* 7.7.1833
and *Caridad* 18.9.1833; on 19.8.1835 under Lieut Henry Joseph
Puget (temp.); took slavers *Isabel* 21.12.1835 and *Diligencia*,
Felix Vescongada and *Eliza*, all on 28.1.1836; paid off 1.6.1836.
Recommissioned 26.12.1836 under Cmdr Henry Edward Coffin;
sailed for Lisbon 21.3.1837 (–1838); paid off 29.5.1841. BU at
Plymouth 7.1841.

Hecate John King, Upnor
   As built: 100ft 2in, 77ft 4in x 30ft 7in x 12ft 9in. 384⁷/₉₄bm. Draught
      6ft 9in/11ft 3in.
   Ord: 5.11.1808. K: 12.1808. L: 30.5.1809. C: 30.5–23.9.1809 at
      Chatham.
   Commissioned 7.1809 under Cmdr William Buchanan; in ?10.1809
      under Cmdr Edward Hoare; sailed for East Indies 31.10.1809. In
      1810 under Lieut George Rennie (acting); at capture of Mauritius
      11–12.1810. In 1811 under Cmdr Thomas Graham, then ?7.1811
      Cmdr Henry Peachey; in 8.1812 under Cmdr William Case, then
      2.1814 under Cmdr John Allen and 11.1815 under Cmdr John
      Reynolds, still in East Indies; paid off 8.1816. Sold to Mr Parkin
      (for £860) 30.10.1817, and thence to the Chileans, being renamed
      Galvarino.

Rifleman John King, Upnor
   As built: 100ft 2in, 77ft 4in x 30ft 8in x 12ft 9in. 386⁸/₉₄bm. Draught
      6ft 8in/11ft 2in.
   Ord: 5.11.1808. K: 1.1809. L: 12.8.1809. C: 12.8–6.12.1809 at
      Chatham.
   First cost: £5,285 to build.
   Commissioned 9.1809 under Cmdr Alexander Innes. In 11.1810
      under Cmdr Joseph Pearce; retook cutter *Alban* (taken by Danes
      12.9.1810) 11.5.1811; sailed for the Leeward Islands 23.3.1813;
      took US 5-gun privateer *Diomede* 28.5.1814. In 6.1814 under
      Cmdr Henry Napier, for North America and the West Indies;
      on 22.8.1815 under Cmdr George Bennett Allen, then 8.1816
      under Cmdr Houston Stewart, at Jamaica, then 6.1817 under
      Cmdr Robert Felix and 9.1817 under Cmdr Norwich Duff; paid
      off 8.1818. Repair at Portsmouth 1–7.1820. Fitted for sea there
      4–7.1823; recommissioned 4.1823 under Cmdr James Montagu,
      for Halifax; in 7.1824 under Cmdr William Webb, then 12.1826
      under Cmdr Frederick Michell, for the Mediterranean; paid off
      26.11.1830. Sold at Portsmouth (for £1,010) 21.1.1836.

Echo John Pelham, Frindsbury
   As built: 100ft 4in, 77ft 6⅛in x 30ft 8in x 12ft 9in. 387⁷/₉₄bm.
      Draught 6ft 2in/10ft 3in.
   Ord: 21.11.1808. K: 12.1808. L: 1.7.1809. C: 1.7–8.11.1809 at
      Chatham.
   Commissioned 9.1809 under Cmdr Robert Keen, for the Downs;
      took 16-gun privateer *Le Capricieux* off Dieppe 12.3.1810. In

10.1810 under Cmdr Arden Adderley; took 16-gun privateer
La Confiance 21.2.1811. In ?5.1811 under Cmdr John Haswell
(died 28.7.1811), then 8.1811 Cmdr Thomas Percival (–1815); on
North American station 1813–14. BU at Chatham 5.1817.

Sophie John Pelham, Frindsbury
   As built: 100ft 3in, 77ft 5⅛in x 30ft 8in x 12ft 9in. 387⁴⁵/₉₄bm.
      Draught 6ft 4in/10ft 6in.
   Ord: 21.11.1808. K: 12.1808. L: 8.9.1809. C: 8.9–23.12.1809 at
      Chatham.
   Commissioned 10.1809 under Cmdr Nicholas Lockyer (–1814),
      for Portsmouth station; sailed for North America 28.8.1812;
      destroyed US 17-gun privateer *Pioneer* in the Chesapeake
      31.12.1812; took US 2-gun privateer *Starks* 24.4.1814; in attack
      on Fort Bowyer (Mobile) 15.9.1814; in New Orleans operations
      12.1814–1.1815; under Lieut James Tattnall (acting) in 12.1814.
      In ?4.1815 under Cmdr Silas Hood, later under Lieut William
      Gilbert Roberts (acting); paid off at Portsmouth 6.9.1815.
      Between Middling and Large Repair at Chatham 12.1815–5.1817.
      Fitted for sea there 8–12.1818; recommissioned 8.1818 under
      Cmdr Sir William Wiseman (–1820), for Jamaica. Fitted for sea
      at Chatham again 12.1820–4.1821; recommissioned 12.1820
      under Cmdr George French, for the East Indies; in 7.1822 under
      Cmdr Robert Dunlop, then under Lieut George Ryves (acting)
      4–10.1823, then under Lieut Edward Blanckley 4/5.1825; in
      Burma War 1824–25; paid off 8.1825. Sold in the East Indies (for
      £3,200) 15.8.1825.

Ordered by Charles Yorke's Board (15)
Of this batch, six (*Wasp*, *Pandora*, *Espiegle*, *Heron*, *Despatch* and
Grasshopper) were altered to ship rig during the 1820s.

Childers Portsmouth Dyd. [M/Shipwright Nicholas Diddams]
   As built: 100ft 1in, 77ft 3⅛in x 30ft 6½in x 12ft 10in. 383⁵¹/₉₄bm.
      Draught 6ft 10in/11ft 6in.
   Ord: 19.1.1811. K: 8.1811. L: 9.7.1812. C: 22.8.1812.
   Commissioned 7.1812 under Cmdr Buckland Bluett. In 8.1812
      under Cmdr John Bedford; sailed for the Leeward Islands
      29.9.1812; took (with *Acasta*, *Maidstone* and *Aeolus*) US 10-
      gun privateer *Snapper* 3.11.1812. In 12.1813 under Cmdr John
      Umfreville (–1815); in attack on Fort Bowyer, Mobile 15.9.1814.
      In 10.1815 under Cmdr Richard Wales, in the Leeward Islands.
      In 9.1816 under Lieut Edward Astley (acting), then 1.1817 Cmdr
      Amos Westropp, still in Leeward Islands; paid off 30.11.1818.
      BU at Chatham 7.3.1822

Curlew (William) Good & Co., Bridport
   As built: 100ft 1in, 77ft 3½in x 30ft 7½in x 12ft 10in. 385⁵¹/₉₄bm.
      Draught 6ft 6in/11ft 6in.
   Ord: 30.8.1811. K: 10.1811. L: 27.5.1812. C: 12.6–6.8.1812 at
      Portsmouth.
   Commissioned 7.1812 under Cmdr Michael Head (–1814);
      sailed for North America 28.8.1812; took (with *Tenedos*) US
      4-gun privateer *Enterprise* 21.5.1813. In 6.1814 under Cmdr
      Hugh Pearson. Large Repair (begun as Small Repair) at
      Chatham 11.1815–1.1817. Fitted for sea at Chatham 2–4.1818;
      recommissioned under Cmdr William Walpole, for the East
      Indies; at Ras al-Khaimah 1819. In 12.1819 under Cmdr George
      Gambier, then 4.1820 Lieut (Cmdr 6.1821) Price Blackwood, still
      in East Indies (–1822). Sold at Bombay to James Matheson (for
      15,100 rupees) 28.12.1822, renamed *Jamesina*.

Wasp Robert Davy, Topsham
   As built: 100ft 1in, 77ft 4in x 30ft 7½in x 12ft 10in. 385⁷¹/₉₄bm.
      Draught 6ft 9in/11ft 1in.

Ord: 30.8.1811. K: 10.1811. L: 9.7.1812. C: 31.7–3.10.1812 at
Plymouth.

First cost: £5,837 to build, plus £7,216 fitting.

Commissioned 8.1812 under Cmdr Thomas Everard (–1814); sailed
for North America 22.4.1813. In 6.1814 under Cmdr John Fisher,
at Halifax, then 8.1815 under Cmdr William Wolrige (–1818), in
the Mediterranean. In 11.1818 under Cmdr Thomas Wren Carter
(–1820), in the Leeward Islands. Large Repair at Portsmouth
(for £6,364) 4.1821–2.1822. Fitted as ship sloop at Portsmouth
(for £6,275) 11.1827–4.1828; recommissioned 4.1828 under
Cmdr William Wellesley, for the Mediterranean. In 1.1829 under
Capt. Richard Dickinson, then 2.1829 Cmdr Thomas Hoste
and 4.1830 Cmdr Brunswick Popham, still in Mediterranean.
Fitted as brig again and Small Repair at Portsmouth (for £4,491)
8.1832–9.1833; recommissioned 19.7.1833 under Cmdr James
Burney, for North America and West Indies; on 29.4.1834 under
Cmdr John Foreman; paid off 27.4.1837. Recommissioned
21.10.1837 under Cmdr Dudley Worsley Anderson Pelham,
for the Mediterranean; at Bombardment of Acre 3.11.1840;
on 26.12.1840 under Cmdr Henry Anthony Murray; aid off
12.5.1842. Recommissioned 6.10.1842 under Cmdr Andrew
Drew, for North America and the West Indies; on 4.6.1844 under
Cmdr Sidney Henry Ussher, for the west coast of Africa; took
slavers *Grande Poder de Dios* 16.9.1844, *Diligencia* 16.11.1844,
Esperanza 8.1.1845, *San Domingo* 20.1.1845, *Isabel* 1.6.1845,
Echo 2.3.1845, *Lobo* 15.1.1846, *Pacquito de Rio* 17.2.1846,
Gaio 5.4.1845 and *Galgo* 20.4.1846; paid off 17.10.1846. BU at
Sheerness 9.1847.

Fairy William Taylor, Bideford

As built: 100ft 1½in, 77ft 4in x 30ft 8in x 13ft 3in. 386⁸⁰⁄₉₄bm.
Draught 6ft 6in/11ft 0in.

Ord: 30.8.1811. K: 1.1812. L: 11.6.1812. C: 17.7–3.10.1812 at
Plymouth.

Commissioned 8.1812 under Cmdr Edward Grey, for South
America. On 18.3.1814 under Cmdr Henry Loraine Baker;
in Potomac operations 1814. In 6.1815 under Cmdr Hugh
Patton; paid off into Ordinary at Plymouth 3.1816. BU there
27.12.1820–1.1821.

Pelican Robert Davy, Topsham

As built: 100ft 0in, 76ft 7⅝in x 30ft 9in x 12ft 9in. 385⁴¹⁄₉₄bm.
Draught 6ft 6in/11ft 1in.

Ord: 30.8.1811. K: 1.1812. L: 8.1812. C: 7.9–11.12.1812 at
Plymouth.

First cost: £5,928 for building, plus £5,903 for fitting.

Commissioned 11.1812 under Cmdr John Fordyce Maples for the
Irish Station; took USS *Argus* (16 guns) on 14.8.1813. In 8.1813
under Cmdr Thomas Mansell; took US 12-gun privateer *Siren*
31.1.1814; at Lisbon in 1814; later under Cmdr William Bamber.
In 12.1814 under Cmdr Thomas Prickett, then 10.1815 Cmdr
Robert Coulson at Portsmouth, then 6.1816 Cmdr Edward
Curzon, at Bermuda; paid off 1817 at Portsmouth. Large Repair
there (for £7,234) 7.1820–3.1821, then laid up. Fitted for sea
at Portsmouth (for £4,730) 8–10.1826; recommissioned 8.1826
under Cmdr Charles Irby, for the Mediterranean; took schooner
Aphrodite in Gulf of Kalamata 3.1.1827. On 2.7.1827 under
Cmdr William Alexander Baillie Hamilton, then 9.8.1828 Cmdr
Francis Deane Hutcheson and 24.4.1830 Cmdr Joseph Gape, still
in Mediterranean; paid off 11.3.1834. Partial Repair and fitted
for sea at Chatham (for £2,981) 8.1834–1.1835; recommissioned
3.12.1834 under Cmdr Brunswick Popham, for the Cape of
Good Hope, then to East Indies (for China War); paid off

5.1839. Between Small and Middling Repair at Chatham (for
£5,624) 6.1839–1.1841. Recommissioned 30.10.1840 under Cmdr
Charles Napier, for China; on 11.11.1841 under Cmdr Philip
Justice, on same station; paid off 14.1.1845 at Portsmouth and
laid up. To coastguard 1850, stationed at Rye Harbour 1850–65
(by 1863 renamed *CGWV 29*). Sold to Mr Fryman, Rye on
7.6.1865.

Bacchus Chatham Dyd [M/Shipwright Robert Seppings]

As built: 100ft 3½in, 77ft 3½in x 30ft 7in x 12ft 9in. 384⁵¹⁄₉₄bm.
Draught 7ft 1in/11ft 0in.

Ord: 30.8.1811. K: 1.1812. L: 17.4.1813. C: 10.6.1813.

Commissioned 5.1813 under Cmdr (Capt. 12.1813) Lewis Hole, for
Irish station. In 2.1814 under Cmdr George Willes, then 5.1814
Cmdr William Slaughter and 6.1814 Cmdr William Hill. In
12.1816 under Cmdr Edward Barnard, then 7.1817 Cmdr John
Parkin. Laid up at Deptford 1.1820. Fitted there to receive coals
8.1826, then at a breakwater 6–8.1829, for Harwich. Towed to
Harwich 13.8.1829.

Pandora Deptford Dyd [M/Shipwright Robert Nelson to 7.1813,
completed by William Stone]

As built: 99ft 10½in, 77ft 1⅞in x 30ft 6½in x 12ft 11in. 382⁷⁷⁄₉₄bm.
Draught 7ft 1in/10ft 11in.

Ord: 30.8.1811. K: 9.1812 (as *Lynx*, renamed 24.9.1812). L:
12.8.1813. C: 28.11.1813.

Commissioned 10.1813 under Cmdr Thomas Stamp, for cruising.
In 6.1814 under Cmdr James Meara, at Cork, and 12.1814
under Cmdr William Popham (temp.) then Cmdr Samuel
Malbon, and in 8.1815 under Cmdr Frederic Noel. In 1.1817
under Cmdr George Jones, still at Cork, then 1.1819 Cmdr
Charles Randolph; paid off 6.1822, but recommissioned same
month under Cmdr Frederick Hunn, and in 1.1823 under
Cmdr William Gordon. Altered to a ship sloop at Portsmouth
11.1824–2.1825; recommissioned 7.1825 under Cmdr William
Jervoise, for the East Indies. In 8.1828 under Cmdr John Gordon
for the East Indies; paid off 2.1830 at Plymouth. Sold there to Mr
Snooks (for £910) 13.4.1831.

Nimrod Jabez Bailey, Ipswich

As built: 100ft 0in, 77ft 2¾in x 30ft 7in x 12ft 10in. 384²²⁄₉₄bm.
Draught 6ft 2in/11ft 4in.

Ord: 26.9.1811. K: 11.1811. L: 25.5.1812. C: 3.6–12.9.1812 at
Sheerness.

Commissioned 8.1812 under Cmdr Nathaniel Mitchell; sailed for
North America 22.9.1812; took (with *Poictiers* and *Maidstone*)
US 20-gun privateer *Yorktown* 17.7.1813. In 6.1814 under Cmdr
George Hilton, then 27.8.1815 Cmdr John M'Pherson Ferguson,
on Leith station. In 1816 under Cmdr John Gedge, then 1.1817
Cmdr John Dalling, still at Leith. In 6.1819 under Cmdr Charles
Nelson, then 7.1822 Cmdr William Rochfort; at Tyne coal strike
1822, later at Cork; paid off 10.1825. In 9.1826 under Cmdr
Samuel Sparshot; bilged in Holyhead Bay in a gale 14.1.1827;
salved 12.2.1827 but sold as unrepairable to Rowland Robert &
Co. (for £510) 22.2.1827.

Saracen (Nicholas) Bools & (William) Good, Bridport

As built: 100ft 1in, 77ft 3½in x 30ft 8in x 12ft 10in. 386⁶⁶⁄₉₄bm.
Draught 6ft 6in/11ft 0in.

Ord: 26.9.1811. K: 11.1811. L: 25.7.1812. C: 10.8–17.9.1812 at
Portsmouth.

Commissioned 8.1812 under Cmdr John Harper; took 14-gun
privateer *Le Courier* off Beachy Head 21.9.1812; sailed for the
Mediterranean 17.11.1812; landing party at Guipanna (Dalmatia)
17.6.1813; took (with *Weazle*) the *Mizzo* 22.7.1813; her boats

(with *Bacchante*'s) took 4 gunboats at Ragusa 13.10.1813; at surrender of Cattaro 5.1.1814. On 7.6.1814 under Cmdr Alexander Dixie (acting), then 18.8.1815 under Cmdr John Gore, at Bermuda; paid off 12.1818. Sold to William Wilkinson at Chatham (for £1,150) 18.8.1819.

Satellite Daniel List, Fishbourne

As built: 100ft 6in, 77ft $7\frac{1}{8}$in x 30ft $6\frac{1}{2}$in x 12ft $9\frac{1}{2}$in. $384\frac{93}{94}$bm. Draught 5ft 10in/11ft 0in.

Ord: 5.10.1811. K: 3.1812. L: 9.10.1812. C: 10.1812–18.12.1812 at Portsmouth.

Commissioned 11.1812 under Cmdr John Porteus, for Jamaica. In 6.1814 under Cmdr Thomas Cecil (died 15.10.1814), then ?10.1814 under Cmdr Charles Samuel White and 19.8.1815 under Cmdr James Murray (–1817), for the Mediterranean. Small Repair and fitted for sea at Portsmouth 11.1820–3.1821; recommissioned 14.10.1820 under Cmdr Armar Lowry Corry, for the East Indies. In 7.1821 under Cmdr Robert Gore, then 9.1.1823 under Cmdr Mark John Currie; paid off 17.3.1824 at Bombay. Sold in the East Indies (for 30,000 rupees) 3.1824.

Arab John Pelham, Frindsbury

As built: 99ft $7\frac{1}{4}$in, 78ft $1\frac{3}{8}$in x 30ft $7\frac{5}{8}$in x 12ft $9\frac{1}{2}$in. $389\frac{92}{94}$bm. Draught 6ft 2in/10ft 6in.

Ord: 24.10.1811. K: 11.1811. L: 22.8.1812. C: 22.8–16.10.1812 at Chatham.

Commissioned 9.1812 under Cmdr John Wilson; sailed for the Leeward Islands 13.12.1812; later under Lieut Robert Standly (acting). In 7.1813 under Cmdr Henry Jane; took US 5-gun privateer *Industry* 3.11.1813. In 1817 at Plymouth; Very Small Repair there 11.1818–2.1819; recommissioned 11.1819 under Cmdr Charles Simeon, for Cork station. In 3.1822 under Cmdr William Holmes, still Cork; wrecked off Belmullet (County Mayo) with loss of all hands 18.12.1823.

Espiegle Jabez Bailey, Ipswich

As built: 100ft 2in, 77ft $4\frac{5}{8}$in x 30ft $7\frac{1}{4}$in x 12ft 9in. $385\frac{50}{94}$bm. Draught 6ft 3in/11ft 6in.

Ord: 2.11.1811. K: 1.1812. L: 10.8.1812. C: 18.8–1.10.1812 at Sheerness.

First cost: £5,666 to build, plus £5,259 fitting.

Commissioned 9.1812 under Cmdr John Taylor; sailed for the Leeward Islands 22.1.1813. In 2.1814 under Cmdr Charles Mitchell, in the West Indies; paid off at Portsmouth 5.1816. Small Repair at Portsmouth (for £4,385) 4–7.1819. Fitted for sea at Portsmouth (for £3,763) 6–9.1822; recommissioned 5.1822 under Cmdr Henry Collier, for Cape of Good Hope. In 1.1823 under Cmdr Isham Chapman, then 12.1824 under Cmdr Luke Wray. Altered to a ship sloop at Portsmouth (for £5,733) 1–6.1826; recommissioned 1.1826 under Cmdr Richard Yates, for Jamaica station, then 3.1827 under Cmdr William Sandom, 2.1828 under Cmdr Henry Gosset, 24.4.1828 under Cmdr Joseph O'Brien, 30.7.1828 under Cmdr Charles Ramsey Drinkwater, and 5.9.1829 under Cmdr Russell Eliott; paid off at Portsmouth 3.1830. Sold to Thomas Ward, Ratcliffe (for £650) 28.11.1832.

Heron (ex-*Rattlesnake*) John King, Upnor

As built: 100ft 2in, 77ft $3\frac{3}{4}$in x 30ft 8in x 12ft 9in. $386\frac{79}{94}$bm. Draught 6ft 2in/10ft 8in.

Ord: 14.11.1811. K: 2.1812. L: 22.10.1812. C: 22.10–24.12.1812 at Chatham.

First cost: £5,845 to build, plus £5,883 fitting.

Commissioned 11.1812 under Cmdr William M'Culloch; sailed for the Leeward Islands 28.3.1813; took US 5-gun privateer *Mary* 7.7.1814. In 7.1814 under Cmdr George Luke, then in same month under Cmdr Francis Annesley, them 1815 Cmdr Timothy Scriven. In 7.1816 under Cmdr George Bentham; at Bombardment of Algiers 27.8.1816. In 10.1816 under Cmdr Henry Powell, on Plymouth station, then in 1817 Cmdr Robert Riddell. In 2.1819 under Cmdr Job Hanmer (–1822), for St Helena; on 19.7.1822 under Cmdr Henry Francis Grenville. Small Repair and altered to a ship sloop at Portsmouth (for £6,589) 9.1823–11.1825; fitted for sea at Portsmouth (for £4,984) 3–8.1826; recommissioned 3.1826 under Cmdr Robert Tait, for South America; from 17.4.1827 under Cmdr Frederick William Grey, then 19.4.1828 Cmdr John Alexander Duntze (–1829), still in South America. BU at Portsmouth 3.1831.

Despatch (ii) John King, Upnor

As built: 100ft 3in, 77ft $3\frac{1}{2}$in x 30ft $8\frac{1}{2}$in x 12ft 9in. $387\frac{65}{94}$bm.

Ord: 14.11.1811. K: 2.1812. L: 7.12.1812. C: 7.12.1812–27.1.1813 at Chatham.

First cost: £5,745 to builder, plus £5,847 fitting.

Commissioned 12.1812 under Cmdr James Galloway (–1814), for Collier's squadron on the north coast of Spain; to North America 1814. In 11.1814 under Cmdr William Cobbe; paid off 1815. Small Repair at Chatham (for £3,747) 10.1818–4.1819. Fitted for sea at Chatham (for £5,382) 1–4.1821; recommissioned 1.1821 under Cmdr William Jervoise (–1822), for the Mediterranean. Small Repair and altered to a ship sloop at Chatham (for £9,204) 4–10.1825; recommissioned 7.1825 under Cmdr Robert Parsons (–1828); at Portsmouth 1827 then Cork 1828. Made good defects and altered the stations of the masts at Plymouth (for £4,025) 10.1828–1.1829; in 10.1828 under Cmdr William Bowyer. In 2.1830 under Cmdr Edward Frankland, at Cork. Altered to a brig and fitted for sea at Chatham (now 16 guns) 2–9.1832; recommissioned 7.6.1832 under Cmdr George Daniell (–1834), for North America and West Indies; paid off 6.10.1835. Sold at Sheerness 5.1836.

Grasshopper (ii) Portsmouth Dyd [M/Shipwright Nicholas Diddams]

As built: 100ft $1\frac{1}{2}$in, 77ft $6\frac{3}{8}$in x 30ft 7in x 12ft $9\frac{1}{2}$in. $385\frac{69}{94}$bm. Draught 7ft 0in/11ft 8in.

Ord: 6.1.1812. K: 8.1812. L: 17.5.1813. C: 28.6.1813.

Commissioned 5.1813 under Cmdr Henry Battersby; sailed with convoy for the Mediterranean 1.1814. In 6.1814 under Cmdr Sir Charles Burrard; paid off 2.1816 into Ordinary. Fitted for sea at Portsmouth 1–5.1818; recommissioned under Cmdr Henry Forbes. In 5.1819 under Cmdr David Buchan, for the Newfoundland station. Altered to a ship sloop at Portsmouth 1–6.1822; in 12.1823 under Cmdr John Aplin (–1826), for Halifax station. On 23.12.1826 under Cmdr Courtenay Edward William Boyle, still at Halifax. Small Repair and fitted for sea at Woolwich 6.1827–2.1828; recommissioned 8.12.1827 under Cmdr Abraham Crawford, for Jamaica, then under Cmdr Charles Deane on 5.1.1829, and Cmdr John Elphinstone Erskine on 3.5.1830; paid off 6.9.1831. Sold to J. Ward at Portsmouth (for £910) 30.5.1832.

Ordered by Melville's Second Board (20)

Of this batch, the *Epervier* (taken by USS *Peacock*), *Halcyon* (wrecked) and *Penguin* (taken by USS *Hornet*) were lost during 1814, and two were later cancelled (*Lynx* and *Samarang*). Eight vessels (*Fly*, *Jaseur*, *Victor*, *Alert*, *Harlequin*, *Harrier*, *Ontario* and *Gannet*) were fitted with ship rig during the 1820s.

Fly Jabez Bailey, Ipswich

As built: 100ft 5in, 77ft 9in x 30ft 7in x 12ft $11\frac{1}{2}$in. $386\frac{77}{94}$bm. Draught 7ft 6in/11ft 6in.

Ord: 23.4.1812. K: 6.1812. L: 16.2.1813. C: 8.3–3.7.1813 at Sheerness.

Commissioned 21.5.1813 under Cmdr Sir William George Parker, for the Channel station; on 6.6.1814 under Cmdr John Baldwin, for Newfoundland station; paid off 4.1815. Post-war, recommissioned under Cmdr James Tomkinson 1818, then 12.8.1819 under Cmdr John Townsend Coffin in the Leeward Islands and later on Cork Station; paid off 12.1821. Altered to a ship sloop at Portsmouth 12.1821–5.1822. Recommissioned 8.1821 under Cmdr George Tayler, for Cape of Good Hope station. Recommissioned 10.10.1822 under Cmdr Edward Curzon, for South America; on 8.2.1823 under Cmdr William Fanshawe Martin and from 20.4.1825 under Cmdr Lord William Paget. On 26.7.1825 under Cmdr Frederick Augustus Wetherall, in the East Indies; on 13.11.1826 under Cmdr Follett Walrond Pennell. Sold at Bombay (for 8,300 Rupees) 10.5.1828 (by AO 20.4.1827).

Jaseur Jabez Bailey, Ipswich

As built: 100ft 5in, 77ft 9¼in x 30ft 7in x 12ft 11½in. 386⁸⁷⁄₉₄bm. Draught 7ft 6in/11ft 6in.

Ord: 6.5.1812. K: 8.1812. L: 2.2.1813. C: 2.3–2.6.1813 at Sheerness. First cost: £5,316 fitting only.

Commissioned 4.1813 under Cmdr George Edward Watts; sailed with convoy for Baltic 29.6.1813; to Halifax station 1814; took US ?14-gun privateer *Grecian* 2.5.1814. On 7.6.1814 under Cmdr Nicholas Lechmere Pateshall, then 18.2.1815 Cmdr Nagle Lock; laid up at Plymouth 7.1816. Very Small Repair and fitted for sea at Plymouth (for £6,648) 12.1820–5.1821; recommissioned 23.1.1821 under Cmdr Henry Edward Napier, for Halifax station; paid off 7.1823. Very Small Repair, rigged as a ship sloop and fitted for sea at Plymouth (for £5,934 + £4,330) 11.1823–9.1824; recommissioned 6.1824 under Cmdr Thomas Martin, for South America; from 8.1826 under Cmdr Edward Handfield. Between Very Small and Small Repair and fitted at Plymouth (for £7,091) 3–10.1828; recommissioned 26.8.1828 under Cmdr John Lyons, for Cape of Good Hope; from 23.7.1830 under Cmdr Francis Harding; on 16.9.1831 under Cmdr Archibald Sinclair; paid off 10.1832. Between Small and Middling Repair and altered to 16-gun brig at Sheerness (for £4,748) 8.1833–2.1834; recommissioned 9.11.1833 under Cmdr John Hackett, for the Mediterranean; paid off 31.5.1837. Recommissioned 11.6.1838 under Cmdr Frederick Moore Boultbee, for the Mediterranean; on 13.2.1841 under Cmdr William Alexander Willis, on same station; paid off 5.11.1842. Sold to Mr Holmes at Portsmouth (for £510) 2.1845.

Argus Thomas Hills, Sandwich

As built: 100ft 1½in, 77ft 3¾in x 30ft 8in x 12ft 9½in. 386⁷⁹⁄₉₄bm. Draught 6ft 4in/10ft 10in.

Ord: 8.6.1812. K: 9.1812. L: 11.9.1813. C: 10–16.10.1813 (for Ordinary) at Sheerness.

Fitted for sea at Chatham 6–9.1821.

Commissioned 2.7.1821 under Cmdr Septimus Arabin, for Halifax. On 20.3.1823 under Cmdr John Dundas; laid up at Portsmouth 11.1824. Sold to Mr Freake (for £2,000) 11.7.1827; resold to John Small Sedger (for £1,110) 26.3.1828, 'Mr Freake having been declared insane'.

Challenger Hobbs & Hellyer, Redbridge

As built: 100ft 1¾in, 77ft 4¼in x 30ft 8in x 12ft 8½in. 386⁸⁹⁄₉₄bm. Draught 6ft 6in/11ft 3in.

Ord: 29.7.1812. K: 8.1812. L: 15.5.1813. C: 17.5–9.7.1813 at Portsmouth.

Commissioned 5.1813 under Cmdr Frederick Vernon; at reduction of San Sebastian 8.9.1813; destroyed (with *Constant* and *Telegraph*) *Le Flibustier* 13.10.1813. In 12.1814 under Cmdr Henry Forbes. Fitted for Channel service 9–10.1815. In 11.1816 under Cmdr Philip Bridges, for East Indies; ordered to be sold in India 26.2.1818 'in consequence of her very defective state'; dismantled at Trincomalee and fitted to receive rice 5.1819; used as mooring tender 1820. Sold at Trincomalee (for 3,000 rupees) 3.1824.

Lynx (ex-*Pandora*, renamed 24.9.1812) Woolwich Dyd [M/Shipwright Edward Sison]

Ord: 7.9.1812. K: — . Cancelled 1818.

Victor East India Co., Bombay Dyd. Teak-built. [M/Shipwright Jamsetjee Bomanjee Wadia]

As built: 100ft 0in, 77ft 3½in x 30ft 6in x 12ft 9in. 382⁴²⁄₉₄bm.

Ord: 2.10.1812. K: 1.1814. L: 29.10.1814. C: 18.3.1815 at Bombay.

First cost: £6,169 to build (paid by HEICo.), plus stores supplied by government £2,838.

Commissioned 2.1814 at Bombay under Cmdr Basil Hall, for passage to England. Arrived Plymouth 30.8.1815 and paid off into Ordinary. Altered to a Ship sloop and fitted for sea there (for £4,993) 7–11.1823; recommissioned 8.1823 under Cmdr Thomas Prickett, for the African station; in 8.1824 under Cmdr George Woollcombe (–1827); at Cork 1826; paid off 8.1827 but recommissioned same month under Cmdr George Lloyd, for Jamaica; in 8.1828 under Cmdr Richard Keane (–1831). Refitted and altered back to a brig at Portsmouth 5–8.1831; recommissioned 12.1831 for North America and West Indies (–1838); recommissioned 8.1841 under Cmdr Charles Otway, for same station; sailed 31.8.1842 from Veracruz for Halifax; lost, presumed foundered with all hands 9.1842.

Zebra East India Co., Bombay Dyd. Teak-built. [M/Shipwright Jamsetjee Bomanjee Wadia]

As built: 100ft 3in, 77ft 5in x 30ft 7in x 12ft 9in. 385¹⁵⁄₉₄bm. Draught 7ft 6in/8ft 6in.

Ord: 2.10.1812. K: 9.11.1814. L: 18.12.1815. C: 1816 at Bombay.

First cost: £6,618 to build.

Commissioned 12.1815 at Bombay under Cmdr Robert Forbes, for passage to England. Reached Portsmouth 14.12.1816 and paid off into Ordinary. Fitted for Sea there (for £5,050) 1.1825–14.5.1825; recommissioned 24.2.1825 under Cmdr Edward Richard Williams, for the Mediterranean. In 4.1827 under Cmdr Charles Cotton (died 1828); in attack on Grabusa 1.1.1828. In 10.1828 under Cmdr Edmund Gilbert, still in the Mediterranean, then 1.1829 under Cmdr Richard Pridham (–1830); paid off 1833. Very Small Repair and fitted for sea at Chatham (for £4,006) 2–9.1834; recommissioned 1.6.1834 under Cmdr Robert Contart M'Crea; sailed for Cape of Good Hope 24.9.1834; in East Indies to 1837; home and paid off 27.10.1838. Recommissioned 3.1.1839 under Cmdr Robert Fanshawe Stopford, for the Mediterranean; on 23.2.1840 under Cmdr James John Stopford; wrecked off Haifa 2.12.1840 (3 drowned).

Carnation (ii) Wm & James Durkin, Northam (Southampton)

As built: 100ft 1½in, 77ft 4⅜in x 30ft 7in x 12ft 9in. 384⁸¹⁄₉₄bm. Draught 6ft 4in/11ft 1in.

Ord: 8.10.1812. K: 11.1812. L: 29.7.1813. C: 9.8–20.12.1813 at Portsmouth.

Commissioned 10.1813 under Cmdr George Bentham (–1815); in attack on US privateer *General Armstrong* at Faial 26–27.9.1814. In 1816 under Cmdr Gregory Grant, then Cmdr Thomas Wren Carter; paid off 1817. Fitted for sea at Portsmouth 2–5.1818;

recommissioned 1818 under Cmdr John Gordon; in 12.1818 under Cmdr William Glasscock, then 5.1819 under Cmdr Henry Shiffner, and 11.1819 Cmdr Roger Hall, at Halifax; paid off 1821. Recommissioned 9.1821 under Cmdr John Walcott; sailed for Jamaica 4.1.1822; in 5.1822 under Cmdr Thomas Herbert, then 10.1824 under Cmdr Rawdon Maclean, still at Jamaica; home to pay off at Plymouth 6.1825. Fitted for the Breakwater Department at Plymouth 5–8.1826. Sold at Plymouth (for £810) 21.1.1836.

Elk (ii) Hobbs and Hellyer, Redbridge (Southampton)

As built: 100ft 3in, 77ft 4½in x 30ft 7½in x 12ft 11½in. 386bm. Draught 6ft 8in/11ft 2in.

Ord: 2.11.1812. K: 12.1812. L: 28.8.1813. C: 29.8.1813–18.4.1814 at Portsmouth.

First cost: £5,865 to build, plus £5,750 fitting.

Commissioned 11.1813 under Cmdr John Bartholomew Hoar Curran (–1815), for the East Indies. In 7.1815 under Lieut (Cmdr 20.11.1815) John Reynolds; paid off into Ordinary at Plymouth 1816. Sold there (for £760) 21.1.1836.

Confiance Mrs Mary Ross, Rochester

As built: 100ft 3½in, 77ft 3⅛in x 30ft 10½in x 12ft 10in. 391 71/94bm. Draught 6ft 4in/11ft 3in.

Ord: 2.11.1812. K: 12.1812. L: 30.8.1813–9.1813 (for Ordinary), 2.1818–26.6.1818 (for sea) at Chatham.

Commissioned 1818 under Cmdr William Morgan; wrecked in a gale on Irish coast (near Crookhaven), with no survivors, 21.8.1822.

Alert Thomas Pitcher, Northfleet

As built: 100ft 4in, 77ft 6¼in x 30ft 8in x 12ft 9in. 387 74/94bm. Draught 6ft 0in/10ft 11in.

Ord: 2.11.1812. K: 1.1813. L: 13.7.1813. C: 27.11.1813 at builder's.

First cost: £5,857 to build, plus £5,146 fitting.

Commissioned under Cmdr George Sartorius (date unknown), but from 10.1813 under Cmdr Joseph Garland, for the Downs; paid off 7.1815. Fitted for Channel service at Woolwich (for £811) 8–10.1815; in 8.1815 under Cmdr John Smith (–1817), for the North Sea; retook Hamburg vessel *Ocean* (taken by Tunisian pirates in North Sea) ?5.1816. In 1818 under Cmdr Henry Leeke, then 2.1819 under Cmdr Charles Farwell, for Home waters. Middling Repair at Chatham (for £6,382) 9.1821–5.1823. Altered to a ship sloop at Chatham (for £3,325) 8.1826–4.1827; recommissioned 1.1827 under Cmdr Samuel Burgess (–1829), for South America; from 11.1829 under Cmdr John Fitzgerald; paid off 9.2.1832. Sold to Crystall, Rotherhithe (for £638) 11.1832.

Harlequin Jabez Bailey, Ipswich

As built: 100ft 2in, 77ft 4½in x 30ft 7in x 12ft 11½in. 384 90/94bm. Draught 7ft 6in/11ft 6in.

Ord: 2.11.1812. K: 2.1813. L: 15.7.1813. C: 19.7.1813–6.6.1814 at Sheerness.

Commissioned 11.1813 under Cmdr William Kempthorne, for the Channel; laid up at Plymouth 11.1815. Fitted there 3–4.1818; recommissioned 2.1818 under Cmdr Alexander Branch, for Queenstown; in 7.1819 under Cmdr Charles Parker, then 6.1822 under Cmdr John Weeks (died 24.10.1824) and 11.1824 under Cmdr James Scott, still at Queenstown. Altered to a ship sloop and fitted for sea at Plymouth 7–10.1825; recommissioned 8.1826 under Cmdr William Sandom for Jamaica; in 4.1827 under Cmdr Charles Elliot. Sold at Jamaica (for £403) 4.9.1829.

Harrier (ii) Jabez Bailey, Ipswich

As built: 100ft 2½in, 77ft 5in x 30ft 7½in x 13ft 0in. 386 20/94bm. Draught 7ft 6in/11ft 6in.

Ord: 2.11.1812. K: 2.1813. L: 28.7.1813. C: 31.7.1813–17.4.1814 at Sheerness.

First cost fitting £5,488.

Commissioned 12.1813 under Cmdr Andrew Green, later under Cmdr Henry Forbes. On 16.6.1814 under Cmdr Charles Thomas Jones, in the Canary Islands, off the French coast, then on Halifax Station; paid off 12.1818. Small Repair at Portsmouth (for £4,714) 12.1818–1.1820. Fitted for sea at Portsmouth 4–5.1823, but did not sail. Altered to a ship sloop at Portsmouth (for £5,042 including previous fitting) 8–12.1823; recommissioned 10.1823 under Cmdr George Gosling, for Cork. In 8.1825 under Cmdr John Pakenham, then 10.1826 under Cmdr William Morier, for the Nore; paid off 26.9.1828. Sold to Tibbetts & Co. (for £810) 8.1.1829.

Ontario (ex-*Mohawk*, renamed 9.4.1813) Richard Chapman, Bideford

As built: 100ft 4in, 77ft 6¼in x 30ft 6½in x 12ft 9½in. 384 59/94bm. Draught 6ft 8½in/11ft 1in.

Ord: 2.11.1812. K: 2.1813. L: 26.10.1813. C: 17–23.12.1813 (for Ordinary) at Plymouth.

Fitted for sea at Plymouth 7.1818–25.9.1818.

Commissioned 7.1818 under Cmdr George Gosling (–1820), for Jamaica; destroyed pirate schooner off Cuba 17.12.1819. In 1821 under Cmdr Jodrell Leigh; paid off 12.1821 and laid up at Plymouth. Re-rigged as a ship 1825. Sold to Mr Stone of Grove Yard (for £760) at Plymouth 11.1832.

Belette Edward Larking & William Spong, King's Lynn

As built: 99ft 10in, 77ft 2⅝in x 30ft 8in x 12ft 9in. 386 26/94bm. Draught 7ft 0in/10ft 6in.

Ord: 14.8.1813. K: 10.1813. L: 18.6.1814. C: 8.1818–27.7.1818 at Sheerness (arrived Sheerness 13.7.1814 and laid up; not finished until 1818).

Commissioned 5.1818 under Cmdr George Pechell, for Halifax station; paid off 12.1821. Very Small Repair and fitted for sea at Plymouth 4–12.1822; recommissioned 9.1822 under Cmdr John Leith, for the West Indies; paid off 1827 at Chatham. Sold to Adam Gordon at Chatham (for £1,210) 26.3.1828.

Gannet Edward Larking & William Spong, King's Lynn

As built: 100ft 2in, 77ft 4½in x 30ft 7½in x 12ft 9½in. 386 (exact)bm. Draught 7ft 0in/10ft 6in.

Ord: 14.8.1813. K: 12.1813. L: 13.11.1814. C: 23.11.1814–28.1.1815 at Sheerness (for Ordinary).

First cost: £5,152 to builder, plus fitting £1,726 (in 1814–15).

Fitted for sea at Sheerness (for £6,019) 4.1821–10.8.1821.

Commissioned 7.1821 under Cmdr William Simpson (–1822), for Cork. On 28.9.1824 under Cmdr Francis Brace (–1827), for the Mediterranean; in 10.1827 under Cmdr William Edwards (acting); paid off at Chatham 24.5.1828. Between Small and Middling Repair and altered to ship rig at Plymouth (for £9,328) 1.1830–2.1831; recommissioned 22.11.1830 under Cmdr Mark Halpen Sweny; to North America and West Indies 1831–32. Refitted as a 16-gun sloop at Sheerness (for £4,969) 3–5.1834. Recommissioned 4.3.1834 under Cmdr John Balfour Maxwell (appointed 6.6.1833), for the West Indies; on 26.1.1837 under Cmdr William George Hyndham Whish; paid off 24.2.1838. Sold to Mr Soames at Sheerness (for £1,140) 16.8.1838.

Samarang Portsmouth Dyd. [M/Shipwright Nicholas Diddams]

Ord: 6.9.1815. K: — . Cancelled 30.9.1820 (unstarted).

***PRIMROSE* Class**. Only one non-standard brig in the 32pdr carronade armed class was built for the RN during the Napoleonic War. The following vessel was designed by Henry Peake on considerably slimmer lines than the *Cruizer* Class, but carried an identical armament.

Dimensions & tons: 108ft 0in, 87ft 0in x 28ft 9in x 13ft 6in. $382\frac{47}{94}$bm.

Men: 121. Guns: 16 x 32pdr carronades, plus 2 x 6pdrs (bow).

Primrose Portsmouth Dyd. [M/Shipwright Nicholas Diddams]

As built: 108ft 0in, 87ft 0⅜in x 28ft 9¼in x 13ft 6in. $383\frac{18}{94}$bm. Draught 7ft 7in/10ft 10in.

Ord: 3.5.1809. K: 5.1809. L: 22.1.1810. Completed fitting 1.5.1810. First cost: £11,715.

Commissioned 3.1810 under Cmdr Thomas Burton for the Baltic station. After Burton was made captain on 21.10.1810, *Primrose* was under Cmdr Charles George Rodney Phillott for the rest of the war. On 12.3.1814, she was in action, due to mistaken signalling, against the British packet *Duke of Marlborough* in the Bay of Biscay; on North American station 1814; captured the US privateer *Pike* on 25.8.1814. Post-war, the *Primrose* was fitted for sea service at Woolwich 9–10.1815, and recommissioned 8.1817 for the Jamaica station, but was paid off 19.12.1818 at Plymouth, and underwent a Middling Repair there until 6.1820. Recommissioned 1.1824 under Cmdr John Stoddart, she was re-rigged as a ship sloop at Plymouth from 2–5.1824, and then assigned to the West Indies. Under Cmdr Octavius Vernon from 5.1825, then under Cmdr Thomas Saville Griffenhoofe from 8.1827 on the African station; took slavers *Nuevo Virgin* 28.7.1828, *Zepherina* 14.9.1828, *Vengador* and *Aurelia* (both on) 15.1.1829. Following Griffenhoofe's death on 9.2.1830, under Cmdr William Broughton, took slavers *Maria de Concepcion* 24.3.1830 and 20-gun *Veloz Passagera* on 7.9.1830, losing 3 killed and 13 wounded in latter action. At Plymouth from 4.1831, paid off 12.10.1831. BU was completed there 25.8.1832.

(ii) 24pdr carronade type

The need during the Napoleonic War for a type of brig sloop somewhat smaller than the 382-ton *Cruizer* Class led to the construction of a series of thirty brigs carrying a primary broadside armament of 24pdr slide-mounted carronades (vice the 32pdr carronades in the *Cruizer* Class).

***SEAGULL* Class** The first of the 24pdr-armed brigs were rated as 16-gun vessels. They were built to a William Rule design, approved 4.1.1805. Five vessels were ordered to this design in December 1804, with a further eight in the following summer. Of this class, *Delight*, *Electra*, *Seagull*, *Satellite* and *Skylark* were wartime losses; *Nightingale* was disposed of in 1815, and *Oberon* and *Paulina* in 1816.

Dimensions & tons: 93ft 0in, 76ft 0⅝in x 26ft 5in x 12ft 0in. $282\frac{26}{94}$bm.

Men: 95. Guns: UD 14 x 24pdr carronades (2 more added later), plus 2 x 6pdrs (bow).

Imogen Jabez Bailey, Ipswich

As built: 93ft 0in, 76ft 0⅝in x 26ft 6in x 12ft 0in. $284\frac{8}{94}$bm.

Ord: 12.12.1804. K: 4.1805. L: 11.7.1805. C: 22.7–25.9.1805 at Chatham.

Commissioned 8.1805 under Cmdr Thomas Garth (–1807), for the North Sea; sailed for the Mediterranean 26.6.1807. In 3.1808 under Cmdr William Stephens, then 2.1813 ?C. Taylor (acting), still in Mediterranean. In 11.1813 under Cmdr William Bamber, on the Clyde, then 1815 Lieut John Gilmore; in Ordinary 7.1815. Sold at Plymouth (for £690) to Mr Ismay 3.4.1817.

Savage Robert Adams, Chapel (Southampton)

As built: 93ft 6in, 76ft 6⅜in x 26ft 7½in x 12ft 0in. $288\frac{54}{94}$bm. Draught 7ft 0in/8ft 7in.

Ord: 12.12.1804. K: 4.1805. L: 30.7.1805. C: 9.11.1805 at Portsmouth.

Commissioned 8.1805 under Cmdr James Maurice (–1808), for the Irish station; sailed with convoy for Jamaica 30.8.1807; took 8-gun privateer *Quixote* 13.12.1807. In 1809 under Cmdr William Robilliard, in the Leeward Islands. In 1810 under Cmdr William Ferrie; sailed for Jamaica again 2.7.1810. Repair at Sheerness 9.1811–3.1812; recommissioned 2.1812 under Cmdr William Bissel; sailed with convoy for Quebec 18.5.1812; stranded on Guernsey 20.1.1814 but salved (Bissel dismissed); paid off 2.2.1814. Sold to John Tibbut (for £950) 3.2.1819.

Orestes Jabez Bailey, Ipswich

As built: 93ft 0in, 76ft 0⅝in x 26ft 6in x 12ft 0in. $284\frac{8}{94}$bm. Draught 7ft 8in/8ft 9in.

Ord: 16.7.1805. K: 8.1805. L: 23.10.1805. C: 3.11.1805–11.3.1806 at Chatham.

Commissioned 1.1806 under Cmdr George Poulett. In 10.1806 under Cmdr John Lapenotière (–1811), in North Sea 1807 and on Plymouth station 1808–10; took privateers – 10-gun *La Dorade* off the Lizard 9.5.1810, and 16-gun *Le Loup Garou* in the Channel 27.10.1810. In 8.1811 under Cmdr John Carter, the 10.1811 Cmdr William Richard Smith (–1815), still in Channel. Sold to Thomas Pittman at Chatham (for £710) 6.3.1817.

Julia Jabez Bailey, Ipswich

As built: 93ft 1in, 76ft 1¼in x 26ft 6in x 12ft 0in. $284\frac{26}{94}$bm. Draught 7ft 1in/8ft 11½in.

Ord: 30.7.1805. K: 10.1805. L: 4.2.1806. C: 23.2–17.4.1806 at Chatham.

Commissioned 2.1806 under Cmdr Robert Yarker, for North Sea; sailed for the West Indies 28.6.1807. In 3.1808 under Cmdr John Ellis Watt, then 4.1808 under Cmdr Charles Warde, in the Leeward Islands; took privateer *Le Petit Décidé* 30.8.1808. In 11.1808 under Cmdr Charles Kerr, then 1809 Cmdr William Dowers. In 10.1809 under Cmdr Henry Coxen, still in Leeward Islands, then 1811 Cmdr Valentine Gardner. In Ordinary at Chatham 1813–15. Between Middling and Large Repair and fitted for Ordinary at Chatham 9.1813–3.1814. Fitted for sea at Chatham 8–11.1815; recommissioned 9.1815 under Cmdr John Watling, for St Helena. In 5.1816 under Cmdr Jenkin Jones, at St Helena; wrecked on Tristan da Cunha 2.10.1817 (55 drowned).

Sheldrake Mark Richards, Hythe

As built: 93ft 1¼in, 76ft 2in x 26ft 6in x 12ft 0in. $284\frac{48}{94}$bm. Draught 7ft 4in/9ft 0in.

Ord: 30.8.1805. K: 10.1805. L: 20.3.1806. C: 21.3–28.5.1806 at Portsmouth.

Commissioned 21.4.1806 under Cmdr John Thicknesse (–7.1810), for the Channel Islands; at destruction of 26-gun (*en flûte*) *La Salamandre* near St Malo 12.10.1806; in the Baltic 1809–12. On 7.7.1810 under Cmdr James Pattison Stewart; at defence of Anholt Island 27.3.1811 (took *Gunboat No. 9* and *Lugger No. 1*); took (with consorts) four Danish gunboats 5.7.1811. On 17.2.1812 under Cmdr James Gifford, then 13.8.1812 under Cmdr George Brine, in North Sea and Baltic; paid off 12.8.1815 into Ordinary at Portsmouth. Sold at Chatham to Mr Manclerk (for £700) 6.3.1817.

***FLY* Class** Following closely on the *Seagull* Class, the contemporary design by Sir John Henslow was also originally built as 16-gun vessels, although an additional pair of carronades was added subsequently,

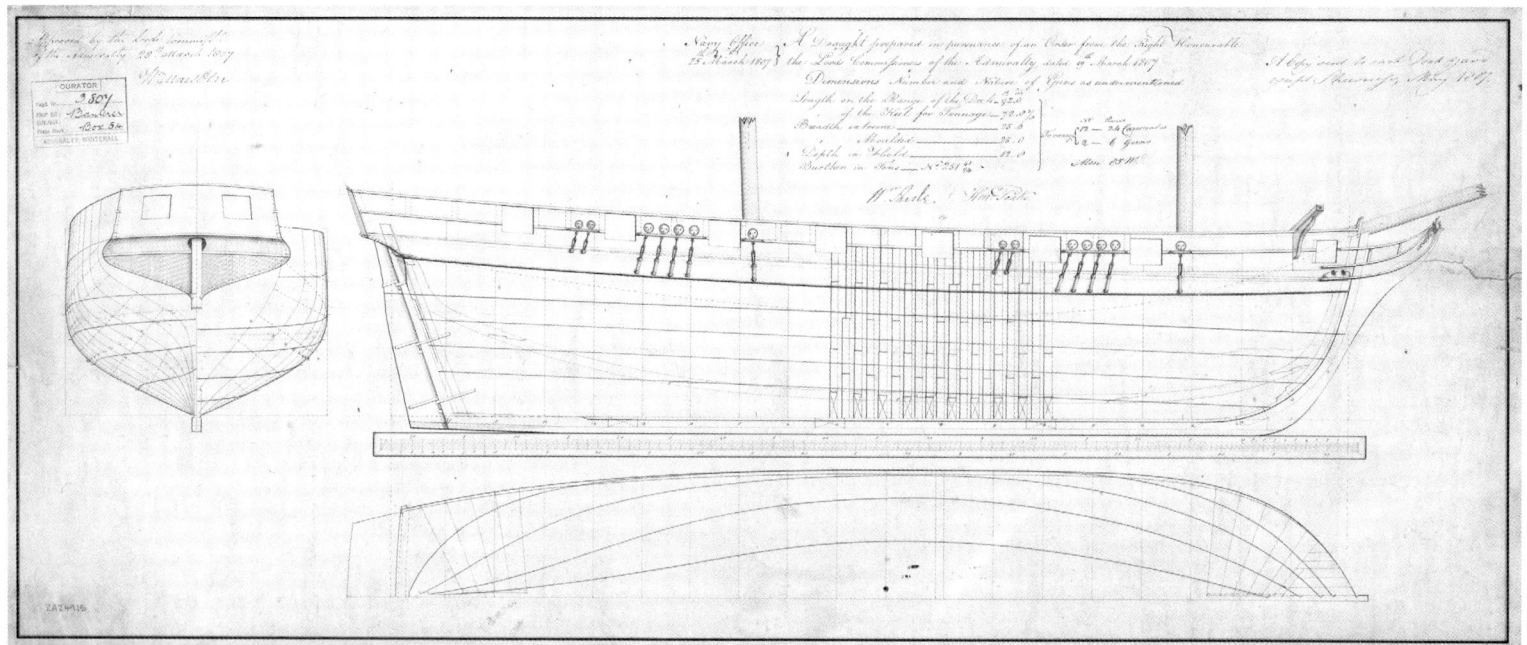

The 25 March 1807 draught for the *Crocus* (or *Banterer*) Class of 14-gun brig sloops. Unlike the larger *Cruiser* class, these ten were all built in the Royal Dockyards. (*NMM J4766*)

raising them to 18 guns. Five vessels had been ordered to this design in January 1805, with a further pair in that summer. However, *Challenger* was taken by the French in 1811, *Fly* was wrecked in 1812 and *Goshawk* in 1813, while *Kite* was sold in 1815 and *Raven*, *Sparrow* and *Wizard* in 1816.

Dimensions & tons: 96ft 0in, 79ft 5in x 25ft 10in x 11ft 6in. 281⁸¹/₉₄bm.

Men: 94. Guns: 14 x 24pdr carronades (2 more added later), plus 2 x 6pdr (bow).

CROCUS Class The only 14-gun brigs to be built for the RN during the Napoleonic Wars, the *Crocus* (sometimes called *Banterer*) Class were designed by the Surveyors of the Navy jointly, approved 28 March 1807. Unusually, all were built in the Royal Dockyards. One was ordered from each dockyard (except Sheerness) on 30 March 1807; four more were ordered in 1808 and a final unit in 1810. The *Crocus* and *Merope* had been sold in 1815, and the *Apelles* and *Prospero* in 1816.

Dimensions & tons: 92ft 0in, 72ft 8³/₈in x 25ft 6in (25ft 0in moulded) x 12ft 8in. 251⁴¹/₉₄bm.

Men: 86. Guns: 12 x 24pdr carronades, plus 2 x 6pdrs (bow).

Podargus Portsmouth Dyd. [M/Shipwright Nicholas Diddams]

As built: 92ft 0in, 72ft 8¼in x 25ft 6¼in x 12ft 8in. 251⁷⁷/₉₄bm. Draught 6ft 1½in/11ft 2½in.

Ord: 30.3.1807. K: 11.1807. L: 26.5.1808. Completed 9.8.1808. First cost: £7,394 for building plus £1,070 for fitting.

Commissioned 9.1808 under Cmdr William Hellard and stationed in the Downs. Recommissioned in 11.1810 under Cmdr John Lloyd, then under Cmdr John Bradley from 10.1811. Recommissioned in 11.1811 under Cmdr William Robilliard for Baltic service, where on 6.7.1812 she participated in the destruction of the Danish 40-gun *Nayaden* off Mardö, Norway, and the capture (temporarily) of the brigs *Laaland*, *Samsö* and *Kiel*. Recommissioned in ?1.1813 under Cmdr George Rennie, she was involved in the Bordeaux operations in 1814; under Cmdr Houston Stewart (temp.) 6–8.1814. Recommissioned

13.8.1814 under Cmdr James Wallis, she was stationed at St. Helena; on 2.8.1817 under Lieut (Acting, Cmdr 26.11.1817) Henry John Rous, still at St. Helena, in 1.1818 under Lieut James Webb Cairnes; paid off 21.4.1820 and laid up in Ordinary at Portsmouth. Sold there to John Small Sedger, Rotherhithe for £510 on 7.8.1833 to BU.

Portia Deptford Dyd. [M/Shipwright Robert Nelson]

As built: 92ft 0in, 72ft 8in x 25ft 7in x 12ft 8in. 251⁷⁷/₉₄bm. Draught 5ft 10in/10ft 8in.

Ord. 30.3.1807. K. 12.1809. L. 30.8.1810. Completed 18.10.1810.

Commissioned 9.1810 under Cmdr Joseph Symes for service in the North Sea. Recommissioned 8.1811 under Cmdr Henry Thompson (–1815.), but in ?8.1813 under Lieut William Adams (temp.). Recommissioned in 4.1815 under Cmdr Silas Hood for Halifax station; in 5.1816 under Cmdr John Wilson, still on Halifax Station, but home in 7.1816 and paid off 8.1816. Sold to Mr Marclark for £800 6.3.1817 to BU.

Muros Chatham Dyd [M/Shipwright Robert Seppings]

As built: 92ft 1in, 72ft 9½in x 25ft 6in x 12ft 8in. 251⁷⁷/₉₄bm. Draught 6ft 10in/8ft 11in.

Ord. 3.5.1808. K. 6.1808. L. 23.10.1809. Completed 13.12.1809.

Commissioned 10.1809 under Cmdr Clement Sneyd. Stranded off Jersey in a gale and paid off 1.1810. Underwent small repair at Portsmouth 11.1811 to 2.1812. Recommissioned 12.1811 under Cmdr James Aberdour; sailed for Newfoundland 24.6.1812. Recommissioned 12.1812 under Cmdr Charles Hobart (died 20.7.1813); sailed for Newfoundland 17.5.1813. Under Lieut Thomas Saville Griffinhoofe (Cmdr 20.6.1813) on South America Station. Recommissioned 4.1815 under Cmdr George Gosling. Laid up at Deptford 6.1816. At Woolwich 1817, then Deptford 1820–22. Sold to Thomas Pitman for £910 (?at Deptford) 18.4.1822 to BU.

Zephyr Portsmouth Dyd [M/Shipwright Nicholas Diddams]

As built: 92ft 5in, 73ft 2in x 25ft 6in x 12ft 9in. 253⁵/₉₄bm. Draught 6ft 4in/10ft 7in.

Ord. 9.6.1808. K. 10.1808. L. 29.4.1809. Completed 17.8.1809.

Commissioned 6.1809 under Cmdr Francis George Dickins for the Downs and Channel Station. Recommissioned 8.1811 under

Cmdr Thomas Cuthbert Hichens; took 16-gun *La Victoire* off Dieppe 11.12.1811; took 10-gun US privateer schooner *Antelope* 10.12.1812; sailed with convoy to Mediterranean 22.4.1813. Recommissioned 7.1814 under Cmdr Richard Creyke. Recommissioned 12.1814 under Cmdr George Frederick Rich; sailed 1815 to St. Helena; paid off 6.1816. Sold to Thomas Pitman for £820 (?at Plymouth) 29.1.1818 to BU.

Banterer Woolwich Dyd [M/Shipwright Edward Sison]
As built: 92ft 0in, 72ft 8¾in x 25ft 6in x 12ft 9in. $251\frac{41}{94}$bm. Draught 6ft 2in/10ft 11in.
Ord. 19.9.1808. K. 12.1809. L: 2.6.1810. Completed 5.7.1810.
Some of her floor timbers, futtocks and top timbers were formed from Holstein oak timber.
Commissioned 6.1810 under Charles Warde for North Sea service. Sold to Gordon & Co. for £850 (?at Deptford) 6.3.1817 to BU.

Wolf Woolwich Dyd [M/Shipwright Edward Sison]
As built: 92ft 0in, 72ft 6⅝in x 25ft 7in x 12ft 8½in. $252\frac{54}{94}$bm. Draught 6ft 3in/11ft 2in.
Ord. 8.8.1810. K. 8.1812. L: 16.9.1814. Not completed in time for end of war. Coppered and sailed 22.9.1814 to Sheerness. Fitted for sea 27.2.1819 and sailed from Sheerness.
Commissioned 5.12.1818 under Cmdr Bernard Yeoman, for the Cork station; paid off 1.1822. Underwent very small repair at Plymouth 4–5.1822. Sold to Thomas S. Benson for £3,100 on 27.1.1825.

(iii) 18pdr carronade type

***CHEROKEE* Class** The only wartime brigs to mount 18pdr carronades on their broadside (other than captured vessels and the one-off vessel *Icarus*) were the numerous but infamous 'coffin brigs'. The contracts all describe these vessels as 'brigantines' rather than brigs. Approval to build the first four vessels was issued by Grenville's Board in early 1807, but the design by Henry Peake was not approved until eight months later on 26 November (by Mulgrave's Board), when the order for these first four was re-affirmed, along with a fresh order for the next pair; none were laid down on the stocks until December 1807. Another twenty were ordered five weeks later, and a final eight during 1808. Of the latter thirty, *Prince Arthur* had been sold to the Sultan of Morocco in 1808 (prior to completion), and six others wrecked in service before 1817: *Achates* and *Wild Boar* in 1810, *Ephira* in 1811, *Sarpedon* and *Rhodian* in 1813, and *Bermuda* in December 1816.

A further two teak-built vessels – *Sphinx* and *Cameleon* – were ordered from Bombay to this design by Melville's Second Board in 1812; neither was completed for service during the War. Another 79 vessels were to be ordered to this design post-war – see Section B – to bring the total number ordered to 115 vessels, the largest group of sailing warships built to a single design. Note that eleven of the names of early *Cherokee* Class brigs were re-used for post-war vessels of the same class after the originals had been lost or disposed of.
Dimensions & tons: 90ft 0in, 73ft 7⅝in x 24ft 6in x 11ft 0in. $235\frac{9}{94}$bm.
Men: 75. Guns: 8 x 18pdr carronades, plus 2 x 6pdrs (bow).

Cherokee (John) Perry, Wells & Green, Blackwall
As built: 90ft 1⅝in, 73ft 8⅝in x 24ft 7in x 11ft 0in. $237\frac{38}{94}$bm. Draught 6ft 6in/9ft 2in.
Ord: 30.3.1807. K: 12.1807. L: 24.2.1808. Coppered by builder to 10.3.1808. C: 10.3–10.7.1808 at Woolwich.
Commissioned 12.4.1808 under Cmdr Richard Arthur for the Downs station; captured 16-gun privateer *L'Aimable Nelly* in Dieppe Harbour 11.1.1810. Arthur posted Captain for this action, and thus from 1.1810 under Cmdr William Ramage for same station. Paid off 1815 after service on North Sea station, but recommissioned (still under Ramage) for service on Leith station; on 26.2.1819 under Cmdr Theobald Jones, on same station; on 7.8.1822 under Cmdr William Keith, for service at Cork; paid off at Deptford 11.3.1826. Sold to Joshua Crystall (?at Deptford) for £610 on 26.3.1828.

Cadmus John Dudman & Co., Deptford
As built: 90ft 3in, 74ft 2⅝in x 24ft 6in x 11ft 0in. $236\frac{90}{94}$bm. Draught 6ft 6in/9ft 3in.
Ord: 30.3.1807. K: 12.1807. L: 26.2.1808. C: 16.3–21.5.1808 at Deptford Dyd.
First cost: £3,912 for building, plus £3,492 for fitting.
Commissioned 3.1808 for Channel station under Cmdr Delamore Wynter (died 10.4.1810), then under Cmdr Thomas Fife; paid off into Ordinary 12.1818. Recommissioned 12.1813 under Cmdr Watkin Evans for North Sea station, then from 7.6.1814 under Cmdr John Gedge for Downs station; paid off 12.1818. Post-war at Chatham 1819–25, then recommissioned 4.1826 under Cmdr Charles Hallowell for South America; on 19.1.1828 under Cmdr Sir Thomas Raikes Trigge Thompson; paid off 7.5.1830 at Portsmouth. Fitted as coastguard watch vessel for Whitstable 2.1835, serving in this role in the Swale until 1860. Renamed *CGWV.24* on 25.5.1863, but sold to Wm Lethbridge for £500 on 12.3.1864 to BU.

Leveret (i) (John) Perry, Wells & Green, Blackwall
As built: 91ft 2in, 73ft 9in x 24ft 7in x 11ft 0in. $237\frac{7}{94}$bm.
Ord: 30.3.1807. K: 12.1807. L: 24.2.1808. C: 7.3–4.5.1808 at Woolwich.
Commissioned 3.1808 under Cmdr Robert Evans, for the Baltic; in 10.1808 under Cmdr Benjamin Crispin, still in Baltic. In 1810 under Cmdr John Worth, then 10.1810 under Cmdr George Willes, in North Sea; took privateers – 3-gun *Le Prospère* 22.8.1811, 14-gun *Le Dunkerquois* in the North Sea 10.11.1811, and 4-gun *Le Brave* in the North Sea 4.7.1812; her boats (with those of *Osprey* and *Britomart*) chased 5-gun privateer *L'Eole* off Heligoland 16.7.1812 (she was taken by the others' boats). In 12.1813 under Cmdr Jonathan Christian. Recommissioned 9.1815 under Cmdr John Theed, for Cape of Good Hope. In 5.1818 under Cmdr Rodney Shannon, for St Helena station. Sold at Portsmouth to Thomas Pitman (for £710) 18.4.1822.

Rolla (i) Thomas Pitcher, Northfleet
As built: 90ft 4in, 74ft 0in x 24ft 7in x 11ft 0in. 238bm. Draught 6ft 7in/9ft 5in.
Ord: 30.3.1807. K: 12.1807. L: 13.2.1808. C: 5.1808 at builder.
Commissioned 3.1808 under Cmdr John Hardy Godby, for the Downs. In 9.1808 under Cmdr Samuel Clarke (–1812); took 16-gun privateer *L'Espoir* in the Channel 6.10.1811. In 10.1812 under Cmdr William Hall, then 6.1814 Cmdr Robert Julyan; laid up at Deptford 6.1816. Sold at Deptford to Thomas Pitman (for £730) 18.4.1822 (still afloat as mercantile until caught fire and burnt in Bonny River [Nigeria] 6.7.1836).

Parthian William Barnard, Deptford
As built: 90ft 3in, 74ft 1¾in x 24ft 6½in x 11ft 0in. $238\frac{33}{94}$bm. Draught 5ft 10in/9ft 5in.
Ord: 26.11.1807. K: 12.1807. L: 13.2.1808. (Coppered by builder to 27.2.1808) C: 5.6.1808 at Deptford Dyd.
Commissioned 3.1808 under Cmdr John Balderston (murdered 12.12.1808). In 1809 under Cmdr Richard Harward; took 14-gun privateer *La Nouvelle Gironde* 5.5.1809. In 9.1809 under Cmdr Henry Dawson, on Texel station. In ?8.1811 under

Cmdr James Tomkinson, then 2.1812 Cmdr James Garrety; laid up at Portsmouth 11.1813. Between Middling and Large Repair at Portsmouth 6.1817–1.1818; fitted for sea 7–10.1818; recommissioned 8.1818 under Cmdr Wilson Biggland, for Jamaica station. Recommissioned 8.1820 at Portsmouth under Cmdr Whitworth Lloyd, for Jamaica, then 2.1823 Cmdr George Barrington, at Plymouth and the Nore. Fitted for sea at Deptford 3–6.1826; in 3.1826 under Cmdr Henry Martin, for the Mediterraneam; on 4.1827 under Cmdr George Hotham; wrecked off Alexandria 15.5.1828.

Briseis (i) John King, Upnor
  As built: 90ft 3in, 73ft 9¾in x 24ft 7½in x 11ft 0in. 238⅞/94bm. Draught 6ft 5½in/9ft 0in.
  Ord: 31.12.1807. K: 2.1808. L: 19.5.1808. C: 30.8.1808 at Chatham.
  Commissioned 6.1808 under Cmdr Robert Pettet; her boats (with *Bruiser*'s) took 1-gun *Comen* in the North Sea 5.1809; operations in the River Elbe 7.1809. In 9.1809 under Cmdr John Adye; took 4-gun Danish privateer *Recipricite* off Heligoland 6.11.1809. In 10.1810 under Lieut George Bentham (acting); took 14-gun privateer *Sans-Souci* in the North Sea 14.10.1810. In 12.1810 under Cmdr Charles T. Smith, then 3.1812 Cmdr John Ross; boats at Pillau 28.6.1812; took 4-gun privateer *Le Petit Poulet* 9.10.1812. In 6.1812 under Lieut (Cmdr, 10.1813) William Rush Jackson (–1815), in the North Sea. In 9.1815 under Cmdr George Dommett; wrecked on Point Pedras, Cuba 5.11.1818.

Jasper Jabez Bailey, Ipswich
  As built: 90ft 0in, 73ft 7in x 24ft 7in x 11ft 0in. 236⅝/94bm. Draught 6ft 6in/8ft 11in.
  Ord: 31.12.1807. K: 2.1808. L: 27.5.1808. C: 18.6–20.8.1808 at Sheerness.
  Commissioned 6.1808 under Cmdr William W. Daniel; sailed for Portugal 3.5.1809. In 11.1810 under Cmdr Thomas Hunloke (died 12.1811), then 12.1811 under Cmdr John Everleigh; sailed for Portugal 9.6.1812. In 8.1812 under Cmdr Henry Jenkinson, then 6.1814 under Cmdr Thomas Carew; wrecked in a gale Plymouth Sound 20.1.1817 (65 drowned, only 2 saved – excluding Carew and his 1st Lieut, both ashore).

Onyx Jabez Bailey, Ipswich
  As built: 90ft 0in, 73ft 7in x 24ft 7in x 11ft 0in. 236⅝/94bm. Draught 6ft 6in/8ft 8in.
  Ord: 31.12.1807. K: 2.1808. L: 8.7.1808. C: 15.7–18.10.1808 at Sheerness.
  Commissioned 9.1808 under Cmdr Charles Gill; recaptured brig *Manly* in the North Sea 1.1.1809. In 2.1809 under Cmdr John Parish, then 8.1810 under Cmdr Gawen Hamilton; sailed for the Mediterranean 30.8.1810. In ?3.1811 under Cmdr William Carrol, then 11.1811 under Cmdr Charles Squire, then ?10.1812 Cmdr Charles Phillips; in 1812–13 at Lisbon. In 12.1812 under Cmdr Smith Cobb; to Jamaica 1814; laid up at Sheerness 7.1816. Sold to Thomas Pittman at Sheerness (for £950) 3.2.1819.

Badger John King, Upnor
  As built: 90ft 0in, 73ft 6⅞in x 24ft 9in x 11ft 0in. 239⁶⁸/94bm. Draught 6ft 0in/9ft 0in.
  Ord: 31.12.1807. K: 2.1808. L: 23.7.1808. C: 17.12.1808 at Chatham.
  First cost: £3,375 to builder, plus £3,725 fitting.
  Commissioned 9.1808 under Cmdr John Lampen Manley (–1812) for North Sea and Downs; in Scheldt operations 1809; sailed for the Mediterranean 20.11.1811; destroyed 11-gun privateer *La CSS d'Emerieau* 2.1812. In 1813 under Cmdr Charles Hole; took 2-gun privateer *L'Aventure* off Minorca 30.10.1813. In 4.1814 under Cmdr Samuel Dickins, for the Downs. In 12.1815 under

Cmdr Charles Bridgeman; laid up at Sheerness 8.1816. Between Middling and Large Repair at Chatham (for £5,035) 1.1820–2.1821; roofed over at Chatham 4.1821 and laid up. Fitted for sea at Chatham (for £3,867) 12.1815–4.1826; recommissioned 12.1825 under Cmdr Charles Crowdy (–1825), for North Sea, then 1826 at Cork, 1828 at Lisbon. In 1.1829 under Cmdr Richard Freeman Rowley, at the Nore, then to Cape of Good Hope; in 2.1830 under Cmdr G. F. Stow (–1833), still at the Cape, then to East Indies 1832; paid off 2.9.1833 at the Cape. Laid up as mooring (?receiving) ship at the Cape 1834–1860, then beached there 22.3.1860 and BU 1864.

Opossum Edward & Jonathan Muddle, Gillingham
  As built: 90ft 2½in, 73ft 7½in x 24ft 7½in x 11ft 0in. 237⁴⁴/94bm. Draught 5ft 10in/8ft 11in.
  Ord: 31.12.1807. K: 3.1808. L: 9.7.1808. C: 15.3.1809 at Chatham.
  Commissioned 11.1808 under Cmdr William Byam (–1810); sailed with West Indies convoy 26.3.1809. In 1.1811 under Cmdr Thomas Wolrige (–1814), in the Leeward Islands. In 3.1815 under Cmdr Lord John Hay (–1817), on the Halifax station. Sold to George Bailey at Portsmouth (for £900) 3.2.1819.

Rinaldo John Dudman & Co., Deptford
  As built: 90ft 3in, 73ft 10⅝in x 24ft 6½in x 11ft 0in. 236⁶⁶/94bm. Draught 6ft 2in/9ft 0in.
  Ord: 31.12.1807. K: 3.1808. L: 13.7.1808. C: 27.7–7.9.1808 at Deptford.
  First cost: £3,695 to builder, plus £3,386 fitting at Deptford.
  Commissioned 9.1808 under Cmdr James Irwin, for the Downs. In 11.1809 under Cmdr James Anderson; took 14-gun privateer *Le Maraudeur* off Dover 7.12.1810; action with four privateer luggers off the Owers, with 16-gun *La Vielle Josephine* sunk 17.12.1810; in action (with *Royalist*) against vessels of the Boulogne Flotilla 3.9.1811; in action (with *Naiad, Castilian, Redpole* and *Viper*) during Napoleon's visit to Boulogne, with 12-gun *Le Vicomte de Lyon* sunk 21.9.1811. In 2.1812 under Cmdr Sir William George Parker, then 4.1812 under Cmdr Edmund Lyons; recaptured (with *Castillian, Bermuda* and *Phipps*) 14-gun *Apelles* off Étaples 4.5.1812. In 6.1814 Cmdr Archibald Tisdall, for Jamaica station. In 7.1815 under Cmdr Thomas Wren Carter; laid up at Sheerness 10.1815. Middling Repair and fitted for Ordinary at Chatham (for £5,596) 3.1819–5.1820. Fitted as packet at Chatham (for £9,099) 11.1823–2.1824; recommissioned as Falmouth packet 11.1823 under Lieut John Arthur Moore, then 3.1829 under Lieut John Hill; paid off 8.1834. Sold at Plymouth (for £610) 6.8.1835.

Chanticleer Daniel List, East Cowes
  As built: 90ft 3in, 74ft 0½in x 24ft 7in x 11ft 0in. 237bm. Draught 6ft 0in/9ft 0in.
  Ord: 31.12.1807. K: 3.1808. L: 26.7.1808. C: 11.8–5.10.1808.
  First cost: £3,254 to builder, plus £3,816 fitting.
  Commissioned 9.1808 under Cmdr Charles Harford (drowned 19.10.1808), for North Sea; in 11.1808 under Cmdr Richard Spear (–1812). In 3.1813 under Cmdr Stewart Blacker; by 1814 under (now Capt.) Spear again. In 7.1814 under Capt. John Thompson, then same month Lieut George Tupman (acting), then 8.1814 Cmdr William Dickson; laid up at Sheerness 7.1816; at Chatham 1817. Middling Repair at Chatham (for £4,910) 12.1818–3.1820. Fitted for sea at Chatham (for £3,361) 1–2.1821; recommissioned 10.1821 under Cmdr Henry Eden, for the Mediterranean. In 7.1822 under Cmdr Burton Macnamara, then Cmdr Charles J. H. Johnstone in 9.1824 and Cmdr John Maxwell in 4.1827. Fitted as Survey vessel (for £6,526) at Portsmouth 12.1827–4.1828;

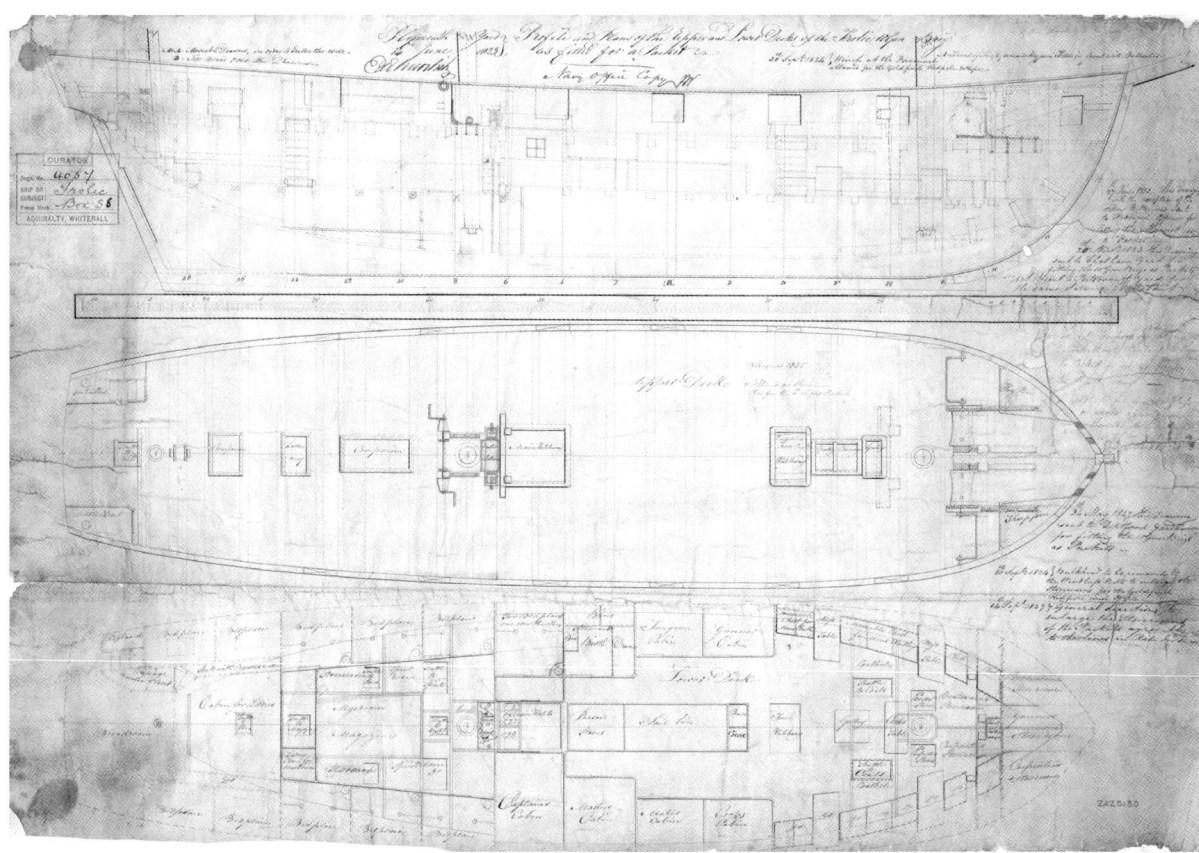

Following the Napoleonic Wars, some of the surviving *Cherokee* Class brig sloops were converted to packets for transporting mails and dispatches (as well as a few passengers), and many of the post-war vessels built to the same design were similarly fitted out. This draught, signed by Edward Churchill (master shipwright at Plymouth from 1815 to 1830) and dated 14 June 1823, is for the fitting of the *Frolic* to be carried out there. It shows the inboard profile, upper deck, lower deck and platforms. Churchill also used this draught for fitting the wartime *Redpole* (of 1808) as a packet, plus the new *Hope*. Copies of the draught were sent to Chatham (to convert the *Goldfinch* of 1808, and for the completion of the *Weazle*, *Procris*, *Espoir*, *Fairy*, *Harpy*, *Hearty*, *Hyaena* and *Lapwing*), to Woolwich (to complete the *Magnet*), to Deptford (for the *Briseis* and *Calypso*) and to Portsmouth (for the *Musquito*, *Myrtle*, *Rapid* and *Recruit*). Other brigs not listed on the draught were similarly adapted, and some of those named above were not modified. *(NMM J0161)*

in 12.1827 under Cmdr Henry Foster (–1831), for survey and exploration in South Atlantic. Lent to Royal Sailing Society Thames 1831–32. Delivered to Customs as watch vessel by AO 30.1.1833; renamed *CGWV No. 5* on 25.5.1863, based on River Crouch; replaced by *Kangaroo* 27.6.1870 and laid up; BU (by AO 9.11.1870) completed 3.6.1871 at Sheerness.

Goldfinch John Warwick, Eling
   As built: 90ft 0in, 73ft 7⅛in x 24ft 8in x 11ft 0in. 237bm. Draught 6ft 0in/9ft 2in.
   Ord: 31.12.1807. K: 3.1808. L: 8.8.1808. C: 27.8–26.12.1808 at Portsmouth.
   First cost: £? To builder, plus fitting £3,076 + £1,079.
   Commissioned 8.1808 under Cmdr Fitzowen Skinner, for the Channel; in action against 16-gun privateer *La Mouche* 17.5.1809. In 10.1809 under Cmdr Arden Adderley, then 11.1810 Cmdr Edmund Waller and 10.1814 Cmdr John Foote; paid off and laid up at Sheerness 12.1815. Between Small and Middling Repair at Chatham (for £4,546) 4.1819–6.1820. Fitted as packet at Chatham 6–11.1824; recommissioned 9.1824 under Lieut John Walkie (–1831), for packet service at Falmouth; in 7.1832 under Lieut Edward Collier (–1838). Sold to R. Willis (for £718.8.0d) 8.11.1838.

Woodlark William Row, Newcastle
   As built: 90ft 0in, 73ft 6¾in x 24ft 7½in x 10ft 10½in. 237²⁶/₉₄bm. Draught 6ft 0in/8ft 6in.
   Ord: 31.12.1807. K: 3.1808. L: 17.11.1808. C: 23.2–2.10.1809 at Chatham.
   Commissioned 5.1809 under Cmdr George Edward Watts (–1813); for the Baltic 1810–12. In 1813 under Cmdr George Anson Byron, later under Cmdr Robert Balfour; in 6.1814 under Cmdr William Cutfield; paid off 9.7.1816. Sold to Mr Grant at

Chatham (for £710) 29.1.1818.

Shearwater William Row, Newcastle
   As built: 90ft 0in, 73ft 6¾in x 24ft 7½in x 11ft 0in. 237²⁶/₉₄bm. Draught 6ft 0in/8ft 6in.
   Ord: 31.12.1807. K: 3.1808. L: 21.11.1808. C: 23.2–1.9.1809 at Chatham.
   First cost: £2,911 to builder, plus £4,371 fitting.
   Commissioned 5.1809 under Cmdr Edward Sibly; sailed for the Mediterranean 27.1.1810. In 1.1811 under Cmdr George William Hughes d'Aeth (acting), then under Cmdr William R. Smith (–1813), in the Mediterranean. In 7.1814 under Cmdr John Coffin; paid off into Ordinary. Middling to Large Repair at Sheerness (for £6,511) 8.1814–11.1815. Fitted for sea at Sheerness (for £3,311) 2–3.1817; recommissioned 1.1817 under Cmdr Edward Rowley, for Jamaica. In 7.1817 under Cmdr Douglas Cox, then 4.1820 under Cmdr John Walter Roberts, for Cape of Good Hope; laid up at Portsmouth 1.1822. Sold to Mr Beatson, Rotherhithe (for £400) 11.1832.

Calliope John Dudman & Co., Deptford
   As built: 90ft 2in, 73ft 9⅛in x 24ft 7in x 11ft 0in. 237²³/₉₄bm. Draught 6ft 5½in/9ft 3in.
   Ord: 31.12.1807. K: 4.1808. L: 8.7.1808. C: 8.9.1808 at Deptford Dyd.
   Commissioned 9.1808 under Cmdr John M'Kerlie (–1813), for the North Sea; took 14-gun privateer *La Comtesse d'Hambourg* in the North Sea 25.10.1810. In 12.1813 under Cmdr John Codd, still in North Sea. In 4.1815 under Cmdr Henry Thompson, then Cmdr Alexander Maconochie in 9.1815; laid up at Portsmouth 9.1815. Middling Repair at Portsmouth 7–12.1820. Fitted at Portsmouth as a Tender to *Apollo* 4–6.1822. Tender to yacht *Royal George* 1825; in 1827 under Lieut John Powney. BU completed there 13.8.1829.

Hope (i) Jabez Bailey, Ipswich
As built: 90ft 0in, 73ft 7in x 24ft 7in x 11ft 0in. $236\frac{50}{94}$bm. Draught 6ft 6in/8ft 9in.
Ord: 31.12.1807. K: 4.1808. L: 22.7.1808. C: 5.8–5.11.1808 at Sheerness.
Commissioned 9.1808 under Cmdr Joseph Pearce, for the Channel. In 11.1810 under Cmdr Edward William Garrett (–1813); took US 6-gun privateer *Lewis* off Halifax 14.8.1812; took a small privateer 24.6.1813. In 7.1813 under Cmdr Edward Saurin, then Cmdr Henry Jauncey in 6.1814. Fitted for Channel service at Woolwich 8–11.1815; paid off 2.10.1818. Sold to Thomas Pittman to BU (for £1,030) 3.2.1819.

Britomart John Dudman & Co., Deptford
As built: 90ft 3in, 73ft $10\frac{5}{8}$in x 24ft 8in x 11ft 0in. 238 ($239\frac{11}{94}$ by calc.)bm. Draught 6ft 4in/9ft 3in.
Ord: 31.12.1807. K: 4.1808. L: 28.7.1808. C: 11.9.1808 at Deptford Dyd.
Commissioned 8.1808 under Cmdr William Buckley Hunt (–1813); in Scheldt operations 1809; her boats (with those of *Desiree*, *Quebec*, and *Bold*) destroyed 6-gun privateer and took two others in the Vlie 29.5.1810; took (with *Osprey* and *Leveret*) 5-gun privateer *L'Éole* off Heligoland 16.7.1812. In 12.1813 under Cmdr Robert Riddell (–1816); on Irish station 1814; at bombardment of Algiers 27.8.1816. In 10.1816 under Cmdr Constantine Moorsom, then 11.1816 under Cmdr George Perceval (–1817), then 1818 (nominally, never actually took command) under Cmdr Bernard Yeoman; paid off 12.10.1818. Sold to G. Bailey at Plymouth (for £900) 3.2.1819.

Cordelia John King, Upnor
As built: 90ft 3in, 73ft $9\frac{1}{2}$in x 24ft 8in x 11ft 0in. $238\frac{77}{94}$bm. Draught 6ft 6in/9ft 1in.
Ord: 31.12.1807. K: 5.1808. L: 26.7.1808. C: 17.11.1808 at Chatham.
Commissioned 9.1808 under Cmdr Thomas Kennedy (–1813), for the North Sea. In 11.1813 under Cmdr Henry Fraser, then 5.1814 under Cmdr William Sargent (–1816); at bombardment of Algiers 27.8.1816. In 9.1816 under Cmdr William Popham; laid up at Sheerness 10.1816. Between Middling and Large Repair at Chatham 4.1820–10.1821, then laid up there. Fitted for sea at Chatham 11.1826–4.1827; in 11.1827 under Cmdr George Mildmay, then 16.8.1828 under Cmdr Courtenay Edmund William Boyle, for the Mediterranean; paid off 1829. On 17.3.1830 under Cmdr Charles Hotham (–1833). Sold to Mr Nixon at Chatham (for £710) 12.12.1833.

Redpole Robert Guillaume, Northam
As built: 89ft 11in, 73ft $8\frac{1}{2}$in x 24ft 8in x 10ft $10\frac{1}{2}$in. $238\frac{52}{94}$bm. Draught 6ft 6in/9ft 3in.
Ord: 31.12.1807. K: 5.1808. L: 29.7.1808. C: 5.8–12.10.1808 at Portsmouth.
Commissioned 9.1808 under Cmdr John Joyce, for the Downs; convoyed fireships to Basque Roads 1809, joining Gambier's fleet 10.4.1809; later to North Sea. In 9.1809 under Cmdr Colin Macdonald; took 16-gun privateer *Le Grand Rôdeur* 10.12.1809; attacked (with *Rinaldo*) vessels of the Boulogne Flotilla 3.9.1811; action (with *Naiad*, *Rinaldo*, *Castilian* and *Viper*) vessels of Boulogne Flotilla 21.9.1811 (taking 12-gun *Le Vicomte de Lyon*). In 10.1812 under Cmdr Alexander Fraser, in the Downs, then 6.1814 under Cmdr Edward Denman (–1815); accompanied *Northumberland* to St Helena 1815. Middling Repair at Plymouth 6.1816–5.1817. Fitted for foreign service at Plymouth 9–11.1817; recommissioned 9.1817 under Cmdr James Pasley, for St Helena. In 10.1819 under Cmdr William Evance, at Portsmouth; then 10.11.1820 under Cmdr Patrick Duff Henry Hay, for the Mediterranean; in 12.1821 under Cmdr Richard Anderson. Fitted as packet at Plymouth 7.1824–2.1825; recommissioned 1.1825 under John Bullock, Master, for Falmouth packet service. Sailed from Rio 10.8.1828 for UK; lost, believed sunk with all hands in action with pirate vessel *Congress* off Cape Frio (Brazil) 8.1828.

Helicon John King, Upnor
As built: 90ft 2in, 74ft 0in x 24ft 7in x 11ft 0in. $237\frac{82}{94}$bm. Draught 6ft 5in/9ft 2in.
Ord: 31.12.1807. K: 5.1808. L: 8.8.1808. C: 21.11.1808 at Chatham.
Commissioned 9.1808 under Cmdr Spelman Swaine, for the Downs. In 6.1810 under Cmdr ?John Campbell, then ?7.1810 Cmdr Harvey Hopkins (–1814), for the Channel; took privateer *Zulma* off Île Bas 25.6.1812; took 14-gun privateer *Le Revenant* off the Eddystone 22.12.1813. In 6.1814 under Cmdr Andrew Mitchell, then 10.1816 under Cmdr Alexander Branch, at Cork. Middling Repair and fitted for sea at Plymouth 12.1817–4.1819. In 1819 under Cmdr William Pettman, then 5.1821 under Cmdr William Dawkins (died 9.1824), to the West Indies; in 9.1824 under Cmdr Thomas Furber, then 7.1825 under Cmdr Charles Dyke Acland, on Cape of Good Hope station, then 1.1828 Cmdr R. H. Stanhope, and finally Cmdr C. Talbot in 5.1829; paid off (as defective) 6.1829. BU at Sheerness 7.1829.

Lyra John Dudman & Co., Deptford
As built: 90ft $6\frac{5}{8}$in, 74ft $1\frac{5}{8}$in x 24ft 8in x 11ft 0in. $239\frac{88}{94}$bm. Draught 6ft 1in/9ft 0in.
Ord: 31.12.1807. K: 5.1808. L: 22.8.1808. C: 3.9–5.10.1808 at Deptford Dyd.
Commissioned 9.1808 under Cmdr William Bevians, for the Channel; guide vessel for fireship attack on the Basque Roads 11.4.1809. In 1.1810 under Cmdr Robert Bloye; operations on North coast of Spain 1812–13. In 1813 under Cmdr John Campbell (acting); took (with *Surveillante*) US 6-gun privateer *Tom* 27.4.1813. In 9.1813 under Cmdr Dowell O'Reilly; passage of the Adour River 23.2.1814. In 10.1815 under Cmdr Basil Hall; fitted at Deptford for foreign service 11–12.1815; sailed 9.2.1816 with Lord Amherst's mission to China; home 10.1817 and paid off 13.11.1817 at Portsmouth. Sold to Thomas Pittman at ?Portsmouth to BU (for £920) 11.7.1818.

Beaver Jabez Bailey, Ipswich
As built: 89ft 7in, 73ft $1\frac{1}{2}$in x 24ft $7\frac{1}{2}$in x 11ft 1in. $235\frac{81}{94}$bm. Draught 6ft 9in/8ft 1in.
Ord: 4.1808. K: 10.1808. L: 16.2.1809. C: 23.2–10.9.1809 at Sheerness (including coppering).
Commissioned 9.1809 under Cmdr Edward O'Brien Drury (–1815), for the North Sea. Middling Repair at Sheerness 1.1815–3.1817; recommissioned 1.1817 under Cmdr Norwich Duff, for Jamaica station; in 9.1817 under Cmdr Robert Felix, then 5.1819 under Cmdr Richard Saumarez. In 6.1820 under Cmdr Frederick Marryat, on St Helena station, then Cmdr Archibald McLean 5.1821 off South America, then Cmdr Thomas Bourchier in 9.1822. Small Repair and fitted for foreign service at Chatham 5–9.1824. Recommissioned 22.7.1824 under Cmdr John James Onslow, on Jamaica station, then Cmdr Sir George Young in 10.1825 and Cmdr Joseph O'Brien in 5.1826; paid off 1827 at Portsmouth. Sold to Joshua Crystall at Portsmouth (for £580) 24.6.1829.

Drake Jabez Bailey, Ipswich
As built: 89ft 3in, 72ft $10\frac{1}{2}$in x 24ft $7\frac{1}{2}$in x 11ft 0in. $235\frac{5}{94}$bm. Draught 6ft 10in/8ft 11in.

Ord: 27.6.1808. K: 8.1808. L: 3.11.1808. C: 6.11.1808–7.3.1809 at
Sheerness.

Commissioned 12.1808 under Cmdr Eyles Mounsher, for the North
Sea; destroyed a privateer schooner off Camperdown 7.3.1810;
took 18-gun privateer *Le Tilsit* off Texel 9.4.1810. In 12.1810
under Cmdr Colin Campbell, then 3.1812 under Cmdr Gregory
Grant (–1815); in 1816 under Lieut Caleb Jackson. Between
Small and Middling Repair at Portsmouth 1–6.1817. Fitted
for sea at Portsmouth 1–3.1818; recommissioned 1.1818 under
Cmdr Henry Shiffner, for Newfoundland, then 5.1819 under
Cmdr William Glasscock and 2.1820 Cmdr Octavious Vernon.
In 12.1820 under Cmdr Charles Baker; wrecked on eastern
end of St Shott's, near Cape Race (Newfoundland) 22.6.1822 (4
drowned including Baker).

Rosario Jabez Bailey, Ipswich
As built: 90ft 0in, 73ft 5½in x 24ft 7in x 11ft 2in. 236¹³⁄₉₄bm. Draught
6ft 6in/9ft 2in.
Ord: 27.6.1808. K: 8.1808. L: 7.12.1808. C: 9.12.1808–29.5.1809 at
Sheerness.
First cost: £3,263 to builder, plus £3,917 fitting.
Commissioned 3.1809 under Cmdr Booty Harvey (–1812); action
with two privateers off Dungeness 10.12.1810, taking 16-gun
Mamelouk; took (with *Griffin*) three armed brigs off Dieppe
27.3.1812. In 4.1812 under Cmdr William Henderson, for the
Downs; in 6.1813 under Cmdr Thomas Peake (–1817); paid off
at Portsmouth 1817. In 1.1819 under Cmdr William Hendry,
for guard duty at St Helena. In 5.1821 under Cmdr Frederick
Marryat, for the Channel; paid off 2.1822 at Portsmouth. Sold to
J. Levy, Rochester (for £380) 11.1832.

Renard John King, Upnor
As built: 90ft 3in, 74ft 1in x 24ft 7in x 11ft 0in. 238¹⁴⁄₉₄bm. Draught
6ft 6in/9ft 2in.
Ord: 9.7.1808. K: 8.1808. L: 5.12.1808. C: 27.6.1809 at Chatham.
Commissioned 2.1809 under Cmdr Hew Steuart (–1812), for
Walcheren operations. In 8.1812 under Cmdr George Brine, for
the Baltic. In 11.1812 under Cmdr David St Clair (–1817), for
the North Sea 1813–14, then Mediterranean 1815–16. Paid off to
Ordinary at Deptford 1817. Sold to Thomas Pittman (for £400)
at Dartford 29.6.1818.

Tyrian (i) Robert Guillaume, Northam
As built: 90ft 2in, 74ft 0⅛in x 24ft 8in x 11ft 0in. 239⁶⁄₉₄bm. Draught
6ft 6in/9ft 4in.
Ord: 18.7.1808. K: 8.1808. L: 16.12.1808. C: 10.4.1809 at
Portsmouth.
Commissioned 1.1809 under Cmdr Henry Thomas Davies, for the
Channel station; on 3.8.1811 under Cmdr Frederick William
Burgoyne, then 7.2.1812 under Cmdr Augustus Baldwin; on
1.1.1817 under Cmdr William Popham, at Jamaica; paid off
10.12.1818. Sold to Revill at Chatham (for £850) 22.7.1819.

Sphinx East India Co., Bombay Dyd [M/Shipwright Jamsetjee
Bomanjee]
As built: 89ft 10¼in, 73ft 6⅛in x 24ft 10½in x 11ft 0in. 241⁸⁸⁄₉₄bm.
Draught 7ft 6in/8ft 9in.
Ord: 2.10.1812. K: 5.1814. L: 25.1.1815. C: 3.4.1815 at Bombay.
First cost: £3,932 for building, plus stores of £1,922.
Commissioned 10.1814 under Cmdr Henry Shiffner (appointed
22.2.1814), in the East Indies; on 12.2.1815 (acting, Cmdr
20.9.1815) under Lieut Arthur Richard Turnour, for voyage
to England. Arrived 19.11.1815 at Chatham where laid up.
Fitted 11.1824 to 2.1825 as a packet (for £11,688) with 33 men,
and 4 x 12pdr carronades; recommissioned 15.12.1824 under

John Watkins, Master, for the Falmouth mail packet service;
on 14.2.1827 under Lieut Augustus R. L. Passenger; paid off
21.4.1832. Sold at Plymouth (for £810) 6.8.1835.

Cameleon East India Co., Bombay Dyd [M/Shipwright Jamsetjee
Bomanjee]
As built: 90ft 2½in, 73ft 9⅛in x 24ft 10½in x 11ft 0in. 242⁷⁷⁄₉₄bm.
Draught 8ft 6in/9ft 5in.
Ord: 2.10.1812. K: 3.1815. L: 15.1.1816. C: 19.1.1816 at Bombay.
First cost: £3,968 for building, plus £2,104 for RN supplies at
Bombay. Fitting at Portsmouth £4,472.
Commissioned 1816 under Lieut John M'Arthur Low (acting), for
passage to England; arrived 14.12.1816 at Portsmouth where
laid up. Fitted 10.1818 to 1.1819; recommissioned 10.1818 under
Cmdr William Mingaye, for Home waters; tender to *Apollo*
in 1822. Recommissioned 23.5.1823 under Cmdr James Ryder
Burton, for the Mediterranean; affair with 20-gun Algerine
Tripoli 31.1.1824. In 2.1824 under Cmdr George Lambert, then
8.1826 Cmdr Charles Cotton, for the Mediterranean, 4.1827
Cmdr Christopher Wyvil and 10.1828 Cmdr Sir Thomas Pasley.
Fitted for sea 8–9.1831 at Portsmouth; recommissioned 12.1834
under Lieut John Bradley, for the Lisbon station; in 11.1838
under Lieut George Hunter, for South America then the East
Indies; paid off into Ordinary at Sheerness 12.1843. BU at
Deptford 4.1849.

ICARUS **Class** As with the 32pdr type, just one non-standard vessel
(again, at Portsmouth) was built of the 18pdr class, other than the very
distinctive *Rapid*. The design was prepared and the work executed by
the Superior Class of shipwright apprentices in the Royal Dockyard.
Dimensions & tons: 90ft 0in, 71ft 5¾in x 24ft 10in x 11ft 0in.
234⁴⁴⁄₉₄bm.
Men: 76. Guns: 8 x 18pdr carronades, plus 2 x 6pdrs (bow).

Icarus Portsmouth Dyd [M/Shipwright Nicholas Diddams]
As built: 90ft 0in, 71ft 5¾in x 24ft 10in x 11ft 0in. 234⁴⁴⁄₉₄bm.
Draught 7ft 0in/9ft 8in.
Ord: 5.9.1812 as *Ephira*, renamed *Icarus* 4.12.1812. K: 3.1813. L:
18.8.1814. Completed fitting 3.1.1815.
First cost: £9,734.
Commissioned 26.9.1814 under Cmdr Thomas Barker Devon;
escorted Napoleon Bonaparte to St Helena; then with despatches
to East Indies; paid off 15.4.1817. Fitted for sea at Portsmouth
(for £3,678) 6–10.1817; recommissioned 24.6.1817 under Cmdr
Charles Orlando Bridgeman, for South America; on 2.9.1819
under Cmdr Henry Algernon Eliot, still in South America;
paid off 6.1821. In 5.1822 under Cmdr Charles Cole, for
Jamaica station, then 6.1823 under Cmdr John Graham; took
pirate *Diaboleto* at Cayo Blanco, Cuba 20.8.1824; paid off into
Ordinary 11.1825. Recommissioned 3.1828 under Cmdr Thomas
Best, for Jamaica, then 7.1830 under Cmdr Thomas Currie; paid
off 10.1831 into Ordinary. Coastguard at Lymington 9.1838.
Sold to Mr Ransom for £450 by AO 4.4.1861.

(iv) Brigs taken from the enemy

Ex-FRENCH PRIZES (1798 and 1808–09)

Considerable numbers of French naval brigs were added by capture
during the Napoleonic Wars and added to the RN, in some cases
amounting almost to entire classes. However, most had been disposed
of before 1817, leaving the five examples below. Sharp-lined and
relatively fast, they tended to be smaller and with poorer sea-keeping
qualities than their British-built equivalents, and few lasted very long

in frontline service. A very few large privateer brigs were also added in this period.

Le BRAVE Class A *canonnière* or *corvette-canonnière*, one of a pair built to a design by Pierre-Alexandre Forfait in 1793 (her sister *La Citoyenne* was deleted in 1804). Initially classed in the RN as a gun-brig under her French name, but renamed 31 August 1798 and re-rated in October 1811 as a brig sloop. Described then as 'nearly the same as the *Crocus* and the other 252-ton brigs'.

Insolent (ex-French gunboat *L'Arrogante*, ex-*Le Brave* renamed 1794, built 1.1793–5.1793 at Le Havre. L: 26.4.1793.), 14 guns
> Dimensions & tons: 91ft 9in, 72ft 1in x 25ft 11⅜in (25ft 6⅜in mld.) x 11ft 2in. 258¹⁴⁄₉₄bm.
> Men: 55 (85 as brig sloop). Guns: 2 x 18pdrs + 10 x 32pdr carronades (2 x 6pdrs + 12 x 24pdr carronades as brig sloop).
> Taken 23.4.1798 by *Jason* and *Naiad* off Brest. Arrived Plymouth 26.4.1798. Retained original name until 31.8.1798, when renamed. Fitted at Plymouth 2–4.1801.
> Commissioned 3.1801 under Lieut William Bevians (drowned 1801), in the Channel Islands. In 1802 under Lieut ?N. Hartwright, at Milford; paid off 1802. Recommissioned 1.1803 under Lieut William Smith, at Guernsey; on 19.10.1803 under Lieut John Row Morris (–1809); coppered at Plymouth 1.1805; sailed for the Mediterranean 25.8.1807. Large repair at Plymouth 10.1809–9.1811; re-rated as a brig sloop and recommissioned 2.8.1811 under Cmdr Edward Brazier, for North Sea and Baltic. In 11.1812 under Cmdr John Forbes, then 6.1814 Cmdr William Kelly, for North America; paid off at Deptford 18.9.1815. Sold to Joshua Crystall there (for £860) 11.6.1818.

Le PALINURE Class Out of twenty brigs built to this 1803 design (by François Pestel), eighteen were captured by the RN (the remaining two were transferred by France to the Italian Navy in June 1810); however, four of these were not added to the RN and twelve of those which were had gone by 1817

Griffon (French *Le Griffon*, built 4.1805–8.1806 at Rochefort. L: 2.6.1806), 16 guns.
> Dimensions & tons: 92ft 6in, 8oft 1oin x 29ft 4in x 8ft 2in. 368bm.
> Men: 100. Guns: 14 x 24pdr carronades, and 2 x 6pdrs.
> Taken 11.5.1808 near Cape St Anthony by *Bacchante*.
> Commissioned 1808 in Jamaica under Lieut Henry Spark Jones; in 12.1808 under Lieut ? Allen. Arrived Sheerness 10.10.1809, then fitted at Chatham 2.1810–12.1811; recommissioned 11.1811 under Cmdr John Tancock. On 1.2.1812 under Cmdr George Barne Trollope; in action (with *Rosario*) against ten brigs off Cherbourg 27.3.1812 (3 captured, 2 driven ashore). In 6.1814 under Cmdr George Hewson, then 5.1816 under Cmdr James Arthur Murray, and on 20.9.1816 under Lieut (acting, made Cmdr 20.8.1817) William Elliott Wright at St Helena; paid off 30.9.1818. Sold to Hill & Co. at Deptford (for £1,400) 11.3.1819.

Achates (French *Le Milan*, built 3.1806–1.1808 at St Malo. L: 6.7.1807), 16 guns
> Dimensions & tons: 97ft 4½in, 76ft 10¾in x 28ft 3½in (27ft 9½in mld.) x 13ft 2in. 327³⁄₉₄ bm.
> Men: 95. Guns: 14 x 24pdr carronades, and 2 x 6pdrs.
> Taken 30.10.1809 in the Atlantic by *Surveillante* and *Seine*. Fitted at Deptford 3.1810–30.6.1810.
> Commissioned 5.1810 under Cmdr John Davies, for the Channel; on 31.5.1813 under Cmdr Isaac Hawkins Morrison; engaged 40-gun *La Trave* 21.10.1813; at capture (with *Eurotas*) of 44-gun *La Clorinde* 25.2.1814. In 6.1814 under Cmdr Thomas Lamb Polden

Laugharne, still in the Channel. Laid up at Plymouth 11.1815. Sold to John Small Sedger (for £1,100) at Plymouth 11.6.1818.

Le SYLPHE Class (Design by Jacques-Noël Sané, 1803), 16 guns. Some thirty-two brigs were built to this design, of which four was added to the RN in 1808–09. Three more of this class were captured in 1814 but were not added to the RN. A sole example remained in 1817.

Sabine (French *Le Requin*, built 10.1805–1.1807 at Rochefort. L: 10.11.1806), 18 guns
> Dimensions & tons: 96ft 0in, 77ft 0⅞in x 28ft 5½in (28ft 0½in mld.) x 13ft 1in. 332¹⁄₉₄bm.
> Men: 100. Guns: 16 x 32pdr carronades, and 2 x 6pdrs.
> Taken 28.7.1808 off Monaco by *Volage*.
> Commissioned ?12.1808 in the Mediterranean under Cmdr James Donnor. Made good defects at Sheerness 11.8–20.11.1809. In 5.1810 under Cmdr Joseph Bott, then 10.1810 under Cmdr Thomas Grove; sailed for the Mediterranean 18.9.1810. In 11.1810 under Cmdr George Price; her boats cut out five gunboats from Sabiona 26.5.1811. In ?1.1812 under Cmdr Edward Wrottesley (died 7.1814); off Portugal 1812–13, then West Indies and Newfoundland 1814. In 7.1814 under Cmdr William Hall (–1815). Sold at Plymouth to Thomas Pitman (for £900) 29.1.1818.

L'ILLYRIEN Class (Design by Jean Tupinier, 1806). Six of this class were built for Napoleon's Italian Navy by Andrea Salvini at Venice, of which four were taken by the RN in 1808 and the last pair (with a further vessel still building) fell into Austrian hands at the surrender of Venice in 4.1814. One remained on the List in 1817.

Tuscan (Franco-Venetian *Le Ronco*, built 6.1807–5.1808 at Venice. L: 4.1808), 16 guns
> Dimensions & tons: 97ft 7in, 76ft 5in x 28ft 8in x 18ft 3in. 334¹⁄₉₄bm.
> Men: 100. Guns: 14 x 24pdr carronades, and 2 x 9pdrs.
> Taken 2.5.1808 off Cape Promontoro by *Unite*.
> Commissioned 8.1808 at Malta under Cmdr John Gourly (acting), for the Barbary coast. In 9.1808 under Cmdr John Wilson; in Rosas Bay 31.10–1.11.1809. In 12.1810 under Cmdr George Matthew Jones (–1812), in the Mediterranean; under Lieut ? Phillips (acting) 1.1811. Recommissioned 5.1813, still under Jones in the Mediterranean; paid off 18.12.1815. Sold at Plymouth to Thomas Pitman (for £800) 29.1.1818.

Ex-SPANISH PRIZE (1798)

Port Mahon (unnamed Spanish brig, building at Port Mahon, Minorca), 18 guns
> Dimensions & tons: 91ft 5½, 82ft 1in x 25ft 2in x 12ft 8in. 276⁵⁄₉₄bm.
> Men: 121. Guns: 18 x 6pdrs.
> Taken on the stocks 15.11.1798. L: 1798.
> Commissioned ?10.1799 under Cmdr William Buchanan (–1802); took 1-gun privateer *L'Enfant Chérie de la Victoire* in the Mediterranean 8.5.1800; too (with *Phoenix*) 12-gun *L'Albanaise* 3.6.1800. Egypt operations 1801. Paid off 1802. Fitted 27.7–27.10.1802 at Portsmouth. Recommissioned 8.1802 under Cmdr Walter Grossett; in 1803 under Capt. Martin Nevile (died 7.1803); sailed for Leeward Islands 5.1803. In ?2.1804 under Cmdr Francis M'Donald (died 6.1804), then Cmdr Thomas Garth, at Jamaica. In 1805 under Cmdr Samuel Chambers (–1809); took 14-gun schooner *Galgo* 30.9.1805; boats took 7-gun privateer *San Josef* at Puerto de Banes 25.6.1806; took 16-gun privateer *Le Furet* 8.2.1808, on Home station; took 3-gun *Le Général Paris* 18.11.1808. Under Cmdr John Lawson in 6.1809, then 3.3.1810

under Cmdr Villiers Francis Hatton, at Portsmouth and then off Spain; and then 22.10.1810 under Cmdr Thomas Everard, in the Channel. On 7.2.1812 under Cmdr Frederick William Burgoyne, on Irish station; to South America 1814; paid off 8.1814. Fitted at Deptford as a receiving ship 5.1815, recommissioned 8.5.1815 for Gravesend under Capt. Charles Carter. Loaned to Thames Police 29.1.1817 (–1837). Sold 8.1837 at Woolwich.

Ex-AMERICAN PRIZES (1812)

Emulous (USN brig *Nautilus*, built 1799 as a schooner by Henry Spencer, Maryland, purchased by the USN 1803, re-rigged as a brig 1810), 14 guns

    Dimensions & tons: 87ft 6in, 71ft 6in x 23ft 8in x 9ft 10in. 213bm.

    Men: …. Guns: 12 x 12pdr carronades, and 2 x 6pdrs.

    Taken 17.7.1812 by *Shannon*, *Aeolus* and *Africa*. Purchase (for £3,252.17.2d) reported 29.8.1812.

    Commissioned 2.1813 under Cmdr William M. Godfrey, on Halifax station. In 6.1815 under Cmdr John Undrell, and in 1816 under Cmdr Thomas Wren Carter, at Jamaica; later under Lieut Caleb Jackson (acting); arrived at Deptford 19.6.1816 to pay off and was laid up there. Sold (for £900) 8.1817.

Ferret (ex-*Nova Scotia*, ex-US privateer brig *Rapid*, built —), 14 guns

    Dimensions & tons: 84ft 0in, 66ft 3in x 24ft 8in x 10ft 4in. 214^{38}⁄₉₄bm.

    Men: 75. Guns: 12 x 12pdr carronades, and 2 x 6pdrs.

    Taken 17.10.1812 by *Maidstone* and *Spartan* off St. George's Bank.

    Commissioned as HMS *Nova Scotia* 11.1812 at Halifax under Lieut Bartholomew Kent. Fitted at Plymouth 7.7–30.9.1813; rated sloop and renamed *Ferret* 12.6.1813; recommissioned 6/7.1813 under Cmdr William Ramsden. In 5.1814 under Cmdr James Stirling; to St Helena (with *Northumberland*) 1815; took slaver *Dolores* (with 275 slaves) 4.4.1815. Fitted for sea at Plymouth 5–6.1815. To Ordinary at Plymouth 6.1816. Fitted at Plymouth 3–4.1817; under Lieut William Pitman. Sold there to Mr Rundle (for £460) 13.1.1820.

Columbia (US privateer brig *Curlew*, built 1812 —), 18 guns

    Dimensions & tons: 94ft 4in, 80ft 3in x 26ft 3in x 13ft 0in. 294^{12}⁄₉₄bm.

    Men: — . Guns: 16 x 18pdr carronades, and 2 x 6pdrs. Purchase (for £3,300.6.7d) reported 25.9.1812.

    Taken 24.7.1812 by *Acasta* off Cape Sable. Fitted at Portsmouth 19.11.1813–18.4.1814.

    Commissioned 3.1813 in North America under Lieut John Kinsman. In 5.1813 under Cmdr Henry Chads; sailed for the Leeward Islands; took US 1-gun privateer *Dolphin* 4.12.1814; paid off into Ordinary at Chatham 11.1815. Sold there 13.1.1820.

(v) Gun-brigs

During the French Revolutionary War, a new type of small warship was developed for coastal patrol and convoy escort purposes. The earliest of these enlarged gunboats were former cutters, re-rigged as small brigs, while a new series of vessels, initially rigged as schooners or brigantines, were soon re-established as brigs, carrying a dozen guns (two long guns as bow chasers and ten carronades on the broadsides) and a complement of 50 men under a lieutenant. The first of these were the twelve vessels of the *Conquest* Class in 1794, of which the name ship was the sole surviving gun-brig from this era surviving into 1817, in harbour service (this *Conquest* was sold in April 1817, and is not covered here in detail). Considerable numbers were built by contract over the following decade; fifteen of the *Acute* class and sixteen of the

Courser class were ordered in 1797; ten of the *Archer* class were ordered in late 1800 and another forty-eight in the first half of 1804; all of these which survived the Napoleonic War were disposed of before 1817.

CONFOUNDER **Class** Sir William Rule design of 1804. The concept of the gun-brig had plainly been re-evaluated by the outbreak of the Napoleonic period. The requirement was for 'small brigs to carry 14 x 18pdr carronades, of nearly the same dimensions and tonnage of the last built gun brigs'; twenty orders were placed on 20 November 1804. A twenty-first vessel, *Richmond*, was ordered 23 August 1805. By the time the vessels were fitted, this had been modified to include two chase guns instead of four of the carronades. The 12pdrs were mounted on traversing carriages, one in the bows and the other in the stern. Nevertheless, these were more clearly seaworthy vessels than the earlier gun-brigs, and more able to face lengthy ocean voyages; compared with the flat-bottomed hulls of previous gun-brig designs, these had a relatively sharp section, with performance under sail now prioritised. They were deployed around the world, being seen as very small brig sloops rather than enlarged gunboats.

    Two vessels (*Indignant* and *Rebuff*) were fitted and recommissioned as mortar brigs in 1809. Eight were wartime losses (*Adder*, *Inveterate*, *Turbulent*, *Bustler*, *Conflict*, *Fancy*, *Exertion* and *Encounter*), while one was BU in 1811 (*Indignant*) and six were sold in 1814 (*Confounder*, *Strenuous*, *Dapper*, *Starling*, *Richmond* and *Rebuff*) and two in 1816 (*Virago* and *Hearty*); the last four (*Martial*, *Intelligent*, *Resolute* and *Havock*) survived in ancillary roles for many years.

    Dimensions & tons: 84ft 0in, 69ft 8¼in x 22ft 0in x 11ft 0in. 179^{48}⁄₉₄bm.

    Men: 50. Guns: originally 10 x 18pdr carronades + 2 x 12pdr (bow & stern chasers); some had 6pdrs or 9pdrs replacing the 12pdrs.

Martial Charles Ross, Rochester

    As built: 84ft 6in, 70ft 5¼in x 22ft 1in x 11ft 0in. 182^{67}⁄₉₄bm. Draught 6ft 6in/7ft 6in.

    Ord: 20.11.1804. K: 1.1805. L: 17.4.1805. C: 17.4–15.5.1805 at Chatham (coppered).

    Commissioned 4.1805 under Lieut Joseph Marrett (–1809), for the Channel; in Basque Roads operations 4.1809. In 1810 under Lieut Joshua Kneeshaw (–1811), for the Baltic, then in Scheldt operations. In 1812 under Lieut Charles Leaver (–1813). In 1814 under Cmdr George Elliott; passage of the Adour 27.2.1814 (Elliott killed); in 2.1814 under Lieut Edward Collins (acting), then Cmdr Henry Forbes, and in 8.1814 Cmdr James Leach. Fitted at Woolwich as Fishery protection vessel 11.1815–4.1816; recommissioned 2.1816 under Lieut Robert M'Kinley (–1830); at Leith 1816–25; fitted again at Deptford 'for the protection of the cod industry' 4–6.1826; at the Nore 1826–30. Fitted as Quarantine vessel at Sheerness 4–8.1831, then to Chatham. Sold at Chatham (for £440) 21.1.1836.

Resolute John King, Dover

    As built: 84ft 1in, 69ft 8in x 22ft 1½in x 11ft 0in. 181^{37}⁄₉₄bm.

    Ord: 20.11.1804. K: 1.1805. L: 17.4.1805. C: 23.4–6.7.1805 at Sheerness (coppered).

    First cost: £3,152 to build.

    Commissioned 4.1805 under Lieut George Higginson. In 1806 under Lieut Edward Harries, then 1810 Lieut Thomas Pettman, both on the Irish station. In 7.1812 under Lieut William Pringle Green (–1815); fitted as a Tender at Plymouth 1–3.1814. Fitted as diving bell ship at Plymouth 11.1815–6.1816. Transferred to Breakwater Dept. 5.1820. Again fitted for diving bell at Plymouth 8–9.1822. Fitted at Plymouth 2–3.1826 to go with diving bell to Bermuda, where she combined the function of

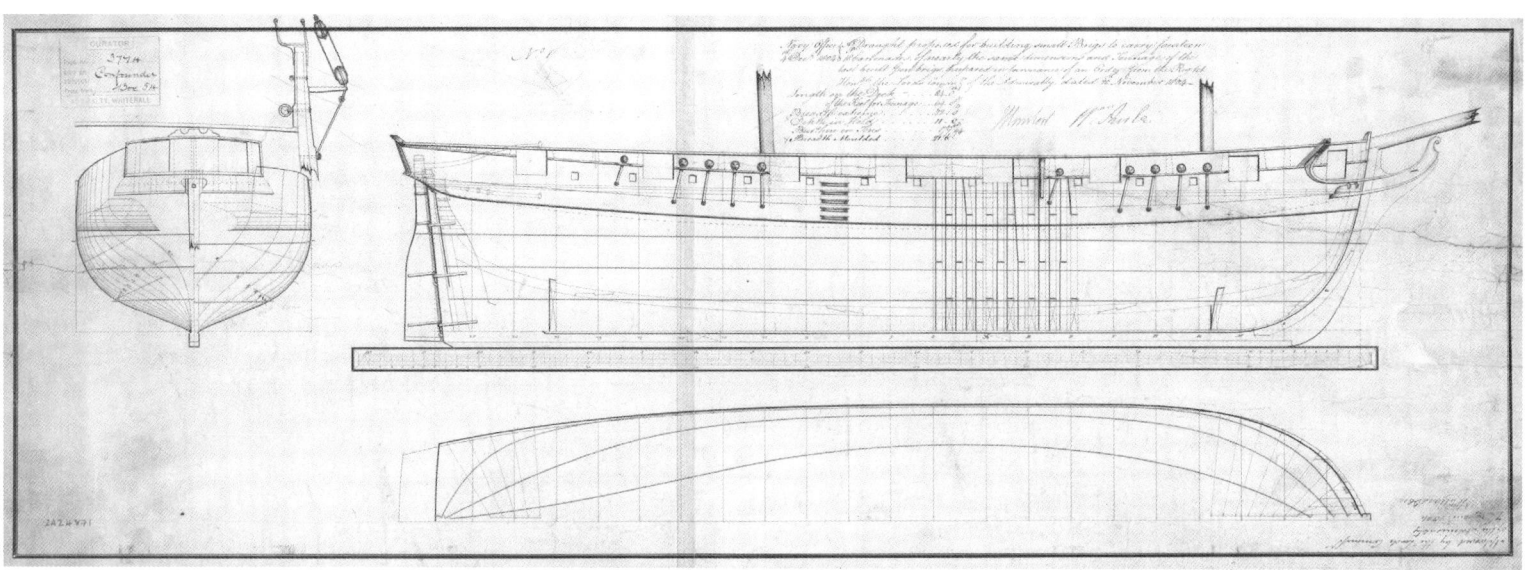

diving bell ship and receiving ship. Fitted as Convict hulk 1844.
BU there 1852.

Intelligent (Nicholas) Bools & (William) Good, Bridport
As built: 84ft 0in, 69ft 9½in x 22ft 1in x 11ft 0in. 181⁴/₉₄bm.
Ord: 20.11.1804. K: 1.1805. L: 26.8.1805. C: 15.9.1805–14.4.1806 at
Plymouth (coppered).
Commissioned 10.1805 under Lieut Nicholas Tucker, in the Downs.
In Ordinary at Plymouth 7.1815. Sold for £610 at Portsmouth
14.10.1815, but the purchaser refused to accept her. Fitted as
mooring lighter at Portsmouth 8–9.1816. By 1864 *Mooring
Lighter No. 4* – fate uncertain.

Havock Stone, Great Yarmouth
As built: 84ft 10in, 70ft 1¾in x 22ft 2¼in x 11ft 0in. 183⁶/₉₄bm.
Draught 6ft 3in/8ft 4in.
Ord: 20.11.1804. K: 2.1805. L: 25.7.1805. C: 7.8–30.10.1805 at
Sheerness.
First cost: £2,940 to build.
Commissioned 10.1805 under Lieut William Bamber (–1810),
for North Sea (based at Great Yarmouth). In 1811 under
Lieut Henry Rowed; in 1812 in Ordinary at Sheerness.
Middling Repair there 8–12.1812, then re-rated as sloop and
recommissioned 1.1813 under Cmdr John Forbes, for the Baltic.
In 10.1813 under Cmdr George Truscott (–1815). In Ordinary
at Portsmouth to 1818, then fitted as a storeship (bullock vessel)
4–7.1818, but does not appear to have served as such. Fitted at
Bembridge as a temporary light vessel 5–9.1821. Transferred to
the Customs as a watch vessel for the Hamble River 3.3.1834
(–1855). BU completed at Portsmouth 25.6.1859.

BOLD (or Modified CONFOUNDER) Class A revival of Sir William
Rule's design of 1804. A batch of twelve to a somewhat modified
design (sometimes called the *Bold* Class) were ordered in 1811, of
which two had been wartime losses in 1813 (*Boxer* and *Bold*) and
one sold in 1815 (*Borer*), and a further group of six followed in 1812.
Unlike earlier brigs of this size, most were re-rated as brig sloops at or
soon after their completion, and were under commanders, at least until
1815–17, when they reverted to gun-brigs, except the *Shamrock* (which
was re-classed as a survey vessel).
Dimensions & tons: 84ft 0in, 69ft 8¾in x 22ft 0in x 11ft 0in.
179⁴⁸/₉₄bm.
Men: 60. Guns: 10 x 18pdr carronades + 2 x 6pdrs (bow).

The 1805 draught for the gun-brigs of the *Confounder* class. This class stood at the end of
a rapid wartime development of the type, which had evolved from a coastal gunboat into
what was essentially the smallest combatant that could be deployed across the world. The
Confounder herself and many of her class had been lost or disposed of by 1817, but four
soldiered on in auxiliary roles for several decades, along with a dozen of more of the later
but very similar *Bold* Class. *(NMM J4681)*

Manly Thomas Hills, Sandwich (rated as sloop)
As built: 84ft 1in, 70ft 0¼in x 22ft 1in x 11ft 0¼in. 181⁶⁰/₉₄bm.
Draught 6ft 8in/8ft 10in.
Ord: 16.11.1811. K: 2.1812. L: 13.7.1812. C: 24.7–24.9.1812 at
Sheerness (coppered).
Commissioned 27.10.1812 under Cmdr Edward Collier, later
under Cmdr Henry Montresor; sailed for North America
23.3.1813; stranded at Halifax 13.11.1813, but salved. In 1814
under Lieut (Cmdr 5.1814) Vincent Newton; on 27.5.1814 under
Cmdr Henry William Bruce; in Potomac operations 8.1814; on
22.10.1814 under Cmdr George Truscott, then ?6.1815 Cmdr
Charles Simeon; paid off 16.9.1815. In Ordinary at Portsmouth
to 1826. Fitted there 2.1824 with iron and zinc on the bottom
'for an experiment to preserve the copper'. Middling Repair and
fitted for sea at Portsmouth 10.1826–5.1827; recommissioned
2.1827 under Lieut William Field, for Halifax station; in 6.1828
under Lieut H. W. Bishop, then 10.1830 Lieut John Wheatley,
still at Halifax; paid off 19.4.1831. Sold at Portsmouth (for £550)
to Sturge 12.12.1833.

Snap Russell & Son, Lyme Regis (rated as sloop)
As built: 84ft 0in, 69ft 9⅛in x 22ft 1½in x 11ft 0in. 181⁶¹/₉₄bm.
Draught 6ft 9in/8ft 6in.
Ord: 16.11.1811. K: 2.1812. L: 25.7.1812. C: 9.8–9.10.1812 at
Portsmouth (coppered).
First cost: £3,212 to builder, plus £3,741 fitting & coppering.
Commissioned 15.8.1812 under Cmdr George Rose Sartorius, for
the Channel and North Sea. On 23.7.1813 under Cmdr William
Bateman Dashwood; took 16-gun privateer *Lion* off St Valery en
Caux 1.11.1813. In 11.1814 under Cmdr George King (–1815).
In Ordinary at Sheerness to 1820. Fitted as surveying vessel
at Sheerness 1–4.1821; recommissioned 3.1821 under Lieut
John Hose, then 10.3.1823 under Lieut Frederick Bullock, for
surveying Newfoundland coast; paid off 18.1.1827. Appropriated
as Powder hulk at Woolwich (by AO 29.9.1827); to Deptford
1829 then back to Woolwich same year. Sold to Levy, Rochester
(for £420) 4.1.1832.

Conflict William Good, Bridport
As built: 84ft 3½in, 70ft 0in x 22ft 0½in x 11ft 1¼in. $180\frac{84}{94}$bm. Draught 6ft 6in/8ft 3in.
Ord: 16.11.1811. K: 2.1812. L: 26.9.1812. C 10.10–14.12.1812 at Portsmouth.
Commissioned 28.10.1812 under Cmdr Henry Loraine Baker; sailed for North America 17.4.1813. On 18.3.1814 under Cmdr Abraham Mills Hawkins, for the Mediterranean; later under Cmdr Valentine. Jones; paid off 1.9.1815. Small Repair at Sheerness 9.1820. Fitted for sea there to 2.1825; recommissioned 11.1824 under Lieut John Christie, for the west coast of Africa; took slaver *Charles* 21.12.1825; on 19.9.1826 under Lieut Arthur Wakefield (–1827), on same station; took slavers *Independencia* 28.2.1827 and *Bahia* 3.4.1827. Between Small and Middling Repair 4–6.1829, became watch vessel at Fowey. Recommissioned 3.1830 under Lieut George Smithers, for African station; fitted for sea 5.1830; took slaver *Ninfa* 24.11.1830; paid off 5.7.1832, and became a receiving ship at Sierra Leone. Sold there 30.12.1840.

Contest William Good, Bridport
As built: 84ft 6½in, 69ft 9in x 22ft 0½in x 11ft 0in. $180\frac{23}{94}$bm. Draught 6ft 8½in/8ft 6in.
Ord: 16.11.1811. K: 2.1812. L: 24.10.1812. C: 6.11.1812–23.2.1813 at Portsmouth.
Commissioned 25.11.1812 under Cmdr James Rattray (to 1815); sailed for North America 22.4.1813; her boats captured (with those of *Mohawk*) 3-gun gunboat *Asp* in Yeocomico Creek (Chesapeake) 11.7.1813, but *Asp* was then retaken by the Americans; paid off 30.8.1815. At Portsmouth 1817–22. Recommissioned 27.3.1826 under Lieut Charles English, for the Halifax station; in 4.1827 under Lieut Edward Plaggenborg; lost with all hands, presumed foundered off Halifax 14.4.1828.

Thistle Mrs Mary Ross, Rochester
As built: 84ft 4½in, 70ft 9½in x 22ft 3in x 11ft 0½in. $186\frac{39}{94}$bm. Draught 6ft 9in/8ft 8in.
Ord: 16.11.1811. K: 3.1812. L: 13.7.1812. C: 13.7–12.9.1812 at Chatham,
Commissioned 8.1812 under Cmdr James Kearney White (–1814); sailed for North America; on 23.6.1814 under Cmdr James Montagu, in North America; paid off 8.1815. Fitted for sea at Portsmouth 2–7.1819; recommissioned 5.1819 under Lieut Robert Hagan, on African station; took 40 slavers over next 4 years; paid off 21.5.1823. BU at Portsmouth 7.1823.

Shamrock Edward Larking, King's Lynn
As built: 84ft 1in, 69ft 9¾in x 22ft 0in x 11ft 0¼in. $179\frac{68}{94}$bm. Draught 6ft 6in/8ft 3in.
Ord: 16.11.1811. K: 3.1812. L: 8.8.1812. C: 18.9.1813–1.1.1813 at Sheerness.
First cost: £3,011 to builders.
Commissioned 11.1812 under Cmdr Andrew Pellet Green; on 11.11.1813 under Cmdr John Marshall, when at Cuxhaven. Under Cmdr Christopher Crackenthorp Askew from 76.1814, on Irish station; paid off into Ordinary at Plymouth 9.10.1815. Fitted at Plymouth as a survey vessel 11.1816–2.1817. Recommissioned 1.1817 under Cmdr Martin White, for surveying in the Channel (to 1828); at Woolwich in 1829. Fitted there for a quarantine service vessel 9.1830–11.1831. Delivered to the coastguard as a watch vessel at Rochester 3.1833; renamed *WV 18* on 25.5.1863. Sale reported 24.1.1867.

Hasty Thomas Hills, Sandwich
As built: 84ft 1in, 69ft 11in x 22ft 1½in x 11ft 0in. $182\frac{4}{94}$bm. Draught 6ft 6in/8ft 9in.
Ord: 16.11.1811. K: 4.1812. L: 26.8.1812. C: 10.9–20.11.1812 at Sheerness.
Commissioned 10.1812 under Cmdr James Dickinson; to Baltic in 1813. From 27.6.1814 under Cmdr John Brenton, on the North Sea and Irish stations; paid off 14.11.1815. Fitted as survey ship at Deptford 4–7.1819. Recommissioned 4.1819 under Lewis Fitzmaurice for surveying (until 1822); at Deptford in 1824. Fitted 'for a mud-engine' (i.e. dredger) at Deptford 2.1826–12.1827 for Port Louis, Mauritius (where still in service in 1870).

Plumper William Good, Bridport
As built: 84ft 1½in, 70ft 0in x 22ft 0½in x 11ft 0in. $180\frac{84}{94}$bm. Draught 6ft 8in/8ft 7½in.
Ord: 16.11.1811. K: 4.1812. L: 9.10.1813. C: 1–25.11.1813 at Portsmouth (for Ordinary), including coppering. Fitted for sea there 5.1815.
Commissioned 4.1815 under Lieut George Dommett (Cmdr 6.1816); paid off 5.9.1815. At Sheerness in 1817. Underwent Small Repair there (for £841) 2.1819. Fitted for sea at Sheerness (for £3,526) 3–11.1820. Recommissioned 9.1820 under Lieut William Hutchinson, for Cork station (to 1824). Recommissioned 7.1827 under Lieut Edward Medley for African station; took slavers *Fanny* 19.5.1828 and *Minerva de Conceicao* 17.10.1828; from 2.1829 under Lieut M. Green; fitted for foreign service at Woolwich (for £2,393) 3–5.1829; took slaver *Ceres* 6.8.1829; from 1.1830 under Lieut John Adams, still on African station; took slavers *Loreto* 12.5.1830, *Maria* 7.11.1830 and *Marie* 26.12.1830; paid off 6.9.1832. Sold to John Levy, Rochester (for £420) 12.12.1833.

Swinger William Good, Bridport
As built: 83ft 11½in, 69ft 7⅞in x 22ft 0¾in x 11ft 0in. $180\frac{32}{94}$bm. Draught 6ft 9in/8ft 7½in.
Ord: 16.11.1811. K: 4.1812. L: 15.5.1813. C: 2.6–2.11.1813 at Portsmouth (and coppered).
Commissioned 23.6.1813 under Cmdr Robert Wauchope; under Cmdr Alexander Branch from 6.1814, on Leeward Islands station. Paid off 9.1815. Recommissioned 12.1815 under Lieut John Mitchell, for Leith station (until 1820); fitted for sea at Deptford 12.1815–7.1816. At Deptford in 1822. Recommissioned 20.6.1823 under Lieut John Scott, for African station (died 1824); from 11.1824 under Lieut Edward Stewart Clerkson (died 5.1825); took slaver *Bon Fim* 14.1.1825; from 6.1825 under Lieut Henry Poingdestre (died 10.1825); from 11.1825 under Lieut George Matson; took slaver *Paqueta de Bahia* 22.11.1825; paid off 7.1826 at Portsmouth. Fitted as mooring lighter at Portsmouth 1–2.1829. BU 3.1877.

Adder Robert Davy, Topsham (Exeter)
As built: 84ft 6in, 70ft 1⅞in x 22ft 1½in x 10ft 11½in. $182\frac{63}{94}$bm. Draught 6ft 11in/8ft 3in.
Ord: 2.11.1812. K: 12.1812. L: 28.6.1813. C: 31.7.1813–19.2.1814 at Portsmouth.
Commissioned 10.1813 under Cmdr Nicholas Pateshall; deployed to Halifax. Under Cmdr James Montagu from 6.1814, then Cmdr Samuel Malbon from 6.1815; paid off 5.9.1815 at Portsmouth. Fitted there 8.1825–1.1826 for the Coast Revenue Blockade service; became a watch vessel on the Sussex-Coast 1.1826. Stationed at Rye in 1827–30. Wrecked near Newhaven 12.1831.

Griper (Mark) Richards & (John) Davidson, Hythe (Southampton)
As built: 84ft 8in, 70ft 5in x 22ft 1in x 11ft 1in. $182\frac{62}{94}$bm. Draught 6ft 8in/8ft 6in.

Ord: 2.11.1812. K: 1.1813. L: 14.7.1813. C: 17.7–7.11.1713 at Portsmouth.

First cost: £3,411 to builders.

Commissioned 7.1813 under Cmdr Charles Mitchell; from 2,1814 under Cmdr Arthur M'Meekan; paid off 14.12.1815 at Chatham. Fitted at Deptford 12.1818–5.1819 for 'a voyage of Discovery in the Arctic Seas' (strengthened for ice navigation); recommissioned under Lieut Matthew Liddon from 1.1819, engaged (with *Hecla*) in exploration of Hudson Bay; paid off 12.1820. Fitted for the Arctic again at Deptford 2–5.1823, then recommissioned 1.4.1823 under Cmdr Douglas Clavering, engaged in magnetic investigations 'in the high latitudes of the Polar Seas' off Norway, Spitzbergen, Greenland and Iceland; paid off 12.1823 at Deptford. Fitted again 1–6.1824; recommissioned 15.1.1824 under Capt. George Lyon; sailed 16.6.1824 for Hudson Bay seeking Northwest Passage; paid off 13.12.1824 at Portsmouth. Fitted for the Coast Blockade service there 8–12.1825. To the coastguard at Blackwall 1825. At Portsmouth 1827–30. later at Chichester 1831–60. Fitted at Portsmouth 7.1860 with a target for gunnery experiments; removed 9.1861; fitted for armour plate experiments 11.1862. BU completed at Portsmouth 11.11.1868.

Clinker Robert Davy, Topsham (Exeter)

As built: 84ft 3½in, 69ft 10½in x 22ft 2½in x 10ft 11in. 183²⁹⁄₉₄ bm. Draught 7ft 3in/8ft 9in.

Ord: 2.11.1812. K: 1.1813. L: 15.7.1813. C: 31.7.1813–19.2.1814 at Portsmouth.

First cost: £3,411 to builders.

Commissioned 10.1813 under Cmdr Joseph Tullidge; paid off 5.12.1815 at Portsmouth. Recommissioned 11.1819 under Lieut Nathaniel Martin, for Newfoundland service; from 7.1821 under Lieut John Eager; from 5.1825 under Lieut Allen Gardiner, then 11.1826 under Lieut George Matson off the African coast; took slavers *Capaioba* 16.5.1827, *Voador* 20.8.1828, *Clementina* 2.9.1829, *Octavio* 9.10.1829, *Emilia* 31.10.1829 and *Altimara* 21.3.1830; paid off 6.10.1830. Fitted for the coastguard at Portsmouth 9.1830–11.1831. To the coastguard 11.1831 at Cuckmere Haven/Yantlet Creek, renamed *WV 12* on 25.5.1863. Sale notified 24.1.1867.

Pelter Henry Tucker, Biddeford

As built: 84ft 0¼in, 69ft 10in x 22ft 2¾in x 1ft 0in. 183⁵¹⁄₉₄ bm. Draught 6ft 6in/8ft 9in.

Ord: 2.11.1812. K: 1.1813. L: 27.8.1813. C: 6.10.1813–17.4.1814 at Plymouth.

First cost: £3,411 to builders.

Commissioned 17.2.1814 under Cmdr George Haye, for North America; to Quebec 1814; paid off 20.9.1815 at Portsmouth. Recommissioned 11.1819 under Lieut William Minchin, for Newfoundland service; from 11.1821 under Lieut Roger Curry; from 11.1824 under Lieut John Adams, on Halifax station; paid off 30.12.1825. Fitted for the Coast Blockade at Portsmouth 2–4.1826, and sent to the Coast Blockade at Folkestone 1826. Sold to J. B. Tolputt for £32 on 8.8.1862.

Mastiff William Taylor, Biddeford

As built: 84ft 2in, 70ft 0½in x 22ft 2¾in x 11ft 0in. 184³⁄₉₄ bm. Draught 6ft 9in/8ft 6in.

Ord: 2.11.1812. K: 1.1813. L: 25.9.1813. C: 17–23.12.1813 for Ordinary at Plymouth.

First cost: £3,707 to builders.

Commissioned 4.1815 under Cmdr Job Hanmer, and fitted for sea (for £3,707) to 27.6.1815; paid off 21.9.1815 at Portsmouth.

Underwent Middling Repair at Portsmouth (for £8,848), then fitted as Survey ship 5.1822–12.1825. Recommissioned 9.1825 under Cmdr Richard Copeland for surveying in the Mediterranean; on 4.2.1830 under Lieut William John Cooling, then 22.11.1830 under Lieut James Wolfe, then 11.5.1832 under Lieut Thomas Graves, all in Mediterranean; paid off 2.1836. Underwent Small Repair, and fitted as a Surveying vessel at Chatham (for £1,320) 3–4.1836. Recommissioned 4.1836 under George Thomas, Master, for surveying in the Orkneys; on 1.4.1847 under Cmdr Alexander Bridport Becher; paid off into Ordinary at Woolwich 13.10.1848. BU at Woolwich 5.1851.

Snapper Hobbs & Hellyer, Redbridge (Southampton)

As built: 84ft 4⅛in, 70ft 1⅜in x 22ft 2½in x 11ft 2in. 183⁸⁸⁄₉₄ bm. Draught 6ft 8in/8ft 6in.

Ord: 2.11.1812. K: 4.1813. L: 27.9.1813. C: 16.5.1815 at Portsmouth.

First cost: £3,411 to builders.

Commissioned 4.1815 under Cmdr Robert Gordon, based at Portsmouth; paid off 18.9.1815. Fitted at Portsmouth 2–7.1819. Recommissioned 5.1819 under Lieut James Henderson, for the west coast of Africa; took slaver *Juanita* 30.9.1819; on 3.1820 under Lieut Richard Nash (acting); from 8.6.1821 under Lieut Christopher Knight; took slaver *Conceicao* 1.8.1821; from 1.1822 under Lieut Thomas Henry Rothery, still off Africa; paid off 31.10.1823. Fitted at Portsmouth for the Coast Blockade 2–5.1824. Sold to Castle & Son for £303 (at Portsmouth?) 3.7.1861.

(vi) Brig sloop on the Great Lakes

Star (ex-*Lord Melville*) Kingston Dyd, Ontario [M/Shipwright George Record]

Dimensions & tons: 71ft 7in, 56ft 9½in x 24ft 8in x 8ft 0in. 186⁴⁹⁄₉₄ bm.

Men: — . Guns: 16 x 32pdr carronades; later 12 x 32pdr carronades, + 2 x 18pdrs.

Ord: 12.1812. K: 1.1813. L: 22.7.1813.

Commissioned 29.12.1813 as *Lord Melville* for Lake Ontario under Cmdr Charles Anthony; renamed *Star* 22.1.1814; at capture of Fort Oswego 6.5.1814. Later in 1814 under Cmdr Alexander Dobbs; took US 3-gun schooners *Ohio* and *Somers* 13.8.1814. On 1.12.1814 under Lieut (Cmdr 28.2.1815) Charles Cunliffe Owen. In 1815 under Lieut Massy Herbert (acting); paid off 6.1816. Sold 1837.

(B) Vessels acquired from 1 January 1817

All post-war brigs were to be flush-decked. With a single exception, production until 1830 concentrated exclusively on the small 10-gun *Cherokee* class, many of which were adapted as brigantines and used for a variety of services, including the packet (mail) routes. Many other brigs of all classes (as well as other British warships) were to spend years patrolling the slaving routes, intercepting thousands of slaving vessels of all nationalities and releasing many hundreds of thousands of slaves from their captivity; only a proportion of these seizures can be included in the rest of this chapter and elsewhere in this volume, but hopefully enough are listed to show the immensity of the achievement.

A highly detailed contemporary model of a *Cherokee* Class brig sloop in the National Maritime Museum at Greenwich. The model, built plank on frame and fully rigged with original masts and spars and modern rigging, is from about 1825, and so probably represents one of the post-war vessels of this class. It is complete with a variety of fittings such as a capstan (rigged with bars), deck hatches and gratings, the large riding bitts in the bows – these are metal-sheathed, which may mean that the brig carries the new-fangled chain cables – and a finely carved wooden figurehead on the stem. The large square gunports are fitted with bottom-hinged half port lids – a post-war development – to allow the lower half of the port to be closed to prevent too much water coming inboard in a seaway. *(NMM F5788-002)*

Later *CHEROKEE* Class 10 guns. Post-war revival of the 1807 Peake design. Part of group of 115 ordered between 1807 and 1830, comprising the 36 included in Section A and 79 others ordered between 1817 and 1826, which were all to be dockyard-built (the original 34 were all contract-built). None of the six vessels finally built as paddlers were laid down as sailing vessels; all were re-ordered as paddlers and built as such *ab initio*.

> Dimensions & tons: 90ft 0in, 73ft 7⅛in x 24ft 8in oa (24ft 6in for tons) x 11ft 0in. 235%94bm. 297 disp.
> Later vessels were built to an amended draught of 72ft 3in x 24ft 8in (231 tons bm).
> Men: 52 in peacetime (33 as packet brigs and 50 as brigantines).
> Guns: 2 x 6pdrs + 8 x 18pdr carronades.
> As brigantine: 2 x 24pdr carronades (10cwt) and 1 x 32pdr (40cwt); but these varied (sometimes 4 x 12pdr carronades only).

***1817* Batch** Twelve vessels were authorised on 16 February 1817

and ordered 13 June 1817, two from each dockyard (except from Sheerness), to which plans were sent on 16 July, modified from the wartime batch by Seppings, with diagonal bracing and by having bulwarks raised (6in at stem, 4in at stern).

Alacrity Deptford Dyd
> As built: 90ft 0½in, 73ft 8⅛in x 24ft 8⅛in x 11ft 0in. 236bm. Draught 7ft 3in/9ft 6in.
> Ord: 13.6.1817. K: 10.1817. (temp. housing built 1.1818) L: 29.12.1818. C: 5.1820–8.8.1820.
> Commissioned 1818 under Cmdr Henry John Leeke, for the Downs. On 8.5.1820 under Cmdr Henry Stanhope, for South America. On 5.3.1821 under Cmdr Frederick Earl Spencer, in South America. On 26.8.1822 under Cmdr Thomas Porter, in South America. In 8.1823 under Cmdr Charles P. Yorke, for the Mediterranean. In 9.1825 under Cmdr George James Hope Johnstone; destroyed three Greek pirate vessels ('misticoes') at Psara 9–10.4.1826. On 8.7.1827 under Cmdr Robert Lambert

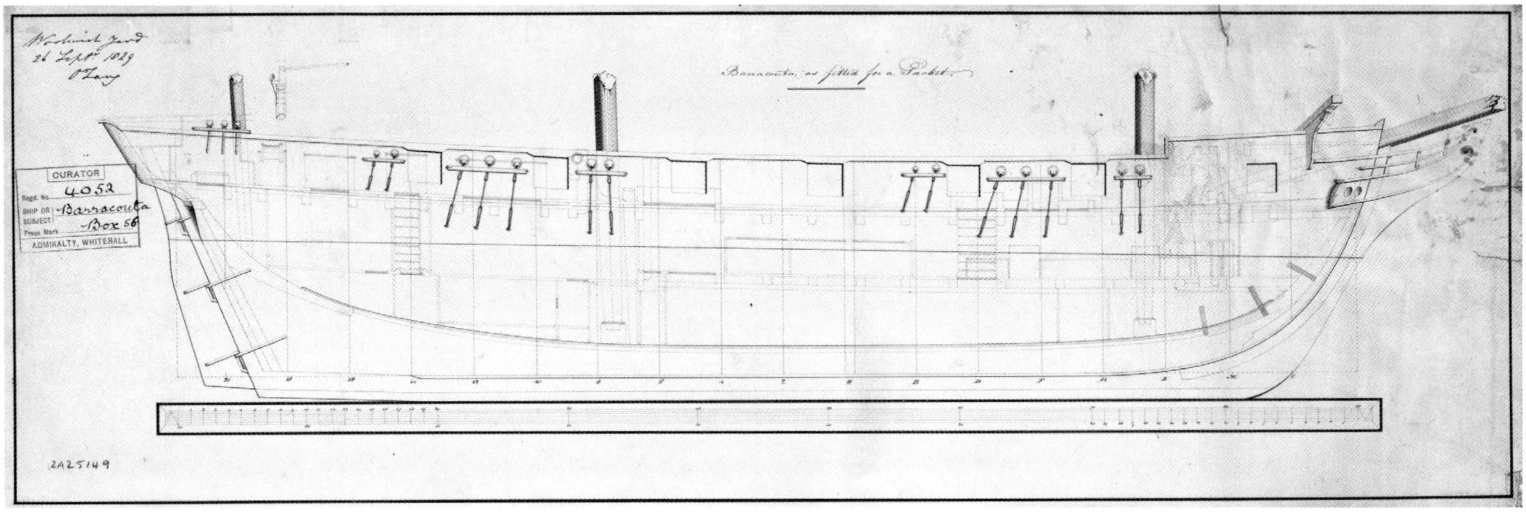

The draught for the post-war *Cherokee* Class brig sloop *Barracouta*; like her celebrated sister ship the *Beagle* (built alongside her at Woolwich) the *Barracouta* was first employed as a survey ship, and – in company with the larger 24-gun *Leven* – surveyed much of the coast and harbours of southern Africa, including Table Bay. The perils of the survey service in tropical waters claimed many lives through disease (two-thirds of the officers succumbed during their service, and the majority of their seamen), including *Barracouta*'s first commander, William Cutfield, who had to be replaced on 15 May 1823 by Lieut Alexander Vidal, seconded from the *Leven*. In 1828–29 the *Barracouta* was fitted as a barque-rigged packet at Woolwich, recorded on this draught of 26 September 1829 by Oliver Lang, then master shipwright at the dockyard there. The fitting of a windlass forward, requiring a shortened forecastle platform, seems to have been a standard modification for the packet service. *(NMM J0192)*

Baynes, then 8.1.1828 under Cmdr Joseph Nias, in the Mediterranean; arrived Portsmouth to pay off 23.12.1829. Sold at Portsmouth (for £660) 28.8.1835.

Ariel Deptford Dyd

As built: 90ft 0½in, 73ft 8⅛in x 24ft 8⅛in x 11ft 0in. 236bm. Draught 6ft 8in/9ft 7in.

Ord: 13.6.1817. K: 2.1819. L: 28.7.1820. Housed over (incomplete): 7.1821. C: 5.1827–1.3.1828.

First cost: £6,879. Completed to a packet brig (6 guns).

Commissioned 12.1827 under Cmdr James Figg; lost, presumed foundered with all (32) hands in North Atlantic 8.12.1828.

Beagle Woolwich Dyd

As built: 90ft 0in, 73ft 7⅝in x 24ft 8in (24ft 6in for tons) x 11ft 0in. $235\frac{10}{94}$bm. Draught 7ft 7in/9ft 5in.

Ord: 13.6.1817. K: 6.1818. L: 11.5.1820. C: by 19.7.1820 (for Coronation of King George IV).

First cost: £7,803 to build, + £5,913 fitting (1825–26) 'to survey Magellan's Streights'.

Completed as a survey ship 27.9.1825–16.3.1826 (sheathed with wood and re-coppered, and rigged as a barque). Commissioned 7.9.1825 under Cmdr Pringle Stokes (suicide 1.8.1828); sailed 22.5.1826 for South America. On 1.8.1828 under Lieut William George Skyring, then 13.11.1828 under Cmdr Robert Fitzroy; returned home 14.10.1830 and paid off 27.10.1830 at Plymouth. Fitted at Plymouth for survey 7–11.1831 (for £7,583); sailed 27.12.1831 for South America (with Charles Darwin as naturalist); returned home 2.10.1836 and paid off 17.11.1836 at Woolwich. Recommissioned 16.2.1837 under Cmdr John Clements Wickham (invalided 24.3.1841); sailed 5.7.1837 for South America and then to Australia; on 24.3.1841 under Lieut (acting, confirmed 16.8.1841) John Lort Stokes; returned home 8.10.1843 and paid off 14.10.1843 at Woolwich. Fitted and transferred to the coastguard 14.6–11.7.1845 at Paglesham. Renamed *WV.5* on 25.5.1863 (still at Paglesham). Sold to Murray & Trainer (for £540 – net £524.19.3d) to BU 13.5.1870.

Barracouta Woolwich Dyd

As built: 90ft 0in, 73ft 7⅝in x 24ft 8in (24ft 6in for tons) x 11ft 0in. $235\frac{10}{94}$bm. Draught 7ft 7½in/9ft 5½in.

Ord: 13.6.1817. K: 6.1818. L: 13.5.1820. C: 9.1821–11.1.1822 as a survey vessel.

First cost: £8,090 to build, + £3,570 fitting for foreign service.

Commissioned 2.10.1821 under Cmdr William Cutfield, for the coast of Africa (died 30.11.1823); on 15.5.1823 under Cmdr Alexander Thomas Emeric Vidal, on same station; paid off 19.9.1826 at Woolwich. Fitted there as a packet brig (for £6,650) 4.1828–7.1829. Recommissioned 30.5.1829 under Lieut Robert Bastard James, for the Falmouth packet service; paid off 29.7.1833 at Plymouth. Sold there (for £600) 21.1.1836.

Bustard Chatham Dyd

As built: 90ft 0in, 73ft 10¾in x 24ft 8½in (24ft 6½in for tons) x 11ft 0in. $236\frac{69}{94}$bm. Draught 7ft 10in/9ft 9in.

Ord: 13.6.1817. K: 11.1817. L: 12.12.1818. C: 1.1819 (for Ordinary). Roofed between the fore and main waist 7.1819.

Commissioned 14.1.1821 under Cmdr James Wigston, for the West Indies; fitted for sea 6.1821–27.9.1821; on 24.9.1822 under Cmdr Edwin Ludlow Rich, then 11.10.1824 under Cmdr William Stephen Arthur, then 10.6.1825 under Cmdr William Sandom, then 14.8.1826 under Cmdr Charles Elliott, then 12.3.1827 under Cmdr George Sidney Smith, all on same station; paid off 13.9.1828 at Portsmouth. Sold to Thomas Surflen (for £630) 24.6.1829.

Brisk Chatham Dyd

As built: 90ft 0in, 73ft 10¼in x 24ft 8½in (24ft 6½in for tons) x 11ft 0in. $236\frac{69}{94}$bm. Draught 7ft 9in/9ft 9in.

Ord: 13.6.1817. K: 11.1817. L: 10.2.1819. C: 22.5.1819 (for sea), then 16.6.1819 for foreign service.

First cost: £9,371.

Commissioned 31.3.1819 under Cmdr John William Montagu, on anti-smuggling service; in 12.1820 under Cmdr Edward Stewart (drowned 23.12.1823); on 26.12.1823 under Cmdr Lord Adolphus Fitzclarence, for the North Sea; on 28.2.1824 under Cmdr Charles Hope, on same station. In 1.1826 under Cmdr William Anson, for the Mediterranean; at Battle of Navarino 20.10.1827; paid off 3.6.1829. Between Small and Middling Repair at Portsmouth (for £3,048, hull only) 8.1829–6.1831, and reduced to 3-gun brigantine. Recommissioned 16.4.1831 under Lieut Edward Harris Butterfield, for Scotland, then on Channel and Irish stations, to the Scheldt and finally to Gambia; on 13.3.1832 under Lieut Josiah Thompson, for the Cape of Good Hope; took slavers *Preuva* (ex-frigate) 3.5.1832 and *Vertude* 23.10.1833; paid off 16.10.1835. Recommissioned 20.12.1837 under Lieut Arthur Kellett, for the coast of Africa; took slavers *Diligente* 15.8.1838, *Ligeira* 16.8.1838, *Veloz* 17.10.1838, *Maria* 8.11.1838, *Constitucion* 30.11.1838 and *Violante* 28.12.1838; paid off 19.8.1842. Sold at Sheerness to John Levy, Rochester (for £500) 7.11.1843.

Delight (i) Portsmouth Dyd

As built: 90ft 1in, 73ft 9in x 24ft 9in (24ft 7in for tons) x 11ft 1in. 237$\frac{7}{94}$bm. Draught 7ft 5in/9ft 1in.

Ord: 13.6.1817. K: 11.1817. L: 10.5.1819. C: 31.8.1822 for sea (housed over in 9.1821).

Commissioned 14.6.1822 under Cmdr Robert Hay, for the East Indies; foundered with all (75) hands off Port Louis, Mauritius 23.2.1824.

Cygnet Portsmouth Dyd

As built: 90ft 1in, 73ft 8$\frac{5}{8}$in x 24ft 9in (24ft 7in for tons) x 11ft 0in. 236$\frac{91}{94}$bm. Draught 7ft 9in/9ft 3in.

Ord: 13.6.1817. K: 11.1817. L: 11.5.1819. C: 10.9.1819 (for sea).

First cost: £10,451.

Commissioned 2.7.1819 under Cmdr Thomas Bennett, for the Irish station, and later off St Helena; paid off 5.1823. Converted at Portsmouth to a packet brig (for £9,754) 12.1823–3.1824; recommissioned 28.11.1823 under Lieut James Glassford Gooding, for the mail packet service at Falmouth; paid off 18.10.1832 at Plymouth. Sold there (for £560) 6.8.1835.

Eclipse Plymouth Dyd

As built: 90ft 0in, 73ft 7$\frac{1}{2}$in x 24ft 8$\frac{1}{4}$in (24ft 6$\frac{1}{4}$in for tons) x 11ft 0in. 235$\frac{44}{94}$bm. Draught 7ft 11in/9ft 1in.

Ord: 13.6.1817. K: 3.1818. L: 23.7.1819. C: 19.8.1819 (for Ordinary).

First cost: £7,647 (£8,377 fitted for Ordinary).

Converted to a packet brig (for £4,260) 3.1823–23.9.1823.

Commissioned 7.1823 under Lieut Augustus Robert Lloyd Passingham, as a Falmouth packet. In 9.1826 under Lieut Charles William Griffith Griffin; in 1.1834 under Lieut William Forrester; paid off 9.12.1836 at Chatham. To coastguard at Woolwich 12.1836, renamed *WV.21* on 25.5.1863. Sold to Messrs Castle (for £365) to BU at Charlton 10.11.1865.

Emulous Plymouth Dyd

As built: 90ft 0in, 73ft 7$\frac{1}{4}$in x 24ft 8$\frac{1}{8}$in (24ft 6$\frac{1}{8}$in for tons) x 11ft 0in. 235$\frac{19}{94}$bm. Draught 8ft 3in/9ft 10in.

Ord: 13.6.1817. K: 6.1818. L: 16.12.1819. C: 14.12.1823 as a packet brig.

First cost: £7,434 (£8,390 fitted for Ordinary).

Converted to a packet brig (for £3,692) 3.1823–14.12.1823.

Commissioned 24.11.1823 under Lieut Wentworth Parsons Croke, as a Falmouth packet; paid off 7.1833. Fitted at Chatham and transferred to coastguard at Haven Hole 8.1834, renamed *WV.13* on 25.5.1863. Sold to Mr T. R. Sargent, Northfleet (for £205) 7.6.1865.

Falcon Pembroke Dyd

As built: 90ft 1in, 73ft 8in x 24ft 9in (24ft 7in for tons) x 11ft 0in. 236$\frac{76}{94}$bm. Draught 7ft 8in/9ft 0in.

Ord: 13.6.1817. K: 5.1818. L: 10. 6.1820. C: 27.6–3.8.1820 at Plymouth.

First cost: £5,308, + £1,529 fitting for Ordinary.

Fitted at Plymouth for sea (for £2,461) to 29.8.1823. Tender to *Windsor Castle* at Plymouth from 9.1823. To Lisbon as tender to *Spartiate* 5.1826–5.1828. Fitted as tender at Plymouth (for £607) 3–4.1827.

Commissioned 30.5.1828 under Cmdr John Pole, for the Cape of Good Hope; fitted at Portsmouth 5–8.1828, then to Cape of Good Hope; on 31.1.1829 under Cmdr Edward Griffith Colpoys, on same station, then to the West Indies; on 4.12.1830 under Lieut Andrew Smith; paid off 30.6.1831. Converted to steam paddle sloop at Sheerness (by AO 6.3.1832), docking there for that purpose 10.5.1833, but the result was unsuccessful and the engine was removed 1834. Sold to Mr Hackwood for mercantile use 16.8.1838, renamed *Waterwitch*.

Frolic Pembroke Dyd

As built: 90ft 1$\frac{1}{2}$in, 73ft 8$\frac{7}{8}$in x 24ft 8$\frac{1}{2}$in (24ft 6$\frac{7}{8}$in for tons) x 11ft 0in. 236$\frac{78}{94}$bm. Draught 7ft 10in/9ft 4in.

Ord: 13.6.1817. K: 8.1818. L: 10. 6.1820. C: 27.6–3.8.1820 at Plymouth (for Ordinary).

First cost: £5,134, +£849 fitting for Ordinary.

Fitted as packet brig at Plymouth (for £4,093) 3.1823–15.6.1823.

Commissioned 4.1823 under Lieut Thomas Charles Barron, for Falmouth packet service. On 3.11.1829 under Lieut William Pringle Green, for same service; paid off 25.11.1832 at Plymouth. Sold to Messrs Dowson (for £690) 16.8.1838.

***1818* Batch** Another batch of twelve vessels were ordered 2 November 1818 (of which one was cancelled), again from each of the same six dockyards, with two more ordered on 8 December (this time from Sheerness Dockyard). Six of these vessels (*Jasper*, *Britomart*, *Lyra*, *Reynard*, *Opossum* and *Onyx*) were given names previously borne by earlier units of this class.

Jasper Portsmouth Dyd

As built: 90ft 1in, 73ft 8$\frac{7}{8}$in x 24ft 9in (24ft 7in) x 11ft 0in. 237$\frac{3}{94}$bm. Draught 7ft 7in/9ft 3in.

Ord: 2.11.1818. K: 5.1819. L: 26.7.1820. C: 8.8.1820 (for Ordinary, housed over).

First cost: £8,152 to build. Fitted for sea 4.1823–2.7.1823.

Commissioned 24.4.1823 under Cmdr Alexander Dundas Young Arbuthnott, for St Petersburg. In 6.2.1824 under Lieut (Cmdr 24.6.1824) George Rolle Walpole Trefusis, for oyster fishery protection. On 24.6.1824 under Cmdr Charles Howe Fremantle, to Mexico. In 6.1826 under Cmdr H. Robert Moorsom in the Mediterranean (died 25.7.1826). On 26.7.1826 under Cmdr Henry Martin Blackwood. On 28.4.1827 under Cmdr Leonard Charles Rooke; in 6.1827 under Lieut Adam Camperdown Duncan (acting), until Rooke resumed command. Wrecked off Santa Maura (Levkas) 13.10.1828. Wreck sold 1.1831 to N. Zumbelli (for £73.6.7d).

Britomart Portsmouth Dyd

As built: 90ft 1$\frac{1}{2}$in, 73ft 9$\frac{1}{2}$in x 24ft 9in (24ft 7in) x 11ft 0in. 237$\frac{19}{94}$bm. Draught 7ft 7in/9ft 0in.

Ord: 2.11.1818. K: 6.1819. L: 24.8.1820. C: 5.9.1820 (for Ordinary).

First cost: £7,607 to build. Fitted for sea 4.1823–15.9.1824 (for £2,877).

Commissioned 6.1824 under Cmdr Q. H. S. V. Vernon. On 20.2.1826 under Cmdr Henry Dundas Trotter, then 9.8.1826 under Cmdr Frederick Chamier. Recommisioned 21.10.1827 under Cmdr Russell Henry Manners, at Plymouth; on 4.3.1829 under Cmdr Edward John Johnson, for Lisbon station; on 22.11.1830 under Cmdr Lord Edward Russell; paid off 22.1.1831 at Plymouth. Fitted for sea at Plymouth (for £3,033) 4–5.1833. Recommissioned 3.1833 under Cmdr H. W. Quin, for Cape of

Good Hope. Very Small Repair and fitted for sea at Plymouth (for £3,176) 8.1837–2.1838. Recommissioned 12.1838 under Cmdr Owen Storey, for Australia; paid off 25.4.1843 when sold by public auction at Singapore (for $4,843) on same day.

Lyra Plymouth Dyd

As built: 90ft 0½in, 73ft 8⅞in x 24ft 8in (24ft 6in) x 11ft 0in. $235^{41}/_{94}$bm. Draught 7ft 11in/10ft 0in.

Ord: 2.11.1818. K: 3.1819. L: 1.6.1821. C: ?8.1821.

First cost: £7,633 to build, plus £3,080 fitting (as tender) 4.1823–23.7.1823.

Tender to *Superb* 6.1823; to *Ocean* 1824, to *Britannia* 1825, then at Lisbon 1826. Fitted as packet at Plymouth (for £5,632) 6.1828–7.1829.

Commissioned as Falmouth packet brig under Lieut James St John 16.6.1829. In 2.1837 under Lieut W. Forrester. Laid up at Plymouth 3.1842. Sold to Hull Dock Co. (for £500) 3.6.1845.

Partridge (i) Plymouth Dyd

As built: 90ft 0⅜in, 73ft 8in x 24ft 8in (24ft 6in) x 11ft 0in. $237^{3}/_{94}$bm. Draught 7ft 11½in/9ft 11½in.

Ord: 2.11.1818. K: 12.1819. L: 22.3.1822. C: 3.1822 (for Ordinary).

Fitted for sea 5.1823–30.8.1823.

Commissioned 1823 under Lieut Edward Yonge, as tender to *Britannia*; wrecked by stranding in the Texel 28.11.1824. Wreck sold (for $7,000) 30.8.1825.

Reynard Pembroke Dyd

As built: 90ft 2in, 73ft 9in x 24ft 9in (24ft 7in for tons) x 11ft 0in. $237^{7}/_{94}$bm. Draught 7ft 6in/9ft 4in.

Ord: 2.11.1818. K: 5.1820. L: 26.10.1821. C: 11–12.1821 at Plymouth (for Ordinary).

First cost: £4,958 to build, + £957 fitting for Ordinary.

Fitted for sea at Plymouth (for £3,295) 5–9.1823. Commissioned 18.12.1824 under Lieut Thomas Hardy, as tender to the *Genoa* 4.1825 at Portsmouth; by AO 24.7.1827 became tender to the *Ocean* at Portsmouth. Name spelt *Renard* from 1828. Recommissioned 6.1829 under Lieut George Dunsford, for the mail packet service at Falmouth; fitted as a packet brig (for £7,988) at Portsmouth 7–8.1829; paid off 20.1.1837. Fitted as mooring lighter at Chatham (for £809) 1.1841. BU at Chatham 8.1857.

Weazle Chatham Dyd

As built: 89ft 11in, 73ft 10in x 24ft 8½in (24ft 6½in for tons) x 10ft 11¾in. $236^{50}/_{94}$bm. Draught 7ft 9in/10ft 2in.

Ord: 2.11.1818. K: 5.1820. L: 26.3.1822. C: 3–4.4.1822 (for Ordinary).

First cost: £6,113 including fitting for Ordinary.

Commissioned 9.1823 under Cmdr Timothy Curtis, for the Mediterranean; fitted for foreign service (for £3,779) 9.1823–10.11.1823; on 6.6.1825 under Cmdr Richard Beaumont; paid off 1827. Recommissioned 9.3.1827 under Cmdr John Burnet Dundas, then 1.1828 under Cmdr William Wellesley, 5.1828 under Cmdr Thomas Edward Hoste, and 17.2.1829 under Cmdr Charles Basden, all on same station; paid off 18.9.1830. Small Repair and fitted for sea at Plymouth (for £5,193) 11.1837–7.1838. Recommissioned 9.7.1838 under Lieut John Simpson, for the Mediterranean; paid off 3.7.1840 at Chatham. Sold to W. Beech (for £540) at Chatham to BU 30.4.1844.

Kingfisher Woolwich Dyd

As built: 90ft 0½in, 73ft 8in x 24ft 9in (24ft 7in for tons) x 11ft 0in. $236^{76}/_{94}$bm. Draught 7ft 7in/9ft 8in.

Ord: 2.11.1818. K: 12.1820. L: 11.3.1823. C: 6.3.1824 as a packet brig.

First cost: £10,026 including fitting.

Commissioned 11.1823 under Lieut James Henderson, for the mail packet service at Falmouth; on 22.12.1825 under Lieut William Poore; on 3.12.1828 under Lieut Bethune James Walker, on same service; paid off 6.4.1830. Sold to Mr Knowland at Plymouth (for £680) 16.8.1838.

Procris Chatham Dyd

As built: 89ft 10¾in, 73ft 8in x 24ft 8in (24ft 6in for tons) x 11ft 0in. $235^{19}/_{94}$bm. Draught 7ft 10in/9ft 11in.

Ord: 2.11.1818. K: 3.1821. L: 21.6.1822. C: 6.1822 (for Ordinary).

First cost: £5,060 to build.

Commissioned 11.1825 under Cmdr William Waldegrave, for the North Sea; fitted for sea (for £3,759) 11.1825–9.3.1826; on 12.8.1828 under Cmdr Charles Henry Paget, for Cork; paid off 1829. Recommissioned 30.11.1829 under Cmdr Sir Thomas Sabine Pasley, for the Mediterranean; on 22.5.1830 under Lieut William Tomlin Griffiths, then 20.5.1831 under Cmdr John Thomas Talbot, on the same station; paid off 19.7.1832. Sold to W.Greenwood, Plymouth (for £660) at Plymouth 12.1.1837.

Algerine (i) Deptford Dyd

As built: 90ft 0in, 72ft 2⅛in x 24ft 8¼in (24ft 6¼in for tons) x 11ft 0in. $230^{79}/_{94}$bm. Draught 8ft 0in/10ft 2in.

Ord: 2.11.1818. K: 4.1821. L: 10.6.1823. C: 28.10.1823–27.5.1824 at Woolwich.

Commissioned 2.1824 under Cmdr Montagu Stopford, for the Mediterranean; on 8.4.1825 under Cmdr Charles Wemyss; lost, presumed foundered with all (75) hands during a storm in the Mediterranean 9.1.1826.

Magnet Woolwich Dyd

As built: 90ft 0½in, 73ft 8in x 24ft 9in (24ft 7in for tons) x 11ft 0in. $236^{76}/_{94}$bm. Draught 7ft 7in/9ft 10in.

Ord: 2.11.1818. K: 7.1821. L: 13.3.1823. C: 14.12.1823 as a packet brig (3 guns).

First cost: £9,860 including fitting

Commissioned 10.1823 under James Porteous, Master, for the mail packet service based at Falmouth; paid off 22.6.1832 at Plymouth. Recommissioned 9.7.1836 under Lieut Smyth Griffith, for the same station; paid off 9.5.1842. Sold at Plymouth (for £630) by AO of 27.7.1847.

Zephyr Pembroke Dyd

As built: 90ft 0in, 71ft 3⅜in x 24ft 8½in (24ft 6½in for tons) x 11ft 0in. $228^{34}/_{94}$bm. Draught 8ft 6in/8ft 8in.

Ord: 2.11.1818. K: 11.1821. L: 1.11.1823. C: 25.11.1823–23.6.1824 at Plymouth as a packet brig (6 guns).

First cost: £4,578.

Commissioned 12.1824 under Lieut Charles Church, for the mail packet service based at Falmouth; paid off 23.5.1832. Sold at Plymouth to Mr Greenwood (for £650) 8.9.1836.

Halcyon Woolwich Dyd

Ord: 2.11.1818. K: — . Cancelled 21.2.1831.

Opossum Sheerness Dyd

As built: 90ft 0in, 73ft 7¾in x 24ft 8½in (24ft 6½in for tons) x 11ft 0in. $236^{88}/_{94}$bm. Draught 7ft 6in/9ft 8in.

Ord: 8.12.1818. K: 11.1819. L: 11.12.1821. C: laid up in Ordinary.

First cost: £6,356 to build.

Fitted (partly) for sea at Sheerness (for £948) 10.1827–21.10.1828, then completed at Chatham (for £4,259) to 3.5.1829 as a packet brig. Commissioned 29.3.1829 under Lieut Thomas Hannam, for the mail packet service at Falmouth; in 11.1830 under Lieut Robert Peter, on same station; paid off 1.7.1840 at Woolwich. Sold to John Levy, Rochester (for £650) 27.5.1841.

Onyx Sheerness Dyd
>As built: 90ft 0in, 73ft 7¼in x 24ft 8¾in (24ft 6¾in for tons) x 11ft 0in. 236¹⁹⁄₉₄bm. Draught 7ft 6in/9ft 8in.
>Ord: 8.12.1818. K: 11.1819. L: 24.1.1822. C: laid up in Ordinary.
>First cost: £6,620 to build.
>Fitted as tender to yacht *Royal George* at Sheerness (for £3,877) 5.1827–29.9.1827. Commissioned 9.1827 under Lieut William John Cole, as tender to yacht *Royal George*; to west coast of Africa 2–5.1828; in 9.1828 under Lieut John Harvey Boteler, then 1.1830 under Lieut William Dawson, in same role. On 16.9.1831 under Lieut Alexander Burgoyne Howe, at Cork; paid off 12.11.1832. Sold to Joshua Crystall (for £600) 12.1.1837 to BU.

1820–21 **Batch** Six vessels were ordered on 23 May 1820, plus another on 6 January 1821 and two more on 19 April 1821. *Hope* and *Tyrian* were given names previously borne by earlier units of this class.

Plover Portsmouth Dyd
>As built: 90ft 1in, 73ft 8in x 24ft 9in (24ft 7in for tons) x 11ft 0in. 236⁶⁹⁄₉₄bm. Draught 7ft 10in/9ft 5in.
>Ord: 23.5.1820. K: 8.1820. L: 30.6.1821. C: 18.7.1821 (for Ordinary). Housed over 9.1821.
>First cost: £6,741 (fitted for Ordinary)
>Commissioned 24.11.1823 under Lieut Edward Jennings, for the mail packet service at Falmouth; fitted as a packet brig (for £10,977) 11.1823–3.1824; on 19.6.1828 under Lieut William Downey, then 11.9.1835 under Lieut William Luce, both on same station; paid off 27.7.1836. Fitted as lazarette at Sheerness 10.1836, then lent to Isambard Brunel as an accommodation hulk for the Thames Tunnel 3.1838; returned to RN at Woolwich 11.1839. Sold to Joshua Crystall (for £640) 27.5.1841.

Ferret Portsmouth Dyd
>As built: 90ft 1in, 73ft 8½in x 24ft 9in (24ft 7in for tons) x 11ft 0in. 236⁸⁸⁄₉₄bm. Draught 7ft 7in/9ft 2in.
>Ord: 23.5.1820. K: 8.1820. L: 12.10.1821. C: 27.12.1821 (for Ordinary).
>First cost: £7,097 (fitted for Ordinary)
>Commissioned 15.1.1825 under Cmdr William Hobson (appointed 1.1.1825), for Jamaica; fitted for sea (for £3,818) 6.1825–11.4.1825; on 10.7.1826 under Cmdr Henry Gosset, then 14.4.1828 under Lieut Charles Ramsey Drinkwater, then 8.1828 under Lieut Charles Deare, all on same station; paid off 3.11.1828 at Portsmouth. Recommissioned 4.11.1828 under Cmdr Thomas Hastings, for the Mediterranean; on 23.7.1830 under Cmdr Edward Thornton Wodehouse, on same station; paid off 9.3.1832 and laid up at Plymouth. Sold to H. Bayly at Plymouth (for £600) 12.1.1837.

Hope (ii) Plymouth Dyd
>As built: 90ft 0in, 72ft 3⅞in x 24ft 8in (24ft 6in for tons) x 11ft 0in. 230⁸⁶⁄₉₄bm. Draught 8ft 0in/9ft 11in.
>Ord: 23.5.1820. K: 3.1822. L: 8.12.1824. C: 1.6.1825 as a packet brig (4 guns).
>First cost: £6,029 to build (+ £2,915 fitting).
>Commissioned 31.5.1825 under Lieut John Wright, for the mail packet service at Falmouth; paid off 3.11.1832. Recommissioned 18.7.1836 under Lieut William Lee Rees, for the mail packet service at Falmouth; on 26.10.1839 under Lieut Thomas Creser; paid off 31.1.1842 at Plymouth. To Milford 5.1846 and fitted as a lazarette at Pembroke 5–8.1846. BU (by AO 18.8.1882) completed at Pembroke 13.10.1882.

Mutine Plymouth Dyd
>As built: 90ft 0in, 72ft 3⅛in x 24ft 8in (24ft 6in for tons) x 11ft 0in.

>230⁷⁹⁄₉₄bm. Draught 8ft 1in/10ft 0in.
>Ord: 23.5.1820. K: 4.1822. L: 19.5.1825. C: 21.10.1826 as a packet brig (4 guns).
>First cost: £6,959 to build (+ £2,890 fitting).
>Commissioned 30.8.1826 under Lieut Richard Pawle, for the mail packet service at Falmouth; paid off 15.9.1840. Sold at Chatham to Tibbett, Stoneman & Spence (for £840) for mercantile use 27.5.1841, renamed *Aladdin*.

Forester Deptford Dyd
>Ord: 23.5.1820. K: not laid down. Re-ordered at Chatham 23.5.1826.

Griffon Deptford Dyd
>Ord: 23.5.1820. K: not laid down. Re-ordered at Chatham 23.5.1826.

Tyrian (ii) Woolwich Dyd (built of winter-felled timber).
>As built: 90ft 0in, 72ft 3in x 24ft 9in (24ft 7in for tons) x 11ft 0in. 232²⁴⁄₉₄bm. Draught 7ft 11in/10ft 1in.
>Ord: 6.1.1821. K: 4.1823. L: 16.9.1826. C: 16.11.1826 as a packet brig.
>First cost: £8,026 to build (£10,813 fitted for sea).
>Commissioned 9.1826 under Lieut Robert Dwyer, for the mail packet service at Falmouth; paid off 23.10.1832. Fitted for sea at Plymouth (for £3,817) 0.1833–8.1834. Recommissioned 6.6.1834 under Lieut Edward Jennings, in same service; paid off 2.1839. Fitted at Plymouth as a Quarantine hulk for the Motherbank (off Portsmouth) 10.1844–4.1845. Fitted as Portsmouth as a receiving ship 2–4.1864. Transferred to coastguard as a depot (receiving) ship by AO 6.11.1866, restored to receiving ship at Portsmouth 1870. Made good defects 1–10.1877 by John White, Cowes. Sold later at Portsmouth.

Philomel Portsmouth Dyd
>As built: 90ft 0in, 72ft 2½in x 24ft 8½in (24ft 6½in for tons) x 11ft 0in. 231³¹⁄₉₄bm. Draught 7ft 4in/9ft 2in.
>Ord: 19.4.1821. K: 6.1821. L: 28.4.1823. C: 23.5.1823 (for Ordinary).
>First cost: £7,351 (fitted for Ordinary)
>Commissioned 23.12.1825 under Cmdr Lord William Paget, for the Mediterranean; fitted for sea 1.1826–22.4.1826; on 18.10.1826 under Cmdr Henry John Chetwynd, Viscount Ingestrie; at Battle of Navarino 20.10.1827; on 14.8.1827 under Cmdr William Keith (nominally from that date, but did not actually take over from Ingestrie until after Navarino); on 18.8.1828 under Cmdr Edward Hawes, on same station; paid off 15.6.1829. Very Small Repair and fitted for sea at Chatham 6–8.1829. Recommissioned 15.6.1829 under Cmdr Charles Graham, for the Mediterranean; on 14.4.1831 under Cmdr William Smith; paid off 16.9.1833. Sold to Capt. Templar (for £690) 12.12.1833.

Royalist Portsmouth Dyd
>As built: 90ft 0in, 72ft 2in x 24ft 8½in (24ft 6½in for tons) x 11ft 0in. 231¹⁸⁄₉₄bm. Draught 7ft 2in/9ft 5in.
>Ord: 19.4.1821. K: 8.1821. L: 12.5.1823. C: 23.5.1823 (for Ordinary).
>First cost: £7,213 (fitted for Ordinary)
>Fitted at Portsmouth as a tender (for £2,623) 3.1826–26.5.1826; served as tender to *Britannia* there 1826, then 1829 as tender to *Britannia* at Plymouth; in 1830 tender to *St Vincent*, later to *Caledonia*, both at Plymouth. Commissioned 15.6.1831 under Lieut Richard Nicholls Williams (–summer 1834), at Lisbon; on 15.12.1834 under Lieut Charles Anstruther Barlow, then 19.1.1837 under Lieut Edward Plunkett; paid off 3.9.1838 at Plymouth. Sold to Mr Lindon (for £804) 8.11.1838.

1823 **Batch** Thirty vessels were ordered on 25 March 1823 (of which two were cancelled and six others replaced by orders for paddle vessels). *Briseis*, *Leveret* and *Rolla* were given names previously borne by earlier units of this class; *Calypso* and *Hyaena* exchanged names during construction in 1826.

Alban Deptford Dyd
Ord: 25.3.1823. Re-ordered as paddle steamer 5.1824 (see Chapter 11).

Briseis (ii) Deptford Dyd
As built: 90ft 0¼in, 72ft 2¼in x 24ft 8in (24ft 6in for tons) x 11ft 0in. $230\frac{45}{94}$bm. Draught 8ft 0in/9ft 11in.
Ord: 25.3.1823. K: 8.1827. L: 3.7.1829. C: 2.11.1829 at Woolwich as packet brig (6 guns).
Commissioned 8.1829 under Lieut John Downey, for the mail packet service at Falmouth; lost, presumed foundered with all (33) hands in North Atlantic 1.1838, on passage from Falmouth to Halifax (Nova Scotia).

Hyaena (ex-*Calypso*) Deptford Dyd
Ord: 25.3.1823. K: – Cancelled 21.2.1831.

Carron Deptford Dyd
Ord: 25.3.1823. Re-ordered as paddle steamer 5.1824 (see Chapter 11).

Columbia Woolwich Dyd
Ord: 25.3.1823. Re-ordered as paddle steamer 5.1824 (see Chapter 11).

Confiance Woolwich Dyd
Ord: 25.3.1823. Re-ordered as paddle steamer 5.1824 (see Chapter 11).

Dee Woolwich Dyd
Ord: 25.3.1823. Re-ordered as paddle steamer 5.1824 (see Chapter 11).

Echo Woolwich Dyd.
Ord: 25.3.1823. Re-ordered as paddle steamer 5.1824 (see Chapter 11).

Espoir Chatham Dyd
As built: 90ft 4in, 72ft 5½in x 24ft 8½in (24ft 6½in for tons) x 11ft 0¼in. $232\frac{12}{94}$bm. Draught 7ft 10in/9ft 11in.
Ord: 25.3.1823. K: 1.1825. L: 9.5.1826. C: 27.5.1826 (for Ordinary).
First cost: £5,941 (fitted for Ordinary).
Commissioned 1.5.1827 under Cmdr Henry Francis Greville, for the Cape of Good Hope; fitted for sea at Chatham (for £4,202) 6.1827–14.8.1827; paid off 15.1.1831. Fitted for sea at Portsmouth (for £3,224) 1–4.1834. Recommissioned 9.1.1834 under Lieut Charles Wilson Riley, for the mail packet service at Falmouth; paid off summer 1837 at Plymouth. Recommissioned 15.1.1838 under Lieut John Thomas Paulson, for Lisbon; paid off 24.12.1841. Between a Very Small and a Small Repair, and fitted for sea at Plymouth (for £3,336) 5.1842–7.1843. Recommissioned 25.5.1843 under Cmdr Arthur Fleming Morell, for the west coast of Africa; took slavers *Agnia* 19.9.1843, *Helena* 29.11.1843 and *El No Sé* 29.9.1844; on 14.12.1844 under Cmdr George Sumner Hand, on same station; took slavers *Karem* 5.1.1845, *Vinte Nova* 27.3.1845 and *Electra* 23.10.1846; paid off 7.4.1847 at Plymouth. Sold 1853 at Plymouth (by AO of 19.3.1853, for £710).

Fairy Chatham Dyd
As built: 90ft 2in, 72ft 5½in x 24ft 9in (24ft 7in for tons) x 10ft 11½in. $232\frac{86}{94}$bm. Draught 7ft 10in/9ft 10in.
Ord: 25.3.1823. K: 7.1824. L: 25.4.1826. C: 22.5.1826 (for Ordinary).
First cost: £5,545 (fitted for Ordinary).
Commissioned 17.4.1827 under Cmdr George William Conway Courtenay, for Jamaica; fitted for sea at Chatham (for £4,212) 4.1827–17.7.1827; on 12.9.1827 under Cmdr David Edwards, for same station; on 24.4.1828 under Cmdr Francis Blair; on 10.3.1830 under Lieut William Molyneux (acting); paid off 30.6.1830 at Chatham. Fitted as survey vessel at Sheerness 12.1831–4.1832. Recommissioned 6.12.1831 under Cmdr William Hewett, for surveying in North Sea; paid off 12.1834. Recommissioned 28.12.1834 and again 31.12.1837, still under Hewett; sailing from Harwich 12.11.1840, wrecked off Suffolk in a storm with all (63) hands 13.11.1840.

Harpy Chatham Dyd
As built: 90ft 2in, 72ft 3½in x 24ft 9in (24ft 7in for tons) x 11ft 0in. $232\frac{36}{94}$bm. Draught 7ft 9in/9ft 11in.
Ord: 25.3.1823. K: 3.1824. L: 16.7.1825. C: 30.7.1825 (for Ordinary).
First cost: £5,854 (fitted for Ordinary).
Commissioned 18.1.1828 under Cmdr Charles Rich, for Jamaica; fitted for sea at Chatham (for £4,065) 1.1828–13.5.1828; on 8.8.1829 under Cmdr Joseph Pafford Dickson Larcom, on same station; paid off 11.6.1831. Fitted for sea at Plymouth (for £3,306) 10.1835–3.1836. Recommissioned 1.1836 under Cmdr G. R. A. Clements, for North America and the West Indies; on 19.1.1838 under Lieut John Spencer Ellman; paid off 5.4.1839. Sold to Greenwood & Clarke at Plymouth (for £710) 27.5.1841.

Hearty Chatham Dyd
As built: 90ft 1¼in, 71ft 0¼in x 24ft 8½in (24ft 6½in for tons) x 11ft 0in. $227\frac{49}{94}$bm. Draught 8ft 0in/9ft 6in.
Ord: 25.3.1823. K: 7.1823. L: 22.10.1824. C: 6.11.1824 (for Ordinary)
First cost: £5,864 (fitted for Ordinary and coppered).
Commissioned 5.1827 under Cmdr Henry Jewry; fitting completed 26.7.1827 as packet brig (10 guns) for the mail packet service at Falmouth; sailed from Falmouth 12.9.1827 bound for Jamaica; lost, presumed foundered with all (32) hands in North Atlantic 9.1827.

Calypso (ex-*Hyaena*) Chatham Dyd
As built: 90ft 3in, 72ft 6in x 24ft 9in (24ft 7in for tons) x 12ft 2in. $233\frac{5}{94}$bm. Draught 8ft 0in/10ft 2in.
Ord: 25.3.1823. K: 3.1825. L: 19.8.1826. C: 29.10.1826 as 2-gun yacht for Governor of Malta (Marquis of Hastings), returned to Navy 5.1829.
First cost: £11,049 (fitted as yacht).
Commissioned 5.1829 under Lieut Richard Payton; fitted at Plymouth 5–10.1829 packet brig (6 guns) for the mail packet service at Falmouth; lost, presumed foundered with all (30) hands off Halifax, Nova Scotia 1.2.1833 on passage for home.

Lapwing Chatham Dyd
As built: 90ft 9¾in, 71ft 1½in x 24ft 8⅝in (24ft 6⅝in for tons) x 11ft 0in. $228\frac{6}{94}$bm. Draught 8ft 0in/9ft 7in.
Ord: 25.3.1823. K: 9.1823. L: 20.2.1825. C: 8.1827–6.1.1829 as packet brig (6 guns).
First cost: £5,878 (fitted for Ordinary and coppered), then £4,737 fitting as packet brig.
Commissioned 14.10.1829 under Lieut George Brooke Forster, for the mail packet service at Falmouth; paid off 7.5.1737. Recommissioned 18.7.1837 under Lieut Francis Rogers Coghlan, for the same service; paid off 7.12.1841. Became breakwater at Keyham (Plymouth) 6.1845; Devonport Dyd lighter 6.1854. Sold to Marshall, Plymouth 22.11.1861.

Leveret (ii) Portsmouth Dyd
As built: 90ft 0in, 72ft 2⅛in x 24ft 8¾in (24ft 6¾in for tons) x 11ft 0in. $231\frac{59}{94}$bm. Draught 7ft 7in/9ft 6in.
Ord: 25.3.1823. K: 5.1823. L: 19.2.1825. C: 27.3.1825 (partly fitted), then 3–5.1826 as tender.

First cost: £6,591, + £3,335 fitting as tender.
Not commissioned until 1832, employed as tender to *Windsor* Castle (1826), then *Warspite* (1829) and *Kent* (1830–32) at Plymouth. Fitted at Plymouth (for £1,547) 1.1832. Commissioned 2.10.1833 under Lieut Gilbert Traill; on 11.9.1835 under Lieut Charles John Bosanquet, for the Cape of Good Hope; took slavers *Atafa Primo* and *Zema* 25.1.1836; at Mozambique 9.1836; took slaver *Diogenes* 12.1836; paid off 6.8.1839. Sold at Plymouth 7.11.1843.

Musquito Portsmouth Dyd
As built: 90ft 0in, 72ft 2⅜in x 24ft 8½in (24ft 6½in for tons) x 11ft 0in. $231\frac{28}{94}$bm. Draught 7ft 6in/9ft 4in.
Ord: 25.3.1823. K: 5.1823. L: 19.2.1825. C: 28.7.1827.
First cost: £6,381 to build, + £3,792 fitting in 1827.
Commissioned 17.4.1827 under Cmdr George Bohun Martin, for the Mediterranean; at Battle of Navarino 20.10.1827; on 19.4.1828 under Cmdr Charles Bentham, on same station; paid off and laid up 10.1830. Sold to Greenwood & Clarke (for £510) at Plymouth 7.11.1843.

Myrtle Portsmouth Dyd
As built: 90ft 0in, 72ft 1⅞in x 24ft 9in (24ft 7in for tons) x 11ft 0in. $231\frac{89}{94}$bm. Draught 8ft 0in/9ft 8in.
Ord: 25.3.1823. K: 7.1823. L: 14.9.1825. C: 1.1827–27.10.1827 as packet brig (6 guns).
First cost: £6,409 to build.
Commissioned 8.1827 under Lieut Samuel Sisson, for the mail packet service based at Falmouth; wrecked off Nova Scotia 3.4.1829; wreck sold 25.6.1829 (for £21).

Rapid Portsmouth Dyd
As built: 90ft 1in, 72ft 3in x 24ft 8½in (24ft 6½in for tons) x 11ft 0in. $231\frac{44}{94}$bm. Draught 7ft 3in/9ft 5in.
Ord: 25.3.1823. K: 1.1824. L: 17.8.1829. C: 23.11.1829.
First cost: £10,288 fitted for sea.
Commissioned 7.9.1829 under Cmdr Charles Henry Swinburne, for the Mediterranean; paid off 15.7.1833, but recommissioned under Lieut Frederick Patten, for South America (–1836). Recommissioned under Lieut Graham Kinnaird, for the Mediterranean; wrecked off Tunisia 12.4.1838 (Kinnaird drowned – the sole fatality).

Recruit Portsmouth Dyd
As built: 90ft 1in, 72ft 2¾in x 24ft 8½in (24ft 6½in for tons) x 11ft 0in. $231\frac{37}{94}$bm. Draught 7ft 3in/9ft 6in.
Ord: 25.3.1823. K: 2.1825. L: 17.8.1829. C: 3.1831–18.8.1831 as tender to *Asia*.
Commissioned 1.7.1831 under Lieut Thomas Hodges, as tender; in 1832 to North America and the West Indies; lost, presumed foundered with all (52) hands in North Atlantic 6.1832 while en route from Halifax for Bermuda.

Reindeer Plymouth Dyd
As built: 90ft 0in, 72ft 1in x 24ft 8¼in (24ft 6¼in for tons) x 11ft 0in. $230\frac{51}{94}$bm. Draught 7ft 7in/9ft 7in.
Ord: 25.3.1823. K: 12.1824. L: 29.9.1829. C: 2.3.1830 as packet brig (4 guns) 1.1830.
First cost: £8,470 (£11,414 fitted for sea).
Commissioned 11.1.1830 under Lieut Henry Perry Dicken, as packet brig based at Falmouth; paid off 26.8.1840 at Chatham. Fitted at Chatham 10.1840–5.1841 for harbour service (guard ship) for Gibraltar. Recommissioned 31.5.1841 under Lieut Philip Bisson, for passage to Gibraltar; paid off 6.1.1843 there. Sold there (for £585.17.5¾d) before 17.1.1857, being replaced by *Samarang*.

Rolla (ii) Plymouth Dyd
As built: 90ft 0in, 72ft 1¼in x 24ft 8¼in (24ft 6¼in for tons) x 11ft 0in. $230\frac{57}{94}$bm. Draught 7ft 8in/9ft 8in.
Ord: 25.3.1823. K: 6.1825. L: 10.12.1829. C: 4.1.1834.
First cost: £7,321 to build (£9,679 fitted for sea).
Commissioned 20.11.1833 under Lieut Frederick Henry Hastings Glasse, for Lisbon, then to Scottish fishery protection; to the west coast of Africa 1836; took slavers *Lusita* 21.11.1836, *San Nicolas* 2.12.1836, and *Experimento* and *Lechuquino* both 27.12.1836; paid off 18.11.1837. Recommissioned 20.9.1838 under Lieut Charles Hall, for the Cape of Good Hope and west coast of Africa; took slavers *Porto Formosa* 16.9.1840 and *Feliz Ventura* 29.11.1840; in action at Gallinas 19–26.11.1840 destroying slave compounds; paid off 18.10.1842 at Chatham. Recommissioned 21.12.1844 under Cmdr John Simpson, for the west coast of Africa; 9.11.1846. Recommissioned 18.11.1846 under Cmdr Hugh Myddelton Ellicombe, for the west coast of Africa; paid off 28.12.1847. Fitted as boys' training brig 2.1848 (reduced to 6 guns). Recommissioned 30.12.1853 under Lieut William Henry Fenwick, as tender to *Victory* at Portsmouth; on 5.9.1856 under Lieut Charles Gudgeon Nelson, in same role; apprentices training brig 1.1.1858. BU completed at Portsmouth 15.9.1868.

Savage Plymouth Dyd
As built: 90ft 7in, 71ft 2⅜in x 24ft 7in (24ft 6in for tons) x 11ft 0in. $227\frac{30}{94}$bm. Draught 8ft 0in/9ft 11in.
Ord: 25.3.1823. K: 10.1829. L: 29.12.1830. C: 19.3.1831.
First cost: £7,654 to build (£10,359 fitted for sea).
Commissioned 10.1.1831 under Cmdr Lord Edward Russell, sailed 17.3.1831 for the coast of Ireland; on 1.11.1832 under Lieut Robert Loney, off Portugal, later to Venezuela. Recommissioned 17.12.1836 under Lieut E. R. Curzon, for the West Indies; paid off 23.7.1836 at Plymouth, then recommissioned (still under Curzon); sailed for Spain 2.1.1838; on 16.7.1838 under Lieut Edward Plunkett, for the coast of Spain; paid off 18.9.1840 at Plymouth. Recommissioned ?9.1840 under Lieut John Harrison Bowker (appointed 31.8.1840), for the Mediterranean; paid off 16.12.1844 at Plymouth. Hulked at Plymouth 4.6.1848 as accommodation for Harbourmaster's staff. Fitted as chain (mooring) lighter at Plymouth (for £2,193) for Malta 9.1852–7.1853. Sold at Malta 10.9.1862 and BU 1866.

Saracen Plymouth Dyd
As built: 90ft 8¾in, 71ft 1¼in x 24ft 7¼in (24ft 6¼in for tons) x 11ft 0in. $227\frac{51}{94}$bm. Draught 7ft 11in/9ft 11in.
Ord: 25.3.1823. K: 12.1829. L: 30.1.1831. C: 30.1.1834.
First cost: £7,617 to build (£10,474 fitted for sea).
Commissioned 20.11.1833 under Lieut Thomas Philip Le Hardy, for the coasts of Portugal and Spain; paid off 4.1837 and recommissioned 14.4.1837 under Lieut Henry Worsley Hill, for the Cape of Good Hope and West coast of Africa; took slavers *Brilhante* 16.10.1839, *Diana* 21.7.1840, *Sirena* 17.8.1840, *Buoao Uniao* 9.12.1840, *San Paolo Leandro* 11.12.1840, *Vracca* 19.1.1841 and *Republicano* 22.2.1841; paid off 20.7.1841. Between Very Small and Small Repair, then fitted for surveying at Plymouth (for £5,028) 12.1851–2.1854. Recommissioned 18.11.1853 under John Richards, Master, for surveying in the China; on 1.9.1858 under William Stanton, Master, on same service; paid off 30.6.1862. Sold at Singapore (by AO 10.9.1862, for $7,050) in late 1862 – a sailing vessel of about 75 tons, *Young Queen*, being purchased in her place for $3,500 and taking her name on 29.12.1862.

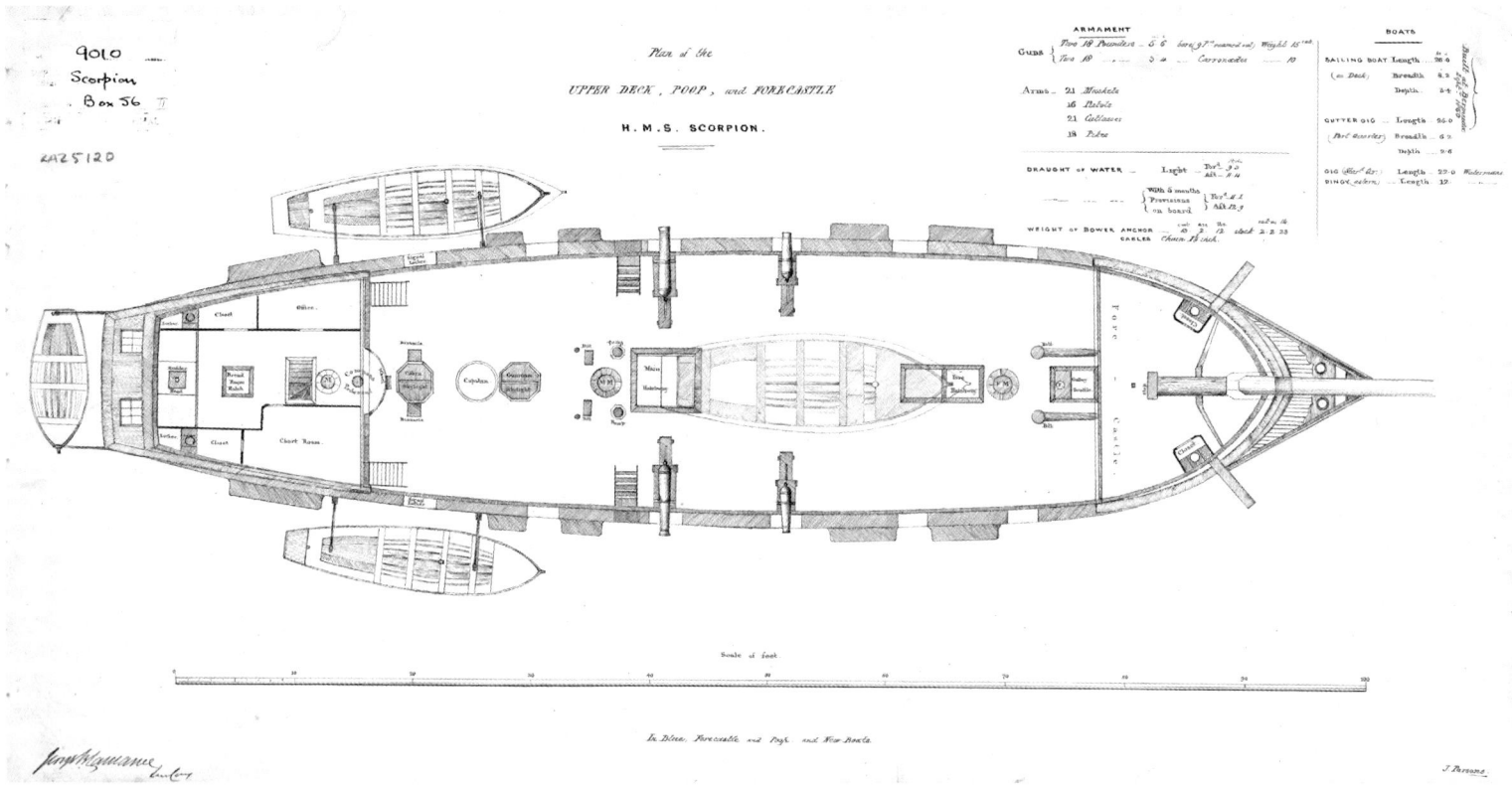

This highly detailed plan view of the upper deck of the brig sloop *Scorpion* reveals many features of her as converted into a three-masted survey vessel at Plymouth in 1849, recorded as schooner-rigged. An extended poop (to house a chart room) and forecastle have been constructed on the upper deck, and the vessel has been re-armed with just two 18pdr long guns (15cwt/5½ft) and two 18pdr carronades in the waist. Even the boats are carefully delineated: a 26ft sailing boat stowed amidships, a 24ft cutter gig on the port quarter (both built in Bermuda), a 22ft gig on the starboard quarter, and a 12ft dinghy on the stern davits. The draught bears the signature of Lieutenant George Bennett Lawrence, who recommissioned her on 2 April 1849 and spent the next four years on survey duties off North America and in the Caribbean. *(NMM J0061)*

Scorpion Plymouth Dyd
As built: 90ft 7⅛in, 71ft 1⅛in x 24ft 7½in (24ft 6½in for tons) x 11ft 0in. 227⁸⁴/₉₄bm. Draught 7ft 11in/9ft 11½in.
Ord: 25.3.1823. K: 6.1830. L: 28.7.1832. C: 22.2.1834.
First cost: £6,219 to build (£9,087 fitted for sea).
Commissioned 12.1833 under Lieut Nicholas Robilliard, as packet brig based at Falmouth; on 8.10.1836 under Lieut Edward Holland; paid off 1.1837. Recommissioned 13.2.1837 under Lieut Charles Gayton, for the Mediterranean; paid off 23.11.1841 at Plymouth. Reduced to 6 guns 1842. Fitted at Plymouth as a survey vessel (for £4,171) 1–5.1849 (rigged as schooner). Recommissioned 2.4.1849 under Lieut George Bennett Lawrence, for surveying North America and the West Indies; paid off 15.1.1853. Recommissioned 14.1.1855 under John Parsons, Master, on the same service; paid off 17.4.1857. Police accommodation hulk at Blackwall 3.1858. BU at Chatham 15.9–17.10.1874.

Sealark Plymouth Dyd
Ord: 25.3.1823. K: 11.1830. L: — . Cancelled 10.1.1831 (frame BU 2.1831).

Sheldrake Pembroke Dyd
As built: 90ft 0in, 71ft 0½in x 24ft 8½in (24ft 6½in for tons) x 11ft 0in. 227¹⁵/₉₄bm. Draught 8ft 2in/9ft 0in.
Ord: 25.3.1823. K: 11.1823. L: 19.5.1825. C: 21.10.1826 at Plymouth as packet brig (4 guns).

First cost: £5,362 to build, + fitting £1,568 (as a packet) + £2,877 (for sea): total £9,807.
Commissioned 8.1826 under Lieut Robert Ede, for the mail packet service based at Falmouth; paid off 28.6.1830. Recommissioned 9.4.1832 under Lieut Augustus Robert Lloyd Passingham, for the same station; paid off 12.1841. Sold 19.3.1853 (by AO of that date, for £710) to Messrs Wilson at Plymouth.

Skylark Pembroke Dyd
As built: 90ft 0in, 72ft 3in x 24ft ?10½in (24ft 8½in for tons) x 11ft 0in. 234⁶⁷/₉₄bm. Draught 8ft 4in/9ft 6in.
Ord: 25.3.1823. K: 5.1825. K: 6.5.1826. C: 2.6.1826–22.2.1827 at Plymouth as packet brig (4 guns).
First cost: £5,151 to build, + £3,544 fitting as a packet
Commissioned 11.1.1827 under Lieut Benjamin Aplin, for the mail packet service based at Falmouth; paid off 31.7.1830. Fitted for sea at Plymouth (for 33,267) 9.1830–21.10.1831. Recommissioned 5.8.1831 under Lieut Charles Pybus Ladd, on same station; paid off 18.12.1840. Recommissioned 14.6.1841 at Sheerness under Lieut John Allan Wright, for the Mediterranean. Recommissioned 25.3.1843 under Lieut George Morris, for the coast of Scotland; wrecked on Kimmeridge Ledge near St Alban's Head, Weymouth 25.4.1845.

Spey Pembroke Dyd
As built: 90ft 0in, 72ft 3in x 24ft 8in (24ft 6in for tons) x 11ft 0in. 230⁶⁴/₉₄bm. Draught 8ft 5in/9ft 6in.
Ord: 25.3.1823. K: 7.1825. L: 6.10.1827. C: 15.10.1827–17.11.1828 at Plymouth as packet brig (4 guns).
First cost: £5,494 to build, + £3,354 fitting as a packet.
Commissioned 10.1828 under William James, Master, for the mail packet service at Falmouth; paid off 26.6.1832. Fitted for sea at Plymouth (for £2,261) 8–9.1833. Recommissioned 25.7.1838 under Lieut Robert Bastard James, for the same service;

A lithograph by A. B. Sturdee of the 10-gun brig *Curlew* coming into Spithead on 1 September 1838 after a notably successful commission off the west coast of Africa, during which she took five slaving vessels. Following 14 years on near-continuous service from 1830 onwards, mostly on the anti-slavery patrols, the brig was paid off at Portsmouth in 1844. During her retirement years, she was seemingly fitted as a mortar brig. The success or otherwise of the experiment is uncertain, as the *Curlew* was taken to pieces in 1849. *(NMM PW8051)*

wrecked on Racoon Cay in the Bahamas 28.11.1840 (wreck later sold for £192.6.0d).

Thais Pembroke Dyd

As built: 90ft 0in, 72ft 3in x 24ft 8in (24ft 6in for tons) x 11ft 0in. 230⁶⁴⁄₉₄bm. Draught 8ft 0in/9ft 0in.

Ord: 25.3.1823. K: 7.1828. L: 12.10.1829. C: 6.1832 at Plymouth as packet brig (6 guns).

Commissioned 25.5.1832 under Lieut Charles Church, for the mail packet service at Falmouth; lost, presumed foundered with all (35) hands off Galway 12.1833, on passage from Falmouth to Halifax (Nova Scotia).

Variable Pembroke Dyd

As built: 231bm (dimensions missing, probably as *Spey* or *Thais*)

Ord: 25.3.1823. K: 5.1826. L: 6.10.1827. Renamed *Pigeon* 2.2.1829. C: 23.2.1829 at Plymouth as packet brig (4 guns).

Commissioned 3.2.1829 (after renaming) under Lieut John Binney, for the mail packet service at Falmouth; on 14.4.1835 under Lieut James Harvey, then 22.12.1836 under Lieut William Luce, then 24.7.1838 under Lieut Thomas James, all on same service; paid off 31.5.1842. Sold by AO 27.7.1847.

1824 **Batch** Two vessels were ordered on 23 November 1824, and another pair a fortnight later on 7 December 1824, although none of these were begun until 1829.

Termagant Portsmouth Dyd

As built: 90ft 10in, 72ft 7⅜in x 24ft 8in (24ft 6in for tons) x 12ft 0in. 231⁷⁸⁄₉₄bm. Draught 7ft 1½in/9ft 7½in.

Ord: 23.11.1824. K: 10.1829. L: 26.3.1838. C: 9.6.1838 as brigantine (3 guns).

First cost: £5,447 (?fitted for sea)

Commissioned 10.4.1838 under Lieut Woodford John Williams, for the west coast of Africa; on 25.7.1838 under Lieut Henry Frowd Seagram, on same station; took slavers *Prova* 9.10.1838, *Braganza* 9.2.1839, *Golfinho* 19.9.1839, *Conceiçao* 28.11.1839, *Julia* 28.11.1839, *Felicidade* 5.10.1840 and *Pacqueta Vera Cruzana* 15.10.1840. Sold to Mr Henry Dunk (for £640) at

Chatham 3.1845.

Lynx Portsmouth Dyd

As built: 90ft 11in, 72ft 5⅜in x 24ft 8½in (24ft 6½in for tons) x 11ft 10¼in. 232⁵⁄₉₄bm. Draught 6ft 8in/9ft 6in.

Ord: 23.11.1824. K: 2.1830. L: 2.9.1833. C: 13.12.1833 as brigantine (3 guns).

First cost: £5,703 (+ fitting £2,551)

Commissioned 23.9.1833 under Lieut Henry Vere Huntley (invalided 1837), for the Cape of Good Hope; took slavers *Arrogante Mayaguesano* 17.9.1834, *Atrevida* 27.12.1834 and *Vandolero* 21.1.1836; on 14.4.1837 under Lieut Thomas Frederick Birch, for the west coast of Africa; paide off 1.9.1837 at Chatham. Recommissioned 24.11.1837 under Lieut Henry Broadhead, for the same station; took slavers *Liberal* 1.11.1838, *Victoria* 24.12.1838, *Simpathia* 27.7.1839, *My Boy* 20.9.1839, *Destimida* 30.10.1839, *Lavandeyra* 27.11.1839, *Octavia* 3.4.1840 and *Olimpa* 18.4.1840; paid off 2.7.1841. Recommissioned 20.8.1841 under Lieut Godolphin James Burslem, for Lisbon and then the coast of Ireland (invalided 8.1843); on 7.9.1843 under Lieut John Thomas Knott, on particular service; paid off 16.10.1845. BU completed at Plymouth (for £125, less proceeds valued £1,454) 3.12.1845.

Nautilus Woolwich Dyd

As built: 90ft 1½in, 72ft 4¾in x 24ft 9½in (24ft 7½in for tons) x 11ft 0in. 233⁴⁸⁄₉₄bm. Draught 7ft 6in/9ft 11in.

Ord: 7.12.1824. K: 4.1829. L: 11.3.1830. C: 28.6.1830.

First cost: £7,173 (£9,801 fitted for sea).

Commissioned 4.3.1830 under Cmdr Lord George Paulet, for the Channel and Irish coast; in 7.1832 to Lisbon; 10.1833 to north coast of Spain; paid off 2.1.1834; recommissioned 3.1.1834 under Lieut William Crooke, for Lisbon, then to the Mediterranean 1836; later to Falmouth packet service?; paid off 9.1837. On 11.1.1838 under Lieut George Beaufoy, for the Cape of Good Hope and west coast of Africa; paid off 4.6.1841. Small Repair and fitted at Portsmouth (for £3,048) 7.1841–3.1842. Tender to yacht *Royal George* at Portsmouth 1.1.1843. Recommissioned

26.12.1844 under Lieut William Robson, for fishery protection in the Channel; on 12.11.1846 under Lieut William Thomas Rivers, still in the Channel. In 1.1.1849 under Lieut Samuel Brooking Dolling, for training apprentices as tender to flagship *Impregnable* at Plymouth; Dolling re-appointed 1.1.1852, then 24.11.1854 under Lieut John Packwood and 1.1.1857 under Lieut William Burley Grant, in the same role until 11.1861, then to harbour service. Hulk 1872. Reported to be in a very rotten and dilapidated state at Plymouth 6.10.1877; BU at Devonport 4–17.10.1878.

Curlew Woolwich Dyd
As built: 90ft 1in, 72ft 4⅜in x 24ft 9¼in (24ft 7¼in for tons) x 11ft 0in. $233\frac{1}{94}$bm. Draught 7ft 5in/9ft 11in.
Ord: 7.12.1824. K: 11.1829. L: 25.2.1830. C: 28.6.1830.
First cost: £10.485 (fitted for sea).
Commissioned 17.3.1830 under Lieut George Woolcombe; on 22.7.1830 under Cmdr Henry Dundas Trotter, for the west coast of Africa; took slaver *Veloz Marianna* 27.4.1833; her boats took pirate schooner *Panda* in Nazareth River 4.6.1833; paid off 6.1834. Recommissioned 17.11.1834 under Lieut Joseph Dedman, at Portsmouth for same station, then 21.4.1835 under Lieut Edmund Norcott; took slavers *Tres Tomases* 19.12.1835, *Rosarita* 2.1.1836, *Esperanza* 4.10.1836, *Quatro de Avril* 19.10.1836 and *Princessa Africano* 26.12.1837; paid off 13.9.1838. Recommissioned 2.5.1839 under Lieut George Rose, for the Cape of Good Hope and west coast of Africa; took slavers *Rapido* 22.6.1840 and *Animo Grande* 3.4.1841; paid off 2.1844. On 8.2.1842 under Lieut John Foote, for South America; on 16.9.1843 under Lieut George Sprigg, on same station; paid off 10.2.1844 (or 30.6.1844?) at Portsmouth. Fitted experimentally there with a mortar 1847, but not recommissioned. BU at Portsmouth 8.1849.

***1826* Batch** Four vessels were ordered on 23 May 1826, plus six more on 28 October 1826 (of which two were cancelled). *Delight*, *Algerine* and *Partridge* were given names previously borne by earlier units of this class, while *Forester* and *Griffon* took the names of units originally ordered in 1820 at Deptford.

Delight (ii) Chatham Dyd
As built: 90ft 2in, 72ft 6¼in x 24ft 9in (24ft 7in for tons) x 11ft 0in. $233\frac{11}{94}$bm. Draught 8ft 0in/10ft 2in.
Ord: 23.5.1826. K: 8.1827. L: 27.11.1829. C: 25.1.1830 as packet brig (4 x 12pdr carronades), but not completed for sea.
First cost: £2,917 recorded
Not commissioned until 1836 (intended to be commissioned under Cmdr Charles Talbot at Chatham in 6.1829, but this was cancelled). Housed over fore and aft in 6.1835, then fitted for sea (for £2,917) 7.1835–27.11.1835. Commissioned 4.11.1835 under Lieut John Moore, for the mail packet service at Falmouth; on 3.5.1837 under Lieut William Lory; paid off 2.1842 at Plymouth. (Possibly briefly used as watch vessel in the Tees?) Sold to Greenwood & Clarke (by AO 10.7.1844, for £510) at Plymouth 30.4.1844.

Algerine (ii) Chatham Dyd
As built: 90ft 1in, 72ft 5in x 24ft 8½in (24ft 6in for tons) x 11ft 0in. $231\frac{20}{94}$bm. Draught 8ft 0in/10ft 1in.
Ord: 23.5.1826. K: 10.1827. L: 1.8.1829. C: 5.10.1829 (for sea).
First cost: £9,993 total
Commissioned 30.6.1829 under Cmdr Charles Talbot, for South America (–25.11.1830); on 14.5.1831 under Cmdr John Frederick Fitzgerald De Ros, on same station; recovered portion of treasure from wreck of HMS *Thetis*; paid off 3.12.1833. Partial repair and fitted for sea at Chatham (for £3,168) 11.1833–8.1834. Recommissioned 6.6.1834 under Lieut George Charles Stovin, for the East Indies; on 26.3.1839 under Lieut Thomas Henry Mason (Acting), on same station; in 1st Anglo-Chinese War; at Amoy 25–26.8.1840; on 16.10.1841 under Lieut William Heriot Maitland; at Woosing 16.6.1842; in Yangtse operations 7.1842; on 20.9.1842 under Lieut Samuel Brooking Dolling, on same station; paid off 2.12.1843. Sold to John Small Sedger, Rotherhithe (for £510) 30.4.1844.

Griffon Chatham Dyd
As built: 90ft 4½in, 72ft 3¾in x 24ft 7¾in (24ft 5¾in for tons) x 11ft 0½in. $230\frac{46}{94}$bm. Draught 7ft 4½in/9ft 8½in.
Ord: 23.5.1826. K: 7.1830. L: 11.9.1832. C: 2.1.1833 as brigantine (with 3 guns – 2 x 24pdr carronades (10cwt) and 1 x 32pdr (40cwt)).
First cost: £4,951 to build, + £3,208 fitting.
Commissioned 11.10.1832 under James Edward Parlby, for the Cape of Good Hope; took slavers *Ingadora* 31.10.1834 and *Clemente* 3.11.1834; paid off 1836. Recommissioned 2.11.1836 under Lieut John Gooch D'Urban, for North America and the West Indies; sailed 28.1.1837 for station; paid off 1841. Small Repair and fitted for sea at Sheerness (for £5,048 + £520) 7.1841–8.1842. Recommissioned 2.7.1842 under Lieut Charles Jenkin, on same station. Recommissioned 9.4.1846 under Lieut Charles Edward Wilmot, for the east coast of South America; on 17.11.1846 under Lieut James Ptolemy Thurburn, on same station; paid off 4.1850 at Portsmouth. Harbour service as watch vessel in Portchester Lake 3.1853. Coal hulk at Portsmouth 1857. Renamed *Griffin* 1858. BU (by AO 6.8.1868) completed at Portsmouth 27.2.1869.

Forester Chatham Dyd
As built: 90ft 4in, 72ft 3⅜in x 24ft 7½in (24ft 5½in for tons) x 11ft 0¼in. 230bm. Draught 7ft 5in/9ft 10½in.
Ord: 23.5.1826. K: 9.1830. L: 28.8.1832. C: 2.1.1833 as brigantine (with 3 guns – 2 x 18pdr carronades (22cwt) and 1 x 18pdr (38cwt)).
First cost: £5,098 to build, + £3,161 fitting.
Commissioned 10.1832 under Lieut William Henry Quin, for the west coast of Africa; grounded off Isles of Scilly in gale 13.2.1833; refloated and (after tow to Plymouth by *Rhadamanthus*) slight repair and fitted for sea (for £2,227) 3–10.1833. Recommissioned 6.9.1833 under Lieut George Gover Miall, for the Cape of Good Hope; took slavers *Manuel* 23.2.1835, *Legitimo Africano* 20.3.1835, *Numero Dos* 21.6.1835, *Luisa* 9.2.1836, *Golondrina* 9.2.1836 and *Victoria* 20.10.1836; on 6.9.1837 under Lieut C. P. Rosenberg, on same station; took slaver *Dos Simaos* 2.4.1838; paid off 5.1838. Recommissioned 27.9.1838 under Lieut Francis Godolphin Bond, for the Cape of Good Hope; during 1839, temp under Lieut Colin Yorke Campbell of *Melville*; took slavers *Rebecca* 11.3.1839, *Serea* 11.3.1839, *Reina dos Anjos* 12.5.1839 and *Recurso* 23.11.1840; paid off 9.1841 at Plymouth Sold to Brabin, Baker & Collins (for £500) 27.11.1843.

Partridge (ii) Pembroke Dyd
As built: 90ft 0in, 72ft 3in x 24ft 8in (24ft 6in for tons) x 11ft 0in. $230\frac{64}{94}$bm. Draught 8ft 0in/9ft 0in.
Ord: 28.10.1826. K: 8.1828. L: 12.10.1829. C: 24.4.1836 at Plymouth as tender.
First cost: £5,790 to build, + £3,019 fitting.
Commissioned ?4.1836 under Lieut Henry Pryce Deschamps, as

tender to *Royal Adelaide*, flagship at Plymouth; in 8.1836 under Lieut Philip Bisson, still as a tender; in 7.1837 to the west coast of Africa ('with presents for native Princes'); home 10.1837. On 8.1.1838 under Lieut William Morris, for the coasts of Scotland, Africa and South America; on 8.12.1841 under Lieut John Thomas Knott, off South America; paid off 6.1843. Fitted as coastguard watch vessel at Portsmouth 6–7.1843, and transferred to coastguard for Southampton Water (based at Netley) 7.1843, renamed *WV.32* on 25.5.1863. Sold to J. Ransome, Southampton (for £550) 2.2.1864.

Wizard Pembroke Dyd
>   As built: 90ft 0in, 72ft 3in x 24ft 8¼in (24ft 6¼in for tons) x 11ft 0in. 231⅞₉₄bm. Draught 7ft 10¼in/9ft 2¼in.
>   Ord: 28.10.1826. K: 10.1829. L: 24.5.1830. C: 15.6.1837 (for Ordinary) at Plymouth.
>   First cost: £6,035 to build, +£641 fitting.
>   Commissioned 4.1837 under Lieut E. L. Harvey, for South America; on 24.4.1838 under Lieut Thomas Frederick Birch, on same station; paid off 5.4.1841. Fitted at Plymouth (for £2,396) 8.1850. Tender to *Ajax* at Queenstown 1850. Recommissioned 30.5.1854 under Lieut Samuel James Brickwell, as tender to *Conway* at Queenstown; on 1.3.1858 under Lieut Alfred Prowse Hasler Helby, as tender to *Nile* at Queenstown; wrecked at Berehaven in Bantry Bay 8.2.1859.

Charybdis Portsmouth Dyd
>   As built: 90ft 3in, 72ft 6in x 24ft 7¼in (24ft 6½in for tons) x 11ft 0in. 232²¹⁄₉₄bm. Draught 7ft 3in/9ft 5in.
>   Ord: 28.10.1826. K: 12.1829. L: 27.2.1831. C: 11.7.1831 as a gun-brig (1 gun), then 9.9.1831 as a brigantine (3 guns).
>   First cost: £3,302 to build.
>   Commissioned 17.5.1831 under Lieut Richard Borough Crawford, for the west coast of Africa and Cape of Good Hope; took slaver *Desengano* 22.2.1833; on 11.1.1834 under Lieut Samuel Mercer, on same stations; took slavers *Tamega* 14.6.1834, *Argos*

Construction of brig sloops during Seppings's tenure as Surveyor was confined to further copies of the Peake's well-tried and ubiquitous *Cherokee* Class. The single exception, the *Columbine*, was an experimental design by William Symonds, the first vessel designed for the Admiralty by this then naval lieutenant (he was promoted commander just four months after *Columbine* was ordered). His naval architectural abilities were first brought to the Admiralty's attention by a number of well-placed individuals for whom he had designed fast-sailing yachts; one of these (Vernon) underwrote the *Columbine*'s design by standing surety to the amount of £20,000. The sloop is here seen barque-rigged on 2 August 1827, returning from her second cruise with the Experimental Squadron. The brig was tried against other small Sixth Rates and sloops in the 1828 competitive sailing trial, and proved satisfactory, although not outstandingly so. Sailing trials became a post-war obsession for the Royal Navy, a process Symonds keenly promoted, although no naval architect was able to draw any objective conclusions from any of them. Nevertheless, despite only moderate success, Symonds's political influence eventually led to his replacing Seppings as Surveyor of the Navy. *(NMM PU8006)*

>   10.10.1835, *Matilde* 5.2.1836, *El Mismo* 4.3.1836, *Cantabra* 21.10.1836, *Lafayette* 12.5.1837 and *General Recaforte* 26.7.1837; paid off 9.1837. Recommissioned 5.2.1838 under Lieut Robert Gore, for North America and the West Indies; on 9.3.1839 under Lieut Edward Burnaby Tinling, in the West Indies; on 4.3.1841 under Lieut Michael De Courcy, for South America; in action at Carbagenda 6.2.1841; on 4.4.1842 under Lieut James Archibald Macdonald, in North America and the West Indies; paid off 27.1.1843 at Sheerness. Sold to Beatson, Rotherhithe (for £500) 7.11.1843.

Buzzard Portsmouth Dyd
>   As built: 90ft 10½in, 73ft 1½in x 24ft 8in (24ft 6in for tons) x 11ft 10in. 233⁴⁴⁄₉₄bm. Draught 6ft 9in/9ft 5in.
>   Ord: 28.10.1826. K: 12.1829. L: 23.3.1834. C: 27.8.1834 as brigantine (3 guns).
>   First cost: £3,411 to build (+ fitting £4,782).
>   Commissioned 14.6.1834 under Lieut William Cave Burbidge, for the west coast of Africa; on 15.12.1834 under Lieut Jeremiah MacNamara, on same station (died 7.1835); took slavers *Formidable* 17.12.1834 and *Bien Venida* 28.3.1835; on 27.7.1835 under Lieut Thomas Lorey Roberts (temp.), then 12.1835 under

Lieut Patrick Campbell (nominal commander from 27.7.1835); took slavers *Semiramis* 2.9.1835, *Norma* 27.11.1835, *Ligera* 24.12.1835, *Mindello* 4.5.1836, *Felicia* 2.7.1836, *Famosa Primiera* 6.7.1836, Portuguese *Joven Carolina* 22.7.1836, *Olympia* 28.10.1836 and *Serea* 12.11.1836(invalided 3.1837); during 1836 under Samuel Otway Wooldridge, Mate (Lieut 6.2.1837); Campbell invalided 3.1837, and 4.1837 under Lieut John Luke Richard Stoll, still on the west coast of Africa; on 26.7.1838 under Lieut Charles Fitzgerald, on same station (invalided 1840); took slavers *Sirse* 17.11.1838, *Empredador* 27.11.1838, *Clara* 13.3.1839, *Adelaide* 1.4.1840 and *Carolina* 10.7.1840; on 25.8.1840 under Lieut Reginald Thomas John Levinge; took slavers *Liberal* 9.2.1841 and *Juliana* 13.2.1841; paid off 19.5.1842. Sold to Greenwood & Clarke at Plymouth (for £520) 7.11.1843.

Foxhound Plymouth Dyd
  Ord: 28.10.1826. K: — . Cancelled 21.2.1831 (unstarted).
Helena Plymouth Dyd
  Ord: 28.10.1826. K: — . Cancelled 21.2.1831 (unstarted).

COLUMBINE **Class** Second Class Brig, 18 guns. Lieut William Symonds design of February 1825; his first ship for the Navy, to demonstrate his radical views on naval architecture. She was, like all Symonds's ships, to combine 'great breadth of beam and extraordinary sharpness . . . with a careful attention to stowage, the stand of the masts, and the cut and setting of the sails'. On 12 March 1825 the Honourable George Vernon was called upon to give a bond of £20,000 'for the fulfilment of a promised engagement to purchase this vessel [from the Navy] if she did not answer'. She did 'answer', but this gesture of support helps to explain why Symonds named a frigate *Vernon* after his patron (and not, as might have been expected, after Admiral Edward Vernon) after becoming Surveyor of the Navy. Originally rigged as a barque, but re-rigged as a brig in the 1834 refit.
  Dimensions & tons (as built): 105ft 0½in, 84ft 0⅛in x 33ft 6¼in oa (33ft 2¼in for tonnage) x 7ft 11in. 492¹⁷⁄₉₄bm.
  Men: 125 (110 as 12-gun). Guns: 2 x 6pdr + 16 x 32pdr carronades; later 2 x 18pdrs + 10 x 32pdr carronades.
Columbine Portsmouth Dyd
  Ord: 9.6.1825. K: 1.1826. L: 1.12.1826. C: 9.3.1827 (for sea).
  First cost: £19,743 including fitting.
  Commissioned 12.1826 at Portsmouth under Lieut William Symonds (her designer), then on 9.11.1827 under Cmdr Charles Crole, for Halifax; on 26.1.1828 under Cmdr John Townshend; paid off 1.6. 1830. Recommissioned 2.6.1830 under Cmdr James Wallace Gabriel, for the West Indies; in 2.7.1831 under Cmdr Henry Ommanney Love; private expedition to the Niger River; paid off 12.3.1834. Small Repair and fitted for sea at Sheerness (for £4,822) 5–8.1834. Recommissioned 2.6.1834 under Cmdr Thomas Henderson, for the Mediterranean; to west coast of Africa 1836; took slavers *Veloz* 14.11.1836, *Latona* 4.2.1837 and *Josephina* 10.2.1837; paid off 4.1838. Recommissioned 29.5.1838 under Cmdr George Augustus Elliot, for the Cape of Good Hope and the west coast of Africa; took slavers *Neptune* and *Angerona* 23.9.1839, *Dos Irmanos* 27.11.1839, *Vigilante* 20.12.1839, *Bom-Fin* 21.12.1839 and *Primo Genito* 14.2.1840; on 3.6.1840 under Cmdr Thomas Jordaine Clarke, for the East Indies; in 1st Anglo-Chinese War; at Amoy 25–26.8.1840; at Tsinghae 1.10.1841; on 16.10.1841 under Cmdr William Henry Anderson Morshead, on same station; destroyed junks at Ningpo 10.3.1842; at Woosung 16.6.1842; paid off 17.3 1843 at Chatham. Recommissioned 25.6.1846 under Cmdr James Richard Booth, for the East Indies and China; on 12.11.1846

under Cmdr Charles Conrad Grey; on 6.12.1847 under Cmdr John Charles Dalrymple Hay; action against pirate junks 24.8.1849; re-rated 12-gun brig from 1849; paid off 15.4.1850. Fitted as coal hulk at Chatham and then Sheerness 3–4.1854. Sold (by AO 27.9.1890) to Henry Castle & Sons to BU at Charlton 12.1.1892.

(C) Vessels acquired from November 1830

Following a thirteen-year period during which brigs had been built exclusively to the wartime *Cherokee* design (the *Columbine* being the sole exception, and the sole large brig built), Symonds's accession to the office of Surveyor saw a variety of new designs, almost all during the 1830s to his own plans, while a more experimental series of designs were built in the 1840s.

SNAKE **Class** Second Class Brigs, 16 guns (12 guns from 1846). Thomas Ditchburn design, 1831. The largest warships to be built in a commercial yard since 1815. At this date, Ditchburn was Fletcher's yard manager.
  Dimensions & tons: 100ft 0in, 78ft 0in x 32ft 0in oa (31ft 9in for tonnage) x 14ft 10in. 418²⁴⁄₉₄bm.
  Men: 110. Guns: 2 x 9pdr (or 18pdrs) + 14 x 32pdr carronades.
Snake Fletcher & Fearnall, Limehouse
  As built: 101ft 10in, 79ft 2¼in x 32ft 4¼in oa (32ft 1¼in for tonnage) x 15ft 0½in. 434¹³⁄₉₄bm. Draught 7ft 11in/12ft 6in.
  Ord: 13.10.1831. K: 12.1831. L: 3.5.1832. C: 15.5–4.8.1832 at Woolwich Dyd.
  First cost: £5,345 to build, + £4,889 fitting.
  Commissioned 30.4.1832 under Cmdr William Robertson; blockade of Dutch coast 1832; sailed for South America 15.6.1833; paid off 22.5.1835. Recommissioned 21.9.1835 under Cmdr Richard Laird Warren; on 26.12.1836 under Cmdr Alexander Milne; sailed 22.3.1837 for the West Indies; took slavers *Arrogante* 23.11.1837 and *Matilda* 5.12.1837; on 30.1.1839 under Cmdr John Baker Porter Hay, in North America and the West Indies; paid off 27.9.1840. Very Small Repair and fitted for sea at Sheerness (for £7,814) 11.1840–12.1841. Recommissioned 6.10.1841 under Cmdr Walter Bourchier Devereux, for the Mediterranean; paid off ?9.1845. Recommissioned 10.1.1846 under Cmdr Thomas Bourmaster Brown, for the Cape of Good Hope; wrecked in the Mozambique Channel 29.8.1847.
Serpent Fletcher & Fearnall, Limehouse
  As built: 102ft 5in, 79ft 10in x 32ft 3in oa (32ft 0in for tonnage) x 15ft 0in. 434⁷⁄₉₄bm. Draught 8ft 6in/12ft 0in.
  Ord: 13.10.1831. K: 2.1832. L: 14.7.1832. C: 6.12.1832 at Woolwich Dyd.
  First cost: £5,326 to build, + £6,052 fitting.
  Commissioned 10.1832 under Cmdr John Charles Symonds, for North America and the West Indies; paid off 13.11.1834. Middling Repair and fitted for sea at Chatham (for £5,595) 3–10.1836. Recommissioned 13.10.1836 under Cmdr Richard Laird Warren, and sailed 21.12.1837 for North America and the West Indies; on 10.5.1839 under Cmdr Robert Gore, on same station. Small Repair and fitted for sea at Chatham (for £5,562) 6–11.1841. Recommissioned 30.8.1841 under Cmdr Charles Henry Seale, at Chatham; on 27.11.1841 under Cmdr William Nevill, for the East Indies; paid off 18.5.1846. Between Small and

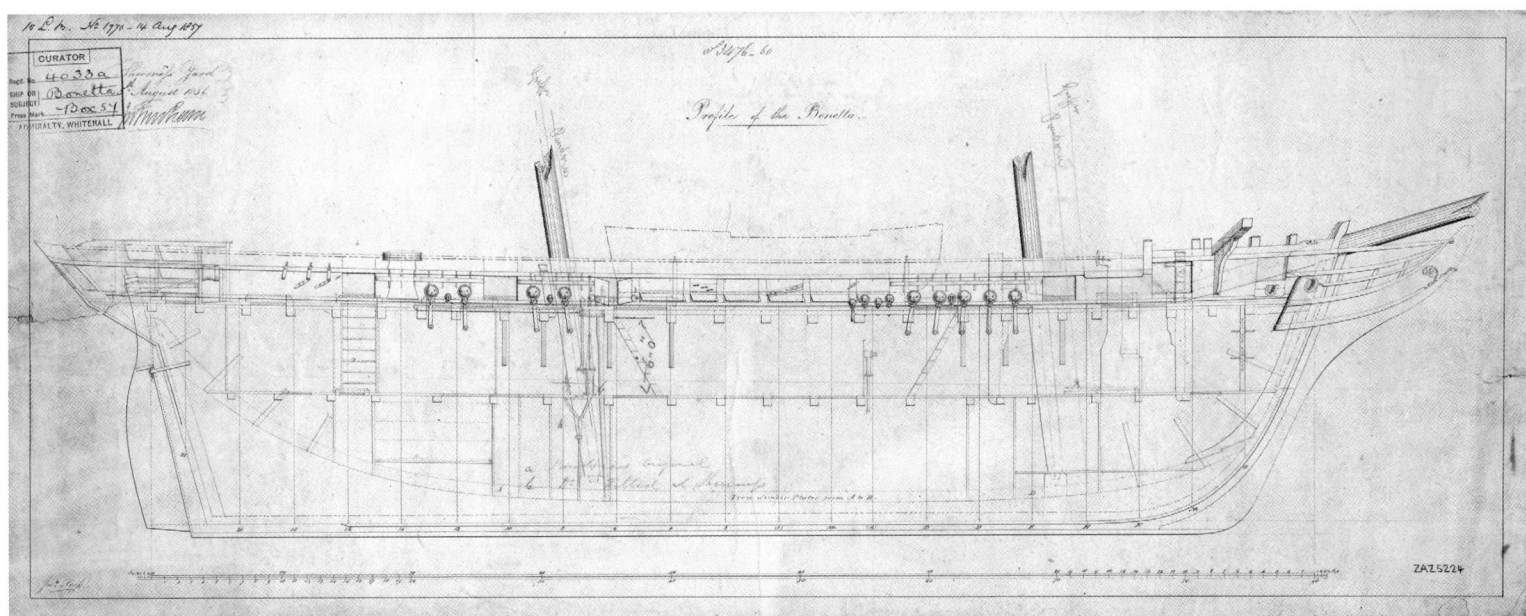

Middling Repair, and fitted for sea at Portsmouth (for £8,532) 9.1846–11.1849. Recommissioned 30.9.1849 under Cmdr Charles Barker, for the East Indies; on 29.9.1850 under Cmdr William Garnham Luard, on same station; in 2nd Anglo-Burmese War; paid off 19.1.1854. Fitted as tender (to *Excellent*) at Portsmouth 12.1857. BU completed 26.7.1861 at Portsmouth.

PANTALOON 10 guns. William Symonds design, 1831. Designed as yacht for Duke of Portland by William Symonds, built in early 1831 and purchase authorised on 27 June; actually bought on 1 October 1831 (effective 5 December 1831).

    Dimensions & tons (as purchased): 90ft 0in, 71ft 4¼in x 29ft 4¼in oa (29ft 2¼in for tonnage) x 12ft 8in. 323bm.

    Men: 68. Guns: 8 x 18pdr carronades (15cwt) + 2 x 6pdrs. From 1848 had 6 x 18pdr carronades + 2 x 18pdrs (20cw5)

Pantaloon Troon Shipyard (owned by Duke of Portland)

    L: 5.1831. Fitted at Portsmouth 18.11.1831–26.7.1832.

    First cost: £8,905 (including masts, rigging, etc.) + £983 fitting.

    Commissioned 5.9.1834 under Lieut Nicholas Cory, on 'particular service', for the Falmouth packet service, as tender to the yacht *Royal George*. Small Repair and fitted for sea at Portsmouth (for £6,611) 11.1840–12.1841. Recommissioned 30.8.1841 under Lieut William Pearson Crozier, at Portsmouth; on 24.11.1841 under Cmdr Charles Horace Lapidge, for the west coast of Africa; took slaver *Desiao* 19.5.1842, paid off 1844. On 7.9.1844 under Lieut (Acting, then Cmdr from 2.1.1845) Edmund Wilson, for the experimental brig squadron, then to the west coast of Africa; took slavers *Graciosa Vingativa* 4.5.1845, 4-gun *Borboleta* off Lagos 26.5.1845, *Frederico* 10.6.1845, *Magico* 9.12.1845 and *Bella Clara* 28.2.1846; on 12.5.1846 under Lieut. Edward Hughes Alston (acting), then on 29.6.1846 under Cmdr Henry John Douglas, for North America and the West Indies; paid off 17.2.1848. Repaired and fitted at Plymouth (for £3,682) 5–8.1848; reduced to 8 guns. Recommissioned 28.6.1848 under Cmdr Lewis de Teissier Prevost, on same station. On 4.7.1849 under Cmdr Hyde Parker, for the Cape of Good Hope; paid off at Portsmouth 22.2.1852. BU there 8.1852.

The inboard profile for the *Pandora* class of 8-gun brigs, the first batch designed by William Symonds. The prototype was built at Woolwich (as a packet brig), but the others of the class were constructed at Sheerness and Woolwich, with slight variations among them. This draught, dated 5 August 1836 at Sheerness, relates specifically to the *Bonetta*. (*NMM J4794*)

PANDORA Class 8 guns. Brigantines to William Symonds design of 1832, rated as brigs. Coastguard watch vessels as shown. Depth in hold varied; prototype *Pandora* was 13ft 10in; Sheerness vessels 14ft 8in (first pair), 14ft 4in (next pair); Portsmouth vessels 13ft 5in only.

    Dimensions & tons: 90ft 0in, 71ft 3in x 29ft 3in (29ft 0in for tonnage) x 13ft 10in (varied – see notes). 318⁶⁸⁄₉₄bm.

    Men: 50/60 (33 as packets). Guns: *Rapid* and *Sealark* carried 8 x 18pdr carronades; others 1 x 32pdr + 2 x 32pdr carronades as packets.

Pandora Woolwich Dyd (packet brig)

    As built: 90ft 0in, 71ft 0¾in x 29ft 2½in (29ft 0½in for tonnage) x 13ft 10in. 318⁷¹⁄₉₄bm. Draught 7ft 2½in/11ft 1½in.

    Ord: 11.12.1831. K: 8.1832. L: 4.7.1833. C: 10.8.1833 at Chatham.

    First cost: £5,397 to build (£7,821 fitted).

    Commissioned 31.7.1833 under Lieut Wentworth Parsons Croke, as packet brig based at Falmouth; on 5.3.1836 under Lieut Robert Wintle Innes, on same service; paid off 14.1.1842. Small Repair and altered to a survey vessel (for £7,777) 10.1842–6.1845. Recommissioned 8.2.1845 under Lieut James Wood, as surveying vessel for the Pacific (along with *Herald*), to conduct preliminary survey of the British Columbia coast (after the Oregon boundary dispute with USA); paid off 27.11.1849 at Plymouth. Recommissioned 20.12.1850 under Cmdr Byron Drury, as surveying vessel for Australia and New Zealand; paid off 29.5.1856 at Plymouth. Became coastguard watch vessel on 29.7.1857. Sold to Marshall to BU (by AO 11.1.1862, for £600).

Sheerness batch

Bonetta Sheerness Dyd (packet brig)

    As built: 90ft 8in, 71ft 3⅛in x 29ft 3½in (29ft 0½in for tonnage) x 14ft 7¾in. 319⁶⁵⁄₉₄bm. Draught 6ft 11½in/11ft 0in.

    Ord: 3.3.1834. K: 10.1834. L: 5.4.1836. C: 1.6.1836.

    First cost: £5,605 (£8,689 fitted).

    Commissioned 1.6.1836 under Lieut Philip Bisson. On 23.7.1836 under Lieut Henry Pryce Deschamps, for the coast of Africa; took slavers *Temerario* 20.1.1837, *Cinco Amigos* 30.3.1837, *Felicidade* 4.11.1837 and *Ligecra* 15.11.1837. On 3.5.1838 under Lieut John Luke Richard Stoll, on the same station; took nine slavers (of which three were taken 50 miles up the Congo River) including *Izabel* 3.12.1838, *Violante* 28.12.1838, *Josephine* and *Liberal* 7.9.1839 (both), *Governado Vidal* 30.12.1839 and *Novo Abismo* 3.1.1840; paid off 6.1840. Recommissioned 1.3.1841 under Lieut Francis William Austen, then 13.1.1842 under Lieut Edward Evans Gray, for same station. Small Repair and fitted for sea at Chatham (for £4,311) 1–6.1844. Recommissioned 5.3.1844 under Cmdr Thomas Saumarez Brock, as surveying vessel in the Mediterranean. Between Small and Middling Repair and fitted for sea at Chatham (for £3,523) 9.1847–1.1848. Recommissioned 24.5.1847 under Lieut Frederick Edwyn Forbes, for the west coast of Africa. On 13.2.1851 under Lieut Charles Wake, for the east coast of South America; paid off at Chatham 6.1.1855. BU completed at Deptford 23.4.1861.

Dolphin Sheerness Dyd (packet brig)
As built: 90ft 7¾in, 71ft 2⅞in x 29ft 3½in (29ft 0½in for tonnage) x 14ft 7¾in. 319⁵⁶/₉₄bm. Draught 6ft 10in/10ft 9½in.
Ord: 3.3.1834. K: 10.1834. L: 14.6.1836. C: 28.9.1836 (as 3-gun brig with 3 x 32pdrs).
First cost: £6,952 (£9,192 fitted).
Commissioned 28.7.1836 under Lieut Thomas Lorey Roberts, for the west coast of Africa; took slavers *Incomprehencivel* 23.12.1836, *Dolores* 19.4.1837, *Cobra d'Africa* 27.5.1837 and *Providencia* 1.6.1837. On 12.9.1837 under Lieut John Macdougall, on same station; took slaver *Primoroza* 25.9.1837; on 2.8.1838 under Lieut Edward Holland; took slavers slavers *Ligeiro*, *Victoria* and *Dous Amigos* 16.11.1838 (all three brigs in Lagos Roads), *Amalia* 26.12.1838, *Jack Wilding* 28.5.1839, *Merced* 18.6.1839, *Carolina* 4.7.1839, *Casualidado* 6.7.1839, *Intrepido* 19.8.1839 and *Dos Amigos* 27.8.1839; on 4.3.1840 under Lieut Edward Littlehales; took slavers *Carolina* 5.1.1841, *Firme* (by boat action off Whydah) 30.5.1841, *Nova Fortuna* 8.6.1841 and *Doris* 3.7.1841; on 5.5.1842 under Lieut (Acting) Octavius Cumberland; took slaver *Minerva* 29.5.1842; on 22.7.1842 under Lieut (Acting) Philip Bisson (died 19.10.1842). Recommissioned 25.5.1843 under Lieut William O'Bryen Hoare, for the east coast of South America; took slaver *Bella Angelina* 7.5.1844; on 7.2.1845 under Lieut Reginald Thomas John Levinge, then 11.1845 under Lieut William Stewart Miller, on same station; at Battle of Obligado 20.11.1845; paid off 4.3.1847 at Sheerness. Recommissioned 19.9.1847 at Portsmouth under Lieut Robert Francis Boyle, for the west coast of Africa; took slavers *Pensamento* 14.6.1848, *Curioso* 24.6.1848 and *Brasiliense* 12.10.1848. Between Very Small and Small Repair and fitted for sea at Sheerness (for 3,269) 2–6.1850. Recommissioned 1.6.1850 at Sheerness under Lieut Henry Temple, for same station; on 3.8.1853 under Lieut Edmund Webber; paid off 14.3.1857 at Chatham. Became coastguard **WV No. 3** on 19.3.1861 at Gravesend. Sold to BU 11.5.1894.

Spy Sheerness Dyd
As built: 90ft 8in, 71ft 3⅛in x 29ft 3½in (29ft 0½in for tonnage) x 14ft 7¾in. 319⁶⁵/₉₄bm. Draught 6ft 11½in/11ft 0in.
Ord: 20.2.1837. K: 9.1838. L: 24.3.1841. C: 23.10.1841.
First cost: £6,411 (£8,387 fitted).
Commissioned 20.8.1841 under Lieut John Moon Potbury (appointed 10.8.1841); on 15.4.1842 under Lieut George Raymond, for the west coast of Africa; took slavers *Clio*

24.1.1843, *Bom Fim* 24.1.1843 and *Esperanza* 29.5.1843; on 1.7.1843 under Lieut Samuel Otway Wooldridge, on same station; took slavers *Furia* 8.8.1843, *Egeria* 25.9.1843 and *Linda* 20.11.1843; paid off 30.7.1844 at Chatham. Repair at Sheerness 8.1844–2.1845. Recommissioned 15.3.1845, still under Wooldridge, for the Pacific; paid off 24.6.1848 at Chatham. Recommissioned 8.8.1848 under Lieut George Western, for the west coast of Africa; took slavers *Amelia* 15.4.1849 and *Estrella* 23.8.1849; paid off 12.1850 at Sheerness. Recommissioned 6.1.1851 under Lieut Edward Hill (died 5.1851), for the west coast of Africa; on 13.5.1851 under Lieut Gorge Morrell (temp.), then 16.6.1851 under Lieut Henry Barré Beresford; paid off 12.1853 at Sheerness. Recommissioned 22.6.1854 under Lieut Robert Francis Boyle, for the east coast of South America; in 5.1855 under Lieut Henry Oakeley; on 15.8.1855 under Lieut Robert Francis Boyle (also in 8.1855 under Lieut Alfred Luckraft), then 23.8.1858 under Lieut Tathwell Benjamin Collinson, then 7.1859 under Lieut John Hanmer, and 5.7.1860 under Lieut Augustus Tabuteau, all on same station; paid off 15.1.1862. Sold to Robert Wilson at Montevideo 20.1.1862.

Dart Sheerness Dyd
As built: 90ft 1in, 71ft 4in x 29ft 3in (29ft 0in for tonnage) x 14ft 4in. 319⁹/₉₄bm. Draught 7ft 4½in/11ft 2½in.
Ord: 20.2.1837. K: 4.1846. L: 17.3.1847. C: 7.6.1847.
First cost: £7,294 (£9,737 fitted).
Commissioned 10.5.1847 under Lieut Edmund Anthony Glynn (appointed 6.5.1847), for the west coast of Africa; took slaver *Maria* 2.10.1847; Glynn invalided 5.1848; on 14.5.1848 under Lieut Edward Hill; took slavers *Nova Rosa* 28.7.1848, *Bom Destino* 19.9.1848, *Nuo Packet* 4.10.1848 and *Despigue* 11.5.1849; paid off 30.11.1849 at Sheerness. Fitted as tender at Sheerness (for £1,765) 12.1849–2.1850, then to Cape of Good Hope, as tender to *Castor*. Recommissioned 10.1854 under Lieut James M'Clure, for the Cape of Good Hope, as tender to *Hydra*; paid off 27.8.1857 at Plymouth. To coastguard by AO 7.12.1857. Renamed as coastguard watch vessel *WV.26* in 5.1863, at Beresford. Completed BU at Chatham 9.1.1875.

Daring Sheerness Dyd
Ord: 14.5.1840. K: — . Cancelled 4.9.1843 (new *Daring* ordered to new design 10.7.1843 – see below).

Portsmouth batch
Two 10-gun brigs from Portsmouth – to be called *Ariel* and *Sealark* – were in the submitted Programme of Work for 1837, but the former was not ordered when the Programme was approved in February. However, another brig (*Rapid*) to this design was added in September, and included in the 1838 Programme of Work.

Sealark Portsmouth Dyd
As built: 90ft 0in, 71ft 2in x 29ft 3½in (29ft 0½in for tonnage) x 13ft 5in. 319²⁵/₉₄bm. Draught 6ft 6in/10ft 8in.
Ord: 20.2.1837. K: 5.1840. L: 27.7.1843. C: 9.12.1843.
First cost: £5,860 (£9,654 fitted).
Commissioned 27.10.1843 under Cmdr Thomas Lewis Gooch, for west coast of Africa (invalided 3.1847); took slaver *Tentacio* 24.9.1846; on 20.3.1847 under Lieut (Cmdr 28.8.1847) Richard Dunning White; took slavers *Gustavo Primiero* 14.4.1847 and *Lebro* 7.8.1847; paid off 25.9.1847. Refitted at Chatham (for £3,118) 9.1847–1.1848. Recommissioned 14.12.1847 under Cmdr William Backhouse Monypenny, on same station; took slavers *San Francisco* 27.8.1848, *Orite* 30.7.1849 and *Francellina* 6.10.1839; on 6.6.1850 under Cmdr Edward Southwell Sotheby,

Symonds followed up the *Pandora* Class brigs with the much larger *Racer* Class, which carried 16 guns. This lithograph by H. John Vernon depicts the *Ringdove* of this class, whose design was largely copied by Symonds from the *Snake* and *Serpent* of 1832, themselves largely an updating (by a commercial shipbuilder, Thomas Ditchburn) of the wartime *Cruiser* Class. *(NMM PW8061)*

on same station; in Lagos river 12.1851; paid off 23.7.1852. Fitted 7.1852–9.1853 as a training brig for naval apprentices (for £1,506), as tender to *Victory*. Recommissioned 11.7.1854 under Lieut Marcus Lowther, then 9.1854 under Abraham J. A. Parks, Master, as training ship and tender to *Illustrious* at Portsmouth; in 1.1859 under Lowther again, as tender to *Britannia* at Portsmouth, then 10.1.1860 under Lieut Charles Edward Stevens, as tender to *Implacable* at Portsmouth; on 1.12.1862 under Lieut James Terence Fitzmaurice, as tender to *St Vincent*, then 5.1865 under Lieut William Henniker Heaton, 6.1868 under Lieut Frederick George Denham Bedford, and 4.1875 under Lieut F. H. Haygarth, all as tender to *Implacable* at Portsmouth. Repair at Plymouth (for £4,143) 9.1876–1.1877. In 5.1878 under Lieut Gerrard R. Bromley, then 1886 under Lieut Powell Cecil Underwood, both as tender to *Ganges*; in 1.1888 under Lieut W. F. G. Clarke, as tender to *Lion*; on 17.5.1894 under Lieut Alfred E. Tizard; paid off 31.3.1898 at Portsmouth. Sold to Mr Harris of Bristol (for £560) 11.11.1898 and BU at Milford Haven.

Rapid Portsmouth Dyd
As built: 90ft 0¾in, 71ft 3½in x 29ft 3in (29ft 0in for tonnage) x 13ft 5½in. 318⁸⁶⁄₉₄bm. Draught 7ft 6in/11ft 3in.
Ord: 19.9.1837. K: 9.1839. L: 3.6.1840. C: 17.3.1841 (for sea).
First cost: £5,693 (£9,081 fitted).
Commissioned 11.12.1841 under Lieut Edward Charles Earl, for the west coast of Africa (died 12.4.1845); took slavers *Quelhe Importa* 24.10.1842, *Temerario* 3.11.1843, another (name unknown) 13.1.1844, *Carlitos* 16.2.1844 and *Santa Anna* 4.4.1844; on 27.12.1845 under Cmdr Henry John Windham Sherbrook Payne Gallwey (invalided 3.1847); in 1.1847 under Cmdr (acting) Edward Hill; took slaver *Constante Amizade* 22.1.1847; on 19.3.1847 under Cmdr Edward Dixon; took slavers *Erpito* 9.7.1847 and (with *Waterwitch*) Brazilian *Romeo Primero*

22.7.1847; paid off 14.5.1848. Small Repair and fitted for sea at Portsmouth (for £5,142) 3.1850–12.1851. Recommissioned 4.11.1851 under Cmdr George Blane, for the East Indies; on 5.4.1855 under Lieut (Cmdr 23.7.1855) John Eglinton Montgomerie, on same station; paid off 20.9.1855. Sold to Messrs Macdonald & Co. (for £2,800) at Singapore 19.1.1856.

RACER Class 16 guns. Sir William Symonds design of 1832, a dockyard-built version of *Snake* Class with slightly modified lines. Brig sloops, rated as Second Class Brigs. The prototype was ordered in March 1832, and six more in July; a further pair ordered in February 1837 were built to the same design.
Dimensions & tons: 100ft 6in, 78ft 7in x 32ft 4in oa (32ft 0in for tonnage) x 15ft 4in (14ft 10in in *Racer* and *Wanderer*). 428²⁄₉₄bm.
Men: 110. Guns: 2 x 9pdr (or 18pdrs) + 14 x 32pdr carronades.

Harlequin Pembroke Dyd
As built: 100ft 6in, 78ft 10½in x 32ft 5½in oa (32ft 1½in for tonnage) x 15ft 2½in. 432⁸¹⁄₉₄bm. Draught 8ft 2in/12ft 6in.
Ord: 28.3.1832. K: 11.1832. L: 18.3.1836. C: 27.4–25.10.1836 at Sheerness.
First cost: £8,231 (fitted £12,299).
Commissioned 16.8.1836 under Cmdr John Elphinstone Erskine, for the Mediterranean; on 27.7.1838 under Cmdr Lord Francis John Russell, in the Mediterranean, then to the Cape of Good Hope and west coast of Africa; took slavers *Constaza* 16.5.1839, *Bella Fiorentina* 20.5.1839, *Victoria de Libertade* 26.6.1839, *Christiano* 27.6.1839, *Emprendador* 28.6.1839, *Sin Gual* 28.6.1839 and *Sociedade Feliz* 24.12.1839; paid off 30.7.1840. Recommissioned 16.8.1841 under Cmdr George Fowler Hastings, for the East Indies; in 1st Anglo-Chinese War; on 1844 under Cmdr George Henry Seymour, on the same station; paid off 31.1.1845. Very Small Repair and fitted for sea at Portsmouth

(for £7,077) 9.1845–4.1846. Recommissioned 30.1.1846 under Cmdr Douglas Curry, for the Mediterranean; on 12.11.1846 under Cmdr John Moore, then 22.11.1848 under Cmdr Frederick Beauchamp Paget Seymour, on same station; paid off 27.3.1849 at Plymouth. Fitted for sea there (for £1,959) 11–12.1850. Recommissioned 6.11.1850 under Cmdr Arthur Parry Eardley Wilmot, for the west coast of Africa; boats in action in Lagos river 25.11.1851; paid off at Plymouth 9.1853. Fitted at Plymouth as a coal depot (for £1,190) 6–7.1859. Sold to Marshall, Plymouth to BU 8.1889; handed over to them 12.9.1889.

Racer Portsmouth Dyd
As built: 100ft 8in, 78ft 9⅛in x 32ft 4¾in oa (32ft 0¼in for tonnage) x 14ft 10½in. 430⁶³⁄₉₄bm. Draught 7ft 10in/12ft 3in.

Ord: 10.7.1832. K: 9.1832. L: 18.7.1833. C: 5.9.1833.

First cost: £8,313 (fitted £12,152).

Commissioned 13.7.1833 under Cmdr James Hope, for North America and the West Indies; paid off 13.6.1838. Recommissioned 19.9.1838 under Cmdr George Byng, on same station; on 6.11.1840 under Cmdr Thomas Harvey; paid off 10.1842 at Portsmouth. Small Repair and fitted at Portsmouth (for £6,920) 10.1842–6.1843. Recommissioned 28.4.1843 under Cmdr Archibald Reed, for South America, later to west coast of Africa; took slaver *Bom Destino* 7.9.1844; paid off 20.11.1847 at Plymouth. Reduced to 12 guns. Recommissioned 24.6.1848 under Lieut Henry Bacon, as tender to *Caledonia*; on 28.2.1849 under Cmdr Charles Henry Beddoes, for the Mediterranean; paid off 8.3.1852 at Plymouth. Sold to Willson & Co. (by AO 17.9.1852, for £820).

Ringdove Plymouth Dyd
As built: 100ft 6in, 78ft 10⅛in x 32ft 4in oa (32ft 0in for tonnage) x 15ft 2in. 429⁴²⁄₉₄bm. Draught 8ft 6in/11ft 7in.

Ord: 10.7.1832. K: 10.1832. L: 18.6.1833. C: 7.2.1834.

First cost: £9,661 (fitted £12,782).

Commissioned 21.11.1833 under Cmdr William Frederick Lapidge, for Lisbon; in 5.1837 to the north coast of Spain (San Sebastian and Bilbao); paid off 4.1837. Recommissioned 4.1837 under Cmdr H. Stopford Nixon, for North America and the West Indies; on 14.10.1838 under Cmdr Keith Stewart, on the same station, then to the west coast of Africa; took slavers *Victoria* 23.1.1840 and *Jesus Maria* 29.12.1840; on 26.1.1842 under Cmdr Sir William Daniell, on same stations (died 15.9.1845). On 16.9.1845 under Lieut Walter Need (acting); paid off 10.11.1845 at Plymouth. Recommissioned 17.12.1845 under Cmdr William Legge George Hoste, and sailed 5.2.1846 for the East Indies; in Brunei river operations; on 12.1846 under Cmdr William John Cavendish Clifford, then 20.8.1847 under Cmdr Edward Augustus Inglefield, on the same station; paid off 28.4.1849 at Plymouth. Towed 7.7.1849 (by *Confiance*) into Yealm River to serve as depot ship for Cholera patients; returned 4.9.1849. BU at Plymouth 8.1850.

Wanderer Chatham Dyd
As built: 100ft 7in, 78ft 7¼in x 32ft 4in oa (32ft 0in for tonnage) x 14ft 10in. 428³⁴⁄₉₄bm. Draught 8ft 10½in/11ft 9in.

Ord: 10.7.1832. K: 2.1833. L: 10.7.1835. C: 22.8.1835.

First cost: £10,233 (fitted £13,817).

Commissioned 4.8.1835 under Cmdr Thomas Dilke, for North America and the West Indies; on 25.1.1837 under Cmdr Thomas Bushby, for the west coast of Africa; took slavers *Flor de Tego* 3.4.1837, another (name unknown) 21.1.1838 and *Scorpio* 26.11.1838; paid off 2.8.1839 at Sheerness. Recommissioned 19.11.1839 under Cmdr Joseph Denman, for the same station;

took slavers *Eliza Davidson* 4.4.1840. *Josephina* 12.5.1840, *Sao Paolo de Loando* 3.6.1840, *Maria Rosario* 9.6.1840, *Pombinha* 3.7.1840, *Vanguardia* 11.11.1840, *Reglano* 10.12.1840 and *Amalia* 3.7.1841; to the East Indies 7.1841; on 23.8.1841 under Cmdr Edward Norwich Troubridge, for the East Indies; in 1st Anglo-Chinese War; on 6.11.1841 under Cmdr Stephen Greville Fremantle, then 8.3.1842 under Cmdr George Henry Seymour, on same station; paid off 24.5.1844. Recommissioned 30.1.1846 under Cmdr Philip Hodge Somerville, for the west coast of Africa; on 14.2.1847 under Cmdr Frederick Byng Montresor, on same station; took slavers *Dez de Outobro* 16.7.1847 and *Subtil* 4.1.1849; paid off 21.1.1849. BU at Chatham 4.3.1850.

Wolverene Chatham Dyd
As built: 100ft 7in, 78ft 8in x 32ft 4in oa (32ft 0in for tonnage) x 15ft 2in. 428⁴⁵⁄₉₄bm. Draught 8ft 9in/11ft 10½in.

Ord: 10.7.1832. K: 2.1833. L: 13.10.1836. C: 1.12.1836.

First cost: £10,085 (fitted £13,307).

Commissioned 13.10.1838 under Cmdr Edward Granville Howard, for the Mediterranean; on 5.1.1839 under Cmdr William Tucker, transferred 3.1839 to the Cape of Good Hope and west coast of Africa; took slavers *Passos* 8.4.1839, *Vigilante* 23.5.1839, *Emprehendedor* 23.6.1839, *Fermeza* 23.7.1839, *Pampeiro* 5.9.1839, *Veloz* 16.11.1839, *Lark* 16.1.1840, *Asp* 17.1.1840, *Santo Antonio Victorioso* 2.4.1840, *Palmira* 15.9.1840, *Gratidao* 14.10.1840, *Emilia* 10.11.1840, *Rapido* 30.1.1841 and another (name unknown) 5.4.1841; paid off 4.9.1841 at Chatham. Re commissioned 18.12.1841 under Cmdr John Samuel Willes Johnson, for the East Indies; on 15.8.1842 under Lieut Henry Gage Morris, then 28.3.1844 under Cmdr Charles Foreman Brown, and 7.9.1844 under Cmdr William John Cavendish Clifford; at Malluda Bay 19.8.1845; on 1.1847 Cmdr John Charles Dalrymple Hay, on same station; paid off 11.5.1847 at Chatham. Small Repair at Chatham (for £3,231) 2–7.1848. Recommissioned 20.9.1849 under Cmdr Maxwell Falcon, for the west coast of Africa; took slavers: name unknown 8.7.1850 and *Flor del Carmamom* 15.8.1850; paid off 14.5.1852. Fitted for sea at Chatham (for £2,296) 10–12.1853. Recommissioned 9.5.1854 under Cmdr John Corbett, for North America and the West Indies; wrecked on the Courtown Bank off Greytown, Mosquito Coast 11.8.1855 (all crew saved); wreck sold 27.5.1857 (for £20).

Sappho Plymouth Dyd
As built: 100ft 6½in, 78ft 8in x 32ft 4in oa (32ft 0in for tonnage) x 15ft 2in. 428⁴⁵⁄₉₄bm. Draught 8ft 7in/11ft 7in.

Ord: 10.7.1832. K: 12.1834. L: 3.2.1837.

First cost: £9,081 (fitted £13,531).

Commissioned 28.2.1837 under Cmdr Thomas Fraser, for North America and the West Indies; took slaving schooner *Rozalia Habaneira* 13.7.1838; paid off 23.11.1841. Very Small Repair at Plymouth (for £3,756, ?+1,059) 1–5.1843. Recommissioned 19.3.1843 under Cmdr George Hope, for the Cape of Good Hope; took slavers *Sociedade* 21.9.1843 and others (names unknown)on 22.1.1844, 9.5.1845, 24.5.1845 and 30.7.1845; on 2.8.1845 under Cmdr Robery Fitzgerald Gambier, on same station; took slaver *Triumphante* 13.12.1845. Small Repair and fitted for sea at Portsmouth (for £7,628) 5.1847–3.1849. Recommissioned 13.2.1849 under Cmdr Reynell Charles Michell, for North America and the West Indies; stranded on coast of Honduras, but re-floated 11.12.1849; on 17.3.1850 under Cmdr (Acting) Arthur Auckland Leopold Pedro Cochrane, on same station. Fitted for commission at Chatham 12.1852–3.1854; completed for sea 1–2.1856. Recommissioned 8.1.1856 under

Cmdr Fairfax Moresby, for the west coast of Africa; disappeared off South Australia while en route from the Cape of Good Hope to Sydney, presumed foundered with all (147) hands 2.1858.

Lily Pembroke Dyd

As built: 100ft 6in, 78ft 8¾in x 32ft 5½in oa (32ft 1½in for tonnage) x 15ft 2½in. $432\frac{17}{94}$bm. Draught 8ft 2in/12ft 4in.

Ord: 10.7.1832. K: 12.1835. L: 28.9.1837. C: 7.10.1837–12.3.1838 at Plymouth.

First cost: £8,563 (fitted £12,591).

Commissioned 15.12.1837 under Cmdr John Reeve, for the west coast of Africa; took slavers *Eagle* 14.1.1839 and *Maria Theresa* 26.1.1839; on 28.5.1839 under Cmdr Charles Deare, for the Cape of Good Hope and the west coast of Africa; took slavers *Jose* 17.4.1840 and *Maria Feliz* 12.11.1840; on 18.12.1841 under Cmdr George Baker, for the Cape of Good Hope; took slavers *Esperanza* 4.3.1843, *Desangano* 14.3.1843 and *Confidencia* 17.3.1843; paid off 2.1.1844. Recommissioned 14.12.1844 under Cmdr Charles James Franklin Newton, for the west coast of Africa; took slaver *Princeza* 17.11.1845. Recommissioned 4.3.1850 under Cmdr Robert Tench Bedford, for the East Indies; on 26.8.1851 under Cmdr John Sanderson, on same station; paid off 7.5.1855 at Portsmouth. Coal depot at Portsmouth 1860, renamed *C.29*; to Chatham as *C.15* in 1871. Sold (for £900) to Castle to BU at Baltic Wharf, Charlton 7.4.1908.

Liberty (ex-*Hearty*, renamed 1837) Pembroke Dyd

As built: 100ft 6in, 78ft 8⅛in x 32ft 3½in oa (31ft 11½in for tonnage) x 15ft 2in. $427\frac{40}{94}$bm. Draught 8ft 2in/12ft 4in.

Ord: 20.2.1837. K: 9.1848. L: 11.6.1850. C: 18.8.1850 into Ordinary at Plymouth.

First cost: £8,768 (unfitted).

Not commissioned until 1866. Fitted at Plymouth as a training ship (for £1,970) 3–7.1861; fitted there again as a training ship (for £3,229 + £614) 2–6.1865. Commissioned 25.5.1866 under Lieut Joshua Reynolds Palmer, as tender to *Ganges* at Falmouth by AO 26.5.1866. Recommissioned 2.1869 under Lieut John James Martin, as tender to *Ganges* at Falmouth; on 14.4.1870 under Lieut Wollaston Comyns Karslake; in 12.1873 under Lieut Charles Johnstone. In 4.1878 under Lieut G. O'Connor, as tender to *Implacable* at Plymouth; in 1.1890 under Lieut Frederick Roope, as tender to *Lion* at Plymouth; in 1.1899 under Lieut C. H. Morgan (–1901). For sale by AO 4.1.1905, and sold at Portsmouth 11.7.1905.

Squirrel Pembroke Dyd

As built: design dimensions quoted

Ord: 20.2.1837. K: 9.1850. L: 8.8.1853. C: 24.8.1853 into Ordinary at Plymouth as a sail training ship.

First cost: £9,202 (unfitted).

Not commissioned until 1862. Fitted at Plymouth as a training ship (for £4,254) 2–5.1862, as tender to *Impregnable*. Commissioned 1.1.1862 under Lieut Arthur Richard Wright in this role; in 7.1864 under Lieut Thomas Keith Hudson, then 4.1868 under Lieut Henry Waller and 7.1874 under Lieut Robert N. Hamond. BU completed at Devonport 11.2.1879.

***ALERT* Class** 6 (later 8) guns. William Symonds design 1834 as packet brigs. Six were ordered to be built by contract in 1834, but one (*Star*) was instead allotted to Woolwich Dockyard; a further six were ordered from the dockyards in 1836, and four more in 1837, but four of these (one pair in each year) were re-ordered to the larger *Acorn* design. The first nine were all fitted as packet brigs, and initially employed on the Falmouth service, but were later deployed to overseas stations.

Dimensions & tons: 95ft 0in, 74ft 10in x 30ft 3in oa (30ft 0in for tonnage) x 14ft 8in (first 8)/14ft 10in (*Crane*)/13ft 6in (rest). $358\frac{18}{94}$bm.

Men: 80 (44 as packet brigs). Guns: 2 x 6pdrs, + 4 x 12pdr carronades; later 6 x 32pdr; later 8 x 18pdr (guns varied).

Star Woolwich Dyd (packet brig)

As built: 95ft 0in, 74ft 11¼in x 30ft 3in oa (30ft 0in for tonnage) x 14ft 8in. $358\frac{70}{94}$bm. Draught 8ft 0in/11ft 7in.

Ord: 29.7.1834. K: 9.1834. L: 29.4.1835. C: 18.6.1835 (sheathed with Grenfell's 'patent yellow metal').

First cost: £10,747 (including fitting).

Commissioned 14.4.1835 under Lieut John Binney, as packet brig based at Falmouth; on 4.4.1836 under Lieut Christopher Smith, on same service; on 5.7.1841 under Lieut Thomas Creser; paid off 21.6.1842. Between Small and Middling Repair and altered to a sloop at Plymouth (for £7,002) 5–11.1843. Recommissioned 15.9.1843 under Cmdr Robert John Wallace Dunlop, for the west coast of Africa (died 1846); took slavers *Nova Cristina* 3.2.1842, *Maria* 1.4.1844, *Cazuza* 30.1.1845, *Diligencia* 8.2.1845, *Vivo* 11.2.1845, *Felicidade* 6.3.1845, *Audaz* 26.3.1845, *Rafael* 27.3.1845, *Minerva* 17.4.1845, *Mariquinhas* 28.6.1845, *Fantasma* 16.7.1845, unknown (brig) 10.10.1845, *Descobridor* 21.10.1845 and (with *Wasp*) *Paquete de Rio* 17.2.1846; on 15.6.1846 under Cmdr (acting) Frederick Leopold Augustus Selwyn, then 5.8.1846 under Cmdr Charles Luxmore Hockin; paid off 11.5.1847 at Sheerness. Recommissioned 24.1.1848 under Cmdr Charles Wilson Riley, for the west coast of Africa; took slaver *Curioso* 24.6.1848; on 23.8.1848 under Selwyn again, on same station; took slavers *Raspate* 17.9.1848, *Minerva* 11.3.1850, *Feliz Lembranca* 14.3.1850, *Pensamento Feliz* 24.4.1850, and *Vengador* 2.5.1850; paid off 1.8.1850 at Chatham. Recommissioned 26.6.1854 under Cmdr William Garnham Luard, for east coast of South America; on 11.8.1855 under Cmdr Alexander Boyle; paid off 21.1.1857. Transferred to coastguard as a watch vessel (by AO 29.4.1857) for Gravesend 9.1857, renamed *WV.11* by AO 25.5.1863. BU *c*.1899.

Ranger Samuel Bottomley, Rotherhithe (packet brig)

As built: 95ft 1⅛in, 75ft 3⅝in x 30ft 4¼in oa (30ft 1¼in for tonnage) x 14ft 8in. $363\frac{34}{94}$bm. Draught 8ft 1in/11ft 3in.

Ord: 29.7.1834. K: 9.1834. L: 25.7.1835. C: 30.7–7.12.1835 at Sheerness.

First cost: £5,656 to build + £195 dockyard expenses; fitting £3,030.

Commissioned 24.10.1835 under Lieut James Howard Turner, as packet brig based at Falmouth; Turner re-appointed 14.5.1838; paid off 21.3.1843 at Plymouth. Between Small and Middling Repair and altered to a sloop at Plymouth (for £8,548) 1.1844–2.1845. Recommissioned 14.12.1844 under Cmdr James Anderson, for the west coast of Africa; took slaver *Emprehendedor* 22.6.1845; on 31.5.1848 under Cmdr Charles Frankland Newland, then 13.6.1849 under Cmdr Thomas Miller, on same station; took slaver Destimeda 15.3.1850; paid off at Plymouth 19.12.1851. Hulked as floating chapel at Kingstown (Dublin) from 6.1859. Sold to John Good of Dublin (for £480) 9.9.1867.

Linnet Joseph White, Cowes (packet brig)

As built: 95ft 2¾in, 75ft 0¾in x 30ft 4in oa (30ft 1in for tonnage) x 14ft 8in. $361\frac{32}{94}$bm. Draught 7ft 9in/11ft 3in.

Ord: 29.7.1834. K: 9.1834. L: 27.7.1835. C: 5.11.1835 at Portsmouth.

First cost: £5,025 to build + £86 dockyard expenses; fitting £3,711.

Commissioned 11.9.1835 under Lieut William Downey, as packet brig based as Falmouth; on 11.9.1839 under Lieut William

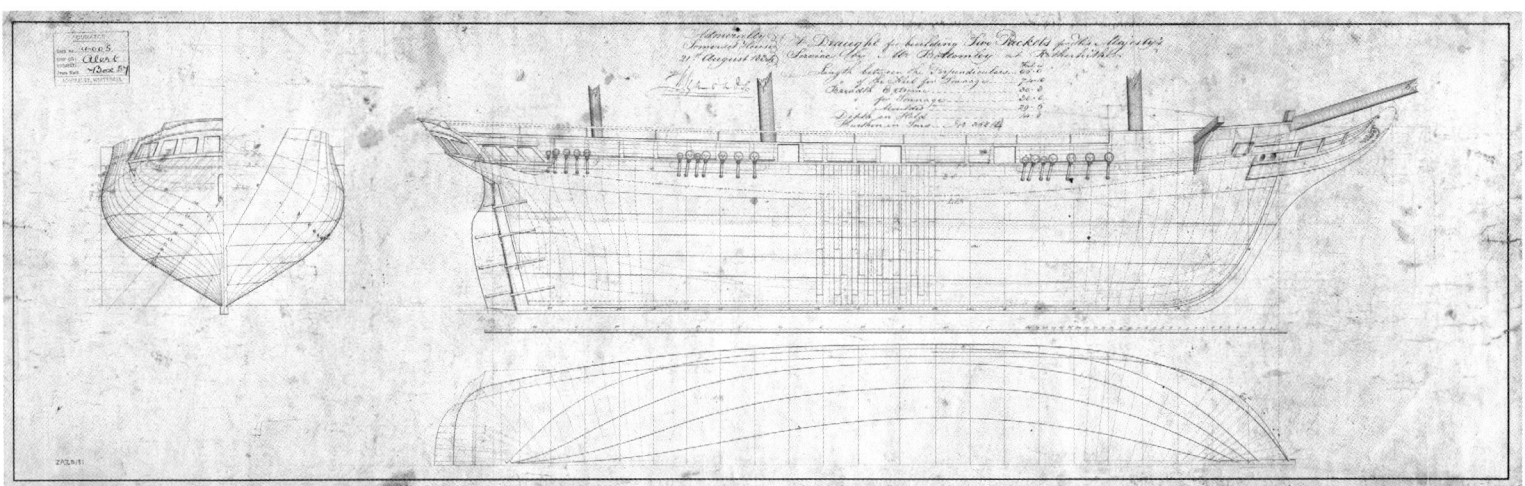

This draught, dated 21 August 1834, is for the *Alert* and *Ranger*, two of a group of brigs designed specifically by Symonds to serve as packet vessels. Both of these were built by Samuel Bottomley at Rotherhithe. (*NMM J4954*)

Forrester; on 13.5.1841 under Lieut Francis Rogers Coghlan, then 18.7.1843 under Lieut Henry Percy Dicken; paid off 19.4.1845. Repaired and fitted at Plymouth (for £4,311) 3.1847–3.1848. Recommissioned 26.2.1848 under Lieut Thomas James, on same service; paid off 18.2.1851. Refitted at Plymouth (for £3,794) 5.1851–11.1852. Recommissioned 4.10.1852 under Cmdr Henry Need, for the west coast of Africa; took slavers *Paulina* 30.4.1853 and *Mellidon* 30.4.1854; paid off 25.9.1856. Transferred to coastguard as a watch vessel (by AO 29.2.1857) in 9.1857, renamed *WV.36* by AO 25.5.1863. Sold to Marshall to BU at Plymouth (by AO 30.10.1866) 1.2.1867.

Alert Samuel Bottomley, Rotherhithe (packet brig)
As built: 95ft 0½in, 74ft 10in x 30ft 3in oa (30ft 0in for tonnage) x 14ft 8in. 358²³/₉₄bm. Draught 7ft 10in/10ft 11in.
Ord: 29.7.1834. K: 9.1834. L: 24.9.1835. C: 25.9.1835–20.1.1836 at Chatham.
First cost: £5,639 to build + £196 dockyard expenses; fitting £3,618.
Commissioned 30.12.1835 under Lieut Charles Harvey Norrington, as packet brig based at Falmouth; on 25.2.1839 under Lieut Edward Jennings, on same station; paid off 3.1842. Very Small Repair at Plymouth (for £373) 9–10.1842; altered to a sloop there (for £5,529) 4–9.1843. Recommissioned 28.7.1843 under Cmdr Charles John Bosanquet, for the west coast of Africa; took slavers: name unknown 4.6.1844, *Triumpho de Inveja* 23.5.1845, *Adelaide* 12.8.1845, *Voador* 3.10.1845, *Bra Sorte* 15.11.1845 and *Emprehendedora* 21.1.1846; on 3.7.1846 under Cmdr William Ellis, Very Small Repair at Plymouth (for £3,900) 7–11.1847. Recommissioned 25.10.1847 under Cmdr Hugh Dunlop, for same station; took slavers *Maria* 7.4.1848, *Lawrence* 25.9.1848 and *Andorinha* 11.10.1848; paid off at Plymouth 4.4.1850. BU at Plymouth 5.1851.

Express Colson, Deptford (packet brig)
As built: 95ft 2⅛in, 74ft 10¾in x 30ft 5in oa (30ft 2in for tonnage) x 14ft 7in. 362⁵/₉₄bm. Draught 7ft 9in/10ft 11in.
Ord: 29.7.1834. K: 9.1834. L: 8.10.1835. C: 9.3.1836 at Woolwich & Sheerness.
First cost: £5,113 to build + £152 dockyard stores; fitting £3,733 (Woolwich £2,042; Sheerness £1,691).
Commissioned 31.1.1836 under Lieut Wentworth Parsons Croke, as packet brig based at Falmouth; on 4.1.1840 under Lieut

Edward Herrick, then 26.6.1846 under Lieut Thomas James, for same service. Repaired and fitted at Plymouth (for £4,294) 5–7.1848. Recommissioned 13.6.1848 under Lieut William Lory, on same service. Recommissioned 3.1.1851 under Cmdr William Frederick Fead, for North America and the West Indies (died 14.5.1853 at Bahia); to east coast of South America 6.1851; paid off 29.1.1853 at Plymouth. Recommissioned 14.5.1853 under Cmdr Henry Boys, for the east coast of South America; paid off 20.9.1856 at Plymouth. Sold to Marshall (for £685) to BU at Plymouth 11.1.1862 (with *Crane*).

Swift Colson, Deptford (packet brig)
As built: 95ft 5in, 75ft 2⅜in x 30ft 3½in oa (30ft 0½in for tonnage) x 14ft 7in. 365⁴⁶/₉₄bm. Draught 7ft 11in/10ft 11in.
Ord: 29.7.1834. K: 9.1834. L: 21.11.1835. C: 26.4.1836 at Woolwich.
First cost: £5,114 to build + £152 dockyard stores; fitting £3,409.
Commissioned 26.2.1836 under Lieut David Welch, as packet brig based at Falmouth; on 5.7.1841 under Lieut John Douglas, then 7.1.1847 under Lieut William Lory, on same service. Between Small and Middling Repair and fitted for sea at Plymouth (for £5,912) 5–7.1849. Recommissioned 25.6.1849 under Cmdr William Cornwallis Aldham, for the Pacific. Fitted at ?Plymouth as a mooring vessel for Cape of Good Hope (for £3,254) 7–11.1860, renamed *YC.3* in 1861. Sold 1866.

Penguin Pembroke Dyd (packet brig)
As built: 95ft 0in, 75ft 0in x 30ft 3½in oa (30ft 0½in for tonnage) x 14ft 8in. 360⁴/₉₄bm. Draught 7ft 11in/10ft 11in.
Ord: 14.4.1836. K: 11.1836. L: 10.4.1838. C: 18.4–21.9.1838 at Plymouth.
First cost: £7,755 to build, + fitting £2,641.
Commissioned 24.7.1838 under Lieut William Luce, as packet brig based at Falmouth; on 5.11.1841 under Lieut Walter Leslie, then 20.11.1846 under Lieut William Swainson and 23.11.1848 under Leslie again, on same service. Recommissioned 6.12.1850 under Cmdr Thomas Etheridge, for Cape of Good Hope; took slavers *Disengano* 8.4.1851, *Isabel* 19.6.1851 and *Presidenta* 3.11.1851; paid off 20.7.1855. Became coastguard watch vessel (by AO 29.9.1857) at Plymouth, to Gamble River and renamed *WV.31* on 25.5.1863. Sold to A. Dockerill of Lloyds (for £750) 5.6.1871.

Peterel Pembroke Dyd (packet brig)
As built: 95ft 1in, 75ft 0in x 30ft 3in oa (30ft 0in for tonnage) x 14ft 8½in. 359⁴/₉₄bm. Draught 7ft 11in/10ft 11in.
Ord: 14.4.1836. K: 4.1837. L: 23.5.1838. C: 31.5–3.10.1838 at Plymouth.

First cost: £7,545 to build, + fitting £2,635.

Commissioned 18.8.1838 under Lieut William Crooke, as packet brig based at Falmouth; on 23.10.1843 under Lieut Thomas Creser, on same service; paid off 13.3.1851 at Plymouth. Coal depot at Portsmouth 1857; became floating workshop at Devonport 1850. Sold to Marshall (along with *Crane* and *Express*, by AO 11.1.1862, for £630) to BU.

Crane Woolwich Dyd (packet brig)

As built: 95ft 0in, 74ft 9½in x 30ft 3½in oa (30ft 0½in for tonnage) x 14ft 10in. 359⁹⁄₉₄bm.

Ord: 20.2.1837. K: 1.1838. L: 28.5.1839. C: 6.10.1839.

First cost: £10,394 (including fitting).

Commissioned 20.8.1839 under Lieut John Hill, as packet brig based at Falmouth; on 8.5.1842 under Lieut Thomas Arundel Lewis, then 28.6.1847 under Lieut John Parsons, on same service. Refitted at Plymouth (for £5,002) 1–11.1851. Recommissioned 18.10.1851 under Cmdr John Julius McDonell, for west coast of Africa; on 19.4.1852 under Cmdr Charles Wright Bonham, then 2.4.1853 under Cmdr Thomas Miller, on same station; took slaver *Oregon* 30.8.1854; paid off 25.4.1855. Sold to Marshall (for £670) to BU at Plymouth 11.1.1862.

Ferret Plymouth Dyd

As built: 95ft 0in, 74ft 10in x 30ft 3in oa (30ft 0in for tonnage) x 13ft 6½in. 358²³⁄₉₄bm. Draught 7ft 6in/10ft 10in.

Ord: 20.2.1837. K: 4.1838. L: 1.6.1840. C: 24.11.1840.

First cost: £8,332 (+ fitting £3,342).

Commissioned 1.12.1841 under Cmdr Josiah Oake, for west coast of Africa; took slavers *Aventura* 28.9.1844, *Oliviera* 2.3.1845, *Uncas* 21.4.1845 and *Conciçao Feliz* 6.5.1845; paid off 7.8.1845 at Sheerness. Recommissioned 3.2.1846 under Cmdr George Sprigg, for same station; took slavers *Forao* 24.6.1847, *San Sebastiao* 2.7.1847, *Faisca* 19.7.1847, *Maria de Gloria* 28.8.1847, *Facerinha* 10.9.1847, *Malaga* 11.12.1847, *Pagaseira* 10.2.1848, *Cazualidade* 22.3.1848, *Flor de Marium* 22.3.1848, *Anna Carolina* 11.5.1848, *Paquette de Cabo* 11.5.1848 and *Castro* 3.6.1848; paid off 8.9.1848 at Plymouth. Middling Repair and fitted for sea at Plymouth (for £6,587) 7.1850–7.1852. Recommissioned 10.6.1852 under Cmdr Reginald John James George Macdonald, for same station; took lavers *Restaurador* 18.3.1853 and *Manuelita* 22.5.1854; on 20.10.1854 under Cmdr Charles Taylor Leckie; took slavor of unknown name 19.6.1855; paid off 4.2.1856. Fitted 4.1859 at Plymouth as a training ship for boys (for £4,591) for Queenstown (Cork). Recommissioned 7.5.1859 under Cmdr William Edward Fisher, for Queenstown; on 16.3.1860 under Lieut Robert Mansel, at Queenstown, as tender to *Sans Pareil*; on 1.7.1853 tender to *Hastings*; on 14.6.1864, still under Mansel, to Southampton Water, as tender to *Boscawen*; on 14.6.1865 under Lieut Hilary Mansell Carré; in 1868 to Portland; wrecked alongside pier at Dover in a gale 29.3.1869.

Cygnet Woolwich Dyd

As built: 95ft 0in, 74ft 6in x 30ft 4in oa (30ft 1in for tonnage) x 13ft 6in. 358⁵⁹⁄₉₄bm. Draught 6ft 11in/10ft 8in.

Ord: 20.2.1837. K: 5.1838. L: 6.4.1840. C: 4.9.1840.

First cost: £11,179 (including fitting).

Commissioned 7.7.1840 under Lieut Edmund Wilson, for west coast of Africa; took slavers *Luisa* 23.1.1841, *Galliana* 2.11.1841, *Resolucao* 4.9.1842, *Pureza de Conceicao* 7.10.1842, *Se Deos Quizer* 10.10.1842 and *Brilhante* 18.3.1843. Refitted at Sheerness as a sloop (for £3,493) 11.1843–6.1844. Recommissioned 19.4.1844 under Cmdr Henry Layton, on same station; took slavers *Ave Maria* 25.10.1844, *Caroline* 17.12.1844, *Alabes*

4.1.1845, *Diligencia* 25.1.1845, *Rosa* 24.9.1845, *Isabella* 3.11.1845, *Quatro de Marco* 18.12.1845 and another (name unknown) 6.1.1846; on 17.2.1846 under Cmdr Frederick Wilmot Horton; took slaver *Clara* 2.3.1845; on 20.4.1846 under Cmdr Frederick Byng Montresor, and 14.2.1847 under Cmdr Philip Hodge Somerville, all on same station; paid off 10.4.1847 at Sheerness. Small Repair and fitted at Sheerness for sea (for £5,816) 4–12.1847. Recommissioned 6.11.1847 under Cmdr George Kenyon, for west coast of Africa; took slavers *Nereide* 19.3.1848 and *Mettre Mao* 1.5.1848; on 4.12.1848 under Cmdr David Robertson; took slavers *Emisades* 22.3.1849, *Volador* 24.5.1849, *Empehendreda* 14.7.1849 and *Maria Jose* 18.7.1849; on 1.7.1850 under Cmdr Richard Durning White, on same station; paid off 5.1853 at Portsmouth. Transferred to coastguard as watch vessel in 9.1857 for Chichester, renamed *WV.30* on 25.5.1863. BU at Portsmouth 11.12.1876–3.1.1877.

Philomel Plymouth Dyd

As built: 95ft 0in, 74ft 9⅛in x 30ft 4¼in oa (30ft 1¼in for tonnage) x 13ft 6in (rest). 360⁴⁵⁄₉₄bm.

Draught 7ft 10in/10ft 1in.

Ord: 20.2.1837. K: 4.1840. L: 28.3.1842. C: 16.7.1842 as surveying vessel 4.1842.

First cost: £7,697 to build (+ fitting £4,286).

Commissioned 2.4.1842 under Cmdr Bartholomew James Sulivan, for surveying Brazil, then Parana River operations; at Battle of Obligado 20.11.1845; paid off 19.6.1846. Refitted as a sloop at Plymouth (for £2,657) 10.1846–5.1847. Recommissioned 19.3.1847 under Cmdr William Cotterell Wood, for west coast of Africa; took slavers *Atrevida* 11.10.1847, *Wandering Jew* 14.1.1848, *Aurora* 2.5.1848, *Igual* 23.1.1849, *Andorinha* 2.2.1849, *Astucio* 11.3.1849 and *Albertina* 3.7.1849. Very Small Repair and fitted for sea at Plymouth (for £2,544) 10.1849–3.1850. Recommissioned 4.2.1850 under Cmdr Thomas George Forbes, for west coast of Africa; took slavers Nova *Espiculaçao* 9.5.1850 and *Condor* 26.6.1850; at repulse at Lagos Bar 25.11.1851. Recommissioned 2.8.1853 under Cmdr John MacDowall Skene, for west coast of Africa; paid off 9.7.1856. Transferred at Southampton to coastguard as watch vessel in 9.1857, renamed **WV.23** on 25.5.1863 at Milton. Foundered in the Swale 1869; wreck sold to Hayhurst & Clasper (for £550) to salve and BU 26.2.1870.

Dispatch Chatham Dyd

Ord: 20.2.1837. K: — . Re-ordered as *Acorn* Class brig 24.5.1839 (see below).

Dove Chatham Dyd

Ord: 20.2.1837. K: — . Re-ordered as *Acorn* Class brig 24.5.1839 (see below).

Heroine Woolwich Dyd

As built: 95ft 0in, 74ft 8⅞in x 30ft 3½in oa (30ft 0½in for tonnage) x 13ft 7in (rest). 358⁷⁴⁄₉₄bm.

Draught 7ft 3in/11ft 2in.

Ord: 20.8.1838. K: 5.1840. L: 16.8.1841. C: 11.12.1841.

First cost: £7,663 to build (+ fitting £3,810).

Commissioned 30.8.1841 under Lieut Thomas Dilnot Stewart, for the west coast of Africa; on 14.2.1843 under Lieut Henry Richard Foote, on same station; took slavers *Maria Louise* 5.4.1843, *Quatro de Setembro* 24.2.1844 and *Atilla* 23.2.1845. Recommissioned 29.8.1845 under Cmdr Charles Edmunds, for west coast of Africa; took slavers *Sierra Leone* 30.1.1846, *Aventureiro* 1.10.1847, *Luisa* 20.11.1847, *Venus* 24.3.1848, *Zephiro* 28.3.1848, *Mercurio* 7.4.1848, *Maria Candida* 5.5.1848 and *Marianna* 20.6.1848. Small Repair at Plymouth (for £2,468)

The influence of yacht-racing on the Navy's small craft was significant in the Symonds era; not only did the Surveyor's reputation stand on his successful work as a yacht designer, but the Admiralty was even persuaded to purchase a few outstanding examples for comparative purposes. One was the *Waterwitch*, designed by the noted yacht-builder Joseph White of Cowes and ordered as a private yacht for the Earl of Belfast, but bought by the Admiralty when nearly complete on the stocks at Cowes in order to participate in competitive sailing trials. Arguably she was the first true clipper ship, as it is claimed that she introduced the clipper's characteristic feature of a hollow bow at the waterline (but no plans exist to confirm this). In 1843 she required a repair, for which Joseph White quoted a contract price that was half of that estimated by Portsmouth Dockyard, and she was hauled up in September on the patent slip at Falcon Yard. Her re-launch in June 1844 was attended by as large and enthusiastic an audience as that present at her first launch in November 1834. As portrayed in this lithograph by H. John Vernon, the brig became one of the most successful anti-slaving vessels in the Navy's history, in a role where speed was the primary requirement. *(NMM X0105)*

2–5.1849. Recommissioned 14.4.1849 under Cmdr John Barling Marsh, for west coast of Africa; paid off at Plymouth 7.1851 into Ordinary. Hulked as blacksmith's shop at Devonport 1865. BU (by AO 31.2.1877, 'she being in a very decayed condition') completed at Plymouth 31.12.1878.

Hound Woolwich Dyd
  As built: 95ft 10in, 74ft 8⅛in x 30ft 4in oa (30ft 1in for tonnage) x 13ft 6in (rest). 359⁵⁵⁄₉₄bm.
  Draught 7ft 4in/11ft 2in.
  Ord: 20.8.1838. K: not begun. Order moved to Deptford Dyd.
  (Re)Ord: 22.6.1844. K: 2.7.1845. L: 23.5.1846. C: 27.8.1846 at Sheerness.
  First cost: £7,497 to build (+ fitting £3,842).
  Commissioned 29.6.1846 under Cmdr Granville Hamilton Wood, for west coast of Africa; took slavers *Bahiano* 11.4.1847, *Brazilieuse* 5.6.1847, *Faisco* 23.7.1847 and *Rey Bango* 27.10.1847; to North America and West Indies 1848. Recommissioned 7.11.1849 under Cmdr Frederick Patten (–6.9.1852), for west coast of Africa; under Cmdr Robert Hall (acting) from 14.12.1850; took slaver *Santissima Trinidad* 27.2.1852; paid off at Plymouth 28.8.1852 (or 4.9.1852). In Ordinary at Devonport 1852–70; became breakwater at Devonport 1872. Sold to Castle & Sons (for £245) 11.1887 to BU at Charlton; left Plymouth Sound for London 3.9.1888.

Mariner Pembroke Dyd
  Ord: 20.8.1838. K: — . Re-ordered as *Acorn* Class brig 24.5.1839 (see above).

Martin Pembroke Dyd
  Ord: 20.8.1838. K: — . Re-ordered as *Acorn* Class brig 24.5.1839 (see above).

WATERWITCH 10 guns. Design by Joseph White, developed from yacht *Louisa* of 1828. Ordered as yacht for Earl of Belfast; but purchased by the RN (by AO 11.10.1834) on stocks.
  Dimensions & tons (as built): 90ft 6in, 71ft 7in x 29ft 4in (29ft 2in for tonnage) x 12ft 8in. 323⁸¹⁄₉₄bm. Draught 7ft 1in/11ft 4in.
  Men: 55–60. Guns: 8 x 18pdr carronades (10cwt, 3½ft long) and 2 x 6pdrs (17cwt, 6ft long).

Waterwitch Joseph White, East Cowes
  L: 15.11.1834. C: 11.11.1834–25.3.1835 at Portsmouth.
  First cost: £3,948 to purchase, + £3,983 fitting (and making good defects).
  Commissioned 17.11.1834 under Lieut John Adams, for packet service; to Spain and then to west coast of Africa; on 3.1837 under Lieut William Dickey, for the weat coast of Africa and Cape of Good Hope; took slavers *El Cazador Santuazanos* 6.2.1836, *Galana Josefa* 13.3.1836, *Soven Maria* 14.3.1836, *Amelia* 6.8.1838 and *Vibora* 28.9.1837; paid off 31.7.1838. Refitted at Portsmouth (for £3,569) 10.1838–4.1839. Recommissioned 2.3.1839 under Lieut Henry James Matson, for the Cape of Good Hope and west coast of Africa; took slavers *Si* (a felucca) 27.5.1839, *Constituacao* 8.7.1839, *Sete de Abril* 27.9.1839, *Calliope* 27.10.1839, *Fortuna* 1.11.1839, *Cabaca* 14.3.1840, *Maria Rita* 17.5.1840, *Andorinha* 28.5.1840, *Doze d'Autobro* 13.11.1840, (with *Fantome* and *Brisk*) *Orozimbo* 8.1.1841, *Flor de Loanda* 2.5.1841, *Donna Elliza Erelinda* (a barque) 27.10.1841, *Donna Francesca* 20.11.1841, *Feliz Triuverato* 31.12.1841, *Himmaleh* 10.2.1842, *Triumfe* 22.7.1842, *Bello Indiana* 7.8.1842, *Nossa Senhora de Juda* 11.8.1842, *Gentil Africano* 28.8.1842, *Duquese de Mendille* 23.9.1842, *Josefina* 12.11.1842 and *Almeida* 27.4.1843; paid off 5.7.1843. Refitted at Portsmouth (for £2,445) 7–9.1843, then repaired at Cowes (for £3,109) 9.1843–6.1844; fitted for sea at Portsmouth (for £4,767) 6–10.1844. Recommissioned 7.9.1844 under Cmdr

The *Acorn* was the lead ship in Symonds's second design for 16-gun brigs, enlarged from the *Racer* Class. The *Acorn* spent her first commission between 1839 and 1843 off the coast of Africa, as part of the British West Africa Squadron (established in 1808 in the wake of the Wilberforce-inspired Act of Parliament to enforce the suppression of activity among British slavers). The squadron had other duties, but it was kept busy seizing foreign slavers (following enactment of anti-slavery legislation by other nations) and protecting legitimate traders. This lithograph after an original by Nicholas Matthew Condy the Younger depicts the *Acorn* in pursuit of the 'piratical' slaver *Gabriel*, which she seized on 6 July 1841. *(NMM PX9195)*

Thomas Francis Birch, for the experimental brig squadron, then to west coast of Africa; took slavers *Venganza* 23.5.1845, *Gabriel* 10.4.1846, *Caxias* 1.5.1846, unknown brig 8.5.1846, *Emprehenadora* 11.6.1846, *Selina* 10.1.1847, *Beulah* 12.7.1847, (with *Rapid*) *Romeo Primero* 22.7.1847, and *Adelaide* 12.8.1847; paid off 25.11.1847. Fitted for sea at Chatham (for £3,532) 2–4.1848. Recommissioned 3.4.1848 under Cmdr Richard Robert Quin, for the same station; took slavers *Despigue* 22.9.1849, another (name unknown) 15.11.1849, *Golfin* 27.11.1849, *Deos te Salve* 26.12.1849, *Encarnacion* 3.3.1850 and *Anna* 14.8.1850; on 28.6.1851 under Cmdr Alan Henry Gardner, for the same station; repulse at Lagos Bar 25.11.1851; took slaver *Gallina* 16.12.1852; paid off 10.5.1854 at Sheerness. Sold to Castle & Sons to BU (by AO of 22.11.1861, for £760).

***ACORN* Class** 16 guns. Second class brigs designed by William Symonds in 1836, enlarged from his *Racer* class, with similar dimensions to *Columbine*. The first six were ordered and named on 31 December 1835. Six more were ordered 20 February 1837, but *Liberty* and *Squirrel* were completed to *Racer* Class design instead, fitted as sail training ships, and in their place *Despatch* and *Dove* (first ordered as 6-gun packets to *Alert* design) were re-ordered to this *Acorn* class. Last six to modified design; a final pair, likewise originally ordered as *Alert* Class brigs in 1838, were also built to this design.

    Dimensions & tons: 105ft 0in, 82ft 2¾in x 33ft 6in oa (33ft 2in for tonnage) x 14ft 10in (15ft 0in in later ships). 481¹¹⁄₉₄bm.

    Men: 110/130. Guns: 2 x 32pdrs (25cwt in bow), + 14 x 32pdr carronades (17cwt); later 4 x 32pdr + 12 x 32pdr carronades (except where otherwise below).

Grecian Pembroke Dyd (16 x 32pdr carronades only)

    As built: 105ft 0in, 82ft 7in x 33ft 6½in oa (33ft 2½in for tonnage) x 14ft 10in. 484⁴⁹⁄₉₄bm. Draught 8ft 11in/11ft 8in.

Ord: 31.12.1835. K: 7.1836. L: 24.4.1838. C: 5.5–10.12.1838 at Plymouth.

First cost: £9,468 (+ fitting £4,057).

Commissioned 19.9.1838 under Cmdr William Smyth, for South America, then to Cape of Good Hope; took slavers *Leal* 11.4.1839, *Ganges* 30.4.1839, *Maria Carlotta* 29.5.1839, *Saudade* 12.2.1841, *Constante* 1.6.1841, *Janaviva* 24.4.1842, *Minerva* 28.4.1842, *Summacca Amizade Feliz* 14.10.1842, *Princesa Dona Francesca* 18.10.1842 and *Sumariva* 23.11.1842; paid off 30.91843. Very Small Repair at Plymouth and fitted for sea (for £4,403) 2.1844–4.1845. Recommissioned 12.3.1845 under Cmdr Alexander Leslie Montgomery, for the east coast of South America; on 16.7.1846 under Cmdr Louis Symonds Tindal, on same station (12 guns only); took slaver *Bello Miguelino* 22.4.1848; paid off 12.3.1849. Very Small Repair at Plymouth and fitted for sea (for £6,324) 10.1849–9.1851. Recommissioned 5.8.1851 under Cmdr George Disney Keane, for the Cape of Good Hope, then to the East Indies & China; on 23.7.1855 under Cmdr George Blane, in the East Indies; paid off 25.4.1856. Sold to J. & E. Marshall, Plymouth 1865; BU completed 1.11.1865.

Fantome Chatham Dyd

    As built: 105ft 1in, 82ft 6⅜in x 33ft 6¼in oa (33ft 2¼in for tonnage) x 14ft 10in. 483⁴⁸⁄₉₄bm. Draught 8ft 3in/11ft 10in.

    Ord: 31.12.1835. K: 5.1837. L: 30.5.1839. C: 17.12.1839.

    First cost: £10,338 (£13,702 fitted for sea).

Commissioned 28.10.1839 under Cmdr Edward Harris Butterfield, for the Cape of Good Hope, then to west coast of Africa; took slavers *Onzede Novembro* 11.10.1840, *Bellona* 14.12.1840, *Aventureiro* 31.12.1840, *Orozimbo* 8.1.1841, *Josephina* 1.5.1841, *Boa Nova* 9.6.1841, *Espardarte* 25.6.1841, *Triumfe* 4.7.1841 and *Conceicao de Maria* 3.10.1841; on 18.12.1841 under Cmdr Philip George Haymes, for South America; took slavers: one (name unknown) 4.1.1842, *Diligentia* 13.2.1842 and *Eugenia* 15.3.1842; paid off 20.10.1843 at Chatham. Very Small Repair there (for £3,718) 10.1843–3.1844; fitted for sea (for £678) 12.1844–2.1845. Recommissioned 14.12.1844 under Cmdr Frederick William Erskine Nicolson, for the Mediterranean; in action with Rif tribesmen at Cape Tres Forcas 12.5.1846; on 4.6.1846 under Cmdr Thomas Philip Le Hardy, on same station. Refitted at Portsmouth (for £5,299) 11.1849–1.1851. Recommissioned 10.12.1850 at Portsmouth under Cmdr John Henn Gennys, for Australia; paid off 6.1856 at Chatham. Sold to Castle & Beech to BU 7.10.1864, completed BU 19.10.1865.

Pilot Plymouth Dyd

As built: 105ft 0½in, 82ft 10½in x 33ft 6in oa (33ft 2in for tonnage) x 14ft 10in. $484\frac{86}{94}$bm. Draught 8ft 11in/11ft 9in.

Ord: 31.12.1835. K: 8.1837. L: 9.6.1838. C: 21.10.1838.

First cost: £10,589 (£14,601 fitted).

Commissioned 3.8.1838 under Cmdr George Ramsay, for North America and the West Indies; on 3.8.1842 under Cmdr Wallace Houston, on same station; on 7.7.1843 under Cmdr William Henry Jervis, for particular service; on 13.10.1845 under Cmdr George Knyvett Wilson, for the East Indies (–1847). Between Very Small and Small Repair at Plymouth, and fitted (for £6,274) 7.1847–7.1848. Recommissioned 7.6.1848 under Cmdr Edmund Moubray Lyons, then 6.11.1849 under Cmdr John Matthew Robert Ince, on same station; paid off 11.2.1852 at Plymouth. Sold to Marshall, Plymouth (for £802) 11.1.1862.

Acorn Plymouth Dyd (12 x 32pdr carronades only)

As built: 105ft 0in, 82ft 10¾in x 33ft 6in oa (33ft 2in for tonnage) x 14ft 10¾in. $485\frac{4}{94}$bm. Draught 8ft 10in/11ft 9in.

Ord: 31.12.1835. K: 11.1837. L: 15.11.1838. C: 21.3.1839.

First cost: £14,317 (fitted).

Commissioned 19.1.1839 under Cmdr John Adams, for the west coast of Africa; took slavers *Jehovah* 28.12.1839, *Pauceo* 9.1.1840, *Rahamana* 23.1.1840, *Quatro de Marco* 8.10.1840, *Amelia* 31.10.1840, *Gabriel* 6.7.1841, *Dous de Fevriero* 15.10.1841, *Minerva* 31.12.1841, another (name unknown) 29.5.1842, *Marianna* 27.6.1842, *Oito de Decembro* 7.7.1842, *Anna* 17.8.1842 and *San Joao Baptista* 27.9.1842; paid off 4.10.1843. Recommissioned 14.12.1844 under Cmdr John Elliot Bingham, for the east coast of South America; to East Indies 1847; paid iff 8.8.1848. Completed (refitting) for sea at Chatham (for £4,986) 5–6.1856. Recommissioned 6.5.1856 under Cmdr Arthur William Acland Hood, for the East Indies and China; in 2nd Anglo-Chinese War; boats in action at Fatshan Creek 1.6.1857; on 1.3.1858 under Cmdr Richard Bulkeley Pearse, on same station; paid off 2.1861 at Hong Kong. Fitted as hospital ship there 6–7.1861; to Shanghai 8.1861. Coal hulk at Shanghai 1861. On 21.2.1863 under Henry Hutchings, Master, as hospital ship at Shanghai; on 26.8.1864 under Denton Hemsworth Speer, Master. Hospital hulk 1867 (complement 26 men + 23 supernumeraries). Sold to Wilkins & Robinson at Yokohama (for £1,500) 15.2.1869.

Arab Chatham Dyd

As built: 105ft 0in, 82ft 2¼in x 33ft 6in oa (33ft 2in for tonnage) x 14ft 10in. $480\frac{84}{94}$bm. Draught 8ft 10in/11ft 9in.

Ord: 31.12.1835. K: 2.1838. L: 31.3.1847. C: 6.10.1847.

First cost: £10,147 to build, + fitting £5,539.

Commissioned 28.8.1847 under Cmdr William Morris, for the Cape of Good Hope, then to the East Indies; paid off 27.2.1851 at Chatham. Fitted for sea at Chatham (for £1,922) 8–9.1853. Recommissioned 18.8.1853 under Cmdr Graham Ogle, for North America and the West Indies; paid off at Chatham 29.6.1857. Became coastguard watch vessel at Queenborough (Swale) 24.5.1863, re-named *WV.18* in 1867. BU (by AO 15.12.1877) at Chatham 12.5–14.6.1879.

Persian Pembroke Dyd

As built: 105ft 1in, 82ft 8½in x 33ft 6in oa (33ft 2in for tonnage) x 14ft 10in. $483\frac{88}{94}$bm. Draught 8ft 10½in/11ft 8½in.

Ord: 31.12.1835. K: 5.1838. L: 7.10.1839. C: 20.10.1839–28.4.1840 at Plymouth.

First cost: £9,069 to build (+ fitting £4,444).

Commissioned 23.2.1840 under Cmdr William Henry Quin, for the west coast of Africa (died 22.11.1840); took slaver *Plant* 7.7.1840; on 20.11.1840 under Cmdr (acting) Thomas Edward Symonds; took slavers *Novo Inveya* 20.1.1841 and *Bom Fin* 21.1.1841; on 4.2.1841 under Cmdr Thomas Rodney Eden, on same station; took slavers *Dous d'Abril* 13.2.1841, *Flor d'America* 29.6.1841, *Cipher* 10.7.1841, *Senhora del Bon Viagen* 22.11.1841 and *Andorinha* 13.6.1843; paid off 13.8.1843 at Devonport. Very Small Repair and fitted at Plymouth (for £4,378) 1.1844–2.1845. Recommissioned 14.12.1844 under Cmdr Henry Coryton, for North America and the West Indies; paid off 14.3.1848. Recommissioned 7.7.1849 at Woolwich under Cmdr Archibald Gibson Bulman, for North America and the West Indies (invalided 24.7.1851); on 22.7.1851 under Cmdr Thomas Mitchell, on same station; paid off 15.3.1853. Recommissioned 14.10.1857 under Cmdr John Hanbury Chads, for Cape of Good Hope; on 27.1.1859 under Cmdr Edward Hardinge, on same station; on 21.7.1861 under Cmdr Cecil William Buckley; paid off 11.7.1861 at Sheerness. Sold to Castle & Beech to BU at Charlton 3.1866; BU 10.5–27.6.1866.

Bittern Portsmouth Dyd

As built: 105ft 1in, 82ft 3½in x 33ft 7in oa (33ft 3in for tonnage) x 14ft 10in. $483\frac{87}{94}$bm. Draught 8ft 1½in/11ft 8½in.

Ord: 20.2.1837. K: 3.1838. L: 18.4.1840. C: 17.8.1841.

First cost: £13,008 (fitted).

Commissioned 5.5.1841 under Cmdr Byron Charles Ferdinand Plantagenet Cary, for the west coast of Africa; took slavers *Sumariva* 23.11.1842, *Ventura* 6.1.1843 and *Furia* 13.1.1843; on 1.7.1843 under Cmdr Edmund Peel, for the Cape of Good Hope; took slavers *Altrevida* 23.9.1843, another (name unknown) 25.9.1843, *Emperador Dom Pedro* 23.6.1844 and *Opio Felice* 23.6.1844; paid off 18.7.1845 at Sheerness. Recommissioned 17.11.1845 under Cmdr Thomas Hope, for the west coast of Africa; took slavers *Adelaide* 21.11.1846, *Phedro* 29.1.1847, *Tebbesero* 11.8.1847 and *Josephina* 4.5.1848; paid off 30.1.1849 at Sheerness. Very Small Repair at Sheerness (for £1,629) 10–12.1849; fitted for sea there (for £2,497) 8–10.1852. Recommissioned 25.8.1852 under Cmdr Edward Westby Vansittart, for the East Indies; on 12.1.1856 under Cmdr William Thornton Bate, on same station; paid off 17.10.1856. For sale 1857, but re-employed on 1.12.1857 under Lieut (Acting) James Graham Goodenough, for Canton operations in 2nd Anglo-Chinese War. Sold to Cheung Sung (for £1,031¼) at Hong Kong 20.2.1860.

Albatross Portsmouth Dyd (16 x 32pdr carronades only)

As built: 105ft 1in, 82ft 3⅝in x 33ft 7in oa (33ft 3in for tonnage) x 14ft 10in. $483\frac{93}{94}$bm. Draught 7ft 10in/11ft 3in.

Ord: 20.2.1837. K: 7.1838. L: 28.3.1842. C: 14.7.1842.

First cost: £8,649 (+ fitting £6,451).

Commissioned 12.5.1842 under Cmdr Reginald Yorke, for North America and the West Indies; then to west coast of Africa 7.1844; took slavers *Constancia* 30.7.1844, *Piedad* 7.10.1844, *San Jose* 2.11.1844, *Albanez* 1.3.1845 and *Beija Flor* 5.7.1845; paid off

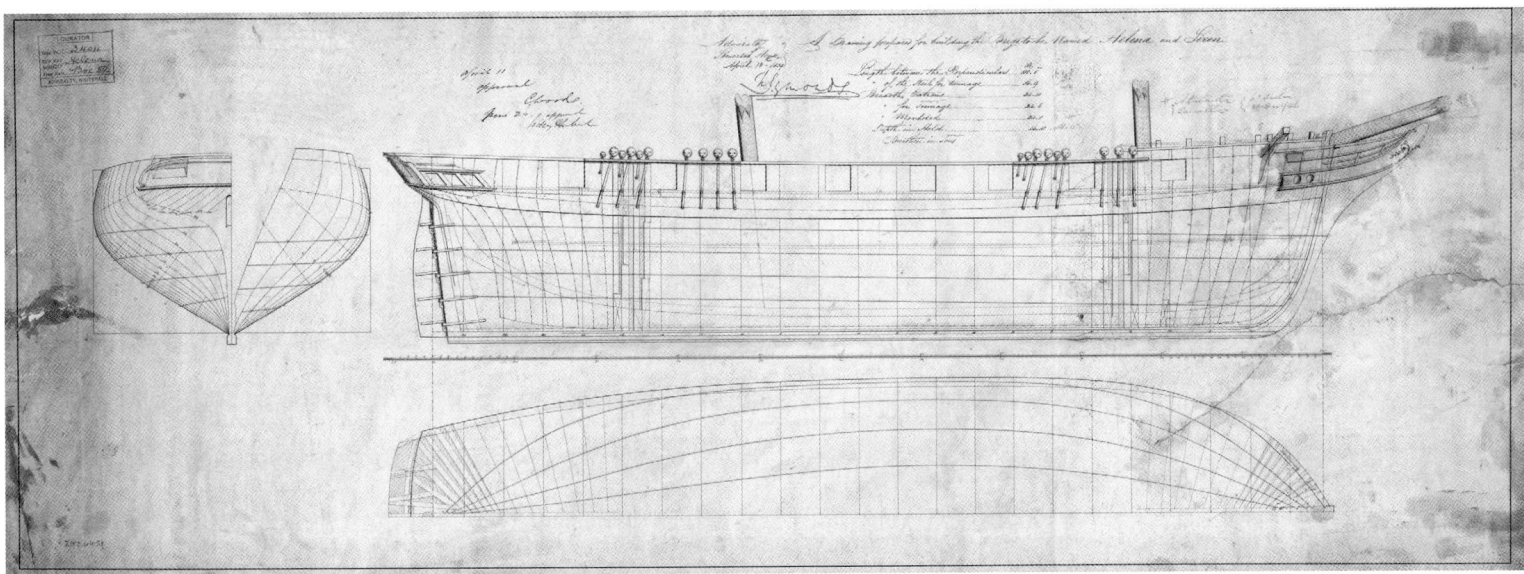

Symonds's third and final group of 16-gun brigs was the *Helena* class (the design is dated 10 April 1839), again somewhat enlarged from the preceding *Acorn* Class. Symonds continued to modify his 16-gun brig design throughout his period in office, and even within a given class there were differences between the early individual ships and those begun later. He was keen on the traditional look for warships, and these brigs carried false quarter galleries, even though there was no cabin behind them. *(NMM J6940)*

9.7.1846. Recommissioned 12.11.1846 at Chatham under Cmdr Arthur Farquhar, for west coast of Africa; took slavers *Rosetta* 19.6.1847, *Prendadore* 22.7.1847, *Imperador* 1.2.1848 and *Constancia* 29.2.1848; to the East Indies 5.1848; in 10.1849 under Lieut Hugh Maximilian Elliott; paid off 26.9.1850 at Chatham into Ordinary. BU completed at Chatham 19.5.1860.

Elk Chatham Dyd
As built: 105ft 0in, 82ft 2⅛in x 33ft 6½in oa (33ft 3in for tonnage) x 15ft 0in. 483²⁴⁄₉₄bm. Draught 8ft 4in/12ft 3in.
Ord: 20.2.1837. K: 11.1846. L (undocked): 27.9.1847. C: 6.1856.
First cost: £9,874 (+ fitting for sea £6,045).
Commissioned 6.5.1856 under Cmdr John Fane Charles Hamilton, for the East Indies and China; in 2nd Anglo-Chinese War; boats action at Fatshan Creek 1.6.1857; at Canton 12.1857; on 1.3.1858 under Cmdr Hubert Campion, on same station, then to Australia; paid off 13.8.1861 into Ordinary. Coastguard watch vessel 3.1863, initially as *WV.13*, re-named *WV.28* on 25.5.1863, stationed at Tilbury. Sold to W. Taylor (by AO 7.4.1893, for £400) 30.5.1893.

Heron Chatham Dyd (built in a dock)
As built: 105ft 0in, 82ft 2⅜in x 33ft 6¼in oa (33ft 2¼in for tonnage) x 15ft 0in. 481¹³⁄₉₄bm. Draught 8ft 3in/12ft 1in.
Ord: 20.2.1837. K: 11.1846. L: 27.9.1847. C: 23.11.1857.
First cost: £9,319 to build.
Commissioned 3.10.1857 under Cmdr William Henderson Truscott, for the west coast of Africa; foundered in tornado off West Africa 9.5.1859 (65 drowned); Truscott survived the sinking, but died (from yellow fever) en route home.

Mariner Pembroke Dyd
As built: 105ft 0in, 82ft 2¾in x 33ft 6in oa (33ft 2in for tonnage) x 14ft 10½in. 481¹³⁄₉₄bm. Draught 8ft 8in/11ft 10in.
Ord: 24.5.1839. K: 5.1845. L: 19.10.1846. C: 15.3.1847 at Plymouth.
First cost: £9,824 to build (£14,376 fitted).
Commissioned 28.1.1847 under Cmdr Charles Mitchell Mathison, for the west coast of Africa, then to the East Indies; paid off 6.11.1850 at Plymouth. Recommissioned 10.5.1854 under Cmdr

Frederick Erskine Johnston; on 29.8.1854 under Cmdr Thomas Cochran, for North America and the West Indies; on 12.1.1856 under Cmdr Wray Richard Glenstanes Palliser, on same station; paid off 12.6.1857 at Chatham. Sold at Chatham to Capt. Henry Frederick McKillop by AO of 12.6.1865.

Despatch Chatham Dyd (12 x 32pdr carronades only)
As built: 105ft 0in, 82ft 7½in x 33ft 6in oa (33ft 2in for tonnage) x 15ft 0in. 483⁴³⁄₉₄bm. Draught 7ft 11in/11ft 10in.
Ord: 24.5.1839. K: 12.1846. L: 25.11.1851. C: 24.12.1851 (for Ordinary).
First cost: £9,267 to build.
Never commissioned. Became coastguard watch vessel 25.5.1863, re-named *WV.24* (or *Cadmus*), based in the East Swale. Sold at Sheerness (by AO 14.12.1900) 13.5.1901.

Martin Pembroke Dyd
As built: 105ft 0in, 82ft 2⅛in x 33ft 6⅛in oa (33ft 2⅛in for tonnage) x 15ft 0¼in. 481³⁹⁄₉₄bm. Draught 8ft 5in/11ft 10in.
Ord: 24.5.1839. K: 9.1848. L: 19.9.1850. C: 3.10.1850 (sailed for Plymouth for Ordinary).
First cost: £9,200 to build.
Fitted as training brig at Plymouth 2–10.1864.
Commissioned 11.6.1864 under Lieut William Henniker Heaton, as tender to *Implacable* at Plymouth; on 19.5.1865 under Lieut James Terence Fitzmaurice, as tender to *St Vincent* at Portsmouth, then 1.1867 under Lieut George Bruce Evans, then 14.4.1870 under Lieut (Cmdr 29.12.1871) Dashwood Goldie Tandy, then 1.1875 under Lieut E. Hicks and finally 4.1878 under Lieut Arthur C. B. Bromley, all in same role. Re-named *Kingfisher* 2.5.1890, as training brig; recommissioned 8.1890 under Lieut Charles Skelton Nicholson. Sold to Collins, Dartmouth 2.10.1907.

Kangaroo (ex-*Dove*, renamed 27.2.1843) Chatham Dyd (4 x 32pdr + 8 x 32pdr carronades)
As built: 105ft 1in, 82ft 7in x 33ft 6in oa (33ft 2in for tonnage) x 15ft 0in. 481¹³⁄₉₄bm. Draught 7ft 11in/11ft 5in.
Ord: 24.5.1839. K: 1.4.1850. L: 31.8.1852. C: 28.2.1853 (for Ordinary).
First cost: £7,806 to build.
Never commissioned. Became coastguard watch vessel 25.5.1863, re-named *WV.20*; towed to Stangate Creek (by *Adder*) 2.7.1863; by AO 27.6.1870 to replace *Chanticleer* in the River Crouch. Sold to Michael Hayhurst (for £360) 10.5.1897.

Four further vessels (of which the last three were to this *Acorn* Class design) were launched after 1863, all at Pembroke Dyd, as training brigs. As these were significantly later than 1863, no service histories are attempted:

Seaflower
　Dimensions & tons: 100ft 6in x 32ft 4in x 15ft 4in. 425bm. 454 disp.
　L: 25.2.1873. Tender to *Boscawen* at Portland from 6.1873.
　　Workshop 1.1904. Sold to Castle to BU at Charlton 7.4.1908.
Nautilus L: 20.5.1879. Tender to *Impregnable*. Sold to Cox, Falmouth 11.7.1905.
Pilot (ii) L: 12.11.1879. Sold to Adrien Merveille, Dunkirk to BU 2.10.1907.
Martin (ii) (ex-*Mayflower*, renamed 4.1888) L: 20.1.1890. Coal hulk *C.23* from 15.11.1907.

HELENA Class First Class Brigs, 16 guns. Sir William Symonds design 1839 (last five to a modified design).). Two were ordered in 1839, one (of teak) in 1843 and two in 1844. Six more were ordered in February 1847, two each at Pembroke (*Musquito* and *Rover*), at Woolwich and at Deptford, but the orders for the latter four were replaced on 27 April by orders for four 'screw schooners', to be named *Pincher*, *Hornet*, *Cracker* and *Reynard* (see Chapter 12).
　Dimensions & tons: 110ft 0in, 86ft 9in x 34ft 10in oa (34ft 6in for tonnage) x 14ft 10in (15ft 0in in last 5). 549²¹⁄₉₄bm.
　Men: 140. Guns: 16 x 32pdr carronades (25cwt).
Siren (or *Syren*) Woolwich Dyd
　As built: 110ft 0in, 86ft 8⅛in x 34ft 10in (34ft 6in for tonnage) x 14ft 10in. 549²⁄₉₄bm. Draught 8ft 5in/11ft 2½in.
　Ord: 18.3.1839. K: 6.1839. L: 23.4.1841. C: 30.10.1841.
　First cost: £9,975 to build (£14,820 including fitting).
　Commissioned 9.8.1841 under Cmdr William Smith, for the East Indies; paid off 23.12.1844. Recommissioned 9.5.1845 under Cmdr Harry Edmund Edgell, for the Mediterranean; reported that her boats took four pirate vessels off Stanchio (Turkey) 1846, but the event was officially denied. On 10.11.1846 under Cmdr Thomas Chaloner, for the coast of Africa; took slavers *Tito* 26.5.1847, *San Jose* 17.1.1848, *San Francisco Boa Fé* 2.4.1848, *Princessa Donna Isabel* 3.4.1848 and *Polka* 5.8.1848; paid off 15.12.1848. Between Very Small and Small Repair at Sheerness (for £4,006) 1–10.1850. Fitted for sea at Sheerness (for £4,009) 8–9.1855. Recommissioned 2.8.1855 under Cmdr Robert Jocelyn Otway, for the east coast of South America; on 6.2.1857 under Cmdr Jasper Henry Selwyn, then on 28.2.1858 under Cmdr George Macintosh Balfour, on same station; paid off 2.4.1860 at Portsmouth. Fitted as target ship at Portsmouth 2–3.1862. BU at Portsmouth 23.11–14.12.1868.
Helena Pembroke Dyd
　As built: 110ft 0in, 86ft 9½in x 34ft 10in oa (34ft 6in for tonnage) x 14ft 10½in. 549⁴⁶⁄₉₄bm. Draught 9ft 0in/11ft 5in.
　Ord: 18.3.1839. K: 9.1839. L: 11.7.1843. C: 23.12.1843 at Portsmouth.
　First cost: £9,541 to build (£15,625 including fitting at Portsmouth).
　Commissioned 23.10.1843 under Cmdr Sir Cornwallis Ricketts, for the Cape of Good Hope; took slaver dhow *Messure Kley* 3.4.1845; paid off 17.7.1847 at Portsmouth. Refitted there for sea (for £5,976) 9.1847–7.1848. Recommissioned 9.6.1848 under Cmdr George Woodberry Smith, for the Channel squadron; on 12.9.1848 under Cmdr Michael De Courcy, for North America and the West Indies; paid off at Portsmouth 30.11.1851. Coal hulk 1861. Police hulk 11.1863. Fitted as floating chapel at Portsmouth 12.1868. To Ipswich on loan as church ship 1.1879. To Chatham

for repairs 7.1880. Fitted as police accommodation ship (replacing *Challenger*) at Chatham 7.11.1883–2.1884, for harbour police. Sold to J. B. Garnham to BU 6.1.1921.
Jumna (ex-*Zebra*, renamed 19.1.1846) Bombay Dyd. Teak-built.
　As built: 110ft 0in, 86ft 9⅝in x 34ft 10in oa (34ft 6in for tonnage) x 15ft 4in. 549¹²⁄₉₄bm. Draught 8ft 6in/11ft 6in.
　Ord: 15.6.1843. K: 12.1843. L: 7.3.1848. C: 12.4.1848 at Bombay.
　First cost: £16,589 (including £837 fitting at Bombay).
　Commissioned 4.1848 under Lieut Matthew Harley Rodney, for voyage to England; arrived 3.10.1848 at Portsmouth, then 7.10.1848 at Chatham and paid off 16.10.1848 there into Ordinary; not otherwise commissioned. Sold (by AO 25.6.1862, for £1,650) to Capt. Lynch.
Atalanta Pembroke Dyd
　As built: 110ft 0in, 86ft 8⅞in x 34ft 11in oa (34ft 7in for tonnage) x 15ft 0½in. 551⁷⁶⁄₉₄bm. Draught 8ft 11in/11ft 6in.
　Ord: 19.2.1844. K: 10.1846. L: 9.10.1847. C: 20.12.1847, then to Plymouth for Ordinary.
　First cost: £9,771 including fitting at Pembroke.
　Fitted for sea at Plymouth (for £6,973) 5–7.1856. Commissioned 20.5.1856 under Cmdr Thomas Malcolm Sabine Pasley, for North America and the West Indies; paid off 14.4.1860 at Plymouth. BU completed at Devonport 12.12.1868.
Camilla Pembroke Dyd
　As built: 110ft 0in, 86ft 9½in x 34ft 10in oa (34ft 6in for tonnage) x 15ft 0½in. 549⁴⁶⁄₉₄bm. Draught 9ft 0in/11ft 6in.
　Ord: 8.7.1844. K: 10.1846. L: 8.9.1847. C: 20.12.1847, then to Plymouth 23.12.1847 for Ordinary.
　First cost: £9,715 including fitting at Pembroke.
　Fitted for sea at Plymouth (for £7,091) 5–8.1856. Commissioned 8.7.1856 under Cmdr George Twisleton Colville, for the East Indies and China; lost off Japan, presumed foundered with all (121) hands 9.9.1860 in a typhoon.
Musquito Pembroke Dyd
　As built: Dimensions as designed.
　Ord: 17.2 & 25.4.1847. K: 5.1849. L: 29.7.1851. C: 25.8.1851, then to Plymouth 27.8.1851 for Ordinary.
　First cost: £9,945.
　Never commissioned for RN. Sold to Prussian Navy (by AO 9.7.1862, for £11,828.16.10) and left Plymouth 28.10.1862 for the Baltic.
Rover Pembroke Dyd
　As built: Dimensions as designed. Draught 8ft 6in/11ft 0in.
　Ord: 17.2 & 25.4.1847. K: 9.1850. L: 21.6.1853. C: 2.7.1853, then to Plymouth for Ordinary.
　First cost: £9,662 including fitting at Pembroke.
　Never commissioned for RN. Sold to Prussian Navy (by AO 9.7.1862, for £11,763.15.10d) and left Plymouth 28.10.1862 for the Baltic.

ROYALIST Teak-built mercantile brig *Mary Gordon*, launched 13 July 1839 at Bombay, and purchased 1841 for use as a survey ship in the Far East.
　Dimensions & tons: 87ft 4in, about 72ft 0in x 25ft 6in x 6ft 8in. 249bm.
　Men: 60. Guns: UD: 4 x 12pdr; QD 2 x 6pdr; Fc 2 x 6pdr.
Royalist Bombay
　Purchased in China by Sir G. Bremner 9.7.1841 (approved under AO 5.1.1842), for £7,200, and renamed.
　Commissioned 22.1.1842 under Lieut Philip Chetwode (died 15.9.1843), then under Gerald Kingsley, 2nd Mate (died 1844),

The 16-gun *Frolic* was a one-off design by Capt. William Hendry. Her construction appears to be an early attempt by the Admiralty to introduce the ideas of other designers by way of comparison with the Surveyor's vessels, which three years later would see a proliferation of competitive designs being ordered. William Hendry was a long-serving sixty-three-year old officer whose last active role had been to bring home from St Helena in 1821 the dispatches announcing the death of Napoleon. When the Naval College opened in 1829 'for the instructions of officers in subjects connected with their profession', Hendry was the senior officer among those admitted. H. John Vernon's lithograph of the *Frolic* at sea shows her crew striking her studding sails off the coast of Brazil, having driven a slaving vessel ashore. *(NMM PW8097)*

then under Eudo Wells, 2nd Mate (died 5.1844); these three allegedly died while in successive command of the brig); then Charles Parkinson, 2nd Master (temp.); on 11.4.1844 (nominally) under Lieut Graham Ogle, for the East Indies, but Ogle did not actually take command until 4.1845; note Lieut Edward Hill was officially appointed w.e.f. 16.9.1843, but seemingly never took up command; on 6.7.1846 under Lieut James Aylmer Dorset Paynter, then 5.8.1846 under Lieut David MacDowall Gordon, and 25.7.1849 under Cmdr William Thornton Bate; paid off 3.6.1854 at Portsmouth. Police hulk on Thames 7.1856. Sold 14.2.1895.

FROLIC First Class Brig, 16 guns. Designed by Captain William Hendry, 1840.

  Dimensions & tons: 108ft 3in, 84ft 3⅛in x 34ft 1in oa (33ft 9in for tonnage) x 15ft 6in. 510⁶⁹⁄₉₄bm.
  Men: 140. Guns: 16 x 32pdr carronades (17cwt).
Frolic Portsmouth Dyd
  As built: 108ft 3in, 84ft 1in x 34ft 1in (33ft 9in for tons) x 15ft 6½in. 509⁴²⁄₉₄bm. Draught 8ft 5½in/10ft 11in.
  Ord: 1.6.1840. K: 2.1841. L: 23.8.1842. C: 11.2.1843.
  First cost: £8,634 to build, plus £6,212 fitting.
  Commissioned 3.11.1842 under Cmdr William Alexander Willis, for South America; on 8.4.1844 under Cmdr Cospatrick Baillie Hamilton, for Brazil and the Pacific; paid off 12.6.1847 at Portsmouth. Refitted and fitted for sea at Portsmouth (for £5,675) 6.1847–7.1848. Recommissioned 7.6.1848 under Cmdr Nicholas Vansittart, for the Mediterranean. Refitted at Portsmouth (for £4,665) 10.1853–3.1854. Recommissioned 25.10.1853 under Cmdr Matthew Stainton Nolloth, for the Cape of Good Hope; on 21.2.1856 under Cmdr John de Courcy Andrew Agnew, then 6.6.1856 under Cmdr (Acting) Lumley Woodyear Peyton, on same station; paid off 23.7.1857 at Chatham. Sold to Castle & Beech 7.10.1864, and BU at Charlton 20.11–29.12.1865.

From mid-1843 a group of experimental brigs were built. The first was ordered from Bombay on 26 June 1843 as *Goshawk*. The next six vessels listed below, although to differing designs, were all ordered nominally under the designation '*Goshawk* Class', and named on 4 September (first three) or 18 September (last three) 1843; excluding *Kingfisher*, the other five formed (with Rule's *Cruizer* of 1828 and Symonds's *Pantaloon* of 1831) an Experimental Squadron to conduct research into improved hull and sailing forms; *Daring*, *Espiegle* and *Flying Fish* were usually the fastest with the wind abeam, *Cruizer* before the wind.

NERBUDDA Second Class Brig, 16 guns (12 guns as completed). Sir William Symonds design, approved 26 June 1843, slightly modified from lines of the *Racer* Class.

  Dimensions & tons: 100ft, 77ft 0⅞in x 32ft 4in oa (32ft 0in for tonnage) x 13ft 6in. 419⁵²⁄₉₄bm.
  Men: 110. Guns: 2 x 32pdr (39cwt) + 14 x 32pdr (25cwt).
Nerbudda (ex-*Goshawk*, renamed 19.1.1846) Bombay Dyd. Teak-built.
  Ord: 26.6.1843. K: 1.11.1843. L: 5.2.1848. C: 6.1853 at Chatham.
  First cost: £12,817 to build (+ £229 fitting) at Bombay, plus £4,001 fitting at Chatham.
  Commissioned 20.2.1848 at Bombay under Cmdr Edward Peirse (appointed 27.10.1847), for voyage to England; sailed from Bombay 28.2.1848; paid off 8.6.1848 at Chatham into Ordinary. Recommissioned 2.7.1853 under Cmdr Henry Ashburton Kerr, for the Cape of Good Hope; lost, presumed foundered with all (126) hands off Algoa Bay 6.1855.

DARING Third Class Brig, 12 guns. This vessel was initially ordered in 1840 to be built at Sheerness to the *Pandora* design, but on 27 July 1841 this design was cancelled and the Sheerness order was to be executed to 'the drawings of *Waterwitch*' (the Joseph White-designed brig purchased in 1832); however, the scantlings and specifications were to be as *Rapid*. Eventually she was re-ordered from Portsmouth to a new

In September 1844 an Experimental Squadron was assembled, comprising five of the newest brigs to conduct sailing trials against the older *Cruiser* (1828) and the purchased *Pantaloon* and *Waterwitch*, which were then regarded as the Navy's best performing designs. Under the overall command and supervision of Captain Armar Lowry Corry, fourteen runs were made under varying conditions, and a detailed report produced. As these were effectively races, where ship handling and seamanship skills were of as much import as design, it is not surprising that different ships excelled in certain conditions and no clear overall winner could be determined – although the newer brigs were all generally an improvement over the elderly *Cruiser*. This lithograph by H. John Vernon shows the squadron with the *Mutine* in the foreground; behind, left to right, are the *Daring*, *Espiegle*, *Osprey*, *Waterwitch* and *Pantaloon*. (*NMM PW8103*)

Joseph White design, approved 23 August 1843; the Sheerness order was discontinued on 4 September. *Osprey* (see below) was originally to have been built to the same design.

Dimensions & tons: 104ft 0in, 83ft 0½in x 31ft 4½in oa (31ft 0½in for tonnage) x 15ft 5½in. 425⁵⁹⁄₉₄bm.

Men: 130. Guns: 2 x 32pdr shell (39cwt) + 10 x 32pdr (25cwt).

Daring Portsmouth Dyd (not built at Cowes as sometimes mis-stated!)
As built: 104ft 0in, 83ft 0½in x 31ft 4½in oa (31ft 0½in for tonnage) x 15ft 5½in. 425⁵⁹⁄₉₄bm.
Ord: 10.7.1843. K: 10.1843. L (undocked): 2.4.1844. C: 22.10.1844.
First cost: £8,036 to build, plus £4,992 fitting.
Commissioned 7.9.1844 under Cmdr Henry James Matson, for North America and the West Indies; from 1.2.1847 under Cmdr William Peel (third son of Sir Robert Peel); paid off 2.10.1748. Fitted for Ordinary at Sheerness 12.1848; fitted at Chatham (for £1,647) 8–12.1849, then to Portsmouth; fitted there for sea (for £2,011) 8–10.1852. Recommissioned 8.1852 under Cmdr Gerard John Napier, for same station; paid off 17.10.1856 at Portsmouth. Sold to Castle & Beech 7.10.1864 and BU at Charlton 3.1865.

MUTINE Third Class Brig, 12 guns. John Fincham design, approved September 1843. *Espiegle* (see below) was originally to have been built to the same design.

Dimensions & tons: 102ft 0in, 81ft 2¼in x 31ft 10in oa (31ft 6in for tonnage) x 13ft 7in. 428⁴⁷⁄₉₄bm.

Men: 130. Guns: 2 x 32pdr shell (39cwt) + 10 x 32pdr (25cwt).

Mutine Chatham Dyd
As built: 101ft 11½in, 81ft 2in x 31ft 11in oa (31ft 6in for tonnage) x 13ft 7in. 428³⁷⁄₉₄bm.
Ord: 10.7.1843. K: 30.10.1843. L: 20.4.1844. C: 21.9.1844.
First cost: £12,983 (including fitting).
Commissioned 7.9.1844 under Cmdr Richard Borough Crawford, for the Cape of Good Hope; took five slaver dhows between

5.1845 and 1.1846; captured Brazilian slavers *Diana* 10.10.1845 and *Amelia* 6.2.1846; suppression of slave traffic in the Mozambique Channel; from 29.9.1846 under Cmdr Robert Tryon, in the Mediterranean, then 26.10.1847 under Cmdr John Jervis Palmer; wrecked at Chioggia, near Venice 20.12.1848, while under temp. command of Lieut Alfred Palmer (5 died).

FLYING FISH Class Third Class Brigs, 12 guns. William Symonds design, approved 4 September 1843.

Dimensions & tons: 103ft 0in, 81ft 8in x 32ft 4in oa (32ft 0in for tonnage) x 14ft 8in. 444⁷⁶⁄₉₄bm.

Men: 130. Guns: 2 x 32pdr shell (39cwt) + 10 x 32pdr (25cwt).

Flying Fish Pembroke Dyd
As built: 103ft 1in, 81ft 7in x 32ft 4½in oa (32ft 0½in for tonnage) x 14ft 4½in. 445⁴⁹⁄₉₄bm.
Ord: 10.7.1843. K: 10.1843. L: 3.4.1844; sailed 18.4.1844. C: 26.4–22.10.1844 at Portsmouth.
First cost: £9,055 to build, plus £4,248 fitting.
Commissioned 7.9.1844 under Cmdr Robert Harris, for the experimental brig squadron, then to the west coast of Africa; took slaver *Eliza* 4.12.1845; on 19.5.1846 under Cmdr Peché Hart Dyke, then 11.6.1846 under Lieut George Oldmixon (acting); took slaver *Jupiter* 18.3.1847; on 3.12.1847 under Lieut Richard Ashmore Powell (acting); paid off 3.3.1848. Recommissioned from 8.1849 under Cmdr George Edward Patey, for the same station; took slavers *Proserpina* 9.11.1849, *Phoenix* 27.2.1850, *Louisa* 19.5.1850; *Constellation* 3.7.1850, *Vingador* 10.7.1850, *Mosquito* 1.11.1850, another (name unknown) 26.12.1850 and *Pepita* 14.7.1851; paid off 6.12.1851 at Portsmouth. BU at Portsmouth 8.1852.

Kingfisher Pembroke Dyd
As built: 103ft 1in, 81ft 7in x 32ft 4½in oa (32ft 0½in for tonnage) x 14ft 8in. 445⁴⁹⁄₉₄bm. Draught 8ft 1in/12ft 1in.

Public interest in the 'experimental brigs' was reflected in a number of published prints, this one by H. John Vernon featuring the *Osprey*, with the *Mutine*, *Pantaloon* and *Espiegle* shown (left to right) behind. *(NMM PW8104)*

Ord: 10.7.1843. K: 4.1844. L: 8.4.1845; sailed 14.5.1845. C: 25.10.1845 at Portsmouth.

First cost: £8,704 to build, plus £4,203 fitting.

Commissioned 28.8.1845 under Cmdr Charles Foreman Brown (invalided 21.3.1846), for the west coast of Africa; took slaver *Ligeiro* 12.3.1846; on 21.3.1846 under Lieut Richard Ashmore Powell (acting); on 24.4.1846 under Cmdr Frederick Wilmot Horton, on the same station; took slavers *Maria* 15.8.1846, *Valerozo* 6.9.1846, *Augusta* 13.9.1846, *Victoria* 16.10.1846, *Genio* 17.10.1846 and another (name unknown) 24.4.1847; paid off 19.7.1848. Recommissioned 11.10.1848 under Cmdr Henry Harvey, for the same station; took slavers *Paquete do Sul* 4.9.1849, *Lusitana* 18.10.1849, *California* 5.11.1849 and *Clio* 30.11.1849; paid off 26.2.1851 at Devonport. Used as tender at Plymouth (to *Impregnable*) 1875–90. Sold to W. Taylor 26.4.1890.

OSPREY Third Class Brig, 12 guns. Richard Blake design, approved 13 November 1843.

　Dimensions & tons: 101ft 5in, 80ft 6¼in x 31ft 10in oa (31ft 6in for tonnage) x 13ft 6in. 424⁹²/₉₄bm.

　Men: 130. Guns: 2 x 32pdr shell (39cwt) + 10 x 32pdr (25cwt).

　First cost: £7,676 (+ fitting £4,822).

Osprey Portsmouth Dyd

　As built: 101ft 6in, 80ft 4in x 31ft 10½in oa (31ft 6½in for tonnage) x 13ft 10in. 425¹¹/₉₄bm.

　Draught 7ft 6in/12ft 9in.

　Ord: 10.7.1843. K: 11.1843. L: 2.4.1844. C: 22.10.1844.

　First cost: £7,676 to build, plus £4,822 fitting.

　Commissioned 7.9.1844 under Cmdr Frederick Patten, for the East Indies; at Ruapekapeka Stockade 11.1.1846; wrecked off Hokianga, New Zealand 11.3.1846.

ESPIEGLE Third Class Brig, 12 guns. Reed, Chatfield and Creuze design, approved 23 November 1843.

　Dimensions & tons: 104ft 8in, 82ft 8in x 31ft 8in oa (31ft 4in for tonnage) x 13ft 1¼in. 431⁸⁶/₉₄bm.

　Men: 130. Guns: 2 x 32pdr shell (39cwt) + 10 x 32pdr (25cwt).

Espiegle Chatham Dyd

　As built: 104ft 8in, 83ft 7¾in x 31ft 9½in oa (31ft 5½in for tonnage) x 13ft 1¼in. 442⁶⁹/₉₄bm.

　Draught 8ft 11in/11ft 9in,

　Ord: 25.10.1843. K: 1.1844. L: 20.4.1844. C: 21.9.1844.

　First cost: £9,722 to build, plus £3,406 fitting.

　Commissioned 7.9.1844 under Cmdr Thomas Pickering Thompson; to East Indies; in attack (with steamer *Pluto*) on Canton River pirates and capture of the Bogue Forts 3–26.4.1847. On 20.8.1847 under Cmdr Frederick Archibald Campbell, still on China Station; paid off 6.1849. Very Small Repair at Sheerness (for £3,309) 9.1849–3.1750. Recommissioned 28.1.1753 under Cmdr George Hancock; refitted at Sheerness (for £3,058) 2–3.1753, for North America and the West Indies. On 19.1.1855 under Cmdr Edward Henry Gage Lambert, for North America and the West Indies; paid off 10.1856 into Ordinary at Sheerness. Sold to Castle & Beech (by AO 22.11.1861, for £805) to BU 22.11.1861.

BRITOMART Third Class Brig, of 8 guns. Design by Sir William Symonds, approved 4 September 1843. A further six ships to this design were ordered 18 June 1844, but they were suspended six days later and never proceeded with.

　Dimensions & tons: 93ft 0in, 73ft 5⅛in x 29ft 3in oa (29ft 0in for tonnage) x 13ft 5in. 328⁴/₉₄bm.

　Men: 80. Guns: 2 x 18pdr + 6 x 18pdr carronades (10cwt).

Britomart Pembroke Dyd

　As built: 93ft 0in, 73ft 5⅛in x 29ft 3¾in oa (29ft 0¼in for tonnage) x 13ft 5½in. 329⁸³/₉₄bm.

　Draught 7ft 9½in/11ft 2in.

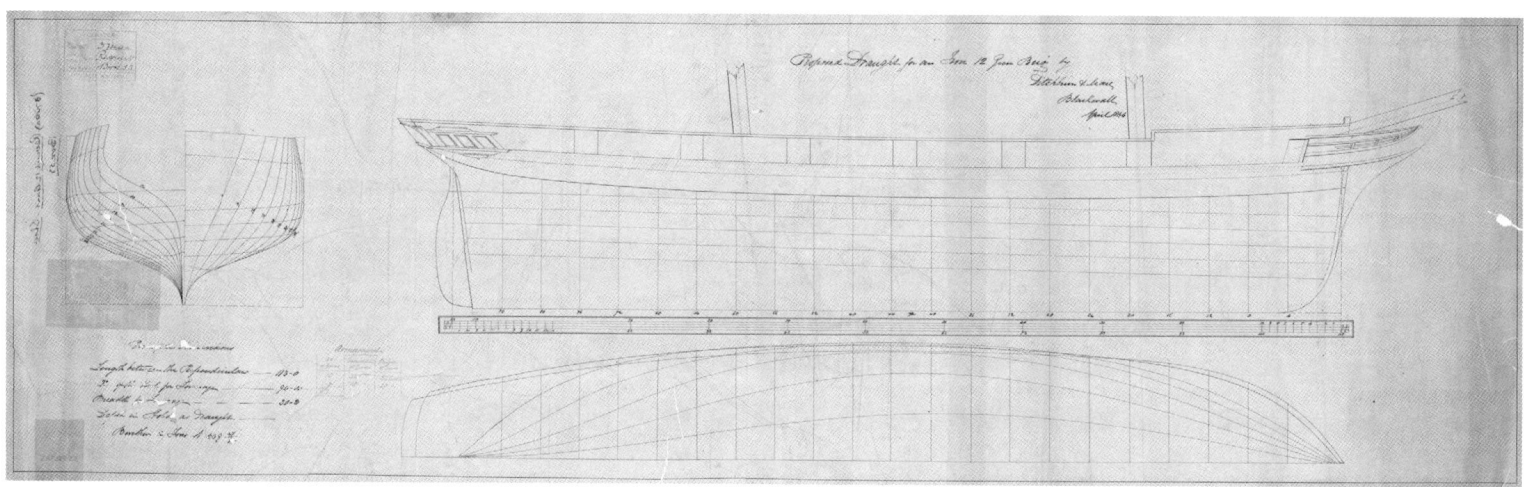

The *Recruit* was the only iron-hulled sailing vessel ever to be built for the Royal Navy. The specialist shipbuilder Thomas Ditchburn drew this draught for her in April 1844, and constructed her under contract at his Blackwall yard. Her first and only naval commander, Adolphus Slade, was highly critical, saying in November 1847 that she was unsuitable for the Navy because the iron suffered severely under cannon fire, while condensation caused health issues for the crew. His report was considered by a Parliamentary Select Committee, and with other evidence this persuaded them to instruct that no further iron-hulled warships should be ordered, a prohibition which lasted until well after the Crimean War, by which time iron-founding methods had improved significantly. *(NMM J6883)*

Ord: 4.9.1843. K: 5.1846. L: 12.6.1847. C: 25.6–24.8.1847 at Plymouth.

First cost: £7,219 to build, plus £3,219 fitting.

Commissioned 3.1852 under Cmdr Albert Heseltine, for the West coast of Africa; paid off 25.10.1855 into Ordinary at Plymouth. To coastguard service 21.10.1857, renamed *WV.25* on 26.5.1863 at Beresford, then to Burnham. BU at Chatham 25.6–25.7.1874.

RECRUIT Third Class iron brig, 12 guns. Design by Thomas Ditchburn, approved 6 May 1844. The RN's only iron-hulled sea-going sailing vessel, for which the tender was accepted on 22 June 1844.

Dimensions & tons: 113ft 0in, 94ft 11in x 30ft 6in x 14ft 0in. 469⁶⁷⁄₉₄bm.

Men: 130. Guns: 2 x 18pdrs + 10 x 32pdr c.

Recruit Ditchburn & Mare, Blackwall

As built: 114ft 5⅛in, 92ft 5in x 30ft 8½in oa (30ft 7¾in for tonnage) x 13ft 10in. 461⁶⁄₉₄bm.

Draught 7ft 11in (fwd) 12ft 3in (aft). Ord: 6.5.1844 (named 20.7.1844). K: — . L: 10.6.1846. C: 17.6–29.8.1846 at Sheerness.

First cost: £10,127 to build, plus £5,306 fitting.

Commissioned 2.7.1746 under Cmdr Adolphus Slade, for particular service; paid off 6.11.1847 as being 'unfit for further service'. Sold back to builders 28.8.1849 for £4,500 (by AO 6.7.1849); subsequently fitted as screw ship and re-sold to become mercantile *Harbinger* in 1850.

CONTEST Third Class Brig, 12 guns. Design by Joseph White (a development of White's *Daring* design), approved 29 April 1845.

Dimensions & tons: 109ft 0in, 88ft 8⅜in x 31ft 4in oa (30ft 9in for tonnage) x 15ft 2in. 446¹¹⁄₉₄bm.

Men: 130. Guns: 2 x 18pdr + 10 x 32pdr carronades.

Contest Joseph White, Cowes

As built: 109ft 0in, 88ft 8⅜in x 31ft 6½in oa (31ft 2½in for tonnage) x 15ft 2in. 459⁴⁸⁄₉₄bm.

Ord: 29.4.1845. K: 6.1845. L: 11.4.1846. C: 11.4–14.8.1846 at Portsmouth.

First cost: £8,301 to build, plus £4,401 fitting.

Commissioned 11.6.1846 under Cmdr Archibald McMurdo, for the African station; took slavers *Esperanza* 30.7.1847, *Phoenix* 29.8.1847, *Sappho* 8.10.1847, *Conceicao* 23.11.1847, *Temerario* 8.5.1848, *Santa Cruz* 14.5.1848, *Oceana* 9.7.1848, *Felix Sociedade* 18.7.1848, another (name unknown) 31.8.1848, *Phoenix* 3.10.1848, *Meteora* 21.11.1848 and *Horizonte* 30.4.1849; in 12.1849 under Cmdr John Welbore Sunderland Spencer, for the East Indies; took slaver *Rosina* 5.4.1850; paid off 3.9.1853. In Ordinary at Portsmouth until 1868. BU (by AO 6.8.1868) completed at Portsmouth 9.9.1868.

9 Cutters and Schooners

(A) Vessels in service or on order at 1 January 1817

Large numbers of fore-and-aft rigged vessels had been employed throughout the Napoleonic Wars, including a number of ex-American merchant and privateer schooners seized during the War of 1812. Most of these small craft had been among the first to be deleted from the post-war surplus of vessels. By the start of 1817, the Royal Navy had disposed of most, retaining just a handful of the newer craft.

On 5 April 1816 the Navy had taken over responsibility for the cruisers of the Revenue and Customs service and over sixty existing R&C craft were taken under Admiralty control during the following year. Separating naval and revenue cutters is often difficult, as the latter were included in the Navy lists (albeit listed in a separate section); these are accordingly omitted except where naval and revenue craft were built to a common design (thus allowing transfer between the two groups). Essentially, the naval craft in 1817 comprised fourteen schooners (including the *Starling*, still building) and just four cutters; a further four schooners (the new *Quail* Class) were rated as tenders.

PICKLE (French *Eclair*, built 1808) Three-masted schooner, 12 guns
> Dimensions & tons: 69ft 3in, about 50ft 6in x 22ft 6in x 9ft 9in. 136bm.
> Men: 50. Guns: 12 x 12pdr carronades.
> Taken 1808, and registered 2.5.1809.
> Commissioned 1809 under Lieut. Uriah Goodwin; in 1810 under Lieut. Andrew Crawford, then 22.4.1812 under Lieut. William Figg; paid off 26.8.1815. Sold to Mr Durkin 11.6.1818.

ARROW Class 14-gun schooner, designed by John Peake in 1805
Arrow Deptford Dyd
> Ord: 3.7.1804. K: 1.1805. L: 7.9.1805.
> Commissioned 9.1805 under Lieut. Richard Hawkes; paid off 1810. Recommissioned 1811 under Lieut. Samuel Knight; in 5.1813 under Lieut. John George Aplin; attacks on coastal batteries near Quimper and Quiberon; paid off 12.3.1814. Became Yard craft at Plymouth 5.1814 (for the Breakwater Dept). BU at Plymouth 5.1828.

CUCKOO (or 'Bird') Class 4-gun schooner, design copied from draught of Bermudan *Fish* Class. The last survivor of a class of twelve, all originally ordered 11 December 1805; the other eleven were wartime losses or disposals, the last being the *Quail* and *Jackdaw* sold in 1816 (as had been the *Bream*, the last of the eighteen similar *Ballahoo* or 'Fish' class).
> Dimensions & tons: 56ft 2in, 42ft 4⅛in x 18ft 3in x 8ft 6in. 75¹/₉₄bm.
> Men: 20. Guns: 4 x 12pdr carronades.
Landrail Thomas Sutton, Ringmore
> As built: 56ft 3½in, 42ft 4¼in x 18ft 3½in x 8ft 7in. 75³¹/₉₄bm. Draught 4ft 6in/7ft 6in.
> Ord: 11.12.1805. K: 1.1806. L: 18.6.1806. C: 7.7–16.7.1806 at Plymouth.

In Ordinary at Plymouth until 1812. Fitted as tender 6–7.1811. Commissioned 26.8.1812 under Lieut John Hill, for the North Sea and Baltic. On 21.6.1814 under Lieut Robert Daniell Lancaster; taken off Ushant by the American privateer schooner *Syren* 12.7.1814, with 5 wounded, but was retaken 9.1814 on her way to the USA, and carried into Halifax, Nova Scotia. At Jamaica in 1817. Paid off 10.1818 and sold 1818?

CHEERFUL Class A 12-gun cutter, designed by Sir John Henslow in 1806. Her sister *Cheerful* had been sold on 31 July 1816.
> Dimensions & tons: 63ft 0in, 46ft 9⅛in x 23ft 6in x 10ft 0in. 137³⁹/₉₄ bm. (no 'as-built' data recorded).
> Men: 50. Guns: 2 x 4pdrs + 10 x 12pdr carronades.
Surly Joseph Johnson, Dover
> Ord: 29.3.1806. K: 7.1806. L: 15.11.1806. C: 23.11–4.5.1807 at Sheerness.
> Commissioned 12.12.1806 under Lieut Peter Crawford, for the North Sea. on 1.2.1809 under Lieut Richard Welch (–1815); took (with *Firm* and *Sharpshooter*) 4-gun privateer *L'Alcide* off Granville 21.4.1810. In 1815 under Lieut Mark Lucas, then 18.5.1816 under Lieut John Hill; paid off 9.1818. Subsequently reduced to 8 guns, and based at the Nore. Between Small and Middling Repair 5–8.1826. On 22.7.1830 under Lieut Horatio James (–6.1831). Fitted as lighter at Chatham 12.1832–2.1833. Sold to Mr Rahn at Chatham (for £400) 1.1837.

PIGMY Class 10-gun schooners, designed by Henry Peake and originally built as cutters. Their sister *Algerine* had been a wartime loss (in 1813).
> Dimensions & tons: 82ft 6in, 70ft 6in x 22ft 8in x 10ft 6in. 192⁶²/₉₄bm.
> Men: 60. Guns: 10 x 12pdr carronades.
Pigmy John King, Upnor
> As built: 82ft 10in, 70ft 8in x 22ft 11in x 10ft 1in. 197³⁸/₉₄bm. Draught 5ft 0in/8ft 3in.
> Ord: 2.10.1809. K: 11.1809. L: 24.2.1810. C: 25.6.1810 at Chatham.
> First cost: £2,770 to build.
> Commissioned 5.1810 under Lieut Edward Moore. Converted to schooner at Sheerness 11–12.1811. In 1812 under Lieut William Hutchinson, in the Downs, then 1813 to Baltic. In 1814 under Lieut Richard Crossman, then Lieut John Little. Fitted for sea 3.1817. In 8.1807 under Lieut Wentworth Parsons Croke. Re-converted to cutter 1819. In 6.1820 under Lieut Thomas Hills. Sold at Plymouth to Joshua Crystall (for £510) 21.5.1823.
Pioneer John King, Upnor
> As built: 82ft 10in, 70ft 7½in x 23ft 0in x 10ft 1in. 197bm (198⁶⁸/₉₄ by calc.). Draught 4ft 8in/8ft 5in.
> Ord: 2.10.1809. K: 12.1809. L: 10.3.1810. C: 21.6.1810 at Chatham.
> First cost: £2,769 to build.
> Commissioned 24.4.1810 under Lieut John Row Morris (–1814); took (with *Decoy*) lugger-privateer *L'Indefatigable* 9.5.1810; took (with *Bermuda* and *Dwarf*) 16-gun privateer *Le Bon Génie* 11.9.1810. Converted to schooner at Sheerness 10–11.1811. In 6.1814 under Lieut John Hill, then 25.8.1815 Lieut John Wood

This model, thought to date from about 1820, cannot be identified with a specific vessel, but exhibits the main features of a naval or revenue cutter of the period – a sharp but deep hull to offset the relatively large area of canvas (gaff main and gaff topsail, with an additional square topsail for running before the wind, plus multiple headsails set from the long bowsprit). Early example of naval cutters tended to be clinker-built (with overlapping strakes), but later ones like this example were carvel (where the strakes butted edge-to-edge, producing a smooth surface). *(NMM F5871-001)*

Rouse, off Scottish coast; paid off 15.4.1818. Recommissioned 16.12.1818 under Lieut William Oldrey; re-converted to cutter 1819; paid off 1821. Fitted at Plymouth for the Coast Blockade 1.1823–4.1824; in coastguard service to 1845. Sold at Plymouth 4.9.1849.

DECOY **Class** 10-gun cutter, designed by William Rule. Her sisters *Decoy* and *Racer* had been wartime losses to the French.
    Dimensions & tons: 74ft 4in, 55ft 10⅛in x 26ft 0in x 11ft 0in. 200⁸⁹⁄₉₄bm.
    Men: 60. Guns 10 x 18pdr carronades.
Dwarf John Lowes, Sandgate
    As built: 74ft 6¾in, 56ft 0⅛in x 26ft 1in x 11ft 0in. 202⁶¹⁄₉₄bm. Draught 4ft 8in/11ft 0in.
    Ord: 2.10.1809. K: 11.1809. L: 24.4.1810. C: 13.4–29.6.1810 at Sheerness.
    Commissioned 5.1810 under Lieut Samuel Gordon (–1817), for the Downs; took (with *Bermuda* and *Pioneer*) 16-gun privateer *Bon Génie* in the Channel 11.9.1812. Between Small and Middling Repair, and fitted for sea at Plymouth 10.1818–1.1819; recommissioned 11.1818 under Lieut Nicholas Chapman, then 11.1821 Lieut George Read, finally 1.1823 Lieut Nicholas Gould; wrecked on the pier at Kingstown, Dublin 3.3.1824 (1 died).

NIMBLE **Class** 10-gun cutter, built 'by the draught of an improved revenue cutter'. Her sister *Nimble* had been a wartime loss in 1812.
    Dimensions & tons: 63ft 0in, 49ft 1¼in x 22ft 9in x 10ft 6in. 140¹⁶⁄₉₄bm.

Men: 50. Guns: 10 x 12pdr carronades.
Swan Thomas Gely, Cowes
    As built: 63ft 5in, 49ft 4½in x 23ft 5in x 10ft 2in. 144bm. Draught 4ft 4in/10ft 0in.
    Ord: 21.1.1811. K: 6.1811. L: 1.11.1811. C: 4.11.1811–16.1.1812 at Portsmouth.
    First cost: £2,453 to build, plus £2,703 fitting.
    Commissioned 11.1811 under Lieut Henry Rowed. In 6.1814 under Lieut James Whitthorn, at the Nore, then 8.1815 under Lieut William Smith, 6.1817 under Lieut James Griffiths and 16.4.1819 under Lieut Thomas Dilnot Stewart then paid off 11.11.1822. Fitted at Sheerness as Fishery protection vessel 7–8.1831. Became Chapel ship for the Seamen's Home Society on the Thames 6–7.1837. BU completed at Sheerness 7.12.1874.

Ex-FRENCH Prize
Algerine (French *Pierre Cézar*, built 1805 Baltimore) Cutter, 14 guns.
    Dimensions & tons: 92ft 9in, 72ft 9¾in x 24ft 4in x 10ft 9in. 229³¹⁄₉₄bm.
    Men: 50. Guns: 14 x 12pdr carronades.
    Taken 29.6.1808 and purchased as *Tigress*.
    Commissioned 10.1811 under Lieut. William Carnegie; renamed *Algerine* 21.4.1814; paid off 1.1817. Sold 29.1.1818.

Ex-AMERICAN PRIZES.
Sealark (American *Fly*, launched 1811 –) Schooner, 10 guns.
    Dimensions & tons: 79ft 6in, 68ft 0⅞in x 22ft 8in x 9ft 10in. 178bm.
    Men: 50. Guns: 10 x 12pdr carronades.
    Taken by *Scylla* 29.12.1811, and purchased 1.5.1812.
    Commissioned 8.1812 under Lieut James Warrend, for the Lisbon station, later to Channel. In 11.1814 under Lieut Philip Helpman; paid off 1.1819. Sold 13.1.1820.
Bermuda (American *Delaware* pilot boat, building details unknown) Schooner/yacht.
    Dimensions & tons: 46ft 2in, c36ft x 15ft 0in x 5ft 6in. 43bm.
    Presented by her captors 1813, and registered in RN 28.4.1813. BU at Bermuda 9.1841.
    Commissioned 11.1814 under Cmdr John Sykes, on the Jamaica station. BU 2.1817.
Telegraph (American privateer *Vengeance*, built 1812 New York) Schooner/gunbrig, 12 guns.
    Dimensions & guns: 83ft 7in, 66ft 5¼in x 22ft 6½in x 10ft 6in. 180bm.
    Men: 60. Guns: 12 x 12pdr carronades.
    Taken by *Phoebe* 1.1.1813. Registered 31.8.1813.
    Commissioned 12.1813 under Lieut Timothy Scriven. In 9.1815 under Lieut Richard Crossman, then 1816 Lieut Jonathan Little; wrecked under Mount Batten, Plymouth Sound 20.1.1817.
Alban (American *William Bayard*, built 1812 by Bergh, New York) Schooner, 14 guns.
    Dimensions & tons: 94ft 4½in, 78ft 6½in x 24ft 7⅛in x 10ft 6in. 252⁷¹⁄₉₄bm.
    Men: 60. Guns: 2 x 6pdr + 12 x 12pdr carronades.
    Taken by the *Warspite* 12.3.1813.
    Commissioned 10.1813 under Lieut Mayson Wright. In 9.1815 under Lieut Hugh Patton; paid off 10.1818. BU 18.2.1822.
Musquidobit (American privateer *Lynx*, built 1812 at Baltimore) Schooner, 10 guns.
    Dimensions & tons: 94ft 7in, 73ft 1¼in x 24ft 0in x 10ft 3in. 223⁹²⁄₉₄bm.
    Men: 50. Guns: 8 x 18pdr carronades, + 2 x 6pdrs.

Taken 16.3.1813 in the Chesapeake by boats of *St Domingo* and
others. Purchased for £1,933.11.5d (amended figure). Arrived
Portsmouth 30.3.1814.

Commissioned 1813 under Lieut John Murray. On 10.9.1815 under
Lieut Joseph Griffiths (–1817); paid off 12.1818. Sold to Mr
Rundle (for £410) at Plymouth 13.1.1820.

Shelburne (American privateer *Racer*, built 1811 in
Maryland) Schooner, 12 guns.

Dimensions & tons: 93ft 8in, 72ft 8in x 23ft 11in x 10ft 8in.
221⁹⁰⁄₉₄bm.

Men: ?50. Guns: 10 x 12pdr carronades, + 2 x 6pdrs.

Taken 16.3.1813 in the Chesapeake by boats of *St Domingo* and
others. Purchased for £1,940.11.5d (amended figure).

Commissioned 6.1813 under Lieut David Hope; took (with
Orpheus) US sloop *Frolic* 20.4.1814. On 1.10.1814 under Lieut
William Hamilton; arrived Deptford 22.6.1816 and paid off. Sold
to Mr Brown (for £600) at Deptford 10.1817.

Pike (American privateer *Dart*, built 1813 at New Orleans) 14-gun
schooner.

Dimensions & tons: 93ft 0in, 78ft 7in x 24ft 8in x 10ft 6in.
254³¹⁄₉₄bm.

Men: 60. Guns: 12 x 12pdr carronades, + 2 x 6pdrs.

Taken 13.11.1813 by *Niger* and purchased (registered) 11.2.1814.
Fitted for sea at Plymouth (for £3,126) 1–4.1814, with coppered
bottom.

Commissioned 2.1814 under Lieut David Buchan (–1817), for
Newfoundland station. Recommissioned 13.8.1818 under Lieut
Simon Hopkinson, for Cork; in 5.1821 under Lieut Stephen
Pain. Middling Repair at Plymouth (for £4,901) 8.1824–1.1825;
recommissioned 9.1824 under Lieut Thomas Pennington, still
for Cork; in 10.1825 under Lieut Edward Kelly, then 9.4.1828
under Lieut John Gwyn Wigley (–1831). In 1831 under Capt.
Alexander Thomas Emeric Vidal, for surveying. Fitted for sea at
Plymouth (for £3,345) 11.1831–3.1832; recommissioned 11.1831
under Lieut Arthur Brooking, for West Indies; paid off 2.1835
but recommissioned 1.1836, still under Brooking; wrecked on
Bare Bush Key, Jamaica 5.2.1836, and wreck sold.

Grecian (American privateer *Grecian*, built 1813 by Kemp,
Baltimore) Schooner, 10 guns.

Dimensions & tons: 95ft 1in, 74ft 3in x 23ft 10in x 10ft 5in.
224³¹⁄₉₄bm.

Men: 50. Guns: 8 x 18pdr carronades, + 2 x 6pdrs.

Taken 2.5.1814 by boats of *Jaseur* (not purchased until 3.11.1815).
Defects made good at Portsmouth 18.8–21.12.1815.

Commissioned 10.1815 under Lieut Henry Jewry (–1817); on
18.8.1818 under Lieut Nathaniel Martin; paid off 1821. Sold to
Joshua Crystall (for £510) at Portsmouth 18.4.1822.

Variable (American *Edward*, built 1812 –) Schooner or brig, 14 guns.

Dimensions & tons: 104ft 6in, 82ft 2in x 27ft 3in x 13ft 0in. 324bm.

Men: 95. Guns: 12 x 24pdr carronades, + 2 x 6pdrs.

Taken and purchased 31.10.1814.

Commissioned 1814 in North America under Lieut Richard
Williams. Reported under Lieut John Sykes at Jamaica on
2.11.1814, then Lieut Robert Baldey, but this may be a different
Variable. Arrived Deptford 25.7.1816 and paid off 8.1816. BU
there 12.1817.

Pictou (ii) (American privateer *Zebra*, built 1812 at New York) 14-gun
schooner.

Dimensions & tons: 101ft 5in, 85ft 6½in x 25ft 7⅜in (25ft 2⅜in mld.)
x 10ft 0½in. 299bm.

Men: — . Guns: 6 x 6pdr + 10 x 12pdr carronades.

Taken 12.4.1813 by *Pyramus* off west coast of France. Made good
defects at Portsmouth 2–4.1815.

Commissioned 6.1815 under Lieut Charles Hare in 9.1815 under
Lieut James Morgan, at Cork; paid off 2.1818. Sold to Mr
Hughes 13.8.1818.

SPEEDWELL Schooner tender

Dimensions & tons: 80ft 0in, 61ft 2in x 25ft 0in x 11ft 3½in.
203³²⁄₉₄bm.

Ex-mercantile *Royal George*, purchased 8.1815.

Sold at Jamaica 1.1834.

QUAIL Class Schooner tenders, 4 guns. Built of fir, to the 'draught
of the *Watchful* revenue cutter, but of increased dimensions'. Eight
were ordered on 23 May 1816, but *Woodlark*, *Hart* and *Highflyer* were
not begun until 1820 and 1821 respectively, while the eighth vessel
(*Arrow*) was never built; their design was approved 15 June 1816. On
31 January 1822 it was directed that *Quail* be renamed *Providence*, but
this AO was rescinded on 11 April 1822. Seemingly some were fitted as
cutters.

Dimensions & tons: 55ft 6in, 44ft 5in x 18ft 7in x 7ft 6in. 81⁵¹⁄₉₄bm.

Men: 30. Guns: 4 x 12pdrs (also 2 x ½pdrs in first four).

Quail Deptford Dyd [M/Shipwright William Stone]

As built: 55ft 8in, 44ft 7⅜in x 18ft 8in x 7ft 6in. 82⁶¹⁄₉₄bm. Draught
4ft 0in/7ft 5in.

Ord: 23.5.1816. K: 8.1816. L: 3.1.1817. C: 5.2.1817.

Renamed *Providence* 31.1.1822, but reverted to *Quail* 11.4.1822.
BU completed 8.4.1829.

Linnet Deptford Dyd [M/Shipwright William Stone]

As built: 55ft 8in, 44ft 7⅜in x 18ft 7in x 7ft 6in. 82⁶¹⁄₉₄bm. Draught
4ft 0in/7ft 5in.

Ord: 23.5.1816. K: 8.1816. L: 3.1.1817.

Built for Revenue Service, transferred 1817 to RN. Defects made
good at Portsmouth 14.11–13.12.1817. Fitted as survey vessel
at Plymouth 11.1827–1.1828; in 2.1828 under Lieut Edward
Barrett, for surveying the Channel Islands and as tender to
Prince Regent; later under Capt. Martin White; tender to
Wellesley 11–12.1830; then in Ordinary 1830–31. Survey vessel
1833. Sold to John Small Sedger, Rotherhithe (for £210) to BU
7.8.1833.

Swift Woolwich Dyd [M/Shipwright Henry Canham]

As built: 55ft 6in, 44ft 5in x 18ft 7in x 7ft 6in. 81⁵¹⁄₉₄bm. Draught 3ft
6in/8ft 0in.

Ord: 23.5.1816. K: 11.1816. L: 15.2.1817. C: 30.3.1817.

Sold to Mr Turner (for £90) 8.1821.

Redbreast Woolwich Dyd [M/Shipwright Henry Canham]

As built: 55ft 6in, 44ft 5in x 18ft 7in oa (18ft 6in for tonnage) x 7ft
6in. 80⁸¹⁄₉₄bm. 80⁸¹⁄₉₄bm

Ord: 23.5.1816. K: 11.1816. L: 18.2.1817.

Quarantine service at Liverpool 1817–1840. Fitted as tender at
Sheerness in 1830s, to Plymouth as lazarette in 1840s. Sold to J.
Brown 1850.

Four more were ordered on the same date as the above, but
Woodlark (at Deptford) and *Hart* and *Highflyer* (at Woolwich)
were not begun until 1820 and 1821 respectively, while *Arrow*
was never built.

Woodlark Deptford Dyd

As built: 55ft 6in, 44ft 5in x 18ft 7in oa (18ft 6in for tonnage) x 7ft
6in. 80⁸¹⁄₉₄bm. Draught 3ft 10in/7ft 1in.

Ord: 23.5.1816. K: 9.1820. L: 2.7.1821. C: 5.9.1821 for sea.

Fitted as a surveying cutter at Deptford 11.1828–4.1829. Tender to

Investigator 1834, then to Woolwich 1837–42. Commissioned 9.4.1845 under Lieut Frederick William Leopold Thomas, as tender to *Mastiff* (–10.1848). Recommissioned 1.1858 under Thomas again, as tender to *Fisgard* at Woolwich.

Hart Woolwich Dyd [M/Shipwright Henry Canham]

As built: 55ft 6in, 44ft 5in x 18ft 8in oa (18ft 6in for tonnage) x 7ft 6½in. $80\frac{89}{94}$bm. Draught 4ft 2in/8ft 1in.

Ord: 23.5.1816. K: 1.1821. L: 12.6.1821. C: 30.7.1822 for sea.

Renamed ***Navy Board*** 1822. Fitted as longboat at Sheerness 6.1825. Fitted as Admiralty tender at Sheerness 3.1833, reverted to *Hart*. Became Sheerness *YC No. 1* in 1860s, then renamed ***Drake*** 17.11.1870. Superintendent's yacht at Sheerness (by AO 13.2.1872) 31.3.1872. Offered for sale to Colonial Office (by AO 12.3.1872) as a mail boat for Heligoland. BU completed at Chatham 6.3.1875.

Highflyer Woolwich Dyd [M/Shipwright Henry Canham]

As built: 55ft 7in, 44ft 6in x 18ft 9in oa (18ft 7in for tonnage) x 7ft 6½in. $81\frac{79}{94}$bm. Draught 4ft 5in/8ft 0in.

Ord: 23.5.1816. K: 1.1821. L: 11.6.1822. C: 14.1.1823 for sea.

Fitted at Chatham for sea 2–4.1825 ; became tender to *Hyperion*, coastguard depot at Newhaven, 4.1825. Sold to John Small Sedger of Rotherhithe (for £240) 7.8.1833.

STARLING One-off cutter, 10 guns, built 'to the draught of the *Griper* (revenue) cutter'. Based at Portsmouth.

Dimensions & tons: 71ft 9in, 60ft 8in x 21ft 7⅞in x 9ft 3in. $151\frac{32}{94}$bm. Draught 4ft 7in/11ft 0in.

Men: 45. Guns: 8 x 6pdr carronades, plus 2 x 6pdrs (brass).

Starling Chatham Dyd [M/Shipwright George Parkin]

Ord: 1816. K: 9.1816. L: 3.5.1817. C: 12.7.1817.

Commissioned 4.6.1817 under Lieut Thomas Sherwin; on 7.11.1818 under Lieut John Reeve, then 10.11.1821 under Lieut Charles Turner, 10.1.1825 under Lieut Richard Stuart and 1.2.1827 under Lieut James Pearl. BU at Portsmouth 11.8.1828.

(B) Vessels acquired from 1 January 1817

COCHIN Schooner/tender for Trincomalee. Design by John Edye

Dimensions & tons: 53ft 6in, 43ft 2in x 15ft 6in x 7ft 10in. $53\frac{92}{94}$bm.

First cost: £550.

Cochin East India Co., Cochin

L: 23.4.1820. Hulked 1840. Sold at Trincomalee 9.4.1850.

EMERALD Cutter tender, built 1819 and purchased 1820.

Dimensions & tons: 57ft 3in, 45ft 8¾in x 18ft 9½in oa (18ft 5½in for tonnage) x 9ft 1in. $85\frac{84}{94}$bm.

Emerald Philip Sainty, Colchester

Purchased at Portsmouth from the Marquis of Anglesea 14.9.1820.

Initially (1821) tender to flagship at Portsmouth; 12.1837 tender to yacht *Royal George*. BU at Portsmouth 12.1847.

GOSSAMER Cutter tender, unarmed

Dimensions & tons: 46ft 4in, 35ft 4⅞in x 16ft 0⅞in x 7ft 11in. 48bm.

Gossamer Gosport

Purchased for £404 on stocks 7.1821, as tender to flagship at Sheerness.

Sold to John Levy, Rochester 22.11.1861 for £230.

***VIGILANT* Class** Cutters, 6 guns. Built to the 'lines of the *Lapwing* revenue cutter, enlarged'. The Pembroke-built pair were for the Revenue Service. The latter two were for the Navy, and were based at Plymouth and Sheerness respectively; they were later ketch-rigged.

Dimensions & tons: 67ft 3in, 51ft 4in x 24ft 3in x 10ft 7in. $160\frac{51}{94}$bm.

Men: 38 (later 31). Guns: 2 x 6pdr + 8 (later 2) x 6pdr carronades.

Skylark Pembroke Dyd

Ord: 3.9.1819. K: 8.1820. L: 17.2.1821.

Revenue vessel. Fate unknown.

Swift Pembroke Dyd

Ord: ?3.9.1819. K: ?8.1820. L: 17.2.1821.

Revenue vessel. Fate unknown (still in service 1846).

Vigilant Deptford Dyd

Ord: 3.9.1819. K: 8.1820. L: 18.4.1821.

Commissioned 14.2.1822 under Lieut George Read; on 25.2.1823 under Lieut Nicholas Colthurst, then 8.4.1825 under Lieut Samuel Meredith, 7.5.1828 under Lieut Charles Jones, and 12.8.1829 under Lieut William Loney. Re-rigged as ketch 5.1826. Sold 11.1832.

Basilisk Chatham Dyd

As built: 67ft 5½in, 51ft 6⅝in x 24ft 3in x 10ft 7in.

Ord: 1819. K: 2.1821. L: 7.5.1822.

Commissioned 16.2.1823 under Lieut David Dickson; on 1.3.1824 under Lieut John James Hough; on 16.3.1826 under Lieut John Chumley, then 1.5.1829 under Lieut William Watts and 22.7.1830 under Lieut John Wright. Converted to a tender 1831. Sold 1.1846.

***BRAMBLE* Class** Cutters, 10 guns. Built 'as the *Diligence* revenue cutter' (built 1799). The *Cheerful* was for the Revenue Service; the other two were for the Navy, and were based at Plymouth and Portsmouth respectively.

Dimensions & tons: 70ft 9in, 52ft 4in x 24ft 2½in x 11ft 0⅜in. $163\frac{13}{94}$bm. Draught (light): *Bramble* 6ft 3in/10ft 3in; *Cheerful* 6ft 1½in/10ft 0in.

Men: 50. Guns: 2 x 6pdr + 8 x 12pdr carronades.

First cost: *Bramble* £6,217 (+ fitting £3,550). *Sparrow* £3,779

Cheerful Plymouth Dyd

Ord: 1818 (for Revenue Service). K: 8.1818. L: 21.8.1819.

Commissioned 1.3.1820 under Lieut William Edwards; paid off 24.11.1821. Recommissioned 1.2.1822 under Lieut Alex Greet. Fate unknown.

Bramble Plymouth Dyd

Ord: 20.3.1819. K: 8.1820. L: 8.4.1822.

Commissioned 21.6.1824 under Lieut Thomas Favell; on 10.8.1827 under Lieut William Haswell, then 24.8.1830 under Lieut James Harvey. Classed as a tender 1831. Fitted as a survey ship with schooner rig 4.1842 (31 men). Recommissioned 2.4.1842 under Lieut Charles Bampfield Yule, on East India Station as tender to *Fly* and *Rattlesnake*. Lent to New South Wales government as diving vessel 31.5.1853. Sold as a lightship 23.12.1876.

Sparrow Pembroke Dyd

Ord: 20.3.1819. K: 10.1827. L: 28.6.1828.

Commissioned 18.7.1828 under Lieut John Moffat. Classed as a tender 1831. Ketch rig 1837 (40 men). Survey ship 1844. BU at Devonport 8.1860.

FLYING FISH Schooner tender (ex-mercantile *Lady Augusta*, built at Leith)

Dimensions & tons: 62ft 0in, 47ft ?10½in x 17ft 6in x 8ft 0in. 78bm.

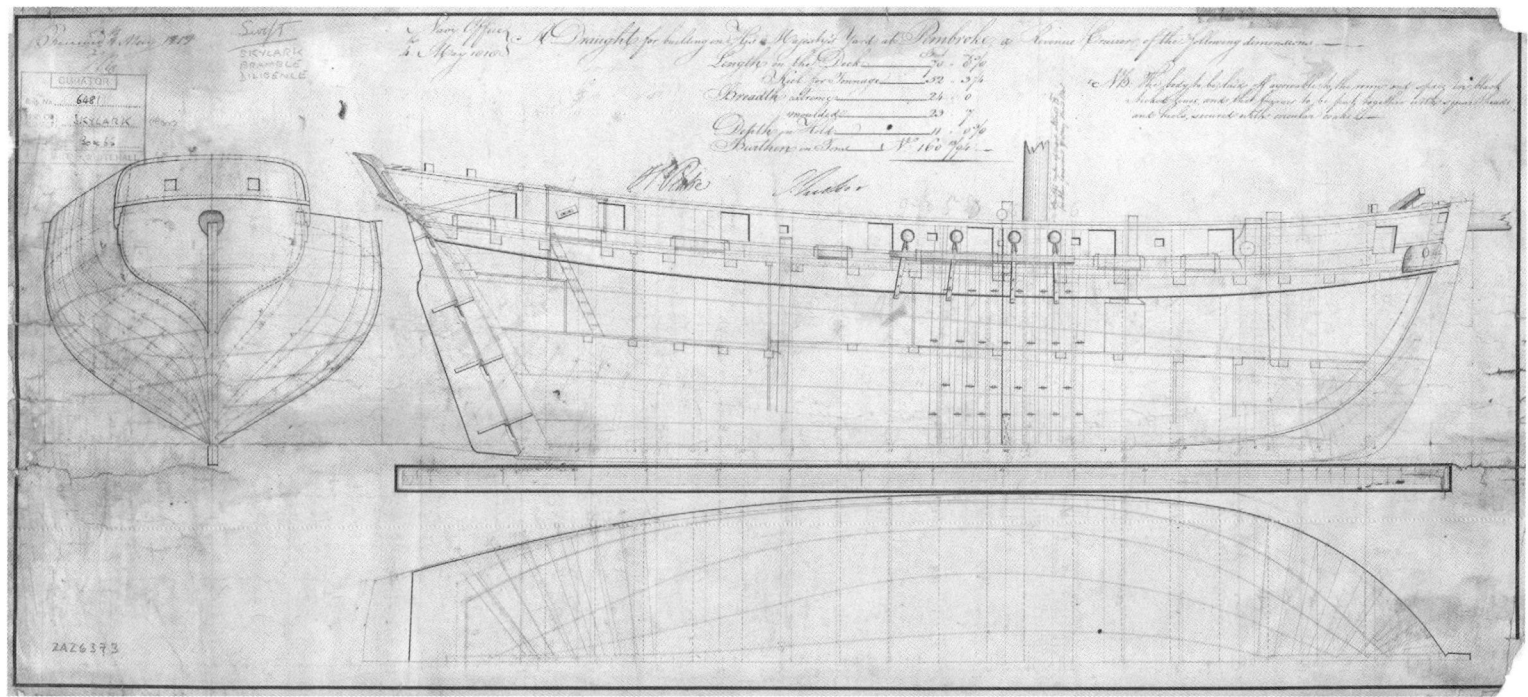

The draught for the cutter *Skylark*, dated 4 May 1818 and signed by Henry Peake and Joseph Tucker, the Surveyors, combines a sheer and profile with details of midship framing (a note instructs the frames to be 'put together with square heads and heels, secured with circular coaks', a recent structural innovation intended to replace the wasteful use of triangular chocks). This plan does not apply to the *Swift* (in theory the two Pembroke-built cutters were supposed to be identical) or to the other cutters pencilled on the draught – *Bramble* and *Diligence*. Both *Skylark* and *Swift* were produced by the dockyard for the Revenue service, and their subsequent histories are unrecorded. *(NMM J0483)*

Purchased 1817, for the Leeward Islands.

Defects made good at Sheerness 7.12.1818–1.1.1819. Sold 31.3.1821 at Antigua.

NIGHTINGALE Class Cutter tenders, 6 guns. 1822 design 'similar to the *Dove* revenue cutter'. Actually built as revenue cutters; to RN on completion. Five of these were built for and served in the Revenue Service as Second Class Revenue cutters, under Admiralty control (but are included to complete class list).

Dimensions & tons: 63ft 9in, 46ft 11½in (first four) 46ft 10in (last three) x 22ft 2in x 9ft 6in. 122bm. (*Speedy*, 22ft 6in beam, 123bm). Draught (*Dove* and *Netley*, light) 5ft 2in/10ft 5in; (*Fancy*, *Kite* and *Nightingale*, light) 5ft 6in/10ft 9in.

Men: 34. Guns: 2 x 6pdr (brass) + 4 x 6pdr carronades.

First cost: *Dove* £3,876. *Snipe* £3,314. *Speedy* £3,329.

Fancy Plymouth Dyd
Ord: 6.1817 (for Revenue Service). K: 9.1817. L: 7.5.1818.
Commissioned 31.12.1818 for Plymouth and Irish stations; paid off 1822. Fate unknown.

Kite Plymouth Dyd
Ord: 6.1817 (for Revenue Service). K: 9.1817. L: 7.5.1818.
Commissioned 23.5.1818. Fate unknown.

Racer Pembroke Dyd
Ord: 1817. K: 8.1817. L: 4.4.1818. Coastguard 1822.
Sold at Malta 4.5.1830.

Sprightly Pembroke Dyd
As built: 67ft 0in, 52ft 7in x 22ft 5in x 9ft 6in. 140⁵¹⁄₉₄bm.
Ord: 1817 (for Revenue Service). K: 10.1817. L: 3.6.1818. Revenue Service 1819. Wrecked off Portland 8.1.1821.

Dove Plymouth Dyd
Ord: 1818 (for Revenue Service). K: 5.1818. L: 29.12.1818. C: 5.8.1819 (for sea).
Commissioned 2.7.1819 under Lieut Francis Little, for the Irish Station; in 1822 to the Portsmouth Station, and in 1830 to Falmouth Station. In 1830 under Lieut Stephen Cocker, then 11.10.1834 under Lieut William George Pearne and 9.1837 under Lieut George Gahan; 'given up' by Customs Service 5.7.1843. Sold to Messrs Castle (for £85) by AO of 12.8.1865.

Netley Plymouth Dyd
Ord: 1821 (for Revenue Service). K: 6.1821. L: 13.3.1823.
Commissioned in RN as tender to flagship at Cork 23.1.1826. Tender to flagship at Plymouth 1829; subsequently receiving ship. Foundered at Spithead 10.1848 but raised; became buoy boat at Halifax 1850. To Bermuda 1850, then Jamaica 1851–52. Sold at St John's, Newfoundland 29.8.1859.

Nightingale Plymouth Dyd
Ord: 1822 (for Revenue Service). K: 4.1822. L: 19.4.1825. Schooner rig 2.1828.
Commissioned 13.4.1825 under Lieut William Hewgill Kitchen, for the Channel. Fitted with schooner rig 2.1828. Recommissioned 18.7.1828 under Lieut George Wood; wrecked on the Shingles 7.2.1829.

Snipe Pembroke Dyd
Ord: 1822. K: 10.1827. L: 28.6.1828.
Tender to flagship at Portsmouth from 1829. Rigged as schooner 1835. BU at Devonport 11.1860.

Speedy Pembroke Dyd
Ord: 1822. K: 11.1827. L: 28.6.1828.
Tender to the guard ship at Cork from 1828 to 1831. Commissioned at Portsmouth 2.11.1833 under Lieut Thomas Henderson; on 16.11.1835 under Lieut John Douglas, then 10.1.1837 under Lieut Joseph Martin Mottley, 25.7.1838 under Lieut John Wright, and 9.8.1841 under Lieut George Bentley; paid off 9.1845 at Sheerness. Mooring lighter 8.1853. To Sheerness 1865, renamed **YC.11** in 1866. BU 1876.

This unusual sail plan, complete with mast and spar dimensions, depicts the schooner-rigged *Nightingale* of 1825, signed on 18 March 1828 by Edward Churchill, master shipwright at Plymouth from 1815 to 1830. The *Nightingale*, nominally an 8-gun cutter, is here labelled a 2-gun schooner (illustrating how both the rig and armament of these small craft were subject to change); she was built on the graving slip in the dockyard, and originally intended for the Revenue service, but was commissioned on 20 April 1825 as a naval cutter, and departed on 1 June for service in the Channel; she was converted to a schooner during February and March 1828.
(NMM J0809)

SYLPH Class Cutter rated as tender
Dimensions & tons: 61ft 7in, about 47ft 10in x 21ft 2in x 10ft 0in. 114bm.
Sylph Woolwich Dyd
Ord: 1820. K: 10.1820. L: 15.6.1821.
Tender to the C-in-C Portsmouth from 1823. Lent to Customs 15.9.1862 as watch vessel. Sold 7.1888. (Note this may not be the same vessel.)

GRECIAN (ex-Revenue cutter *Dolphin*, built 1799 at Cowes)
Dimensions & tons: 68ft 9¼in, 50ft 9in x 23ft 2⅛in x 9ft 6½in. 145bm.
Men: 50. Guns: 2 x 6pdr + 8 x 6pdr carronades.
Purchased 20.11.1821.
Commissioned 20.11.1821 under Lieut John Cawley, for the West Indies; took pirate schooner *La Gata* 20.3.1823 (see *Lion* below); paid off 11.5.1825 at Portsmouth. Sold 11.7.1827 to Mr Freake but sale cancelled due to latter being declared insane; resold to Samuel Phillips for £300.

ARROW Class Cutter, 10 guns. Design by Captain John Hayes
Dimensions & tons: 64ft 0in, 48ft 8⅛in x 24ft 9in x 9ft 2½in. 157bm.
Men: 50. Guns: 2 x 6pdr (brass) + 8 x 6pdr carronades.
First cost: £3,916.
Arrow Portsmouth Dyd
Ord: 11.5.1822. K: 9.1822. L: 14.3.1823.
Commissioned 20.3.1823 under Lieut John Powney (at Capt. Hayes's request), for oyster fisheries protection and anti-smuggling off Jersey; paid off 10.1825. On 7.4.1826 under Lieut Arthur Brooking, then on 30.4.1830 under Lieut Edwin Thrackston (–1832). Customs tender 1831. Ketch-rigged 1838. BU 1.1852.

ASSIDUOUS (ex-pirate vessel *Jackal*, built about 1820) Schooner/tender, 2 guns. Captured 1823, and employed on Jamaica station, in suppression of piracy.
Dimensions & tons: dimensions unknown. 54 bm.
Men: 31. Guns: 1 x 12pdr + 1 x 32pdr carronade.
Commissioned 11.1824 under Lieut Richard Dowse; paid off and sold 5.5.1825.

LION (ex-pirate vessel *La Gata*, built 1821 at Baltimore). Employed on Jamaica station.
Dimensions & tons: 61ft 8in, 43ft 0¼in. x 19ft ?7in x 5ft 9in. 88 bm.
Men: 31. Guns: 1 x 12pdr.
Captured 20.3.1823 by *Grecian*.
Commissioned 3.1823 under Lieut William Hobson; on 18.3.1824 under Lieut Francis Liardet; took 9 pirate vessels and a slaver, and recaptured French ship *Calypso*. Sold 15.5.1826.

RENEGADE (ex-pirate slaver Zaragozana, built 1820 in America). Schooner, employed on Jamaica station.
Dimensions & tons: 76ft 3in, 59ft 10¾in x 19ft 2in oa (19ft 0in for tonnage) x 7ft 4in. 115 bm.
Men: 35. Guns: 2 x 12pdr + 2 x 32pdr carronades.
Captured 31.3.1823 by *Tyne* and *Thracian*, and purchased 7.1823.
Commissioned 10.10.1824 under Lieut Joseph O'Brien; in 10.1824 under Lieut Henry Ommanney Love, then on 30.8.1825 under Lieut Charles Elliot. Sold 8.1.1826.

UNION (ex-mercantile *City of Kingston*, built 1821 at Kingston, Jamaica)
Dimensions & tons: 59ft 9in, 44ft 6¼ x 18ft ?10in x 6ft 10in. 84bm.
Men: 31. Guns: — .
Purchased 1823.

The aquatint by William John Huggins shows the capture of the Spanish slaving brig *Midas* (on the left) by the schooner *Monkey* on the Grand Bahama Bank on 27 June 1829. In the label of the print the *Monkey* is credited with only a single 12pdr pivot gun to tackle the slaver's four long 18pdrs and four long 12pdrs. *(NMM PX9093)*

Commissioned 30.1.1823 under Lieut Thomas Marriott, later under Lieut William Henderson and then under Lieut Francis Liardet; on 27.9.1824 under Lieut Henry Love, then 26.10.1824 under Lieut James Bennett, 19.6.1825 under Lieut Charles Elliot, 25.8.1826 under Lieut Berry Haines, 29.3.1827 under Lieut Charles Madden, and 1828 under Lieut John Wills; wrecked on a reef in the West Indies 17.5.1828.

MAGPIE **Class** Schooners, 4 guns. 1825 design 'similar to the *Assiduous* tender'. Only two were accepted on delivery; the cost of the two amounted to £2,438.

Dimensions & tons: 53ft 3in, 40ft 8½in x 18ft 0in x 7ft 3in. 70bm.
Men: 35. Guns: (*Magpie*) 2 x 9pdrs + 2 x 18pdr carronades.

Magpie McLean, Jamaica
Ord: 16.7.1825 K: — . L: 6.1826.
Commissioned 1826 under Lieut Edward Smith; capsized in a squall off western Cuba 27.8.1826 (33 died including Smith, with 2 survivors).

Monkey McLean, Jamaica
Ord: 16.7.1825 K: — . L: 6.1826.
Commissioned 26.7.1826 under Lieut Edward Holland, for Jamaica; in 7.1827 under Lieut James Beckford Lewis Hay, then 20.9.1828 under Lieut Martin Cole (appointment); but in 10.1828 under Lieut Joseph Sherer; took slavers – Spanish schooner *Josepha* 4.1829 and 4-gun brig *Midas* 27.6.1829. Tender to *Blossom* in 1830. In 1831 under Thomas Downes, mate (possibly the later Lieut Thomas Hardwicke Downes); stranded and wrecked 13.5.1831 off Tampico; remains sold 24.5.1831.

Nimble McLean, Jamaica
Ord: 16.7.1825 K: — . L: 5.1826.
Found to be unsatisfactory on delivery and not accepted by RN.

NIMBLE Schooner, *Bolivar* (built 1822), 5 guns
Dimensions & tons: 83ft 7in, 64ft 7⅛in x 22ft 2in x 9ft 5in. 168bm.
Men: 41. Guns: 4 x 18pdr carronades + 1 x 18pdr.

Nimble Jamaica
Purchased 26.7.1826 and renamed as a replacement when the smaller (70bm) schooner *Nimble* (see previous entry) was found to be unsatisfactory and not accepted, being returned to that builder 5.1826.

Commissioned 1.8.1826 under Lieut Edward Holland; in 8.1829 under Lieut Joseph Sherer; took slaver *Gallito* 11.1829; on 29.4.1830 under Lieut John McDonnell, 26.1.1831 under Lieut John Putbury and 24.2.1833 under Lieut Charles Bolton; wrecked on Key Verde in the Bahamas Channel 4.12.1834.

ASP **Class** Cutters rated as tenders, 2 guns. Design by Joseph Tucker, they were intended to be employed during the oyster season on the coast of Jersey, but on delivery they proved very unsatisfactory; these craft 'failed to answer the purpose intended' and never saw active service, being paid off in September and ordered to be converted into pitch vessels for Portsmouth Dyd. Symonds's autobiography reported that *Cracker* was found so poor on delivery that she was immediately replaced by a smaller (47bm) vessel purchased at Gosport, while he then designed and built a second replacement (*Sylvia*) for the other pair. Note all three are recorded as being ordered 12 June 1826, but this is plainly erroneous and may reflect the date on which it was ordered that they be replaced.

Dimensions & tons: 50ft 2in, 37ft 3in x 16ft 5in x 7ft 11½in. ?53³⁷/₉₄bm (by calculation).
Men: 30. Guns: 2 x 12pdr carronades.

Asp Joseph Tucker, Gosport.
L: 3.1826.
Commissioned 21.3.1826 under Lieut Nicholas Colthurst; paid off 9.1826 and employed as a dockyard craft; sold 2.1829.

Cracker Joseph Tucker, Gosport.
L: 3.1826.
Commissioned 21.3.1826 under Lieut John James Hough; paid off 9.1826 and converted to dockyard craft. Later pitch boat at Pembroke. Sold 11.1842 (or this may have been her replacement of the same name).

Dwarf Joseph Tucker, Gosport.
L: 3.1826.
Commissioned 12.3.1826 under Lieut James St John (–10.4.1826); paid off 9.1826 and converted to dockyard lighter 1827. Coastguard brig 1855. Sold to Messrs Hood at Rye 1862.

CRACKER Purchased cutter in 1826. One of two replacements (see also *Sylvia* below) for the unsuccessful *Asp* Class cutters. She may have been built by White at Cowes, but was purchased at Gosport.

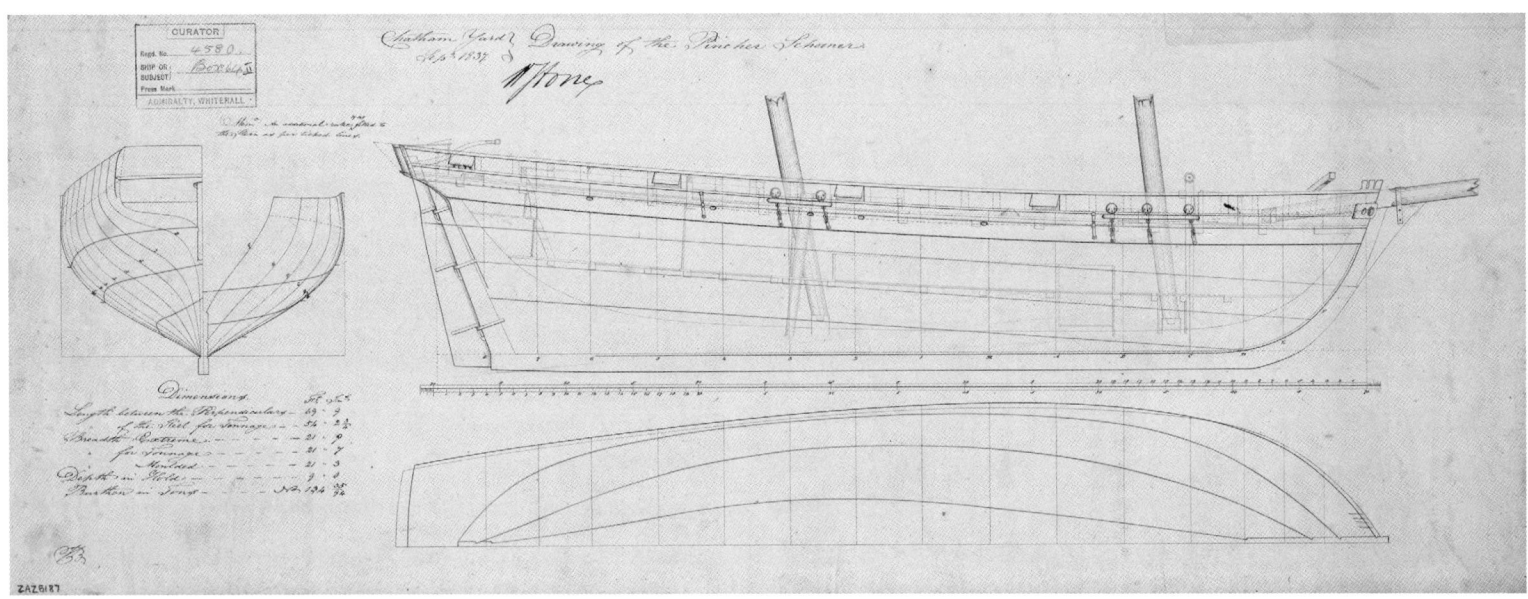

A draught showing the body plan with transom outline, sheer lines with inboard detail, and longitudinal half-breadth for the 3-gun schooner *Pincher*. It is signed by William Stone (master shipwright at Chatham Dockyard from 1830 to 1839) and dated September 1837. Both schooners and cutters tended to exhibit a lot of drag aft (i.e. they floated much deeper at the stern than at the bow), although this is disguised by the draught being oriented on the keel line instead of the load waterline, which consequently runs down the drawing from left to right; the feature was intended to preserve weather helm (the tendency for the vessel to turn up into the wind), when the craft was steeply heeled under a lot of sail. *(NMM J8069)*

Dimensions & tons: 52ft 4in (41ft 10in keel) x 15ft 0in x 6ft 1½in (50%₉₄tons).

Commissioned 28.10.1826 under Lieut John James Hough (taking over from previous *Cracker*). Became dockyard training hulk for artificers 11.1832. Her subsequent history is difficult to separate from that of the earlier *Cracker*.

SYLVIA Cutter, 1 gun. Design by William Symonds for oyster fishery service at Jersey, as a replacement for the unsuccessful *Asp* Class.
Dimensions & tons: 52ft 7in, 40ft 8½in x 18ft 1in x 8ft 6in. 70bm.
Men: 25. Guns: 1 x 3pdr (brass).
First cost: £2,351.
Sylvia Portsmouth Dyd
Ord: 9.9.1826. K: 1.1827. L: 24.3.1827.
Commissioned 14.3.1827 under Lieut David Dickson; on 2.10.1827 under Lieut John Morgan, then 12.5.1830 under Lieut Thomas Spark. Yard craft at Portsmouth 1833. Survey vessel 3.1842, as tender to *Seaflower*. Sold at Londonderry 9.1859.

FANNY Tender, unarmed. Employed at Deptford and Woolwich.
Dimensions & tons: 34ft 0in x 11ft 6in. 20bm.
Fanny Deptford Dyd
L: 5.1827.
BU 8.1835 at Woolwich.

PICKLE Class Schooners, 3 guns. Designed 'by a draught prepared in Jamaica'.
Dimensions & tons: 62ft 8in, 54ft 0in x 21ft 2in x 9ft 9in. 118bm.
Men: 36 (including 6 boys). Guns: 2 x 18pdr carronades + 1 x 18pdr.
Pickle Bermuda
Ord: 16.7.1825. L: 8.1827.
Commissioned 1.1.1828 under Lieut John Bunch Bonnemaison

McHardy; took slaver *Bolodera* 6.6.1829; arrived England 9.1830 and paid off 10.1830 at Plymouth. Recommissioned 16.10.1830 under Lieut Thomas Taplen, for Jamaica; on 15.1.1832 under Lieut Christopher Bagot, then 1.5.1834 under Lieut Archibald Gibson Bulman, then 18.7.1837 under Lieut Philip Hast, 9.3.1839 under Lieut Frederick Holland, 12.12.1840 under Lieut Frederick Montresor, 28.1.1843 under Lieut Joseph Bainbridge and 27.2.1846 under Lieut Henry Bernard; paid off 3.6.1847 at Bermuda. BU 1847.

Pincher Bermuda
Ord: 16.7.1825. L: 8.1827.
Commissioned 19.12.1826 under Lieut Richard Fegen; on 7.3.1828 under Lieut Justus Oxenham, then 5.9.1828 under Lieut William Tulloh; paid off 1833. Tender to the flagship on the North American and West Indies station 1833–34. Recommissioned 12.12.1835 under Lieut George Byng, then 2.1837 under Lieut Edward Bevan; home to pay off at Portsmouth 23.8.1837. Recommissioned 30.11.1837 under Lieut Bartholomew Sulivan; on 27.1.1838 under Lieut Thomas Hope; capsized and lost with all (40) hands off the Owers 6.3.1838 but raised 6.1838. Sold 31.8.1838.

Skipjack (ex-*Skylark*) Bermuda
Ord: 16.7.1825. L: 8.1827.
Commissioned 15.12.1826 under Lieut James Pulling; on 8.9.1829 under Lieut John Roche, then 21.5.1831 under Lieut Willoughby Shortland, 25.6.1833 under Lieut William Willis, 4.9.1834 under Lieut Sidney Ussher, 24.3.1836 under Lieut John Robinson, 31.8.1839 under Lieut Henry Wright, and 2.3.1841 under Lieut Augustus May; wrecked on Point Pedro, off Grand Cayman 1.6.1841; wreck sold.

ADELAIDE (ex-slaver *Bella Josephina*, built ?1823 in Sardinia)
Dimensions & tons: 66ft 8in x 18ft 5in x — . 95bm.
Slaver captured 1827 and used as tender to flagship at Rio de Janeiro. Sold 1833 at Rio de Janeiro.

BLACK JOKE (ex-slaver *Henriquetta*, built — at Baltimore)
Dimensions & tons: 90ft 10in x 26ft 5in x 12ft 3in. 260bm.
Men: 55. Guns: 1 x 18pdr + 1 x 18pdr carronade (on pivot).
Tender to the *Sybille* and later to *Dryad*; in 1.1828 under Lieut William Turner; took 14-gun privateer *Providencia* 2.4.1828

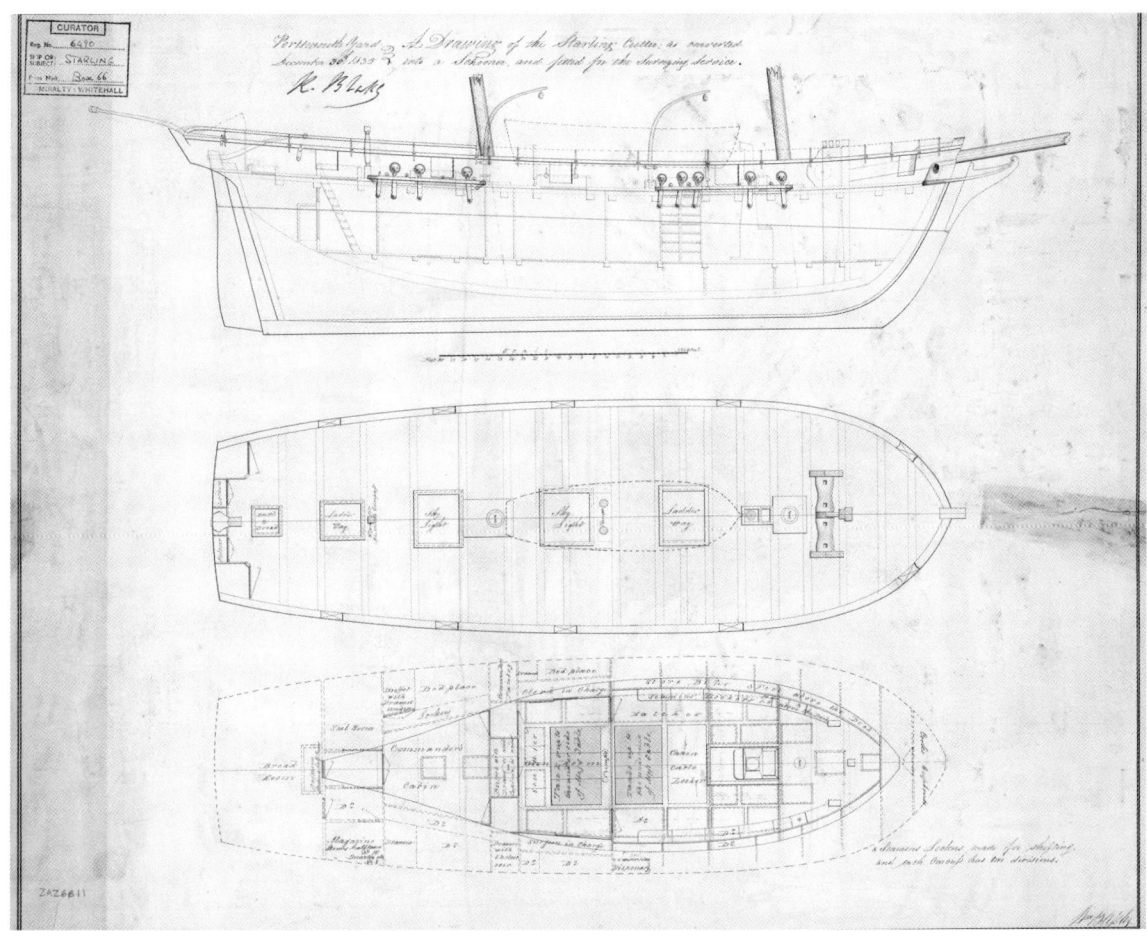

This draught, dated 30 December 1835 and signed by Robert Blake (master shipwright at Portsmouth Dockyard from 1830 to 1839) shows the inboard profile, upper deck and lower deck with fore and aft platforms of the *Starling* cutter of 1829 as converted to a two-masted schooner and fitted for surveying duties. Intended for ancillary duties from the onset (as tenders), this class did not possess either the extreme proportions or the sharp lines of the 'cruising' cutters. Under her commander for eight years, Henry Kellett, the *Starling* served in the survey role both in South America and subsequently at the start of 1841 in China, where she took part in several operations during the First China War. In late 1842 she was re-classed as a sloop-of-war, so that Kellett, who had been promoted Commander in May 1841, could be further advanced to post rank while remaining in command of the *Starling*.

and slaver *Vengador* 16.5.1828; in 11.1828 under Lieut Henry Downes; took slaver *Almirante* 1.2.1829; in 5.1829 under Lieut Edward Iggulden Parrey, then 9.1830 under Lieut William Ramsay; took slavers *Primeira* 22.2.1831, *Marinerito* 25.4.1831 off Calabar, *Regulo* and *Rapido* 10.9.1831, and *Frasquita* 15.2.1832. Coal depot at the Cape 1832, then stripped 4.1832 and remains burnt 3.5.1832.

LARK Class Cutter/tenders, 4 guns. Designed by Sir Robert Seppings, 1828. All were initially employed as tenders on home stations, but from 1832 were commissioned for sea duty overseas, usually for surveying.

Dimensions & tons: 60ft 9in, 49ft 5in x 20ft 3in x 9ft 0in. 107^{7}⁄$_{94}$bm.

Men: 35. Guns: 2 x 6pdr + 2 x 6pdr carronades.

First cost: *Raven* £2,804.

Raven Pembroke Dyd

As built: 60ft 9in, 49ft 5in x 20ft 5in (20ft 4in for tonnage) x 9ft 0in. 108^{6}⁄$_{94}$bm. Draught 5ft 3in/9ft 5in.

Ord: 8.11.1828. K: 6.1829. L: 21.10.1829.

C: 11.1829–29.12.1829 (as tender) at Plymouth.

First cost: £2,809, + £377 (hull only) fitting as a tender.

Tender to the *Hyperion* at Newhaven 1829–32. Commissioned 1832 at Portsmouth under Lieut William Arlett, for surveying in the Mediterranean; in 12.1833 under Lieut Henry Kellett (acting), for surveying on the coast of Africa, Madeira, etc.; paid off 1835. Fitted as survey cutter (yawl rig) by AO 9.9.1835; recommissioned 9.12.1835 under Lieut George Augustus Bedford, in same role; paid off end 1838. Recommissioned 9.5.1839 under Lieut David Robert Bunbury Mapleton, on

particular service; on 10.6.1841 under Lieut Juste Peter Roepel, in the Channel and North Sea; on 26.2.1842 under Lieut James Waldegrave Ludlow Sheils, then 26.1.1844 under Lieut John Stephen, stationed at Sheerness; paid off 16.11.1846. Fitted at Sheerness (for £146) as a quarantine ship 12.1847–1.1848. Coastguard service on Humber 1850. Sold (by AO 28.10.1859, for £160) at Sheerness to Messrs Hunt of Aldborough.

Starling Pembroke Dyd

As built: 60ft 9in, 49ft 5in x 20ft 5½in (20ft 4½in for tonnage) x 9ft 0in. 109^{1}⁄$_{94}$bm. Draught 5ft 3in/9ft 5in.

Ord: 8.11.1828. K: 6.1829. L: 31.10.1829. C: 6.11.1829–14.1.1830 (as tender) at Plymouth.

First cost: £3,009, + £1,643 fitting as tender.

Tender 1830–2.1832 (when laid up at Portsmouth). Fitted there as survey cutter (schooner rig) (for £1,901) 7–12.1835. Commissioned 29.10.1835 under Lieut (Cmdr 6.5.1841) Henry Kellett, for South America; later to East Indies (including 1st China War); re-classed as a sloop-of-war 23.12.1842 (to allow Kellett to receive post-rank of capt.). Sold in China (for $1,300) by AO 15.2.1844.

Lark Chatham Dyd

As built: 60ft 11½in, 49ft 6¼in x 20ft 5¾in (20ft 4¾in for tons) x 8ft 11¾in. 109^{15}⁄$_{94}$bm. Draught 5ft 0in/9ft 0in.

Ord: 8.11.1828. K: 6.1829. L: 23.6.1830. C: 15.10.1830 (as a tender).

First cost: £3,903 (including fitting as tender).

Tender 1830–1835 at Sheerness. Fitted at Sheerness for sea as a schooner (for £1,526) 7–11.1835. Commissioned 22.9.1835 under Lieut Edward Barnett, for North America and the West Indies; on 30.11.1837 under Lieut Thomas Smith, on same station; on

15.2.1843 under Lieut George Bennett Lawrence, on same station; paid off 13.10.1848. Fitted at Sheerness as a colonial cutter with 2 guns (for £2,045) 8.1848–3.1849; loaned to Liberian government. Returned by Liberia 7.1858 (to be replaced by *Quail*) and surveyed. BU at Plymouth 26.6.1860.

Jackdaw Chatham Dyd
As built: 60ft 9½in, 49ft 6⅝in x 20ft 5in (20ft 4in for tons) x 9ft 0¼in. 108$\frac{91}{94}$ bm. Draught 5ft 0in/8ft 11in.
Ord: 8.11.1828. K: 8.1829. L: 4.8.1830.
First cost: £2,727, + £1,721 fitting as a tender).
Tender to the *Thunderer* at Chatham 1830–32. Fitted as a schooner 12.1832–4.1833. Commissioned 20.2.1833 under Lieut Edward Barnett, for surveying Costa Rica; wrecked on a reef off Isla Providencia (Honduras/Nicaragua) 11.3.1835; hull sold for £30.6.8d.

Magpie Sheerness Dyd
As built: 60ft 9in, 49ft 4⅝in x 20ft 5½in (20ft 4½in for tons) x 9ft 0in. 10$\frac{95}{94}$bm. Draught 5ft 0in/8ft 6½in.
Ord: 8.11.1828. K: 12.1829. L: 30.9.1830. C: 26.1.1831 (as tender) at Sheerness
First cost: £3,006 to build.
Fitted for sea at Sheerness 1.1833. Commissioned 28.8.1833 under Lieut Frederick Henry Hastings Glasse, for home and Lisbon stations. Fitted at Plymouth as a survey vessel (for £631) 5–7.1836. Recommissioned 11.5.1836 under Lieut Thomas Saumarez Brock, for surveying in the Mediterranean; paid off ?7.3.1842. Fitted at Chatham as a harbour water tank (for £1,259) 10.1845–2.1846. Sold about 1908.

Quail Sheerness Dyd
As built: 60ft 9in, 49ft 5in x 20ft 5in oa (20ft 4in) x 9ft 0in. 108$\frac{63}{94}$bm. Draught 5ft 0in/8ft 7in.
Ord: 8.11.1828. K: 12.1829. L: 30.9.1830. C: 4.1831
First cost: £3,060 to build + £67 fitting as tender (at Pembroke).
Fitted for sea at Plymouth (for £1,053) 8–10.1835. Commissioned 11.9.1835 under Lieut Philip Bisson for Falmouth (to Lisbon) packet service; swamped in gale off Ushant 27–30.3.1836, with 16 men drowned, but vessel was saved although heavily damaged. Fitted as Harbour lighter at Sheerness 6–9.1838. Fitted for transfer to Liberian Government 10.1858–5.1859 (replacing *Lark*).

***MINX* Class** Schooners, 3 guns. Designed to lines of schooner *Union* of 1820. A third schooner to this design was also ordered in 1828, but was cancelled in the same year.
Dimensions & tons: 59ft 9in, 44ft 6½in x 18ft 3½in x 6ft 10in. 84bm.
Men: 32. Guns: 2 x 12pdr carronades and 1 x 5½in howitzer.

Firefly Bermuda
Ord: 10.1.1828. K: — . L: 1828.
Commissioned 25.10.1828 under Lieut Edward Holland, for the West Indies. On 8.12.1830 under Lieut John Julius McDonell, on same station; wrecked off the coast of British Honduras 27.2.1835.

Minx Bermuda
Ord: 10.1.1828. K: — . L: 23.12.1829.
Commissioned 22.6.1929 under Lieut John Simpson, in the West Indies and South America; paid off 8.1832. Recommissioned 1.11.1832 under Lieut. George Gover Miall; paid off 8.1833. Sold at Jamaica (for £523.5.0d) 6.1833.

KANGAROO Schooner tender (ex-*Las Damas Argentinas*, built at Baltimore), purchased 1829 at Jamaica.
Guns: 2 x 12pdrs, 1 x 5in howitzer.
Commissioned early 1831 under Lieut Willoughby Shortland, for the West Indies; on 21.3.1831 under Lieut James Hookey; home and paid off 8.1833. Sold 1.1834 (for £162½).

ALBATROSS Schooner tender, built in Scotland
Dimensions & tons: 58ft 5in, 42ft 0in x 16ft 11in. 64 bm.
Purchased 1826; arrived Deptford 6.9.1826.
Fitted as tender to Hecla at Deptford 6–27.3.1828.
Sold to John Small Sedger (for £155) 30.8.1833.

SEAFLOWER Cutter, 4 guns. Design by Captain John Hayes.
Dimensions & tons (as built): 60ft 0in, 46ft 10⅜in x 21ft 8½in (21ft 6½in for tonnage) x 10ft 6in. 1156$\frac{3}{94}$bm. Draught 5ft 11in/7ft 1in.
Men: 35. Guns: 2 x 6pdr (brass) + 2 x 6pdr carronades.
First cost: £2,907.

Seaflower Portsmouth Dyd
Ord: 5.12.1829. K: 2.1830. L: 20.5.1830. C: 21.7.1830 (for sea).
Commissioned 28.8.1832 under Lieut John Morgan, for Home Waters (probably Channel Islands); on 6.9.1835 under Lieut Joseph Roche, for protection of the Jersey fisheries, then 7.1838 (or 30.1.1839) under Lieut (Cmdr, 23.11.1841) Nicholas Robilliard, still on the Home station; on 25.7.1845 under Cmdr Juste Peter Roepel, in Jersey (–6.11.1846); on 12.11.1846 under Cmdr Henry Dumaresq, at Portsmouth; paid off 15.9.1847. Tender to the Cuckoo at Portsmouth in 1850. Fitted as survey ship 1855. Recommissioned 1856 under Edward Kilwick Calver, Master, for surveying in North Sea; paid off 1862. BU 21.9–8.10.1866 (under AO 12.12.1866) by Castle & Beech, Charlton.

CERUS Cutter tender, ?unarmed
Dimensions & tons: 38ft 5½in, — x 12ft 8½in x 7ft 8in. 25bm.
Purchased 8.9.1830 for £430.
Used as buoy tender at Portsmouth. Renamed ***Ceres*** 5.6.1833, then ***YC.3*** in 3.1860. Sold at Greenock 26.7.1877.

FANNY Cutter yacht, rated as tender. Designed by Captain John Hayes as 'the Admiralty yacht'.
Dimensions & tons: 68ft 6in, ?55ft 4in x 21ft 6in x 10ft 4in. 136bm.
First cost: £3,842 (fitted for sea).

Fanny Portsmouth Dyd
Ord: 8.1830. K: 8.1830. L: 28.2.1831.
Tender to *Britannia* at Portsmouth; on 4.8.1855 under George Clark, Acting Master, as tender to *Victory* at Portsmouth. Fitted as coastguard vessel 5.1862. Sold about 1863. Sunk in collision off the Tuskar Rock 31.10.1878.

***COCKATRICE* Class** Schooners, 6 guns. Design by Sir Robert Seppings, approved 11 September 1828. *Hornet* and *Viper* were completed as packets, with brigantine rig, while *Spider* and *Cockatrice* were schooners.
Dimensions & tons: 80ft 0in, 64ft 5½in x 23ft 0in x 9ft 10in. 181$\frac{35}{94}$bm.
Men: 33–42. Guns: 2 x 6pdr + 4 x 12pdr carronades.

Hornet Chatham Dyd
As built: 80ft 3in, 64ft 1in x 23ft 3⅝in oa (23ft 0⅝in) x 9ft 10in. 181$\frac{13}{94}$bm. Draught 6ft 1in/9ft 0in.
Ord: 11.9.1828. K: 12.1829. L: 24.8.1831. C: 21.11.1831–10.10.1832.
First cost: £2,576 to build.
Commissioned 10.7.1832 under Lieut Francis Rogers Coghlan (–1837) for South America. On 18.7.1837 under Lieut Henry

Baillie, for the West Indies. On 9.3.1839 under Lieut Robert
Boyle Miller, for North America and the West Indies; paid off
early 1845). BU 7.1845.

Spider Chatham Dyd
As built: 80ft 2in, 64ft 8⅜in x 23ft 3in oa (23ft 0in) x 9ft 10in.
182⁴/₉₄bm. Draught 6ft 0in/9ft 4in.
Ord: 11.9.1828. K: 3.1830. L: 23.9.1835. C: 14.12.1835.
First cost: £2,887 (fitting).
Commissioned 4.11.1835 under Lieut John O'Reilly (–1841) for
South America. On 5.4.1842 under Lieut Richard Elsworthy Pym
(–1847), for Brazilian station. Repaired and fitted as a Packet (for
£2,744) at Plymouth 2–10.1847. Recommissioned 28.8.1847 under
Lieut Charles Haydon, then 3.4.1849 under Lieut James Ward
Tomlinson, for South America; paid off 30.4.1851. Sold to S.
Clark (for £375) 22.11.1861.

Viper Pembroke Dyd
As built: 80ft 0in, 64ft 5½in x 23ft 3in oa (23ft 0in) x 9ft 10½in.
181¹³/₉₄bm. Draught 6ft 4in/9ft 5in.
Ord: 11.9.1828. K: 6.1830. L: 12.5.1831. C: 31.8.1831 at Plymouth.
First cost: £4,077 to build (+ fitting £2,270).
Commissioned 15.6.1831 under Lieut Horatio James, for the Tagus.
On 3.4.1834 under Lieut Louis Augustus Robinson (–3.1837), off
coasts of Portugal and Spain. On 17.3.1847 under Lieut William
Robert Wolseley Winniett, for Cape of Good Hope and coast
of Africa; in 7.1838 under Lieut Godolphin James Burslem, in
South America; paid off 1.1841. Between a Small and a Middling
Repair and fitted for sea at Portsmouth (for £3,245) 3–12.1841.
Recommissioned 1.9.1841 under Lieut James Carter, for South
America. On 8.12.1845 under Lieut Edward Evans Gray, for
North America & the West Indies; in 10.1846 under Lieut E. G.
Hare; paid off 9.11.1846. Recommissioned 21.11.1847 under Lieut
Henry Bernard, for same station; home in 9.1847. In Ordinary at
Sheerness, then BU 5.1851 at Chatham.

Cockatrice Pembroke Dyd
As built: 80ft 0in, 64ft 1⅞in x 23ft 4in oa (23ft 1in) x 9ft 10in.
181⁷⁸/₉₄bm. Draught 6ft 5in/9ft 5in.
Ord: 11.9.1828. K: 7.1831. L: 14.5.1832. C: 15.9.1832 at Plymouth.
First cost: £3,893 to build (+ fitting £2,365).
Commissioned 7.1832 under Lieut William Lee Rees, for South
America. On 24.9.1836 under Lieut John Douglas, then 27.2.1841
Lieut Justus Oxenham, on same station; paid off 21.1.1845 at
Chatham. Fitted at Chatham as a victualling transport (for £3,418)
1.1845–5.1846. Fitted at Sheerness as a stores tender for the Pacific
(for £376) 10–11.1847. Tender in the Pacific to the *Asia* 1850–51,
and to the *Portland* 1855. Sold at Callao (for £983) 9.1858.

MONKEY Schooner rated as tender, 6 guns, ex-mercantile *Courier*
built 1827.
Dimensions & tons: 56ft 0in, 40ft 3½in x 17ft 10in x 9ft 4in. 68¹⁶/₉₄bm.
Men: 35. Guns: 2 x 12pdrs + 1 x 5½in howitzer.

Monkey Bermuda
Purchased at Bermuda (for £818) 25.10.1831.
Sold at Jamaica (for £403) 8.8.1833.

SEAGULL **Class** Three-masted schooners, 6 guns. Designed by
Seppings. *Seagull* re-rigged as a brigantine in 1834, but a schooner again
in 1837.
Dimensions & tons: 95ft 0in, 77ft 9¼in x 26ft 3in x 11ft 2in.
279⁷⁴/₉₄bm.
Men: 29. Guns: 2 x 6pdr + 4 x 12pdr carronades (*Seagull* later 2 x
6pdr + 6 x 18pdr carronades).

Seagull Chatham Dyd
Ord: 18.7.1829. K: 8.1830. L: 21.11.1831.
Commissioned 6.6.1834 under Lieut John Parsons, as packet at
Falmouth; paid off 15.3.1842. On 8.4.1845 under Lieut Henry
Percy Dicken, then 21.3.1848 under Lieut James Smail, again at
Falmouth. Employed as a packet again 1850–51. BU 10.1852.

Peterel Woolwich Dyd
Ord: 18.7.1829. K: — . Cancelled 28.2.1831.

(C) Vessels acquired from November 1830

FAIR ROSAMOND (ex-Spanish slaver *Dos Amigos*, built ? at
Baltimore).
Dimensions & tons: 74ft 11½in, 64ft 1½in x 23ft 2½in x 10ft 4½in.
172 bm.
Men: 40. Guns: 1 x 18pdr carronade + 1 x 18pdr ('bored up').
Slaver schooner captured 9.11.1830 by *Atholl*, then purchased (for
£609) 22.2.1831 and used as tender.
Middling repair at fitted for sea at Portsmouth (for £3,904)
30.7.1832–9.7.1833. Commissioned 6.1833 under Lieut George
Rose, for the Cape of Good Hope; paid off 2.1837 at Portsmouth.
Recommissioned 3.4.1837 under Lieut William Browne Oliver,
for the west coast of Africa; took slavers *Constitucao* 21.8.1838,
Matilde 22.1.1839, *Tejo* 31.1.1839, *Pomba da Africa* and *Sedo
ou Tarde* 25.6.1839, and (barque) *Augusto* 5.9.1839; paid off

Many slaving vessels were schooners, which tended to have an advantage to windward
over the Navy's brigs, their usual hunters. On the old country notion of a poacher-turned-
gamekeeper, some of the best of the former slaving schooners were brought into the Navy
when captured and turned against their own kind. The *Fair Rosamond*, shown in this drawing
by Alfred Basil Lubbock done in 1936, was previously the slaving schooner *Dos Amigos*. The
rig depicted might be more accurately described as a brigantine. (*NMM PY5778*)

1.10.1840. Recommissioned 8.4.1841 under Lieut Archibald Gibson Bulman, for North America and the West Indies; paid off 1842 at Portsmouth. BU there 20.11.1845.

FAWN (ex-slaver *Caroline*, built 1835 in ?Brazil).
Dimensions & tons: 75ft 0in, 60ft 0in x 22ft 10in x 7ft 9in. 169 bm.
Men: 40. Guns: 6.
Ex-slaver brigantine purchased 27.5.1840 at Rio de Janeiro (by AO of 4.11.1839) for £834.
Commissioned 4.11.1839 under Lieut John Foote; took slavers *Sandade* 30.1.1840, *Asseiceiro* 30.12.1840, *Dois de Fevereiro* 18.2.1841 and *Venus* 29.4.1841. On 8.2.1842 under Lieut Joseph Nourse; took slavers *Boa Harmonia* 18.3.1842, another (name unknown) 12.4.1842 and *San Antonio* 7.5.1842. Became tank vessel at Cape of Good Hope 12.1842. Sold to Natal Colonial Government 5.1847.

LOUISA Cutter rated as tender, no guns. Built 1827 at Northfleet, and purchased 1835 at Canton for use as tender.
Dimensions & tons: dimensions unknown. 83 tons.
In service as tender 1839 under Thomas Carmichael, Mate (Lieut 8.6.1841); in attack on Canton 21.3.1841; wrecked on an island between Hong Kong and Macau 21.7.1841.

GIPSY Class Cutters rated as tenders, 2 guns. Designed by Sir William Symonds.
Dimensions & tons: 54ft 0in, 41ft 7⅛in x 18ft 0in oa (17ft 10in) x 8ft 7in. 70³⁴⁄₉₄bm.
Men: 6. Guns: 2 x 3pdrs.
First cost: *Gipsy* £1,647 (fitted for sea). *Mercury* £1,724 or £2,076 (fitted for sea).
Gipsy Sheerness Dyd
Ord: 18.4.1836. K: 4.1836. L: 27.10.1836.
Tender at Sheerness 1837–45. Tender at *Ajax* at Queenstown 1850. In 7.1856 under Richard Reed, 2nd Master, as tender to *Conway*. In 1860 tender to *Amphion*, later to *Sans Pareil*, (still under Reed, reappointed 10.3.1860); in 1865–70 tender to *Hastings*, 1875–80 to *Revenge*, and 1890 to *Triumph*, all at Queenstown. Sold 12.8.1892.
Mercury Chatham Dyd
Ord: 18.4.1836. K: 5.1836. L: 7.2.1837. C: 3.3.1837.
Tender at Chatham on completion. In 10.1841 under James Scarlett, as tender at Portsmouth (–1846). In 1848 under John Roskilly, for tidal observations. Tender to *Dasher* at Portsmouth 1860–65. Renamed *YC.6* as yard craft at Plymouth 1866. Renamed *Plymouth* 8.2.1876. Sold there 17.5.1904.

PROMPT (i) (ex-Portuguese slaver *Josephine*, built ? at ?Portugal).
61 bm slaver captured 10.1840, and purchased 22.1.1842.
Tender on coast of Africa. Sold 1845.

FELICIDADE Schooner. Captured as Brazilian slaver 27 February 1845 by *Wasp* off West Africa.
Dimensions & tons: unknown.
In service under Lieut Robert Douglas Stupart 27.2.1845, then 1.3.1845 under Midshipman Harmer with reduced prize crew of 9; all overpowered and murdered 1.3.1845 by prisoners. Retaken 7.3.1845 by *Star*, with new prize crew put aboard under Lieut John Wilson; overset in squall and foundered 16.3.1845.

PROMPT (ii) ?
Slaver captured early 1845, and purchased 17.7.1845.

Sold about 1847.

BERMUDA Schooner, 3 guns. Designed by Sir William Symonds.
Dimensions & tons: 80ft 0in, 60ft 1⅜in x 23ft 3in (23ft 0in for tonnage) x 11ft 0in. 180³⁸⁄₉₄bm. Draught 4ft 10½in/9ft 10in.
Men: 35 (later 52). Guns: 1 x 32pdr and 2 x 12pdr carronades.
Bermuda Outerbridge & Hollis, Bermuda.
Ord: 24.12.1845. K: — L: 3.1848.
First cost: £6,187.
Commissioned 16.7.1847 under Lieut Archibald Douglas Jolly, for North America and the West Indies. In 1855 under Lieut William Cashman, still in West Indies; wrecked on Caicos Reef, near Turks Island 20.1.1855.

RENIRA Schooner tender, purchased 27 June 1846 (for £1,000).
Dimensions & tons: dimensions unknown. 86 bm.
Sold 1850.

ADELAIDE Cutter/tender (not to be confused with paddle tug of same name; see Chapter 11). Captured as slaver in early 1848, and purchased on 3 May (for £263) as tender to *Penelope* for anti-slavery patrol off West Africa.
Dimensions & tons: dimensions unknown. 140 bm.
Commissioned in 1849 under Lieut John L. MacLeod; wrecked on Banana Island, West Africa 9.10.1850; remains sold or £31.

AUSTRALIAN Schooner. Teak-built, designed by John Edye for the South Australian government. Apparently ordered by the colony's Lieutenant Governor 'without clearance for use of public funds'. Note the name is not certain, and may be a description instead.
Dimensions & tons: 60ft 0in, 47ft 6⅛in x 18ft 4in x 8ft. 82bm.
Australian Bombay Dyd.
Ord: 1.6.1847. L: 24.3.1849.
Commanded by Lieut Henry Wandesford Comber, sailed 4.4.1849 from Bombay and arrived Sydney 11.9.1849. Sold 24.4.1850 at Adelaide (for £810) to Mr J. F. Bennett.

AZOV Class Schooner-rigged gunboats, 2 guns. Ordered (and possibly designed) by Rear-Admiral Montagu Stopford, the Superintendent at Malta, originally planned as bomb vessels for the Crimea, but not completed as such. Their poor sailing qualities (slow and leewardly, steering badly in light winds) were attributed to their rounded bottom and very scant run.
Dimensions & tons: 64ft 0in, 52ft 7⅛in x 18ft 8in x 6ft 11in. 93⁵⁄₉₄bm.
Men: — . Guns: 2 x 32pdr (pivot).
First cost: *Azov* £591. *Kerch* £604.
Azov Guman, Malta
Ord: 1855. K: 4.1855. L: 14.7.1855.
Used as Superintendent's yacht at Malta, then as tender to *Hibernia* 1860–80. Powder hoy 1871. Sold at Malta to Mr Puolo Elliel (for £255, by AO 26.10.1897) 9.6.1899.
Kerch Guman, Malta
Ord: 1855. K: 4.1855. L: 10.7.1855.
Water tank at Gibraltar 10.1860, renamed *YC.1*. Deleted about 1875.

JERSEY Cutter/tender to *Dasher* in the Channel Islands
Dimensions & tons: 55ft 10in x 17ft 0in x 9ft 0in. 70bm.
Jersey White, Cowes.
L: 22.3.1860.
Sold to E. A. S. Mignon 8.1873.

In addition to the battlefleet and the various categories of cruising (patrol and escort) vessels, the Royal Navy also operated a variety of specialised and auxiliary units. The combatant vessels in this category by 1817 consisted chiefly of the bomb vessels – now an obsolescent type – although during the 1850s their direct equivalent – the mortar vessels and mortar floats – would be built in large numbers for the Russian War.

Other essentially non-combatant types reflected new or evolved duties of the Navy. While some of the former bomb vessels found subsequent role in exploring the polar regions, other were specially acquired for this service or for the Navy's growing hydrological and surveying responsibilities, mapping seas and harbours around the world. While the mail packet service, with its need for reliability for routine voyages, was an early candidate for the adoption of steam power, pure sailing vessels were originally used in this role (in addition to the sailing brigs often adapted to this service). Another function to be fulfilled was the provision of vessels to transport royal passengers or other dignitaries, and a number of specialised 'yachts' were built for this role.

Considerable numbers of elderly or obsolescent warships, on retiring from active status, were utilised for various ancillary duties. In a seagoing capacity, many of these served as troop transports or as storeships to handle British naval and military commitments worldwide, although the latter were augmented by the non-combatant ships of the separate Transport Board until that body and its responsibilities were merged with the Navy Board in 1817. Other former warships fulfilled a more sedentary role as harbour hulks, employed as hospital, prison and receiving ships. These vessels and their full careers are listed in this book under their original classification, and to avoid duplication they are not included here.

BOMB VESSELS

(A) Vessels in service or on order at 1 January 1817

The ship-rigged bomb vessels coming into service at the end of the war were a new design and the first purpose-built bomb vessels produced since the War of American Independence. They were also the last of their kind to be built for the RN, as the subsequent mortar vessels of the Russian ('Crimean') War of 1854–56 were much smaller, simpler and less powerful. The stout framing of the bomb vessels, intended to resist the effect of the recoil of their mortars, suited them for polar exploration where strength to resist the crushing effect of the ice was a major requirement.

VESUVIUS **Class** Designed by Henry Peake, approved 21 May 1812. Ship-rigged bomb vessels, with hulls designed in frigate/sloop style instead of the pink-sterned type used for earlier purpose-built bombs.

> Dimensions & tons: 102ft 0in, 83ft 10in x 27ft 0in x 12ft 6in. 325⁷⁄₉₄bm.
> Men: 67. Guns: UD 8 x 24pdr carronades and 2 x 6pdrs. Mortars: 2 (1 x 13in + 1 x 10in).

Vesuvius Robert Davy, Topsham
> As built: 102ft 4in, 84ft 2in x 27ft 0in x 13ft 0in. 326³⁵⁄₉₄bm. Draught 8ft 10in/10ft 1in.
> Ord: 30.3.1812. K: 7.1812. L: 1.5.1813. C: 22.5–27.9.1813 at Portsmouth.
> Commissioned 6.1813 under Cmdr William Hart (–1814). Laid up at Deptford 6.1816–4.1819 (handed over 7.1818). Sold to Mr Garratt (for £980) 22.7.1819.

Terror Robert Davy, Topsham
> As built: 102ft 4½in, 84ft 2½in x 27ft 3⅜in x 12ft 11½in. 333³⁵⁄₉₄bm. Draught 8ft 10in/10ft 0in.
> Ord: 30.3.1812. K: 9.1812. L: 29.6.1813. C: 31.7.1813 at Portsmouth.
> First cost: £7,495 to builder, plus £7,416 fitting.
> Commissioned 7.10.1813 under Cmdr John Sheridan; to North America 1814; on 16.6.1815 under Cmdr Constantine Richard Moorsom; paid off 8 or 9.1815 into Ordinary at Portsmouth. Large Repair at Portsmouth (for £12,487) 3.1821–11.1822. Fitted for sea at Portsmouth (for £4,626) 4.1824. Fitted for sea at Portsmouth (for £6,861) 12.1827–6.1828. Small Repair at Plymouth (for £7,839) 11.1828–6.1829; housed over fore and aft. Housing removed and used as tender to *Howe* 12.1835. Converted to Arctic discovery vessel at Chatham (for £6,940) 1–6.1836. Recommissioned 11.5.1836 under Capt. George Back, for discovery voyage to Hudson's Bay; returned home 3.9.1837. Fitted for Capt. James Clark Ross's expedition to South Pole (for £6,765) at Chatham, and coppered 4–9.1839; recommissioned 11.5.1839 under Cmdr. (Capt., 16.8.1841) Francis Rawdon Moira Crozier, for Antarctic exploration. Refitted for expedition to North Pole 2–5.1845, fitted with auxiliary steam engine (ex railway locomotive) and hoisting screw propeller. Recommissioned 8.3.1845, still under Crozier; sailed 19.5.1845 from Greenhithe. Abandoned in the Arctic ice 22.4.1848 during the final and fatal Franklin expedition to find the Northwest Passage (deleted from Navy List 16.6.1854).

Beelzebub William Taylor, Bideford
> As built: 102ft 8in, 84ft 2⅜in x 27ft 3¾in x 12ft 11¾in. 334¹⁄₉₄bm. Draught 8ft 8in/9ft 5in.
> Ord: 30.3.1812. (contract 3.8.1812) K: 11.1812. L: 30.7.1813. C: 1813 at Plymouth for Ordinary.
> Commissioned 7.1816 under Cmdr (Capt. 9.1816) William Kempthorne; at Bombardment of Algiers 27.8.1816. In Ordinary at Plymouth 1817. Was to have been repaired under AO of 28.6.1820, but instead BU completed 23.9.1820.

Fury William Good, Bridport
> Ord: 30.3.1812. (contract 20.8.1812) K: 1812. Cancelled 3.2.1813.

Also reported was *Thunder*, to have been built by Josiah and Thomas

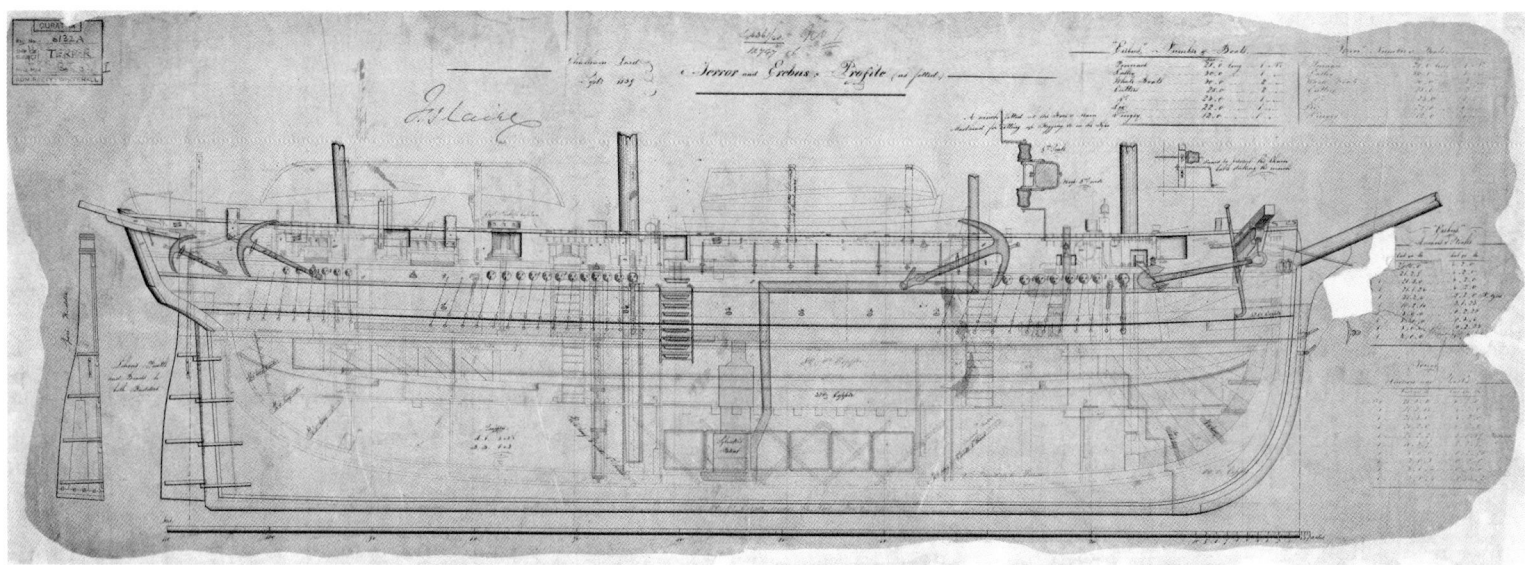

An inboard profile of the *Terror* (also used for the *Erebus*) as drawn in 1839, showing the modifications for her new role of Arctic exploration ship. Most significant features included a heating system, and a large increase in the establishment both of anchors (ten of differing weights) and boats (nine, ranging in size from a pair of 30ft whaleboats to a 12ft dinghy). She had been converted to an Arctic/Antarctic exploration vessel at Chatham in 1836, and employed in Captain George Back's expedition to Hudson's Bay. The structural strength of the bomb vessels, designed to withstand the concussion of repeated firings of their 13in. mortars, made them ideal vessels to withstand the crushing caused by polar ice. After her return she was adapted for a fresh expedition to the other Pole, led by Captain James Clark Ross in the *Erebus*, 'for the purpose of magnetic research and geographical discovery in the Antarctic Seas'. For a later voyage the two ships were fitted with screw propulsion. *(NMM J1407)*

Brindley, Frindsbury. However, there is no trace of an order being placed, and no such vessel was built.

FURY Class Designed by Henry Peake, modified while building from *Vesuvius* Class. The three vessels ordered in 1813 were completed too late for service in the Napoleonic conflict, but all (plus *Beelzebub*, above) were commissioned for active service in the Algiers campaign in 1816.

    Dimensions & tons: 105ft 0in, 86ft 1¼in x 28ft 6in x 13ft 10in. 372¹/₉₄bm.
    Men: 67. Guns: UD 10 x 24pdr carronades and 2 x 6pdrs. Mortars: 2 (1 x 13in + 1 x 10in).

Fury Mrs Mary Ross, Rochester
    As built: 105ft 8in, 86ft 6⅛in x 28ft 7½in x 14ft 11in. 377bm.
        Draught 8ft 6in/11ft 2in.
    Ord: 5.6.1813. K: 9.1813. L: 4.4.1814. C: 4.1814 (for Ordinary), then 7.1816 (foreign service) at Chatham.
    Commissioned 2.7.1816 under Cmdr Constantine Richard Moorsom; at Bombardment of Algiers 27.8.1816. In Ordinary at Deptford 1817; housed over fore and waist at Woolwich 4.1817–12.1818. Fitted at Deptford as Arctic Seas discovery vessel (re-registered as a sloop) 11.1820–4.1821; recommissioned 30.12.1820 under Cmdr (Capt. 11.1821) William Edward Parry; sailed to explore Northwest Passage 8.5.1821 (with *Hecla*); paid off 14.11.1823. Fitted at Deptford for same again 10.1823–5.1824; recommissioned under Cmdr Henry Parkyns Hoppner, sailed to explore Northwest Passage 8.5.1824 (with *Hecla* again). Bilged in Prince Regent Inlet 1.8.1825, and abandoned in Arctic 25.8.1825.

Hecla Barkworth & Hawkes, North Barton (Hull)
    As built: 105ft 0in, 86ft 1¼in x 28ft 7½in x 13ft 10½in. 375²⁶/₉₄bm.

        Draught 8ft 0in/10ft 0in.
    Ord: 5.6.1813. K: 7.1813. L: 22.7.1815. C: 6.9.1815–14.7.1816 at Sheerness.
First cost: £9,028 to builder.
    Commissioned 27.6.1816 under Cmdr William Popham, for the Mediterranean; at Bombardment of Algiers 27.8.1816. Fitted at Deptford as Arctic discovery vessel 1.1818–5.1819, re-registered as a sloop; recommissioned 16.1.1819 under Lieut William Edward Parry, for Northwest Passage Exploration; sailed 5.1819; paid off 21.12.1820. Refitted at Deptford for Arctic discovery 11.1820–4.1821; recommissioned 1.1821 under Cmdr George Lyon and sailed 8.5.1821 (with *Fury*); paid off 14.11.1823. Refitted again at Deptford for Arctic discovery 10.1823–5.1824; recommissioned 17.1.1824 under Parry (now Capt.), and sailed 8.5.1824 (with *Fury* again); paid off 10.1825. Fitted for Ordinary at Woolwich 11–12.1825. Fitted at Deptford for voyage to Spitzbergen 8.1826–3.1827; recommissioned 11.1826 under Parry, and sailed 4.4.1827 (this time reaching 82¾ degrees N); paid off 1.11.1827. Fitted at Deptford to survey the coast of Africa 10.1827–3.1828; recommissioned 12.1827 under Cmdr Thomas Boteler (died 28.11.1829 from fever, along with most of his crew); from 28.11.1829 under Lieut Francis Harding (acting). Sold to Sir E. Banks (for £1,990) 13.4.1831.

Infernal Barkworth & Hawkes, North Barton (Hull)
    As built: 105ft 0in, 86ft 1in x 28ft 4in x 13ft 10in. 374bm.
    Ord: 5.6.1813. K: 7.1813. L: 26.7.1815. C: 7.1816 at Sheerness.
    Commissioned 7.1816 under Cmdr George James Perceval, for the Mediterranean; at Bombardment of Algiers 27.8.1816, losing 2 killed and 27 wounded. At Deptford in Ordinary 1817–1824. Fitted for sea at Deptford 3–6.1824; recommissioned 4.1824 under Cmdr (Capt. 10.1824) Robert Heriot Barclay; paid off Autumn 1824. Recommissioned 8.12.1827 under Cmdr Edmund Williams Gilbert, then 30.10.1828 under Cmdr Brunswick Popham, for the Mediterranean. Sold to Mr Snook (for £1,910) 13.4.1831.

In addition to the above vessels, the Navy List in January 1817 also included the elderly bomb vessel (1789) *Discovery* of 337bm (in harbour service).

François-Étienne Musin's depiction of the *Erebus* in 1846 during the Arctic venture under the command of Sir John Franklin. Both she and the similar *Terror* were fitted with a steam engine and screw propeller for this voyage. Here she is still in open water, albeit surrounded by icebergs; but in 1848 both vessels were icebound and had to be abandoned by their crews. Notwithstanding various clues discovered in later years, the remains of the two vessels have still to be ascertained. *(NMM BHC3225)*

(B) Vessels acquired from 1 January 1817

A further eight units of the *Fury* Class were ordered in 1819–20, and a final vessel in 1823, but only five of these nine were completed. The new type were developed from the wartime *Terror* Class, but were squarer in cross-section, with shallow floors and flat sides. Designed with a continuous main deck, a full square stern with quarter galleries, and a traditional stem with figurehead. No swivel guns or sweep ports. On 10 January 1831 the Admiralty cancelled the four ships still on order and directed that no further bomb vessels were to be built, the type being now obsolete and the best of the existing vessels having already been converted into exploration and survey ships.

Meteor Pembroke Dyd
> As built: 106ft 0in, 87ft 1in x 28ft 11in (28ft 7in for tons) x 13ft 10in. 378bm. Draught 8ft 10in/10ft 9in.
> Ord: 18.5.1819. K: 5.1820. L: 25.6.1823. C: 19–26.7.1823 (for Ordinary) at Plymouth.
> First cost: £12,754 to build, + £1,440 fitting for Ordinary.
> Commissioned 4.5.1824 under Cmdr James Scott, for the Mediterranean; fitted for sea at Plymouth (for £3,929) 17.6.1824. Fitted for sea at Plymouth (for £10,635) 7–9.1828. Recommissioned 26.7.1828 under Cmdr David Hope; sailed for the Mediterranean 13.9.1828; blockade of Tangier; on 4.2.1830 under Cmdr Richard Copeland. Fitted as survey vessel at Portsmouth 5–7.1832, renamed **Beacon** 6.1832. Commissioned 2.8.1836 under Lieut (Cmdr 22.2.1841) Thomas Graves, for surveying in the Mediterranean; paid off 2.4.1846. Sold (by AO 17.8.1846, for £1,500) at Malta.

Aetna Chatham Dyd
> As built: 105ft 2¼in, 86ft 1in x 28ft 11¼in (28ft 7¼in for tons) x 13ft 10in. 375bm. Draught 9ft 1in/11ft 0in.
> Ord: 18.5.1819. K: 9.1821. L: 14.5.1824. C: 5.1824 (for sea).
> First cost: £20,433 fitted for sea.

Commissioned 4.5.1824 under Cmdr Williams Sandom, for the Mediterranean; reached Gibraltar on 28.5.1824 (on the fourteenth day after launch, having been fully manned and equipped for sea service before sailing). Fitted at Woolwich (for £6,097) 12.1827. Recommissioned 12.1827 under Cmdr Thomas Edward Hoste, for the Mediterranean; on 13.5.1828 under Stephen Lushington, on same station; operations against Morea Castle 10.1828; on 28.10.1829 under Robert Ingram, for the west coast of Africa; paid off 26.5.1830. Recommissioned 27.5.1830 under Cmdr Edward Belcher, on same service; refitted at Portsmouth (for £8,905) 6–10.1830, and fitted as a Survey vessel at Portsmouth 8–12.1831; to River Douro at end 1832 (during Portuguese civil war), then 1833 to the Mediterranean; paid off 10.9.1833. Recommissioned 21.9.1833 under Cmdr William George Skyring, for surveying East African coast (killed 23.12.1833 on the Gambia River); on 23.12.1833 under Lieut William Arlett (acting), surveying on west coast of Africa; paid off 6.10.1834. Recommissioned 15.9.1835 under Capt. Alexander Thomas Emeric Vidal, on same station; paid off end 1838. Fitted as receiving ship ('for the reception of newly-raised men') at Portsmouth (for £1,069) 1–4.1839; recommissioned 31.1.1839 under Lieut John Willson, for the Mersey; paid off 25.4.1842. 'Defects made good for Liverpool' (for £1,127) 4–7.1842. Recommissioned 23.5.1842 under Lieut Charles George Butler, for Liverpool; paid off 2.1844. Sold to the Bristol Seamens Friendly Society (for £843) 20.2.1846.

Sulphur Chatham Dyd
> As built: 105ft 5¼in, 86ft 2½in x 28ft 11in (28ft 7in for tons) x 13ft 10¼in. 375bm. Draught 8ft 8in/10ft 2in.
> Ord: 18.5.1819. K: 5.1824. L: 26.1.1826. C: 21.2.1826 (for sea).
> First cost: £14,738 fitted for sea.
> Commissioned 25.11.1828 under Cmdr William Townsend Dance; fitted at Chatham (for £8,619) 11.1828–1.1829 to carry settlers to New Holland (Western Australia), being re-registered as a sloop

for this particular service; paid off 12.1832. Fitted at Woolwich as a Survey vessel (for £7,790) 7–12.1835. Recommissioned 9.11.1836 under Cmdr Edward Belcher, for surveying in South American waters; in 1841 to China; in Canton River 1841; paid off 2.8.1842. Fitted as receiving ship and convict ship at Woolwich (for £1,020) 7.1842–5.1843. BU completed 20.11.1857 (the last bomb ship in RN service).

Thunder Deptford Dyd
>As built: 105ft 0⅞in, 86ft 2⅛in x 28ft 10in (28ft 6in for tons) x 13ft 10in. 372bm. Draught 8ft 6in/10ft 5in.
>Ord: 18.5.1819. K: 11.1826. L: 4.8.1829. C: 26.10.1829 (for Ordinary).
>First cost: £14,919 to build.
>In Ordinary until 1832; to Sheerness 4.1831; to Chatham 11.1832 and fitted there as a Survey vessel (for £9,142) 11.1832–6.1833. Commissioned 25.3.1833 under Cmdr Richard Owen, for the West Indies; in 1837 under Lieut B. Allen; paid off 5.1837. Fitted for the West Indies 9.1837. Recommissioned 30.11.1837 under Lieut (Cmdr 28.6.1838) Edward Barnett, for surveying North America and the West Indies; refitted at Chatham (for £5,254) 7–10.1841; refitted at Chatham (for £3,880) 8–11.1844 (still under Barnett); paid off 31.8.1848. BU completed at Chatham 14.5.1851.

Vesuvius Deptford Dyd
>Ord: 18.5.1819. K: — . Order transferred to Chatham Dyd. Re-ord: 30.8.1828. K: 8.1830. L: — . Cancelled 10.1.1831.

Devastation Plymouth Dyd
>Ord: 18.5.1819. K: 1820. L: — . Cancelled 10.1.1831.

Volcano Plymouth Dyd
>Ord: 18.5.1819. K: 1821. L: — . Cancelled 10.1.1831.

Beelzebub Plymouth Dyd
>Ord: 31.10.1820. K: — . Cancelled 10.1.1831 (unstarted).

Erebus Pembroke Dyd
>As built: 105ft 0in, 86ft 1¼in x 28ft 10in (28ft 6in for tons) x 13ft 10in. 372bm. Draught 9ft 6in/10ft 7in.
>Ord: 9.1.1823. K: 10.1824. L: 7.6.1826. C: 15.6.1826–2.1828 (for sea) at Plymouth.
>First cost: £14,603 to build, + £9.339 fitting.
>Commissioned 8.12.1827 under Cmdr George Haye, for the Mediterranean; on 8.7.1828 under Cmdr Sir Philip Broke, on same station; paid off 7.1830 at Portsmouth. Fitted as Antarctic discovery vessel at Chatham (for £14,316) 4–9.1839. Recommissioned 8.4.1839 under Capt. James Clark Ross, for expedition (with earlier conversion *Terror*) towards the South Pole; returned home 9.1839. Fitted at Woolwich for Arctic exploration 2–5.1845, with steam engine (converted railway locomotive) and screw propeller. Recommissioned 3.3.1845 under Capt. Sir John Franklin, for expedition to explore for Northwest Passage (again in company with *Terror*); on 6.1847 (upon Franklin's death) under Capt. James Fitzjames; abandoned in Arctic 22.4.1848 along with *Terror*; both were retained on Navy List until 1854.

PACKET VESSELS

In addition to the following purchased vessels, numerous naval-built brigs (primarily of the *Cherokee* Class – see Chapter 8) were employed as Falmouth packets.

Dove (ex-mercantile *Manchester*, built 1805 at Falmouth) Brig packet, 6 guns.
>Dimensions & tons: 187bm.
>Purchased 1823.
>Commissioned 12.1823 under Lieut James Crosby, for the Falmouth packet service. In 11.1826 under Lieut George Brooke Forster, on same service. Sold to Joshua Crystall (for £720) 31.1.1829.

Swallow (ex-Post Office packet *Marquis of Salisbury*, built 1819 at Falmouth) Brig sloop, 10 guns.
>Dimensions & tons: —, 68ft 7in x 25ft 1in x — 236bm.
>Men: 28. Guns: 4 x 12pdr carronades.
>Purchased 7.1824.
>Commissioned 13.9.1824 under Lieut Thomas Baldock (who built the vessel and commanded her under Post Office authority from 1819 until her purchase by the Admiralty), for the Falmouth packet service. On 25.11.1831 under Lieut Smyth Griffith, on same service. Sold 8.9.1836.

Marchioness of Queensbury (ex-Post Office packet of same name)
>Purchased 1829.
>To Police service at Gibraltar 1830; fate and other details unknown.

Cynthia (ex-mercantile *Prince Regent*, built 1821 at Falmouth) Brig packet, 6 guns.
>Dimensions & tons: 87ft 2in x 25ft 0in x — 232bm.
>Guns: 2 x 9pdrs, 4 x 9pdr carronades.
>Purchased 1826.
>Commissioned 20.7.1826 under Lieut John White, for the Falmouth packet service; sailed from Falmouth 7.5.1827 with mail for West Indies, but wrecked near Barbados 6.6.1827.

Nightingale (ex-Post Office packet *Marchioness of Salisbury*) Brig sloop, 6 guns.
>Dimensions & tons: 82ft 1½in, 66ft 5⅛in x 23ft 10in (23ft 8in for tonnage) x 14ft 11in. 198³⁄₉₄bm.
>Men: 33 as packet. Guns: 2 x 6pdrs, 4 x 12pdr carronades.
>Purchased 28.7.1829 (after survey 5.1829). Between Small and Middling Repair at Plymouth and fitted as a packet 24.6.1829–30.4.1830 (for £7,749).
>Commissioned 15.3.1830 under Lieut George Brooke Forster, for the Falmouth packet service (–1838). Fitted as depot at Plymouth 10.1838–1.1839. Sold to Greenwood & Clarke (for £610) 24.11.1842.

Kestrel (ex-private yacht *Kestrel*, built 1837 by White, Cowes) Packet brigantine.
>Dimensions & tons: —, 62ft 6in x 23ft 7in x 11ft 3in. 202bm.
>Purchased 4.12.1846 from the Earl of Yarborough (for £2,727) as brigantine). Fitted as packet at Portsmouth (for £1,505) 23.2.1847–4.5.1847.
>Commissioned 3.1847 under Lieut Horace Mann Baker, for East coast of South America; paid off 5.8.1850 at Plymouth. Re-rigged as a schooner 1851, but BU at Plymouth 11.1852.

SURVEYING AND EXPLORATION VESSELS

Vessels acquired from 1 January 1817

A number of specialised vessels were commissioned by the Royal Navy for Arctic exploration around mid-century. The key event at this time was the 1845–48 ill-fated expedition led by Sir John Franklin in search of the Northwest Passage, aboard the adapted bomb ships *Erebus* and *Terror* (for which see under Bomb Vessels above). Much subsequent activity over the next decade originated in attempts to learn of Franklin's fate. In addition to the vessels below, acquired for the purposes of exploration and surveying, a considerable number of redundant warships were converted for the purpose, details of which appear in earlier chapters.

Mermaid (ex-mercantile — , built Calcutta 1816) Survey cutter
> Dimensions & tons: 55ft 0in, 46ft 0in x 18ft 6in x — . 84bm.
> Men: 14. Guns: 1 x 6pdr.
> Purchased 16.10.1817 (for £2,000) at Sydney.
> Commissioned 10.1817 under Lieut Philip Parker King, for surveying the coast of Australia; paid off 12.1820 and given to the New South Wales government. Sold 1823.

Kangaroo (origin unknown) Survey brig
> Dimensions & tons: 87ft 0in, 75ft 5⅞in x 22in 7in x 7ft 10in. 204⁷⁴/₉₄bm.
> as 'doubled' in 1823: 87ft 0in, 75ft 5⅞in x 22ft 11in x 7ft 10in. 210⁸⁰/₉₄bm.
> Purchased in West Indies 11.1.1819, registered 11.1.1819.
> Commissioned 23.1.1819 under Lieut Frank Hastings; on 19.6.1819 under Anthony de Mayne, Master. Hull doubled and re-rigged as a ship at Deptford Dyd 1823. Wrecked 18.12.1828 off Cuba.

Bathurst (ex-mercantile, built at Port Jackson, Australia) Survey brig.
> Dimensions & tons: dimensions unknown. 170bm.
> Purchased 7.7.1821 for survey use.
> Commissioned 7.7.1821 under Lieut Philip Parker King, for surveying the coast of Australia (–1822); arrived Portsmouth 24.3.1823 and paid off. To coastguard 2.1824. Sold to Castle 11.4.1858 to BU.

Albatross (ex-mercantile, built in Scotland 1826) Survey schooner
> Dimensions & tons: 58ft 5in x 16ft 11in. 64bm.
> Purchased 1828.
> Tender to *Hecla* 1828–30, for surveying the coast of West Africa. Laid up at Woolwich 6.1830. Sold to John Small Sedger, Rotherhithe 30.8.1833.

Wasp (ex-mercantile — , built 1801 by Thomas White, Cowes)
> Dimensions & tons: dimensions unknown. 81bm.
> Purchased 1821.
> BU 1829.

Violet (Dutch fishing boat, building at Scheveningen) Survey tender
> Dimensions & tons: 40ft 2in, 29ft 8½in x 17ft 2in x 6ft 5⅛in. 46¹³/₉₄bm.
> Purchased 20.6.1835 on stocks as tender to HMS *Fairy*. Sold 24.11.1842.

Speedwell (ex-mercantile *Speedwell*, built Hull) Survey cutter
> Dimensions & tons: 55ft 11½in, 45ft 6⅛in x 17ft 7½in x 8ft 3in. 73bm.
> Purchased 12.7.1841 (for £900).
> Tender to *Blazer* 1843–47. Commissioned 27.10.1847 under Edward Calver, master, as independent survey vessel. Dockyard lighter at Woolwich from 8.1853. Sold 28.5.1855 (for £480).

Plover (ex-East India Co. *Lady William Bentinck*, building details unknown) Arctic survey cutter. Teak-built.
> Dimensions & tons: 82ft 2in, 64ft 0½in x 25ft 0in x 14ft 7½in. 213bm. Draught 7ft 6in/8ft 0in.
> Men: 50. Guns: 4 x 18pdr carronades (10cwt).
> Purchased 14.2.1842 at Canton for £5,951 including masts, rigging, etc. (vessel hired from 1841) and renamed 29.11.1842.
> Commissioned 18.5.1842 (appointed 19.2.1842) under Cmdr (Capt., 23.12.1842) Richard Collinson, for surveying in China and the East Indies; in Yangtze operation 7.1842; paid off ?9.1846 at Woolwich, as receiving ship. Fitted at Sheerness for Arctic service (for £10,182) 10.1847–3.1.1848. Recommissioned 17.11.1847 under Lieut (Cmdr, 11.1.1848) Thomas Edward Laws Moore, for surveying in North Pacific; employed as depot ship 1850 for Arctic search for Sir John Franklin; in 1852 under Cmdr Rochfort Maguire, on same service. Sold 24.11.1854 (for £1,300.16.8d) at San Francisco to Moore & Folyer.

Young Hebe (ex-mercantile) Survey schooner, no guns
> Dimensions & tons: 42ft 0in, 27ft 9in x 12ft 0in x 5ft 0in. 22bm.
> Purchased 1843 for surveying duties on China station. 16 men.
> Commissioned 1.1842 under Cmdr William Cotterell Wood; paid off 5.1842. Recommissioned 8.7.1843 under Lieut William Thornton Bate, still in the East Indies. Paid off and sold 7.1.1847 at Hong Kong.

Research (ex-mercantile *Research*) Survey tender
> Dimensions & tons: dimensions unknown. 40bm.
> Purchased 12.3.1846 at Malta as tender to HMS *Beacon*.
> BU 6.1859 at Malta.

Castlereagh (ex-mercantile, origins unknown) Survey schooner
> Dimensions & tons: dimensions unknown. 93bm.
> Men: 50. Guns: 4 x 18pdr carronades.
> Hired 11.1845, then purchased 3.4.1846 at Sydney.
> Tender to *Bramble* 11.1845, for surveying coast of Australia. Paid off 3.8.1847 and sold 23.8.1847 (for £600) at Sydney to Mr Fotheringham.

The next four vessels were all purchased by the Admiralty and outfitted for the protracted search to locate Sir John Franklin's missing expedition (which had sailed in May 1845 with the former bomb vessels *Erebus* and *Terror* – for which see earlier in this chapter). An initial expedition comprising the *Enterprise* and *Investigator* sailed in 1848, and both ships were recommissioned in 1850 for subsequent searches, during which the *Investigator* itself was crushed by the ice. A further expedition under Capt. Horatio Thomas Austin was despatched in 1850, comprising the *Resolute* and *Assistance*, together with the screw steamers *Pioneer* and *Intrepid* (see Chapter 15); the same four ships were recommissioned in 1852 for a further expedition under Sir Edward Belcher, in the course of which all four had to be abandoned.

Enterprise (ex-mercantile, building by Wigram, Blackwall) Arctic discovery ship
> Dimensions & tons: 125ft 7½in, 108ft 6in x 28ft 8½in (?28ft 7in for tonnage) x 20ft 0in. 471bm.
> Purchased 2.1848 on stocks. L: 5.4.1848. Fitted for Arctic expedition.
> Total first cost: £24,545 including masts, rigging, etc.
> Commissioned 2.1848 under Capt. Sir James Clark Ross (appointed 31.1.1848). Recommissioned ?1850 under Capt. Richard Collison. Coal depot 1860, lent to Board of Trade. Sold 15.9.1903.

Following the disappearance of the Franklin Expedition in *Erebus* and *Terror* during the search for the fabled Northwest Passage, numerous searches were conducted to seek an explanation for the mystery and to locate any survivors. In this painting by Thomas Sewell Robins is the Assistance, commanded by Captain Erasmus Ommanney, which was forced to winter in the Canadian ice at the close of 1850 and returned home in 1851. *(NMM BHC4239)*

Investigator (ex-mercantile, built by Scott's, Greenock) Arctic discovery ship
> Dimensions & tons: 118ft 0in, 101ft 2½in x 28ft 3in (28ft 0in for tonnage) x 18ft 11in. 422bm.
> Purchased 2.1848 and fitted for Arctic service by Green's, Blackwall.
> Total first cost: £25,337 including masts, rigging, etc.
> Commissioned 29.2.1848 under Capt. Edward Joseph Bird. Recommissioned ?1850 under Robert John Le Mesurier McClure. Abandoned in ice off Banks Island 3.6.1853 but rediscoverd in the same location 25.7.2010.

Resolute (ex-mercantile *Ptarmigan*, built 1848 by Smith's Dock, South Shields) Arctic discovery barque
> Dimensions & tons: 115ft 0in, about 99ft 3in x 28ft 4in x 11ft 6in. 424bm.
> Purchased 21.2.1850 from builders (for £10,777) 'for the Arctic Seas' and renamed *Refuge*, then renamed *Resolute* 7.3.1850. Fitted at Woolwich 23–26.2.1850, then fitted for Arctic service by Green's, Blackwall (for £8,045) 25.2–5.4.1850. Completed fitting at Woolwich (for £4,831) 5–25.4.1850.
> Commissioned 2.1850 under Capt. Horatio Thomas Austen, for exploration in Baffin's Bay. Refitted for the Arctic at Woolwich (for £1,763) 10.1851–4.1852. Recommissioned 1852 under Capt. Henry Kellett. Abandoned in ice off Bathurst Island 15.5.1854 (along with steam tender *Intrepid*). Floated free and recovered 12.12.1856 by American whaler. Purchased by US government and presented to Queen Victoria by President Buchanan; arrived Portsmouth 12.12.1856. Remained in Ordinary at Chatham as discovery ship 1856–75. BU (by AO 11.6.1879) 30.6–16.8.1879 at Chatham.

Assistance (ex-mercantile *Baboo*, built 1835 at Howrah, Calcutta) Teak-hulled Arctic discovery barque
> Dimensions & tons: 117ft 4in, about 98ft 6in x 28ft 5in x 13ft 7in. 423bm.

> Men: 58. Guns: 2.
> Purchased 28.2.1850 from Mr Kincade (or Kincaid) (for £5,751). Fitted by Wigram, Blackwall (for £4,900) 8–25.4.1850.
> Commissioned 2.1850 under Capt. Erasmus Ommaney, for Arctic service. Refitted for the Arctic at Woolwich (for £1,982) 10.1851–4.1852. Recommissioned 1852 under Cmdr George Henry Richards, for Arctic exploration. Recommissioned 1854 under Sir Edward Belcher; abandoned in the ice off Bathurst Island 25.8.1854 (along with steam tender *Pioneer*).

Sophia (ex-mercantile *Sophia*, built 1850 at Dundee) Survey vessel
> Dimensions & tons: 83ft 0in, about 70ft 6in x 20ft 0in x 11ft 9in. 150bm.
> Purchased 11.5.1850 for Arctic service.
> At Woolwich 22.9.1851 – 10.1851, then to Chatham. Not employed, but sold (by AO of 20.4.1853, for £903.12.0d) 5.5.1853 to W. H. Rayden.

Rose (ex-mercantile *Lord Star*, built 1846 at Cowes) Survey cutter
> Dimensions & tons: 70ft 2in, ?59ft 4in x 20ft 10in x — '37bm' (probably 137bm).
> Purchased 20.5.1857 for surveying in home waters.
> Commissioned 1857 under Cmdr Henry Laird Cox, as tender to *Impregnable*. On 4.9.1860 under Capt. John Lort Stokes, as tender to *Impregnable*. Stranded 20.7.1864, wreck sold 9.8.1864.

YACHTS

This section lists the yachts (the term does not denote a particular form of sailing rig; most yachts were ship-rigged, i.e. with three masts carrying spars and sails) used by the royal family or by local dignitaries, often the superintendents of the Royal Dockyards

The painting by William Anderson shows the *Royal George* yacht at anchor off Greenwich Hospital on 10 August 1822, on the occasion of HM George IV's return from his visit to Scotland (the first by any monarch since Charles I). The king has disembarked and is being rowed to the Watergates, while in the foreground a ceremonial barge is being manoeuvred by bargemen dressed in scarlet tunics. *(NMM BHC0620)*

at which they were based. While the royal yachts were prestige commands, and were Rated vessels under permanent captains, the dockyard yachts did not generally have fixed commanding officers, as the lists below will indicate.

Royal Charlotte Deptford Dyd [M/Shipwright John Hollond]
(A ship-rigged royal yacht, designed by Sir Joseph Allen. Carried 10 x 3pdr guns)
> Dimensions & tons: 90ft 1in, 72ft 2½in x 24ft 7in x 11ft 0in. 232¹¹⁄₉₄bm.
> Men: — . Guns (org): 8 x 4pdrs, + 8 x ½pdr swivels.
> L: 29.1.1750.
> Recommissioned 5.1801 (her 18th-century service prior to paying off in 1793 is here omitted) under Capt. Sir Harry Neale. In 2.1804 under Capt. George Grey, then under Capt. George Henry Towry, then 1805 under Capt. Edward James Foote. In 1813 under Capt. Thomas Eyles, then 6.1814 under Capt. George Scott (–1820, when he transferred to the new *Royal Charlotte*). BU 7.1820.

Princess Augusta Deptford Dyd [M/Shipwright Adam Hayes]
(Built as *Augusta*, but renamed 23.7.1773)
> Dimensions & tons: 80ft 9in, 65ft 2½in x 23ft 1¼in x 10ft 10in. 185¹⁵⁄₉₄bm.

> Men: 40. Guns: 6 x 2pdrs.
> L: 1771.
> Sold to Thomas Pittman (for £500) 13.8.1818.

Medina Portsmouth Dyd [M/Shipwright Thomas Bucknall]
(This royal yacht was theoretically a Great Repair of the yacht *Portsmouth* built at that port in 1703)
> Dimensions & tons: 52ft 10in, 42ft 10in x 17ft 0in x 8ft 6½in. 65⁷⁹⁄₉₄bm.
> Men: 15. Guns: 6 x 2pdrs.
> L: 8.1772.
> From 1803 to 1813 under Cmdr Peter Baskerville, based on the Isle of White; then briefly under Cmdr James Hills (died 1804), then 6.1814 under Cmdr William Love; paid off 1817. Thereafter based at Portsmouth, with no permanent commander appointed. BU at Portsmouth 8.1832.

Chatham Chatham Dyd [M/Shipwright Thomas Pollard]
(A dockyard yacht, a rebuilding of the former *Chatham* yacht of 1741)
> Dimensions & tons: 59ft 6½in, 48ft 5¼in x 19ft 0in x 9ft 8in. 93bm.
> Men: 10. Guns: 6 x 2pdrs.
> L: 1793. C: 1793.
> Rebuilt in 1828, then again in 1842 (when lengthened to 104bm). BU at Chatham 1867.

Portsmouth Portsmouth Dyd [M/Shipwright Edward Tippett]
(A dockyard yacht, a rebuilding of the former *Portsmouth* yacht of 1742)
> Dimensions & tons: 70ft 4in, 53ft 9in x 18ft 11in x 11ft 8in. 102²⁹⁄₉₄ bm.
> Men: 10. Guns: 6 x 2pdrs.

L: 1794. C: 1794.
BU completed 4.9.1869.

Royal Sovereign Deptford Dyd
(A ship-rigged royal yacht, designed by Sir John Henslow. Carried 8 x ½pdr swivels.)

Dimensions & tons: 96ft 0in, 80ft 5in x 25ft 8in (25ft 6in for tonnage) x 10ft 6in. 278bm.

Ord: 1800. K: 11.1801. L: 12.5.1804. C: 21.8.1804.

First cost: £17,824.

Commissioned 6.1804 under Capt. Sir Harry Burrard Neale (–1808, but briefly under Capt. Mark Robinson in 1805); in 8.1810 under Capt. Thomas Byam Martin, then 8.1811 under Capt. Graham Moore, 9.1811 under Capt. Henry William Bayntun, 8.1812 under Capt. William Hotham, 6.12.1813 under Capt. Sir Edward Berry, and 15.12.1814 under Capt. Charles Adam (Acting); paid off 7.2.1816. Middling Repair (for £12,515) 1815–16. Recommissioned 2.1816 under Capt. Sir Edward William Campbell Rich Owen, then 20.7.1821 under Capt. Charles Adam again, 27.5.1825 under Capt. Sir William Hoste, 25.6.1828 under Capt. Robert Cavendish Spencer, 9.10.1828 under Capt. Sir Christopher Cole, 22.7.1830 under Capt. Charles Bullen (–10.1.1837), 29.2.1838 under Capt. Samuel Jackson (–23.11.1841), 17.12.1841 under Capt. Sir Watkin Owen Pell, 17.2.1845 under Capt. Gordon Thomas Falcon and 3.8.1848 under Capt. Peter Richards, all based at Pembroke (the captains from 1832 being concurrently Superintendent of Pembroke Dockyard); paid off 5.5.1849. BU 11.1849 at Pembroke.

William and Mary Deptford Dyd
(A ship-rigged royal yacht, designed by Sir John Henslow. Carried 8 x 3pdr guns.)

Dimensions & tons: 85ft 0½in, 70ft 3½in x 23ft 2½in (23ft 0½ ft for tonnage) x 11ft 1in. 199bm.

Ord: 1800. K: ?1801. L: 14.11.1807.

Commissioned 4.1807 under Capt. Thomas Fremantle; in 8.1810 under Capt. Alexander Fraser, then 8.1811 under Capt. Graham Moore (or Capt. Thomas Surridge), 8.1812 under Capt. Charles Dudley Pater, 27.1.1814 under Capt. Andrew Sproule, 13.4.1820 under Capt. John Philimore, 23.1.1822 under Capt. William Bowles, 8.7.1822 under Capt. Sir Charles Malcolm, 26.7.1827 under Capt. John Chambers White, 22.7.1830 under Capt. Samuel Warren; paid off 1836 at Woolwich. Recommissioned 2.1837 under Capt. Sir John Louis, 6.1.1838 under Capt. Phipps Hornby, 17.12.1841 under Capt. (Commodore 16.1.1844) Sir Francis Augustus Collier, 1.5.1846 under Capt. Lord John Hay, 17.7.1846 under Capt. Houston Stewart and 24.11.1846 under Commodore Sir James John Gordon Bremer, all based at Woolwich (the captains from 1832 being concurrently Superintendent of Woolwich Dockyard). BU 4.1849.

Admiralty Woolwich Dyd
(A dockyard yacht, believed to be a rebuilding of the former *Plymouth*)

Dimensions & tons: 68ft 7in, 54ft 5⅝in x 30ft 1in oa (19ft 11in for tonnage) x 12ft 8in. 114⁸⁷⁄₉₄bm.

Men: 8. Guns: none.

L: 21.5.1814.

Renamed *Plymouth* 7.1830, as dockyard yacht for that port; renamed *Dockyard YC.1* in 1866. Sold 10.5.1870.

Royal George Deptford Dyd
(A ship-rigged royal yacht, designed by Peake)

Dimensions & tons: 103ft 0in, 88ft 4¼in x 26ft 8in (26ft 6in for tonnage) x 11ft 6in. 330bm.

Ord: 1813. K: 5.1814. L: 17.7.1817. C: 9.8.1817.

First cost: £24,258.

Commissioned 4.9.1815 under Capt. Sir Edward Berry; on 12.8.1819 under Capt. Charles Paget. Registered as a Third Rate 15.12.1821. On 15.12.1821 under Capt. Thomas Bladen Capel; on 27.5.1825 under Capt. Sir Michael Seymour, then 29.12.1828 under Capt. George Mundy, 22.7.1830 under Capt. Lord Adolphus Fitzclarence. Fitted for accommodation of crews of royal yachts at Portsmouth 1843. BU 9.1905 at Deptford.

Prince Regent Portsmouth Dyd
(Royal yacht, designed by the Portsmouth Academy apprentices)

Dimensions & tons: 96ft 0½in, 81ft 3in x 25ft 8¼in (25ft 6¼in for tonnage) x 10ft 0¼in. 282bm.

Ord: 1815. K: 9.1815. L: 12.6.1820. C: 23.8.1820.

First cost: £28,936.

Commissioned 12.8.1819 at Deptford under Capt. Sir Michael Seymour (nominally Capt. Sir Edward Hamilton was CO from 12.1815, then Capt. Charles Paget from 1.1819 while the yacht was building); on 27.5.1825 under Capt. George Mundy, then on 29.12.1828 under Capt. William Parker and 22.7.1830 under Capt. George Tobin; paid off 23.8.1837 at Deptford. Fitted there (for £3,207, + £210 for 'defects') for transfer to the Imaum of Muscat; sailed Plymouth 29.9.1837 for Muscat. BU 1847?

Navy Board Woolwich Dyd
(Navy Board yacht, formerly revenue cutter *Hart*. Carried 2 x 6pdr guns.)

Dimensions & tons: 53ft 6in, 44ft 5in x 18ft 8in (18ft 6in for tonnage) x 6ft 6½in. 80⁸⁰⁄₉₄bm. Draught 4ft 2in/8ft 1in.

Ord: 1821. K: 1.1821. L: 16.2.1822. C: 30.7.1822 (for sea).

Fitted as yard craft ('longboat') at Sheerness 6.1825. Fitted as Admiralty tender at Sheerness 5.1833, known as 'Sheerness YC No. 1'. Renamed *Drake* 17.11.1870. Employed as Superintendent's yacht at Sheerness (by AO 13.2.1872) from 31.3.1872. BU completed 6.3.1875 at Chatham.

Royal Charlotte Woolwich Dyd
(A ship-rigged royal yacht, designed by Seppings. Carried 6 x 1pdr swivels.)

Dimensions & tons: 85ft 8in, 72ft 8⅜in x 23ft 0½in x 8ft 2in. 202bm.

Ord: 1819. K: 4.1820. L: 22.11.1824.

Commissioned 20.5.1820 under Capt. George Scott; on 20.7.1821 under Capt. Jahleel Brenton, then 27.5.1825 under Capt. John White, 6.11.1827 under Capt. Lord William Paget, 27.1.1829 under Capt. Josceline Percy, and 6.12.1830 under Capt. Edward Galway. BU 10.1832.

The sail-powered royal yachts gave way early in the Victorian period to steam-driven vessels, which will be found later in this volume.

VESSELS FOR AMPHIBIOUS WARFARE (FROM 1854) — (MORTAR VESSELS AND MORTAR FLOATS)

A large programme of specialised vessels was developed for the Russian War of 1854–56, which – apart from the famous Crimean gunboats (see Chapter 5) – are usually unreported in naval compilations. The conflict faced the Navy with an urgent need for large numbers of vessels capable of coastal bombardment. They reverted to a type long considered obsolete, the bomb vessel, but in miniature form. The mortar vessels carried about 17 or 18 men (generally a mate to command, a dozen seamen and four or five marines). They carried no guns, but simply a single 13in mortar, capable of projecting a 196¼lb bomb or an incendiary 'carcass' far inland. Sitting on (and thereby trainable through 360°) a circular iron race that was fixed in its a 4½ft deep octagonal well, the 5½-ton mortar was fired at a fixed 45° elevation, the range being varied simply by altering the size of the charge.

Charge	Range	Time of flight	Fuse length
2lb	690 yds	13 secs	2.70in
8lb	2,575 yds	24¾ secs	5.09in
12lb	3,500 yds	29 secs	6.02in
20lb	4,200 yds	31 secs	6.44in

The vessels resembled shortened Thames sailing barges, complete with lee-boards, and stepped a single mast, located forward to improve the angle of fire of the mortar, with a light mizzen seated in a tabernacle, well aft, and carried a rudimentary rig – presumably fore-and-aft, as there is no evidence of spars fitted aloft. They saw service not only in the Black Sea and Sea of Azov, but also in the Baltic. There were also a larger screw mortar frigate, the *Horatio*, and four similar conversions ordered but never completed and armed (see under 'screw frigates' in Chapters 5).

MORTAR VESSELS

The first pair of these vessels were converted in October 1854 from existing harbour tank vessels/lighters (see below for original details). They were followed by 54 new-built vessels; originally launched with names, the first two batches (each of 10 vessels) were given numbers *MV.3–MV.22* instead of names on 19 October 1855, with most of the withdrawn names being then re-issued to new screw gunboats; the last 34 never received names. None of these vessels were commissioned, and were placed under the command of warrant officers only (thus no commanders' names are recorded below). The 60-foot, 65-foot and 70-foot types were deployed overseas, the majority being towed to the Black or Baltic Seas by other warships; none of the later 75-foot type was completed in time to deploy overseas.

The six vessels sent to the Black Sea were the *Flamer*, *Firm*, *Hardy*, *Raven*, *Magnet* and *Beacon*, each commanded by a Mate. The senior of these was Henry Knox Leet (*Firm*), the other five being John Brasier Creagh, Thomas Livingstone Pearson, Harry Woodfall Brent, Henry Vaughan and Albert Frederick Hurt; all six were commissioned as Lieuts within a year of their bombardments of Sebastopol (on 9 September 1855) and Kinburn (on 17 October 1855). The remaining sixteen vessels were sent to the Baltic, where they all took part in the bombardment of Sveaborg (on 9–10 August 1855).

60-FOOT TYPE Both vessels were converted from sailing lighters, but saw no service and were swiftly reconverted just over a year later. Both were deployed to the Baltic with Dundas's fleet in 1855, and took part in the bombardment of Sveaborg (and were on that occasion commanded by Acting-Gunners John Dew and Henry Wallace respectively).

Drake Portsmouth Dyd
As built: 60ft 0in, 47ft 5¼in x 20ft 9¼in (20ft 6¼in for tons) x 8ft 11in. 105¹¹/₉₄bm. Draught 4ft 1in/5ft 4in.

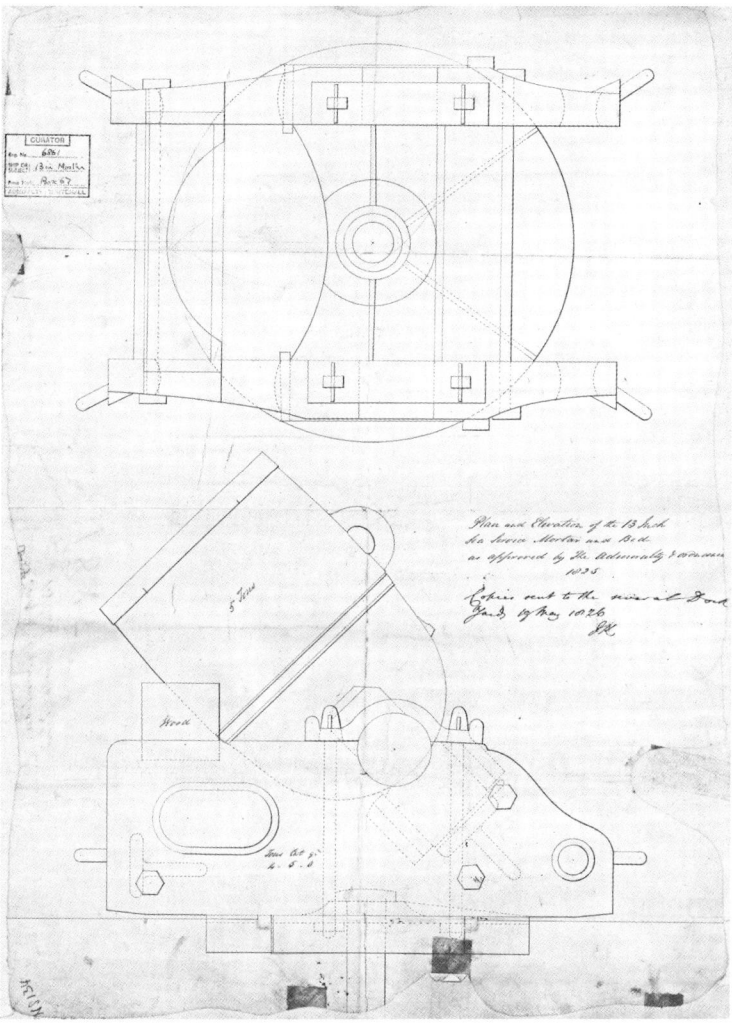

An 1825 plan and elevation of the latest model 13in 5-ton sea service mortar and its bed. A single weapon of this kind armed all the mortar boats and floats built for the Russian War. (*NMM J0718*)

K: 5.1833. L: 25.3.1834. C: 5.1834. Converted 7–11.1854 (by AO 6.10.1854) at Portsmouth Dyd.
First cost: £3,233 to build and fit (1834); £1,750 fitting (1854).
Renamed *MV.1* on 19.10.1855. Refitted as a lighter again at Woolwich 4–5.1856 (by AO 28.6.1856), renamed *Sheppey* by AO 7.7.1856. Harbour service (renamed *YC.1*) at Pembroke Dyd in 4.1865. Sold 1884.

Sinbad Pembroke Dyd
As built: 60ft 1in, 47ft 6in x 20ft 9in (20ft 5in for tons) x 9ft 0in. 105bm. Draught 3ft 11in/5ft 0in.
K: 11.1832. L: 27.2.1834. C: 30.6.1834. Converted 10.1854–6.1855 (by AO 5.10.1854) at Woolwich Dyd.
First cost: £3,066 to build (1834); £1,046 fitting (1854).
Renamed *MV.2* on 19.10.1855. Refitted as a lighter again at Woolwich 10.1856 (by AO 26.6.1856), renamed *YC.3* at Woolwich on 3.7.1856. BU at Woolwich 10.1866–10.11.1866.

65-FOOT TYPE (designed by Capt. Roberts) Ordered 6 October 1854 (on the same day as the 20 *Dapper* class gunboats). *Flamer*, *Firm* and *Hardy* were deployed to the Black Sea in 1855, and took part in the bombardment of Kinburn, while the other seven of this class were all deployed to the Baltic with Dundas's fleet in 1855, and took part in the bombardment of Sveaborg (and were on that occasion commanded by the petty officers mentioned below).

    Dimensions & tons: 65ft 0in, 52ft 5¼in x 20ft 10in x 7ft 6in. 117²³⁄₉₄bm.

Flamer Money Wigram & Son, Blackwall
    Ord: 6.10.1854. K: 9.10.1854. L: 1.3.1855.
    First cost (contractor paid + Admiralty supplies/fitting costs):
      £2,360 + 280/937.
    In 1855 in the Black Sea; took part in the bombardment of Kinburn. Renamed *MV.10* on 19.10.1855. Malta yard craft (mooring lighter) *YC.5* in 1858. For disposal about 1901.
Firm Money Wigram & Son, Blackwall
    Ord: 6.10.1854. K: 9.10.1854. L: 1.3.1855.
    First cost (contractor paid + Admiralty supplies/fitting costs):
      £2,360 + 280/302.
    In 1855 in the Black Sea, under Lieut Henry Knox Leet; took part in the bombardment of Kinburn. Renamed *MV.11* on 19.10.1855. Sold at Malta 1858.
Havock C. J. Mare & Co., Blackwall
    Ord: 6.10.1854. K: 9.10.1854. L: 14.3.1855.
    First cost (contractor paid + Admiralty supplies/fitting costs):
      £2,360 + 280/874.
    In 1855 to the Baltic; took part in bombardment of Sveaborg (under Boatswain Thomas Foreman). Renamed *MV.5* on 19.10.1855. Became Customs watch vessel *WV.27*. BU at Chatham 7.1874.
Hardy Money Wigram & Son, Blackwall
    Ord: 6.10.1854. K: 9.10.1854. L: 14.3.1855.
    First cost (contractor paid + Admiralty supplies/fitting costs):
      £2,359 + 280/835.
    In 1855 in the Black Sea; took part in the bombardment of Kinburn. Renamed *MV.12* on 19.10.1855. Sold at Malta 21.4.1858.
Growler Money Wigram & Son, Blackwall
    Ord: 6.10.1854. K: 9.10.1854. L: 31.3.1855.
    First cost (contractor paid + Admiralty supplies/fitting costs):
      £2,360 + 280/843.
    In 1855 to the Baltic; took part in bombardment of Sveaborg. Renamed *MV.4* on 19.10.1855. Completed as landing stage at Chatham 16.12.1863.
Surly Money Wigram & Son, Blackwall
    Ord: 6.10.1854. K: 9.10.1854. L: 31.3.1855.
    First cost (contractor paid + Admiralty supplies/fitting costs):
      £2,360 + 280/825.
    In 1855 to the Baltic; took part in bombardment of Sveaborg. Renamed *MV.9* on 19.10.1855. BU 11.1863.
Blazer C. J. Mare & Co., Blackwall
    Ord: 6.10.1854. K: 14.10.1854. L: 5.5.1855.
    First cost (contractor paid + Admiralty supplies/fitting costs):
      £2,206 + 139/962.
    In 1855 to the Baltic; took part in bombardment of Sveaborg (under Acting-Gunner Josiah Hunt). Renamed *MV.3* on 19.10.1855. Became yard craft *YC.12* at Chatham. Transferred to Thames Conservancy Board 10.1867.
Mastiff C. J. Mare & Co., Blackwall
    Ord: 6.10.1854. K: 14.10.1854. L: 5.5.1855.
    First cost (contractor paid + Admiralty supplies/fitting costs):
      £2,206 + 139/937.
    In 1855 to the Baltic; took part in bombardment of Sveaborg (under Acting-Gunner Richard Fowell). Renamed *MV.7* on 19.10.1855. Became Customs watch vessel *WV.37* in 11.1864. BU at Chatham 9.1875.

Manly John Jenkins Thompson, Rotherhithe
    Ord: 6.10.1854. K: 14.10.1854. L: 16.5.1855.
    First cost (contractor paid + Admiralty supplies/fitting costs):
      £1,481 + 448/1,095.
    In 1855 to the Baltic; took part in bombardment of Sveaborg (under Acting-Boatswain John Bosanquet). Renamed *MV.6* on 19.10.1855. Hulked 7.1866.
Porpoise John Jenkins Thompson, Millwall
    Ord: 6.10.1854. K: 14.10.1854. L: 26.5.1855.
    First cost (contractor paid + Admiralty supplies/fitting costs):
      £1,497 + 448/–
    In 1855 to the Baltic; took part in bombardment of Sveaborg (under Acting-Boatswain Charles Haydon). Renamed *MV.8* on 19.10.1855. Hulked as a bathing place 7.1866. Sold to Castle 25.6.1885.

70-FOOT TYPE *Raven*, *Magnet* and *Camel* were deployed to the Black Sea in 1855, and took part in the bombardment of Kinburn, while the other seven of this class were all deployed to the Baltic with Dundas's fleet in 1855, and took part in the bombardment of Sveaborg (and were on that occasion commanded by the petty officers mentioned below).

    Dimensions & tons: 70ft 0in, 55ft 7½in x 23ft 4in x 9ft 4in. Draught 8ft 6in. 156⁴⁸⁄₉₄bm.

Raven R. & H. Green, Blackwall (No. 316)
    Ord: 22.12.1854. K: 24.12.1854. L: 19.4.1855.
    First cost (contractor paid + Admiralty supplies/fitting costs):
      £3,053 + 273/1,178.
    In 1855 in the Black Sea; took part in the bombardment of Kinburn. Renamed *MV.13* on 19.10.1855. Ordered to be sold at Constantinople 11.7.1856.
Rocket R. & H. Green, Blackwall (No. 317)
    Ord: 22.12.1854. K: 24.12.1854. L: 5.5.1855.
    First cost (contractor paid + Admiralty supplies/fitting costs):
      £3,054 + 841/1,026.
    In 1855 to the Baltic; took part in bombardment of Sveaborg (under Boatswain John Thoms). Renamed *MV.20* on 19.10.1855. BU completed by Marshall at Plymouth 27.9.1865.
Redbreast R. & H. Green, Blackwall (No. 318)
    Ord: 22.12.1854. K: 24.12.1854. L: 5.5.1855.
    First cost (contractor paid + Admiralty supplies/fitting costs):
      £3,054 + 273/1,109.
    In 1855 to the Baltic; took part in bombardment of Sveaborg (under Acting-Gunner G. Taylor). Renamed *MV.19* on 19.10.1855. BU completed by Marshall at Plymouth 27.9.1865.
Prompt C. J. Mare & Co., Blackwall
    Ord: 22.12.1854. K: 24.12.1854. L: 23.5.1855.
    First cost (contractor paid + Admiralty supplies/fitting costs):
      £3,148 + 139/1,164.
    In 1855 to the Baltic; took part in bombardment of Sveaborg (under Acting-Boatswain Charles Ford). Renamed *MV.21* on 19.10.1855. BU completed by Marshall at Plymouth 27.9.1865.
Pickle C. J. Mare & Co., Blackwall
    Ord: 22.12.1854. K: 24.12.1854. L: 23.5.1855.
    First cost (contractor paid + Admiralty supplies/fitting costs):
      £3,148 + 139/1,025.
    In 1855 to the Baltic; took part in bombardment of Sveaborg (under Acting-Boatswain Richard Jones). Renamed *MV.22* on 19.10.1855. BU completed by Marshall at Plymouth 20.10.1865.
Magnet Money Wigram & Co., Northam
    Ord: 22.12.1854. K: 27.12.1854. L: 2.5.1855.

A watercolour by Colonel Durnford depicting *Mortar Vessel 22* (ex-*Pickle*) cleared for action. The boat is anchored with much of the rigging struck, and the artillerymen are loading the 13in mortar. Among the interesting details are the Thames barge-style leeboards and the spars lashed to the unengaged side of the hull. (*NMM PU9636*)

First cost (contractor paid + Admiralty supplies/fitting costs): £3,125 + 280/1,781.

In 1855 in the Black Sea; took part in the bombardment of Kinburn. Renamed *MV.15* on 19.10.1855. Dockyard craft at Malta 2.1857. BU 6.1867.

Camel Money Wigram & Son, Blackwall
Ord: 22.12.1854. K: 1.1855. L: 21.4.1855.
First cost (contractor paid + Admiralty supplies/fitting costs): £3,125 + 280/1,051.

In 1855 in the Black Sea; took part in the bombardment of Kinburn. Renamed *MV.14* on 19.10.1855. Became a crane lighter *YC.6* at Malta in 1871. For disposal 1901.

Beacon Money Wigram & Son, Blackwall
Ord: 22.12.1854. K: 1.1855. L: 21.4.1855.
First cost (contractor paid + Admiralty supplies/fitting costs): £3,125 + 280/1,008.

In 1855 to the Baltic; took part in bombardment of Sveaborg (under Acting-Boatswain Richard Broad). Renamed *MV.16* on 19.10.1855. Became a dockyard lighter 5.1862.

Carron Money Wigram & Son, Blackwall
Ord: 22.12.1854. K: 1.1855. L: 28.4.1855.
First cost (contractor paid + Admiralty supplies/fitting costs): £3,125 + 280/837.

In 1855 to the Baltic; took part in bombardment of Sveaborg (under Acting-Boatswain J. Terdre). Renamed *MV.17* on 19.10.1855. Hulked 7.1866. BU at Devonport 11.1884.

Grappler Money Wigram & Son, Blackwall
Ord: 22.12.1854. K: 1.1855. L: 1.5.1855.
First cost (contractor paid + Admiralty supplies/fitting costs): £3,125 + 280/969.

In 1855 to the Baltic; took part in bombardment of Sveaborg (under Acting-Boatswain Thomas Hawkins). Renamed *MV.18* on 19.10.1855. Hulked 7.1866. Sold 24.4.1896.

75-FOOT TYPE This final batch received numbers only, and are listed in order of that number rather than chronologically. Thirty were ordered in October 1855 and a final four in December 1855. To late to see service in the Russian War, most were employed as harbour craft or as other auxiliaries.

Dimensions & tons: 75ft 0in, 59ft 2in x 23ft 4in x 9ft 4in. Draught ?8ft 6in. 166⁴⁵/₉₄bm.

First cost (contractor paid + Admiralty supplies/fitting costs):
MV.23 £3,547 = 489/— . *MV.24* £3,547 + 489/291. *MV.25* £3,539 + 497/308. *MV.26* £3,539 + 497/— . *MV.27* £3,245 + 501/12. *MV.28* £3,155 + 591/— . *MV.29* £3,110 + 552/382. *MV.30* £3,129 + 533/— . *MV.31* £3,130 + 532/— . *MV.32* £3,131 + 531/— . *MV.33* £2,684 + 978/— . *MV.34* £2,684 + 978/— . *MV.35* £3,700 + 128/1,194. *MV.36* £3,753 + 75/1,013. *MV.37* £3,310 + 477/994. *MV.38* £4,267 + 19/508. *MV.39* £4,267 + 19/269. *MV.40* £4,267 + 19/132. *MV.41* £4,267 + 19/— . *MV.42* £3,976 + 19/— . *MV.43* £3,976 + 19/— . *MV.44* £3,709 + 120/914. *MV.45* £3,849 + 187/961. *MV.46* £3,897 + 139/641. *MV.47* £3,832 + 204/1,087. *MV.48* £3,839 + 197/165. *MV.49* £4,082 + 71/700. *MV.50* £4,120 + 28/— . *MV.51* £4,028 + 60/— . *MV.52* £3,801 + 57/— . *MV.53* £4,295 + 67/1,300. *MV.54* £4,306 + 50/1,242. *MV.55* £3,378 + 456/— . *MV.56* £3,534 + 300/— .

MV.23 Charles Lungley, Rotherhithe
Ord: 2.10.1855. K: 9.10.1855. L: 24.4.1856.
Became mooring lighter (Luggage Lighter No. 4) at Sheerness 3.1865, later renamed *YC.15*.

MV.24 Charles Lungley, Rotherhithe
Ord: 2.10.1855. K: 9.10.1855. L: 24.4.1856.
Lent to HM Customs as a revenue vessel 12.3.1861, renamed *Harpy*. BU at Chatham 12.10.1872.

MV.25 Charles Lungley, Rotherhithe
Ord: 2.10.1855. K: 9.10.1855. L: 22.4.1856.
Became mooring lighter (No. 11) at Chatham 1862, later renamed *YC.9*. BU about 1901.

MV.26 Charles Lungley, Rotherhithe

Ord: 2.10.1855. K: 9.10.1855. L: 22.4.1856.
Became mooring lighter 10.1862. BU 1.1866.

MV.27 Charles Lungley, Rotherhithe
Ord: 2.10.1855. K: 9.10.1855. L: 22.5.1856.
Became mooring lighter (No. 9) at Chatham 11.1857, later renamed
YC.8. BU about 1901.

MV.28 Charles Lungley, Rotherhithe
Ord: 2.10.1855. K: 9.10.1855. L: 22.5.1856.
Became coastguard *CGWV 38* in 1864. Sold 18.10.1878. BU at
Poole.
Note that at least one of the vessels attributed to Lungley was built
at Northam, apparently subcontracted to Wigram.

MV.29 John Jenkins Thompson, Rotherhithe
Ord: 2.10.1855. K: 9.10.1855. L: 3.5.1856.
Became coastguard *CGWV 21* on 25.5.1863. Sold to BU 12.1905.

MV.30 John Jenkins Thompson, Rotherhithe
Ord: 2.10.1855. K: 9.10.1855. L: 19.7.1856.
Became coal pontoon at Portland in 4.1868. To Portsmouth as
YC.44 in 7.1879. BU 1860.

MV.31 John Jenkins Thompson, Rotherhithe
Ord: 2.10.1855. K: 9.10.1855. L: 6.10.1856.
Became coastguard *CGWV 9* in 1864. To War Dept for torpedo
experiments 6.1870. Sold to BU 1887.

MV.32 John Jenkins Thompson, Rotherhithe
Ord: 2.10.1855. K: 9.10.1855. L: 3.9.1856.
BU at Chatham 16.5.1860.

MV.33 John Jenkins Thompson, Rotherhithe
Ord: 2.10.1855. K: 9.10.1855. L: 15.11.1856.
Became harbour craft *YC.9* at Woolwich in 11.1856. To Sheerness as
YC.27 in 1872. BU about 1901.

MV.34 John Jenkins Thompson, Rotherhithe
Ord: 2.10.1855. K: 9.10.1855. L: 1.11.1856.
Became harbour craft *YC.10* at Woolwich in 11.1856. To Sheerness
as *YC.28* in 1872. BU at Sheerness 3.1873.

MV.35 Hoad Brothers, Rye
Ord: 2.10.1855. K: 15.10.1855. L: 25.2.1856.
BU 10.1865.

MV.36 Hoad Brothers, Rye
Ord: 2.10.1855. K: 15.10.1855. L: 25.3.1856.
BU 9.1865.

MV.37 George Inman, Lymington
Ord: 2.10.1855. K: 15.10.1855. L: 22.3.1856.
Became harbour craft *YC.24* at Devonport in 1864. BU 1876.

MV.38 Scott & Long, Greenock
Ord: 3.10.1855. K: 16.10.1855. L: 23.2.1856.
Became coastguard *CGWV 4* on 28.5.1864. BU 1901.

MV.39 Scott & Long, Greenock
Ord: 3.10.1855. K: 16.10.1855. L: 8.3.1856.
Became coastguard *CGWV 8* in 1866. Sold to BU 23.6.1897.

MV.40 Scott & Long, Greenock
Ord: 3.10.1855. K: 29.10.1855. L: 8.4.1856.
BU at Haslar Creek, Portsmouth 7.1867.

MV.41 Scott & Long, Greenock
Ord: 3.10.1855. K: 8.11.1855. L: 22.4.1856.
Became coastguard *CGWV 32* in 7.1863, but reverted to *MV.41* in
1871. Sold 19.2.1881.

MV.42 Scott & Long, Greenock
Ord: 3.10.1855. K: 3.12.1855. L: 16.6.1856.
Became coastguard *CGWV 14* on 6.6.1867, but reverted to *MV.42* in
1870. Sold to BU at Southampton 12.1877.

MV.43 Scott & Long, Greenock

Ord: 3.10.1855. K: 7.12.1855. L: 23.6.1856.
Became coastguard *CGWV 12* on 6.2.1864. Sold 11.1898.

MV.44 Hassall & Holmes, Rye
Ord: 2.10.1855. K: 15.10.1855. L: 11.3.1856.
Became coastguard *CGWV 21* in 1866. Later fate unknown.

MV.45 Thomas Harvey, Ipswich
Ord: 9.10.1855. K: 12.10.1855. L: 21.3.1856.
Became coastguard *CGWV 22* in 1866. Later fate unknown.

MV.46 Thomas Harvey, Ipswich
Ord: 9.10.1855. K: 12.10.1855. L: 22.4.1856.
Became coastguard *CGWV 10* on 25.5.1863, renamed *CGWV.31* in
1865. Sold to BU 30.5.1893.

MV.47 Thomas Harvey, Wivenhoe
Ord: 9.10.1855. K: 12.10.1855. L: 7.4.1856.
Became bathing 'place' for Boscawen in 7.1866. Sold 21.12.1892.

MV.48 Thomas Harvey, Wivenhoe
Ord: 9.10.1855. K: 12.10.1855. L: 28.5.1856.
Became mooring lighter (No. 10) in 10.1862. Became harbour craft
YC.11 at Chatham (beached as accommodation) 1864. BU 1868.

MV.49 William Patterson & Son, Bristol
Ord: 11.10.1855. K: 15.10.1855. L: 28.2.1856.
Became coastguard *CGWV 19* in 1864, renamed *CGWV.22* in
5.1867. Sold 11.11.1898.

MV.50 William Patterson & Son, Bristol
Ord: 11.10.1855. K: 15.10.1855. L: 5.4.1856.
To coastguard 1864; renamed *CGWV.17* in 1866. Sold to BU
12.1887.

MV.51 William Patterson & Son, Bristol
Ord: 11.10.1855. K: 15.10.1855. L: 18.6.1856.
Became coastguard *CGWV 13* in 1864. Sold to BU 4.8.1883.

MV.52 William Patterson & Son, Bristol
Ord: 11.10.1855. K: 8.11.1855. L: 14.7.1856.
Became coastguard *CGWV 16* on 6.2.1864. Lent to HM Customs in
10.1871. BU 5.1894.

MV.53 Briggs & Co., Sunderland
Ord: 14.12.1855. K: 1.1856. L: 24.3.1856.
BU 8.1865.

MV.54 Briggs & Co., Sunderland
Ord: 14.12.1855. K: 1.1856. L: 24.3.1856.
Became harbour craft *YC.12* at Devonport 1866. BU 1893.

MV.55 J. & R. White, East Cowes
Ord: 14.12.1855. K: 15.12.1855. L: 15.6.1856.
Became coastguard *CGWV 6* on 9.10.1865. Sold 18.6.1881.

MV.56 J. & R. White, East Cowes
Ord: 14.12.1855. K: 15.12.1855. L: 21.6.1856.
Became coastguard *CGWV 22* on 8.10.1864, renamed *CGWV 19* in
1868. BU at Sheerness 31.8.1874.

MORTAR FLOATS

Unlike the wooden-hulled mortar vessels, the mortar floats were
built of iron (and, as Stuart Rankin has observed, must have rung like
a cathedral bell whenever the mortar was fired!). 50 were ordered in
1855 and (except the earliest three) launched in 1856. The only one to
receive a name in RN service was *MF.103*, which was named *Cupid*
(until 1865); otherwise all bore numbers only, and appear in order of
that number rather than chronologically.

None of the 50 vessels vessels were ready in time for participation
in the Russian War – indeed the main batch of 47 units were not begun

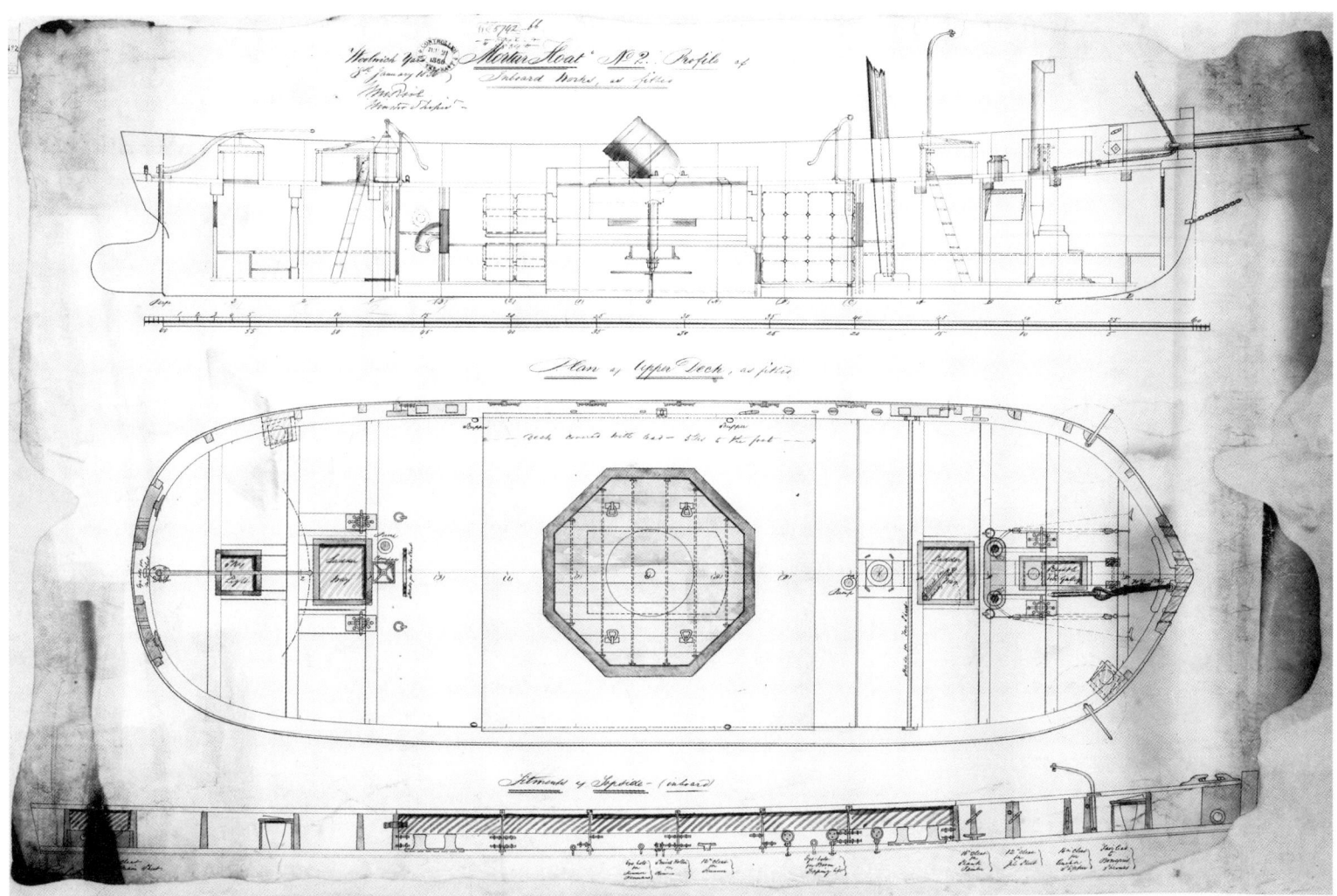

A draught marked 'Mortar Float No. 2. Profile of Inboard Works, as fitted' (presumably for *MF.102*) and dated January 1856 at Woolwich. Among the interesting details annotated on the plan is the fact that the deck around the mortar was protected from blast (and from catching alight) by a covering of lead, at 5lb to the foot. Although this class is generally described as 'dumb' (without any means of propulsion), the draught clearly shows provision for a sailing rig; this may have been confined to the prototypes only. *(NMM DR8042)*

until the start of 1856. None therefore saw any active service, or were ever deployed outside the UK, most being towed to naval ports and immediately laid up. It was reported in November 1856 that it had been decided that 'all such vessels shall be converted into Admiralty lighters, and thus be available for useful purposes' (*Morning Post*, 6 November 1856).

60-FOOT TYPE. The first three in November 1855 were largely experimental; *MF.101* carried out firing trials in December 1855 at Shoeburyness, and was reported to have been 'totally disabled'; the mortar beds were unable to withstand the concussion as well as wooden beds, and much 'concussion' (or shock) damage resulted; she was returned to Samuda's yard in January 1856 for repairs. *MF.102* was modified as a consequence of these trials, and she carried out firing trials in January 1856 without receiving damage.

MF.101 Samuda Brothers, Poplar

    Dimensions & tons: 60ft 3in x 20ft 3in x 5ft 11in. 102bm.

    K: 10.1855. L: 24.11.1855.

    Became a coal pontoon at Queenstown 1871.

MF.102 J. Scott Russell, Millwall

    Dimensions & tons: 60ft 0in x 20ft 1in x 6ft 7in. 102bm.

    K: 10.1855. L: 24.11.1855.

    Became a dredger at Chatham 29.9.1865, renamed *YC.53*.

MF.103 Laird Brothers, Birkenhead

    Dimensions & tons: 59ft 10in x 20ft 1in x 6ft 7in. 102bm.

    K: 10.1855. L: 13.11.1855.

    Became harbour vessel *YC.3* at Haulbowline in 1865.

For the remaining 47 vessels, *MF.104* to *MF.123* were built by C. J. Mare & Co., Blackwall; *MF.124* to *MF.135* were built by John Scott Russell, Millwall; and *MF.136* to *MF.150* were built by Laird Brothers, Birkenhead. Note that Mare became bankrupt on 30.7.1856, and some of these MFs may have been delivered by the shipyard's new owners (Thames Iron Sbdg Co); most of Laird's MFs were built across the Mersey at their yard at Dingle, Liverpool. *MF.141* to 150 were all hauled up on slipway at Haslar in 1858, pending disposal, but were subsequently used as harbour craft. The ultimate fate of many harbour craft is unknown where not stated.

    Dimensions & tons: 60ft 0in, 47ft 2¼in x 20ft 0in x 6ft 6in. Draught 5ft 7in. 100⁴³/₉₄bm.

    First cost (contractor paid + Admiralty supplies/fitting costs):

        MF.104 £2,311 + 4/424; *MF.105* £2,311 + 4/437; *MF.106* £2,311 + 4/372; *MF.107* £2,311 + 4/395; *MF.108* £2,311 + 4/426; *MF.109* £2,311 + 4/438; *MF.110* £2,311 + 4/449; *MF.111* £2,311 + 4/475; *MF.112* £2,311 + 4/501; *MF.113* £2,311 + 4/457; *MF.114* £2,311

+ 4/221; *MF.115* £2,311 + 4/220; *MF.116* £2,311 + 4/159; *MF.117* £2,311 + 4/162; *MF.118* £2,311 + 4/69; *MF.119* £2,311 + 4/65; *MF.120* £2,311 + 4/90; *MF.121* £2,311 + 4/—; *MF.122* £2,311 + 4/—; *MF.123* £2,311 + 4/—; *MF.124* £2,599 + 1/37; *MF.125* £2,599 + 1/— (next 10 the same); *MF.136* £2,396 + 4/255; *MF.137* £2,397 + 3/276; *MF.138* £2,397 + 3/252; *MF.139* £2,395 + 5/247; *MF.140* £2,396 + 4/243; *MF.141* £2,397 + 3/194; *MF.142* £2,393 + 7/— (next 3 the same); *MF.146* £2,389 + 11/—; *MF.147* £2,387 + 13/—; *MF.148* £2,388 + 12/—; *MF.149* £2,389 + 11/—; *MF.150* £2,389 + 11/— .

MF.104
K: 1.1.1856. L: 23.2.1856.
Became water tank vessel at Woolwich from 12.1856, then harbour vessel *YC.1* at Deptford in 1865, renamed *Steady* in 9.1865 as store carrier. Later became harbour vessel *YC.1* for Victualling Yard at Deptford. BU about 1873.

MF.105
K: 1.1.1856. L: 23.2.1856.
Coast defence 1864? BU about 1893.

MF.106
K: 1.1856. L: 23.2.1856.
Became mooring lighter *YC.6* at Woolwich in 1.1866. To Sheerness as YC.23 in 1872. BU about 1901.

MF.107
K: 1.1.1856. L: 23.2.1856.
Became coastguard *CGWV 42* in 1876, later renamed *CGWV 7*. Sold 5.1.1891.

MF.108
K: 1.1.1856. L: 20.3.1856.
Became mooring lighter *YC.7* at Woolwich in 6.1866. To Sheerness as *YC.24* in 1872.

MF.109
K: 1.1.1856. L: 20.3.1856.
Became pontoon to *Thunderbolt* at Chatham.

MF.110
K: 1.1.1856. L: 20.3.1856.
Became harbour vessel *YC.4* at Deptford in 1865, renamed *Trusty* as store carrier in 1866. BU about 1874.

MF.111
K: 1.1.1856. L: 5.4.1856.
Became harbour vessel *YC.7* for Victualling Yard at Deptford in 1863. Later to Pembroke as *YC.18*. Wrecked 1892.

MF.112
K: 1.1.1856. L: 5.4.1856.
Became harbour vessel *YC.17* at Pembroke in 1867. Became pontoon to police hulk at Chatham in 1880. BU about 1901.

MF.113
K: 1.1.1856. L: 5.4.1856.
Became a lighter in 1856; renamed *Iron Lighter No. 1* in 1867, then *YC.24* at Chatham in 1872. BU about 1901.

MF.114
K: 1.1.1856. L: 19.4.1856.
Became a barge at Chatham in 5.1872.

MF.115
K: 1.1.1856. L: 19.4.1856.
Became torpedo lighter (No. 2) on 19.7.1876.

MF.116
K: 1.1.1856. L: 24.4.1856.
Lent to War Dept for torpedo (i.e. mine) experiments 1872. Towed (by *Discovery*) to Devonport by AO 26.2.1889 to replace *Torpedo Lighter No. 5*.

MF.117
K: 1.1.1856. L: 24.4.1856.
Became a coal pontoon at Queenstown 5.1872 (by AO 11.11.1871).

MF.118
K: 1.1.1856. L: 8.5.1856.
Became pontoon to *Thunderbolt* at Chatham by 1875.

MF.119
K: 1.1.1856. L: 8.5.1856.
Became a lighter 7.1856, renamed *Iron Luggage Lighter No. 1*. Later to Sheerness as *YC.12*. BU about 1901.

MF.120
K: 2.1.1856. L: 17.5.1856.
Became a lighter 7.1856, renamed *Iron Luggage Lighter No. 2*. Later to Sheerness as *YC.13*. BU about 1901.

MF.121
K: 2.1.1856. L: 17.5.1856.
Became water tank vessel in 7.1856. To Woolwich as *YC.11* in 1866. To Sheerness as *YC.25* in 1872. BU about 1901.

MF.122
K: 2.1.1856. L: 21.5.1856.
Became a lighter 7.1856. Dummy pier at Gillingham in 8.1872. BU 1876.

MF.123
K: 2.1.1856. L: 4.6.1856.
Became iron luggage lighter *YC.8* at Woolwich in 1866; hulked in Chatham basin by 1879. To Pembroke as *YC.2* in 1886. Sold 1894.

MF.124
K: 2.1.1856. L: 8.5.1856.
To War Dept. on 15.7.1871.

MF.125
K: 2.1.1856. L: 5.7.1856.
Became torpedo lighter (*No. 1*) on 19.7.1876.

MF.126
K: 3.1.1856. L: 14.6.1856.
Became coal pontoon at Portland 7.1869. To Portsmouth as *YC.47* in 7.1879. BU about 1901.

MF.127
K: 3.1.1856. L: 14.6.1856.
To Chatham as *YC.53* in 5.1875. BU about 1901.

MF.128
K: 3.1.1856. L: 4.6.1856.
Became pontoon to police hulk at Chatham. BU about 1870.

MF.129
K: 3.1.1856. L: 30.8.1856.
Became coal pontoon at Portland in 1869.

MF.130
K: 3.1.1856. L: 5.7.1856.
Became a lighter, renamed *Iron Lighter No. 2* (date unknown). Later to Chatham as *YC.25*. BU about 1901.

MF.131
K: 3.1.1856. L: 5.7.1856.
Became coal pontoon at Portland in 7.1869. To Portsmouth as *YC.49* in 7.1879. BU about 1901.

MF.132
K: 6.1.1856. L: 5.7.1856.
Became pontoon at Chatham in 2.1868. BU about 1898.

MF.133
K: 6.1.1856. L: 5.9.1856.
Became a lighter, renamed *Iron Lighter No. 3* (date unknown). Later to Chatham as *YC.26*. BU about 1901.

MF.134

K: 6.1.1856. L: 5.9.1856.

'appropriated for the conveyance of steam machinery' at Woolwich 11.1856. Became coal pontoon at Portland in 7.1869. To Portsmouth as *YC.50* in 7.1879. BU about 1901.

MF.135

K: 6.1.1856. L: 10.9.1856.

Became a lighter, renamed ***Iron Lighter No. 4*** (date unknown). Later to Chatham as *YC.27*. BU about 1901.

MF.136

K: 1.1.1856. L: 6.3.1856.

Became tank vessel *YC.49* at Chatham in 5.1872. Later to Devonport as *YC.37*. BU about 1901.

MF.137

K: 1.1.1856. L: 6.3.1856.

Became harbour vessel *YC.12* at Woolwich in 8.1866. To Sheerness as *YC.26* in 1872.

MF.138

K: 1.1.1856. L: 11.3.1856.

Became a lighter, renamed ***Iron Lighter No. 5*** in 12.1856. Later to Chatham as *YC.28*. Became torpedo lighter (***No. 3***) on 19.7.1876. To War Dept. in 1.1886.

MF.139

K: 1.1.1856. L: 11.3.1856.

Became pontoon to *Thunderbolt* at Chatham in 1870. To War Dept. for torpedo experiments in 8.1871.

MF.140

K: 1.1.1856. L: 25.3.1856.

Became coal pontoon at Portland in 12.1869. To Portsmouth as *YC.43* in 7.1879.

MF.141

K: 1.1.1856. L: 25.3.1856.

Became harbour vessel *YC.34* at Malta in 10.1867. To Portsmouth as a barge *YC.42* in 10.1867.

MF.142

K: 1.1.1856. L: 26.3.1856.

Became a lighter, renamed *YC.29* at Portsmouth on 4.2.1867.

MF.143

K: 1.1.1856. L: 26.3.1856.

Became a lighter, renamed *YC.23* at Portsmouth on 4.2.1867.

MF.144

K: 1.1.1856. L: 5.4.1856.

Became mooring lighter *YC.4* at Malta in 9.1867.

MF.145

K: 12.2.1856. L: 9.4.1856.

Became a yard craft in 1864. Coastal defence in 1866. To Portsmouth as *YC.34* in 1866.

MF.146

K: 12.2.1856. L: 9.4.1856.

Became coal pontoon at Portland in 4.1868. To Portsmouth as *YC.44* in 7.1879.

MF.147

K: 12.2.1856. L: 14.4.1856.

Became a gunnery tender to HMS *Excellent* in 9.1866. BU about 1896.

MF.148

K: 12.2.1856. L: 17.4.1856.

Became a hopper barge 12.1861. To Portsmouth as *YC.13*.

MF.149

K: 12.2.1856. L: 17.4.1856.

Became coal pontoon at Portland in 4.1868 (by AO 14.9.1867). To Portsmouth as *YC.45* by AO 9.7.1879.

MF.150

K: 12.2.1856. L: 17.4.1856.

Became coal pontoon at Portland in 4.1868 (by AO 14.9.1867). To Portsmouth as *YC.46* by AO 9.7.1879.

11 Steam Paddle Vessels

James Watt patented the steam engine in 1769, and by 1815 Britain already had an estimated total of 210,000hp of steam engine capacity (according to French observer Baron Dupin), but attempts to install steam power afloat were limited, and the few early steamships were quite small. Following Admiralty experiments in 1794 with the prototype *Kent Ambinavigator*, little further was attempted until 1815 when a beam engine was ordered for installation in the river exploration gunboat *Congo*, designed by Sir Robert Seppings. The first steam-propelled vessel ordered for the RN, 'to be built for the purpose of exploring the interior of Africa by the River Congo', she never served as such. On trials under steam in the Thames, she was found to be too weak to carry the steam engine which was intended for her, so in 1816 the engine was removed and she was fitted out as a sailing vessel (a three-masted schooner).

From the earliest days of the steam era, it was recognised by the Navy Board and the Admiralty that the constant vibration of a ship's engines would need much more robust construction to cope with the shaking that steam ships experienced. Consequently all paddle steamships (and later screw-driven steamships) were built with exceedingly strong construction. A result of this was that such vessels lasted, on average, longer than their pure sailing equivalents. As this was a period of rapid evolution in ordnance, most vessels underwent multiple re-arming, and the data below is thus not exhaustive but an attempt to illustrate changes for each type's weaponry.

(B) Vessels acquired from 1 January 1817

In 1821 the Admiralty ordered its first purpose-built steam vessel – the *Comet* – for use as a tender for towing its warships out of becalmed harbours. Further vessels followed, and their use soon expanded to take on a variety of tasks, including the employment of paddle vessels to explore African and Asian rivers. During the First Anglo-Chinese (or 'Opium') War, Adm. Sir William Parker used Indian naval steam ships to tow British warships into Chinese rivers, where they could bombard major forts and land British troops to cut the Grand Canal.

Steam engines became more powerful and reliable over the following decades, but a major problem was that paddle boxes along a ship's sides, and the paddlewheels within them, were vulnerable to disabling cannon fire in battle, while the space they occupied prevented the placing of guns in a broadside battery. In 1836 the first screw propellers were patented independently by Francis Smith and John Ericsson, and following expensive trials by the Admiralty, the first British screw-propelled warships were ordered (see later chapters of this book). These quickly spelled the demise of the paddle warship.

***COMET* Class 1821** Oliver Lang design; Lang was the master shipwright at Woolwich, and he adapted the lines of a sailing sloop.

Lang was responsible for the vast majority of the early designs for steam paddle vessels. The *Comet* was the first effective steamer built for the RN 'to be employed in towing HM Ships in the Thames and Medway' (although in practice her towing duties took her much farther afield). Carried a two-masted schooner rig. Not on the Navy List until 9 October 1828. In 1835 she was fitted with iron plates containing quicksilver in the boilers. From January 1837 classed as a Fifth Class steam vessel (SV5), but re-classed as a tender from 1844.

> Dimensions & tons: 115ft 0in, 101ft 2⅞in x 21ft 3in (21ft 0in for tonnage) x 11ft 11in. Draught 7¼ft (design), 9¼ft (actual). 238bm. 239 disp.
> Men: 16 (1827). Guns: (orig.) 2 x 6pdrs (6cwt) brass; (later) 4 x 9pdrs (13½cwt).
> Machinery: 2-cylinder side lever engine (operating at 4lb/sq.in.). 80nhp. 7.6kts.

Comet Deptford Dyd/Boulton, Watt & Co.
> Ord: 10.11.1819. K: 21.11.1821. L: 23.5.1822. C: 13.7.1822 (for sea).
> First cost: hull £8,052, fitting (masts & yards, stores) £1,313, machinery £5,050.
> At Deptford from 1822, and at Woolwich 1834. In 23.4.1836 under Lieut (Cmdr 6.1.1837) Robert Otway, off the coasts of Spain and Portugal; on 1.7.1837 under Lieut George Thomas Gordon. Refitted at Deptford as a Surveying vessel (for £3,195) 1842, to serve on the coast of Ireland; on 15.7.1842 under Cmdr George Alexander Frazer; then 12.7.1844 Lieut William Pretyman, on 29.6.1846 under Lieut Charles Richardson Johnson, then 25.10.1847 Lieut Charles Gray Rigge; surveying west coast of Scotland 1849–53; on 7.5.1850 under Cmdr Henry Charles Otter (–1853), at Cork, then back to Scottish survey. Fitted as surveying vessel, then as tug at Portsmouth 1854. BU from 10.12.1868 at Portsmouth Dyd (No. 6 Dock).

***LIGHTNING* Class 1822** Oliver Lang design. Ordered to meet Admiralty need for steam tugs for Plymouth and Portsmouth respectively. From January 1837 classed as Fifth Class steam vessels (SV5), but *Meteor* was re-classed as a tender from 1844. *Lightning* was used as a surveying vessel during the Crimean War, and by the 1860s was classed as a Second Class paddle gunvessel. In 1830 *Meteor* began the steam mail service to Malta and Corfu. Her re-engining was with *Tartarus*'s engine (see below).

> Dimensions & tons: 126ft 0in, 111ft 9¾in x 22ft 8in (22ft 4in for tonnage) x 13ft 8in. 296bm. 349 disp. Draught 9ft 0in (fwd), 9ft 6in (aft).
> Men: 20 (16 from 1827). Guns: 2 (later 3) x 6pdrs (6cwt) brass.
> Machinery: 2-cylinder side lever. 100nhp (2 engines of 50nhp each). 11kts.

Lightning Deptford Dyd/Maudslay, Sons & Field.
> Ord: 26.12.1821. K: 2.1823. L.19.9.1823. C.18.12.1823 (for sea).
> First cost: £15,744 (including £5,350 for machinery).
> Commissioned 4.12.1827 under Lieut George Evans. Reboiled at Woolwich 1828. On 4.6.1828 under Lieut George Hutchings, then 22.8.1828 under Lieut George Bissett. Reboiled and refitted at Woolwich (for £3,146) 1833. On 23.4.1834 under Thomas Allen, Master, then 1835? under Cmdr Edward

Like many of the early paddle vessels, the *Lightning* was extremely robust in construction, as the Admiralty recognised that the side-lever engines would cause excessive vibration that would quickly weaken a ship built with normal scantlings. Consequently she had a long and active life, for much of which she was used for surveying. In 1854 she was deployed to the Baltic as part of Napier's fleet, and undertook a survey of the defences of Bomarsund, in the Aland Islands, during early June; she returned to the Baltic with Dundas's fleet in 1855, and participated in the attack on Sveaborg. This modern model depicts the ship as she appeared in 1855–56. (*NMM F2938-002*)

Belcher, surveying Irish Channel. Reboiled at Woolwich 1836. On 4.3.1836 under Lieut James Shambler. On 30.11.1838 recommissioned under Lieut Richard Nicholls Williams, paid off 9.1840. Some time in 1842 under G. Snell, Master. Recommissioned 5.11.1842 under Lieut William Robert Wolseley Winniett, at Woolwich. Re-classed as a tender and employed as a tug at Woolwich and on the Thames from 10.1843, under William Roberts, Master, from 29.10.1843. Reboiled at Woolwich (for £3,854) 1845. On 11.1.1845 under John Eaton Petley, Master, then 7.2.1851 Henry William Allen, Master. Refitted at Woolwich (for £8,297) in 1853–54 and 1855. On 25.2.1854 under Capt. Bartholomew James Sulivan, in the Baltic, for survey of the Lumper Channel at Bomarsund. On 12.3.1855 under Capt. James Carter Campbell, still in the Baltic, as tender to *Duke of Wellington*. From 1858 under George Williams, Master, as tender to *Saturn* at Pembroke Dock. Reboiled (with *Viper*'s boilers) 1864, re-classed as 2nd Class steam gunvessel. Recommissioned 5.4.1864 under Staff Cmdr Timothy W. Sulivan, as tender to Saturn at Pembroke Dock. On 2.1.1865 under Capt. Edward James Brooker, as tender to *Fisgard*, for surveying West coast of England; in 1868 under Daniel J. May, Master, for Atlantic soundings. In 4.1870 under John Richards, Master, as tender to *Nankin*, for surveying West coast of England. BU 1872 at Devonport.

Meteor Deptford Dyd/Boulton & Watt.
Ord: 26.12.1821. K: 2.1823. L.17.2.1824. C.10.6.1824.
First cost: hull & fitting £10,427, machinery £5,500.

Commissioned 4.12.1827 under Lieut George James Hay. On 20.9.1828 under Lieut William Henry Symonds; to the Mediterranean 1832. On 16.12.1834 under Lieut John Duffill. On 3.7.1835 under Lieut George Woodberry Smith, in the West Indies; paid off 7.1837 at Woolwich. Re-engined there (for £3,144, with engines ex-*Tartarus* and new boilers) 1837. Recommissioned 13.2.1838 under Lieut Richard Davison Pritchard, as tender at Falmouth; paid off 14.9.1840 at Woolwich. Refitted there as a tug 1841, and recommissioned 14.2.1841 under Lieut John Waugh. In 9.1841 under Cmdr Michael Atwell Stater, for surveying North Sea and Scottish waters; paid off 12.1841 at Woolwich. Recommissioned 13.7.1842 under Lieut George Buttler, at Sheerness, for the Mediterranean. Reboiled and refitted at Woolwich (for £4,328) 1845, then back to Mediterranean (still under Buttler); paid off 2.11.1848 at Sheerness. BU there 8.1849.

DIANA East India Company vessel, the ex-mercantile *Diana*, launched 12 July 1823 by Kyds, Kidderpore (Calcutta), was operated (and possibly purchased) at the suggestion of Commander Frederick Marryat (the author) and put to riverine use in the First Anglo-Burmese War – the RN's first employment of steam power in combat.
Dimensions & tons (approx): 100ft 0in, 90ft 0in x 16ft 8in x — . 133bm.
Machinery: 64nhp.
Transferred to Burmese government 1826 (following the end of that War); BU 1836.

ALBAN **Group 1824** All six of these vessels were originally ordered 25 March 1823 as *Cherokee* Class sailing brigs (see Chapter 8), but not begun as such. They were re-ordered as paddle steamers in May 1824 and so completed, but with three variations in basic design as noted below. Initially rated Steam Vessels (SV5 Class from 1837), from 1844 rated tenders.

First design:
Dimensions & tons: 109ft 8in, 91ft 11⅛in x 24ft 10in oa (24ft 6in for tonnage) x 13ft 6in. Draught 11ft 2in (fwd) 12ft 4in (aft). 293⁶⁰/₉₄bm. 487 disp.
(*Alban* lengthened 1839–40: 145ft 0in, 127ft 6in x 24ft 10in oa (24ft 6in for tonnage) x 13ft 6in. 407bm.)
Men: 16–25 (1827), later 33. Guns: (orig) 2 x 6pdrs (6cwt) brass; (later) 4 x 18pdrs (10cwt) + 4 x 18pdr carronades.
Machinery: 2-cylinder side lever. 100nhp. 7.6kts.

Alban Deptford Dyd/Boulton & Watt (initially 80nhp)
Ord: 5.1824. K: 7.1824. L: 27.12.1826. C: 6.5.1827.
First cost: £12,293, +£4,950 for engine.
Tender at Deptford from 1827, for towing and transport duties. On 11.6.1830 under Lieut Thomas John James Davis, at Woolwich (died 27.8.1831); on 29.8.1831 under Lieut Henry Walker. Re-engined at Woolwich (for £4,403) with new B&W 100nhp engine installed 3–4.1831. On 27.11.1832 under Lieut Andrew Kennedy, for the Mediterranean and later to West Indies; navigated Orinoco upriver as far as Angostura. On 17.6.1834 under Lieut Juste Peter Roepel, in the Mediterranean, then to North America and West Indies, then 17.6.1836 under Lieut Edward Burnaby Tinling; paid off 25.6.1839. Rebuilt at Chatham (for £6,908) 1839–40, lengthened to 145ft, 127½ft (= 407bm), then to Woolwich where fitted with *Carron*'s engine and new boilers (for £8,319), then back to Chatham where fitted for trooping (for £1,127). On 1.6.1840 at Portsmouth under John King, Acting Master, as transport. Recommissioned 8.3.1843 under Lieut John Jeayes, for particular service, then 16.8.1844 under Lieut Frederick Lowe, and 5.9.1845 under Lieut ?White. On 17.10.1845 under Jonathan Aylen, Master, for trooping; on 29.6.1846 under Henry Burney, Master, then on 6.11.1846 under Manser Bradshaw, Master; paid off 16.2.1849 at Woolwich. From 8.1850 at Jamaica, as tender to *Imaum*; returned to Woolwich 8.1853 and paid off 13.8.1853. Refitted and re-coppered 9.1853–3.1854. Recommissioned 6.3.1854 under Lieut Edward Burstal, then 25.4.1854 under Cmdr Henry Charles Otter, for surveying in the Baltic; on 1.1.1855 under Cmdr Lacon Usher Hammett, at Woolwich; on 12.3.1855 under Lieut William Edward Fisher, at Sheerness. Repairs at Sheerness 1–2.1856 following ice damage to paddle wheels; on 19.5.1856 under Lieut William Barnard De Blaquiere, as tender to Port Admiral at Portsmouth; paid off 3.4.1858 into Steam Reserve at Portsmouth. BU there 5.1860.

Carron Deptford Dyd/Boulton & Watt (100nhp)
Ord: 5.1824. K: 7.1824. L: 9.1.1827. C: 31.5.1827.
First cost: £10,853 including fitting (plus machinery £5,648).
Tender at Portsmouth from 1828, for towing and transport duties. Paddle boxes lengthened at Portsmouth 4.1829. Fitted there as a packet 3–5.1830. Commissioned 4.6.1830 under Lieut William Frederick Lapidge, for Home waters as packet; paid off 11.1831 at Woolwich for repairs. Recommissioned 17.10.1832 under Lieut John Duffill, as a packet; to the Mediterranean 1834. On 11.3.1835 under Lieut William Dow, for Spain and then to the West Indies; on 25.11.1836 under Lieut Edward E. Owen (died 2.1838); at Woolwich 1838. Recommissioned 7.3.1838 under Lieut John Bettinson Cragg; to Woolwich 6.1838, paid off

6.7.1838. Fitted with new paddle-box (designed by Capt. George Smith) 1839, used as tender/tug at Plymouth. Fitted at Woolwich as a tug (for £1,028) for Bermuda 8–9.1841. At Bermuda as local tender 10.1841–5.1843. Fitted as coal hulk 2.1844; lent as coal hulk at Woolwich 1846. Breakwater 1848 at Harwich. Deleted 6.1877. BU 1.1885 at Devonport.

African (ex-*Dee*, renamed 5.1825) Woolwich Dyd/Boulton & Watt (80nhp)
Ord: 5.1824. K: 9.1824. L: 30.8.1825. C: 24.9.1825.
First cost: £10,029 including fitting (plus machinery £8,290).
In 5.1825 it was ordered that she be fitted for colonial service in Africa (and was renamed accordingly) and in 2.1826 she was delivered to the Colonial Department, sailing for Sierra Leone under Lieut ?Horatio Austen; home to Portsmouth 1.5.1828. In 9.1828 under Lieut Andrew Kennedy, as the first government steamer stationed between Corfu and Ancona. At Deptford in 1829; believed to be returned to Navy in 1.1831. Commissioned 3.2.1831 under Lieut James Harvey, as Mediterranean packet. On 4.6.1834 under Lieut Joseph West, as a Falmouth packet. On 18.7.1837 under Capt. Frederick William Beechey, as survey ship for Irish coast (?in 8.1837 under James Hogg); paid off 1840. Fitted as tug at Woolwich (for £3,069) 4.1840–12.1842. On 14.7.1842 under Alex-Mackey, Master, as tug at Woolwich; on 8.3.1843 under John King, Master, as tug at Sheerness in 1843–49. BU at Sheerness 12.1862.

Second design:
Dimensions & tons: 111ft 8in, 92ft 3⅜in x 24ft 10in oa (24ft 6in for tonnage) x 13ft 6in. 294⁷¹/₉₄bm.]

Echo Woolwich Dyd/Boulton & Watt (100nhp)
Ord: 5.1824. K: 12.1824. L: 28.5.1827. C: 22.12.1827.
First cost: £12,468 including fitting (plus machinery £7,047).
Commissioned 4.12.1827 under Lieut Frederick Bullock, for surveying Thames. In 10.1829 under Lieut George Bissett. On 22.7.1830 under Lieut Robert Otway (–end 1833); boilers (ex-*Confiance*) installed at Woolwich (for £2,065) 12.1830–2.1831; at Oporto 1832–33. Engines and boilers removed at Woolwich 10.1833, but she was repaired and new machinery fitted during hull refit there (for £3,129 + £1,966) 7.1835–1.1836. Recommissioned 25.11.1836 under Lieut William James; at Sheerness 2.1837, then to North America and the West Indies; paid off 1839. Fitted as tug at Woolwich (for £4,157) 7.1839–5.1840; used as tug at Portsmouth from 1841. Sold to Castle & Son, and towed 25.6.1885 to BU.

Confiance Woolwich Dyd/Maudslay, Sons & Field (100nhp)
Ord: 5.1824. K: 2.1825. L: 28.3.1827. C: 3.6–12.1827.
First cost: £11,354 including fitting (plus machinery £7,047)
Commissioned 3.1828 under Lieut Robert James, at Chatham. On 30.5.1829 under Lieut William Richardson. Repaired and fitted with Morgan's paddle wheels at Woolwich (for £2,214) 10.1829–2.1830. On 30.8.1830 under Lieut Henry Fage Belson (–4.1832), as a Falmouth packet. On 12.4.1834 under Lieut John Middleton Waugh, again as Falmouth packet; on 25.11.1836 under Lieut William Arlett, for the Mediterranean. On 20.2.1838 under Lieut Edward Stopford, on same station; later in 1838 under Lieut John Henn Gennys; actions off Syrian coast 1840. Fitted as tug at Woolwich (for £4,354) 6.1841–4.1842, and employed as such at Plymouth. In 4.1842 under J. Jagos, Master; on 19.1.1849 under William Martin, Master. BU 6.1873 at Devonport.

Third design:
Originally as *Alban*, but lengthened during construction. Initially rated Steam Vessel (SV4 Class from 1837).

Dimensions & tons: 130ft 6in, 111ft 3½in x 24ft 9¼in x 13ft 7½in. 361bm. 504 disp. Draught 10ft 5in (fwd) 11ft 7in (aft).

Men: — . Guns: 4 x 18pdrs; 4 x 18pdr carronades.

Machinery: 2-cylinder side lever. 120nhp. 8kts.

Columbia Woolwich Dyd/Maudslay, Sons & Field

Ord: 5.1824. K: 3.1827. L: 1.7.1829. C: 29.7.1830 (as a packet).

First cost: £14,928 to build, plus £2,541 fitting, plus £7,158 machinery

Commissioned 7.1830 under Lieut Robert Ede, as a Falmouth packet. Engine removed at Woolwich 12.1832; new Boulton & Watt 100nhp engine (cost £6,080) and Morgan's paddle wheels (£1,360) installed at Woolwich (for £2,087) 1–5.1833. On 4.2.1834 under Lieut Benjamin Aplin, as Falmouth packet. Fitted as troop steam vessel at Woolwich (for £3,297) 8–11.1834. Recommissioned 20.8.1834 under James Henderson, Master; sailed for the West Indies 25.12.1834. On 24.3.1838 under Alexander Thompson, Master; trooping in North America and West Indies 1838. Refitted at Woolwich (cycloidal paddle wheels fitted) 9.1839–1.1840. On 2.4.1841 under Benjamin Robinson, Master, on same station. Fitted as survey vessel 1842; recommissioned 4.5.1842 under Lieut Alfred Kortright, then 14.12.1842 under Lieut John Harding, for North America. On 3.12.1844 under Lieut Peter Frederick Shortland; on 22.4.1847 under Capt. William Frederick Owen. Between small and middling repair at Chatham (for £5,955) 8.1848–5.1849, then 2.4.1849 under Shortland (now Cmdr) again, all surveying on the North American station; to St Johns, New Brunswick 1852; paid off 16.10.1857 at Halifax, Nova Scotia. Coal hulk 1857. Engines sold (for £500) 5.1858, boilers to England aboard *Gorgon* (value £2,000) 9.1858; hull sold (for £500) 29.10.1859 at Halifax, Nova Scotia to Lewis R. Fairbanks.

DEE Built as new (not converted on stocks from sailing vessel, as mistakenly recorded in some sources). Design by Seppings, modified by Oliver Lang.

Dimensions & tons: 166ft 7in, 146ft 6in x 30ft 4½in (30ft 0½in for tonnage) x 16ft 4in. Draught 11ft 6in (fwd & aft). 704bm. 907 disp.

Men: — . Guns: 2 x 18pdrs (22cwt); later 6 x 32pdrs (4 x 63cwt; 2 x 56cwt), all pivot-mounted; the 2 x 56cwt guns were later replaced by 1 x 10in (86cwt) gun.

Machinery: 2-cylinder side lever. 200nhp. 272ihp = 8kts. (new 220nhp engine in 1866)

Dee Woolwich Dyd/Maudslay, Sons & Field

Ord: 4.4.1827. K: 10.1829. L: 5.4.1832. C: 26.8.1832.

First cost: £19,275 including fitting; machinery cost £11,261.

Commissioned 9.6.1832 under Cmdr Robert Oliver, for the blockade of the Dutch coast. On 5.11.1833 under Cmdr Edward Stanley, on the Home station, then 28.5.1834 under Cmdr William Ramsay, for the West Indies; paid off 4.1837. Recommissioned 2.1838 under Cmdr Joseph Shearer, for South America; paid off 5.1841. Fitted as transport at Sheerness (for £6,939) 12.1841–6.1842, then at Woolwich (for £5,461) 6–9.1842. Recommissioned 26.5.1842 under Thomas Driver, Master; on 18.5.1848 under George Filmer, Master, for Cape of Good Hope; on 14.9.1852 under Lieut George Thomas Cleather Smith, on same station. Fitted as storeship at Portsmouth (for £860) 11–12.1854. Recommissioned 23.11.1854 under Thomas C. Pullen, Master; paid off 1860. On 12.6.1863 under George Raymond, Master, as a storeship. On 14.10.1869 under George Waters, Master, at Woolwich; paid off 17.6.1871 at Sheerness. BU there 10.1871.

Until the end of 1827 none of the steam vessels were commissioned into the Navy. The Admiralty issued instructions that the above vessels be registered on the Navy List and commissioned on 4 December 1827 (*Lightning*, *Meteor* and *Echo*) and 9 October 1828 (*Comet*, *Alban*, *Carron*, *Confiance*, *Columbia* and *Dee*). *African*, being under control of the Colonial Department, was not registered until 28 January 1831.

VICTORY Paddle packet, purchased 1828

Dimensions & tons: — . 85bm. 150 disp.

No record of service. Abandoned 1832.

MESSENGER Class Purchased vessels. Originally built for the General Steam Navigation Company, these two vessels were purchased from Joliffe & Banks as wooden paddle packets for Corfu route, but quickly re-rated as sloops. *Messenger* was lengthened by 3½ft and re-engined with a Maudslay 200nhp engine (costing £12,560) in 1831; her old engine was then installed in the new *Hermes* (by AO 20 April 1840).

Dimensions & tons: 155ft 6in, 133ft 4in x 32ft 9in oa (32ft 3in for tonnage) x 12ft 0in. Draught 10ft 3in (fwd), 10ft 9in (aft). 733⁴⁴/₉₄bm. 912 disp. *Messenger* as lengthened 159ft 0in, 137ft 3¼in (759³³/₉₄bm).

Men: 30. Guns: 1 x 12pdr carronade.

Machinery: 2-cylinder side lever. 80nhp.

Messenger [mercantile *Duke of York*, built by Benjamin Wallis at Blackwall, 1824]

Authorised (to purchase): 21.5.1830. Fitted at Woolwich as a packet 13–28.5.1830. Purchased 20.8.1830. Became sloop 1831.

First cost (purchase): £12,489.

Commissioned 1830 under Lieut William Frederick Lapidge. On 20.5.1830 under Lieut Benjamin Aplin, as a Falmouth packet. In 5.1834 under John King, Master (–1839), as a transport. Fitted as coal depot by AO 27.7.1840 (engine removed) at Woolwich 5–12.1840. Sold to Henry Castle & Son (for £360, by AO 22.11.1861) to BU at Charlton.

Hermes [mercantile *George the Fourth*, built by Benjamin Wallis at Blackwall, 1824, as *Courier*]

Authorised (to purchase): 24.6.1830. Purchased 20.8.1830. Became sloop 1831.

Commissioned 24.6.1830 under Lieut Andrew Kennedy, as a Falmouth packet. On 19.10.1832 under Lieut John Wright, at Portsmouth; paid off 30.11.1833. Renamed *Charger* 22.1.1834, as a coal hulk at Woolwich. BU 6.1854.

PLUTO Class Steam vessel (from 1837 rated SV4 as a First Class gunvessel). Oliver Lang design, approved 19 February 1831. Shallow-draught vessel intended to combat piracy in the Caribbean, but she served chiefly in West African anti-slavery duties.

Dimensions & tons: 135ft 0½in, 118ft 8½in x 24ft 2½in oa (24ft 0½in for tonnage) x 11ft 10in. Draught 6½ft. 365bm. 386 disp.

Men: 80. Guns: 2 x 18pdrs (22cwt); later 1 x 32pdr (42cwt) pivot-mounted + 1 x 18pdr (22cwt).

Machinery: 2-cyliunder side lever. 100nhp. 7½kts.

Pluto Woolwich Dyd/Boulton, Watt & Co.

Ord: 9.11.1830. K: 2.1831. L: 28.4.1831. C: 9.9.1831 (for sea).

First cost: £17,432 (including £5,465 for the machinery).

Commissioned 20.5.1831 under Cmdr George Buchanan (died 1833), for the Cape of Good Hope. On 18.3.1833 under Cmdr Thomas Ross Sulivan; paid off 12.12.1834. Refitted at Woolwich with *Columbia*'s repaired boilers from store; fitted to carry British Ambassador to Persia, disarmed. Recommissioned

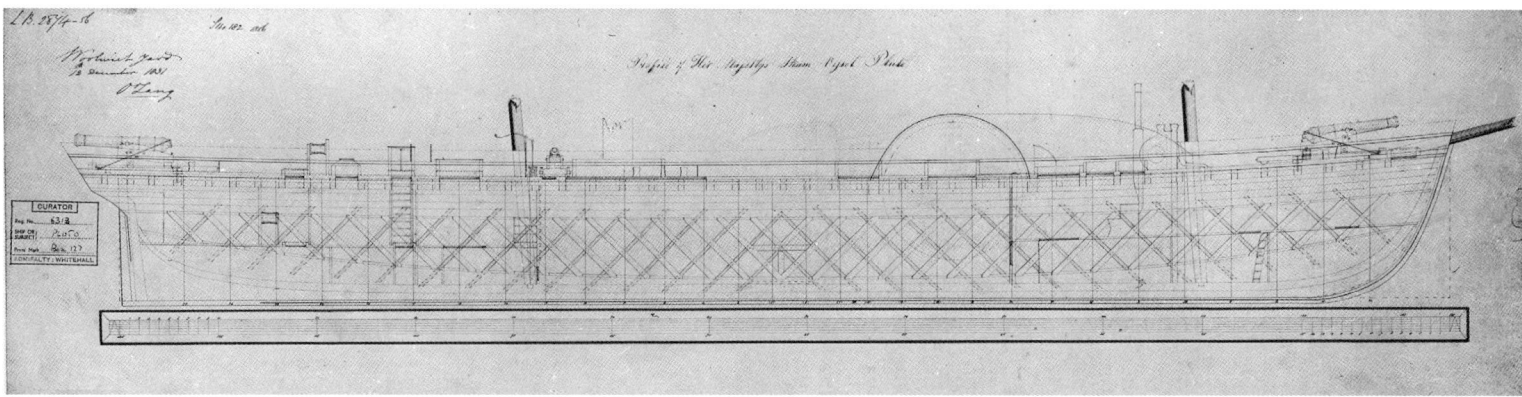

This profile plan, dated 18 December 1831 at Woolwich, is for the *Pluto*, ordered in 1831 to a design by Oliver Lang as a shallow-draught (6½ feet) vessel intended to combat piracy in the West Indies, although she spent much of her career off the coasts of Africa and South East Asia. Such a long, low hull – especially one that had to withstand the vibration of steam machinery – would have been impossible before the advent of Seppings's structural innovations, so it is fitting that the diagonal iron strapping forms such an eye-catching element of the draught. *(NMM DR6313)*

1.4.1835 under Cmdr John Duffill (–15.4.1835), delivering Ambassador and his suite to Trebizond, then surveying Crimea and Black Sea. Recommissioned as a packet 21.6.1836; at Lisbon 2.1837. On 10.8.1838 under Lieut John Lunn, on North American and West Indies station. Refitted at Woolwich (for £6,417) 1840–41. Recommissioned 1.1.1841 under Lieut William Simpson Blount, for the west coast of Africa; took slavers *Paz* 27.10.1841 and *Balurca* 20.2.1842; paid off 17.6.1842 at Woolwich. Refitted at Woolwich 1842–43. Recommissioned 19.12.1843 under Lieut William Pearson Crozier, for Irish waters; on the 'West India and Home' stations. On 21.6.1845 under Lieut Frederick Lowe, in Irish waters; famine relief 1845–46; paid off 1.9.1847 at Woolwich. Recommissioned 29.2.1848 under Lieut Richard McKinley Richardson, for west coast of Africa (died 2.11.1848); on 3.11.1838 under Lieut William Kynaston Jolliffe (temp. at first, confirmed 16.1.1849, on same station; took slavers *Quatro Andhorina* 5.11.1848, *Merea* 22.12.1848, *Tentativa Feliz* 12.4.1849, *Casco* 28.11.1849, *Rowena* 11.1.1850, *Anne D. Richardson* 14.2.1850, and *J. W. Huntingdon* 16.3.1850; paid off 19.11.1850 at Woolwich. Refitted there 1851. Recommissioned 4.3.1852 under Lieut Henry West, for the East Indies; in Second Burmese War, as flagship of Rear-Adm. Charles John Austen (who died 7.10.1852); on 10.2.1853 under Lieut Norman Bernard Bedingfeld (invalided 8.1854), then 12.1854 under Lieut Henry Augustus Clavering; paid off 11.8.1855. Refitted at Deptford 1855–56. Recommissioned 3.12.1856 under Lieut William Swinburn, for west coast of Africa; on 19.4.1858 under Lieut Cortland Herbert Simpson, all on same station; paid off 12.5.1860 at Woolwich. BU completed 26.3.1861 at Sheerness.

(C) Vessels acquired from November 1830

On 10 January 1831, within two months of taking office, the new First Lord Sir James Graham gave orders for four paddle vessels to be built in the dockyards to competitive designs, one (*Phoenix*) by the Surveyor and the others by different master shipwrights. Among the largest and most powerful steamers of that date, all four were to have Maudslay engines of 220 HP, carry a schooner rig (later changed to barque or barquentine) and mount one or two 10in shell guns of 80cwt (in fact, 84cwt was the model fitted). They were named on 22 March 1831. With the existing *Dee* they provided the Navy with unrivalled early experience in operating steam warships. All were re-rated Second Class Sloops by AO of 31 May 1844 except the *Rhadamanthus* and *Dee*, which became transports until 1846 when they likewise became Second Class sloops.

SALAMANDER Steam vessel, later Second Class sloop. Joseph Seaton design, approved 1831. Originally to have been built at Portsmouth (Seaton was the master shipwright at Sheerness 1826–30).
   Dimensions & tons: 175ft 5in, 151ft 8¼in x 32ft 2in oa (31ft 10in for tonnage) x 17ft 0in. Draught 12ft 6in (fwd), 13ft 6in (aft). 818bm. 1,014 disp.
   Men: 135. Guns: 2 x 10in (84cwt) shell guns on pivots, 2 (later 4) x 32pdrs (25cwt). In 1862 one of the 10in shell guns was replaced by a 110pdr rifled BL gun, also pivot-mounted.
   Machinery: 2-cylinder side lever 220nhp. 506ihp = 7.2kts.
Salamander Sheerness Dyd/Maudslay, Sons & Field
   Ord: 12.1.1831. K: 4.1831. L: 14.5.1832. C.12.2.1833.
   First cost: £34,224 (builder £20,429, machinery £11,201, fitting £2,704).
   Commissioned 27.11.1832 under Cmdr Horatio Thomas Austin, at Woolwich. On 15.2.1834 under Cmdr William Langford Castle, for the Channel. On 15.4.1836 under Cmdr John Duffill, then on 16.8.1836 under Cmdr Sidney Colpoys Dacres, for particular service on the north coast of Spain; on 16.9.1840 under Cmdr Hastings Reginald Henry; paid off 11.8.1841. Recommissioned 24.6.1842 under Cmdr Andrew Snape Hamond, for South America, then the Pacific in 1842; repaired at Jamaica 2.1847 (new mainmast and bowsprit); paid off 11.1847 at Woolwich. Refit at Sheerness, then to Woolwich and 1.1849 to Plymouth for Steam Reserve (used as tug/tender). Recommissioned 17.7.1850 under Cmdr John Spencer Ellman, for the East Indies; home in 8.1854, then 19.8.1854 under Cmdr Benjamin Portland Priest, in the Mediterranean; paid off 23.11.1854 at Portsmouth into Steam Reserve. Recommissioned 6.11.1855 under Cmdr George Frederick Mecham, for the west coast of Africa; home in 6.1856; used as transport, and sent to 64° N to search for missing merchantmen from Archangel; to Sheerness 2.1857 for repairs to ice damage, but found unfit and paid off 4.2.1857. Repair and refit (reboilered) at Chatham 1857–58, then to Medway Steam Reserve. At Woolwich in 1860. 'Very extensive repair' at Chatham 1863 (machinery overhauled, new poop and masts). Recommissioned 8.12.1863 under Cmdr John Carnegie, for

A photograph of *Salamander* in later life, with a barque rig. One of the first true paddle fighting ships, *Salamander*'s long career included commissions on some of the most distant stations – the East Indies, South America and the Pacific – which demonstrates more confidence in the new technology that the Admiralty is usually credited with. She, and most early paddle sloops, initially carried two 10in shell guns on pivot mountings, but in 1862 one of these was replaced by a 110pdr rifled breech-loader in most of the surviving early vessels. In April 1852 the *Salamander* took part in operations in the Rangoon River during the Second Burma War, and in 1873, by now used as a tug, it was she who towed the ancient frigate *Unicorn* to what was to become the latter's final home in Dundee. *(NMM A0501)*

Australia; on 11.7.1865 under Cmdr George Strong Nares, in Australia; sailed for home 4.7.1867 and paid off 12.1867 into Steam Reserve. At Sheerness 1869. During 1870 under Cmdr George B. F. Swain and in 1872 Cmdr John C. Solfleet (both intermittently upon employment as transport). Recommissioned 9.4.1872 as transport under Cmdr Edward Youel, then 23.12.1872 under Cmdr James Kiddle; paid off 20.7.1875 into Steam Reserve at Portsmouth. Sold to Castle 10.1883 and towed away 19.12.1883 to BU.

PHOENIX Steam vessel, later Second Class sloop. Robert Seppings design, approved 1831. The projected name was altered on 22 March 1831 to *Charon*, but restored to *Phoenix* on 8 April (before being laid down). Seemingly a second vessel was ordered in May 1831, to have taken the name *Charon*, but this was cancelled in November 1832 without being laid down. The *Phoenix* was built in a dry dock. In 1844–45 she was converted to a screw sloop (later re-classed as corvette); the conversion to screw at Deptford cost £18,663 (the new Penn engine cost £13,576); her old engine and boilers were fitted to *Firefly*. In 1855 she was fitted for service in the White Sea (against the Russian Arctic coast).

Dimensions & tons: 174ft 7in, 153ft 10⅜in x 31ft 10in oa (31ft 6in for tonnage) x 16ft 9in. Draught 12ft 0in (fwd), 12ft 6in (aft). 812bm. 1,024 disp.

Men: 135. Guns: 1 x 10in (84cwt) shell gun on pivot, 1 x 8in (52cwt) on pivot, 4 x 32pdrs (17cwt).

Machinery: 2-cylinder side lever 220nhp.

Phoenix Chatham Dyd/Maudslay, Sons & Field.

Ord: 12.1.1831. K: 5.1831. L (undocked): 25.9.1832. C: 7.1.1834 (at Chatham), 27.9.1835 (at Woolwich).

First cost: Building £14,714, machinery £10,781, fitting £7,968.

Commissioned 6.11.1833 under Cmdr Robert Oliver, for the Channel; to Rotterdam with Queen Adelaide in summer 1834; on 9.9.1835 under Cmdr William Honyman Henderson, for the north coast of Spain. On 19.11.1836 under Capt. Lord John Hay,

off the north coast of Spain. In 7.1838 under Cmdr Anthony W. Milward; on 20.7.1839 under Cmdr Robert Spencer Robinson, in the Mediterranean. On 1.3.1840 under Cmdr Robert Fanshawe Stopford, for the Mediterranean; at bombardment of Acre 3.11.1840; on 26.12.1840 under Cmdr John Richardson; paid off 1.11.1842. Altered to screw sloop 4.1844–2.1845 (see Chapter 12 for later history).

Fitted as Arctic storeship 1851–53. Sold to Castle 26.1.1864 to BU at Charlton.

RHADAMANTHUS Steam vessel, later Second Class sloop. Thomas Roberts design, approved 1831 (Roberts was the master shipwright at Plymouth 1813–15 and 1830–37). On launch, the ship was sailed (jury-rigged) to Woolwich where machinery was fitted and completed. The first RN steamer to cross the Atlantic. Poop removed during 1836 refit.

Dimensions & tons: 164ft 7in, 143ft 2in x 32ft 10in oa (32ft 8in for tonnage) x 17ft 10in. Draught 11ft 0in (fwd) 13ft 0in (aft). 813bm. 1,086 disp.

Men: 135. Guns: 1 x 10in (84cwt) on pivot, 2 x 32pdrs (25cwt), 2 x 6pdrs (6cwt) brass. Later the brass guns were removed and a 32pdr (42cwt) pivot-mounted gun added.

Machinery: 2-cylinder side lever. 220nhp. 385ihp = 10kts.

Rhadamanthus Plymouth Dyd/Maudslay, Sons & Field

Ord: 12.1.1831. K: 9.1831. L.16.4.1832. C: 2.11.1832.

First Cost: £31,919 (building £18,534, fitting £2,197, machinery £11,188).

Commissioned 4.10.1832 under Cmdr George Evans, for blockade of the Dutch coast, then to North America and the West Indies; paid off 21.4.1835 at Woolwich. Major refit at Woolwich 1836. Recommissioned 23.10.1836 under Cmdr John Duffill, as a packet vessel for the coast of Spain. On 13.7.1837 under Cmdr Arthur Wakefield, for the Mediterranean; paid off 20.10.1840. Recommissioned 28.8.1841 under Thomas Laen, Master, at Woolwich after fitting as a transport at Sheerness. On 29.6.1846 under Jonathan Aylen, Master; paid off 13.2.1849 at Woolwich.

This full hull 1:48 scale model of the paddle sloop *Rhadamanthus* was made in 1832 by the ship's designer, Thomas Roberts, but may have been altered later as it does not show the original poop, which was removed in 1836. More deck detail is depicted than is usual with the official models of this era, and it is finished in the authentic colour scheme adopted by the Royal Navy in this period. Her masts and spars indicate that she carried a barquentine rig. *(NMM L2456-003)*

Fitted as troopship 3.1851; recommissioned 7.3.1851 under John Belam, Master, for particular service. On 17.11.1855 under John E. Petley, Master (temp.), then 25.1.1856 under Edmund P. Cole, Master, on 29.10.1857 under Frederick R. Sturdee, Master, and on 8.12.1862 under George Raymond, Master; paid off 11.6.1863 at Sheerness. BU completed 8.2.1864 at Sheerness (machinery transferred to *Virago*).

MEDEA – Steam vessel, later Second Class sloop. Oliver Lang design, approved 1831 (Lang was the master shipwright at Woolwich 1826–32). Re-engined 1846 with a Maudslay 4-cylinder Siamese engine of 220nhp (900ihp = 10.6kts).
  Dimensions & tons: 179ft 4½in, 157ft 4⅞in x 31ft 11in oa (31ft 7in for tonnage) x 20ft 0in. Draught 13ft 10in (fwd) 14ft 6in (aft). 835bm. 1,142 disp.
  Men: 135. Guns: 2 x 10in (84cwt) on pivots, 2 (later 4) x 32pdrs (25cwt). In 1862 one of the 10in shell guns was replaced by a 110pdr rifled BL gun, also pivot-mounted.
  Machinery: 2-cylinder side lever 220nhp.
Medea Woolwich Dyd/John Penn & Sons
  Ord: 12.1.1831. K: 4.1832. L: 2.9.1833. C: 3.4.1834.
  First cost: £35,961 (including machinery £13,080).
  Commissioned 30.1.1834 under Cmdr Horatio Thomas Austin, for the Mediterranean; paid off 10.1837. Recommissioned 14.2.1838 under Cmdr John Neale Nott, for North America and the West Indies (including the St Lawrence); paid off 11.1839. Recommissioned 12.8.1840 under Cmdr Frederick Warden, for the Mediterranean; in operations on the Syrian coast 1840; brought 'Xanthian Marbles' from Greece to UK; paid off 15.5.1845. Recommissioned 2.11.1846 under Cmdr Francis Thomas Brown, then 5.11.1846 under Cmdr Graham Eden William Hamond (died 23.1.1847); from 25.1.1847 under Cmdr Thomas Henry Mason; on 20.2.1849 under Cmdr William Nicholas Lockyer (temp., confirmed 7.12.1849), for the East Indies and China; in action against five pirate junks 8.9.1849; home on 1.7.1850 (carrying the Koh-i-Noor diamond) and paid off 8.7.1850 at Portsmouth. Refit 1851–52. Recommissioned

18.12.1852 under Cmdr John Crawshay Bailey, for North America and the West Indies; home 11.1853, then 16.11.1853 under Cmdr Augustus Phillimore, for home waters; grounded 30.12.1853 off Spurn Point, then docked 1.1854 at Woolwich for repairs; to North America and West Indies again 5.1854; on 10.10.1855 under Cmdr Edward Peirse, on same station; paid off 17.6.1856 at Portsmouth into Steam Reserve. Repair and refit 1860. Recommissioned 18.9.1861 under Cmdr D'Arcy Spense Preston, on same station; paid off 22.6.1865 at Portsmouth. Condemned 12.1966 and BU from 1.1867.

FIREBRAND Class Steam vessels (SV3 from 1837), re-classed 1844 as First Class SGVs. Thomas Ditchburn design (for *Flamer*; presumably used also for *Firebrand*). The first steamers built for the RN by private builders instead of the Royal Dockyards. *Firebrand* had her Butterley engine removed in December 1832 (fitted into *Spitfire* instead) and replaced by a Maudslay 120nhp engine in July 1833 (cost £2,606) with Morgan's paddle wheels (£1,375). In 1843 she was lengthened and a new 260nhp Penn engine with 'tubular boilers' was fitted (machinery cost £14,200; entire cost of this rebuild was £22,587). *Flamer* was originally fitted with Morgan's experimental 140nhp high pressure engine and paddle wheels, but the former 'did not answer' and was replaced in 7.1832 before completion by a Boulton & Watt 120nhp engine.
  Dimensions & tons: 155ft 3in, ?136ft 9in x 26ft 5in x 14ft 10in. Draught 9¾ft. 495bm. 510 disp.
  Men: 80. Guns: Initially 6 x 9pdrs (13½cwt); later 1 x 32pdr (25cwt) on pivot, 2 x 32pdr (17cwt) carronades.
  Machinery: 2-cylinder side lever. *Firebrand*: 140nhp, *Flamer*: 120nhp
Firebrand Curling, Young & Co., Limehouse/Butterley Co.
  Ord: 28.1.1831. K: 4.1831. L: 11.7.1831. C: 11.7.1831.
  First cost: £19,964 (builder paid £7,474, machinery £7,543?).
  Commissioned 25.11.1831 under Lieut Thomas Baldock, for the Mediterranean. On 9.11.1832 under Lieut William George Buchanan, at Falmouth. Re-engined 1833. Recommissioned 23.4.1834 under John Allen, Master, for service as a tender at

A lithograph by Thomas Goldsworth Sutton depicting the paddle gunvessel *Firefly* in action on 8 August 1855 at Brandon, the harbour and shipbuilding centre for the town of Vasa on the Gulf of Bothnia. Commander Otter's 4 guns destroyed a 4-gun shore battery and then set fire to the port's magazines and storehouses, as well as sinking four small vessels. *(NMM 8089)*

Woolwich. In 1837 under Luke Smithett, Master, then 22.5.1839 under Joseph Sanders, Master. Refitted at Woolwich (for £6,848) 7–8.1840, with engine power was raised to 132nhp. Recommissioned 18.7.1840 under Samuel Cook, Master; from 1.1.1842 under Cmdr John Neale Nott, as tender to *William and Mary*; renamed *Black Eagle* 5.2.1842, re-rated 'steam yacht' to convey King of Prussia from Ostend to England (and later return); Cook resumed command after (–1851). Lengthened forward by 13ft (becoming 540bm) at Deptford 6–9.1843, re-rated paddle yacht, completing service as Admiralty yacht until 1857; on 12.6.1844 under Capt. Charles Philip Yorke, Earl of Hardwicke; on 7.2.1851 under John Eaton Petley, Master, as tender to *Fisgard* at Woolwich. On 16.8.1860 under Augustus G. Whichelo, Master. On 25.8.1862 under Thomas J. Whillier, Master (Cmdr, 11.6.1863). On 31.3.1868 under Charles R. P. Forbes, as tender to *Duke of Wellington*, at Portsmouth. BU (by AO 22.11.1873) at Portsmouth in 3.1876.

Flamer Fletcher & Fearnall, Limehouse/Boulton, Watt & Co.
Ord: 28.1.1831. K: 4.1831. L: 11.8.1831. C: 27.10.1832 (for sea).
First cost: £22,865 (builder paid £7,489, original machinery £2,293, Boulton & Watt machinery cost £6,145).
Commissioned 19.11.1831 under Lieut Richard Bastard, for the Mediterranean (–20.4.1832). On 6.9.1832 (after an absence) under Bastard again, as Falmouth packet. On 26.2.1834 under Lieut Charles William Griffith Griffin, for North America and the West Indies; sailed 18.12.1834 for the West Indies; on 13.11.1835 under Lieut John Moon Potbury, for same station; reboiled by Boulton & Watt at Woolwich 8–12.1836; paid off 1839. Reboiled and 'cycloidal' paddle wheels were fitted during refit at Woolwich (for £9,023 total) 7.1839–1.1840. Recommissioned 21.11.1839 under Lieut William Robson, for the West Indies; paid off end 1842. Reboiled and refitted at Deptford and Woolwich (for £8,288) 2–10.1843. Recommissioned 6.10.1843 under Lieut (Cmdr 12.12.1845) Charles James Postle, as a packet; on 28.12.1845 under Lieut George Lavie (reduced to 3 guns), for the Mediterranean; paid off 19.6.1848. Refitted 'for the Holyhead

Station' at Deptford and Woolwich (for £6,404) 10.1848–8.1849. Recommissioned 15.8.1850 under Cmdr James Aldworth St Leger; wrecked on the coast of West Africa on 22.11.1850–12 miles east of Monrovia (Liberia).

On 6 March 1832 an AO was issued to convert the brig sloop *Falcon* (of 1820 – see Chapter 8) to steam power. This was effected at Sheerness Dyd, using a Mills (?) high pressure engine; however, the conversion was deemed a failure, and the engine was removed in 1834.

FIREFLY Class Steam vessels (SV3 from 1837), re-classed 1844 as First Class SGVs. Oliver Lang design, approved 28.3.1832. *Firefly* re-engining in 1844 was done with the engine and boilers from *Phoenix*. *Spitfire* was fitted with the engine taken from *Firebrand* and with Morgan's paddle wheels – at one stage she was classed as a packet.
　Dimensions & tons: 155ft 0in, 136ft 8in x 27ft 8½in (27ft 6½in for tonnage) x 16ft 7in. 550bm.
　Men: 80. Guns: 1 x 32pdr (50cwt) on pivot; plus 2 x 32pdrs (17cwt) in *Firefly* or 4 x 32pdrs (25cwt) in *Spitfire*.
　Machinery: 2-cylinder side lever. *Firefly*: 220nhp, *Spitfire*: 140nhp; 380ihp.

Firefly Woolwich Dyd/Maudslay, Sons & Field
Ord: 28.3.1832. K: 5.1832. L: 29.9.1832. C: 31.1.1833.
First cost: £16,057 + machinery £5,369.
Commissioned 9.11.1832 under Lieut Thomas Baldock, for Falmouth packet service. On 14.6.1836 under Lieut Joseph Pearse, on same service. Refit at Woolwich (for £10,152) 6–11.1839 including reboiling and fitting with 'cycloidal' paddle wheels. Recommissioned 9.10.1839 under Lieut William Robert Wolseley Winniett, for North America and the West Indies. Refitted and re-engined (for £10,569) 11.1843–6.1844, then recommissioned 16.3.1844 under Capt. Frederick William Beechey, as survey ship for Irish coast; paid off 16.10.1847. Recommissioned 24.11.1847 under Cmdr Thomas Ponsonby, for west coast of Africa; took Brazilian slavers *Mercurio* 7.4.1848 and *Guaybo* 24.4.1848; on 15.6.1848 under Cmdr

John Tudor, on same station; took slavers *Gerardo* 5.10.1848, *Assombro* 15.1.1849, *So* 29.6.1849, *Juliet* 6.12.1849 and *Navarre* 19.3.1850; on 3.9.1850 under Cmdr George Alexander Seymour, on same station; in Congo River 22.6.1852; paid off 30.7.1853 at Woolwich. Recommissioned 15.2.1855 under Capt. Henry Charles Otter, for the Baltic; struck submerged mine ('infernal machine') off Kronstadt 9.6.1855; attacked harbour at Brandon 2.8.1855; home to Woolwich 12.1855. On 26.5.1856 under Cmdr George Fiott Day, for the west coast of Africa; took US slaver *William Clark* 22.8.1857 and Spanish slaver *Conchita* 27.8.1857; paid off 5.8.1858 at Woolwich. Recommissioned 1858 under Cmdr Frederick William Sidney, for surveying the Channel Islands. Recommissioned 21.2.1860 under Cmdr Arthur Lukis Mansell, for surveying duties; on 6.11.1863 under Lieut (Cmdr, 1.1.1865) George Robinson Wilkinson, on same station; paid off 7.2.1866 at Malta. BU there 3.1866.

Spitfire Woolwich Dyd/Butterley Co.
    Ord: 29.11.1832. K: 12.1832. L: 26.3.1834. C: 23.7.1834.
    Commissioned 4.6.1834 under Lieut Andrew Kennedy, as a Falmouth packet; paid off at Woolwich 13.8.1838. Recommissioned 30.11.1838 under Lieut James Shambler, for North America and the West Indies; on 12.11.1839 under Lieut Edward Hoile Kennett (temp.), then 20.12.1839 under Lieut John Evans; paid off 19.12.1840 at Woolwich. Reboilered at Woolwich (for £5,407 plus £1,501 for the boilers) 1841. Recommissioned 19.4.1841 under Lieut William Tringham, for the West Indies; on 24.2.1842 under Lieut Hay Erskine Shipley Winthrop, for North America and the West Indies; wrecked 10.9.1842 while trooping on Half Moon Cay lighthouse reef, Belize.

TARTARUS Class Steam vessels (SV4 from 1837), re-classed 1844 as First Class SGVs. Oliver Lang design, 1833. Originally Boulton & Watt engines of 100nhp from store was fitted to *Tartarus*; she was re-engined 1837–38. *Blazer* was fitted with the engine taken from *Echo*, and new boilers.
    Dimensions & tons: 145ft 0in, 125ft 6in x 28ft 4in (28ft 0in for tonnage) x 14ft 9in. 523²⁴⁄₉₄bm. 560 disp. Draught 10½ft.
    Men: 80. Guns: Initially 2 x 9pdrs (13½cwt); later 1 x 32pdr (25cwt) on pivot, 2 x 32pdr (17cwt) carronades. Machinery: 2-cylinder side lever 136nhp.

Tartarus Pembroke Dyd/Boulton & Watt
    Ord: 2.7.1833. K: 9.1833. L: 23.6.1834. C: 3.10.1834 (for sea) at Woolwich.
    First cost: £22,012 (including hull £10,552, machinery £7,644).
    Commissioned 27.8.1834 under Lieut Horatio James, for the packet service; paid off 27.2.1837 at Woolwich. Re-engined at Woolwich (by AO 4.5.1837) with Miller & Ravenhill engines of 136nhp (which cost £7,897) 1837–38. Recommissioned 2.2.1838 under Lieut George Woodberry Smith, for North America and the West Indies; paid off 3.5.1842 at Woolwich. On 1.1.1843 under Capt. Frederick Bullock, for surveying English coast; on 8.11.1843 (briefly) under Capt. Horatio Thomas Austin, then on 26.12.1843 under Capt. James Wolfe, surveying in Irish waters. Reboilered at Woolwich (for £3,867) 1846–47. Recommissioned 9.10.1847 under Lieut Sir Godfrey Vassal Webster, for the Mediterranean; paid off 3.9.1850 at Woolwich. Recommissioned 2.4.1851 under Lieut Richard Hawkins Risk, on particular service (fishery protection off Scottish coast); paid off 27.7.1854 at Woolwich. Recommissioned 7.1.1856 under Cmdr Arthur Lukis Mansell, for surveying in the Mediterranean; paid off

28.2.1860. BU completed 6.11.1860 at Malta.

Blazer Chatham Dyd/Boulton & Watt
    Ord: 2.10.1833. K: 11.1833. L: 5.1834. C: 12.2.1835 (for sea) at Woolwich.
    First cost: £19,994 (including £8,967 for hull).
    Commissioned 17.11.1834 under Lieut Joseph Pearse, at Woolwich. On 7.11.1836 under Lieut John Middleton Waugh; to the Mediterranean 6.1840. On 6.10.1840 under Lieut John Steane, for North America and the West Indies; paid off Woolwich in March 1842. Fitted as survey ship 1.1843. Recommissioned 29.1.1843 under Capt. John Washington, for surveying the east coast of England; on 7.12.1846 under Cmdr Henry Edward Wingrove, for famine relief in Ireland; paid off at Harwich 1.10.1847. Recommissioned 20.10.1847 under Lieut George Thomas Cleather Smith, on the west coast of Africa; took Brazilian slavers *Atrevida* 29.12.1848, *Esperanza* 23.2.1849 and *Finale* 29.3.1849; paid off at Portsmouth 7.8.1849. In 7–8.1851 fitted 'to take telegraph to Dover', during which the hold 'was gutted to make room for this gigantic and gutta-percha tube'; towed 25.9.1851 to eastern Channel to lay Dover–Calais cable. Became store-hulk 10.1852, towed to ferry Napier steam engines from Glasgow to Portsmouth. BU completed 8.1853 at Portsmouth.

HERMES Class Steam vessels (SV3), later Second Class sloops. Symonds specification but John Edye design, approved 1834. *Hermes* initially fitted with Morgan's paddle wheels (ex-the previous *Hermes* – the ex-*Courier*). *Acheron* ordered first at Chatham, moved 18.9.1837 to Sheerness (since moulds were there).
    Dimensions & tons: 150ft 0in, 128ft 0in x 32ft 9in (32ft 5in for tonnage) x 17ft 0in. 715⁴³⁄₉₄bm. 1,006 disp. Draught 11½ft (fwd), 12ft (aft).
    Hermes (lengthened design 1842): 170ft 0in, 148ft 1¼in x 32ft 9in (32ft 5in for tons) x 18ft 2in. 827⁸⁸⁄₉₄bm.
    Men: 135. Guns: 2 x 9pdrs (13½cwt) brass; from 1842–43 – 1 x 8in (52cwt) on pivot, 2 x 32pdrs (17cwt).
    Machinery: 2-cylinder side lever. 140nhp. (*Acheron* 160nhp)

Hermes Portsmouth Dyd/Butterley Co.
    As lengthened 1842: 170ft 0in, 147ft 9½in x 32ft 10in (32ft 6in for tonnage) x 18ft 2in. 830³²⁄₉₄bm.
    Ord: 22.1.1834. K: 4.1834. L: 25.6.1835. C: 9.1835–25.11.1835 at Woolwich.
    First cost: £24,452 total.
    Commissioned 11.1835 under Lieut William Simpson Blount, for the Mediterranean packet service. New cycloidal wheels fitted at Woolwich 4–7.1838, then resumed Mediterranean packet service (still under Blount); home (for refit) 30.3.1840; engines and boilers removed at Woolwich 3–5.1840; paid off there 5.10.1840. Lengthened (forward) at Chatham (for £9,037) 11.1841–11.1842. New engines (Maudslay 220nhp machinery of 'Siamese' type = 8½kts) and boilers (combined cost £11,225) fitted at Woolwich 12.1842–4.1843; completed fitting for sea at Chatham (for £3,303) 4–7.1843. Recommissioned 30.5.1843 under Lieut Washington Carr, for North America and the West Indies; paid off 26.10.1847 at Chatham. Recommissioned 7.1.1850 under Cmdr Edmund Gardiner Fishbourne, for the Cape of Good Hope, then to the East Indies and China; in 2nd Anglo-Burmese War 1852; in Bassein River 5.1852; in 4.1853 at Nanling; paid off 10.6.1854 into Steam Reserve at Woolwich. Recommissioned 13.6.1855 under Cmdr Henry Coryton, for North America and the West Indies. On 21.2.1856 under Cmdr William Everard Alphonso Gordon, for the Cape of Good Hope; paid off

8.6.1860 at Sheerness. Sold to Castle & Beech 10.1864 to BU at
Charlton.

Volcano Portsmouth Dyd/Seaward & Capel

As built: 150ft 8in, 128ft 7⅞in x 32ft 9⅜in (32ft 5⅜in for tonnage) x
18ft 0½in. 720⁴⁹⁄₉₄bm. 1,006 disp.

Ord: 19.11.1834. K: 7.1835. L: 30.6.1836. C: 8.10.1836–17.1.1837 at
Woolwich.

First cost: £27,884 total (including £17,011 for hull, £8,875 for
machinery).

Commissioned 12.1836 under Lieut William M'Uwaine, for
the Mediterranean packet service. On 14.5.1838 under Lieut
Joseph West, on same service, then 1840 to the North America
and West Indies station; paid off 16.3.1841 at Woolwich, but
recommissioned 23.8.1841, still under West. On 2.12.1841
under Lieut Craven John Featherstone; on 6.4.1844 under
Lieut Edward Charles Miller (died 29.9.1845). On 1.10.1845
under Lieut John Hay Crang, in the Mediterranean; paid off
16.2.1849. Repaired and fitted for sea at Woolwich and Deptford
(for £9,628) 1849–50. Recommissioned 6.12.1850 under Cmdr
William Thomas Rivers, at Devonport, for the west coast of
Africa; on 21.4.1851 under Cmdr Robert Coote, on same station;
at Lagos 25.11.1851; paid off 30.1.1854 at Woolwich. Fitted as
depot ship (Engineer's Workshop, also listed as 'Steam Smithery'
and 'Floating Factory') at Woolwich (for £6,054 + £2,565) 1854.
Recommissioned 16.6.1854 under Martin Roberts, Master, for
the Baltic; on 21.11.1854 under Richard Dyer, Master, then
27.2.1855 under James H. Ryan, 2nd Master, still in the Baltic;
paid off 1856. Repair and refit at Portsmouth (reboilered, re-
coppered, forepart 'nearly all new', re-rigged as a brig) 1857.
Recommissioned 17.3.1857 under John M. Hockly, Master,
for the East Indies and China; at Canton and Hong Kong
1857–59; paid off 25.5.1859 at Portsmouth into Steam Reserve.
Harbour craft from 1862. Sold to Castle & Son 11.1894 to BU at
Charlton.

Megaera Sheerness Dyd/Seaward & Capel

Ord: 19.11.1834. K: 8.1836. L: 17.8.1837. C: 30.3.1838 (for sea) at
Sheerness (after being at Limehouse 6.10.1837–13.2.1838 for
machinery fitting).

First cost: £27,778 total (including £15,161 for hull, £8,983 for
machinery).

Commissioned 13.12.1837 under Lieut Hugh Colville Goldsmith,
for the Mediterranean; paid off 10.1841 (Goldsmith dead
8.10.1841). Refit at Limehouse, Woolwich and Deptford –
reboilered and smoke-consuming apparatus fitted (for £6,763)
3–12.1842. Recommissioned 10.11.1842 under Lieut George
Oldmixon, for the West Indies; wrecked on Barc Bush Key (off
Port Royal), Jamaica, on 4.3.1843.

Acheron Sheerness Dyd/Seaward & Capel

Ord: 15.9.1837 (named 27.9.1837). K: 10.1837. L: 23.8.1838. C:
8.1.1839 (for sea) at Sheerness (after being at Limehouse to
11.10.1838 for machinery fitting).

First cost: £25,509 total (including £16,819 for hull & rigging, £8,690
for machinery).

Commissioned 27.11.1838 under Lieut Andrew Kennedy,
for the Mediterranean; paid off 12.1841. Recommissioned
3.12.1842 under Lieut Benjamin Aplin, for the Mediterranean.
On 10.9.1846 under Lieut Andrew Robert Dunlap; paid off
13.10.1847 at Woolwich. Re-classed as survey ship 1847.
Recommissioned 14.10.1847 under Capt. John Lort Stokes, for
surveying New Zealand; paid off 13.8.1851 at Sydney, became
tender to *Calliope* at Sydney. Sold 23.4.1855 to G. A. Lloyd &

Co. at Sydney by Captain Fremantle for £2,067.16s.

FURY A steam paddle vessel was purchased (for £530) at Bermuda on
2 May 1834 (approval was given by AO on 31 July), and the vessel was
named *Fury* on 27 August.

Dimensions & tons: 90ft 0in, 78ft 0in x 20ft 0in x 10ft 0in. 166 tons.
No record of service; probably employed as a local tender. Taken to
pieces at Bermuda 8.1842.

GORGON Class Steam vessel (SV1), later First Class sloop. Symonds
and Eyde design, conceived in late 1833 as a slightly enlarged *Medea*
with two 110nhp engines; following an offer by Seaward & Capel of
300nhp engines, and increased coal capacity of 380 tons, an enlarged
design was finally approved 29 April 1836. Originally allotted to
Plymouth, to use the teak frames, originally cut in Bombay for frigate
Tigris, but no slips were available to build this ship at Plymouth, and
on 31 January 1835 the order was moved to Pembroke. Initially carried
a schooner rig (later brig). The original design was for both *Gorgon*
and *Cyclops*.

Dimensions & tons: 178ft 0in, 152ft 3in x 37ft 6in oa (37ft 0in for
tonnage) x 23ft 0in. 1,108⁶⁄₉₄bm.

Men: 160. Guns (UD): 2 x 84pdrs (10in/84cwt) shell guns on pivots
fwd and aft; 4 x 32pdrs (41cwt) broadside.

Machinery: 2-cylinder direct-acting ('Gorgon type'). 4 tubular
boilers. 320nhp. 800ihp = 9½kts. 380 tons coal.

Gorgon Pembroke Dyd/Seaward & Capel

As built: 178ft 0in, 152ft 2½in x 37ft 6¼in oa (37ft 0½in for tonnage)
x 23ft 0in. 1,110⁸¹⁄₉₄bm. 1,610 disp.

Ord: 10.7.1834. K: 7.1836. L: 31.8.1837. C: 30.8.1838 (for sea) at
Deptford Dyd.

First cost: £54,306 (including machinery £22,662).

Commissioned 19.6.1839 under Capt. William Honeyman
Henderson, for the Mediterranean; at bombardment of St
Jean d'Acre 3.11.1840. Recommissioned 25.11.1842 under
Capt. Charles Hotham, for the east coast of South America;
stranded in River Plate 11.5.1844; in Anglo-French operation
in Parana River 11.1845; at Battle of Obligado 20.11.1845. On
13.5.1846 under Cmdr Edward Crouch, on same station; paid
off 28.11.1846 at Woolwich. On 29.3.1848 under Cmdr James
Aylmer Dorset Paynter, for the Pacific; paid off 2.2.1852 at
Portsmouth into Steam Reserve. Recommissioned 25.2.1854
under Cmdr (Capt., 19.4.1854) Arthur Cuming; on 8.5.1854
under Cmdr Peter Cracroft, for the Baltic. On 28.11.1854 under
Cmdr Richard Borough Crawford, for the Baltic. Employed as
transport in 1856. On 13.5.1856 under Cmdr George William
Towsey; to the Mediterranean 11.1856; home in 7.1857 (carrying
marbles from the Mausoleum at Helicarnassus); paid off 1.8.1857
at Woolwich. On 31.3.1858 under Cmdr Joseph Dayman; in
April 1858 completed fitting (for £4,389) to assist *Agamemnon*
lay the Atlantic cable; paid off 5.11.1858 at Woolwich. On
27.4.1859 under Cmdr Bedford Clapperton Trevelyan Pim, for
the Cape of Good Hope; home in 7.1860. Repairs and refit at
Portsmouth 1860 (reboilered); to Cape of Good Hope again
12.1860, still under Pim (towing mooring brig *Swift* to the
Cape). On 25.3.1861 under Cmdr John Crawford Wilson, on
same station; paid off 11.2.1864 at Woolwich. Sold to Castle
17.10.1864 to BU at Charlton.

CYCLOPS Class Steam vessel (SV1), later Second Class frigate, 6
guns. Sir William Symonds design. Originally ordered as sister to
Gorgon, the design was extended on 14 July 1838 by an additional
12¼ft mid-section. The original engine (ordered 4 September 1837) was

This lithograph by William Adolphus Knell, showing the *Cyclops* off Spithead on first commissioning, celebrated what at that time was the largest steam frigate in the world. She was the first steam vessel to be rated as a frigate, a rather surprising categorisation given the traditional understanding of the term, since she was rigged as a brig and carried all her guns on the upper deck. On her upper deck she carried two 98pdr guns on pivots fore and aft, and four 68pdrs, two on each broadside; she had been designed to carry sixteen 32pdrs on the main deck below, but the ship was overweight and the loss of freeboard allegedly meant that she could not fit main deck guns. *(NMM X0628)*

diverted to the *Gorgon* on 17 October 1837, and a replacement ordered from Seaward on 20 October 1838. *Cyclops* fitted 1856/7 to assist with the laying of the Atlantic cable.

Dimensions & tons: 190ft 3in, 163ft 6in x 37ft 6in oa (37ft 0in for tonnage) x 23ft 0in. 1,190⁵¹⁄₉₄bm.

Men: 175. Guns (UD): 2 x 84pdrs (10in/84cwt) shell guns on pivots fwd and aft; 4 x 68pdrs (8in/65cwt) broadside.

Machinery: 2-cylinder (64in diameter, 5½ft stroke) vertical direct-acting ('Gorgon' type). 4 flue boilers. 320nhp. 1,100ihp = 9½kts. 420 tons coal.

First cost: £53,931 (including machinery £22,103).

Cyclops Pembroke Dyd/Seaward & Capel

As built: 190ft 3in, 164ft 1in x 37ft 6in oa (37ft 0in for tonnage) x 23ft 0in. 1,194⁷⁹⁄₉₄bm. 1,862 disp.

Ord: 25.6.1836. K: 8.1838. L: 10.7.1839. C: 4.2.1840 at Sheerness (after fitting machinery at Limehouse).

First cost: £53,931 (including machinery £22,103).

Commissioned 19.11.1839 under Capt. Horatio Thomas Austin, for the Mediterranean; in operations on the Syrian coast 1840; paid off 16.5.1843 at Woolwich. Recommissioned 27.5.1843, still under Austin, for particular service (including escorting HM the Queen in *Victoria and Albert* to France and Belgium); paid off 29.9.1843 at Woolwich. Recommissioned 23.11.1843 under Capt. William Frederick Lapidge, for particular service, including coast of Ireland, then to east coast of South America 8.1845, home in 2.1846 (joining Squadron of Evolution for sailing trials), and then to Channel Squadron 9.1846; paid off 1.1847 at Woolwich. To Portsmouth 1.1848 for Steam Reserve. Recommissioned 29.7.1848 under Capt. George Fowler Hastings, for troop transport to Ireland; paid off 15.8.1848, but recommissioned 8.9.1848, still under Hastings, for the African station; took slavers *Bomm Successo* 25.12.1848, *Esperanza* 10.5.1849, *Sophia* 11.8.1849, *Apollo* 29.10.1849 and *Sociedade* 17.6.1850; paid off 28.1.1851 at Woolwich. Recommissioned 17.6.1851 under George Hoffmeister, Master, as a transport;

paid off 12.1851. Repair and refit at Sheerness 1852, then to Steam Reserve. Recommissioned 16.12.1853 under Robert Wilson Roberts, Master, for the Black Sea; at bombardment of Sebastopol 11.10.1854. On 20.6.1855 under John F. Rees, Master, for the Mediterranean; in Black Sea 1854–55; on 19.6.1855 under John F. Rees, Master; paid off 15.9.1856 at Sheerness. Fitted to assist with the Atlantic cable (for £13,181) 1856–57 and re-armed as a frigate (with 2 x 10in and 44 x 8in guns). Recommissioned 28.4.1857 under Lieut Joseph Dayman, for sounding to assist *Agamemnon* in laying the Atlantic Telegraph Cable. On 15.9.1857 under Capt. William John Samuel Pullen, for the Red Sea (surveying for laying of a telegraph cable) 7–8.1858, then to China; paid off 22.5.1861 at Sheerness into Steam Reserve. Sold to Castle 26.1.1864 to BU at Charlton (for £5,000).

The year 1837 was when steam propulsion finally came of age within the Royal Navy. Before then, the Admiralty had defined just a single category of 'Steam Vessels'. An Admiralty committee was set up to consider this, and its report on 8 October 1836 recommended that this single category be divided into five classes, in order to establish standard dimensions of masts and yards to enable steam vessels to make voyages under sail as well as under steam. The change would also enable rigging warrants to be established, along with sea store establishments for boatswains', carpenters' and engineers' stores. This report was implemented by AO on 21 January 1837. The five classes (and the vessels grouped within them) were as follows:

Steam Vessel 1 (SV1) *Gorgon, Cyclops*
Steam Vessel 2 (SV2) *Dee; Messenger; Salamander, Phoenix, Rhadamanthus, Medea.*
Steam Vessel 3 (SV3) *Hermes, Volcano, Megaera* (*Acheron* ordered later); *Firebrand, Flamer; Firefly, Spitfire.*
Steam Vessel 4 (SV4) *Columbia; Pluto; Tartarus, Blazer.*
Steam Vessel 5 (SV5) *Comet; Lightning, Meteor, Alban, Carron, African, Echo, Confiance.*

The Admiralty Board formed a new Packet Department on 16 January 1837, and a new Steam Department on 19 April 1837, and on the latter date Capt. Sir Edward Parry was appointed as Comptroller of the Steam Machinery and Packet Department, with its offices in Somerset House. By the end of 1836 the Navy had procured 24 steam paddle vessels; but in 1837 the number was more than doubled when the Admiralty took over the 26 steam paddle vessels previously operated by the Post Office. All but one of these 50 vessels had side lever engines (a modification of Watt's beam engine); the exception – the ex-PO tender *Redwing* – had a steeple engine built by David Napier.

Three months later, on 19 July 1837, an Order in Council approved the creation of a new group of warrant officers, that of Engineers, to compare with the long-established ranks of Boatswains, Carpenters and Gunners. It fixed monthly pay scales for three grades of Engineers and four grades of Apprentices (Engineer Boys), and regulations for their qualification, appointment, superannuation and uniforms; thus was established the engineering branch of the Navy.

The remainder of this chapter, after listing the packets acquired from the Post Office in 1837, groups into sections the subsequent categories of steam paddle vessels according to the 1844 classification of vessels:

Steam (paddle) frigates:
> The former SV1 class (excluding the *Gorgon*, which became a steam sloop); these were to be commanded by captains; there were 2 classes:
> First Class – new frigates including *Penelope*, *Retribution*, *Terrible* and *Avenger*
> Second Class – *Cyclops*, and the six new frigates ordered on 18 March 1841

Steam (paddle) sloops:
> The former SV2 and SV3 classes; these were to be commanded by commanders.

Steam (paddle) gunvessels:
> The former SV4 and SV5 classes; these were to be commanded by lieutenants.

PADDLE PACKETS ACQUIRED FROM THE POST OFFICE IN 1837

While the paddle vessels *Messenger* and *Courier* were purchased in 1830 for the Corfu service, they were quickly re-classed as sloops and appear earlier in this chapter. However, when in 1837 the Admiralty took over responsibility for the packet service from the Post Office as part of a policy of rationalising services, the following 26 paddle steamers were all transferred by the Post Office to the Admiralty's jurisdiction. Note that two other Post Office paddle packets which would undoubtedly had been part of this transfer arrangement – the 1821–built *Meteor* and *Dasher* – had been wrecked on 23 February 1830 and 19 December 1830 respectively.

Note that for some of the former PO vessels the keel lengths were not recorded; these figures for these have been calculated by the author from the quoted bm tonnage and ships' breadths. Service histories for the period before transfer in 1837 are not shown.

Large Type Built for the prestige Liverpool–Kingstown (now Dun Laoghaire) service. Four vessels were contracted for on 1 November 1824 (*Thetis* and *Dolphin*) and 23 November 1824 (*Aetna* and *Comet*), all with 140nhp engines, they commenced service in August 1826. *Thetis* (in 1834) and *Comet* (in 1836) were lengthened by Humble &

Hurry, Liverpool and fitted with new engine cylinders (new power of 170nhp and 190nhp respectively). With these must be included the Liverpool-based tender *Richmond*, which had been purchased in 1834 to replace an earlier smaller tender *Jonathan Hulls* (which had been disposed of in March 1834). All were transferred to Admiralty control at Liverpool on 1 June 1837 and were renamed on that date. In 1839–40 all four of the large packets (the *Redwing* excluded) were replaced on the Liverpool service by the new *Merlin* Class packets (see later page) and were re-classed as paddle gunvessels (SV4) – *Avon* and *Kite* on 6 May 1839, *Lucifer* on 31 March 1840 and *Shearwater* on 7 August 1840.

AVON 1837–63 [ex-PO *Thetis*, built 1825 by George Graham, Harwich]
> Dimensions & tons: 143ft 11¼in, 131ft 1½in x 23ft 0in (22ft 9in for tonnage) x 14ft 7in. 361bm.
> Men: 22. Guns: 2 x 6pdr carronades.
> Machinery: Boulton & Watt 2-cylinder side lever. 160nhp.
> (The *Thetis* was originally of 301bm tons, with 140nhp engine; she was lengthened and re-engined 1835)
> Fitted and repaired at Liverpool (for £7,085) 1837. Commissioned 27.6.1837 under Lieut Charles Haswell Townley, as mail steamer there until 6.1839. Fitted as transport at Woolwich 1839, then Second Class paddle gunvessel 6.5.1839. In 6.1839 under John King, Master, as a transport. Reboilered and refitted at Woolwich (for £16,230) 1839–40. Recommissioned 10.1840 (confirmed 14.3.1841) under Lieut Richard Davison Pritchard, for Falmouth packet service; on 9.9.1841 under Lieut Charles Jenkin, at Woolwich; on 11.6.1842 under Lieut Henry Byng, for North America and the West Indies (Byng invalided 4.1843); on 31.5.1843 under Lieut David Robert Bunbury Mapleton, on same station; paid off 7.9.1843. Reboilered and fitted for sea at Woolwich (for £9,565) 1845. Recommissioned 30.7.1845 under Cmdr Henry Mangles Denham, for surveying west coast of Africa (Niger Delta and coast of Guinea); paid off 28.11.1846 at Woolwich. Recommissioned 1.1.1847 under Cmdr Henry Charles Otter, for surveying west coast of Scotland; paid off 3.4.1849 at Plymouth. From 9.1849 tender to the Flagship at Devonport. In 1852 under Alfred Joseph Veitch, Master, as tender to flagship at Devonport. On 27.8.1855 became tender to *Impregnable* at Plymouth. Sold to Marshall 19.1.1863 to BU at Plymouth.

SHEARWATER 1837–57 [ex-PO *Dolphin*, built 1826 by George Graham, Harwich]
> Dimensions & tons: 136ft 10¾in, 123ft 3in x 22ft 10½in x 14ft 9in. 343bm.
> Men: 12. Guns: 2 x 6pdr carronades.
> Machinery: Boulton & Watt 2-cylinder side lever. 160nhp.
> First cost: £20,542 (including £8,500 for machinery and £4,609 for fitting).
> Commissioned 27.6.1837 under Lieut William Smithett, as mail steamer until 1840 Fitted at Woolwich for the West Indies (for £1,418) 1840, then as a survey ship (for £5,673) 3.1841. Recommissioned 16.3.1841 as survey vessel under Cmdr (Capt., 16.3.1842) John Washington at Woolwich, to survey east coast of England; paid off 28.1.1843 at Woolwich. Refitted at Deptford and Woolwich and reboilered (for £4,739) 1842–43. Recommissioned 31.1.1843 under Cmdr Charles Gepp Robinson, for surveying coast of Scotland (–end 1847). On 2.3.1848 under Lieut Edmund Edward Turnour, for particular

In May 1821 the Post Office introduced the first two steam mail packets – the *Lightning* and *Meteor* – to carry mail and passengers between Holyhead and Dublin. HM King George IV paid his first state visit to Ireland in that year; he travelled to Holyhead under sail aboard the royal yacht *Royal George*, but then – facing contrary headwinds – transferred to the *Lightning* for the sea crossing; this portrait by William John Huggins shows the packet departing for Dublin on 7 August. *(NMM BHC0619)*

service; paid off 5.3.1851. Refitted at Woolwich (for £2,536), then under 9.8.1851 under Lieut William Horton, for the Mediterranean; paid off 30.9.1854 at Malta; in 1855 tender to *Ceylon* at Malta. Sold 2.7.1857 at Malta for £1,025.

KITE 1837–65 [ex-PO *Aetna*, built 1825 by Humble & Hurry, Liverpool]

> Dimensions & tons: 124ft 11in, 109ft 9in x 22ft 8in x 14ft 5in. 300bm. Draught 5ft 11in/7ft 10in.
> Men: 12. Guns: 2 x 6pdr brass.
> Machinery: Fawcett, Preston & Co. side lever. 170nhp.
> Refitted at Woolwich and Deptford (for £3,672) 1837–39. Commissioned 1.7.1838 under Lieut George Evan Davis, as mail steamer. Reboilered and refitted again at Woolwich (for £6,350) and deployed to overseas stations. Recommissioned 1.4.1839 under Lieut George Snell, for the West Indies. Refitted at Woolwich (for £3,560) 1842, with a new plan for disconnecting paddle wheels. Recommissioned 21.2.1842 under Lieut Thomas Lewis Gooch; on 18.5.1842 under Lieut William Montagu Isaacson George Pasco, for the coast of Africa; paid off 30.8.1843 at Woolwich. Repaired and fitted for sea at Deptford and Woolwich (for £9,085) 1846–47. Recommissioned 5.1847 under George Filmer, Master, as tender/transport at Portsmouth; on 15.9.1847 under Cmdr Henry Dumaresq (–17.12.1847), for service at Jersey; paid off 17.12.1847. Fitted at Portsmouth as a tug (for £2,147) 12.1847–1.1848, and stationed 2.1848 at Bermuda; home in 7.1855. Repair/refit at Woolwich 1855–56; returned to Bermuda 5.1856; home again in 3.1863, and condemned by survey at Woolwich 16.3.1863. Sold to T. Sargent 11.1864 and BU 1865 by Castle & Beech at Charlton.

LUCIFER 1837–75 [ex-PO *Comet*, built 1825 by Humble & Hurry, Liverpool]

> Dimensions & tons: 155ft 3in, 140ft 0in x 22ft 9½in x 14ft 1in. 387bm.
> Men: 12. Guns: 2 x 6pdr carronades.
> Machinery: Fawcett, Preston & Co. 2-cylinder side lever. 170nhp.

(The *Comet* was originally of 300bm tons, with 140nhp Maudslay engine; lengthened and re-engined 1836.)

> Based at Liverpool and Pembroke. Later First Class paddle gunvessel. Commissioned 27.6.1837 under Lieut John Philipps Philipps; paid off 8.8.1839. Refitted at Woolwich 1840 (for £8,807) for survey work. Recommissioned 7.5.1840 under Capt. Frederick William Beechey, for survey of Irish coast. In 1.1842 under Cmdr Henry Mangles Denham. Refitted and reboilered at Woolwich (for £5,901) 1843–44. Recommissioned 12.2.1844 under Cmdr George Alexander Frazer, for surveying coast of Ireland. On 13.1.1848 under Lieut Richard Hawkins Risk; in 9.1848 under Lieut Edward Alexander Tylden Lloyd, and on 17.10.1850 under Lieut George Melville Jackson, on particular service; paid off 30.10.1851 at Woolwich. Moved to Portsmouth 7.1853 for use as harbour tender/tug. Reboilered 1855. Harbour service at Portsmouth 1870. Engines removed by AO 12.5.1875 and transferred to coastguard (became *CGWV.41* at Cowes). Sold 1893 to BU.

REDWING 1837–48 [ex-PO *Richmond*, built 1834 by Hunter & Dow, Glasgow] Although the hull was built by Hunter & Dow, this was a speculative venture by David Napier, and was bought by the PO following trials. Tender for Liverpool packet service (rather than a packet herself).

> Dimensions & tons: 144ft 4½in, 105ft 4in x 16ft 0in (15ft 9in for tonnage) x 10ft 8in. 139bm.
> Men: 11. Guns: 2 x 6pdr carronades.
> Machinery: David Napier & Co. steeple engine (the only ex-PO packet not fitted with a side lever engine). 60nhp.
> First cost: estimated value £5,660.
> Based at Liverpool. On 27.6.1837 under Cmdr Edward Chappell, on 27.12.1838 under Lieut John Tudor (temp.), and on 5.2.1839 under Cmdr Thomas Bevis (as superintendent of the packet service at Liverpool). Sold to Messrs Sutherland 17.1.1849 for £650.

Medium Type Built for the Holyhead–Howth, Milford–Dunmore and Weymouth–Channel Islands services. The first steam packet, *Lightning*, was contracted for on 17 July 1820 and the second, *Meteor* (wrecked 1830), on 23 August 1820; their construction was supervised by Oliver Lang on behalf of the Admiralty. The *Ivanhoe* was purchased in December 1821 and the *Vixen* ordered from Deptford Dyd in May 1822. These four vessels were stationed on the Holyhead–Howth service until June 1824, then transferred to the Milford station.

Ten further packets were contracted for on 26 February 1823 (*Aladdin*), 15 April 1823 (*Harlequin* and *Cinderella*), 1 July 1824 (*Crocodile*), 22 March 1825 (*Watersprite*, *Escape* and *Wizard*), 10 March 1827 (*Sibyl*), 17 March 1827 (*Dragon*) and August 1830 (*Flamer*, to replace the lost *Meteor*); a final ship, *Gulnare*, was ordered from Chatham Dyd in March 1833. Four of these vessels were re-classed as paddle gunvessels – *Boxer* and *Fearless* (as SV5) in early 1839, *Gleaner* (as SV4) on 20 November 1839, and *Monkey* (as SV5) on 21 April 1841.

MONKEY 1837–87 [ex-PO *Royal Sovereign* ex-*Lightning*, built 1821 by William Evans, Rotherhithe] This is the vessel on which was based the tenacious legend that 'Their Lordships of the Admiralty so despised steam that they named the first steamer they built *Monkey*, which was a tug'. As will seen above, the vessel in question was built as *Lightning* – changed on 12 August 1821 at the wish of George IV to *Royal Sovereign King George the Fourth* (on the occasion of his visit to Ireland in 1821) which was quickly shortened to *Royal Sovereign*. She was not built for the Admiralty, but for another government department, and she was not named *Monkey* till 1837 – and was not a tug until 1844. The legend presumably comes from some publicist looking at a navy list and seeing that the oldest steamer there was called *Monkey*. The legend was already being propagated in Fincham's *History of Naval Architecture* of 1851, written by an author who, as an aspirant for Surveyor, should have known better.

    Dimensions & tons: 106ft 6in, 92ft 0¼in x 21ft 0in x 11ft 2in. 212bm.

    Men: 12. Guns: 1 x 12pdr carronade, travelling carriage.

    Machinery: Boulton, Watt & Co. side lever. 80nhp. 373ihp.

    First cost: Hull £4,861; engine £5,608.

    L: 18.1.1821.

    Transferred 3.1837. Based at Pembroke. Commissioned 13.7.1838 under Robert Roberts, Master, for Pembroke Dock; to Woolwich 11.1839; used as a tug from 1.1841. On 18.6.1842 under William Bryant, Master; at Chatham 1844. In 1850 machinery of 130nhp (taken from the *Gleaner* – see below) was fitted at Chatham. Recommissioned 16.7.1850 under Robert Sallenger, Master, then 19.11.1853 under George Syndercombe, 2nd Master (re-appointed 1.1.1858). In 1860 became harbour craft at Woolwich (later to Chatham). Sold 9.1887.

BOXER 1837–41 [ex-PO *Ivanhoe*, launched 17.2.1820 by Scott & Co., Greenock for the City of Dublin Steam Packet Co. – first purchased in December 1821 by the PO for the Holyhead Station – later moved to the Weymouth Station.

    Dimensions & tons: 101ft 2in, 91ft 9in x 18ft 8in x 11ft 0in. 170bm.

    Men: — . Guns: — .

    Machinery: Originally David Napier side lever, replaced by Maudslay side lever 1826. 60nhp.

    Transferred 1.4.1837. Commissioned 8.6.1837 under Cmdr Frederick Bullock (–1841), for surveying the Thames Estuary. Sold 27.5.1841 to BU.

ADVICE 1837–70 [ex-PO *Vixen*, built 1822 at Deptford Dyd. – Oliver Lang design of 6.5.1822]

    Dimensions & tons: 107ft 8in, 93ft 11in x 20ft 0in oa (19ft 10in for tonnage) x 11ft 11in. 197bm.

    Men: 12. Guns: 1 x 12pdr carronade, travelling carriage.

    Machinery: Boulton, Watt & Co. side lever. 80nhp.

    Transferred 4.1837. New cylinders installed 1837. Commissioned 13.4.1837 under William Evans, Master, for mail packet service at Pembroke; on 11.2.1840 under Lieut Abraham Darby, then on 26.1.1843 under Lieut Charles Adolphus Petch, on same station; paid off 8.1848 at Woolwich. Used as a tug/tender at Woolwich/Deptford from 1849; to Cork 12.1850 – then paid off for conversion to a dredger (?not done?). New 100nhp engine fitted 1851. On 27.8.1855 under T. H. Laity, 2nd Master, as tender at Cork. New boilers fitted at Keyham 1856. Recommissioned 1.7.17856 under Michael C. Raymond, 2nd Master, as tender to *Sans Pareil* at Queenstown (Cobh) until the 1860s. Sold to J. J. Stark 12.5.1870.

JASPER 1837–54 [ex-PO *Aladdin*, built 1824 by Richard Symonds, Falmouth for Simmonds – later purchased by PO]

    Dimensions & tons: 112ft 6in, about 95ft 0in x 21ft 4in x 12ft 5in. 223bm.

    Men: 12. Guns: 1 x 12pdr carronade, travelling carriage.

    Machinery: Fawcett, Preston & Co. side lever. 100nhp.

    Transferred 4.1837. Commissioned 13.4.1837 under Edward Rose, Master, for mail packet service at Pembroke; paid off 11.1848 at Woolwich. Recommissioned 9.5.1849 under Cmdr Richard Brydges Beechey, for surveying in Bristol Channel, then paid off 12.1849 at Woolwich. Used as a tug there from 2.1850. Repair and refit at Woolwich in 1853 (engine overhauled, reboilered), then to Plymouth 1853 as a tug. Fitted as gunboat at Plymouth 4.1854 (with 1 x 32pdr on pivot and 2 x 18pdr carronades). Recommissioned 21.4.1854 under Lieut Charles Gibbs Crawley, for the Baltic, but caught fire 15.5.1854, exploded and sank off Beachy Head.

SPRIGHTLY 1837–89 [ex-PO *Harlequin*, built 1823 by Wigram & Green, Blackwall (No. 186)]

    Dimensions & tons: 119ft 9in, 104ft 2in x 21ft 1in x 12ft 8in. 234bm.

    Men: — . Guns: 1 x 6pdr (long).

    Machinery: Boulton, Watt & Co. side lever (2 cylinders of 39½in diameter, 3½ft stroke). 100nhp (orig. Maudslay 80nhp, re-engined 1831).

    Transferred 1837. Commissioned 14.7.1837 under James Moon, Master, for mail packet service at Holyhead; paid off 8.1848 at Portsmouth. Tender to *Victory* at Portsmouth 1850. Sold 1.1889 (after being advertised 30.4.1888).

CUCKOO 1837–64 [ex-PO *Cinderella*, built 1822 by Wigram & Green, Blackwall. (No. 187)]

    Dimensions & tons: 120ft 6in, 100ft 0in x 20ft 0in x 12ft 6in. 234bm.

    Men: 12. Guns: 1 x 6pdr (4cwt) brass.

    Machinery: Boulton, Watt & Co. side lever. 100nhp.

    Transferred 1837. Commissioned 14.7.1837 under William Wadling, Master, for mail packet service at Holyhead; on 23.7.1839 under William Comben, Master, for mail packet service at Weymouth; on 3.9.1841 under Lieut Abraham Parks, on same station; at Sheerness 28.5.1845; fishery protection duties off Scotland 8.1845–10.1847. On 17.12.1847 under Cmdr Henry Dumaresq, for fishery protection off Channel Islands (based at

Jersey); grounded 14.5.1850 on Huitres Rocks off Jersey, sank at jetty but salved; paid off 19.5.1850 at Portsmouth for repairs. Recommissioned 25.6.1850 under Cmdr Nicholas Lefebvre, on same station; paid off 3.7.1851 at Sheerness into Steam Reserve. From 6.1852 tender to *Waterloo* at Sheerness. In 11.1853 under George Brockman, Master, as a transport. Fitted as gunvessel at Sheerness 4.1854 (with 1 x 32pdr on pivot and 2 x 18pdr carronades). Recommissioned 2.5.1854 under Lieut Augustus George Ernest Murray, for the Baltic; home in 10.1854; repair and refit at Sheerness 10.1854–3.1855; back to Baltic 5.1855; in action at Raumo 24.7.1855 and at Bjorneborg 17.8.1855, destroying shipping; home in 11.1855; repair and refit at Sheerness 11.1855–2.1856; based at Sheerness 1856–57; grounded in the Medway 1.12.1857 and waterlogged, but refloated 6.12.1857; paid off 22.12.1857 at Chatham for repairs, then into Steam Reserve. Fitted for a tug 6.1861 for Gibraltar; home in 3.1862 and laid up at Portsmouth. Sold 1864.

ADDER 1837–70 [ex-PO *Crocodile*, built 1826 by George Graham, Harwich]
  Dimensions & tons: 116ft 4in, 102ft 10⅛in x 21ft 2in (21ft 0in for tonnage) x 12ft 6in. 241²³⁄₉₄bm. 272 disp.
  Men: 12. Guns: 1 x 12pdr carronade, travelling carriage.
  Machinery: Boulton, Watt & Co. side lever. 100nhp.
  First cost: £10,960 total (hull £4,859, machinery £4,750, fitting £1,351).
  Transferred 5.1837. Commissioned 13.4.1837 under John Hammond, Master, for mail packet service at Pembroke; paid off 8.1848 at Chatham. Used as a tug 1849. Reboilered at Woolwich (for £2,406) 1855. On 1.5.1858 recommissioned under James Minchin, 2nd Master, as tender to Wellesley at Chatham; on 1.1.1862 under William Blakey, Master, still tender to *Wellesley*, then 4.1865 tender to *Cumberland*; reboilered 1864 (with boilers ex-*Pluto*). Sold to Wilson, Maclay & Co. 13.5.1870.

WILDFIRE 1837–56 [ex-PO *Watersprite*, built 1826 by George Graham, Harwich]. Originally of 162bm tons, with 60nhp engine; lengthened by White at Gosport and re-engined in 1836.
  Dimensions & tons (as lengthened): 115ft 9in, 102ft 7in x 18ft 8in oa (18ft 5in for tonnage) x 11ft 7¼in. 185½⁄₉₄bm. Draught 7ft 4in/8ft 6in.
  Men: 12. Guns: 1 x 6pdr (4cwt) brass.
  Machinery: Boulton, Watt & Co. side lever. 75nhp. 198ihp.
  Transferred 1.1838. Commissioned 12.5.1837 under Robert White, Master, for mail packet service at Weymouth. Refitted and reboilered at Woolwich and Deptford (for £3,597) 1838. On 23.7.1838 under William Roberts, Master, on same station; on 28.8.1841 under Lieut Charles Adolphus Petch, then 26.1.1843 under Lieut Abraham Darby, on same station; paid off 3.5.1845 at Woolwich. Recommissioned 1.11.1845 under George Brockman, and used as a tug/tender at Sheerness for Thames and Medway; from 1851 tender to the Flagship at the Nore; in 1857 (still under Brockman) tender to *Waterloo*, in 1.1865 to *Formidable*, then to *Agincourt*. In 4.1870 under Charles G. Johnston, as tender to *Pembroke*; in 4.1875 under Robert L. Cleveland, as tender to *Duncan*. Sold to W. Walker 12.1888.

DOTEREL 1837–50 [ex-PO *Escape*, built 1826 by George Graham, Harwich]
  Dimensions & tons: 118ft 7in, 101ft 9in x 20ft 9in x 12ft 7in. 233bm.
  Men: — . Guns: 1 x 6pdr (long).

Machinery: Boulton, Watt & Co. side lever. 100nhp.
First cost: £12,136 total (hull £5,214, machinery £5,800, fittings 1,722)
Transferred 1837. Commissioned 14.7.1837 under John Grey, Master, for mail packet service at Holyhead; paid off 1846. Recommissioned 19.5.1848 under Lieut Edward Wylde, on same station; paid off 3.1.1849 at Woolwich. Sold to H. H. Hull (by AO 4.12.1850, for £1,417) 17.7.1850.

OTTER 1837–78 [ex-PO *Wizard*, built 1827 by George Graham, Harwich]
  Dimensions & tons: 120ft 0in, 102ft 0in x 20ft 9in x 12ft 8in. 237bm. Draught 8ft 3in/8ft 5in.
  Men: — . Guns: 1 x 6pdr (long).
  Machinery: Boulton, Watt & Co. side lever. 120nhp. 272ihp.
  First cost: £12,160 total (hull £5,214, machinery £5,800, fitting £1,156).
  Transferred 1837. Commissioned 24.8.1837 under Lieut Henry Paget Jones (who had actually been in command, of *Wizard* as a PO vessel, since 1826), for the mail packet service at Holyhead. On 9.6.1844 under Lieut Edward Wylde; paid off 26.8.1848 at Portsmouth; moved to Woolwich 3.1849, then back to Holyhead 7.1849 under Charles Pulfer, Master; paid off 5.1850 at Chatham. Fitted (for £4,761) as a gunvessel in 3.1854 at Woolwich (with 1 x 32pdr on pivot and 2 x 18pdr carronades). Recommissioned 23.5.1854 under Lieut William Andrew James Heath, for the Baltic; home in 10.1854, had repair and refit at Woolwich 11.1854–2.1855; on 12.3.1855 under Lieut John Hawley Glover, for the Baltic 5–9.1855; at Sheerness 1856, then on transport duties. On 25.2.1857 under Lieut Cortland Herbert Simpson, for fishery protection off Scotland; paid off 8.4.1858 at Sheerness into Steam Reserve. Fitted as tug 1860 at Sheerness; reboilered 1863. Coal 'haulabout' at Chatham 1878, then coal hulk at Sheerness 1884. Sold 1893.

PIGMY 1837–79 [ex-PO *Sibyl*, built 1827 by Humble & Hurry, Liverpool]
  Dimensions & tons: 114ft 6in, 101ft 6in x 20ft 9in oa (20ft 6in for tonnage) x 13ft 2in. 227bm.
  Men: 12. Guns: 1 x 12pdr carronade, on a travelling carriage.
  Machinery: Fawcett, Preston & Co. side lever (2 cylinders of 40in diameter, 3½ft stroke). 86nhp. 178ihp.
  First cost: £10,307 total (hull £4,977, machinery £4,850, fitting £490).
  Transferred 4.1837. Commissioned 31.5.1837 under Lieut William Hoseason, for the mail packet service at Holyhead; on 21.8.1837 under Henry Williams, Master, on same station; on 29.10.1839 under Lieut Juste Peter Roepel, for the mail packet service at Pembroke; on 19.6.1841 under Lieut Charles Autridge, then 26.6.1846 under Lieut Abraham Darby, on same station; paid off 9.1848 at Woolwich at Chatham 7.1849. Repair and refit (reboilered) at Woolwich 7–10.1853, and fitted as a tug 1853–54 at Portsmouth and Woolwich (for £5,164). Recommissioned 21.4.1854 under Lieut James Hunt, as a tender to the Baltic fleet; home in 9.1854; on 25.10.1854 under William Johnston, Master, as a tender at Portsmouth; reboilered 1857 (with boilers ex-*Dasher*); on 26.5.1862 under William Vine, Master. On 14.6.1865 under Cmdr William H. Petch, as tender to the Flagship at Portsmouth; on 27.10.1870 under Cmdr William H. Drysdale; reboilered 1871; on 5.1.1874 under Cmdr George Christie and 27.9.1875 under Cmdr William Hewlett; paid off 1877 and

used as a floating workshop at Portsmouth. BU there (by AO 14.5.1877) 2.1879.

ZEPHYR 1837–65 [ex-PO *Dragon*, built 1827 by George Graham, Harwich]

> Dimensions & tons: 116ft 0in, 103ft 0in x 20ft 8in x 12ft 3in. 234bm. 272 disp. Draught 8ft 1in/8ft 3in.
> Men: 12. Guns: 1 x 6pdr (long).
> Machinery: Boulton, Watt & Co. side lever. 100nhp (orig. 80nhp, re-engined 1830).
> First cost: £10,507 total (hull £4,977, machinery £5,400, fitting £130).
> Transferred 1.4.1837. Commissioned 22.7.1837 under Lieut James Duncan, for the mail packet service at Holyhead; on 18.12.1837 under Lieut James Smail, then 21.5.1845 under Lieut Charles Pybus Ladd, then 7.7.1848 under Lieut Andrew Robert Dunlap, all on same station; paid off 24.11.1848 at Woolwich. Fitted (for £3,185) as a Gunvessel in 3.1854 at Woolwich (with 1 x 32pdr on pivot and 2 x 18pdr carronades). Recommissioned 23.5.1854 under Lieut Charles Gibbs Crawley, for the Baltic; paid off at Devonport 11.11.1854. Used as a tug at Devonport from 1860. Sold to Marshall 6.1865 to BU at Plymouth.

FEARLESS 1837–75 [ex-PO *Flamer*, built 1831 by Fletcher & Fearnall, Limehouse]. Built as a replacement for the lost *Meteor* on Weymouth-Channel Islands service.

> Dimensions & tons: 112ft 6in, 96ft 0in x 18ft 5in x 11ft 1in. 165bm.
> Men: 12. Guns: 1 x 6pdr (4cwt) brass.
> Machinery: Boulton, Watt & Co. side lever. 60nhp (increased to 76nhp in 1838 refit).
> Transferred 8.1837. Commissioned 22.8.1837 under William Comben, Master, for the Weymouth mail service. Reboilered and re-engined at Woolwich (for £2,857) 1838. Recommissioned 19.11.1839 under George Bower, Master, as a tender at Woolwich. Refitted at Woolwich as surveying vessel (for £525) 1841. On 2.3.1841 under Capt. Frederick Bullock, for surveying east coast of England. Refitted and reboilered at Woolwich and Deptford (for £3,582) 1842. On 1.1.1843 under Cmdr William Louis Sheringham, for survey on south coast; paid off 1.8.1845 at Woolwich. Refitted at Woolwich as a tug in 1845; reboilered again at Woolwich (for £748) 1846. To Sheerness 1846 then Plymouth 1849, and Woolwich/Sheerness again in 1851. From 1860 tender to guard ship at Chatham and Sheerness. BU 11.1875 at Chatham.

GLEANER 1837–49 [ex-PO *Gulnare*, built 1833 at Chatham Dyd.] Design approved 20 May 1833.

> Dimensions & tons: 120ft 0in, 106ft 10in x 23ft 3in oa (23ft 0in for tonnage) x 13ft 0in. 300⁵⁷⁄₉₄bm.
> As lengthened 1839: 138ft 0in, 124ft 10in x 23ft 3in oa (23ft 0in for tonnage) x 13ft 0in. 351²¹⁄₉₄bm.
> Men: — . Guns: 1 x 6pdr (long).
> Machinery: Boulton, Watt & Co. side lever. 100nhp. (1838, increased to 130nhp.)
> Based at Holyhead from 1834 (replacing *Aladdin*). Transferred 29.6.1837. Commissioned 14.7.1837 under Lieut George Evan Davis (having previously been employed in this role with the Post Office); paid off 31.3.1839. Machinery replaced by Boulton, Watt & Co. 1838 at Woolwich. Lengthened by 18ft at Chatham 1839 (plans submitted 6.7.1839). Recommissioned 4.2.1840 under Lieut John Jeayes, for the West Indies; home tp

pay off 20.9.1842. Repaired and refitted at Woolwich, and fitted as a tug. On 13.12.1842 under Cmdr Charles Gepp Robinson (–31.1.1843). On 11.4.1843 under Henry Hill, Master, for Bermuda (–7.1844); in 10.1843 tender on North America and West Indies; home and paid off 8.1848 at Woolwich. BU 8.1849 at Deptford (her engines being transferred to the *Monkey*).

Small Type: Built for and employed on the shortest routes: Dover–Calais, Margate–Ostend (until 1830) and Portpatrick–Donaghadee services. Seven vessels had been contracted for on 11 June 1821 (*Arrow* and *Dasher*), in April 1823 (*Fury* and *Spitfire*), May 1826 (*Crusader* and *Salamander*), and August 1830 (*Firefly*). Of these *Dasher* had been wrecked on 19 December 1830, and a further vessel (*Ferret*) built as her replacement. Several were lengthened during the 1830s – all at Fletcher & Fearnall, Limehouse (*Arrow* in 1833, *Ferret* in 1835, *Crusader* in 1836, *Beaver* ex-*Salamander* in 1837, and *Firefly* ex-*Myrtle* after transfer from the Post Office) – and their engine capacity raised.

ARIEL 1837–50 [ex-PO *Arrow*, built 1822 by William Evans, Rotherhithe]

> Dimensions & tons: 107ft 11½in, 95ft 10⅛in x 17ft 5in (17ft 3in for tonnage) x 10ft 3in. 151⁶⁵⁄₉₄bm.
> (also recorded as 107ft 11in, 96ft 2in x 17ft 3in. 149bm.)
> Men: 12. Guns: 1 x 4pdr carronade.
> Machinery: Boulton, Watt & Co. side lever (2 cylinders of 31½in diameter, 3ft stroke). 60nhp.
> Originally of 130bm tons, with 40nhp engine; lengthened by Fletcher & Fearnall, Limehouse and re-engined 1833. Transferred 1.2.1837 (nominally) 12.4.1837 (actual, and renamed). Completed alterations and fitting out at Woolwich (for £4,266) 17.6.1837.
> Commissioned 3.6.1837 under Luke Smithett, Master, and based at Dover; refitted at Woolwich (for £3,290) 1840–41; paid off 30.1.1846 at Woolwich. Sold to T. Marston (by AO 17.5.1850) for commercial service.

PIKE 1837–68 [ex-PO *Spitfire*, built 1824 by George Graham, Harwich]

> Dimensions & tons: 89ft 0in, 79ft 3in x 16ft 3in x 8ft 10½in. 111bm. Draught 6ft 6in/5ft 7½in.
> Men: — . Guns: 1 x 6pdr (long).
> Machinery: Boulton, Watt & Co. side lever. 50nhp.
> Transferred 1837 (estimated value of hull £2,120; machinery £2,500). Based at Portpatrick and Holyhead. Commissioned 24.7.1839 under Lieut Abraham Parks; on 4.8.1842 under Lieut Alexander Boyter; paid off 6.11.1849 at Portsmouth. In 6.1850 became tug at Devonport. BU 7.1868.

ASP 1837–69 [ex-PO *Fury*, built 1825 by George Graham, Harwich]

> Dimensions & tons: 89ft 7in, 79ft 10in x 16ft 3in x 8ft 11in. 112bm. Draught 6ft 2in/6ft 4in.
> Men: 16. Guns: 1 x 6pdr (long).
> Machinery: Boulton & Watt side lever. 50nhp.
> Transferred 1837 (estimated value of hull £2,120; machinery £2,500). Based at Portpatrick and Holyhead from 1837 – later at Pembroke. Commissioned 1.4.1839 under Lieut William Henry; on 3.5.1841 under Lieut William Walter Oke; paid off 6.11.1849 at Portsmouth. Recommissioned 6.1850 under Lieut George Alldridge, to assist *Comet* with surveying; in 1851 tender to *Saturn* at Devonport, to Pembroke 1865 under George H. Blakey, Master. Replaced *Widgeon* at Deptford 1868. Transferred

A painting by Samuel Walters dated 1831 shows two views of a paddle steamer off the Wirral, usually identified as the small packet *Ariel* (although at the time the painting was done she would still have been called *Arrow*). This vessel normally operated out of Dover, but she did visit Liverpool on a number of occasions around this time. *(NMM BHC3204)*

to the Director of Works 1869 for use (as a tug?) in the extension of Chatham Dyd. BU 7.1881 at Chatham.

BEAVER 1837–50 [ex-PO *Salamander*, built 1827 by George Graham, Harwich]

   Dimensions & tons: 102ft 2in, about 92ft 0in x 16ft 2in x 8ft 8in. 128bm.
   Men: 12. Guns: 1 x 4pdr carronade.
   Machinery: Maudslay, Sons & Field side lever. 62nhp.
   Originally of 110bm tons, with 50nhp engine; lengthened by Fletcher & Fearnall, Limehouse and re-engined 1836–37. Transferred 1837.
   Commissioned 3.6.1837 under Lieut Robert Mudge, for the Dover station (–1845). Refitted at Woolwich (for £1,751) 1840, but was a dockyard craft by 1845. Hulked (as 'lump' at Woolwich Dyd) by AO 22.2.1850. BU completed 25.3.1858 at Woolwich.

CHARON 1837–49 [ex-PO *Crusader*, built 1827 by George Graham, Harwich]

   Dimensions & tons: 100ft 4in, about 90ft 0in x 16ft 2in x 9ft 11in. 125bm.
   Men: 12. Guns: 1 x 4pdr carronade.
   Machinery: Maudslay, Sons & Field side lever. 60nhp.
   Originally of 110bm tons, with 50nhp engine; lengthened by Fletcher & Fearnall, Limehouse and re-engined 1836. Transferred 1.4.1837.
   Commissioned .6.1837 under Ethell Lyne, Acting Master; stationed at Dover (except for brief transfer to Holyhead in 1839–40). On 1.9.1840 under Edward C. Rutter, Master, on same service; refitted and reboiled at Woolwich (for £4,581) 1843–44; paid off 11.1846 at Woolwich. Sold to the 'Trinity Board' (Trinity House?) (for £1,010) 15.7.1849.

MYRTLE 1837–68 [ex-PO *Firefly*, built 1831 by Fletcher & Fearnall, Limehouse]

   Dimensions & tons: 96ft 3in, 84ft 0in x 16ft 1in x 9ft 7½in. 116bm.
   Men: 12. Guns: 1 x 4pdr carronade.
   Machinery: Boulton, Watt & Co. side lever. 50nhp.
   Transferred 9.1837. Refitted and reboiled at Woolwich (for £3,288) 1837–38.
   Commissioned 3.6.1838 under Robert Sherlock, Master, stationed

at Dover; on 1.7.1839 under Edward Rutter, 2nd Master; on same service; paid off 8.1840. From 1840 based at Woolwich, used as a relief steamer for Dover packet service whenever required. In 9.1846 under Alfred Balliston, 2nd Master, as a tender at Portsmouth; Repair at Woolwich (for £3,621) 12.1847–8.1848; returned 8.1848 as a tender to Portsmouth; from 4.1849 under Samuel Braddon, Master; from 1850 at Sheerness. Recommissioned 1.1.1858 under William S. Bourchier, Master, at Sheerness, as tender to *Formidable* and later (3.7.1860) to *Monarch*. Converted to a tug 1861 at Portsmouth. BU 5.1868 at Portsmouth Dyd by Israel Marks.

SWALLOW 1837–48 [ex-PO *Ferret*, built 1831 by Wm & Henry Pitcher, Northfleet]

   Dimensions & tons: 107ft 6in, 97ft 8in x 16ft 0in x 9ft 9in. 133bm.
   Men: 12. Guns: 1 x 4pdr carronade.
   Machinery: Boulton, Watt & Co. side lever. 50nhp.
   Lengthened by Fletcher & Fearnall, Limehouse 1835. Transferred 1.4.1837.
   Commissioned 3.6.1837 under John Hamilton, Master; stationed at Dover. Refitted and new 70nhp engine and boilers fitted at Woolwich (for £5,529 including machinery £4,095) 1838. Recommissioned 29.9.1838 under Robert Sherlock, Master, for same service (–1846). Refitted at Woolwich (for £2,034) 1842. Tender at Sheerness (or Woolwich?) 3.1846; to Portsmouth 12.1847. BU there 3.1848.

PADDLE FRIGATES

Initially classed as First Class steam vessels, SV1, the category of 'steam frigate' was introduced on 31 May 1844, with all existing SV1 vessels except one being classed as either First Class or Second Class steam frigates; the exception among the SV1 group was the *Gorgon*, which was classed as a First Class sloop (and *never* as a frigate); she is thus listed with the sloops in the following section. There was little difference between the *Gorgon* and the *Cyclops*, and it is unclear why the different classification of 'frigate' was chosen for the latter. First Class frigates, with guns on both UD and MD, had a complement of 200 men; Second Class frigates, with guns on UD only, had a complement of 175 men.

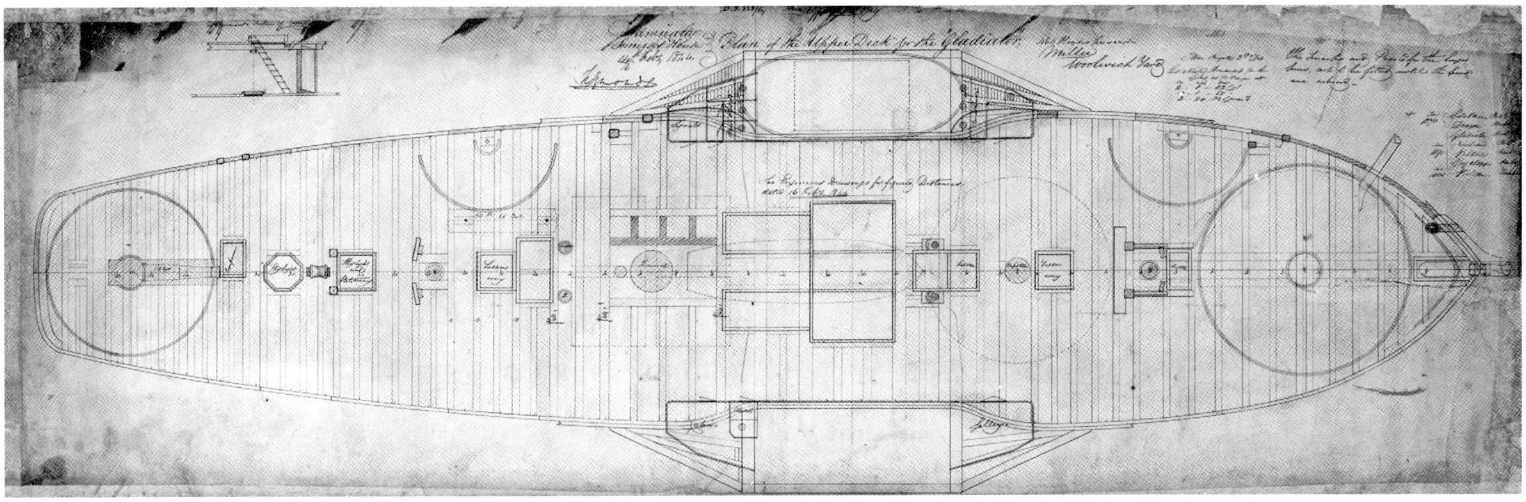

The upper deck plan of the *Gladiator*, dated 16 February 1844, a Second Class frigate of the *Firebrand* Class, whose design derived from that of the *Cyclops*. The draught indicates that the intended gun establishment was two 8in (112cwt) guns on pivots fore and aft, four 8in (65cwt) guns on broadside trucks, and two 24pdr carronades on slides, but like most of the early paddle warships the armament was changed frequently; at a later date she carried two 68pdr guns on pivots fore and aft, and four 10in shell guns, two on each broadside. The pivot mountings were on the centreline, while those on the broadside were fitted to traverse through a large angle, the wheels of both types of mounting requiring the prominent circular iron tracks, called 'racers', let into the deck. Because of the space taken up by their paddle boxes, steam frigates could never carry as many guns as their sailing (or screw-propelled) equivalents, so a different tactical role was evolved for them, emphasising their power to manoeuvre and keep their distance – hence an armament of few but heavy long-range shell guns. *(NMM DR7580)*

On 6 March 1841 the Navy Board approved the construction of six 'First Class Steamers' (SV1) under the 1841 Programme, based on the prototype *Cyclops*. Tenders were called for on 20 October 1841, with the first two to be begun as soon as possible, but the last three were to be delayed, resulting in the six ships forming three distinct classes. All six were re-classed as Second Class frigates in 1844; then, during the 1850s, all six – plus the *Cyclops* – were to be re-classed as corvettes.

Later *CYCLOPS* Class Steam vessels (SV1), later Second Class frigates, 6 guns. Sir William Symonds design. *Firebrand* was originally ordered as *Beelzebub* but the name was changed by AO of 5 February 1842. *Vulture*'s cylinders were 80⅝in diameter, 5¾ft stroke; *Firebrand*'s 75½in diameter, 5ft stroke; *Gladiator*'s 78¾in diameter, 6¾ft stroke.

Dimensions & tons: 190ft 0½in, 163ft 6in x 37ft 6in oa (37ft 0in for tonnage) x 23ft 0in. 1,190$^{55}/_{94}$bm. 1,960 disp.

Men: 175–195. Guns: UD only 2 x 68pdrs (95cwt) on pivots, 4 x 8in (65cwt).

Machinery: 2-cylinder vertical direct-acting. 4 tubular boilers (flue boilers in *Vulture*). See below for nhp. 415 tons coal.

Vulture Pembroke Dyd/William Fairbairn (470nhp; 9½kts)

As built: 190ft 1in, 163ft 7in x 37ft 6in oa (37ft 0in for tonnage) x 23ft 0in. 1,191$^{18}/_{94}$bm. 1,960 disp.

Ord: 18.3.1841. K: 9.1841. L: 21.9.1843. C: 7.6.1845 at Sheerness (engines fitted in East India Docks to 23.1.1844).

First cost: Hull £24,323, machinery £21,429, fitting £10,139.

Commissioned 15.2.1845 under Capt. John M'Dougall, for the East Indies; home in 5.1848 and paid off 17.6.1848 at Woolwich. Small repair at Sheerness and Woolwich (for £17,334) 1848–49; moved to Plymouth 11.1849. Recommissioned 25.11.1852 under Capt. Frederick Henry Hastings Glasse; to the Baltic 1854; attacked shipping and shore facilities in the Gulf of Finland mid-1854; incident at Gamla-Karleby (Kokkola) 7.6.1854. Recommissioned 3.6.1856 under Capt. Frederick Archibald Campbell, for the Mediterranean. On 7.12.1859 under Capt. George Parker, for the Mediterranean; paid off 5.4.1860 at Portsmouth into Steam Reserve. Sold to Castle 10.1863, left Portsmouth 20.10.1863 under tow to BU at Charlton.

Firebrand Portsmouth Dyd/Seaward & Capel (410nhp)

As built: 190ft 0½in, 163ft 5⅝in x 37ft 6in oa (37ft 0in for tonnage) x 23ft 0in. 1,190$^{34}/_{94}$bm. 1,960 disp.

Ord: 18.3.1841. K: 12.1841. L: 5.9.1842. C: 22.10.1844 (for sea).

First cost: Hull £23,310, machinery £18,747, fitting £13,034.

Commissioned 7.9.1844 under Capt. Armar Lowry Corry, in command of 1844 experimental squadron of brigs. On 13.12.1844 under Capt. James Hope, for east coast of South America; at Battle of Obligado 20.11.1845; paid off 17.6.1848 at Woolwich. Small repair at Woolwich (for £22,186) 10.1848–10.1849. Recommissioned 6.10.1849 under Capt. Thomas Owen Knox, for the Mediterranean (died 30.4.1851). Recommissioned 18.12.1852 under Capt. Hyde Parker, for the Mediterranean; in 1854 to the Black Sea (killed 7.7.1854 in action at Sulina); on 8.7.1854 under Capt. William Houston Stewart (temp). In 10.1854 under Capt. William Moorsom; at bombardment of Sebastopol 11.10.1854. On 7.7.1855 under Capt. Edward Augustus Inglefield, in the Black Sea; on 7.3.1856 under Capt. John Welbore Sunderland Spencer; paid off 6.8.1856 at Sheerness into Steam Reserve. Reboilered at Woolwich (for £8,230) 11.1858–6.1859. Recommissioned 27.4.1859 under Cmdr James Minchin Bruce, for North America and the West Indies. On 14.5.1859 under Cmdr Joseph Dayman, for cable-laying to Gibraltar; Cmdr Bruce resumed command 15.11.1859, for Channel fleet; to the West Indies 12.1860; home in 9.1861, then to Mediterranean 11.1861; on 22.4.1862 under Cmdr Shute Barrington Piers, in the Mediterranean; paid off 9.4.1863. Sold to Castle 7.10.1864 to BU at Charlton.

Gladiator Woolwich Dyd/Miller, Ravenhill & Col (430nhp/943ihp).

As built: 190ft 0in, 163ft 8⅝in x 37ft 8in oa (37ft 2in for tonnage) x 23ft 0in. 1,210$^{28}/_{94}$bm. 1,960 disp.

Ord: 18.3.1841. K: 2.1842. L: 15.10.1844. C: 25.4.1846 at Woolwich (after engines fitted in East India Docks)

First cost: Hull £21,535, machinery £23,579, fitting £12,659.

Commissioned 12.12.1845 under Capt. John Robb, for the Channel and later the Mediterranean; paid off 10.3.1849. Recommissioned 4.1.1850 under Capt. John Adams, for the west coast of Africa;

took slavers *Bom Fim* 24.5.1850 and *Gira Sol* 4.9.1850; paid off 16.5.1852 at Portsmouth. Repair and refit there 1852–53. Recommissioned 3.4.1854 under Capt. Sir George Nathaniel Broke, for the Mediterranean and Black Sea. In 5.1855 under Capt. Robert Hall, then 7.6.1855 under Capt. Charles Farrell Hillyar, for the Black Sea; paid off 22.5.1857 at Plymouth into Steam Reserve. Recommissioned 22.6.1859 under Cmdr Henry Dennis Hickley, for North America and the West Indies; paid off 29.10.1861 at Plymouth into Steam Reserve. Recommissioned 9.3.1864 under Capt. Francis Henry Shortt, for particular service. On 15.6.1865 under Capt. Elphinstone D'Oyley D'Auverne Aplin. On 19.8.1869 under Capt. Norman Bernard Bedingfeld, for the east coast of South America; paid off 20.2.1872. Sold to Castle (by AO 25.1.1877) and BU 3.1879.

SAMPSON Class Steam vessel (SV1), later Second Class frigate, 6 guns. Sir William Symonds design, approved 4 April 1843. Lengthened by 13½ft from the *Cyclops* design, with much less rise of floor. Barque-rigged, with two funnels. Her machinery was fitted at East India Dock.

  Dimensions & tons: 203ft 6in, 178ft 1¼in x 37ft 6in oa (37ft 0in for tonnage) x 23ft 0in. 1,296⁸/₉₄bm.
  Men: 175. Guns: UD only 2 x 68pdrs (95cwt) on pivots, 4 x 8in (65cwt). Also (temp.) 2 x 24pdrs (13cwt) carronades.
  Later 2 x 8in (112cwt), 2 x 10in (85cwt), 2 x 32pdrs (56cwt), 2 x 24pdr carronades.
  Machinery: 2-cylinder (81in diameter, 68in stroke) vertical direct-acting. 4 tubular boilers. 467nhp. 9½kts.

Sampson Woolwich Dyd/G. & J. Rennie
  As built: 203ft 6in, 178ft 5½in x 37ft 6in oa (37ft 0in for tonnage) x 23ft 0in. 1,299⁴⁸/₉₄bm. Draught 7ft 6in/8ft 9in.
  Ord: 18.3.1841. K: 11.1843. L: 1.10.1844. C: 5.2.1846 at Woolwich and Sheerness.
  First cost: Total £56,764 (hull £23,931, machinery £21,043, fitting £11,790).
  Commissioned 8.12.1845 under Capt. Thomas Henderson, for the Pacific; paid off at Woolwich 9.12.1848; to Plymouth 6.1849 for Steam Reserve, then to Woolwich 12.1849 for repairs and refit (reboilered) before returning 6.1850 to Plymouth. Recommissioned 24.12.1850 under Capt. Lewis Tobias Jones, for the west coast of Africa; took slavers *Deseada* 7.7.1851 and *Purisima* 10.10.1851; in Lagos operations 12.1851; home in 2.1852, joined Channel Fleet, then to Mediterranean 9.1852; at bombardment of Odessa 22.4.1854, then bombardment of Sebastopol 11.10.1854; severe storm damage off Balaklava 14.11.1854 (losing all three masts, bowsprit and figurehead); on 18.11.1854 under Capt. Thomas Saumarez Brock, in the Black Sea; home and paid off 31.1.1855 at Portsmouth for repairs. Recommissioned 22.10.1855 under Capt. George Sumner Hand; for the East Indies and China; in 10.1856 and 12.1857 at Canton; in 8.1860 at Peiho; paid off 4.5.1861 at Portsmouth. Sold to Castle & Beech 15.7.1864 to BU at Charlton.

CENTAUR/DRAGON Class Steam vessels (SV1), later Second Class frigates, 6 guns (but re-rated sloops 1859). Sir William Symonds design, approved 12 February 1844. Ordered to *Firebrand* design, then amended by AO of 26 December 1843 to lengthen them by 10ft, with the engine-room enlarged by 6ft. *Dragon* name altered from *Janus* on 31 July 1843. Both vessels towed to East India Dock for installation of machinery, in April 1846 and November 1846 respectively. *Dragon*'s cylinders were 88in diameter, 5¾ft stroke; *Centaur*'s cylinders were 85½in diameter, 6ft stroke. Brig-rigged on completion (mizzens added).

  Dimensions & tons: 200ft 0in, 174ft 4¼in x 37ft 6in oa (37ft 0in for tonnage) x 23ft 0in. 1,269⁵/₉₄bm.
  Men: 175. Guns: UD only: 2 x 68pdrs (95cwt) on pivots, 4 x 10in (84cwt) as completed; later 2 x 32pdr carronades or 2 x 8in (112cwt), 4 x 8in (65cwt), 2 x 24pdr carronades (*Dragon*).
  Machinery: 2-cylinder vertical direct-acting. 4 tubular boilers. *Centaur* 540nhp, 10½kts. *Dragon* 560nhp, 11½kts. 430 tons coal.

Dragon (ex-*Janus*) Pembroke Dyd/William Fairbairn
  As built: 200ft 1in, 174ft 4¼in x 37ft 6in oa (37ft 0in for tonnage) x 23ft 0in. 1,269⁵/₉₄bm.
  Ord: 18.3.1841. K: 1.1844. L: 17.7.1845. C: 7.1847.
  First cost: Hull £26,401, machinery £25,005, fitting £16,024 (for relief service at Sheerness for £9,256, then at Woolwich for sea for £5,828).
  Commissioned 12.1846 under Capt. William Ramsay (temp., being captain of *Terrible* at same time) for famine relief to Ireland 1–5.1847. On 3.5.1847 under Capt. William Hutcheon Hall, for the Channel fleet; to the Mediterranean 9.1849; paid off 8.6.1850 at Portsmouth. Recommissioned 3.8.1852 under Capt. Henry Wells Giffard, for the Mediterranean; paid off 19.6.1852 at Portsmouth. Repair and refit 1852–53. Recommissioned 23.1.1854 under Capt. James Wilcox, for the Baltic; on 2.2.1855 under Capt. William Houston Stewart, still in the Baltic; home 12.1855, then served at transport 1856; paid off 20.2.1857 at Sheerness. Repair and refit at Chatham 1858, then into Steam Reserve. Sold 7.10.1864 to Castle and BU 2.1865 at Charlton.

Centaur Portsmouth Dyd/Boulton, Watt & Co.
  As built: 200ft 0in, 174ft 4⅛in x 37ft 6in oa (37ft 0in for tonnage) x 23ft 0in. 1,269⁸/₉₄bm.
  Ord: 18.3.1841. K: 12.1844. L: 6.10.1845. C: 3.2.1849 at Portsmouth (after engines fitted at East India Dock).
  First cost: £64,107 total (hull £27,080, machinery £26,050, fitting £10,977).
  Commissioned 1.1.1849 under Capt. Claude Henry Mason Buckle, for the west coast of Africa, with pennant of Commodore Arthur Fanshawe; took slavers *Sirena* 31.5.1849 and *Veloz* 2.10.1849. Buckle invalided end 1849, and on 1.1.1850 under Cmdr Frederick Patten, still Fanshawe's flag; took slavers *Esperanza* 2.8.1850 and *Feliz Aurora* 19.10.1850; paid off 2.7.1851 at Portsmouth. Recommissioned 23.7.1851 under Capt. Edward St Leger Cannon, for east coast of South America, as flag of Rear-Adm. William Wilmott Henderson; in 9.1853 under Capt. Thomas Harvey; paid off 6.5.1854 at Portsmouth. Repaired and refitted 1854. Recommissioned 2.2.1855 under Capt. William John Cavendish Clifford, for the Channel fleet and later the Baltic; home in 12.1855; to the Mediterranean and Black Sea 8.1856; paid off 17.11.1858 at Plymouth into Steam Reserve. Recommissioned 20.9.1859 under Cmdr Elphinstone D'Oyley D'Auvergne Aplin, for the East Indies and China; at Peiho forts 20.8.1860. On 6.2.1861 under Cmdr John Eglinton Montgomerie, still in China; in Hankow expedition 1861. On 15.1.1863 under Cmdr John Zell Creasy, on same station; paid off at Devonport 11.6.1864. BU 12.1864 at Devonport.

PENELOPE Converted from sailing frigate to Steam Vessel (SV1), later First Class paddle frigate, 22 guns. Originally launched October 1829 as a sailing frigate (see Chapter 5). Conversion designed by John Edye (nominally by Symonds), approved in May 1842; she was lengthened with a 63ft 4in mid-section added and rebuilt as a paddle vessel 1842–43. Barque-rigged, and with wire rigging. Edye proposed at least fifteen similar conversions to follow, but the Admiralty decided that converting frigates to screw propulsion was more effective and probably easier.

> Dimensions & tons: 215ft 2in, 184ft 4in x 40ft 9in oa (40ft 4in for tonnage) x 26ft 8in. 1,595bm. 2,766 disp.
>
> Men: 200. Guns: UD (spar deck) 2 x 42pdrs (84cwt) on pivots, 10 x 42pdr (23cwt) carronades; MD 8 x 8in /68pdrs (65cwt), 2 x 8in /68pdr (36cwt) carronades. Reduced to 16 guns by 1856 and 14 guns in 1862.
>
> Machinery: 2-cylinder (91½in diameter, 80in stroke) vertical direct-acting. 4 tubular boilers. 650nhp. About 11kts. 500tons coal.

Penelope Chatham Dyd/Seaward & Capel

> As lengthened: 215ft 2in, 186ft 10⅜in x 40ft 9in oa (40ft 4in for tonnage) x 26ft 8in. 1,616⁷²⁄₉₄bm. 2,766 disp. Draught 10ft 7in/13ft 1in (light), 19ft 3in/20ft 3in (loaded).
>
> Conversion ord: 26.3.1842. Conversion begun (docked): 11.6.1842. L: (undocked) 1.4.1843. C (unpowered): 6.1843, (powered) 7 -9.1843. At Limehouse 4–7.1843 to fit machinery.
>
> First cost (for 1842–43) £66,267 total (hull £18,337, machinery £34,192, fitting £13,738)
>
> Commissioned 27.6.1843 under Capt. (later Commodore) William Jones, for the west coast of Africa (died 24.5.1846); took slavers *Maria Louisa* 3.4.1844, *Virginia* 20.10.1844 and *Legeira* 30.9.1845; paid off 5.1846 at Portsmouth. Refitted there (for £12,440) 6.1846–1.1847. Recommissioned 13.10.1846 under Capt. Henry Wells Giffard, for the west coast of Africa, with broad pennant of Commodore Sir Charles Hotham; took slavers *Saron* 18.3.1847, *Joanito* 4.4.1847, *Felicidade* 30.4.1847 and *Sylphide* 17.10.1847; on 18.12.1847 under Capt. Lewis Tobias Jones, still Hotham's pennant; took slavers *Theresa* 22.6.1848 and *Jacinto* 3.1.1849; paid off 16.4.1849. Refitted at Portsmouth (for £28,853) 12.1849–4.1851. Recommissioned 7.3.1851 under Capt. Henry Lyster for west coast of Africa, with broad pennant of Commodore Henry William Bruce; in Lagos operations 12.1851; paid off 24.3.1854. Recommissioned 27.3.1854 under Capt. James Crawford Caffin; stranded 10.8.1854 at Bomarsund (jettisoned guns to float off); home to pay off 12.1854. Repair and refit at Portsmouth 12.1854–3.1855. Recommissioned 2.2.1855 under Capt. Sir William Saltonstall Wiseman, for transport duties; in 11.1855 to the Cape of Good Hope; paid off 7.7.1858 at Portsmouth into Steam Reserve. Sold to Castle & Beech 15.7.1864 to BU.

RETRIBUTION Steam vessel (SV1), later First Class frigate, 22 guns (28 guns from 1856). John Edye design, approved 13 July 1842. Originally assigned the name *Dragon* on 13 July, but this was changed to *Watt* by AO of 1 August 1842 and to *Retribution* by AO of 26 April 1844. Originally no MD guns were carried. She was overweight and her first year's commission was exceedingly discouraging, and it was intended that she be either 'dismantled' or converted to a tender to the gunnery hulk *Excellent*, but nevertheless she was retained for another 18 years.

> Dimensions & tons: 220ft 0in, 192ft 10¼in x 40ft 6in oa (40ft 0in for tonnage) x 26ft 4in. 1,641¹⁹⁄₉₄bm. Draught 10ft 3½in/9ft 9¼in.
>
> Men: 200. Guns: (1843) UD 2 x 68pdrs (112cwt) + 4 x 8in/68pdrs (65cwt) + 2 x 32pdrs (40cwt). (1845) UD 2 x 68pdrs (112cwt) + 6 x 8in/68pdrs (65cwt) + 2 x 32pdrs (34cwt) + 2 x 24pdrs (13cwt). (1849) MD 12 (later 18) x 32pdrs (50cwt); UD 1 x 8in/68pdr (112cwt) on pivot, 9 x 8in/68pdr (65cwt/9ft) – 8 on broadside and 1 aft on pivot.
>
> Machinery: 4-cylinder (72in diameter, 8ft stroke) direct-acting 'Siamese' type. 4 boilers. 800nhp. About 10kts. 530 tons coal.

Retribution Chatham Dyd/Maudslay, Sons & Field

> Ord: 26.3.1842. K: 8.1842. L: 2.7.1844. C: 23.12.1845 at Chatham (after fitting machinery at East India Dock).
>
> First cost: £91,540 total (hull £34,456, machinery £43.210, fitting £13,874).
>
> Commissioned 14.10.1845 under Capt. Stephen Lushington, for the Channel; paid off 16.10.1846 at Portsmouth. Re-engined at Woolwich in 1849–50 with Penn 2-cylinder (72in diameter, 7ft stroke) 400nhp oscillating engine, new tubular boilers and paddlewheels; converted from brig rig by addition of a mizzen mast; fitted for sea at Portsmouth (for £36,076) 7–10.1850. Recommissioned 7.8.1850 under Capt. Frederick Warden, on particular service. Recommissioned 10.6.1852 under Capt. James Robert Drummond, for the Mediterranean and Black Sea; bombardment of Odessa 22.4.1854; bombardment of Sebastopol 11.10.1854; paid off 2.2.1855 at Portsmouth. Recommissioned 3.1855 under Capt. Thomas Fisher, as flag of Rear-Adm. Robert Lambert Baynes, for the Baltic; home in 12.1855 and paid off 22.8.1856 at Portsmouth. Recommissioned 24.8.1856 under Capt. Charles Barker, for the Channel fleet; in 3.1857 to East Indies and China (invalided 26.1.1859); on 26.1.1859 under Capt. Peter Cracroft, then on 15.2.1859 under Capt. (Commodore, 27.4.1859) Harry Edmund Edgell; at Hankow 11.1859; paid off 22.12.1860 at Portsmouth. Sold to Castle & Beech 15.7.1864 to BU at Charlton.

TERRIBLE Class Steam vessel (SV1), later First Class frigate, 19 guns. Oliver Lang design, approved 28 October 1842. The first RN ship with four funnels, and the largest paddle warship ever built for the RN. When built the most powerful steam warship of her day. Her name was changed from *Simoom* to *Terrible* on 23 December 1842. She underwent very expensive repairs in 1862–64 (see below), amounting to a rebuild, with new boilers, other machinery repaired or replaced, hull timbers replaced, and masts and rigging replaced; re-armed with 9 x 110pdr Armstrong guns, 10 x 8in (60cwt) and 2 x 68pdrs (96cwt).

> Dimensions & tons: 226ft 2in, 196ft 10¼in x 42ft 6in oa (42ft 0in for tonnage) x 27ft 4in. 1,847⁷⁄₉₄bm.
>
> Men: 300. Guns: MD 4 x 68pdrs (95cwt), 4 x 56pdrs (98cwt), 3 x 12pdrs (probably 6cwt boat guns, deleted 1847); UD 4 x 68pdrs (95cwt), 4 x 56pdrs (98cwt).
>
> Machinery: 4-cylinder (72in diameter, 8ft stroke) direct-acting 'Siamese' type. 4 tubular boilers. 800nhp. 2,059ihp = 10.9kts.

Terrible (ex-*Simoom*) Woolwich Dyd/Maudslay, Sons & Field

> As built: 226ft 2in, 197ft 1½in x 42ft 6in oa (42ft 0in for tonnage) x 27ft 4in. 1,849¹⁸⁄₉₄bm. 3,189 disp. Draught 8ft 10in/11ft 4in.
>
> Ord: 22.7.1842. Order transferred to Deptford Dyd on 24.2.1843. K: 11.1843. L: 6.2.1845. C: 25.3.1846 at Woolwich (after engines fitted at East India Dock).
>
> First cost: £101,842 total (hull £38,346, machinery £41,820, fitting £21,676).
>
> Commissioned 5.12.1845 under Capt. William Ramsay, for the Channel, later to the Mediterranean. In 1.1847 under Capt. William Hutcheon Hall, then 2.5.1847 under Capt. William Ramsay again; paid off 27.9.1849 at Woolwich. Recommissioned 23.11.1849 under Capt. James Hope, for the Mediterranean; paid off 13.10.1852 at Woolwich. Repair and refit (reboilered; engines overhauled) 1852–53. Recommissioned 7.6.1853 under Capt. James Johnstone McCleverty, for the Mediterranean; to the Black Sea 1854–55; bombardment of Odessa 22.4.1854; bombardment of Sebastopol 11.10.1854; paid off 1.9.1856 at Sheerness into Steam Reserve. On 2.6.1858 under Capt.

Terrible, 'the largest war steamer in the world', as she appeared in this portrait by John T. Wood. She was built with four boilers working at 9lb/sq in, each boiler having its own funnel – so there were two abreast between the main mast and the paddle boxes, and two abreast further forward – but two funnels and their associated boilers were removed in about 1850 as they were producing more steam than the 800 nhp engines could cope with. Her lower frames were so close together that they actually touched, so that the ship was watertight even before she was planked. Her armament of eight 56pdrs and eight 68pdrs was divided equally between the main and upper decks, but in 1852 the main deck 68pdrs and the upper deck 56pdrs were all replaced by equal numbers of 8in shell guns, and in 1862 the entire ordnance was superseded by nine 110pdr rifled breechloaders and ten 8in shell guns. *(NMM PU6189)*

Frederick Henry Hastings Glasse, for the Mediterranean; paid off 4.2.1862 at Sheerness. 'Very extensive repairs' (see above) there 6.1862–12.1864. Recommissioned 27.4.1865 under Capt. Gerard John Napier, for the Channel fleet; assisted *Great Eastern* in laying the Atlantic cable 6–8.1865 (reduced to 19 guns); to the Mediterranean 10.1865; home in 4.1866; on 3.5.1866 under Capt. John Edmund Commerell, again for Channel fleet; assisted *Great Eastern* in Atlantic cable laying 6–9.1866; to the Mediterranean 9.1867; home in 3.1868. On 14.5.1868 under Capt. Trevenen Penrose Coode, for Channel fleet; paid off 10.5.1869 at Plymouth. Recommissioned 11.6.1869 under Cmdr Hamilton Edward George Earle; on 9.1869 under Capt. John Adolphus Pope Price; assisted *Black Prince* and *Warrior* towing floating dock to Bermuda 1869 (the other warships towed the dock, while *Terrible* was lashed astern to act as a rudder); escorted turret ship Scorpion to Bermuda; paid off 12.2.1870 at Plymouth. Sold 7.7.1879 to BU at Devonport.

BIRKENHEAD Class Second Class iron-hulled frigate, later troopship. Builders design by John Laird, approved 23 August 1843. The only iron-hulled paddle frigate built for the RN, her hull was divided into 8 watertight compartments by transverse bulkheads, while the engine-room had divided by 2 longitudinal bulkheads into 4 compartments, making 12 in total. Her name, first approved as *Vulcan* on 12 June 1843, changed to *Birkenhead* by AO on 1 February 1845. Rated Second Class frigate 26 February 1845, but re-rated 14 October 1846 as a troopship as a result of official unease about iron as a material for fighting ships. Her original brig rig was altered in 1851 from a brig rig to barquentine.

Dimensions & tons: 210ft 0in, 187ft 9in x 37ft 6in (50ft 6in over paddle boxes) x 22ft 11in. 1,405bm. 1,918 disp.
Men: 250 (as frigate), 100 (as troopship). Guns: as frigate, UD 2 x 10in (84 cwt) on pivots, 4 x 68pdrs (64 cwt) on broadside; as troopship, re-armed with 4 x 32pdrs (25 cwt) only.
Machinery: 2-cylinder direct-acting engine. Tubular boilers. 536nhp. 12–13kts. 500 tons coal.
Birkenhead (ex-*Vulcan*, renamed 26.2.1845) Laird & Co., Birkenhead. (No. 51)/George Forrester & Co.

Ord: 13.4.1843. K: 9.1843. L: 30.12.1845. C: 17.1.1847 (for sea) at Sheerness, after engines fitted 7.1846.
First cost: £98,844 total (hull £30,685, machinery £57,678, fitting £10,481).
Commissioned 5.11.1846 under Cmdr Augustus Henry Ingram; to Fort George 25.2.1847 with troops; ordered to Cork 5.1847 carrying Indian meal (probably famine relief); assisted attempts to refloat the stranded Great Eastern in Dundrum Bay 9.1847; collion with merchant brig 15.9.1847 (starboard paddle box smashed); paid off 14.1.1848 at Portsmouth. Fitted there as troopship (for £6,756) 21.6.1850–15.2.1851, with a forecastle and poop deck added to increase her accommodation Recommissioned 8.2.1851 under Robert Salmond, Master, as a troopship; wrecked off Point Danger, South Africa, 27.2.1852. She was lost off South Africa while carrying troops, who behaved most gallantly in standing in formation while the ship sank and the women and children were saved (438 men died). Portions of the hull and masts were later salved and sold at the Cape for £164.13.6. Attempts at salvage continue to the present day.

AVENGER Class First (from 31 July 1846, Second) Class Frigate. Sir William Symonds design, approved 25 March 1844. Originally ordered as Steam vessel (SV1) from Deptford, the order was moved 22 June 1844 to Devonport. On launch, the ship sailed (jury-rigged) to Deptford where machinery fitted and completed.

Dimensions & tons: 210ft 0in, 183ft 2⅛in x 39ft 0in oa (38ft 6in for tonnage) x 25ft 8in. 1,444³⁶⁄₉₄bm.
Men: 200 (later 250). Guns: 2 x 68pdr/8in (95 cwt) on pivots, 4 x 8in (65cwt) on pivots, 4 x 32pdr (25 cwt) gunnades.
Machinery: 2-cylinder direct-acting. 650nhp. 9½kts.
Avenger Devonport Dyd/Seaward & Capel

Ord: 19.2.1844. K: 27.8.1844. L: 5.8.1845. C: 21.6.1846 at Devonport (after engines fitted at Limehouse and Deptford).
First cost: £89,147 (hull £44,777, machinery £32,740, fitting £11,630).
Commissioned 28.4.1846 under Capt. Woodford John Williams, for the Channel; tender to *Caledonia* in 1847. On 9.4.1847 under

A lithograph of the paddle frigate *Sidon* by Thomas Goldsworth Dutton, showing a row of main deck gun ports (including false ports painted on the paddle box). Although based on the hull of the *Odin*, the designed is credited to Sir Charles Napier, one of a couple of famous naval officers allowed to try their hand at designing steam warships (the other being Lord Cochrane, who produced the *Janus*). Although the *Sidon* had a reputation for being crank, her captain reported her to be a good, easy seaboat. *(NMM A2469)*

Capt. Sidney Colpoys Dacres, then 20.11.1847 under Capt. Charles George Elers Napier; wrecked 20.12.1847 – while on passage for Malta from Gibraltar – on the Sorelli Rocks off the North African coast (246 of 250 aboard drowned, including Napier).

Under the 1844 Programme of Work, besides the one-off *Avenger* (see above), five frigates 'similar to the *Sampson*' were ordered on 19 February 1844 – two to be built at Woolwich (*Niger* and *Odin*), two at Pembroke (*Conflict* and *Desperate*) and one at Portsmouth (*Dauntless*). Later that year the *Odin* was transferred to Deptford and re-ordered to a new design by Fincham, and the other four were subsequently ordered to be built as screw vessels (see relevant chapters).

ODIN Class Second Class frigate, 16 guns. John Fincham design, approved 12 February 1845. Initially intended to carry 16 guns with 8 x 32pdr carronades on the middle deck – but this idea was later discarded and these guns never fitted. 1850–51 fitted for an Ambassador.
    Dimensions & tons: 208ft 0in, 183ft 8½in x 37ft 0in oa (36ft 6in for tonnage) x 24ft 2in. 1,301⁷⁄₉₄bm.
    Men: 175. Guns: MD 10 x 32pdrs (56cwt); UD 2 x 68pdrs (112cwt) + 4 x 10in (84cwt).
    Machinery: 2-cylinder (88½in diameter, 5¾ft stroke) vertical direct-acting. 5 tubular boilers. 560nhp. 10kts. 500 tons coal.
Odin Deptford Dyd/William Fairbairn
    As built: 208ft 0in, 183ft 2⅜in x 37ft 2in oa (36ft 8in for tonnage) x 24ft 3in. 1,310⁹⁄₉₄bm. Draught 8ft 0in/9ft 1½in.
    Ord: 19.2.1844. K: 2.1845. L: 24.7.1846. C: 12.8.1847 at Woolwich and Deptford (after machinery fitted at East India Dock).
    First cost: £76,661 total (hull £35,056, machinery £25,810, fitting £15,795).
    Commissioned 24.5.1847 under Capt. Frederick Thomas Pelham, for the Mediterranean; paid off 1.6.1850 at Portsmouth. Fitted at Portsmouth (for £13,060) for an Ambassador 10.1850–8.1851. Recommissioned 1.8.1851 under Cmdr William Saltonstall Wiseman (–10.1851), to convey Britsh Ambassador to St

Petersburg; paid off 8.10.1851 into Steam Reserve at Portsmouth. Recommissioned 16.7.1852 under Capt. William Henderson, to escort royal yacht; paid off 20.8.1852. Recommissioned 18.12.1852 under Capt. Francis Scott, for the Baltic; in the Gulf of Finland mid-1853 attacking enemy shipping and shore facilities; incident at Gamla Karleby 7.6.1854; on 7.9.1855 under Capt. James Willcox, for the Mediterranean; paid off 7.8.1856. Recommissioned 20.9.1859 under Capt. Lord John Hay, for the East Indies and China; at Peiho forts 8.1860; paid off 28.8.1863. Sold to Castle & Beech 7.1864 and left Portsmouth 1.3.1865 under tow to BU at Charlton.

SIDON Class Second Class frigate, 22 guns. Design by Admiral Sir Charles Napier, approved 30 May 1845. Design based on *Odin* with increased depth in hold and to carry 400 additional tons of coal as suggested by Napier.
    Dimensions & tons: 210ft 8in, 185ft 3in x 37ft 0½in oa (36ft 6½in for tonnage) x 27ft 0in. 1,315⁶³⁄₉₄bm. Draught 8ft 7in/9ft 4in.
    Men: 270. Guns: MD 2 x 68pdrs (88wt) on pivots (fwd and aft) + 14 x 8in (60cwt); QD 4 x 8in (52cwt) on slides; Fc 2 x 8in (52cwt) on slides.
    Machinery: 2-cylinder (86½in diameter, 6ft stroke) vertical direct-acting. 4 tubular boilers. 560nhp. 10kts. 650 tons coal.
Sidon Deptford Dyd/Seaward & Capel
    Ord: 7.4.1845. K: 26.5.1845. L: 26.5.1846. C: 26.11.1846 at Deptford (after machinery fitted at East India Dock). Fitted at Portsmouth for the Governor-General of India 10–11.1847.
    First cost: £81,290 (including machinery £25,178) +£3,100 for 1847 fitting.
    Commissioned 25.8.1846 under Capt. William Honeyman Henderson, for the Mediterranean; rescued crew of sinking P&O ship *Ariel* 28.5.1848; later voyage up the Nile; paid off 30.3.1849 at Portsmouth. Refitted for troops at Portsmouth 6.1851. Recommissioned 23.6.1851 under Cmdr George Henry Gardner, to take troops to South Africa; paid off 12.1851 at Portsmouth. Recommissioned 26.11.1852 under Capt. George Goldsmith, for the Mediterranean and Black Sea; blockade of

coast from Bulgaria to the Danube estuary 4.1854; in attack on Fort Nicolaiev 4.10.1854; on 7.3.1856 under Capt. Edward Augustus Inglefield, for the Mediterranean; paid off 30.7.1856 at Portsmouth into Steam Reserve. Recommissioned 11.6.1859 under Capt. Richard Borough Crawford, for Cape of Good Hope; took seven slaver dhows between 20.2.1861 and 22.4.1861, one on 13.8.1861 and three more 2–3.10.1861; paid off 1.3.1862 at Portsmouth. Sold to Castle & Beech 15.7.1864 to BU at Charlton.

LEOPARD Class Second Class frigate, 18 guns. Committee of Reference (John Fincham) design, approved 11 February 1848. Originally ordered to *Odin* design, but on 9 September 1846 (and again on 13 January 1847) work was suspended pending the result of trials of the *Odin*. The altered plan of 2 February 1848 was lengthened by 10ft. In 1852 *Leopard* was re-armed with – MD 8 x 32pdrs (56cwt) + 4 x 40pdr Armstrong BLs (32cwt); UD 5 x 100pdr Armstrong BLs (82cwt) + 1 x 68pdr (95cwt).

> Dimensions & tons: 218ft 0in, 194ft 0in x 37ft 6in oa (37ft 0in for tonnage) x 25ft 2in. 1,412⁶⁹/₉₄bm.
> Men: 200. Guns: MD 12 x 32pdrs (56cwt), UD 4 x 10in (84cwt) + 2 x 68pdrs (95cwt) on pivots.
> Machinery: 2-cylinder (91½in diameter, 80in stroke) vertical direct-acting. 4 tubular boilers. 560nhp. 11.2kts.

Leopard Deptford Dyd/Seaward & Capel

> As built: 218ft 0in, 193ft 5¾in x 37ft 5½in oa (36ft 11½in for tonnage) x 25ft 1½in. 1,405⁶⁸/₉₄bm. Draught 7ft 10in/9ft 1in.
> Ord: 26.3.1846. K: 8.1846 (but suspended 9.9.1846 to 2.1848). L: 5.11.1850. Engines fitted at East India Dock and then Woolwich, 21.11.1850–5.1851. C: 1–3.1853 (for sea) at Portsmouth.
> First cost: £67,850 (+ 1853 fitting £14,400).
> Commissioned 6.1.1853 under Capt. George Giffard, for the Channel fleet; flag of Rear-Adm. James Hanway Plumridge in the Baltic 1854, then 8.1854 flag of Rear-Adm. Henry Byam Martin; home in 10.1854; to the Mediterranean and Black Sea 12.1854. On 6.3.1856 under Capt. Courtenay Osborne Hayes, on same station; paid off 13.8.1856 at Plymouth into Steam Reserve. Recommissioned 28.4.1857 under Capt. James Francis Ballard Wainwright, for Atlantic Telegraph cable; flag of Rear-Adm. Sir Stephen Lushington 30.1.1858, for east coast of South America, paid off 18.7.1861 at Plymouth. Recommissioned 10.10.1862 under Capt. Charles Tayler Leckie, for China; at Shimonoseki 5–8.9.1864; paid off 22.9.1866 at Plymouth. Sold to Marshall 8.4.1867 to BU at Plymouth.

In 1847 it was proposed that the three 560 nhp paddle frigates (*Sidon*, *Odin* and *Leopard*) should be armed with just 12 guns – MD 6 x 32pdrs (56cwt/9½ft); UD 2 x 68pdrs (95cwt/10ft) + 4 x 10in (85cwt/9ft 4in). The seven older Second Class paddle frigates (the *Cyclops* and the six ordered on 18 March 1841) should carry the same guns on the UD, but without the MD ordnance.

Under the 1847 Programme of Works, five First Class sloops of the *Sphinx* Class, with engines of 400nhp, were submitted for construction, and orders placed with the dockyards on 25 April 1847; one was cancelled in 1850, but the other four were all re-graded to Second Class (paddle) frigates in August 1847 (first 2) and 1852 (last 2), prior to their completion in 1852 and 1853. When finished, these constituted the last group of paddle warships built for the RN.

MAGICIENNE Class Second Class frigates, 16 (later 14) guns. John Edye design, approved 12 August 1847. Originally ordered on 25 April 1847 as First Class sloops, but on 5 August they were re-ordered to be lengthened to 210ft and were subsequently re-classed as frigates.

> Dimensions & tons: 210ft 0in, 185ft 6¼in x 36ft 0in (35ft 8in for tonnage) x 24ft 6in. 1,255³¹/₉₄bm. 2,300 disp.
> Men: 175. Guns: MD 8 (later 6) x 32pdrs (50cwt) aft + 2 x 32pdrs (50cwt) fwd; UD 1 x 68pdr (95cwt) + 1 x 10in (85cwt) + 4 x 32pdrs (56cwt). After 1861–62 refit had UD guns as 2 x 110-pdr Armstrongs and 4 x 32pdrs.
> Machinery: 2-cylinder (72in diameter, 7ft stroke) oscillating. 400nhp. 1,300ihp. 9–10kts.

Magicienne Pembroke Dyd/John Penn & Sons

> As built: 210ft 0in, 185ft 6in x 36ft 0½in (35ft 8½in for tonnage) x 24ft 6¼in. 1,258¹²/₉₄bm. Draught 9ft 0in/8ft 4in.
> Ord: 25.4.1847. K: 9.1847. L: 7.3.1849. C: 20.2.1853 at Devonport (after engine fitted at East India Dock, then to Chatham).
> First cost: £71,303.
> Commissioned 25.11.1852 under Capt. Thomas Fisher, for particular service; in 1854 to the Baltic; on 26.1.1855 under Capt. Nicholas Vansittart; home in 12.1855; to the Mediterranean and Black Sea 7.1856; to China 7.1857 (via South America); destruction of pirate junks at Coulan 8–9.1858; at Peiho forts 6.1859 (Vansittart mortally wounded). On 28.9.1859 under Capt. John Edmund Commerell; at Peiho forts 8.1860; paid off 7.4.1861 at Plymouth (having circumnavigated the globe). Repair and refit 1861–62. Recommissioned 13.3.1862 under Capt. Ernest Leopold Victor Charles Auguste Joseph Enrick, Prince of Leiningen, for the Mediterranean; on 1.4.1863 under Capt. William Armytage, for same station; paid off 12.12.1864. BU by Marshall, Plymouth 9.1866.

Valorous Pembroke Dyd/Miller & Ravenhill

> As built: 210ft 0in, 185ft 6⅛in x 36ft 0¼in (35ft 8¼in for tonnage) x 24ft 6in. 1,256⁶⁹/₉₄bm. Draught 9ft 7in/8ft 8in.
> Ord: 25.4.1847. K: 3.1849. L: 30.4.1851. C: 7.7.1853 at Devonport (after fitting at Woolwich and engine fitted at East India Dock).
> First cost: £69,064 (hull £31,366, machinery £24,329, fitting £13,369).
> Commissioned 11.12.1852 under Capt. Claude Henry Mason Buckle, for the Channel fleet; to the Baltic 3.1854, in Gulf of Finland mid 1854 attacking enemy shipping; grounded and lost her false keel, but repaired from 9.1854 at Sheerness; thence to the Mediterranean and Black Sea 11.1854; paid off 19.8.1856 at Plymouth. Recommissioned 26.8.1857 under Capt. William Cornwallis Aldham, for North America and the West Indies; escort to Atlantic cable layers 1858; paid off 14.8.1861 at Plymouth into Steam Reserve. Recommissioned 13.4.1863 under Capt. Charles Codrington Forsyth, for the Cape of Good Hope; paid off 13.9.1867 at Plymouth into Steam Reserve. Recommissioned 28.11.1869 under Capt. Francisco Saugro Tremlett, then 8.3.1870 under Cmdr Edward D. Hardinge, for North America and the West Indies; returned home in 10.1870, then to Channel fleet. On 27.2.1872 under Capt. Arthur Thomas Thrupp; fishery protection off northern Scotland 6–9.1872; paid off 10.9.1873. Refitted at Devonport (for £19,966) 1874–75. Recommissioned 25.2.1875 under Capt. Loftus Francis Jones, for home waters. Recommissioned 25.2.1878 at Portsmouth under Capt. Archibald George Bogle; on 1.10.1878 under Capt. George Digby Morant, for particular service; to west coast of Ireland 1880–81; on 7.10.1882 under Capt. Walter Stewart, then 14.11.1882 under Capt. Richard Hare, for particular service; on 7.8.1886 under Capt. John Hugh Bainbridge. Sold to E. Marshall, Plymouth 27.2.1891 to BU.

The modern steam frigate *Tiger* was the Royal Navy's most significant loss during the Russian War. This lithograph is based on an eyewitness sketch by Lieutenant Montagu Buccleuch Dunn of HMS *Niger*, seen in the foreground with the *Vesuvius* to port. Having run aground in fog off Odessa on 12 May 1854, the *Tiger* is shown beached and on fire under cliffs lined by Russian troops. Her commander, Captain Henry Gifford, had his left leg shot away as he stood next to the pivot gun, and died later of his wounds in Odessa after the ship was surrendered. (*NMM PW4896*)

TIGER **Class** Second Class frigate, 16 guns. Originally intended to be a sloop of the *Sphinx* type, but the order was altered to a John Edye design, approved 29 July 1847; re-classed as a frigate in 1852.

> Dimensions & tons: 205ft 0in, 180ft 6¼in x 35ft 11⅞in (35ft 8in for tonnage) x 24ft 6in. Draught 17ft (mean). 1,221⁴⁹⁄₉₄bm.
> Men: 175. Guns: Designed for MD 4 x 32pdrs (56cwt); UD 1 x 68pdr (95cwt) + 1 x 10in (84cwt) + 4 x 32pdrs (42cwt). Completed with MD 8 x 32pdrs (56cwt); UD 2 x 10in (85cwt) + 6 x 32pdrs (56cwt).
> Machinery: 2-cylinder (72in diameter) oscillating. 400nhp. 1,300ihp. 9–10kts.
> First cost: £64,900.

Tiger Chatham Dyd/John Penn & Sons
> Ord: 25.4.1847. K: 11.1847. L (undocked): 1.12.1849. C: 21.8.1852 at Portsmouth (after fitting engine in East India Dock by 9.12.1850).
> First cost: £60,398 total (+ £4,502 for fitting in 1852).
> Commissioned 19.6.1852 under Capt. Henry Wells Gifford, for the Mediterranean; to Black Sea 1854; bombardment of Odessa 22.4.1854; grounded off Odessa in fog 12.5.1854 and was set on fire by Russian bombardment (Gifford mortally wounded; he died of gangrene in captivity 1.6.1854); surrendered to Russians, then blew up after further bombardment from shore when British ships tried to retake her. Some sources suggest she was later salved by the Russians and commissioned by them as *Tigr* but this is impossible in the circumstances, and due to a misreading of Russian navy lists; in fact, her *engines* were salvaged and installed in the Russian royal yacht *Tigr*.

FURIOUS **Class** – Second Class frigates, 16 guns. John Fincham design, approved 30 August 1848. Originally ordered as sloops of the *Sphinx* type, with a designed ordnance of 14 guns: MD 8 x 32pdrs (56cwt); UD 1 x 68pdr (95cwt) on pivot, 1 x 10in (84cwt), and 4 x 32pdrs (42cwt).

> Dimensions & tons: 206ft 0in, 181ft 7in x 37ft 0in (36ft 6in for tonnage) x 23ft 3in. 1,286⁷⁄₉₄bm.
> Men: 175. Guns: MD 10 x 32pdrs (50cwt); UD 2 x 10in/110pdrs (84cwt) on pivots, 4 x 32pdrs (50cwt);

Machinery: 2-cylinder oscillating. 400nhp.

Furious Portsmouth Dyd/Miller & Ravenhill
> As built: 206ft 0in, 181ft 7in x 37ft 0in (36ft 6in for tonnage) x 23ft ?7in. 1,286⁷⁄₉₄bm. Draught 8ft 7in/9ft 9in.
> Ord: 25.4.1847. K: 6.1848. L: 26.8.1850. C: 18.2.1853 at Portsmouth (after engine fitted at East India Dock).
> First cost: £64,794 total (hull £32,420, machinery £24,577, fitting £7,797).
> Commissioned 25.11.1852 under Capt. William Loring, for the Channel fleet; to the Mediterranean 6.1853; to the Black Sea 1854; bombardment of Odessa 22.4.1854; bombardment of Sebastopol 11.10.1854; in the Mediterranean 1855; paid off 20.8.1856 at Portsmouth. Refitted there (for £8,105) 1856–57. Recommissioned 11.3.1857 under Capt. Sherard Osborn, for China; at Canton 12.1857; to Hankow 11.1858. On 17.3.1859 under Capt. Oliver John Jones, still in East Indies and China; paid off 30.8.1859 at Portsmouth into Steam reserve. Became a coal hulk at Portsmouth 3.1867 (by AO 6.1865). Sold to Castle 1884, towed from Portsmouth 10.7.1885 to BU at Charlton.

Resolute Portsmouth Dyd/John Penn & Sons.
> Ord: 25.4.1847. K: – Order moved to Sheerness Dyd 8.1.1849. Cancelled 23.3.1850.

Further paddle frigates were contemplated as late as 1852, when a design of 2,540bm tons and 260ft length (with 1,000nhp engines) was considered, to have carried 36 x 68pdrs (2 of 113cwt and 34 of 95cwt); but none were ordered.

PADDLE SLOOPS

In December 1834 the various paddle vessels were divided into three groups (as listed in Adm 222/5) of which the smallest comprised what would later be classed as gunvessels, and the other two groups comprised the sloops:

SV 145ft length, 100 HP – *Tartarus*, *Blazer*, *Alban*, *African*, *Confiance*,

Carron, *Meteor*; *Colombia* (130ft 6in)
SV 155ft length, 140 HP – *Spitfire*, *Firefly*, *Flamer*; *Hermes* (150ft);
SV 220 HP – *Phoenix*, *Salamander*, *Rhadamanthus*, *Medea*; *Dee*

On 21 January 1837, as mentioned under 'Paddle Frigates' earlier, the former single category of Steam Vessels was split into five classes, in accordance with an Admiralty Committee report of 8 October 1836, in order to establish dimensions of masts and yards, and to provide rigging warrants and sea store establishments of boatswains', carpenters' and engineers' stores. The first category (SV1) covered the two new-building vessels *Gorgon* (see above) and *Cyclops* (the first steam frigate), while the last two categories (SV4 and SV5) covered the smaller ships (later defined as steam gunvessels). The sloops were divided as follows:

SV2 – *Phoenix*, *Salamander*, *Rhadamanthus*, *Medea* ; *Dee*, *Messenger*. (the following *Merlin* Class packets were added to this class in 1838–40: *Merlin*, *Medusa* and *Medina*.)
SV3 – *Hermes*, *Volcano*, *Megaera*, *Spitfire*, *Firefly* ; *Firebrand*, *Flamer*.

***HYDRA* Class** Originally Steam vessels (SV2), from 1844 Second Class sloops. Sir William Symonds design, approved 3 November 1837.
Dimensions & tons: 165ft 0in, 143ft 7¼in x 32ft 10in oa (32ft 8in for tonnage) x 20ft 4in. Draught 12ft 1in (fwd), 13ft (aft). 814$\frac{91}{94}$bm. 1,096 disp.
Men: 135. Guns: 2 x 68pdrs/8in (65cwt) on pivots, fwd & aft; 2 x 32pdrs (50cwt). In 1842 had 2 x 10in (84cwt) guns on pivots instead of the 8in guns.
Machinery: 2-cylinder side lever 220nhp (Boulton, Watt & Co.) or 240nhp (Scott & Sinclair). About 9kts.

Hydra Chatham Dyd/Boulton, Watt & Co.
As built: 165ft 0in, 143ft 8⅝in x 32ft 10½in oa (32ft 8½in for tonnage) x 20ft 4in. 817$\frac{79}{94}$bm. Draught 6ft 7in/7ft 1in (light).
Ord: 18.9.1837. K: 1.1838. L: 13.6.1838. C: 23.3.1839 at Woolwich, Deptford and Chatham.
First cost: £37,239 total (hull £13,154, machinery £18,289, fitting £5,796).
Commissioned 19.1.1839 under Cmdr Anthony William Milward, for the Mediterranean (died 2.1840); on 1.3.1840 under Cmdr Robert Spencer Robinson, for the Mediterranean; operations on the Syrian Coast 1840; on 26.12.1840 under Cmdr Alexander Murray, for North America and the West Indies; paid off 21.8.1842. Recommissioned 7.8.1843 under Cmdr Horatio Beauman Young, for the west coast of Africa; took slavers *Cyrus* 6.6.1844, *Huracan* 24.2.1845, *Pepito* 4.3.1845, *Africano* 30.7.1845, *Amelia* 13.9.1846 and *Isabel* 30.9.1846; on 12.11.1846 under Cmdr Arthur Fleming Morrell, on the same station; paid off 21.4.1847 at Woolwich. Recommissioned 2.12.1847 under Cmdr Grey Skipwith, for the east coast of South America; her boats took the slaver *Uniao* in 1849. Recommissioned 17.1.1852 under Cmdr Thomas Belgrave, for Cape of Good Hope (invalided 10.8.1852); on 11.8.1852 under Cmdr William Everard Alphonso Gordon (acting), then 22.1.1853 under Cmdr Henry Gage Morris, on same station; paid off 10.5.1856 at Sheerness. Recommissioned 4.6.1858 under Cmdr (Capt. 27.1.1862) Richard Vesey Hamilton, for the west coast of Africa, then to North America and the West Indies; paid off 23.8.1862 at Woolwich. On 10.11.1863 under Capt. Arthur Lukis Mansell, for surveying in the Mediterranean (–1866). In 1866 under Capt. Peter Frederick Shortland, in same role; to the Arabian Gulf and Indian Ocean in 11.1867; paid off 8.1868 at Sheerness. Sold for BU (for £3,200) to Castle, Charlton on 13.5.1870.

Hecla Chatham Dyd/Scott & Sinclair
As built: 165ft 0in, 143ft 8¾in x 32ft 10¼in oa (32ft 8¼in for tonnage) x 20ft 5in. 816$\frac{81}{94}$bm. Draught 6ft 0½in/7ft 7½in (light)
Ord: 12.3.1838. K: 6.1838. L: 14.1.1839. C: 28.8.1839 at Chatham (after machinery fitted at Glasgow).
First cost: £31,404 (including £10,888 for machinery).
Commissioned 18.7.1839 under Lieut John Bettinson Cragg, for the West Indies; paid off 31.3.1842 at Woolwich. Recommissioned 21.1.1843 under Cmdr John Duffill, for the Mediterranean; on 3.7.1846 under Cmdr Charles Starmer, on same station; paid off 18.1.1848 at Sheerness. Engines removed and replaced in a refit (total £15,110) of 1848–49 at Woolwich and Sheerness. Recommissioned 5.9.1849 at Plymouth under Cmdr Edward Halhead Beauchamp-Proctor, for the west coast of Africa; took slavers *Rosita* 18.1.1850, *Assombra* 20.1.1850, *Andorinha* 8.4.1850, *Leao* 19.4.1850, *Nova Andorinha* 9.5.1850, *Flor de Maria* 11.6.1850 and *Caramara* 19.6.1850; paid off 31.1.1852 at Portsmouth. Recommissioned 4.2.1854 under Peter Tucker, Master, for reconnoitring the Baltic. On 16.3.1854 under Cmdr William Hutcheon Hall, in the Baltic; at bombardment of Bomarsund 6.1854 (Mate Charles Lucas winning first ever VC); on 4.11.1854 under Cmdr Henry Samuel Hawker, for Gibraltar; grounded 23.1.1855 and returned to Plymouth for repairs; paid off 23.5.1855. Recommissioned 22.10.1855 at Plymouth under Cmdr Elphinstone D'Oyly D'Auvergne Aplin, for the west coast of Africa; paid off 16.2.1859 at Plymouth. Later torpedo depot ship at Portsmouth. Sold to Williams & Co. 15.6.1863 (for £2,550) for commercial service, renamed *Typhoon*.

Hecate Chatham Dyd/Scott & Sinclair
As built: 165ft 0½in, 143ft 9¼in x 32ft 10¼in oa (32ft 8¼in for tonnage) x 20ft 5in. 817$\frac{9}{94}$bm. Draught 6ft 2½in/7ft 6½in (light)
Ord: 12.3.1838. K: 6.1838. L: 30.3.1839. C: 7.12.1839 at Chatham (including machinery).
First cost: £27,742 (including £11,113 for machinery).
Commissioned 14.9.1839 under Cmdr James Hamilton Ward, for the Mediterranean; operations on the Syrian Coast 1840; paid off 23.8.1843. Recommissioned 11.6.1845 under Cmdr Joseph West, for the African coast; took slavers *Uniao* 30.10.1845 and *Relampago* 8.7.1846; on 14.9.1847 under Cmdr Richard Moorman; paid off 25.9.1848 at Portsmouth. Recommissioned 5.5.1849 under Cmdr William Cornwallis Aldham (temp) for the Baltic, to safeguard British interests during the 1st Schleswig War; paid off 20.6.1849. Recommissioned 30.10.1849 under Cmdr Charles Frederick Alexander Shadwell, to transport British Ambassador (with family and suite) to the USA; paid off 1.2.1850. Recommissioned 20.9.1850 under Cmdr George Sumner Hand, for particular service. Recommissioned 1.3.1855 under Cmdr Henry Carr Glyn, for the West coast of Africa; on 10.9.1855 under Cmdr Alexander Duff Gordon, on same station (died 1.1857); took slaver *Chatsworth* 1.1.1856; then on 13.2.1857 under Cmdr George Foster Burgess; took slaver *Niagara* 7.12.1857; paid off 3.7.1858 at Woolwich. Fitted as survey ship 6.1860; recommissioned 15.5.1860 under Cmdr Anthony Hiley Hoskins, for the Pacific; on 1.9.1860 under Capt. George Henry Richards; paid off 1.1864 at Woolwich. Sold to Castle & Beech, Charlton 11.1864 for BU; towed from Woolwich 1.1865.

***STROMBOLI* Class** Steam vessels (SV2), later First Class sloops. Sir William Symonds design, approved 29 August 1838. Originally to be of

A coloured drawing by Oswald Walters Briery depicting the paddle sloop *Vesuvius* destroying Russian stores in the Sea of Azov, 1 September 1855. *(NMM PW5680)*

Medea design, altered to new draught derived from *Gorgon*.

Dimensions & tons: 180ft 0in, 157ft 0¾in x 34ft 4in oa (34ft 0in for tonnage) x 21ft 0in. Draught 13ft (fwd), 13½ft (aft). 965⁷⁹⁄₉₄bm. 1,283 disp.

Men: 149 (later 160). Guns: 2 x 10in/42pdrs (84cwt) on pivot, 2 x 68pdrs (64cwt), 2 x 42pdrs (22cwt). 1 x 68pdr (95cwt) replaced 1 x 10in in 1856, and 4 x 32pdrs (42cwt) replaced the 4 smaller guns. A 110pdr later replaced the 68pdr.

Machinery: 2-cylinder side lever. 280nhp. About 9kts.

Stromboli Portsmouth Dyd/Robert Napier & Son

As built: 180ft 1⅛in, 157ft 2¾in x 34ft 4in oa (34ft 0in for tonnage) x 21ft 0in. Draught 7ft 1in/8ft 7in (light). 966⁷¹⁄₉₄bm.

Ord: 12.3.1838. K: 9.1838. L: 27.8.1839. C: 6.9.1840.

First cost: £41,240 total (hull £19,248, machinery £13,280, fitting £8,712).

Commissioned 18.7.1840 under Cmdr Woodford John Williams, for the Mediterranean; at bombardment of Acre 3.11.1840. On 11.6.1841 under Cmdr William Louis, on same station; paid off 6.1843 at Woolwich. Recommissioned 13.10.1843 under Cmdr Edward Plunkett, on Irish station, then on particular service; on 13.6.1845 under Cmdr Thomas Fisher; on 11.11.1847 under Cmdr Lord Amelius Wentworth Beauclerk; paid off 17.9.1850 at Portsmouth. Recommissioned 18.8.1853 under Cmdr Robert Hall, for the Mediterranean; in 8.1855 under Cmdr Cowper Phipps Coles, for Azov operations; on 27.2.1856 under Cmdr George Foster Burgess. Recommissioned 9.12.1861 at Portsmouth under Cmdr William Buller Fullerton Elphinstone, at Portsmouth; on 21.12.1861 under Cmdr Arthur Robert Henry, for east coast of South America (invalided 4.6.1863); on 4.6.1863 under Cmdr Alexander Philips, on same station; paid off 8.6.1866. Sold 8.1866; towed away 24.8.1866 to BU by White, Cowes.

Vesuvius Sheerness Dyd/Robert Napier & Son

As built: 180ft 0in, 157ft 1⅜in x 34ft 5in oa (34ft 1in for tonnage) x 20ft 11½in. Draught 7ft 0in/8ft 10½in (light). 970⁷⁸⁄₉₄bm.

Ord: 12.3.1838. K: 9.1838. L: 11.7.1839. C: 20.4.1840 at Chatham.

First cost: £39,505 (hull 21,707, machinery £13,309, fitting £4,389).

Commissioned 5.1840 under Lieut William Simpson Blount (temp), at Sheerness, for trials, then used as transport. Recommissioned 31.8.1840 under Cmdr Thomas Henderson, for the Mediterranean; at bombardment of Acre 3.11.1840. On 23.8.1841 under Cmdr Erasmus Ommanney; paid off 15.11.1844 at Woolwich. Recommissioned 22.3.1845 under Cmdr George William Douglas O'Callaghan, for North America and the West Indies; on 19.2.1847 under Lieut Herbert Grey Austen; paid off 14.9.1848. Recommissioned 1852 under Cmdr Frederick Lamport Barnard (–22.3.1852). Recommissioned 17.8.1853 under Cmdr Richard Ashmore Powell; at bombardment of Sebastopol 11.10.1854; on 16.2.1855 under Cmdr Sherard Osborn, for Azov operations; on 6.6.1855 under Cmdr Francis Marten, as tender to *Royal Albert*, still in Black Sea; on 29.10.1855 under Cmdr Edward George Hore, in the Mediterranean; paid off 17.1.1856 at Woolwich. Recommissioned 20.4.1857 under Commodore Charles Wise, for the west coast of Africa; paid off 22.2.1860 at Deptford. Fitted with Armstrong guns 1862; on 16.7.1862 under Capt. Richard Vesey Hamilton, in the West Indies; paid off 8.11.1864 at Portsmouth. Sold 6.1865 to BU, and towed away 10.11.1865 to White, Cowes.

ALECTO Class Steam vessels (SV3), later Third Class sloops. Sir William Symonds design, approved 12 April 1839, to which four vessels were built. A fifth vessel – *Rattler* – was ordered from Sheerness Dyd, but a year later was re-ordered as a screw sloop (see Chapter 12); she was the first true warship fitted with a screw in the RN and took part in a famous tug of war with her original sister *Alecto* in March 1845. A sixth vessel, *Argus*, was ordered on 25 April 1847 from Portsmouth Dyd to the *Alecto* design, but on 18 August 1847 a new design was requested for her (see below).

Dimensions & tons: 164ft 0in, 142ft 6in x 32ft 8in oa (32ft 6in for tonnage) x 18ft 7in. 800⁵⁸/₉₄bm. 878 disp.

Men: — . Guns: 2 x 32pdrs (45cwt) on pivots; 2 x 32pdr (25cwt).

Machinery: 2-cylinder direct acting. 200nhp. 370ihp = 8½–9½kts.

Alecto Chatham Dyd/Seaward & Capel

As built: 164ft 0in, 141ft 8in x 32ft 8in oa (32ft 6in for tons) x 18ft 7in. 795⁸⁷/₉₄bm. Draught 6ft 0in/7ft 5in.

Ord: 25.2.1839. K: 7.1839. L: 7.9.1839. C: 12.12.1839 at Chatham (after machinery fitted at Limehouse).

First cost: £27,268 total (including machinery £10,700).

Commissioned 26.10.1839 under Lieut William Hoseason, for the Mediterranean; paid off 24.4.1845 at Woolwich. Trials against *Rattler* 1845. Refitted and reboiled at Woolwich (for £18,110 + £10,073) 1845. Recommissioned 10.11.1845 under Lieut Francis William Austen, for east coast of South America. On 17.11.1846 under Cmdr Vincent Amcotts Massingberd, on same station, later to North America and the West Indies; paid off 22.6.1849 at Woolwich. Re-engined at Woolwich (for £10,073) 1850–1. Recommissioned 27.1.1852 under Cmdr Stephen Smith Lowther Crofton, for west coast of Africa; paid off 11.5.1854. Refitted at Woolwich and Deptford (for £10,143) 1855. Recommissioned 2.8.1855 under Cmdr Robert Phillips, for the west coast of Africa; on 7.5.1856 under Cmdr James Hunt; took slavers *Eliza Jane* 27.8.1857, *Onward* 13.9.1857, *Lewis McLane* 15.10.1857, *Clara Williams* 26.10.1857, and *Windward* 4.11.1857; paid off 15.1.1859. Refitted at Woolwich (for £13,469) 1859–60. Recommissioned 27.1.1860 under Cmdr Henry James Raby, for same station; took slaver *Constancia* 15.8.1860 and another slaver (name unknown) 11.7.1861; boat action at Porto Novo (Niger River) 26.4.1861); paid off 25.6.1862. On 23.1.1863 under Cmdr William Hans Blake, for east coast of South America; paid off 3.1865. BU 11.1865 by Castle, Charlton.

Prometheus Sheerness Dyd/Seaward & Capel

As built: 164ft 0in, 141ft 8in x 32ft 8in oa (32ft 6in for tonnage) x 18ft 7in. 795⁸⁷/₉₄bm. Draught 6ft 0in/7ft 7in.

Ord: 25.2.1839. K: 7.1839. L: 21.9.1839. C: 20.2.1840 (after engines fitted at Limehouse).

First cost: £29,433 (including machinery £10,700).

Commissioned 21.11.1839 under Lieut Thomas Spark, for the Mediterranean; on 23.2.1843 under Lieut Frederick Lowe, on particular service; paid off 27.3.1844. Refitted and reboiled at Woolwich and Limehouse (for £9,248) 1843–44. Recommissioned 15.5.1844 under Lieut John Hay, for the west coast of Africa; took slavers *Marinero* 13.9.1844, *Audaz* 23.10.1844, *Albeira* 18.4.1845, *Tentador* 3.6.1845, *Suspiro* 6.7.1845, *Belmira* 11.11.1845, *Recuperador* 27.11.1845, *San Domingo* and *Eugrazia* 25.12.1845, and *San Lorenzo* 26.12.1845; paid off 2.9.1847. Between Small and Middling repair at Woolwich and Deptford (for £14,010) 1848–50. Recommissioned 21.5.1850 under Cmdr Henry Richard Foote, for same station; paid off 3.1.1853 at Woolwich. Moved to Devonport, where recommissioned 25.2.1854 under Cmdr Edward Bridges Rice; action with Rif tribesmen off Cape Treforeas 26.6.1854; on 12.9.1854 under Cmdr Jasper Henry Selwyn; on 19.1.1856 under Cmdr Charles Webley Hope; took slaver *Adams Gray* 16.4.1857; paid off 15.9.1857 at Woolwich. Refitted at Woolwich (for £14,520) 1858–59. Recommissioned 19.10.1859 under Cmdr Sidmouth Stowell Skipwith, for the west coast of Africa; on 13.11.1860 under Cmdr Norman Bernard Bedingfeld, for the same station; took slaver *Jacinta* 27.5.1861; paid off 21.6.1862 at Woolwich. To Chatham 11.1862 for repair, but found to be

'exceedingly rotten'. Sold to Castle & Sons (for £1,525) 5.3.1863 to BU at Charlton.

Polyphemus Chatham Dyd/Seaward & Capel

As built: 164ft 0in, 142ft 6in x 32ft 8in oa (32ft 6in for tonnage) x 18ft 7in. 800⁵⁸/₉₄bm.

Ord: 25.2.1839. K: 2.1840. L: 28.9.1840. C: 24.4.1841 at Chatham (after machinery fitted at Limehouse).

First cost: £27,596 total (hull £13,198, machinery £10,700, fitting £3,698).

Commissioned 25.2.1841 under Lieut John Evans, for the Mediterranean (died 21.2.1843). On 23.2.1843 under Lieut Thomas Spark. On 23.4.1846 under Cmdr James Johnstone M'Cleverty, for the Channel squadron, later to the Mediterranean; rescue of brig *Three Sisters* from Rif tribesmen 8.11.1846. On 28.12.1848 under Cmdr Richard Borough Crawford; paid off 11.4.1849. Repaired at Deptford and Woolwich (for £16,601) 1850–52. Recommissioned 16.2.1852 under Cmdr Charles Gerrans Phillips, for the west coast of Africa; paid off 23.9.1854 at Portsmouth. Recommissioned 22.10.1855 under Cmdr Frederick Pelham Warren, for the Baltic; wrecked on a steep sandy beach 7 miles from Hansholme on the Baltic coast of Jutland on 29.1.1856 (27 drowned); proceeds of the wreck sold for £467.

Ardent Chatham Dyd/Seaward & Capel

As built: 164ft 0in, 142ft 6½in x 32ft 8in oa (32ft 6in for tonnage) x 18ft 7in. 800⁸/₉₄bm. Draught 6ft 5in/7ft 6in.

Ord: 25.2.1839. K: 2.1840. L: 12.2.1841. C: 16.9.1842 at Chatham (after engines fitted at Limehouse).

First cost: £28,593 total (hull £13,385, machinery £10,700, fitting £4,061).

Commissioned 24.7.1841 under Cmdr John Russell, for the South American and African stations; paid off 11.1845. Repaired and reboiled at Woolwich (for £10,340) 1845–46. Recommissioned 23.4.1846 under Cmdr George Spong, for the Mediterranean; in 17.11.1846 under Lieut John Robinett Baker, then 9.10.1847 under Lieut William Calmady Nowell. Recommissioned 1.10.1854 under Lieut William Horton, for the Mediterranean and the Black Sea; on 18.8.1855 under Lieut Hubert Campion. Refitted at Woolwich (for £15,641) 1856–57. Commissioned 12.11.1857 under Cmdr John Halliday Cave, for the west coast of Africa; in action in Great Scarcies River 1–2.1858; grounded 24.12.1858 off Benin; home in 3.1859 and paid off 4.4.1859 at Woolwich. Recommissioned 19.10.1859 under Cmdr John Edward Parish, for the east coast of South America; paid off 5.8.1863. BU 2.3.1865 (arrived) at Castle, Charlton.

Rattler Sheerness Dyd/— .

Ord: 18.3.1841. K: — . On 24.2.1842 re-ordered as a screw sloop.

DRIVER Class Steam vessels (SV2), later First Class sloops. Sir William Symonds design, 1840. The first six were ordered on 12 March 1840 and named on 14 May 1840. Ten more were ordered on 18 March 1841, but only six of the latter were built to this plan; one (*Rattler*) was initially amended to the *Alecto* Class design (see above), and was then swiftly re-ordered as a screw sloop; for the other three (*Bulldog*, *Inflexible* and *Scourge*), the design was lengthened (at the bow) to provide the basis for the *Bulldog* Class (see below). A further five (*Furious*, *Magicienne*, *Resolute*, *Tiger* and *Valorous*) were included in the 1847 Programme of Work, ordered on 25 April 1847, but four of these were then re-ordered as paddle frigates (which see), the fifth (*Resolute*) being cancelled. The Maudslay-engined *Devastation* had a 4-cylinder 'Siamese' engine (400nhp; 785ihp = 10kts) and the Penn-

An annotated reference drawing of the paddle sloop *Driver*, as shown in almost photographic detail by Sir Oswald Walters Brierly at Ledsund in the Aland Island during the first season's campaign in the Baltic, on 4 August 1854, prior to the attack on Bomarsund. The *Driver* was back at Harwich by the end of 1854, but returned to the Baltic during the 1855 campaign, where on 2 July she was engaged (along with the *Harrier*) in the bombardment of the port of Raumo. *(NMM PW8488)*

engined *Sphynx* had a 2-cylinder oscillating engine (500nhp; 1,169ihp = 12kts).

Dimensions & tons: 180ft 0in, 156ft 0¾in x 36ft 0in oa (35ft 8in for tonnage) x 21ft 0in. 1,055⁶²⁄₉₄bm. 1,590 disp.

Men: 149 (later 160). Guns: 2 x 10in/42pdrs (84cwt) on pivot, 2 x 68pdrs (64cwt), 2 x 42pdrs (22cwt). 1 x 68pdr (95cwt) replaced 1 x 10in in 1856, and 4 x 32pdrs (42cwt) replaced the 4 smaller guns; a 110pdr later replaced the 68pdr.

Machinery: 2-cylinder direct-acting (side lever in *Spiteful*). 280nhp in first six (except *Devastation*); others as stated individually below. *Devastation* and *Sphynx* were differently engined (see above).

Driver Portsmouth Dyd/Seaward & Capel

As built: 180ft 0½in, 156ft 1⅜in x 36ft 0in oa (35ft 8in for tonnage) x 21ft 0in. 1,056³¹⁄₉₄bm. Draught 7ft 1in/8ft 6in.

Ord: 12.3.1840. K: 6.1840. L: 24.12.1840. C: 5.11.1841 at Portsmouth (after machinery fitted at Limehouse).

First cost: £39,707 total (hull £19,433, machinery £13,866, fitting £6,408).

Commissioned 30.8.1841 under Cmdr Samuel Fielding Harmer, for the East Indies (died 16.4.1843). On 6.5.1843 under Cmdr Courtenay Osborn Hayes, in the East Indies; New Zealand operations 1846; paid off 25.5.1847 at Woolwich (the first steam vessel to complete a circumnavigation). Recommissioned 8.1848 under Lieut Edward King Barnard; on 16.9.1848 under Charles Richardson Johnson, for the Pacific; paid off 9.5.1852. In 3.1853 at Chatham under Cornelius Fox, Master. On 25.1.1854 under Cmdr Arthur Auckland Leopold Pedro Cochrane, for the Baltic. On 12.9.1854 under Cmdr Edward Bridges Rice, for particular service; on 26.2.1855 under Cmdr Alan Henry Gardner, in the Baltic, then 23.3.1856 under Cmdr Ennis Chambers, on particular service; paid off 12.2.1857 into Steam Reserve at Sheerness. Recommissioned 3.5.1861 under Cmdr Horatio Nelson; wrecked 3.8.1861 on Mayaguana Island in the West Indies.

Styx Sheerness Dyd/Seaward & Capel

As built: 180ft 0½in, 154ft 10in x 36ft 2in oa (35ft 10in for tonnage) x 21ft 0in. 1,057⁴⁷⁄₉₄bm. Draught 7ft 3in/8ft 3in.

Ord: 12.3.1840. K: 22.6.40. L: 26.1.1841. C: 4.10.1841 at Sheerness (after machinery fitted at Limehouse).

First cost: £39,805 total (hull £20,521, machinery £13,866, fitting £5,418).

Commissioned 12.8.1841 under Cmdr Hastings Reginald Henry. On 15.9.1841 under Capt. Alexander Thomas Emeric Vidal, for surveying in the Azores; paid off 1.1845. Recommissioned 7.2.1845 under Cmdr William Wyndham Hornby, for west coast of Africa; took slavers *Regenerador* 22.10.1845, *Izabel* 27.10.1845 and *Espiga* 1.10.1845; on 20.4.1846 under Cmdr Henry Chads, for the same station; on 30.4.1846 under Cmdr George Oldmixon, on same station; took slavers *Rolla* 17.9.1846, *Anna & Constanca* 19.1.1847, *Nictheroi* 26.2.1847, *Nero* 18.8.1847, *Atrevida* 14.9.1847, another (name unknown) 10.10.1847, *Quatorze de Novembre* 19.10.1847, *Heroina* 24.10.1847, another *Izabel* 7.11.1847, *Flamingo* and *Boa Fe* 15.11.1847, *Santa Anna* 11.12.1847, *Gaio* 16.12.1847, *Cidade d'Angra* 27.12.1847, *Umbelina* 1.1.1848, *Pedreira* 4.1.1848, *Maria Constanca* 12.1.1848, *Leopoldina* 14.1.1848, *Sylphide* 24.1.1848, *Gentil Africano* 18.2.1848, and *Flora* 13.3.1848; paid off 7.6.1848. Repaired at Woolwich and Deptford (for £22,197) 1848–51. Recommissioned 15.7.1851 under Cmdr William King Hall, for the Cape of Good Hope; on 7.6.1854 under Cmdr Frederick Woollcombe, for the East Indies and China (died 3.4.1855). On 3.4.1855 under Cmdr James Minchin Bruce, in the East Indies; at Goulan Bay 13.11.1855. On 26.8.1857 under Cmdr Charles Vesey, for North America and the West Indies. On 31.7.1860 under Cmdr John Halliday Cave, on same station. On 26.2.1862 under Cmdr William John Ward, on same station; in 7.1864 under Cmdr William Brabazon Urmston, on same station; paid off 25.10.1865 at Plymouth. BU 4.1866.

Vixen Pembroke Dyd/Seaward & Capel.

As built: 180ft 0in, 155ft 9¼in x 36ft 0in oa (35ft 8in for tonnage) x 20ft 11½in. 1,054³⁄₉₄bm. Draught 7ft 9in/8ft 5in.

Ord: 12.3.1840. K: 6.1840. L: 4.2.1841. C: 28.12.1841 at Chatham and Woolwich (after engines fitted at Limehouse).

First cost: £39,098 total (hull £18,206, machinery £13,866, fitting £7,026).

Commissioned 30.8.1841 under Cmdr Henry Boyes, for China; in Yangtze operations 7.1842. On 5.1.1843 under Cmdr George Giffard, in the East Indies; at Brunei 11.8.1845; flag of Rear-Adm. Sir Thomas John Cochrane at Malluda Bay 18.8.1845; paid off 10.7.1846 at Woolwich. Refit at Woolwich (for £16,625) 1846–47. Recommissioned 26.5.1847 under Cmdr Alfred Phillipps Ryder, for North America and the West Indies; at Fort Serapaqui (Nicaragua) 12.2.1848; on 10.5.1848

under Cmdr Robert Jenner; paid off 22.7.1850 at Portsmouth.
Recommissioned 22.3.1852 under Cmdr Frederick Lamport
Barnard, for east coast of South America; paid off 9.1855. Refit
at Woolwich (for £10,683) 1856–57. Recommissioned 4.2.1857
under Cmdr George Frederick Mecham, for the Pacific (died
17.2.1858 at Honolulu). On 18.2.1858 under Cmdr Lionel
Lambert, in the Pacific (murdered 8.2.1860); on 9.2.1860
under Cmdr Frederick William Richards; paid off 20.4.1861 at
Woolwich. Sold to Castle & Son (by AO 12.11.1862, for £3,740)
to BU at Charlton.

Devastation Woolwich Dyd/Maudslay, Sons & Field
As built: 180ft 0in, 156ft 4¾in x 36ft 0in oa (35ft 8in for tonnage) x
21ft 0in. 1,058²⁴⁄₉₄bm. Draught 7ft 6in/8ft 3in.
Ord: 12.3.1840. K: 27.7.40. L: 3.7.1841. C: 30.11.1841 at Woolwich.
First cost: £45,744 total (including machinery £14,300).
Commissioned 15.9.1841 under Cmdr Hastings Reginald Henry,
for the Mediterranean; in 5.1842 under Lieut John James
Robinson (acting), for the Mediterranean; on 14.2.1843 under
Cmdr Henry again; in 11.1843 under Cmdr Swynfen Thomas
Carnegie, then 17.2.1844 under Cmdr William Hewgill Kitchen;
paid off 8.10.1845 at Woolwich. Recommissioned 2.3.1846 under
Cmdr Edward Crouch, for 1846 Squadron of Evolution, then
west coast of Africa; on 13.5.1846 broad pennant of Commodore
Sir Charles Hotham, for Cape of Good Hope; took slavers
Tres Amigos 19.3.1837, *Rey de Aquiton* 25.3.1847 and *Rosetta*
19.6.1847; on 4.7.1847 Crouch was invalided, command passed
to Cmdr Richard Thomas John Levinge; took slaver *Voador*
9.7.1847; on 19.9.1847 under Cmdr Reynell Charles Michell;
took slavers *Eolo* 6.11.1847, *Agonogro* 24.12.1847 and *Adelaide*
21.1.1848; paid off 18.9.1848 at Portsmouth. Recommissioned
13.5.1851 under Cmdr Colin Yorke Campbell, for North
America and the West Indies; on 10.6.1853 under Cmdr
Algernon Frederick Rous De Horsey, for same station, later to
the Mediterranean; paid off 9.1855. Recommissioned 20.12.1855
under Cmdr Edward Marshall, at Portsmouth; on 17.10.1857
under Cmdr Leveson Eliot Henry Somerset, then 20.11.1857
under Cmdr Charles Wake, for North America and the West
Indies; in 1858 under Cmdr John Kennedy Erskine Baird.
Recommissioned 9.12.1861 under Cmdr John Dobree McCrea,
for the Pacific; on 9.5.1862 under Cmdr John William Poke, then
12.1864 under Cmdr William Kynaston Joliffe, in the Pacific;
paid off 8.8.1866 at Woolwich. BU 9.1866 by Castle at Charlton.

Geyser Pembroke Dyd/Seaward & Capel
As built: 180ft 0in, 155ft 9¾in x 36ft 0in oa (35ft 8in for tonnage) x
21ft 1in. 1,054²⁹⁄₉₄bm. Draught 7ft 7in/8ft 6in.
Ord: 12.3.1840. K: 8.1840. L: 6.4.1841. C: 8.3.1842 at Sheerness and
Chatham (after engines fitted at Limehouse).
First cost: £39,853 total (hull £17,293, machinery £13,633, fitting
£8,922).
Commissioned 13.12.1841 under Cmdr Edward John Carpenter,
for the Mediterranean; paid off 2.7.1846. Recommissioned
28.10.1846 under Cmdr Francis Thomas Brown, for the Cape
of Good Hope; on 9.10.1850 under Cmdr Edward Tatham, for
east coast of South America; grounded near Rio de Janeiro, and
paid off for repair at Woolwich 4.12.1851. Recommissioned
8.9.1852 under Cmdr Thomas Wilson, for North America and
the West Indies; paid off 17.6.1853. Recommissioned 5.12.1854
under Cmdr Roderick Dew, for the Baltic. On 12.1.1856
under Cmdr Arthur Tower, for the Cape of Good Hope; paid
off 12.1857. Recommissioned 23.6.1860 under Cmdr George
Melville Jackson, for the Channel squadron; on 15.4.1862 under

Cmdr Colin Andrew Campbell, then 31.3.1863 under Cmdr
Mark Robert Pechell. On 24.2.1864 under Cmdr Arthur Thomas
Thrupp, at Portsmouth; paid off 1.6.1865. Towed away 2.4.1866
to BU by White at Cowes.

Growler Chatham Dyd/Seaward & Capel
Ord: 12.3.1840. K: 1.1841. L: 20.7.1841. C: 9.3.1842 at Chatham
(after engines fitted at Limehouse).
First cost: £39,461 total (hull £19,559, machinery £14,108, fitting
£5,804).
Commissioned 7.12.1841 under Cmdr Claude Henry Mason
Buckle, for South America; to west coast of Africa 7.1844; took
slavers *Veterano* 21.7.1844, *Concepcion* 23.9.1844, *Enganador*
4.11.1844 and *El Cayman* 11.1.1845; took part in destruction
of barracoons at Dombocorro 2.1845; paid off 11.1845. Repair
at Deptford and Woolwich (for £16,764) 9.1846–1.1847;
fitted for the 'transport of free blacks [i.e. former slaves] from
America to the West Indies'. Recommissioned 30.3.1847 under
Cmdr John Moon Potbury; paid off 23.3.1848 at Portsmouth.
Recommissioned 4.7.1849 under Cmdr James Stoddart, for
the Mediterranean; paid off 25.9.1852 at Portsmouth. Docked
7.1853 for a 'thorough' refit, but found to be 'quite rotten'.
BU commenced 19.11.1853 and completed by 17.1.1854 at
Portsmouth.

Thunderbolt Portsmouth Dyd/Robert Napier (314nhp)
As built: 180ft 0in, 156ft 0in x 36ft 0in oa (35ft 8in for tonnage) x
21ft 0in. 1,055⁵¹⁄₉₄bm.
Ord: 18.3.1841. K: 4.1841. L: 13.1.1842. C: 8.2.1843 at Portsmouth
(after machinery fitted at Pembroke).
First cost: £44,567 total (hull £24,454, machinery £13,600, fitting
£5,113).
Commissioned 28.11.1842 under Cmdr George Nathaniel Broke,
for the Cape of Good Hope. On 27.12.1845 under Cmdr
Alexander Boyle, on same station; wrecked on Cape Recife,
Algoa Bay (South Africa) on 3.2.1847.

Cormorant Sheerness Dyd/William Fairbairn (291nhp)
Ord: 18.3.1841. K: 17.5.1841. L: 29.3.1842. C: 28.6.1843 at Deptford
and Sheerness (after engines fitted at Limehouse).
First cost: Hull £22,695, machinery £14,976, fitting £7,227.
Commissioned 11.4.1843 under Cmdr George Thomas Gordon, for
the Pacific coast of South America (–2.1847). On 5.6.1847 under
Cmdr Frederick Beauchamp Paget Seymour, for the Pacific;
paid off 13.12.1847 at Portsmouth. Refitted at Portsmouth
(for £16,644) 9.1849. Recommissioned 29.8.1849 under Cmdr
Herbert Schomberg, for east coast of South America; took
slavers *Astrea* 12.12.1849, *Rival* off Cape Frio 26.6.1850, and
Campadora, *Serea* and *Donna Anna* at Paragua 1.7.1850; on
13.1.1851 under Cmdr William Charles Chamberlain, on same
station; paid off 28.2.1852 at Woolwich. Docked 18.8.1853 for
repairs/refit, but found to be in poor condition and condemned
12.1853. Engine removed and hull towed to Deptford for BU.

Spiteful Pembroke Dyd/Scott & Sinclair (294nhp) beam engine
As built: 180ft 0in, 155ft 10in x 36ft 0in oa (35ft 8in for tonnage) x
21ft 1in. 1,054⁴²⁄₉₄bm. Draught 7ft 8in/8ft 7in.
Ord: 18.3.1841. K: 8.1841. L: 24.3.1842. C: 24.3.1843 at Plymouth.
First cost: £40,367 (hull £18,580, machinery £13,094, fitting £8,693).
Commissioned 14.12.1842 under Cmdr William Maitland, for the
East Indies (died 10.1846); in Brunei River operations 1846;
on 29.10.1846 under Cmdr William Legge George Hoste; paid
off 30.7.1847 at Woolwich. Refitted at Woolwich (for £11,560)
3–8.1848. Recommissioned 4.10.1849 under Cmdr Thomas
Carmichael, for the Mediterranean (–3.6.1851); in 1851 under

Cmdr Thomas Abel Brimage Spratt, then 3.6.1851 under Cmdr George Parker; aid off 7.12.1852 at Plymouth. Recommissioned 12.7.1854 at Woolwich under Cmdr Augustus Frederick Kynaston, for the Baltic; in bombardment of Sebastopol 11.10.1854 (Kynaston severely wounded); on 19.1.1855 under Cmdr Francis Henry Shortt, in the Mediterranean; paid off 23.5.1857 at Woolwich. Recommissioned 30.8.1860 under Cmdr William Charles Fahie Wilson, for the east coast of South America (invalided 15.1.1867); on 8.3.1867 under Cmdr Benjamin Langlois Lefroy, on same station; in 1868 in punitive expedition against the Emperor of Ethiopia; paid off 24.12.1868. Recommissioned 7.1873 under Cmdr Mervyn Bradford Medlycott, for the Cape of Good Hope and then west coast of Africa; in Congo River 8.1875; in 11.1875 under Cmdr Armand Temple Powlett; at blockade of Dahomey; paid off 6.7.1877. Sold to Castle 9.1883.

Infernal Woolwich Dyd/Miller, Ravenhill & Salkeld (296nhp)
As built: 180ft 0in, 156ft 6in x 36ft 0in oa (35ft 8in for tonnage) x 21ft 1in. 1,058⁹⁹⁄₉₄bm.
Ord: 18.3.1841. K: 8.1841. L: 31.5.1843. C: 9.9.1844 at Woolwich (after machinery fitted at East India Dock).
First cost: £42,651 total (hull £18,225, machinery £14,079, fitting £7,357).
Commissioned 26.8.1844 (and renamed *Eclair*) under Cmdr Walter Grimston Bucknall Estcourt, for the west coast of Africa (died 16.9.1845); took slaver *Mariana* 15.5.1845; on 16.9.1845 under Cmdr Henry Cooke Harston (acting); fever killed 71 of the 146 men aboard; paid off 13.11.1845 at Sheerness. Renamed *Rosamond* 14.10.1846. Recommissioned 5.11.1846 under Cmdr John Foote, for Cape of Good Hope; later to the Mediterranean; paid off 12.4.1850. Recommissioned 20.10.1854 under Cmdr Stephen Smith Lowther Crofton, for particular service; paid off 6.2.1856. Fitted as steam floating factory at Portsmouth 5.1861–7.1862. BU at Portsmouth from 6.9.1864.

Virago Chatham Dyd/Boulton, Watt & Co. (294nhp)
As built: 180ft 0in, 156ft 2in x 36ft 0½in oa (35ft 8½in for tonnage) x 21ft 0in. 1,059¹⁷⁄₉₄bm. Draught 7ft 2in/8ft 6in.
Ord: 18.3.1841. K: 15.11.1841. L: 25.7.1842. C: 29.7.1843 at Chatham (after machinery fitted at Woolwich).
First cost: £42,714 total (hull £19,650, machinery £15,865, fitting £7,199).
Commissioned 25.5.1843 under Cmdr George Graham Otway, for the Mediterranean; on 20.5.1846 under Cmdr John Lunn, on same station; paid off 16.11.1847 at Woolwich. Recommissioned 5.8.1851 under Cmdr William Houston Stewart, for the Pacific; recaptured rebel Chilean colony of Punta Arenas; in 12.1852 under Cmdr James Charles Prevost, surveying on west coast of Canada; on 5.4.1853 under Cmdr Edward Marshall, for the Pacific; with Anglo-French squadron; in action at Petropavlovsk (Kamchatka) 31.8.1854; paid off 31.7.1855. In 5.5.1856 under Cmdr Henry Vachell Haggard, for east coast of South America. On 15.3.1858 under Cmdr Montagu Buccleuch Dunn, at Devonport; paid off 9.6.1860. At Sheerness in 1860. Recommissioned 31.7.1861 under Cmdr William George Hope Johnstone, for the Channel; to the West Indies in 8.1863; paid off 1.1865. Recommissioned 25.8.1869 under Cmdr Elibank Harley Murray, for Australia; paid off 25.11.1871. BU 11.1876 at Chatham.

Sphinx Woolwich Dyd/John Penn & Sons (500nhp)
As built: 180ft 0in, 156ft 5½in x 36ft 0½in oa (35ft 8½in for tonnage) x 21ft 0in. 1,061¹¹⁄₉₄bm. Draught 6ft 6in/7ft 11in.

Ord: 18.3.1841. K: 5.1844. L: 17.2.1846. C: 3.11.1846 (for sea).
First cost: Total £54,843 (hull £16,802, machinery £25,578, fitting £12,463).
Commissioned 29.6.1846 under Cmdr John Bettinson Cragg; paid off 20.2.1847. Recommissioned 2.2.1850 under Cmdr Charles Frederick Alexander Shadwell, for the East Indies; in 2nd Anglo-Burmese War; repulse of landing parties near Rangoon 4.2.1852; paid off 22.7.1853. Recommissioned 15.9.1854 under Capt. Arthur Parry Eardley Wilmot, for the Mediterranean and Black Sea; at bombardment of Sebastopol 11.10.1854; paid off 2.5.1857. Recommissioned 20.9.1859 at Portsmouth under Cmdr George Fiott Day, for the East Indies and China; on 1.1.1862 under Cmdr Theodore Morton Jones; paid 9.7.1863 at Woolwich. In 4.1865 under Cmdr Richard Vesey Hamilton, for the West Indies; on 17.4.1868 under Capt. John Edward Parish; paid off 5.12.1868. In 5.1870 under Cmdr Henry Bouchier Phillimore, for North America and the West Indies; paid off 3.9.1874. At Plymouth from 1875. BU 9.1881 at Devonport.

BULLDOG Class Steam vessels (SV2), later First Class sloops. Sir William Symonds design, 1844. Developed from the *Driver* Class by AO of 26 December 1843, with the bow re-designed and lengthened by 10ft (mainly to provide an engine-room lengthened by 6ft). *Scourge* had 2 funnels, and was completed as a bomb vessel, with a 13in mortar plus 2 x 68pdrs (95cwt/10ft) on the UD; the others each had a single funnel. In July 1844 it was queried whether *Fury* should be completed as a screw ship, but it was decided to complete her with paddles as planned as she was too far advanced for conversion.
Dimensions & tons: 190ft 0in, 166ft 0¾in x 36ft 0in oa (35ft 8in for tonnage) x 21ft 0in. 1,122¹¹⁄₉₄bm.
Men: 160. Guns: (original) 2 x 42pdrs (84cwt/10ft) on pivots, 2 x 68pdrs (64cwt/9ft), 2 x 42pdr carronades (22cwt). By 1862 had 1 x 10in (84cwt/10ft) and either 1 x 68pdr (95cwt/10ft) or 1 x 110pdr (82cwt), plus 4 x 32pdrs (42cwt) or (*Inflexible*) 4 x 8in (52cwt).
Machinery: 2-cylinder direct-acting, nhp as given individually below.

Inflexible Pembroke Dyd/Fawcett, Preston & Co. (378nhp. 680ihp = 9½kts)
As built: 190ft 0in, 165ft 10in x 36ft 0in oa (35ft 8in for tonnage) x 21ft 1in. 1,122¹¹⁄₉₄bm. Draught 7ft 8in/8ft 1in.
Ord: 18.3.1841. K: 1.1844. L: 22.5.1845. C: 9.8.1846 at Devonport (after engine fitted at Liverpool).
First cost: Total £50,114 (hull £22,338, machinery £18,458, fitting £9,418).
Commissioned 10.6.1846 under Cmdr John Cochrane Hoseason, for the East Indies and China; in New Zealand operations 1847; in action against pirate junks at Lemma Island 30.5.1849; on 12.7.1850 under Cmdr Peché Hart Dyke; to Channel Fleet 6.1852; on 13.7.1852 under Cmdr George Rhodes Wolrige, for particular service. Recommissioned 7.1853 under Cmdr George Otway Popplewell, for the Mediterranean and Black Sea; in attack on Fort Nicolaiev 4.10.1854; paid off 15.8.1855. Recommissioned 28.7.1856 under Cmdr John Corbett, for the East Indies and China; boats at Fatshan 1.6.1857; on 15.8.1857 under Cmdr George Augustus Cooke Brooker; at destruction of pirate junks at Coulan 8–9.1858; paid off 27.4.1861. Sold to Castle & Beech 7.1864 and towed from Portsmouth 8.9.1864 to BU at Charlton.

Scourge Portsmouth Dyd/Maudslay, Sons & Field (420nhp)
As built: 190ft 0in, 166ft 0¾in x 36ft 0in oa (35ft 8in for tonnage) x

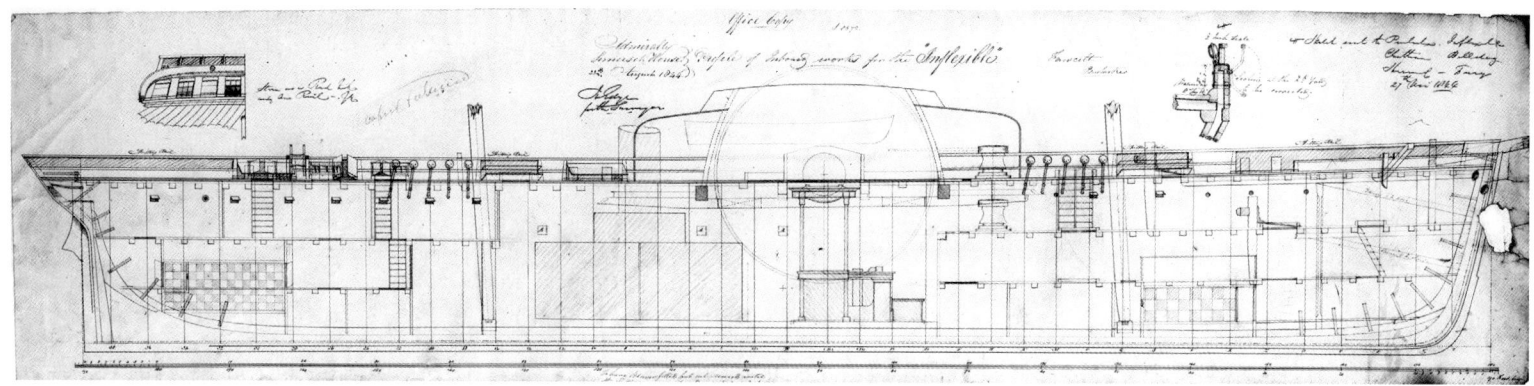

A profile of the inboard works for the *Inflexible*, dated 23 August 1844. The hatched areas of the bulwarks – referred to on the draught as 'shifting rails' – were removable. The *Inflexible* was one of the *Bulldog* Class of paddle sloops designed by Symonds as a development of his *Driver* Class of 1840. The *Inflexible*, *Scourge* and *Bulldog* were originally ordered under the 1841 Programme of Works as units of the *Driver* Class, but in December 1843 it was ordered that these three ships (which had not been started at that date) should have their engine-room extended by 6ft to accommodate their more powerful engines. (*NMM DR6290*)

21ft 0in. 1,123⁶²⁄₉₄bm. Draught 7ft 9¼in/8ft 4in.
Ord: 18.3.1841. K: 2.1844. L: 9.11.1844. C: 13.5.1846 at Portsmouth (after engine fitted in East India Docks).
First cost: Total £55,002 (hull £21,328, machinery £20,390, fitting £13,284).
Commissioned 26.11.1845 under Cmdr James Crawford Caffin, for the Channel squadron; on 1.10.1847 under Cmdr Henry Edward Wingrove, for North America and the West Indies; in 10.1849 under Cmdr Lord Frederick Herbert Kerr, for the Mediterranean; paid off 27.11.1852. Recommissioned 4.1854 under Commodore John Adams, for the west coast of Africa; paid off 28.11.1857. Recommissioned 28.6.1858 under Cmdr Prince Victor Ferdinand Franz Eugen Gustaf Adolf Constantin Friedrich of Hohenlobe-Langenbourg, for the Mediterranean; on 13.12.1859 under Cmdr William Gore Jones; paid off 8.2.1862. BU 1865.

Bulldog Chatham Dyd/Rennie (orig. 420nhp intended, revised to 500nhp. 10.2kts)
As built: 190ft 0in, 166ft 2⅛in x 36ft 0¼in oa (35ft 8¼in for tonnage) x 21ft 0in. 1,125⁷¹⁄₉₄bm. Draught 7ft 11½in/8ft 5in.
Ord: 18.3.1841. K: 7.7.1844. L: 2.10.1845. C: 7.9.1846 at Chatham (after engine fitted in East India Dock).
First cost: Total £58,122 (hull £23,342, machinery £24,892, fitting £8,338.
Commissioned 25.6.1846 at Devonport under Cmdr George Evan Davis, for the Cape of Good Hope; on 4.5.1847 under Cmdr Astley Cooper Key, off Portugal; paid off 16.4.1850. Recommissioned 23.1.1854 under Cmdr William King Hall, for the Baltic; flag of Sir Charles Napier at bombardment of Bomarsund 16.8.1854; in 2.1855 under Cmdr Alexander Crombie Gordon; paid off 25.3.1857. Recommissioned 2.6.1860 under Sir Francis Leopold McClintock, ocean sounding for Atlantic Telegraph; on 3.12.1860 under Cmdr Henry Frederick McKillop. Fitted with Armstrong guns in 1861. In 3.1864 under Capt. Charles Wake, for North America and the West Indies; in action at Cape Haytien with Haitian ships and forts 23.10.1865, she sank the Haitian *Valorogue* and a schooner before running aground and then being deliberately blown up.

Fury Sheerness Dyd/Rigby (515nhp. 10.5kts)
As built: 190ft 0in, 166ft 0¼in x 36ft 0in oa (35ft 8in for tonnage) x 21ft 0in. 1,123⁶²⁄₉₄bm. Draught 8ft 5in/8ft 5½in.
Ord: 19.2.1844. K: 6.1845. L: 31.12.1845. C: 6.7.1847 at Sheerness (after engine fitted at Liverpool).
First cost: Total £51,688 (hull £24,764, machinery £22,142, fitting £4,782).
Commissioned 20.7.1847 under Cmdr James Willcox, for the East Indies and China; destroyed (with *Columbine*) 23 pirate junks at Tysami 29.9.1849 and pirate fleet at Haiphong on 20–21.10.1849. Refitted at Woolwich (for £12,987) 1851–52. Recommissioned 4.12.1851 under Cmdr Edward Tatham; in 1854 to the Black Sea; in 8.1854 under Cmdr Ennis Chambers. Refitted at Portsmouth (for £23,838) 1855–56. Recommissioned 1.8.1856 under Cmdr Charles Taylor Leckie, for the East Indies and China; boats at Fatsham 1.6.1857. In 7.1859 under Cmdr William Andrew James Heath, for East Indies and China; at Peiho forts 26.6.1859; on 2.1.1860 under Cmdr John Crawford Wilson; paid off 19.6.1861. Sold to Castle & Beech 7.1864 to BU at Charlton.

TRIDENT Second Class iron-hulled sloop. Thomas Ditchburn design, approved 23 August 1843. This design was commissioned 2 August 1842 for a steam yacht to replace *Black Eagle* (ex-*Firebrand*), rated Steam Vessel (SV3); re-rated Second Class sloop 31 May 1844. Originally intended to be fitted with Maudslay 200nhp side lever engine.
Dimensions & tons: 180ft 0in, 161ft 1in x 31ft 6in x 17ft 3in. Draught 10¼ft (mean). 850²⁶⁄₉₄bm. 903 disp.
Men: 135. Guns: 2 x 10in (85cwt), 2 (later 4) x 32pdr (25cwt/6ft) gunnades.
Machinery: Boulton, Watt & Co. 2-cylinder (70¼in diameter, 5ft stroke) oscillating. 350nhp. 9½kts.

Trident Ditchburn & Mare, Blackwall/Boulton, Watt & Co.
Ord: 30.4.1843. K: 1845. L: 16.12.1845. C: 8.8.1846 at Chatham (after engines fitted at East India Dock).
First cost: £41,446 total (hull £17,000, machinery £17,502, fitting £6,864).
Commissioned 28.6.1846 under Lieut Charles Gray Rigge, for the Mediterranean; on 10.5.1848 under Lieut Richard Hawkins Risk, at Sheerness (–1850). Recommissioned 19.7.1852 under Lieut Robert Beazley Harvey, for east coast of South America (–1855). On 24.11.1856 under Cmdr Francis Arden Close, for the west coast of Africa; took slaver *Lydia Gibbs* 29.5.1858. Recommissioned 5.1.1861 at Woolwich under Cmdr Beville Granville Wyndham Nicolas, for the Mediterranean; on 11.12.1861 Nicolas was dismissed from the Navy for excessive cruelty. On 1.1.1862 under Cmdr Charles John Balfour, on same station; paid off 20.12.1864 at Woolwich. Sold 9.1865 and BU 1.1866 by Castle at Charlton.

The *Oberon* was one of three iron-hulled packets built for the mail service between the Ionian Islands, Malta and Greece. They were re-classed as Third Class sloops in 1847 and subsequently as gunvessels. This watercolour portrait is by William Edward Atkins. *(NMM PW5616)*

JANUS Steam vessel (SV2), later First Class sloop. Thomas Cochrane, Lord Dundonald double-ended design, 1843. An unsatisfactory experimental design by a distinguished naval officer.

    Dimensions & tons: 180ft 0in, 159ft 5in x 30ft 6in (30ft 0in for tonnage) x 17ft 6in. 763¹¹/₉₄bm.

    Men: — . Guns: 2 x 10in (84 cwt) on pivots. Re-armed 1855: 1 x 68pdr (95cwt) replaced 1 x 10in; 4 x 32pdrs (17cwt) added.

    Machinery: Rotary engine, with watertube boilers of Cochrane's own design.

Janus Chatham Dyd/— .

    Ord: 31.7.1843. K: 6.9.1843. L: 6.2.1844. C: 1.1846 at Chatham.

    First cost: £37,133 (including machinery £10,948).

    Commissioned 12.2.1849 under Lieut Richard Ashmore Powell, for Gibraltar; completed fitting at Woolwich as a tug for Gibraltar (for £14,397) 3.1849; in action against Rif tribesmen at Cape Treforcas 19.10.1851; paid off 6.8.1852. Completed fitting at Woolwich as a gunvessel (for £7,378) 4.1854. Recommissioned 20.1.1854 under Lieut Colin Campbell Kane, as gunner training vessel at Sheerness; paid off 7.7.1855. Re-engined 1855 with the 2-cylinder 220nhp side lever engine ex-*Sydenham*. Sold to Henry Castle (for £3,300 including engines) 4.1856 to BU.

ANTELOPE Class Third Class iron-hulled sloops, also listed as packets. Thomas Ditchburn design, approved 30.9.1845. For the mail service between the Ionian Islands, Malta and Greece. Later re-classed as gunvessels. The original order (on 12 August 1845) for two wooden packets of 600bm and 200nhp (with 1 x 40cwt and 2 x 17cwt 32pdrs) was replaced on 5 September 1845 by below order (third ship added 20 September 1845). Rennies appear on some records as contractor for *Oberon* (built 'at Deptford'), but they did not own a shipyard until 1858, and Ditchburn & Mare seem to have constructed all three hulls. Thomas Ditchburn retiring in 1847, *Triton* was delivered by the successor firm of C. J. Mare & Co.

    Dimensions & tons: 170ft 0in, 155ft 8¼in x 28ft 0in x 18ft 0in. 649²⁹/₉₄bm. 1,055 disp. Draught about 10½ft (mean).

    Men: 100. Guns: 1 x 68pdr (95cwt) on pivot, 3 x 32pdrs (17cwt). By 1856 – 1 x 32pdr (45cwt) on pivot + 2 x 32pdrs (25cwt).

    Machinery: 2-cylinder oscillating, 260nhp. 10½–11kts.

Antelope Ditchburn & Mare, Blackwall/John Penn & Sons

    As built: 170ft 7¾in, 153ft 8¾in x 28ft 3½in x 18ft 6in. 654⁴⁸/₉₄bm. 1,055 disp. Draught 5ft 11in/7ft 1in.

    Ord: 5.9.1845. K: 10.12.1845. L: 20.7.1846. C: 4.9.1847 at Chatham (after machinery fitted at East India Dock).

    First cost: £36,201 total (hull £15,037, machinery £14,705, fitting £6,459).

    Commissioned 27.7.1847 under Lieut Francis Smyth, for the Mediterranean (–1850), paid off 17.5.1852 at Woolwich. Refitted at Woolwich and Portsmouth (for £7,832) 1851–52. Recommissioned 24.2.1852 under Lieut Charles Henry Young, for the Mediterranean, later to the west coast of Africa; paid off 12.4.1856 at Woolwich. Refitted at Woolwich (for £12,385) 1856–57. Recommissioned 1.12.1856 under Lieut John William Pike, for the west coast of Africa; took slavers *Jupiter* 29.6.1857 and *Joseph Record* 3.9.1857; paid off 12.10.1859 at Woolwich. At Devonport 1860, then to Woolwich where recommissioned 5.8.1861 under Lieut Constantine O'Donnel Allingham, for same station (died 11.5.1865); on 11.5.1865 under Alfred Thomas, 2nd Master; paid off 8.7.1865 at Woolwich. Recommissioned 4.7.1866 under Lieut John Bruce, for same station (invalided 9.1867); on 20.9.1867 under Lieut James Buchanan; to the Mediterranean 7.1868 (Bruce invalided 6.1870); on 1.7.1870 under Lieut Frederick Canning Lascelles (acting); on 12.10.1870 under Lieut Charles Sheldon Pearse Woodruffe; fitted 1870–71 at Malta for the service of the Ambassador at Constantinople; paid off 11.2.1874 at Malta. Recommissioned 5.12.1873 under Lieut John Coke Burnell, for the Mediterranean; paid off 2.1874, but immediately recommissioned, still under Burnell but with a new crew. On 23.2.1877 under Lieut Edward John Wingfield, then on 25.2.1880 under Lieut Walter Hylton Joliffe, still in the Mediterranean; paid off 8.5.1883. Sold (by AO 16.5.1883) 20.9.1883 at Malta.

Oberon Ditchburn & Mare, Blackwall/G. & J. Rennie

    As built: 172ft 3⅞in, 155ft 6⅛in x 28ft 0in x 18ft 0in. 648⁵⁶/₉₄bm. 1,055 disp. Draught 5ft 9in/6ft 6in.

    Ord: 5.9.1845. K: 7.1.1846. L: 2.1.1847. C: 13.11.1847 at Chatham (after machinery fitted at East India Dock).

    First cost: £28,088 total (hull £14,669, machinery £4,177, fitting £7,442).

    Commissioned 27.8.1847 under Lieut George Johnson Gardner, for the Mediterranean; on 23.10.1850 under Lieut Robert Beazley Harvey. Recommissioned 72.1855 under Lieut John Osmond Freeland, again for the Mediterranean; on 2.6.1856 under Lieut

The paddle sloop *Basilisk* is shown in 1872 on the Australian Station, en route from Brisbane to Cape York on 5 February 1872. Her captain, John Moresby, had previously commanded the paddle sloop *Argus* during the western allies' attack on the Strait of Shimonoseki in September 1864 in the so-called 'Satsuma War', earning promotion to Captain in that action. The drawing shows the *Basilisk* stopping the 'blackbirding' (slaving) schooner *Peri*, whose Portuguese crew had kidnapped 50 Polynesians from Fiji in the preceding December; the gunvessel discovered that the Portuguese had been either killed or had abandoned the schooner, and that barely a dozen Fijians remained alive. The graphite drawing is anonymous, but looks like the amateur work of somebody present at this shocking incident. (*NMM PW8116*)

John Milbourne Jackson; paid off 10.1.1857. Recommissioned 7.6.1858 under Lieut Frederick George Charles Paget, for east coast of South America; paid off 1.11.1862. Recommissioned 14.11.1865 under Lieut Edmund Hope Verney, for the west coast of Africa; on 29.6.1865 under Lieut Henry Hand; on 25.10.1867 under Lieut John Shortt; paid off 20.7.1869. Lent to the War Office for torpedo experiments from 1870; sunk by 'Whitehead torpedo' (actually a mine with a 500lb gun cotton charge) 20.5.1875, but raised 11.1875 and repaired. For sale on 27.10.1880, and sold 15.11.1880 to Isaacs.

Triton Ditchburn & Mare, Blackwall/Ravenhill & Miller
 As built: 172ft 11½in, 155ft 1½in x 28ft 2in (28ft 0¾in for tonnage) x 18ft 0½in. 648⁵⁶/₉₄bm. 1,055 disp. Draught 6ft 1in/7ft 0in.
 Ord: 5.9.1845. K: 3.12.1845. L: 24.10.1846. C: 13.4.1848 at Chatham and Woolwich (after machinery fitted at East India Dock).
 First cost: £35,671 total (hull £15,328, machinery £14,525, fitting £5,818).
 Commissioned 21.3.1848 under Lieut Charles James Price Glinn, for the Mediterranean (–1850). Refitted for sea at Woolwich (for £6,517) to 25.6.1852. Recommissioned 24.5.1852 under Lieut Henry Lloyd; at bombardment of Sebastopol 11.10.1854. On 26.12.1854 under Lieut Archibald Douglas William Fletcher, in the Mediterranean and Black Sea; paid off 23.9.1856 at Woolwich. Refitted at Woolwich (for £13,347) to 12.11.1857. Recommissioned 14.10.1857 under Lieut Robert Heron Burton, for the west coast of Africa; paid off 19.1.1861 at Woolwich. Recommissioned 13.1.1863 under Lieut Edward Francis Kerby, for the east coast of South America; on 16.2.1864 under Lieut Richard Henry Napier; paid off 25.10.1867. Sold 17.2.1872 to Isaacs (for £2,525).

BASILISK First Class sloop. Oliver Lang design, 1846 (with similar lines to the screw sloop *Niger*). At one time during 1847, it was

intended that this vessel be built as a First Class screw sloop of 400 ihp, with 8 guns on the UD (1 x 68pdr and 1 x 10in on pivots; 6 x 8in on broadside slides), but she was completed as shown below. When built, measured 5in beamier than design (at 1,031bm). Featured in a tug-of-war trial with her screw-powered near sister *Niger*.

 Dimensions & tons: 190ft 0in, 166ft 1in x 34ft 0in (33ft 8in for tonnage) x 21ft 6in. 1,001²⁵/₉₄bm. 1,710 disp.
 Men: 145. Guns: 1 x 68pdr (95cwt) on pivot, 1 x 10in (84cwt), 4 x 32pdrs (42cwt).
 Machinery: 2-cylinder oscillating. 400nhp. 1,033ihp.
 First cost: £54,745.

Basilisk Woolwich Dyd/Miller, Ravenhill & Salkeld
 As built: 190ft 0in, 166ft 9¾in x 34ft 5in (34ft 1in for tonnage) x 21ft 5in. 1,030⁷¹/₉₄bm. 1,691 disp. Draught 8ft 1in/9ft 3in.
 Ord: 26.3.1846. K: 11.1846. L: 22.8.1848. C: 7.1852 at Portsmouth (after engines fitted in East India Dock).
 First cost: £54,745 total.
 Commissioned 25.5.1852 under Cmdr Francis Egerton, for particular service; to the Baltic 1854. On 8.2.1855 under Cmdr Robert Jenner, still in the Baltic. Recommissioned 15.2.1856 under Cmdr Stephen Smith Lowther Crofton, for the West Indies; on 28.7.1856 under Cmdr George Annesley Phayre; paid off 18.1 1860. 'Thorough' repair at Woolwich 1863–64. In 1865 under William G. Atkinson, Master, at Sheerness; recommissioned 16.9.1865 under Capt. William Nathan Wright Hewett, for China; paid off 13.4.1869. Recommissioned 23.1.1871 under Capt. John Moresby, for Australia; sailed 13.2.1871; at Melbourne 4.6.1871; took slaver *Crispna* 14.1.1873; paid off 12.1874. BU 1882 at Chatham.

The 1847 Programme of Work (submitted on 17 February) provided for twelve additional steam sloops – five First Class sloops of 400nhp 'as *Sphinx* class' (*Tiger*, *Furious*, *Resolute*, *Magicienne* and *Valorous*), two Second Class of 300nhp 'as *Ardent* class' (*Argus* and *Buzzard*) and five Third Class 'duplicates of *Rattler*' (*Archer*, *Parthian*, *Highflyer*, *Brisk* and *Grinder*). All were ordered from the dockyards on 25 April, together with a third Second Class sloop (*Barracouta*), on which date the Surveyor was instructed to submit a drawing for the First Class sloops 'if he thinks it advisable to propose any alterations', and to submit drawings for the Third Class (now described as '*Rattler* enlarged') for consideration to carry a heavier armament than the *Rattler*.

However, it was noted that these ships were not to be built until their plans had been reported on by the Committee of Reference, and a decision made by the Admiralty Board. Only the three Second Class sloops were eventually built as paddle sloops – to separate designs, and all were delayed several years. The *Barracouta* was the last sea-going paddle warship to be built for the Royal Navy. The intended First Class sloops were all completed as paddle frigates (except the *Resolute*, which was cancelled), while the Third Class sloops were all confirmed as screw vessels (see Chapter 12).

BUZZARD Second Class sloop. John Edye design, approved 29 July 1847.

Dimensions & tons: 185ft 0in, 162ft 2in x 34ft 0in oa (33ft 8in for tonnage) x 20ft 0in. $977\frac{61}{94}$bm. 1,530 disp.

Men: 100. Guns: 2 x 10in (84cwt) on pivots, 4 x 32pdrs (25cwt).

Machinery: 2-cylinder oscillating 300nhp. 853ihp = 10kts.

Buzzard Pembroke Dyd/Miller, Ravenhill & Salkeld

As built: 185ft 0in, 162ft $1\frac{5}{8}$in x 34ft $0\frac{1}{2}$in oa (33ft $8\frac{1}{2}$in for tonnage) x 20ft 0in. $979\frac{87}{94}$bm. Draught 8ft 6in/8ft $6\frac{1}{2}$in.

Ord: 25.4.1847. K: 10.1847. L: 24.3.1849. C: 8.11.1850 at Deptford, East India Docks and Woolwich. Fitted for sea at Deptford 23.6.1852.

First cost: £45,699 (+ £2,311 fitting for sea in 1852).

Commissioned 7.5.1852 under Cmdr William Hugh Dobbie, for North America and the West Indies; paid off 15.7.1856. On 5.8.1857 under Cmdr Francis Peel, for east coast of South America; on 20.8.1860 under Cmdr John Trevenen (Acting); paid off 5.4.1861. Recommissioned 21.1.1863 under Cmdr Herbert Frederick Winnington Ingram, for North America and the West Indies; on 14.5.1863 under Cmdr Thomas Hutchinson Mangles Martin, then 21.12.1865 under Cmdr Charles Gowan Lindsay, both on same station; paid off 27.11.1866. Reboilered 1867. Recommissioned 10.12.1867 under Cmdr James George Hobbs Thain, for home serve; on 6.4.1870 under Cmdr Alexander Brown; paid off 29.7.1872. BU 9.1883 at Devonport (by AO 17.6.1881).

ARGUS Second Class sloop. John Fincham design, approved 19 August 1847. Originally planned as an *Alecto* Class sloop, but her design was lengthened by 6ft during 1847. Barque-rigged.

Dimensions & tons: 190ft 0in, 168ft 4in x 33ft 6in oa (33ft 0in for tonnage) x 20ft 11in. $975\frac{3}{94}$bm. 1,630 disp.

Men: 100. Guns: 2 x 10in (84cwt) on pivots, 4 x 32pdrs (25cwt).

Machinery: 2-cylinder oscillating 300nhp. 764ihp = 10kts.

Argus Portsmouth Dyd/John Penn & Sons

As built: 190ft $2\frac{1}{2}$in, 168ft $2\frac{1}{4}$in x 33ft $5\frac{1}{2}$in oa (33ft $1\frac{1}{4}$in for tonnage) x 20ft 11in. $980\frac{37}{94}$bm. Draught 7ft 7in/8ft 3in.

Ord: 25.4.1847. K: 6.1848. L: 15.12.1849. C: 2.5.1853 at Woolwich, Blackwall, Deptford and Devonport.

First cost: £49,188 total (including machinery £18,419).

Commissioned 20.12.1852 at Plymouth under Cmdr Richard Purvis, for North America and the West Indies; paid off 26.8.1856 at Sheerness. Refitted for sea at Sheerness (for £6,684) 1858. Recommissioned 3.6.1858 under Cmdr Herbert Frederick Winnington Ingram, for the Channel, then to the Mediterranean; paid off 6.1861 at Sheerness. Recommissioned 7.8.1862 under Cmdr Lewis James Moore, for the East Indies and China; at bombardment of Kagoshima 15.8.1863; on 12.11.1863 under Cmdr John Moresby; at bombardment of Shimonoseki 5–6.9.1864. On 5.12.1864 under Cmdr Henry Lewis Round, off China; on 30.3.1867 under Cmdr Frederick William Hallowes, still off China; to Indian Ocean 1867–68; in Abyssinian War 1868; paid off 3.12.1869. Recommissioned 26.3.1873 under Cmdr Percy Putt Luxmoore, for west coast of Africa; in Ashanti War 1873; on 10.4.1874 under Cmdr Edward John Jermain, on same station; in 7.1874 to North America and the West Indies; on 6.4.1877 under Cmdr Robert Hastings Harris; paid off 12.12.1879 at Chatham. BU 10.9.1881–10.1881 at Chatham.

BARRACOUTA Second Class sloop. Committee of Reference design, approved 14 January 1848. In 1856 a 68pdr (95cwt) pivot-mounted gun replaced 1 x10in, and the 25cwt guns were replaced by 42cwt; in 1862 the 68pdr was replaced by a 110pdr (82cwt) BLR pivot-mounted gun.

Dimensions & tons: 190ft 2in, 166ft $4\frac{1}{2}$in x 35ft 0in (34ft 6in for tonnage) x 20ft 5in. $1,048\frac{40}{94}$bm. 1,676 disp.

Men: 100. Guns: 2 x 10in (84cwt) on pivots, 4 x 32pdrs (25cwt).

Machinery: 2-cylinder direct-acting 300nhp. 881ihp = $10\frac{1}{2}$kts.

Barracouta Pembroke Dyd/Miller, Ravenhill & Salkeld

As built: 190ft 2in, 166ft $6\frac{1}{4}$ x 34ft $11\frac{3}{4}$in (34ft $5\frac{3}{4}$in for tonnage) x 20ft $5\frac{1}{4}$in. $1,052\frac{93}{94}$bm. Draught 7ft 4in/8ft 4in.

Ord: 25.4.1847. K: 5.1849. L: 31.3.1851. C: 30.7.1853 at Woolwich (after fitting engine at West India Dock).

First cost: £50,042 total (hull £21,653, machinery £18,228, fitting £10,161).

Commissioned 6.1852 (for trials) under Capt. Swynfen Thomas Carnegie; paid off 9.1852. Recommissioned 18.12.1852 under Capt. George Parker, for the East Indies and China; at Petropavlovsk (Kamchatka) 5.1855; action against pirates in Coulan Bay 18.11.1854. In 2.1855 under Cmdr Frederick Henry Stirling, for East Indies and China; on 23.2.1855 under Cmdr Thomas Dyke Acland Fortescue; at Canton 10.1856; at French Folly Fort 4.11.1856; paid off 21.8.1857. On 28.5.1860 under Cmdr William Wood, for North America and the West Indies; on 24.8.1861 under Cmdr George John Malcolm; at Veracruz 1.1862; on 13.11.1863 under Cmdr John D'Arcy, on same station; paid off 23.11.1864 at Sheerness, for repair and then to Steam Reserve. Recommissioned 23.3.1866 under Cmdr William Brabazon Urmston; on 14.4.1866 under Cmdr George Dacres Bevan, for North America and the West Indies; paid off 2.4.1870. Recommissioned 12.3.1873 under Capt. Edmund Robert Fremantle, for west coast of Africa; in Ashanti War 1873; home in 5.1874; in 6.1874 under Capt. Charles Edward Stevens, for the Pacific; at Samoa 3.1876; Stevens dismissed by Court Martial 4.1877, and paid off 14.4.1877. BU 12.1881 at Chatham.

PADDLE GUNVESSELS

On 21 January 1837, as mentioned under 'Paddle Frigates', the former single category of Steam Vessels was split into five classes, the first three categories covering the steam frigates and sloops (see above). The last two categories (SV4 and SV5) covered the smaller ships which were later to be defined as steam gunvessels. These were divided as follows:

> SV4 – *Tartarus*, *Blazer* ; *Colombia*, *Pluto*.
> SV5 – *Lightning*, *Meteor*, *Confiance*, *Echo*, *Alban*, *Carron* ; *African*, *Comet*.

During 1839–41, the following former packets were added to these classes:

> to SV4 – *Kite*, *Avon* (6.5.1839); *Gleaner* (20.11.1839); *Lucifer* (21.3.1840); *Shearwater* (7.8.1840).
> to SV5 – *Boxer*, *Fearless* (early 1839); *Dover* (1840); *Monkey* (21.4.1841); *Dasher* (1841?).

The category of 'steam gunvessel' (or SGV) was introduced on 31 May 1844, with most existing SV4 vessels classed as First Class SGVs (excluding the *Colombia*, which had become a survey vessel in 1842). In theory, existing SV5 vessels became Second Class SGVs, but most of the early ships in the SV5 category were, however, instead re-classed as tenders at this time: the *Comet*, *Meteor*, *Alban* (becoming a transport), *Carron*, *Echo*, *African*, *Confiance*; as also did the *Fearless*, *Monkey* and *Dwarf*. First Class SGVs had a complement of 80 men, and hypothetically would carry 1 x 32pdr (26cwt) and 2 x 32pdr (17cwt) carronades; Second Class SGVs had a complement of 60 men, and were to carry 1 x 18pdr (22cwt) and 2 x 18pdr (15cwt) carronades.

LIZARD Class Steam vessels (SV5), re-classed 1844 as Second Class SGVs. William Symonds design, approved 22 May 1839.

> Dimensions & tons: 120ft 0in, 104ft 10¾in x 22ft 8in (22ft 6in for tonnage) x 13ft 0in. 282⁴⁹⁄₉₄bm.
> Men: 60. Guns: 1 x 18pdr (22cwt) on pivot, 2 x 18pdrs (15cwt) amidships.
> Machinery: 2-cylinder side lever. 100nhp. 157ihp (*Locust*).

Lizard Woolwich Dyd/Maudslay, Sons & Field

> Ord: 18.3.1839. K: 7.1.1839. L: 7.1.1840. C: 22.5.1840 (for surveying), 26.1.1841 (for sea).
> First cost: £15,085 (hull £7,180, machinery £6,453, fitting £1,452).
> Commissioned 12.1.1842 under Lieut Charles James Postle, for the Mediterranean, fitted for surveying; lost in collision with French armed paddle sloop *Véloce* between Gibraltar and Cadiz on 24.7.1843 (no casualties).

Locust Woolwich Dyd/Maudslay, Sons & Field

> As built: 120ft 0in, 105ft 0⅛in x 22ft 8½in (22ft 6½in for tonnage) x 13ft 0in. 283⁷⁷⁄₉₄bm. Draught 4ft 11in/5ft 5in.
> Ord: 18.3.1839. K: 7.1.1839. L: 18.4.1840. C: 30.11.1840.
> First cost: £15,244 (including machinery £6,424).
> Commissioned 12.10.1840 under Lieut John Lunn, for the Mediterranean (–1842). Recommissioned 19.12.1844 under Lieut Henry Eden, for same station; on 1.7.1846 under Lieut Edward Roche Power; paid off 11.1.1849. Refitted at Woolwich (for £5,338) 1850. Recommissioned 19.11.1850 under Lieut Roger Lucius Curtis; on 12.11.1851 under Lieut George Fiott Day, for east coast of South America; paid off 1854. Recommissioned 12.3.1855 under Lieut John Blythesea, for the Baltic. On 20.5.1856 under Lieut John Bousquet Field; paid off 13.7.1860 at Sheerness. Fitted as tug at Chatham 2.1863, for Sheerness. Sold 2.1895 at Sheerness.

PORCUPINE Class Steam vessel (SV5), re-classed 1844 as First Class SGV. 'Admiralty' design, approved 11 November 1843. : The engine was ex-*Black Eagle*, fitted with new tubular boilers.

> Dimensions & tons: 141ft 0in, 124ft 7½in x 24ft 1¼in (24ft 0in for tonnage) x 13ft 0in. 381⁷⁸⁄₉₄bm.
> Men: 80. Guns: 1 x 32pdr (26cwt) on pivot, 2 x 32pdr (17cwt) carronades.
> Machinery: 2-cylinder side lever. 132nhp. 285ihp.

Porcupine Deptford Dyd/Maudslay, Sons & Field

> As built: 141ft 0in, 124ft 7⅛in x 24ft 1¼in (24ft 0in for tonnage) x 13ft 6in. 381⁶⁹⁄₉₄bm. Draught 4ft 11in/5ft 11in.
> Ord: 11.11.1843. K: 1.1844. L: 17.6.1844. C: 19.8.1844 at Woolwich (for sea).
> First cost: £19,604 total (hull £7,997, machinery £4,557, fitting £7,050).
> Commissioned 1.7.1844 under Capt. Frederick Bullock, for surveying in the Thames. Refitted at Woolwich (for £2,028) 1847. Recommissioned 9.11.1847 under Lieut Edward Forward Roberts, for the Mediterranean; paid off 3.8.1850. Refitted at Portsmouth (for £4,623) 1851–52. Recommissioned 29.4.1852 under Lieut George Melville Jackson, for Scotland; to the Baltic 1854. Recommissioned 17.1.1856 under Lieut George Mackintosh Balfour, for surveying duties; on 13.5.1856 under Capt. Henry Charles Otter, for Scotland and Newfoundland ('for particular service'); paid off 12.1861. Refitted for survey duties at Devonport, and recommissioned 2.6.1862, under Richard Hoskyn, Master, to survey route for Atlantic telegraph. On 8.8.1863 under Cmdr Edward K. Calver, then 2.1.1865 under Cmdr Edward Wolfe Brooker, for surveying, as tender to *Fisgard*; in 4.1872 under John Parsons, Master, still tender to *Fisgard*, later to *Duncan*; paid off 11.11.1881. Sold 7.1883 to Castle & Son to BU.

SPITFIRE Class Steam vessel (SV5), re-classed 1844 as First Class SGV. 'Admiralty' design, approved 11 November 1843. Fitted with the engine that had come from *Hermes*, together with new boilers.

> Dimensions & tons: 147ft 2in, 130ft 0in x 25ft 1in (25ft 0in for tonnage) x 14ft 6in. 432¹⁷⁄₉₄bm. Draught 5ft 0in/6ft 1in.
> Men: 80. Guns: 1 x 32pdr (26cwt) on pivot, 2 x 32pdrs (17cwt) carronades.
> Machinery: 2-cylinder side lever. 140nhp.

Spitfire Deptford Dyd/Butterley Co.

> Ord: 11.11.1843. K: 6.1844. L: 26.3.1845. C: 30.6.1845 at Woolwich (for sea)
> First cost: Hull £7,963, fitting & machinery £9,320.
> Commissioned 12.6.1845 under Lieut Cmdr James Archibald Macdonald, for the Mediterranean; on 3.4.1848 under Lieut George Ommaney Willes; paid off 16.5.1848 at Malta. Fitted as survey ship 1851; recommissioned 21.4.1851 under Cmdr (Capt., 3.1.1855) Thomas Abel Brimage Sprat, for surveying in Crete; in 1854 to the Black Sea; at bombardment of Sebastopol 11.10.1854; later surveying in the Mediterranean and Black Sea; paid off 7.2.1856 at Woolwich. Repair/refit 1856–57. Recommissioned 22.10.1857 under Lieut (Cmdr, 1.4.1858) James Carter Campbell, for the west coast of Africa; on 8.4.1858 under Lieut (Cmdr 17.6.1859) William Cox Chapman; on 18.11.1859 under Lieut Constantine O'Donnel Allingham, for same station; took slavers (names unknown) 26.9.1859 and 16.8.1860; paid off 23.2.1861. Fitted at Woolwich as a tug 1861–62; to Bermuda 12.1862. BU from 7.5.1888 at Bermuda.

On 16 January 1844, the Admiralty accepted tenders from Robert Napier and from Ditchburn & Mare to each built three iron-hulled steam vessels 'of the *Locust* class'; the plans were approved on 29

January. The two builders were each informed on the following day that they could proceed with construction of one of the three vessels on their own plans, and the six names were assigned on 13 February 1844. They were ordered as steam vessels (SV5) but re-rated as gunvessels on 31 May 1844.

JACKAL **Class** Second Class iron-hulled gunvessels. Symonds design, approved 17 April 1844. After launchings, both sailed from Govan in March 1845 for Devonport

> Dimensions & tons: 142ft 7¼in, 126ft 10½in x 22ft 6in x 12ft 9½in. 340bm. 492 disp.
> Men: 60. Guns: 1 x 18pdr (22cwt) on pivot; 2 x 24pdrs (13cwt).
> Machinery: 2-cylinder side lever. 150nhp. 455ihp.

Jackal Robert Napier & Sons, Govan (No. 8)/Robert Napier & Sons

> Ord: 16.1.1844. K: 1844. L: 28.10.1844. C: 22.9.1845 at Woolwich and Devonport.
> Sold 11.1887 to BU.
> First cost: £14,665 (hull £5,680, machinery £6,000; fitting £2,985).
> Commissioned 7.4.1845 under Lieut William Montagu Isaacson George Pasco, for the Mediterranean; on 17.11.1846 under Lieut George Western, for the Home station, then to Lisbon and the Mediterranean; paid off 24.4.1848 at Portsmouth. To Steam Reserve at Devonport 1.1849. Recommissioned 26.1.1850 under Thomas Yeatman, Master, for west coast of Africa, as tender to *Gladiator*; boats attack in Benin River 28.3.1850; later tender to *Tortoise*, storeship at Ascension; took slavers 21.8.1850 (name unknown) and *Gira Sol* 4.9.1850; on 22.5.1851 under Lieut Norman Bernard Bedington, on west coast of Africa; paid off 12.6.1852. Refitted at Woolwich (for £4,612) 1854. Recommissioned 30.6.1854 under Lieut William Travers Forbes Jackson, for particular service; on 13.9.1855 under Lieut Charles Thomas William George Cerjat; paid off 29.10.1858 at Sheerness. Refitted at Woolwich (for £7,847) 1858–59. Recommissioned 21.5.1859 under Lieut Augustus George Ernest Murray, then 17.12.1859 under Lieut James Simpson, for fishery protection on west coast of Scotland; on 7.8.1860 under Lieut Edward Francis Loder. On 19.5.1863 under Lieut Hugh McNeile Dyer, then 11.12.1868 under Lieut Arthur Edward Dupuis, 12.4.1870 under Lieut John Bruce, 2.1872 under Henry Thomas Clanchy, 9.1875 under H. A. Digby, 12.1878 under Richard Bingham, all on same station. Recommissioned 5.1.1885 under Lieut John Warde Osborne; paid off 30.4.1886. For sale 8.1887, sold 11.1887.

Lizard Robert Napier & Sons, Govan (No. 9)/Robert Napier & Sons

> Ord: 16.1.1844. K: 1844. L: 28.11.1844. C: 27.11.1845 at Woolwich.
> First cost: £14,338 (hull £5,680, machinery £6,000; fitting £2,658).
> Commissioned 15.3.1845 under Lieut James Archibald Macdonald, for the Mediterranean; on 31.10.1845 under Lieut Henry Manby Tylden, for east coast South America; in Parana River operations 11.1845 and 1846; on 8.6.1847 under Lieut William Alfred Rumbulow Pearse, for the same station; in 12.1848 under Lieut Henry Augustus Hollinworth; paid off 24.8.1849. Recommissioned 12.1851, as tender to the *Waterloo* at Sheerness; in 1852–53 under George Brockman, 2nd Master, then 1854–55 under Samuel Winnecot, 2nd Master, and 1855–56 under Lieut Thomas Borradaile Christopher. On 1.1.1858 under Lieut Edward Eyre Maunsell, for fishery protection duties as tender to the *Monarch*, later (3.7.1860) to the *Formidable*. On 30.8.1861 under Lieut William Spratt, and 28.3.1863 under Lieut Henry Joseph Challis, still as tender to the *Formidable* at Sheerness. Reboilered 1865, but 1.1865 still under Challis, then 9.1865 under Lieut John Buchan Telfer, and 23.4.1866 under Lieut

Stanhope Grove Price; paid off 31.10.1868. On sales list in 2.1869, but no satisfactory offers were received, and BU 4.1869 in dock at Chatham.

TORCH **Class** Second Class iron-hulled gunvessels. Symonds design, approved 17 April 1844.

> Dimensions & tons: 141ft 0in, 127ft 6in x 22ft 6in x 13ft 6in. 343bm. 505 disp.
> Men: 60. Guns: 1 x 18pdr (22cwt) on pivot; 2 x 24pdrs (13cwt).
> Machinery: 2-cylinder side lever. 150nhp. (*Harpy* cylinders were 55in diameter, 4ft stroke, oscillating. 200nhp 520ihp = 9kts).

Torch Ditchburn & Mare, Blackwall/Seaward & Capel

> As built: 140ft 10in, 127ft 4in x 22ft 6in (22ft 6in for tonnage) x 13ft 6in. 343bm. Draught 3ft 10in/5ft 1½in.
> Ord: 16.1.1844. K: 1844. L: 25.2.1845. C: 28.5.1846 at Woolwich.
> First cost: £15,483 total (hull £5,305, machinery £7,620; fitting £2,558).
> Commissioned 9.4.1846 under Lieut William Hoseason; on 1.9.1846 under Lieut David Robert Bunbury Mapleton, then on 23.11.1846 under Lieut George Morris; paid off 19.10.1848. Fitted as tender for *Herald* at Woolwich (for £5,244) in 1852; recommissioned 18.2.1852 under Lieut William Chimmo, for surveying Fiji, as tender to Herald; paid off 25.1.1856. Sold 15.5.1856 to D. Egan at Sydney (per AO 6.12.1854) for £2,400.

Harpy Ditchburn & Mare, Blackwall/John Penn & Sons

> As built: 141ft 1½in, 127ft 7⅜in x 22ft 6¼in (22ft 6¼in for tonnage) x 13ft 7in. 344bm. Draught 3ft 9in/5ft 0in.
> Ord: 16.1.1844. K: 1844. L: 4.3.1845. C: 1.2.1846 at Woolwich (after machinery fitted at East India Dock).
> First cost: £17,249 total (hull £5,148, machinery £9,636; fitting £2,465).
> Commissioned 26.11.1845 under Lieut Edward Halhed Beauchamp-Proctor, for east coast of South America; in Anglo-French operations in Parana River 5.1846; on 24.11.1846 under Lieut James Ward Tomlinson. In 1850 tender to the *Southampton*, on same station; paid off 1.3.1851. Refitted at Portsmouth (for £5,435) 1853–54. Recommissioned 12.4.1854 under Lieut Charles Gowan Lindsay, as flag of Rear-Adm. Montagu Stopford, at Constantinople; on 11.8.1855 under Lieut George Augustus Brine, as flag of Rear-Adm. Frederick William Grey, on same station; paid off 25.11.1858. Refitted for sea at Woolwich (for £5,497) 1859–60. Recommissioned 28.1.1860 under Lieut Alexander Henderson, as tender to *Hogue* at Greenock; on 25.8.1860 under Lieut Francis Hewson, on same station. On 1.7.1864 under Lieut William Howarth, as tender to *Lion* at Greenock. Repaired 1876–77 (for £13,843). In 1880 tender to *Royal Adelaide* at Plymouth. She was sold to the War Office 26.10.1892 as target for use at Milford Haven for experiments with a pneumatic gun. Wreck sold 1909 for £20.

BLOODHOUND **Class** Second Class iron-hulled gunvessel. Napier's design, approved 26 April 1844. After launching, sailed April 1845 from Govan for Devonport.

> Dimensions & tons: 146ft 0in, 134ft 4½in x 23ft 0in x 13ft 6in. 378¹⁵⁄₉₄bm.
> Men: 60. Guns: 1 x 18pdr (22cwt) on pivot; 2 x 24pdrs (13cwt).
> Machinery: 2-cylinder side lever. 150nhp.

Bloodhound Robert Napier & Sons, Govan (No. 10)/Robert Napier & Sons

> Ord: 16.1.1844. K: 1844. L: 9.1.1845. C: 26.9.1845.
> First cost: £14,984 (hull £5,992, machinery £6,000; fitting £2,992).

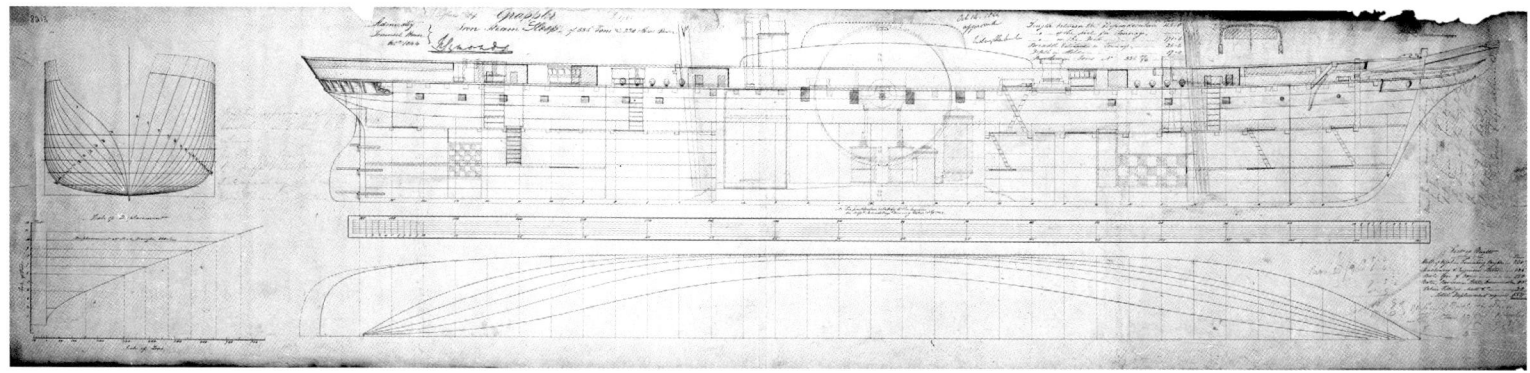

An inboard profile for the iron-hulled gunvessel *Grappler*, dated October 1844; designed and built by William Fairbairn. The rapid corrosion of this hull, coupled with Admiralty reservations about the battle-worthiness of iron, led the Navy to abandon the building of iron warships until 1855, when the availability of better quality iron led to its reintroduction. *(NMM DR8213)*

Commissioned 18.7.1845 under Lieut Robert Phillipps, for the Mediterranean; paid off 28.12.1848. Fitted as tender to *Sampson* at Portsmouth 1849–51. Recommissioned 8.1.1851 under Lewis Fisher, 2nd Master, as tender to Sampson, for west coast of Africa; on 28.1.1851 under John Waye, 2nd Master; took slaver Deseada 7.7.1851; on 12.4.1852 under Lieut Henry Christian, still on west coast of Africa; paid off 6.8.1853. Recommissioned 27.8.1855 under Lieut George Thomas Ceather Smith; on 14.9.1855 under Lieut George Bell Williams, for the west coast of Africa; on 13.12.1856 under Lieut Charles Refus Robson; paid off 3.7.1858. Recommissioned 25.6.1860 under Lieut Francis William Bennett, for same station; on 8.9.1861 under Lieut John Edward Stokes; paid off 15.9.1863. Boilers and machinery for sale 12.4.1866, hull BU 1866.

MYRMIDON Class Second Class iron-hulled gunvessel. Thomas Ditchburn design, approved 1 June 1844.
    Dimensions & tons: 151ft 0in, 137ft 6in x 22ft 6in x 13ft 8in. 370²⁴/₉₄bm.
    Men: 60. Guns: 1 x 18pdr (22cwt) on pivot; 2 x 24pdrs (13cwt).
    Machinery: 2-cylinder (48in diameter, 4ft stroke) side lever. 150nhp. 10kts.
Myrmidon Ditchburn & Mare, Blackwall/Boulton, Watt & Co.
    Ord: 16.1.1844. K: 1844. L: 2.1.1845. C: 16.3.1846.
    First cost: £14,381 (hull £5,148, machinery £6,910; fitting £2,323).
    Commissioned 23.12.1845 under Lieut Charles Jenkin, for particular service; on 23.11.1846 under Lieut Edward Forward Roberts, for same role. On 22.6.1848 under Lieut Jasper Henry Selwyn, as tender to *Ganges* at Sheerness; paid off 19.10.1848. Recommissioned 30.6.1851 under Lieut William Kynaston Joliffe, for the west coast of Africa; took slaver *Maria* 31.5.1853; paid off 22.6.1854. Recommissioned 13.6.1855 under Lieut Edward Eyre Maunsell, for west coast of Africa, then 6.12.1855 under Lieut Edward Wingfield Shaw, and 22.4.1857 under Lieut Charles Aylmer Pembroke Vallancey Robinson, all on same station; took slavers (names unknown) on 19.5.1856 and 4.10.1857; paid off 5.2.1858. Sold 1.12.1858 at Fernando Po (for £120).

GRAPPLER Class First Class iron-hulled gunvessel, to builders' design, which was commissioned 11 May 1844, and approved 15 October. A contract with Fairbairn was signed on 21 November. Initially rated as steam vessel (SV3) but re-rated as gunvessel 31 May

1844. Built to receive the 220nhp engine originally intended for *Trident* (qv). Her hull was beset with severe early corrosion and thus required early disposal.
    Dimensions & tons: 165ft 0in, 149ft 1in x 26ft 6in x 16ft 7in. 556⁸⁸/₉₄bm.
    Men: 80. Guns: 2 x 32pdrs (26cwt/6ft) on pivots + 2 x 32pdr (17cwt/4ft) carronades.
    Machinery: 4-cylinder 'Siamese' engine. 220nhp.
Grappler William Fairbairn & Co., Millwall/Maudslay, Sons & Field
    As built: 165ft 2in, 149ft 3in x 26ft 7in x 16ft 7in. 561¹/₉₄bm.
    Ord: 27.7.1844. K: 1844. L: 30.12.1845. C: 26.11.1846.
    First cost: £35,161 total (hull £13,863, machinery £16,830, fitting £4,468).
    Commissioned 10.11.1846 under Lieut Thomas Henry Lysaght, for the west coast of Africa; took slavers *Taglioni* 30.3.1847, *Pampa* 26.11.1847 and *Aguia* 8.12.1847; paid off 10.1.1849. Sold to W. P. Beach 2.2.1850 to BU (for £560).

RECRUIT Class First Class iron-hulled gunvessels, to builder's design. Ordered 1849 by the Prussian Navy (with a third sister ship, which was subsequently retained by Prussia) and commissioned by them in July 1851, these two double-ended gunboats (with rudders both forward and aft) were handed over to the RN on the outbreak of the Crimean War (by agreement, the sailing frigate *Thetis* was given to Prussia in exchange on 12 January 1855). Both arrived at Devonport on 21 December 1854 and were renamed by AO of 22 December. British colours were hoisted and the ships handed over to the RN on 12 January 1855; the exchange was considered complete on 5 February, and both vessels completed fitting for sea on 4 April (for £1,553 and £1,948 respectively). They were iron-hulled with 8in teak backing, but were considered 'sad failures', as they were overweight and their performance was disappointing.
    Dimensions & tons: 181ft 0in (178ft wl), 162ft 9⅞in x 26ft 1½in x 11ft 0in. 591¹⁰/₉₄bm.
    Men: 74–80. Guns: 4 x 8in (65cwt) shell, 2 x 32pdr carronades (25cwt); later 4 x 32pdrs (56cwt), 2 x 12pdr (10cwt) howitzers.
    Machinery: 2-cylinder (48in diameter, 4½ft stroke) oscillating direct-acting. 4 boilers. 160nhp. 754ihp = 11½kts. 2,500 miles radius @ 10kts.
Recruit (ex-Prussian *Salamander*) J. Scott Russell & Robinson, Millwall/same for engines
    L: 1850. C: 1.7.1851 (for Prussia).
    Commissioned 2.1855 under Lieut George Fiott Day, for the Black Sea; in Azov operations 1855; on 30.11.1855 under Cmdr Henry Frederick McKillop, in the Mediterranean. Recommissioned 26.2.1859 under Cmdr David Spain, for the Mediterranean; paid off 11.10.1861. Sold to E. Bates 23.9.1869.

Weser (ex-Prussian *Nix*) J. Scott Russell & Robinson, Millwall/same for engines.

  L: 1850. C: 29.7.1851 (for Prussia).

  Commissioned 21.2.1855 under Lieut John Edmund Commerell, for the Black Sea; in Azov operations 1855 (for which Commerell awarded VC); in 11.1855 under Cmdr John Francis Ross; on 10.5.1856 under Cmdr Charles Arthur Wise, in the Mediterranean (died 18.2.1859); on 19.2.1859 under Cmdr Arthur Henry John Johnstone, still in the Mediterranean; paid off 1.6.1859 at Woolwich. Refitted for sea at Woolwich (for £8,167) 1859. Recommissioned 12.10.1859 under Cmdr James Raby; intended for South America, but found unfit for Atlantic crossing and paid off 26.1.1860 at Woolwich. Recommissioned 2.3.1862 again under Johnstone), for the Mediterranean; paid off 19.7.1865 at Malta. Harbour service 1866; converted to water distilling ship at Malta 1867 (also used as tug). Sold 29.10.1873 at Malta for £1,500.

PADDLE PACKETS

Further purchases and new construction directly by the Admiralty from 1837 until 1847 follow. A number of these were built with iron hulls, by contract with specialist builders who also designed them. In 1850 the British government decided that future mail should be carried by commercial steam liners operating with reliable timetables on scheduled routes, and contracted with shipping lines for this purpose. The Admiralty transferred certain of its paddle packets to such commercial operators and sold the remaining vessels.

WIDGEON Class 1837   William Symonds design, approved 22 June 1837. Stationed at Dover for packet service.

  Dimensions & tons: 108ft 0in, 98ft 1in x 18ft 1in (17ft 11in for tonnage) x 10ft 7in. 167⁴⁶/₉₄bm. 200 disp.

  Men: — . Guns: 1 x 6pdr.

  Machinery: 90nhp. 191ihp.

  First cost: £10,121 total (including machinery £5,065).

Widgeon   Chatham Dyd/Seaward & Capel

  As built: 108ft 0in, 96ft 3½in x 18ft 0½in (17ft 10½in for tonnage) x 10ft 10in. 163⁶¹/₉₄bm. Daught 4ft 5in/4ft 5½in.

  Ord: 24.5.1837. K: 6.1837. L: 12.9.1837. C: 26.10.1837 at Chatham (after machinery fitted at Limehouse).

  Commissioned 1.7.1837 under John Hamilton, Master, for mail packet service at Dover. Reboilered at Woolwich (for £1,588) 1841. Recommissioned 1.9.1841 under Lieut Thomas Swain Scriven for the packet service. Reboilered at Woolwich (for £1,807) 1846. Recommissioned 18.10.1847 under Lieut George Raymond; paid off 11.1848. From 3.1849 re-classed as a Tender, first to *Ocean* at the Nore, then to *Monarch* at Sheerness, then to survey Thames estuary under command of Capt. Frederick Bullock of *Fisgard* (to 12.1850). To Pembroke 8.1852 as a tug, then to Deptford 1854 as tender (tug) to *Fisgard*. Reboilered again at Woolwich 1855. Recommissioned 3.1859 under George Brockman, Master, as tender to *Monarch* at Sheerness. Tug at Deptford 1865. Lent to the Royal Victoria Victualling Yard 1866 (replacing *Asp*). In 1869 tender to *Nankin* at Pembroke (–1880); reboilered again 1874. For sale 7.2.1884, and sold to Castle & Sons 3.1884 to BU at Charlton.

URGENT Purchased 1837 [ex-mercantile *Collonsay*, built 1826 at Liverpool]. Purchased 26.7.1837 for £26,000.

  Dimensions & tons: 173ft 3in, 155ft 0in x 26ft 1in x 17ft 8½in. 560⁸⁴/₉₄bm.

  Men: 12. Guns: 2 x 6pdr carronades.

  Machinery: — . 280nhp.

  Commissioned 26.7.1837 under John Emerson, Master, for the mail packet service at Liverpool. Engines and boilers replaced at Woolwich and Chatham (for £10,696 including £4,516 for machinery) 1842–43, then returned to Liverpool. On 19.6.1846 under Lieut Aaron Stark Symes, for the Liverpool packet service; paid off 10.1848 at Portsmouth. Sold to Mr Hall (by AO 14.12.1850, for £7,261) 1850, and left Portsmouth 19.12.1850 for delivery to buyer, becoming mercantile *Levant*.

PROSPERO   Purchased 1837 [ex-mercantile *Belfast*, built 1829 by Coates & Young, Belfast]. Purchased 4.8.1837 for £11,000.

  Dimensions & tons: 129ft 1in, 115ft 1in x 20ft 4in oa (20ft 2in for tonnage) x 11ft 5in. 248⁹⁰/₉₄bm.

  Men: 12. Guns: 1 x 12pdr carronade, travelling carriage.

  Machinery: — . 144nhp.

  Commissioned 4.8.1837 under Lieut William Hoseason, for the mail packet service at Pembroke; paid off 1838. Refitted at Woolwich (for £7,164) 1838–39. Recommissioned 26.10.1839 under Lieut William Viner Read, then 30.3.1840 under Lieut Edward Keane, then in 12.1841 under Peter Rundle, Master, all for the same station. Reboilered and refitted at Woolwich (for £3,752) 1843. Recommissioned 17.10.1846, still under Rundle, for the same station; paid off 6.1850. Fitted as tender at Woolwich (for £3,107) 1853; to Pembroke 12.1853 as tug/tender. BU 1866 by Marshall, Plymouth.

DASHER 1837   Sir William Symonds design, approved 26 September 1837.

  Dimensions & tons: 120ft 2in, 105ft 8¾in x 21ft 8in oa (21ft 6in for tonnage) x 13ft 0in. 259⁹³/₉₄bm. 357 disp.

  Men: — . Guns: 1 x 6pdr (brass).

  Machinery: 100nhp. 234ihp.

Dasher   Chatham Dyd/Seaward & Capel

  Ord: 5.9.1837. K: 10.1837. L: 5.12.1837. C: 8.4.1838 for sea.

  First cost: £12,519 (including machinery £5,276).

  Commissioned 27.7.1838 under Lieut John Bettinson Cragg, for the mail packet service at Weymouth; in 7.1839 under Robert White. Reboilered at Woolwich (for £1,987); reboilered again and fitted as a survey ship at Woolwich (for £4,800) 1845. Recommissioned 1.8.1845 under Cmdr William Louis Sheringham, for surveying south coast of England ship 1845. In 10.1847 under Lieut Abraham Parks, at Sheerness. On 5.7.1851 under Cmdr (Capt. 31.12.1853) Nicholas Le Febvre; in 1854 on fishery protection service in the Channel Islands, as tender to *Victory*; on 12.12.1856 under Cmdr Edward George Hore. Recommissioned 2.1.1860 under Cmdr Philip de Saumarez, for the Channel Islands; on 1.1.1866 under Cmdr George John Malcolm, then 14.4.1867 under Cmdr James Henry Bushell, then 1.6.1869 under Cmdr Edward Bond Harrison Franklin, then 8.4.1870 under Cmdr William Frederick Johnson, then 10.4.1873 under Cmdr John Patton, then 15.4.1876 under Cmdr Edmund Charles Drummond, then 23.2.1877 under Cmdr Adolphus Augustus Frederick Fitzgeorge, then 8.10.1879 under Cmdr Charles Vernon Anson, and finally 21.9.1882 under Cmdr Charles E. Reade, all on same station; paid off 31.7.1884. For sale 27.2.1885, and sold to Castle & Sons 23.3.1885 to BU at Charlton.

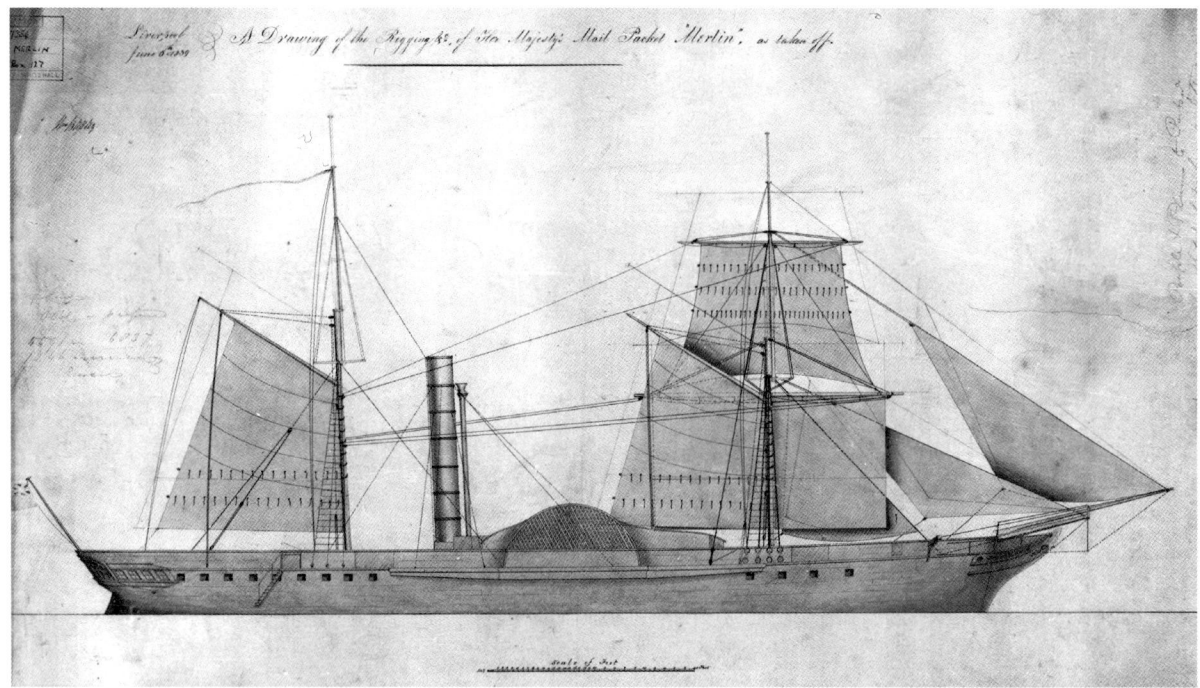

The combination of sail and paddle propulsion was problematic because any heel caused by wind action on the sails made one paddle wheel dig deeper than the other, with a consequent pull to one side which made steering a straight course difficult. Eventually seamen worked out how to use them most efficiently in combination, but it was sometimes found necessary to produce bespoke sail plans for specific ships because the rig was no longer remotely standard, even on ships being to the same design. This plan for the *Merlin* was drawn at Liverpool (the ship's home port as a packet) in June 1839, and has the position of larger sails outlined, so was presumably a planned increase in sail area based on her early passages. *(NMM DR7354)*

MERLIN Class 1838 Sir William Symonds design, approved 2 April 1838. All were built as steam mail packets for the Liverpool Station, but were fitted as Mediterranean packets in 1848.

Dimensions & tons: 175ft 0in, 153ft 6in x 33ft 2in oa (33ft 0in for tonnage) x 16ft 5in. 889¹⁴/₉₄bm.

Men: — . Guns: 2 x 6pdr carronades.

Machinery: 312nhp.

Merlin Pembroke Dyd/Fawcett, Preston & Co.

Ord: 10.3.1838. K: 4.1838. L: 18.9.1838. C: 4.1839 (commission).

First cost: £34,285 (including machinery £14,510).

Commissioned 20.4.1839 under Lieut Charles Haswell Townley, based at Liverpool; on 5.2.1842 under Lieut Edward Keane, on same service; paid off 6.2.1847. Recommissioned 22.5.1848 under Lieut James Howard Turner, for the Mediterranean. Converted 1854 to survey ship. On 6.1.1855 under Capt. Bartholomew James Sulivan, for the Baltic. Converted 1856 as gunvessel. Recommissioned 24.5.1856 under Cmdr Cecil William Buckley, for the west coast of Africa; paid off 23.4.1858. On sales list 18.9.1961, and sold to A. E. Williams & Co. 18.5.1863 for commercial service, renamed *Sea Hawk*.

Medusa Pembroke Dyd/Fawcett, Preston & Co.

Ord: 10.3.1838. K: 5.1838. L: 31.10.1838. C: 12.8.1839.

First cost: £33,822 (including machinery £14,510).

Commissioned 8.8.1839 under Lieut John Phillips Philipps, based in Liverpool; on 13.7.1844 under Lieut James Grant Raymond, on same service. Recommissioned 17.5.1848 under Lieut John Oldenshaw Bathurst, for the Mediterranean; paid off 28.5.1852. On 15.2.1855 under Lieut William Andrew James Heath, for the Black Sea; on 18.8.1855 under Capt. Sherard Osborn, in the Sea of Azov. Recommissioned 1.8.1856 under Cmdr Henry James Raby, for the west coast of Africa; on 18.3.1857 under Cmdr William Bowden, on same station; paid off 17.5.1860. Converted at Woolwich to a tug 1861–62. Recommissioned 3.2.1862 under Jabez Loane, Master; on 1.9.1863 under John Hillary Allard, Master, at Sheerness, then 11.11.1865 under Thomas Potter, Master (Staff Cmdr from 11.11.1866), then 1.6.1868 under Staff Cmdr George B. F. Swain, and 26.9.1870 under Staff Cmdr

Charles J. Polkinghorne; paid off 15.12.1871. Sold to Castle 17.2.1872 to BU at Charlton.

Medina Pembroke Dyd/Fawcett, Preston & Co.

Ord: 30.3.1839. K: 6.1839. L: 18.3.1840. C: 4.1840 (commission).

First cost: Hull £16,411, fitting £21,311 (including machinery £14,767).

Commissioned 19.10.1848 under Lieut Abraham Darby, for the Mediterranean; on 23.1.1851 under Lieut Louis Rivett Reynolds, then 4.5.1854 under Lieut Henry Barré Beresford, then 7.1.1856 as survey vessel under Capt. Thomas Abel Brimage Sprat, all in the Mediterranean; paid off 10.11.1863 at Malta. BU there 3.1864.

DOVER Class 1839 Built of iron, to builders design. The first iron ship ordered and built for the RN.

Dimensions & tons: 110ft 5in, 95ft 8in x 21ft 0in x 10ft 6in. Draught 5ft 2½in. 224⁴³/₉₄bm.

Machinery: 90nhp.

First cost: £10,153 (including machinery £5,300).

Dover William Laird & Son, Birkenhead (No. 31)/Fawcett, Preston & Co.

Ord: 4.2.1839. (named 30.3.1839) K: 1839. L: 30.5.1840. C: — .

Commissioned 7.1840 under Edmund Lyne, Master, for the mail packet service at Dover; on 4.3.1847 under Lieut George Raymond; paid off 18.10.1847. Refitted for the River Gambia 1847–48, recommissioned 24.8.1848 under John Boulton, Master, and sailed for there 10.1848. In 1849 under Lieut John R. Harward. Loaned to the Colonial Office 1850 for use on the Gambia; returned home 10.1852 and refitted at Woolwich 1853, then to the Gambia again; returned home again 10.1859 and refitted at Woolwich 1859–60; then to Gambia again. Sold 1866.

PRINCESS ALICE Iron paddle vessel, purchased 1844. Designed by Thomas Ditchburn and built by Ditchburn & Mare, Blackwall 1843.

Dimensions & tons: 140ft 4in, 126in 4in x 20ft 4in x 10ft 10¼in. 270bm.

Machinery: Maudslay 120nhp. 431ihp.

Purchased from Maudslay by AO 23.1.1844.

Commissioned 5.2.1844 under Luke Smithett, Master, for the mail packet service at Dover; on 11.1.1848 under Lieut Thomas Swain Scriven, then 11.1848 under R. L. Canny, Master, and then 4.11.1852 under John Warma, 2nd Master; paid off 4.1854. Loaned to Churchward, Jenkins & Co. 1854 for mail packet service. In 2.1855 tender to *Duke of Wellington*, as despatch vessel in the Baltic (various Cmdrs.). Lent to Mr Chads for the Dover packet service 1857. In 1.1858 tender to *Fisgard* at Woolwich. On 25.4.1862 under Richard Cossantine Dyer, Master, as tender to *Royal Adelaide* at Plymouth. BU completed 16.7.1878 at Devonport.

ONYX Class 1845 Iron paddle vessels, to Thomas Ditchburn design, approved 14 April 1845. The original design as approved was 130ft length, 275 tons; it was lengthened by 7ft by AO on 9 July 1845. Both actually measured 139ft length, 298bm as built (see below). Both were sold (together with *Undine*) for £13,000 under AO 16 February 1854.

Dimensions & tons: 137ft 0in, 124ft 2²/₃in x 21ft 0in x 10ft 8in. 291³⁷/₉₄bm.

Men: — . Guns: — .

Machinery: 124nhp.

Onyx Ditchburn & Mare, Blackwall/John Penn & Son

As built: 139ft 3in, 126ft 8in x 21ft 0in x 10ft 8in. 297¹²/₉₄bm.

Ord: 16.4.1845. K: — . L: 11.1845. C: 1.1846 at Deptford Dyd.

First cost: £12,174 total (hull £6,673, fitting £6,673 incl. machinery £6,376).

Commissioned 12.1.1846 under Lieut Robert Pepperel Mudge, for the mail packet service at Dover; on 11.1.1848 under Lieut George Raymond. Sold to Jenkins & Churchward by AO 16.2.1854.

Violet Ditchburn & Mare, Blackwall/John Penn & Son

As built: 139ft 0in, 126ft 5in x 21ft 1in x 10ft 2in. 298⁸⁴/₉₄bm.

Ord: 9.7.1845. K: — . L: 1.12.1845. C: — .

First cost: Hull £5,409, machinery £6,360, fitting — .

Commissioned 1.6.1850 under Lieut Henry Paget Jones (died 2.1854), for the mail packet service at Dover. Sold to Jenkins & Churchward by AO 16.2.1854 (subsequently lost on the Goodwin Sands in 1.1857).

GARLAND Class 1845 Oliver Lang design, proposed 8 August 1845 ('to build a steamer of three thicknesses of diagonal planks') and approved 13 September. Fitted out by builders.

Dimensions & tons: 140ft 0in, 126ft 11¼in x 21ft 0in x 10ft 9in. 298bm.

Men: — . Guns: — .

Machinery: 120nhp.

Garland Fletcher & Fearnall, Limehouse/John Penn & Sons.

As built: 141ft 5¼in, 126ft 11⅛in x 21ft 3in (21ft 2in for tonnage) x 10ft 8in. 302⁵⁴/₉₄bm. Draught 3ft 5in/3ft 6in.

Ord: 13.9.1845. K: 10.1845. L: 26.2.1846. C: 18.5.1846 at builders (after machinery fitted at Deptford).

First cost: Hull £5,312, machinery £6,710.

Commissioned 18.5.1846 under Luke Smithett, Master, for the mail packet service at Dover; on 7.2.1848 under Lieut Benjamin Aplin; on 14.3.1849 under Lieut Edward Wylde. Sold to Jenkins & Churchward 1.2.1856 (for £4,800 'which has been deducted from Bills to them for the Packet Service').

On 29 September 1846 the Admiralty ordered a design competition to be held for four new paddle packets of 350nhp for the Holyhead–Kingston service, to compare with the Surveyor's plans, which were drawn by Lang. As a result, they ordered the following four vessels (including the iron-hulled *Caradoc*), which were all based initially at Liverpool, then at Holyhead. In 1850, the packet service was sold to commercial users; the *Caradoc* and *Banshee* were then fitted as gunboats for the Mediterranean, while the *Saint Columba* and *Llewelyn* were sold to the City of Dublin Steam Packet Company.

CARADOC Sir William Symonds design, approved 6 November 1846. Based at Liverpool, then Holyhead. The packet service was sold to commercial users soon after, and in 1850–51 the *Caradoc* was fitted for the Mediterranean and commissioned as a gunboat/dispatch vessel.

Dimensions & tons: 190ft 0in, 174ft 0in x 26ft 9in x 14ft 9in. 662²¹/₉₄bm.

Men: 65 (from 1851). Guns: (from 1850) 3 x 18pdrs.

Machinery: 350nhp. ?997ihp.

First cost: £35,859 total (hull £16,041, machinery £18,322, fitting £1,496).

Caradoc C. J. Mare & Co., Blackwall/Seaward & Capel

Ord: 9.11.1846. K: 12.1846. L: 3.7.1847. C: 6.6.1848.

Commissioned 4.4.1848 under Lieut Charles Pybus Ladd, for the mail packet service; paid off 1.10.1850. Refitted as a gunboat 1850–51.

Recommissioned 14.4.1851 under Lieut Samuel Hosk Derriman, as a despatch vessel for the Mediterranean; to the Black Sea 1854–55; home 7.1855 (bearing the body of the deceased Lord Raglan); paid off 10.8.1855. Repairs and refit at Woolwich (including new boilers) 1855. Recommissioned as a despatch vessel 26.9.1855 under Lieut Chandos Scudamore Scudamore Stanhope, for the Mediterranean; on 18.3.1858 under Lieut Charles Matthew Buckle, then 10.8.1860 under Lieut John Binney Scott, then 12.1.1863 under Lieut Edward Henry Wilkinson, then 19.9.1867 under Lieut John Maxwell Dalrymple Elphinstone, and then 3.4.1868 under Lieut Henry Holmes A'Court, all on same station; paid off 9.6.1869 at Portsmouth. Sold to E. Bates 12.5.1870.

BANSHEE Oliver Lang design, approved 7 November 1846. Designed for the Holyhead Station. As Thompson's own yard at Horseferry was too small, it is highly probable that *Banshee* was built by Thompson in the huge and under-utilised Barnard's Wharf site in Rotherhithe. On completion, based at Liverpool, then at Holyhead (and was the fastest of the four new packets). The packet service was sold to commercial users soon after, and in 1850–51 the *Banshee* was fitted for the Mediterranean and commissioned as a gunboat/dispatch vessel.

Dimensions & tons: 189ft 0in, 172ft 9½in x 27ft 2in oa (27ft 0in for tonnage) x 14ft 4in. 670⁵/₉₄bm.

Men: 65 (from 1851). Guns: 3 x 18pdrs (from 1851).

Machinery: 2-cylinder oscillating (72in diameter, 5ft stroke). 350nhp. 16.13kts (trials).

First cost: £37,980 total (hull £15,034, machinery £21,350, fitting £1,596).

Banshee John Jenkins Thompson, Rotherhithe/John Penn & Sons.

Ord: 9.11.1846. K: 12.1846. L: 13.10.1847. C: 28.2.1848.

Commissioned 5.1.1848 under William Smithett, Master, for the mail packet service; paid off 1.10.1850. Refitted as a gunboat 850–51. Recommissioned 14.2.1851 under Lieut James Hosken, as a despatch vessel for the Mediterranean; on 6.12.1853 under Lieut Louis Rivett Reynolds; to the Bklack Sea 1854–55; in 8.1855 under Lieut John William Pike; paid off 27.8.1856. Repair and refit at Woolwich 1856–57. Recommissioned 14.2.1857

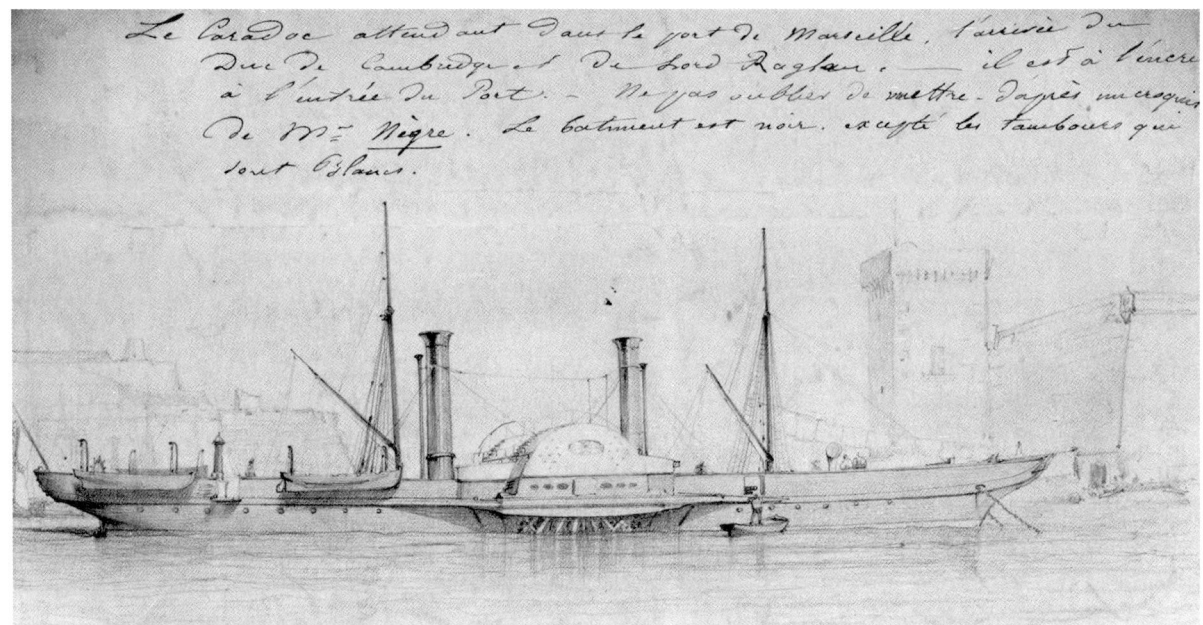

The *Caradoc* (shown here) and similar *Banshee* were originally built by the Admiralty as paddle-driven packets for the Holyhead–Kingston mail service, designed by Symonds and Lang respectively. At the same time, the Admiralty purchased on the stocks two other ships – *Saint Columba* and *Llewelyn* – which were commercially designed by their builders. All four entered service in 1848, but in 1850 the packet service was privatised and sold to commercial operators, along with the *Saint Columba* and *Llewelyn*. The *Caradoc* and *Banshee* were ordered on 30 November to be converted to paddle gunboats at Woolwich, and were deployed to the Mediterranean, although both were actively involved in the Black Sea during the Russian War (reduced from 3 to 2 guns). *Caradoc* is shown in this French drawing at Marseille, according to the rubric 'awaiting the arrival of the Duke of Cambridge and Lord Raglan'. *(NMM PU9008)*

under Henry W Allen, Master, as an Admiralty Yacht, as tender to Fisgard at Woolwich; on 5.1.1858 under Vincent Williams, Master, on same service; paid off 10.11.1858. Recommissioned 7.5.1859 as a despatch vessel under Cmdr Colin Andrew Campbell, for the Mediterranean; on 8.1.1861 under Cmdr Edward Madden, on same station; paid off 31.5.1862 at Woolwich. Listed for BU 11.1864 and BU there 8.1865.

SAINT COLUMBA Builders' design. Originally named *Kathleen*, then *Columba*, finally *Saint Columba* on on 13 February 1847. She was purchased from Lairds on 5 July 1847. Initially based at Liverpool, then at Holyhead. Her sister ship, the Holyhead packet *Cambria* (launched in January 1848 by Lairds), was not a Navy vessel.
   Dimensions & tons: 198ft 6½in, about 184ft 3in x 27ft 3in x 15ft 5in. 719bm.
   Men: — . Guns: — .
   Machinery: 350nhp.
   First cost: £34,997 (hull £15,292, machinery £19,705) excluding fitting.
Saint Columba William Laird & Son, Birkenhead (No. 65)/George Forrester & Co.
   Ord: 9.11.1846. K: 11.1846. L: 5.7.1847. C: — .
   Commissioned 5.1.1848 under Lieut Aaron Stark Symes. Sold to the City of Dublin Steam Packet Company 1850 (by AO 14.6.1850) for commercial service.

LLEWELLYN Class 1846 Builders' design. Based at Holyhead 1848–50.
   Dimensions & tons: 190ft 5in, 174ft 4in x 27ft 0in x 14ft 0in. 671bm.
   Men: — . Guns: — .
   Machinery: 2-cylinder oscillating (68in diameter, 6ft stroke). 350nhp.
   First cost: £38,962 total (hull £16,355, machinery £21,200, fitting £1,407).
Llewellyn Miller & Ravenhill, Blackwall/Miller & Ravenhill
   Ord: 9.11.1846. K: 4.1847. L: 22.1.1848. C: 15.5.1848 at builders (after machinery fitted at East India Dock).
   Commissioned 3.5.1848 under John Grey, Master. Sold to the City of Dublin Steam Packet Company 1850 (by AO 14.6.1850) for commercial service, renamed *Saint Patrick*.

VIVID Class 1847 Oliver Lang design, approved 6 May 1847. Improved version of *Garland*.
   Dimensions & tons: 150ft 0in, 136ft 9⅝in x 22ft 1in oa (22ft 0in for tonnage) x 11ft 4in. 352¹⁷⁄₉₄bm. 350 disp. Draught 3ft 9in/4ft 7in.
   Men: — . Guns: 2.
   Machinery: 2-cylinder oscillating (49½in diameter, 4ft stroke). 160nhp. 832ihp.
   First cost: £19,437 total (hull £7,685, machinery £10,400, fitting £1,352).
Vivid Chatham Dyd/John Penn & Sons
   Ord: 4.2.1847 (named 13.2.1847). K: 5.1.1847. L: 7.2.1848. C: 7.4.1848 at Chatham (after machinery fitted at Deptford).
   Commissioned 2.1848 under Luke Smithett, Master, for the mail packet service at Dover. In 5.1854 under Henry W. Allen, Master, as tender to *Fisguard* at Woolwich (–1860); in 2.1865 under Timothy W. Sulivan, Master, on same service. Recommissioned 10.1872 under John Thompson, Master as tender to *Royal Adelaide* and Port Admiral's yacht at Devonport (–1875); in 11.1878 under John R. Ryan, Master, on same service. On 14.8.1882 under Cmdr John Gravener, then on 1886 under ?Thomas Robinson, as tender to *Royal Albert*. On 8.10.1888 under Cmdr George James Tomlin, as flagship from 11.1888 of Adm. Sir William Montagu Dowell at Devonport; in 4.1890 flagship of HRH (and Adm.) Prince Alfred Ernest Albert, the Duke of Edinburgh. Sold to G. Cowen & Sons 5.1894.

UNDINE Purchased 1847 [ex-mercantile *Ondine*, built 1845 by Miller, Ravenhill & Co., Blackwall]. Iron paddle packet.
   Dimensions & tons: 145ft 0in, about 133ft 6in x 20ft 0in x 9ft 10in. 284bm.
   Men: — . Guns: — .
   Machinery: Miller, Ravenhill & Co. 2-cylinder oscillating (34in diameter, 3ft stroke). 110nhp.
   Purchased 13.2.1847 from Edward Baldwin (for £10,936).
   Commissioned 9.5.1848 under Lieut Henry Paget Jones, for the mail packet service at Dover; on 6.2.1851 under John Warman, Master, then 4.11.1852 under Edmund Lyne, Master. Sold to

Jenkins & Churchward for the Dover Mail Service (with *Onyx* and *Violet*, for £13,000) by AO 16.2.1854.

FIREQUEEN Purchased 1847 [yacht *Fire Queen* launched 27.7.1844 by Robert Napier, Glasgow]. Iron paddle packet, to have been based on Holyhead, and fitted as such 1850–51, but never used in this role.

> Dimensions & tons: 163ft 9in, 147ft 1½in x 20ft 0in x 9ft 8in. Draught 4¾ft–5ft. 313²/₉₄bm. 251 disp.
>
> Men: — . Guns: — .
>
> Machinery: Napier 120nhp. 466ihp.
>
> Purchased from Mr Assheton Smith 23.7.1847 (by AO 10.7.1847, for £5,000).
>
> Commissioned 7.1847 under Lieut Charles Richardson Johnson, as tender to *Victory* at Portsmouth; on 1.12.1847 under George Allen, Master, then 30.7.1852 under Frederick W. Paul, Master, in same role (re-appointed 1.1.1858). Reboiled 1860. In 4.1870 under William H. Harris, Master, as tender to *Duke of Wellington*; repaired and refitted at Portsmouth (for £6,690) 1871–72; in 9.1873 under Thomas Pounds, Master. Port Admiral's yacht at Portsmouth 1874. In 10.1878 under William H. Fawckner, Master, on same service (–1880). Sold to Castle 4.8.1883 (for £1,100) for BU at Charlton.

PADDLE DESPATCH VESSELS

COROMANDEL Purchased 1855 [ex-mercantile *Tartar*, built 1853 by Thomas & John White, West Cowes]

> Dimensions & tons: 172ft 9in x 22ft 5in x 11ft 4in. 303bm. (these figures are clearly in error, and the *keel* length is probably around 113ft)
>
> Men: — . Guns: — .
>
> Machinery: Maudslay, Sons & Field. 150nhp. 550ihp.
>
> Purchased by AO of 8.1.1855 from the Peninsular & Oriental Steam Navigation Co. (for £26,000). Renamed *Coromandel* by AO 22.8.1855.
>
> Commissioned 10.1856 under Lieut Sholto Douglas, for Canton River operations; at Fatshan Creek 1.6.1857, as flag of Rear-Adm. Sir Michael Seymour. On 22.9.1857 under William H. Vine, Master, as tender to *Chesapeake*; on 23.8.1859 at Taku Forts, as flag of Vice-Adm. James Hope. On 25.10.1861 under Lieut Duncan George Davidson, as tender to *Imperieuse*; on 29.10.1862 under Lieut Robert Peel Dennistoun, as tender to *Euryalus*; on 17.8.1863 under Lieut George Poole. Sold 17.8.1866 to Bonnel, Hong Kong; became the Japanese *Naruto*, BU 1876.

BANN Class Iron-hulled despatch vessels/gunboats to builder's design, 1855. These were diminutive versions of the ex-Prussian iron paddle gunvessels acquired in 1855. Originally designed 'for exploring the river Nile', with a shallow draught, but when this scheme fell through they were taken over by the Admiralty.

> Dimensions & tons: 140ft 0in, 122ft 10in x 20ft 2in x 8ft 6in. Draught 4ft 0in. 267bm. 291 disp.
>
> Men: 40. Guns: 2 x 8in (65cwt) shell; 2 x 32pdr carronades (25cwt).
>
> Machinery: 2-cylinder (40in diameter, 3½ft stroke) oscillating. 2 boilers. 80nhp. *Bann* 341ihp. *Brune* 364ihp. 10½kts.

Bann J. Scott Russell, Millwall/Russell

> As built: 135ft 6in, 122ft 10⅞in x 20ft 2in x 9ft 1½in. Draught 4ft 0in. 265⁸²/₉₄bm. 291 disp.

Ord: 11.1855. K: 2.1.1856. L: 5.7.1856. C: 9.1856 & purchased for £10,495.

> Commissioned 13.3.1857 under Lieut Joseph Samuel Hudson; on 1.1.1858 under James S. Wells, Master, for surveying on coast of Cornwall, as tender to *Fisgard*. Reported fitting for sea 23.3.1859 but paid off; refitted at Woolwich (for £1,836) 17.5.1859 and recommissioned, still under Wells (–1865); in 1866 tender to *Royal Adelaide* at Devonport. Sold 18.2.1873 to BU.

Brune J. Scott Russell, Millwall/Russell

> As built: 135ft 6in, 122ft 10⅞in x 20ft 2in x 9ft 1½in. Draught 4ft 0in. 265⁸²/₉₄bm. 291 disp.
>
> Ord: 11.1855. K: 2.1.1856. L: 3.6.1856. Purchased 30.8.1856 for £10,495. C: 7.5.1857 (completed for sea at Woolwich for £1,437).
>
> Commissioned 4.5.1857 under Lieut (Charles *or* William) Walker; on 13.11.1857 under Lieut Edward Francis Lodder, for Cape of Good Hope and west coast of Africa, as tender to *Vesuvius*. On 20.12.1860 tender to *Arrogant*, on same station; in 1861 under Lieut John Edward Stokes; at Porto Novo on the Niger River 24.2.1861. Sold 19.5.1863 at Lagos.

PSYCHE Class Isaac Watts design, 1860, to serve as despatch vessels and yachts for Commanders-in-Chief. The first two were built with conventional knee bows; the design of the second pair was modified by Edward Reed; the *Helicon*'s original clipper bow being modified to a ram bow. Two additional vessels were built to this design in 1870–71 and are included to complete the class listing.

> Dimensions & tons: 220ft 0in, about 200ft 0in x 28ft 2in x 14ft 6in. 835bm. 985 disp.
>
> Men: 65 . Guns: 1 x 40pdr (32cwt/10ft) Armstrong gun; later 2 x 20pdrs.
>
> Machinery: 2-cylinder oscillating (62in diameter, 4½ft stroke in *Lively*). 250nhp. See individual ships for ihp and speed.
>
> First cost: av c.£43,000.

Psyche Pembroke Dyd/John Penn & Sons (1,440ihp)

> As built: 220ft 0in, about 197ft 0in x 28ft 2in x 14ft 6in. 835bm. 952 disp.
>
> Ord: 1861. K: 8.1.1861. L: 29.3.1862. C: 1862.
>
> Commissioned 20.10.1862 under Lieut Robert Sterne, for the Mediterranean. In 6.1865 under Lieut Arthur Rodney Blane, then 6.1869 under Lieut Armand Temple Powlett, and 9.4.1870 under Lieut John Fellowes, all in the Mediterranean; wrecked near Catania 15.12.1870, wreck blown up in 2.1871.

Enchantress Pembroke Dyd/John Penn & Sons (1,318ihp = 14kts)

> As built: 220ft 0in, about 197ft 0in x 28ft 2in x 14ft 6in. 835bm. 952 disp.
>
> Ord: 1861. K: 8.2.1861. L: 2.8.1862. C: 1.1863.
>
> Commissioned 16.10.1862 under John E. Petley, Master; in 4.1870 under George L. Carr, Master; in 9.1873 under William H Harris, Master; in 5.1876 under Edward H. Hills, Master, as tender to *Asia*. Recommissioned 4.1877. In 1886 under Staff Cmdr William Wallis Vine, at Portsmouth. For sale 12.1888, and sold to Read, Portsmouth, to BU; delivered 8.5.1889.

Helicon Portsmouth Dyd/Miller, Ravenhill & Salkeld (1,610ihp = 14½kts)

> As built: 220ft 0in, about 197ft 0in x 28ft 3in x 14ft 6in. 837bm. 945 disp. Draught 9ft 11in/10ft 3in.
>
> Ord: 1861. K: 30.5.1861. L: 31.1.1865. C: 1886.
>
> Commissioned 9.1865 under Cmdr Morgan Singer, for the Channel Squadron; on 10.8.1866 under Cmdr Edward Field, then 6.1869 under Cmdr Henry Edward Crozier, on same station. In 7.1872 under Lieut Frank A. Rougemont, for the

The elegant *Enchantress*, one of the first two *Psyche* Class dispatch vessels that were built with a traditional knee bow; the second pair had a form of ram bow that was more hydrodynamically efficient, giving them a slightly higher sustained speed. *(NMM C5855)*

Mediterranean; in 7.1875 under Lieut Attwell Peregrine Macleod Lake, then 1.1879 under Lieut Sir Baldwin Wake Walker, on same station. In 1882 under Lieut William Llewellyn Morrison, then 16.8.1882 under Lieut Alfred Leigh Winsloe, for Suez patrol; on 8.1.1884 under Staff Capt. William Wallis Vine, but Winsloe was back in command for Nile operations in 1885. Repair and refit at Woolwich 1887. Renamed *Enchantress* 1.4.1888, and recommissioned under Staff Capt. Henry Emilius Wood, as tender at Portsmouth. Sold to Laidler, Sunderland 11.7.1905.

Salamis Chatham Dyd/Miller, Ravenhill & Salkeld (1,440ihp = 13¼kts)
  As built: 220ft 0in, about 197ft 0in x 28ft 2in x 14ft 7in. 835bm. 985 disp. Draught 10ft 3in/10ft 8in.
  Ord: 1861. K: 10.8.1861. L: 19.5.1863. C: 1865.
  Commissioned 1.1.1866 under Cmdr Francis Grant Suttie, for China. In 10.1868 under Cmdr Henry Matthew Miller, then 12.4.1870 under Lieut Herbert Dolphin, then 10.2.1872 under Lieut Seymour Spencer Smith, on same station, and 19.12.1872 under Lieut Algernon Charles Littleton; in Malaya operations 2.1874; paid off 21.5.1874. Repair and refit at Devonport 1875–76. Recommissioned 29.8.1876 under Cmdr Frederick Wilbraham Egerton, for the Mediterranean; flag of Vice-Adm. Geoffrey Thomas Phipps Hornby (C-in-C Mediterranean), for passage of Dardanelles 2.1878; in 10.1879 under Adolphus Fitzgeorge, in the Channel; on 15.10.1881 under Cmdr Frederick Ross Boardman, for Egypt operations; paid off 31.12.1882. BU 9.1883 at Sheerness.

Lively Sheerness Dyd/John Penn & Sons (1,757ihp)
  As built: 220ft 0in, 200ft 2⅛in x 28ft 3in (28ft 1½in for tonnage) x 14ft 6½in. 842³/₉₄bm. 940 disp.
  Ord: — . K: 4.1870. L: 10.12.1870. C: 1871.
  Commissioned 15.5.1872 under Cmdr Edward Hobert Seymour; in 10.1873 under Cmdr Henry St Leger Bury Palliser. Repair and refit at Sheerness 1876–77, then to Steam Reserve. Recommissioned 9.4.1878 under Lieut Baldwin Wake Walker; in 1.1880 under Cmdr C. Le Strange, as tender to *Penelope* at Harwich. In 1883 under Cmdr Alfred Arthur Chase Parr; wrecked 7.6.1883 near Stornoway, wreck sold 7.1910.

Vigilant Devonport Dyd/James Watt & Co. (1,815ihp = 13.2kts)
  As built: 220ft 0in, 200ft 3⅜in x 28ft 2in (28ft 0in for tonnage) x 28ft 3in x 14ft 6in. 835²¹/₉₄bm. 951 disp.
  Ord: — . K: 11.4.1870. L: 17.2.1871. C: 12.1871.
  Commissioned 3.1.1872 under Cmdr Edward Hobart Seymour; paid off 14.5.1872. Recommissioned 30.6.1872 under Staff Cmdr George L. Carr, as temporary Admiralty Board yacht;

on 14.1.1873 under Staff Cmdr Robert L. Cleveland; paid off 30.6.1874. Recommissioned 1.9.1874 under Cmdr Hugh Cuthbert Dudley Ryder, for China; in 9.1877 under Cmdr William Martin Annesley, still in China; paid off 1880. Recommissioned 8.1884 at Hong Kong under Lieut Peyton Hoskyns. Sold 7.10.1886 at Hong Kong.

PADDLE ROYAL YACHTS

VICTORIA & ALBERT [i] 1842 Sir William Symonds design. Initially known simply as *Royal Yacht*, the name *Victoria & Albert* was assigned 21 December 1842. Fitted at Chatham and Deptford, using internal fittings from old yacht *Royal George* of 1817. The engine used was that intended for *Firebrand*.
  Dimensions & tons: 200ft 1in, 178ft 6in x 33ft 0in x 22ft 1in. 1,034bm.
  As lengthened 1853: 260ft 0in, 238ft 6in x 33ft 0in x 22ft 1in. 1,381⁴⁸/₉₄bm.
  Machinery: 430nhp. 15kts. As lengthened 1853: 600nhp.
Victoria & Albert Pembroke Dyd/Maudslay, Sons & Field.
  Ord: 31.10.1842. K: 9.11.1842. L: 26.4.1843. C: 23.8.1843.
  First cost: Hull & machinery + initial fitting £32,265, completion of fitting £13,391.
  Commissioned 1.7.1843 under Capt. Lord Adolphus Fitzclarence; paid off 1.1.1845, but recommissioned, still under Fitzclarence. On 31.12.1853 under Capt. Joseph Denman; renamed *Osborne* by AO 22.12.1854; paid off 2.1855. Fitted and then recommissioned as an Admiralty yacht 14.7.1856 under George Henry Kerr Bower, Master, as tender to *Asia*. On 20.1.1865 under Cmdr John D'Arcy. BU 7.1868 at Portsmouth.

ELFIN 'Steam dispatch boat', to serve as tender to *Victoria & Albert*. Oliver Lang design, approved 24 August 1848. Built of Spanish mahogany. Very high speed. Unarmed, but 'to carry 1 x32pdr (25cwt) if needed'.
  Dimensions & tons: 103ft 6in, 93ft 11in x 14ft 0in x 7ft 3in. 98bm. 93 disp.
  Machinery: 40nhp. 181ihp.
Elfin Chatham Dyd/John Rennie
  Ord: 13.7.1848. K: 7.8.1848. L: 8.2.1849, then to Woolwich 12.3.1849 for machinery. C: 30.4.1849 (fitted for sea at Chatham).
  First cost: £5,510 (including hull £2,873, machinery £2,358).

HRH Prince Albert was a great supporter of steam propulsion, and presumably by association this enthusiasm was shared by his wife, Queen Victoria. The first of three royal yachts to be called *Victoria & Albert* was launched in 1843, and when the second was ordered in 1855, the first was renamed *Osborne* after the royal residence on the Isle of Wight. This contemporary model of the first *Victoria & Albert* includes a complete complement of boats, finished in a high-gloss royal blue. *(NMM F2939-001)*

Commissioned 25.4.1849 under Alfred Balliston, Master (Staff Cmdr from 1864), her commander for almost 29 years; reboilered 1873; on 1.4.1878 under Cmdr William Burgess Goldsmith, then 13.7.1883 under Cmdr William Patterson Haynes, then 18.6.1892 under Cmdr George Alexander Broad, and 11.10.1897 under Cmdr James Edward Tully; paid off 9.3.1901. BU 1901.

VICTORIA & ALBERT [ii] 1854 Surveyor's Department design, approved 12 January 1854. The name was changed from *Windsor Castle* to *Victoria & Albert* by AO on 22 December 1854.

Dimensions & tons: 300ft 1in, 275ft 3in x 40ft 3¼in x 24ft 6in. 2,342⁵³/₉₄bm. 2,470 disp.

Machinery: 600nhp. 2,980ihp. 18kts.

First cost: £137,656 (including hull £52,579, machinery £35,493).

Victoria & Albert [ex-*Windsor Castle*] Pembroke Dyd/John Penn & Sons

Ord: 20.10.1855. K: 2.1854. L: 16.1.1855. C: 2.7.1855.

Commissioned 3.3.1855 under Capt. Joseph Denman; on 1.4.1863 under Capt. Prince Ernest Leopold Victor Charles Auguste Joseph Enrick of Leiningen; on 15.3.1877 under Capt. Frank Tourle Thomson, then 15.10.1844 under Capt. John Reginald Thomas Fullerton. BU and burnt 1904 at Portsmouth.

ALBERTA Built to replace *Fairy* as tender to *Victoria & Albert* as 'Her Majesty's passage boat . . . between the Mainland and the Isle of Wight'.

Dimensions & tons: 160ft 0in, 143ft 0in x 22ft 8in x 12ft 6in. 391bm. 370 disp.

Machinery: 2-cylinder Penn oscillating (49½in diameter, 4ft stroke). 160nhp. 1,208ihp.

Alberta Pembroke Dyd

Ord: 1863 at Pembroke Dyd. L: 3.10.1863.

Commissioned 20.8.1864 under Staff Cmdr (Capt., 8.6.1871) David Nairne Welch; damaged in collision with *Mistletoe* 1875; on 1.4.1878 under Capt. Alfred Balliston, then 12.7.1883 under Capt. William Burgess Goldsmith; on 14.10.1897 under Capt. (retd) George Alexander Broad. BU 1912 at Portsmouth.

EXPLORATION VESSELS

The Royal Navy played a key role in the exploration of the River Niger during the early years of Victoria's reign. Following the journey of Commander Hugh Clapperton from Tripoli to Kano in 1827, two expeditions were led by Lieut William Allen in 1832 and by Lieut William Bullock some years later. A third naval expedition under Capt.

Henry Dundas Trotter was despatched in 1841, and for this expedition three iron-hulled paddle vessels (rigged as two-masted topsail schooners) were specially built by Lairds. Not ordered by the RN, but they were registered in the Royal Navy by AO on 19 August 1840 while already under construction. Two wooden-hulled paddle vessels were built in 1860–61.

SOUDAN Class 1840 Builders design.

Dimensions & tons: 110ft 0in, 96ft 10in x 22ft 0in x 8ft 2in. Draught 4½ft. 249²⁷/₉₄bm. 253 disp.

Men: — . Guns: 1 x 12pdr howitzer.

Machinery: 35nhp.

Soudan William Laird & Son, Birkenhead. (No. 34)/Forrester, Liverpool

K: 1840. L: 8.6.1840. C: 10.1840.

Commissioned 9.1840 under Cmdr Bird Allen, for 1841 Niger Expedition (died 25.10.1841 at Fernando Po); on 19.9.1841 under Lieut (Cmdr, 1.11.1841) Edmund Gardiner Fishborne (acting); wrecked in the Sierra Leone River 1.1844. By AO 22.11.1844 the engines were to be removed from the wreck and the stores to be sold locally.

ALBERT Class 1840 Builders design.

Dimensions & tons: 136ft 0in, about 118ft 4in x 27ft 0ft x 10ft 6in. 459bm. 340 disp.

Men: — . Guns: 3 x 12pdrs, 4 x 1pdrs.

Machinery: 70nhp.

Albert William Laird & Son, Birkenhead. (No. 35)/Fawcett, Preston & Co., Liverpool (457 t)

As built: 134ft 0, 117ft 8in x 27ft 0in x 10ft 6in. Draught 6ft. 456²⁵/₉₄bm. 340 disp.

K: 8.1840. L: 29.8.1840. C: 10.1840.

Commissioned 8.9.1840 under Capt. Henry Dundas Trotter, for the west coast of Africa (invalided 12.11.1841); on 12.7.1843 under Lieut Daniel Woodruffe; on 1.2.1845 under Lieut Andrew Robert Dunlap; took slavers *Sua Majestade* 3.2.1845 and *Triumfo* 11.2.1845; wrecked on 13.7.1843 but then salved. Transferred to the Colonial Government of Gambia (in lieu of *Wilberforce*) by AO 1.3.1845. BU 1850.

Wilberforce William Laird & Son, Birkenhead. (No. 36)/George Forrester & Co., Liverpool

As built: 134ft 0, 117ft 9in x 27ft 0in x 10ft 6in. 456¹⁵/₉₄bm. 340 disp.

K: 8.1840. L: 10.10.1840. C: 10.1840.

Commissioned 10.10.1840 under Cmdr William Allen, for the 1841 Niger expedition (–29.6.1841). Refitted at Woolwich (for £3,671) 1842–43, with machinery ex-*Carron* fitted. Recommissioned

27.2.1843 under Lieut Robert Seppings Moore. By AO 4.3.1845 she was to be removed from the List of the Navy, with a note to the effect that she had been transferred to the Colonial Office, although this hardly squares with the contemporary order for her sister. BU 6.1850 in Gambia and materials sold for £59.8.4d.

PIONEER Controller's Department design, 1860. While the builder was recorded as being John Jenkins Thompson, he had actually died in 1859, and the building should really be ascribed to Charlotte Jenkins, who ran the business after his death. A 'vessel for Zambesi River' – built by the Admiralty for the Colonial Office, and transferred by them to the RN in April 1864, for use as a tender on the Congo river. Not to be confused with the 295bm stern-wheel gunboat of the same name, built 1863 in Sydney for use in the New Zealand Wars.

> Dimensions & tons: 115ft 0in, 104ft 1⅛in x 16ft 0in x 7ft 3in. 141⁷⁹⁄₉₄bm.
> Men: — . Guns: 1 x 20pdr., 1 x howitzer.
> Machinery: 34nhp. 171ihp.
> *Pioneer* Thompson, Rotherhithe (see above)/John Penn & Sons.
> Ord: 18.4.1860. K: 30.4.1860. L: 4.8.1860. C: 1.9.1860.
> Commissioned 1866 under Lieut Harry Farr Yeatman, as tender to Flagship of West Africa squadron; on 9.11.1866 under Lieut William F. E. Freeland, then 7.2.1871 under Lieut Thomas Larcom. Sold in 8.1873 at Fernando Po (by AO 14.2.1873) for £500.

INVESTIGATOR Controller's Department design, 1861. Built in dry dock as 'exploring vessel for Gabon River'.

> Dimensions & tons: 121ft 0in, 110ft 1¼in x 16ft 0in x 7ft 5¼in. 149bm.
> Men: 21. Guns: 2?
> Machinery: 34nhp.
> First cost: — .
> *Investigator* Deptford Dyd/—
> Ord: — . K: 15.6.1861. L: 16.11.1861. C: 14.12.1861, 'to be put at the service of Dr Livingstone'.
> Commissioned 19.10.1861 under Lieut Benjamin Langlois Lefroy, for exploration. On 1.4.1863 under Lieut Philip Ruffle Sharpe, as tender at Lagos; on 19.10.1863 under Lieut James Nias Croke, then 18.1.1864 under Lieut Charles George Frederick Knowles, then 25.2.1865 under Lieut John George Graham McHardy, then 1866 under Lieut John William Jones, and on 5.4.1867 under Lieut Albert Elms Kay. Sold 2.8.1869 to the local authorities at Lagos.

PADDLE TUGS/TENDERS

FURY [ex-mercantile — . — . — .] Purchased at Bermuda on 2 May 1834 for £530 for use as gunboat.

> Dimensions & tons: 90ft 0in, 78ft 0in x 20ft 0in x 10ft 0in. 166bm.
> Men: — . Guns: 2 x 9pdrs.
> Machinery: —nhp.
> BU 8.1843 at Bermuda.

BEE Combined paddle/screw tender. William Symonds design, 1840. Tender for the 'Royal Academy, Portsmouth' for the use of the students. Intended to be fitted for both paddle and screw propulsion. Began life as a paddle steamer, but screw fitted in 1844 (and replaced in 1846). As such she was the first screw steamer built for the RN.

> Dimensions & tons: as built: 60ft 0in, 52ft 5¾in x 12ft 4in x 6ft 0in. 42bm.
> Lengthened 1844: 63ft 0in, 55ft 8¼in x 12ft 2in x 6ft 0in. 43⁸⁰⁄₉₄bm.
> Machinery: 10nhp.
> *Bee* Chatham Dyd/Maudslay, Sons & Field.
> Ord: 1.9.1840. K: 14.6.1841. L: 28.2.1842. C: 3.1843?
> First cost: Hull £737, machinery £990.
> Screw fitted 1844. BU 9.1874 at Sheerness.

RUBY Class 1841–Builder's design, approved 17 September 1841. Iron-hulled tender.

> Dimensions & tons: 93ft 0in, about 84ft 0in x 12ft 9½in x 7ft 1in. 73bm.
> Men: — . Guns: nil.
> Machinery: 2-cylinder 20nhp.
> *Ruby* Acramans, Morgan & Co., Bristol/—
> Ord: 1.9.1841. K: 1841. L: 7.1842. C: 25.9.1842 (arrived at Devonport).
> Used in gunnery trials 1846. Her remains were sold to Mr Barnard 2.11.1846 (for £20).

ROCKET Class 1841 Builder's design, approved 15 November 1841. Iron-hulled tender, initially described as a 'steam packet'.

> Dimensions & tons: 90ft 0in, about 82ft 0in x 12ft 8in x 7ft 4in. 70³¹⁄₉₄bm.
> Men: — . Guns: nil.
> Machinery: 2-cylinder 20nhp.
> First cost: fitting £481.
> *Rocket* William Fairbairn & Co., Millwall/William Fairbairn
> Ord: 1.9.1841. K: 1841. L: 7.1842. C: 27.2.1843.
> Commissioned 19.8.1842 under Cmdr William Louis Sheringham (–31.12.1842). BU completed 6.1850 at Woolwich.

ADELAIDE 1848 Built for the Colonial Government of South Australia, under Admiralty supervision. Only in RN control for delivery to Australia.

> Dimensions & tons: 140bm.
> Men: — . Machinery: — .
> *Adelaide* C. J. Mare & Co., Blackwall/—
> Ord: — . K: 9.1847. L: 26.2.1848. C: 4.1848.
> Commissioned at Woolwich Dyd 3.7.1848 for passage to South Australia. Handed over to colonial government at Port Adelaide 7.3.1849.

No further seagoing paddle tugs/tenders were built for the Navy, but from the outbreak of the Crimean War a number of vessels were purchased by the Navy. Some of these known to have iron hulls are so stated below, but it should be noted that of the following the *Steady* may possibly also be of iron.

CIRCASSIAN Purchased 1854 [ex-mercantile *Swan*, built 1852 by T. D. Marshall at South Shields]. Wooden(?) paddle tender of 90nhp, purchased 21 July 1854 by Admiral Dundas at Constantinople for £4,800.

> Dimensions & tons: 115ft 0in x 19ft 0in x — . 170bm.
> Commissioned 1854 under Edward Codrington Ball, Second Master (Acting); at bombardment of Sebastopol 11.10.1854.
> Sold 28.7.1856 to Dr Glascote for £1,037.

CRESCENT Purchased 1854 [ex-mercantile, origins unknown]. Wooden paddle tender of 50nhp, purchased 21 July 1854 by Admiral Dundas at Constantinople for £3,000.

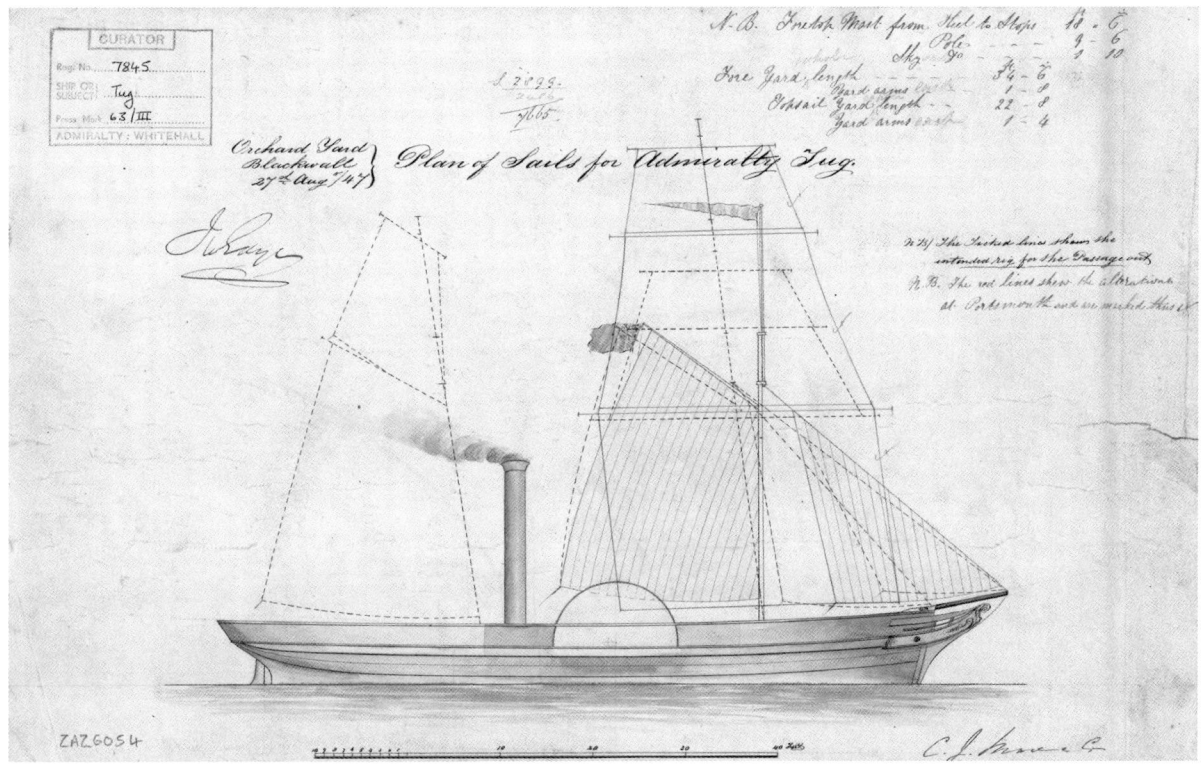

A sail plan for the Admiralty paddle tug *Adelaide*, dated 27 August 1847 at the Orchard Yard in Blackwall, and signed by John Edye, the Assistant Surveyor. The *Adelaide* was built at Blackwall for the South Australian Colonial Government, and originally had only a foremast with a gaff-headed sail and a single jib, but was re-rigged at Portsmouth for the passage to Australia – the ticked lines show the first proposal for a brigantine rig with heavily raked masts, but the larger sail plan outlined on the drawing was the one adopted. *(NMM J7979)*

Dimensions & tons: 80ft 0in, — x 16ft 0in x — . 90bm.
Sold 7.7.1856 for £1,700 to Mr Wood at Constantinople.

DANUBE Purchased 1854 [ex-mercantile *Tuna*, built 1854 by Money Wigram & Sons at Blackwall]. Wooden paddle tug of 90nhp, purchased 13 August 1854 for £13,000 at Constantinople.
Dimensions & tons: 138ft 0in x 18ft 3in x 9ft 0in (regd). 110bm. 170 disp.
Commissioned 6.1855 under Lieut. Ralph Peter Cator; in action at Taganrog 6.1855. Sold 23.7.1856 at Constantinople to Aali Hilmi for £4,013.

PIGEON/PIDGEON Purchased 1854 [ex-mercantile *Brothers* — . — .]. Wooden paddle tug of 53nhp, purchased 15 August 1854 at Constantinople for £4,420.
Dimensions & tons: 92ft 0in, — . x 16ft 0in x 9ft 7in. 33bm.
Sold 14.7.1856 to Hamilton & Hodge at Constantinople for £2,005, becoming mercantile *Shark*.

MALTA Purchased 1854 [ex-mercantile *Britannia*, building details unknown]. Wooden paddle tug of 53nhp, purchased 22 August 1854 (?) at Constantinople.
Dimensions & tons: 97ft 0in, — . x 17ft 5in x 10ft 5in. 42bm.
Sold 14.7.1856 at Constantinople for £1,500 to Mr Benzolos (?).

HELEN FAWCIT Purchased 1855 [ex-mercantile *Helen Fawcit*, built Middlesborough 1849]. Wooden paddle tug of 70nhp, purchased 3.2.1855 at Constantinople by Admiral Fred. Grey from Pisani & Co. for £7,200.
Dimensions & tons: — — . 45bm.
Sold 1856.

MULLET Purchased 1855 [ex-mercantile *Express*, built North Shields 1846]. Wooden paddle tug of 100nhp, purchased 9 March 1855 at Constantinople by Admiral Fred. Grey from Whitall (?).

Fitted by Admiral Lyons off Sebastopol as a gunboat.
Dimensions & tons: — —. 99bm.
Men: — . Guns: 1 x 32pdr (42 cwt).
Sold 21.7.1856 to C. Jacob for £1,400 at Constantinople.

STEADY Purchased 1855 [ex-mercantile *Powerful*, building details unknown]. Wooden (?or iron) paddle tug of 75nhp, purchased for Constantinople from T. D. Marshall at Cork for £6,000 by AO of 26 March 1855.
Dimensions & tons: 108ft 0in, 84ft 9in x 18ft 0in x 10ft 0in. 146bm.
Sold 21.7.1856 at Constantinople to Glavanay, Sons & Co. for £1,850.

BUSTLER Purchased 1855 [ex-mercantile *Merry Andrew*, built 1852 by T. & W. Smith, at North Shields]. Wooden paddle tug of 100nhp, purchased by AO of 28 March 1855 from T. & W. Smith for £5,500 or £5,800.
Dimensions & tons: 111ft 6in, 97ft 0in x 20ft 6in x 10ft 6in. 217bm.
Employed at Chatham. Sold 11.1894.

HEARTY Purchased 1855 [ex-mercantile *Merry Monarch*, built 1854 by T. & W. Smith, North Shields]. Wooden paddle tug of 100nhp (379ihp), purchased by AO of 28 March 1855 from T. & W. Smith for £7,500.
Dimensions & tons: 112ft 0in, 99ft 0in x 20ft 6in x 10ft 6in. 221bm. 346 disp.
BU 1876 at Malta.

LANDRAIL Purchased 1855 [ex-mercantile *Gipsy Queen*, built North Shields 1854]. Wooden paddle tug of 50nhp, purchased 24 May 1855 at Constantinople by Admiral Fred. Grey for £7,500 from Mr Ward (or Wood?).
Dimensions & tons: — — . 36bm.
Sold 21.7.1856 at Constantinople by Admiral Grey to J.Parry for £2,400.

REDPOLE Purchased 1855 [ex-mercantile *Racehorse*, launched 9 May 1855 by W. C. Miller at Liverpool]. Wooden paddle tug of 160nhp, purchased from W. Jack for £11,000 or £12,078 by AO of 2 March 1855.
> Commissioned 29.10.1855 under Cmdr. Edward Knatchbull Hughes Hallett, as tender to Hibernia at Malta; on 13.12.1856 under Capt. Frederick Warden, as tender to *Hibernia* at Gibraltar.
> Dimensions & tons: 142ft 0in, — x 23ft 0in x 12ft 0in. 360bm.
> BU 27.1.1872.

SULTANA Purchased 1855 [ex-mercantile — , built 1855 North Shields]. Wooden paddle tug of 50nhp, purchased from Octavius Swan at Constantinople for service at Balaclava for £6,250.
> Dimensions & tons: — — . 98bm. (61.84 register)
> Sold 22.9.1856 at Pera to H.Riddley for £2,700.

TIVERE Purchased 1855 [ex-mercantile *Tivere*, building details unknown]. Wooden (coppered) paddle tug of 60nhp, purchased at Constantinople for £5,100 on 13.11.1855.
> Dimensions & tons: — — . unknown.
> Wrecked 3.12.1855, remains sold to Arajgi Vasil.

MINNA Purchased 1854 [ex-mercantile *Pomona* — —]. Iron-hulled paddle tug of 120nhp, purchased 19 August 1854 at Constantinople for £9,500.
> Dimensions & tons (approx data only): ?180ft x 23ft x 9ft. ?170bm.
> Sold 28.7.1856 for £2,600 at Constantinople to Swan Brothers.
>> There seems to be some confusion as to whether this vessel was also known as *Sulina* – or whether *Sulina* was a separate vessel purchased and sold at the same times. Certainly *Sulina* was in operation at Taganrog in 6.1855, under Charles Henry Williams, mate.

MOSLEM Purchased 1854 [ex-mercantile *Towk Beira* — —]. Iron-hulled paddle tug of 120nhp, purchased at Constantinople for £9,500.
> Dimensions & tons (approx data only): ?180ft x 23ft x 9ft. ?170bm.
> Commissioned 1854 under Lieut. J. Simpson; in Sea of Azov operation 5–6.1855. Sold (as *Brenda*) to H. Henchfield 8.1856 for £2,100 at Constantinople. There is some confusion as to whether this vessel was also known as *Brenda* (which seems likely) or whether *Brenda* was a separate vessel with a similar history.

BOURNABAT Purchased 1854 [ex-mercantile *Bournabat*, builder uncertain but possibly built 1844 by 'the Assignees of Acramans, Morgan & Co., Bristol' following that firm's liquidation in 1842.]. Iron-hulled paddle tug, purchased by AO of 9 September 1854 at Constantinople for £2,200.
> Dimensions & tons: 79ft 0in x 13ft 0in x 7ft 0in.
> Ran ashore and lost at Scutari 29.3.1855. Wreck sold for £150.

NIMBLE Purchased 1855 [ex-mercantile *Morecambe's Queen*, built 1853 by Alexander Denny, Dumbarton (No. 27).]. Iron-hulled paddle tug, purchased from Jas Ward at Constantinople for £9,600 by Admiral Frederick Grey 1 May 1855.
> Dimensions & tons: — — . 91bm.
> Machinery: Tulloch & Denny 2-cylinder (34in diameter, 4ft stroke). 90nhp.
> Sold 22.7.1856 to the Turkish government at Constantinople for £4,050.

WALLACE Purchased 1855 [ex-mercantile – possibly *Lyons*, built 1854 by J. Scott Russell, Millwall]. Iron-hulled paddle tug of 100nhp, purchased (at Malta?) for £9,500 by Admiral Lyons in July 1855.
> Dimensions & tons: 112ft 3in, 99ft 3in x 19ft 10½in x 10ft 0in. 209⁷⁄₉₄bm.
> To Portsmouth 1857 and attached to the Clarence Victualling Yard. Sold 1869.

THAIS Builders design of 1856. Double ended iron-hulled tug/ferry. Built (or purchased?) for use at Constantinople. Alternative dimensions reported may indicate possible lengthening.
> Dimensions & tons: 125ft 0in, 108ft 0in x 22ft 0in x 9ft 6in. 278⁴⁄₉₄bm. also reported as: 134ft 0in, about 117ft 0in x 22ft 0in x 8ft 11in. 302bm.
> Men: — . Machinery: Laird. 80nhp.
> First cost: £10,250.

Thais John Laird, Birkenhead. (No. 129)/Laird
> Ord: 16.1.1856. K: — . L: 2.1856. C: 22.2.1856.
> Turned over to Devonport Victualling Yard by AO 16.3.1857. In 5.1857 under Lieut Edward Dent, at Battle of Fanshan Creek. Sold to Mr E. Bates 2.12.1869 for mercantile service. Re-sold (?to BU) at Buenos Aires 6.1878.

HONG KONG Purchased 1856 [previous history unknown]
> No further details. Sold 1858.

DRAGON Purchased 1860 [ex-mercantile *Dragon*, built 1855 at Newcastle]. Wooden paddle tug of 40nhp, purchased 11 July 1860 at Malta for £1,200 for towing mud barges.
> Dimensions & tons: 85ft 0in, 74ft 10¼in x 16ft 10in x 8ft 7in. 113bm.
> Sold 1867.

COWPER Purchased 1860 [ex-mercantile *Hei Sean*, building details unknown]. Wooden paddle tender of 130nhp, purchased 17.4.1860 at Hong Kong by Admiral Sir Henry Hope from Lyall, Still & Co. of Hong Kong for 120,000 Mexican pesos (= £25,000) as she lay in harbour with everything aboard (AO 20.10.1860 confirmed purchase)
> Dimensions & tons: 175ft 0in, — x 27ft 0in x — 342bm.
> Sold 1861 at Hong Kong to become mercantile *Fei Seen*.

WATERMAN Purchased 1861 [ex-mercantile *Sir Charles Forbes*, launched 3.2.1846 at Bombay]. wooden paddle tender of 60nhp, purchased 10 January 1861 at Hong Kong by Admiral Sir Henry Hope (for £9,375). Teak-built.
> Dimensions & tons: 130ft 1in, — x 18ft 5in x 10ft 2in. 141bm.
> Employed as distilling vessel. Sold 1862 to become mercantile *Nagasaki Maru*.

WARDS Purchased 1861 [ex-mercantile *Wards*, built 1861] wooden paddle tug, purchased by the Director of Works Department for use at Gibraltar.
> Dimensions & tons: 73ft 3in, 64ft 3¼in x 14ft 11in x 7ft 6in. 76bm.
> Machinery: 27nhp.
> First cost: £3,765 (hull £2,280 + machinery £1,485).
> BU 1889.

PADDLE VESSELS ON THE CANADIAN GREAT LAKES

The Royal Navy acquired nine small paddle vessels (eight wooden-hulled, and the iron-hulled *Mohawk*) during the 1830s and 1840s for patrol duties on the Canadian Great Lakes, to prevent gunrunning from the United States into Canada. A further vessel (*Montreal*) is included although it is not certain that she had an engine.

TORONTO Purchased 1838 [ex-mercantile *Sir Charles Adam*, built 1834 in USA]. Paddle gunboat, purchased 7 July 1838 for use on Lake Erie.

Dimensions & tons: 147ft 1in, 136ft 0in x 27ft 4in x 7ft 10in. 342bm.
Men: 40. Guns: 4 x 18pdr carronades.
Machinery: Gibson & Co., Black Rock? 100nhp.
Commissioned 21.4.1838 under Lieut John Duffill; in 1841 under Lieut William Rowlatt. Sold 1843.

EXPERIMENT Purchased 1838 [ex-mercantile — — , built 1838 by Niagara Dock Co.]. Former sailing ship, purchased 21 July 1838 and converted to paddle gunboat for use on Lake Huron.

Dimensions & tons: 96ft 6in, 86ft 0in x 14ft 6in x 7ft 2½in. 220bm.
Men: 40? Guns: 1 x 6pdr brass on pivot + 1 x 12pdr carronade.
Machinery: Builders? 25nhp.
Commissioned 18.4.1843 under Lieut James Boxer; on 16.12.1843 under Lieut James Harper. Sold 1847 or 1848.

TRAVELLER Purchased 1839 [ex-mercantile — —, built 1838 by Niagara Dock Co.]. Paddle gunboat, purchased 30 April 1839 at Niagara for use on Lake Ontario.

Dimension & tons: 138ft 0in, 132ft 0in x 23ft 9in x 10ft 5in. 335bm.
Men: 42. Guns: 1 x 12pdr carronade.
Machinery: Ward & Co., Montreal 90nhp.
Sold 1844 for mercantile use.

MONTREAL Purchased 1839 [ex-mercantile, built 1836 in Canada]. A two-masted schooner (perhaps not a paddler), purchased 18 October 1839 for use on Lake Erie.

Dimensions & tons: 82ft 0in, 65ft 0in x 20ft 6in x 9ft 0in. 145bm.
Men: 35. Guns: 1 x 18pdr SBML.
Machinery: ?
Commissioned 3.4.1843 under Cmdr William Newton Fowell; on 10.8.1843 under Lieut John Tyssen; paid off 8.1848 and sold 1848.

MINOS 1840 Symonds design. Wooden paddle gunboat, for use on Lake Erie.

Dimensions & tons (design): 136ft 0in, 121ft 9½in x 24ft 2in x 13ft 0in. 363bm.
Men: 52. Guns: 1 x 12pdr on the forecastle.
Machinery: Ward & Co., Montreal 90nhp.
Minos Chippawa
As built: 152ft 8in, 132ft 7½in x 24ft 2in x 13ft 0in. 406bm.
Ord: — . K: — . L: 6.1840? C: — .
Commissioned 1.8.1846 under Lieut James Harper. Sold 3.1852 to Mr Weston for '£1,200 Halifax currency equal to £986/6/0 Sterling'.

MOHAWK 1841 Symonds design. Iron paddle gunboat for patrol service on Lake Ontario, later on Lake Huron. Lengthened in 1846 with additional 25ft 1½in section constructed at Millwall (probably shipped out to Canada and section inserted at Kingston Yard).

Dimensions & tons (as built): 99ft 1½in, 87ft 0in x 19ft 6in x 9ft 10in. 176bm.
Dimensions & tons (as lengthened 1846): Lengthened to 124ft 3in, 111ft 3in. 225bm.
Men: 20. Guns: 1.
Machinery: — 6onhp.
First cost: (in pieces) £1,446 (hull?).
Mohawk William Fairbairn & Co., Millwall/William Fairbairn
Ord: 6.1841. K: 1841. Delivered in pieces 11.1841 for shipment to Canada. L: 21.2.1843 after assembly at Kingston Yard, Lake Ontario.
Commissioned 3.4.1843 under Lieut John William Bedford; on 26.7.1843 under Cmdr William Newton Fowell, then 1.7.1846 under Lieut John Tyssen and 26.8.1848 under Lieut Frederick Charles Herbert. Sold 21.6.1852 to J.F.Parke for £1,643.

SYDENHAM Purchased 1841 [ex-mercantile, built 1841 at Montreal] First Class paddle gunvessel, purchased while building at Montreal per AO of 27 November 1841 for £26,940 (for hull & rigging – machinery appears to have been £13,200 extra).

Dimensions & tons: 170ft 0in, 153ft 10in x 27ft 2in x 16ft 9in. 596bm.
Men: 80. Guns: 1 x 32pdr (26cwt) + 2 x 32pdr (17cwt); later 2 x 68pdrs.
Machinery: — , Montreal. 220nhp.
Commissioned at Montreal 23.9.1842 under Lieut William Pearson Crozier (appointed 21.6.1842); to England 5.1843 and paid off 15.6.1843 at Woolwich for repairs. Refitted and reboilered 1843–44 at Woolwich (for £17,380). Recommissioned 27.4.1844 under Lieut David Robert Bunbury Mapleton, for the Mediterranean packet service; paid off 7.5.1846 at Malta. Sold there 11.7.1846 (for £1,060).

CHEROKEE Class 1841 Symonds design. Wooden paddle gunvessel, formerly a brig-sloop.

Dimensions & tons: 170ft 0in, 150ft 1¾in x 30ft 10in x 16ft 3in. 750bm.
Men: 36. Guns: 1 x 8in, 1 x 32pdr.
Machinery: Maudslay, Sons & Field 200nhp.
Cherokee Kingston Navy Yard, Ontario.
Ord: — . K: 1842. L: 22.9.1842. C: — .
Commissioned 1.7.1846 under Cmdr William Newton Fowell. Sold by AO 30.10.1851 to Messrs. Campbell, Forsyth, Yarwood & Gaskin (for £3,287.13.5d).

MAGNET Paddle steamer built at Hamilton, Canada, about 1846. Built to carry up to 1,000 troops with arms and accoutrements 'for the Inland Navy'. The British government advanced £6,112 towards the estimated value when completed (which was a total of £15,260) – retaining the right to assume possession of the vessel on payment of the remaining portion of the value – right relinquished by AO 12 August 1864 (or 1866?).

CANADA Purchased 1847 – paddle gunvessel (possibly stern-wheeler).

Dimensions & tons: 198ft 0in x 29ft 0in x 10ft 0in.
Men: — . Guns: 2 x 68pdrs.
Machinery: 120nhp.

Vessels ordered or re-ordered as steam screw sloops (from 1845)

During the first decade of the screw-propeller era (from the *Rattler/Alecto* trials of 1845 to the outbreak of the Crimean War), no type of vessel was more affected by rapidly evolving changes in design and designation than the fluctuating corvette/sloop/gunvessel categories. While the detail given below reflects the chief stages for every vessel ordered, some intermediate changes have been omitted to simplify the picture. The number of guns carried by each of these early screw sloops rose rapidly from between 5 and 8 initially to virtually double within a few years.

RATTLER Sloop, 9 guns. Sir William Symonds design, 1844 (with Brunel acting as consulting engineer), based on *Alecto* Class paddle sloops but modified for screw. Originally a paddle sloop (SV2) of the *Alecto* Class had been ordered on 12 March 1841 and named *Rattler*, but was re-ordered as a new vessel (not, as sometimes stated, a conversion while building) using 'such parts of the frame of the original *Rattler* as was feasible'. Towed to Maudslay's yard on completion, to fit engine, then coppered at Woolwich Dyd. While full 'completion' was as stated above, *Rattler* first put to sea for propeller trials on 30 October 1843. *Rattler* had her famous tug of war with her paddle-powered sister *Alecto* on 3 April 1845.

    Dimensions & tons: (195ft 0in oa) 176ft 6in, 157ft 9½in x 32ft 8in (32ft 6in for tonnage) x 18ft 7½in. 888⁸⁹⁄₉₄bm. 1,112 disp.
    Men: 120, later 180. Guns: 1 x 8in/68pdr (65cwt/10ft) on pivot + 8 x 32pdrs (25cwt).
    Machinery: 4-cylinder vertical single expansion, double cylinders, driving screw through straps rather than gearing. Single screw. 200nhp. 428ihp = 10.074kts.
Rattler Sheerness Dyd/Maudslay, Sons & Field

Ord: 24.2.1842. K: 4.1842. L: 12.4.1843. Machinery and screw fitted at East India Dock 1.1844. C: 30.1.1845.
First cost: Machinery £9,400, fitting £17,413.
Commissioned 12.12.1844 at Woolwich under Cmdr Henry Smith (Capt. 27.6.1846); to Portsmouth 30.1.1845; to Squadron of Evolution 1846; on 17.11.1846 under Cmdr Richard Moorman, for South America; paid off 13.9.1847 at Woolwich. Recommissioned 12.2.1849 at Plymouth under Cmdr Arthur Cumming, for west coast of Africa; paid off 15.4.1851 at Woolwich. Recommissioned 28.8.1851 at Woolwich under Cmdr Arthur Mellersh, for the East Indies; in action against junks near Namguan 11.5.1853; in 2nd Burma War; on 5.4.1855 under Cmdr William Abdy Gellowes, for China and the East Indies; paid off 17.5.1856 at Woolwich. BU 1856 (22.7–26.11) at Woolwich (by contract with Mr Fulcher).

PHOENIX 6 (later 12) guns. Converted from an 1832 paddle sloop (see Chapter 9) in accordance with an Oliver Lang design, 1843. Docked at Deptford Dyd to fit 'Mr Steinman's' screw propeller (by AO 1 April 1844); towed down to East India Docks 11 December 1844 to complete fitting her machinery.

    Dimensions & tons (as converted): 174ft 7in, 153ft 5in x 31ft 10in oa (31ft 6in for tonnage) x 16ft 9in. 809⁶⁷⁄₉₄bm. 1,460 displacement.
    Men: 120. Guns: 1 x 10in (84cwt) on pivot + 1 x 8in (52cwt) on pivot + 8 x 32pdr carronades (17cwt).
    Machinery: 2-cylinder vertical single expansion, oscillating, geared. Mr Steinman's screw propeller (fitted by AO 9.8.1843). 260nhp. 524ihp = 7.67kts.
Phoenix [Conversion by] Curling & Young's, Limehouse/John Penn & Son.
    Ord: 25.5.1843. Altered 4.1844–2.1845. Undocked 2.1845 at Limehouse.

This depiction shows the celebrated steam sloop trials between the screw-propelled *Rattler* and the paddle-driven *Alecto*, which took place on 3 April 1845 in the North Sea in perfectly calm weather. The two were near sisters (although the aft part of the *Rattler* was lengthened by 12½ft to allow the screw to be inserted, and accordingly the *Rattler* carries a three-masted rig, whereas the *Alecto* has just two masts) and were linked stern to stern by a cable. With both ships' 200nhp engines working, *Rattler* developed 300 ihp while *Alecto* could only reach 141 ihp, and the former succeeded in towing the latter backwards at a speed of 2.8 knots. In reality, the trial was academic, as the Admiralty had already decided that the future of steam propulsion lay with the screw propeller, and the trial was little more than a public relations exercise designed to convince sceptics. *(NMM PY0923)*

Recommissioned 27.12.1845 at Woolwich under Cmdr James
Samuel Aked Dennis, on Home station, then sailed 13.4.1846
to the Mediterranean; to Squadron of Evolution 1846; paid off
4.9.1847 at Woolwich. Recommissioned 16.7.1849 at Portsmouth
under Cmdr George Wodehouse, for west coast of Africa;
captured slaver 13.4.1850; on 3.8.1850 under Cmdr Thomas
Henry Lysaght, on same station; paid off 15.8.1851. Fitted to
carry stores to Beachey Island (in Barrow Strait, Arctic) (for
£13,655) 12.1851–5.1853. Recommissioned 2.1853 at Deptford
under Capt. Edward Augustus Inglefield, as supply ship for
Sir Edward Belcher's expedition to Barrow Strait; returned
with news of the discovery of the Northwest Passage; paid off
25.10.1853 at Woolwich. Recommissioned 21.2.1854, still under
Inglefield, for further supply voyage to Belcher's expedition;
paid off 25.10.1854 at Woolwich. Fitted there for service in the
White Sea (for £8,312) in early 1855. Recommissioned 10.2.1855
under Cmdr (Capt., 9.7.1855) John Montagu Hayes, for the
White Sea during the Russian War; sailed for home 9.10.1855 and
paid off at Sheerness. Sold to Castle 26.1.1864 to BU at Charlton.

The 1844 Programme of Works as amended by the Admiralty on 19
February provided for seven steam vessels, of which five were initially
intended to be built to the lines of the paddle-driven *Sampson* (see
Chapter 9); the five comprised the *Niger* and *Odin* at Woolwich, the
Dauntless at Portsmouth, and the *Conflict* and *Desperate* at Pembroke;
four of these (see below) would eventually be completed with screw
propulsion, three as screw sloops as set out below (the fourth was
Dauntless – see Chapter 6). Ironically the sole steam vessel proposed in
this year's programme for screw propulsion – the *Fury* at Sheerness –
would be completed as a paddle-driven vessel.

Sir George Cockburn, the First Naval Lord, directed John Edye
on 29 February to produce a new classification system for the Navy's
steam vessels. By 7 March Edye had compiled a new list of vessels in
accordance with Cockburn's order, and on 31 May a new Admiralty
Order set out the detail. There would be three grades of steam sloops,
each under a Commander; the largest (First Class, or S1) additionally
carried 3 lieutenants among its established complement of 145 men,
and was hypothetically to be armed with 6 guns; the Second Class
(S2) and Third Class (S3) would each have just 2 lieutenants among
its established 100 men; the Second Class would carry 4 guns – 2 x
10in (84cwt) and 2 x 32pdr carronades (25cwt/6ft), while the Third
Class would carry 1 x 68pdr and 4 x 32pdr carronades (17cwt). Both
the *Rattler* and the *Phoenix* were included in the List as Second Class
sloops.

On 13 August the Surveyor was instructed to prepare two sets of
draughts for the five *Sampson* class vessels of the 1844 Programme –
one set for paddle propulsion, the other as screw sloops.

ENCOUNTER Class 6 (later 14) guns. On 12 December 1844 the
Surveyor was ordered to report on a drawing by John Fincham (M/
Shipwright at Portsmouth) for a steam sloop, to carry 8 guns (2 x
85cwt, 4 x 65cwt and 2 x 25cwt); Symonds was directed to propose
the lines for a steam sloop with screw propulsion and the same
armament as the Fincham design. The Surveyor on 4 February 1845
submitted a design based on the paddle-driven *Devastation* Class, but
on the following day the Admiralty ordered this ship to be built at
Pembroke according to Fincham's plans, as a First Class sloop with
screw propulsion. On 11 October 1845 the Admiralty accepted a
tender from Penn for a 360nhp engine, to be of a new type in lieu of
the old oscillating cylinder type. The *Encounter* was 'altered abaft' and
lengthened at Deptford Dyd 1848 before completion. A very successful

vessel, sailing very fast and capable of 10½ knots under steam. Re-
classed from sloop to corvette in 1862.

Dimensions & tons: Original design 180ft 0in, 157ft 7in x 33ft 2in
oa (32ft 8in for tonnage) x 20ft 10in. 894⁴⁹⁄₉₄bm. As lengthened
1848: 190ft 0in, 167ft 10in (breadth and depth unchanged).
953bm. 1,194 disp.
Men: 145. Guns: 1 x 68pdr (95cwt/10ft), 1 x 10in (85cwt/9ft 4in), 4
x 8in (65cwt/9ft), later 2 x 8in/56pdr (85cwt), 4 x 68pdr (65cwt)
and 2 x 32pdr (25cwt). Re-armed 1850 with 12 guns, increased
1856 to 14 guns (probably all 32pdr MLSB).
Machinery: 2-cylinder (55in diameter, 2¼ft stroke) horizontal single
expansion, trunk. Single screw (12ft diameter). 360nhp. 673ihp =
10¼kts.
Encounter Pembroke Dyd/John Penn & Co.
Ord: 5.2.1845. K: 6.1845. L: 5.9.1846. C: 12.10.1849.
First cost: £52,587 total (hull £19,734, machinery £20,192, fitting
£12,661).
Commissioned 7.9.1849 at Portsmouth under Capt. George Thomas
Gordon, for particular service; paid off 15.9.1852 at Portsmouth.
Recommissioned next day under Capt. George William Douglas
O'Callaghan, for the East Indies and China; off Russian Pacific
coast during the Russian War, the 2nd Anglo-Chinese War; paid
off 19.2.1858 at Plymouth. Refitted there (for £20,993) 1858–59.
Recommissioned 22.9.1859 under Capt. Roderick Dew, for
East Indies and China; escorted gunboats *Bouncer* and *Snap* to
China; in action during Taiping rebellion, then to Japan; paid off
3.8.1864 at Plymouth. BU 5.1866 at Devonport.
Harrier Pembroke Dyd/— .
Ord: 26.3.1846. K: —.1846. Suspended 9.9.1846. Cancelled 4.4.1851.

NIGER Class – 8 (later 14) guns. On 20 February 1845 the Admiralty
decided that the *Niger*, originally scheduled for building at Woolwich
to the *Sampson* design, should instead be built as a First Class screw
sloop along the lines of plans drawn up by Oliver Lang, with a hull
similar to paddle sloop *Basilisk*. At this date the design was for a sloop
of 185ft length, to carry the same power (engine) and armament as the
Encounter. Woolwich's Slip 3 was allocated on 7 April, with the *Niger*
scheduled to be completed before the end of 1845. On 25 October
the Admiralty accepted a tender from Maudslay for a 400hp engine
for her. The *Niger* took part in three series of trials (including a noted
tug of war) with her paddle-powered sister *Basilisk* over the summer
of 1849 which, while less well-known than the earlier *Rattler/Alecto*
trial of 1845, effectively sealed the fate of paddle propulsion. Like the
Encounter, she was found to sail and steam very well. The *Niger* was
re-classed from a sloop to a corvette in 1862, and re-engined in 1864
with a 350nhp 2-cylinder engine by Miller, Ravenhill & Salkeld.

Dimensions & tons: Original design 185ft 0in, about 162ft (est.) x
32ft 8in x 21ft 6in. 911bm.
As lengthened 1848: 194ft 4in, 170ft 11¾in x 34ft 8in oa (34ft 4in for
tonnage) x 21ft 5½in. Draught 15¼ft -16¼ft. 1,072⁶⁹⁄₉₄bm. 1,496
disp.
Men: 145, later 175. Guns: 1 x 56pdr (87cwt), 1 x 10in (85cwt), 4
x 8in (65cwt), 2 x 32pdrs (17cwt) OR 2 x 56pdrs (85cwt), 4 x
68pdrs (65cwt), 2 x 32pdrs (25cwt). Re-armed 1850 with 14 guns
– probably all 32pdr MLSB (32cwt/6ft) on broadside trucks; 1 x
68pdr on a pivot was added in 1856.
Machinery: 4-cylinder horizontal single expansion. Single screw
(12½ft diameter). 400nhp. 1,002ihp. 10¼kts.
Niger Woolwich Dyd/Maudslay, Sons & Field.
Ord: 20.2.1845. K: 5.1845. L: 18.11.1846. Lengthened at Deptford
Dyd before completion. C: 16.8.1850.

First cost: £57,597 (including hull £14,830, machinery £21,057).
Commissioned 9.7.1850 at Portsmouth under Cmdr Leopold
George Heath, for the west coast of Africa, then at end 1852
to the Mediterranean, and in 1854 to the Black Sea (during
the Russian War). On 2.2.1855 under Cmdr Henry William
Hire, in the Black and Mediterranean Seas; paid off 31.1.1856
at Woolwich. Recommissioned 14.5.1856 under Capt. Arthur
Auckland Leopold Pedro Cochrane, for the East Indies and
China; in 2nd Anglo-Chinese War; on 8.9.1858 under Capt.
Peter Cracroft (when Cochrane invalided), on same station, then
to Australian station; in New Zealand Wars; paid off 13.9.1861 at
Woolwich. Recommissioned 3.2.1865 under Capt. John Clarke
Byng, for North America and West Indies; on 9.10.1865 under
Capt. James Minchin Bruce (when Byng invalided), on same
station; paid off 9.12.1868 at Woolwich. Sold 2.12.1869 to Castle
for BU at Charlton.

Florentia Woolwich Dyd/— .
   Ord: 26.3.1846. K: —.1846. Suspended 6.10.1846. Cancelled
   22.5.1849.

CONFLICT Class – 8 guns. On 9 May 1845 the Admiralty directed
that the *Conflict* and *Desperate*, originally intended for building at
Pembroke to the *Sampson* design, should now be built as First Class
screw sloops according to a design by the Surveyor, Sir William
Symonds. On 11 October the Admiralty accepted tenders for 400nhp
engines for these two sloops from Maudslay and Seaward respectively
(although the engines for the two were swapped over on 22 October).
The cylinders in *Conflict* were 45in diameter, 2ft stroke, and the screw
was 13½ft diameter; the cylinders in *Desperate* were of 55in diameter,
2½ft stroke, and the screw was 13ft diameter. *Conflict* was lengthened
(by altering the stern) at Wigram's Yard, Blackwall in early 1848, and
Desperate was also lengthened before completion under AO of 13 July
1848. Unlike the other First Class sloops built at this time, the pair
were not to be re-classed as corvettes in 1862.
   Dimensions & tons: original design 185ft 0in, 164ft 6¾in x 34ft 4in
      (33ft 8in for tonnage) x 22ft 8½in. 992¹³/₉₄bm.
   as lengthened 1848: 192ft 6½in, 172ft 3½in x 34ft 4in (33ft 8in for
      tonnage) x 22ft 8½in. 1,038⁶⁹/₉₄bm. 1,628 disp.
   Men: 145, later 170. Guns: 2 x 56pdrs (85cwt), 6 x 8in/68pdrs
      (65cwt), 2 x 32pdrs (25cwt). 1857: 1 x 68pdr (95cwt), 1 x 10in
      (85cwt), 6 x 8in (65cwt).
   Machinery: 4-cylinder horizontal single expansion (geared in
      Desperate). Single screw. 400nhp.

Conflict Pembroke Dyd/Seaward & Capel. (772ihp = 9.378kts)
   Ord: 9.5.1845. K: 7.1845. L: 5.8.1846. C: 20.11.1849.
   First cost: Hull £20,496, machinery £21,514, lengthening £5,410,
      fitting £11,088.
   Commissioned 4.10.1849 at Plymouth under Cmdr Thomas George
      Drake, for east coast of South America; on 15.12.1851 under
      Acting Cmdr Robert Jenner, on same station; paid off 4.6.1852
      at (?)Portsmouth. Recommissioned 25.2.1854 at Plymouth
      under Capt. John Foote (drowned 18.4.1854 off Memel), for
      the Baltic (during the Russian War); on 9.5.1854 under Cmdr
      Arthur Cuming, still in the Baltic, including Liepaja and Riga;
      on 6.2.1855 under Cmdr Stephen Smith Lowther Crofton,
      in the Baltic, then 9.5.1855 under Cmdr William Charles
      Chamberlain, at Devonport. On 21.2.1856 under Cmdr Thomas
      Cochran, for the Mediterranean; paid off 24.2.1857 at Plymouth.
      Recommissioned 29.8.1857 under Cmdr Richard William
      Courtenay, for the west coast of Africa; paid off 13.12.1859 at
      Plymouth. Sold 1863.

Desperate Pembroke Dyd/Maudslay, Sons & Field. (699ihp =
9.432kts)
   Ord: 9.5.1845. K: 10.1845. L: 23.5.1849. C: 9.5.1853.
   First cost: £57,740 (including machinery £21,007).
   Commissioned 12.4.1852 under Capt. ?William Knighton Stephens,
      to accompany Sir Edward Belcher's squadron to search for lost
      Franklin expedition. Recommissioned 18.12.1852 at Plymouth
      under Capt. William Wylly Chambers, to accompany *Phoenix*
      (Capt. Inglefield) in taking supplies to Belcher's expedition at
      Barrow Strait; to Lisbon 10.1853; on 29.3.1854 under Capt.
      Edwin Claton Tennyson d'Eyncourt, for the Baltic (during
      the Russian War); on 6.1.1855 under Cmdr Richard Dunning
      White, still in the Baltic; on 10.5.1856 under Cmdr George
      Melville Jackson, for the Mediterranean; on 11.11.1857 under
      Cmdr Robert George Craigie, still in the Mediterranean; paid off
      1.1.1859 at Plymouth. Recommissioned 19.10.1860 under Cmdr
      John Francis Ross, at Devonport, then to Vera Cruz (Mexico);
      on 31.7.1862 under Cmdr Arthur Thomas Thrupp, on North
      America and West Indies station; paid off 7.11.1863 at Plymouth.
      BU at Devonport Dyd 8.1865.

Enchantress Pembroke Dyd/—
   Ord: 26.3.1846. K: —.1846. Suspended 18.9.1846. Cancelled
   4.4.1851.
Falcon Pembroke Dyd /—
   Ord: 26.3.1846. K: —.1846. Suspended 9.9.1846. Cancelled 4.4.1851.

ARCHER Class 12, later 15 guns. John Edye design, approved 25
August 1847. Originally ordered as gunvessels (of the *Rifleman* Class),
but orders for both vessels were suspended on 9 September 1846 and
they were re-ordered as sloops on 25 April 1847. *Archer* was fitted
with the 202nhp engine removed from the screw gunvessel *Rifleman*;
on later trials, this raised her speed to 8.242 kts. *Wasp* was sheathed
with Muntz metal. *Archer* was re-classed as a corvette 1862, *Wasp*
remaining a sloop.
   Dimensions & tons: 186ft 4in, 162ft 6¼in x 33ft 10in oa (33ft 6in for
      tonnage) x 19ft 0in. 970⁴⁵/₉₄bm.
   Men: 170 (later 175). Guns: 2 x 68pdrs (87cwt/9½ft), 10 x 32pdrs
      (42cwt/6½ft).
   Machinery: 2-cylinder (*Archer* – horizontal, geared – 202nhp; *Wasp*
      – vertical, oscillating – 100nhp). Single screw.

Archer Deptford Dyd/Miller, Ravenhill & Co. (347ihp = 7.818kts)
   As built: 186ft 4½in, 162ft 7¼in x 33ft 10½in oa (33ft 6½in for
      tonnage) x 18ft 11in. 973²²/₉₄bm. Draught 14ft 7in/14ft 10in.
   Ord: 26.3.1846. K: 18.10.1847. L: 27.3.1849. C: 9.3.1850.
   First cost: £41,404 (including hull £20,785).
   Commissioned 2.4.1850 at Plymouth under Cmdr James
      Newburgh Strange, for west coast of Africa, then to Leith for
      fishery protection duties; paid off 15.11.1853 at Woolwich.
      Recommissioned 25.2.1854 under Capt. Edmund Heathcote,
      for the Baltic (during the Russian War), then to North
      America and the West Indies; paid off 11.6.1857 at Woolwich.
      Recommissioned 21.5.1858 under Capt. John Sanderson
      (died 1859), for west coast of Africa; on 17.8.1859 under
      Cmdr Richard William Courtenay, then 4.4.1860 under
      Capt. Frederick Augustus Buchanan Crauford (Courtenay
      invalided), on same station; paid off 5.10.1861 at Woolwich.
      Recommissioned 30.3.1863 under Capt. John Bythesea, for same
      station; on 11.4.1864 under Capt. Francis Marten (Bythesea
      being invalided), on same station; in action against (Congo) river
      pirates 1865; paid off 30.1.1866 at Woolwich. Sold to Castle to
      BU, arriving at Charlton on 15.3.1866.

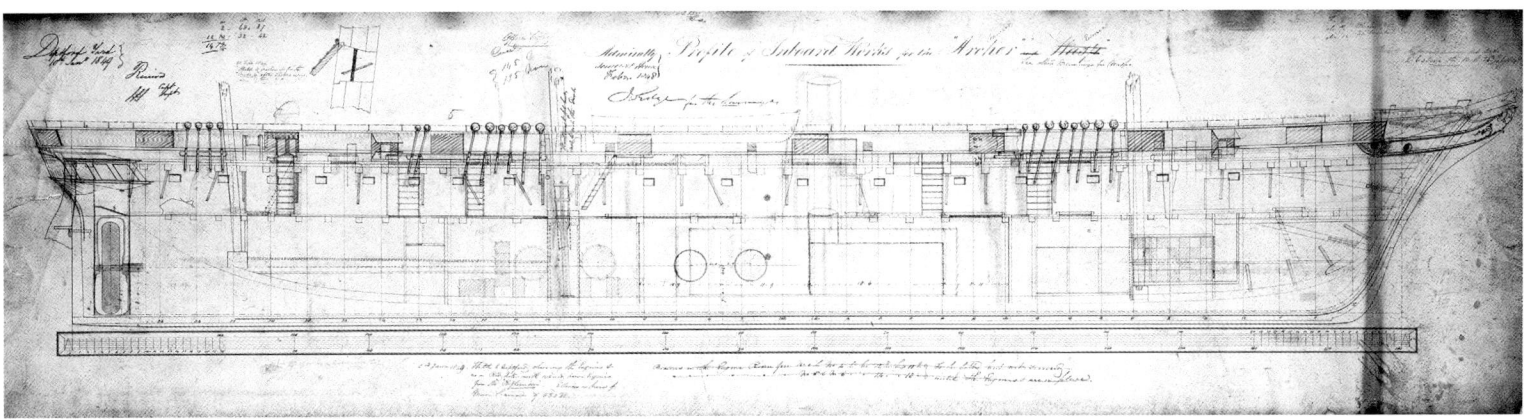

Dated February 1848, this is the profile of inboard works for the *Archer* and *Wasp*, although the latter name was later crossed out, with the note 'See other drawings for *Wasp*'. The drawing is covered in annotations reflecting a degree of uncertainty about the fittings for these early screw vessels. One notable problem with screw ships was caused by the run of the propeller shaft, which prevented the after masts from being stepped on the keel in the traditional fashion; in this drawing the mainmast is shown fitted into an iron yoke at lower deck level that straddles the propeller shaft. *(NMM DR66666)*

Wasp (ex-*Parthian*, renamed 25.8.1847) Deptford Dyd/Miller, Ravenhill & Co. (280ihp = 8.178kts)

Ord: 26.3.1846. K: 10.1847. L: 28.5.1850. C: 26.10.1850 at Woolwich (for sea).

First cost: £33,521 (including hull £21,326).

Commissioned 5.10.1850 at Woolwich under Cmdr William Pearson Crozier (invalided 13.3.1852), for west coast of Africa; on 16.4.1852 under Cmdr Charles Wright Bonham (Acting), then 8.1852 under Lieut Samuel Pritchard (Acting), for return from Africa. On 20.8.1852 under Cmdr Lord John Hay, in the Mediterranean and Black Sea (for the Russian War); on 2.2.1855 under Cmdr Henry Lloyd, still in the Mediterranean; paid off 7.1.1856 at Sheerness. Recommissioned 18.7.1856 under Cmdr Frederick Henry Stirling, for east coast of South America; paid off at Sheerness. Recommissioned 10.4.1860 under Cmdr Charles Stirling, for Cape of Good Hope; paid off 10.12.1861 at Portsmouth. Recommissioned 16.11.1863 under Capt. William Bowden, for the East Indies; took Arab slaver off Zanzibar 2.5.1865; on 29.1.1866 under Capt. Norman Bernard Bedingfeld, in the East Indies; paid off 22.4.1868 at Portsmouth. Sold to Charles Marshall on 2.12.1869 to BU at Plymouth.

1847 Programme On 17 February 1847, five screw sloops were ordered to the original *Rattler* design: *Brisk* and *Highflyer* from Woolwich Dyd, *Archer* and *Parthian* (previously on order as gunvessels) from Deptford Dyd, and *Grinder* from Sheerness Dyd. On 25 April 1847 they were re-ordered to an enlarged design, to carry a heavier armament. *Highflyer* was eventually built to a larger design, and rated as a corvette (which see); the other four vessels were built to three different designs.

Four 'steam gun schooners' of new design were ordered on 25 April 1847 'in lieu of sailing brigs originally planned': *Cracker* and *Plumper* at Deptford, and *Pincher* and *Hornet* at Woolwich. They were re-designated as 'gunvessels' in 1849, and as sloops in 1850 before completion. Four further screw schooners of this type were in the 1848 Programme of Works proposed on 23 March, being ordered on 30 March from Deptford and Chatham Dockyards (two at each) but were not named or proceeded with under this description.

MIRANDA sloop (re-classed as a corvette 1862), 14 (later 15) guns. John Fincham design, 'Improved *Rattler* Class', approved 3 November 1847. Although her officers and crew received the Baltic medal in 1854, she never entered that theatre of action; they justly received the Crimea Medal in 1855 with clasps for both Sebastopol and Azov operations.

Dimensions & tons: 196ft 0½in, 169ft 0in x 34ft 0in (for tonnage) x 20ft 9in. 1,039¹⁶/₉₄bm (as built 1,062 bm). 1,350 disp.

Men: 170 (later 175). Guns: 14 x 32pdrs (42cwt), later 10 x 32pdrs and 4 x 20pdrs; 1 x 68pdr (87cwt) added 1856.

Machinery: 2-cylinder horizontal single expansion, geared. Single screw. 250nhp. 613ihp = 10¾kts.

Miranda (ex-*Grinder*, renamed 3.11.1847) Sheerness Dyd/Robert Napier & Sons

Ord: 25.4.1847. Re-ord: 3.11.1847. K: 9.1848. L: 18.3.1851. C: 9.3.1854 at Sheerness (for sea).

First cost: £48,393 (hull £24,232, machinery £14,235, fitting £9,926).

Commissioned 25.2.1854 at Sheerness under Capt. Edmund Moubray Lyons (died 23.6.1855), for the Baltic, but never served there, being sent instead to the White Sea; joined Capt. Erasmus Ommanney's squadron, then returned to Portsmouth 9.1854; deployed 1855 to the Black Sea (during the Russian War); in Kerch operations 17.6.1855 when Lyons mortally wounded (died 23.6.1855); on 24.6.1855 under Capt. Robert Hall, in the Mediterranean and Black Seas (Hall was Senior Officer in the Strait of Kerch); paid off 21.4.1857 at Sheerness. Recommissioned 4.10.1860 under Cmdr Henry Carr Glyn, for the Australian station; on 29.8.1861 under Capt. Robert Jenkins, on same station; in New Zealand War; paid off 3.6.1865 at Sheerness. Sold to C. Lewis 2.12.1869 for BU.

BRISK Sloop (re-classed as a corvette 1862), 10 (later 14, then 15) guns. Committee of Reference design, 1847. Enlarged *Rattler* Class. *Brisk* completed fitting to carry the Governor to Port Royal [Jamaica]. Re-engined 1864 by Miller, Ravenhill & Salkeld with new 200nhp engine (837ihp = 9.989kts)

Dimensions & tons: 193ft 7¼in, 169ft 9¼in x 35ft 0in oa (34ft 6in for tonnage) x 20ft 5½in. 1,074⁸¹/₉₄bm (1,087 as completed).

Men: 170, later 175. Guns: 2 x 68pdrs (87cwt), 8 (later 12) x 32pdrs (42cwt); later 1 x 68pdr (87cwt/9½ft) on pivot and 14 x 32pdrs (32cwt/6½ft) on broadsides.

Machinery: 2-cylinder horizontal single expansion. Single screw. 250nhp. 505ihp = 7.35kts.

Brisk Woolwich Dyd/Scott, Sinclair & Co.

As built: 193ft 7¼in, 169ft 10in x 35ft 2¼in oa (34ft 8¼in for tonnage) x 20ft 5¼in. 1,086⁹⁰/₉₄bm. Draught 14ft 8in/16ft 8in.

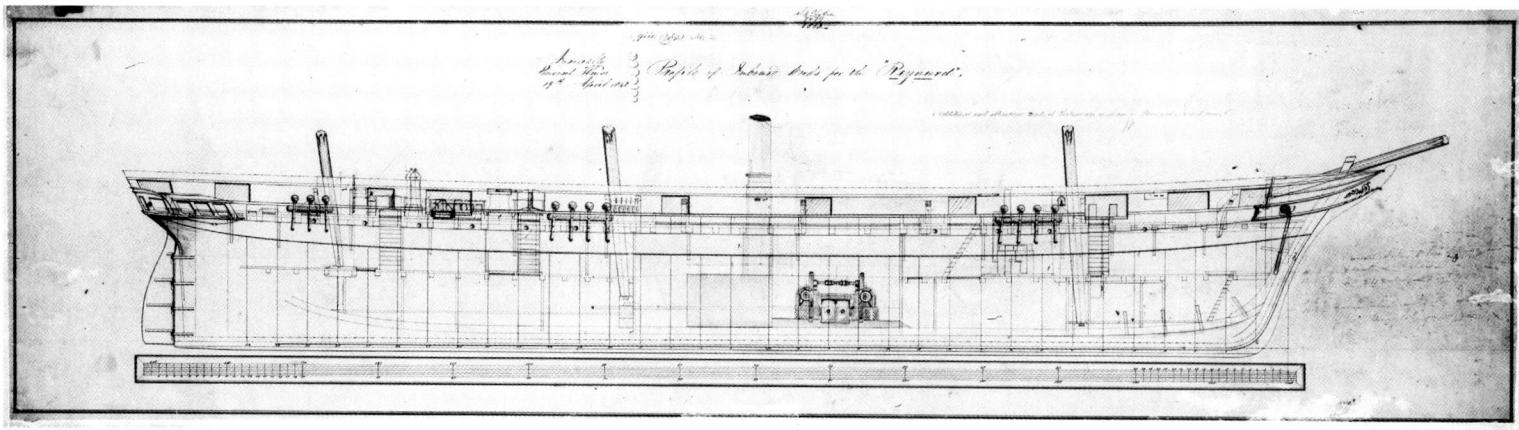

A profile of inboard works for the *Reynard*, dated 19 April 1848. The two-cylinder horizontal engine is reproduced in more detail than is usual on profile draughts. This type of machinery was favoured by the Navy as it fitted entirely below the waterline, reducing vulnerability in action. *(NMM DR6279-001)*

Ord: 25.4.1847. K: 1.1849. L: 2.6.1851. C: 24.8.1853.
First cost: £47,482 (including hull £20,677).
Commissioned 24.5.1853 under Cmdr Frederick Beauchamp Paget Seymour, for North America and the West Indies, then to White Sea to join Capt. Erasmus Ommanney's squadron, then to Russian Pacific coast (during Russian War); on 20.10.1854 under Cmdr Alfred John Curtis, in the Pacific; paid off 13.6.1857 at Plymouth. Recommissioned 4.5.1859 under Capt. Algernon Frederick Rous de Horsey, for Cape of Good Hope; on 24.2.1862 under Capt. John Proctor Luce (de Horsey invalided), to west coast of Africa; paid off 22.8.1863 at Plymouth. Recommissioned 30.8.1864 under Capt. Charles Webley Hope, for Australian station; paid off 19.1.1869 at Plymouth. Sold 31.1.1870 to mercantile owners.

CRACKER Class 1847 Originated as four 'steam gun schooners' of new design ordered on 25 April 1847, 'in lieu of sailing brigs originally planned' in the 1847 Programme of Work. These 516-ton vessels were to carry an armament of 2 x 32pdrs of 56cwt and 6 x 32pdrs of 25cwt. They were re-designated as 'gunvessels' in the 1849 Programme, and as sloops on 1 November 1850. Five further screw schooners were in the programme proposed on 23 March 1848, and these were ordered 30 March from Deptford Dyd (3) and Chatham Dyd (2), but not proceeded with under this description.
Cracker Deptford Dyd/George & John Rennie
  Ord: 25.4.1847. Suspended 12.8.1847, replaced by new sloop design 1.11.1850 (see *Cruizer* below).
Hornet Woolwich Dyd/Maudslay, Sons & Field.
  Ord: 25.4.1847. Suspended 12.8.1847, replaced by new sloop design 1.11.1850 (see below).

PLUMPER Sloop, 8 guns. Design by John Fincham. Originally ordered 25 April 1847 from Woolwich Dyd as a steam schooner under the name of *Pincher* (to have been engined by Rennie), she was re-ordered to a new design in August 1847.
  Dimensions & tons: 140ft 0in, 121ft 10½in x 27ft 10in oa (27ft 6in for tonnage) x 14ft 6in. 490²¹⁄₉₄bm. 577 disp.
  Men: 100. Guns: 2 x 32pdrs (56cwt), 6 x 32pdrs (25cwt).
  Machinery: 2-cylinder (27in diameter, 2ft stroke) vertical single expansion, oscillating, geared. Single screw (9ft diameter). 60nhp. 148ihp = 7.4kts.

Plumper Portsmouth Dyd/Miller, Ravenhill & Co.
  Ord: 25.4.1847. Re-ord: 12.8.1847. K: 10.1847. L: 5.4.1848. C: 17.12.1848.
  First cost: £20,446.
  Commissioned 6.11.1848 under Cmdr Matthew Stainton Nolloth, for Sir Charles Napier's Western Squadron, then 1.1849 to North America and West Indies, then 6.1851 to east coast of South America; paid off 6.1.1853 at Portsmouth. Recommissioned 1.8.1853 under Cmdr John Anthony Lawrence Wharton, for west coast of Africa; on 5.4.1855 under Cmdr William Henry Haswell, on same station; paid off 9.12.1856 at Portsmouth. Recommissioned next day under Capt. George Henry Richards as a surveying vessel for service on the Canadian Pacific coast (including the Fraser River, Burrard Inlet, Victoria and Esquimalt); in 1.1861 under Cmdr Anthony Hiley Hoskins, for return to England; paid off 2.7.1861 at Portsmouth. Survey vessel by 1860. Sold 2.6.1865 to White, Cowes to BU.

REYNARD Sloop, 8 guns. Design by John Edye. Originally ordered 25 April 1847 as a steam schooner under the name of *Plumper* (to have been engined by Maudslay), she was re-ordered to a new design in August 1847.
  Dimensions & tons: 147ft 0in, 128ft 4½in x 27ft 10in oa (27ft 6in for tonnage) x 14ft 6in. 516³⁷⁄₉₄bm. 656 disp. Draught 10ft 1½in/11ft 8in.
  Men: 100. Guns: 2 x 32pdrs (56cwt) on pivots, 6 x 32pdrs (25cwt) on broadsides.
  Machinery: 2-cylinder (28in diameter, 2ft stroke) horizontal single expansion. Single screw (8¾ft diameter). 60nhp. 165ihp = 8.238kts.
Reynard Deptford Dyd/George & John Rennie
  Ord: 25.4.1847. Re-ord: 12.8.1847. K: 8.1847. L: 21.3.1848. C: 1.8.1848.
  First cost: Hull £10,262, machinery & fitting £8,625.
  Commissioned 4.7.1848 at Woolwich under Cmdr Peter Cracroft, for Sir Charles Napier's Western Squadron, then to the East Indies for anti-pirate services; wrecked 31.5.1851 on unknown rock off the Pratas Shoal on the China Station (no casualties)

MALACCA 17 guns. Surveyor's Department design, approved 7 December 1848, similar to *Conflict*, but built of teak. The screw aperture was not opened until she was fitted with her engine in England. Departed under sail for Britain 16 May 1853 and arrived at Chatham Dyd 16 September 1853 with her vacant engine-room loaded with enough teak for Chatham to build a similar vessel, where she was engined and fitted from 31 January (order) to 8 August 1854 (undocked). Re-engined and re-classed as a corvette by 1858.

> Dimensions & tons: 192ft 0in, 168ft 2½in x 34ft 4in oa (34ft 0in for tonnage) x 22ft 8in. 1,034²⁸⁄₉₄bm. Draught 15ft 10in/18ft 10in.
>
> Men: 180 . Guns: 1 x 10in (85cwt/9ft 4in),) or 1 x 8in (65cwt); 16 x 32pdrs (32cwt/6ft 4in).
>
> Machinery: 2-cylinder inclined single expansion, trunk, high pressure. Single screw. 200nhp. 692ihp = 9.2kts.

Malacca Moulmein, Burma – begun by Mr Mould (who died, his estate being declared bankrupt) – completed by Mr Ladd, the Government Inspector/John Penn & Son.

> Ord: 9.11.1847. K: 29.5.1849. L: 9.4.1853. C: 17.8.1854.
>
> First cost: £49,980 (including hull £23,857, machinery £11,516).
>
> Commissioned 16.5.1853 at Moulmein under Lieut John Adolphus Pope Price for voyage to England; paid off 28.9.1853 at Chatham. Recommissioned 19.6.1854 under Capt. Arthur Farquhar, for the Mediterranean; to North America and the West Indies 11.1855; paid off 16.6.1857 at Sheerness. Recommissioned 3.9.1861 at Sheerness under Capt. Gerard John Napier, for the Mediterranean; paid off 8.12.1863 at Portsmouth. Recommissioned 10.11.1865 under Capt. Radulphus Bryce Oldfield, for the Pacific; paid off 9.9.1869 at Portsmouth. Re-engined 1862, the new engine producing 707ihp = 9.5kts Sold 6.1869 to E. Bates who, later the same year, sold her to the Japanese Navy as *Tsukuba* (or *Tsukuba Kan*). In Japanese service she was armed with 6 x 4.5in breech loading rifled guns + 2 x 30pdrs + 2 x 24pdrs. From 1892 re-armed with 4 x 6in QF guns. Became stationary training ship about 1900. BU 1906.

By the start of 1850 the experimental stage had substantially ended, and two series of wooden-hulled sloops were built during the next decade which experienced relatively few changes in armament. The larger were the Second Class sloops, of which sixteen were completed with 17 guns each (to three designs); the smaller were the Third Class sloops, also sixteen in the number completed with either 9 (first four vessels) or 11 guns (again to three designs). Others of both classes on order in 1863 were cancelled. Each carried one pivot-mounted gun along the centreline on the forecastle, the balance initially being 32pdr carronades mounted on slides or trucks along the broadsides.

CRUIZER Class 17 guns. *Cruizer* was a Lord John Hay design of 1850; others by Walker (Surveyor). 17 guns. First pair originally ordered 1847 at Deptford as screw schooners (later gunvessels) *Cracker* and *Hornet*; they are taken here to be new vessels. The *Mutine* (see below) was originally to have been of this design. *Alert* re-engined 1876 with R. & W. Hawthorn compound engine of similar power (312ihp = 7.684kts).

> Dimensions & tons: 160ft 0in, 140ft 1¾in x 31ft 10in oa (31ft 8in for tonnage) x 17ft 5in. 747⁵¹⁄₉₄bm. 1,045 displacement. Draught 12ft 8in/14ft 5in.
>
> Men: 160. Guns: 1 x 32pdr (56cwt/9ft 4in) on pivot, 16 x 32pdrs (32cwt/6ft 4in) on broadside trucks.
>
> Machinery: 2-cylinder horizonal single expansion. Single screw. 60nhp (*Cruizer*) 100nhp (others). (See below for individual ihp/speed.)

Cruizer [later *Cruiser*] Deptford Dyd/George & John Rennie (132ihp = 6.608kts)

> Ord: 1.11.1850. K: 4.1851. L: 19.6.1852. C: 3.2.1853.
>
> First cost: £25,213 (£15,239 for hull, £2,805 machinery, £7,169 fitting)
>
> Commissioned 18.12.1852 at Woolwich under Cmdr George

Henry Douglas; to the Baltic (during the Russian War); paid off 21.5.1856 at Portsmouth. Renamed *Cruiser* 1856. Recommissioned 2.8.1856 under Cmdr Charles Fellowes, for East Indies and China; accompanied gunboats *Haughty*, *Staunch* and *Forester* out to China (for 2nd Anglo-Chinese War); in action at Fatshan Creek 1.6.1857 and at Canton 12.1857; on 4.3.1858 under Cmdr John Bythesea, on same station; to Hankow 9.1858; at Peiho 26.6.1859; paid off 1.5.1861 at Portsmouth. Recommissioned 23.7.1866 under Cmdr Edward Field, then 10.8.1866 under Cmdr Morgan Singer, for the Mediterranean; on 2.6.1869 under Cmdr Thomas Anthony Swinburne, then 28.9.1869 under Cmdr George Graham Duff, on same station; paid off 13.12.1870. Recommissioned 15.11.1872 under Cmdr Alfred Taylor Dale, as training vessel for the Mediterranean; on 27.7.1875 under Cmdr John Hext, then 17.8.1878 under Cmdr Frederick Augustus Wetherall and 1.1.1890 under Cmdr John Rolleston Prickett, in same role. Renamed *Lark* in 5.1893 as a training ship. Became tender to *Hibernia* at Malta. Sold 1912 at Malta.

Hornet Deptford Dyd/James Watt & Co. (233ihp = 7.75kts)

> Ord: 1.11.1850. K: 6.1851. L: 13.4.1854. C: 14.7.1854.
>
> First cost: £29,142 (£15,889 for hull, £5,450 machinery, £7,803 fitting)
>
> Commissioned 15.5.1854 at Woolwich under Cmdr Frederick Archibald Campbell, for the Baltic (during the Russian War); on 13.9.1854 under Cmdr Charles Codrington Forsyth, to the East Indies and China; in 2nd Anglo-Chinese War; at Canton 11.1856; in 1.1857 under Cmdr William Montagu Dowell; at Fatshan Creek 1.6.1857 and at Canton 12.1857; on 26.2.1858 under Cmdr Viscount (Richard James) Gilford, on same station; paid off 14.7.1859 at Portsmouth. Recommissioned 20.7.1860 under Cmdr William Butler Fullerton Elphinstone, for East Indies and China; on 3.5.1861 under Cmdr Joseph Dayman, then 1864 under Acting Cmdr Richard Hare, on same station; paid off 12.9.1864 at Portsmouth. BU by White, Cowes 1868.

Harrier Pembroke Dyd/Humphrys, Tennant & Dykes (360ihp = 8.32kts)

> Ord: 18.7.1851. K: 11.1851. L: 13.5.1854. C: 3.11.1854.
>
> First cost: £32,705 (including £5,615 for machinery)
>
> Commissioned 16.8.1854 at Portsmouth under Cmdr Henry Alexander Story, for the Baltic (during the Russian War); on 4.10.855 under Cmdr Samuel Hoskins Derriman, on same station, then to east coast of South America; on 16.9.1858 under Cmdr Malcolm MacGregor, on same station; paid off 7.9.1859 at Portsmouth. Recommissioned 29.10.1860, still under MacGregor, for Australia; on 24.6.1862 under Cmdr Francis William Sullivan, on same station; in New Zealand War; on 9.11.1863 under Cmdr Edward Hay (killed 30.4.1864), on same station; on 15.7.1864 under Cmdr William Henry Fenwick, still in Australia; paid off 31.3.1865 at Portsmouth. BU completed 12.1866 at Portsmouth.

Fawn Deptford Dyd/Miller, Ravenhill & Salkeld (434ihp = 8.77kts)

> Ord: 27.3.1852. K: 4.5.1854. L: 30.9.1856. C: 26.11.1859.
>
> First cost: £32,798 (£18,207 for hull, £6,317 machinery, £8,274 fitting)
>
> Commissioned 30.10.1859 at Sheerness under Cmdr Ralph Peter Cator, for Australia; paid off 11.4.1863 at Sheerness. Recommissioned 20.6.1864 under Cmdr Charles Joseph Wrey, for the West Indies; on 2.3.1865 under Cmdr Walter Cecil Talbot, then 14.4.1866 under Cmdr Basil Sidmouth de Ros Hall, for North America and West Indies, then 20.4.1867 under

While it is widely appreciated that most early steam warships spent much of their sea-time under sail, there are surprisingly few photographs of such ships under canvas. This is one of the Second Class sloop *Cruiser* taken in her later days as a Mediterranean Fleet training ship in 1884. The watch are setting royals and flying jib in very light winds. *(NMM B8803)*

Cmdr Charles Augustus John Heysham, on same station; paid off 11.6.1868 at Sheerness. Recommissioned 28.2.1870 under Cmdr Herbert Price Knevitt, for the Pacific; in 1874 under Cmdr Frederick Augustus Wetherall, on same station; paid off 19.3.1875 at Sheerness. Converted to survey ship 1876. Recommissioned 1.6.1876 at Chatham under Cmdr William James Lloyd Wharton, for surveying on east coast of Africa; to Sea of Marmara 1879; on 1.1.1880 under Cmdr Pelham Aldrich, for surveying in the Mediterranean; paid off 6.4.1883. Sold 1884.

Falcon Pembroke Dyd/Miller, Ravenhill & Salkeld (312ihp = 7.87kts)
    Ord: 2.4.1853. K: 11.1853. L: 10.8.1854. C: 30.3.1855.
    First cost: £34,423 (£19,059 for hull, £6,200 machinery, £9,164 fitting)
    Commissioned 7.2.1855 at Portsmouth under Cmdr William John Samuel Pullen, then to the Baltic Sea (during the Russian War); on 10.5.1856 under Cmdr Hubert Campion, to North America and the West Indies; paid off at Portsmouth 25.8.1857. Recommissioned 4.5.1859 under Cmdr Arthur George Fitzroy (died 9.1.1861), for west coast of Africa; on 10.1.1861 under Cmdr Algernon Charles Fieschi Heneage, on same station; paid off 6.10.1862 at Portsmouth. Recommissioned 22.10.1863 under Cmdr George Henry Parkin, for Australia; in New Zealand Wars; on 16.4.1866 under Cmdr Arthur Rodney Owen, then 12.6.1866 Cmdr William Hans Blake and 19.9.1867 Cmdr Henry Legge Perceval, on same station; paid off 3.10.1868 at Woolwich. Sold to Marshall 27.9.1869 to BU at Plymouth.

Alert Pembroke Dyd/Miller, Ravenhill & Salkeld (383ihp = 8.786kts)
    Ord: 2.4.1853. K: 1.1855. L: 20.5.1856. C: 24.1.1858.
    First cost: £36,743 (£18,893 for hull, £6,052 machinery, £11,798 fitting)
    Commissioned 12.11.1857 at Plymouth under Cmdr William Alfred Rumbulow Pearse, for the Pacific; paid off 8.10.1861. Recommissioned 15.5.1863 under Cmdr Henry Cholmeley Majendie, for Scottish fishery protection; to the Pacific 10.1863; on 21.9.1865 under Cmdr Arthur John Innes, then 20.4.1867 under Cmdr Hugh Horation Knocker, on same station; paid off 30.5.1868 at Plymouth. Reduced to 5 guns in 1871. Converted 1874 for Arctic exploration with strengthened hull (re-engined and re-armed). Recommissioned 15.4.1875 at Portsmouth under Capt. George Strong Nares, for British Arctic Expedition of 1875–76; paid off 5.12.1876 at Portsmouth. Converted to survey ship 1878; recommissioned 20.8.1878, still under Nares, for survey of Strait of Magellan; on 12.3.1879 under Capt. John Fiot Lee Pearse Maclear, for survey of Canadian and Australian waters; paid off 20.9.1882 at Sheerness. Transferred to the US Navy 20.2.1884 for use by American Research Society, used for polar expedition under USN Capt. George W. Coffin, then to Canadian government 5.1885 for survey of Hudson Bay under Lieut Andrew R. Gordon (RN), then used as lighthouse supply vessel and buoy tender in Nova Scotia and subsequently in Gulf of St Lawrence. Laid up 11.1894, being formally returned to RN and sold to BU 1895.

SWALLOW **Class** 9 guns. Surveyor's Department design of 1852. Built in response to an Admiralty requirement for 'a type of screw vessel below the *Cruizer* (design)', the first three were ordered as 'gunvessels, class 1', but while still on the stocks they were re-rated as Third Class sloops under the reclassifications of early 1854. *Curlew*'s engines were those taken out of the *Hornet*. The *Icarus* (as ordered 2.4.1853 – and named 3.4.1854), *Cordelia* and *Gannet* (see below) were originally to have been of this design, but were built as units of the succeeding *Racer* Class.

Dimensions & tons: 139ft 0in, 120ft 6in x 27ft 10in oa (27ft 6in for tonnage) x 13ft 5in. 484⁶⁸⁄₉₄bm. 625 disp.

Men: 100. Guns: 1 x 32pdr (56cwt/9ft 6in) on pivot forward, 8 x 32pdrs (25cwt/5ft 4in) broadsides.

(in 1854, the 5ft 4in guns were ordered to be replaced by 6ft weapons of same weight).

Machinery: 2-cylinder horizontal single expansion. Single screw. 60nhp. (See below for individual ihp/speed.)

Curlew Deptford Dyd/Maudslay, Sons & Field (ihp/speed unknown)
Ord: 5.7.1852. (named same day). K: 19.10.1852. L: 31.5.1854. C: 18.10.1854.
First cost: £21,003 (£10,430 for hull, £4,193 machinery, £6,380 fitting)
Commissioned 16.8.1854 at Woolwich under Cmdr Rowley Lambert, for the Mediterranean and Black Seas (during the Russian War); on 1.10.1855 under Cmdr John James Kennedy, in the Mediterranean; on 1.2.1856 under Cmdr William Horton, still in the Mediterranean; paid off 11.12.1858 at Plymouth. Recommissioned 20.8.1860 under Cmdr Edward Wingfield Shaw, for the east coast of South America; on 9.8.1861 under Cmdr Charles Stuart Forbes, then 17.11.1862 under Cmdr Thomas Alexander Packenham, on same station; paid off 18.5.1865 at Plymouth. Sold 29.8.1865 to Marshall, Plymouth to BU.

Swallow Pembroke Dyd/Miller, Ravenhill & Salkeld. (182ihp = 6.585kts)
Ord: 5.7.1852. (named same day). K: 30.8.1853. L: 12.6.1854. C: 23.10.1854.
First cost: 'more than' £22,000 (including £12,735 for hull, £3,800 machinery)
Commissioned 18.8.1854 at Portsmouth under Cmdr Frederick Augustus Buchanan Crauford, for the Mediterranean and Black Seas (during the Russian War); on 10.5.1856 under Cmdr Charles Lodowick Darley Waddilove, in the Mediterranean; paid off 15.2.1859 at Sheerness. Converted to survey ship 1861; recommissioned 21.11.1861 under Master Edward Wilds, for surveying in Japan and China; paid off 1.10.1866 at Sheerness. Sold 12.1866 to BU.

Ariel Pembroke Dyd/Miller, Ravenhill & Salkeld (193ihp = 6.823kts)
Ord: 2.4.1853. (named 8.6.1853). K: 11.1853. L: 11.7.1854. C: 12.4.1855.
First cost: unknown.
Commissioned 27.1.1855 at Portsmouth under Cmdr John Proctor Luce, for the White Sea (during the Russian War); sailed for home 9.10.1855. Recommissioned 5.12.1855 under Cmdr Frederick Augustus Maxse, to the Mediterranean; on 25.11.1857 under Cmdr Charles Bromley, still in the Mediterranean; paid off 7.7.1859 at Portsmouth. Recommissioned 30.8.1860 under Cmdr John Hobhouse Inglis Alexander, for the Cape of Good Hope; in 10.1861 under Cmdr Radolphus Bryce Oldfield, on same station (note the *Ariel* and *Lyra* exchanged commanders, as ordered on 17.8.1861); on 22.4.1862 under Cmdr William Cox Chapman, on same station; paid off 1.12.1864 at Portsmouth.

Sold 23.5.1865 to Shaw & Thompson to BU.

Lyra Deptford Dyd/Miller, Ravenhill & Salkeld (224ihp = 7.48kts)
Ord: 3.4.1854. K: 8.7.1854. L: 26.3.1857. C: 11.12.1857.
First cost: £21,917 (£13,024 for hull, £3,583 machinery, £5,310 fitting)
Commissioned 12.11.1857 at Woolwich under Cmdr Radulphus Bryce Oldfield, for Cape of Good Hope; in 10.1861 under Cmdr John Hobhouse Inglis Alexander, on same station; paid off 3.1.1862 at Portsmouth. Recommissioned 19.12.1863 under Cmdr Robert Augustus Parr, for the East Indies; paid off 21.4.1868. Reduced to 7 guns in 1868. BU 1876 at Portsmouth Dyd.

RACER **Class** 11 guns. Surveyor's Department design of 1855. Lengthened version of the *Swallow* design, to obtain higher speed.

Dimensions & tons: 151ft 0in, 131ft 3¾in x 29ft 1in (28ft 9in for tonnage) x 15ft 10in. 577³⁹⁄₉₄bm. 861 disp. Draught 13ft 8in/14ft 3in.

Men: 120. Guns: 1 x 32pdr (58cwt/9½ft) MLSB on pivot; 10 x 32pdr carronades (25cwt/6ft) MLSB on broadside trucks.

Machinery: 2-cylinder horizontal single expansion. Single screw. 150nhp. (See below for individual ihp/speed.)

Cordelia Pembroke Dyd/James Watt & Co. (461ihp = 9.912kts)
Ord: 3.4.1854. K: 10.1855. L: 3.7.1856. C: 1856.
First cost: £33,428 (£15,592 for hull, £9,014 machinery, £8,822 fitting)
Commissioned 11.4.1857 at Plymouth under Cmdr Charles Egerton Harcourt-Vernon, for Australia; in June 1861 under Cmdr Francis Alexander Hume, on same station; paid off 2.4.1862 at Plymouth. Recommissioned 24.6.1864 under Cmdr John Binney Scott, for North America and the West Indies; on 3.3.1865 under Cmdr Thomas Alexis de Wahl (Scott invalided), then on 16.9.1867 under Cmdr Charles Parry, on same station; paid off 9.7.1868 at Plymouth. Sold to Marshall, Plymouth on 12.5.1870.

Racer Deptford Dyd/Humphrys, Tennant & Dykes (522ihp = 9.519kts)
Ord: 3.4.1854. (named 31.3.1855). K: 1.10.1856. L: 4.11.1857. C: 22.7.1858.
First cost: £31,167 (£17,082 for hull, £9,142 machinery, £4,943 fitting)
Commissioned 3.6.1858 at Woolwich under Cmdr Thomas Alexander Pakenham, for the Channel Squadron; to North America and the West Indies 2.1859; on 15.5.1860 under Cmdr Algernon McLennan Lyons, on same station; paid off 9.12.1862 at Woolwich. Recommissioned 22.6.1864 under Cmdr Isaac Newton Thomas Saulez, for the Mediterranean; on 27.5.1865 under Cmdr Lindesay Brine, on same station; paid off 17.6.1868 at Woolwich. Commissioned 2.3.1870 under Lieut Arthur Knyvet Wilson, as tender to Boscawen at Portland; paid off 24.4.1873. BU 1876 at Portsmouth.

Gannet Pembroke Dyd/Miller, Ravenhill & Salkeld (617ihp = 10.817kts)
Ord: 3.4.1854. (named 31.3.1855). K: 12.1856. L: 29.12.1857. C: 9.3.1859.
First cost: £33,259 (£15,934 for hull, £9,138 machinery, £8,287 fitting)
Commissioned 31.1.1859 at Plymouth under Cmdr Edward Henry Gage Lambert, for the Mediterranean; on 11.7.1861 under Cmdr Maurice Horatio Nelson, on same station; paid off 9.1.1863 at Plymouth. Recommissioned 3.7.1865 under Cmdr William Chimmo, for the West Indies; paid off 5.10.1868 at Plymouth.

A watercolour painting of the Third Class sloop *Icarus*, portrayed on the China Station during the late 1860s. This was one of a series produced by William Frederick Mitchell, depicting the ships of the Royal Navy 'then and now'; some of these were painted long after the event (in this case in 1880, five years after the ship had been broken up), but his research was generally good and the detail is authentic. *(NMM PU9502)*

BU 2.1877 at Devonport.

Icarus Deptford Dyd/George Rennie & Sons. (608ihp = 10.145kts)
Ord: 3.2.1855. K: 7.4.1857. L: 22.10.1858. C: 28.12.1859.
First cost: £31,928 (£17,452 for hull, £8,633 machinery, £5,843 fitting)
Commissioned 1.11.1859 at Sheerness under Cmdr Nowell Salmon, for the West Indies; to Channel Squadron 4.1861; to the Mediterranean 2.1862; paid off 24.5.1864 at Sheerness. Recommissioned 26.2.1866 under Cmdr Samuel Philip Townsend, for the China station; in 1868 under Cmdr Charles Thomas Montague Douglas Scott, on same station; paid off 5.12.1871 at Sheerness. Sold to Castle 23.1.1875 to BU at Charlton.

Pantaloon Devonport Dyd/Greenock Foundry Co. (484ihp = 8.934kts)
Ord: 1.4.1857. K: 2.11.1857. L: 26.9.1860. C: 17.9.1861.
First cost: unknown.
Commissioned 24.8.1861 at Plymouth under Cmdr William Robert Hobson, for the East Indies. Recommissioned in the East Indies under Cmdr Francis Reginald Purvis; on 22.4.1866 under Cmdr George Lydiard Sulivan; paid off 6.3.1867 at Plymouth. Sold to C. Marshall, Plymouth 18.9.1867 (for £1,610).

GREYHOUND Class 17 guns. Surveyor's Department design of 1855 ordered by AO of 3 February 1855. This was a lengthened version of the *Cruizer* design, to obtain higher speed and be fitted with 200nhp engines with two funnels. For this and the following *Camelion* Class a raised forecastle was introduced to give protection to the pivot-mounted bow-chaser. *Mutine* had originally been ordered to *Cruizer* design, changed 1856. These two sloops were later re-classed as corvettes.

Dimensions & tons: 172ft 6in, 151ft 6in x 33ft 2in oa (33ft 0in for tonnage) x 17ft 5in. 877⁵³⁄₉₄bm. 1,260 disp. Draught 14ft 0in/15ft 0in.
Men: 165. Guns: 1 x 32pdr (58cwt/9½ft) on pivot, 16 x 32pdrs (32cwt/6½ft) on broadside trucks. Later 5 x 40pdrs (58cwt/9½ft)

MLSB, 12 x 32pdrs (32cwt/6½ft) MLSB.
Machinery: 2-cylinder horizontal single expansion. Single screw. 200nhp. (See below for individual ihp/speed.)

Mutine Deptford Dyd/Maudslay, Sons & Field (786ihp = 10.25kts)
Ord: 27.3.1852. K: 1.6.1857. L: 30.7.1859. C: 26.11.1859.
First cost: £30,874 (including £23,632 for hull, £5,900 machinery)
Commissioned 21.10.1859 at Woolwich under Cmdr William Graham, for the Pacific; paid off at Woolwich 5.1.1864. Recommissioned 7.4.1865 under Cmdr William Hans Blake, for the Pacific again; on 2.11.1867 under Cmdr William Swinburn; in 1.1868 under Cmdr Henry McClintock Alexander, on same station; paid off 30.3.1869 at Sheerness. Sold 26.2.1870 for commercial service, renamed *Chieftain*.

Greyhound Pembroke Dyd/Miller, Ravenhill & Salkeld (743ihp = 9.77kts)
Ord: 3.4.1854. K: 12.1856. L: 15.6.1859. C: 6.2.1860.
First cost: £41,394 (£23,630 for hull, £?12,475 machinery, £?5,289 fitting)
Commissioned 14.12.1859 at Plymouth under Cmdr Francis William Sullivan, for the Mediterranean; paid off 30.10.1861 at Plymouth. Recommissioned 9.11.1861 under Cmdr Henry Dennis Hickley, for North America and the West Indies; on 16.2.1864 under Cmdr Henry Rushworth Wratislaw, on same station; paid off 23.11.1864 at Plymouth. Recommissioned 16.9.1865 under Capt. Charles Stirling, for west coast of Africa; in 3.1868 to east coast of South America; paid off 11.9.1869 at Plymouth for harbour service. Employed for William Froude's hydrodynamic tests 1871–72. Breakwater at Devonport 1879. Sold on 3.4.1906.

At the start of 1854 in a series of re-classifications the large gunvessels *Rifleman* and *Sharpshooter* (see Chapter 11 for details) were re-rated as Third Class sloops, as were the newer *Swallow* Class vessels (see above). At the same time, the larger *Archer*, *Wasp*, *Miranda* and *Brisk* was classed with the First Class sloops.

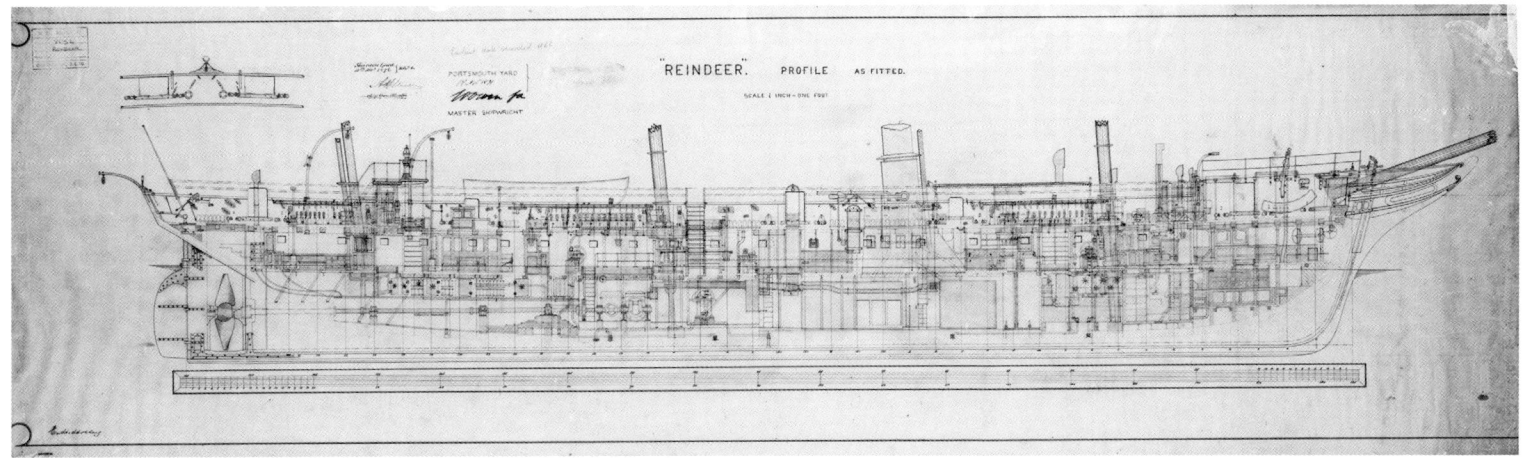

This superbly executed 'as fitted' profile of the *Reindeer* shows a remarkable level of internal detail and appears to have been drawn following the ship's refit at Portsmouth in 1871 to prepare her for a Pacific commission. The *Reindeer* belonged to the *Camelion* Class of wooden screw sloops, later reclassed as corvettes. This was a very heterogeneous class, with considerable variations in interior construction and in deployment of armament. Work on the *Reindeer* was suspended in 1862, but resumed later. *(NMM DR8052)*

CAMELION Class Second Class sloops (later re-classed as corvettes), 17 guns. Isaac Watts design, approved 22 February 1858. A stretched version of the *Greyhound* design (lengthened by 8ft at the bow and 4½ft at the stern). *Camelion* and *Pelican* were first ordered to the *Greyhound* design (this pair were referred to as '*Camelion* Class altered') and their frames had to be altered by contract. *Zebra* and *Rattler* also had their frames converted by contract. In 1858 *Pelican*, *Rinaldo*, *Zebra*, *Chanticleer*, *Reindeer* and *Rattler* were listed as a new class, and *Camelion* and *Perseus* were added in 1860. *Rinaldo* was built in dry dock. *Reindeer* completed (after suspension) with 1 x 110pdr + 5 x 64pdrs.

Dimensions & tons: 185ft 0in, 164ft 0¼in x 33ft 2in oa (33ft 0in for tonnage) x 17ft 5in. 950%₉₄bm. 1,365 disp. Draught 13ft 8in/15ft 7in.

Men: 170, later 180. Guns: : 1 x 32pdr (58cwt/9½ft) MLSB on pivot, 16 x 32pdrs (32cwt/6½ft) MLSB on broadside trucks. Later 5 x 40pdrs (58cwt/9½ft), 12 x 32pdrs MLSB.

Machinery: 2-cylinder horizontal single expansion. Single screw. 200nhp. (See below for individual ihp/speed.)

Camelion Deptford Dyd/Maudslay, Sons & Field (678ihp = 10.004kts)
Ord: 3.4.1854. K: 8.11.1858. L: 23.2.1860. C: 30.7.1861.
First cost: Hull £23,328, machinery £12,000.
Commissioned 9.7.1861 at Sheerness under Cmdr Edward Hardinge, for the Pacific; on 14.12.1863 under Cmdr Theodore Morton Jones, on the same station; paid off 28.4.1866 at Sheerness. Recommissioned 1.5.1867 under Cmdr William Henry Annesley, for the Pacific; paid off 1.5.1871 at Panama. Recommissioned 5.1871 under Cmdr Andrew James Kennedy, for the Pacific; paid off 10.6.1875 at Sheerness. Sold 1883.

Pelican Pembroke Dyd/Miller, Ravenhill & Salkeld (671ihp = 10.045kts)
Ord: 3.4.1854. K: 16.6.1859. L: 19.7.1860. C: 25.9.1861.
First cost: Machinery £12,000.
Commissioned 6.9.1861 at Portsmouth under Cmdr Philip Brock, for the Mediterranean; on 1.10.1863 under Cmdr Henry Wandesford Comber (when Brock invalided), then 7.6.1865 under Cmdr Ralph Peter Cator (when Comber invalided), on same station; paid off 7.5.1866 at Portsmouth. Sold to Arthur &

Co. 2.1867, renamed *Hawk* then resold to the Portuguese Navy 24.5.1868 and renamed *Infanta Dom Henrique*; deleted from Portuguese Navy 1.12.1879 and sold.

Rinaldo Portsmouth Dyd/Humphrys & Tennant (752ihp = 10.588kts)
Ord: 1.4.1857. K: 1.3.1858. L: 26.3.1860. C: 8.6.1861.
First cost: Hull £23,369.
Commissioned 3.5.1861 under Cmdr William Nathan Wright Hewett, for North America and the West Indies; on 26.11.1862 under Cmdr James Andrew Robert Dunlop, then 2.6.1864 under Cmdr Maurice Horatio Nelson (when Dunlop invalided), on same station; paid off 10.2.1865 at Portsmouth. Recommissioned 1.11.1866 under Cmdr William Kemptown Bush, for the China station; on 28.11.1868 under Cmdr Frederick Charles Bryan Robinson (when Bush invalided), then 1.1871 under Cmdr George Robinson (when F. C. B. Robinson invalided), then 5.6.1872 under Cmdr George Parsons, all on same station; paid off 10.7.1874 at Portsmouth. Sold 4.1884.

Zebra Deptford Dyd/Humphrys & Tennant (772ihp = 9.874kts)
Ord: 1.4.1857. K: 4.7.1859. L: 13.11.1860. C: 23.5.1861(completed for reserve).
Commissioned 7.4.1862 at Sheerness under Cmdr Anthony Hiley Hoskins, for west coast of Africa; in 2.1864 under Cmdr Charles Gowan Lindsay, on same station; paid off 31.10.1865 at Sheerness. Recommissioned 13.2.1867 at Woolwich under Cmdr Edwin John Pollard, for China station; on 6.4.1868 under Cmdr Henry Anthony Trollope, then 1870 under Cmdr Albert Denison Somerville Denison on same station; paid off 30.4.1873 at Hong Kong. Sold 20.8.1873 in the Far East.

Perseus Pembroke Dyd/George Rennie & Sons (614ihp = 10.148kts)
Ord: 1.4.1857. K: 20.7.1860. L: 21.8.1861. C: 9.1862.
Commissioned 28.9.1862 at Plymouth under Cmdr Augustus John Kingston, for China station; at bombardments of Kagoshima and Simonoseki; on 21.11.1864 under Cmdr Charles Edward Stevens, on same station; paid off 22.1.1866. Recommissioned 23.1.1866, still under Stevens; paid off 24.10.1869. Training ship at Devonport 1886, attached 1890 to Defiance for torpedo instruction, renamed *Defiance II* in 3.1904. Sold 26.6.1931.

Chanticleer Portsmouth Dyd/A. & J. Inglis. (694ihp = 10.132kts)
Ord: 27.3.1858. K: 2.2.1860. L: 9.2.1861. C: 12.1861.
Commissioned 11.12.1861 at Portsmouth under Cmdr Charles Stirling, for the Channel Squadron; to the Mediterranean 10.1862; on 14.12.1863 under Cmdr John Erskine Field Risk, then 6.6.1865 under Cmdr William Henry Fenwick, on same station; paid off 20.2.1866 at Woolwich. Recommissioned

4.4.1867 under Cmdr William Wilson Somerset Bridges, for the Pacific; paid off 12.3.1872 at Sheerness. Sold to Castle 23.1.1875 to BU at Charlton.

Reindeer Chatham Dyd/George & J. Rennie. (722ihp = 10.002kts)
Ord: 27.3.1858. K: 1.5.1860. (Suspended 1862, but resumed 5.1863) L: 29.3.66. C: 10.1866.
Commissioned 30.10.1866 at Chatham under Cmdr Edward Nares, for the Pacific; paid off 15.2.1871 at Portsmouth. Recommissioned 10.8.1871 under Cmdr William Robert Kennedy, for the Pacific; in 1875 under Cmdr Charles Vernon Anson, on same station; paid off 4.11.1875 at Sheerness. BU completed 22.12.1876 at Chatham.

Rattler Deptford Dyd/Maudslay, Sons & Field (843ihp = 10.006kts)
Ord: 27.3.1858. K: 28.8.1860. L: 18.3.1862. C: 7.1862.
Commissioned 5.7.1862 at Sheerness under Cmdr Edward Henry Howard, for China station; on 16.2.1864 under Cmdr John Whitmarch Webb, then 8.7.1867 under Cmdr John Thomlinson Swann (died 25.4.1868), on same station; by 9.1868 under Cmdr Henry Frederick Stephenson, on same station; wrecked on 24.9.1868 on an uncharted reef in the La Perouse Straits, Japan (no casualties).

Harlequin Portsmouth Dyd/—.
Ord: 5.3.1860. K. 13.2.1861. L: –. Cancelled 16.12.1864 (after costs of £5,123); completed taking to pieces 2.2.1865.

Tees Chatham Dyd. ['converted frame only']/ —.
Ord: 5.3.1860. K: 2.1861. L: –. Cancelled 12.12.1863 (after costs of £1,135).

Sappho Deptford Dyd/—.
Ord: 5.3.1860. K: 1.5.1861. L: –. Cancelled 12.12.1863 (after costs of £2,921).

Trent Pembroke Dyd/—.
Ord: 5.3.1860. K: 3.9.1861. Completed as ironclad *Research* (which see).

Circassian Deptford Dyd/—.
Ord: 25.3.1861. K: 5.5.1862. Completed as ironclad *Enterprise* (which see).

Diligence Chatham Dyd/—.
Ord: 25.3.1861. K: 1862. L: –. Cancelled 12.12.1863.

Imogene Portsmouth Dyd/—.
Ord: 25.3.1861. K: –. Cancelled 12.12.1863 (unstarted)

Success Pembroke Dyd/—.
Ord: 25.3.1861. K: –. Cancelled 12.12.1863 (unstarted).

ROSARIO Class Third Class sloops, 11 guns. Isaac Watts design, approved 12 August 1858. An intended modification to this design (which would have lengthened the design to 162ft) would have affected the vessels from *Acheron* onwards, but in the event these later ships were all cancelled. At the time of cancellation, *Bittern* had been laid down, but for *Circassian*, *Acheron*, *Sabrina*, *Cynthia* and *Fame* the conversion of their frames had begun on the 'K' dates below, although none had been laid down.

Dimensions & tons: 160ft 0in, 139ft 8½in x 30ft 4in oa (30ft 0in for tonnage) x 18ft 11in. 15ft 10in draught. 668⁷⁄₉₄bm. 913 disp. Draught 12ft 8in/13ft 10in.
Men: 130, later 150. Guns: 1 x 32pdr (58cwt/9½ft) MLSB on pivot; 10 x 32pdr carronades (30cwt/6ft 4in) on trucks on broadsides. Later 1 x 40pdr (58cwt/9½ft) Armstrong BL (slide-mounted), 6 x 32pdrs (30cwt/6ft 4in) MLSB (truck-mounted), 4 x 20pdr Armstong BL (pivot-mounted). By 1869 the first six ships all had 3 guns (1 x 7in ML + 2 x 40pdrs).
Machinery: 2-cylinder horizontal single expansion. Single screw.

150nhp. (See below for individual ihp/speed.)

Rosario Deptford Dyd/Greenock Foundry Co. (436ihp = 9.236kts)
Ord: 1.4.1857. K: 13.6.1859. L: 17.10.1860. C: 7.1862.
Commissioned 20.6.1862 under Cmdr James Stanley Graham, for fishery protection in the North Sea; in 10.1862 to North America and the West Indies; on 14.5.1863 under Cmdr Henry Duncan Grant, then 9.1864 under Cmdr Louis Hutton Versturme, on same station; paid off 13.10.1866 at Chatham. Recommissioned 28.9.1867 at Woolwich, still under Versturme; on 4.10.1867 under Cmdr George Palmer, to Australia; on 8.4.1870 under Cmdr Henry Joseph Challis, then 10.1871 under Acting Cmdr Alfred H. Markham, then 22.1.1874 under Cmdr Arthur Edward Dupuis, on same station; to government of South Australia 1874, as hulk for young criminals; paid off 12.10.1875 at Sheerness. Sold to Castle 31.1.1884 to BU at Charlton.

Peterel Devonport Dyd/Greenock Foundry Co. (478ihp = 8.982kts)
Ord: 1.4.1857. K: 5.12.1859. L: 10.11.1860. C: 3.1862.
Commissioned 2.2.1862 at Plymouth under Cmdr George Willes Watson, for North America and the West Indies; on 20.2.1864 under Cmdr Edward Madden. on same station; paid off 21.8.1865 at Plymouth. Recommissioned 2.6.1866 under Cmdr William Ebrington Gordon, for Cape of Good Hope and east coast of Africa; on 16.4.1868 under Cmdr Ernest Grey Lambton Cochrane, on same station; in 7.1869 to west coast of Africa; paid off 7.4.1870 at Plymouth. Recommissioned 1.3.1872 under Cmdr Cecil George Sloane Stanley, for the Pacific; in 5.1874 under Cmdr William Edgar de Crackenthorpe Cookson, on same station; paid off 23.6.1876 at Plymouth. Converted to light vessel 12.1877; in 1879 to coast of Ireland as lightship to mark the wreck of the *Vanguard*. Coal hulk 12.1885. Sold 10.1901.

Rapid Deptford Dyd/Greenock Foundry Co. (460ihp = 9.117kts)
Ord: 27.3.1858. K: 18.8.1859. L: 29.11.1860. C: 6.1862.
Commissioned 12.6.1862 at Woolwich under Cmdr Charles Trelawney Jago, for the Cape of Good Hope; on 10.6.1866 under Acting Cmdr William Belford Stubbs, on same station; paid off 24.1.1867 at Woolwich. Recommissioned 14.5.1868 under Cmdr Francis Lindley Wood, for the Mediterranean; paid off 1871 at Malta. Recommissioned 9.8.1871 under Cmdr Victor Alexander Montagu, for the Mediterranean; on 20.9.1874 under Cmdr Seymour Spencer Smith, then 4.9.1875 under Cmdr Adolphus Augustus Frederick Fitzgeorge (when Smith invalided), then 4.1.1878 under Cmdr Charles Cooper Penrose-Fitzgerald, then 3.4.1880 Cmdr William Frederick Stanley Mann, all in the Mediterranean; paid off 14.1.1881 at Malta. BU 9.1881 at Malta.

Shearwater Pembroke Dyd/R. & W. Hawthorn. (532ihp = 8.957kts)
Ord: 27.3.1858. K: 2.4.1860. L: 17.10.1861. C: 12.1862.
Commissioned 13.12.1862 at Plymouth under Cmdr Robert Gordon Douglas, for the Pacific; on 14.4.1866 under Cmdr Thomas Edward Smith, on same station; paid off 31.1.1868 at Woolwich. Recommissioned 21.7.1871 at Sheerness under Capt. George Strong Nares, for surveying in the Mediterranean; in 5.1872 under Cmdr William James Lloyd Wharton, still in the Mediterranean, then to east coast of Africa for surveying; paid off 4.8.1875 at Sheerness. BU completed 8.2.1877 at Sheerness.

Royalist Devonport Dyd/Greenock Foundry Co. (627ihp = 9.269kts)
Ord: 27.3.1858. K: 20.10.1860. L: 14.12.1861. C: 10.1863.
Commissioned 5.11.1863 at Plymouth under Cmdr Edwin John Pollard, for North America and the West Indies; on 7.7.1865 under Cmdr Maurice Horatio Nelson (when Pollard invalided), then on 14.4.1866 under Cmdr Hamilton Edward George Earle, on same station; paid off 13.4.1867 at Woolwich.

The final class of wooden screw sloops to be designed by Isaac Watts were the *Rosario* Class, of which the *Peterel* appears in this photograph taken in the 1870s. They had a single hoisting screw (of 10 feet in diameter) and a telescopic funnel, for retracting when under sail power, and were barque-rigged. These Third Class sloops were a development from the *Racer* Class, and the first five vessels were actually ordered to that design. They were actually shorter in length than the contemporary *Cormorant* Class gunvessels, but with more guns; the difference between the sloops and gunvessels was that the former were designed primarily for oceanic cruising, whereas the gunvessels were designed with coastal and inshore work in mind. *(NMM N5438)*

Recommissioned 9.1.1868 under Cmdr Loftus Francis Jones, for North America and the West Indies; on 11.4.1870 under Cmdr Richard Sacheverell Bateman, on same station; paid off 15.8.1871 at Sheerness. BU 9.1875 at Chatham.

Columbine Deptford Dyd/Greenock Foundry Co. (598ihp = 9.504kts)
Ord: 8.4.1859. K: 16.5.1860. L: 2.4.1862. C: 5.1863.
Commissioned 13.5.1863 at Sheerness under Cmdr Thomas le Hunte Ward, for fishery protection in Scotland; to the Pacific 10.1863; in 1865 under Cmdr James Elphinstone Erskine, in the Pacific; paid off 22.1.1868 at Sheerness. Recommissioned 16.6.1870 under Cmdr John Collier Tucker, for the East Indies; on 22.11.1872 under Cmdr Edward William Hereford, on same station; paid off 15.5.1874 at Sheerness. BU 6.1875 at Chatham.

Africa Devonport Dyd/Greenock Foundry Co. (530ihp = 9.5kts)
Ord: 5.3.1860. K: 22.12.1860. L: 14.2.1862. C: 13.8.1862.
Not commissioned in British Navy. Sold to Chinese Imperial Customs 13.8.1862, renamed *China* and sailed 4.1863 (to join Capt. Sherard Osborn's 'Vampire Fleet'), but returned to England

when agreement with the Chinese government collapsed. Resold 30.12.1865 to the Egyptian government.

Circassian (ex-*Enterprise* – name changed by AO of 22.7.1862) Deptford Dyd/— .
Ord: 5.3.1860. K: 1.5.1861. Cancelled 12.12.1863 (after costs of £3,112).

Acheron Deptford Dyd/— .
Ord: 5.3.1860. K: 14.10.1861. Cancelled 12.12.1863 (after costs of £1,947).

Bittern Devonport Dyd/— .
Ord: 5.3.1860. K: 17.12.1861. Cancelled 12.12.1863 (after costs of £2,607). BU on stocks completed 11.2.1865.

Sabrina Pembroke Dyd/— .
Ord: 25.3.1861. K: 28.5.1861. Cancelled 12.12.1863.

Cynthia Devonport Dyd/— .
Ord: 25.3.1861. K: 2.12.1861. Cancelled 12.12.1863 (after costs of £3,677).

Fame Deptford Dyd/— .
Ord: 25.3.1861. K: 2.12.1861. Suspended 5.2.1863. Cancelled 12.12.1863.

In a series of re-classifications which took place in March 1862 among the older sloops, the *Encounter*, *Niger*, *Archer*, *Miranda*, *Brisk* and *Malacca* were re-rated as corvettes (Sixth Rates) under captains, while the *Phoenix*, *Conflict*, *Desperate* and *Wasp* remained rated as First Class sloops under commanders.

(A) Vessels ordered as steam screw gunvessels (from 1845)

On 16 May 1845, the Admiralty instructed the Surveyor to report on designs for a screw version of the First Class steam gunvessel, and on 22 May made a similar request for a Second Class gunvessel design. Two versions of each Class were ordered, with one of each pair being built of wood, and the second of that pair contracted to have an iron hull. The lines of these hulls differed around the stern for comparative purposes, the two wooden vessels (below) having full lines around the stern, and the two iron hulls being drawn much finer. After trials in 1847 had demonstrated the significantly higher speeds with the finer stern lines, the two wooden vessels were modified with similar fine lines. All four were initially completed with engines by Miller, Ravenhill & Co., but three of them were quickly re-engined. The *Rifleman* and *Sharpshooter* were re-classed as Third Class sloops in 1854, and the *Teazer* and *Minx* as gunboats, but the smaller pair became tenders in 1855.

RIFLEMAN First Class gunvessel. John Fincham design, approved 22 May 1845. Similar to iron-hulled *Sharpshooter*, but with fuller stern section. Built in a dry dock. Rebuilt by AO of 7 January 1848 with finer stern lines and re-engined with a Miller, Ravenhill & Co. vertical oscillating geared engine of 100nhp. 10lb. 188ihp = 8kts. The old engine was installed in the *Archer* under AO of 26 October 1848.

   Dimensions & tons: 150ft 0in, 132ft 7in x 26ft 6in oa (26ft 2in for tonnage) x 15ft 6in. 482⁸¹/₉₄bm. 565 disp.
   Men: 80. Guns: 2 x 10in (85cwt/9ft), 4 x 32pdrs (25cwt/6ft); also reported as 1 x 56pdr (85cwt), 1 x 10in (85cwt), 4 x 32pdr (25cwt).
   Machinery: 2-cylinder (46in diameter, 3ft stroke) horizontal geared 202nhp. 7lb 350ihp = 9½kts. 100ND/166FD
Rifleman Portsmouth Dyd/Miller, Ravenhill & Co.
   Ord: 23.5.1845. K: 7.1845. L: 10.8.1846. C: 1.7.1848 at Portsmouth (for sea).
   First cost: Hull £11,776, machinery £6,301, fitting £3,460.
   Commissioned 14.7.1848 (appointed 5.7.1848) under Lieut Cmdr Stephen Smith Lowther Crofton, for Sir Charles Napier's Western Squadron; in 5.1849 to east coast of South America; on 25.4.1850 under Lieut Cmdr John Powell Branch, for same station; on 5.11.1850 under Lieut Cmdr Richard Henry Dalton, then 24.10.1853 under Lieut Cmdr Henry Christian, for same station; paid off at Woolwich 14.9.1857. Docked there 10.1858 and extensive rot found; engines removed and vessel repaired 11.1858–12.1859. Fitted as survey ship 11.1861–1.1862. Recommissioned 14.11.1861 under John William Reid, Master, for surveying in China seas; in 3.1864 under John Ward, for same role; paid off 1865. On 1.4.1866 under Navigating Lieut John Reed; paid off 9.1867. Sold 18.11.1869 at Hong Kong.

SHARPSHOOTER First Class Gunvessel. John Fincham design, approved 22 May 1845, contract dated 30 August. Half-sister to wooden *Rifleman* built at Portsmouth Dyd. Engines ex-*Minx*? Re-engined 1864 by John Penn & Son with 2-cylinder (35⅛in diameter, 20in stroke) 160nhp engine (ex-HMS *Lynx* of 1854); 438ihp = 8.4kts. Re-engined again 1870 (after sale) by Laird.

   Dimensions & tons: 150ft 5in, 134ft 6in x 26ft 7½in x 15ft 9in. 503bm. 658 disp.
   Men: 80. Guns: 2 x 10in (85cwt/9ft) + 4 x 32pdrs (25cwt/6ft 8in).
   Machinery: 2-cylinder (46in diameter, 3ft stroke) horizontal single expansion, geared. Single screw. 200nhp. 348ihp = 9.327kts.
Sharpshooter Ditchburn & Mare., Blackwall/Miller, Ravenhill & Co.
   Ord: 23.5.1845. K: 8.1845. L: 25.7.1846. C: 16.5.1848.
   First cost: Hull £11,030, machinery £10,734, fitting £3,539?
   Commissioned 4.4.1848 (appointed 29.3.1848) at Portsmouth under Lieut Cmdr John Crawshay Bailey, for the Channel Squadron; in 1.1849 to the Mediterranean, then 6.1850 to east coast of South America; took slaver *Polke* at Macahe (Brazil) 23.6.1849; paid off 19.11.1851 at Portsmouth. Fitted there (for £5,509) 1852; recommissioned 27.9.1852 under Lieut Cmdr John Edward Parish, for same station; paid off 16.10.1856 at Portsmouth. Refitted there (for £9,938) 1856–58; recommissioned 12.11.1857 under Lieut Cmdr Charles Gibbons, for west coast of Africa. Recommissioned 28.9.1864 under Lieut Cmdr Henry Augustus Clavering, for east coast of South America; on 3.6.1865 under Lieut Cmdr Richard Hare, on same station. On 19.9.1867 under Lieut Blair Skeffington Hamilton; paid off 3.9.1868 at Portsmouth. Sold 2.12.1869 for commercial service (name unchanged).

TEAZER Second Class Gunvessel. Sir William Symonds design, approved 31 May 1845. Similar to iron-hulled *Minx*, but with a fuller stern section. Rebuilt by AO of 7 January 1848; the original Miller machinery (described below) was in February removed from *Teazer*, new machinery (John Penn & Sons – vertical, oscillating, 40nhp, 27in diameter, 30in stroke) and screw (5ft diameter) were installed and the stern section rebuilt with finer lines by 21 December 1848; in trials on 6 October 1848, the new machinery recorded 128.2ihp = 7.685kts; it cost £2,530.

   Dimensions & tons: 130ft 0in, 118ft 10¼in x 22ft 0in oa (21ft 10in for tonnage) x 9ft 2in. 301⁴⁴/₉₄bm. 205 disp.
   Men: 30. Guns: 1 x 10in (86cwt); later 1 x 8in (65cwt/9ft) on pivot, + 1 brass 2pdr.
   Machinery: 2-cylinder vertical, oscillating (34in diameter, 2¾ft stroke). 100nhp. Single 4½ft diameter screw. Trials (8.6.1848) 175.6ihp = 6.315kts.
Teazer Chatham Dyd/Miller, Ravenhill & Co.
   Ord: 31.5.1845. K: 6.8.1845. L: 25.6.1846. C: — .
   First cost: Hull £5,974, machinery £5,259.
   Commissioned 19.10.1848 under Lieut Cmdr Jasper Henry Selwyn, for west coast of Africa (as tender to *Penelope*); paid off 4.1851. Recommissioned 8.3.1855 under Lieut Beville Granville Wyndham Nicolas, then 28.5.1855 under Lieut Cmdr Walter James Hunt-Grubbe, for same station; on 14.3.1857 under Lieut Cmdr William Henry Whyte, on same station; took

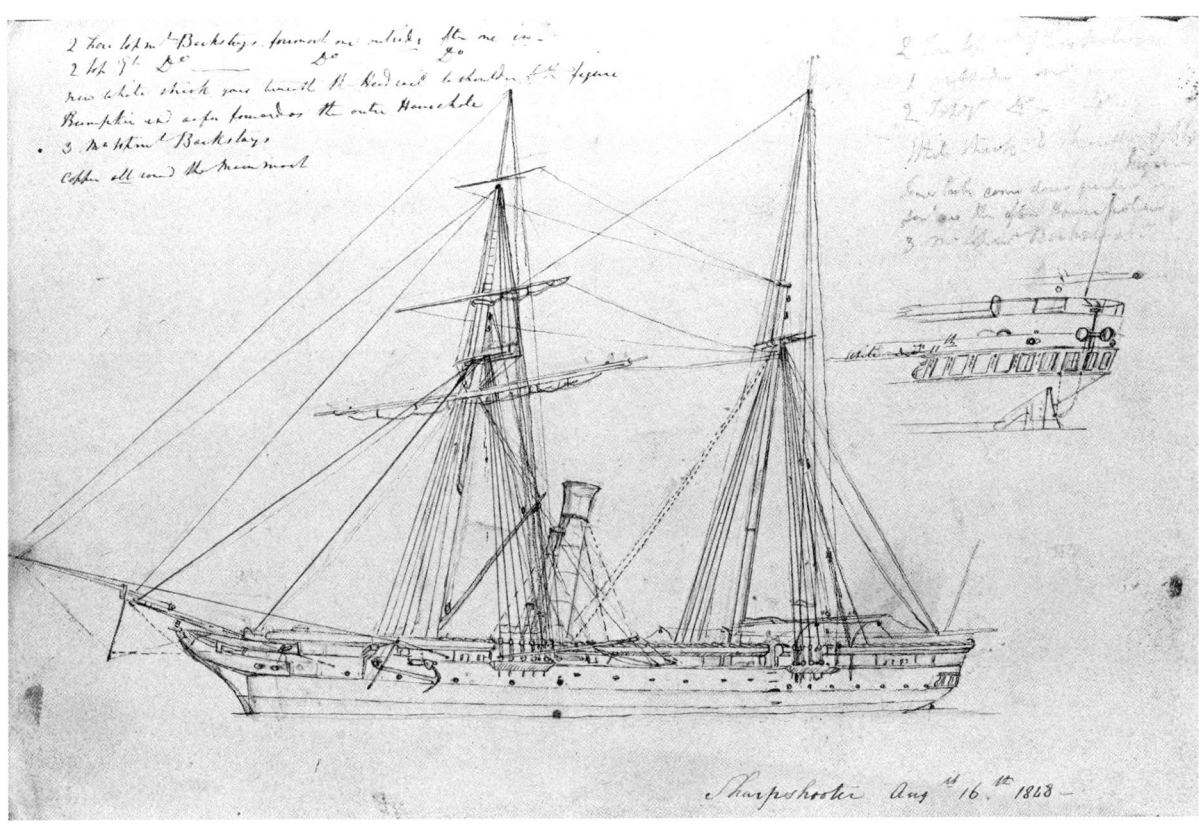

An anonymous artist's highly detailed reference sketch of the First Class gunvessel *Sharpshooter*, dated 16 August 1848. The *Sharpshooter* and her near-sister *Rifleman* were built to the same design, but the former was constructed of iron and the latter with a traditional wood hull, initially with fuller stern lines. In 1847-48 the two gunvessels were evaluated together in a series of trials, as a result of which *Rifleman*'s stern was rebuilt with finer lines. Both were later deployed overseas as standard small cruisers, and were reclassified as sloops in 1854. *(NMM X1325)*

slaver schooner *Abbot Devereux* 1.8.1857; on 8.4.1858 under Lieut Cmdr Charles Joseph Wrey, on same station; took slaver *General Scott* 5.5.1858; paid off 22.10.1858 at Woolwich. Sold to Castle & Son by AO 20.6.1862 for BU at Charlton.

MINX Second Class gunvessel. Sir William Symonds design, approved 31 May 1845, contract dated 5 July. Half-sister to wooden *Teazer* built at Chatham Dyd. Machinery fitted at East India Dock by December 1846, but following initial sea trials in December, *Minx* was laid up before completion. In May 1847 she became trials ship for the new screw propeller, under control of Engineering Dept at Woolwich Dyd.

  Dimensions & tons: 131ft 0in, 117ft 10in x 22ft 1in x 8ft 4in. 301bm. 250 disp.
  Men: 30. Guns: 1 x 10in (86cwt); later 1 x 8in (65cwt/9ft) on pivot, + 1 brass 2pdr.
  Machinery: 2-cylinder (34in diameter, 2¼ft stroke) vertical single expansion, oscillating. Single 4½ft diameter screw. 100nhp. 234ihp = 9.137kts.
Minx Miller, Ravenhill & Co., Blackwall/Miller, Ravenhill & Co.
  Ord: 22.5.1845. K: 5.9.1845. L: 5.9.1846. C: 2.1854–4.6.1854 at Deptford & Woolwich (for sea).
  First cost: £13,677 (including hull £5,182; machinery £4,518).
  Commanded from 5.1847 by Lieut Robert Robertson. Fitted between 6.1847 and 9.1848 with a variety of screws to designs of Woodcroft and Atherton, achieving speeds of between 7.44 and 8.84 knots. Re-engined (by AO of 3.1.1849) 1–6.1849 at Woolwich with a Seaward & Capel 2-cylinder (9³/₁₆in diameter, 9in stroke) horizontal, oscillating engine of 10nhp (36ihp = 5.44kts). After re-engining, continued with screw experiments until 1.1850 then laid up.
  Commissioned 9.5.1854 under Lieut Cmdr Herbert Thomas Ryves, for west coast of Africa; on 16.3.1855 under Lieut Cmdr Richard Henry Roe, on same station; paid off 25.4.1857 at Woolwich

and laid up. Converted at Woolwich to water tank steamer (for £2,601) 1859, for Portland; by 1870 to Plymouth. Sold 15.12.1899.

The 1846 Programme of Works, submitted for approval on 10 March and formally approved on 26 March, included six further wooden-hulled screw gunvessels, four to be First Class as the *Rifleman*, and two to be Second Class like the *Teazer*. The larger four were to be named the *Archer* and *Parthian* (both to be built at Deptford), and the *Sepoy* and *Cossack* (both assigned to Portsmouth); the smaller pair were to be named the *Boxer* and *Biter* (both for Chatham).

Chatham Dockyard on 20 June proposed to launch the *Teazer* on 25 June (in fact she was not launched until six weeks later) as they intended to lay down the *Boxer* on the same slip; the *Biter* would in turn follow her on the same slip. On 22 July they requested the plans for the *Boxer*. However on 6 October the Admiralty ordered work on the *Boxer* and *Biter* suspended (the former's frames were taken down in June 1847), and their names were removed from the Navy List (i.e. they were cancelled) on 22 May 1849.

On 9 September 1846 the Admiralty ordered the dockyards to suspend work on the four First Class gunvessels. In the next year's Programme, as proposed on 17 February 1847, the *Archer* and *Parthian* at Deptford were re-envisaged as screw sloops to the design of the *Rattler*, along with a similar pair (*Highflyer* and *Brisk*) at Deptford; on 25 April these four screw sloops were ordered to an enlarged design (see Chapter 10). The *Sepoy* and *Cossack* remained on the Navy List as intended construction until 22 May 1849, when they were cancelled.

In 1852 the Admiralty decided that they had need of a new class of screw vessels below the size of the *Cruiser* class sloops (which had by now been re-classed as screw sloops), and for this purpose revived the classification of First Class Steam Gunvessel (GV1). Two vessels of this rating (*Swallow* and *Curlew*) were ordered on 5 July 1852, and on 8 July were allocated for building to Pembroke and

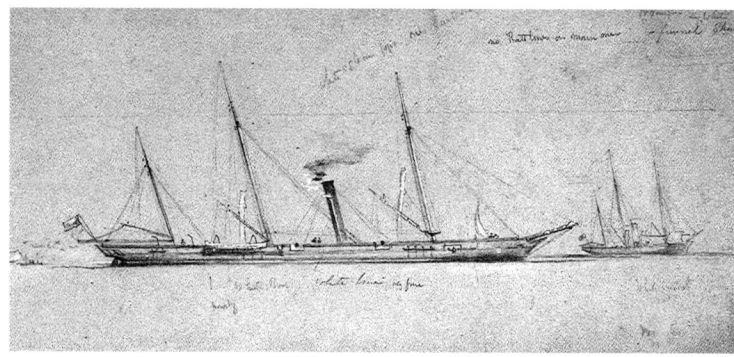

A coloured sketch drawn in 1855 by Oswald Walters Brierly of the screw dispatch vessel *Beagle* (with her sister *Wrangler* in the background). The *Beagle* and four of her five sisters were in the Black Sea from 1854, and the *Wrangler* joined them there in 1855. All six took part in the expeditionary force to the Sea of Azov in May–June 1855, and in the assault on Kinburn at the mouth of the Dnieper in October; they (and the following *Vigilant* and *Intrepid* classes) were all re-classed as gunvessels in 1856. *(NMM PU9293)*

Deptford respectively; these were to be vessels of 485bm, with 50 HP engines and 8 guns. Two more (*Ariel* and *Icarus*) were ordered on 2 April 1853, and allocated to the same dockyards. However, in a major reclassification at the start of 1854 all four – together with the *Rifleman* and *Sharpshooter* – were re-classed from GV1 to Third Class sloops (S3). The *Teazer* and *Minx* were re-classed from GV2 to the new category of Gunboat (GB), although in 1855 they were again re-classed, this time as tenders.

The outbreak of the Russian War demonstrated the need for numerous shallow-draught, manoeuvrable vessels for coastal and inshore operations in the Baltic and Black Seas. Six small screw steamers of the *Arrow* Class were approved in early 1854 to be built by contract in the Thames (8 builders tendered by 6 April 1854), initially classed as despatch vessels. Twenty further vessels, to two designs rated as respectively First Class (the six *Intrepid* Class) and Second Class (the fourteen *Vigilant* Class) were ordered in 1855. A further Second Class design (the *Philomel* Class, of which twenty vessels were completed) was ordered from 1857, derived from the wartime gunboat designs but with dimensions enlarged. The next design (the *Cormorant* Class, of which nine were completed) were enlarged again, but their draught as built considerably exceeded the planned 8ft, and further units were cancelled.

ARROW Class Despatch vessels, re-classed as Second Class Gunvessels 1856. Surveyor's Department design 1854. Authorised 28 March 1854, all were ordered either 10 or 11 April, and named 2 May 1854. On completion, *Wrangler* posted to the Baltic, and the other five to the Black Sea, but *Wrangler* later joined her sisters in the Black Sea, and all six were deployed into the Sea of Azov during 1855. Due to problems with the Lancaster guns, most were quickly disarmed and used as despatch vessels. Most reduced to 80nhp by 1860.

  Dimensions & tons: 160ft 0in, 143ft 4¾in x 25ft 4in oa (25ft 0in for tonnage) x 13ft 3in. Draught 10ft 0in (fwd), 11ft 8in (aft). 476⁶⁸⁄₉₄bm. 586 disp.
  Men: 65. Guns: 2 x 68pdrs (95cwt/10ft) Lancaster MLR on pivots, 4 x 32pdrs (25cwt/6ft).
  Machinery: 2-cylinder horizontal single expansion. Single screw. 160nhp. (See below for individual ihp/speed.)

Arrow C. J. Mare & Co., Blackwall/Humphrys, Tennant & Dykes (594ihp = 11.0kts)
  Ord: 10.4.1854. K: 15.4.1854. L: 26.6.1854. C: 17.8.1854.
  First cost: £21,234 (including £8,307 for hull and £9,724 for machinery).

Commissioned 13.7.1854 under Lieut William Kynaston Jolliffe, for the Black Sea (via the Mediterranean) and Sea of Azov. On 10.5.1856 under Cmdr Samuel Hood Henderson, for the Mediterranean station; paid off 29.8.1857 at Plymouth. Hull sold to Marshall (by AO of 19.5.1862, for £482), machinery returned to store.

Beagle C. J. Mare & Co., Blackwall/Humphrys, Tennant & Dykes (ihp/speed not tried when built)
  Ord: 10.4.1854. K: 15.4.1854. L: 20.7.1854. C: 3.9.1854.
  First cost: £23,091 (including £8,302 for hull and £9,725 for machinery).
  Commissioned 7.1854 under Lieut Edward George Hore, for the Black Sea (via the Mediterranean). On 15.12.1854 under (acting) Lieut William Nathan Wrighte Hewett, in the Black Sea and Sea of Azov. Recommissioned 20.9.1859 under Cmdr Edward Hay, for the East Indies and China; paid off 13.1.1863. Sold at Hong Kong by AO of 16.7.1863, became the Japanese Navy *Kanko* in 1865. BU 1889.

Lynx C. J. Mare & Co., Blackwall/John Penn & Son (ihp/speed not tried when built)
  Ord: 10.4.1854. K: 15.4.1854. L: 22.7.1854. C: 14.9.1854.
  First cost: £23,443 (including £8,364 for hull and £10,745 for machinery).
  Commissioned 7.1854 under Lieut John Proctor Luce, for the Black Sea (via the Mediterranean). On 23.11.1854 under Lieut Charles Murray Aynsley, in Black Sea and Sea of Azov, then 10.5.1856 under Cmdr Radolphus Bryce Oldfield, still in the Black Sea. Recommissioned 12.11.1857 under Lieut Henry Berkeley, for the Cape of Good Hope, then to west coast of Africa; paid off 12.1860. Sold to Marshall (by AO of 19.5.1862 for £482) to BU at Plymouth.

Snake C. J. Mare & Co., Blackwall/John Penn & Son (460ihp = 10.302kts)
  Ord: 10.4.1854. K: 9.5.1854. L: 6.9.1854. C: 4.11.1854.
  First cost: £24,075 (including £8,459 for hull and £10,747 for machinery).
  Commissioned 14.9.1854 under Lieut Henry Frederick McKillop, for the Black Sea (via the Mediterranean) and Sea of Azov. On 15.6.1855 under Lieut Cecil William Buckley, then 27.2.1856 under Cmdr John Edmund Commerell, still in the Black Sea; paid off 22.7.1857 at Sheerness. Recommissioned 20.9.1859 under Cmdr Henry Harvey, for the East Indies and China; on 4.12.1861 under Cmdr John Moresby (Harvey invalided 1.12.1861), on same station; paid off 6.1863. Sold to Marshall, Plymouth 1864, and BU at Keyham 1.1865.

Viper R. & H. Green, Blackwall/Maudslay, Sons & Field (673ihp = 11.86kts)
  Ord: 10.4.1854. K: 12.4.1854. L: 23.7.1854. C: 25.9.1854.
  First cost: £23,927 (including £9,081 for hull and £10,070 for machinery).
  Commissioned 5.8.1854 under Lieut Charles Arthur Lodder, for the Black Sea (via the Mediterranean). In 3.1854 under Lieut William Armytage, in the Black Sea and Sea of Azov; on 16.5.1855 under Lieut Henry Wandesford Comber, in the Black Sea. Recommissioned 12.11.1857 under Lieut Arthur Bissell Hodgkinson, for west coast of Africa; in 1.1859 under Cmdr William Nathan Wrighte Hewett, on same station; paid off at Plymouth 14.7.1860. Sold to Marshall (by AO 19.5.1862, for £482) to BU at Plymouth.

Wrangler R. & H. Green, Blackwall/Maudslay, Sons & Field (551ihp = 10.101kts)

Ord: 10.4.1854. K: 19.4.1854. L: 19.6.1854. C: 2.9.1854.

First cost: £24,860 (including £8,610 for hull and £9,920 for machinery).

Commissioned 18.7.1854 under Lieut Richard Hawkins Risk, for the Baltic Sea; redeployed 1855 to the Black Sea (via the Mediterranean) and Sea of Azov. On 15.6.1855 under Lieut Hugh Talbot Burgoyne, in the Black Sea. On 10.5.1856 under Cmdr Joseph Henry Marryat, in the Black Sea. Recommissioned 15.5.1860 under Cmdr Henry Hamilton Beamish, for west coast of Africa; took slaver felucca 31.8.1861, and slaver *Island Queen* 28.12.1862; paid off 21.9.1863 at Sheerness. BU 5.1866 by Castle at Charlton.

VIGILANT Class Despatch vessels, re-classed as Second Class Gunvessels by 16 November 1855. Surveyor's Department design, approved 18 April 1855, to be 'despatch vessels of *Arrow* Class but of increased dimensions'. The first four were ordered on 18 April 1855 to be built by contract. The contracts were placed on 15 May, and they were given names on 30 May. The next ten, ordered on 27 July 1855, were named on 21 August. Another four ordered 9 April 1856 (two from Deptford Dyd and two from Portsmouth Dyd) were not named or built (cancellation date unrecorded). Most were re-rated as sloops in 1859 or 1860, but all these reverted to gunvessels in 1862.

Dimensions & tons: 180ft 0in, 160ft 7½in x 28ft 4in oa (28ft 0in for tonnage) x 14ft 0in. 669⁷⁹⁄₉₄bm. 860 disp.

Men: 80. Guns: As designed – 2 x 68pdrs (95cwt/10ft) Lancaster MLR on pivots, 2 x 12pdr howitzers. As completed – 1 x 110pdr, 1 x 68pdr, 2 x 20pdrs.

Machinery: 2-cylinder horizontal single expansion. Single screw. 200nhp. (See below for individual ihp/speed.)

Coquette R. & H. Green, Blackwall/Miller, Ravenhill & Co. (690ihp = 10.853kts)

Ord: 15.5.1855. K: 18.5.1855. L: 25.10.1855. C: 8.3.1856.

First cost: £31,460 (including £13,205 for hull and £12,999? for machinery).

Commissioned 30.11.1855 at Woolwich under Cmdr Richard Hawkins Risk. On 10.7.1856 under Cmdr Henry Carr Glyn, at Portsmouth. On 10.7.1858 under Cmdr Fitzgerald Algernon Charles Foley, for the Mediterranean. Recommissioned 5.4.1862 under Cmdr John Hobhouse Inglkis Alexander, for China; at bombardment of Kagoshima (in Anglo-Satsuma War) 15.8.1863; on 17.8.1863 under Cmdr Arthur George Robertson Roe, in China; at bombardment of Simonoseki 5–8.9.1864; paid off 22.1.1867 at Portsmouth. Arrived Cowes 10.5.1867 for BU by White.

Wanderer R. & H. Green, Blackwall/Maudslay, Sons & Field (707ihp = 10.733kts)

Ord: 15.5.1855. K: 18.5.1855. L: 22.11.1855. C: 19.3.1856.

First cost: £32,108 (including £13,192 for hull and £12,529 for machinery).

Commissioned 5.12.1855 under Cmdr John Proctor Muce, for the Mediterranean. On 24.2.1858 under Cmdr Mark Robert Pechell, on same station; paid off at Sheerness. Recommissioned 20.6.1861 under Cmdr Michael Culme-Seymour, for the Mediterranean; paid off 16.8.1865 at Chatham, later to Sheerness. BU 31.8.1866 (arrival date) by Castle at Charlton.

Alacrity C. J. Mare & Co., Blackwall/Miller, Ravenhill & Co. (724ihp = 1.87kts)

Ord: 15.5.1855. K: 21.5.1855. L: 20.3.1856. C: 20.4.1856.

First cost: £31,220 (including £12,689 for hull and £12,049 for machinery).

Commissioned 3.1856 under Cmdr Henry Cholmeley Majendie; on 23.2.1857 under Cmdr Armine Wodehouse; paid off 20.2.1858. Recommissioned 23.12.1859 under Cmdr John Kennedy Erskine Baird, for the Mediterranean; paid off 8.1863. Sold to Castle 7.10.1864 to BU at Charlton.

Vigilant C. J. Mare & Co., Blackwall/Maudslay, Sons & Field (630ihp = 9.763kts)

Ord: 15.5.1855. K: 21.5.1855. L: 20.3.1856. C: 20.8.1856.

First cost: £32,514 (including £13,043 for hull and £12,533 for machinery).

Commissioned 19.3.1856 under Cmdr William Armytage, for the Mediterranean. Recommissioned 21.12.1861 under Cmdr John William Pike, for the Channel Squadron; on 10.5.1862 under Cmdr William Buller Fullerton Elphinstone, on same station. Recommissioned 9.2.1863 at Chatham under Cmdr William Robert Hodson, for the East Indies; on 21.8.1865 under Cmdr W. H. Jones Byron, on same station. In 1868 under Cmdr Ralph Abercrombie Otho Brown, for the Persian Gulf. Ordered 25.2.1869 to be sold at Bombay.

Lapwing T. & J. White, West Cowes/Miller, Ravenhill & Co. (689ihp = 11.021kts)

Ord: 26.7.1855. K: 30.7.1855. L: 26.1.1856. C: 31.5.1856.

First cost: £31,744 (including £13,285 for hull and £12,180 for machinery).

Commissioned 5.2.1856 under Cmdr Thomas Saumarez, for East Indies and China; in 2nd Anglo-Chinese War. Recommissioned 19.11.1857 under Cmdr Montagu Frederic O'Reilly, for the Mediterranean; paid off 23.1.1862. Sold to Marshall, Plymouth 1864 (BU 1865).

Ringdove T. & J. White, West Cowes/Miller, Ravenhill & Co. (677ihp = 10.824kts)

Ord: 26.7.1855. K: 30.7.1855. L: 22.2.1856. C: 31.5.1856.

First cost: £31,748 (including £13,313 for hull and £12,161 for machinery).

Commissioned 2.1856 under Cmdr Isaac Newton Thomas Saulez. Recommissioned 20.9.1859 under Cmdr Robert George Craigie, for the East Indies and China; in Taiping Rebellion; on 16.9.1862 under Cmdr Ralph Abercrombie Otho Brown, on same station; paid off 10.12.1864 at Portsmouth. Sold 2.6.1865 and arrived Cowes 26.9.1866 to BU 11.1866 by White.

Surprise Wigram & Son, Blackwall/Miller, Ravenhill & Salkeld (778ihp = 11.148kts)

Ord: 26.7.1855. K: 30.7.1855. L: 6.3.1856. C: 12.4.1856.

First cost: £33,356 (including £14,656 for hull and £12,096 for machinery).

Commissioned 18.3.1856 under Cmdr Charles Egerton Harcourt Vernon. Recommissioned 11.3.1857 under Cmdr Samuel Gurney Cresswell, for the East Indies and China; in 2nd Anglo-Chinese War. Recommissioned 2.8.1861 under Cmdr William Henry Whyte, for the Mediterranean; in 8.1864 under Cmdr George Tryon, on same station; paid off 24.4.1866 at Plymouth. BU 11.1866 by Marshall, Plymouth.

Renard C. J. Mare & Co., Blackwall/?George Rennie & Sons

Ord: 26.7.1855. K: 1.8.1855. L: 24.4.1856. C: 12.10.1859.

First cost: £33,906 (including £13,717 for hull and £12,000 for machinery).

Commissioned 20.9.1859 under Cmdr James Graham Goodenough, for the East Indies and China; on 1.5.1862 under Comdr. Charles John Rowley (when Goodenough invalided), on same station; paid off 6.7.1863 at Sheerness. BU 3.1866 by Castle at Charlton.

Foxhound C. J. Mare & Co., Blackwall/Humphrys, Tennant & Dykes

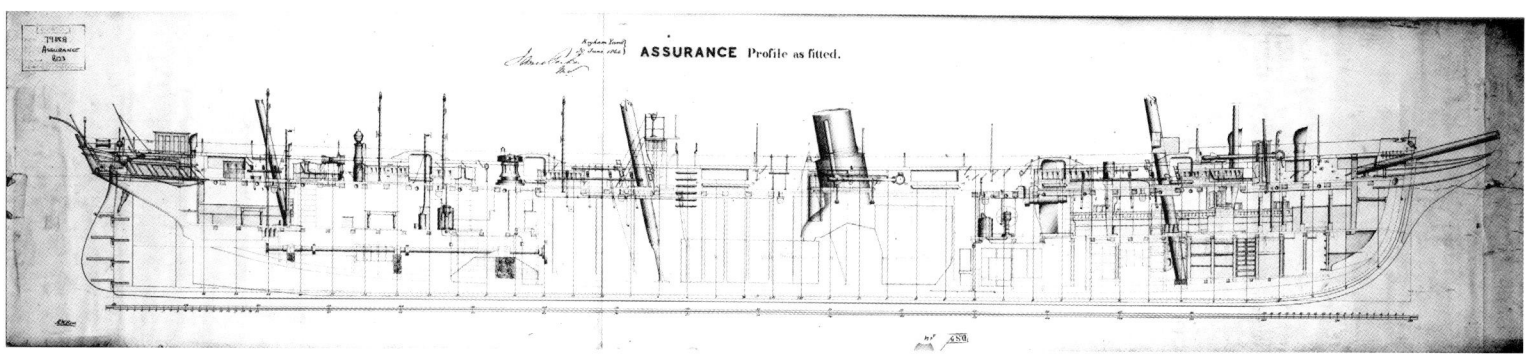

ASSURANCE Profile as fitted.

A profile of the *Assurance* as refitted at Keyham Yard, dated 29 June 1864. The *Assurance* was one of twenty ships to be ordered as 'dispatch vessels' in 1855, all being re-labelled as 'gun vessels' by the end of that year. Sixteen were completed by the middle of 1856, but were too late for employment in the Russian War; the *Roebuck* and *Nimrod* of the larger (First Class) *Intrepid* type were not completed until 1857, while *Renard* and *Foxhound* of the smaller (Second Class) *Vigilant* type were delayed to 1859 and 1860 respectively. (*NMM DR79881*)

(600ihp = 11.6kts)
   Ord: 26.7.1855. K: 1.8.1855. L: 16.8.1856. C: 6.10.1860.
   First cost: £27,437 (including £13,643 for hull and £11,999 for machinery).
   Commissioned 7.9.1860 under Cmdr Alfred Mitchell, for the Mediterranean; on 15.8.1861 under Cmdr Augustus Charles Hobart, in the Mediterranean; on 10.4.1863 under Cmdr Warren Hastings Anderson; paid off 22.10.1864 at Chatham. BU 8.1866 by Castle at Charlton.

Mohawk Young, Magnay & Co., Limehouse/Humphrys, Tennant & Dykes (642ihp = 10.721kts)
   Ord: 26.7.1855. K: 6.8.1855. L: 11.1.1856. C: 23.3.1856.
   First cost: £33,095 (including £14,304 for hull and £11,996 for machinery).
   Commissioned 26.1.1856 under Cmdr Francis Arden Close. Recommssioned 16.7.1857 under Cmdr Patrick Charles Campbell McDougall, for the East Indies and China. Recommissioned 30.5.1860 under Cmdr Edward Henry Howard, for the Mediterranean; paid off 25.10.1861 at Plymouth. Sold 20.9.1862 to the Chinese Imperial Customs, renamed *Peking* and sailed 4.1863 (to join Capt. Sherard Osborn's 'Vampire Fleet'), but returned to England when agreement with Chinese government collapsed Resold 30.12.1865 to the Egyptian government.

Sparrowhawk Young, Magnay & Co., Limehouse/Humphrys, Tennant & Dykes (726ihp = 11.065kts)
   Ord: 26.7.1855. K: 6.8.1855. L: 9.2.1856. C: 7.4.1856.
   First cost: £33,167 (including £14,464 for hull and £11,984 for machinery).
   Commissioned 5.4.1856 under Cmdr Samuel Gurney Cresswell. Recommissioned 18.7.1857 at Portsmouth under Cmdr John Clarke Byng, for the East Indies and China; paid off 13.5.1861 at Portsmouth. Repaired there 1864 (having been reported to be in 'a very dilapidated state'). Recommissioned 4.3.1865 under Cmdr Edwin Augustus Porcher, for the Pacific; on 18.8.1868 under Cmdr Henry Wentworth Mist, on same station; paid off 5.10.1872. Sold 26.11.1872 at Esquimault (for £4,000) to Corbitt & Co., Portland, Oregon.

Osprey Fletcher, Limehouse/Maudslay, Sons & Field (593ihp = 10.15kts)
   Ord: 26.7.1855. K: 2.8.1855. L: 22.3.1856. C: 30.7.1856.
   First cost: £31,633 (including £12,570 for hull and £12,400 for machinery).
   Commissioned 18.3.1856 under Cmdr Henry John Blomfield, for the Mediterranean; paid off 23.3.1860 at Woolwich. Commanded (not actually commissioned) from 3.1.1863 by Cmdr William Robert Hobson, at Woolwich; actually recommissioned 12.1.1863 under Cmdr Arthur John Innes, for the East Indies and China; on 14.12.1864 under Cmdr William Menzies, in China; wrecked on the Southeast African coast near Klippen Point (Algoa Bay) on 30.5.1867 en route to England, with one man lost.

Cormorant Fletcher, Limehouse/Robert Napier & Sons (727ihp = 11.155kts)
   Ord: 26.7.1855. K: 9.8.1855. L: 19.5.1856. C: 20.6.1856.
   First cost: £33,095 (including £12,728 for hull and £12,728 for machinery).
   Commissioned 1.2.1856 under Cmdr William Bowden. Recommissioned 20.5.1858 under Cmdr Thomas Saumarez, for China; in 2nd Anglo-Chinese War; in attack on Taku forts 20.5.1858; on 16.8.1858 under Cmdr Armine Wodehouse; sunk in action with the Peiho River forts (China) on 28.6.1859.

Assurance R. & H. Green, Blackwall/Miller, Ravenhill & Co. (745ihp = 11.142kts)
   Ord: 26.7.1855. K: 26.8.1855. L: 13.3.1856. C: 16.4.1856 at Chatham (for sea).
   First cost: £32,517 (including £14,711 for hull and £12,042 for machinery).
   Commissioned 3.1856 at Chatham under Cmdr William Gore Jones. Recommissioned 17.7.1857 under Cmdr Charles Murray Aynsley, for the East Indies and China. Recommissioned 9.5.1864 under Cmdr Henry Bedford Woolcombe, for the Mediterranean. In 5.1866 under Cmdr William Henry Pym; to West Africa station 4.1867. On 4.9.1867 under Cmdr John Binney Scott. Sold to Marshall, Plymouth, on 8.3.1870 to BU.

INTREPID Class Despatch vessels, re-classed as First Class Gunvessels by 16 November 1855. Surveyor's Department design, approved 27 January 1855. They had twin funnels and three masts. The first four named 30 May 1855, last two on 21 August 1855. *Flying Fish* and *Pioneer* were 'built on the diagonal principle'. *Flying Fish* was selected for hull tests after her completion, with a false bow fitted that added 18ft to her length; revised speed on trials was 12.43 knots. Building of *Roebuck* and *Nimrod* completed by workmen from Deptford Dockyard (by AO 18.2.1856), possibly because Scott Russell's experience had been with iron ships; for these two vessels the second figure under hull costs is dockyard expenses, the first being what was paid the builder. Two more, ordered from Pembroke Dockyard on 9 April 1856, were not built (cancellation date unrecorded). All except *Victor* were re-rated as sloops in 1859 (*Pioneer* in 1860), but reverted to being gunvessels in 1862.

Dimensions & tons: 200ft 0in, 179ft 5¼in x 30ft 4in oa (30ft 2in for tonnage) x 14ft 6in. 868⁴⁹⁄₉₄bm. 1042 disp. Draught 10½ft (normal) – 12½ft (full load).

Men: 100. Guns: 1 x 68pdr (95cwt/10ft) MLR; 4 x 32pdrs (25cwt/6ft). (Later 1 x 110pdr BL; 1 x 60pdr BL; 4 x 20pdrs (BL).)

Machinery: 2-cylinder horizontal single expansion. Single screw. 350nhp. (See individual vessels below for ihp/speed.)

Flying Fish Pembroke Dyd/Maudslay, Sons & Field (1,302ihp = 11.585kts)

Ord: 18.4.1855. K: 6.1855. L: 20.12.1855. C: 29.1.1856.

First cost: £54,901 (£20,373 for hull, £22,660 machinery, £11,868 fitting)

Commissioned 1.1856 at Portsmouth under Cmdr Roderick Dew; paid off 2.1857. Recommissioned 4.5.1859 under Cmdr Charles Webley Hope, for the Channel Squadron; in 7.1860 escort to HRH Prince of Wales for his journey to America; on 16.5.1861 under Cmdr Warren Hastings Anderson for Channel Squadron, then to west coast of Africa; paid off 29.11.1862. Sold to Castle to BU, arriving 8.1866 at Charlton.

Pioneer Pembroke Dyd/Miller, Ravenhill & Salkeld (1,150ihp = 11.332kts)

Ord: 18.4.1855. K: 7.1855. L: 19.1.1856. C: 31.5.1856.

First cost: £48,966 (£20,407 for hull, £21,350 machinery, £7,209 fitting)

Commissioned 8.1.1856 at Portsmouth under Cmdr George Pechell Mends, for North America and the West Indies. Recommissioned 7.5.1859 at Portsmouth under Cmdr Charles Henry May, for the Channel squadron, then to China; on 27.6.1859 under Cmdr Hugh Arthur Reilly, in the East Indies and China; in action for Taiping Rebellion; to Australia 1862; on 17.8.1862 under Acting Cmdr Frederick Charles Bryan Robinson, in Australia; paid off 22.9.1863 at Plymouth. Sold to Marshall 10.1864 to BU 1865 at Plymouth.

Victor Wigram & Son, Blackwall/Miller, Ravenhill & Co. (1,166ihp = 11.583kts)

Ord: 15.5.1855. K: 24.5.1855. L: 2.11.1855. C: 1.4.1856.

First cost: £45,450 (£16,205 for hull, £20,960 machinery, £8,225 fitting)

Commissioned 30.11.1855 under Cmdr Algernon Frederick Rous de Horsey, for North America and the West Indies; paid off 1.7.1857 at Sheerness. Sold 29.9.1863 as *Scylla* to Gordon Coleman & Co., who resold her later the same month to the Confederacy as the raider CSS *Rappahannock*. She was interned by the French at Cherbourg (or Calais?) and never served the Confederacy at sea. At the conclusion of the Civil War she was handed over to the US government.

Intrepid Wigram & Son, Blackwall/Maudslay, Sons & Field (930ihp = 10.35kts)

Ord: 15.5.1855. K: 24.5.1855. L: 13.11.1855. C: 4.4.1856.

First cost: £46,114 (£16,382 for hull, £24,339 machinery, £5,393 fitting)

Commissioned 6.12.1855 under Commander William Wood, for North America and the West Indies; on 25.9.1857 under John Waye, Master, at Devonport. Recommissioned 4.5.1859 under Cmdr Joseph Henry Marryat, for the Mediterranean; paid off 25.6.1862 at Plymouth. Sold to Marshall 7.10.1864 to BU 1865 at Plymouth.

Roebuck J. Scott Russell, Millwall/Miller, Ravenhill & Salkeld (1,099ihp = 11.377kts)

Ord: 25.7.1855. K: 1.8.1855. L: 22.3.1856. C: 25.8.1857 at Woolwich and Portsmouth (for sea).

First cost: £48,993 (£10,101 to Russell + £7,772 dockyard expenditure for hull, £20,977 machinery, £10,143 fitting).

Commissioned 18.5.1857 at Portsmouth under Cmdr Francis Marten, for the East Indies and China; from 17.5.1858 until 7.1859 (Acting Cmdr) under Lieut Edwin C. Symons, in the Red Sea, before reverting to Marten; paid off 16.7.1861 at Portsmouth. Sold to Castle, Charlton 1864 to BU.

Nimrod J. Scott Russell, Millwall/Maudslay, Sons & Field (?363ihp = 8.812kts)

Ord: 25.7.1855. K: 1.8.1855. L: 21.4.1856. C: 23.2.1857 at Sheerness.

First cost: £47,544 (including £11,811 to Russell + £5,982 dockyard expenditure for hull)

Commissioned 2.1857 under Cmdr Roderick Dew, for the East Indies and China; in action for 2nd Anglo-Chinese War; at Bombardment of Canton 12.1857. On 1.3.1858 under Cmdr George Pechell Mends, on same station; at Bombardment of the Taku Forts 20.5.1858; in 6.1859 under Acting Cmdr Robert James Wyniatt; in attack on the Peiho River Forts 26.6.1859; later under Lieut William Arthur (temp.) for voyage home; paid off 1.8.1861 at Portsmouth. Sold to White, Cowes 2.6.1865 to BU.

PHILOMEL (*RANGER*) **Class** Second Class gunvessels, 5 guns. This design derived not from the wartime gunvessels, but from an enlargement of the Algerine Class gunboats. The first batch of six were ordered from the dockyards as 'new type steam schooners', but were re-classed as gunvessels on 8 June 1859. This type having been found 'very useful and much required on foreign stations', another twelve were then ordered to be contract built, with fitting out at these dockyards: Deptford (*Lee* and *Dart*), Woolwich (*Snipe* and *Sparrow*), Chatham (*Torch* and *Plover*), Devonport (*Penguin* and *Steady*), Portsmouth (*Cygnet* and *Philomel*) and Sheerness (*Griffon*), to which was added *Mullet* (Deptford). A final eight were later ordered from the dockyards, but six of these were cancelled.

The first eighteen were all named on 24 September 1859, the next six on 4 October 1860, and the last pair on 21 March 1861. *Griffon*, *Mullet* and *Philomel* were originally ordered as follow-on *Algerine* Class gunboats, but were 'reintegrated with gunvessels' in 1860. *Sparrow* fitted for Armstrongs in 1861 and then for service on the West Coast of Africa (65 men).

Dimensions & tons: 145ft 0in, 127ft 10¼in x 25ft 4in oa (25ft 0in for tonnage) x 13ft 0in. 425¹⁴⁄₉₄bm. 570 displacement.

Men: 60. Guns: 1 x 68pdr (later replaced by 1 x 110pdr BL), 2 x 24pdr howitzers, 2 x 20pdr BL.

Machinery: 2-cylinder horizontal single expansion (single trunk in *Ranger*). Single screw. 80nhp (originally to have been 60nhp). (see individual vessels below for ihp/speed.)

First cost: *Alban*: before cancellation £951. *Undine*: before cancellation £868. *Humber*: before cancellation £1,149. *Rye*: before cancellation £1,293.

First batch (1857 to 4.1859 Orders)

Ranger Deptford Dyd/George Rennie & Sons (284ihp = 9.006kts)

Ord: 1.4.1857. K: 16.11.1857. L: 26.11.1859. C: 6.2–10.5.1860 at Woolwich.

First cost: £12,537 for hull.

Commissioned 16.4.1860 at Woolwich under Cmdr Henry Rushworth Wratislaw, for the west coast of Africa; at Porto Novo (Niger River) 26.4.1861; paid off end 1863 at Woolwich. Recommissioned 27.2.1864 under Cmdr William Ebrington

The first group of post-war gunvessels were the *Philomel* Class, of which the *Pandora* is seen in this photograph. At first classed as 'new type steam schooners', their design was an enlargement of the *Algerine* Class gunboats, with increased length and draught to improve their seaworthiness, and carrying a square rig on both fore and main masts. In 1875 the *Pandora* was sold to Allen Young, who fitted her for polar exploration, partly with funding from Sir John Franklin's widow, and tried to get through the Northwest Passage by way of Peel Sound. Young later sold the *Pandora*, but purchased her original sister ship *Newport*, which he renamed *Pandora II*; both vessels were subsequently lost during Arctic exploration voyages. *(NMM 5386)*

Gordon, then 19.5.1665 under Cmdr Charles Gudgeon Nelson, and 10.8.1866 under Cmdr William Alfred Gambier, on same station; paid off 24.1.1868 at Woolwich. Sold to Messrs. Moss Isaacs on 3.11.1869.

Espoir Pembroke Dyd/A. & J. Inglis (286ihp = 9.658kts)
Ord: 1.4.1857. K: 25.4.1859. L: 7.1.1860. C: 27.9.1860 at Plymouth. First cost: £12,067 for hull.
Commissioned 21.8.1860 at Plymouth under Cmdr Sholto Douglas, for the west coast of Africa; took slaver bark 22.7.1862 and slaver *Haydee* 12.8.1863; paid off 4.1864 at Plymouth. Recommissioned 1.11.1864 under Cmdr Mountford Stephen Lovielle Peille, for west coast of Africa (Peille being Senior Officer in the Bight of Benin division); paid off 24.12.1867 at Sheerness. Converted to dredging vessel *Y.C.19* at Sheerness 10.1868–5.1869. Sailed 26.10.1869 to Bermuda under Navigating Lieut William Hutton, and BU there in 6.1881.

Landrail Deptford Dyd/Humphrys & Tennant (253ihp = 8.432kts)
Ord: 27.3.1858. K: 3.8.1859. L: 27.3.1860. C: 27.9.1860 at Woolwich.
Commissioned 7.9.1860 at Woolwich under Cmdr Thomas Hutchinson Mangles Martin, for North America and the West Indies; on 14.5.1863 under Lieut William Arthur, in the West Indies; ordered home 12.1863 and paid off 15.3.1864 at Woolwich. Recommissioned 14.1.1865 under Cmdr Horatio Laurence Arthur Lennox Maitland, for the west coast of Africa; paid off 15.2.1868 at Sheerness. Sold 23.9.1869 (for £1,328) for mercantile use, renamed *Walrus*.

Nimble Pembroke Dyd/A. & J. Inglis (334ihp = 9.933kts)
Ord: 27.3.1858. K: 30.10.1859. L: 15.9.1860. C: 27.9.1860–8.4.1861 at Devonport.
Commissioned 12.2.1861 at Plymouth under Lieut John D'Arcy, for North America and the West Indies, as tender to *Nile*. No captain 9.1863–1.1864. Recommissioned 4.2.1864 under Lieut Frederick William Hallowes, as tender to *Duncan*. No captain 9.1865. Recommissioned 17.10.1865 under Cmdr Alfred John Chatfield, for North America and the West Indies; paid off 1868 at Devonport. Recommissioned 19.10.1870 under Cmdr

William Frederick Lee, for the East Indies; on 11.6.1872 under Cmdr Richard Hastings Harington, then 10.10.1873 under Cmdr Henry Compton Best. Recommissioned at Bombay 3.3.1874, still under Best; on 1.12.1874 under Cmdr William Derenzy Donaldson Selby; ordered home 3.1876 and paid off 1.8.1876 at Chatham. To Hull as RNR drill ship, as tender 5.1878 to *Endymion*, *Audacious*, then *Repulse* at Hull (engines removed 4.1878). RNR drill ship at Hull for most of remainder of century. Sold to W. R. Jones on 10.7.1906.

Speedwell Deptford Dyd/C. A. Day & Co. (322ihp = 8.883kts)
Ord: 27.3.1858. K: 2.4.1860. L: 12.2.1861. C: 19.10.1861 at Woolwich (for Reserve).
Commissioned 8.9.1863 under Cmdr George Frederick Cottam, for the west coast of Africa; on 25.2.1865 under Cmdr James Elphinstone Erskine; paid off 16.2.1867 at Woolwich. Recommissioned 19.7.1867 under Cmdr John Parry Jones Parry, for east coast of South America; ordered home 6.1871; paid off 21.11.1871. BU 7.1876 at Chatham.

Pandora Pembroke Dyd/C. A. Day & Co. (346ihp = 9.387kts)
Ord: 8.4.1859. K: 30.3.1860. L: 7.2.1861. C: — .
Commissioned 19.3.1863 at Portsmouth under Cmdr William Fitzherbert Ruxton, for the west coast of Africa; on 18.12.1865 under Cmdr Ernest Augustus Travers Stubbs; paid off 21.2.1867 at Portsmouth. Recommissioned 24.8.1868 under Cmdr John Burgess, for the same station; to the Mediterranean 4.1869; paid off 36.7.1872. Sold 13.1.1875 to Allen Young for polar exploration; she was re-sold to James Gordon Bennett Jnr in 1878, and renamed *Jeanette*, departing for a further polar voyage in 7.1879, but was crushed by the Arctic ice and sank on 13.6.1881.

Second batch (6.1859 Orders)

Lee Wigram & Son, Blackwall/Robert Napier & Sons (352ihp = 9.691kts)
Ord: 14.6.1859. K: 20.6.1859. L: 25.1.1860. C: 24.4.1862 at Woolwich and Sheerness.
Commissioned 13.3.1862 at Sheerness under Cmdr Edwin Charles

Symons, for the west coast of Africa; on 30.6.1863 under Lieut
Philip Ruffle Sharpe, as tender to *Rattlesnake* on same station;
on 12.1.1854 under Lieut Charles Edward Foot, then 2.1.1865
under Lieut Oliver Thomas Lang; on 24.12.1865 under Lieut
Herbert Holden Edwards, as tender to *Bristol* on same station;
paid off at Sheerness. Recommissioned 8.6.1867 under Cmdr
Charles William Andrew, for the west coast of Africa; ordered
home by 3.1869. On 20.6.1870 under Cmdr Charles Sedgwick
Fitton, for the Mediterranean; ordered home by 6.1871, paid
off at Sheerness. To Chatham 1873. BU completed 20.3.1875 at
Sheerness.

Dart C. J. Mare & Co., Millwall/Robert Napier & Sons (281ihp =
8.708kts)
    Ord: 14.6.1859. K: 27.6.1859. L: 10.3.1860. C: 1860 at Woolwich,
        then to Chatham (for Reserve).
    Commissioned 8.3.1862 under Cmdr Frederick William Richards,
        for the west coast of Africa; paid off 5.1.1866 at Portsmouth.
        Recommissioned 25.8.1866 under Cmdr Marcus Lowther,
        on same station; later to West Indies; on 26.10.1868 under
        Cmdr John Carnegie; paid off 8.11.1870 at Portsmouth.
        Recommissioned 9.5.1872 under Cmdr D'Arcy Anthony
        Denny, for east coast of South America; paid off 31.5.1876 at
        Portsmouth. Renamed *Kangaroo* 1.4.1882. BU 12.1884.

Snipe J. Scott Russell, Millwall/Robert Napier & Sons (363ihp =
10.32kts)
    Ord: 14.6.1859. K: 27.6.1859. L: 5.5.1860. C: 1860 at Woolwich,
        then to Sheerness (for Reserve).
    Commissioned 6.8.1863 under Cmdr Albert Henry William
        Battiscombe, for the west coast of Africa; on 22.9.1865 under
        Cmdr Henry Anthony Trollope, on same station; ordered
        home by 4.1867; paid off 30.7.1867 at Portsmouth. BU 5.1868 at
        Sheerness.

Sparrow J. Scott Russell, Millwall/Robert Napier & Sons (392ihp =
10.872kts)
    Ord: 14.6.1859. K: 27.6.1859. L: 7.7.1860. C: 16.3.1863 at Woolwich,
        then to Chatham (for Reserve).
    Commissioned 24.3.1863 under Cmdr Ernest Grey Lambton
        Cochrane, for the west coast of Africa; on 13.7.1864 under Cmdr
        Loftus Francis Jones; paid off 17.8.1866 at Plymouth. BU by
        Marshall 1868 at Plymouth.

Torch R. & H. Green, Blackwall/Robert Napier & Sons (281ihp =
8.843kts)
    Ord: 14.6.1859. K: 29.6.1859. L: 24.12.1859. C: 14.7.1860.
    Commissioned 14.6.1860 under Cmdr Frederick Harrison Smith,
        for the west coast of Africa; at Saba 21.2.1861; ordered home
        9.1863 and paid off at Chatham. Recommissioned 22.2.1865
        under Cmdr George Amelius Douglas, for the west coast
        of Africa; paid off 1869 at Woolwich, then to Sheerness.
        Recommissioned 5.1.1871 under Cmdr Louis Geneste, for same
        station; on 11.9.1871 under Cmdr Joseph Edward Maitland
        Wilson then on 17.9.1871 under Cmdr Hugh McNeile Dyer,
        on same station, then 3.1873 to the Mediterranean; on 8.10.1873
        under Cmdr Richard Henry Napier; paid off 30.10.1873.
        Recommissioned 27.1.1876 under Cmdr Richard Henry
        Hamond, for the Mediterranean. Recommissioned 13.6.1878 at
        Malta, still under Hamond; on 30.1.1880 under Cmdr William
        Charles Selby. BU 9.1881 at Malta.

Plover R. & H. Green, Blackwall/Robert Napier & Sons (341ihp =
10.507kts)
    Ord: 14.6.1859. K: 29.6.1859. L: 19.1.1860. C: 5.9.1861 at Chatham.
    Commissioned 9.8.1861 at Chatham under Cmdr Armar Lowry

Corry, for North America and the West Indies; paid off
13.4.1865 at Chatham. Sold 12.9.1865 (for £2,000) for mercantile
use, renamed *Hawk*; wrecked 1876.

Penguin W. C. Miller, Toxteth Dock, Liverpool/Robert Napier & Sons
(365ihp = 11.078kts)
    Ord: 14.6.1859. K: 29.6.1859. L: 8.2.1860. C: 13.3.1861 at
        Devonport.
    Commissioned 26.12.1860 at Plymouth under Lieut John George
        Graham McHady, for the Cape of Good Hope, as tender to
        Narcissus. Recommissioned 17.2.1864 under Lieut Edmund St
        John Garforth, for the East Indies and Cape (later East Indies) as
        tender to *Princess Royal*; paid off 13.8.1868 at Devonport. Sold
        to Lethbridge & Drew 26.2.1870 to BU.

Steady W. C. Miller, Toxteth Dock, Liverpool/Robert Napier & Sons
(360ihp = 11.053kts)
    Ord: 14.6.1859. K: 29.6.1859. L: 8.2.1860. C: 12.6.1861 at
        Devonport.
    Commissioned 18.5.1861 at Plymouth under Cmdr Henry Duncan
        Grant, for North America and the West Indies; on 14.5.1863
        under Cmdr Frederick Harvey, for same station; on 18.11.1864
        under Cmdr Walter Cecil Talbot (when Harvey invalided), then
        3.3.1865 under Cmdr Thomas Thelwall Bullock, on same station;
        invalided 20.4.1867. On 21.4.1867 under Cmdr John Parry Jones
        Parry; paid off 5.7.1867 at Plymouth. Sold to W. & T. Jolliffe
        12.5.1870.

Cygnet Wigram & Co., Northam/Robert Napier & Sons (354ihp =
11.233kts)
    Ord: 14.6.1859. K: 28.7.1859. L: 6.6.1860. C: 13.6.1861 at
        Portsmouth.
    Commissioned 18.5.1861 at Portsmouth under Cmdr Arthur
        Thomas Thrupp, for North America and the West Indies; on
        30.7.1872 (acting) 28.1.1863 (confirmed) under Cmdr Walter
        Sidney de Kantzow, for same station. Recommissioned 10.1864.
        On 29.4.1865 under Cmdr Gover Rose Miall, for same station;
        on 14.8.1866 under Cmdr Henry Wayland Chetwynd-Gore;
        paid off 17.8.1867. BU 8.1868 at Portsmouth.

Griffon W. & H. Pitcher, Northfleet/Robert Napier & Sons (326ihp =
10.119kts)
    Ord: 14.6.1859. K: 28.6.1859. L: 25.2.1860. C: 28.7.1861 at
        Sheerness.
    Commissioned 27.6.1861 at Sheerness under Cmdr John Laisné
        Perry, for the west coast of Africa; paid off 25.6.1865 at
        Sheerness. Recommissioned 1866 under Cmdr Duncan George
        Davidson, for west coast of Africa; collided with *Pandora* off
        Little Popo in West Africa 2.10.1866 and stranded.

Mullet Charles Lungley, Rotherhithe/Robert Napier & Sons (355ihp
= 10.067kts)
    Ord: 14.6.1859. K: 29.6.1859. L: 3.2.1860. C: 30.4.1862 at Deptford,
        later to Sheerness (for Reserve).
    Commissioned 20.3.1862 at Sheerness under Cmdr Corland
        Herbert Simpson, for the west coast of Africa; paid off 1865
        at Sheerness. Recommissioned 30.4.1866 under Cmdr Charles
        Aylmer Pembroke Vallancey Robinson, for west coast of Africa;
        on 7.11.1867 under Cmdr Edward Kelly, for North America and
        the West Indies; paid off 19.7.1870 at Sheerness. Believed sold in
        UK for mercantile use, then resold 25.4.1872 at Hong Kong for
        mercantile use, renamed *Formosa*.

Philomel J. & R. White, Cowes/Robert Napier & Sons (327ihp =
10.85kts)
    Ord: 14.6.1859. K: 9.7.1859. L: 10.3.1860. C: 11.2.1861 at
        Portsmouth.

Commissioned 19.12.1860 at Portsmouth under Cmdr Leveson Widman, for the west coast of Africa; ordered home 7.1864; paid off 11.8.1864. Sold back to White, Cowes, 2.6.1865 to BU.

Third batch (1860 and 1861 Orders)

Newport Pembroke Dyd/Laird Brothers (325ihp = 9.25kts)
> Ord: 5.3.1860. K: 17.9.1860. Suspended 1862/63. L: 20.7.1867. C: 4.1868 as survey ship.
>
> Commissioned 31.3.1868 under Cmdr George Strong Nares, for hydrographic work in the Mediterranean; by 1871 surveying in Red Sea; paid off 8.6.1871 at Sheerness, later to Chatham. Sold 5.1881 to Sir Allen Young for polar expedition, and renamed *Pandora II*. Re-sold in 1890 and renamed *Blencathra*; subsequently sold again in 1912 and employed in Russian expeditions to open up a Northeast Passage, but was lost in the Arctic ice.

Alban Deptford Dyd /—
> Ord: 5.3.1860. K: 1.10.1860. Suspended 1862/63. L: – . Cancelled 12.12.1863

Jaseur Deptford Dyd /—
> Ord: 5.3.1860. K: 7.1.1861. L: 15.5.1862. C: — .
>
> Commissioned 6.4.1863 at Sheerness under Cmdr Walter James Hunt-Grubbe, for the west coast of Africa; paid off 13.2.1867 at Sheerness. Recommissioned 16.8.1867 under Cmdr Charles Frederick Hotham, for the Mediterranean; ordered home 6.1871 and paid off at Sheerness, later to Chatham. Sold to the Commissioners of the Irish Lighthouses in 12.1874.

Humber Pembroke Dyd /—
> Ord: 5.3.1860. K: 8.2.1861. L: – . Cancelled 12.12.1863.

Undine Deptford Dyd /—
> Ord: 5.3.1860. K: 31.12.1861. L: – . Cancelled 12.12.1863.

Rye Pembroke Dyd /—
> Ord: 5.3.1860. K: – Cancelled 12.12.1863 (not laid down).

Portia Deptford Dyd /—
> Ord: 3.1861. K: – Cancelled 12.12.1863 (unstarted).

Discovery Deptford Dyd /—
> Ord: 3.1861. K: – Cancelled 12.12.1863 (unstarted).

CORMORANT (*ECLIPSE*) Class First Class gunvessels. A return to the larger dimensions of the *Intrepid* Class in 1859 came as the result of the Admiralty's wish to deploy overseas gunvessels with greater firepower than the *Philomel* design, which necessitated not only increased draught (mitigating against inshore operations) but also requiring greater engine power.

The first six were ordered under the 1859 Programme to be contract built, with fitting out at these dockyards: Chatham (*Cormorant* & *Racehorse*) and Woolwich (*Serpent*, *Star*, *Eclipse* & *Lily*); these six were all named on 24 September 1859; initially planned to carry 2 x 68pdr MLSB and 2 x 32pdr MLSB, they were completed with the armament detailed below, but subsequently re-armed with 1 x 7in (6½ton) MLR and 2 x 64pdr MLR. The next four were ordered to be built at the dockyards under the 1860 Programme, and named on 4 October 1860; the construction of *Sylvia*, *Nassau* and *Myrmidon* was suspended 1862/63, but later resumed; 150nhp engines fitted to *Sylvia* and *Nassau* for service as survey ships. The last three were ordered under the 1860 Programme, also for building in the dockyards, and named on 21 March 1861, but were cancelled in 1863.

> Dimensions & tons: 185ft 0in, 165ft 7¼in x 28ft 4in oa (28ft 1in for tonnage) x 14ft 0in. Draught 11ft – 12ft. 694⁶⁶/₉₄bm. 877 disp.
> Men: 90. Guns (as built): 1 x 110pdr MSLB; 1 x 68pdr MLSB; 2 x 20pdrs.
> Machinery: 2-cylinder horizontal single expansion. Single screw.

200nhp. (see individual vessels below for ihp/speed.)

Cormorant Wigram & Son, Blackwall/Robert Napier & Sons (727ihp = 11.155kts)
> Ord: 14.6.1859. K: 20.6.1859. L: 9.2.1860. C: — .
>
> Commissioned 14.1.1862 at Chatham under William S. Bourchier, Master. Recommissioned 16.5.1863 under Cmdr Charles Matthew Buckle, for the East Indies and China. In 1866 under Cmdr George Doherty Broad, then 22.6.1869 under Cmdr Albert Denison Somerville Denison, on same station. Sold 7.6.1870 at Hong Kong.

Racehorse Wigram & Son, Blackwall/Robert Napier & Sons (732ihp = 10.937kts)
> Ord: 14.6.1859. K: 20.6.1859. L: 19.3.1860. C: —
>
> Commissioned 16.5.1862 at Sheerness under Cmdr Charles Richard Fox Boxer, for China; at Bombardment of Kagoshima 15.8.1863 (during Anglo-Satsuma War); wrecked near Chefoo, China, on 4.11.1864 with loss of 99 lives (9 saved including Boxer).

Serpent C. J. Mare & Co., Millwall/Robert Napier & Sons (774ihp = 10.042kts)
> Ord: 14.6.1859. K: 27.6.1859. L: 23.6.1860. C: — .
>
> Commissioned 17.4.1865 under Cmdr Charles James Bullock, for surveying in China Seas; paid off 10.8.1869 at Plymouth and laid up. Sold to Castle 13.4.1875 to BU at Charlton.

Star C. J. Mare & Co., Millwall/Robert Napier & Sons (892ihp = 11.1kts)
> Ord: 14.6.1859. K: 27.6.1859. L: 15.12.1860. C: — .
>
> Commissioned 5.1865 at Woolwich under Cmdr George Patterson, then in 7.1865 under Cmdr C. Wooton. In 6.1868 under Cmdr Walter Sidney de Kantzow, for the East Indies; paid off 4.8.1870 at Plymouth. Completed BU 14.3.1877 at Plymouth.

Eclipse J. Scott Russell, Millwall/Robert Napier & Sons (838ihp = 11.0kts)
> Ord: 14.6.1859. K: 8.8.1859. L: 18.9.1860. C: — .
>
> Commanded (but not commissioned) 16.2.1861 at Sheerness under Cmdr Edmund Robert Fremantle; in 10.1861 to Steam Reserve at Sheerness. Commissioned 24.9.1862 at Sheerness under Cmdr Richard Charles Mayne, for Australia; in operations during New Zealand War; in 2.1864 under Fremantle again; paid off 9.2.1867 at Sheerness. BU there 7.1867.

Lily J. Scott Russell, Millwall/Robert Napier & Sons (799ihp = 11.48kts)
> Ord: 14.6.1859. K: 8.8.1859. L: 27.2.1861. C: — .
>
> Commissioned 10.11.1862 at Sheerness under Cmdr Henry Harvey, for North America and the West Indies; on 14.12.1863 under Cmdr Algernon Charles Fieschi Heneage, on same station; paid off 7.11.1866 at Sheerness. BU 10.1867 at Sheerness.

Sylvia Woolwich Dyd/Humphrys & Tennant (689ihp = 10.343kts)
> Ord: 5.3.1860. K: 7.5.1860. L: 20.3.1866. C: 10.1866 (as survey ship).
>
> Commissioned 12.10.1866 at Woolwich under Cmdr Edward Wolfe Brooker, for surveying on the China Station; on 7.10.1869 under Cmdr Henry Craven St John (when Brooker invalided), surveying in Japan; paid off 9.4.1873 at Sheerness. Recommissioned 18.11.1873, still under St John (now Capt.), in same role; paid off 1877 at Hong Kong. Recommissioned 9.5.1877 at Hong Kong under Capt. Bonham Ward Bax (died 15.7.1877 at Nagasaki from dysentery), surveying in Japan; in 7.1877 (acting) 27.11.1877 (confirmed) under Cmdr Pelham Aldrich, in same role; paid off 17.11.1880 at Sheerness. Recommissioned 27.3.1882 under Capt. William James Lloyd Wharton, for South America, surveying Straits of Magellan. On 8.4.1884 under Aldrich again (now Capt.), for surveying Cape

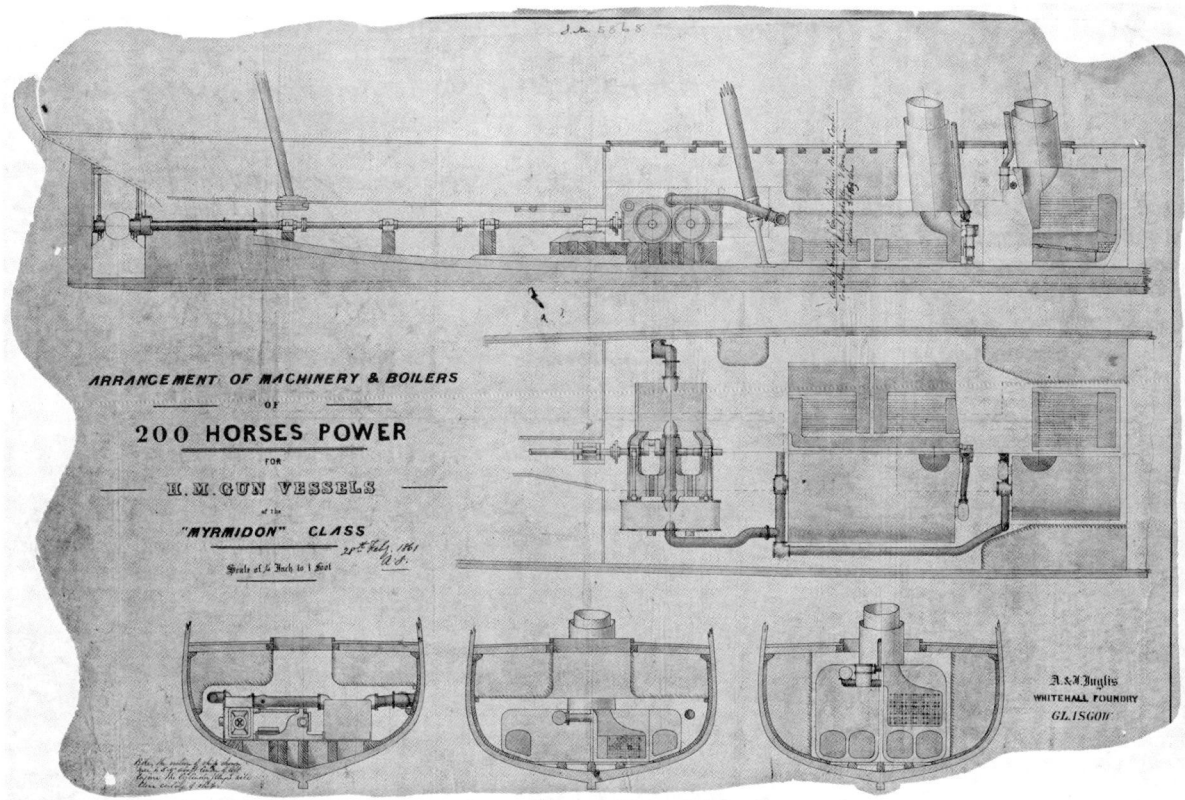

ARRANGEMENT OF MACHINERY & BOILERS
OF
200 HORSES POWER
FOR
H.M. GUN VESSELS
of the
"MYRMIDON" CLASS

The arrangement of machinery and boilers for the *Cormorant/Eclipse* Class, although this drawing dated 28 February 1861 refers to them as the *Myrmidon* Class (in fact the *Myrmidon* was one of the second – 1860 Programme – batch ordered). The drawing is labelled 'A & J Inglis, Whitehall Foundry, Glasgow'; the firm was not one of the contracted engine-builders for any of the completed ships, so this may apply to one of the cancelled vessels or indicates a subcontract from one of the nominated engineering companies. The increase to 200 nhp reciprocating engines certainly placed a strain on the engine contractors, who in many cases only carried out the final assembly and installation of these engines themselves.
(NMM DR05868)

of Good Hope, then to Gibraltar; on 28.10.1885 under Cmdr Llewellyn S. Dawson, in the Mediterranean. Recommissioned 12.1885 at Malta, still under Dawson; paid off 22.2.1889 at Sheerness. Sold to George Cohen 8.1889 to BU.

Nassau Pembroke Dyd/Humphrys & Tennant (735ihp = 10.585kts)
Ord: 5.3.1860. K: 20.6.1860. L: 20.2.1866. C: 7.1866 (as survey ship).
Commissioned 6.7.1866 at Woolwich under Cmdr Richard Charles Mayne, for South America, surveying Straits of Magellan; paid off 12.8.1869 at Plymouth. Recommissioned 22.4.1870 under Cmdr William Chimmo, for surveying in China Seas; in action with pirates off Sulu 11.5.1872; paid off at Malta 17.3.1873, and refitted there. Recommissioned 5.10.1873 under Lieut Francis John Gray (died 12.12.1875), for the East Indies; in attack at Mombasa 19.1.1875; on 22.1.1876 under Cmdr Richard Henry Napier; recommissioned (still under Napier) at Cape Town 3.1876, for China; paid off 15.4.1879 at Sheerness. BU 4.1880 at Sheerness.

Myrmidon Chatham Dyd/Humphrys & Tennant (782ihp = 10.838kts)
Ord: 5.3.1860. K: 24.7.1860. L: 5.6.1867. C: 10.1867.
Commissioned 29.10.1867 under Cmdr Henry Boys Johnstone, for the west coast of Africa. In 8.1869 under Cmdr Henry Lowe Holder, for North America and the West Indies; paid off 2.12.1871 at Sheerness. Recommissioned 24.10.1872 under Cmdr Richard Hare, for the Pacific; paid off 10.5.1877 at Sheerness, and laid up. Rebuilt (for £14,750) as a survey ship at Sheerness 6–12.1883 (timbers reported very rotten and were replaced, as were boilers); recommissioned 10.1.1884 under Cmdr Alfred Carpenter, for surveying in the Mediterranean and Red Sea. On 30.8.1884 under Cmdr Richard Frazer Hoskyn; transferred to Australian station 5.1885; damaged in collision with *Tyne* in the Bass Strait 25.6.1887 (Hoskyn reprimanded), and repaired at Sydney; on 1.11.1887 under Cmdr F. P. Vereker, for surveying in Australia; paid off

31.1.1889 at Hong Kong. Sold 4.1889 at Hong Kong.
Tartarus Pembroke Dyd/—
Ord: 5.3.1860. K: 25.10.1860. L: — . Cancelled 16.12.1864. BU 4.11.1865.
First cost: *Tartarus*: £6,268 before cancellation.
Pegasus Woolwich Dyd/—
Ord: 25.3.1861. K: 13.1.1862. L: — . Cancelled 12.12.1863 (cost £339)
Albatross Chatham Dyd/—
Ord: 25.3.1861. K: — .1862. Cancelled 12.12.1863.
Guernsey Pembroke Dyd/—
Ord: 25.3.1861. K: — . Cancelled 12.12.1863 (not laid down).

(B) Vessels ordered or re-ordered as steam screw gunboats (from 1854)

The numerous wooden-hulled small gunboats that were mass-produced during the Russian War of 1854–58 resulted from the difficulties encountered by the Royal Navy during their initial 1854 campaigns in the Baltic and Black Seas. As the Russian Navy declined to leave the security of its ports, the British Admiralty – in response to public pressure – required a massive increase in its coastal bombardment forces. Besides the ironclad floating batteries detailed in Chapter 6, the Admiralty ordered a large flotilla of small craft, designed to operate in very shallow waters. Collectively called the 'Great Armament', this programme was to build 156 new screw gunboats (as well as 56 sail-powered mortar vessels and 50 mortar floats – for details see Chapter 10) by 1 March 1856. As even the

During the Russian War, to cope with fitting out the numerous gunvessels and gunboats, the Admiralty established a purpose-built Gunboat Yard at Haslar Farm, near Gosport on the western side of Portsmouth Harbour. Once acceptance trials had been conducted, each new vessel arriving from its machinery contractor would be hauled up on a slipway, hoisted onto a steam-driven 'traverser' or transporting carriage (popularly known as 'The Elephant') designed by Isambard Brunel, and moved into one of a series of ten large sheds where they could be coppered and armed, and then re-launched and rigged. The Gunboat Yard also provided storage for those boats which were not fitted out before the end of the Russian War, and which in some cases were never to be completed; by 1867 some 36 of the unused boats, which had rotted beyond repair, were broken up and their engines installed (in sets of two engines) into 18 new composite-hulled gunvessels of the *Beacon* Class. Besides its repair and maintenance role, the Gunboat Yard also became the construction site for the later gunboats of the *Britomart* Class, ordered in 1860 and 1861.

Navy's gunvessels drew too much water, the new gunboats were flat-bottomed, and designed to draw only 6½ft. Some sources record these vessels as being masters' commands, but the Navy List (1860) makes clear that each officer in command received a lieutenant's commission on the dates shown below.

This massive industrial programme (only six of these gunboats were built by the dockyards) brought to warship building several small firms without previous experience of naval work. These contractors built only the bare hulls, with the propulsion machinery being installed by the engine-builders. The 150 boats were then taken to a naval dockyard to be coppered, rigged, armed and stored. To aid the process, a new specialist gunboat yard was built at Haslar Creek, Gosport (opposite Portsmouth), where many contract-built gunboats were fitted out (and a number of later gunboats built). Note comprehensive trials data is deficient for most Crimean era gunboats (unlike all larger vessels); individual trials ihp, but not speeds, are shown in parenthesis below

where known. Most Crimean gunboats were quickly built using green or unseasoned timber, resulting in the short careers of many.

Following on from these 156 vessels (comprising five classes with little variation in design), the lessons learnt were incorporated into six new *Algerine* Class gunboats, considerably larger and more seaworthy than their predecessors. After the war's end, 20 improved versions of the Crimean type were built with hoisting screws and a barquentine rig.

GLEANER (or ***PELTER***) **Class** Wooden 'Crimean' gunboats. Design by W. H. Walker in the Surveyor's Office. All were ordered on 16 June 1854 (two awarded to Deptford Dyd and the rest built by contract with Pitcher at Northfleet), initially called gunvessels, and named on 24 June. At the outset it was planned that each would carry 2 x 68pdrs, but the one forward was substituted by a 32pdr. They carried a light three-masted schooner rig, with a sail area of 2,426 sq.ft.

Dimensions & tons: 100ft 0in, 86ft 4in x 22ft 0in oa (21ft 8in for tonnage) x 7ft 10in. 215⁵³⁄₉₄bm. 290 disp.

Men: 36. Guns: 1 x 68pdr (95cwt/10ft) MLSB shell (on pivot, aft); 1 x 32pdr (56cwt/9½ft) MLSB (on pivot, fwd); 2 x 24pdr howitzers (on trucks).

Machinery: 1 set horizontal single expansion, trunk. 2 cylindrical boilers. Single (non-hoisting screw) 60nhp. 270ihp = 7½kts. 25 tons coal.

Pelter W. & H. Pitcher, Northfleet/John Penn & Sons
Ord: 16.6.1854. K: 29.6.1854. L: 26.8.1854. C: 20.10.1854.
First cost: £8,720 (including £2,738 for hull, £3,390 for machinery).
Commissioned 20.4.1855 under Lieut William Frederick Lee, for the Baltic (during the Russian War) as tender to *Cressy*; in

2.1856 under Lieut Henry Lewis Round; paid off 10.9.1856 at Sheerness. Sold for BU 1.1864, BU completed 1.2.1864 by Tolpult.

Pincher W. & H. Pitcher, Northfleet/John Penn & Sons (215ihp)
  Ord: 16.6.1854. K: 29.6.1854. L: 5.9.1854. C: 28.3.1855.
  First cost: £8,730 (including £3,397 for hull, £3,481 for machinery).
  Commissioned 3.1855 under Lieut Keith Stewart, for the Baltic (during the Russian War) as tender to *Blenheim*; on 28.11.1855 under Lieut Alfred Frederik Marescaux, at Portsmouth, still tender to *Blenheim*; paid off there 2.7.1856. BU completed 17.2.1864 there.

Badger [ex-*Ranger*, renamed before launching] W. & H. Pitcher, Northfleet/John Penn & Sons (215ihp)
  Ord: 16.6.1854. K: 6.1854. L: 23.9.1854. C: 26.3.1855.
  First cost: £7,526 (including £2,761 for hull, £3,363 for machinery).
  Commissioned 6.8.1855 under Lieut William Henry Cuming, for the Baltic (during the Russian War) as tender to *Hogue*; on 12.1.1856 under Lieut Thomas Thelwall Bullock; paid off 18.7.1856. In 3.1858 at Kingstown (Dublin) for coastguard (reserve), as tender to *Ajax*; stored in Gunboat Shed at Haslar from 9.1860. BU 6.1864 at Portsmouth.

Snapper W. & H. Pitcher, Northfleet/John Penn & Sons
  Ord: 16.6.1854. K: 6.1854. L: 4.10.1854. C: 27.3.1855.
  First cost: £8,462 (including £3,304 for hull, £3,481 for machinery).
  Commissioned 20.4.1855 under Lieut Arthur Julian Villiers, for the Baltic (during the Russian War) as tender to *Edinburgh*; paid off 8.1856 at Portsmouth into Steam Reserve. Fitted as coal hulk at Portsmouth 9.1863, sold 1906.

Ruby Deptford Dyd/John Penn & Sons
  Ord: 16.6.1854. K: 29.6.1854. L: 7.10.1854. C: 8.3.1855.
  First cost: £9,840 (including £5,679 for hull, £3,277 for machinery).
  Commissioned 24.7.1855 under Lieut Henry George Hale, for the Baltic (during the Russian War) as tender to *Majestic*; paid off 10.9.1856 at Sheerness. Recommissioned 26.6.1858 under Lieut Robert John Stotherd, as tender to *Cornwallis*, on the Humber; in 2.1859 under Lieut William Vicary, as tender to *Dauntless*, on the Humber. BU 10.1868 at Devonport.

Gleaner Deptford Dyd/John Penn & Sons
  Ord: 16.6.1854. K: 29.6.1854. L: 7.10.1854. C: 28.3.1855.
  First cost: £10,285 (including £5,328 for hull, £3,481 for machinery).
  Commissioned 22.9.1855 under Lieut Archibald George Bogle, for the Baltic (during the Russian War) as tender to *Ajax*; paid off 29.8.1856 at Portsmouth. Recommissioned 21.4.1865 under Lieut Frederick Hardy, for South America; in 9.5.1866 under Lieut Charles Frederick Hill; paid off 31.3.1868. Sold 4.1868 at Montevideo.

DAPPER Class Wooden 'Crimean' gunboats. W. H. Walker design, slightly enlarged from the *Gleaner* Class. The design was approved and twenty vessels were authorised on 6 October 1854. Two were awarded to Deptford Dyd and the rest were to be built by contract, authorised on 9 October, of which twelve were by Pitcher at Northfleet. Names were issued on 14 October. Half (ten vessels) were engined by Maudslay, Sons & Field; and half by John Penn & Sons. On completion, fourteen were deployed to the Baltic campaign; the remaining six went to the Black Sea, and all these took part in the Kertch campaign and in the Sea of Azov.
  Dimensions & tons: 106ft 0in, 93ft 2½in x 22ft 0in oa (21ft 8in for tonnage) x 8ft 0in. 6¾ft draught. 232⁶⁸/₉₄bm. 284 displacement.
  Men: 36. Guns: 1 x 68pdr (95cwt/10ft) MLSB shell (on pivot, aft); 1 x 32pdr (56cwt/9½ft) MLSB (on pivot, fwd, replacing planned

second 68pdr); 2 x 24pdr howitzers (on broadside trucks).
  Machinery: 1 set horizontal single expansion, direct-acting (trunk in Penn-engined vessels). 3 cylindrical boilers (2 in Penn vessels). Single (non-hoisting) screw. 60nhp. 270ihp = 7½kts. 25 tons coal.

Lark Deptford Dyd/Maudslay, Sons & Field (210ihp)
  Ord: 6.10.1854. K: 5.10.1854. L: 15.3.1855. C: 3.4.1855.
  First cost: £10,497 (including £6,334 for hull, £3,450 for machinery).
  Commissioned 3.1855 under Lieut George Teal Sabor Winthorp, as tender to *Exmouth*. On 9.4.1855 under Lieut Mark Robert Pechell, to the Baltic as tender to *Exmouth*. On 12.1.1856 under Lieut William Henry Cuming; paid off 18.7.1856 at Plymouth. Became tender to *Wellington* (guard ship of the Reserve) 1859–61; at Queenstown (Ireland) from 11.1861 for coastguard (reserve) as tender to *Hawke* until 1865, then tender to *Mersey* 1866–71. From 1871 in Steam Reserve at Plymouth. Sold there to Marshall 18.7.1878 to BU.

Magpie Deptford Dyd/Maudslay, Sons & Field (210ihp)
  Ord: 6.10.1854. K: 5.10.1854. L: 15.3.1855. C: 3.4.1855.
  First cost: £10,300 (including £6,245 for hull, £3,450 for machinery).
  Commissioned 16.3.1855 under Lieut Bedford Clapperton Tryvellion Pim, to the Baltic as tender to *Orion*; paid off 3.9.1856. From 3.1858 at Queenstown (Cork) for coastguard reserve, as tender to *Hawke*. Wrecked in Galway Bay 8.4.1864.

Dapper R. & H. Green, Blackwall/John Penn & Sons (272ihp)
  Ord: 6.10.1854. K: 10.1854. L: 31.3.1855. C: 19.4.1855.
  First cost: £9,990 (including £4,179 for hull, £3,567 for machinery).
  Commissioned 3.1855 under Lieut Augustus Thomas John Bullock, to the Baltic as tender to *Colossus*. On 22.3.1855 under Lieut Henry Lewis Round, then Lieut Henry James Grant, to the Baltic; then 10.1855 under Lieut Edward Henry Cecil, in the Baltic, then 7.1.1856 under Lieut Hugh McNeile Dyer, still tender to *Colossus*; paid off 4.9.1856. From 3.1858 coastguard (reserve) vessel at Liverpool, as tender to *Hastings*, later to *Majestic*. Became tender to training ship *Britannia* at Dartmouth from 1864. Cooking depot 1897. Renamed *YC.37* in 1909. Sold to Perry 10.5.1922 to BU.

Fancy R. & H. Green, Blackwall/John Penn & Sons (262ihp)
  Ord: 6.10.1854. K: 10.1854. L: 31.3.1855. C: 16.4.1855.
  First cost: £10,617 (including £4,679 for hull, £3,887 for machinery).
  Commissioned 3.1855 under Lieut Charles Gerveys Grylls, as tender to *Cornwallis*; on 11.5.1855 to the Black Sea, as tender to *Royal Albert*; paid off 16.5.1867 into Steam Reserve at Portsmouth. Major refit at Portsmouth 1861–62 ('scarcely one of her original timbers will remain in her'). From 1865–76 tender to *Asia*, guard ship of the Reserve at Portsmouth (used as trials ship for new propeller designs in 1867). Engines removed 2.1876 and converted to laundry/drying room as part of the *St Vincent* training establishment 1876. Sold 11.7.1905 at Portsmouth.

Grinder T. & J. White, West Cowes/Maudslay, Sons & Field
  Ord: 6.10.1854. K: 13.10.1854. L: 7.3.1855. C: 17.5.1855.
  First cost: £10,624 (including £4,084 for hull, £3,567 for machinery).
  Commissioned 11.5.1855 under Lieut Francis Trevor Hamilton, to the Black Sea as tender to *Royal Albert*. On 16.7.1855 under Lieut Hugh Talbot Burgoyne, in the Black Sea. Steam Reserve at Portsmouth 1856–64. BU 15.7.1864 at Haslar.

Jasper T. & J. White, West Cowes/Maudslay, Sons & Field (205ihp)
  Ord: 6.10.1854. K: 13.10.1854. L: 2.4.1855. C: 7.5.1855.
  First cost: £10,326 (including £3,963 for hull, £3,440 for machinery).
  Commissioned 3.1855 under Lieut William Henderson Truscott, intended for the Baltic as tender to *Hastings*; but went to Black Sea instead. On 16.7.1855 under Lieut Joseph Samuel Hudson, in

the Black Sea; grounded in action and lost 23.7.1855 at Taganrog, Sea of Azov.

Hind John Jenkins Thompson, Rotherhithe/Maudslay, Sons & Field (216ihp)
> Ord: 6.10.1854. K: 18.10.1854. L: 3.5.1855. C: 4.6.1855.
> First cost: £10,324 (including £4,471 for hull, £3,475 for machinery).
> Commissioned 25.5.1855 under Lieut John Ward, to the Baltic as tender to *Cornwallis*; paid off 29.10.1855. From 2.1858 coastguard reserve vessel at Falmouth, as tender to *Russell* 1858–64, then to *St George* 1864–69. Move to Liverpool 1870, as tender to *Mersey* to 1871. BU 10.1872 at Devonport.

Jackdaw John Jenkins Thompson, Rotherhithe/Maudslay, Sons & Field (207ihp)
> Ord: 6.10.1854. K: 18.10.1854. L: 18.5.1855. C: 7.6.1855.
> First cost: £10,016 (including £4,189 for hull, £3,475 for machinery).
> Commissioned 10.9.1855 under Lieut Joshua Berkeley, to the Baltic as tender to *Duke of Wellington*. On 8.2.1856 under Lieut William Swinburn. From 2.1858 coastguard vessel at Greenock, as tender to *Hogue* 1858–62. Became a cooking depot in 1868. Sold to C. Wort 11.1888.

Thistle W. & H. Pitcher, Northfleet/John Penn & Sons
> Ord: 6.10.1854. K: 25.10.1854. L: 3.2.1855. C: 7.4.1855.
> First cost: £?9,136 (including £4,264 for hull, £3,065 for machinery).
> Commissioned 16.3.1855 under Lieut David Spain, to the Baltic as tender to *Royal George*; paid off 29.12.1856 into Steam Reserve. Completed BU 11.11.1863 at Deptford.

Starling W. & H. Pitcher, Northfleet/John Penn & Sons
> Ord: 6.10.1854. K: 28.10.1854. L: 1.2.1855. C: 7.4.1855.
> First cost: £9,820 (including £4,264 for hull, £3,575 for machinery).
> Commissioned 16.3.1855 under Lieut Shute Barrington Piers, to the Baltic as tender to *Duke of Wellington*. Recommissioned 9.9.1856 under Lieut Arthur Julian Villiers, for China; wrecked in attack on Peiho River forts, but salved. Recommissioned 7.12.1858 under Lieut James Hawkins Whitshed, for the East Indies and China; paid off 2.1.1861 and laid up at Hong Kong. Recommissioned 4.5.1867 under Lieut Francis Dent; in 9.1870 under Lieut Augustus H. B. Bradshaw; paid off 23.12.1870. Sold 1.12.1871 at Hong Kong.

Snap W. & H. Pitcher, Northfleet/John Penn & Sons
> Ord: 6.10.1854. K: 3.11.1854. L: 3.2.1855. C: 7.4.1855.
> First cost: £9,926 (including £4,264 for hull, £3,569 for machinery).
> Commissioned 3.1855 under Lieut Charles Arthur Wise, to the Baltic as tender to *Edinburgh* and *James Watt.*; in 1856 under Lieut Claude Augustus Champion De Crespigny; paid off 9.6.1856. On 1.3.1858 under Lieut William Frederick Johnson, as tender to *Hogue* at Greenock. Recommissioned 9.1859 under Lieut William Ormonde Butler, for East Indies and China. In 1861 under Lieut Horatio Pack; paid off 1.1863 and laid up at Hong Kong. Recommissioned 23.1.1866 under Lieut George Powys; paid off 19.7.1867. Sold 1868 at Hong Kong and then resold to the Japanese as the warship *Kaku-ten-shan* (presumably for one of the rebellious clans, as no such vessel appears in the main lists of the Imperial Navy), then 1872 became *Snap* again as a merchantman.

Redwing W. & H. Pitcher, Northfleet/Maudslay, Sons & Field (210ihp)
> Ord: 6.10.1854. K: 3.11.1854. L: 19.3.1855. C: 11.4.1855.
> First cost: £10,046 (including £4,264 for hull, £3,497 for machinery).
> Commissioned 26.3.1855 under Lieut Charles Stuart Forbes, to the Baltic as tender to *Nile*; briefly under William Greenhill Silverlock (mate) for bombardment of Bomarsund 8.1855.

Became tender to the training ship *Cambridge* at Devonport 1857. Sold 2.12.1878 at Devonport.

Weazel W. & H. Pitcher, Northfleet/John Penn & Sons
> Ord: 6.10.1854. K: 3.11.1854. L: 19.3.1855. C: 14.4.1855.
> First cost: £10,267 (including £4,264 for hull, £3,568 for machinery).
> Commissioned 16.3.1855 under Lieut Robert George Craigie, to the Baltic as tender to *Caesar*; paid off 7.1856 and laid up at Plymouth. Recommissioned 20.9.1859 under Lieut William Howarth, for the East Indies and China; on 1.1.1863 under Lieut Henry Hale, then 10.4.1867 under Lieut William Richards; paid off 30.6.1869. Sold 18.11.1869 at Hong Kong.

Clinker W. & H. Pitcher, Northfleet/John Penn & Sons
> Ord: 6.10.1854. K: 4.12.1854. L: 2.4.1855. C: 21.4.1855.
> First cost: £9,217 (including £4,264 for hull, £3,575 for machinery).
> Commissioned 28.3.1855 under Lieut Marcus Smithett, to the Black Sea as tender to *Royal Albert*; on 25.8.1855 under Lieut Joseph Samuel Hudson, still in the Black Sea and Sea of Azov; in the Mediterranean in 1856; paid off 1.1857 at Woolwich. Moved to Plymouth 5.1858. From 11.1861 tender to *Hibernia*, flag of Admiral Superintendent at Malta Dockyard. Sold to Castle 6.6.1871 for BU at Charlton.

Cracker W. & H. Pitcher, Northfleet/Maudslay, Sons & Field (207ihp)
> Ord: 6.10.1854. K: 16.12.1854. L: 2.4.1855. C: 21.4.1855.
> First cost: £10,159 (including £4,264 for hull, £3,628 for machinery).
> Commissioned 26.3.1855 under Lieut Joseph Henry Marryat, to the Black Sea as tender to *Royal Albert*; in the Mediterranean in 1856; paid off 14.3.1857 at Portsmouth and laid up. BU completed 12.4.1864.

Boxer W. & H. Pitcher, Northfleet/John Penn & Sons
> Ord: 6.10.1854. K: 5.1.1855. L: 7.4.1855. C: 27.4.1855.
> First cost: £9,423 (including £4,264 for hull, £3,575 for machinery).
> Commissioned 3.4.1855 under Lieut Samuel Philip Townsend, to the Black Sea as tender to *Royal Albert*; Mediterranean fleet in 1856; in 1.1.1859 under Lieut James Graham. From 11.1861 tender to *Hibernia*, flag of Admiral Superintendent at Malta Dockyard. BU 10.1865 at Malta.

Stork W. & H. Pitcher, Northfleet/Maudslay, Sons & Field (233ihp)
> Ord: 6.10.1854. K: 19.1.1855. L: 7.4.1855. C: 20.5.1855.
> First cost: £9,709 (including £4,264 for hull, £3,497 for machinery).
> Commissioned 24.5.1855 under Lieut George John Malcolm, to the Baltic as tender to *Pembroke*; tender to *Excellent*, as gunnery training ship from 6.1856. Coal hulk 1874. Sold 4.1884 to BU.

Skylark W. & H. Pitcher, Northfleet/John Penn & Sons (225ihp)
> Ord: 6.10.1854. K: 19.1.1855. L: 3.5.1855. C: 25.5.1855.
> First cost: £9,925 (including £4,264 for hull, £3,497 for machinery).
> Commissioned 25.5.1855 under Lieut Frederick Whiteford Pym, to the Baltic as tender to *Russell*; by 4.1856 under Lieut Francis Alexander Hume; paid off 16.6.1856 and stored in Gunboat Shed at Haslar from 30.1.1857. Refitted at Portsmouth in 1861 to replace unseasoned timber. Recommissioned 14.6.1865 under Lieut Robert Swinton for the Mediterranean (stationed at Gibraltar). On 12.9.1865 under Lieut James Prevost, then 9.1868 under Lieut Aretas Collins; paid off 4.9.1869 at Portsmouth. From 8.1870 tender to *Excellent* as gunner training ship. Gunnery tender 1884. Sold to Garnham 10.7.1906 for BU.

Biter W. & H. Pitcher, Northfleet/John Penn & Sons
> Ord: 6.10.1854. K: 5.2.1855. L: 5.5.1855. C: 25.5.1855.
> First cost: £9,882 (including £4,264 for hull, £3,561 for machinery).
> Commissioned 28.5.1855 under Lieut Warren Hastings Anderson, to the Baltic as tender to *Hawke*: paid off 8.1856 and laid up at Portsmouth. From 3.1858 coastguard (reserve) vessel at

The *Dapper* Class screw gunboat *Cracker* passing within 600 yards of the Kinburn shore batteries on the morning of 15 October 1855, as drawn by Edward Wolfe Brooker, Master aboard the gunboat at the time. *Cracker* and her sisters *Fancy*, *Boxer* and *Clinker*, together with four French gunvessels (*Alerte* Class) *Tirailleuse*, *Stridente*, *Mutine* and *Meurtrière*, entered the estuary of the Dnieper on the previous day, and *Cracker* laid a series of white marker buoys. On 17 October the expedition commander, Rear-Adm. Sir Houston Stewart, transferred his flag to the *Cracker* to supervise the bombardment of the coastal fort, which surrendered later that day. *(NMM PW4907)*

Weymouth, initially as tender to *Blenheim*, later to *Colossus*. Became a coal hulk 21.4.1865, later renamed *C.16*. Sold to Castle, Woolwich 12.4.1904 to BU.

Swinger W. & H. Pitcher, Northfleet/Maudslay, Sons & Field (217ihp)
Ord: 6.10.1854. K: 5.2.1855. L: 10.5.1855. C: 7.6.1855.
First cost: £10,104 (including £4,264 for hull, £3,497 for machinery).
Commissioned 25.5.1855 under Lieut Maurice Horatio Nelson, to the Baltic as tender to *Hastings*; paid off 13.9.1856 into Steam Reserve at Portsmouth. BU 6.9.1864.

ALBACORE Class Wooden 'Crimean' gunboats. W. H. Walker design, approved 18 April 1855. Four vessels were ordered on 18 April, six on 2 May, fourteen on 30 June and twenty-four on 26 July; a second batch of fifty was ordered during early October; of the latter, the original *Surge* and *Hound* were renamed *Lively* and *Mastiff* in 1855. Exactly half (49 vessels) were engined by Maudslay, Sons & Field; and half by John Penn & Sons. Of the October batch (50 vessels), all vessels from Smith, Lairds, Hill and Patterson yards (also *Cochin* from Green's) were built under cover.
Dimensions & tons: 106ft 0in, 93ft 2½in x 22ft 0in oa (21ft 8in for tonnage) x 8ft 0in. 6½ft draught. 232⁶⁸⁄₉₄bm. 284 displacement.
Men: 36–40. Guns: 1 x 68pdr (95cwt/10ft) MLSB shell (on pivot, aft); 1 x 32pdr (56cwt/9½ft) MLSB (on pivot, fwd, replacing planned second 68pdr); 2 x 24pdr howitzers (on broadside trucks).
Machinery: 1 set horizontal single expansion, direct-acting (trunk in Penn-engined vessels). 3 cylindrical boilers (2 in Penn vessels). Single (non-hoisting) screw. 60nhp. 203–233ihp = 7½kts. 25 tons coal.
Beaver Wigram & Son, Northam/John Penn & Sons
Ord: 18.4.1855. K: 28.5.1855. L: 28.11.1855. C: 31.5.1856.
First cost: £10,629 (£4,159 for hull, £3,611 for machinery, £2,859 fitting).

Commissioned 10.1.1856 under Lieut Anthony Hiley Hoskins; paid off 5.9.1856 at Portsmouth into Steam Reserve. BU 1864 at Portsmouth.
Whiting Wigram & Son, Northam/John Penn & Sons (183ihp)
Ord: 18.4.1855. K: 28.5.1855. L: 9.1.1856. C: 8.2.1856.
First cost: £9,538 (£4,140 for hull, £3,611 for machinery, £1,787 fitting).
Commissioned 14.2.1856 under Lieut Iltid Thomas Mansel Nicholl; paid off 2.7.1856 at Portsmouth into Steam Reserve. Stored in Gunboat Shed at Haslar from 10.3.1857. A survey in 7.1860 reported her 'sap and rot, fore and aft; not a sound timber in her'; refitted (re-timbered and planked) at Portsmouth 1861. In 3.1866 to Queenstown, as tender initially to *Black Prince*, later to *Mersey*; paid off 24.9.1869. In 8.1870 to Devonport, as tender to *Indus*. BU 12.1881.
Nightingale C. J. Mare & Co., Blackwall/Maudslay, Sons & Field (200ihp)
Ord: 18.4.1855. K: 1.6.1855. L: 22.12.1855. C: 11.3.1856.
First cost: £10,521 (including £4,310 for hull, £3,537 for machinery).
Commissioned 2.1856 under Lieut Tathwell Benjamin Collinson; paid off 12.6.1856 at Plymouth into Steam Reserve. Sold to W. Lethbridge 16.7.1867 to BU.
Violet C. J. Mare & Co., Blackwall/Maudslay, Sons & Field (206ihp)
Ord: 18.4.1855. K: 1.6.1855. L: 9.1.1856. C: 18.3.1856.
First cost: £10,374 (including £4,236 for hull, £3,537 for machinery).
Commissioned 15.2.1856 under Lieut Henry Bedford Woollcombe, as tender to *Brunswick*; paid off 6.1856 at Plymouth into Steam Reserve. In 3.1858 under Lieut William Bush, as tender at Pembroke to *Eagle*, for coastguard (reserve). On 12.6.1860 under Lieut Eugene Chambers Batty, as tender to *Blenheim* at Pembroke Dock; paid off 8.10.1861 at Plymouth. Sold to Marshall 7.10.1864 to BU at Plymouth.
Seagull W. & H. Pitcher, Northfleet/Maudslay, Sons & Field (212ihp)
Ord: 2.5.1855. K: 1.5.1855. L: 4.6.1855. C: 22.12.1855.
First cost: £10,454 (including £4,369 for hull, £3,573 for machinery).
Commissioned 30.11.1855 under Lieut Montagu Frederic O'Reilly, as tender to *Porcupine* for surveying duties from 16.5.1856; paid off 25.1.1857. Refitted at Sheerness for surveying duties 2–4.1857. In 4.1857 under William Stanton, Master. On 24.5.1858 under Lieut William Chimmo, for surveying duties, as tender to

Porcupine; paid off 15.1.1862. Sold to Marshall 7.10.1864 to BU at Plymouth.

Skipjack W. & H. Pitcher, Northfleet/John Penn & Sons (265ihp)
 Ord: 2.5.1855. K: 8.5.1855. L: 4.8.1855. C: 10.12.1855.
 First cost: £10,382 (including £4,396 for hull, £3,516 for machinery).
 Commissioned 30.11.1855 under Lieut Henry Weyland Chetwynd (–1856); paid off 1.8.1856 at Plymouth into Steam Reserve. Recommissioned 13.8.1857 under Lieut John Murray, for North America and the West Indies; in 1861 under Lieut Horatio Packe; paid off 6.11.1861 at Plymouth. Boilers removed in 1862, became a cooking depot in 1874. BU at Devonport completed 4.2.1879.

Sandfly W. & H. Pitcher, Northfleet/Maudslay, Sons & Field (205ihp)
 Ord: 2.5.1855. K: 16.5.1855. L: 1.9.1855. C: 4.1.1856.
 First cost: £10,327 (including £4,492 for hull, £3,507 for machinery).
 Commissioned 30.11.1855 under Lieut Beville Granville Wyndham Nicolas; paid off 10.9.1856 at Sheerness. Tender to *Cornwallis* 121.1858, coastguard and reserve depot at Hull 12.1858; moved to Queenstown 4.1862, as tender to *Hawke*. On 6.6.1866 under Lieut Charles Edward Foot, as tender to *Black Prince* at Queenstown; paid off 10.1867 at Plymouth. Sold to W.Lethbridge 5.11.1867 to BU.

Sheldrake W. & H. Pitcher, Northfleet/John Penn & Sons
 Ord: 2.5.1855. K: 18.5.1855. L: 1.9.1855. C: 16.12.1855.
 First cost: £10,378 (including £4,527 for hull, £3,566 for machinery).
 Commissioned 12.1855 under Lieut Cortland Herbert Simpson; paid off 12.6.1856, to Steam Reserve at Plymouth. Recommissioned 29.11.1861 under Lieut Edward Spencer Meara, for South America; on 12.5.1863 under Lieut John Nott; paid off 20.6.1865. Sold 30.6.1865 at Montevideo.

Plover W. & H. Pitcher, Northfleet/Maudslay, Sons & Field (214ihp)
 Ord: 2.5.1855. K: 16.6.1855. L: 8.9.1855. C: 3.1.1856.
 First cost: £10,106 (including £4,500 for hull, £3,587 for machinery).
 Commissioned 30.11.1855 under Lieut Keith Stewart, for the East Indies (including 2nd Anglo-Chinese War); on 26.9.1857 under Lieut Robert James Wynniatt, then 25.6.1859 Lieut William Hector Rason; sunk in action with the Taku forts in the Peiho River 26.6.1859.

Tickler W. & H. Pitcher, Northfleet/John Penn & Sons
 Ord: 2.5.1855. K: 20.6.1855. L: 8.9.1855. C: 26.12.1855.
 First cost: £10,394 (including £4,533 for hull, £3,566 for machinery).
 Commissioned 30.11.1855 under Lieut Charles John Balfour; paid off 12.6.1856 into Steam Reserve at Sheerness. BU at Deptford completed 21.11.1863.

Banterer W. & H. Pitcher, Northfleet/Maudslay, Sons & Field (218ihp)
 Ord: 30.6.1855. K: 12.7.1855. L: 29.9.1855. C: 9.1.1856.
 First cost: £10,469 (including £4,789 for hull, £3,534 for machinery).
 Commissioned 24.12.1855 under Lieut Edward Hardinge; later under Lieut James Hawkins Whitshed; paid off 4.9.1856 into Steam Reserve at Portsmouth. Recommissioned 9.10.1857 under Lieut Bedford Clapperton Tryvellion Pim, for China (including 2nd Anglo-Chinese War). On 10.2.1858 under Lieut John Jenkins, still in China; grounded in action with the Taku forts in the Peiho River 25.6.1859, but refloated. On 20.7.1860 under Lieut Edward Bond Harrison Franklin, for East Indies and China. On 8.9.1864 under Lieut James Stephen Tonkin; on 31.8.1866 under Lieut James Thomas Pringle; paid off 31.12.1870 at Hong Kong. Sold there 30.12.1872.

Bullfrog W. & H. Pitcher, Northfleet/John Penn & Sons (263ihp)
 Ord: 30.6.1855. K: 20.7.1855. L: 6.10.1855. C: 6.1.1856.

First cost: £10,648 (including £4,800 for hull, £3,408 for machinery).
 Commissioned 5.1.1856 under Lieut Cornwallis Wykeham Martin (actually appointed 24.12.1855); paid off 12.6.1856 into Steam Reserve at Sheerness. *Possibly* recommissioned 1.2.1857 under Lieut Frederick William Hallowes, for East Indies and China; if so was soon paid off. Moved to Chatham 8.1860, returned 11.1862 to Sheerness. To North Shields 7.1865 for coastguard reserves, as tender to *Castor*; paid off 3.1875 at Sheerness. BU there 8.6.1875.

Bustard W. & H. Pitcher, Northfleet/Maudslay, Sons & Field (212ihp)
 Ord: 30.6.1855. K: 1.8.1855. L: 20.10.1855. C: 9.1.1856.
 First cost: £10,485 (including £4,796 for hull, £3,224 for machinery).
 Commissioned 1.1856 under Lieut (acting) William Nathan Wrighte Hewett. Later in 1856 under Lieut Richard James Meade. On 4.9.1856 under Lieut Tathwell Benjamin Collinson, for East Indies and China (including 2nd Anglo-Chinese War); fitted for foreign service 1856. On 1.9.1857 under Lieut Frederick William Hallowes, on same station. On 2.12.1862 under Lieut John Collier Tucker, them 30.5.1867 Lieut Cecil Johnson; paid off 30.6.1869. Sold to Cheeong Loong 18.11.1869 at Hong Kong.

Carnation W. & H. Pitcher, Northfleet/John Penn & Sons
 Ord: 30.6.1855. K: 7.8.1855. L: 20.10.1855. C: 16.1.1856.
 First cost: £10,398 (including £4,802 for hull, £3,408 for machinery).
 Commissioned 24.12.1855 (appointment) under Lieut Philip Saumarez; paid off 11.9.1856 into Steam Reserve at Sheerness. Moved to Chatham 8.1860. Condemned 4.1863 as 'rotten' and BU same year at Sheerness.

Charger W. & H. Pitcher, Northfleet/Maudslay, Sons & Field (215ihp)
 Ord: 30.6.1855. K: 9.8.1855. L: 13.11.1855. C: 24.1.1856.
 First cost: £?9,315 (including £4,808 for hull, £3,334 for machinery).
 Commissioned 12.1.1856 under Lieut Edwin Charles Symons; paid off at Portsmouth 2.7.1856. Stored in gunboat shed at Haslar from 28.2.1857. Survey in 6.1860 condemned her as largely rotten – 'as bad a state from sap as can e … planking badly worked . . . fingers may be freely passed between her bilge streak and timbers'. Large refit ('almost a rebuild') at Portsmouth 1860–61, then back to reserve in gunboat shed. Fitted with experimental Curtis steering screw at Young & Mangay, Limehouse 6.1863; trials conducted 9.1863, then to Steam Reserve at Portsmouth. Towed by *Barracouta* 1.5.1866 from Plymouth to Halifax; became a buoy boat 6.1866 at Halifax – later renamed *YC.3* on 24.6.1866, then *YC.6* in 1869. Sold mercantile 7.1887 as s.s. *Rescue*. BU 1921.

Cockchafer W. & H. Pitcher, Northfleet/John Penn & Sons (288ihp)
 Ord: 30.6.1855. K: 6.9.1855? L: 24.11.1855. C: 26.1.1856.
 First cost: £10,552 (including £4,804 for hull, £3,408 for machinery).
 Commissioned 8.1.1856 under Lieut Edwin Augustus Porcher; paid off 11.6.1856 into Steam Reserve at Plymouth On 20.9.1859 under Lieut Henry Lowe Holder, for East Indies and China; fitted for service in China 1859. On 2.12.1862 under Lieut Edmund Marmaduke Dayrell, then 1.12.1866 under Lieut Edmund Kerr, 5.3.1869 under Lieut Seymour Spencer Smith, and 10.2.1872 under Lieut Henry Clancy (or Clanchy?), all in China; paid off 13.5.1872. Sold to Telge Northing Company 1872 at Shanghai.

Dove W. & H. Pitcher, Northfleet/Maudslay, Sons & Field (215ihp)
 Ord: 30.6.1855. K: 11.9.1855. L: 24.11.1855. C: 24.1.1856.
 First cost: £10,709 (including £4,797 for hull, £3,334 for machinery).
 Commissioned 1.1856 under Lieut Frederick Anstruther Herbert. On 7.10.1856 under Lieut Charles James Bullock, as tender to *Actaeon*, for East Indies and China; recommissioned 20.11.1858,

still under Bullock. On 27.1.1864 under Lieut ?, them 10.6.1867 under Lieut Martin J. Dunlop; paid off 29.1.1873. Sold to P. & O. Steam Navigation Co. 14.4.1873 at Shanghai.

Forward W. & H. Pitcher, Northfleet/John Penn & Sons
Ord: 30.6.1855. K: 13.9.1855. L: 8.12.1855. C: 9.2.1856.
First cost: £10,554 (including £4,798 for hull, £3,408 for machinery).
Commissioned 22.1.1856 under Lieut Charles Gudgeon Nelson (–4.9.1856). On 24.10.1857 under Lieut Henry Davies, for Cuba and the West Indies; paid off 12.11.1858 at Plymouth. On 29.6.1859 under Lieut Charles Rufus Robson, for the Pacific; fitted in 1859 for service in British Columbia; in 12.1860, proceeded from Nootka Sound to the wreck of the US brig *Consort* in San Josef Bay (a feat praised by US President James Buchanan). On 7.11.1861 under Lieut Horace Douglas Lascelles, then 24.4.1866 under Lieut D'Arcy Anthony Denny; paid off 19.8.1869. Sold 28.9.1869 to Hill & Ready, Esquimault.

Grasshopper W. & H. Pitcher, Northfleet/Maudslay, Sons & Field (223ihp)
Ord: 30.6.1855. K: 14.9.1855. L: 12.1.1856. C: 5.2.1856.
First cost: £10,397 (including £4,802 for hull, £3,374 for machinery).
Commissioned 2.1856 under Lieut William Frederick Lee; paid off 2.7.1856 at Portsmouth. On 27.9.1859 under Lieut John Collier Tucker, for the East Indies and China. On 2.12.1862 under Lieut Francis William Bennett, then 23.11.1864 under Lieut George Digby Morant, 17.2.1866 under Lieut John Conyngham Patterson and 5.4.1868 under Lieut Robert E. Stopford, all in China; paid off 2.5.1871. Sold 5.1871 at Newchang.

Hasty W. & H. Pitcher, Northfleet/John Penn & Sons
Ord: 30.6.1855. K: 8.10.1855. L: 10.1.1856. C: 20.2.1856.
First cost: £10,120 (including £4,805 for hull, £3,408 for machinery).
Commissioned 1.1856 under Lieut William Menzies; paid off 12.6.1856 into Steam Reserve at Sheerness. To coastguard reserve 4.1861 at North Shields, as tender to *Castor*. Sold to Castle 11.1865 for BU at Charlton.

Herring W. & H. Pitcher, Northfleet/Maudslay, Sons & Field (225ihp)
Ord: 30.6.1855. K: 8.10.1855. L: 10.1.1856. C: 29.2.1856.
First cost: £10,369 (including £4,791 for hull, £3,224 for machinery).
Commissioned 1.1856 at Woolwich under Lieut Louis Geneste; paid off 28.5.1856 at Sheerness. To coastguard reserve 7.1862 at Queensferry, as tender to *Edinburgh*. Damaged by grounding in Firth of Forth and paid off 4.7.1865. BU 8.1865 at Sheerness.

Insolent W. & H. Pitcher, Northfleet/John Penn & Sons
Ord: 30.6.1855. K: 24.10.1855. L: 26.1.1856. C: 8.3.1856.
First cost: £10,610 (including £4,804 for hull, £3,408 for machinery).
Commissioned 2.1856 under Lieut Charles Harcourt Smith; paid off 19.7.1856. Landed guns 8.1856 and transferred to control of Board of Trade 'to be employed in the construction of a lighthouse on the Great Basses rocks off Ceylon'. Returned at Hong Kong 8.1861 and refitted for Naval service. Recommissioned 8.12.1861 under Lieut George Parsons; on 11.12.1862 under Lieut Granville Toup Nicholas (re-appointed 22.11.1864); on 20.4.1867 under Lieut Leicester Chantrey Keppel; paid off 31.3.1869. Sold 1.5.1869 at Chefoo in China.

Mayflower W. & H. Pitcher, Northfleet/Maudslay, Sons & Field (203ihp)
Ord: 30.6.1855. K: 1.11.1855. L: 31.1.1856. C: 17.3.1856.
First cost: £10,372 (including £4,746 for hull, £3,334 for machinery).
Commissioned 2.1856 under Lieut Grenville Marshall Temple; paid off 10.9.1856 into Steam Reserve at Sheerness. Became tender to *President* in 8.1862 at Barking Creek. BU 8/9.1867 at Sheerness.

Staunch W. & H. Pitcher, Northfleet/John Penn & Sons

Ord: 30.6.1855. K: 2.11.1855. L: 31.1.1856. C: 12.3.1856.
First cost: £10,722 (including £4,783 for hull, £3,410 for machinery).
Commissioned 15.2.1856 under Lieut Leveson Wildman, for the East Indies; fitted for foreign service 9.1856. On 14.11.1858 under Lieut Edwin John Pollard, for the East Indies and China (including 2nd Anglo-Chinese War): between 30.7.1862 and 3.8.1862 stuck on a rock at Canton but managed to get off. Recommissioned 15.5.1863 under Lieut John Smith Keats; paid off 31.12.1865. Sold 12.1866 at Hong Kong and BU.

Charon W. & H. Pitcher, Northfleet/John Penn & Sons
Ord: 26.7.1855. K: 14.11.1855. L: 9.2.1856. C: 19.3.1856.
First cost: £10,566 (including £4,790 for hull, £3,560 for machinery).
Commissioned 2.1856 under Lieut Edwin John Pollard; paid off 9.6.1856 into Steam Reserve at Plymouth. To Pembroke Dock 1.1.1862, as tender to *Blenheim*, coastguard reserve. BU 10.1865 by Marshall at Plymouth.

Haughty W. & H. Pitcher, Northfleet/John Penn & Sons
Ord: 26.7.1855. K: 20.11.1855. L: 9.2.1856. C: 18.3.1856.
First cost: £10,810 (including £4,805 for hull, £3,560 for machinery).
Commissioned 28.2.1856 under Lieut Richard Vesey Hamilton, for China. On 30.7.1858 under Lieut George Doherty Broad, for East Indies and China (including 2nd Anglo-Chinese War). On 2.12.1862 under Lieut John Campion Wells, on same station; paid off 13.5.1867. Sold 23.5.1867 at Hong Kong.

Leveret W. & H. Pitcher, Northfleet/John Penn & Sons
Ord: 26.7.1855. K: 10.11.1855. L: 8.3.1856. C: 8.4.1856.
First cost: £10,363 (including £4,797 for hull, £3,560 for machinery).
Commissioned 16.3.1856 under Lieut William Codrington; paid off 13.9.1856. To Southampton 3.1857 for coastguard reserve, as tender to *Arrogant*, then 8.1859 tender to *Dauntless*; but in 9–10.1858 under Lieut Annesley Turner Denham, for Fishery Protection in North Sea. BU 10.1867 at Portsmouth.

Mackerel W. & H. Pitcher, Northfleet/John Penn & Sons
Ord: 26.7.1855. K: 10.11.1855. L: 8.3.1856. C: 8.4.1856.
First cost: £10,437 (including £4,799 for hull, £3,560 for machinery).
Commissioned 26.3.1856 under Lieut Edmund Francis Weld; paid off 2.7.1856 at Portsmouth. At Haslar in Gunboat Shed from 12.2.1857. Condemned by survey 1.1862 as rotten; BU completed at Portsmouth on 18.7.1862.

Procris W. & H. Pitcher, Northfleet/Maudslay, Sons & Field (219ihp)
Ord: 26.7.1855. K: 19.12.1855. L: 13.3.1856. C: 12.4.1856.
First cost: £10,311 (including £4,806 for hull, £3,315 for machinery).
Commissioned 31.3.1856 under Lieut St George Caulfield D'Arcy Irvine; paid off 14.6.1856 into Steam Reserve at Plymouth. Recommissioned 9.3.1861 under Lieut John Carnegie at Plymouth, 'for the service of the Governor of Gibraltar'. On 11.7.1861 under Lieut George Palmer, then 24.6.1862 under Lieut John Brabazon Vivian; paid off 9.9.1865 at Plymouth. Cooking depot ship at Devonport 30.6.1869. Sold to T. Hockling 31.5.1893 and BU at Stonehouse.

Shamrock W. & H. Pitcher, Northfleet/John Penn & Sons
Ord: 26.7.1855. K: 19.12.1855. L: 13.3.1856. C: 9.4.1856.
First cost: £10,770 (including £4,774 for hull, £3,560 for machinery).
Commissioned 5.1856 under Lieut William Andrew James Heath; in 6.1856 under Lieut Charles James Bullock, for surveying on the coast of Ireland. On 1.4.1858 under William B. Carver, Master, as tender to *Fisgard* at Woolwich; paid off 6.11.1858. Recommissioned 13.4.1859 at Plymouth under Cmdr Joseph Edge, for surveying (–30.5.1859). On 26.7.1860 under Lieut Edward Wolfe Brooker, for surveying, as tender to *Fisgard* at Woolwich. On 1.1.1863 under Capt. Henry Charles Otter, for

surveying, 'carried on the books of *Fisgard*'; paid off 22.9.1866. Sold to Marshall 4.1867 to BU at Plymouth.

Spey W. & H. Pitcher, Northfleet/John Penn & Sons
Ord: 26.7.1855. K: 1.2.1856. L: 29.3.1856. C: 14.4.1856.
First cost: £10,399 (including £4,792 for hull, £3,560 for machinery).
Commissioned 7.4.1856 under Lieut William Gore (or William Henry?) Annesley; paid off 6.1856 into Steam Reserve at Sheerness. BU 12.1863 at Deptford.

Tilbury W. & H. Pitcher, Northfleet/John Penn & Sons
Ord: 26.7.1855. K: 1.2.1856. L: 29.3.1856. C: 14.4.1856.
First cost: £10,448 (including £4,779 for hull, £3,569 for machinery).
Commissioned 7.4.1856 under Lieut ? Wells; paid off 7.6.1856 into Steam Reserve at Plymouth. BU 2.8.1865 by Marshall at Plymouth.

Fervent R. & H. Green, Blackwall/Maudslay, Sons & Field (209ihp)
Ord: 26.7.1855. K: 26.8.1855. L: 23.1.1856. C: 24.3.1856.
First cost: £10,551 (including £4,942 for hull, £3,283 for machinery).
Commissioned 6.1856 by Lieut Ernest Augustus Travers Stubbs (appointed 18.2.1856). Also during 1856 (?before Stubbs was available) under Lieut Alfred Mitchell, but paid off 16.6.1856 into Steam Reserve at Portsmouth; stored in Haslar gunboat shed from 24.2.1857. Survey in 6.1860 showed her to be in very poor condition, with 'timbers showing sap and rot fore and aft'; large refit at Portsmouth 1861. To Plymouth 11.1862, then 11.1864 to Bristol, for reserves training, as tender to *Daedalus*. BU 2.1879 at Devonport.

Forester R. & H. Green, Blackwall/Maudslay, Sons & Field (223ihp)
Ord: 26.7.1855. K: 26.8.1855. L: 23.1.1856. C: 24.3.1856.
First cost: £10,327 (including £4,894 for hull, £3,283 for machinery).
Commissioned 20.2.1856 under Lieut Arthur John Innes, for East Indies & China (including 2nd Anglo-Chinese War); believed laid up 1863 at Hong Kong. Recommissioned 16.4.1866 under Lieut John Edward Stokes, for China; paid off 30.9.1867. Became yard craft *YC.7* at Hong Kong in 1868 and was lost in a hurricane at Hong Kong on 2.9.1871.

Griper R. & H. Green, Blackwall/Maudslay, Sons & Field (219ihp)
Ord: 26.7.1855. K: 26.8.1855. L: 11.12.1855. C: 22.2.1856.
First cost: £10,363 (including £4,851 for hull, £3,283 for machinery).
Commissioned 4.3.1856 under Lieut Morgan Singer; paid off 6.1856 into Steam Reserve at Sheerness. In 6.1861 under Lieut John Smith Keats, as tender to *Cornwallis* on the Humber and was fitted for Armstrong guns in 1862. BU 18.3.1869 at Devonport.

Spanker R. & H. Green, Blackwall/Maudslay, Sons & Field (210ihp)
Ord: 26.7.1855. K: 26.8.1855. L: 22.3.1856. C: 12.4.1856.
First cost: £10,302 (including £4,625 for hull, £3,270 for machinery).
Commissioned 19.3.1856 under Lieut George Stanley Bosanquet; paid off 10.9.1856 at Sheerness. On 1.3.1858 under Lieut Alfred John Chatfield, for coastguard reserve training, as tender to *Cornwallis* on the Humber; on 1.7.1859 under Lieut John Smith Keats. To Steam Reserve at Chatham 1860. BU 8.1874 at Chatham.

Bouncer C. J. Mare & Co., Blackwall/John Penn & Sons
Ord: 26.7.1855. K: 15.8.1855. L: 23.2.1856. C: 3.1856.
First cost: £? (including £? for hull, £3,524 for machinery, £4,631 for fitting).
Commissioned 20.3.1856 under Lieut George Arthur Tyrhitt Drake; paid off 7.1856 into Steam Reserve at Plymouth. Recommissioned 29.9.1859 under Lieut Arthur Rodney Owen, for East Indies & China. On 15.1.1863 under Lieut Henry Lowe Holder; at bombardment of Shimonoseki, then 24.11.1864 under Lieut Frederick William Lewis. On 1.1.1868 under Lieut Rodney Macleane Lloyd, as tender to *Princess Charlotte* at Hong Kong.

Sold 1.2.1871 at Hong Kong.

Hyaena C. J. Mare & Co., Blackwall/John Penn & Sons
Ord: 26.7.1855. K: 15.8.1855. L: 3.4.1856. C: 12.4.1856.
First cost: £10,929 (including £5,207 for hull, £3,504 for machinery).
Commissioned 20.3.1856 under Lieut William Filmer Gregory; paid off 10.6.1856 into Steam Reserve at Plymouth. From 1.1865 to Milford, for coastguard reserve, as tender to *Blenheim*; paid off 11.3.1869. Sold to W. E. Joliffe as a salvage vessel 8.3.1870; BU 1894.

Savage C. J. Mare & Co., Blackwall/John Penn & Sons
Ord: 26.7.1855. K: 15.8.1855. L: 5.5.1856. C: 13.10.1856.
First cost: £?9,100 (including £4,840 for hull, £3,450 for machinery).
Never commissioned; towed to Portsmouth 10.1856 after launch; stored in Haslar gunboat shed from 9.1.1857. Reported as condemned 9.1863, but became mooring lighter *YC.3* at Malta 1864. BU 9.1888 at Malta.

Wolf C. J. Mare & Co., Blackwall/Maudslay, Sons & Field (210ihp)
Ord: 26.7.1855. K: 15.8.1855. L: 5.7.1856. C: 8.2.1856.
First cost: £?9,192 (including £4,879 for hull, £3,285 for machinery).
Never commissioned; towed to Portsmouth 10.1856 after launch; stored in Haslar gunboat shed from 2.2.1857. Completed BU there 8.7.1864.

Thrasher R. & H. Green, Blackwall/Maudslay, Sons & Field (212ihp)
Ord: 26.7.1855. K: 16.8.1855. L: 22.3.1856. C: 12.4.1856.
First cost: £10,281 (including £4,708 for hull, £3,260 for machinery).
Commissioned 19.3.1856 under Lieut George Lydiard Sulivan; paid off 12.6.1856 into Steam Reserve at Sheerness. From 9.1863 tender to *Cumberland* (guard ship of Steam Reserve) at Sheerness. To Southampton 11.1874, for reserves training, as tender to *Zealous*, then 1875 tender to *Hector*, and then 3.1877 tender to *Duke of Wellington* (flagship of C-in-C Portsmouth). Sold by AO of 9.5.1883.

Traveller R. & H. Green, Blackwall/Maudslay, Sons & Field (213ihp)
Ord: 26.7.1855. K: 26.8.1855. L: 13.3.1856. C: 11.4.1856.
First cost: £10,389 (including £4,810 for hull, £3,270 for machinery).
Commissioned 20.3.1856 under Lieut Prince Victor of Hohenlohe-Langenburg; paid off 7.1856 into Steam Reserve at Sheerness. From 2.1858 to Portsmouth, as tender to *Asia*. In 3.1860 to Steam Reserve there. Fitted for Armstrong guns in 1861. Completed BU 28.12.1863 at Portsmouth.

Goldfinch Wigram & Son, Blackwall/Maudslay, Sons & Field (205ihp)
Ord: 26.7.1855. K: 30.7.1855. L: 2.2.1856. C: 15.3.1856.
First cost: £10,552 (including £4,639 for hull, £3,283 for machinery).
Commissioned 3.1856 under Lieut Charles Richard Fox Boxer; paid off 12.6.1856 into Steam Reserve at Plymouth. From 11.1861 to Greenock, for coastguard reserve, as tender to *Hogue*, then 1864 tender to *Lion* (–1868). BU 22.6.1869 at Pembroke.

Goshawk Wigram & Son, Blackwall/Maudslay, Sons & Field (210ihp)
Ord: 26.7.1855. K: 30.7.1855. L: 9.2.1856. C: 17.3.1856.
First cost: £10,459 (including £4,620 for hull, £3,318 for machinery).
Commissioned 2.1856 under Lieut James Graham Goodenough; paid off 6.1856 into Steam Reserve at Plymouth. From 5.1858 to Greenock, for coastguard reserve, as tender to *Hogue*. From 1.1861 to Liverpool, for coastguard reserve, as tender to *Majestic*, then 1864 tender to *Donegal* (–12.1868). BU 18.3.1869 at Devonport.

Opossum Wigram & Son, Northam/Maudslay, Sons & Field (210ihp)
Ord: 26.7.1855. K: 10.9.1855. L: 26.2.1856. C: 31.8.1856.
First cost: £11,376 (including £4,954 for hull, £3,297 for machinery).
Commissioned 28.2.1856 under Lieut Colin Andrew Campbell, for East Indies & China (including 2nd Anglo-Chinese War). On

4.6.1858 under Lieut William Henry Jones; on 28.6.1859 under Lieut Charles John Balfour, including unsuccessful attack on Peiho forts; on 9.8.1859 under Lieut George Dacres Bevan; on 1.5.1860 under Lieut Lindesay Brine, on same station; in 1864 under Lieut John Edward Stokes, then 18.6.1869 under Lieut Alfred Caldcleugh May, and 1.1871 Lieut John Hope; paid off 25.6.1874. Became a hospital hulk in 1876, a mooring vessel in 1891, was renamed *Siren* in 1895 and sold at Hong Kong in 1896.

Partridge Wigram & Son, Northam/Maudslay, Sons & Field (210ihp)
  Ord: 26.7.1855. K: 10.9.1855. L: 29.3.1856. C: 4.1856.
  First cost: £10,904 (including £4,946 for hull, £3,303 for machinery).
  Commissioned 20.3.1856 under Lieut William Henry Jones; paid off 19.7.1856 at Portsmouth. Fitted for reserve, 6.1858 became tender to *Royal Albert*, flagship of the Channel Squadron; in 6.1860 to Steam Reserve at Plymouth. Sold to Messrs. Habgood 8.9.1864 to BU.

Julia Fletcher, Union Dock, Limehouse/John Penn & Sons
  Ord: 26.7.1855. K: 2.8.1855. L: 27.11.1855. C: 25.2.1856.
  First cost: £9,901 (including £3,910 for hull, £3,019 for machinery).
  Commissioned 8.2.1856 under Lieut Charles Henry Clutterbuck; paid off 7.1856 into Steam Reserve at Sheerness. From 3.1862 at Pembroke Dock, for coastguard reserve, as tender to *Blenheim*. BU by Marshall 2.1866 at Plymouth.

Louisa Fletcher, Union Dock, Limehouse/John Penn & Sons
  Ord: 26.7.1855. K: 9.8.1855. L: 12.1855. C: 17.3.1856.
  First cost: £9,868 (including £3,920 for hull, £3,528 for machinery).
  Commissioned 2.1856 under Lieut William Berjew Pauli; paid off 19.7.1856 into Steam Reserve at Sheerness. On 1.3.1858 under Lieut Thomas William Olivier, for coastguard reserve, as tender to *Edinburgh* at Leith. To Steam Reserve at Chatham 6.1862. From 6.1864 at Queensferry (Cork), for coastguard reserve, as tender to *Trafalgar*. Sold to W. Lethbridge 27.8.1867 to BU.

October batch

Peacock W. & H. Pitcher, Northfleet/Maudslay, Sons & Field (215ihp)
  Ord: 3.10.1855. K: 5.2.1856. L: 12.4.1856. C: 22.4.1856.
  First cost: £10,779 (including £4,910 for hull, £3,308 for machinery).
  Commissioned 9.1.1856 under Lieut Henry Barré Beresford, at Portsmouth; paid off there 2.7.1856; stored in Haslar gunboat shed from 23.2.1857. In 1861 stripped of outer planking and timberwork, being kept in frame. BU 25.3.1869 at Portsmouth.

Pheasant W. & H. Pitcher, Northfleet/Maudslay, Sons & Field (215ihp)
  Ord: 3.10.1855. K: 15.2.1856. L: 1.5.1856. C: 31.1.1857.
  First cost: £9,288 (including £4,904 for hull, £3,327 for machinery).
  Never commissioned; towed to Portsmouth after launch; stored in Haslar gunboat shed from 9.2.1857; re-launched 11.1861 and became tender to *Asia*, guard ship of the Reserve at Portsmouth. From 1866 at Southampton, for coastguard reserve, as tender to *Irresistible*, later to *Hector*. BU completed 31.8.1877 at Sheerness.

Primrose W. & H. Pitcher, Northfleet/Maudslay, Sons & Field (215ihp)
  Ord: 3.10.1855. K: 15.2.1856. L: 3.5.1856. C: 3.1.1857.
  First cost: £9,205 (including £4,924 for hull, £3,327 for machinery).
  Never commissioned; towed to Portsmouth after launch; stored in Haslar gunboat shed from 25.2.1857. BU completed 25.5.1864.

Pickle W. & H. Pitcher, Northfleet/Maudslay, Sons & Field (215ihp)
  Ord: 3.10.1855. K: 20.2.1856. L: 3.5.1856. C: 3.1.1857.
  First cost: £8,809 (including £4,500 for hull, £? for machinery).
  Never commissioned; towed to Portsmouth after launch; stored in Haslar gunboat shed from 14.2.1857. BU completed 12.4.1864.

Prompt W. & H. Pitcher, Northfleet/Maudslay, Sons & Field (215ihp)
  Ord: 3.10.1855. K: 25.2.1856. L: 21.5.1856. C: 3.1.1857.
  First cost: £8,840 (including £4,519 for hull, £3,327 for machinery).
  Never commissioned; towed to Portsmouth after launch; stored in Haslar gunboat shed from 11.2.1857. BU completed 6.5.1864.

Porpoise W. & H. Pitcher, Northfleet/Maudslay, Sons & Field (215ihp)
  Ord: 3.10.1855. K: 8.3.1856. L: 7.6.1856. C: 29.9.1860.
  First cost: £8,632 (including £4,505 for hull, £3,027 for machinery).
  Never commissioned; moved to Woolwich after launch, for Steam Reserve. Moved to Plymouth from 5.1858; from 6.1860 tender to flagship of Channel Squadron. BU completed 22.2.1864.

Firm Fletcher & Co., Limehouse/John Penn & Sons (260ihp)
  Ord: 4.10.1855. K: 9.10.1855. L: 22.3.1856. C: 12.4.1856.
  First cost: £10,627 (including £4,626 for hull, £4,052 for machinery).
  Commissioned 20.3.1856 under Lieut Thomas Sherlock Gooch; paid off 19.7.1856 at Portsmouth. Recommissioned 17.3.1857 under Lieut Beville Granville Wyndham Nicolas, for the East Indies and China; on 20.11.1858 under Lieut William Raymond Boulton, on 30.4.1867 under Lieut Horace W. Rochfort, and on 1.4.1869 under Lieut John Hext, all in China; paid off 13.5.1872. Sold 1872 at Shanghai.

Flamer Fletcher & Co., Limehouse/John Penn & Sons (257ihp)
  Ord: 4.10.1855. K: 26.11.1855. L: 10.4.1856. C: 10.11.1859.
  First cost: £12,299 (including £4,623 for hull, £3,512 for machinery).
  Not commissioned until 1859; towed to Portsmouth 10.1856 after launch, for the Steam Reserve. Commissioned 22.9.1859 under Lieut Henry Maynard Bingham, for the East Indies and China (including British involvement in the Taiping Rebellion); on 1.6.1863 under Lieut George Stanley Bosanquet, then 13.11.1863 under Lieut Thomas Sherlock Gooch; paid off 22.1.1866. Coastal defence in 1868. Hospital ship 1871. Blown ashore during a typhoon 22.9.1871 at Hong Kong and the wreck then sold.

Fly Fletcher & Co., Limehouse/John Penn & Sons
  Ord: 4.10.1855. K: 10.12.1855. L: 5.4.1856. C: 5.7.1856.
  First cost: £9,844 (including £4,629 for hull, £3,525 for machinery).
  Never commissioned; moved to Sheerness 7.1856 after launch, for Steam Reserve. From 9.1860 to Hull, for coastguard reserve, as tender to *Cornwallis*. To Sheerness 2.1862 for refit, but found to be 'so exceedingly rotten' that she was BU there 8.1862.

Sepoy T. & W. Smith, Newcastle/Maudslay, Sons & Field (217ihp)
  Ord: 4.10.1855. K: 8.10.1855. L: 13.2.1856. C: 5.4.1856.
  First cost: £10,725 (including £5,667 for hull, £3,298 for machinery).
  Commissioned 4.1856 under Lieut Henry Needham Knox; paid off 9.6.1856 into Steam Reserve at Sheerness. Fitted for Armstrong guns 1861. In 2.1864 to Hull, for coastguard reserve, as tender to *Cornwallis*, later to *Dauntless*. BU 5.1868 at Devonport.

Erne T. & W. Smith, Newcastle/Maudslay, Sons & Field (212ihp)
  Ord: 4.10.1855. K: 8.10.1855. L: 18.2.1856. C: 3.4.1856.
  First cost: £10,743 (including £5,668 for hull, £3,298 for machinery).
  Commissioned 21.5.1856 under Lieut John D'Arcy; paid off 10.9.1856 at Sheerness. In 3.1858 to Leith, for coastguard reserve, as tender to *Edinburgh*. From 6.1862 to Steam Reserve at Chatham. In 7.1865 to Queensferry (Cork), for coastguard reserve, as tender to *Trafalgar*, later to *Duncan*, *Favourite* and finally *Repulse*. BU 1874 at Chatham.

Spider T. & W. Smith, Newcastle/Maudslay, Sons & Field (218ihp)
  Ord: 4.10.1855. K: 13.10.1855. L: 23.2.1856. C: 11.4.1856.
  First cost: £10,385 (including £5,669 for hull, £3,298 for machinery).
  Commissioned 20.3.1856 under Lieut Henry Davies; paid off 14.6.18565 into Steam Reserve at Plymouth. Fitted for

Armstrong guns 1861, and recommissioned 23.11.1861 under
Lieut Ernest Augustus Travers Stubbs, for South America; on
7.6.1865 under Lieut Andrew James Kennedy, then 8.5.1866
under Lieut John Bullock Michell and 29.10.1867 under Lieut
Francis William Prosser; paid off 1.5.1869 at Plymouth. Sold to
Castle 12.5.1870 to BU at Charlton.

Lively T. & W. Smith, Newcastle/Maudslay, Sons & Field (216ihp)
  Ord: 4.10.1855. K: 15.10.1855. L: 23.2.1856. C: 11.4.1856.
  First cost: £10,660 (including £5,666 for hull, £3,298 for machinery).
  Commissioned 20.3.1856 under Lieut John Clarke Byng; paid
    off 6.1856 into Steam Reserve at Sheerness. Recommissioned
    19.2.1859 under Lieut William Vicary, as tender to *Cornwallis* on
    the Humber; wrecked 23.12.1863 on the Dutch coast, later salved
    and became the German mail steamer *Helogolanderin*.

Surly T. & W. Smith, Newcastle/Maudslay, Sons & Field (217ihp)
  Ord: 4.10.1855. K: 15.10.1855. L: 18.3.1856. C: 5.7.1856.
  First cost: £9,867 (including £5,656 for hull, £3,298 for machinery).
  To Steam Reserve 7.1856 at Sheerness. In 4.1861 to Hull, for
    coastguard reserve, as tender to *Cornwallis*, later to *Dauntless*;
    paid off 4.1869. Sold (by public auction, for £1,025) to Thomas J.
    Begbie 21.10.1869.

Swan T. & W. Smith, Newcastle/Maudslay, Sons & Field (215ihp)
  Ord: 4.10.1855. K: 14.11.1855. L: 12.4.1856. C: 2.10.1856.
  First cost: £9,848 (including £5,667 for hull, £3,298 for machinery).
  Never commissioned; towed to Portsmouth after launch; stored in
    Haslar gunboat shed from 25.2.1857. Became a coal hulk in 1869.
    Sold 1906.

Delight Wigram & Son, Blackwall/Maudslay, Sons & Field (218ihp)
  Ord: 4.10.1855. K: 12.10.1855? L: 15.3.1856. C: 5.4.1856.
  First cost: £10,559 (including £5,233 for hull, £3,312 for machinery).
  Commissioned 19.3.1856 under Lieut Henry Maynard
    Bingham; paid off 12.6.1856 into Steam Reserve at Plymouth.
    Recommissioned 7.1864 under Lieut Charles Harford, for North
    America and the West Indies; on 2.3.1866 under Lieut Thomas
    Lacom. Sold at Halifax as mercantile *M. A. Starr* 11.1867.

Grappler Wigram & Son, Blackwall/Maudslay, Sons & Field (211ihp)
  Ord: 4.10.1855. K: 12.10.1855. L: 29.3.1856. C: 12.4.1856.
  First cost: £10,559 (including £5,238 for hull, £3,283 for machinery).
  Commissioned 20.3.1856 under Lieut William Greenhill
    Silverlock; paid off 14.6.1856 into Steam Reserve at Plymouth.
    Recommissioned 29.6.1859 under Lieut Alfred Prowse Hasler
    Helby, for the Pacific; fitted for service in British Columbia
    1859. On 27.2.1862 under Lieut Edmund Hope Verney,
    stationed at Esquimalt (British Columbia); paid off 31.10.1864
    but recommissioned 1.11.1864, still under Verney; paid off
    28.7.1865. Sold for mercantile service at Esqimault 6.1.1868.
    Burnt 3.5.1883. BU 1884.

Growler Wigram & Son, Blackwall/Maudslay, Sons & Field (215ihp)
  Ord: 4.10.1855. K: 12.10.1855. L: 8.5.1856. C: 30.7.1858.
  First cost: £11,067 (including £5,256 for hull, £3,312 for machinery).
  Not commissioned until 1858; at Woolwich from 6.1856; to
    Portsmouth 4.1858 for Steam Reserve. Commissioned 10.5.1858
    under Lieut Henry Edward Crozier, for the Mediterranean; paid
    off 19.12.1861 at Malta; from 1.1862 tender to *Hibernia*, flagship
    there. BU there 8.1864.

Parthian Wigram & Son, Blackwall/Maudslay, Sons & Field (215ihp)
  Ord: 4.10.1855. K: 12.10.1855. L: 8.5.1856. C: 8.12.1856.
  First cost: £9,592 (including £5,250 for hull, £3,314 for machinery).
  Never commissioned; at Woolwich from 6.1856; moved to
    Portsmouth 12.1858; stored in Haslar gunboat shed from
    6.2.1857. BU completed 14.9.1864.

Quail Wigram & Son, Blackwall/Maudslay, Sons & Field (215ihp)
  Ord: 4.10.1855. K: 12.10.1855. L: 2.6.1856. C: 30.7.1858.
  First cost: £11,812 (including £4,858 for hull, £3,311 for machinery).
  Not commissioned until 1858; at Woolwich from 6.1856; to
    Chatham 11.1856 for Steam Reserve; in 4.1858 moved to
    Portsmouth. Commissioned 5.5.1858 under Lieut Noel Osborn,
    for the Mediterranean. BU 9.1861 at Malta.

Ripple Wigram & Son, Blackwall/Maudslay, Sons & Field (215ihp)
  Ord: 4.10.1855. K: 12.10.1855. L: 2.6.1856. C: 24.4.1857.
  First cost: £8,950 (including £4,866 for hull, £3,011 for machinery).
  Never commissioned; at Woolwich from 6.1856 for Steam Reserve;
    moved to Plymouth 8.1858 for Steam Reserve. BU 4.1866 by
    Marshall at Plymouth.

Cochin R. & H. Green, Blackwall/John Penn & Sons
  Ord: 7.10.1855. K: 1.12.1855. L: 8.4.1856. C: 20.4.1856.
  First cost: £? (including £5,437 for hull, £3,524 for machinery).
  Never commissioned; at Sheerness from 6.1856 for Steam Reserve;
    towed to at Chatham from 8.1860. BU 3.1863 at Sheerness.

Cherokee R. & H. Green, Blackwall/John Penn & Sons
  Ord: 7.10.1855. K: 1.12.1855. L: 30.4.1856. C: 3.10.1856.
  First cost: £9,636 (including £5,468 for hull, £3,512 for machinery).
  Never commissioned; towed to Portsmouth 11.1856; stored in
    Haslar gunboat shed from 27.2.1857, stripped of timbers and left
    in frame. BU 25.3.1869 at Portsmouth.

Camel R. & H. Green, Blackwall/John Penn & Sons
  Ord: 7.10.1855. K: 1.12.1855. L: 3.5.1856. C: — .
  First cost: £? (including £5,477 for hull, £3,506 for machinery).
  Never commissioned; at Woolwich from 8.1856; towed to
    Portsmouth 10.1856; stored in Haslar gunboat shed from
    30.1.1857. BU 30.6.1864.

Caroline R. & H. Green, Blackwall/John Penn & Sons
  Ord: 7.10.1855. K: 1.12.1855. L: 9.5.1856. C: 2.10.1856.
  First cost: £9,612 (including £5,484 for hull, £3,512 for machinery).
  Never commissioned; towed to Portsmouth 10.1856; stored in
    Haslar gunboat shed from 21.3.1857. A survey in 1860 revealed
    her as poorly built; her timbers were very sappy and rotten, and
    'her planks were not worked close to the timbers, and not well
    fastened; very few bolts in the bottom were clinched, many of
    them being too short'. BU 19.2.1862 at Portsmouth.

Confounder R. & H. Green, Blackwall/John Penn & Sons
  Ord: 7.10.1855. K: 12.1855. L: 21.5.1856. C: — .
  First cost: £? (including £5,477 for hull, £3,517 for machinery).
  Never commissioned; at Woolwich from 8.1856; towed to
    Portsmouth 10.1856; stored in Haslar gunboat shed from
    28.1.1857. BU 4.10.1864.

Crocus R. & H. Green, Blackwall/John Penn & Sons
  Ord: 7.10.1855. K: 1.12.1855. L: 4.6.1856. C: — .
  First cost: £8,846 (including £5,069 for hull, £3,511 for machinery).
  Never commissioned; at Deptford from 6.1856; towed to
    Portsmouth 10.1856; stored in Haslar gunboat shed from
    7.2.1857. BU 27.7.1864.

Beacon Laird & Co., Birkenhead/John Penn & Sons
  Ord: 8.10.1855. K: 17.10.1855. L: 11.2.1856. C: 29.3.1856.
  First cost: £10,233 (including £5,660 for hull, £3,593 for machinery).
  Commissioned 10.3.1856 under Lieut Ernest Augustus Travers
    Stubbs; paid off 21.6.1856 at Portsmouth; stored in Haslar
    gunboat shed from 9.3.1857. BU 27.8.1864.

Brave Laird & Co., Birkenhead/John Penn & Sons
  Ord: 8.10.1855. K: 17.10.1855. L: 11.2.1856. C: 3.1856 (not
    completed).
  First cost: £? (including £5,657 for hull, £3,593 for machinery).

Commissioned 7.3.1856 under Lieut Edward Hardinge; paid off 16.6.1856 at Portsmouth; stored in Haslar gunboat shed from 12.3.1857; relaunched as an experiment 9.2.1859. BU 25.3.1869 at Portsmouth.

Brazen Laird & Co., Birkenhead/John Penn & Sons
Ord: 8.10.1855. K: 17.10.1855. L: 23.2.1856. C: 20.3–7.4.1856 (for sea – not completed).
First cost: £10,953 (£5,682 for hull, £3,593 for machinery, £1,678 for fitting).
Commissioned 7.3.1856 under Lieut William Wilson Somerset Bridges; paid off 13.9.1856 into Steam Reserve at Portsmouth. BU 8.1864.

Blazer Laird & Co., Birkenhead/John Penn & Sons
Ord: 8.10.1855. K: 17.10.1855. L: 23.2.1856. C: 7.4.1856.
First cost: £11,159 (including £5,658 for hull, £3,586 for machinery).
Commissioned 11.3.1856 under Lieut Loftus Christopher Hawker Robinson; paid off 8.1856 into Steam Reserve at Portsmouth. From 1.1862 to coastguard reserve at Queenstown (Cork), as tender to *Hawke*, later to *Frederick William*; paid off 9.12.1867 Became steam dredger *YC.29* at Portsmouth in 6.1868, later *YC.4* at Gibraltar. Sold 4.5.1877 at Gibraltar.

Bullfinch Laird & Co., Birkenhead/John Penn & Sons
Ord: 8.10.1855. K: 17.10.1855. L: 25.2.1856. C: 6.1857.
First cost: £11,899 (including £9,275 for combined hull and machinery).
Commissioned 19.3.1856 under Lieut Francis Stafford Thompson. On 28.6.1858 under Lieut John William James, at Portsmouth as tender to *Illustrious*, later to *Britannia*. BU 8.1864.

Rainbow Laird & Co., Birkenhead/John Penn & Sons (156ihp)
Ord: 8.10.1855. K: 17.10.1855. L: 8.3.1856. C: 14.4.1856.
First cost: £11,130 (including £5,658 for hull, £3,591 for machinery).
Commissioned 20.3.1856 under Lieut James Blair Grove; paid off 19.7.1856 at Portsmouth into Steam Reserve. From 5.1860 to coastguard reserve at Kingstown (Dublin), as tender to *Ajax*. In 3.1864 to Steam Reserve at Plymouth. From 7.1868 to coastguard reserve at Hull, as tender to *Dauntless*. In 4.1872 to Steam Reserve at Chatham. Rebuilt at Wivenhoe as RNVR training ship 1874, then fitted at Woolwich Dyd; from 11.1874 moored off Somerset House (London), as drill ship for RN Artillery Volunteers. Sold to Castle 11.1888 to BU at Charlton.

Raven Laird & Co., Birkenhead/John Penn & Sons
Ord: 8.10.1855. K: 17.10.1855. L: 8.3.1856. C: — .
First cost: £11,116 (including £5,659 for hull, £3,591 for machinery).
Commissioned 20.3.1856 under Lieut Charles George Frederick Knowles; paid off 2.7.1856; stored in Haslar gunboat shed from 16.2.1857. In 6.1860 to Steam Reserve at Portsmouth. From 2.1862 to coastguard reserve at Pembroke Dock, as tender to *Blenheim*. From 5.1864 to coastguard reserve at Kingstown (Dublin), as tender to *Royal George*, later to *Pallas*. In 8.1870 to Steam Reserve at Plymouth. Sold to Castle 13.4.1875 to BU at Charlton.

Redbreast Laird & Co., Birkenhead/John Penn & Sons
Ord: 8.10.1855. K: 17.10.1855. L: 11.3.1856. C: 14.4.1856.
First cost: £10,376 (£5,656 for hull, £3,593 for machinery, £1,127 fitting).
Commissioned 20.3.1856 under Lieut Henry Rushworth Wratislaw; paid off 16.6.1856 at Portsmouth; stored in Haslar gunboat shed from 27.2.1857. BU completed 24.9.1864.

Rose Laird & Co., Birkenhead/John Penn & Sons
Ord: 8.10.1855. K: 6.12.1855. L: 21.4.1856. C: 18.5.1856.
First cost: £9,975 (including £5,450 for hull, £3,592 for machinery).

Never commissioned; at Portsmouth from 8.1856, in Steam Reserve; fitted with Armstrong guns in 1862?; to coastguard reserve at Queenstown (Cork) 1.1862, as tender to *Hawke*. BU 8.1868 at Devonport. (I have seen a note that *Rose* was commanded from 1.1.1862 by Lieut John Lort Stokes, for surveying service, then tender to *Royal Adelaide* at Plymouth; but other evidence indicates this to be incorrect.)

Rocket Laird & Co., Birkenhead/John Penn & Sons
Ord: 8.10.1855. K: 8.12.1855. L: 21.4.1856. C: — .
First cost: £9,275 (including £5,660 for hull, £3,327 for machinery).
Never commissioned; at Portsmouth from 8.1856; stored in Haslar gunboat shed from 13.1.1857. BU 10.1864.

Hardy Charles Hill & Sons, Bristol/Maudslay, Sons & Field (215ihp)
Ord: 8.10.1855. K: 19.10.1855. L: 1.3.1856. C: 10.11.1859.
First cost: £12,424 (including £5,658 for hull, £3,591 for machinery).
Commissioned 7.3.1856 under Lieut William Leyland Wilson (died 25.6.1856); paid off 7.1856 at Portsmouth into Steam Reserve. On 20.9.1859 under Lieut Archibald George Bogle, for the East Indies and China (including British involvement in the Taiping Rebellion); on 2.2.1863 under Lieut Henry John Fletcher Campbell, then 14.4.1864 under Lieut George Morice; paid off 31.3.1867. Sold 9.2.1869 at Hong Kong.

Havock Charles Hill & Sons, Bristol/Maudslay, Sons & Field (215ihp)
Ord: 8.10.1855. K: 20.10.1855. L: 20.3.1856. C: 10.11.1859.
First cost: £12,306 (including £5,675 for hull, £3,321 for machinery).
Commissioned 20.3.1856 under Lieut Henry Beverley; paid off 2.7.1856 at Portsmouth into Steam Reserve. Recommissioned 23.9.1859 under Lieut Charles Fairholme, for East Indies and China; on 1.1.1863 under Lieut George Poole; at bombardment of Kagoshima 8.1863 (during Anglo-Satsuma War); in 11.1863 under Lieut Edward Barkley; on 20.9.1865 under Lieut Philip Edward Luard, then 10.6.1867 under Lieut Yelverton O'Keefe; in action (with *Bouncer*) against pirate junks 26.6.1867; paid off 10.3.1870. Sold 31.3.1870 at Yokohama.

Highlander Charles Hill & Sons, Bristol/Maudslay, Sons & Field
Ord: 8.10.1855. K: 22.10.1855. L: 29.4.1856. C: — .
First cost: £8,999 (including £5,630 for hull, £3,325 for machinery).
Never commissioned; stored in Haslar gunboat shed from 31.1.1857; re-launched 12.1861 to join Steam Reserve at Portsmouth. In 1.1862 to Queenstown (Cork), for coastguard reserve, as tender to *Hawke*, then to *Black Prince*; paid off 5.8.1867. Became steam dredger *YC.51* at Chatham in 1868. Sold 5.1884.

Albacore T. & J. White, West Cowes/Maudslay, Sons & Field (215ihp)
Ord: 9 or 10.10.1855. K: 12.10.1855. L: 3.4.1856. C: — .
First cost: £8,978 (including £4,808 for hull, £3,284 for machinery).
Not commissioned until 1867; stored in Haslar gunboat shed from 28.1.1857; re-launched 9.1862 to join Steam Reserve at Portsmouth. In 10.1864 to Bermuda, as tender to flagship there. Commissioned 1.1.1867 under Lieut Charles H. C. Langdon; paid off 28.10.1869. Tender at Bermuda 1870; tank vessel 1874. Hulked in 1882. BU 6.1885 at Bermuda.

Amelia T. & J. White, West Cowes/Maudslay, Sons & Field (215ihp)
Ord: 9 or 10.10.1855. K: 12.10.1855. L: 19.5.1856. C: — .
First cost: £8,627 (including £4,675 for hull, £3,291 for machinery).
Never commissioned; stored in Haslar gunboat shed from 23.1.1857; re-launched 1.5.1861 to join Steam Reserve at Portsmouth. To Pembroke Dock 4.1864, for coastguard reserve, as tender to *Blenheim*. BU 29.9.1865 at Pembroke.

Foam Wigram & Son, Northam/John Penn & Sons
Ord: 9 or 10.10.1855. K: 29.10.1855. L: 8.5.1856. C: — .

The *Albacore* Class gunboat *Magnet* during the 1860s during her service as a tender. Forty-eight were ordered to this design between April and July 1855, virtually identical to the preceding *Dapper* Class. Following the successful performance of the *Dappers* in the assault on Sveaborg in early August, orders for another fifty units were placed during the autumn with English contractors; the design was also shared with French yards, who built a further eight *Albacore* Class in Rochefort (two) and Nantes (six) for the French Navy. (*NMM 5385*)

First cost: £8,655 (£5,053 for hull, £3,602 for machinery) excluding fitting.

Never commissioned; stored in Haslar gunboat shed from 29.1.1857; re-launched 12.1861 to join Steam Reserve at Portsmouth. Sold 6.1867 to BU.

Wave Wigram & Son, Northam/John Penn & Sons
Ord: 9 or 10.10.1855. K: 29.10.1855. L: 25.6.1856. C: — .
First cost: £9,130 (£4,682 for hull, £3,602 for machinery, £946 fitting).
Never commissioned; stored in Haslar gunboat shed from 12.1.1857. Coal hulk at Portsmouth 1869. Hulk, renamed *Clinker* 30.12.1882. Sold 1890.

Magnet Briggs & Co., Sunderland/John Penn & Sons (267ihp)
Ord: 10.10.1855. K: 16.10.1855. L: 29.1.1856. C: 27.3.1856.
First cost: £11,324 (£5,600 for hull, £3,638 for machinery, £2,086 fitting).
Commissioned 7.3.1856 under Lieut Charles John Rowley; paid off 10.9.1856 at Sheerness into Steam Reserve. In 3.1858 to Harwich, for coastguard reserve, as tender to *Pembroke*; paid off at Sheerness 10.7.1869. BU 1874 at Chatham.

Manly Briggs & Co., Sunderland/John Penn & Sons
Ord: 10.10.1855. K: 16.10.1855. L: 29.1.1856. C: 27.3.1856.
First cost: £11,389 (£5,583 for hull, £3,638 for machinery, £2,162 fitting).
Commissioned 7.3.1856 under Lieut William Arthur; paid off 7.1856 at Sheerness into Steam Reserve; to Chatham 1860. BU 1.1864 at Deptford.

Mastiff (ex-*Hound*) Briggs & Co., Sunderland/John Penn & Sons
Ord: 10.10.1855. K: 16.10.1855. L: 22.2.1856. C: 1.4.1856.
First cost: £11,228 (£5,575 for hull, £3,638 for machinery, £2,015 fitting).
Commissioned 7.3.1856 under Lieut William Frederick Johnson; paid off 10.6.1856 at Sheerness into Steam Reserve; to Chatham 1860. BU 10.1863 at Deptford.

Mistletoe Briggs & Co., Sunderland/John Penn & Sons
Ord: 10.10.1855. K: 16.10.1855. L: 22.2.1856. C: 1.4.1856.
First cost: £11,314 (£5,569 for hull, £3,638 for machinery, £2,107 fitting).
Commissioned 7.3.1856 under Lieut Frederick Harvey; paid off 7.1856 at Sheerness into Steam Reserve; to Chatham 1860; in

7.1862 to Leith, for coastguard reserve, as tender to *Edinburgh*. BU completed 28.9.1864 at Sheerness.

Earnest William Patterson & Son, Bristol/Maudslay, Sons & Field (215ihp)
Ord: 11.10.1855. K: 15.10.1855. L: 29.3.1856. C: — .
First cost: £8,804 (£5,592 for hull, £3,212 for machinery) excluding fitting.
Never commissioned; stored in Haslar gunboat shed from 30.1.1857; re-launched 7.7.1861 to join Steam Reserve at Portsmouth. In 1867 to Falmouth, for coastguard reserve, as tender to *St George*; in 9.1870 to Portsmouth, as tender to the *Duke of Wellington*, harbour guard ship and depot. Sold to Castle 17.1.1885 to BU at Charlton.

Escort William Patterson & Son, Bristol/Maudslay, Sons & Field (215ihp)
Ord: 11.10.1855. K: 15.10.1855. L: 26.5.1856. C: — .
First cost: £8,511 (£5,179 for hull, £3,310 for machinery, £22 fitting).
Never commissioned; stored in Haslar gunboat shed from 28.1.1857; re-launched 5.1861 to join Steam Reserve at Portsmouth. In 2.1862 to Pembroke Dock, for coastguard reserve, as tender to *Blenheim*. BU 10.1865 at Pembroke.

CHEERFUL Class 'Crimean' gunboat. W. H. Walker design, 1855, with only 4ft draught. The first pair were approved on 2 May 1855 to be built in HM Dockyards, and were named on 30 May, although the order was not confirmed until 2 July. Two to the same design were approved on 12 November to be built by contract (order placed 2 days later) and orders for a further ten were approved on 21 and 23 November to be built by contract. The ordering of a final six vessels to the same design from HM Dockyards was approved on 30 November. The Laird and Joyce craft (also *Pert*) were built under cover. Original *Careful* and *Ramble* renamed 1855 (while building). All were engined by John Penn & Sons, although the Penn engines, unusually, were single piston-rod type instead of Penn's usual trunked variety. The sail area was only 1,919 sq. ft.

These were an unsuccessful design; nine were never commissioned, and of those that were another nine were paid off into the Steam Reserve within a few months, leaving just *Nettle* and *Onyx* deployed to Bermuda.

Dimensions & tons: 100ft 0in, 85ft 5½in x 21ft 10in oa (21ft 7in for tonnage) x 6ft 7in. Draught 6ft 6in. 211⁶⁴/₉₄bm.

Men: 30. Guns: 2 x 32pdrs (56cwt/9½ft) MLSB. Most re-armed with 2 x 40pdr Armstrong BL.

Machinery: 1 set 1-cylinder horizontal direct-acting single expansion. 2 cylindrical boilers. Single (non-hoisting) screw. 20nhp. *Onyx* 92ihp = 6¾kts.

Cheerful Deptford Dyd/John Penn & Sons
Ord: 2.7.1855. K: 2.7.1855. L: 6.10.1855. C: 11.12.1855.
First cost: £7,072 (including £5,148 for hull, £1,137 for machinery).
Commissioned 30.11.1855 under Lieut William Hector Rason; paid off 7.1856 at Portsmouth into Steam Reserve. BU completed 16.1.1869 at Haslar.

Chub Sheerness Dyd/John Penn & Sons
Ord: 2.7.1855. K: 6.7.1855. L: 15.10.1855. C: 20.2.1856.
First cost: £8,243 (including £5,475 for hull, £1,137 for machinery).
Commissioned 14.1.1856 under Lieut Ernest Grey Lambton Cochrane; paid off 11.9.1856 at Portsmouth into Steam Reserve. BU completed 29.1.1869 at Haslar.

Daisy Thomas Westbrook, Blackwall/John Penn & Sons
Ord: 14.11.1855. K: 14.11.1855. L: 20.3.1856. C: 29.4.1856.
First cost: £7,681 (including £4,509 for hull, £1,187 for machinery).
Commissioned 19.3.1856 under Lieut Harry Woodfall Brent; paid off 11.9.1856 at Portsmouth into Steam Reserve. BU 7.1.1869 at Haslar.

Dwarf Thomas Westbrook, Blackwall/John Penn & Sons
Ord: 14.11.1855. K: 19.11.1855. L: 8.4.1856. C: 3.7.1756.
First cost: £7,166 (including £4,487 for hull, £1,187 for machinery).
Never commissioned. In 7.1856 at Sheerness, for the Steam Reserve; in 8.1860 at Chatham, also Steam Reserve. BU 8.1863 at Deptford Dyd.

Blossom (ex-*Careful*) Laird, Birkenhead/John Penn & Sons
Ord: 21.11.1855. K: 11.1855? L: 21.4.1856. C: — .
First cost: £? (including £5,114 for hull, £1,287 for machinery).
Never commissioned. In 6.1856 at Portsmouth, for the Steam Reserve; stored in Haslar gunboat shed from 2.1.1857. BU 21.10.1864 at Haslar.

Gadfly Laird, Birkenhead/John Penn & Sons
Ord: 21.11.1855. K: 1.12.1855. L: 21.4.1856. C: — .
First cost: £6,448 (including £5,146 for hull, £1,288 for machinery).
Never commissioned. In 6.1856 at Portsmouth, for the Steam Reserve; stored in Haslar gunboat shed from 8.1.1857. BU 11.1864.

Garland Laird, Birkenhead/John Penn & Sons
Ord: 21.11.1855. K: 12.1855. L: 7.5.1856. C: — .
First cost: £? (including £5,153 for hull, £1,290 for machinery).
Never commissioned. In 7.1856 at Portsmouth, for the Steam Reserve; stored in Haslar gunboat shed from 7.1.1857. BU 6.1864.

Gnat Laird, Birkenhead/John Penn & Sons
Ord: 21.11.1855. K: 12.1855? L: 10.5.1856. C: — .
First cost: £? (including £5,153 for hull, £1,290 for machinery).
Never commissioned. In 7.1856 at Portsmouth, for the Steam Reserve; stored in Haslar gunboat shed from 1.1.1857. BU 10.8.1864.

Fidget Joyce, Greenwich/John Penn & Sons
Ord: 23.11.1855. K: 27.11.1855. L: 7.4.1856. C: 5.7.1856.
First cost: £7,138 (including £4,399 for hull, £1,147 for machinery).
Never commissioned. In 7.1856 at Sheerness, for the Steam Reserve; in 8.1860 at Chatham, also Steam Reserve. BU 8.1863 at Deptford Dyd.

Flirt Joyce, Greenwich/John Penn & Sons

Ord: 23.11.1855. K: 27.11.1855. L: 7.6.1856. C: 29.4.1858.
First cost: £6,043 (including £4,130 for hull, £1,147 for machinery).
Never commissioned. In 8.1856 at Deptford, for the Steam Reserve; towed 5.1858 to Portsmouth; stored in Haslar gunboat shed from 14.5.1858. BU 30.4.1864 at Haslar.

Onyx Young, Magnay & Co., Limehouse/John Penn & Sons
Ord: 23.11.1855. K: 15.12.1855. L: 3.4.1856. C: 15.4.1856.
First cost: £7,869 (including £4,618 for hull, £1,201 for machinery).
Commissioned 19.3.1856 by Lieut William Nathan Wrighte Hewett, for North America. Became tender to *Terror* at Bermuda 9.1857. In 1865 under Lieut Herbert Charles A. Brand, as tender at Port Royal, Jamaica. Dockyard craft (steam lump) 1869. Sold 8.7.1873 in Jamaica.

Pert Young, Magnay & Co., Limehouse/John Penn & Sons
Ord: 23.11.1855. K: 15.12.1855. L: 3.4.1856. C: 15.4.1856.
First cost: £7,953 (including £4,616 for hull, £1,201 for machinery).
Commissioned 19.3.1856 by Lieut Frederick Proby Doughty; paid off 8.1856 at Portsmouth into Steam Reserve; stored in Haslar gunboat shed from 7.7.1861. BU completed 12.3.1864.

Midge Young, Magnay & Co., Limehouse/John Penn & Sons
Ord: 23.11.1855. K: 15.12.1855. L: 8.5.1856. C: 26.7.1856.
First cost: £7,353 (including £4,717 for hull, £1,206 for machinery).
Never commissioned. In 7.1856 at Sheerness, for the Steam Reserve; towed 4.1858 to Portsmouth; stored in Haslar gunboat shed from 18.5.1858. BU 10.1864.

Tiny Young, Magnay & Co., Limehouse/John Penn & Sons
Ord: 23.11.1855. K: 15.12.1855. L: 8.5.1856. C: — .
First cost: £7,359 (including £4,723 for hull, £1,208 for machinery).
Never commissioned. In 7.1856 at Sheerness, for the Steam Reserve; towed 4.1858 to Portsmouth; stored in Haslar gunboat shed from 17.5.1858. Completed BU 28.1.1864 at Plymouth.

Angler Devonport Dyd/John Penn & Sons
Ord: 1.12.1855. K: 4.12.1855. L: 8.3.1856. C: 7.4.1856 (for sea).
First cost: £6,729 (including £3,069 for hull, £1,265 for machinery).
Commmissioned 7.3.1856 under Lieut William Howorth; paid off 7.1856 at Portsmouth into Steam Reserve; stored in Haslar gunboat shed from 20.4.1860. BU completed 21.1.1869 at Haslar.

Ant Devonport Dyd/John Penn & Sons
Ord: 1.12.1855. K: 4.12.1855. L: 22.3.1856. C: 28.6.1858.
First cost: £7,346 (including £4,211 for hull, £1,265 for machinery).
Commissioned 7.3.1856 under Lieut Nowell Salmon; paid off 7.1856 at Portsmouth into the Steam Reserve. BU completed 23.2.1869 at Haslar.

Nettle Pembroke Dyd/John Penn & Sons
Ord: 4.12.1855. K: 1.12.1855. L: 9.2.1856. C: 17.4.1856.
First cost: £8,047 (including £5,001 for hull, £1,256 for machinery).
Commissioned 18.2.1856 under Lieut Gilbert Troward Key, for North America; from 9.1857 at Bermuda. Became tender to *Terror* at Bermuda 1860; from 1865 tender at Port Royal (Jamaica). BU 10.1867 at Bermuda.

Pet Pembroke Dyd/John Penn & Sons
Ord: 4.12.1855. K: 12.1855. L: 9.2.1856. C: 17.4.1856.
First cost: £8,065 (including £5,082 for hull, £1,285 for machinery).
Commissioned 2.1856 under Lieut William Belford Stubbs; paid off 7.1856 at Portsmouth into Steam Reserve; stored in Haslar gunboat shed from 1861. Fitted as coal hulk at Portsmouth 9.1863. Renamed *C.17* from about 1900. Sold to Castle 12.4.1904 to BU.

Decoy Pembroke Dyd/John Penn & Sons
Ord: 4.12.1855. K: 12.1855. L: 21.2.1856. C: 17.4.1856.
First cost: £7,508 (including £? for hull, £1,265 for machinery).
Commissioned 2.1856 under Lieut Reginald Treby Clark; paid off

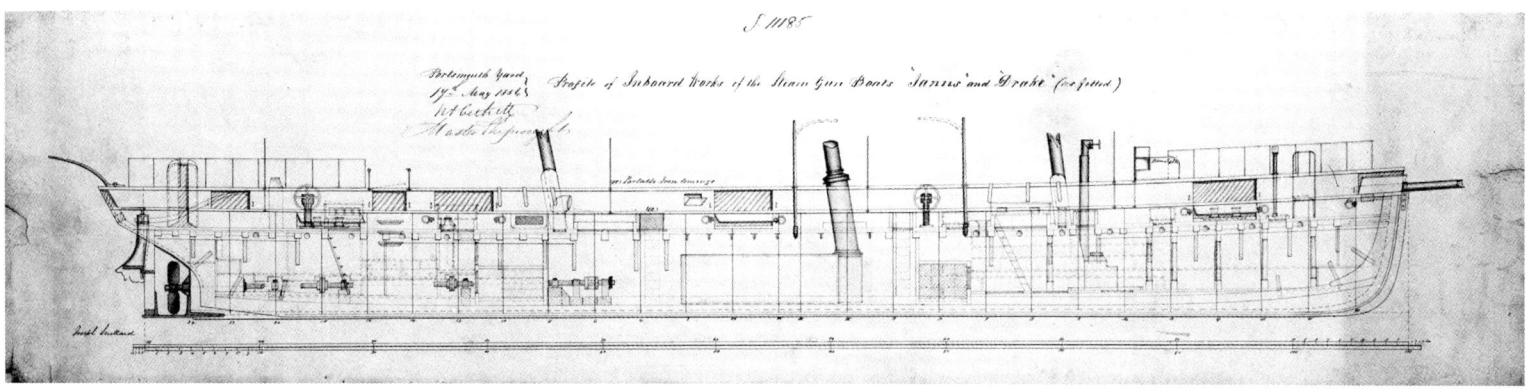

An inboard profile of the *Clown* Class gunboats *Janus* and *Drake*, as fitted, dated 17
May 1856 at Portsmouth. The extremely shallow draught of these 'Crimean' gunboats is
apparent. After the disappointing result of the *Cheerful* Class, whose frail hulls and tiny
engines mitigated against their sole advantage of being able to float in four feet of water,
the *Clown* Class signalled a return to stronger scantlings, a more powerful 68pdr main
gun, and a more effective (if still single-cylinder) 40 nhp engine. Moreover their sail area
was virtually doubled to 4,889 square feet, with square sails now carried on the fore mast,
improving their suitability for overseas service. Six were not commissioned, and were
towed to Haslar for storage in the sheds there; the other six were deployed to China.
(NMM DR11185)

> 11.9.1856 at Portsmouth into Steam Reserve; stored in Haslar
> gunboat shed from 1858; launched from yard 20.4.1861 for
> Steam Reserve. BU completed 8.2.1869 at Haslar.

Rambler (ex-*Ramble*) Pembroke Dyd/John Penn & Sons
> Ord: 4.12.1855. K: 12.1855. L: 21.2.1856. C: 17.4.1856.
> First cost: £7,868 (including £4,803 for hull, £1,265 for machinery).
> Commissioned 2.1856 under Lieut George Rivington; paid off
> 11.9.1856 at Portsmouth into Steam Reserve. BU 7.1.1869 at
> Haslar.

CLOWN Class 1855 'Improved *Cheerful*' design. All twelve were
named on 19 January 1856. The Miller and Smith vessels were built
under cover. All were engined by John Penn & Sons, although the
Penn engines, unusually, were single piston-rod type instead of Penn's
usual trunked variety. Six were not brought into service, while the
other six were adapted for service in China, where they spent their
service lives. The sail area was increased to 4,889 sq.ft compared with
their predecessors.
> Dimensions & tons: 110ft 0in, 95ft 5¼in x 21ft 10in oa (21ft 5in for
>   tonnage) x 6ft 7in. Draught 4ft 0in. 232⁸⁹⁄₉₄bm. 249 disp.
> Men: 36. Guns: 1 x 68pdr (95cwt/10ft) MLSB + 1 x 32pdr
>   (56cwt/9½ft).
> Machinery: 1 set 1-cylinder horizontal direct-acting single
>   expansion. 2 cylindrical boilers. Single (non-hoisting) screw.
>   40nhp. *Woodcock* 145ihp = 7½kts.

Fenella W. & H. Pitcher, Northfleet/John Penn & Sons
> Ord: 31.12.1855. K: 5.3.1856. L: 19.5.1856. C: 26.1.1857.
> First cost: £8,204 (including £4,759 for hull, £2,374 for machinery).
> Never commissioned; towed 1.1857 to Portsmouth, where remained
>   incomplete; stored in Haslar gunboat shed from 11.1.1857;
>   launched from yard 1861 for Steam Reserve. Became a steam
>   dredger 3.1867 at Woolwich (–1876), renamed *YC.3* in 1868. BU
>   14.11.1878.

Garnet W. & H. Pitcher, Northfleet/John Penn & Sons
> Ord: 31.12.1855. K: 5.3.1856. L: 31.5.1856. C: 26.1.1857.
> First cost: £8,264 (including £4,769 for hull, £2,374 for machinery).
> Never commissioned; towed 1.1857 to Portsmouth, where remained
>   incomplete; stored in Haslar gunboat shed from 13.2.1857. BU
>   completed 25.5.1864.

Handy W. & H. Pitcher, Northfleet/John Penn & Sons
> Ord: 31.12.1855. K: 5.3.1856. L: 31.5.1856. C: 26.1.1857.
> First cost: £8,373 (including £4,766 for hull, £2,374 for machinery).
> Not commissioned until 1861; towed 1.1857 to Portsmouth, where
>   remained incomplete; stored in Haslar gunboat shed from
>   10.2.1857; launched from yard 8.1861 and fitted for tropical
>   service. Commissioned 6.9.1861 under Lieut John Glover, for
>   west coast of Africa, as tender to *Arrogant*, station flagship at
>   Sierra Leone; on 23.12.1862 under Lieut William Dolben, on
>   same station, as tender to *Investigator*; paid off 7.3.1867. Sold
>   5.1868 at Lagos.

Hunter W. & H. Pitcher, Northfleet/John Penn & Sons
> Ord: 31.12.1855. K: 12.3.1856. L: 7.6.1856. C: 26.1.1857. Only
>   partially fitted, never completed for sea.
> First cost: £8,443 (including £4,798 for hull, £2,374 for machinery).
> Never commissioned; towed 1.1857 to Portsmouth, where remained
>   incomplete; stored in Haslar gunboat shed from 13.2.1857;
>   launched from yard 1865 for Steam Reserve. On the sale list
>   6.1883 and sold 1884.

Watchful T. & W. Smith, Newcastle/John Penn & Sons
> Ord: 31.12.1855. K: 25.2.1856. L: 4.6.1856. C: 11.4.1857.
> First cost: £10,502 (including £5,572 for hull, £2,448 for machinery).
> Commissioned 11.3.1857 under Lieut James Hawkins Whitshed,
>   for the East Indies & China; on 2.3.1859 under Lieut Frederick
>   Warren Inglefield; paid off 30.11.1861 at Hong Kong. Tender
>   1863. Sold 1.2.1871 at Hong Kong.

Woodcock T. & W. Smith, Newcastle/John Penn & Sons
> Ord: 31.12.1855. K: 25.2.1856. L: 6.6.1856. C: 11.4.1857.
> First cost: £10,512 (including £5,572 for hull, £2,487 for machinery).
> Commissioned 11.3.1857 under Lieut Edwin John Pollock, for the
>   East Indies & China; on 14.11.1858 under Lieut George Stanley
>   Bosanquet; at attack on Peiho northern forts; paid off 15.12.1861.
>   Sold there 1.2.1871.

Drake Pembroke Dyd/John Penn & Sons
> Ord: 3.1.1856. K: — . L: 8.3.1856. C: 31.5.1856.
> First cost: £9,747 (including £? for hull, £? for machinery).
> Commissioned 22.3.1856 under Lieut Mountford Stephen Lovide
>   Peile; paid off 6.9.1856 into Steam Reserve at Portsmouth.
>   Recommissioned 14.3.1857 under Lieut William Arthur, for
>   the East Indies & China; on 4.6.1858 under Lieut Charles John
>   Balfour; then on 26.10.1858 under Lieut Cmdr Arthur Rodney
>   Blane; paid off 28.11.1861 at Hong Kong. Recommissioned
>   24.5.1867 under Lieut Charles Crowdy; paid off 31.10.1868 at
>   Hong Kong. Sold there 9.2.1869.

Janus Pembroke Dyd/John Penn & Sons
> Ord: 3.1.1856. K: 25.1.1856. L: 8.3.1856. C: 31.5.1856.
> First cost: £9,737 (including £5,104 for hull, £2,432 for machinery).

Commissioned 21.3.1856 under Lieut Charles Rufus Robson; paid off 6.9.1856 into Steam Reserve at Portsmouth. Recommissioned 13.3.1857 under Lieut William Henry Jones, for the East Indies and China; on 4.6.1858 under Lieut Herbert Price Knevitt; in unsuccessful attack on the Peiho River forts during 2nd Anglo-Chinese War, 28.6.1859. On 18.11.1859 under Lieut Duncan George Davidson; in attack on Taku forts 21.8.1860. On 2.12.1862 under Lieut Richard Adams; in 10.1864 under Lieut Charles Edward Powys (or George Powys?), still in China. In 1865 under Lieut Cecil Frederick William Johnson, then 30.5.1867 under Lieut Rodney Lloyd; in 1869 under Lieut Leicester Chantrey Keppel; paid off 28.2.1869. Became Coal lighter 12.1869 at Hong Kong, renamed *YC.6*. Sold 1871 (possibly shared fate of *Clown*?).

Clown William Cowley Miller, Toxteth Dock, Liverpool/John Penn & Sons
> Ord: 3.1.1856. K: 14.1.1856. L: 20.5.1856. C: 4.5.1857.
> First cost: £10,196 (including £5,162 for hull, £2,429 for machinery).
> Commissioned 12.3.1857 under Lieut William Frederick Lee, for the East Indies & China; paid off 5.12.1861 at Hong Kong. Became a coal lighter *YC.1* at Hong Kong 1867. Renamed *YC.6* in 12.1869. Wrecked at Hong Kong in the typhoon of 2.9.1871.

Kestrel William Cowley Miller, Toxteth Dock, Liverpool/John Penn & Sons
> Ord: 3.1.1856. K: 14.1.1856. L: 26.5.1856. C: 4.5.1857.
> First cost: £10,273 (including £5,162 for hull, £2,428 for machinery).
> Commissioned 11.3.1857 under Lieut William Hector Rason, for the East Indies and China; on 28.6.1859 under Lieut George Dacres Bevan; sunk during unsuccessful attack on the Peiho River forts during 2nd Anglo-Chinese War, 28.6.1859, but salved and returned to service. On 7.1.1860 under Lieut Henry Huxham, still in China; in Taiping Rebellion. On 10.12.1862 under Lieut Hamilton Dunlop, then in 10.1864 under Lieut Duncan E. K. Grant, in China. Sold 16.3.1866 to Glover & Co., Yokohama then resold to Japanese owners (said to be the navy but there is no sign of this ship in listings of the Japanese Navy – perhaps it was to one of the rebellious clans that she was sold?).

Ready Briggs & Co., Sunderland/John Penn & Sons
> Ord: 3.1.1856. K: 2.1856. L: 12.5.1856. C: 29.4.1857 (for Reserve).
> First cost: £8,693 (including £6,011 for hull, £2,064 for machinery).
> Never commissioned, held in reserve at Deptford; towed 5.1858 to Portsmouth and stored in Haslar gunboat shed from 3.6.1858. BU completed 25.1.1864.

Thrush Briggs & Co., Sunderland/John Penn & Sons
> Ord: 3.1.1856. K: 2.1856. L: 12.5.1856. C: 29.4.1857 (for Reserve).
> First cost: £8,690 (including £6,011 for hull, £2,464 for machinery).
> Never commissioned, held in reserve at Deptford; towed 5.1858 to Portsmouth and stored in Haslar gunboat shed from 2.6.1858. BU completed 14.3.1864.

ALGERINE Class 1856 Enlarged *Albacore* Class, designed by the Surveyor's Department to afford more accommodation and stowage for coals and provisions for service on a distant station (the last four spent their entire service lives in the Far East). All six were ordered on 22 September 1856 and were named on 16 February 1857. They were re-classed as gunvessels on 8 June 1859. Three more were ordered in the same month, but were completed as gunvessels of the *Philomel* Class. Two similar vessels (*Clyde* and *Sir Hugh Rose*) were built for the Bombay Marine in Bombay Dyd in 1859.
> Dimensions & tons: 125ft 0in, 110ft 1½in x 23ft 0in oa (22ft 8in for tonnage) x 9ft 3in. 300⁸⁸⁄₉₄bm. 370 disp.

Men: 50. Guns: 1 x 10in/68pdr (87cwt/9½) on pivot, + 2 x 24pdr howitzers.
Machinery: 1 set horizontal direct-acting single expansion. Single screw. 80nhp. 294ihp = 9kts.

Jaseur R. & H. Green, Blackwall
> Ord: 22.9.1856. K: 2.9.1856. L: 7.3.1857. C: 9.3–5.9.1857 (for sea) at Deptford Dyd.
> First cost: £14,178 (including £5,668 for hull, £4,350 for machinery).
> Commissioned 7.1857 under Lieut John Binney Scott, for the West Indies; wrecked on the Bajo Nuevo Shoal 26.2.1859, while on passage from Port Royal (Jamaica) to Greytown on the Mosquito Coast.

Jasper R. & H. Green, Blackwall
> Ord: 22.9.1856. K: 26.9.1856. L: 18.3.1857. C: 19.3–5.9.1857 at Deptford Dyd.
> First cost: £14,195 (including £5,665 for hull, £4,350 for machinery).
> Commissioned 28.7.1857 under Lieut William Henry Pym, for North America and the West Indies; paid off 3.5.1861. Sold 2.8.1862 to the Chinese Imperial Customs, renamed *Amoy*, and sailed 4.1863 (to join Capt. Sherard Osborn's 'Vampire Fleet'), but returned to England when agreement with Chinese government collapsed. Resold 30.12.1865 to the Egyptian government.

Algerine W. & H. Pitcher, Northfleet
> Ord: 22.9.1856. K: 1.10.1856. L: 24.2.1857. C: 27.2–9.4.1857 (for sea) at Woolwich Dyd.
> First cost: £13,530 (including £5,641 for hull, £4,350 for machinery).
> Commissioned 4.6.1858 under Lieut William Arthur, for East Indies and China; action against pirate junks at Coulan 8–9.1858; action against Taku forts 21.8.1860. On 12.5.1862 under Lieut Arthur Rodney Blane, then 10.1864 under Lieut John Collier Tucker, on China station, then 14.4.1866 under Lieut Compton Edward Domville, on 1.10.1868 under Lieut Henry Rowland Ellison Gray, and in 1869 under Lieut Thornhaugh Philip Gurdon, all in China; various actions against pirate forces; paid off 7.5.1871. Sold 2.4.1872 at Hong Kong, became the mercantile *Algerine*. BU 1894.

Lee W. & H. Pitcher, Northfleet
> Ord: 22.9.1856. K: 1.10.1856. L: 28.2.1857. C: 7.3–13.4.1857 (for sea) at Woolwich Dyd.
> First cost: £13,424 (including £5,541 for hull, £4,350 for machinery).
> Commissioned 14.3.1857 under Lieut William Graham; in 1858 under Lieut William Henry Jones, for East Indies and China; to Hankow 11.1858; sunk 25.6.1859 in action with the Taku forts, Peiho River, China.

Leven W. & H. Pitcher, Northfleet
> Ord: 22.9.1856. K: 1.10.1856. L: 7.3.1857. C: 10.3–24.4.1857 (for sea) at Woolwich Dyd.
> First cost: £13,994 (including £5,541 for hull, £4,350 for machinery).
> Commissioned 4.1857 under Lieut Joseph Samuel Hudson, for the East Indies and China; in action with Taku forts 20.5.1858. On 2.12.1862 under Lieut Herbert Price Knevitt. On 25.5.1867 under Lieut Orford Somerville Cameron; to Outingpoi Creek 28.1.1869; on 25.2.1869 under Lieut Herbert Holden Edwards, then 25.5.1870 Lieut Albert W. Whish; paid off 24.7.1873. Sold 21.7.1873 at Shanghai.

Slaney W. & H. Pitcher, Northfleet
> Ord: 22.9.1856. K: 1.10.1856. L: 17.3.1857. C: 18.3–27.4.1857 (for sea) at Woolwich Dyd.
> First cost: £13,970 (including £5,549 for hull, £4,350 for machinery).
> Commissioned 26.3.1857 under Lieut Anthony Hiley Hoskins, for

The *Algerine* Class gunboat *Lee* leads a line comprising the gunboat *Dove* (*Albacore* Class), the paddle frigates *Retribution* and *Furious* and the screw sloop *Cruiser* in an engagement with Taiping forces on 20 November 1858 at Nankin on the Yangtse river. This coloured lithograph by T. G. Dutton is from a sketch by Frederick le Breton Bedwell. The *Lee* was to be sunk the following year in a similar operation, this time against the Taku forts on the Peiho River. Although often characterised purely as a 'rebellion' in the West, this struggle during the 1850s and early 1860s was an even more bloody civil war than the contemporary conflict between the States in North America, with an estimated 20 million Chinese killed during these climactic events. (*NMM PZ6561*)

East Indies and China; on 2.1859 under Lieut Henry Knox Leet, then 1.5.1860 Lieut George Dacres Bevan. In 10.1864 under Lieut John Smith Keats, on China station, then 4.1866 Lieut William Francis Leoline Elwyn; wrecked in a typhoon 9.5.1870 on the Paracel Islands (south of Hong Kong).

BRITOMART Class 1859 Improved version of the *Dapper* Class. The first batch of ten vessels was approved on 19 October 1859, and were all contracted on 11 November and given names on 17 January 1860; the first three were moved from Newcastle to the Steam Reserve in January 1861. Another six were ordered on 5 March 1860, all to be built at Portsmouth Dyd (they were actually built across the harbour at Haslar Creek, Gosport) and were named on 4 October; construction of these six vessels was suspended 1862/63, but later resumed. A final four were ordered in February 1861, again to be built at Portsmouth; these were named on 21 March; all four were cancelled at the close of 1863, by which time work had begun on the first three (not on *Danube*) which were BU on the stocks. They carried a barquentine rig (three masts with square sails on foremast only). *Heron* was originally fitted for reserve and carried Armstrong guns. Three were moved via the Canadian canal system onto the Great Lakes in 1866, to intercept Fenian raiders into Canada.

    Dimensions & tons: 120ft 0in, 105ft 7in x 22ft 0in oa (21ft 10in for tonnage) x 9ft 0in. 267^{82}/$_{94}$bm. 330 disp.

    Men: 36–40. Guns: 2 x 68pdr MLR (later 2 x 64pdr MLR).

    Machinery: Single expansion reciprocating. Single (hoisting) screw. 60nhp. 260ihp = 9kts.

Britomart T. & W. Smith, Newcastle/John Penn & Sons

Ord: 19.10.1859. K: 11.11.1859. L: 7.5.1860. C: 3.1.1861 for Steam Reserve at Chatham. In 4.1863 at Pembroke Dock, coastguard reserve, as tender to *Blenheim*.

Commissioned 31.3.1866 under Lieut Arthur Hildebrand Alington, for North America and the West Indies; ; patrolled Lake Erie in 1866–67 (based at Dunville). On 11.12.1869 under Lieut Basil Edward Cochrane, then 29.11.1872 under Lieut William Henry Richards; paid off 27.9.1873 at Plymouth into Steam Reserve. Recommissioned 1.11.1876 under Lieut James Sulivan at Southampton, coastguard reserve, as tender to *Hector*; on 29.10.1879 under Lieut William Edward Breeks Atkinson; to coast of Ireland 1881 (–1891); on 5.1.1884 under Lieut Frederick Isaac, then 5.1.1887 under Lieut Henry Oldfield; paid off 19.1.1891. Sold 12.1.1892 to Castle who resold her to S. Williams of Dagenham as a mooring hulk. BU 6.1946.

Cockatrice T. & W. Smith, Newcastle/John Penn & Sons

Ord: 19.10.1859. K: 11.11.1859. L: 24.5.1860. C: 7.4.1861 for Steam Reserve at Portsmouth.

Commissioned 6.3.1861 under Lieut Robert Moore Gillson, for the Mediterranean; on 14.4.1866 under Lieut Spencer Robert Huntley, then on 17.4.1868 under Cmdr James Ferris Prowse, 15.2.1872 under Cmdr George Digby Morant, 13.2.1873 under Cmdr William Home Chisholme St Clair, 5.2.1876 under Cmdr Archer John William Musgrave, and 10.3.1879 under Cmdr Harry Tremenheere Grenfell; paid off 31.12.1880 and became a luggage lighter as *YC.10* at Malta in 1882. Sold 1885.

Wizard T. & W. Smith, Newcastle/John Penn & Sons

Ord: 19.10.1859. K: 11.11.1859. L: 3.8.1860. C: 3.1.1861 for Steam Reserve at Sheerness.

Commissioned 22.7.1864 under Lieut Tom Falcon, for passage to the Mediterranean; paid off 10.9.1864. Recommissioned 12.1864 under Lieut George H Lawson; on 10.3.1866 under Lieut Patrick James Murray, then 1.12.1869 under Lieut Augustus Phillimore, 5.3.1872 under Lieut Herbert Holden Edwards, 9.11.1874 under Lieut Hugh Berners, 26.1.1876 under Lieut William Neilson, and

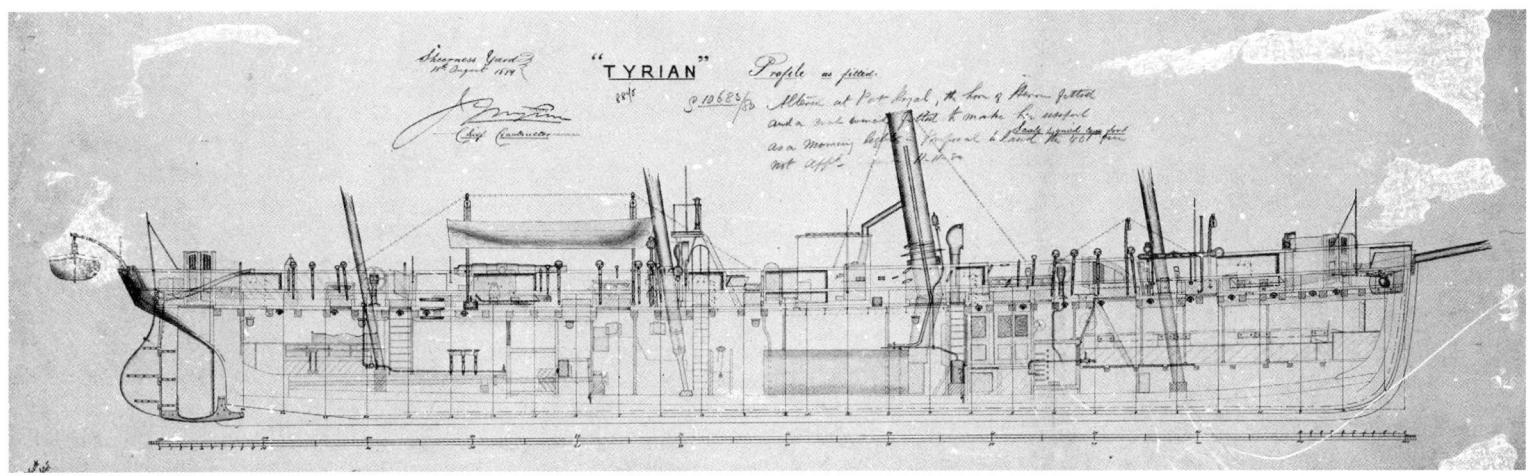

A profile of the *Britomart* Class gunboat *Tyrian* as fitted at Sheerness, dated 15 August 1879, before she was sent to Jamaica. An endorsement on the draught notes that she was modified at Port Royal to make her useful as a mooring lighter, but that a proposal to remove her guns was turned down in November 1880. This class was a marked improvement over the wartime gunboats: the contract-built vessels were built at an easier pace than their predecessors and were more sturdily built, and the later batch, all built at the Haslar Gunboat Yard, proved equally capable, although financial constraints led to the cancellation of the last four. Three of the class were deployed to the Canadian Great Lakes to patrol the waters and protect Canada from growing attacks by Fenian rebels. (*NMM DR8005*)

5.1877 under Lieut Charles Doxat; paid off 17.8.1878. BU 10.1878 at Malta.

Speedy Charles Lamport, Workington
Ord: 19.10.1859. K: 11.11.1859. L: 18.7.1860. C: 2.9.1861 for Steam Reserve at Plymouth.
Commissioned 10.1861 under Charles Burney, Master, for the Channel Islands, as tender to *Dasher*; on 1.1.1867 under Lieut William Clifton; paid off 24.6.1869. In 1870 at Portsmouth, as tender to *Asia*, harbour flagship. Sold to Castle 8.1889 to BU at Charlton.

Dotterel William Cowley Miller, Toxteth Dock, Liverpool
Ord: 19.10.1859. K: 15.11.1859. L: 5.7.1860. C: — ..
Commissioned 25.9.1861 under Lieut Benjamin Langlois Lefroy; on 19.10.1861 under Lieut William Frederick Irwin, for South America; on 22.4.1862 under Lieut William ?Frederick Johnson, then 12.4.1865 under Lieut Robert Elliot, 25.4.1866 under Lieut Robert Henry Thompson, and 8.6.1867 under Lieut William Scott; home and paid off 16.9.1868. Sold to Marshall, Plymouth on 6.6.1871.

Heron William Cowley Miller, Toxteth Dock, Liverpool
Ord: 19.10.1859. K: 15.11.1859. L: 5.7.1860. C: 4.6.1862 for Steam Reserve at Plymouth.
In 4.1862 at Pembroke Dock, coastguard reserve, as tender to *Blenheim*. Commissioned 3.4.1866 under Lieut Henry F. Stephenson, for North America; patrolled Lake Ontario in 1866–67 (based at Toronto); on 24.2.1867 under Lieut Lewis Barrett Solly; paid off 4.9.1869 as 'unfit for further service'. BU 1881 in Jamaica.

Pigeon Briggs & Co., Sunderland
Ord: 19.10.1859. K: 10.11.1859. L: 7.6.1860. C: 6.1862. for Steam Reserve at Chatham.
In 6.1866 at Plymouth, as tender to the flagship of Channel Squadron (*Caledonia* in 1866–67, then *Minotaur* 1868–70), then tender to *Cambridge* 1870 for gunnery/torpedo training. **Recommissioned** 13.9.1870 under Lieut Ralph Lancelot Turton for Senior Officer at Gibraltar; on 10.2.1872 under Lieut Francis

G. Crofton, then 15.2.1873 under Lieut John A. H. Trotter, and 10.10.1873 under Lieut Tom Falcon; paid off 24.11.1873 at Plymouth and BU there 29.9.1876.

Linnet Briggs & Co., Sunderland
Ord: 19.10.1859. K: 18.11.1859. L: 7.6.1860. C: — . for Steam Reserve at Chatham.
Commissioned 8.8.1865 under Lieut Arthur Salwey, for South America; on 20.4.1867 under Lieut Charles Percy Bushe, then 19.10.1868 under Lieut Henry John Carr; paid off 3.7.1869 at Sheerness. Fitted as torpedo trials vessel at Chatham 12.1871. BU at Chatham, completed 15.7.1872.

Tyrian Courtenay, Newhaven
Ord: 19.10.1859. K: 21.11.1859. L: 7.9.1861. C: — . for Steam Reserve at Portsmouth.
Commissioned 22.7.1864 under Lieut Patrick James Murray, for the Mediterranean; in 1865 under Lieut William Derenzy Donaldson Selby, then 12.1865 under Murray again (temp), 1.1866 under Lieut Henry Francis Hovendon (died 8.1866), 8.1866 under Lieut Victor Alexander Montagu, and 19.9.1867 Lieut Edmund Church; paid off 30.9.1868 at Plymouth into Steam Reserve; to Portsmouth Steam Reserve 1870. Recommissioned 22.3.1873 under Lieut William C. J. Blount, for service at Queensferry as tender to *Favorite*; on 27.10.1873 under Lieut Joshua Cole, then 7.1.1875 under Lieut John Eliot Pringle. Refit at Sheerness 1878–79. In 9.1879 at Jamaica, as tender to *Urgent* at Port Royal; tug there in 1883. Sold to John Wort (for £275) at Jamaica 11.8.1892.

Trinculo Joseph Banks, Plymouth
Ord: 19.10.1859. K: 1.12.1859. L: 15.9.1860. C: — . for Steam Reserve at Plymouth.
Commissioned 14.7.1862 under Lieut John Brasier Creagh at Plymouth, as tender to the flagship of Channel Squadron (*Revenge* in 1862, *Edgar* in 1863–65, and *Caledonia* in 1866). Recommissioned 19.3.1867 under Lieut Frederick de Veullé Sanders, for the Mediterranean; on 7.4.1870 under Lieut Francis G. Crofton; wrecked on 5.9.1870 after collision with ss *Moratin* off Gibraltar.

1860 Programme (6 dockyard-built)
Cherub Portsmouth Dyd
Ord: 5.3.1860. K: 3.12.1860. L: 29.3.1865. C: 1866.
Commissioned 14.4.1866 under Lieut Spencer Robert Huntley, for North America and the West Indies; patrolled Lake Huron in 1866–67 (based at Goderich); on 30.7.1869 under Lieut Noel Stephen Fox Digby, then 4.6.1872 under Lieut Francis C. Baker;

paid off 9.1874 at Sheerness into Steam Reserve. In 3.1876 to
Harwich, coastguard reserve, as tender to *Penelope*; from 1886
tender to *Hercules* at Portland. Sold to Castle 5.5.1890 to BU at
Charlton.

Netley Portsmouth Dyd
Ord: 5.3.1860. K: 3.12.1860. L: 22.7.1866. C: 1867.
Commissioned 6.1867 at Queensferry, coastguard reserve, as tender
to *Duncan* (1867–70), then to *Repulse* 1871, to *Favorite* 1872–76
and *Lord Warden* 18786–82. To Steam Reserve at Chatham 1884,
and sold there to Castle 9.1885 to BU at Charlton.

Minstrel Portsmouth Dyd
Ord: 5.3.1860. K: 17.12.1860. L: 16.2.1865. C: 1866.
Commissioned 14.4.1866 under Lieut Mervyn Bradford Medlycott,
for North America and the West Indies; on 20.2.1869 under Lieut
Noel Stephen Fox Digby, then 27.4.1869 under Lieut Harry F.
Yeatman, and 4.7.1872 under Lieut William Parsons; paid off
30.6.1873 at Bermuda. Coal hulk in 1874. Sold 1903 (but still
listed to 1906).

Orwell Portsmouth Dyd
Ord: 5.3.1860. K: 24.12.1860. L: 27.12.1866. C: 1867.
Commissioned 15.6.1867 under Lieut Alfred Frederick Marescaux,
for Ireland; on 5.11.1868 under Lieut Henry T. Price, then
30.10.1871 under Lieut Francis C. H. Dent, 11.1874 under Lieut
Bouverie Francis Clark, 10.11.1875 under Lieut Robert Cornwall,
1.1.1879 under Lieut Frederick Rendell, 26.9.1881 under Lieut
R. B. Needham, 5.7.1883 under Lieut Richard Bingham, and
11.8.1888 under Lieut Powell Cecil Underwood; paid off 7.8.1889
at Plymouth. Sold to the Customs Board 20.12.1890.

Cromer Portsmouth Dyd
Ord: 5.3.1860. K: 4.7.1861. L: 20.11.1867. C: ?1969.
Commissioned 9.1869 under Lieut George Robert Bell, at
Birkenhead, coastguard reserve, as tender to *Resistance* there

1869–74. On 16.2.1872 under Lieut Thomas Searle Dickinson,
then 10.10.1873 under Lieut William F. Boughey, and 19.10.1875
under Lieut Victor Von Donop. In 1874 tender to *Caledonia*
at Birkenhead; in 1875 tender to *Achilles* at Birkenhead.
On5.11.1877 under Lieut Edward Bolitho, as tender to *Resistance*
again, then on 1.1.1880 under Lieut George L. Poe, as tender
to *Defence* at Rockferry; on 1.1.1882 under Lieut Edward
Hodgkinson. Paid off 21.4.1882 at Plymouth and sold 24.8.1886
to BU.

Bruiser [*Bruizer*] Portsmouth Dyd
Ord: 5.3.1860. K: 14.8.1861. L: 23.4.1867. C: — .
Commissioned 25.7.1867 under Lieut Arthur C. H. Paget, at
Queenstown, coastguard reserve, as tender to *Mersey* there 1867–
72, then to *Revenge* 1872–73. Onn 14.1.1870 under Lieut John
Bruce, then on 16.2.1872 under Lieut Charles S. Shuckburgh;
paid off 22.7.1873 into Steam Reserve at Plymouth. In 1874–79
tender to *Cambridge* at Plymouth for gunnery training.
Recommissioned 6.1879 under Lieut Thomas E. Maxwell, as
tender to *Royal Adelaide*, harbour flagship at Devonport. On
29.11.1880 under Lieut Lionel Fanshawe, then 26.9.1881 under
Lieut William M. Annesley; paid off 24.3.1882 into Steam Reserve
at Plymouth. BU 5.1886.

1861 Programme (all 4 cancelled)

Bramble Portsmouth Dyd
Ord: 2.1861. Cancelled 12.12.1863.

Crown Portsmouth Dyd
Ord: 2.1861. Cancelled 12.12.1863.

Protector Portsmouth Dyd
Ord: 2.1861. Cancelled 12.12.1863.

Danube Portsmouth Dyd
Ord: 2.1861. Cancelled 12.12.1863 (unstarted).

The first vessel with screw propulsion to be ordered for the Admiralty was the instructional tender *Bee* in 1840. However, she was uniquely fitted with both screw and paddles, and her details and history are found in Chapter 11. The next vessel to be procured for screw trials was the *Dwarf*, a steam yacht designed by and built for Rennie.

DWARF The first screw vessel to serve in the Royal Navy; ex-mercantile *Mermaid*, built 1842 and purchased from G. & J. Rennie, her designers (for £5,350) by AO of 22 June 1843.
> Dimensions & tons: 130ft 0in, 120ft 4in x 16ft 6in (16ft 0in for tons) x 8ft 0in. Draught 5ft 8in (mean). 164bm. 98 disp.
> Machinery: 2-cylinder (40in diameter, 32in stroke) Rennie vertical single expansion, geared. Single screw (5ft 8in diameter). 90nhp. 216ihp = 10.537 knots.

Dwarf Ditchburn & Mare, Blackwall/G. & J. Rennie (originally Galloway).
> L: 1842. C: 10.1.1843 (for sea) at Woolwich (for £2,138) as *Mermaid*. Renamed by AO of 3 November 1843.
> Commissioned 11.11.1843 under Lieut Edward Nicolls (drowned 11.3.1833), for the Irish station; on 12.3.1844 under Lieut William Charles Chamberlain, at Portsmouth, then 14.2.1845 under Lieut Edward Halhead Beauchamp Proctor, at Sheerness. Experimental alteration 1846 to after body at Woolwich, later removed. In 1846 became trials vessel for different types of screw propeller. Employed as a tender at Woolwich and Sheerness from 1847. Sold to John Broughton by AO of 5.9.1853 for £500.

The iron-hulled *Dwarf* was the former yacht *Mermaid*, designed by Rennie to try out the Galloway rotary engine. Sir George Cockburn was persuaded by Rennie to purchase the yacht for the Navy as a trials vessel, and a variety of speed trials were held in 1845 to test out various proportions for screw propellers, varying the lines of the vessel's stern to minimise the drag. This model is thought to have been made in the G. & J. Rennie yard. *(NMM D4391-1)*

IRON SCREW TROOPSHIPS (originally ordered as frigates and sloops)

The success of a number of commercial shipyards in Britain in building ships with iron hulls led the Admiralty in 1844 to embark on a programme of four large iron frigates, to be fitted with screw propulsion, each from a specialist merchant builder. However, the Admiralty's confidence in iron ships was quickly shaken, following adverse reports of damage to iron hulls under fire. On 1 December 1846 they enquired as to the costs of cancelling the programme, and in April 1847 decided to complete the four as troopships. One (*Greenock*) was sold prior to completion and the other three were converted to troopships (in which role they served with distinction throughout the Crimean War), fitted with engines intended for blockship frigates instead of those originally ordered. No further iron warships were to be ordered until after the Crimean War. (See Table 1)

GREENOCK Iron screw troopship. Originally ordered as a 10-gun Second Class iron screw frigate, to Builders' design, approved 28.6.1845. Sold off before completion. Tender submitted 9 November 1844; design approved 10 June 1845 for vessel of 198ft x 37ft x 21½ft dimensions. By AO on 12 January 1846 to be lengthened 12ft amidships from the original 198ft, and by another AO of 1 June 1847 to be lengthened a further 3ft aft. Her name was changed from *Pegasus* to *Greenock* by AO of 8 November 1846. It was ordered on 23 April 1847 to complete her as a transport.
> Dimensions & tons: 213ft 0in, 190ft 7in x 37ft 4in x 23ft 0in. Draught 15½ft. 1,412⁷⁶⁄₉₄bm. 2,065 disp.
> Men: 175? (intended as frigate) Guns: 2 x 8in (112cwt), 4 x 68pdrs/8in (65cwt), 4 x 32pdrs (25cwt).
> Machinery: 2-cylinder (71in diameter, 4ft stroke) horizontal single expansion, geared. 4 rectangular (box) boilers. Single screw. 564nhp. 707ihp = 9.63kts.

Table 1 Tenders for Second Class Iron Screw Frigates and First Class Iron Screw Sloops, November 1844

The ten tenders received in November 1844 for such ships were (from each builder) of two types, the larger vessel to be rated as Second Class Frigates, and the smaller as First Class Sloops, showing the intended hull dimensions the builders designed:

Contractor	Second Class Frigates		First Class Sloops	
Scott & Sinclair, Greenock	204ft 0in x 36ft 6in x 24ft.	1,290 tons.	192ft 0in x 35ft 6in x 21ft 3in.	1,144 tons.
Fairbairn & Co., Millwall	208ft 0in x 36ft 9in x 23ft.	1,335 tons.	196ft 0in x 36ft 0in x 20ft 6in.	1,202 tons.
Ditchburn & Mare, Blackwall	215ft 0in x 38ft 6in x 24ft.	1,509 tons.	210ft 0in x 37ft 6in x 21ft 6in.	1,399 tons.
Robert Napier, Govan	230ft 0in x 38ft 0in x 22ft 6in.	1,591 tons.	215ft 0in x 35ft 6in x 20ft 0in.	1,298 tons.
Laird, Birkenhead	221ft 9in x 39ft 6in x 23ft.	1,616 tons.	(not tendered)	
Guppy, Bristol	205ft 6in x 42ft 4in x 23ft.	1,716 tons.	192ft 0in x 39ft 6in x 20ft 8in.	1,396 tons.
Rennie, Deptford	220ft 0in x 40ft 9in x 23ft.	1,726 tons.	210ft 0in x 38ft 6in x 22ft 6in.	1,438 tons.
Rigby, Hawarden	220ft 0in x 42ft 0in x 25ft 6in.	1,828 tons.	200ft 0in x 39ft 6in x 22ft 6in.	1,463 tons.
Bury, Curtis & Co., Liverpool	225ft 0in x 42ft 0in x 16ft.	1,874 tons.	216ft 0in x 40ft 0in x 20ft 6in.	1,634 tons.
Miller & Ravenhill, Blackwall	(not tendered)		207ft 0in x 41ft 0in x 22ft 5in.	1,608 tons.

Greenock [ex-*Pegasus*] Scott, Sinclair & Co., Greenock/Scott, Sinclair & Co.

> Ord: 19.2.1845. K: 3.9.1845. L: 30.4.1849. C: 19.5.1850 (sailed from Greenock).
> First Cost: Hull £41,261, machinery £29,306.
> By AO of 18.8.1850 sold to J. Scott Russell (the shipbuilder) on behalf of the Australian Royal Mail Co. – 'A wood screw vessel complete of *Highflyer*'s [Class] in exchange when built' (the vessel she was exchanged for was the *Esk*). Re-engined 1852 with 250nhp engine by George & John Rennie; her original Scott, Sinclair engine was fitted into the 91-gun battleship *Hannibal*. Sailed 18.9.1852, having been fitted for sea at Woolwich. Renamed *Melbourne* 25.11.1852 for the Australian RMCo. Finally sold to BU 1873.

VULCAN Iron screw troopship. Originally ordered as a 10-gun Second Class iron screw frigate, to Builders' design, approved 30 September 1845. Named by AO of 14 April 1845 (after the paddle frigate then building by Laird as *Vulcan* had been renamed *Birkenhead* on 26 February 1845). It was ordered on 23 April 1847 to complete her as a transport (with a capacity of 677 troops). During construction, Thomas Ditchburn retired and the ship was completed by the new firm of C. J. Mare & Co.

> Dimensions & tons: 220ft 0in, 195ft 4¾in x 41ft 5in oa (41ft 0in for tonnage) x 26ft 10in. Draught 17½ft. 1,747¹⁵/₉₄bm. 2,474 disp.
> Men: 175 (intended as frigate). Guns: (intended, as frigate, same as *Pegasus* above); (as troopship) 14 x 6in.
> Machinery: 4-cylinder (49½in diameter, 2ft stroke) horizontal single expansion. Single screw. 350nhp. 837ihp = 9.511kts.

Vulcan Ditchburn & Mare, Blackwall/George & John Rennie

> Ord: 4.3.1845. K: 12.3.1846. L: 27.1.1849. C: 3.3.1851 (for troops) at

Chatham and Sheerness.

> First cost: £98,927 (hull £53,797, machinery £22,853, fitting £22,277).
> Commissioned 17.4.1852 under Cmdr Edward Pelham Brenton Von Donop, for particular service; to the Mediterranean 1855; on 2.10.1855 under Cmdr George Le Geyt Bowyear, then 10.5.1856 under Cmdr James Holmes Furneaux, on particular service. Refitted for sea 1858–59 and re-engined by Maudslay, Sons & Field with the 2-cylinder (62in diameter, 3ft stroke) 400nhp engine taken from the frigate-blockship *Seahorse* and installed for £10,704 (new rating 837ihp = 8.936kts). Recommissioned 28.6.1859 under Cmdr Augustus Chetham Strode, for the East Indies and China; in Taiping rebellion; paid off 3.11.1864 at Portsmouth. Sold 1.2.1867 for mercantile service, converted to barque *Jorawur*.

MEGAERA Iron screw troopship. Originally ordered as a 20-gun First Class iron screw sloop, to Builders' design. The design was lengthened 11ft by AO of 25.11.1845. It was ordered 23 April 1847 to complete her as a screw-driven troop transport (capacity 418 troops).

> Dimensions & tons: 207ft 0in, 184ft 4¼in x 37ft 8in x 24ft 3in. 1,391³¹/₉₄bm. 2,025 disp. Draught 14ft 4in/16ft 6in.
> Men: 167. Guns: 6 x 32pdrs (25cwt/6ft).
> Machinery: 4-cylinder (49½in diameter, 2ft stroke) horizontal single expansion. Single screw. 350nhp. 843ihp = 9.8kts.

Megaera William Fairbairn & Co., Millwall/George & John Rennie

> Ord: 4.3.1845. K: 8.1845. L: 22.5.1849. C: 29.12.1851 (completed fitting engines in East India Dock 3.12.1849).

A profile of inboard works for the *Vulcan*, dated 24 June 1846. At this stage the ship was still envisaged as an iron screw frigate. *(NMM DR06397)*

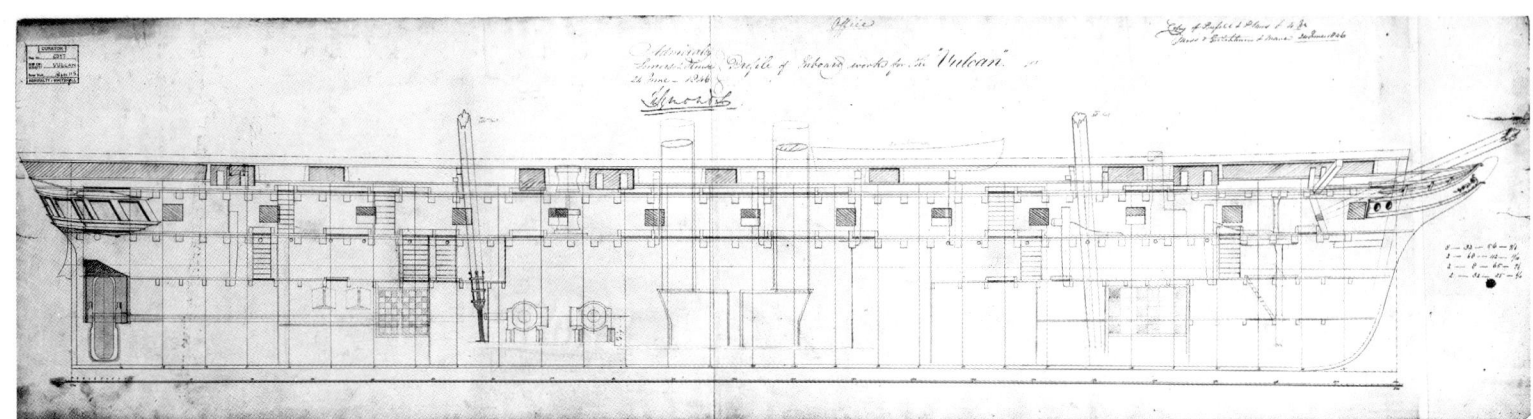

First cost: £83,942 (hull £40,415, machinery £22,349, fitting £21,178 at Woolwich and Sheerness).
Commissioned 17.5.1851 under Capt. George Hoffmeister, to carry 2 artillery companies to Cape of Good Hope, but broke down and had to be towed back to Sheerness (replaced as transport by *Cyclops*); on 18.12.1851 (after repairs) under John Clark Barlow, Master, to take Buller's rifle battalion to Cape of Good Hope; completed fitting for troops at Sheerness 29.12.1851; on 2.7.1852 under Cmdr John Ormsby Johnson; to the Mediterranean 1855, then to Black Sea; paid off 1856 at Portsmouth. Recommissioned 20.12.1856 under Cmdr George Thomas Maitland Purvis; refitted at Portsmouth for sea (for £19,408) 5.2.1857; paid off at Portsmouth. Recommissioned 10.3.1860 under Cmdr James Hunt; on 2.4.1860 under Cmdr Samuel Hood Henderson (–25.3.1863); on 20.4.1863 under Cmdr Edward Madden; paid off at Plymouth. In 1864 fitted with a Miller, Ravenhill & Salkeld 2-cylinder (55in diameter, 2½ft stroke) 400nhp engine (achieving 1,431ihp) which had been taken from *Pioneer*. Recommissioned 18.1.1865 under Capt. Montagu Buccleuch Dunn; on 19.11.1866 under Capt. James Simpson; paid off 21.12.1867 at Woolwich. Commissioned 12.2.1868 under Staff Cmdr Henry D. Sarratt; on 22.12.1868 under Staff Cmdr Jabez Loane, and 11.1869 under William F. Harris. On 31.1.1871 under Capt. Arthur Thomas Thrupp, for the East Indies; beached on Saint Paul's Island, Indian Ocean 16.6.1871 because of major leaks rendering her unseaworthy due to decay of her plating.

SIMOOM Iron screw troopship. Originally ordered as a 20-gun Second Class iron screw frigate, to builders' design. On 23 April 1847 in was ordered to complete her as a screw-driven troopship (capacity 769 troops).

> Dimensions & tons: 246ft 0in, 221ft 4¾in x 44ft 0in oa (41ft 0in for tonnage) x 26ft 7in. 1,979⁵⁷⁄₉₄bm. 2,920 disp. Draught 15ft 3in/10ft 2in.
> Men: 174. Guns: Intended as frigate: MD 12 x 32pdrs (56cwt); UD 2 x 8in/68pdrs (112cwt) + 4 x 8in/32pdrs (56cwt) + 2 x 32pdrs (25cwt). As troopship 2 x 32pdrs (56cwt) + 2 x 32pdrs (25cwt).
> Machinery: 4-cylinder (45⅔in diameter, 2½ft stroke) horizontal single expansion, oscillating. Single screw. 350nhp. 604ihp = 8.696kts.
> *Simoom* Robert Napier & Sons, Govan (No. 7)/Boulton, Watt & Co.
> Ord: 4.3.1845. K: 10.1845. L: 24.5.1849. C: 2.5.1850 (sailed from Govan).
> First cost: £105,787 (hull £59,400, machinery £23,359, fitting £23,028 at Portsmouth).
> Commissioned 9.2.1852 under Capt. John Kingcome; sailed 20.3.1852 after fitting for troops at Portsmouth. Recommissioned 2.12.1854 under Capt. Thomas Ross Sulivan, for the Mediterranean; to Black Sea and Sea of Azov 1855. Refitted and re-engined in 1855 by Portsmouth Dyd with a 2-cylinder (62½in diameter, 3ft stroke) 400nhp engine (achieving 1,699ihp = 11.589kts). On 18.12.1856 under Cmdr John Murray Cooke, for the East Indies and China; at Peiho River 1859. Reduced to 4 guns 1863. Recommissioned at Portsmouth 2.2.1866 under Capt. Thomas Bridgeman Lethbridge; paid off 9.4.1870 at Portsmouth. Recommissioned 5.12.1871 under Capt. Edward Madden. Recommissioned 5.5.1873 at Plymouth under Capt. Mountford Stephen Loviele Peile; in 10.1876 under Capt. Russell Graves Sabine Pasley; on 23.10.1878 under Capt. John Frederick George Grant; paid off 31.1.1879 at Devonport. Sold 6.1887 to Collings, Dartmouth to BU.

OTHER SCREW-DRIVEN TRANSPORTS/TROOPSHIPS

The advent of the Russian War in 1854 witnessed the first large-scale taking up of large merchant vessels to serve as auxiliaries, instead of the traditional practice of using elderly warships; the majority of these were iron-hulled screw ships, which gave the Navy considerable experience of metal hulls prior to their wholesale adoption for large warships.

PERSEVERANCE Purchased 1854 (was building for Russia as *Sobraon*) iron screw troopship. Purchased on stocks from builders for £19 per ton, value of machinery seized from Russians £11,800, builder paid £44,901, by AO of 1 May 1854. The *Urgent* (see below) was a near sister but not identical. Manned and operated by P & O Steam Navigation Co. as a troopship for the Russian War campaign (2nd Baltic expedition in 1855).

> Dimensions & tons: 270ft 7in, 247ft 0in x 38ft 4in x 17ft 10in. Draught 25ft 9in. 1,929⁷⁸⁄₉₄bm. 2,299 disp.
> Machinery: John Penn & Son 2-cylinder (55in diameter, 3ft stroke) horizontal single expansion, trunk. 360nhp. 912ihp = 11.297kts.
> *Perseverance* C. J. Mare, Blackwall./John Penn & Son
> L: 13.7.1854. Completed for sea 19.6.1855 at Sheerness (previously at Woolwich).
> First cost: £87,188.
> Commissioned at Woolwich 23.11.1854 under Cmdr William John Samuel Pullen (–6.2.1855). Completed for sea at Woolwich and then at Sheerness 19.6.1855. Commanded (not commissioned) 9.6.1855 by Capt. Henry Harris (–21.11.1855); to the Mediterranean 9.1855. Recommissioned 28.11.1855 at Portsmouth under Cmdr John Hay Crang, for particular service; grounded on Portland Bill (Crang superseded 26.2.1856); on 27.2.1856 under Cmdr John Wallace Douglas McDonnald, on particular service; paid off 30.9.1857. Recommissioned 21.7.1858 under Cmdr Edward Roche Power, on particular service; wrecked 21.10.1860 on the island of Maio, Cape Verde Islands (all saved).

PRINCE Purchased 1854 (ex-mercantile *Prince*, launched 12 April 1854). Never commissioned as a Navy vessel, but taken over for use by the Transport Service as *Transport Number 107*.

> Dimensions & tons: 291ft 6in x 41ft 6in x 30ft 8in. 2,710bm.
> Machinery: 2-cylinder horizontal single expansion. 300nhp. 815ihp.
> *Prince* C. J. Mare, Blackwall/Maudslay, Sons & Field
> Purchased from General Screw Steam Shipping Co. (for £105,000) by AO of 7.1854.
> Commanded 1854 by John Goodall, (civilian) master, with a merchant crew (and Cmdr Benjamin Baynton, RN, aboard as Transport Agent, not as CO), for the Black Sea; wrecked 14.11.1854 in a hurricane-force storm off Balaklava (all 150 aboard died, including Baynton, except for 6 men).

HIMALAYA Purchased (ex-P & O *Himalaya*). Was the largest vessel in the world when completed as a passenger liner for the P & O.

> Dimensions & tons: 340ft 5in, about 304ft 10in x 46ft 1¼in x 24ft 0in. 3,452⁶⁹⁄₉₄bm. 4,490 disp. Draught 18ft 10in/21ft 3in.
> Men: 213. Guns: 66 x 32pdrs (25cwt/6ft).
> Machinery: 2-cylinder (78in diameter, 3½ft stroke) horizontal single expansion, trunk. 700nhp. 2,556ihp = 12.8kts.
> *Himalaya* C. J. Mare, Blackwall/John Penn & Sons

The iron-hulled screw troopship *Orontes* in the Suez Canal, shown in this 1880 lithograph following her 1873-76 reconstruction and fitting with a compound engine by her builders, Lairds at Birkenhead, in which she was lengthened by 50ft. She was defensively armed with 3 x 4pdr guns. The white hull was designed to reflect as much heat as possible in the tropics, and with the associated buff funnels made up a livery that was still used by British warships on the Far Eastern Station as late as the 1930s. *(NMM PU6233)*

Purchased from the Peninsular and Oriental S. N. Co. for £130,000 by AOs of 10.7.1854 & 26.7.1854. L: 24.5.1853.

Commissioned 5.1.1855 under Cmdr Benjamin Portland Priest; on 24.3.1857 under Cmdr William Henry Haswell (–31.5.1858); on 9.6.1858 under Cmdr Shute Barrington Piers, then 21.7.1858 under Cmdr (Capt. 13.12.1859) John Seccombe; on 25.7.1862 under Capt. Edward Lacy. Re-engined 1863 by Keyham Factory with 2-cylinder (77⅛in diameter, 3½ft stroke) horizontal single expansion engine (re-rated 2,609ihp). On 30.6.1865 under Capt. Thomas Bridgeman Lethbridge; on 22.3.1870 under Capt. Edward Madden, then 1.7.1872 under Capt. William Burley Grant, 28.7.1876 Capt. Edward White, and 12.8.1879 under Capt. Harry Woodfall Brent; paid off 2.3.1881 at Devonport into Steam Reserve. Recommissioned there 7.12.1882 under Capt. Henry St Leger Bury Palliser; on 2.1.1886 under Capt. John Edward Stokes, then 1.3.1889 under Capt. Robert Hammick, and 20.4.1892 under Capt. Edward Chichester; paid off 28.9.1894 at Devonport. Converted to coal hulk 11.1894 at Devonport, renamed *C.60* in 12.1895. Sold to E. W. Payne 28.9.1920 (became a hulk at Portland, and was sunk there by dive-bombers 6.1940 during a German air raid).

TRANSIT Purchased on stocks from builder (building for Russia?) by AO of 13 June 1854 at £21 per ton.

Dimensions & tons: 302ft 6in, 277ft 8¾in x 41ft 8½in x 24ft 1½in. 2,569⁷⁴⁄₉₄bm.

Machinery: 2-cylinder (58⅛in diameter, 3¼ft stroke) horizontal single expansion, trunk. 400nhp. 1,234ihp = 11.9kts.

Transit C. J. Mare, Blackwall/John Penn & Sons

Purchased from builder by AO of 13.6.1854. L: 20.3.1855. Completed for sea 20.7.1855 at Sheerness.

First cost: £109,261 (hull £62,293, machinery £14,689, fitting £32,279).

Commissioned at Portsmouth 24.7.1855 under Cmdr Charles Richardson Johnson; wrecked 10.7.1857 on a sunken rock in the Straits of Banca while carrying troops to the East.

URGENT Purchased on stocks from builder (building for Russia as *Assaye*) by AO of 13 June 1854. A near, but not identical, sister to *Perseverance* (see above).

Dimensions & tons: 272ft 3½in, 250ft 2½in x 38ft 5in x 26ft 8½in. 1,964³⁸⁄₉₄bm. 2,801 disp.

Proposed for lengthening: 300ft 6in, 277ft 3½in x 38ft 5in x 26ft 8½in. 2,204⁴⁹⁄₉₄bm.

Machinery: 2-cylinder (64in diameter, 3ft stroke) horizontal single expansion. 400nhp. 1,483ihp = 11.72kts.

Urgent C. J. Mare, Blackwall/Maudslay, Sons & Field

Purchased from builder by AO of 13.6.1854. L: 2.4.1855. Completed fitting for sea 29.9.1855 at Sheerness.

First cost: £89,936 (hull £48,194, machinery £18,826, fitting £22,916).

Commissioned 25.7.1855 under Cmdr Charles Gerrans Phillips, on particular service; embarked 1,114 troops from Portsmouth 30.9.1855 for Malta; sailed 7.11.1855 from Malta for home; paid off 6.6.1857. Refitted 1857 at Portsmouth (for £17,700, plus £12,280 to Maudslay for machinery). Recommissioned 1.10.1857 under Cmdr John Wallace Douglas McDonnald, on particular service; on 31.3.1859 under Cmdr Henry William Hire, for the East Indies and China; at Peiho forts 20.8.1859; paid off 27.11.1862 at Portsmouth. Recommissioned 22.7.1864 under Capt. Samuel Hood Henderson, on particular service; paid off 20.8.1869 at Portsmouth into the Steam Reserve. Recommissioned 28.11.1870 (still under Henderson) to take scientific party to the Mediterranean to observe solar eclipse; paid off 13.1.1971 into Steam Reserve again. Fitted as depot ship at Portsmouth (engines and armament removed, and broad, thick planking fitted along her sides at water level) for Jamaica 1875–3.1876. Recommissioned 1.5.1877 under Cmdr Robert Boyle; arrived 20.7.1877 at Port Royal and recommissioned next day under Commodore Algernon McLennan Lyons as local deport ship and Commodore's flagship for Jamaica; on 8.2.1878 under Commodore William John Ward, then 3.8.1880 under Commodore William Samuel Brown and 10.3.1882 under Commodore Edward White. Recommissioned 10.1883, still under White; on 2.6.1883 under Commodore Francis Mowbray Prattent, then 29.6.1886 under Commodore Henry Hand, 19.9.1889 under Commodore Rodney Maclaine Lloyd, 20.9.1892 under Commodore Thomas Sturges Jackson, 4.9.1895 under Commodore Herbert Ward Dowding, 31.3.1898 under Commodore William Hannam Henderson, 13.1.1900 under Commodore Edward Henry Megga Davis, and 3.1901 under Commodore Daniel MacNab. Riddel; paid off 13.3.1903 at Port Royal. Sold locally to Butler & Co. 6.1903.

ORONTES Class. Designed by the Surveyor's Dept.
>   Dimensions & tons: 300ft 1in x 44ft 8in. 2,812bm. 4,857 disp. Draught 22ft 0in/22ft 6in.
>   Machinery: 2-cylinder (71in diameter, 3ft stroke) horizontal single expansion. 500nhp.
>
> *Orontes* Laird Brothers, Birkenhead (No. 286)/James Watt & Co. 2,143ihp = 12.354kts.
>   As built: 300ft 1in x 44ft 8in. 2,823 bm.
>   Guns: 2 x 40pdrs (32cwt/10ft) Armstrong guns.
>   Ord: 1861. K: 7.1861. L: 22.22.1862. C: 3.1863.
>   Commissioned 20.4.1863 under Capt. Henry William Hire; on 4.1.1867 under Capt. Henry Phelps; on 23.3.1870 under Capt. John Laisné Perry; paid off 2.5.1873. Refit at Lairds, Birkenhead 1873–76 – lengthened to 350ft 0in (becoming 5,330 disp.) and re-engined with Laird 2-cylinder inverted compound engine (high pressure cylinder 54in, low pressure cylinder 94½in; stroke 3½ft), 2,569ihp = 13.017kts. Recommissioned 7.2.1876 under Capt. Edward Hobart Seymour; on 21.3.1879 under Capt. Richard George Kinahan, then 10.6.1882 under Capt. Hilary Gustavus Andoe, 2.4.1885 under Capt. Charles Barstow Theobald, 4.4.1888 under Capt. Alexander George McKechnie, and 28.7.1891 under Capt. Oswald Peploe Tudor; paid off 13.5.1892 at Portsmouth. Sold to Messrs Crawford (for £7,255) and sailed thence 3.7.1893 to BU on the Thames.
>
> *Tamar* Samuda Brothers, Millwall/Ravenhill & Salkeld [1,869ihp = 10.778kts]
>   As built: 300ft 0in, 267ft 0in x 44ft 7in x 23ft 7in. 2,812bm.
>   Guns: 3 x 6pdrs.
>   Ord: 1861. L: 5.1.1863. C: ?10.1863.
>   Commissioned 10.1863 under Capt. Frederick Henry Stirling. On 4.7.1866 under Capt. Francis William Sullivan. In 9.1869 under Capt. Henry Dennis Hickley. Recommissioned 8.5.1873 under Capt. Walter James Hunt-Grubbe, for Ashanti coast 1873–74; paid off 21.5.1874 at Plymouth. Recommissioned 19.4.1878 under (briefly) Capt. Charles James Brownrigg, then 22.4.1878 under Capt. William Henry Liddell; in 5.1880 under Capt. Thomas Harvey Royse; landing troops at Alexandria 17.7.1882. In 1886 under Capt. John Borlase Warren, then 11.1887 under Capt. Basil Edward Cochrane. Became receiving ship at Hong Kong; recommissioned 10.1897 in that role; finally scuttled 12.12.1941 at Hong Kong during Japanese occupation.

STORESHIPS

INDUSTRY Iron-hulled storeship, purchased 1854 on stocks from her builders.
>   Dimensions & tons: 179ft 5in, 162ft 6in x 27ft 2in x 14ft 9in. 637⁸/₉₄bm. 1,126 disp.
>   Machinery: 2-cylinder (36½in diameter, 3ft stroke) vertical single expansion, geared, oscillating. 100nhp. 318ihp = 9.12kts.
>   Purchased from the builders (for £19,250) by AO of 19.4.1854. L: 1854. C: 18.7.1854 (for sea) at Woolwich (for £2,408).
>
> *Industry* C. J. Mare, Blackwall/James Watt & Co.
>   Commissioned 31.5.1854 under George Lee Bradley, Master, for the Mediterranean; in 4.1855 under George H? K. Bower, Master. On 28.1.1857 under George James Hodges, Master, at Woolwich, for particular service; on 16.12.1862 under Edward Charles Taylor Youel, Master (from 15.9.1865, Staff Cmdr); paid off 1.1868. On 1.1.1868 under Staff Cmdr Charles James Polkinghorne, for the west coast of Africa; on 2.3.1869 under Staff Cmdr Robert Lawcay Cleveland, for the Cape of Good Hope and west coast of Africa; paid off 10.6.1871 at Devonport into Steam Reserve. Recommissioned 14.3.1877 under Staff Cmdr Richard Cossantine Dyer, for the west coast of Africa and the Cape of Good Hope (in the Zulu War); took possession of Walfisch (Walvis) Bay 12.3.1878; paid off 19.6.1880 at Devonport, becoming tender to *Asia*. Converted to boom defence vessel at Southampton 1896–97, and stationed at Southampton from 11.1897. Sold to Ward, Preston 10.10.1911 to BU.

SUPPLY Iron-hulled storeship, purchased 1854 on stocks from her builders. Sister ship to *Industry* above.
>   Dimensions & tons: 178ft 0in, 161ft 8in x 27ft 0in x 19ft 9in. 627³⁸/₉₄bm. 1,126 disp.
>   Machinery: 2-cylinder (34½in diameter, 2¼ft stroke) inclined single expansion, trunk, geared. 80nhp. 265ihp = 8.42kts.
>   Purchased from the builders for (£19,250) by AO of 19.4.1854. L: 3.6.1854. C: 18.8.1854 (for sea) at Woolwich (for £2,019).
>
> *Supply* C. J. Mare, Blackwall/G. & J. Rennie
>   Commissioned 5.5.1854 under John Penn, Master, as tender to the *Fisgard* at Woolwich. Recommissioned 26.4.1856 under William Henry Balliston, Master, for particular service; re-appointed 8.12.1860. On 24.4.1863 under Charles Bawden, Master; Bawden re-appointed (now as Staff Cmdr) 18.11.1865; paid off 3.3.1868 at Woolwich. Recommissioned 25.10.1870 under Staff Cmdr Narcissus George Arguimbeau, for the west coast of Africa and Cape of Good Hope; paid off 6.5.1873. Recommissioned 13.12.1873 under Staff Cmdr Frank Inglis, for same station; at the blockade of Dahomey 1876–77; paid off 29.6.1877. BU completed 8.2.1879 at Chatham.

RESOLUTE Purchased 1855 on stocks from builders. Not at first commissioned as a Navy vessel, but taken over for use by the Transport Service as *Transport Number 214* during 1855–57.
>   Dimensions & tons: 282ft 10½in, ?254ft 7in x 36ft 4¾in x 24ft 1in. 1,794bm. 2,510 disp. Draught 17ft 3in/17ft 6in.
>   Men: — . Guns: 2 x 18pdr carronades.
>   Machinery: 4-cylinder (45⅜in diameter, 3ft stroke) horizontal single expansion, trunk. 400nhp. 825ihp = 10.316kts.
>
> *Resolute* Laird Bros, Birkenhead (No. 115)/Fawcett, Preston & Co.
>   Purchased from Laird for £66,000 by AO of 16.1.1855 as storeship. L: 17.2.1855.

Renamed *Adventure* as troopship by AO of 16.2.1857.

Commanded 1855–57 by Nicholas Pentreath, (civilian) master, with merchant crew. After renaming, commissioned 20.4.1857 under Cmdr Edward Lacy, for the East Indies and China; paid off 10.5.1861 at Portsmouth. Re-engined 1862 with Ravenhill, Salkeld & Co. 2-cylinder (64in diameter, 3ft stroke) horizontal 400nhp engine (rating 1,227ihp = 11.447kts). Recommissioned 22.4.1862 under Cmdr Thomas Bridgeman Lethbridge; on 17.11.1863 under Cmdr Charles Thomas Curme, then 25.2.1864 under Capt. Charles Lodowick Darley Waddilove; paid off 1867. On 22.6.1868 under Capt. Henry James Raby; on 23.12.1871 under Cmdr Joseph Edward Maitland Wilson. Recommissioned 29.5.1874 at Chatham under Capt. John D'Arcy; paid off 11.8.1875 at Chatham. BU 2.1877 at Chatham.

ASSISTANCE Purchased 1855 on stocks from builders. Not at first commissioned as a Navy vessel, but taken over for use by the Transport Service as *Transport Number 215* during 1855–56.

    Dimensions & tons: 282ft 10in, ?254ft 2in x 36ft 5in x 24ft 1in. 1,793bm. 2,260 disp.

    Machinery: 4-cylinder (45⅛in diameter, 3ft stroke) horizontal single expansion, trunk. 400nhp. 879ihp = 10.663kts.

Assistance Laird Bros, Birkenhead. (No. 116)/Fawcett, Preston & Co

    Purchased by AO of 16.1.1855 (for £66,000). L: 5.4.1855. C: 19.6.1855 at Birkenhead.

    Commanded 1855–56 by James Blow, (civilian) master, with merchant crew, for the Mediterranean. Refitted at Portsmouth (for £11,612) 1856–57. Recommissioned 20.4.1857 under Cmdr William Andrew James Heath, for the East Indies and China; in 11.1859 under Cmdr Charles J. Balfour; wrecked 1.6.1860 near Hong Kong, wreck sold by auction (for £1,010.8.4d) 30.7.1860.

BUFFALO Purchased 1855 (ex-mercantile *Baron von Humbolt*, built by C. J. Mare, Blackwall). Not at first commissioned as a Navy vessel, but taken over for use by the Transport Service as *Transport Number 243* during 1855–56.

    Dimensions & tons: 137ft 2in, — . x 26ft 2in x 14ft 4in. 440bm. 620 disp.

    Men: — . Guns: 2.

    Machinery: — . 60nhp. 252ihp. (re-engined 1860 with 80nhp engine).

    Purchased by the Treasury for the Crimean Railway Company for £11,965, then transferred to the Admiralty 5.9.1855. Name altered to *Buffalo* by AO of 26.11.1855.

    Commanded 1855–56 by a civilian master, with merchant crew. Commissioned 9.12.1856 under Theophilus Jones, 2nd Master; in 4.1857 under John Eaton, Master, then 4.12.1857 under George Moore, Master, for the west coast of Africa; on 6.7.1859 under Alexander Brown, Master, on the west coast of Africa; paid off 25.1.1862 at Woolwich. Became victualling tender there 17.1.1864. For sale 4.1888, but not sold and instead BU at Chatham 11.1888.

HESPER Purchased 1855 [ex-mercantile *Hesperus*, built 1854 by Charles Mitchell, Walker. (No. 7)]. Not at first commissioned as a Navy vessel, but taken over for use by the Transport Service as *Transport Number 47* during 1855–56.

    Dimensions & tons: 203ft 7in, — . x 28ft 7in x 17ft 11in. 808bm.

    Men: — . Guns: 4.

    Machinery: — . 120nhp. (re-engined with 150nhp engine in 1860).

    Transferred from the Treasury 24.5.1855 – from 1858 on the China Station.

    Commanded 1855–56 by Mr Cruikshank, (civilian) master, with merchant crew; later under Mr Walling. Commissioned 16.11.1857 under Henry Hill, Master, for the East Indies and China; on 15.3.1860 under William Henry Harris, Master, then 6.6.1863 under Alexander Fraser Boxer, Master, and 23.6.1865 under Staff Cmdr James George Hobbs Thain, on same station; paid off 14.9.1867 at Woolwich. Sold 10.1868, retaking original name.

WYE Purchased 1855 (mercantile *Hecla*) iron steam screw storeship/tank ship. Fitted with condensers to produce 40 tons of fresh water per day from sea water.

    Dimensions & tons: 206ft 0in, — . x 26ft 8in x 17ft 2in. 700bm. 1,161 disp.

    Machinery: Blair & Co. 2-cylinder inverted compound engine. 130nhp. 629ihp = 10.928kts.

    Transferred to the Admiralty on 24.5.1855 by the Treasury who had purchased her for £24,000 for the Crimean Railway. Renamed 6.6.1855 and arrived Portsmouth 16.6.1855 to be fitted as a tank vessel for the Crimea .

    Commissioned 11.10.1855 under Philip Charles Durham Bean, Master, for the Mediterranean; in 12.1855 under Lieut James Boucher Ballard, then 6.1856 under George Moore, Master; paid off 28.2.1857 at Woolwich into Steam Reserve. In 1860 described as a 'distilling ship'. Fitted for a storeship at Chatham 1860–61. On 6.4.1861 under Valentine G. Roberts, Master, for Sierra Leone; paid off 30.10.1865 at Woolwich. Sold to Mr Royden 3.7.1866.

MANILLA Purchased 1860 [ex-mercantile *Atalante*, built 1860 by Pile, West Hartlepool].

    Dimensions & tons: 160ft 0in, — . x 23ft 0in x 12ft 4in. 295bm.

    Machinery: — . 70nhp.

    Purchased at Hong Kong by the authority of the Commander in Chief for £13,000 on or before 6.7.1860. Name changed to *Manilla* by AO of 19.7.1861.

    Commissioned 11.11.1862 under Henry William Burnett, Master, for the East Indies and China; on 9.9.1865 under Frederick William Rea, Master, then 24.5.1866 under Lieut John Rowe Ryan. Exchanged 28.2.1870 for a sailing barque (mercantile *Ingeberg*, renamed **Manilla**).

DEPOT SHIPS

BRUISER Iron screw storeship, purchased 1854 while building (ex-mercantile *Robert Stephenson*) as a floating flour mill for the Black Sea. Fitted by Fairbairn & Sons to grind up to 800 bushels of wheat daily.

    Dimensions & tons: 183ft 0in, — . x 25ft 6in x 15ft 0in. 580bm.

    Men: — . Guns: — .

    Machinery: — . 80nhp.

Bruiser Richardson & Duck, Stockton-on-Tees

    L: 27.7.1854. Purchased by AO of 8.12.1854 (for £16,000). C: 24.4.1855 at Deptford, then Woolwich (still as *Robert Stephenson*, name changed to *Bruiser* by AO of 10.6.1855) for £876.

    Stationed at Balaklava 1855. Sold to Charles Colman by AO of 12.1.1857 for £8,500.

A contemporary full hull and rigged model of the screw yacht *Fairy* of 1845 The model is fully decked, rigged with three masts and spars, and equipped with a variety of fittings, including a capstan and binnacle, wheel, pumps, a complete set of boats slung from davits, and several skylight along the deck to illuminate the accommodation below decks. *(NMM F2856-1)*

ABUNDANCE Iron screw storeship, purchased January 1855 (ex-mercantile *Alfred*, built 1854) as a floating bakery ('flour mill') for the Black Sea, and renamed February 1855. Fitted by Swaine & Bevill of Millwall to produce 20,000 lb of bread per day.

> Dimensions & tons: 170ft 0in, ?153ft 4in x 27ft 6in x 15ft 8in. 617bm.
> Machinery: Horizontal single expansion. 70nhp.
> *Abundance* James Laing, Sunderland (No. 214)/—
> > Purchased from Mr Parnal? for £14,400 by AO of 6.1.1855. C: 7.4.1855 at Deptford (for £1,189).
> > Stationed at Balaklava 1855. Sold 3.1.1856 (by AO 31.10.1855) to Messrs Hambro & Sons (for £8,500).

CHASSEUR Purchased 1855 on stocks from builders (ex-mercantile) as a floating factory for repairing material for the Army in the Crimea.

> Dimensions & tons: 159ft 5½in, 143ft 5½in x 26ft 9in (26ft 8in for tons) x 17ft 2in. 543bm.
> Machinery: — — 70nhp (engine later removed). 350ihp.
> *Chasseur* T. & W. Smith, North Shields/—
> > Purchased by AO of 28.5.1855 (for £12,500).
> > Stationed at Balaklava 1855, later as steam factory at Sheerness 11.1856–?1870, then at Chatham. Sold to Ward, Preston 25.5.1901.

VICTORIA Purchased 1856 [ex-mercantile *Water Rail* (?) — ..] iron steam screw tank vessel, based at Malta 1857.

> Dimensions & tons: 90ft 0in x 15ft 0in x 6ft 10in. 98bm. 60 disp.
> Machinery: — . 30nhp.

Purchased at Constantinople from Henry Aspinall for £7,000 by Admiral Sir Houston Stewart; purchase approved by AO of 12.1.1856.
By 7.1859 known as *Humber*. Sold 1866 as mercantile *Virginia*.

ROYAL YACHTS

FAIRY 1844 design. Iron screw tender for the royal yacht *Victoria & Albert*, at Portsmouth. In many ways the forerunner of the Crimean gunboat designs. Used for trials with anti-fouling compounds.

> Dimensions & tons: 144ft 8in, 133ft 6in x 21ft 1½in x 9ft 10in. 317³/₉₄bm.
> Machinery: 2-cylinder (42in diameter, 3ft stroke) vertical single expansion, oscillating, geared. Single screw. 128nhp. 364ihp = 13.324kts.
> *Fairy* Ditchburn & Mare, Blackwall/John Penn & Co.
> > Ord: 22.11.1844. (named 19.12.1844) K: 1844. L: 3.1845. C: 6.8.1846.
> > First cost: £20,984 (hull £12,268, machinery £7,110, fitting £1,606).
> > Commissioned 21.5.1845 under George Henry Forster, Master; on 15.2.1848 under David Nairn Welch, Master (as Staff Cmdr from 8.8.1863). HM Queen Victoria was aboard during the Spithead Review on 11.3.1854, and she led the Baltic Fleet (under Sir Charles Napier) out to sea as far as St Helens. Stripped and 'internally demolished' from 11.1867 at Portsmouth, then sold 1.1868.

The Arctic exploration barque *Assistance* (see Chapter 9) is shown under tow in ice-studded Arctic waters by the screw tender *Pioneer* in this watercolour by Walter William May. In the autumn of 1851 the two ships proceeded up the Wellington Channel (in the Parry Islands) as far as 77 degrees N, and wintered in Northumberland Inlet, while the other 'team' in this expedition – the barque *Resolute* and screw tender *Intrepid* – proceeded further west to Melville Island. *(NMM PW7061)*

EMPEROR Iron steam screw yacht, 1856. Built as a present for the Emperor of Japan.

>Dimensions & tons: 136ft 9in, 123ft 5¼in x 22ft 1in x 11ft 6in. 318⁷⁄₉₄bm.
>
>Machinery: 60nhp.
>
>First cost: £14,849 (hull £6,370, machinery £3,840, fitting £4,639).

Emperor R. & H. Green, Blackwall. (No. 339)/John Penn & Sons.

>Ord: — . K: 24.5.1856. L: 29.11.1856. C: 22.4.1857. Presented to Japan 7.1858.
>
>Named *Emperor* on commissioning by AO of 27.2.1857; commissioned under Lieut John Ward. Transferred at Shinagawa in 7.1858 and named *Banryu*. In 10.1868 went over to the rebels and was sunk in 5.1869. Salvaged by an American firm who renamed her s.s. *Emperor*. Returned to Japan 1873 as *Raiden Maru* and in 1877 transferred to the Imperial Navy as *Raiden*; served as a gunboat and then a transport. Sold 1888 for conversion to whaler *Raiden Maru*, broken up at Osaka about 1899.

TENDERS, etc.

SIREN Purchased 1855 on stocks (building as mercantile *Mengin*) for the Colonial Government of Bermuda; builder's design?

>Dimensions & tons: 107ft 0in, — . x 17ft 1in x 6ft 7in. 146bm. 99 disp.
>
>Men: — . Guns: — .
>
>Machinery: J. Key 25nhp.
>
>First cost: £5,235.

Siren Laird & Co., Birkenhead. (No. 132)/J. Key.

>Ord: — . K: 21.5.1855. L: 21.10.1855. C: — .
>
>Never on the Navy List, as intended solely for use by the Royal family. On 19.11.1863 the Admiralty, having been asked by the Colonial Department whether an application by Attwood & Co. to purchase the vessel for £2,000 should be approved, agreed to the proposal. Sold 12.1863.

DISCOVERY SHIPS

Purchased Arctic schooners

While most of the ships used for Arctic exploration in this era were sailing vessels, in February 1850 the Admiralty purchased from the Continental Cattle Steam Ship Conveyance Company, Southampton (for £9,000 each, by AOs of 16 & 25 February) two mercantile vessels for use in Capt. Austin's expedition to search for the missing Franklin expedition. These sharp-bowed wooden-hulled screw steamers were strengthened for ice navigation, and were rated as screw tenders to the sailing barques *Assistance* and *Resolute* respectively (see Chapter 9).

>Dimensions & tons: 143ft 6in, about 124ft 3in x 22ft 9in x 14ft 7in. 342bm. (for each vessel)
>
>Machinery: James Watt & Co. 2-cylinder oscillating. 60nhp.

Pioneer [ex-mercantile *Eider*] R. & H. Green, Blackwall (No. 277)/ James Watt & Co.

>Purchased 16.2.1850, then fitted at Woolwich for Arctic Seas (for £9,600) 17.2–27.4.1850.
>
>Commissioned 28.2.1850 under Lieut Sherard Osborn. Refitted at

Woolwich for the Arctic (for £1,871) 1851–52. Recommissioned 30.10.1852, still under Osborn; abandoned in the ice off Bathurst Island 25.8.1854 (written off the Navy List by AO of 25.10.1854).

Intrepid [ex-mercantile *Free Trade*] R. & H. Green, Blackwall (No. 274)/James Watt & Co.

>Purchased 28.2.1850, then fitted at Woolwich for Arctic Seas (for £9,111) 21.3–30.4.1850. Initially named *Perseverance*, but renamed *Intrepid* 3.1850.
>
>Commissioned 20.2.1850 under Lieut John Bertie Cator. Refitted at Woolwich for the Arctic (for £2,151) 1851–52. Recommissioned 10.2.1852 under Lieut Francis Leopold M'Clintock. Abandoned in the ice 15.6.1854 (written off the Navy List by AO of 25.10.1854).

Dredgers

MALTA Class 1859: Builder's design? Both served at Malta.

>Dimensions & tons: dimensions unknown. 280bm.
>
>Machinery: — .

Malta Summers, Day & Baldock, Northam

>Ord: 1.1859. K: 17.2.1859. L: 23.7.1859. C: 29.8–11.10.1859 at Portsmouth.
>
>Sold 6.1870.

Valetta Summers, Day & Baldock, Northam.

>Ord: 1.1859. K: 17.2.1859. L: 3.8.1859. C: 5.9–24.11.1859 at Portsmouth.
>
>Became *YC.27* at Malta.

Screw Lighters

PROGRESSO ex-slaver, captured 12 April 1843 by *Cleopatra*, and purchased 23 April 1844 for use as a tank vessel.

>Dimensions & tons: dimensions unknown. 140bm. 230 displ.
>
>Stationed at the Cape of Good Hope. BU 3.1869.

PERA Purchased 1855 [ex-mercantile, previous details unknown] iron screw lighter

>Dimensions & tons: 80ft 0in x 20ft 0in x 9ft 6in. 126bm (88.9 register)
>
>Machinery — . 25nhp.
>
>Purchased from Joseph Aspinall at Constantinople by Admiral Grey for £4,250 on 30.10.1855.
>
>Sold 10.10.1856 at Constantinople to the Peninsular & Oriental SNCo. for £800.

THISTLE Purchased 1857 [ex-mercantile —] iron screw coal lighter

>Hull purchased at Hong Kong from Curry for £312 on 12.6.1857 after being seized and set fire to by the Chinese on 11.12.1856, became *Hong Kong Yard Craft No. 1*. Sold at Hong Kong 1869.

CHESTER Iron screw tank vessel

>Dimensions & tons: 104ft 8in x 18ft 3in x 8ft 2in. 164bm. 234 disp.
>
>Machinery: — . 30nhp. 90ihp.

Chester Laird Bros, Birkenhead (No. 274)/—

>K: 5.11.1860. L: 14.3.1861. C: 20.5.1861 (with machinery, at Birkenhead), then 24.8.1861 (at Portsmouth).
>
>Sold 12.1925.

Postscript: The First Ironclads

While nominally included in the Rating System of the Navy, the ironclad warships of the early 1860s require a separate treatment. Under the rating system, ships carrying a complement of between 600 and 800 men – such as the *Warrior* – were technically classed as Third Rates, while ironclads with a complement of between 400 and 600 men were classed as Fourth Rates.

The marriage of iron hulls with steam screw propulsion – in retrospect so obvious – was accomplished in three phases during the first half of Victoria's reign. In the mid-1840s a series of iron frigates and gunvessels was initiated; but almost before any of these took to the water, Admiralty jitters over the vulnerability of iron hulls to shellfire led to a quick about-face and the cancellation or conversion of most of these ships; their fear was at least partly justified, given the quality of early iron-making.

During the Russian War, a second phase began with the construction of a series of floating batteries for coastal bombardment; in addition a number of iron-hulled screw-driven auxiliaries, particularly troopships and storeships, were acquired which, while not themselves warships, gave the Navy considerable experience in the practical operation of vessels of such construction. However, no battlefleet units or cruising vessels were built at this time.

Finally, in 1859, the Navy, now reassured as regards iron construction, began the building of genuine ironclad frigates commencing with the *Warrior*, supplemented by the fitting of iron armour to wooden hulls as a means of making up for a shortage of units (compared with the French).

Armoured Iron Frigates (Broadside Ironclads)

Between 1859 and 1861 inclusive, ten new armoured frigates were laid down, all carrying their weapons in traditional broadside fashion. The first six originally mounted 7in/110pdr (82cwt) Armstrong breech-loaders (BLs) and 68pdr (95cwt) muzzle-loaders, but the Armstrong guns were quickly superseded by muzzle-loading rifled guns (MLRs).

WARRIOR **Class** 41 guns. Designed by Isaac Watts, 1859. Built under the 1859 Programme; tenders were sought for the first ship on 29 April, and that from Thames Iron Works was accepted on 11 May; she was named *Warrior* on 24 September. Originally planned to carry 48 x 68pdrs (95cwt) MLSB (MD 38, UD 10), but completed as below. Tenders for the second ship were sought on 23 September, and the contract with Napier was agreed on 6 October; her name was first intended to be *Invincible*, but was renamed *Black Prince* on 17 January 1860. Ship-rigged, with 48,400 sq.ft sail area. Armour belt was 213ft long over midships section and 22ft deep. The hull was subdivided into 92 water-tight compartments. Both re-armed 1867 with 4 x 8in (9ton) MLR, on MD; 28 x 7in (6½ton) MLR (*Black Prince* only 24 x 7in), 20 on MD, 8 on UD; 4 x 20pdrs (16cwt) BL (saluting guns). Both vessels were re-classed as armoured cruisers in 1881 in view of the comparatively thin armour of these early ironclads.

> Dimensions & tons: (420ft 0in oa) 380ft 2in pp x 58ft 4in. Draught 26ft 4in (fwd), 26ft 10in (aft). 6,039bm. 9,180 disp.
> Men: 635 (later 705). Guns: MD 8 x 7in/110pdr (6½ton) Armstrong BL, 26 x 68pdr (95cwt) MLSB; UD 3 x 7in/110pdr (6½ton) BL, 4 x 40pdr BL on broadside trucks.
> Armour: Belt 4½in (with 18in teak backing); bulkheads 4½in.
> Machinery: 2-cylinder (104in diameter, 4ft stroke) horizontal single expansion, trunk. 10 rectangular boilers (20 lb/sq.in). Single screw (hoisting in *Warrior*, not in *Black Prince*). 1,250nhp (see below for trial ihp and speed). 850 tons coal = 2,100 miles @ 11kts. Under sail *Warrior* 13kts, *Black Prince* 11kts.

Warrior Thames Iron Sbdg Co., Blackwall (No. 76E)/John Penn & Son. (5,267ihp = 14.079kts)
> Ord: 29.4.1859 & 11.5.1859. K: 25.5.1859. L: 29.12.1860. C: 24.10.1861.
> First cost: £377,292.
> Commissioned 1.8.1861 under Capt. Arthur Auckland Leopold Pedro Cochrane, for the Channel squadron; escort of HRH Princess Alexandra; paid off 22.11.1864 at Portsmouth. Recommissioned 7.7.1867 under Capt. John Corbett, for the Channel squadron; on 25.7.1867 under Capt. Henry Boys; collision with *Royal Oak* 1867. Recommissioned 21.8.1869

The 'as fitted' Admiralty draught of the ironclad frigate *Warrior*, dated 21 December 1861 and signed by George Turner (the master shipwright at Woolwich from 12 July 1859). This inboard profile included minor alterations carried out at Portsmouth, which on the original draught are picked out in red. The radical features of this ship – the iron hull and armour protection – are not apparent in this draught, which serves to underline the fact that in many ways the *Warrior* stood at the culmination on an evolutionary process rather than marking a revolutionary breaking of the mould. (*NMM J8603*)

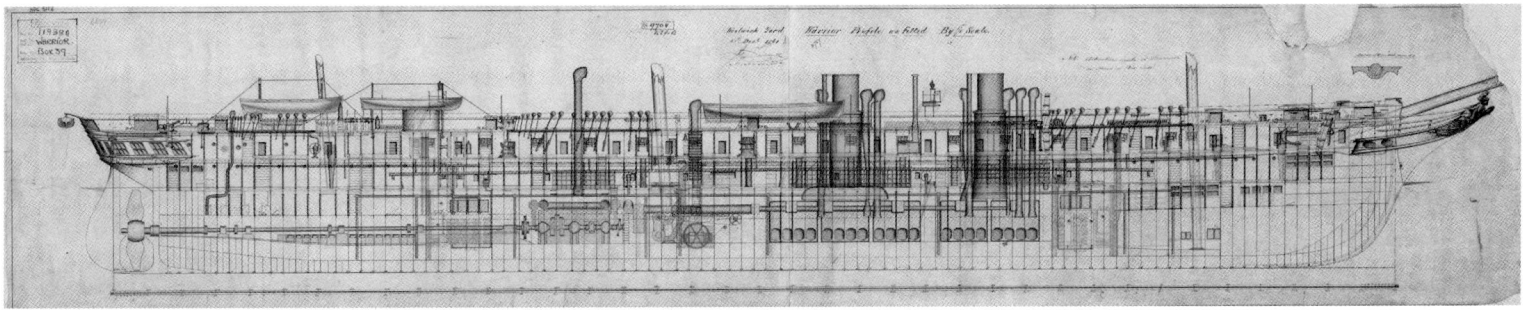

A coloured lithograph, published in August 1872, of the *Defence* in a seaway, showing off the new form of ram bow. Originally built with double topsails, the ship is seen here with single topsail yards, but was later converted from barque to ship rig, when a striking bowsprit (one than could be retracted) was also fitted. She and her sister *Resistance* were regarded as handy vessels under sail, but slow, never exceeding 10½ knots in even the most favourable conditions. *(NMM PU6229)*

under Capt. Frederick Henry Stirling, for the Channel squadron; on 22.2.1870 under Capt. Henry Carr Glyn, on same station; paid off 15.9.1871. Refitted, with a poop and a steam capstan added. Recommissioned 11.6.1875 under Capt. William Henry White, for the Channel squadron; in 1876 to the coastguard at Portland; on 15.3.1878 under Capt. Robert Gordon Douglas, then 7.1.1881 under Capt. Algernon Charles Fieschi Heneage, in same role; paid off 30.4.1881. Recommissioned 1.5.1881 under Capt. Samuel Philip Townsend, for the coastguard at Greenock; on 27.11.1882 under Capt. Edward Stanley Adeane, on same station; at Portsmouth 6.1883, as tender to *Asia*; paid off 1864. Harbour service there as destroyer depot ship 7.1902. Renamed *Vernon III* in 3.1904. Hulked as oil jetty at Pembroke 1923, reverting to *Warrior*. Renamed *C.77* on 27.8.1942. Towed to West Hartlepool for conversion to a museum ship 1979, then to Portsmouth for exhibition – preserved there (1997).

Black Prince Robert Napier & Sons, Govan (No. 98)/John Penn & Son. (5,772ihp = 13.6kts)

> Ord: 6.10.1859. K: 12.10.1859. L: 27.2.1861. C: 12.9.1862.
>> First cost: £377,954.
>> Commissioned 17.6.1862 under Capt. James Francis Ballard Wainwright, for the Channel squadron; on 20.9.1864 under Capt. Frederick Herbert Kerr, on same station; on 7.5.1866 under Capt. John Corbett, as flagship at Queenstown (Cork). Recommissioned 9.9.1868 under Capt. Alexander Crombie Gordon, for the coastguard at Greenock (replacing *Lion*); cruise of the Reserve Fleet 5.1869; on 6.9.1871 under Capt. Edward Lacy, on same station; cruise of the Reserve squadron 1873. Recommissioned 29.7.1875 under Capt. Thomas Bridgeman Lethbridge, as flagship of Rear-Adm. Lord John Hay; on 1.1.1877 under Rear-Adm. Augustus Phillimore, Channel squadron; on 27.4.1878 under Capt. the Duke of Edinburgh, for the Mediterranean; paid off 23.12.1878 at Portsmouth. Training ship at Queenstown from 2.1896. Renamed *Emerald* in 3.1903, then to Devonport as *Impregnable III* in 6.1910. Sold 21.3.1923 to BU.

***DEFENCE* Class** 22 (*Defence*) or 18 (*Resistance*) guns. Designed by Isaac Watts, 1859. Built under the 1859 Programme, as smaller editions of the *Warrior*; tenders were accepted and contracts issued on 14 December, and names were given to the two ships on 17 January 1860. Both barque-rigged, with 24,500 sq. ft sail area. *Resistance*'s machinery was fitted in Victoria Docks. Both re-armed (*Defence* 1866–69; *Resistance* 1866–68) with 2 x 8in (9ton) MLR and 14 x 7in (6½ton) MLR.

> Dimensions & tons: (302ft 0in oa) 280ft 0in pp x 54ft 2in. Draught 24ft 6in (fwd), 26ft 0in (aft). *Defence* 3,720bm, 6,150 disp. *Resistance* 3,710bm, 6,070 disp.
> Men: 450. Guns: *Defence* 8 x 7in/110pdr (6½ton) BL, 10 x 68pdr (95cwt) MLSB, 4 x 5in BL; *Resistance* 6 x 7in/110pdr (6½tpn) BL, 10 x 68pdr (95cwt) MLSB, 2 x 32pdr MLSB.
> Armour: Belt 4½in (with 18in teak backing); bulkheads 4½in.
> Machinery: 2-cylinder (70¾in diameter 3½ft stroke) horizontal single expansion, trunk. 4 rectangular boilers (20 lb/sq.in). Single (hoisting) screw. 600nhp. 2,540ihp = 10¾kts. 460 tons coal = 1,200 miles radius. Under sail 10½ knots.

Defence Palmer Bros, Jarrow. (No. 91)/John Penn & Son. (2,537ihp = 11.618kts)

> Ord: 14.12.1859. K: 14.12.1859. L: 24.4.1861. C: 12.2.1862.
> First cost: £252,422.
>> Commissioned 5.12.1861 under Capt. Richard Ashmore Powell, for the Channel squadron; on 1.10.1862 under Capt. Augustus Phillimore. Recommissioned 13.1.1868 under Capt. Charles Henry May; on 25.3.1869 under Capt. Nowell Salmon, for the West Indies, later to the Mediterranean; paid off 2.8.1872 into Reserve at Devonport. Coastguard depot ship in River Shannon 2.1874. Recommissioned 25.2.1876 under Capt. Frederick Anstruther Herbert, for the Channel squadron, later to the Mediterranean; paid off 5.12.1879. Coastguard depot ship on the Mersey 6.1880; paid off into Reserve at Devonport 8.1885. Became floating workshop 1890; training ship, renamed *Indus I* in 6.1898. Hulked 1922. BU 8.1935 at Plymouth (arrived 16.8.1935 at Cattedown to BU).

Resistance Westwood & Baillie, Poplar/John Penn & Son (2,428ihp = 11.834kts)

Ord: 14.12.1859. K: 21.12.1859. L: 11.4.1861. C: 5.10.1862.
First cost: £258,120.

Commissioned 6.7.1862 at Chatham under Capt. William Charles
Chamberlain, for the Channel squadron; in 12.1863 to the
Mediterranean; in 4.1866 under Capt. Henry Shank Hillyard, in
the Mediterranean; paid off 14.6.1867. Recommissioned 1.7.1869
under Capt. Edward Winterton Turnour, for the coastguard
at Liverpool (replacing *Donegal*); on 7.5.1870 under Capt.
William Henry Haswell, on same station; paid off 23.3.1872.
Recommissioned 8.1873 under Capt. William Graham, for the
Channel squadron; paid off 5.1877. Recommissioned 16.5.1877
under Capt. Sholto Douglas, for the coastguard at Liverpool;
on 1.4.1878 under Capt. Arthur Thomas Thrupp, on the same
station; paid off 6.1880. Used as target 1885 for gunnery &
torpedo experiments. Sold 11.11.1898. Foundered in Holyhead
Bay 4.3.1899; raised and BU 1900 at Garston.

HECTOR Class 18 guns. Designed by Isaac Watts, 1861. Similar to
the *Defence* Class, but with straight stems (no rams) and increased
speed, protection and armament. Tenders were sought on 12 January
1861, and two were accepted on 25 January. Like the *Warrior*, the hull
was subdivided into 92 water-tight compartments. Originally designed
to carry 24 x 68pdr MLSB, *Hector* was completed with 4 x 7in BL
on the UD and 20 x 68pdr MLSB on the MD; she was re-armed in
1867 with the ordnance listed above, and *Valiant* was completed with
similar armament. *Valiant* was completed by Thames Iron Sbdg, after
Westwood & Baillie became bankrupt in 1861. Both were barque-
rigged, with 24,500 sq.ft of sail area.

Dimensions & tons: 280ft 2in pp, 241ft 6½in x 56ft 3in (for
tonnage). Draught 24ft 0in (fwd) 25ft 7in (aft). 4,063bm. 6,713
load disp.
Men: 530. Guns: MainD ('battery deck') 12 x 7in (6½ton) MLR; UD
2 x 8in (9ton) MLR + 4 x 7in (6½ton) MLR.
Armour: Battery and belt 4½in (with 18in teak backing); bulkheads
2½in.
Machinery: 2-cylinder (82in diameter, 4ft stroke) horizontal return
connecting-rod. 6 rectangular boilers (20 lb/sq.in). Single
2-bladed, variable pitch (hoisting) screw. 800nhp. 450 tons coal =
1,600 miles radius. Under sail 10kts.

Valiant Westwood & Baillie, Poplar/Maudslay, Sons & Field (3,560ihp
= 12.65kts)
As built: 280ft 2in pp, 241ft 6½in x 56ft 4in (for tonnage). 4,077bm,
6,123 disp.
Ord: 12.1.1861. K: 1.2.1861. L: 14.10.1863. C: 15.9.1868.
First cost: £325,000.

Commissioned 27.11.1868 under Cmdr Frederick Proby Doughty,
for the coastguard at Tarbert (on the River Shannon); on
24.3.1869 under Capt. William Henry Haswell, then 27.5.1869
under Capt. William John Samuel Pullen, on the same station;
cruise of the Reserve Fleet 5.1869, then back to Tarbert; on
7.5.1870 under Capt. Arthur Wilmshurst, then 12.1871 under
Capt. Cecil William Buckley (invalided 10.1872), then in
4.1874 under Capt. Nowell Salmon; on 10.11.1874 under Cmdr
Edmund Hope Verney (–17.11.1874). Recommissioned 1.1.1876
under Capt. William Cox Chapman, on same station; to the
Channel squadron 1879, then back to Tarbert; in 4.1874 under
Capt. Nowell Salmon; on 1.1.1881 under Capt. James Augustus
Poland; on 18.9.1883 under Capt. Charles George Frederick
Knowles; laid up 8.1885 at Devonport in Dockyard Reserve.
Depot ship for TBDs 1897. Renamed *Indus IV* in 1904, then
Valiant (old) in 1916, *Valiant III* on 3.1.1918, as store hulk at

Devonport. Floating oil tank 1924. Arrived to BU 9.12.1956 at
Zeebrugge.

Hector Robert Napier & Sons, Govan (No. 104)/Robert Napier &
Sons. (3,258ihp = 12.36kts)
As built: 280ft 2in pp, 241ft 6½in x 56ft 5in (for tonnage). 4,089bm,
6,455 disp.
Ord: 25.1.1861. K: 8.3.1861. L: 26.9.1862. C: 22.2.1864.
First cost: £294,000.

Commissioned 12.1.1864 at Portsmouth under Capt. George
William Preedy, for the Channel squadron; on 20.4.1866 under
Capt. William Garnham Luard, on same station; paid off
19.3.1867 at Portsmouth. Recommissioned 1.5.1868 under Capt.
George Le Geyt Bowyear, for the coastguard on Southampton
Water (as ship of First Reserve); on 25.5.1868 under Capt.
Algernon Rous De Horsey, then 12.5.1871 under Capt. Thomas
Cochran, then 21.5.1874 under Capt. John Hobhouse Inglis
Alexander, then 26.4.1875 under Capt. Anthony Hiley Hoskins,
then 7.9.1875 under Capt. Edward Madden (died 11.12.1876),
all on same station. On 1.4.1877 under Capt. Cortland
Herbert Simpson, then 1.4.1880 under Capt. Richard Carter,
then 29.11.1882 under Capt. William Arthur, then 30.3.1885
under Capt. William Elrington Gordon, then 11.7.1885 under
Capt. George Fane, all on same station; paid off 22.4.1886
at Portsmouth. Hulked 1900, as part of *Vernon* torpedo
establishment. Sold 11.7.1905 to BU.

ACHILLES 26 guns (later 16). Designed by Isaac Watts, 1861. The
first iron-hulled ship to be built in a Royal Dockyard. Completed
with four-masted rig – the only British warship since the seventeenth
century to have 4 masts, carrying 44,000 sq.ft of canvas excluding
stunsails; reduced to three-masted in 1865, and re-rigged as a barque
in 1877 with 30,133 sq.ft sail area. Re-armed in 1868 with 4 x 8in MLR
and 22 x 7in MLR; re-armed in 1874 with 14 x 9in (12ton) MLR and 2
x 7in (6½ton) MLR; again re-armed 1889, with 7in removed and 2 x 6in
BL, 8 x 3pdr QF and 16 MGs added.

Dimensions & tons: 380ft 0in pp, 338ft 9in x 58ft 3½in (for tonnage)
x 27ft 3in. Draught 25ft 10in (fwd) 27ft 2in (aft). 6,121bm. 9,820
(load) disp.
Men: 705. Guns: 4 x 110pdr BL, 16 x 100pdr SB; 6 x 68pdr SB added
1865. Later (1874) 14 x 9in MLR, 2 x 7in MLR.
Armour: Battery and belt 4½in (with 18in teak backing); bulkheads
4½in.
Machinery: 2-cylinder (104¼in diameter, 4ft stroke) horizontal
single expansion, trunk. 10 rectangular boilers (25 lb/sq.in).
Single screw. 1,250nhp. 740 tons coal = 1,800 miles radius @
6½kts. Under sail 13kts.

Achilles Chatham Dyd/John Penn & Son (5,720ihp = 14.3kts)
Ord: 24.10.1860 & 10.4.1861. K: 1.8.1861. L: 23.12.1863. C:
26.11.1864.
First cost: £469,572.

Commissioned 14.9.1864 under Capt. Edward Westby Vansittart,
for the Channel squadron; paid off 2.11.1868 at Plymouth.
Re-armed 1868–69. Recommissioned 12.4.1870 under Capt.
Richard Vesey Hamilton, for the coastguard at Portland; on
21.4.1873 under Capt. Radulphus Bryce Oldfield, on same
station; paid off 12.2.1874 and re-armed. Recommissioned
10.3.1875 under Capt. William Henry Whyte, for the
coastguard at Liverpool; on 1.4.1875 under Capt. Sholto
Douglas, on same station; paid off 16.5.1877. Recommissioned
next day under Capt. William Nathan Wright Hewett, for the
Channel squadron; later to the Mediterranean and the Sea of

A contemporary waterline model of the broadside ironclad *Achilles*, its relatively simple hull detail suggesting that it was probably intended to demonstrate features of the novel four-masted square rig (this was the first four-masted warship since the seventeenth century). From this she set the largest area of canvas (44,000 square feet, excluding studding sails) ever carried by a warship. Although based on the *Warrior*, the appearance of the ship was very different, with a near-vertical stem and a new form of round stern. The ship was regarded as a superb seaboat and a steady gun platform. *(NMM F8906-003)*

Marmara (during the Russo-Turkish War). Recommissioned 8.9.1880 under Capt. Edward Kelly, for the Channel squadron; on 6.10.1882 under Capt. George Digby Morant, then 1.1884 under Capt. Alexander Buller, on same station; paid off 28.5.1885. Became base ship, renamed *Hibernia* 1902. Renamed *Egmont* in 3.1904, then *Egremont* in 6.1916, and *Pembroke* in 6.1919. Sold to Granton Shipbreaking Co. 26.1.1923 to BU in 1925.

MINOTAUR class 36 guns. Designed by Isaac Watts, 1861. Six ironclads of the *Achilles* type were included in the 1861 Programme, and tenders were sought for all six on 31 May 1861 but only three were accepted on 2 September. The first two were ordered as *Elephant* and *Captain* respectively, but were renamed on 18 October 1861. Five-masted, with square canvas on the first four (later the first three) masts, and gaffs on all but the second mast (where the sail would have fouled the funnels).

    Dimensions & tons: (407ft 0in oa) 400ft 3in pp x 59ft 6in x 27ft 9in. 6,621bm. 10,627 disp.
    Men: 705 (later 800). Guns: 4 x 9in (12ton), 24 x 7in (6½ton), 8 x 24pdr MLSB. Later 17 x 9in (12 ton) MLR.
    Armour: Battery and belt 5¾in amidships and 4½in at ends (with 10in teak backing); bulkheads 5½in.
    Machinery: 2-cylinder (104¼in diameter; 52in stroke except *Achilles* 48in) horizontal single expansion (trunk in *Minotaur*, double piston rods in *Agincourt*). 10 rectangular boilers. Single screw. 1,350nhp. See below for individual ihp/speed.

Minotaur Thames Iron Sbdg Co., Blackwall (No. 81E)/John Penn & Son. (6,702ihp = 14.411kts).
    As built: 400ft 0in pp x 59ft 4¾in. 6,643bm, 10,690 disp.
    Ord: 2.9.1861. K: 12.9.1861. L: 12.12.1863. C: 1.6.1867.
    First cost: £478,855.
    Commissioned 8.4.1867 under Capt. James Graham Goodenough, for the Channel squadron. On 6.9.1871 under Capt. Robert Gibson, for the Channel squadron, as flagship of Rear-Adm. Geoffrey Thomas Phipps Hornby. On 6.8.1875 under Capt.

Lord Walter Talbot Kerr, for the Channel squadron, as flagship of Rear-Adm. Frederick Beauchamp Paget Seymour; paid off 9.8.1877. Recommissioned 10.11.1877 under Capt. Harry Holdsworth Rawson, for the Channel squadron. On 10.3.1884 under Capt. Richard Frederick Britten, on same station; paid off 23.11.1887 into Reserve at Portsmouth; attached to Boscawen at Portland as accommodation ship 1.1893; training ship there 1895, renamed *Boscawen* in 3.1904. Renamed *Ganges* 21.6.1906, then *Ganges II* on 25.4.1908. Sold 30.1.1922 and BU at Swansea.

Agincourt Laird Brothers, Birkenhead (No. 291)/Maudslay, Sons & Field. (6,867ihp = 15.433kts).
    As built: 400ft 0in, 354ft 0⅞in x 59ft 3½in. 6,638bm, 10,600 disp.
    Ord: 2.9.1861. K: 30.10.1861. L: 27.3.1865. C: 19.12.1868.
    First cost: £483,003.
    Commissioned 1.1.1869 under Capt. Donald McLeod MacKenzie, as flagship of Reserve; in 5.1869 under Capt. Thomas Miller; towed (with *Northumberland*) the Bermuda Floating Dock to Madeira 6.1869; on 1.10.1869 under Capt. Henry Carr Glyn, then 22.2.1870 under Capt. Henry Hamilton Beamish, for the Channel squadron; grounded on Pearl Rock (off Gibraltar) 1871, but salved; on 7.8.1872 under Capt. Edward Stanley Adeane, for the Channel squadron, as flagship of Rear-Adm. Reginald John James George Macdonald; on 1.10.1874 under Capt. Lord Walter Talbot Kerr, as flagship of Rear-Adm. Frederick Beauchamp Paget Seymour; paid off 5.8.1875. Extensive refit at Devonport 1875–77. Recommissioned 10.7.1877 under Capt. Richard Wells, for the Mediterranean; on 16.8.1880 under Capt. Elibank Harley Murray, for the Channel squadron; on 22.12.1882 under Capt. James George Mead; paid off ?. Recommissioned 17.5.1885 under Capt. Frederick Charles Bryan Robinson, for the Channel squadron; on 15.8.1885 under Capt. Charles Thomas Montague Douglas Scott; paid off 1886. Recommissioned 1.12.1888 under Capt. Richard Frederick Britten; paid off 27.5.1889. Training ship at Portland 1895, renamed *Boscawen III* in 3.1904. Renamed *Ganges II* on 21.6.1906. Coal hulk *C.109* in 9.1908. Sold to Ward, Swansea and arrived to BU 21.10.1960.

The *Agincourt* (seen here), *Minotaur* and *Northumberland* were the only five-masted warships ever built for the Royal Navy. Despite the impressive appearance, they were poor performers under sail – 'no ships ever carried so much dress to so little purpose', according to Admiral George Ballard, who knew them well. The masts were originally designated fore, second, main, fourth and mizzen, but square canvas was later confined to the first three masts, after which the aftermost pair were renamed mizzen and jigger (they carried spare square yards but set no sails from them). This photograph was taken in the 1880s and a careful examination will reveal light quick-firing guns and searchlights along the topsides. *(NMM NL4759)*

Modified *MINOTAUR* class 36 guns. *Northumberland* was, strictly speaking, a half-sister to the *Minotaur* and *Agincourt*, ordered at the same date but was modified by Reed from the Watts design; construction completed by Millwall Iron Works; her armament was always different, with heavier but fewer guns, so she had need of only 184ft 6in of her sides protected, rather than the 213ft of her near sisters; she also had an armoured conning tower. The first ironclad to be completed with steam steering gear. When built, rigged like the *Minotaur* class but with gaffs on all five masts; re-rigged as a three-masted barque in 1875–79.

    Dimensions & tons: (407ft 0in oa) 400ft 3in pp, 352ft 7in keel x 59ft 5in x 27ft 9in. 6,621bm. 10,584 disp.
    Men: 705 (later 800). Guns: Main deck 4 x 9in (12½ton) MLR, 18 x 8in (9ton) MLR, 2 x 7in (6½ton) MLR; UD 4 x 8in (9ton) MLR. Re-armed in 1875–79 with 7 x 9in (12ton) MLR, 20 x 8in (9ton) MLR and 2 x 20pdr (15cwt) BL and 4 Whitehead 16in torpedo launchers; re-armed in 1885–87 with 1 x 6in BL and 1 x 5in BL replacing 2 x 8in.
    Armour: Battery and belt 5¼in amidships and 4½in at ends (with 10in teak backing); conning tower 4½in; bulkheads 5½in.
    Machinery: 2-cylinder (108.9in diameter; 52in stroke) horizontal single expansion (trunked). 10 rectangular boilers. Single screw. 1,350nhp. 6,558ihp = 14.132kts.
Northumberland Thames Iron Sbdg Co., Blackwall/John Penn & Son
    As built: 6,631bm. 10,784 disp.
    Ord: 2.9.1861. K: 10.10.1861. L: 17.4.1866. C: 8.10.1868 at Devonport.
    First cost: £444,256.
    Commissioned 14.8.1868 at Devonport under Capt. Roderick Dew (died 24.3.1869), for the Channel squadron; on 25.3.1869 under Capt. Charles Henry May; towed (with *Agincourt*) the Bermuda Floating Dock to Madeira 6.1869; badly damaged 25.12.1872 in collision with *Hercules*; repaired at Malta. Recommissioned 2.10.1873 under Capt. Thomas Bridgeman Lethbridge, for the Channel squadron, as flagship of Rear-Adm. George Hancock; later flagship of Rear-Adm. Lord John Hay; paid off 28.7.1875. Re-rigged 1875–79, and re-armed with MLR guns. Recommissioned 30.10.1879 under Capt. Henry

Rushworth Wratislaw, for the Channel squadron; on 31.12.1880 under Capt. George Stanley Bosanquet, on same station, later to the Mediterranean; in Egyptian war. On 15.8.1883 under Capt. Frederick Charles Bryan Robinson, for the Channel squadron; paid off 16.5.1885. Refitted 1885–97 and partly re-armed. Recommissioned 26.11.1887 under Capt. James Andrew Thomas Bruce, for the Channel squadron; on 17.4.1888 under Capt. Sachaveral Darwin; paid off 7.3.1891. Depot ship 1898. Training ship for stokers, renamed *Acheron* in 3.1904. Coal hulk *C.8* in 1909 at Invergordon; renamed *C.68* in 1926. Sold to Ward 6.1927, resold as *Stedmound*, as hulk at Dakar; finally BU 1935.

Armoured Wooden Cruisers (Broadside or Central Battery Ironclads)

From 1861 seven wooden screw ships of the line, all of the 91-gun *Bulwark* Class (see Chapter 2) in early stages of construction on the stocks, were converted into ironclad frigates. Four of these two-deckers, more recently laid down (during 1860), were in frame only, and these were converted to designs by Isaac Watts with full-length side armour. The other three had been laid down in 1859 and work on them had already advanced to a significant extent; work on these was initially suspended and then, after some delays, they were converted according to designs by Edward Reed with an armoured central battery. Their upper decks were removed, and they were fitted with side armour, ranging from 4½in in the earliest treated to 5½in in the later conversions. Those completed with BL and MLSB guns were re-armed from 1867 onwards with MLR guns. None of these expensive conversions was very successful, and all were removed from the list at an early date.

Three smaller conversions from wooden hulls in the same era were the armoured sloops *Research* and *Enterprise*, and the somewhat larger (and more successful) corvette *Favorite*; all were central-battery designs by Edward Reed. Note that additional dates below are those of conversion order being approved ('App') and conversion work beginning ('Began').

The first of a crash programme to build up the numbers of the armoured battlefleet by utilising the hulls of wooden screw ships of the line in the early stages of construction, the wooden-hulled ironclad *Royal Oak* is shown in a watercolour portrait by William Frederick Mitchell. She is depicted after conversion from double topsails in 1864, but before the adoption of square sails on the mizzen turned her from a barque to full-rigged ship; in the condition portrayed, the ship set a record on 9 February 1863, logging 13.5 knots, the only armoured vessel ever to exceed her steam trial speed while under canvas. *(NMM PU9519)*

ROYAL OAK Wooden-hulled ironclad frigate. Design by Isaac Watts, 1861. The first British wooden-hulled armoured ship, she was approved for conversion from a 91-gun Second Rate to a 36-gun (36 x 68pdr MLSB) ironclad frigate on 14 May 1861. Conversion involved lengthening by 21ft. Barque-rigged (25,000 sq.ft sail area) on completion, but re-rigged as ship in June 1866. Re-armed 1867 with 4 x 8in (9ton) MLR and 16 x 7in (6½ton) MLR on the main deck, and another 4 x 7in (6½ton) MLR on the upper deck.

> Dimensions & tons: 273ft 0in pp, 232ft 8½in x 58ft 3in (57ft 3in for tonnage) x 24ft 0in. 4,057bm. 6,366 disp.
> Men: 585–605. Guns: Main deck: 8 x 7in (82cwt) BL and 24 x 68pdr (95cwt) MLSB; UD 3 x 7in (82cwt) BL.
> Armour: Belt and battery 4½in (with 28in oak backing), extending to ends of ship at 2½in.
> Machinery: 2-cylinder (82in diameter, 4ft stroke) horizontal single expansion, return connecting-rod. 6 rectangular boilers. Single 19ft (hoisting) screw. 800nhp. 3,708ihp = 12.529kts. Coal 550 tons = 2,200 miles @ 5kts. Under sail 13½kts (the fastest ironclad under sail alone).

Royal Oak Chatham Dyd/Maudslay, Sons & Field
> Ord: 8.4.1859. K: 1.5.1860. App: 14.5.1861. Began 3.6.1861. L: 10.9.1862. C: 28.5.1863.
> First cost: £254,537.
> Commissioned 27.4.1863 under Capt. Frederick Archibald Campbell, for the Channel squadron, later to the Mediterranean. On 21.3.1866 under Capt. George Disney Keane, in the Mediterranean; paid off 13.12.1867 at Portsmouth. Recommissioned next day under Capt. Henry Shank Hillyar, for the Channel squadron; later to the Mediterranean; at opening of the Suez Canal 11.1869, when grounded off Port Said, but was refloated; on 5.1.1871 under Capt. Leveson Eliot Henry Somerset; paid off 9.1.1872 at Portsmouth. Sold 30.9.1885 to Castle to BU.

CALEDONIA Class Wooden-hulled ironclad frigates. Design by Isaac Watts, 1861. All approved for conversion from 91-gun Second

Rates to ironclad frigates on 27 May 1861, with a 21ft section inserted when already partially in frame. *Prince Consort* begun as *Triumph*, but was renamed 14 February 1862; she was re-armed a second time in 1871 with 7 x 9in (12ton) MLR + 8 x 8in (9ton) MLR. More powerful engines than the prototype, *Royal Oak*, but the extras weight cancelled this out and they were slower. Initially all were barque-rigged, converted in 1866 to ship rig. All BU at Charlton.

> Men: 605. Guns: (initially all different) *Prince Consort* 7 x 7in (82cwt) BL, 8 x 100pdr MLSB, 16 x 68pdr MLSB; *Ocean* 24 x 7in (6½ton) MLR; *Caledonia* 10 x 7in (82cwt) BL, 8 x 100pdr MLSB, 12 x 68pdr MLSB. All 3 were re-armed in 1867 with 4 x 8in (9ton) MLR + 16 x 7in (6½ton) MLR on the main deck, and another 4 x 7in (6½ton) on the upper deck.
> Armour: Belt and battery 4½in extending to 3in (with 29½in teak backing).
> Machinery: 2-cylinder (92in diameter, 4ft stroke) horizontal single expansion, return connecting-rod (double piston rods in last 2). 8 rectangular boilers. Single 21ft screw. 1,000nhp. See below for ihp/speed. Coal 550 tons = 2,000 miles @ 5kts. Under sail 10–11½kts.

Prince Consort Pembroke Dyd/Maudslay, Sons & Field (4,234ihp = 13,119kts).
> Dimensions & tons: 273ft 1in pp, 232ft 8¾in x 58ft 7in (57ft 2in for tonnage). 4,045bm. 6,430 disp.
> Ord: 8.4.1859. K: 13.8.1860. App: 31.5.1861. Began: 6.6.1861. L: 26.6.1862. C: 6.2.1864.
> First cost: £266,173.
> Commissioned 27.10.1863 under Capt. Charles Vesey, for trials; paid off 10.11.1863. Recommissioned 14.1.1864 under Capt. George Ommanney Willes (appointed 13.11.1863), for the Channel squadron, on 10.4.1866 under Capt. Edward Augustus Inglefield; to the Mediterranean 1867; on 3.3.1868 under Capt. William Armytage, still in the Mediterranean; back to the Channel 1869; paid off 1871. Sold to Castle 3.1882 to BU.

Ocean Devonport Dyd/Maudslay, Sons & Field (4,244ihp = 12.896kts)

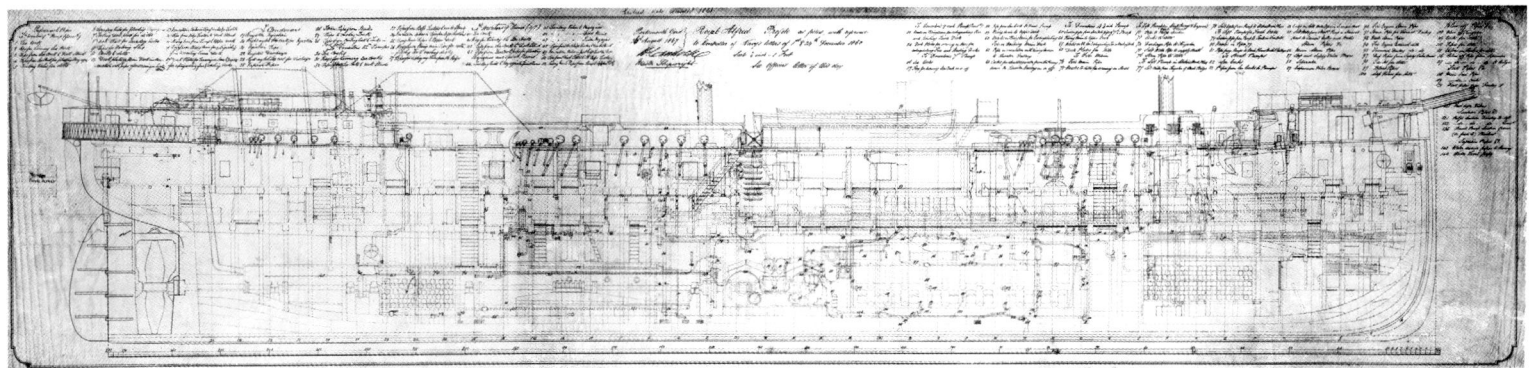

A profile draught of the *Royal Alfred* as fitted, dated 26 August 1867 at Portsmouth. As an indication of the unfamiliar complexity of these vessels, the draught is keyed with 144 items relating to the ship's pipework, done in response to a specific request from the Controller of the Navy, Vice-Admiral Robert Spencer Robinson. *(NMM DR8046)*

Dimensions & tons: 273ft 1in pp, 232ft 9⅛in x 58ft 5in (57ft 2in for tonnage). 4,047bm. 6,535 disp.

Ord: 5.3.1860. K: 23.8.1860. App: 5.6.1861. Began: 3.6.1861. L: 19.3.1863. C: 6.9.1866.

First cost: £298,451.

Commissioned 27.7.1866 under Capt. Chandos Scudamore Scudamore Stanhope, for the Mediterranean; later to China, as flagship of Vice-Adm. Sir Henry Kellett; in 1870 under Capt. William Nathan Wright Hewett, still Kellett's flag in China; paid off 22.6.1872 into Reserve at Devonport. Sold to Castle 11.5.1882 to BU.

Caledonia Woolwich Dyd/Maudslay, Sons & Field (4,092ihp = 12.94kts)

Dimensions & tons: 273ft 0in pp, 231ft 3¼in x 59ft 2in (57ft 11in for tonnage). 4,125bm. 6,753 disp.

Ord: 5.3.1860. K: 1.10.1860. App: 31.5.1861. Began: 6.6.1861. L: 24.10.1862. C: 6.7.1865.

First cost: £312,034.

Commissioned 27.4.1865 at Plymouth under Capt. Fitzgerald Algernon Charles Foley, for the Mediterranean, as flagship of Rear-Adm. Hastings Reginald Yelverton; on 10.6.1867 under Capt. Alan Henry Gardner, as flagship of Rear-Adm. Lord Clarence Edward Paget, still in the Mediterranean; paid off 31.5.1869 at Malta. Recommissioned next day under Capt. Thomas Cochran, in the Mediterranean; on 11.5.1871 under Capt. Chandos Scudamore Scudamore Stanhope (died 7.7.1871), then 8.7.1871 under Capt. Charles Pringle; on 11.8.1871 under Capt. Edward Henry Gage Lambert; paid off 24.9.1872 at Devonport. Recommissioned 1.10.1872 under Capt. John Eglinton Montgomerie, as coastguard deport ship at Birkenhead. Recommissioned 18.11.1874 under Cmdr Edmund Hope Verney. Sold to Castle 30.9.1886 to BU.

ROYAL ALFRED Wooden-hulled ironclad frigate, central battery. Edward Reed design, 1861. Approved for conversion from a 91-gun Second Rate to an ironclad frigate on 27 May 1861, with a 21ft section inserted. Intended to be of *Caledonia* Class, but delayed to evaluate that Class's outcome, and instead completed with a 115ft long central box battery. Intended to be barque-rigged, but in August 1866 was re-rigged as a ship, with 29,200 sq. ft of sail area.

Dimensions & tons: 273ft 0in pp, 232ft 8⅛in x 58ft 7in (57ft 2in for tonnage). Draught 23ft 7in (fwd) – 27ft 2in (aft). 4,068bm. 6,720 disp.

Men: 605. Guns: MD (battery) 10 x 9in (12½ton) MLR + 4 x 7in (6½ton) MLR; UD 4 x 7in (6½ton) MLR (2 fwd, 2 aft); 6 saluting (howitzers).

Armour: Belt and battery 6in, bulkheads 4½in (all with 29½in teak backing).

Machinery: 2-cylinder (82in diameter, 4ft stroke) horizontal single expansion, return connecting-rod (double piston rods). 6 rectangular boilers. Single screw. 800nhp. 3,434ihp = 12.36kts. 550 tons coal = 2,200 miles @ 5kts. Under sail 12½kts.

Royal Alfred Portsmouth Dyd/Maudslay, Sons & Field

Ord: 8.4.1859. K: 1.12.1859. App: 5.6.1861. Began: 22.6.1861. L: 15.10.1864. C: 23.3.1867.

First cost: £282,803.

Commissioned 14.1.1867 under Capt. Frederick Anstruther Herbert, for North America and the West Indies, as flagship of Vice-Adm. George Robert Mundy; on 1.9.1869 under Capt. Richard Wells, on same station, as flagship of Vice-Adm. George Greville Wellesley; in 9.1870 under Capt. Henry Frederick Nicholson, on same station, as flagship of Vice-Adm. Edward Gennys Fanshawe; paid off 15.1.1874 into Reserve. Recommissioned 4.2.1874 as guard ship for Reserve at Portland; paid off 30.3.1875 into Reserve at Portsmouth. Sold to Castle 12.1885 (for £5,562) to BU.

RESEARCH Wooden-hulled ironclad sloop, central battery. Designed by Edward Reed, 1862. Approved for conversion from a wooden screw sloop (*Camelion* Class) to an ironclad sloop in 1862 (lengthened by 10ft). She had a full-length armour belt reaching a depth of 10ft down from the main deck, with a box battery 34ft long on the MD. Barque-rigged, with 18,250 sq. ft sail area. Re-armed 1869–70 with 4 x 7in (6½ton) MLRs replacing the 100pdrs.

Dimensions & tons: 195ft 0in pp, 168ft 3½in x 38ft 6in (37ft 5in for tonnage). Draught 12ft 4in – 15ft 3in. 1,253bm. 1,743 disp.

Men: 139–150. Guns: 4 x 100pdr MLSB.

Armour: Belt, box-battery and bulkheads 4½in (with 19½in teak backing).

Machinery: 2-cylinder (50in diameter, 2ft stroke) horizontal single expansion, direct-acting. 2 tubular boilers. Single (hoisting) screw, 12ft diameter. 200nhp. 937ihp = 10.176kts (after refit in 1969, 1,042ihp = 10.331kts). 130 tons coal. Only 6kts under sail.

Research [ex-*Trent*, renamed 9.1862] Pembroke Dyd/James Watt & Co.

Ord: 5.3.1860. K: 3.9.1861. App: 1.9.1862. Began: —1862. L: 15.8.1863. C: 6.4.1864 at Devonport.

First cost: £71,287.

Commissioned 9.3.1864 under Capt. Arthur Wilmhurst, for the

Channel squadron; on 19.9.1865 under Cmdr Robert Anthony Edwards Scott, on same station; paid off 5.1866 at Portsmouth. Recommissioned 30.11.1866 under Cmdr William Burley Grant; on 18.4.1867 under Cmdr Arthur Morrell; paid off 3.5.1868 into the Steam Reserve at Devonport. Refitted and re-armed 1869–70. Recommissioned 10.7.18871 under Capt. William John Ward, for the Mediterranean; on 21.5.1872 under Capt. Charles Matthew Buckle; paid off 16.8.1875 at Malta. Recommissioned next day under Capt. Hamilton Edward George Earle; paid off 2.8.1879 and laid up at Chatham. Sold 1884 to BU.

ENTERPRISE Wooden-hulled ironclad sloop, central battery. Designed by Edward Reed, 1862. Approved for conversion from a wooden screw sloop (*Camelion* Class) to an ironclad sloop in 1862. Full-length armour belt as in the *Research*, but with the box battery on the UD. Barque-rigged, with 18,250 sq.ft sail area. Re-armed 1868 with 4 x 7in MLRs replacing 110pdrs and 100pdrs.

Dimensions & tons: 180ft 0in pp, 152ft 11½in x 36ft 0in (34ft 11½in for tonnage). Draught 11ft 10in – 15ft 8in. 994bm. 1,350 disp.

Men: 130. Guns: 2 x 110pdr MLR; 2 x 100pdr MLSB.

Armour: Belt, box-battery and bulkheads 4½in (with 19½in teak backing).

Machinery: 2-cylinder (45in diameter, 1½ft stroke) horizontal single expansion, direct-acting. 2 tubular boilers. Single screw. 160nhp. 692ihp = 9.944kts. 95 tons coal. 9¼kts under sail.

Enterprise [ex-*Circassian*, renamed 22.7.1862] Deptford Dyd/ Ravenhill, Salkeld & Co.

Ord: 21.3.1861. App: 17.4.1862. K: 5.5.1862. L: 9.2.1864. C: 3.6.1864.

First cost: £62,474.

Commissioned 26.3.1864 under Cmdr Charles John Rowley, for the Channel squadron, later to the Mediterranean; on 7.7.1866 under Cmdr George Stanley Bosanquet, in the Mediterranean; on 11.4.1870 under Cmdr George Digby Morant, in the Mediterranean; paid off 30.8.1871 at Sheerness and laid up. Sold 11.1886 to BU.

FAVORITE Wooden-hulled ironclad corvette, central battery. Designed by Edward Reed and the Controller's Department, 1862. Approved for conversion from a wooden screw corvette to an ironclad corvette in May 1862. Fitted with engines originally intended for cancelled *North Star*. She had a full-length waterline armour belt reaching from the upper deck down to 3ft below the waterline, with a box battery 66ft long on the UD. Ship-rigged, with 18,250 sq.ft sail area. Re-armed 1869 with 8 x 7in (6½ton) MLR in battery, plus 2 x 68pdr /6in (64cwt) MLR as UD chase guns.

Dimensions & tons: 225ft 0in pp, 195ft 6¾in x 46ft 9in (44ft 10½in for tonnage). Draught 19ft 7in (fwd), 22ft 7in (aft). 2,081bm. 3,169 disp.

Men: 250-270. Guns: 8 x 100pdr /8in (9ton) MLSB in central battery.

Armour: Belt 4½in (with 26in teak backing), battery 4½in (with 19in teak backing), bulkheads 4½in.

Machinery: 2-cylinder (64in diameter, 32in stroke) horizontal single expansion, direct-acting. 4 boilers. Single Mangin screw. 400nhp. 1,773ihp = 11.825kts. 350 tons coal. Under sail 10½kts only (as her screw could neither hoist nor disconnect).

Favorite Deptford Dyd/Humphrys & Tennant

Ord. 8.4.1859. K: 23.8.1860. App: 29.5.1862. L: 5.7.1864. C: 17.3.1866.

First cost: £152,374.

Commissioned 8.2.1866 under Capt. Francis Henry Shortt, for North America and the West Indies; paid off 25.7.1869 at Portsmouth. Recommissioned 22.3.1872 under Capt. Leverson Eliot Henry Somerset, as first reserve coastguard ship at Queenferry; in 9.1875 under Capt. Norman Bernard Beddingfeld; paid off 20.12.1876 at Portsmouth and laid up in reserve there. Sold there 30.3.1886 to BU.

ZEALOUS Wooden-hulled ironclad central battery frigate. Designed by Edward Reed, 1862. Approved for conversion from a 91-gun Second Rate to an ironclad frigate in 1862, without lengthening. Ship-rigged, with 29,200 sq.ft sail area.

Dimensions & tons: 252ft 0in pp, 213ft 9¼in x 58ft 7in (57ft 2in for tonnage). Draught 24ft 10in (fwd), 25ft 11in (aft). 3,716bm. 6,102 disp.

Men: 510. Guns: 20 x 7in (6½ton) MLR – 16 in central battery on MD, 4 on UD (2 under fo'csle, 2 on QD).

Armour: Belt 4½in-2½in (with 30in teak backing), box battery 4½in, bulkheads 3in, pilot tower 3in.

Machinery: 2-cylinder (82in diameter, 4ft stroke) horizontal single expansion, return connecting-rod. 6 rectangular boilers. Single screw. 800nhp. 3,623ihp = 11.71kts. 660 tons coal. Under sail 10kts.

Zealous Pembroke Dyd/Maudslay, Sons & Field

Ord: 1.4.1857. K: 26.10.1859. App: 2.7.1862. Began: —1862. L: 7.3.1864. C: 4.10.1866.

First cost: £239,258.

Commissioned 13.9.1866 under Capt. Richard Dawkins, as flagship of Rear-Adm. George Fowler Hastings, for the Pacific; on 19.1.1870 under Capt. Francis Alexander Hume, as flagship of Rear-Adm. Arthur Farquhar, in the Pacific; on 15.9.1873 under Capt. Thomas Cochran, as coastguard depot ship at Southampton; paid off 18.6.1875 into Reserve at Portsmouth. Sold to Castle 9.1886 to BU at Charlton.

REPULSE Wooden-hulled ironclad central battery frigate. Designed by Edward Reed. Originally approved in 1861 for conversion from a 91-gun Second Rate to an ironclad frigate (without lengthening), but work was suspended in 1861 and not resumed until 1866 (under Reed). The engine she received had previously been installed the three-decker *Prince of Wales*. In 1868 towed to Sheerness Dyd for completion, as Woolwich Dyd closed in October 1869. Ship-rigged, with sail area of 29,200 sq.ft. Later re-armed with 10 x 9in (12ton) MLR, with 4 x 16in TC added in 1880.

Dimensions & tons: 252ft 0in pp, about ?212ft 4in x 59ft 0in oa (ca. 57ft 6in for tonnage). Draught 24ft 4in (fwd), 25ft 10in (aft). 3,734bm. 6,010 disp.

Men: 515. Guns: 12 x 8in (9ton) MLR – 8 in central battery on MD, 4 on UD (2 under fo'csle, 2 under poop); 2 x 20pdr (16cwt) BL (saluting).

Armour: Belt 6in-4½in, box battery 6in, bulkheads 4½in.

Machinery: 2-cylinder (82in diameter, 4ft stroke) horizontal single expansion, trunk. 6 rectangular boilers. Single screw. 800nhp. 3,347ihp = 12.28kts. 460 tons coal. Under sail 10½kts.

Repulse Woolwich Dyd/John Penn & Son

Ord: 1.4.1857. K: 29.4.1859. Approved: 14.5.1861, but then suspended. Re-ord: 9.10.1866. L: 25.4.1868. C: (at Sheerness) 31.1.1870.

First cost: £183,640.

Commissioned 28.2.1870 under Capt. William Rae Rolland, for the coastguard at Queensferry (replaced *Duncan*); paid off

A vision of the future, and a nightmare for traditionalists – the *Royal Sovereign* as originally converted from a three-decker to a turret ship. Not only were the Cowper Coles turrets revolutionary, but the ship was effectively mastless, with a range dictated by the size of her coal bunkers. While cruising warships retained masts and spar for a couple of decades more, in any meaningful sense, for the Royal Navy the Age of Sail was over. *(NMM 5253)*

21.3.1872 at Portsmouth. Recommissioned 9.7.1872 under Capt. Charles Thomas Curme, as flagship of Rear-Adm. Charles Farrell Hillyar, for the Pacific; on 7.6.1873 under Capt. Joseph Edward Maitland Wilson, as flagship of Rear-Adm. Arthur Auckland Leopold Pedro Cochrane, in the Pacific; on 5.7.1875 under Capt Richard Carter; on 24.9.1876 under Capt. Frederick William Wilson; paid off 12.6.1877 into Reserve at Portsmouth. Recommissioned 15.4.1881 under Capt. George Lydiard Sulivan, as coastguard deport ship at Hull; on 15.4.1884 under Capt. Henry Craven St John; paid off 26.8.1885 at Portsmouth. Sold 2.1889 to BU.

Armoured Coastal Defence Ships (Turret Ironclads)

Finally in the same period, two armoured vessels for coastal defence were built with guns mounted in revolving turrets, the initial decision to order these two being in January 1862. The first was a conversion from a 121-gun three-decker; this has additional dates below which are those of conversion order being approved ('App') and conversion work beginning ('Began'). The second was a new-built ship.

ROYAL SOVEREIGN Wooden-hulled turret ironclad, coast defence. Redesigned at behest of Cowper Coles, 1862. Virtually complete as a 121-gun three-decker when she was approved for conversion to an ironclad frigate on 3 April 1862 (just 25 days after the duel in Hampton Roads demonstrated the effectiveness of the revolving turret). The original MLSB guns were replaced by 9in (12ton) MLR guns in 1867.

Dimensions & tons: 240ft 6in pp, 201ft 11½in x 62ft 0½in (59ft 2½in for tonnage). Draught 21ft 2in (fwd), 24ft 8in (aft). 3,963bm. 4,955 disp.

Men: 296. Guns: 5 x 10½in/150pdr MLSB in 4 turrets (one twin fwd, 3 single).

Armour: Belt 5½in-4½in (with 36in oak hull behind), turrets 10in-5½in, conning tower 5½in, deck 1in.

Machinery: 2-cylinder (82in diameter, 4ft stroke) horizontal single expansion, return connecting-rod. Single screw. 800nhp. 2,463ihp = 11kts. 280 tons coal.

Royal Sovereign Portsmouth Dyd/Maudslay, Sons & Field

Ord: 29.6.1848 (see Chapter 2). K: 17.12.1849. L: 25.4.1857. Re-ord: 3.4.1862. C: 20.8.1864.

First cost: £180,572.

Commissioned 7.7.1864 under Capt. Sherard Osborn, for the Channel Squadron; paid off 15.10.1864 as tender to *Excellent*. On 1.7.1865 under Capt. Frederick Anstruther Herbert, for the Channel squadron; paid off 9.10.1865 as tender to *Excellent* again. Recommissioned 7.1867 under Capt. Cowper Phipps Coles, for the 17.6.1867 Naval Review; paid off 3.9.1869. Sold 5.1885 to BU.

PRINCE ALBERT Iron-hulled turret ironclad, coast defence. Design by Isaac Watts, 1862. The twelfth ship under the 1861 ironclad programme, on 16 April 1862 she was given a name personally chosen by Queen Victoria. Completed with four hand-turned iron turrets (six were originally planned), each weighing 11 tons and containing a single 9in gun. She was reboilered in 1878 and 6 x MGs added during the 1880s. At the Queen's request, this vessel – named after her late husband – remained on the active list long after ceasing to have any naval role.

Dimensions & tons: 240ft 7in pp, 206ft 4½in x 48ft 1in (47ft 10in for tonnage) x 19ft 0in. Draught 18ft 8in (fwd) 20ft 4in (aft). 2,537bm. 3,687 disp (3,880 load).

Men: 200. Guns: 4 x 9in (12ton) MLR – in single centreline UD turrets, 2 fwd and 2 aft.

Armour: Belt 4½in (with 18in teak backing); turrets 10½in -5½in (with 14in teak backing); UD 1⅛in - ¾in.

Machinery: 2-cylinder (72in diameter, 3ft stroke) horizontal single expansion, direct acting. 4 rectangular boilers. Single screw. 500nhp. 2,130ihp = 11¼kts. 230 tons coal.

Prince Albert Samuda Bros, Poplar/Humphrys & Tennant

Ord: 8.4.1862. K: 29.4.1862. L: 23.5.1864. C: 23.2.1866.

First cost: £208,345.

Commissioned 30.1.1866 at Woolwich under Capt. Arthur Wilmhurst (for trials); paid off 6.1866. Gunnery tender (uncommissioned) at Devonport 1866–90; in Spithead Review 17.7.1867; in Particular Reserve Squadron formed 8.1878. Recommissioned 1887 under Cmdr Charles Inglis for Jubilee Review, and participated in naval manoeuvres in 1889. To Dockyard Reserve 1898. Sold to Ward, Preston 16.3.1899 for BU.

Appendix A   The Wooden Steam Battlefleet

The following shows the construction of screw-driven wooden-hulled ships of the line between 1848 and 1861; the number of guns in the first column is that with which the ships in question were initially established; the date following each ship name is the year of launch.

Rating	New Construction	Lengthened on stocks	Not lengthened
First Rates			
131 guns		*Duke of Wellington*[*] (1852)	
		Royal Sovereign (1857)	
		Prince of Wales (1860)	
121 guns	*Victoria* (1859)	*Royal Albert* (1854)	
	Howe (1860)	*Marlborough* (1855)	
102 guns			*Windsor Castle* (1858)
Second Rates			
101 guns	*Saint Jean d'Acre* (1853)		
	Conqueror (1855)		
	Donegal (1858)		
	Duncan (1859)		
	Gibraltar (1860)		
91 guns	*Agamemnon* (1852)	*Princess Royal* (1853)	*Nile* (1854)[‡]
	James Watt (1853)	*Caesar* (1853)	*Exmouth* (1854)
	Victor Emmanuel[†] (1855)	*Algiers* (1854)	*Aboukir* (1858)[‡]
	Renown (1860)	*Hannibal* (1854)	*Albion* (1861)[‡]
	Edgar (1858)	*Orion* (1854)	
	Hero (1858)	*London* (1859)[‡]	
	Revenge (1859)	*Hood* (1859)	
	Atlas (1860)	*Rodney* (1860)[‡]	
	Anson (1860)		
	Defiance (1861)		
86 guns			*Queen* (1859)[‡]
			Frederick William (1860)
81 guns		*Bombay* (1861)[‡]	
80 guns			*Majestic* (1853)
			Cressy (1853)
			Brunswick (1855)
			Colossus (1855)[‡]
			Centurion (1855)[‡]
			Mars (1856)[‡]
			Meeanee (1857)[‡]
			Goliath (1857)[‡]
			Lion (1859)[‡]
			Irresistible (1859)
			Collingwood (1861)[‡]

Notes: [*] renamed from *Windsor Castle* soon after launch; [†] renamed from *Repulse* soon after launch; [‡] year of re-launch (originally launched as a sailing warship)

List of the Steam Reserve, September 1860

The following shows the four classes of the Steam Reserve at Portsmouth (sourced from *The Times*, 11 September 1860); the number of guns with which the vessel was established follow in brackets:

First Class   *Duke of Wellington* (131), *Princess Royal* (91); *Shannon* (51), *Immortalité* (51); *Volcano* (6), *Philomel* (6), gunboats *Beaver*, *Blazer*, *Brazen*, *Grinder*, *Snapper* and *Traveller* (each 2 guns).

Second Class   *Royal Sovereign* (131), *Prince of Wales* (131), *Victoria* (121), *Duncan* (101), *Nelson* (91); *Sutlej* (51); *Harrier* (17), *Rinaldo* (17), *Medea* (6), *Stromboli* (6), *Coquette* (6), gunboats *Badger*, *Cracker*, *Fancy*, *Pincher* and *Swinger* (each 2 guns).

Third Class   *Tribune* (31); *Rosamund* (6), *Vulture* (6), *Cygnet* (5), *Vigilant* (4), gunboats *Angler*, *Ant*, *Cheerful*, *Chub*, *Daisy*, *Decoy*, *Pert*, *Pet* and *Rambler* (each 2 guns).

Fourth Class   *Fox* (troopship 42), *Erebus* (16), *Glatton* (14) and *Meteor* (14).

Appendix B   Principal Officers of the Navy

The Royal Navy was administered by the Board of Admiralty, led initially in the post-Napoleonic period (under the Earl of Liverpool's premiership) by Robert Dundas, Viscount Melville, as (civilian) First Lord of the Admiralty from 25 March 1812 until 2 May 1827. During this period the senior Naval member (or First Naval Lord) was initially Sir Joseph Yorke until 2 April 1818, then Sir Graham Moore until 13 March 1820, and subsequently Sir William Johnstone Hope.

In May 1827 the Board was suspended, with the appointment of HRH Prince William Henry, Duke of Clarence, as Lord Admiral. During the next 16½ months, the Lord Admiral was assisted by an appointed council, on which Sir William Johnstone Hope continued as the senior Naval member (replaced in 12 March 1828 by Sir George Cockburn).

The Board of Admiralty was restored on 19 September 1828, again with Viscount Melville as First Lord of the Admiralty and with Cockburn as its First Naval Lord. Following the General Election of 1830, a new Board was constituted on 25 November with Sir James Graham as First Lord and with Sir Thomas Masterman Hardy (Nelson's Captain of the *Victory* at Trafalgar) as First Naval Lord.

The separate Navy Board and Victualling Board were both abolished with effect from 1 June 1832, and their responsibilities absorbed into those of the Board of Admiralty. The leading members of the Board over the next forty years were as follows:

First Lord of the Admiralty (civilians/political appointees)

25 Nov 1830	Sir James Robert George Graham, MP
11 June 1834	George Eden, Lord Auckland
23 Dec 1834	Thomas, Earl de Grey
25 Apr 1835	George Eden, Lord Auckland
19 Sep 1835	Gilbert Elliot, Earl of Minto
8 Sep 1841	Thomas Hamilton, Earl of Haddington
13 Jan 1846	Edward Law, Earl of Ellenborough
13 July 1846	George Eden, Earl of Auckland
18 Jan 1849	Sir Francis Thomas Baring, MP
2 Mar 1852	Algernon Percy, Duke of Northumberland
5 Jan 1853	Sir James Robert George Graham, MP
8 Mar 1855	Sir Charles Wood, MP
8 Mar 1858	Sir John Somerset Pakington, MP
28 June 1859	Edward Seymour, Duke of Somerset

(The Duke of Somerset remained in post until 13 July 1866)

First Naval Lord (the senior serving naval officer on the Board)

25 Nov 1830	Sir Thomas Masterman Hardy
1 Aug 1834	George Dundas
1 Nov 1834	Charles Adam
23 Dec 1834	Sir George Cockburn
25 Apr 1835	Sir Charles Adam
8 Sep 1841	Sir George Cockburn
13 July 1846	Sir William Parker
24 July 1846	Sir Charles Adam
20 July 1847	James Deans Dundas
13 Feb 1852	Maurice Berkeley
2 Mar 1852	Hyde Parker
3 June 1854	Maurice Berkeley
24 Nov 1857	Sir Richard Saunders Dundas
8 Mar 1858	William Martin
28 June 1859	Sir Richard Saunders Dundas
15 June 1861	Rear-Adm. Sir Frederick William Grey

(Sir Frederick Grey remained in post until 13 July 1866)

First Secretary of the Navy (a political appointment)

12 Oct 1809	John Wilson Croker, MP
29 Nov 1830	Capt. George Elliot
24 Dec 1834	George R. Dawson
27 Apr 1835	Charles Wood, MP
4 Oct 1839	R. More O'Ferrall, MP
9 June 1841	John Parker, MP
10 Sep 1841	Sidney Herbert, MP
Feb 1845	Henry Thomas Lowry Corry, MP
13 July 1846	Henry G Ward, MP
21 May 1849	John Parker, MP
3 Mar 1852	Augustus Stafford, MP
6 Jan 1853	Ralph Bernal Osborne, MP
9 Mar 1858	Henry Thomas Lowry Corry, MP
30 June 1859	Rear-Adm. Lord Clarence Edward Paget, MP

Second Secretary of the Navy (a non-political appointee)

9 Apr 1807	John Barrow (Sir John from 1835)
28 Jan 1845	Capt. William Alexander Baillie Hamilton
22 May 1855	Thomas Phinn
7 May 1857	William Govett Romaine

(The office was subsequently re-labelled as Permanent Secretary in 1870.)

Controller of the Navy

9 Feb 1816	Rear-Adm. Sir Thomas Byam Martin
2 Nov 1831	Rear-Adm. George Heneage Lawrence Dundas.

(The office was abolished from 9 June 1832 until it was revived in 1860.)

Surveyor of the Navy

26 May 1813	initially jointly Sir Henry Peake (retired 1822), Joseph Tucker (retired 1831) and Robert Seppings (Sir Robert from 1819), but Seppings soon established himself as the principal source of warship design.
9 June 1832	Capt. William Symonds (Sir William from 1836)
5 Feb 1848	Capt. Sir Baldwin Wake Walker (Rear-Adm. 1858)

(The office was reconstituted as *Controller of the Navy* in 1860, still under Wake Walker. Design responsibilities passed to the *Chief Constructor*, a post created on 4 May 1848 and filled by Isaac Watts from that date until 9 July 1863.)

Admirals of the Fleet

The pinnacle of the naval rank structure was the Admiral of the Fleet. Traditionally the rank had been awarded only to a single individual at any one time, and each recipient held the rank until his death. From 1821 there was able to be more than one holder of the rank:

24 Dec 1811	HRH The Duke of Clarence (later HM King William IV, died 20 June 1837)
19 July 1821	Sir John Jervis, Earl of St Vincent (died 14 Mar 1823)
28 June 1830	William Peere Williams Freeman (died 10 Feb 1832)
22 July 1830	James Gambier, Baron Gambier (died 19 Apr 1833)
22 July 1830	Sir Charles Morice Pole (died 31 Aug 1830)
24 Apr 1833	Sir Charles Edmund Nugent (died 7 Jan 1844)
8 Jan 1844	Sir James Hawkins Whitshed (died 28 Oct 1849)
9 Nov 1846	Sir George Martin (died 28 July 1847)
13 Oct 1849	Sir Thomas Byam Martin (died 21 Oct 1854)
1 July 1851	Sir George Cockburn (died 19 Aug 1853)
8 Dec 1857	Sir Charles Ogle (died 16 June 1858)
25 June 1858	Sir John West (died 18 Apr 1862)
20 May 1862	Sir William Hall Gage (died 5 Jan 1864)
10 Nov 1862	Sir Graham Eden Hamond (died 20 Dec 1862)
27 Apr 1863	Sir Francis William Austin (died 10 Mar 1865)
27 Apr 1863	Sir William Parker (died 13 Nov 1866)

Appendix C   Annual Expenditure on the Navy and Manning Levels

Year	Expenditure voted	Seamen and boys	Marines	Total authorised	Actual carried	Year	Expenditure voted	Seamen and boys	Marines	Total authorised	Actual carried
1816	£13,114,345	24,000	9,000	33,000	35,196	1840	£6,182,247	26,165*	9,000	35,165*	37,655
1817	£7,645,422	13,000	6,000	19,000	22,944	1841	£6,772,969	32,500	10,500	43,000	41,389
1818	£6,547,809	14,000	6,000	20,000	23,026	1842	£7,000,442	32,500	10,500	43,000	43,105
1819	£6,527,781	14,000	6,000	20,000	23,230	1843	£6,579,960	28,500	10,500	39,000	40,229
1820	£6,691,345	15,000	8,000	23,000	23,985	1844	£6,466,019	25,500	10,500	36,000	38,343
1821	£6,391,902	14,000	8,000	22,000	24,937	1845	£7,344,363	29,500	10,500	40,000	40,084
1822	£6,480,325	13,000	8,000	21,000	23,806	1846	£7,920,324	29,500	10,500	40,000	43,314
1823	£5,442,540	16,000	8,700	24,700	26,314	1847	£8,068,985	29,500	11,000†	40,500†	44,969
1824	£5,762,893	20,000	9,000	29,000	30,502	1848	£7,955,001	29,500	12,500†	42,000†	43,978
1825	£5,983,126	20,000	9,000	29,000	31,456	1849	£7,021,724	28,000	12,000	40,000	39,535
1826	£6,135,004	21,000	9,000	30,000	32,519	1850	£6,672,588	28,000	11,000	39,000	39,093
1827	£6,125,850	21,000	9,000	30,000	33,106	1851	£6,543,255	28,000	11,000	39,000	38,957
1828	£6,395,965	21,000	9,000	30,000	31,818	1852	£6,705,746	28,000**	11,000**	39,000**	40,451
1829	£5,878,794	21,000	9,000	30,000	32,458	1853	£7,197,804	33,000	12,500	45,500	45,885
1830	£5,594,955	20,000	9,000	29,000	31,160	1854	£15,017,591	48,000	15,500	63,500	61,457
1831	£7,221,797	22,000	10,000	32,000	29,336	1855	£19,590,833	54,000	16,000	70,000	67,791
1832	£5,045,827	18,000	9,000	27,000	27,328	1856	£16,568,614	60,000‡	16,000	76,000‡	60,659
1833	£4,803,647	18,000	9,000	27,000	27,701	1857	£9,962,840	38,700‡	15,000	53,700‡	54,291
1834	£4,716,894	18,500	9,000	27,500	28,066	1858	£9,878,859	44,380	15,000	59,380	n/a
1835	£4,434,783	17,500	9,000	26,500	26,041	1859	£11,775,718	47,400	15,000	52,400	n/a
1836	£4,689,651	24,700	9,000	33,700	30,195	1860	£11,836,100	66,100	18,000	84,100	n/a
1837	£4,930,736	25,165	9,000	34,165	31,289	1861	£12,640,588	59,000	18,000	77,000	n/a
1838	£4,960,911	25,165	9,000	34,165	32,028	1862	£11,794,305	56,850	18,000	74,850	n/a
1839	£5,532,724	25,165	9,000	34,165	34,857	1863	£10,708,651	53,000	18,000	71,000	n/a

Source: Laird Clowes (see Bibliography).

The periods were calendar years until 1830; 1831 ran from 1 Jan 1831 to 31 March 1832, and thus comprised 15 months expenditure; thereafter years ran from 1 April to 31 March. Months were calendar months from 1832 (until then, lunar months were used).

*   in 1840, the authorised total was raised by 2,000 seamen after 2 months, and by a further 2,500 after another 7 months.

†   in 1847, the authorised total was raised by 1,000 marines after 6 months, and an equal increase of 1,000 marines was made after 6 months in 1848.

**   in 1852, the authorised total was raised by 5,000 seamen and 1,500 marines after 8 months.

‡   in 1856, the authorised total was reduced by 20,000 seamen after 3 months; in 1857, the authorised total was raised by 2,000 seamen after 3 months.

Appendix D   Dockyard Launchings 1817 to 1869

The following table gives a chronological list of naval vessels launched in each of HM Dockyards (including Bombay, actually operated by the HEICo.) over the period of this volume, with their rated number of guns. For those ships which were re-rated at the start of 1817, the rated armament following that change in given. Vessels' names are those at the date of launch. Those names affixed by an asterisk (*) indicates vessels which when *launched* were fitted with steam propulsion (either paddle or screw).

A scheme to concentrate on fewer shipbuilding yards was announced in July 1863. Deptford Dockyard (which had temporarily closed in 1832 – with the frigate *Worcester* still on the stocks – but re-opened in 1844) was listed for closure in March 1868 and formally closed in March 1869, while Woolwich Dockyard was listed for closure in January 1869 and actually closed in October 1969. So that the final vessels constructed at these yards can be included for the sake of completeness (as well as to cover vessels built at other dockyards which had been ordered before mid 1863 but were not completed until after that date) the lists of launchings at all of the dockyards have been extended to the close of 1869.

DEPTFORD

Linnet 4 (schooner)	1817, 3 Jan.
Quail 4 (schooner)	1817, 3 Jan.
Royal George (yacht)	1817, 17 July
Alacrity 10 (brig)	1818, 29 Dec.
Blonde 46	1819, 12 Jan.
Ariel 10 (brig)	1820, 28 July.
Venus 46	1820, 10 Aug.
Amazon 46	1821, 15 Aug.
Southampton 52	1820, 7 Nov.
Vigilant 10 (cutter)	1821, 18 April
Woodlark 2 (cutter)	1821, 2 July
Russell 74	1822, 22 May
Comet 2 (sloop) *	1822, 23 May
Algerine 10 (brig)	1823, 10 June
Lightning 2 (sloop) *	1823, 19 Sept.
Meteor 2 (sloop) *	1824, 17 Feb.
Aeolus 46	1825, 17 June
Alban 2 (sloop) *	1826, 27 Dec.
Carron 2 (sloop) *	1827, 9 Jan.
Fanny — (tender)	1827, May
Nimrod 20	1828, 26 Aug.
Briseis 10 (brig)	1829, 3 July
Thunder 10 (bomb)	1829, 4 Aug.
Worcester 50	1843, 10 Oct.
Porcupine 3 (gunvessel) *	1844, 17 June
Terrible 22 (frigate) *	1845, 6 Feb.
Hound 6 (sloop)	1846, 23 May
Sidon 12 (frigate) *	1846, 26 May
Odin 12 (frigate) *	1846, 24 July
Oberon 3 (gunvessel) *	1847, 2 Jan.
Termagant 24 (frigate) *	1847, 25 Sept.
Renard 8 (sloop) *	1848, 21 March
Phaeton 50	1848, 25 Nov.
Archer 4 (sloop) *	1849, 27 March
Wasp 14 (sloop) *	1850, 28 May
Leopard 12 (frigate)	1850, 5 Nov.
Cruizer 16 (sloop) *	1852, 19 June
Imperieuse 51 (frigate) *	1852, 15 Sept.
Hannibal 90 *	1854, 31 Jan.
Hornet 17 (sloop) *	1854, 13 April
Curlew 9 (sloop) *	1854, 31 May
Gleaner 2 (gunboat) *	1854, 7 Oct.
Ruby 2 (gunboat) *	1854, 7 Oct.
Lark 2 (gunboat) *	1855, 15 March
Magpie 2 (gunboat) *	1855, 15 March
Cheerful 2 (gunboat) *	1855, 6 Oct.
Emerald 51 (frigate) *	1856, 19 July
Fawn 17 (sloop) *	1856, 30 Sept.
Lyra 9 (sloop) *	1857, 26 March
Racer 11 (sloop) *	1857, 4 Nov.
Forte 51 (frigate) *	1858, 29 May
Icarus 11 (sloop) *	1858, 22 Oct.
Ariadne 26 (frigate) *	1859, 4 June
Mutine 17 (sloop) *	1859, 30 July
Ranger 5 (gunvessel) *	1859, 26 Nov.
Cameleon 17 (sloop) *	1860, 23 Feb.
Landrail 5 (gunvessel) *	1860, 28 March
Newcastle 51 (frigate) *	1860, 16 Oct.
Rosario 11 (sloop) *	1860, 17 Oct.
Liverpool 51 (frigate) *	1860, 30 Oct.
Zebra 17 (sloop) *	1860, 13 Nov.
Rapid 11 (sloop) *	1860, 29 Nov.
Speedwell 5 (gunvessel) *	1861, 12 Feb.
Investigator 2 (survey) *	1861, 16 Nov.
Columbine 11 (sloop) *	1862, 2 April
Rattler 17 (sloop) *	1862, 18 March
Jaseur 5 (gunvessel) *	1862, 15 May
Enterprise 4 (sloop) *	1864, 9 Feb.
Favorite 10 (corvette) *	1864, 5 July
Endymion 36 (frigate) *	1865, 18 Nov.
Niobe 4 (sloop) *	1866, 31 May
Nymphe 4 (sloop) *	1866, 24 Nov.
Plover 3 (gunvessel) *	1867, 20 Feb.
Philomel 3 (gunvessel) *	1867, 20 Oct.
Juno 6 (corvette) *	1867, 28 Nov.
Boxer 2 (gunvessel) *	1868, 25 Jan.
Thistle 2 (gunvessel) *	1868, 25 Jan.
Curlew 3 (gunvessel) *	1868, 20 Aug.
Spartan 6 (sloop) *	1868, 14 Nov.
Druid 10 (corvette) *	1869, 13 March

WOOLWICH

Swift 4 (schooner)	1817, 15 Feb.
Redbreast 4 (schooner)	1817, 18 Feb.
Talavera 74	1818, 15 Oct.
Isis 58	1819, 5 Oct.
Hawke 74	1820, 16 March
Beagle 10 (brig)	1820, 11 May
Barracouta 10 (brig)	1820, 13 May
Atholl 28	1820, 23 Nov.
Niemen 28	1820, 23 Nov.
Sylph 2 (cutter)	1821, 15 June
Highflyer 2 (cutter)	1822, 11 June
Navy Board 2 (cutter)	1822, 12 June
Winchester 52	1822, 21 June
Kingfisher 10 (brig)	1823, 11 March
Magnet 10 (brig)	1823, 13 March
Pylades 18	1824, 29 June
Royal Charlotte 6 (yacht)	1824, 22 Nov.
North Star 28	1824, 7 Dec.
African 2 (sloop) *	1825, 30 Aug.
Tyrian 10 (brig)	1826, 16 Sept.
Tyne 28	1826, 30 Nov.
Hebe 46	1826, 14 Dec.
Confiance 2 (sloop) *	1827, 28 March
Echo 2 (sloop) *	1827, 28 May
Clyde 46	1828, 9 Oct.
Columbia 2 (sloop) *	1829, 1 July
Curlew 10 (brig)	1830, 25 Feb.
Nautilus 10 (brig)	1830, 11 March
Pluto 2 (gunboat) *	1831, 28 April
Thunderer 84	1831, 22 Sept.
Dee 4 (sloop) *	1832, 5 April
Vernon 50	1832, 1 May
Firefly 2 (sloop) *	1832, 29 Sept.
Pandora 4 (brig)	1833, 4 July
Medea 4 (sloop) *	1833, 2 Sept.
Spitfire 2 (gunboat) *	1834, 26 March
Star 6 (brig)	1835, 29 April
Modeste 18	1837, 31 Oct.
Crane 6 (brig)	1839, 8 May
Lizard 3 (gunvessel) *	1840, 7 January
Cygnet 10 (brig)	1840, 6 April
Locust 3 (gunvessel) *	1840, 8 April
Siren 16 (brig)	1841, 23 April
Trafalgar 120	1841, 21 June
Devastation 6 (sloop) *	1841, 3 July
Heroine 10 (brig)	1841, 16 Aug.
Infernal 6 (sloop) *	1843, 31 May
Chichester 50	1843, 12 July
Boscawen 70	1844, 3 April
Éclair 6 (sloop) *	1844, 31 May
Sampson 6 (frigate) *	1844, 1 Oct.
Gladiator 6 (frigate) *	1844, 15 Oct.
Amphion 36 (frigate) *	1846, 14 Jan.
Sphinx 6 (sloop) *	1846, 17 Feb.
Niger 8 (sloop) *	1846, 18 Nov.
Basilisk 8 (sloop) *	1848, 22 Aug.
Nankin 50	1850, 16 March
Brisk 14 (sloop) *	1851, 2 June
Agamemnon 91 *	1852, 22 May
Royal Albert 121 *	1854, 13 May
Pearl 21 (corvette) *	1855, 13 Sept.
Scout 21 (corvette) *	1856, 31 Dec.
Challenger 21 (corvette) *	1858, 13 Feb.
Edgar 91 *	1858, 23 Oct.

Galatea 26 (frigate) *	1859, 14 Sept.
Barrosa 21 (corvette) *	1860, 10 March
Blonde 36 (frigate) *	1860, 10 Sept.
Anson 91 *	1860, 15 Sept.
Bristol 51 (frigate) *	1861, 12 Feb.
Caledonia 30 (ironclad) *	1862, 24 Oct.
Wolverine 21 (corvette) *	1863, 29 Aug.
Pallas 6 (corvette) *	1865, 14 March
Sylvia 4 (gunvessel) *	1866, 20 March
Dwarf 2 (gunvessel) *	1867, 28 Nov.
Repulse 12 (ironclad) *	1868, 25 April
Thalia 6 (corvette) *	1869, 14 July

CHATHAM

Starling 10 (cutter)	1817, 3 May
Bustard 10 (brig)	1818, 12 Dec.
Brisk 10 (brig)	1819, 10 Feb.
Blanche 46	1819, 26 May
Trafalgar 106	1820, 26 July
Latona 46	1821, 16 June
Renard 10 (brig)	1821, 26 Oct.
Diana 46	1822, 8 Jan.
Rattlesnake 28	1822, 26 March
Weazle 10 (brig)	1822, 26 March
Basilisk 10 (cutter)	1822, 7 May
Procris 10 (brig)	1822, 21 June
Prince Regent 120	1823, 12 April
Thames 46	1823, 21 Aug.
Rainbow 28	1823, 20 Nov.
Unicorn 46	1824, 30 March
Aetna 10 (bomb)	1824, 14 April
Hearty 10 (brig)	1824, 22 Oct.
Lapwing 10 (brig)	1825, 20 Feb.
Formidable 84	1825, 19 May
Harpy 10 (brig)	1825, 16 July
Mermaid 46	1825, 30 July
Crocodile 28	1825, 28 Oct.
Sulphur 8 (bomb)	1826, 26 Jan.
Fairy 10 (brig)	1826, 25 April
Espoir 10 (brig)	1826, 9 May
Powerful 84	1826, 21 June
Calypso 10 (brig)	1826, 19 Aug.
Acorn 18	1826, 16 Nov.
Mercury 46	1826, 16 Nov.
Childers 18 (brig)	1827, 23 Aug.
Royal George 120	1827, 22 Sept.
Africaine 46	1827, 20 Dec.
Cruizer 18 (brig)	1828, 19 Jan.
Eurotas 46	1829, 19 Feb.
Algerine 10 (brig)	1829, 1 Aug.
Penelope 46	1829, 13 Aug.
Delight 10 (brig)	1829, 27 Nov.
Thalia 46	1830, 12 Jan.
Lark 4 (cutter)	1830, 4 Aug.
Jackdaw 4 (cutter)	1830, 4 Aug.
Hornet 6 (brigantine)	1831, 24 Aug.
Seagull 8 (schooner)	1831, 21 Nov.
Conway 28	1832, 2 Feb.
Castor 36	1832, 2 May
Scout 18	1832, 15 June
Rover 18	1832, 17 July
Forester 3 (brigantine)	1832, 28 Aug.
Griffon 3 (brigantine)	1832, 11 Sept.
Phoenix 4 (sloop) *	1832, 25 Sept.
Monarch 84	1832, 8 Dec.
Waterloo 120	1833, 10 June
Gulnare 1 (packet) *	1833, 30 Sept.
Rochester – (lighter)	1833, 26 Nov.

Blazer 2 (gunboat) *	1834, May
Wanderer 16 (brig)	1835, 10 July
Spider 6 (brigantine)	1835, 23 Sept.
Devon – (lighter)	1835, 24 Sept.
Bat – (victualling hoy)	1836, 29 Aug.
Wolverine 16 (brig)	1836, 13 Oct.
Mercury – (tender)	1837, 7 Feb.
No.1 tank vessel	1837, 19 May
Widgeon 1 (packet) *	1837, 12 Sept.
Dasher 1 (packet) *	1837, 5 Dec.
Hydra 4 (sloop) *	1838, 13 June
Aid – (lighter)	1838, 30 Nov.
Hecla 4 (sloop) *	1839, 14 Jan.
Hecate 4 (sloop) *	1839, 30 March
Fantome 16 (brig)	1839, 30 May
Alecto 1 (sloop) *	1839, 7 Sept.
Maeander 44	1840, 5 May
London 92	1840, 28 Sept.
Polyphemus 1 (sloop) *	1840, 28 Sept.
Ardent 3 (sloop) *	1841, 12 Feb.
Growler 6 (sloop) *	1841, 20 July
Bee – (tender) *	1842, 28 Feb.
Goliath 80	1842, 25 July
Virago 6 (sloop) *	1842, 25 July
Cumberland 70	1842, 21 Oct.
Janus 6 (sloop) *	1844, 6 Feb.
Espiegle 12 (brig)	1844, 20 April
Mutine 12 (brig)	1844, 20 April
Retribution 6 (frigate) *	1844, 2 July
Raleigh 50	1845, 8 May
Calypso 20	1845, May
Active 36	1845, 19 July
Bulldog 6 (sloop) *	1845, 2 Oct.
Teazer 2 (gunvessel) *	1846, 25 June
Arab 12 (brig)	1847, 31 March
Elk 12 (brig)	1847, 27 Sept.
Heron 12 (brig)	1847, 27 Sept.
Vivid 2 (packet) *	1848, 7 Feb.
Mars 80	1848, 1 July
Tiger 16 (frigate) *	1849, 1 Feb.
Elfin 1 (yacht) *	1849, 8 Feb.
Despatch 12 (brig)	1851, 25 Nov.
Kangaroo 12 (brig)	1852, 31 Aug.
Cressy 80 *	1853, 21 July
Euryalus 51 (frigate) *	1853, 5 Oct.
Majestic 80 *	1853, 1 Dec.
Orion 91 *	1854, 6 Nov.
Chesapeake 51 (frigate) *	1855, 27 Sept.
Severn 50 (frigate) *	1856, 24 Jan.
Aetna 16 (floating battery) *	1856, 5 April
Cadmus 21 (corvette) *	1856, 20 May
Renown 91 *	1857, 28 March
Racoon 21 (corvette) *	1857, 25 April
Hero 91 *	1858, 15 April
Mersey 40 (frigate) *	1858, 13 Aug.
Hood 90 *	1859, 4 May
Charybdis 21 (corvette) *	1859, 1 June
Irresistible 80 *	1859, 27 Oct.
Orpheus 21 (corvette) *	1860, 23 June
Atlas 91 *	1860, 21 July
Undaunted 51 (frigate) *	1861, 1 Jan.
Rattlesnake 21 (corvette) *	1861, 9 July
Royal Oak 35 (ironclad) *	1862, 10 Sept.
Salamis 2 (dispatch) *	1863, 19 May
Achilles 20 (ironclad) *	1863, 23 Dec.
Bellerophon 14 (ironclad) *	1865, 26 May
Lord Warden 20 (ironclad) *	1865, 27 May
Reindeer 7 (sloop) *	1866, 29 March

Myrmidon 4 (gunvessel) *	1867, 5 June
Beacon 2 (gunvessel) *	1867, 17 Aug.
Blanche 6 (sloop) *	1867, 17 Aug.
Hercules 12 (ironclad) *	1868, 10 Feb.
Monarch 7 (ironclad) *	1868, 25 May

SHEERNESS

Opossum 10 (brig)	1821, 11 Dec.
Onyx 10 (brig)	1822, 24 Jan.
Daedalus 46	1826, 2 May
Magpie 4 (cutter)	1830, 30 Sept.
Quail 4 (cutter)	1830, 30 Sept.
Medway – (lighter)	1830, 30 Sept.
Salamander 4 (sloop) *	1832, 16 May
Vestal 26	1833, 6 April
Royal Adelaide – (yacht)	1833, Dec.
Bonetta 10 (brigantine)	1836, 5 April
Dolphin 10 (brigantine)	1836, 14 June
Gipsy 2 (cutter)	1836, 28 Oct.
Megaera 2 (sloop) *	1837, 17 Aug.
Calliope 28	1837, 5 Oct.
Acheron 2 (sloop) *	1838, 23 Aug.
Vesuvius 4 (sloop) *	1839, 11 July
Prometheus 1 (sloop) *	1839, 21 Sept.
Styx 6 (sloop) *	1841, 26 Jan.
Spy 3 (brigantine)	1841, 24 March
Cormorant 6 (sloop) *	1842, 29 March
Rattler 6 (sloop) *	1843, 12 April
Alarm 26	1845, 22 April
Fury 6 (sloop) *	1845, 31 Dec.
Dart 3 (brigantine)	1847, 17 March
Diamond 28	1848, 29 Aug.
Miranda 14 (sloop) *	1851, 18 March
Tribune 31 (frigate) *	1853, 21 Jan.
Staunch – (lighter)	1854, 10 July
Pylades 20 (corvette) *	1854, 23 Nov.
Chub 2 (gunboat) *	1855, 15 Oct.
Scylla 21 (corvette) *	1856, 19 June
Clio 21 (corvette) *	1858, 28 Aug.
Orestes 21 (corvette) *	1860, 18 Aug.
Eclipse 6 (sloop) *	1867, 14 Nov.
Bullfinch 3 (gunvessel) *	1868, 13 Feb.
Briton 10 (corvette) *	1869, 6 Nov.
Vulture 3 (gunvessel) *	1869, 6 Nov.

PORTSMOUTH

Waterloo 80	1818, 16 Oct.
Delight 10 (brig)	1819, 10 May
Cygnet 10 (brig)	1819, 11 May
Prince Regent (yacht)	1820, 12 June
Minerva 46	1820, 13 June
Jasper 10 (brig)	1820, 26 July
Britomart 10 (brig)	1820, 24 Aug.
Ranger 28	1820, 7 Dec.
Martin 20	1821, 18 May
Rose 18	1821, 1 June
Plover 10 (brig)	1821, 30 June
Ferret 10 (brig)	1821, 12 Oct.
Arrow 10 (cutter)	1823, 14 March
Tweed 28	1823, 14 April
Philomel 10 (brig)	1823, 28 April
Royalist 10 (brig)	1823, 12 May
Carnatic 74	1823, 21 Oct.
Champion 18	1824, 31 May
Orestes 18	1824, 31 May
Leveret 10 (brig)	1825, 19 Feb.
Musquito 10 (brig)	1825, 20 Feb.
Volage 28	1825, 20 Feb.

Myrtle 10 (brig)	1825, 14 Sept.
Princess Charlotte 110	1825, 14 Sept.
Challenger 28	1826, 14 Nov.
Columbine 18	1826, 1 Dec.
Wolf 18	1826, 1 Dec.
Sapphire 28	1827, 31 Jan.
Sylvia 1 (cutter)	1827, 24 March
President 52	1829, 20 April
Favourite 18	1829, 21 April
Fox 46	1829, 17 Aug.
Rapid 10 (brig)	1829, 17 Aug.
Recruit 10 (brig)	1829, 17 Aug.
Seaflower 4 (cutter)	1830, 20 May
Actaeon 26	1831, 31 Jan.
Charybdis 10 (brig)	1831, 27 Feb.
Admiralty 1 (yacht)	1831, 28 Feb.
Neptune 120	1832, 27 Sept.
Racer 16 (brig)	1833, 18 July
Lynx 10 (brigantine)	1833, 2 Sept.
Buzzard 10 (brigantine)	1834, 25 March
Drake – (lighter)	1834, 25 March
Hermes 2 (sloop) *	1835, 26 June
Inconstant 36	1836, 10 June
Volcano 2 (sloop) *	1836, 29 June
Hazard 18	1837, 21 April
Electra 18	1837, 28 Sept.
Termagant 10 (brigantine)	1838, 25 March
Indus 80	1839, 16 March
Queen 110	1839, 15 May
Stromboli 4 (sloop) *	1839, 27 Aug.
Bittern 16 (brig)	1840, 18 April
Rapid 10 (brig)	1840, 3 June
Driver 6 (sloop) *	1840, 24 Dec.
Thunderbolt 6 (sloop) *	1842, 13 Jan.
Albatross 16 (brig)	1842, 28 March
Frolic 16 (brig)	1842, 23 Aug.
Firebrand 5 (frigate) *	1842, 6 Sept.
Eurydice 26	1843, 16 May
Sealark 10 (brig)	1843, 27 July
Daring 12 (brig)	1844, 2 April
Osprey 12 (brig)	1844, 2 April
Scourge 6 (sloop) *	1844, 8 Nov.
Centaur 6 (frigate) *	1845, 6 Oct.
Rifleman 4 (gunvessel) *	1846, 10 Aug.
Dauntless 24 (frigate) *	1847, 5 Jan.
Leander 50	1848, 8 March
Arrogant 46 (frigate) *	1848, 5 April
Plumper 8 (sloop) *	1848, 5 April
Buzzard 6 (sloop) *	1849, 24 March
Desperate 8 (sloop) *	1849, 23 April
Argus 6 (sloop) *	1849, 15 Dec.
Furious 16 (sloop) *	1850, 26 Aug.
Princess Royal 90 *	1853, 23 June
Marlborough 131 *	1855, 31 July
Shannon 51 (frigate) *	1855, 24 Nov.
Royal Sovereign 121 *	1857, 25 April
Bacchante 51 (frigate) *	1859, 30 July
Victoria 121 *	1859, 12 Nov.
Duncan 101 *	1859, 13 Dec.
Prince of Wales 131 *	1860, 25 Jan.
Frederick William 110 *	1860, 24 March
Rinaldo 17 (sloop) *	1860, 26 March
Chanticleer 17 (sloop) *	1861, 9 Feb.
Glasgow 51 (frigate) *	1861, 28 March
Mooring lighter A	1861, 9 July
Mooring lighter B	1861, 21 Aug.
Royal Albert 18 (ironclad) *	1864, 15 Oct.
Helicon 2 (dispatch) *	1865, 31 Jan.

Minstrel 2 (gunboat) *	1865, 16 Feb.
Cherub 2 (gunboat) *	1865, 29 March
Netley 2 (gunboat) *	1866, 2 July
Orwell 2 (gunboat) *	1866, 27 Dec.
Bruiser 2 (gunboat) *	1867, 23 April
Danae 6 (sloop) *	1867, 21 May
Cromer 2 (gunboat) *	1867, 20 Aug.
Ringdove 3 (gunvessel) *	1867, 4 Sept.
Avon 2 (gunvessel) *	1867, 2 Oct.
Cracker 2 (gunvessel) *	1867, 27 Nov.
Elk 2 (gunvessel) *	1868, 10 Jan.
Magpie 3 (gunvessel) *	1868, 12 Feb.
Sirius 6 (sloop) *	1868, 24 April
Swallow 3 (gunvessel) *	1868, 16 Nov.
Dido 6 (sloop) *	1869, 23 Oct.

PLYMOUTH/DEVONPORT

Agincourt 74	1817, 19 March
Eclipse 10 (brig)	1819, 23 July
Emulous 10 (brig)	1819, 16 Dec.
Britannia 120	1820, 20 Oct.
Lyra 10 (brig)	1821, 1 June
Partridge 10 (brig)	1822, 22 March
Bramble 10 (cutter)	1822, 8 April
Portland 52	1822 , 8 May
Netley 6 (cutter)	1823, 13 March
Lancaster 52	1823, 23 Aug.
Hope 10 (brig)	1824, 8 Dec.
Nightingale 8 (cutter)	1825, 19 April
Mutine 10 (brig)	1825, 19 May
Cerberus 46	1827, 30 March
Circe 46	1827, 22 Sept.
Royal Adelaide 110	1828, 28 July
Hyacinth 18	1829, 6 May
Reindeer 10 (brig)	1829, 29 Sept.
Rolla 10 (brig)	1829, 10 Dec.
Racehorse 18	1830, 24 May
Proserpine 46	1830, 1 Dec.
Savage 10 (brig)	1830, 29 Dec.
Saracen 10 (brig)	1831, 10 Jan.
Rhadamanthus 4 (sloop) *	1832, 16 April
Scorpion 10 (brig)	1832, 28 July
Ringdove 16 (brig)	1833, June
Pique 36	1834, 21 July
Sappho 16 (brig)	1837, 3 Feb.
Pilot 16 (brig)	1838, 9 June
Acorn 16 (brig)	1838, 15 Nov.
Nile 92	1839, 28 June
Ferret 10 (brig)	1840, 1 June
St George 120	1840, 27 Aug.
Hindostan 80	1841, 2 Aug.
Spartan 26	1841, 16 Aug.
Philomel 6 (brig)	1842, 28 March
Albion 90	1842, 6 Sept.
From this point, the dockyard became 'Devonport'	
Flora 36	1844, 11 Sept.
Amethyst 26	1844, 7 Dec.
Avenger 6 (frigate) *	1845, 5 Aug.
Creole 26	1845, 1 Oct.
Thetis 36	1846, 21 Aug.
Arachne 18	1847, 30 March
Aboukir 90	1848, 4 April
Indefatigable 50	1848, 27 July
Niobe 28	1849, 18 Sept.
Sans Pareil 81 *	1851, 18 March
St Jean d'Acre 101 *	1853, 23 March
Algiers 90 *	1854, 26 Jan.
Phoebe 50	1854, 12 April

Exmouth 90 *	1854, 12 July
Conqueror 100 *	1855, 2 May
Satellite 21 (corvette) *	1855, 26 Sept.
Angler 2 (gunboat) *	1856, 8 March
Ant 2 (gunboat) *	1856, 22 March
Liffey 51 (frigate) *	1856, 6 May
Pelorus 21 (corvette) *	1857, 5 Feb.
Robust 91 *	1857, 1 April
Ister 36 (frigate) *	1857, 8 Nov.
Topaze 51 (frigate) *	1858, 28 May
Donegal 101 *	1858, 23 Sept.
Narcissus 51 (frigate) *	1859, 26 Oct.
Jason 21 (corvette) *	1859, 10 Nov.
Gibraltar 101 *	1860, 16 Aug.
Pantaloon 11 (sloop) *	1860, 26 Sept.
Peterel 11 (sloop) *	1860, 10 Nov.
Royalist 11 (sloop) *	1861, 14 Dec.
Africa 11 (sloop) *	1862, 14 Feb.
Ocean 24 (ironclad) *	1862, 19 March
Dryad 4 (sloop) *	1866, 25 Sept.
Carron – (tug) *	1867, 31 May
Lapwing 3 (gunvessel) *	1867, 8 Nov.
Flirt 2 (gunvessel) *	1867, 20 Dec.
Fly 2 (gunvessel) *	1867, 20 Dec.
Seagull 3 (gunvessel) *	1868, 6 March

PEMBROKE

Thetis 46	1817, 1 Feb.
Arethusa 46	1817, 29 July
Racer 6 (cutter)	1818, 4 April
Sprightly 6 (cutter)	1818, 3 June
Belleisle 74	1819, 26 April
Fisgard 46	1819, 8 July
Falcon 10 (brig)	1820, 10 June
Frolic 10 (brig)	1820, 10 June
Melampus 46	1820, 10 Aug.
Nereus 46	1821, 30 July
Meteor 10 (bomb)	1823, 25 June
Hamadryad 46	1823, 25 July
Zephyr 10 (brig)	1823, 1 Nov.
Vengeance 84	1824, 27 July
Thisbe 46	1824, 9 Sept.
Talbot 28	1824, 9 Oct.
Sheldrake 10 (brig)	1825, 17 May
Druid 46	1825, 1 July
Success 28	1825, 30 Aug.
Erebus 12 (bomb)	1826, 7 June
Nemesis 46	1826, 19 Aug.
Satellite 18	1826, 2 Oct.
Mooring Lighter	1826, 27 Dec.
Clarence 84	1827, 25 July
Spey 10 (brig)	1827, 6 Oct.
Variable 10 (brig)	1827, 6 Oct.
Leda 46	1828, 15 April
Snipe 8 (cutter)	1828, 28 June
Sparrow 8 (cutter)	1828, 28 June
Speedy 8 (cutter)	1828, 28 June
Comet 18	1828, 14 Aug.
Hotspur 46	1828, 9 Oct.
Lightning 18	1829, 2 June
Pigeon 10 (brig)	1829, 6 Oct.
Partridge 10 (brig)	1829, 12 Oct.
Thais 10 (brig)	1829, 12 Oct.
Raven 4 (cutter)	1829, 21 Oct.
Starling 4 (cutter)	1829, 31 Oct.
Seahorse 46	1830, 20 May
Wizard 10 (brig)	1830, 24 May
Stag 46	1830, 2 Oct.

Viper 6 (brigantine)	1831, 12 May	*Encounter* 6 (sloop) *	1846, 24 Sept.	*Greyhound* 17 (sloop) *	1859, 15 June
Imogene 28	1831, 24 June	*Mariner* 16 (brig)	1846, 19 Oct.	*Immortalite* 51 (frigate) *	1859, 25 Oct.
Fly 18	1831, 25 Aug.	*Sybille* 36	1847, 15 April	*Espoir* 5 (gunvessel) *	1860, 7 Jan.
Harrier 18	1831, 8 Nov.	*Britomart* 10 (brig)	1847, 12 June	*Howe* 121 *	1860, 7 March
Cockatrice 6 (brigantine)	1832, 14 May	*Lion* 80	1847, 29 July	*Pelican* 17 (sloop) *	1860, 19 July
Comus 18	1832, 14 Aug.	*Camilla* 16 (brig)	1847, 8 Sept.	*Nimble* 5 (gunvessel) *	1860, 15 Sept.
Andromache 28	1832, 27 Aug.	*Atalanta* 16 (brig)	1847, 9 Oct.	*Pandora* 5 (gunvessel) *	1861, 7 Feb.
Royal William 120	1833, 2 April	*Colossus* 80	1848, 1 June	*Defiance* 91 *	1861, 27 March
Rodney 92	1833, 18 June	*Magicienne* 16 (frigate) *	1849, 2 March	*Aurora* 51 (frigate) *	1861, 22 June
Forth 46	1833, 1 Aug.	*Arethusa* 50	1849, 20 June	*Perseus* 17 (sloop) *	1861, 21 Aug.
Sinbad – (lighter)	1834, 27 Feb.	*Octavia* 50	1849, 18 Aug.	*Shearwater* 11 (sloop) *	1861, 17 Oct.
Tartarus 2 (gunboat) *	1834, 23 June	*Liberty* 12 (brig)	1850, 11 June	*Psyche* 2 (dispatch) *	1862, 29 March
Cleopatra 26	1835, 28 April	*Martin* 12 (brig)	1850, 19 Sept.	*Prince Consort* 35 (ironclad) *	1862, 26 June
Vanguard 80	1835, 25 Aug.	*Barracouta* 6 (sloop) *	1851, 31 March	*Enchantress* 2 (dispatch) *	1862, 2 Aug.
Harlequin 16 (brig)	1836, 18 March	*Valorous* 16 (frigate) *	1851, 30 April	*Research* 4 (ironclad) *	1863, 15 Aug.
Dido 18	1836, 13 June	*Musquito* 16 (brig)	1851, 29 July	*Alberta* – (yacht) *	1863, 2 Oct.
Carysfort 26	1836, 12 Aug.	*Duke of Wellington* 131 *	1852, 14 Sept.	*Zealous* 20 (ironclad) *	1864, 7 March
Cremill – (victualling hoy)	1836, 29 Aug.	*James Watt* 90 *	1853, 23 April	*Lord Clyde* 23 (ironclad) *	1864, 13 Oct.
Gorgon 6 (sloop) *	1837, 31 Aug.	*Rover* 16 (brig)	1853, 21 Feb.	*Amazon* 4 (sloop) *	1865, 23 May
Lily 16 (brig)	1837, 28 Sept.	*Squirrel* 12 (brig)	1853, 8 Aug.	*Vestal* 4 (gunvessel) *	1865, 16 Nov.
Penguin 6 (brig)	1838, 10 April	*Caesar* 90 *	1853, 17 Aug.	*Nassau* 4 (sloop) *	1866, 20 Feb.
Grecian 16 (brig)	1838, 24 April	*Curacoa* 31 (frigate) *	1854, 13 April	*Daphne* 4 (sloop) *	1866, 23 Oct.
Peterel 6 (brig)	1838, 23 May	*Harrier* 17 (sloop) *	1854, 13 May	*Penelope* 10 (corvette) *	1867, 18 June
Daphne 18	1838, 6 Aug.	*Swallow* 9 (sloop) *	1854, 12 June	*Newport* 5 (gunvessel) *	1867, 20 July
Merlin 2 (sloop) *	1838, 18 Sept.	*Ariel* 9 (sloop) *	1854, 11 July	*Gnat* 2 (gunvessel) *	1867, 26 Nov.
Medusa 2 (sloop) *	1838, 31 Oct.	*Falcon* 17 (sloop) *	1854, 10 Aug.	*Inconstant* 16 (frigate) *	1868, 12 Nov.
Cyclops 6 (frigate) *	1839, 10 July	*Victoria & Albert* (yacht) *	1855, 16 Jan.	*Bittern* 3 (gunvessel) *	1869, 20 Sept.
Persian 16 (brig)	1839, 7 Oct.	*Sutlej* 50	1855, 17 April		
Medina 2 (sloop) *	1840, 18 March	*Brunswick* 80 *	1855, 1 June	**BOMBAY**	
Iris 26	1840, 14 July	*Victor Emmanuel* 90 *	1855, 27 Sept.	*Melville* 74	1817, 17 Feb.
Vixen 6 (sloop) *	1841, 4 Feb.	*Flying Fish* 6 (gunvessel) *	1855, 20 Dec.	*Trincomalee* 46	1817, 12 Oct.
Cambrian 36	1841, 5 July	*Pioneer* 6 (gunvessel) *	1856, 19 Jan.	*Malabar* 74	1818, 28 Dec.
Geyser 6 (sloop) *	1841, 20 July	*Nettle* 2 (gunboat) *	1856, 9 Feb.	*Seringapatam* 48	1819, 5 Sept.
Collingwood 80	1841, 17 Aug.	*Pet* 2 (gunboat) *	1856, 9 Feb.	*Ganges* 84	1821, 10 Nov.
Spiteful 6 (sloop) *	1842, 24 March	*Decoy* 2 (gunboat) *	1856, 21 Feb.	*Madagascar* 46	1822, 15 Nov.
Superb 80	1842, 6 Sept.	*Rambler* 2 (gunboat) *	1856, 21 Feb.	*Asia* 84	1824, 19 Jan.
Victoria & Albert (yacht) *	1843, 26 April	*Drake* 2 (gunboat) *	1856, 8 March	*Bombay* 84	1828, 17 Feb.
Helena 16 (brig)	1843, 11 July	*Janus* 2 (gunboat) *	1856, 8 March	*Andromeda* 46	1829, 6 Jan.
Vulture 6 (frigate) *	1843, 21 Sept.	*Alert* 17 (sloop) *	1856, 20 May	*Calcutta* 84	1831, 14 March
Flying Fish 12 (brig)	1844, 3 April	*Cordelia* 11 (sloop) *	1856, 3 July	*Nerbudda* 12 (brig)	1848, 5 Feb.
Centurion 80	1844, 2 May	*Diadem* 32 (frigate) *	1856, 14 Oct.	*Jumna* 16 (brig)	1848, 7 March
Juno 26	1844, 1 July	*Doris* 32 (frigate) *	1857, 25 March	*Meeanee* 80	1848, 11 Nov.
Kingfisher 12 (brig)	1845, 8 April	*Melpomene* 51 (frigate) *	1857, 8 Aug.		
Inflexible 6 (sloop) *	1845, 12 April	*Gannet* 11 (sloop) *	1857, 29 Dec.		
Dragon 6 (frigate) *	1845, 17 June	*Orlando* 50 (frigate) *	1858, 12 June		
Constance 50	1846, 12 March	*Windsor Castle* 116 *	1858, 26 Aug.		
Conflict 8 (sloop) *	1846, 5 Aug.	*Revenge* 91 *	1859, 16 April		

Please note that while the *Sylvia* (Woolwich, 1866), *Myrmidon* (Chatham, 1867) and *Nassau* (Pembroke, 1866) are listed above as 4-gun sloops, the three were actually completed as survey ships.

Index to Named Vessels

This index lists alphabetically all named vessels referenced in this book, but only for their primary inclusion (mention elsewhere in the book is not noted). Each is noted with its date of launch (or of acquisition for ships not ordered for the Navy) and the date on which it ceased to be known under that name, its original type, and the appropriate page number.

Vessels which were numbered only are not included. These included the mortar vessels and mortar floats covered in Chapter 10, although the 22 mortar vessels initially named are included below. Other vessels which ceased to have names were mainly those relegated to harbour hulks (which were often numbered with a YC or C prefix) or as coastguard watch vessels (which were numbered with a WV prefix from 1863 onwards). Please note that the term 'cancelled' below includes cancellation by virtue of the ship being re-ordered to a fresh design or type.